Simulation Analysis of the Ground-Water Flow System in the Portland Basin, Oregon and Washington

By DAVID S. MORGAN and WILLIAM D. MCFARLAND

Prepared in cooperation with
Oregon Water Resources Department,
City of Portland Bureau of Water Works, and
Intergovernmental Resource Center

U.S. GEOLOGICAL SURVEY WATER-SUPPLY PAPER 2470–B

U.S. DEPARTMENT OF THE INTERIOR
BRUCE BABBITT, Secretary

U.S. GEOLOGICAL SURVEY
Gordon P. Eaton, Director

For sale by the
U.S. Geological Survey
Branch of Information Services
Box 25286
Federal Center
Denver, CO 80225

Library of Congress Cataloging in Publication Data

Morgan, David S.
Simulation analysis of the ground-water flow system in the Portland Basin, Oregon and Washington / by David S. Morgan and William D. McFarland.
p. 83 cm. — (U.S. Geological Survey water-supply paper ; 2470–B
"Prepared in cooperation with the Oregon Water Resources Department, City of Portland Bureau of Water Works, and Intergovernmental Resource Center [of Clark County, Wash]."
Includes bibliographical references.
Supt. of Docs. no. : I 19.13:2470–B
1. Groundwater flow — Oregon —Portland Region — Simulation Methods.
I. McFarland, William D. II. Oregon. Water Resources Dept. III. Portland (Or.). Water Bureau IV. Intergovernmental Resource Center of Clark County (Wash.) V. Title. VI. Series.
GB1197.7.M67 1996 96–52463
551.49′09795′49—dc21CIP

ISBN 0-607-87040-0

CONTENTS

PLATES

[Plates are in pocket]

FIGURES

TABLES

CONVERSION FACTORS AND VERTICAL DATUM

Multiply	By	To obtain
	Length	
inch (in)	25.4	millimeter
foot (ft)	0.3048	meter
mile (mi)	1.609	kilometer
	Area	
acre	4,047	square meter
acre	0.407	hectare
square mile (mi^2)	259.0	hectare
square mile (mi^2)	2.590	square kilometer
	Volume	
cubic foot (ft^3)	0.02832	cubic meter
	Flow	
cubic feet per second (ft^3/s)	0.02832	cubic meter per second
gallons per minute (gal/min)	0.06308	liter per second
gallons per minute (gal/min)	0.002228	cubic foot per second

Sea level: In this report "sea level" refers to the National Geodetic Vertical Datum of 1929 (NGVD of 1929)—a geodetic datum derived from a general adjustment of the first-order level nets of both the United States and Canada, formerly called Sea Level Datum of 1929.

Simulation Analysis of the Ground-Water Flow System in the Portland Basin, Oregon and Washington

By David S. Morgan *and* William D. McFarland

Abstract

A numerical model of the ground-water flow system in the Portland Basin, Oregon and Washington, was used to (1) test and refine the conceptual understanding of the flow system, (2) estimate the effects of past and future human-caused changes to ground-water recharge and discharge on ground-water levels and streamflow, and (3) determine priorities for ground-water monitoring and data collection that would facilitate improvements in the utility and accuracy of the model. The model covered an area of 981 square miles that includes most of Multnomah County, Oregon, and Clark County, Washington, as well as parts of Clackamas, Washington, and Columbia Counties in Oregon, and Skamania County in Washington.

Model results were most sensitive to the horizontal hydraulic conductivity of the Troutdale gravel aquifer and the Troutdale sandstone aquifer and to the vertical hydraulic conductivity of the confining units and undifferentiated fine-grained sediments. The model was insensitive to the hydraulic characteristics of the unconsolidated sedimentary aquifer because of the strong control that streams and rivers have on water levels in the unit.

The model was calibrated using time-averaged data for 1987–88. Horizontal hydraulic conductivity, vertical hydraulic conductivity, and streambed and riverbed conductances were adjusted until a reasonable fit was obtained between simulated and observed water levels and seepage to streams. Final horizontal hydraulic conductivities had median values ranging from 11 to 110 ft/d (feet per day) in the four aquifer units and from 1 to 8 ft/d in finer grained units. Horizontal to vertical anisotropy ratios ranged from 100:1 in the aquifer units to 1,000:1 in the finer grained units. Approximately 410 observed water levels were available for comparison with simulated water levels in the four aquifer units. The root-mean-squared error ranged from 54 feet in the unconsolidated sedimentary aquifer to 74 feet in the Troutdale sandstone aquifer. Simulated seepage to streams agreed closely with observed seepage at 67 measurement sites.

Recharge to the basin in 1987–88 consisted primarily of 1,440 ft^3/s (cubic feet per second) from direct infiltration of precipitation, but was augmented in unsewered urban areas by 62 ft^3/s from runoff to drywells and 27 ft^3/s from on-site waste-disposal systems. Forty-nine percent of the recharge in the basin enters the system through the Troutdale gravel aquifer and 22 percent enters through the unconsolidated sedimentary aquifer. Simulated discharge in 1987–88 was chiefly by seepage to small rivers and streams (971 ft^3/s), followed by seepage to the Columbia and Willamette Rivers and other large water bodies (457 ft^3/s) and by pumpage (166 ft^3/s). The unconsolidated sedimentary aquifer and Troutdale gravel aquifer supplied 85 percent of 1987–88 pumpage. Most of the ground-water flux through the aquifer system occurred within these two units with 75 percent of total recharge to the system entering through them and 80 percent of total discharge leaving through them in 1987–88.

Recharge under predevelopment conditions was 180 ft^3/s (12 percent) more than in 1987–88 due to the absence of impervious surfaces associated with urbanization. Simulation of the effects of the increased recharge and no well discharge indicates that water levels may have declined as much as 50 feet in the Troutdale gravel aquifer in southern Clark County in response to municipal pumping. Declines similar in magnitude were simulated in the gravel aquifer in the Troutdale gravel aquifer and Troutdale sandstone aquifer in the vicinity of Sandy, Oregon, in response to pumping for irrigation. Since predevelopment conditions, the combination of reduced recharge and increased pumpage may have reduced discharge to large rivers by 25 percent and reduced discharge to small rivers and streams by 16 percent.

One hypothetical condition for future ground-water development was simulated to test the effects of additional pumping stress on the ground-water system. Pumpage estimates in this condition were based on projected municipal supply demands in Clark County through the year 2010 and on limited use of the city of Portland Columbia South Shore well field. The resulting pumpage was 55 percent (92 ft^3/s) greater than 1987–88 pumpage. Equilibrium water-level declines were as much as 20–40 ft within the Troutdale gravel aquifer in Clark County and in the Troutdale sandstone aquifer underlying the Columbia South Shore well field. At equilibrium, much of the additional pumpage was supplied by capture of discharge to the Columbia River (14 ft^3/s) and by additional recharge induced from the Columbia River (26 ft^3/s).

Further data collection through ground-water-level monitoring and periodic updating of well discharge and recharge from on-site waste-disposal systems and drywells would facilitate future improvements in the model developed in this study.

INTRODUCTION

The Portland Basin study area is located in northwestern Oregon and southwestern Washington (fig. 1). The study area, referred to in this report as the "Portland Basin," is defined by geologic, hydrologic, and political boundaries that encompass an area of about 1,310 mi^2 (square miles). It is referred to as a basin because most of the area is underlain by a sediment-filled structural basin and the related streams and rivers all flow to the Columbia River. Population of the basin is approximately 1 million (1991), of which about 238,000 reside in Clark County, Washington. A large part of the population lives within the city of Portland (440,000) and about 46,000 people live within the neighboring city of Vancouver.

Historically, most of the industrial and municipal water needs of Portland and surrounding Oregon communities have been met by surface-water supplies from the Bull Run watershed, located approximately 25 miles southeast of downtown Portland, on the western flank of the Cascade Range (see fig. 1). In the late 1970's, ground water was identified as the source that would supply Portland's emergency and peak water needs; therefore, a 100 million gallon per day (155 ft^3/s [cubic feet per second]) capacity well field was developed on the south shore of the Columbia River, east of the city of Portland (Willis, 1977; Willis, 1978). Ground water is used extensively in rural and agricultural areas to supply domestic and agricultural water needs.

In Clark County, ground water is the primary source for municipal and industrial uses and is an important source for agricultural use. Also, a large percentage of rural Clark County residents rely on individual private wells for their domestic water needs.

Adequacy of ground-water supplies in the basin has not been a major issue in the past; however, an increasing reliance on ground water has emphasized the need for an improved quantitative understanding of the resource. Ground-water-level declines that have occurred in response to ground-water withdrawals have become a problem in some areas and are anticipated to be a continuing problem as more ground-water supplies are developed to accommodate future growth in the basin. Contamination of ground water in the basin also has reinforced the need for a quantitative description of the rate and direction of ground-water movement in the basin.

In 1987, the U.S. Geological Survey began a cooperative study with the Oregon Water Resources Department, the city of Portland Bureau of Water Works, and the Intergovernmental Resource Center located in Clark County, Washington, with the objective of describing and quantifying the ground-water resource in the basin. This report is the seventh and final report resulting from this cooperative study. Previous reports are described in the following sections of this report.

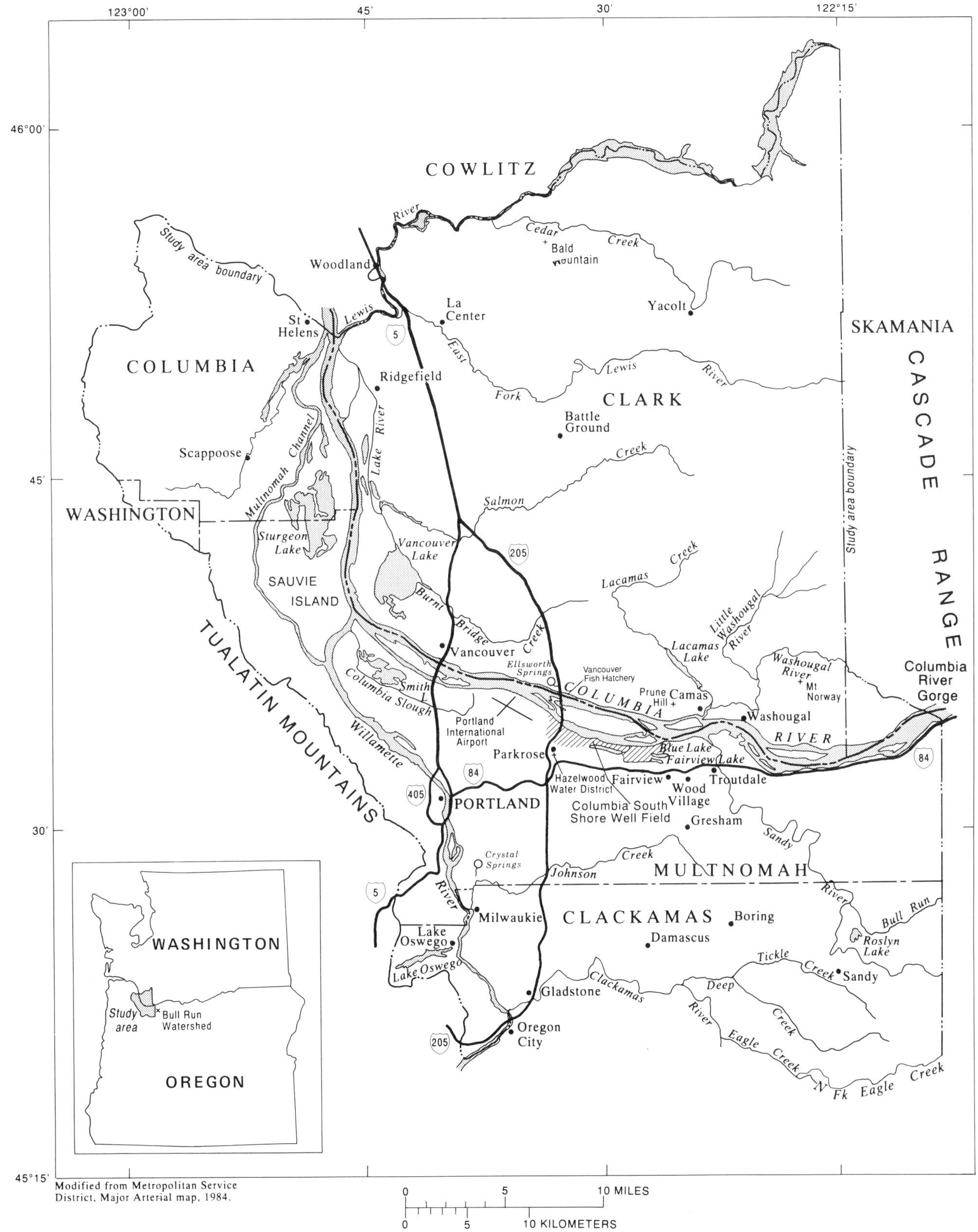

Figure 1. Location and general features of the Portland Basin study area.

Purpose and Scope

The purpose of this report is to describe the results of a numerical model analysis of the regional ground-water flow system of the Portland Basin.

A numerical model of the ground-water flow system in the Portland Basin, Oregon and Washington, was used to (1) test and refine the conceptual understanding of the flow system, (2) estimate the effects of past and future manmade changes to ground-water recharge and discharge on ground-water levels and streamflow, and (3) determine priorities for ground-water monitoring and data collection that would facilitate improvements in the utility and accuracy of the model.

The numerical model described in this report was constructed using data and information published in several other reports produced during the study; these reports are referenced frequently within this report and are summarized briefly below. Basic ground-water data used in the study are listed by McCarthy and Anderson (1990) and include (1) information on the location, ownership, construction, and yield of 1,586 field-located wells; (2) hydrographs for selected wells; and (3) information on the location, ownership, and discharge of 41 springs. A software interface for data exchange between the Geographic Information System and the ground-water flow model used in the study was documented by Orzol and McGrath (1992). The thickness and extent of hydrogeologic units defined in this study were described by Swanson and others (1993). Estimated ground-water withdrawals for municipal, industrial, and agricultural uses during 1987–88 were listed by Collins and Broad (1993). Recharge to the ground-water system from infiltration of precipitation, on-site waste-disposal systems, and urban runoff to drywells were described by Snyder and others (1994). Finally, the ground-water flow model used in the numerical model analysis described in this report is based on the conceptual model of the ground-water flow system presented by McFarland and Morgan (1996).

This report describes (1) the assumptions made to represent the conceptual model of the ground-water flow system within the framework of the mathematical model, including justification of the boundary conditions; (2) the hydraulic characteristics of the ground-water flow system; (3) the model calibration procedure and results; and (4) the results of simulations of predevelopment steady-state conditions and of a hypothetical future pumping condition.

Description of the Study Area

The study area is in the Willamette-Puget Trough, a north-south structural basin between the Coast Range on the west and the Cascade Range on the east. The study area encompasses 1,310 mi^2 and is located in the lower Columbia River Basin at the confluence of the Columbia and Willamette Rivers (fig. 1). In this report, the study area is referred to as the "Portland Basin" and is bounded approximately by the Lewis River on the north, the Clackamas River on the south, the western Cascade Range on the east, and the Tualatin Mountains on the west. The study area includes all or part of six counties in two States: Multnomah, Clackamas, Washington, and Columbia Counties in Oregon; and Clark and Skamania Counties in Washington (fig. 1). The model described in this report covers all of the study area except for the eastern and northeastern parts of Clark County, which are underlain by low-permeability volcanic rocks, and small areas in the extreme northwestern and south-western parts of the study area (fig. 2). The total area covered by the model was 981 mi^2.

Land-surface altitude in the basin ranges from about 10 ft (feet) at the Columbia River to more than 3,000 ft in the western Cascade Range. The basin floor is elongated southeast to northwest in accordance with the structural configuration of the bedrock. In Clark County, a series of nearly flat, step-like plains and benches rise to the north and east from the Columbia River, finally giving way to the steeper western slopes of the Cascade Range. South of the Columbia River, smooth alluvial plains near the river contrast with the hilly topography formed by volcanic buttes such as Mount Scott and Powell Butte.

The climate of the basin is temperate, with moderately warm, dry summers and mild, wet winters. Precipitation ranges from 36 in/yr (inches per year) near Portland to more than 80 in/yr in the western Cascade Range.

The Columbia River occupies the lowest part of the basin and is the ultimate sink for all water entering the basin that is not lost to evapotranspiration. With a drainage area of nearly 240,000 mi^2 upstream of the study area and a mean annual discharge of about 195,000 ft^3/s (period of record, 1878–1958) (Mundorff, 1964), the river directly or indirectly controls all surface- and ground-water movement in the basin. The other major rivers in the basin are the Lewis, East Fork Lewis, and Washougal Rivers in Washington, and the Willamette, Sandy, and Clackamas Rivers in Oregon.

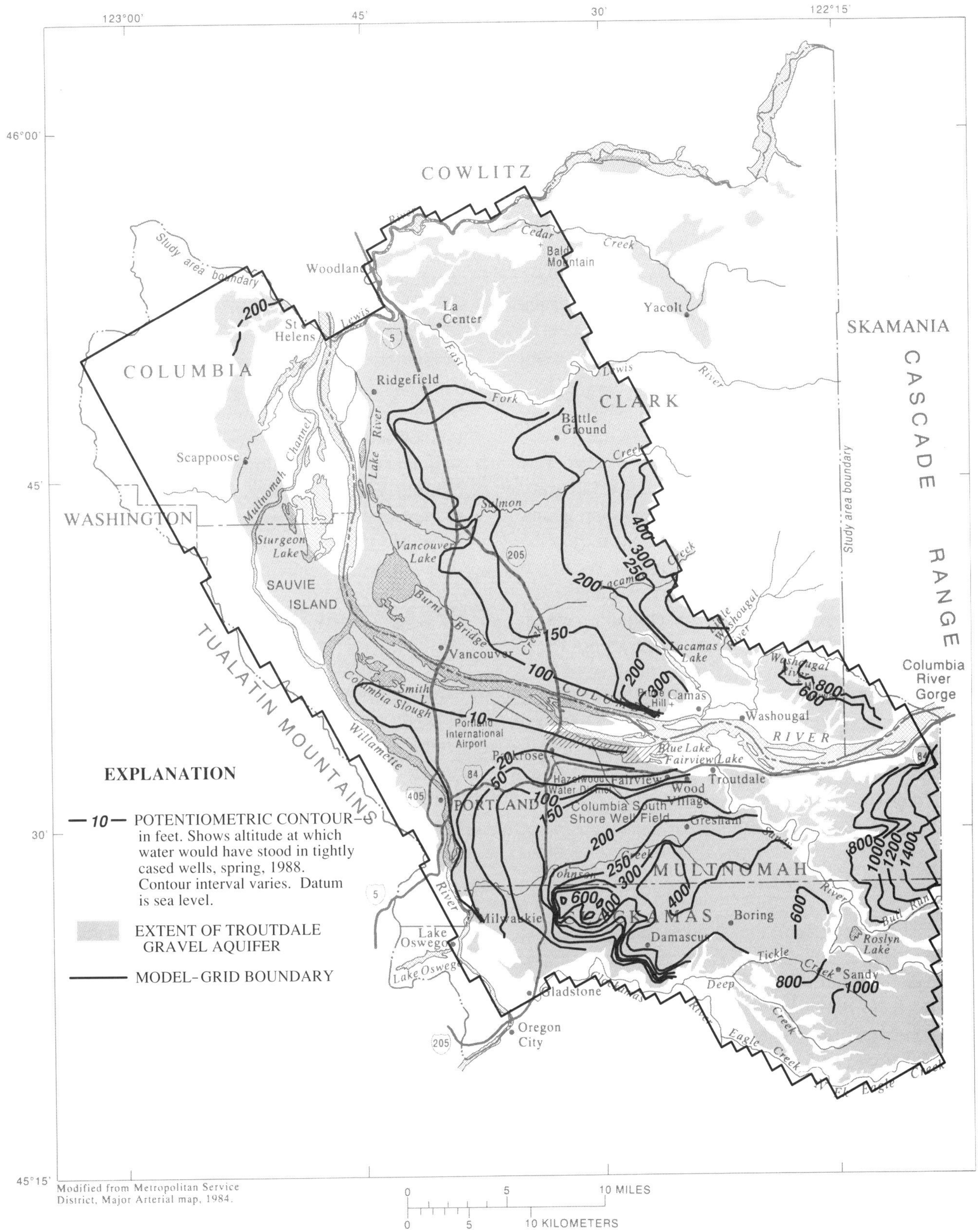

Figure 2. Potentiometric surface in the Troutdale gravel aquifer, spring 1988 (modified from McFarland and Morgan, plate 3, 1996).

Principal streams include Cedar, Salmon, Burnt Bridge, and Lacamas Creeks in Washington and Johnson and Deep Creeks in Oregon (fig. 1).

Acknowledgments

The authors extend their appreciation to the many staff members from cooperating agencies who provided valuable data and consultation throughout the study. In particular, we would like to thank Rodney Swanson (formerly of the Intergovernmental Resource Center located in Clark County, Washington); Jeffery Leighton (city of Portland Bureau of Water Works); Ken Lite (Oregon Water Resources Department); and Terry McClure and Patrick Easley (city of Vancouver).

Special acknowledgment is made to Daniel T. Snyder, U.S. Geological Survey, Portland, Oregon, for his efforts in the preparation of model input data and in making runs and compiling results during the sensitivity analysis of the model. David G. Weatherby, formerly of the U.S. Geological Survey, Eugene, Oregon, deserves special credit for compiling extensive statistical and graphical summaries of model results. Finally, Timothy McGrath, formerly of the U.S. Geological Survey, Eugene, Oregon, is gratefully acknowledged for his outstanding contributions to the project through development of many useful geographic-information software tools that facilitated construction and calibration of the model.

HYDROGEOLOGY

The overviews of the geology and hydrology of the Portland Basin in the following sections summarize more detailed descriptions in reports by (1) Swanson and others (1993), who discussed the thickness, extent and lithology of hydrogeologic units in the basin; and (2) McFarland and Morgan (1996), who describe the ground-water flow system of the basin, including its boundaries, hydraulic characteristics, and components of recharge and discharge.

Geologic Setting

The Portland Basin was formed by structural deformation of the underlying Eocene and Miocene volcanic and marine sedimentary rocks. As much as 1,800 ft of fluvial and lacustrine sediment has accumulated in the basin. The youngest deposits in the basin are primarily unconsolidated Pleistocene sediments brought into the basin during catastrophic flooding of the Columbia River. These sediment fill the center of the basin (pl. 1) and tend to be coarsest and thickest near the channels of the Columbia and Willamette Rivers. Most ground water developed in the basin is extracted from these deposits. The most extensive sedimentary geologic unit in the basin is the Troutdale Formation (Trimble, 1963), which, where it is not exposed, directly underlies the Pleistocene sediments. The Sandy River Mudstone (Trimble, 1963) underlies and is interleaved with the Troutdale Formation (Swanson and others, 1993). The Sandy River Mudstone is composed primarily of fine-grained sediments that overlie the volcanic and marine sedimentary bedrock throughout most of the basin. Mundorff (1964) referred to the Troutdale Formation as the Upper Troutdale Formation and to the Sandy River Mudstone as the Lower Troutdale Formation.

Hydrogeologic Units

Hydrogeologic units, as defined in this report, may comprise one or more geologic units; if a hydrogeologic unit contains more than one geologic unit, the geologic units must be adjacent to each other and have similar hydraulic characteristics so as to form a hydraulically connected continuum of similar materials. These hydrogeologic units can function as aquifers, or confining beds, or, where heterogeneous, both.

The nine hydrogeologic units delineated within the basin were the:

(1) unconsolidated sedimentary aquifer (US);

(2) Troutdale gravel aquifer (TG);

(3) confining unit 1 (C1);

(4) Troutdale sandstone aquifer (TS);

(5) confining unit 2 (C2);

(6) sand and gravel aquifer, upper coarse-grained subunit (SC);

(7) sand and gravel aquifer, lower fine-grained subunit (SF);

(8) undifferentiated fine-grained sediments (UF); and

(9) older rocks (OR).

The lithologic characteristics of these units, along with the geologic formations each unit comprises, are listed in plate 1.

In a detailed study of the hydrogeology of the city of Portland's Columbia South Shore well field in east Multnomah County (fig. 1), Hartford and McFarland (1989) defined the thickness and nature of several hydrogeologic units in the immediate vicinity of the well field. Swanson and others (1993) mapped the regional thickness and extent of these hydrogeologic units by using field mapping, well logs, and geophysical data. These units are described briefly in the following paragraphs; for more complete descriptions of the thickness, nature and extent of these units the reader is referred to Swanson and others (1993) and McFarland and Morgan (1996). The surface exposure of the units and representative sections are shown on plate 1.

Swanson and others (1993) grouped the hydrogeologic units into three subsystems on the basis of regionally continuous contacts between units with distinctly different lithologic and hydrogeologic characteristics: the upper sedimentary subsystem, lower sedimentary subsystem, and older rocks (pl. 1).

The unconsolidated sedimentary aquifer (US), a member of the upper sedimentary subsystem, contains Pleistocene catastrophic flood deposits, Holocene Columbia River alluvium, and to a lesser extent, flood-plain and terrace deposits along major tributaries and glacial outwash deposits in small basins in northern Clark County (McFarland and Morgan, 1996). The average thickness of the saturated part of the aquifer is about 75 ft; however, local accumulations of catastrophic-flood deposits are as much as 240 ft thick. Some parts of the unit are saturated only during the winter and spring or during prolonged periods of normal or above-normal recharge. The thickness and extent of the saturated part of the unconsolidated sedimentary aquifer is shown in Appendix A, figure A–1.

The Troutdale gravel aquifer (TG) is also a part of the upper sedimentary subsystem and includes sandy and cemented conglomerates, Boring Lava and Cascade lavas, and mantling soil horizons (McFarland and Morgan, 1996). Conglomerates within the unit extend basinwide and include the upper part of the Troutdale Formation of Trimble (1963) and Mundorff (1964), Cascade Range volcaniclastics of the Springwater and Gresham Formations (Trimble, 1963), and the informal upper member of the Troutdale Formation (Mundorff, 1964). Boring Lava and vent rocks are included in the Troutdale gravel aquifer. These rocks form many of the isolated hills in the southern part of the study area, including Kelly Butte, Mount Tabor, and the Boring Hills; occurrences in Clark County include Mount Norway and Prune Hill. Throughout much of the basin, the Boring Lava lies above the regional water table (Hogenson and Foxworthy, 1965), and ground water contained within them is perched above old soil zones and dense, low-permeability flow centers. In the southeastern parts of the study area and east of the Sandy River, the unit is composed of a thickening sheet of Cascade lavas (McFarland and Morgan, 1996). The unit has a maximum saturated thickness of approximately 900 ft and thins to zero at the margins of the basin and where it has been eroded by the Columbia, Sandy, Clackamas, and East Fork Lewis Rivers (Appendix A, fig. A–2); average thickness of the saturated part of the unit is 180 ft.

Confining unit 1 (C1) is the uppermost member of the lower sedimentary subsystem and consists of silt, clay, and discontinuous beds of medium- to fine-grained sand. Some low-discharge wells develop water from the interbedded sands, but overall the unit is not an important source of supply. The unit is overlain by the Troutdale gravel aquifer and underlain by the Troutdale sandstone aquifer. Confining unit 1 has a maximum thickness of nearly 300 ft in T.3N., R.2E. and in T.1N., R.2E. (Appendix A, fig. A–3); the average thickness of the saturated part of the unit is 110 ft.

The Troutdale sandstone aquifer (TS) underlies confining unit 1, except in the area east of the Sandy River and Lacamas Creek where the unit directly underlies the Troutdale gravel aquifer. The unit consists of two distinctive lithologic subunits: an upper, vitric sandstone and a lower, basaltic conglomerate (McFarland and Morgan, 1996). Both are highly productive. Thickness of the unit ranges from zero where it interfingers with the undifferentiated fine-grained sediments or has been eroded by the Columbia River to 320 ft in the eastern part of the basin (Appendix A, fig. A–4); the average thickness of the saturated part of the unit is 120 ft.

Confining unit 2 (C2) underlies the Troutdale sandstone aquifer and overlies either the upper, coarse-grained subunit of the sand and gravel aquifer or older rocks. The principal lithology of the unit is clay and silt with some lenses of silt and fine- to medium-grained sand (McFarland and Morgan, 1996). The extent of confining unit 2 is defined by the extent of the Troutdale sandstone aquifer. Where the Troutdale sandstone aquifer is absent, confining beds 1 and 2 are indistinguishable and are mapped as the undifferentiated fine-grained sediments. This definition results in large variations in the thickness of the unit, as shown in Appendix A, figure A–5. The maximum thickness of the unit, 1,100 ft, occurs in Clark County, where the sand and gravel aquifer is absent and confining unit 2 directly overlies older rocks. The unit thins to 50 ft where it overlies the upper, coarse-grained subunit of the sand and gravel aquifer; the average thickness of the saturated part of the unit is 450 ft.

The sand and gravel aquifer was mapped as a single hydrogeologic unit by Hartford and McFarland (1989) and Swanson and others (1993). McFarland and Morgan (1996) noted that the unit contained two distinct lithologic subunits: a relatively coarse upper subunit composed mostly of sandy gravel, and a predominantly fine-grained lower subunit composed of silty sand, clay, and fine sand. The coarse-grained subunit is an important aquifer, being tapped by several public-supply systems on the south shore of the Columbia River and being actively explored for development in southern Clark County (McFarland and Morgan, 1996). The fine-grained subunit is not used as an aquifer. To permit a more accurate simulation of the contrast of their hydraulic properties, the sand and gravel aquifer was subdivided in the model into an upper, coarse-grained subunit (SC) and a lower, fine-grained subunit (SF). Well logs show that the upper subunit has an average thickness of about 100–150 ft. Where the sand and gravel aquifer was more than 125 ft thick, the upper subunit was assigned a constant thickness of 125 ft (Appendix A, fig. A–6) and the lower subunit was assigned the remaining thickness of the unit. Where the thickness of the sand and gravel unit was less than 125 ft, the entire thickness was assigned to the upper subunit (Appendix A, fig. A–7). The average thickness of the saturated part of the lower subunit is 500 ft.

The undifferentiated fine-grained sediments (UF) occur where the Troutdale sandstone aquifer is missing and confining units 1 and 2 cannot be differentiated. Throughout most of the basin, undifferentiated fine-grained sediments underlie the Troutdale gravel aquifer and overlie the older rocks. Regionally, the undifferentiated fine-grained sediments are not a good aquifer; however, coarse facies within the unit are an important source of ground water in central and northern Clark County. Thickness of the undifferentiated fine-grained sediments ranges from zero at the margins of the basin to 1,400 ft near Vancouver (Appendix A, fig. A–8). The average thickness of the saturated part of the undifferentiated fine-grained sediments is 580 ft.

The older rock subsystem (OR) includes Miocene and older volcanic and marine sedimentary rocks that underlie and, to a large extent, form a low-permeability boundary to the basin-fill sediments. The lower permeability members of this subsystem include the dense, altered basalts of the Skamania Volcanics, Goble Volcanics, basalts of Waverly Heights, and fine-grained marine sedimentary rocks such as the Scappoose Formation. The Columbia River Basalt Group is a member of the older rock subsystem and can have high permeability interflow zones that yield large quantities of water to wells (McFarland and Morgan, 1996). The thickness of older rocks in the basin is not well defined; however, on the basis of well-log data, they are known to extend to at least 2,000 ft below sea level (Swanson and others, 1993). The thickness of the older rocks above this altitude is shown in Appendix A, figure A–9.

Occurrence and Movement of Ground Water

The ground-water system in the basin is recharged primarily by direct infiltration of precipitation, but recharge from urban runoff shunted to drywells and from on-site waste-disposal systems also are significant sources in densely populated, unsewered, urban areas (Snyder and others, 1994). The distribution of infiltration of precipitation is closely related to annual precipitation, altitude, and the percentage of the area covered by impervious surfaces. Infiltration from precipitation ranges from zero in highly developed urban areas to nearly 50 in/yr in the upland forested areas of the basin (Snyder and others, 1994). Streams may constitute locally important sources of seasonal recharge during periods of high flow; however, much of this recharge is lost to base flow during stream recession. High discharge wells near the Columbia River also can induce recharge from the river to the shallow alluvial aquifers (Mundorff, 1964).

Movement and discharge of ground water is controlled primarily by the topography of the basin. Regional, intermediate, and local ground-water flow systems exist in the basin. The Willamette, Lewis, Clackamas, and Columbia Rivers represent the discharge areas for the regional ground-water flow system. Most of the ground water discharging to them enters the system in upland recharge areas, moves downward through hundreds of feet of sediment, moves horizontally, and finally moves upward to discharge to the rivers. The East Fork Lewis River, Salmon Creek, and the Sandy River are examples of discharge areas for intermediate ground-water flow systems. The general directions of ground-water flow in the Troutdale gravel aquifer are shown in figure 2. Local ground-water flow systems are much smaller and ground water within them has much shorter residence times; commonly, recharge and discharge points are within hundreds of feet of each other. Discharge can occur as direct seepage to a stream, hillslope seepage, or as evapotranspiration by riparian vegetation. All the conditions discussed above are based on the assumption of natural flow conditions. Manmade changes, such as the addition of wells, drywells, on-site waste-disposal systems, and paved surfaces in the basin, disrupt these natural flow paths. One of the goals of this study was to develop a quantitative understanding of the influence of manmade changes on the natural flow system.

SIMULATION ANALYSIS OF THE GROUND-WATER FLOW SYSTEM

A steady-state numerical model of the ground-water system was constructed and calibrated to time-averaged data for the period April 1987 through March 1988. Time-averaged data for the 1987–88 period were used in the calibration because data needed to define long-term historical changes in ground-water levels, pumpage, and recharge within the basin were not available. The April 1987 through March 1988 period was used because this was the time during which most of the water-level and water-use data were collected in the study.

Estimates of all model-parameter values, fluxes, and boundary conditions were made to construct the initial model. A systematic sensitivity analysis of model response to changes in each parameter value and boundary conditions was then made. Calibration of the model consisted of making changes to parameters and boundary conditions, focusing on the most sensitive parameters, until the best fit between model simulated and measured ground-water levels and fluxes was obtained.

The model was then used to simulate past and future changes to the ground-water system. First, historical changes that have occurred due to the loss of pervious surfaces in urban areas, ground-water pumpage, and diversion of surface runoff were simulated. Second, the future changes that may occur due to increased ground-water pumpage under one hypothetical condition were simulated.

General Features of the Model

A three-dimensional, steady-state, finite-difference, ground-water flow model of the Portland Basin was constructed using the McDonald and Harbaugh (1988) modular code (MODFLOW). Since MODFLOW was published, several additional modules, or packages, have been developed that facilitate the simulation of more complex hydrologic systems. One example (Prudic, 1989) is a streamflow-routing package that was used in this study to simulate stream-aquifer relations in the basin. Another package, written by Orzol and McGrath (1992) for use in this study, allows MODFLOW to read and write data directly from and to the ARC/INFO geographic information system file format.

The data manipulation and analysis tasks related to construction, calibration, and use of the ground-water model were enormous. The complex and varied sedimentary environment of the Portland Basin presented significant challenges when trying to represent that structure within the framework of a numerical model. The use of a geographic information system to store and manage spatial data and the software interface to MODFLOW by Orzol and McGrath (1992) made it possible to include a great deal of that complexity in the model and facilitated graphical presentation of simulation results.

MODFLOW simulates ground-water flow through three-dimensional, heterogeneous, anisotropic aquifer systems by iteratively solving a finite-difference approximation of the partial differential equation for ground-water flow (McDonald and Harbaugh, 1988). The flow equation may be solved using one of several iterative techniques; the strongly implicit procedure (SIP) was used for the Portland Basin model.

The general equation for three-dimensional ground-water flow through porous material is:

$$\frac{\partial}{\partial x}\left(K_{xx}\frac{\partial h}{\partial x}\right)+\frac{\partial}{\partial y}\left(K_{yy}\frac{\partial h}{\partial y}\right)+\frac{\partial h}{\partial z}\left(K_{zz}\frac{\partial h}{\partial z}\right)-W=S_s\frac{\partial h}{\partial t}\ , \qquad (1)$$

where

K_{xx}, K_{yy}, K_{zz} are values of hydraulic conductivity along the x, y, and z axes (LT^{-1});

h is the potentiometric head (L);

W is a volumetric flux per unit volume and represents sources and or sinks of water (T^{-1});

S_s is the specific storage of the porous material (L^{-1}); and

t is time (T).

Equation (1) can be used to simulate either transient or steady-state ground-water flow systems. In transient analyses, the storage term on the right-hand side of equation (1) represents the rate of change in storage in the system; for steady-state simulations, the rate of change in storage is zero. The general flow equation may also be time averaged (Prych, 1983) by integrating over time and dividing by the time interval, resulting in the following equation:

$$\frac{\partial}{\partial x}\left(K_{xx}\frac{\partial \bar{h}}{\partial x}\right) + \frac{\partial}{\partial y}\left(K_{yy}\frac{\partial \bar{h}}{\partial y}\right) + \frac{\partial}{\partial z}\left(K_{zz}\frac{\partial \bar{h}}{\partial z}\right) - \overline{W} = S_s\frac{\left(h_{t_2} - h_{t_1}\right)}{t_2 - t_1}, \quad (2)$$

where

$\bar{h}$ is average potentiometric head for the time interval, t_2-t_1 (L);

$\overline{W}$ is the average volumetric flux per unit volume for the time interval, t_2-t_1 (T^{-1}); and

t_1, t_2 are times at the beginning and end of the time interval, respectively (t).

Time averaging is a method of converting certain transient flow problems into steady-state flow problems. If the heads in the system at times t_1 and t_2 are the same, then the storage term on the right hand side of equation (2) is zero. However, if there is a change in head and thus a change in storage between times t_1 and t_2, then the storage term represents the average rate of change in storage during the time interval. The time-averaged flow equation (2) was used to model the Portland Basin for the calibration period of April 1987 through March 1988; the rationale for this approach is described in the section, "Calibration of the Model".

The finite-difference grid used to subdivide the ground-water flow system in the Portland Basin consisted of 91 rows and 50 columns. This grid formed 4,550 cells, 3,040 of which were "active" in each of the 8 model layers. The dimensions of each cell were 3,000 ft per side, but the thicknesses were variable (fig. 3).

The complex history of deposition and erosion in the basin has created hydrogeologic units that are discontinuous and that vary greatly in thickness. A hydrogeologic section illustrating the stratigraphic complexity of the basin is shown in figure 4. Discontinuous hydrogeologic units present special problems to the construction of ground-water models. These problems stem from the inability of most models (including MODFLOW) to allow vertical flow across inactive model cells that must separate nonadjacent layers; thus, if a hydrogeologic unit has been truncated or has pinched out, vertical flow cannot be simulated between the overlying and underlying units. This problem has historically been addressed by either leaving "pseudo-cells" in missing layers with very large vertical conductances and very low horizontal hydraulic conductivity (Morgan and Dettinger, 1994) or by modifying the model to allow vertical flow between nonadjacent layers (Hansen, 1993).

The primary goal in construction of the Portland Basin model was to maintain as much of the stratigraphic complexity of the real ground-water flow system as possible and yet avoid making changes to the basic MODFLOW code or incorporating "pseudo-cells" into the system. The approach used to meet these goals was a "unit-code" method in which hydrogeologic units are not restricted to a single model layer, but may span multiple layers. Hydraulic characteristics and thicknesses vary within model layers to reflect the character and thickness of the unit represented by individual cells within the layer.

Although nine hydrogeologic units were defined in the basin, the maximum number of units that were found to occur in any vertical section was eight; thus, eight model layers were used to simulate the ground-water flow system. A vertical column of model cells could simulate from one to eight hydrogeologic units. If eight units were present in the section, each would be simulated by one of the cells in the column. Where the section consisted of only one unit, all eight cells in the column were used to represent the unit. The vertical subdivision of the hydrogeologic section with the model section is shown in figure 4.

For cases in which sections contained more than one but fewer than eight hydrogeologic units, hydrogeologic units were allocated to the eight cells on the basis of stratigraphic relations between units (table 1).

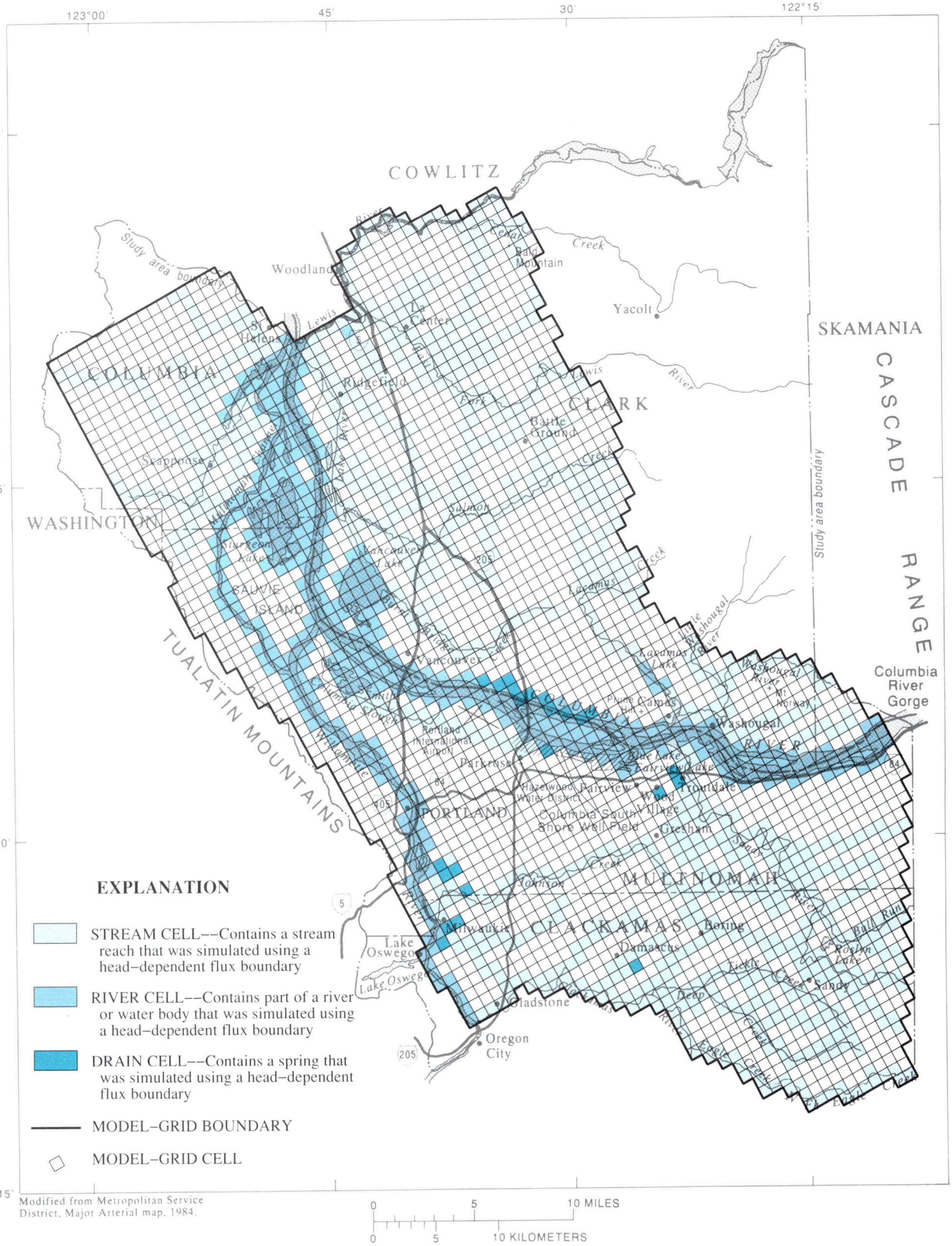

Figure 3. Boundary conditions and extent of grid used in the ground-water flow model.

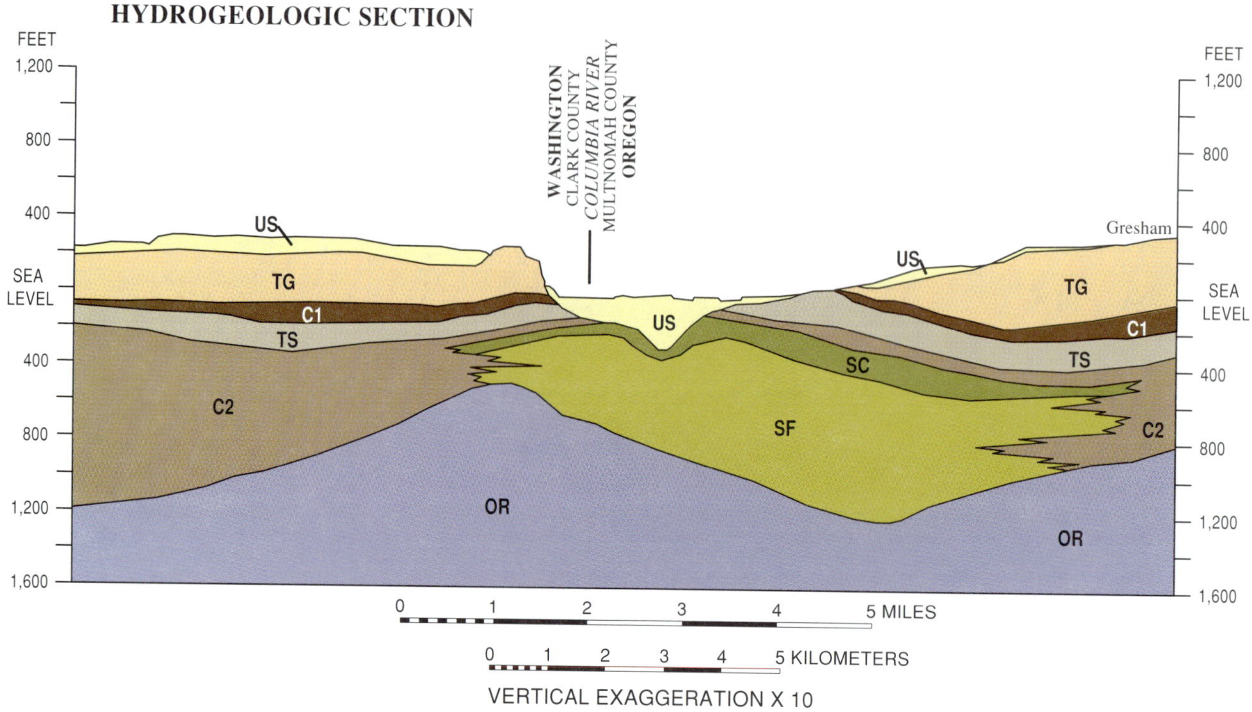

MODEL SECTION

MODEL LAYER NUMBER	42	43	44	45	46	47	48	49	50	51	52	53	54	55	56	57	58	59	60	61	62
1	44	49	74	44	26	45	44	62	52	66	29	19	11	12	22	27	28	27	26	40	71
2	275	246	294	276	262	205	205	252	149	129	29	19	11	12	23	73	222	362	398	355	256
3	100	100	100	118	119	117	115	110	101	65	29	19	11	12	67	100	100	100	120	149	150
4	150	150	150	150	150	150	150	150	150	85	104	19	11	120	85	150	141	125	125	125	125
5	380	340	303	261	219	176	124	75	75	56	50	19	50	68	80	76	83	88	77	75	108
6	185	171	162	151	138	139	130	125	125	125	125	125	125	125	125	125	125	125	125	125	112
7	482	450	414	371	312	295	254	193	197	287	398	523	651	753	804	789	603	522	438	425	332
8	610	685	759	879	1025	1098	1193	1273	1161	1188	1245	1269	1139	907	804	743	852	895	992	1021	1157

MODEL ROW NUMBER

EXPLANATION

HYDROGEOLOGIC UNIT

- **US** Unconsolidated sedimentary aquifer
- **TG** Troutdale gravel aquifer
- **C1** Confining unit 1
- **TS** Troutdale sandstone aquifer
- **C2** Confining unit 2
- NOT SHOWN Undifferentiated fine-grained sediments
- **SC** Sand and gravel (upper coarse subunit)
- **SF** Sand and gravel (lower fine subunit)
- **OR** Older rocks

71 MODEL GRID CELL--Number in cell is thickness of layer, in feet.

——— CONTACT--Approximately located.

NOTE: Complete sections are shown on Plate 1 as C–C' and F–F' for (A) hydrogeologic and (B) model sections respectively.

Figure 4. Hydrogeologic and model sections showing correlation of model layers with hydrogeologic units along column 28 of model. (Location of sections do not match exactly and hydrogeologic section shows total thickness [saturated and unsaturated]; model section shows saturated thickness only.)

Table 1. Assignment of hydrogeologic units to model layers

[US, unconsolidated sedimentary aquifer; TG, Troutdale gravel aquifer; C1, confining unit 1; TS, Troutdale sandstone aquifer; C2, confining unit 2; SC, sand and gravel aquifer, upper coarse-grained subunit; SF, sand and gravel aquifer, lower fine-grained subunit; UF, undifferentiated fine-grained sediments; OR, older rocks; ●, unit occurs most frequently in layer; 0, unit occurs in layer; --, unit does not occur in layer]

	Hydrogeologic unit								
	Upper sedimentary subsystem		Lower sedimentary subsystem					Older rock subsystem	
	Unit code								
	1	2	3	4	5	7	8	6	9
	Unit name								
Model layer	US	TG	C1	TS	C2	SC	SF	UF	OR
1	●	0	0	0	0	0	--	0	0
2	0	●	0	0	0	0	--	0	0
3	0	--	●	0	0	0	--	0	0
4	0	--	--	●	0	0	--	0	0
5	0	--	--	--	●	0	--	0	0
6	0	--	--	--	0	●	--	0	0
7	--	--	--	--	0	0	●	0	0
8	--	--	--	--	--	--	--	--	●

If a unit was represented by only one layer, the entire saturated thickness of the unit was assigned to that layer. If multiple layers were assigned to the unit, the unit's thickness was apportioned among the layers to minimize the changes in thickness between adjacent cells in the same layer. A computer program was written to automate the allocation of model layers to units and calculation of unit thickness in each model cell. The assignment of units to model layers is summarized in table 1.

Data arrays, containing integer unit codes to indicate the hydrogeologic unit represented by each cell, were constructed for each model layer. Input arrays for hydraulic characteristics were prepared for each model layer from arrays for each hydrogeologic unit and the unit-code arrays that link the layer arrays to the units arrays. The unit code files were used to prepare the input data files for several of the model packages.

Boundary Conditions

The boundaries of the model were selected to coincide with either geologic or hydrologic boundaries of the ground-water flow system. Three types of boundary conditions were used to represent the individual hydrogeologic conditions: no-flow, specified flux, and head-dependent flux (fig. 3).

No-Flow Boundaries

The model grid was extended horizontally to boundaries that were assumed to allow very little or no ground-water flow across them into or out of the basin. These boundaries coincide with the edge of the model grid (fig. 3). The following description of no-flow boundaries applies to each layer of the model. This type of boundary condition was used to represent three types of physical boundaries:

(1) nearly impermeable rocks,

(2) ground-water flow divides, and

(3) streams that function as sinks to the regional ground-water system.

The northern boundary of the model is formed by the Lewis River from Lake Merwin to the Columbia River. Because older rocks that underlie the Lewis River have very low hydraulic conductivity and because the river acts as a regional ground-water sink, the outer edge of cells underlying the river was designated a no-flow boundary. The uppermost layer of cells containing the Lewis River were designated head-dependent flux cells to allow flux between the river and the ground-water system.

The eastern boundary of the model, from the Lewis River south to the Columbia River, follows the contact between the basin-fill sediments and the older rocks. The older rocks in this area are basalts of low hydraulic conductivity and were designated a no-flow boundary to the model. The no-flow boundary extends southward from the Columbia River across the western flank of the Cascade Range to the North Fork of Eagle Creek, a tributary of the Clackamas River. The Troutdale gravel aquifer underlying this part of the no-flow boundary is composed of low-permeability Cascade lavas and volcaniclastic conglomerates (McFarland and Morgan, 1996).

The southern boundary of the model is coincident with North Fork Eagle Creek to its confluence with the Clackamas River and with the Clackamas River to its confluence with the Willamette River. The channels of Eagle Creek and the Clackamas River have been cut down into the lower-permeability older rocks or fine-grained sediments throughout most of their length. There is no evidence to indicate that ground-water flow occurs beneath the rivers, and they were assumed to function as regional sinks that intercept all ground-water flow.

The western boundary of the model follows several faults in the older rocks from the confluence of the Clackamas and Willamette Rivers northwest to the axis of the Tualatin Mountains anticline (pl. 1). The anticlinal axis coincides with the drainage divide for the Portland Basin and the boundary follows the divide northward to Scappoose Creek, where the boundary of the model grid coincides approximately with the drainage divide of the creek. The western boundary of the model was assumed to be a no-flow boundary because of the structural barrier to flow created by the anticline and because a ground-water divide probably exists that coincides with the surface-water divide.

The no-flow boundaries shown in figure 3 extend through all eight model layers. The lower boundary of the model is also a no-flow boundary, and it is located at an altitude of 2,000 ft below sea level throughout the model. This altitude was selected so that older, less permeable rocks would form the basal unit of the model throughout the area and have a minimum thickness of about 400 ft. Where the unit has been used as an aquifer, or is likely to be, its top is generally at or above sea level. At least the uppermost 2,000 ft of the unit is simulated wherever it might be subjected to pumping stress that would induce upward vertical flow and large changes in storage.

Specified Flux Boundaries

Specified flux boundaries were used to simulate recharge and discharge processes that are not a function of head. Two types of flux specified in the model were discharge by wells and recharge. During calibration of the model to time averaged conditions for 1987–88, a third type of flux was specified to account for changes in ground-water storage during the calibration period; this flux is described in the section, "Calibration of the Model." Rates of recharge and change in storage were specified using the "RECHARGE" package of MODFLOW; well-discharge rates were specified using the "WELL" package.

Recharge

The annual recharge to the ground-water system is principally derived from three sources: (1) infiltration of precipitation, (2) runoff to drywells, and (3) on-site waste-disposal systems (Snyder and others, 1994).

Snyder and others (1994) used a deep percolation model to estimate recharge from direct infiltration of precipitation in three small drainages within the basin. Regression analysis of the results revealed that recharge from direct infiltration of precipitation is most dependent on mean annual rainfall, altitude, and the percentage of impervious area. They found that the regression equation:

$$R = -4.79 + 0.482\,P - 0.354\,I + 0.0097\,E, \qquad (3)$$

where

R = mean annual recharge, in/yr;

P = mean annual precipitation, in/yr;

I = impervious area, in percent; and

E = altitude of land surface, in feet;

has an r-squared value of 0.91. The recharge rates estimated by Snyder and others (1994) are average rates for land areas within each cell. The mean annual rate for the study area is about 21 in/yr. The recharge rate for model cells containing substantial areas of surface water was adjusted by multiplying by the ratio of land area to the total area of the cell. For example, a model cell with a recharge rate of 10 in/yr, with one-half of its area covered by surface water, would have a recharge rate of 5 in/yr in the model.

Application of the regression equation resulted in estimated recharge rates that range from 0 to 49 in/yr within the model area. The mean annual rate for the basin is about 21 in/yr. Recharge rates generally are greatest in the higher altitudes of the Tualatin Mountains and the western Cascade Range. Recharge rates are least in the most urbanized parts of the basin, where impervious surfaces, such as streets and parking lots, intercept water that would otherwise infiltrate to the ground-water system. In some urban areas, however, runoff from impervious areas is routed to the subsurface through drywells, making them a locally important source of recharge, with areally averaged rates within model cells of from 1 to 26 in/yr. Effluent from on-site waste-disposal systems also is an important source of recharge in densely populated, unsewered areas. In east Multnomah County where densities are highest, Snyder and others (1994) estimated areally averaged recharge rates within model cells of as much as 26 in/yr from on-site waste-disposal systems. The areal distribution of total mean annual recharge is shown on plate 2.

Well Discharge

Average well-discharge rates in the model were based on data published by Collins and Broad (1993). Discharge rates for 1988 were used for calibration of the model because they were more representative of average conditions during the 1987–88 calibration period. The discharge rates for 1987 were unusually high due to operation of wells in the city of Portland's Columbia South Shore well field from August through October of that year.

In 1988, the average rate of well discharge for the basin was 166 ft^3/s; 40 percent (66 ft^3/s) of this total was used for public supply, 50 percent (83 ft^3/s) for industrial process water, and 10 percent (17 ft^3/s) for irrigation (Collins and Broad, 1993). The distribution of well discharge, by model cell, is shown in figure 5. Fifty-eight percent of 1988 well discharge was from the unconsolidated sedimentary aquifer, 27 percent from the Troutdale gravel aquifer, and 7 percent from the Troutdale sandstone aquifer. The areal distributions of well discharge from these units are shown in Appendix B. The remaining 8 percent of well discharge was from the undifferentiated fine-grained sediments (3 percent), the upper, coarse-grained subunit of the sand and gravel aquifer (3 percent), and the older rocks (2 percent).

Head-Dependent Flux Boundaries

Ground-water interaction with surface-water bodies was simulated using three separate packages in MODFLOW. Lakes and large rivers were simulated using the RIVER package. Streams and medium-sized rivers were simulated using the STREAM package (Prudic, 1989). Springs were simulated using the DRAIN package.

Lakes and Large Rivers

The Willamette and Columbia Rivers, Multnomah Channel, Vancouver and Sturgeon Lakes, and other low-lying water bodies are regional discharge points for ground-water flow and were simulated in the model as head-dependent flux boundaries using the RIVER package of MODFLOW (fig. 3). The rates of ground-water discharge to, or recharge from RIVER cells are simulated as a piecewise linear function of the hydraulic conductance of the riverbed and the difference between river stage and the ground-water level in the cell. The mean annual stage at RIVER cells on the Willamette and Columbia Rivers was estimated by extrapolation of data from the gaging stations at Portland and Vancouver. The mean altitude of the riverbed in each RIVER cell was estimated from 7.5-minute USGS topographic maps. The 434 model cells designated as RIVER boundaries are shown in figure 3.

Streams and Small Rivers

All other rivers and major streams were designated as head-dependent flux boundaries in the model using the STREAM package developed by Prudic (1989) for use with MODFLOW (fig. 3). The STREAM package calculates leakage between the stream and the aquifer in a manner identical to that used in the RIVER package, and it also calculates flow into and out of each reach of the stream.

Use of the STREAM package requires that streams in the model be divided into segments and reaches (Prudic, 1989). A segment is a length of stream that has no tributaries or diversions; a reach is a part of a segment that lies within one model cell. A segment can be made up of one or more reaches. Segment and reach numbers are assigned in downstream order to facilitate the flow accounting by the model. A second input file defines the tributary and diversion relations between segments. In the Portland Basin model, 1,253 stream reaches were specified in 150 segments. Model cells containing STREAM boundaries are shown in figure 3.

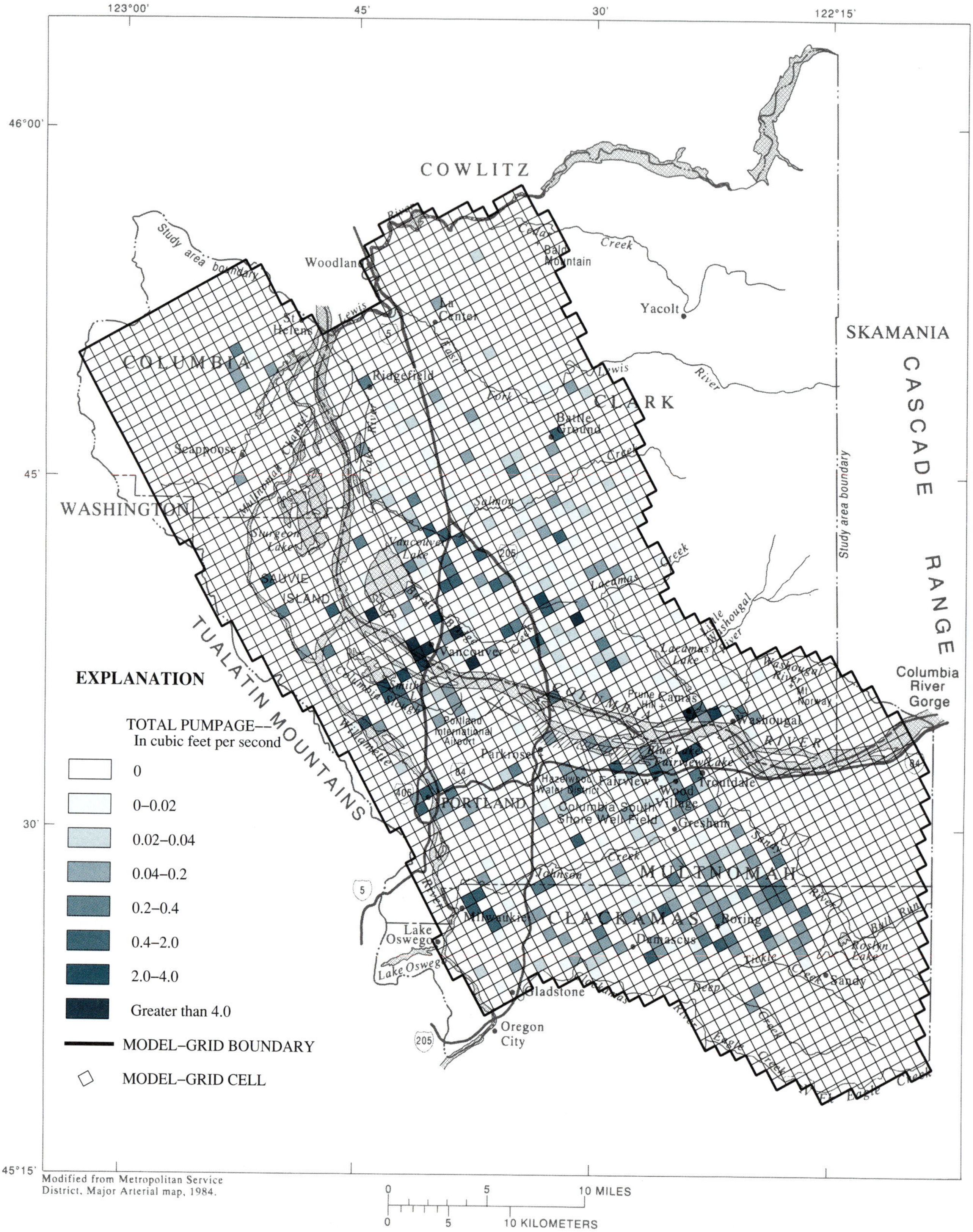

Figure 5. Estimated rates and distribution of ground-water discharge by wells, 1988 (modified from Collins and Broad, 1993).

The mean streambed altitude of each reach was determined from 7.5-minute USGS topographic maps. The mean stream stage in each reach was assumed to be 2 ft above the streambed, and the streambed thickness was assumed to be 2 ft.

Springs

All major springs in the basin were represented as head-dependent flux boundaries in the model using the DRAIN package of MODFLOW. In contrast to the STREAM and RIVER packages, the DRAIN package allows only discharge from the system. Thus, when the calculated head in the drain cell is above the altitude of the spring, the discharge rate of the spring is calculated as a piecewise linear function of the hydraulic conductance of the spring and the difference between the spring altitude and the head in the cell. The 23 model cells containing springs simulated with the DRAIN package are shown in figure 3.

Hydraulic Characteristics

The hydraulic characteristics of the rocks and sediment that form aquifers and confining beds of the ground-water system are major controlling factors on the direction and velocity of ground-water movement within the system.

Horizontal Hydraulic Conductivity

Estimates of horizontal hydraulic conductivity were made by McFarland and Morgan (1996) from approximately 500 single-well specific-capacity tests and 55 multiple-well aquifer tests. These estimates were used to delineate the distributions of hydraulic conductivity for the unconsolidated sedimentary aquifer, Troutdale gravel aquifer, Troutdale sandstone aquifer, and the upper, coarse-grained subunit of the sand and gravel aquifer. These distributions were used as initial values that were subsequently modified during calibration of the numerical model. The statistical distributions of the precalibration and postcalibration hydraulic conductivity estimates are compared for each hydrogeologic unit in figure 6.

Few wells in the basin have openings adjacent to fine-grained units; therefore, specific-capacity or aquifer-test data were insufficient to directly estimate the areal distribution of hydraulic conductivity in these units. There are, however, many wells that penetrate the confining units (1 and 2) and the undifferentiated fine-grained sediments. This made it possible to delineate variations in the lithology of the fine-grained units that might be related to variations in the hydraulic characteristics of the units. Lithologic logs from drillers' reports on these wells were used to construct a map of the average percentage of fine-grained sediments within the units. For wells with specific- capacity tests, the percentage of fine-grained sediments was compared with the estimated hydraulic conductivity, but no relation was found. The reasons for this lack of correlation are probably twofold. First, differences in the fraction of the hydrogeologic unit penetrated by the wells or well-screen placement probably bias the estimated hydraulic conductivities toward high values, because wells typically are selectively completed in coarser-grained sediments. Second, the percentage of fine-grained sediments at each well was estimated on the basis of subjective interpretation of drillers' descriptions of lithology.

The percentage of fine-grained sediments nonetheless was considered an indicator of the hydraulic conductivity of the fine-grained units. A simplistic, but reasonable, relation between the percentage of fine-grained sediments and the hydraulic conductivity of the fine-grained units was derived using the distribution of hydraulic-conductivity estimates for the undifferentiated fine-grained sediments. It was assumed that the maximum hydraulic conductivity of 37 ft/d (feet per day) would occur where there were no fine-grained sediments in the unit, and the minimum (0.1 ft/d) would occur where the unit was composed entirely of fine-grained sediments. Assuming a linear relation between percentage of fine-grained sediments and hydraulic conductivity, the following relation was used to estimate the distribution of hydraulic conductivity in the fine-grained units:

$$Kh = 37.0 - 0.369\,Pf, \qquad (4)$$

where

Kh is horizontal hydraulic conductivity (LT^{-1}),

and

Pf is percentage of fine-grained sediments (dimensionless).

Not enough wells penetrated the fine-grained units to map the percentage of fine-grained sediments for each unit. A composite map of the percentage of fine-grained sediments was constructed using data from all of the wells that penetrated fine-grained units.

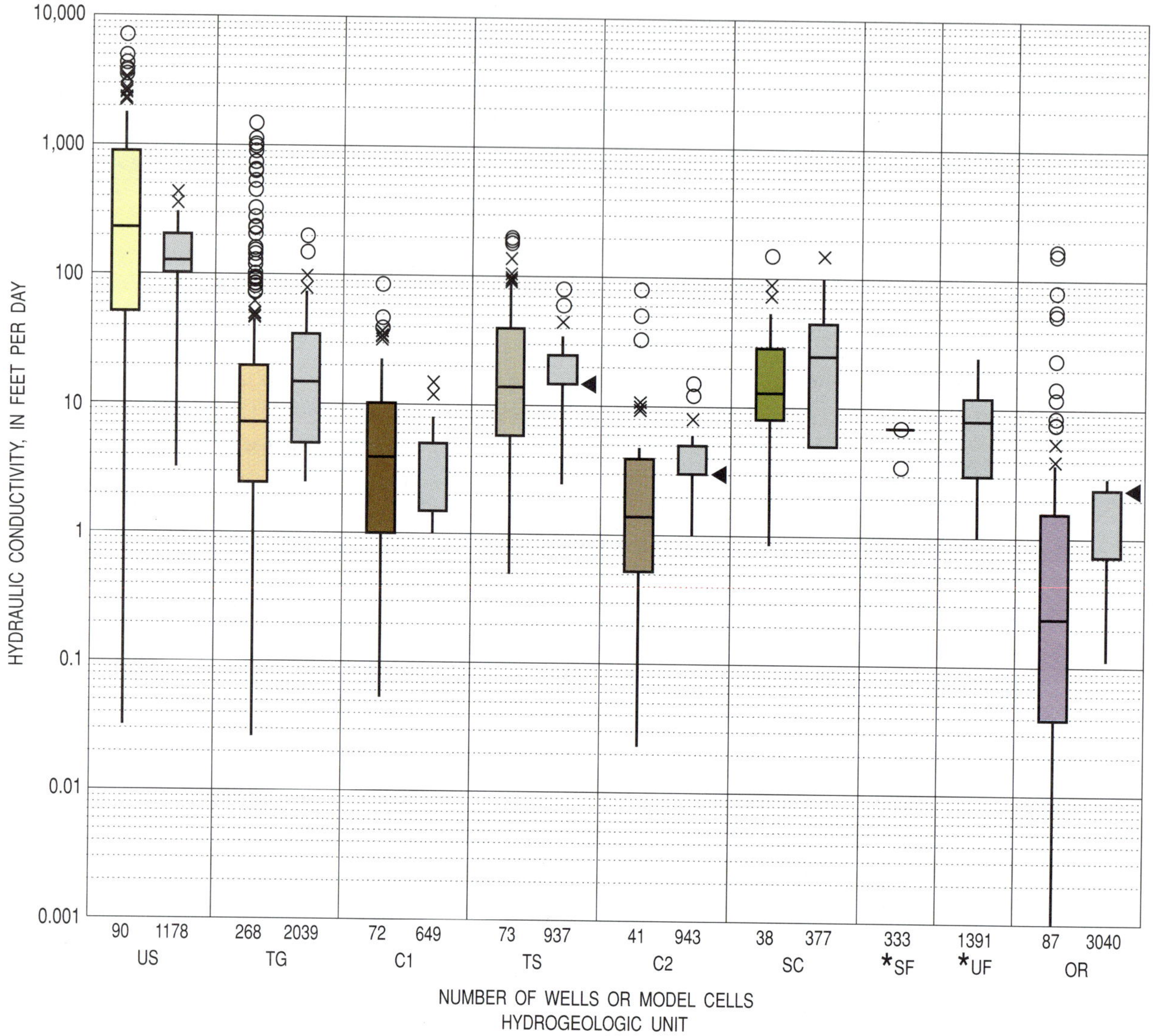

EXPLANATION

Interquartile range equals the value of the 75th percentile minus the value of the 25th percentile.

- ○ More than 3 times the interquartile range from the 75th percentile value
- × 1.5 to 3 times the interquartile range from the 75th percentile value
- Less than 1.5 times the interquartile range from the 75th percentile value
- 75th percentile value
- ◀ Median value
- 25th percentile value
- Less than 1.5 times the interquartile range from the 25th percentile value

Data from wells — Data from model calibration

HYDROGEOLOGIC UNIT

US	Unconsolidated sedimentary aquifer
TG	Troutdale gravel aquifer
C1	Confining unit 1
TS	Troutdale sandstone aquifer
C2	Confining unit 2
SC	Sand and gravel aquifer (upper subunit)
SF	Sand and gravel aquifer (lower subunit)
UF	Undifferentiated fine-grained sediments
OR	Older rocks

* Note: Data from model calibration only

Figure 6. Distributions of hydraulic conductivity for hydrogeologic units. (Data are from specific-capacity and aquifer tests on wells and from model calibration.)

Therefore, the initial areal distribution of hydraulic conductivity, derived by determining the percentage of fine-grained sediments at each model cell and applying the above relation, was the same for each of the fine-grained units (confining units 1 and 2, the lower sub-unit of the sand and gravel aquifer, and the undifferentiated fine-grained sediments).

Hydraulic conductivity of the older rocks also was estimated from specific-capacity data; however, the areal distribution of data was insufficient to allow contouring of hydraulic conductivity. A subsurface geologic map of the older rocks subsystem (McFarland and Morgan, 1996) was used to delineate zones of differing hydraulic conductivity within the older rocks. First, the mean hydraulic conductivity of each geologic unit composing the older rocks subsystem was computed from available specific-capacity and aquifer-test data. Geologic units included within the older rocks subsystem are the Columbia River Basalt Group, Rhododendron Formation, Skamania Volcanics, Goble Volcanics, and the basalts of Waverly Heights (McFarland and Morgan, 1996). Second, the predominant geologic unit within each model cell was determined. Finally, each cell was assigned the mean hydraulic conductivity of the predominant geologic unit in the cell.

Estimates of hydraulic conductivity for some units were modified during the calibration process to achieve a best fit between simulated and observed data. The postcalibration distributions of hydraulic conductivity for all units are shown on plates 3–8.

Vertical Hydraulic Conductivity

In an aquifer system, the hydraulic conductivity varies with direction. In sedimentary aquifer systems, the horizontal and vertical hydraulic conductivity can differ by orders of magnitude. Hydraulic conductivity is typically greatest in the horizontal direction because that is the usual orientation of sediment particles and layers during deposition.

Field measurement of vertical hydraulic conductivity is difficult and expensive, requiring carefully designed multiple-well aquifer tests using observation wells completed at multiple depths. This type of test was not within the scope of this study, and no other test data were found to allow this type of analysis. Laboratory measurement of vertical hydraulic conductivity can be made on core samples; however, the potential for sample disturbance and the very small part of the sediment represented severely limit the utility of these data. Because of these problems, vertical hydraulic conductivity generally is estimated through the model calibration process.

The hydraulic characteristic required by MODFLOW to define the capability of adjacent hydrogeologic units to transmit ground water vertically is called the vertical conductance (*VCONT*). Vertical conductance values must be specified between all adjacent model layers and typically are calculated as:

$$VCONT_{k,k+1} = \frac{2\left(\frac{Kh_k}{An_k}\right)\left(\frac{Kh_{k+1}}{An_{k+1}}\right)}{b_{k+1}\left(\frac{Kh_k}{An_k}\right) + b_k\left(\frac{Kh_{k+1}}{An_{k+1}}\right)}, \quad (5)$$

where

$VCONT$ = vertical conductance T^{-1};

Kh = horizontal hydraulic conductivity (LT^{-1});

An = vertical anisotropy ratio, horizontal to vertical (dimensionless);

b = thickness (L); and

k = model layer number.

In the model constructed for this study, vertical anisotropy ratios of hydraulic conductivities (horizontal to vertical) were estimated for each hydrogeologic unit from published values for similar classes of sediments. These ratios were modified during calibration. The initial ratios were 100:1 for all sediments and 1,000:1 for older rocks. Vertical anisotropy ratios were assumed to be constant within a unit; however, vertical conductance values varied because of variations in thickness and horizontal hydraulic conductivity within each unit.

Conductances of Head-Dependent Flux Boundaries

Ground-water flux across each of the head-dependent boundaries (rivers, streams, and drains) is calculated as a piecewise linear function of the difference between the water level in the aquifer and the water level at the boundary, and the conductance of that boundary. The conductance represents the ease with which ground water can flow across riverbeds, streambeds, or seepage faces (springs) (McDonald and Harbaugh, 1988).

Head-dependent boundary conductances are analogous to the vertical conductance between model layers described above; however, boundary conductances are adjusted external to the model to account for the area of a cell that is covered by the boundary, whereas vertical conductances are adjusted internally. The generalized boundary conductance equation is:

$$COND = \frac{KA}{b} , \tag{6}$$

where

$COND$ = boundary conductance (T^{-1});

K = hydraulic conductivity of the boundary normal to the direction of flow (LT^{-1});

b = length of the flow path (thickness) (L); and

A = area of boundary normal to the flow path within the cell (L^2).

Initial values of riverbed and streambed conductance were determined by assuming that the beds had the same hydraulic conductivity (K) as the underlying hydrogeologic units. The area (A) of each river or stream cell that was covered by water was estimated from 7.5-minute USGS topographic maps. Riverbed and streambed sediments were assumed to have thickness (b) of 5 ft and 2 ft, respectively, throughout the model.

Initial values of drain conductance also were determined by assuming that drains had the same hydraulic conductivity as the hydrogeologic units within the cells. The area (A) of the boundary normal to the flow path was estimated to be 3,000 ft (one cell width) wide by one-half the saturated thickness of the cell high. The length of the flow path was assumed to be 1,500 ft (one-half cell width). Initial estimates of boundary conductance were modified during calibration.

Calibration of the Model

Procedure

The first step in the calibration process was to analyze model sensitivity to changes in parameters and boundary conditions. The purpose of the analysis was to gain an understanding of the effects of uncertainty in hydraulic characteristics and boundary conditions on the heads and fluxes simulated by the model.

Using results of the sensitivity analysis as a guide, the model was calibrated to time-averaged conditions for the period April 1987 through March 1988. Time-averaged data were used for the calibration because there were no data available for any period in which the ground-water system was in a true steady state. Calibration to transient conditions was ruled out because of the lack of information on long-term or seasonal changes in stresses such as pumpage and recharge and the problem of specifying initial conditions.

When using the time-averaged, ground-water-flow equation (equation 2), if changes in storage occur in the ground-water flow system during the time-averaging interval, the rate of change in storage must be added to the right-hand side of the equation. Changes in ground-water storage occurred in some areas of the basin during the 1987–88 calibration period, as evidenced by changes in water levels mapped by McFarland and Morgan (1996, pl. 7). Water-level changes in unconfined aquifers can cause large changes in ground-water storage because water is lost from or taken into storage at a rate proportionate to the specific yield of the aquifer. Specific-yield values are similar in magnitude to porosity and generally range between 0.1 and 0.3. Thus, for a 5 ft/yr (feet per year) decline in ground-water level in an unconfined aquifer with a specific yield of 0.2, the rate of storage depletion would be 1 ft/yr, or about 0.3 ft^3/s within the area of a model cell. This rate is similar in magnitude to estimated recharge rates in the basin (pl. 2).

The rate of change in storage in the unconfined aquifers of the basin was estimated for the calibration period by determining the mean water-level change at each model cell by using maps of changes in the unconsolidated sedimentary aquifer and the Troutdale gravel aquifer compiled by McFarland and Morgan (1996, pl. 7). Estimates of specific yield for lithologies similar to those of the unconsolidated sedimentary aquifer and Troutdale gravel aquifer were taken from Johnson (1967). A value of 0.2 was selected for the unconsolidated sedimentary aquifer for sediments ranging from gravel and sand to silty sand and sandy silt. A value of 0.1 was selected for the Troutdale gravel aquifer due to the abundance of cemented sands and gravels in the unit. Ground-water storage was depleted by 12,000 acre-feet within the areas where water levels declined and was replenished by 50,000 acre-feet within areas where water levels increased. The net volume of change in storage was a gain of 38,000 acre-feet, which amounts to only 3 percent of total recharge to the modeled area.

Storage change estimates could be made only for the areas of the basin where water-level changes were mapped by McFarland and Morgan (1996). The areas mapped included approximately 80 percent of the modeled region and all of the areas where ground-water levels would be most affected by pumping. The estimated rates of change in storage in the unconfined parts of the aquifer system are shown in figure 7; these rates were specified fluxes during model calibration. Errors in the estimated water-level changes or in the specific yield of the aquifers would contribute to errors in estimated rates of storage change. These errors could be large if insufficient water-level data were available to define changes or if specific yields varied greatly from estimated values.

Storage changes in confined parts of the aquifer system were negligible because the storage coefficients of confined aquifers are at least two orders of magnitude less than the specific yield of unconfined aquifers. For example, in a confined aquifer with a storage coefficient of 0.001, a water-level decline of 5 ft/yr would result in a rate of storage depletion of only 0.005 ft/yr, or about 0.001 ft^3/s within the area of a model cell. This rate is insignificant compared with typical recharge rates of 1.5 ft/yr (0.428 ft^3/s) or typical well-discharge rates of 200 gallons per minute (0.446 ft^3/s).

Model calibration also includes the process of adjusting the hydraulic characteristics and boundary conditions of the model until the head and flux conditions simulated by the model agree, within acceptable limits, with observations in the real system. Parameter adjustment is an iterative process that can be accomplished either manually, by using a trial-and-error procedure, or automatically, by using regression techniques (Cooley and Naff, 1990). Parameters in the model constructed for this study were adjusted manually by systematically adjusting parameters, making simulations, and comparing simulated with observed ground-water levels and flow conditions. Approximately 70 model simulations were made in which horizontal hydraulic conductivity and vertical anisotropy ratios were modified from the original estimates. Streambed, riverbed, and drain-conductance values also were recomputed whenever values of horizontal and vertical hydraulic conductivity were modified. Recharge, rates of storage change, and well pumping rates were not modified during the calibration process.

Transmissivity, the product of horizontal hydraulic conductivity and saturated thickness, was read directly by MODFLOW after being computed externally. Only horizontal hydraulic conductivity values were varied during the calibration process; saturated thickness was held constant. Transmissivity also was assumed to remain constant within each simulation and did not change as a function of saturated thickness.

Parameter Sensitivity

A sensitivity analysis of the model was made in order to determine the relative sensitivity of model simulated heads and fluxes to uncertainty in hydraulic characteristics and boundary conditions. The relative sensitivities of these parameters were used as a guide to calibration of the model and later to help determine which types of data would be most useful to improve the ability of the model to simulate the ground-water flow system and thus should have priority for future collection.

The analysis was performed on the uncalibrated model with initial estimates of hydraulic characteristics and initial boundary conditions as a base condition. Individual parameters were changed and effects of the changes on model results were calculated after each simulation. The parameters that were changed included horizontal hydraulic conductivity (*Kh*), vertical hydraulic conductivity (*Kv*), streambed and riverbed conductance, recharge, and boundary conditions.

Although the eastern edge of the modeled area is a no-flow boundary in the calibrated model, this boundary was simulated as a specified-head boundary during the initial stages of model development and during the sensitivity analysis. Most of this boundary coincides with the surface contact between low-permeability older rocks (Skamania Volcanics) and basin-fill sediments; the remaining part follows the eastern extent of the more permeable part of the Troutdale gravel aquifer. Later in the calibration process, this boundary condition was changed to the no-flow condition described in the section on boundary conditions. The justification supporting this change is discussed later in the report.

The sensitivity of the model to parameter changes was measured by using the root-mean square (RMS) of the difference in simulated heads. RMS values were calculated for the entire model and for each hydrogeologic unit so that the relative sensitivities of units could be compared.

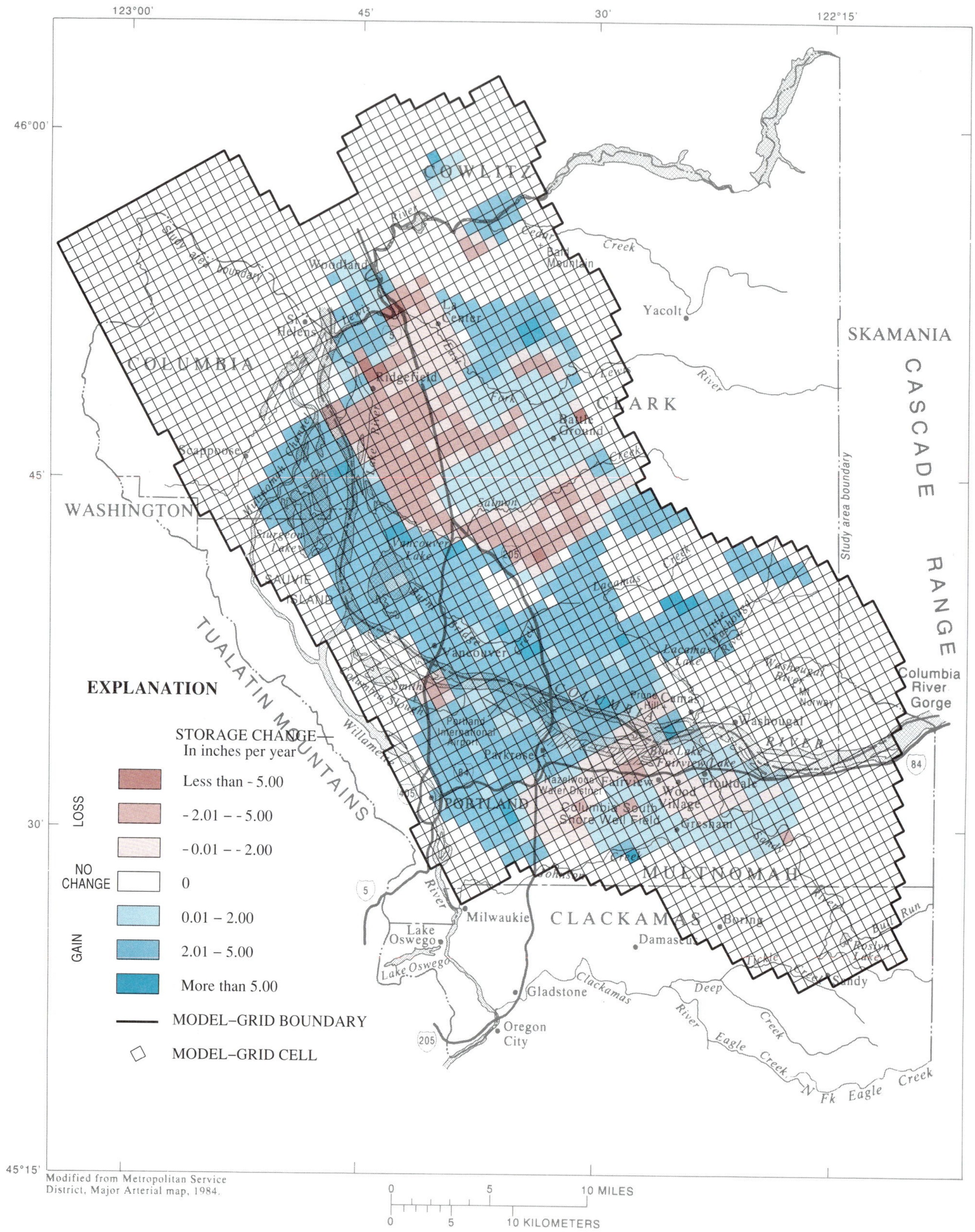

Figure 7. Estimated change in storage in unconfined parts of the ground-water flow system, 1987–88.

Horizontal Hydraulic Conductivity

Sixteen simulations were made, using four multipliers (0.01, 0.1, 10, and 100) to modify the horizontal hydraulic conductivity values in the four primary aquifer units (unconsolidated sedimentary aquifer, Troutdale gravel aquifer, Troutdale sandstone aquifer, and upper subunit of the sand and gravel aquifer). The results of these simulations, showing the sensitivity of simulated water levels to changes in horizontal hydraulic conductivity of the primary aquifer units, are shown in figure 8. The term "cumulative RMS," which is used to compare the sensitivities of units in the following discussion, is the sum of the RMS values for each of the four simulations for an individual unit.

The sensitivity of simulated water levels to changes in the horizontal hydraulic conductivity of the unconsolidated sedimentary aquifer was low (cumulative RMS values less than 100) to moderate (cumulative RMS values of 100 to 200) (fig. 8). Increases to the already high values of horizontal hydraulic conductivity in the unit had less effect on water levels than did decreases, and units at depth were less affected than shallow units. Increasing hydraulic conductivity by a factor of 100 resulted in physically unreasonable values in some areas, and the model was unable to reach a solution. The low sensitivity of simulated water levels in other units to changes in the horizontal hydraulic conductivity of the unconsolidated sedimentary aquifer and the low sensitivity of simulated water levels in the unconsolidated sedimentary aquifer to changes in the horizontal hydraulic conductivity of other units (fig. 8) are explained by the high hydraulic conductivity of the unit and the control on water levels imposed by surface-water features (stream and river boundaries) within the unit.

Changes to the horizontal hydraulic conductivity of the Troutdale gravel aquifer produced the most significant and widespread effects of any of the variations in this parameter. Simulated water levels in most units had moderate, high (cumulative RMS values of 200–300), and very high (cumulative RMS values greater than 300) sensitivity to changes in the horizontal hydraulic conductivity of the Troutdale gravel aquifer; the exception being the unconsolidated sedimentary aquifer, which had low sensitivity (fig. 8). Decreasing the horizontal hydraulic conductivity of the unit by a factor of 0.01 produced some of the largest RMS values in the entire analysis, both within the Troutdale gravel aquifer itself and in the underlying undifferentiated fine-grained sediments. The dominant control that horizontal hydraulic conductivity of the Troutdale gravel aquifer has on ground-water flow in the basin is due to the unit's broad areal distribution and exposure to recharge, and to its high permeability.

The horizontal hydraulic conductivity of the Troutdale sandstone aquifer also exerts important controls on ground-water movement in the basin as evidenced by the high sensitivity of simulated water levels in the Troutdale gravel aquifer, confining unit 1, Troutdale sandstone aquifer, and confining unit 2 (fig. 8). Simulated water levels were also disproportionately more sensitive to increases in the horizontal hydraulic conductivity in the Troutdale sandstone aquifer than they were to decreases.

The sensitivity of simulated water levels to changes in hydraulic conductivity of the upper, coarse-grained subunit of the sand and gravel aquifer was low in all units except the underlying fine-grained subunit of the sand and gravel aquifer, which had moderate sensitivity, and the coarse-grained subunit itself, which had high sensitivity (fig. 8). The depth and limited areal extent of the unit restrict the influence that changes to horizontal hydraulic conductivity have on the overall flow system.

Vertical Hydraulic Conductivity

The vertical anisotropy ratio (*Kh*/*Kv*) was varied to determine the sensitivity of the model to vertical hydraulic conductivity. Initial estimates of vertical anisotropy ratio for the uncalibrated model were 1,000:1 for the older rocks and 100:1 for all other units. The multipliers used to change vertical anisotropy ratios were 0.1, 0.5, 5, and 10, yielding ratios ranging from 100:1 to 10,000:1 in the older rocks and from 10:1 to 1,000:1 in all other units. The sensitivity of simulated water levels was calculated for all of the hydrogeologic units; however, the vertical hydraulic conductivities of confining units 1 and 2 and the undifferentiated fine-grained sediments were varied together, whereas vertical hydraulic conductivities were varied separately for each of the other units. The results of these simulations are shown for the Troutdale gravel aquifer, Troutdale sandstone aquifer, confining units and undifferentiated sediments, and the lower, fine-grained subunit of the sand and gravel aquifer in figure 9.

Within the ranges tested, the sensitivities of simulated heads to changes in vertical hydraulic conductivity were low (cumulative RMS values rarely exceeding 50) compared with those calculated for changes in horizontal hydraulic conductivity.

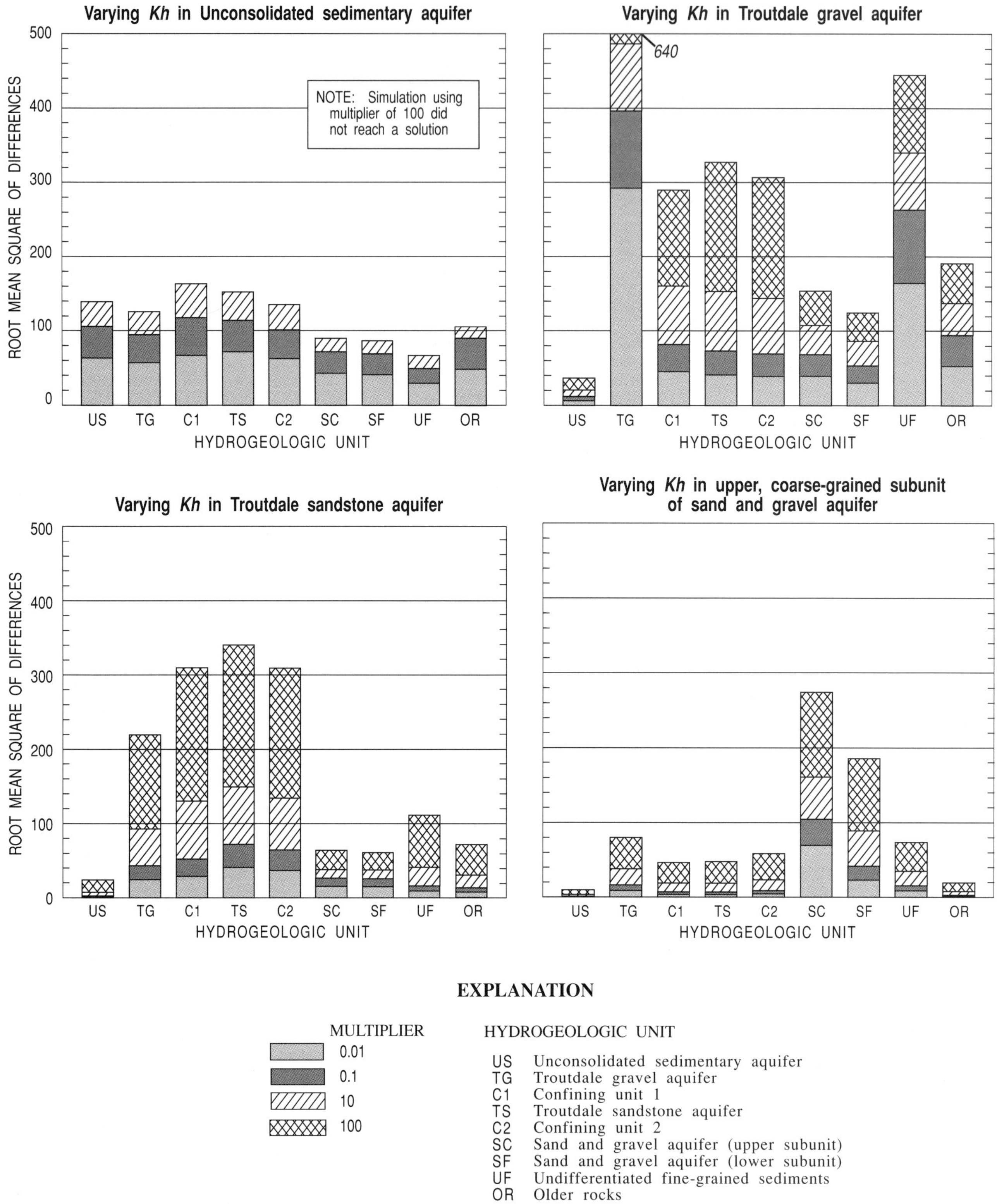

Figure 8. Sensitivity of simulated heads in different hydrogeologic units to variation in horizontal hydraulic conductivity in the unconsolidated sedimentary aquifer; Troutdale gravel aquifer; Troutdale sandstone aquifer; and upper, coarse-grained subunit of the sand and gravel aquifer.

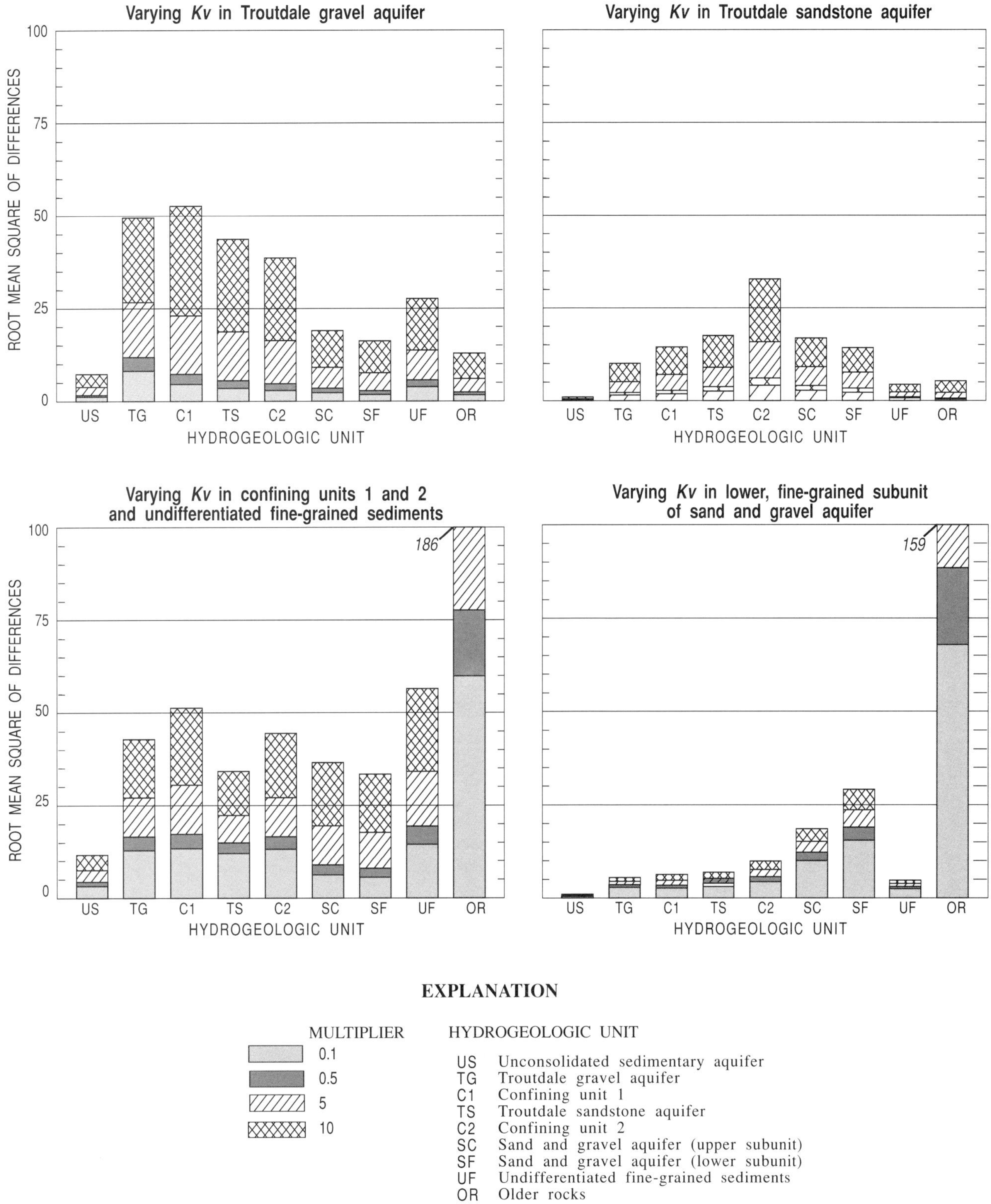

Figure 9. Sensitivity of simulated heads in different hydrogeologic units to variation in vertical anisotropy ratios in the Troutdale gravel aquifer; Troutdale sandstone aquifer; confining units 1 and 2 and undifferentiated fine-grained sediments; and lower, fine-grained subunit of the sand and gravel aquifer.

In the following discussion, the terms low, moderate, and high are used to describe the relative sensitivity of simulated water levels to changes in vertical hydraulic conductivity and correspond to cumulative RMS values of 0 to 25, 25 to 50, and greater than 50, respectively.

The sensitivity of simulated water levels to changes in the vertical hydraulic conductivity of the Troutdale gravel aquifer was high in the unit itself and in the underlying confining unit 1 (fig. 9). Moderate sensitivities were calculated for the Troutdale sandstone aquifer and undifferentiated fine-grained sediments, which also directly underlie the Troutdale gravel aquifer in some places, and confining unit 2, which underlies the Troutdale sandstone aquifer. Low sensitivity was calculated for the unconsolidated sedimentary aquifer due to the strong controls on water level imposed by surface-water features.

Decreasing the vertical hydraulic conductivity of the Troutdale sandstone aquifer had the greatest effect on simulated heads in confining unit 2 (fig. 9); this indicates that restricting the vertical conductance between the Troutdale sandstone and confining unit 2 requires much larger vertical head gradients between the two units in order to maintain the same net vertical flux. Nonetheless, sensitivity to changes in this parameter were comparatively low and showed a tendency to attenuate with vertical distance from the unit.

The vertical hydraulic conductivities of confining units 1 and 2 and the undifferentiated fine-grained sediments strongly influence water levels in nearly all units (fig. 9). Calculated sensitivities were moderate to high in all units except the unconsolidated sedimentary aquifer. This result can be attributed to the wide extent of these units and their relatively low vertical hydraulic conductivities. The extremely high cumulative RMS of the older rocks (186) resulted because that unit receives large quantities of recharge where it is exposed in the higher altitudes of the basin, and discharges by upward leakage in the lower parts of the basin, where it is overlain by the undifferentiated fine-grained sediments. Changes to the vertical hydraulic conductivity of the overlying unit in a discharge area greatly influence the water-level gradient required to maintain the same flux.

The effects of changes in the vertical hydraulic conductivity of the lower, fine-grained subunit of the sand and gravel aquifer were less pronounced for most units (fig. 9) because of the limited extent of the unit (Appendix A, fig. A–7). However, effects on water levels in the older rocks were again large (RMS of 159) and support the idea that water levels within discharge areas of the model are very sensitive to changes in vertical hydraulic conductivity within discharge areas.

Recharge

The sensitivity of simulated heads to changes in recharge rate was determined by multiplying each of the three components of total recharge (infiltration of precipitation, runoff into drywells, and on-site waste-disposal systems) by a constant for each sensitivity simulation. The RMS differences for the drywell and on-site waste-disposal system recharge simulations were relatively low (less than 5 ft) because these are important sources in only a small part of the modeled area. This makes it difficult to compare the sensitivity of these parameters with the sensitivities of regionally distributed parameters. Snyder and others (1994) determined that drywells and on-site waste-disposal systems contribute a large part of the recharge to the ground-water system in highly urbanized, unsewered areas. In those areas, simulated heads were more sensitive to variations in recharge from these sources.

Recharge from infiltration of precipitation occurs over nearly the entire modeled area, and thus variation of this parameter provides a measure of model sensitivity that can be compared with sensitivities related to aquifer properties. In three simulations, recharge was decreased by 50 percent and increased by 50 and 100 percent, respectively. Changes in recharge of this magnitude had significant effects on simulated heads in all units except the unconsolidated sedimentary aquifer (fig. 10). These effects were notably greater in the shallower units through which most of the recharge moves before being discharged and in the lower permeability units where heads are more sensitive to flux.

Streambed and Riverbed Conductance

Streambed and riverbed conductance values were changed together because each was estimated using the assumption that the conductance of the bed was a function of the vertical hydraulic conductivity of the underlying hydrogeologic unit. Increases in conductance values had little effect on simulated heads (fig. 10), indicating that initial values were high enough to not limit the flux across these boundaries.

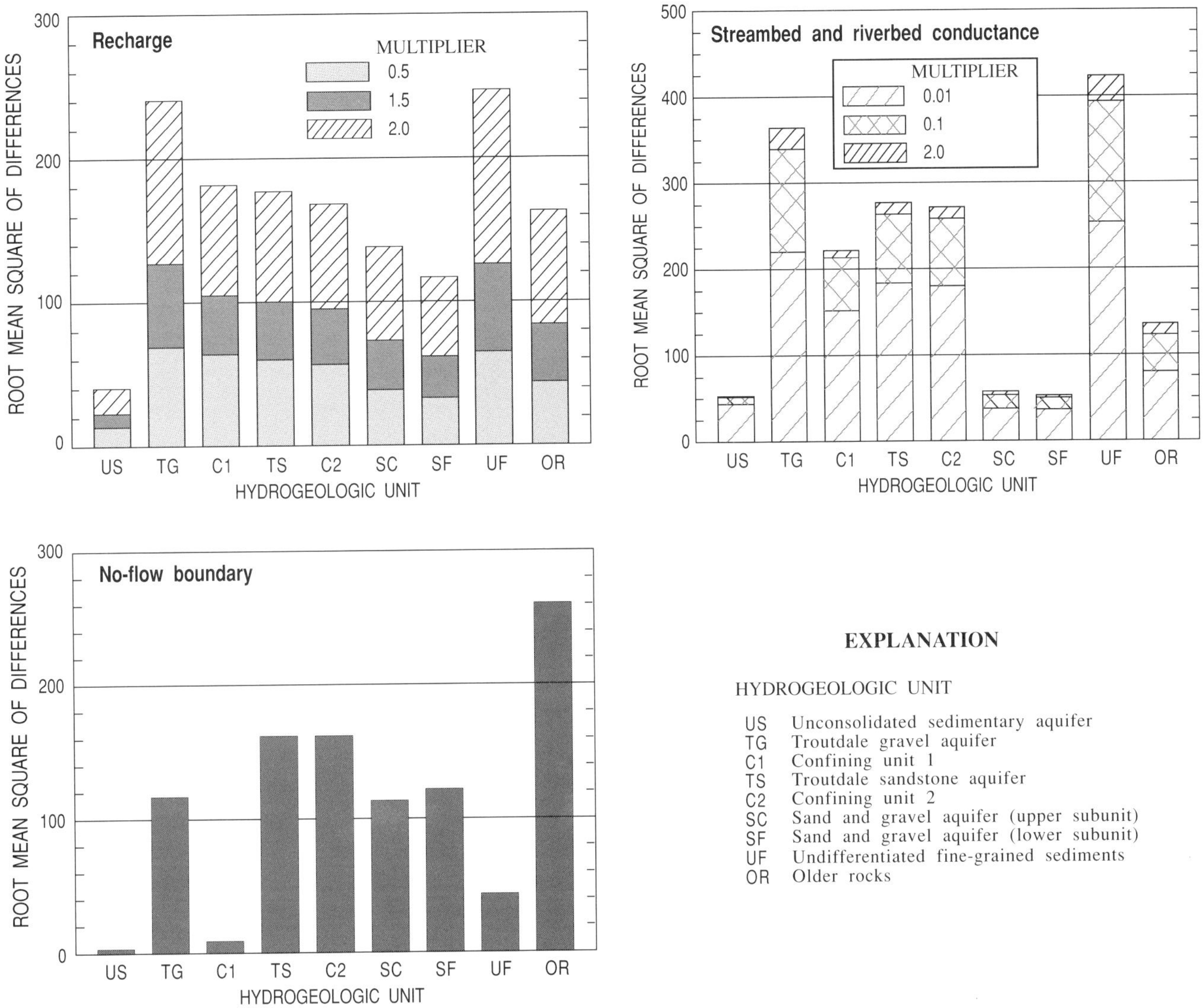

Figure 10. Sensitivity of simulated heads in different hydrogeologic units to variation in recharge from infiltration, variation in streambed and riverbed conductance, and conversion from constant-head to no-flow boundary condition on eastern edge of model.

Simulated heads did show sensitivity to reductions in conductance of one and two orders of magnitude. These conditions forced the model to produce larger head gradients near streams and rivers in order to generate the same discharge to these head-dependent flux boundaries.

Boundary Conditions

Model sensitivity to the boundary condition specified at the eastern edge of the model was evaluated by changing the boundary condition from specified-head to no-flow. Units that had specified-head cells had high RMS differences in this sensitivity simulation (fig. 10). There was much uncertainty in the values of the specified heads because there were few wells in the east boundary area to use for control; thus, large head differences between the specified-head and no-flow simulations do not necessarily indicate that use of a no-flow boundary resulted in a less accurate simulation of the ground-water flow system.

Parameter Adjustment

Initial estimates of hydraulic conductivity were adjusted during calibration in three significant ways: (1) distributions were smoothed within zones to reflect the average cell hydraulic conductivities rather than point values at wells, (2) hydraulic conductivity values of the unconsolidated sedimentary aquifer were reduced by 85 percent modelwide, and (3) hydraulic conductivity values of the sedimentary units overlying a northwest-southeast trending anticline in Clark County were reduced. The distributions of hydraulic conductivity values obtained by calibration are shown on plates 3–8.

Statistical distributions of the calibrated hydraulic conductivity values in the model were compared with the estimates of hydraulic conductivity values at wells by hydrogeologic unit, as shown in figure 6. The estimates, based on well data, have ranges of three to five orders of magnitude within a hydrogeologic unit. Smoothing of the initial distributions and interpolation of the well data to the model grid resulted in narrowing the ranges of hydraulic conductivity distributions for most units to two to three orders of magnitude. This reduction is reasonable considering (1) that lithologies can vary widely within small areas in natural geologic environments and (2) that the model-cell values must represent the mean value for a very large volume of aquifer material compared to the volume represented by an individual well.

Initial estimates of the hydraulic conductivity of the unconsolidated sedimentary aquifer were biased toward high values by estimating them using well-test data. The aquifer units in the basin consist of heterogeneous sequences of coarse- and fine-grained sediment. Wells typically are screened in the most permeable parts of the unit; therefore, the hydraulic conductivity values estimated from these tests are not accurate estimates of the average hydraulic conductivity of the entire unit. Additionally, many of the wells for which test data were available were near the Columbia River and completed in permeable sands and gravels. The median of initial estimates of hydraulic conductivity in the unconsolidated sedimentary aquifer was 232 ft/d. Reduction of hydraulic conductivity in this unit resulted in more reasonable values with a median of 125 ft/d.

An especially difficult area for the model to simulate was southwestern Clark County. The difficulty stemmed from the sharp change in horizontal hydraulic gradient (fig. 2) that occurs along a northwest-southeast trending line extending from near Ridgefield to Prune Hill (McFarland and Morgan, 1996). Gradients along this trend average 50–100 ft/mi (feet per mile) in contrast to gradients of only 15–25 ft/mi to the northeast and southwest. The trend of this line is coincident with an anticline in the underlying older rocks (pl. 1). Swanson and others (1993) suggested that this deformation in the older rocks might have occurred after deposition of the younger sediments, thus displacing them upward in the section. If the younger sediments were deformed along the axis of this anticline, movement of ground water through this section could be along paths that are not parallel to the bedding planes and the principal axis of horizontal hydraulic conductivity. The effective hydraulic conductivity might lie between the horizontal and vertical values. On the basis of this concept, horizontal hydraulic conductivities of the younger sediments were decreased to one-half their initial values along a zone coincident with the anticlinal structure. This modification produced a much improved representation of the water-level distribution in this area. The effect of this modification on the median hydraulic conductivities of each unit were minor.

The general relations between the initial estimates of hydraulic conductivity values were preserved in the calibrated distributions (fig 6). As discussed above, the median hydraulic conductivity of the unconsolidated sedimentary aquifer unit was reduced from about 230 ft/d to about 125 ft/d. The median values of hydraulic conductivity of the Troutdale gravel aquifer, Troutdale sandstone aquifer, and upper, coarse-grained subunit of the sand and gravel aquifer increased slightly from precalibration values and remained the most permeable units, aside from the unconsolidated sedimentary aquifer. Hydraulic conductivity distributions are typically log-normally distributed (Domenico and Schwartz, 1990); thus, mean and median values can be quite different.

The vertical anisotropy ratio was initially estimated to be 1,000:1 in older rocks and 100:1 in all other units. This parameter was varied during calibration until a best fit was obtained between observed and simulated direction and magnitude of vertical hydraulic gradient. The vertical anisotropy ratios determined from calibration were 1,000:1 in the older rocks and all of the fine-grained units (confining units 1 and 2, the lower fine-grained subunit of the sand and gravel aquifer, and the undifferentiated fine-grained sediments). In the primary aquifer units (unconsolidated sedimentary aquifer, Troutdale gravel aquifer, Troutdale sandstone aquifer, and upper, coarse-grained subunit of the sand and gravel aquifer), which have significant interbedding, the initial ratio of 100:1 yielded the best results.

Boundary conditions on the eastern margin of the model also underwent modification during the calibration process. In the early conceptualization of the ground-water flow system, it was hypothesized that some quantity of subsurface inflow occurs from the older rocks to the sedimentary units along this boundary. The amount of inflow was uncertain, however, because of limited data on the hydraulic gradient in the older rocks and their hydraulic conductivity. A rough estimate of inflow was made using Darcy's law:

$$Q = -KhIA, \qquad (7)$$

where

Q is inflow ($L^3 T^{-1}$);

Kh is horizontal hydraulic conductivity (LT^{-1});

I is hydraulic gradient (dimensionless); and

A is the cross-sectional area of flow (L^2).

Assuming a hydraulic gradient of 100 ft/mi (0.02), a hydraulic conductivity of 1.0 ft/d (1.2×10^{-5} ft/s [feet per second]), and a 2,500-ft-thick flow section 50 miles long (6.6×10^{8} ft^2), the inflow across the boundary would be:

$$Q = (1.2 \times 10^{-5})\,(0.02)\,(6.6 \times 10^{8}) = 160\ ft^3/s. \qquad (8)$$

This estimate is uncertain and is presented here only to indicate the potential magnitude of inflow.

Heads were initially specified in cells on the eastern boundary so that inflow across the boundary could be estimated using the model. This approach assumed that the values of the specified heads could be estimated accurately and that simulated heads in the interior of the model approximate observed heads. Estimating heads in the boundary cells proved to be very difficult because of the small number of wells in the remote, upland areas; available head data had to extrapolated horizontally to assign heads to all cells on the boundary and vertically to assign heads to cells in the underlying layers of the model.

During parameter adjustment, it was found that the model was insensitive to relatively large changes in some parameters. Numerical experiments by Franke and Reilly (1987) on the effects of boundary conditions on steady-state models show that specified-head boundaries can make models insensitive to the hydraulic characteristics of the ground-water flow system. This can lead to introduction of large errors in the hydraulic characteristics of the aquifer system if they are modified during model calibration when attempting to make simulated heads match observed heads.

A no-flow boundary condition was adopted on the eastern boundary because of the potentially small amount of inflow across the boundary (160 ft^3/s, which is approximately 10 percent of total recharge), the large uncertainty in estimating heads on the boundary, and the potential for introducing errors into the hydraulic characteristics through calibration.

Model Evaluation and Results

The ability of the model to simulate the ground-water flow system was evaluated by comparing simulated water levels and fluxes with measured water levels and fluxes.

Ground-Water Levels

Observed ground-water levels used for comparison with simulated water levels were from 412 wells completed in the four primary aquifer units in the basin (unconsolidated sedimentary aquifer, Troutdale gravel aquifer, Troutdale sandstone aquifer, and coarse-grained upper subunit of the sand and gravel aquifer). Spring 1988 water levels were used as an approximation of the mean water levels for the calibration period. The validity of using spring 1988 water levels for comparison with simulated mean water levels was checked by comparing annual mean water levels (1987–88) and spring 1988 water levels from the 156 observation wells in the basin that were measured bimonthly between spring 1987 and 1988 (McCarthy and Anderson, 1990). It was found that the mean difference between annual mean and spring 1988 water levels was only 0.4 ft. Most observation wells were screened in only one hydrogeologic unit. Because observation wells were rarely located at the center of a model cell (where the simulated water level is most applicable), the model-simulated water level at each observation well was estimated by interpolation from the simulated water levels in the four adjacent model cells. The goodness of fit between observed and simulated water levels was statistically summarized for each calibration run by computing the RMS of the difference between observed and simulated water levels for each of the four primary aquifer units.

The simulated water level at each cell represents the water level at the center of the cell. Some model cells represent thick sequences of the ground-water flow system in which there are vertical hydraulic gradients. Observation wells are commonly water-supply wells and are open to only a part of the hydrogeologic unit; water levels from these wells do not represent the average water level in the unit. Although vertical interpolation of simulated water levels is possible, it was not done in this analysis because of the small number of wells where it would have been appropriate.

The general directions of ground-water movement and magnitude of horizontal hydraulic gradients simulated within the unconsolidated sedimentary aquifer (pl. 3) agreed well with observed water levels. The most notable exception to this agreement was in a 15 mi^2 area of Clark County south of Salmon Creek and north of Burnt Bridge Creek in the northwest part of T.2N., R.2E. and the northeast part of T.2N., R.1E., where anomalously high heads occur in the unconsolidated sedimentary aquifer (McFarland and Morgan, 1996, pl. 5). Observed water levels in wells in this area are at 200–250 ft above sea level, which is 50–100 ft higher than the regional water table in the area. McFarland and Morgan (1996) suggested that the apparent "mound" in the water table might be caused by high recharge rates from drywells and on-site waste-disposal systems or low permeability sediments within the unconsolidated sedimentary aquifer that impede downward leakage in this area. Snyder and others (1994) estimate recharge from drywells in the area of about 7–9 in/yr, recharge from on-site waste-disposal systems of 1–2 in/yr, and total recharge of about 18–22 in/yr; the total recharge rates are slightly more than in the surrounding area, but are not high enough to explain such a localized anomaly.

Well logs from the area describe beds of "clay" and "sandy clay" within the unconsolidated sedimentary aquifer, and in Mundorff's (1964) description of ground-water occurrences in the area, he reported that these low-permeability beds trap recharge in perched aquifers above the regional water table. It is likely that the water-level anomaly in the area is not a recharge-induced mound, but is a perched aquifer or zone of high downward vertical head gradient caused by low-permeability beds within the unconsolidated sedimentary aquifer. Further evidence that the head anomaly in the unconsolidated sedimentary aquifer does not represent a mound in the regional water table is offered by the fact that there is no similar feature in the underlying Troutdale gravel aquifer.

The heads simulated by the regional model represent the average head within the unconsolidated sedimentary aquifer. The difference between simulated and observed heads in the unconsolidated sedimentary aquifer was less than 30 ft at 86 percent of the 63 observation wells used for comparison (fig. 11). The RMS difference for the unit was 54 ft and the mean error was 11 ft. All of the observation wells used by McFarland and Morgan (1996) to construct head maps for the primary aquifers were used to compute differences of simulated and observed head presented in figures 11. A small group of differences for the unconsolidated sedimentary aquifer shown in figure 11 fall in the 110 to 210 ft range; these differences are for wells in the area where the existence of large vertical gradients or perched zones within the unconsolidated sedimentary aquifer are hypothesized. The observations from these wells were retained in the group used for comparison with model results in order to maintain consistency between this and previous reports (McFarland and Morgan, 1996; Swanson and others, 1993).

Ground-water movement within the Troutdale gravel aquifer, as indicated from the simulated water-level contours on plate 4, is generally from the major recharge areas in the upper Sandy and Clackamas River Basins in the southeast and the western slope of the Cascade Range in Clark County toward the principal discharge areas in the basin. East of the Sandy River, ground water moves to the west and northwest toward the Sandy River canyon and north toward the Columbia River. From upland recharge areas southeast of Sandy, movement is generally to the northwest; however, flowlines diverge as they move downgradient toward the lower Clackamas, Willamette, Columbia, and Sandy Rivers. In Clark County, ground-water movement is strongly influenced by streams and rivers such as Salmon Creek and the East Fork Lewis River. Simulated flow directions and gradients (pl. 4) agree well with those mapped by McFarland and Morgan (1996, pl. 3).

In the Troutdale gravel aquifer, simulated heads were within 30 ft of observed heads at 52 percent of the 243 wells and within 50 ft at 73 percent of the observation wells (fig. 11). The RMS difference for the unit was 63 ft and the mean difference was –3.3 ft. The largest head differences in the Troutdale gravel aquifer were for observation wells located in the southern part of the study area at altitudes above 500 ft. Many of the wells with large differences are completed in the Boring Lava, an intrusive basalt included in the Troutdale gravel aquifer (McFarland and Morgan, 1996).

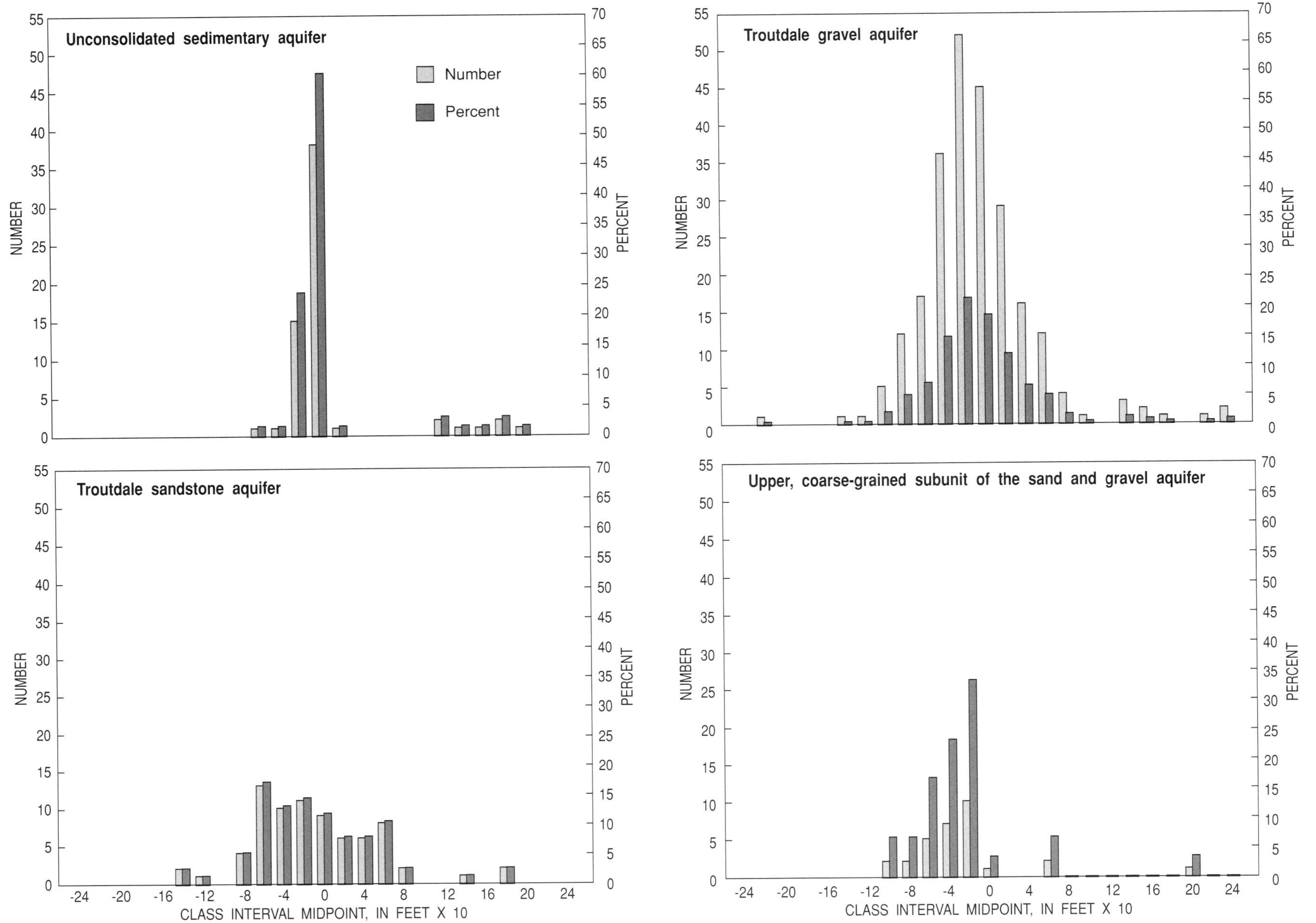

Figure 11. Frequency of the difference between observed and simulated water levels at selected observation wells in the unconsolidated sedimentary aquifer; Troutdale gravel aquifer; Troutdale sandstone aquifer; and upper, coarse-grained subunit of the sand and gravel aquifer.

According to Hogenson and Foxworthy (1965), the regional water table lies below the base of most of the Boring Lava, and any saturated zones within it are perched above old soil horizons or dense, low-permeability flow centers.

The Troutdale sandstone aquifer receives most of its recharge by downward leakage from the Troutdale gravel aquifer through confining unit 1; the principal recharge areas for the Troutdale gravel aquifer are along its updip edge between Camas and Battle Ground in Clark County, in the western Cascade Range east of the Sandy River, and in the upper Sandy and Clackamas River drainages near Sandy. Movement through the unit is northwestward from Sandy, with flow diverging downgradient westward toward the Willamette River and northward toward the Columbia River. Simulated water-level contours on plate 5, from which horizontal directions of flow can be inferred, indicate that much of the discharge from the Troutdale sandstone aquifer is by horizontal flow into the unconsolidated sedimentary aquifer above Blue Lake, where the Columbia River has truncated the Troutdale sandstone aquifer (pl. 1), and by upward flow into the unconsolidated sedimentary aquifer between Blue Lake and Portland International Airport, where the Troutdale sandstone aquifer is continuous beneath the Columbia River. There were 76 wells in the Troutdale sandstone aquifer for which observed heads were available for comparison with simulated heads. Thirty-four percent of the head differences in the Troutdale sandstone aquifer were less than 30 ft, 55 percent were less than 50 ft, and 83 percent were less than 70 ft (fig. 11). The RMS difference for the unit was 74 ft and the mean difference was –2.8 ft.

Observed head data were available from 30 wells in the upper, coarse-grained subunit of the sand and gravel aquifer for comparison with simulated heads (pl. 6). Of the differences between simulated and observed heads, 37 percent were less than 30 ft, 60 percent were less than 50 ft, and 84 percent were less than 70 ft; the RMS difference for the unit was 61 ft and the mean difference was –27 ft (fig. 11).

The RMS and mean differences for the older rocks were 138 and 31 ft for 50 observation wells. The undifferentiated fine-grained sediments had RMS and mean differences of 92 and –46 for 50 observation wells.

Vertical Head Gradients

Vertical gradients in simulated heads showed the presence of local and regional ground-water flow systems in the basin. Local ground-water flow systems generally coincide with small surface-water drainages, where recharge occurs in the upland parts of the drainage, moves through the upper 100–200 ft of the saturated sediments, and is discharged to small streams, seepage faces, or springs at low altitudes in the drainage. The extent of these local ground-water flow systems within the simulated ground-water flow system of the basin was analyzed by computing the vertical water-level gradient between the water-table unit (model layer 1) and the underlying unit (model layer 2). The resulting gradients were used to create a map in which each model cell was shaded according to the direction and magnitude of the simulated vertical head gradient in the shallow part of the system. The relation of ground-water discharge areas to surface-water drainages can readily be seen from the patterns of upward flow (fig. 12).

The extent of the regional ground-water flow systems and the discharge areas for these systems was analyzed by computing the vertical water-level gradient between the water-table unit (model layer 1) and deeper units (model layer 5). The shaded model-cell map in figure 13, which was created using the simulated vertical gradients, shows the location of recharge and discharge areas for the regional ground-water flow system. Comparison of figures 12 and 13 reveals several important features of the ground-water flow system: (1) compared with regional ground-water flow systems, local systems are smaller and more dispersed within the basin, (2) local systems are controlled by local topography and have shorter flow paths from recharge areas to discharge areas than do regional systems, (3) regional systems are broad and the boundaries between their recharge and discharge areas are well defined, and (4) regional discharge areas are controlled by major drainages in the basin, such as the Columbia, Willamette, Lewis, Clackamas, Sandy, and East Fork Lewis Rivers.

Discharge areas for local ground-water flow systems may overlie recharge areas for regional systems, as they do in the upper parts of the Lacamas, Deep, and Tickle Creek Basins (figs. 12 and 13). In the lower parts of the regional ground-water flow system, downward gradients often occur in the shallow part of the system and upward gradients in the deeper system, with gradient reversals occurring in the intermediate zone. In some areas, such as the northern half of T.1N., R.2E. and southern one-half of T.2N., R.2E., high downward gradients exist between the shallow aquifers overlying deeper aquifers that have upward gradients.

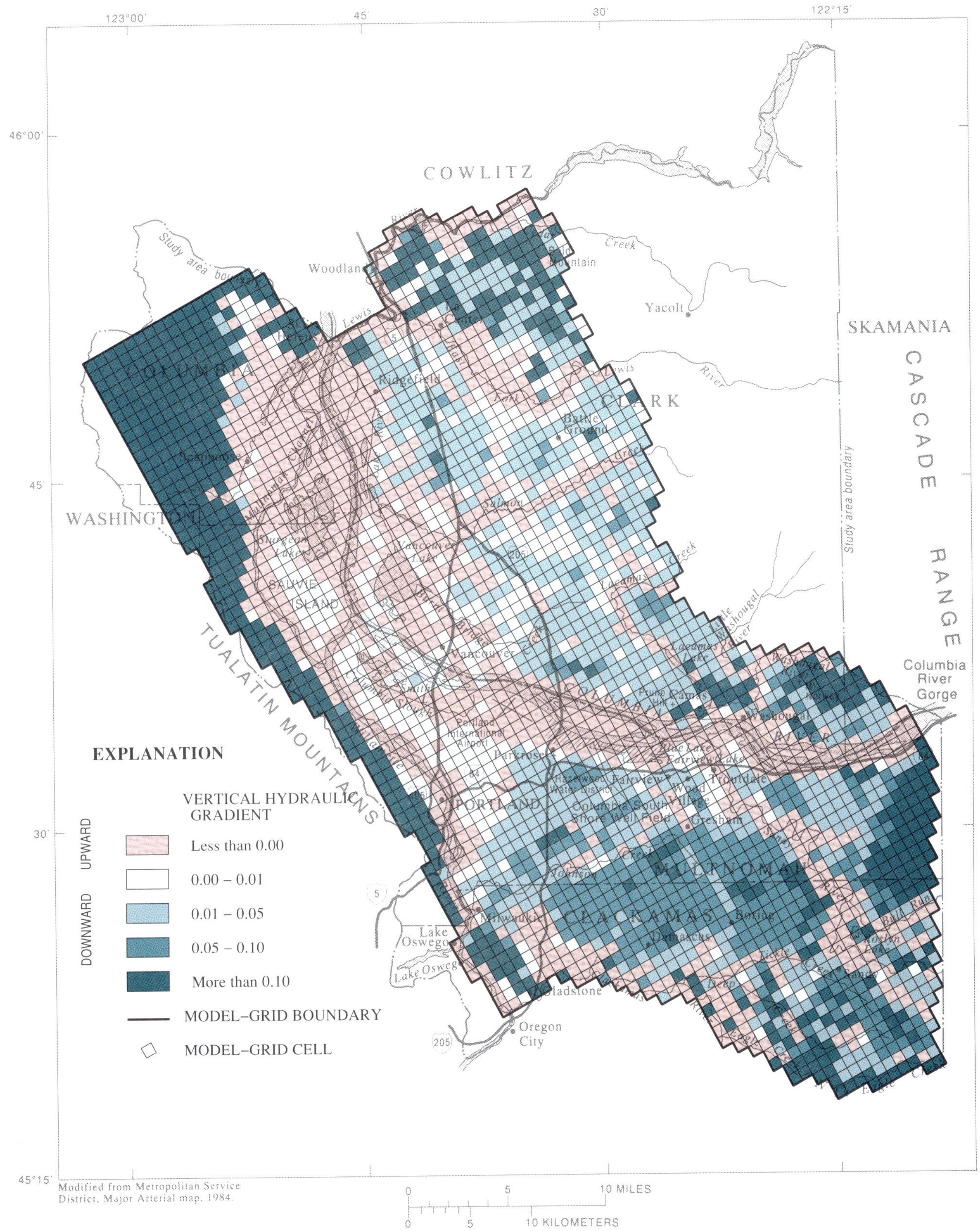

Figure 12. Distribution and extent of shallow, local ground-water flow systems as indicated by simulated vertical head gradients between the water table and underlying unit.

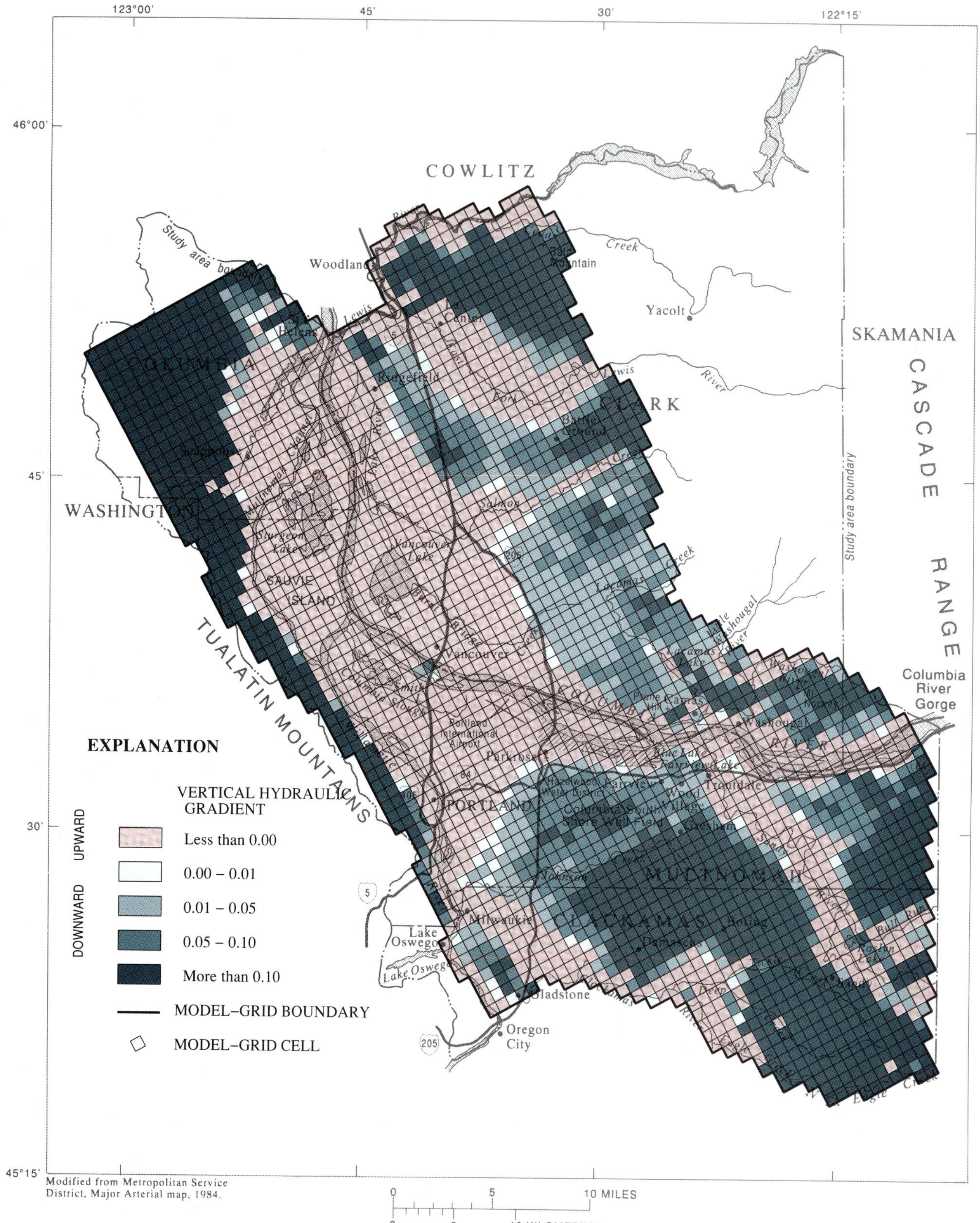

Figure 13. Distribution and extent of deep, regional ground-water flow systems as indicated by simulated vertical head gradients between the water table and deeper units.

These conditions probably result from either high recharge rates to the water table from drywells and on-site waste-disposal systems, or reductions in head in the Troutdale gravel aquifer caused by pumping.

Simulated flux rates from head-dependent flux boundaries (fig. 14) are closely related to the vertical head gradients discussed. Most cells specified as head-dependent flux boundaries, including rivers, streams, and springs, are discharge points in the ground-water system. Independent estimates of ground-water discharge to the Columbia and Willamette Rivers are not available for comparison with simulated ground-water discharge rates to river cells; however, the direction (discharge in most areas) and magnitude (1–2 ft^3/s) of the simulated discharge rates agree with the conceptual model of the system. Recharge to the ground-water system was simulated at head-dependent boundary cells in the headwater areas of some streams and upper reaches of streams that have headwaters outside the model area; although conceptually reasonable, there were few seepage measurements available to verify the simulated flux rates in these areas. Recharge to the shallow aquifers that underlie the Columbia River south of Vancouver Lake also was simulated by the model, indicating that pumping in this area has reversed natural gradients near the river. Additional water-level and chemical data, such as stable isotope concentration, from wells adjacent to the river would be needed to verify this hypothesis.

Discharge to Streams

Estimates of mean stream leakage and discharge rates at selected reaches were used for comparison with simulated leakage and discharge to provide another indicator of the model's ability to simulate the real ground-water flow system. Simulated stream leakage rates represent the mean flux between the stream and the ground-water system during the April 1987 to March 1988 calibration period. Stream discharge simulated by the model represents the mean ground-water contribution to streamflow (base flow) during the 1987–88 calibration period.

Low-flow discharge was measured at 126 sites (pl. 1) on 24 streams and tributaries in September of 1988 (McFarland and Morgan, 1996) and used to estimate mean base flow by assuming a relation between the observed discharge and mean base-flow discharge. Mean base-flow estimates made by hydrograph separation for Cedar and Salmon Creeks in Washington and Johnson Creek in Oregon were compared with low-flow measurements made during September 1988. September 1988 discharges on the three streams equaled 19, 23, and 27 percent of mean base-flow estimates for Johnson, Cedar, and Salmon Creeks, respectively. Despite the wide range of basin characteristics for the three streams, their respective base-flow to low-flow ratios agreed reasonably well. Mean base flow and stream leakage for the 1987–88 period were estimated by assuming the observed low flow was 25 percent of the mean base flow at each of the measurement sites; estimates are listed in Appendix C.

Simulated stream leakage rates were compared with estimated mean leakage rates at 67 sites in the basin (fig. 15; Appendix C). Ground-water discharge occurs at most measurement sites (streams are gaining) and the model simulated this condition well in most parts of the basin. Stream leakage estimates have uncertainties that are proportionate to the discharge of the stream. Leakage estimates on high discharge streams, such as the Sandy River, have large uncertainties because the discharge measurements from which they were estimated have uncertainties of approximately 5 percent. Simulated leakage rates shown in figure 15 lie within the uncertainty range of the estimated leakage rates at 37 (55 percent) of the 67 measurement sites.

Simulated base flow was compared with estimated 1987–88 mean base flow at 126 sites (fig. 16; Appendix C). Low-flow measurements from September 1988 also were used for comparison with model results, but were considered an estimate of the minimum ground-water contribution to streamflow. On streams that have headwaters outside the model boundaries, such as the East Fork Lewis, Sandy, and Bull Run Rivers, the estimated mean base flow was specified at the first stream reach inside the model. Estimated and simulated discharge at all stream reaches where low-flow measurements were made is shown in figure 16. Agreement between the simulated discharge and estimated discharge was generally good. For example, discharge measurements for September 1988 and estimated mean base-flow discharge were compared with simulated base-flow discharge at several sites on Salmon and Tickle Creeks (fig. 17). Simulated base-flow discharge on both creeks closely approximated estimates of 1987–88 base flow.

Water Budgets

The overall 1987–88 water-budget for the model is summarized in table 2. Several components of the budget were specified-flux boundaries to the model.

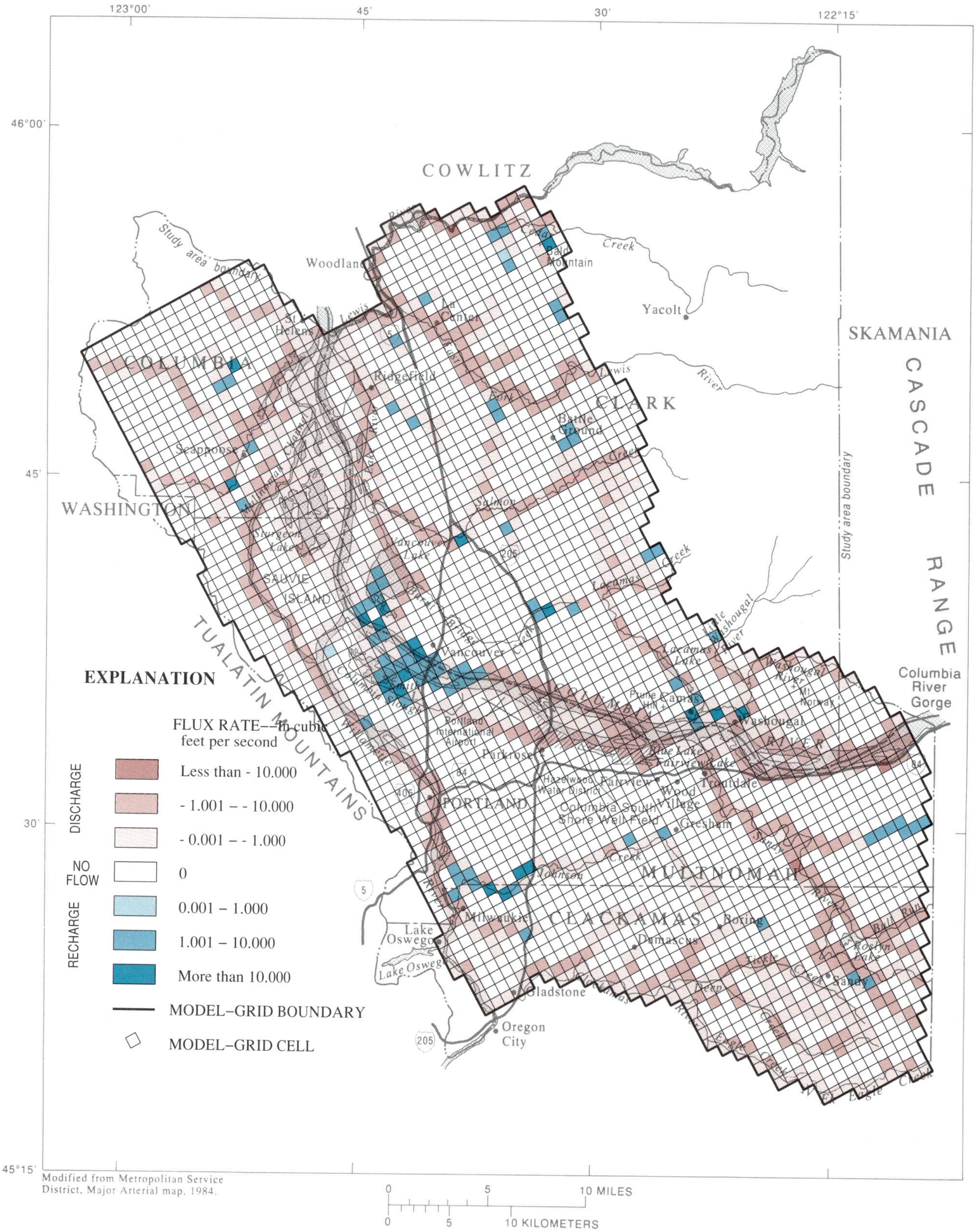

Figure 14. Simulated flux rates at head-dependent flux boundaries (rivers, streams, and springs), 1987–88. (Negative values indicate ground-water discharge.)

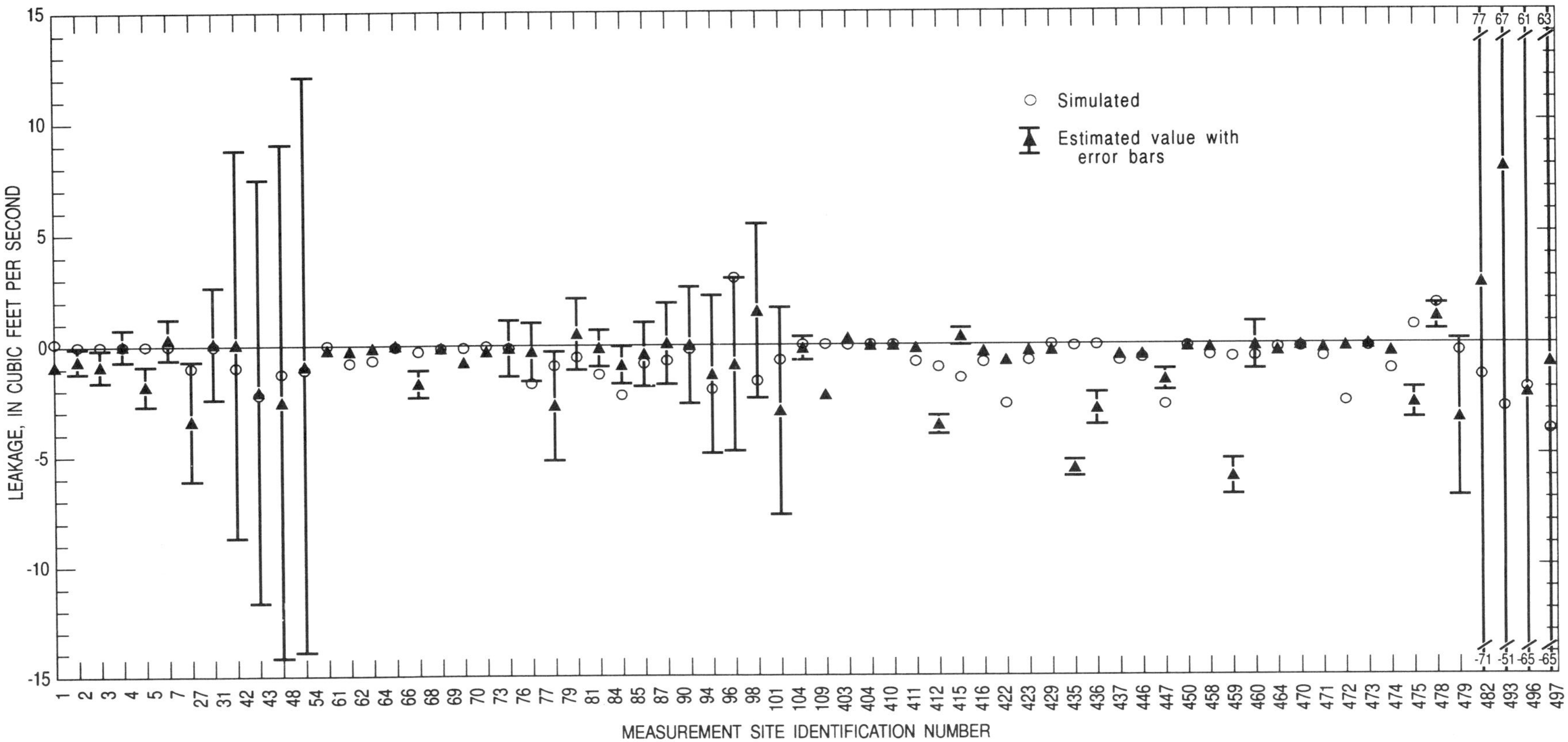

Figure 15. Simulated and estimated stream leakage at measurement sites. (Locations of sites are shown on plate 1.)

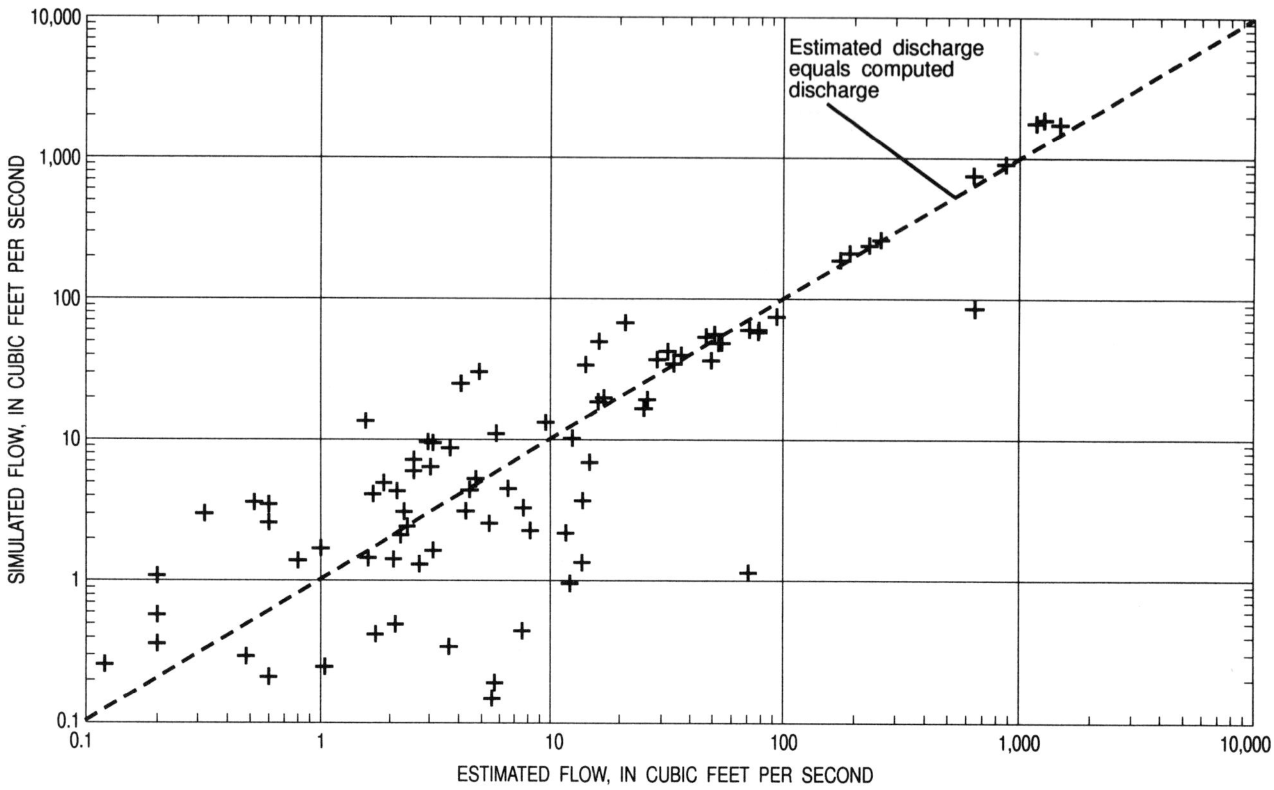

Figure 16. Simulated and estimated mean base flow at measurement sites. (Locations of sites are shown on plate 1.)

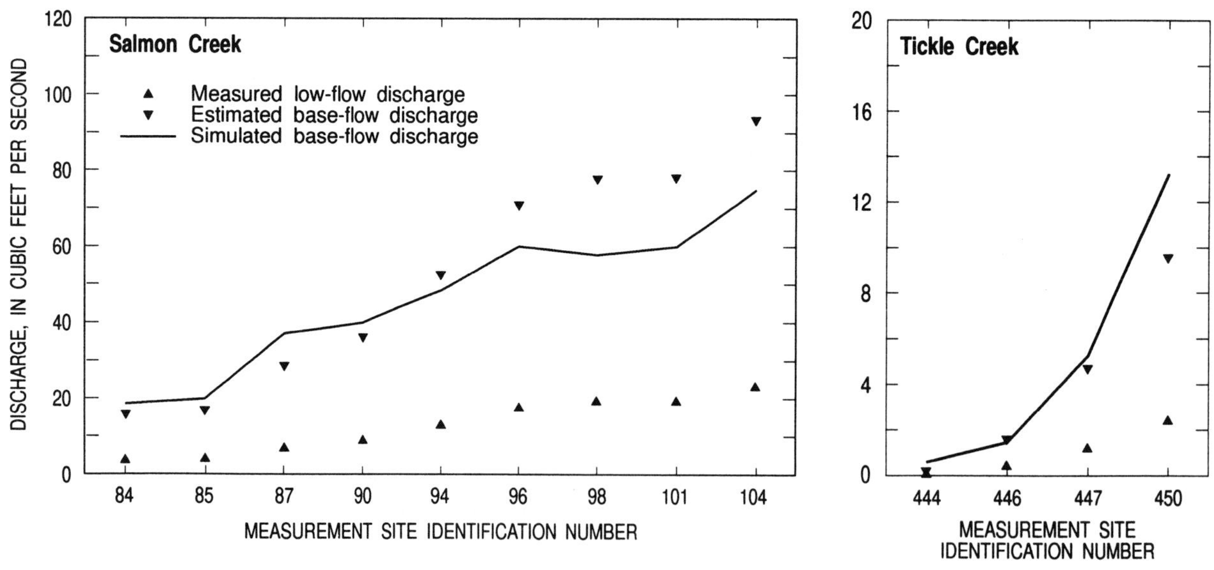

Figure 17. Simulated mean base-flow discharge and estimated discharge at measurement sites on Salmon Creek and Tickle Creek.

Table 2. Ground-water budgets for the Portland Basin, under predevelopment, 1987–88, and hypothetical conditions
[*, budget items simulated by model]

Budget items	Rate (in cubic feet per second)		
	Predevelopment	1987–88	Hypothetical conditions
	Inflow		
Infiltration of precipitation	1,710	1,440	1,440
Drywells	0	62	62
On-site waste-disposal systems	0	27	27
*Seepage from rivers	1	36	62
*Seepage from streams	48	88	87
Storage change (decrease)	0	16	0
Total	1,760	1,670	1,679
	Outflow		
Well discharge	0	166	258
*Seepage to rivers	606	457	443
*Seepage to streams	1,150	971	972
*Spring discharge	7	6	6
Storage change (increase)	0	70	0
Total	1,760	1,670	1,679

These items included recharge from infiltration of precipitation, drywells, and on-site waste-disposal systems; compensation for estimated changes in storage during the 1987–88 calibration period; and discharge by wells. The components of the budget calculated by the model as head-dependent boundary fluxes included seepage to and from rivers, seepage to and from streams, and spring discharge. Regionally, infiltration of precipitation (1,440 ft^3/s) was by far the most important source of recharge to the system; locally, recharge from drywells (62 ft^3/s) and on-site waste-disposal systems (27 ft^3/s) comprised a large part of recharge to shallow aquifers (Snyder and others, 1994). The storage changes listed as inflow and outflow items in table 2 offset estimated losses and gains in storage in the unconfined aquifers during the calibration period.

The total simulated discharge to streams of 971 ft^3/s (table 2) includes all streams and small rivers simulated with the STREAM package (Prudic, 1989), as described in the section on boundary conditions. The next largest outflow from the system was discharged to rivers (457 ft^3/s); most of this total discharged to the Columbia River; however, seepage to the Willamette River, Multnomah Channel, Vancouver Lake, Sturgeon Lake, and other surface-water bodies also occurred. Well discharge accounted for 166 ft^3/s of total outflow, spring discharge 6 ft^3/s, and storage change about 70 ft^3/s.

In order to develop a more complete understanding of the ground-water flow system represented by the model, information from the model was used to prepare a table of fluxes between hydrogeologic units (table 3) and individual water budgets for each of the units (table 4). The principal components of flux to and from the units are shown diagrammatically in figure 18.

The primary inflows to the unconsolidated sedimentary aquifer (US) are areal recharge (341 ft^3/s) and ground-water flux from the Troutdale gravel aquifer (277 ft^3/s) and older rocks (113 ft^3/s). A large part of the areal recharge to the unconsolidated sedimentary aquifer is from drywells (62 ft^3/s) and on-site waste-disposal systems (27 ft^3/s). Flux from the Troutdale gravel aquifer occurs horizontally along the upgradient contact between the two units and vertically by upward leakage in discharge areas near major rivers and streams. Ground-water movement from the older rocks to the unconsolidated sedimentary aquifer takes place where the younger sediments abut the older rocks at the margins of the basin; much of this flux occurs along the base of the Tualatin Mountains. Outflow from the unconsolidated sedimentary aquifer is chiefly to the Columbia and Willamette Rivers (416 ft^3/s) and to other rivers and streams (290 ft^3/s).

Table 3. Simulated horizontal and vertical fluxes between hydrogeologic units in the Portland Basin, 1987–88
[Values are in units of cubic feet per second. To read fluxes from a unit, read vertically down columns; to read fluxes to a unit, read horizontally along rows. US, unconsolidated sedimentary aquifer; TG, Troutdale gravel aquifer; C1, confining unit 1; TS, Troutdale sandstone aquifer; C2, confining unit 2; UF, undifferentiated fine-grained sediments; SC, sand and gravel aquifer, upper coarse-grained subunit; SF, sand and gravel aquifer, lower fine-grained subunit; OR, older rocks]

	Hydrogeologic unit	Outflow from unit									Total inflow
		US	TG	C1	TS	C2	UF	SC	SF	OR	
Inflow to unit	US	--	277	4	53	7	60	38	0	113	552
	TG	150	--	10	27	0	105	3	0	67	362
	C1	0	66	--	13	0	4	0	0	0	83
	TS	5	129	63	--	19	11	0	0	4	231
	C2	0	1	0	60	--	23	11	0	3	98
	UF	13	158	6	19	25	--	3	1	43	268
	SC	1	10	0	1	17	10	--	23	2	64
	SF	0	0	0	0	10	1	10	--	8	29
	OR	2	15	0	6	18	20	2	3	--	66
Total outflow		171	656	83	179	96	234	67	27	240	1,753
Total inflow minus total outflow		381	−294	0	52	2	34	−3	2	−174	0

Table 4. Simulated water budgets for hydrogeologic units, 1987–88
[All values are in cubic feet per second. Total inflow that does not equal total outflow is because of minor mass balance errors in the model. US, unconsolidated sedimentary aquifer; TG, Troutdale gravel aquifer; C1, confining unit 1; TS, Troutdale sandstone aquifer; C2, confining unit 2; UF, undifferentiated fine-grained sediments; SC, sand and gravel aquifer, upper coarse-grained subunit; SF, sand and gravel aquifer, lower fine-grained subunit; OR, older rocks; GW, ground water (from table 3); RIV, rivers; RCH, areal recharge (includes infiltration of precipitation, drywells, on-site waste-disposal systems, and change in storage); STR, streams; DRN, drains (springs); WEL, well discharge]

Hydro-logic unit	Inflow					Outflow						
	GW	RIV	RCH	STR	Total	GW	RIV	RCH	STR	DRN	WEL	Total
US	552	35	341	58	986	171	416	7	290	6	96	986
TG	362	1	743	8	1,114	656	22	0	391	0	44	1,113
C1	83	0	0	0	83	83	0	0	0	0	1	84
TS	231	0	35	0	266	179	6	0	68	0	12	265
C2	98	0	1	0	99	96	0	0	2	0	1	99
UF	268	0	75	0	343	234	0	0	105	0	4	343
SC	64	0	10	0	74	67	0	0	5	0	3	75
SF	29	0	0	0	29	27	0	0	0	0	2	29
OR	66	0	278	0	344	240	13	0	89	0	3	345
Total	1,753	36	1,483	[1]66	3,338	1,753	457	7	[1]950	6	166	3,339

[1] Totals do not equal those in table 2 for seepage to and from streams, because of the method used to compute the budget totals by hydrogeologic unit. See Harbaugh (1990, p. 3) for a detailed discussion. Net flux (inflow minus outflow) is the same in tables 2 and 4.

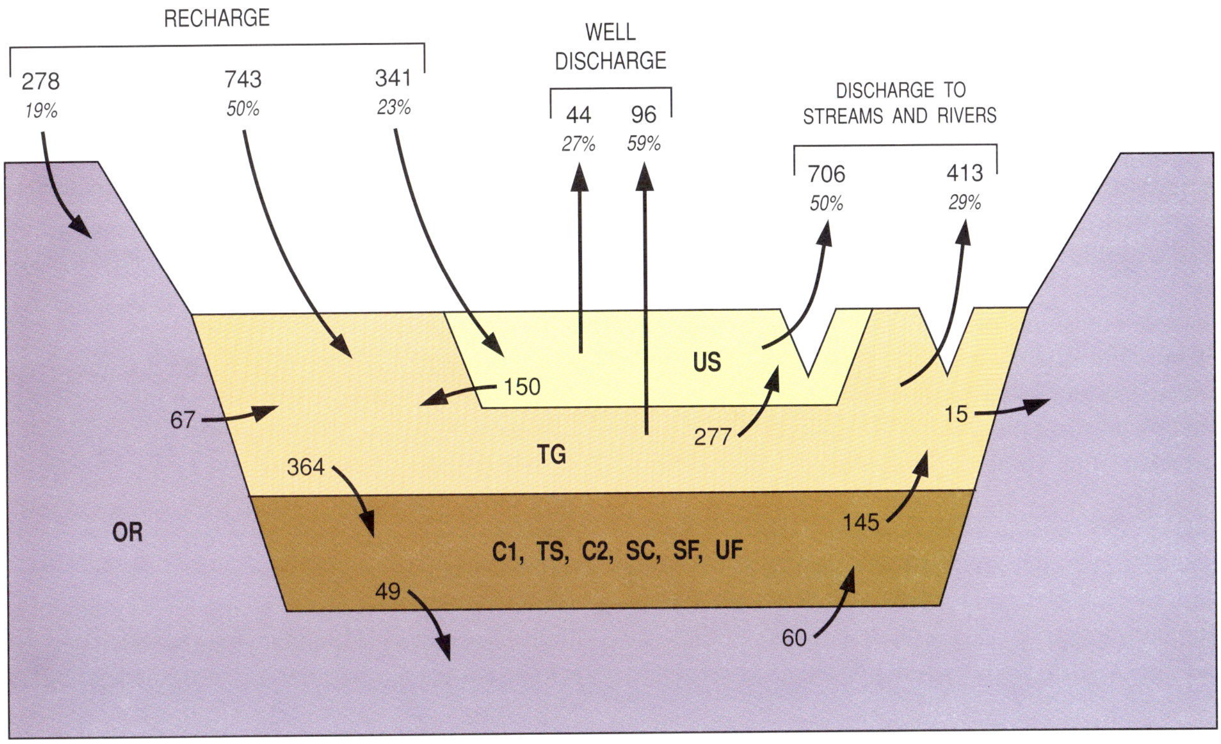

EXPLANATION

Hydrogeologic unit

US	Unconsolidated sedimentary aquifer	**SC**	Sand and gravel aquifer (upper subunit)
TG	Troutdale gravel aquifer	**SF**	Sand and gravel aquifer (lower subunit)
C1	Confining unit 1	**UF**	Undifferentiated fine-grained sediments
TS	Troutdale sandstone aquifer	**OR**	Older rocks
C2	Confining unit 2		

Figure 18. Principal components of the simulated 1987–88 ground-water budget. (Values are from tables 3 and 4 and are in units of cubic feet per second. Numbers in italics are the percentage of each component of the total budget. For example, discharge to streams and rivers from the unconsolidated sedimentary aquifer is 706 ft^3/s, or 50 percent of the total discharge to streams and rivers.

The Troutdale gravel aquifer receives the majority of its inflow from areal recharge (743 ft^3/s), but also receives significant flux from the unconsolidated sedimentary aquifer (150 ft^3/s) and the undifferentiated fine-grained sediments (105 ft^3/s). Direct areal recharge to the unit is principally from infiltration of precipitation where the unit is exposed at land surface (pl. 1) in the upper parts of the Sandy and Clackamas River drainages and in eastern Clark County. Flux from the unconsolidated sedimentary aquifer occurs primarily by downward leakage. Flux from the undifferentiated fine-grained sediments is by upward leakage where the unit underlies the Troutdale gravel aquifer. Discharge to other units (656 ft^3/s) and to streams (391 ft^3/s) comprise the major outflows from the Troutdale gravel aquifer. Most of the discharge to other units is by upward leakage to the unconsolidated sedimentary aquifer (277 ft^3/s) or by downward leakage to the Troutdale sandstone aquifer (129 ft^3/s) and the undifferentiated fine-grained sediments (158 ft^3/s). The importance of the unconsolidated sedimentary aquifer and Troutdale gravel aquifer in the regional flow system is readily apparent from the magnitude of fluxes through them (table 3; fig. 18); 75 percent of recharge to the system and 80 percent of discharge from the system enters and leaves through these units.

The Troutdale sandstone aquifer is not as extensive as the unconsolidated sedimentary aquifer or Troutdale gravel aquifer and, therefore, does not have inflows or outflows of the same magnitude. The Troutdale sandstone aquifer has relatively few exposures in the basin and thus receives only 35 ft^3/s

from areal recharge; most inflow to the unit is by flux from other units (231 ft^3/s), and most of that enters by way of downward leakage from the Troutdale gravel aquifer (129 ft^3/s) where it directly overlies the unit in the upper parts of the Sandy and Clackamas River Basins. Most outflow from the Troutdale sandstone aquifer is by downward leakage to confining unit 1 (60 ft^3/s) and by horizontal flow into the unconsolidated sedimentary aquifer (53 ft^3/s) where the Troutdale sandstone aquifer has been truncated by the Columbia River and abuts the permeable channel sands (see section C-C', pl. 1). Significant outflow also occurs by discharge to streams (68 ft^3/s) such as Deep and Tickle Creeks and the Sandy River.

Inflows to the undifferentiated fine-grained sediments are leakage from other units (268 ft^3/s) and areal recharge (75 ft^3/s). Outflow from the unit is principally by leakage to other units (234 ft^3/s), discharge to streams (105 ft^3/s), and a small amount of well discharge (4 ft^3/s) from lenses of coarser sediments in north-central Clark County.

Virtually all inflow and outflow to and from the confining units 1 and 2 and the lower, fine-grained subunit of the sand and gravel aquifer occurs by leakage to and from adjacent units. A few wells in the basin tap localized permeable strata within the units and withdraw a total of 4 ft^3/s from all three units. In addition, about 2 ft^3/s of discharge is to streams.

Older rocks receive about 19 percent (278 ft^3/s) of the total areal recharge within the modeled area (fig. 18); most of this occurs where older rocks are exposed in the upland areas such as the Tualatin Mountains. Some discharge from the unit is to streams (89 ft^3/s), rivers (13 ft^3/s), and wells (3 ft^3/s), but the primary outflow is by leakage to other units (240 ft^3/s). Most of the leakage is to the unconsolidated sedimentary aquifer (113 ft^3/s), with lesser amounts to the Troutdale gravel aquifer (67 ft^3/s) and the undifferentiated fine-grained sediments (43 ft^3/s) where they overlie older rocks.

Model Usage and Limitations

Many assumptions are necessary to simplify a real hydrogeologic system to the extent that it can be represented by a mathematical model. The assumptions used to construct the model described by this report have been described fully in previous sections. Some of these assumptions, however, limit the scope of application of the model and the hydrologic questions that can reasonably be addressed. The major simplifying assumptions and the limitations they impose are discussed below.

The model developed as part of the Portland Basin study, simulated time-averaged conditions for the period 1987–88. Construction and calibration of a model for simulating transient ground-water flow was prevented by the lack of data documenting historical changes in stress (recharge and pumpage) on the ground-water system and hydrologic response of the system (changes in water levels and discharge to streams). Because the model has not been calibrated to transient conditions, the model cannot be used to predict the transient response of the system. The limitation imposed by this is that intermediate heads and fluxes in the system, between the time a new stress is applied and the time the system reaches a new steady state, cannot be predicted using the model. The model can, however, be used to simulate steady-state conditions for various stress conditions, and the steady-state water levels and fluxes under various ground-water-management conditions can be compared and evaluated on the basis of the eventual effect they would have on the system.

A second limitation on the use of the model is that, as constructed, transmissivities of hydrogeologic units do not change when the saturated thickness of the units change. This is not a serious limitation unless new stresses on the system are great enough to cause significant change to the saturated thickness of any unit. In the most severe case, a model cell, or group of cells, might become completely desaturated; less severe errors might occur if a smaller change in saturated thickness occurs. Model results should be checked for these conditions if large water-level changes are simulated in the uppermost hydrogeologic units.

Finally, boundary conditions involve considerable simplification of the hydrologic system and can have substantial effects on model results; thus, they must be clearly understood to avoid serious errors in model application (Franke and others, 1987). The lateral boundary of the Portland Basin model was specified as a "no-flow" boundary on the basis of assumptions that it coincided with either ground-water flow divides (for example, the Clackamas and Lewis Rivers) or low-permeability rocks (Skamania Volcanics). These assumptions were considered valid for the stress conditions in the basin during the 1987–88 simulation period; however, they should be evaluated carefully when simulating other stress conditions.

An example of a simulation in which model boundary conditions would affect simulated response would be placement of new stresses (such as pumpage from new wells) near a ground-water divide that is simulated as a no-flow boundary. In the real hydrologic system, flux would be induced across that boundary as the location of the divide changed. In the simulated system, the additional flux would not be available and the simulated water-level declines would be greater than those in the real system. Similarly, simulation of addition of pumping stresses near a low-permeability rock contact represented as a no-flow boundary would yield predicted drawdowns greater than actual drawdowns if pumping near the boundary in the real system could induce additional recharge from the lo permeability rocks. For many purposes, these problems do not impose serious limitations on the use of the model as long as the potential effects of the boundary conditions on the results are understood.

SIMULATION OF PREDEVELOPMENT STEADY-STATE CONDITIONS

The calibrated model was used to simulate predevelopment ground-water conditions in the basin to determine if a reasonable match could be obtained between observed and simulated heads for another time period. Predevelopment conditions are defined as those that existed in the basin prior to the drilling of the first well, construction of the first drywell, installation of the first on-site waste-disposal system, first diversion of surface water, and prior to paving of the first street—in other words, conditions that existed before any significant human influence was exerted on the hydrologic system. If this strict definition were imposed, predevelopment conditions have not existed in the basin for over 100 years. Predevelopment conditions are usually assumed to represent true steady-state (equilibrium) conditions, in which recharge and discharge are balanced and the volume of ground water in storage is nearly constant except for changes caused by normal fluctuations in precipitation and evapotranspiration.

The predevelopment period was chosen to test the model because, by definition, human-caused stresses that are difficult to estimate, such as well discharge and drywell recharge, did not exist. The disadvantage of using the predevelopment period was that there is little information on which to base a quantitative comparison of simulated and real changes in the ground-water system. Under ideal circumstances, long-term records of water-level change would be available for wells in the major aquifers in various parts of the basin; in reality, only about 15 wells in the basin have water-level records prior to 1970 and the oldest records in the basin only date back to the early 1950's. Data from these wells were used to make qualitative comparisons of simulated and real water-level changes since predevelopment, by assuming that water-level changes in these wells represented the minimum water-level changes since predevelopment.

Stresses on the system were modified by eliminating all well discharge and recharge from drywell and on-site waste-disposal systems and by modifying the average annual rate of infiltration by precipitation to reflect the absence of impervious surfaces. Eliminating well discharge reduced total discharge by 166 ft^3/s (table 2). Snyder and others (1994) showed that rates of infiltration by precipitation in the basin were closely related to average annual rainfall, altitude, and the percentage of impervious area. Their regression equation (see equation 3) showed the importance of impervious area to recharge in the basin; for example, recharge in an area where 25 percent of the surface is impervious would be nearly 9 in/yr less than in the same area with no impervious surface. Recharge from infiltration of precipitation during the predevelopment period was estimated by using equation 3 with the same values of mean annual precipitation (1968–82) and land surface altitude as was used by Snyder and others (1994); however, the percentage of impervious area was assumed to be zero for each model cell for predevelopment conditions.

Estimates made using these assumptions show that annual recharge from infiltration of precipitation decreased by 270 ft^3/s (16 percent) from 1,710 ft^3/s to 1,440 ft^3/s between the predevelopment period and 1987–88. Most of this decrease occurred in high density residential, commercial, and industrial areas of Portland and Vancouver (fig. 19), where recharge from infiltration of precipitation has been reduced by up to 25 in/yr. In spite of increases in impervious surfaces, areas such as east Multnomah County and the Burnt Bridge Creek area of Clark County, where drywells and on-site waste-disposal systems are used extensively, received more recharge in 1987–88 than under predevelopment conditions (fig. 19).

Water-level changes in the basin since predevelopment were calculated as the difference between simulated predevelopment and 1987–88 water levels.

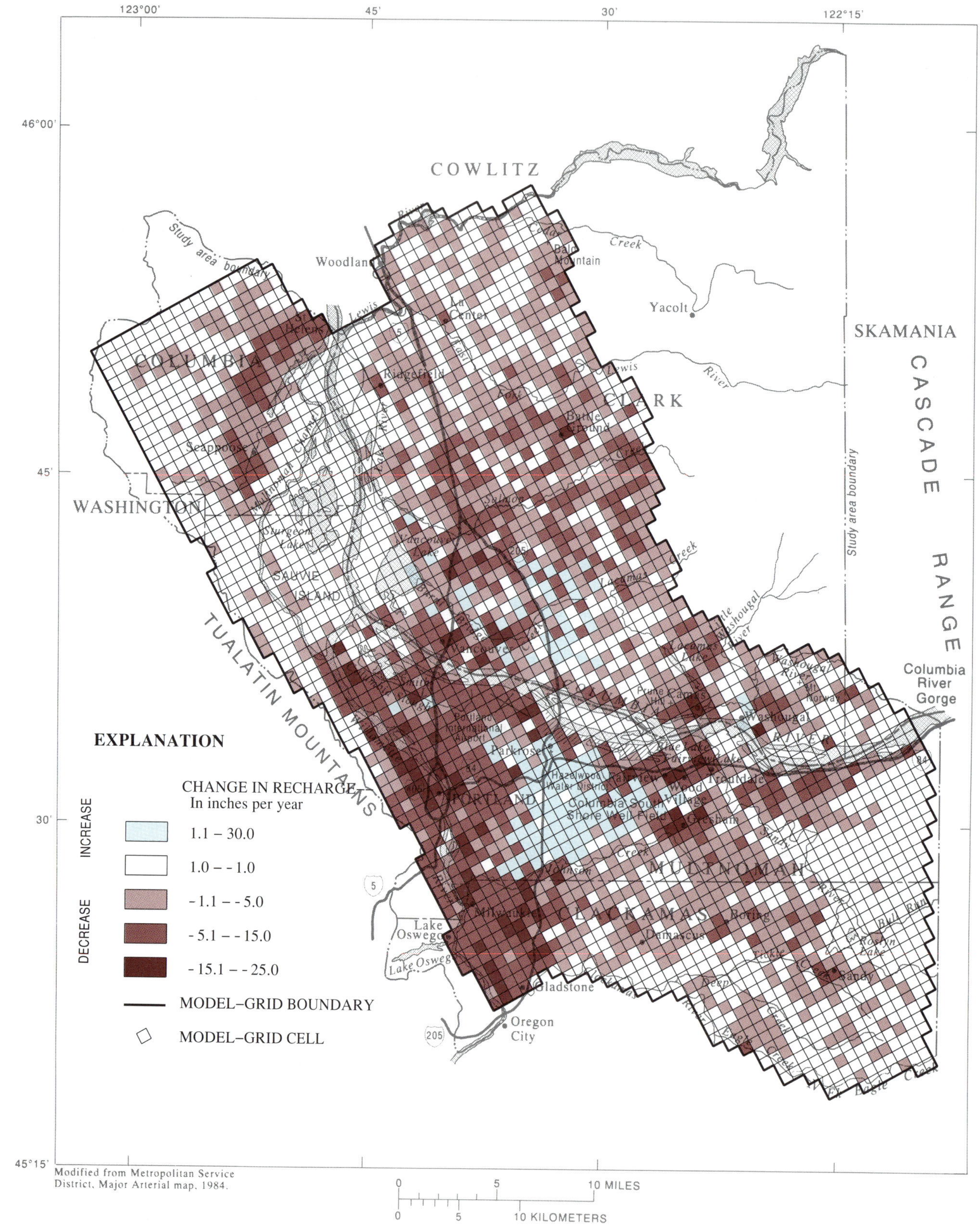

Figure 19. Estimated change in recharge distribution from predevelopment to 1987–88 conditions.

These changes were compared to estimates of water-level changes in selected long-term observation wells in the basin. Changes in simulated water levels reflect only the changes in recharge and well discharge from predevelopment to 1987–88; no other model parameters or boundary conditions were modified. Maps of simulated water-level change for the unconsolidated sedimentary aquifer, Troutdale gravel aquifer, Troutdale sandstone aquifer, and undifferentiated fine-grained sediments are shown in figures 20, 21, 22, and 23, respectively.

Results indicate that water levels have declined moderately (less than 10 ft) throughout most of the basin, primarily in response to reduced recharge in urban areas. Larger declines occurred where pumping is concentrated for public-supply and industrial use in southern Clark County and for agricultural use in the Sandy-Boring area of Oregon. Simulated declines in Clark County were most extensive in the Troutdale gravel aquifer, with a maximum of 40–50 ft and a minimum of 20–30 ft over a broad area within T.2E. bounded by the Columbia River on the south and by Salmon Creek on the north (fig. 21). These simulated declines result from both pumping and reduction in recharge due to urbanization. Concentrated pumping for agricultural use in the area surrounding Sandy and Boring resulted in simulated declines of as much as 40–50 ft in both the Troutdale gravel aquifer (fig. 21) and the Troutdale sandstone aquifer (fig. 22).

Increases in ground-water level were simulated in localized areas of the unconsolidated sedimentary aquifer and Troutdale gravel aquifer in the northeastern part of T.1S., R.2E., where drywell and on-site waste-disposal systems provide large quantities of secondary recharge (figs. 20 and 21).

One long-term observation well in Clackamas County, locally known as the "Issacs" well (T.2S., R.3E., section 6), is completed in the Troutdale gravel aquifer and has been monitored since 1962. During the period 1962–88, water levels in the well declined approximately 25 ft (McCarthy and Anderson, 1990). It is not known how much additional change occurred between predevelopment conditions and 1962; however, 25 ft could be assumed to be the minimum amount of change for the aquifer in this area. Simulated water-level change in the Troutdale gravel aquifer in this area was 20–30 ft (fig. 21). The "Meirs" well, located approximately 7 miles east (T.2S., R.4E., section 5) and completed in the same unit, has been monitored since 1958 and shows a decline of about 20 ft (McCarthy and Anderson, 1990). Simulated declines in this area were 30–40 ft (fig. 21). A well in T.1N., R.2E., section 23 has been monitored since 1967 and shows no long-term trend of decline or rise; simulated water-level changes within the unconsolidated sedimentary aquifer and Troutdale gravel aquifer in this area were 0 to 10 ft.

Two observation wells in Clark County also gave indications of the minimum water-level changes in the Troutdale gravel aquifer since predevelopment. The first well, located in T.2N., R.2E., section 30, showed a 5 ft decline between 1955 and 1988; declines of 10–20 ft were simulated by the model in this area. The second well indicated that declines of at least 12 ft have occurred in T.3N., R.1E., section 21; declines of 0–10 ft were simulated in this area (fig. 21).

Greater recharge from infiltration of precipitation and the absence of pumping left more ground water available to discharge to rivers and streams during predevelopment conditions. Discharge to rivers was 33 percent greater and discharge to streams was 18 percent greater than under 1987–88 conditions (table 2). Less recharge as a result of more impervious surfaces has decreased the flow rate through the shallow aquifers and decreased discharge to rivers and streams. The addition of pumpage near rivers and streams has increased the quantity of seepage from rivers (from 1 to 36 ft^3/s) and streams (from 48 to 88 ft^3/s) to the ground-water system. Streams and rivers throughout the basin received less ground-water discharge in 1987–88 than during predevelopment conditions. In some areas where the simulation shows that pumping has lowered the water table, streams may have stopped receiving discharge completely, or if they have sufficient flow, have come a source of recharge to the shallow aquifers (pl. 9). Some river and stream reaches that contributed recharge to the ground-water system under predevelopment conditions also contributed more recharge after the water table was lowered by nearby pumping.

SIMULATION OF HYPOTHETICAL PUMPING CONDITIONS

The most important use of the model described in this report, beyond the present study, will be to simulate the response of the ground-water system to hypothetical management plans or conditions. Therefore, a set of conditions typical of those that could be simulated was defined and used to test the response of the model to pumping stresses greater than those in effect during the 1987–88 calibration period.

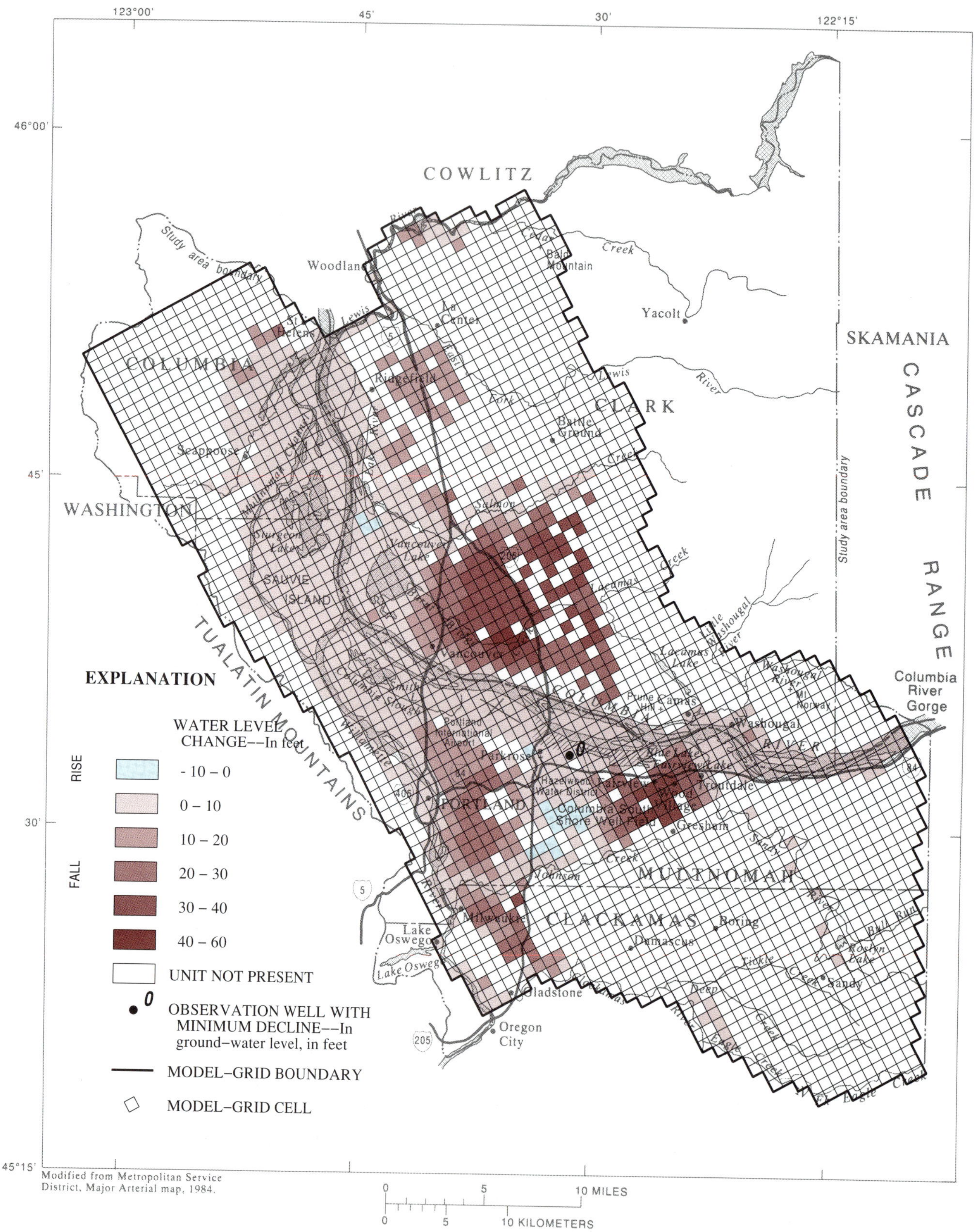

Figure 20. Simulated water-level change between predevelopment and 1987–88 conditions in the unconsolidated sedimentary aquifer. (A negative change indicates rising water level.)

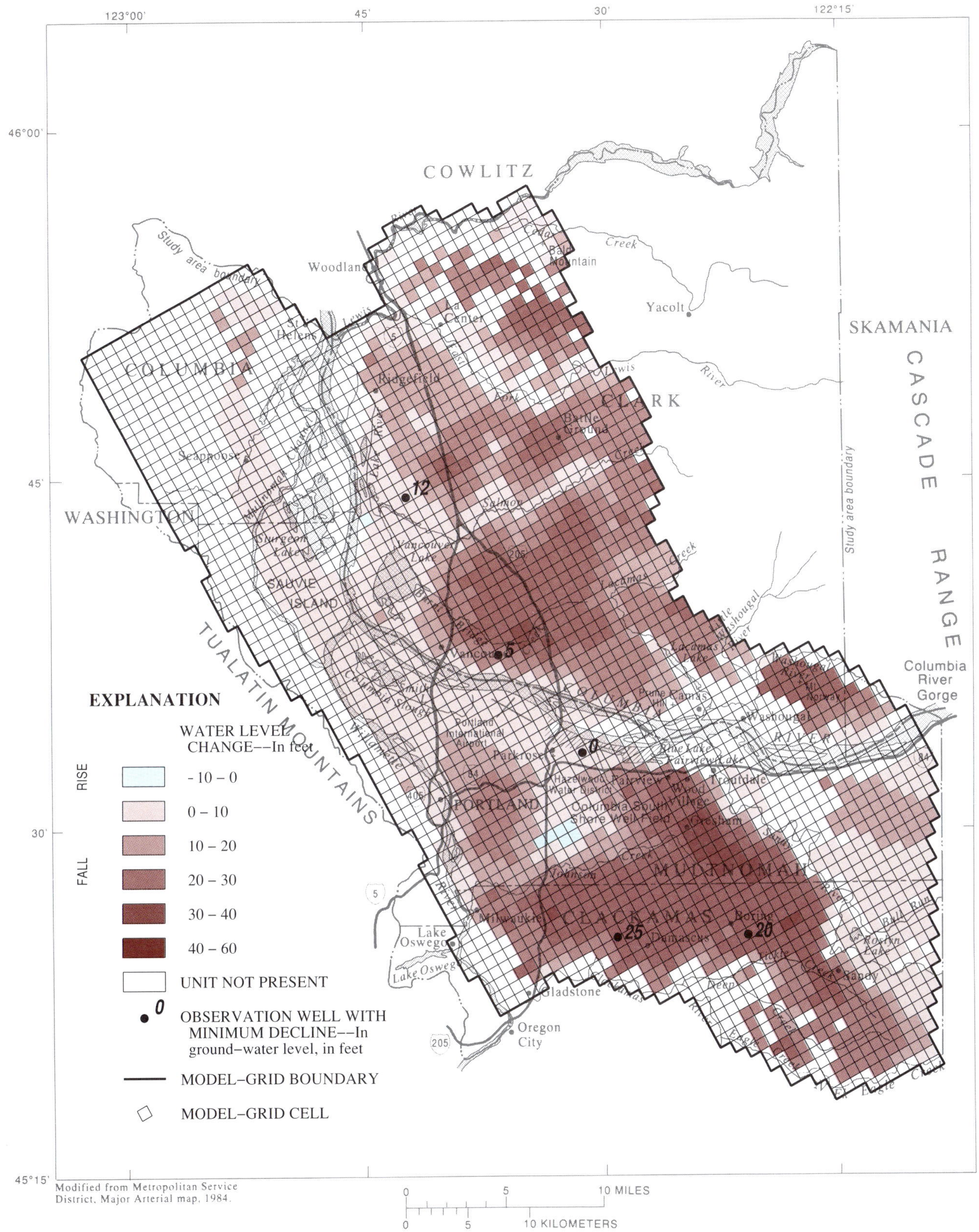

Figure 21. Simulated water-level change between predevelopment and 1987–88 conditions in the Troutdale gravel aquifer. (A negative change indicates rising water level.)

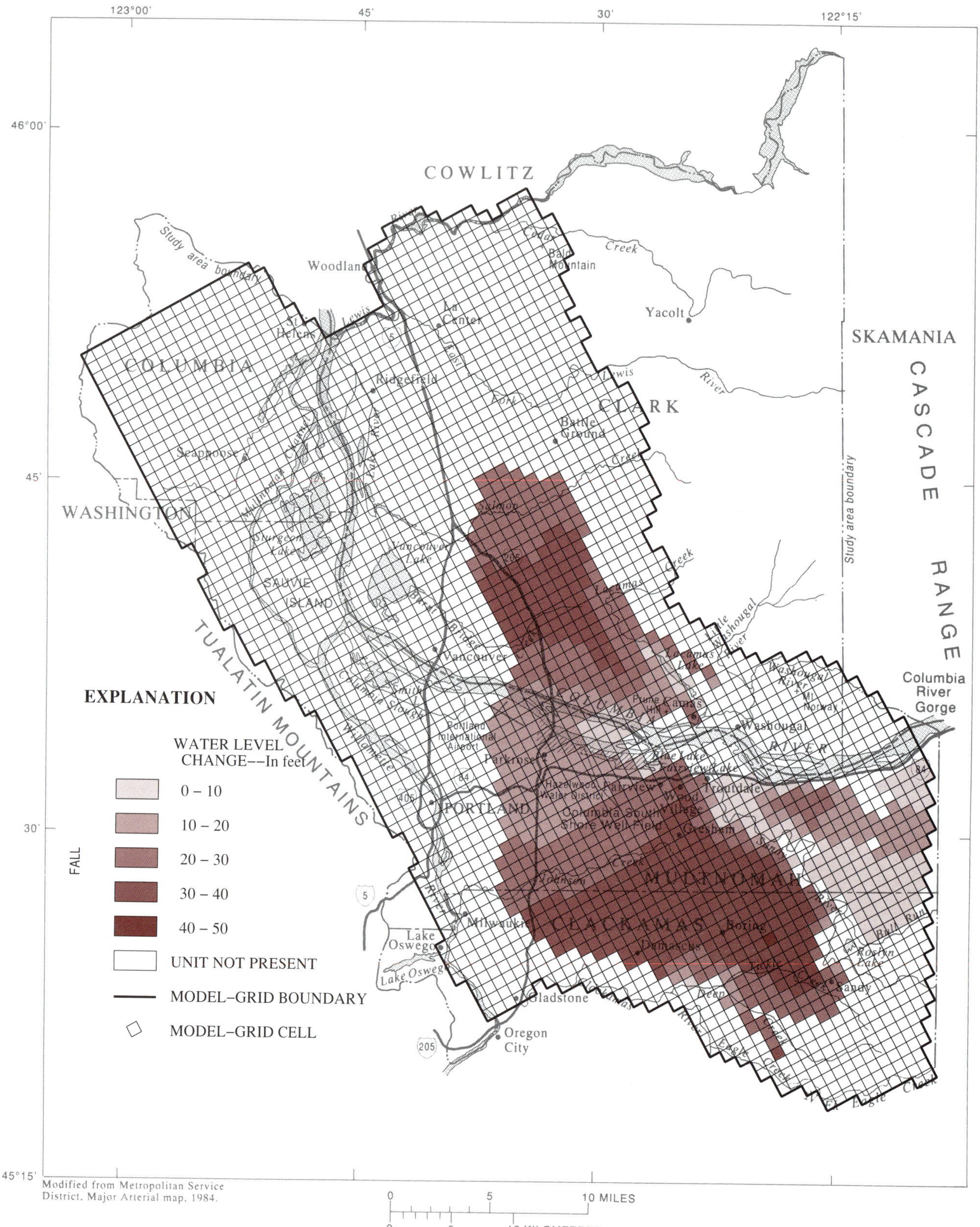

Figure 22. Simulated water-level change between predevelopment and 1987–88 conditions in the Troutdale sandstone aquifer. (A negative change indicates rising water level.)

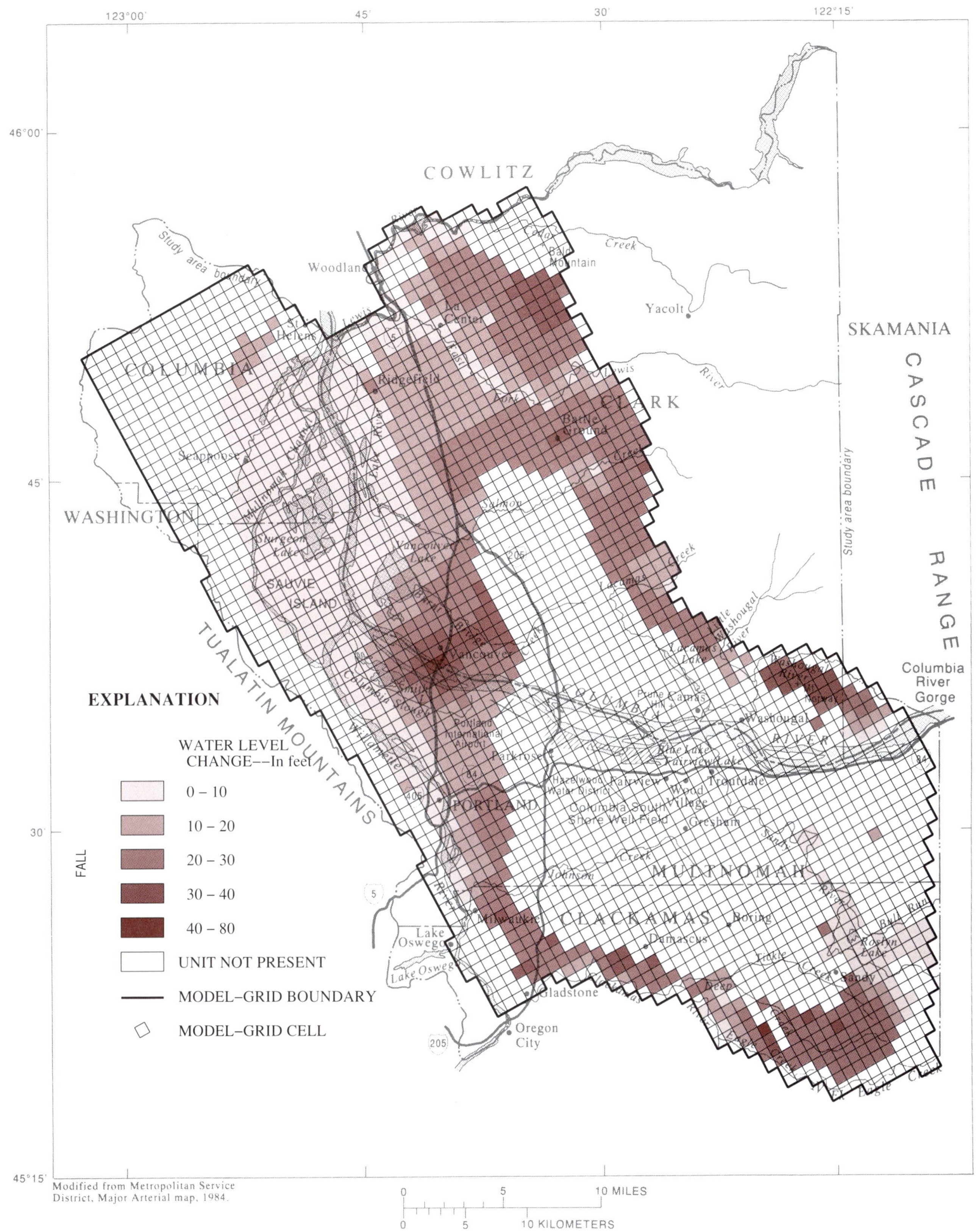

Figure 23. Simulated water-level change between predevelopment and 1987–88 conditions in the undifferentiated fine-grained sediments. (A negative change indicates rising water level.)

The hypothetical management condition defined for this simulation was one in which increased demands for public water supply in Clark County and the city of Portland are met by increasing ground-water withdrawals. The projected ithdrawals in Clark County were based on estimates of increased water needs for the year 2010 (R.D. Swanson, Intergovernmental Resource Center, Clark County, Washington, oral commun., April 1991); withdrawals by the city of Portland were based on the assumption that the Columbia South Shore well field, presently used only rarely, would be operated at 25 percent of its 150 ft^3/s capacity. The condition under which the well field would be operated at 25 percent of capacity would be when the wells are used to meet late summer and peak demand during a 3-month period each year. As a result, total withdrawals in Clark County were increased by 54 ft^3/s and in the Columbia South Shore well field by 38 ft^3/s (fig. 24); the total, 92 ft^3/s, represented a 55 percent increase from 1987–88. About 48 percent (44 ft^3/s) of the increase in pumping was from the unconsolidated sedimentary aquifer, 11 percent (10 ft^3/s) from the Troutdale gravel aquifer, 13 percent (12 ft^3/s) from the Troutdale sandstone aquifer, and 17 percent (16 ft^3/s) from the upper, coarse-grained subunit of the sand and gravel aquifer. In Clark County, most of the new pumpage was from the Troutdale gravel aquifer and Troutdale sandstone aquifer, whereas in the Columbia South Shore well field, most of the increased withdrawals were from the unconsolidated sedimentary aquifer near Blue Lake and Troutdale sandstone aquifer and upper, coarse-grained subunit of the sand and gravel aquifer. Recharge was maintained at the same rates used in the 1987–88 simulation.

Changes in simulated water levels in the unconsolidated sedimentary aquifer, Troutdale gravel aquifer, Troutdale sandstone aquifer, and undifferentiated fine-grained sediments are shown in figures 25, 26, 27, and 28, respectively. Declines in simulated water levels in the unconsolidated sedimentary aquifer were less than 5 ft throughout most of the basin except in Clark County, where declines were as much as 20 ft (fig. 25). The large declines in Clark County probably were caused by increased downward leakage from the unconsolidated sedimentary aquifer that was induced by the increased pumping in the underlying Troutdale gravel aquifer. Declines of 20–40 ft were simulated in the Troutdale gravel aquifer in response to increased pumping from that unit south of Salmon Creek and near Camas (fig. 26). A small area of decline in the Troutdale gravel aquifer also was simulated near Battle Ground. The broadest area of large water-level declines was within the Troutdale sandstone aquifer (fig. 27); simulated declines of 20–40 ft accompanied withdrawals in the Portland well field area and similar declines were simulated near Camas. Declines in the Troutdale sandstone aquifer affected the undifferentiated fine-grained sediments where the two units are in contact and resulted in declines of 10–20 ft within a small area north of Vancouver (fig. 28).

The simulated water budget for the hypothetical conditions (table 2) indicates that, at equilibrium, much of the additional 92 ft^3/s of pumpage is supplied by increased recharge from, and decreased discharge to, the Columbia River. Placement of new pumping near the Columbia River, such as that in the Columbia South Shore well field near Blue Lake and near Camas, induced an additional 26 ft^3/s of recharge from the river (pl. 9). A multiwell aquifer test conducted on wells in the unconsolidated sedimentary aquifer near Blue Lake and subsequent analysis of stable isotope concentrations (oxygen-18 and deuterium) in the well discharge indicates that substantial quantities of river water recharge the aquifer under pumping stress (McCarthy and others, 1992). Areas adjacent to the Columbia River near Vancouver and northeast Portland (southeast T.2W., R.1E.), where flux between the river and the aquifer reversed from discharge to recharge between predevelopment and 1987–88, supplied even more recharge under the hypothetical conditions (pl. 9). Another source of ground water to the new pumping stress was the simulated reduction, or capture, of an additional 14 ft^3/s that discharged from the aquifer to rivers in 1987–88.

The total recharge from streams and discharge to streams listed in table 2 indicates that the hypothetical pumping conditions had little effect on simulated stream leakage from the small streams. Comparison of simulated base-flow discharge at selected sites on Salmon Creek for 1987–88 and under hypothetical pumping conditions shows, however, that within the area of increased pumpage, the ground-water discharge to Salmon Creek would be reduced under these pumping conditions (fig. 29). Mean base-flow discharge at the lower end of Salmon Creek would be reduced by approximately 8 percent. In contrast, Tickle Creek is located outside the area influenced by additional pumpage under these conditions, and simulated base flow was virtually unaffected (fig. 29).

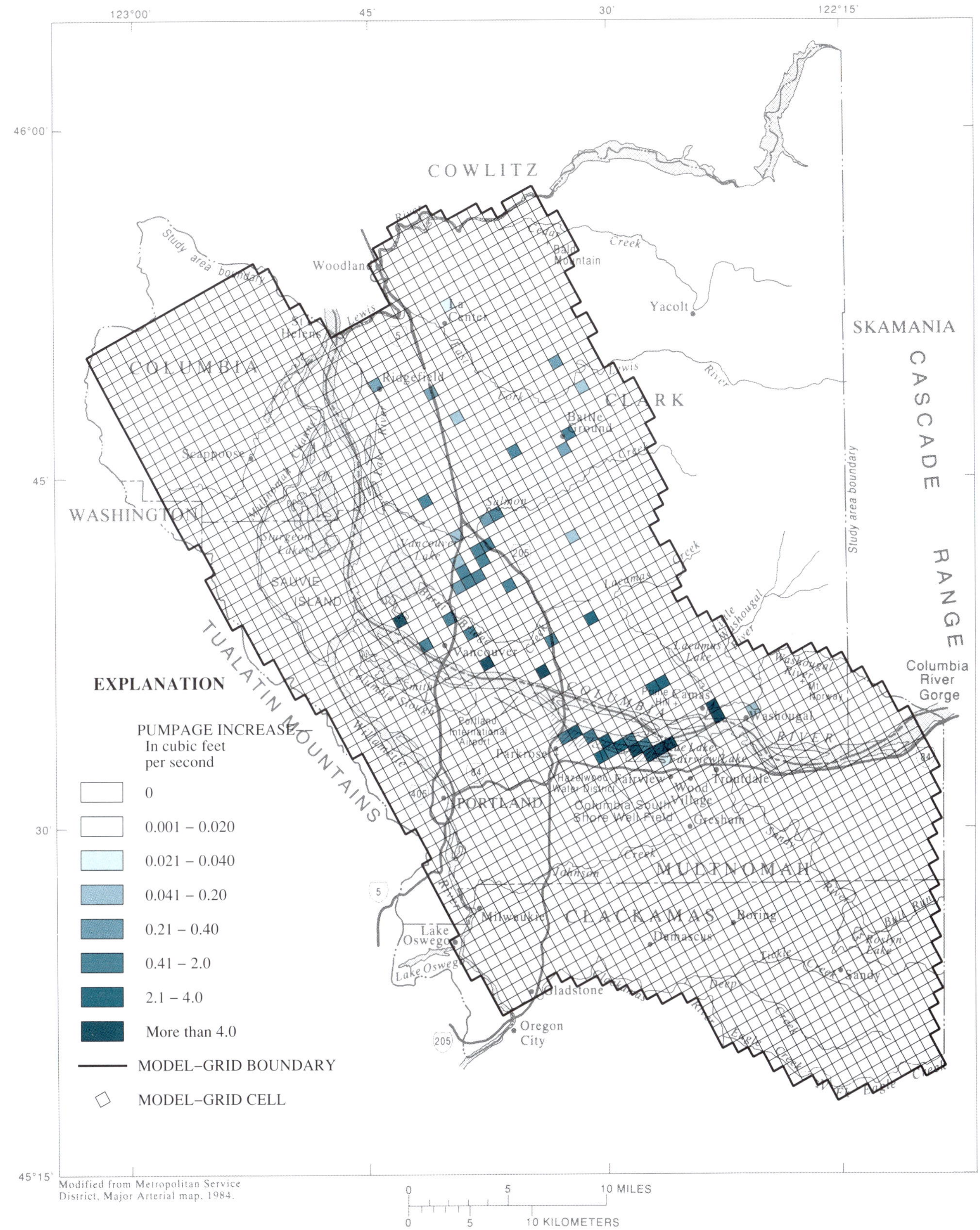

Figure 24. Pumpage increase from 1987–88 conditions to hypothetical pumping conditions.

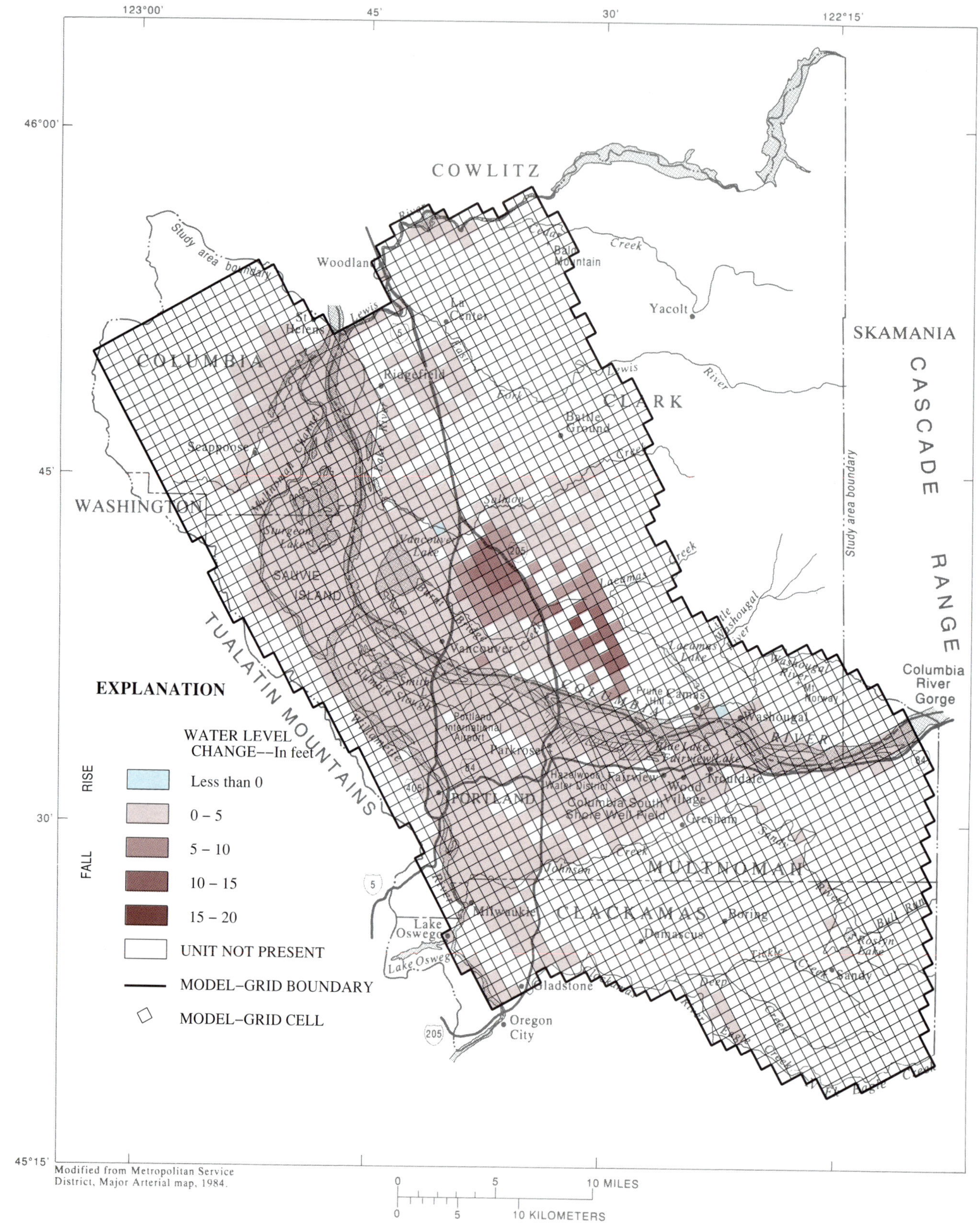

Figure 25. Simulated water-level change from 1987–88 to steady state under hypothetical pumping conditions in the unconsolidated sedimentary aquifer.

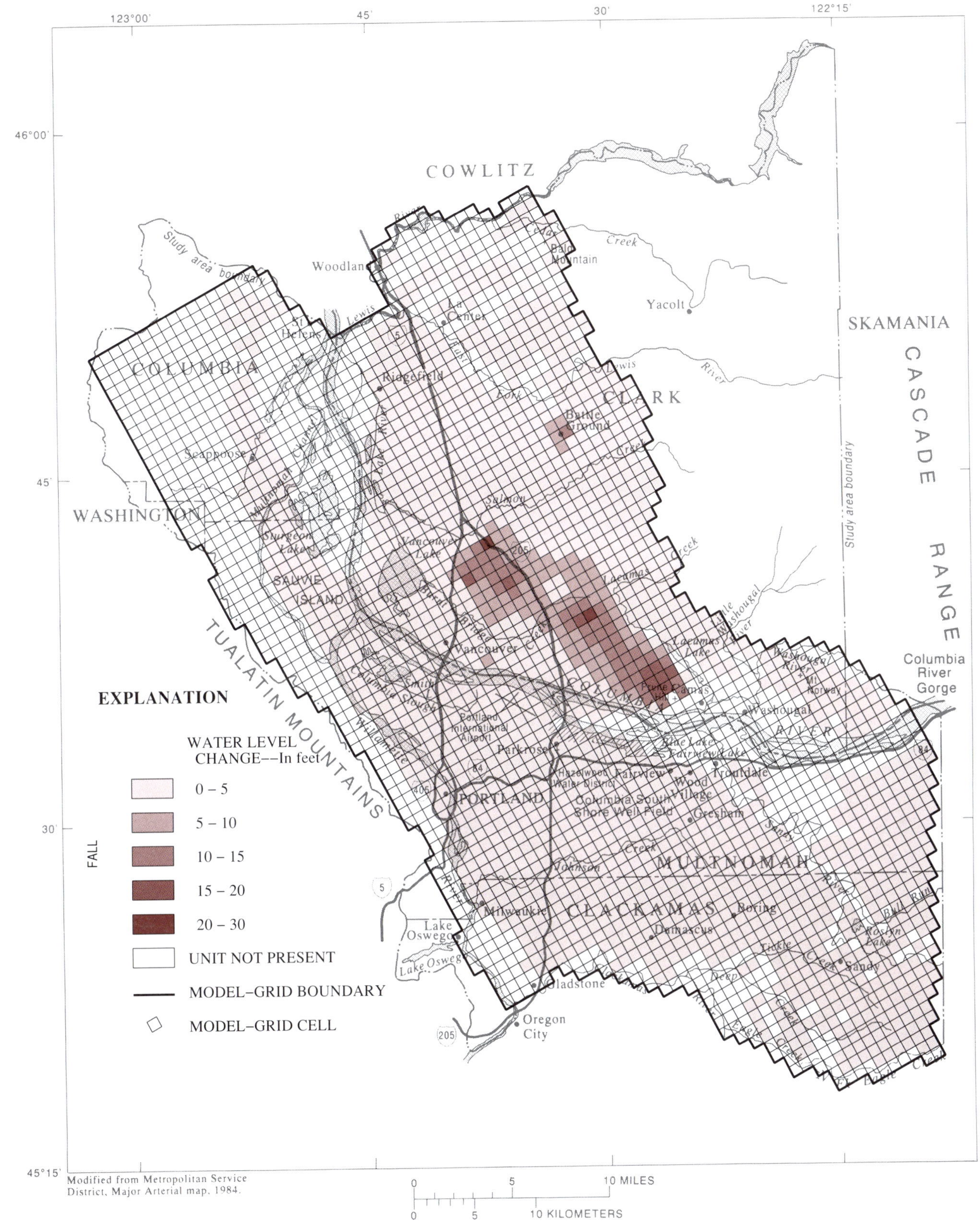

Figure 26. Simulated water-level change from 1987–88 to steady state under hypothetical pumping conditions in the Troutdale gravel aquifer.

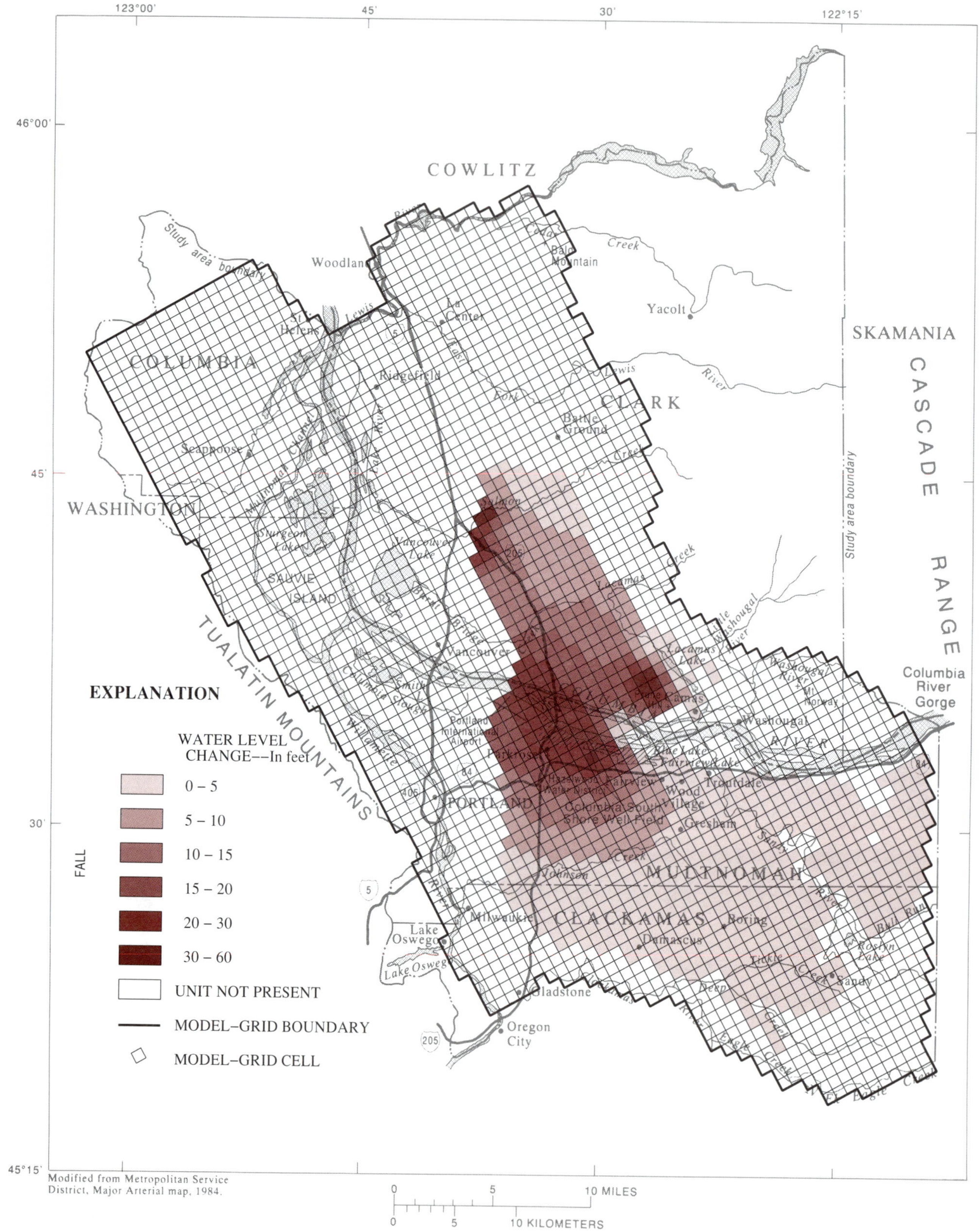

Figure 27. Simulated water-level change from 1987–88 to steady state under hypothetical pumping conditions in the Troutdale sandstone aquifer.

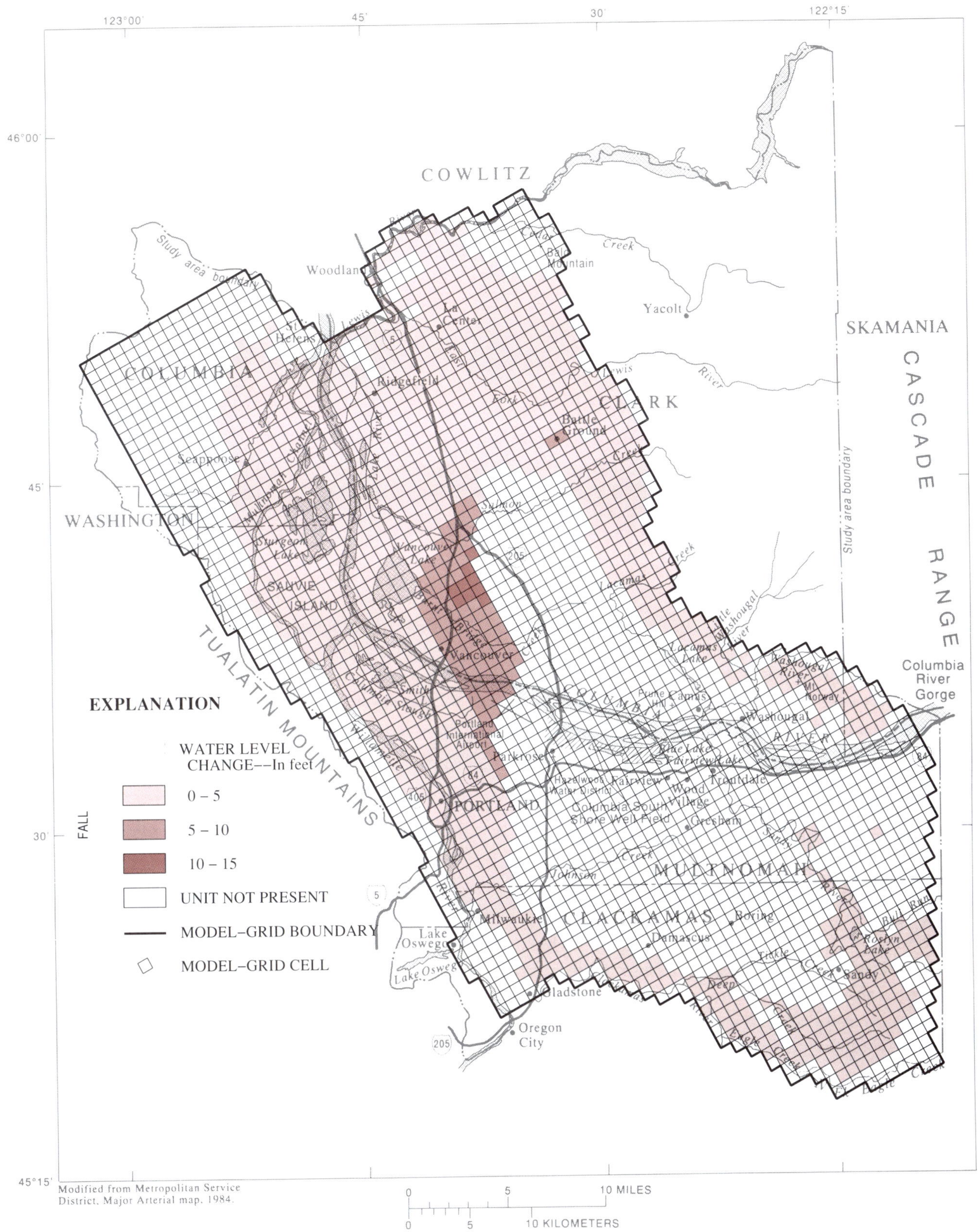

Figure 28. Simulated water-level change from 1987–88 to steady state under hypothetical pumping conditions in the undifferentiated fine-grained sediments.

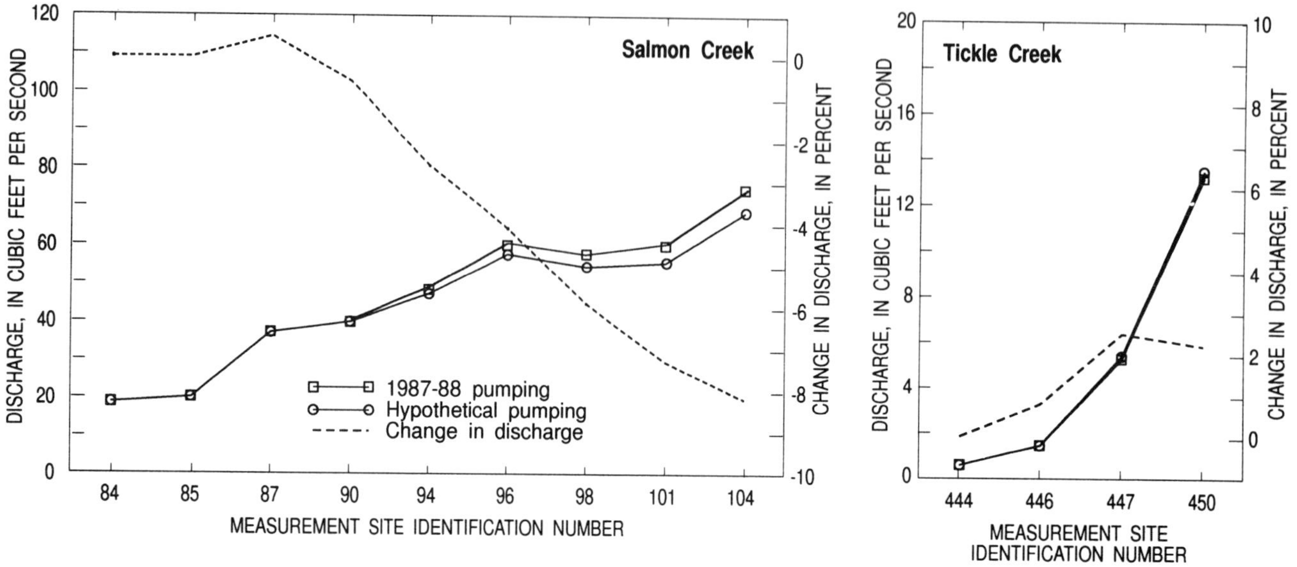

Figure 29. Simulated mean base-flow discharge for 1987–88 and hypothetical pumping conditions on Salmon Creek and Tickle Creek. (Simulated change in base-flow discharge between 1987–88 and hypothetical pumping conditions is shown as percentage of 1987–88 base-flow discharge.)

PRIORITIES FOR FUTURE DATA COLLECTION

A program of data collection and further model calibration would improve the understanding of the hydrologic system and aid in development of a model that could be used to simulate changes in the system over time. A data-collection program for the Portland Basin that would allow calibration of a transient model would include at least the following:

(1) Monitoring of long-term ground-water levels: Many of the monitoring wells established for this study could be retained in a long-term monitoring network. Wells could be selected that would give good areal coverage in each of the major aquifer units. Higher density coverage near pumping centers and, if possible, the capability to measure water level at different depths at the same locations would help to define the vertical hydraulic gradients.

(2) Inventory of ground-water withdrawals: Most withdrawals in the basin are for public-supply and industrial uses. These withdrawals would be inventoried each year; for highly seasonal uses, monthly or weekly withdrawals would be inventoried.

(3) Periodic updating of recharge rates: The distribution and rate of recharge will continue to change as urbanization decreases pervious areas where infiltration can occur and sanitary-sewer installation reduces on-site waste-disposal systems. Elimination of combined sanitary and storm sewers might increase the number of drywells in the basin. To accommodate these changes in the model, recharge rates would be updated on the basis of current land-use and sewering practices.

This model was constructed to simulate the ground-water system at a regional scale. Many situations in the basin occur at scales that do not allow them to be addressed directly with the "regional" model. For example, determination of the effects of pumping a single well on flow in a stream 500 ft from the well is not a reasonable application of a regional-scale model in which the minimum cell dimension is 3,000 ft per side. This type of problem would be best addressed with a more detailed local analysis, possibly a "subregional" model that included only the area immediately surrounding the stream and well and used smaller model cells.

The boundaries of a subregional model could be arbitrarily located because the fluxes across them would be calculated by the regional model described in this report. This approach has been used to allow more detailed modeling of local areas within a regional ground-water flow system; examples include a study of the ground-water system on Long Island, New York (Buxton and Reilly, 1986) and a study of the Owens

Valley in California (W.R. Danskin, U.S. Geological Survey, oral commun., December 1991). Development of subregional models in the basin will require data-collection programs designed to specifically address the problems for which each subregional model is developed.

SUMMARY AND CONCLUSIONS

In order to provide an improved understanding of the regional hydrogeologic framework of the Portland Basin, the U.S. Geological Survey conducted a 3-year cooperative study of the basin with the Oregon Water Resources Department, the city of Portland Bureau of Water Works, and the Intergovernmental Resource Center located in Clark County, Washington. This report, the seventh completed as part of the cooperative study, describes a numerical model of the regional ground-water flow system and the analysis that was made using the model.

The numerical model of the ground-water flow system in the Portland Basin, Oregon and Washington, was used to (1) test and refine the conceptual understanding of the flow system, (2) estimate the effects of past and future human-caused changes to ground-water recharge and discharge on ground-water levels and streamflow, and (3) determine priorities for ground-water monitoring and data collection that would facilitate improvements in the utility and accuracy of the model. The model covered an area of 981 square miles and included most of Multnomah County, Oregon, and Clark County, Washington, as well as parts of Clackamas, Washington, and Columbia Counties in Oregon and Skamania County in Washington.

Volcanic and marine sedimentary rocks form a structural basin that has been filled with up to 1,800 ft of fluvial and lacustrine sediments. Nine hydrogeologic units have been defined on the basis of their regional hydraulic characteristics. The primary aquifers in the basin are the unconsolidated sedimentary aquifer and the Troutdale gravel aquifer; secondary aquifers include the Troutdale sandstone aquifer, the upper, coarse-grained subunit of the sand and gravel aquifer, and the older rocks where they consist of Columbia River basalts. The four remaining units function primarily as confining beds, but might locally contain coarse-grained beds used as aquifers.

Principal components of recharge to the system include infiltration of precipitation, effluent from on-site waste-disposal systems, and runoff to drywells. Recharge from direct infiltration of precipitation was estimated using a regression model relating mean annual recharge to mean annual precipitation, altitude, and impervious area. Recharge rates within the model area ranged from zero where impervious area was near 100 percent to 49 in/yr in the higher altitude, forested parts of the basin. Recharge from on-site waste-disposal systems and drywells was a significant source (as much as 26 in/yr) in unsewered areas of east Multnomah County, Oregon, and southern Clark County, Washington.

Natural discharge from the ground-water system is by seepage to rivers, streams, and, to a lesser extent, springs. Discharge also occurs from wells, and in 1988 the mean withdrawal rate was 166 ft^3/s. Ninety percent of the ground water pumped in 1988 was used for public-supply and industrial needs. Agricultural needs accounted for only 10 percent of pumpage. Eighty-five percent of the total withdrawals were from the unconsolidated sedimentary aquifer (58 percent) and the Troutdale gravel aquifer (27 percent).

A three-dimensional finite-difference model of the basin was constructed by dividing the nine hydrogeologic units into eight model layers, each having 3,040 square cells with dimensions of 3,000 ft per side. The sensitivity of simulated water levels and fluxes to uncertainty in input data, such as hydraulic conductivity and recharge, was assessed by observing the variation of simulated heads and fluxes in response to changes in hydraulic characteristics and boundary conditions. Model results were found to be most sensitive to horizontal hydraulic conductivity. A steady-state model was calibrated using time-averaged data for the 1987–88 period. Horizontal hydraulic conductivities, streambed and riverbed conductances, and vertical anisotropy ratios were the principal parameters changed during calibration. Median horizontal hydraulic conductivity ranged from 130 ft/d in the unconsolidated sedimentary aquifer to about 1 ft/d in the older rocks. Vertical anisotropy ratios ranged from 1,000:1 in the older rocks and fine-grained units to 100:1 in the primary aquifer units.

Simulated heads were compared to measured heads in 417 observation wells, and parameter adjustment proceeded until close agreement between the observed and simulated heads was obtained. Simulated stream leakage and base-flow discharge also were compared with observed data during the calibration process. The root-mean square of the difference of observed and simulated heads in the primary aquifers (unconsolidated sedimentary, Troutdale gravel, Troutdale sandstone, and sand and gravel) was used to measure the goodness of fit of the model; root-mean-square values ranged from 54 ft (mean error 11 ft) in the unconsolidated sedimentary aquifer to 74 ft (mean error –2.8 ft) in the Troutdale sandstone aquifer. Many of the larger differences in the unconsolidated sedimentary aquifer and the Troutdale gravel aquifer were in an area where wells are believed to be completed in perched water zones.

The regional-scale flow systems within the basin have their recharge areas in the upland exposures of older rocks and the Troutdale gravel aquifer and are characterized by long ground-water flow paths, which ultimately discharge to one of the basin's large rivers, such as the Columbia, Willamette, Lewis, or Clackamas. Superimposed on these regional-scale flow systems are many local-scale systems in which ground-water flow is controlled by local topographic features and discharge is to smaller rivers and streams. Most ground water in the basin moves through these local flow systems within the unconsolidated sedimentary aquifer and the Troutdale gravel aquifer before being discharged by seepage to streams or rivers. Fifty-eight percent of ground-water discharge in the basin is to smaller rivers and streams, 27 percent is to the Columbia and Willamette Rivers, and 10 percent is to wells and springs.

As a semiquantitative check on the validity of model-predicted response under another set of stresses on the system, the model was used to simulate predevelopment (no human-caused influences) steady-state conditions in the basin. A predevelopment recharge distribution was simulated, in which areas that now have drywells and on-site waste-disposal systems had reduced recharge and areas now covered by impervious surfaces had increased recharge. Recharge under predevelopment conditions was estimated to be 180 ft^3/s greater than under 1987–88 conditions. Ground-water discharge by wells was assumed to be zero. In Clark County, model-simulated predevelopment water levels in the Troutdale gravel aquifer were as much as 50 ft higher than 1987–88 water levels in areas where the aquifer has been used most heavily for public supply. Similar declines were simulated in the Troutdale sandstone aquifer near Sandy, where agricultural pumpage is concentrated. Data from long-term observation wells in the basin indicate that simulated changes in water level between predevelopment and present conditions were reasonable.

Another simulation was made to estimate the steady-state response of the ground-water flow system to a hypothetical future condition in which increased pumpage was used to meet the public-supply requirements within the basin in 2010. Total pumpage in the hypothetical condition was 55 percent greater than the 1987–88 pumpage of 166 ft^3/s. The additional pumpage was allocated within Clark County (54 ft^3/s) and in the area of the Portland well field (38 ft^3/s). Simulated water-level declines under this condition were greatest in the Troutdale sandstone aquifer on both sides of the Columbia River. Most of the additional 92 ft^3/s of pumpage was supplied by increased recharge from, and decreased discharge to, the Columbia River. Simulated base-flow discharge to Salmon Creek was decreased by as much as 8 percent in the lowermost reaches of the stream.

REFERENCES CITED

Buxton, H.T., and Reilly, T.E., 1986, A technique for analysis of ground-water systems at regional and subregional scales applied on Long Island, New York: U.S. Geological Survey Water-Supply Paper 2310, p. 129–142.

Collins, C.A., and Broad, T.M., 1993, Estimated average annual ground-water pumpage in the Portland Basin, Oregon and Washington: U.S. Geological Survey Water-Resources Investigations Report 91–4018, 26 p., 4 pls.

Cooley, R.L., and Naff, R.L., 1990, Regression modeling of ground-water flow: U.S. Geological Survey Techniques of Water-Resources Investigations, book 3, chap. B4, 232 p.

Domenico, P.A., and Schwartz, F.W., 1990, Physical and chemical hydrogeology: New York, John Wiley and Sons, 824 p.

Franke, O.L., Reilly, T.E., and Bennett, G.D., 1987, Definition of boundary and initial conditions in the analysis of saturated ground-water flow systems—an introduction: U.S. Geological Survey Techniques of Water-Resources Investigations, book 3, chap. B5, 15 p.

Franke, O.L., Reilly, T.E., 1987, The effects of boundary conditions on steady-state response of three hypothetical ground-water flow systems—Results and implications of numerical experiments: U.S. Geological Survey Water-Supply Paper 2315, 12 p.

Hansen, A.J., Jr., 1993, Modifications to the modular three-dimensional finite-difference ground-water flow model used for the Columbia Plateau Regional Aquifer System Analysis, Washington, Oregon and Idaho: U.S. Geological Survey Open-File Report 91–532, 162 p.

Harbaugh, A.W., 1990, A computer program for calculating subregional water budgets using results from the U.S. Geological Survey modular three-dimensional finite-difference ground-water flow model: U.S. Geological Survey Open-File Report 90–392, 46 p.

Hartford, S.V., and McFarland, W.D., 1989, Lithology, thickness, and extent of hydrogeologic units underlying the east Portland area, Oregon: U.S. Geological Survey Water-Resources Investigations Report 88–4110, 23 p., 6 pls.

Hogenson, G.M., and Foxworthy, B.L., 1965, Ground water in the east Portland area, Oregon: U.S. Geological Survey Water-Supply Paper 1793, 78 p., 2 pls.

Johnson, A.I., 1967, Specific yield—Compilation of specific yields for various materials: U.S. Geological Survey Water-Supply Paper 1662–D, 74 p.

McCarthy, K.A., and Anderson, D.B., 1990, Ground-water data for the Portland Basin, Oregon and Washington: U.S. Geological Survey Open-File Report 90–126.

McCarthy, K.A., McFarland, W.D., Wilkinson, J.M., and White, L.D., 1992, The dynamic relationship between ground water and the Columbia River—Using deuterium and oxygen-18 as tracers: Journal of Hydrology, v. 135, no. 1/4, p. 1–12.

McDonald, M.G., and Harbaugh, A.W., 1988, A modular three-dimensional finite-difference ground-water flow model: U.S. Geological Survey Techniques of Water-Resources Investigations, book 6, chap. A1, 576 p.

McFarland, W.D., and Morgan, D.S., 1996, Description of the ground-water flow system in the Portland Basin, Oregon and Washington: U.S. Geological Survey Water-Supply Paper 2470–A, 58 p., 7 pls.

Morgan, D.S., and Dettinger, M.D., 1994, Ground-water flow conditions in Las Vegas Valley, Nevada, Part II—Simulation analysis: U.S. Geological Survey Open-File Report 90–179, 151 p.

Mundorff, M.J., 1964, Geology and ground-water conditions of Clark County, Washington, with a description of a major alluvial aquifer along the Columbia River: U.S. Geological Survey Water-Supply Paper 1600, 268 p., 3 pls.

Orzol, L.L., and McGrath, T.S., 1992, Documentation of a software interface between a modular, finite-difference, ground-water flow model and a geographic information system data base: U.S. Geological Survey Water-Resources Investigations Report 92–50, 202 p.

Prudic, D.E., 1989, Documentation of a computer program to simulate stream-aquifer relations using a modular, finite-difference, ground-water flow model: U.S. Geological Survey Open-File Report 88–729, 113 p.

Prych, E.A., 1983, Numerical simulation of ground-water flow in Lower Satus Creek Basin, Yakima Indian Reservation, Washington: U.S. Geological Survey Water-Resources Investigations Report 82–4065, 78 p.

Snyder, D.T., Morgan, D.S., and McGrath, T., 1994, Estimates of recharge from infiltration of precipitation, on-site sewage-disposal systems, and dry wells in the Portland Basin, Oregon and Washington: U.S. Geological Survey Water-Resources Investigations Report 92–4010, 34 p.

Swanson, R.D., McFarland, W.D., Gonthier, J.B., and Wilkinson, J.W., 1993, A description of hydrogeologic units in the Portland Basin, Oregon and Washington: U.S. Geological Survey Water-Resources Investigations Report 91–4196, 56 p.

Trimble, D.E., 1963, Geology of Portland, Oregon and adjacent areas: U.S. Geological Survey Bulletin 1119, 119 p., 1 pl.

Willis, R.F., 1977, Ground water exploratory program: Portland, Oregon, Bureau of Water Works, 284 p., 17 pls.

———1978, Pilot well study: Portland, Oregon, Bureau of Water Works, 150 p., 23 pls.

APPENDIX A

Maps showing saturated thickness of hydrogeologic units

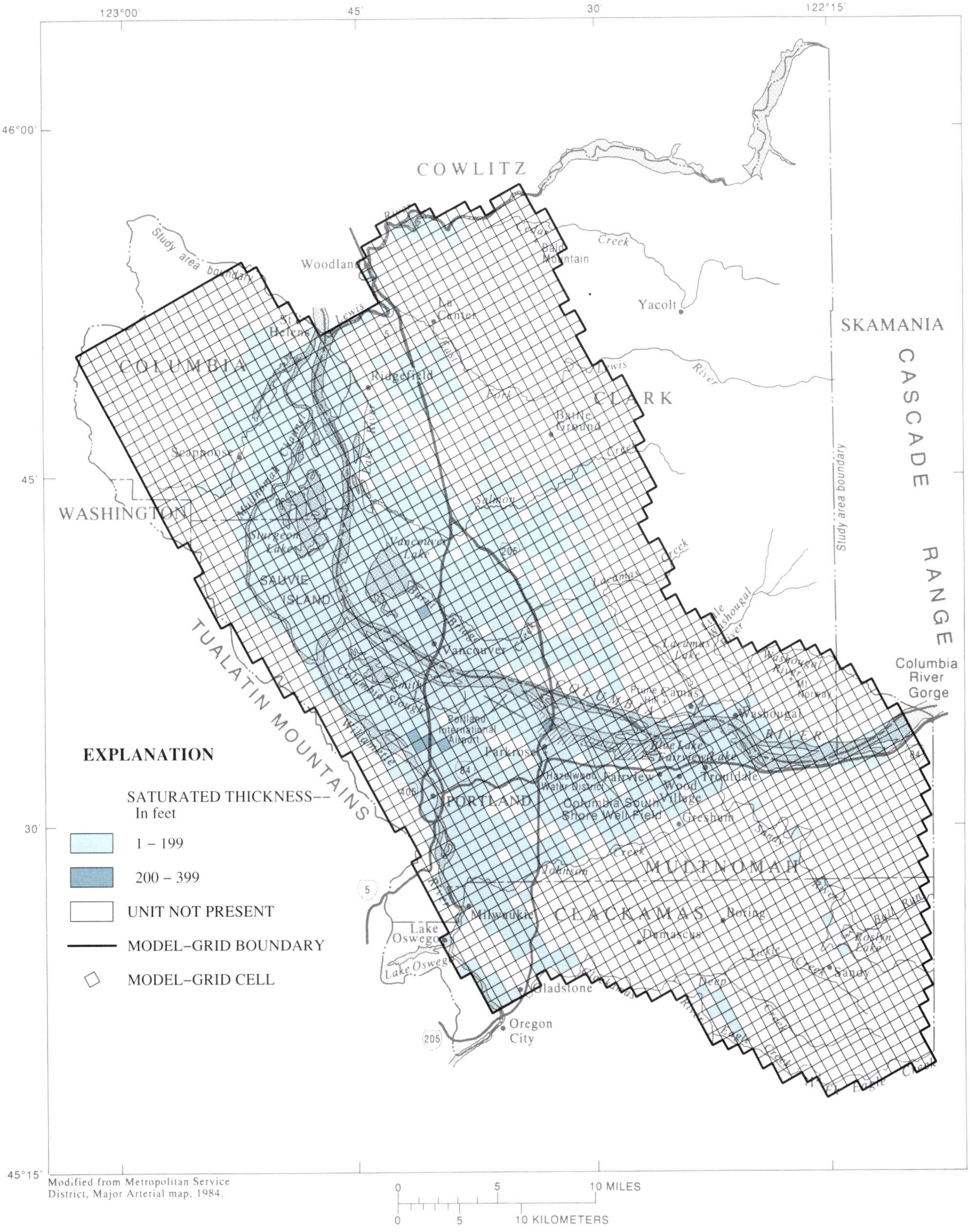

Figure A–1. Saturated thickness of unconsolidated sedimentary aquifer.

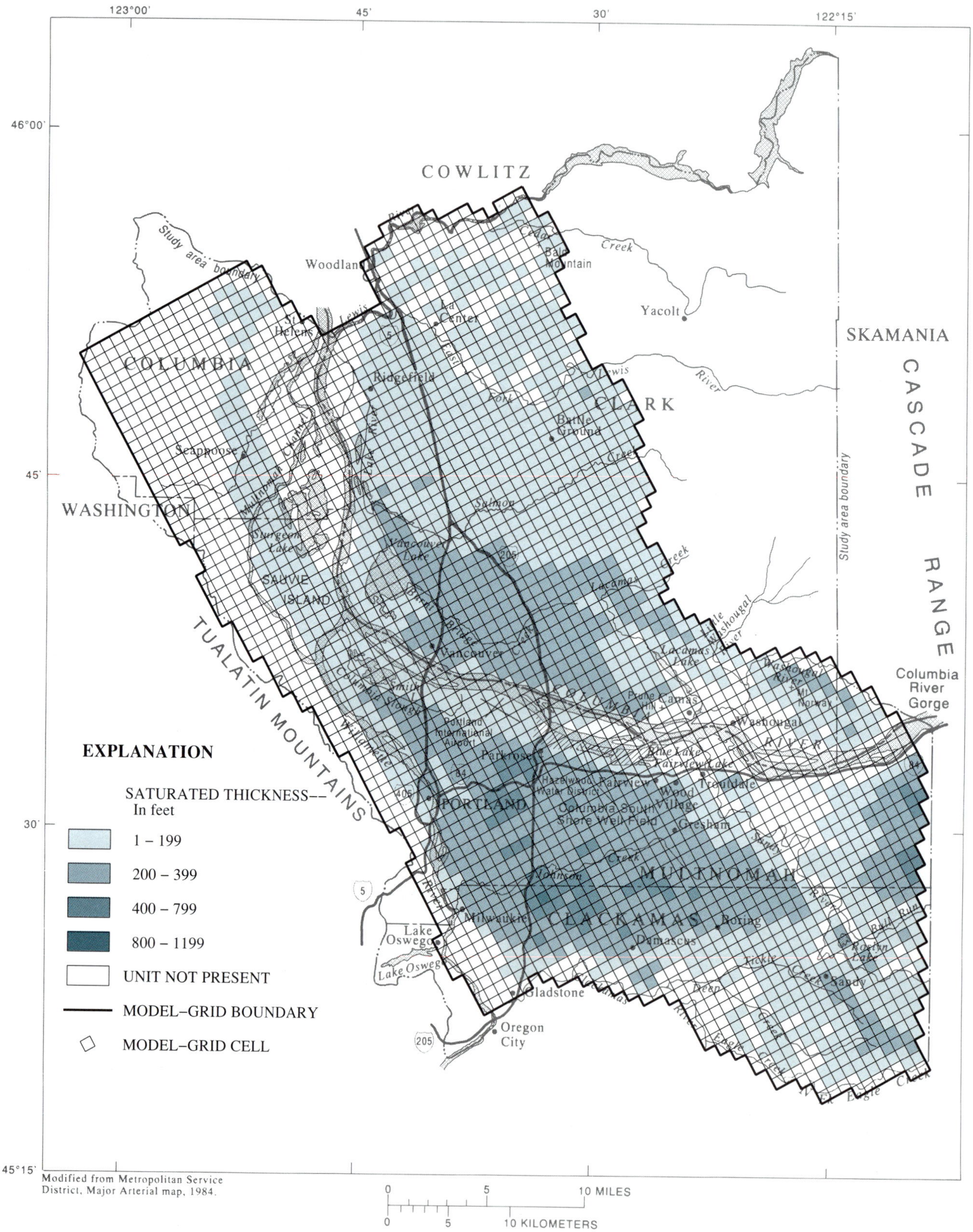

Figure A–2. Saturated thickness of Troutdale gravel aquifer.

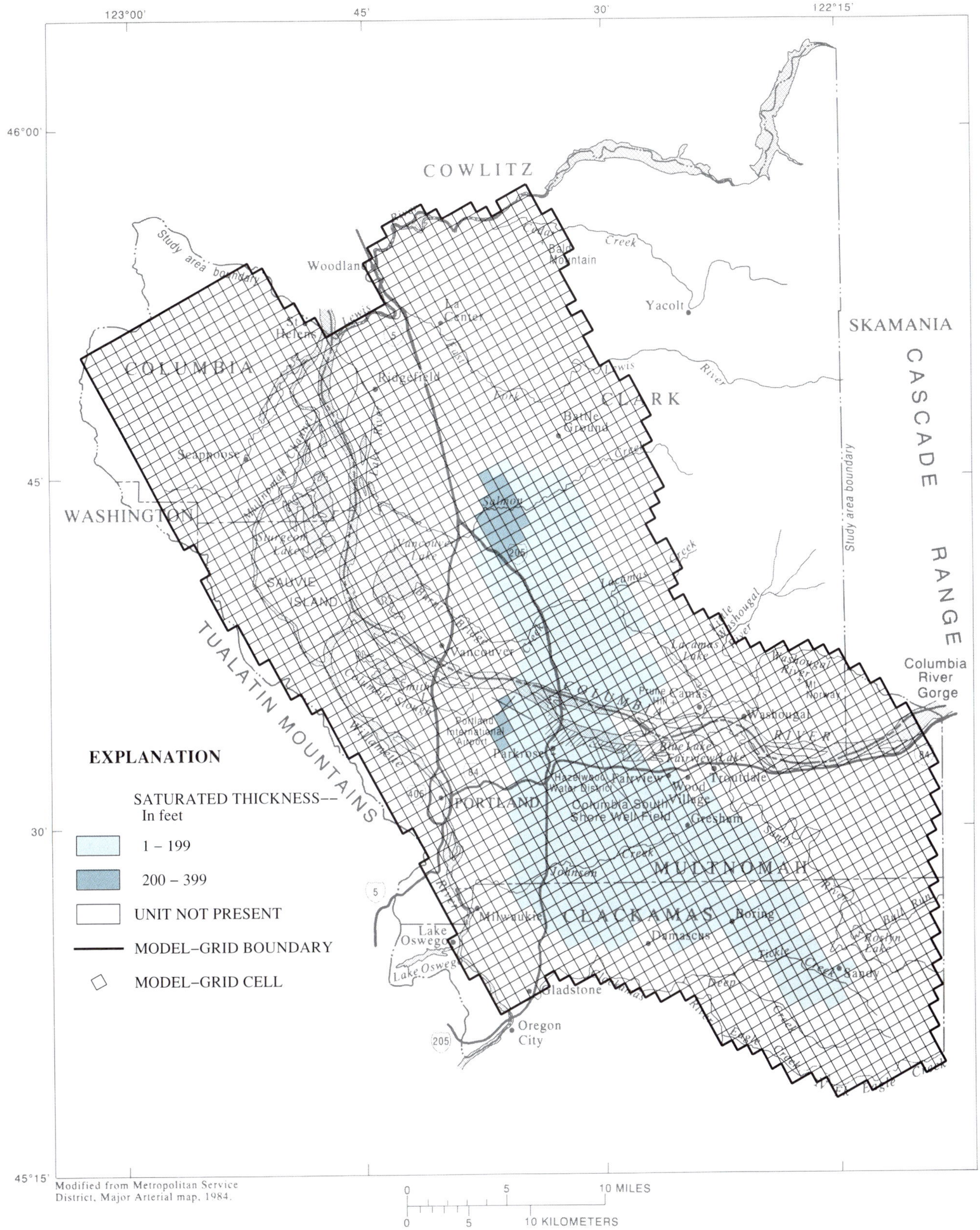

Figure A–3. Saturated thickness of confining unit 1.

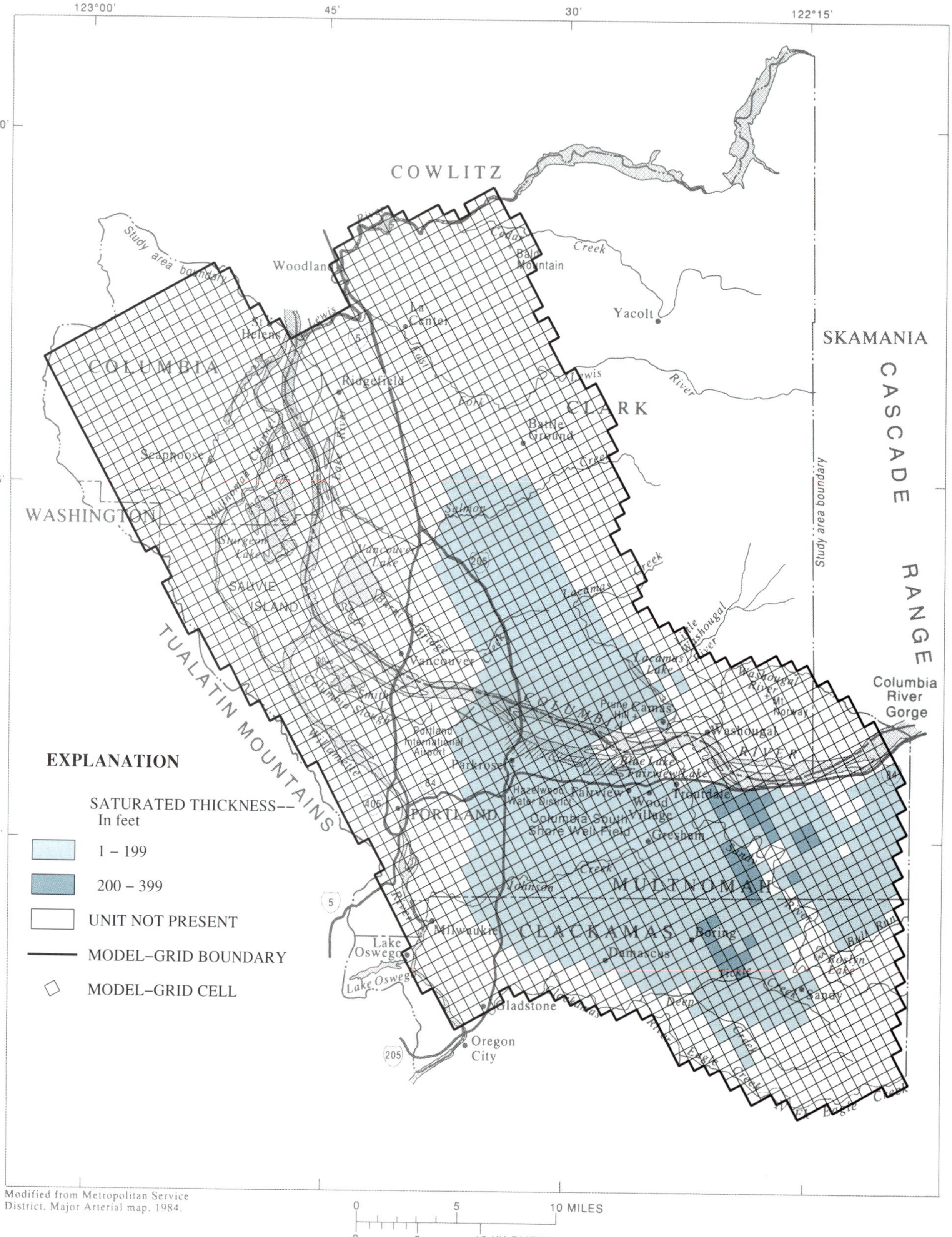

Figure A–4. Saturated thickness of Troutdale sandstone aquifer.

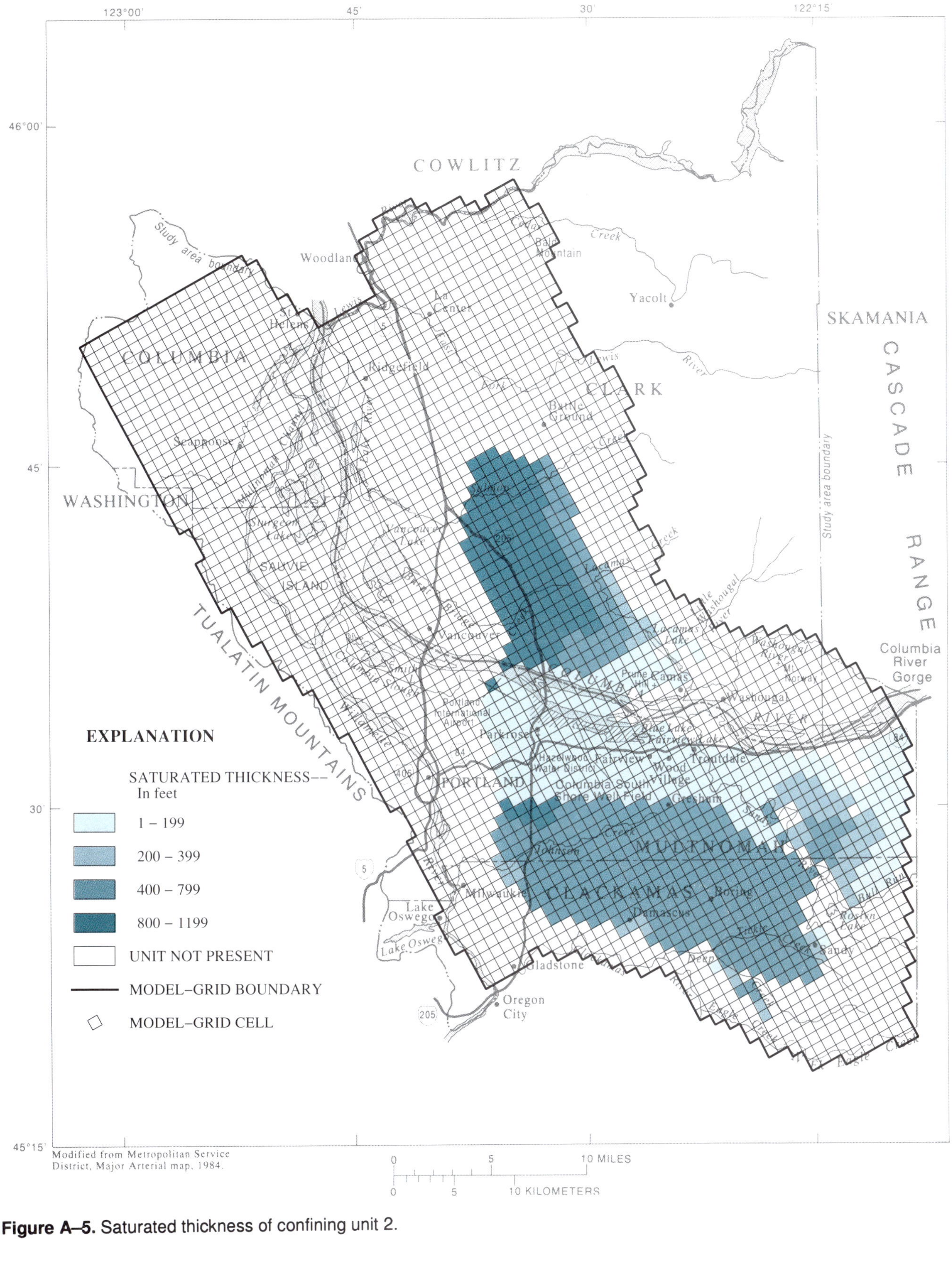

Figure A–5. Saturated thickness of confining unit 2.

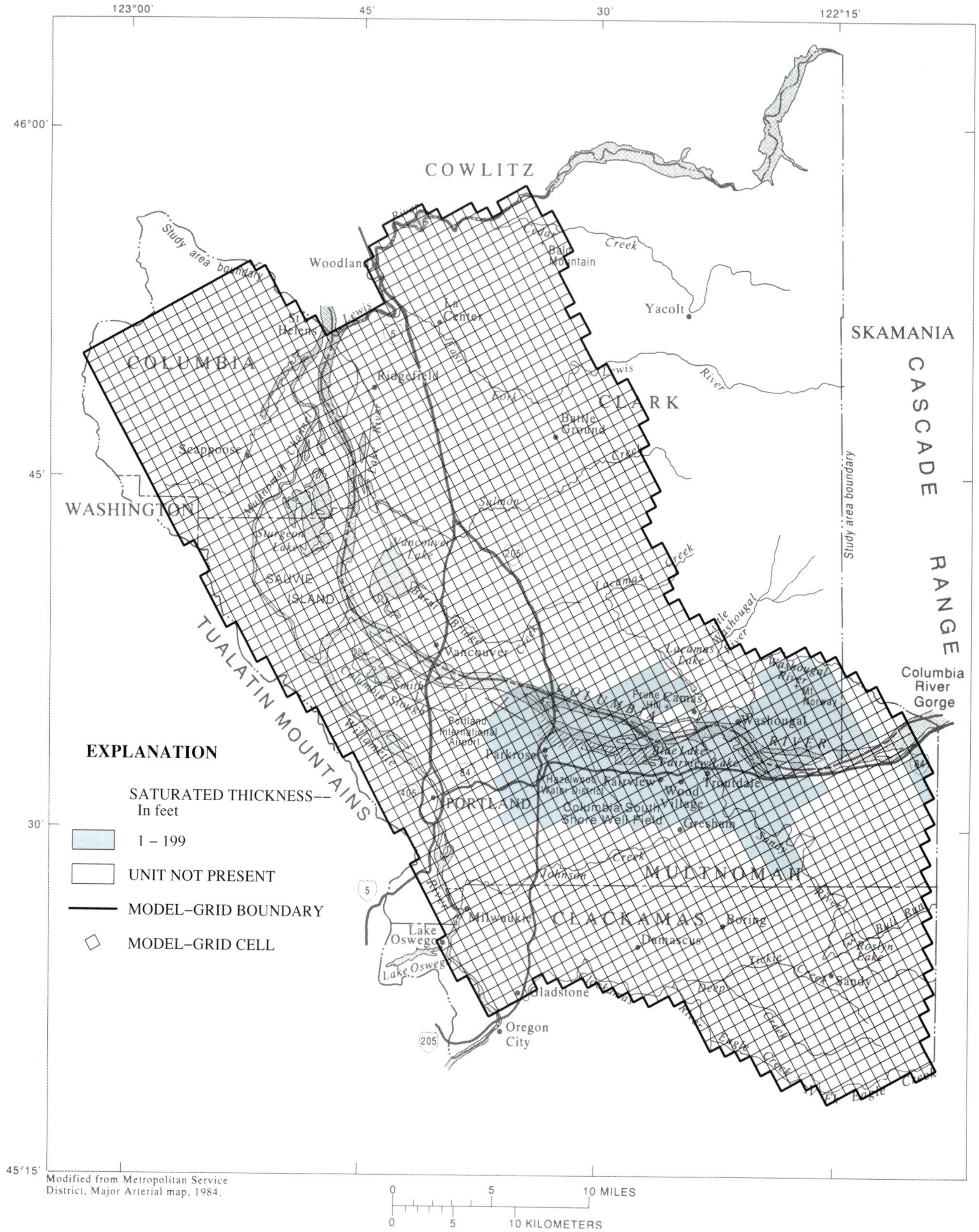

Figure A–6. Saturated thickness of the upper, coarse-grained subunit of the sand and gravel aquifer.

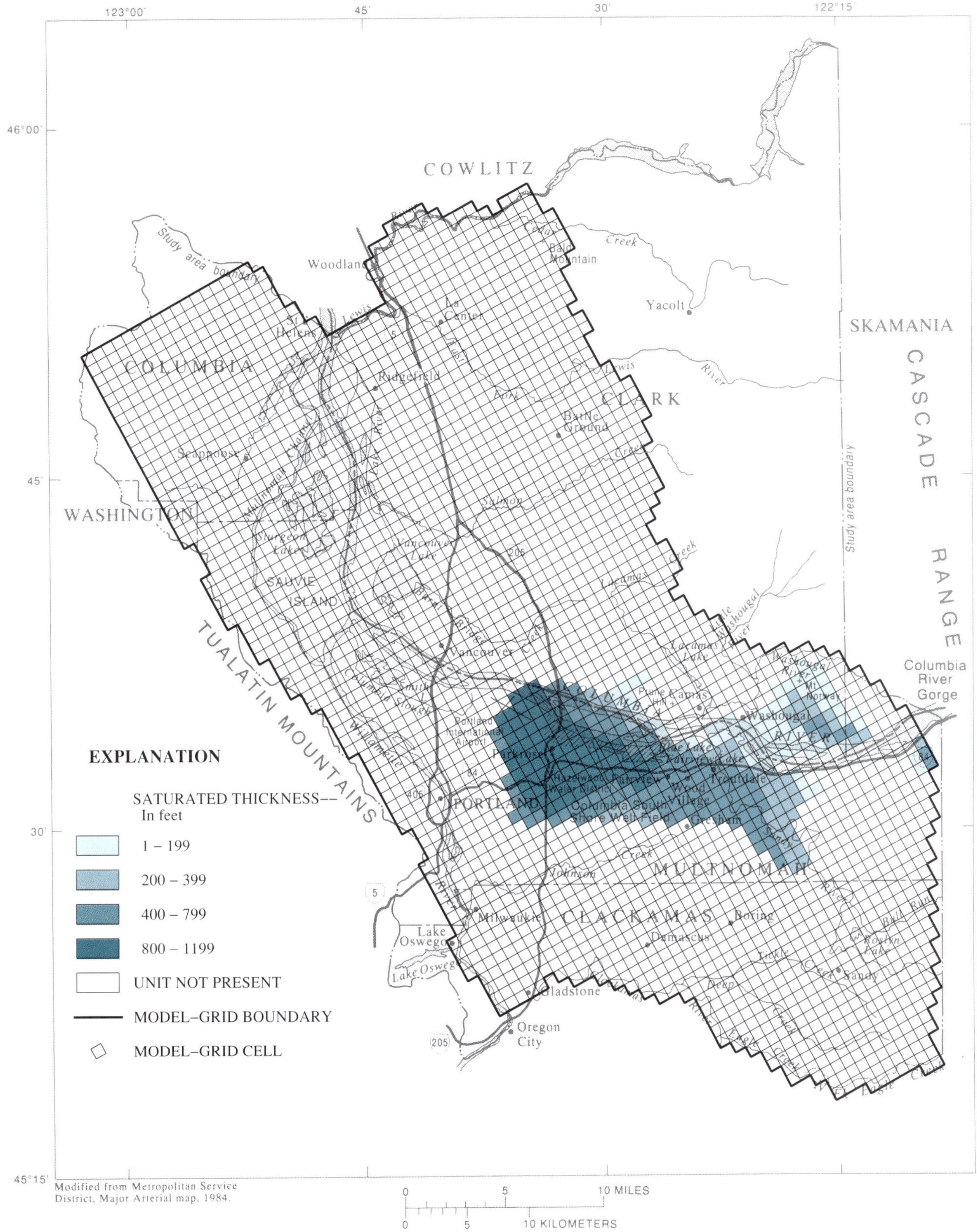

Figure A–7. Saturated thickness of the lower, fine-grained subunit of the sand and gravel aquifer.

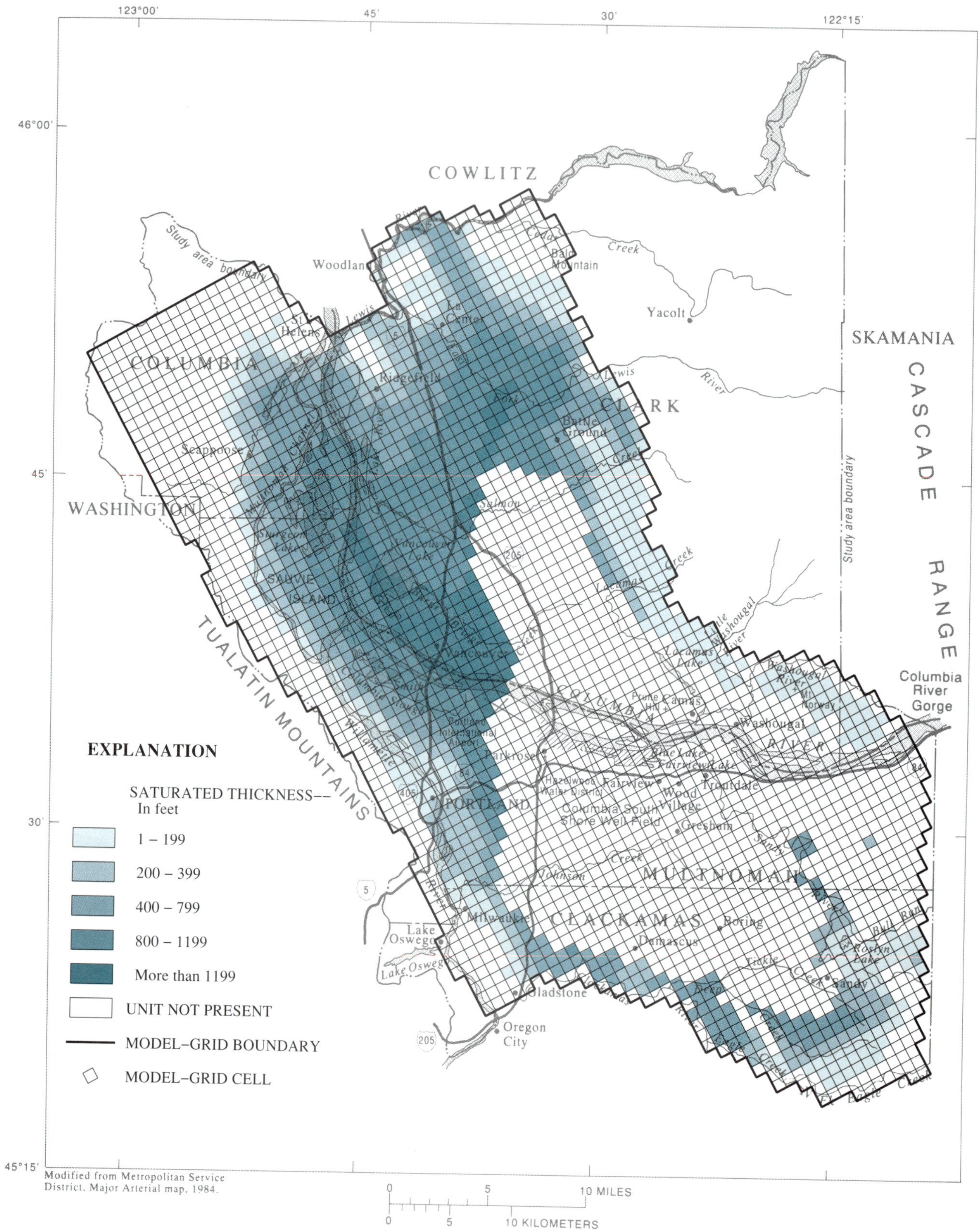

Figure A–8. Saturated thickness of the undifferentiated fine-grained sediments.

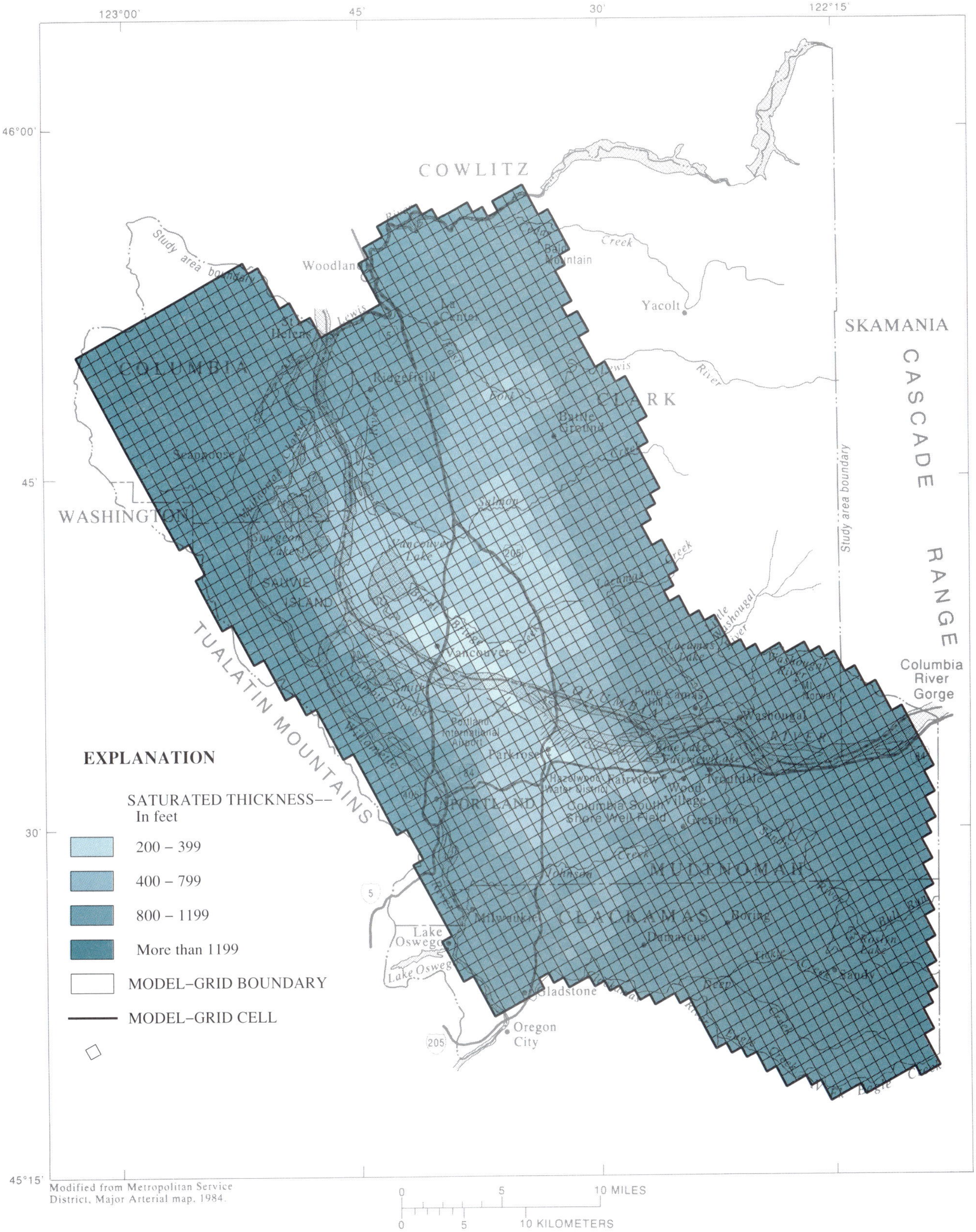

Figure A–9. Saturated thickness of older rocks above altitude of 2,000 feet below sea level.

APPENDIX B

Maps showing the areal distribution of well discharge within hydrogeologic units, 1988

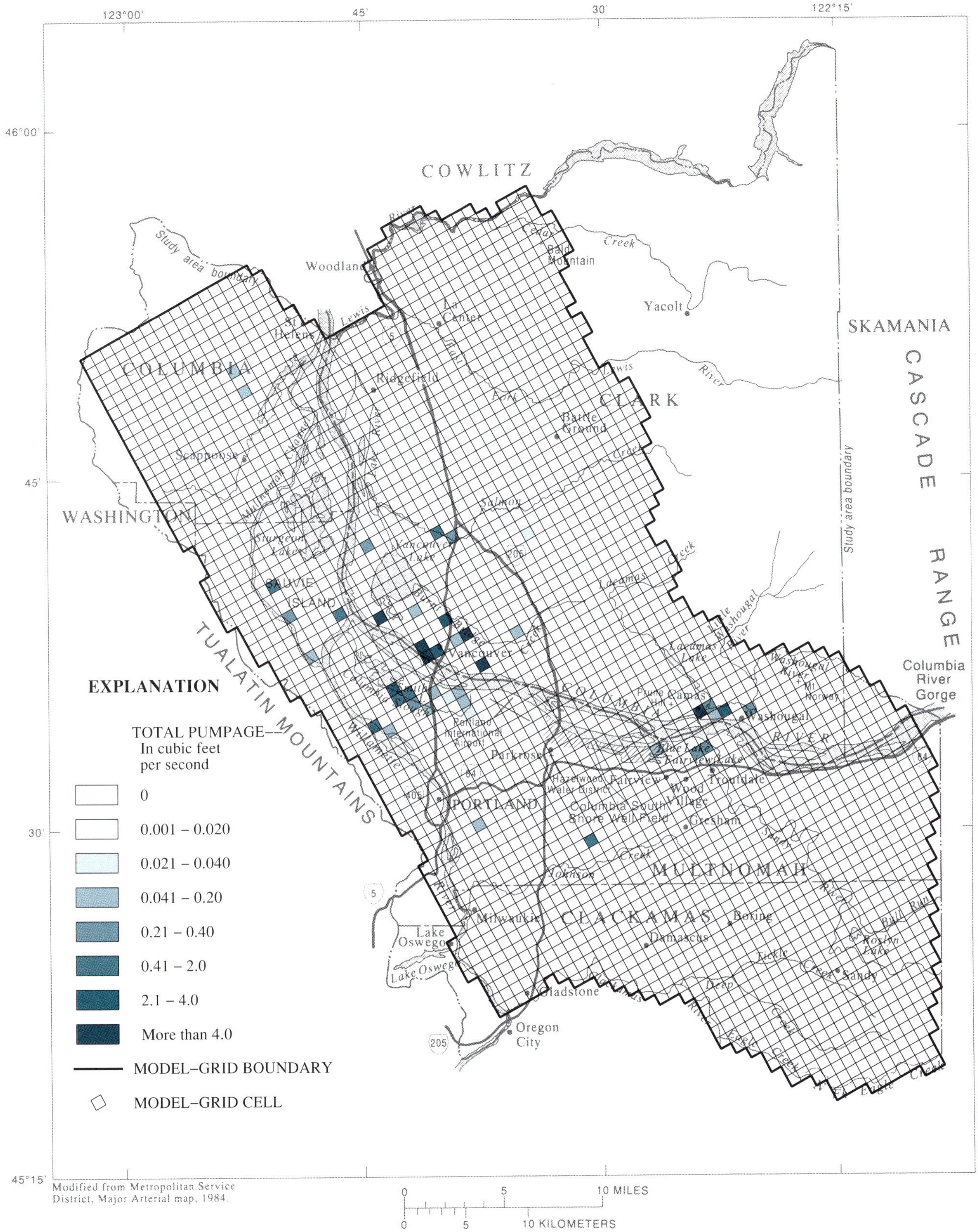

Figure B–1. Estimated rates and distribution of ground-water discharge by wells in the unconsolidated sedimentary aquifer, 1988.

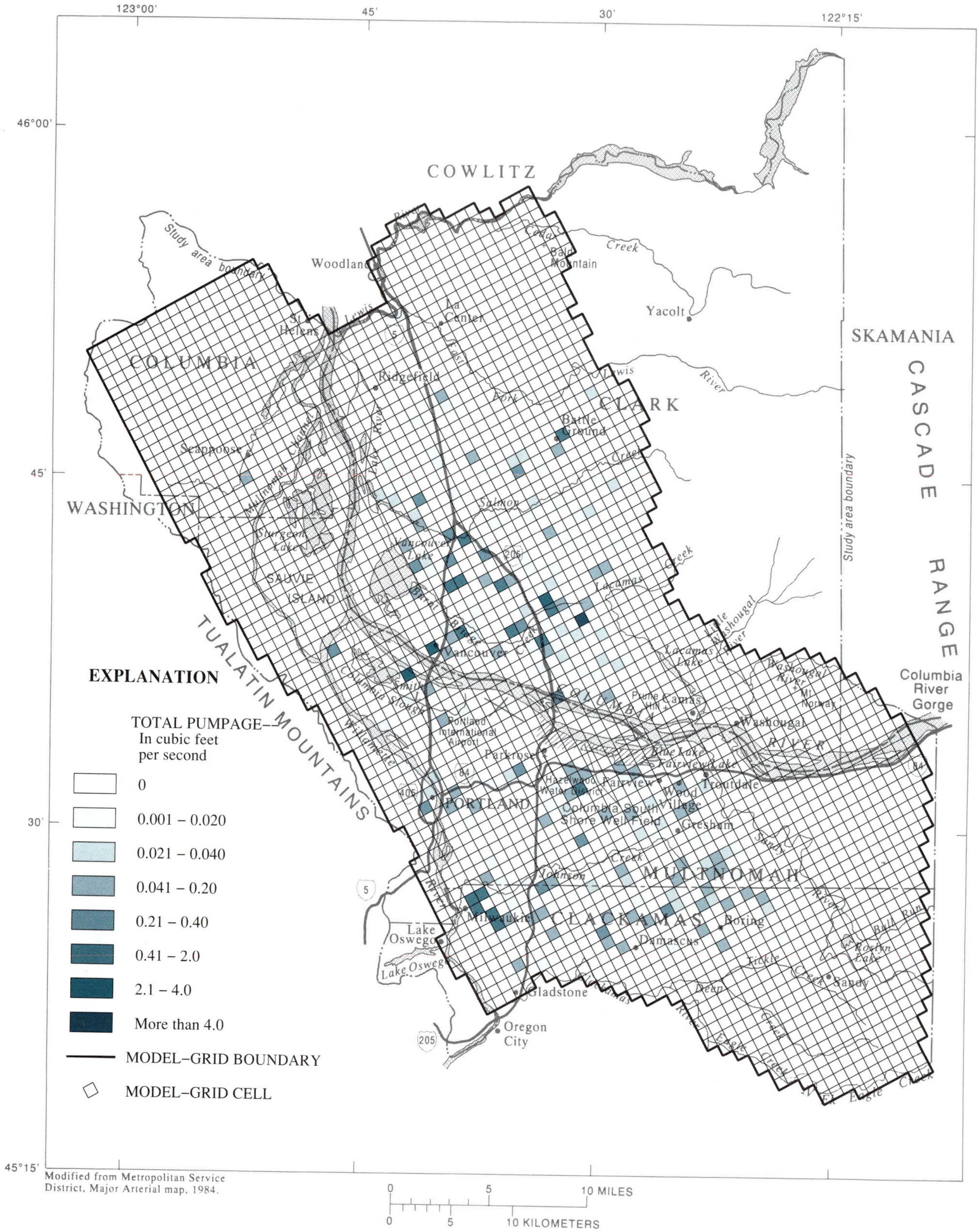

Figure B–2. Estimated rates and distribution of ground-water discharge by wells in the Troutdale gravel aquifer, 1988.

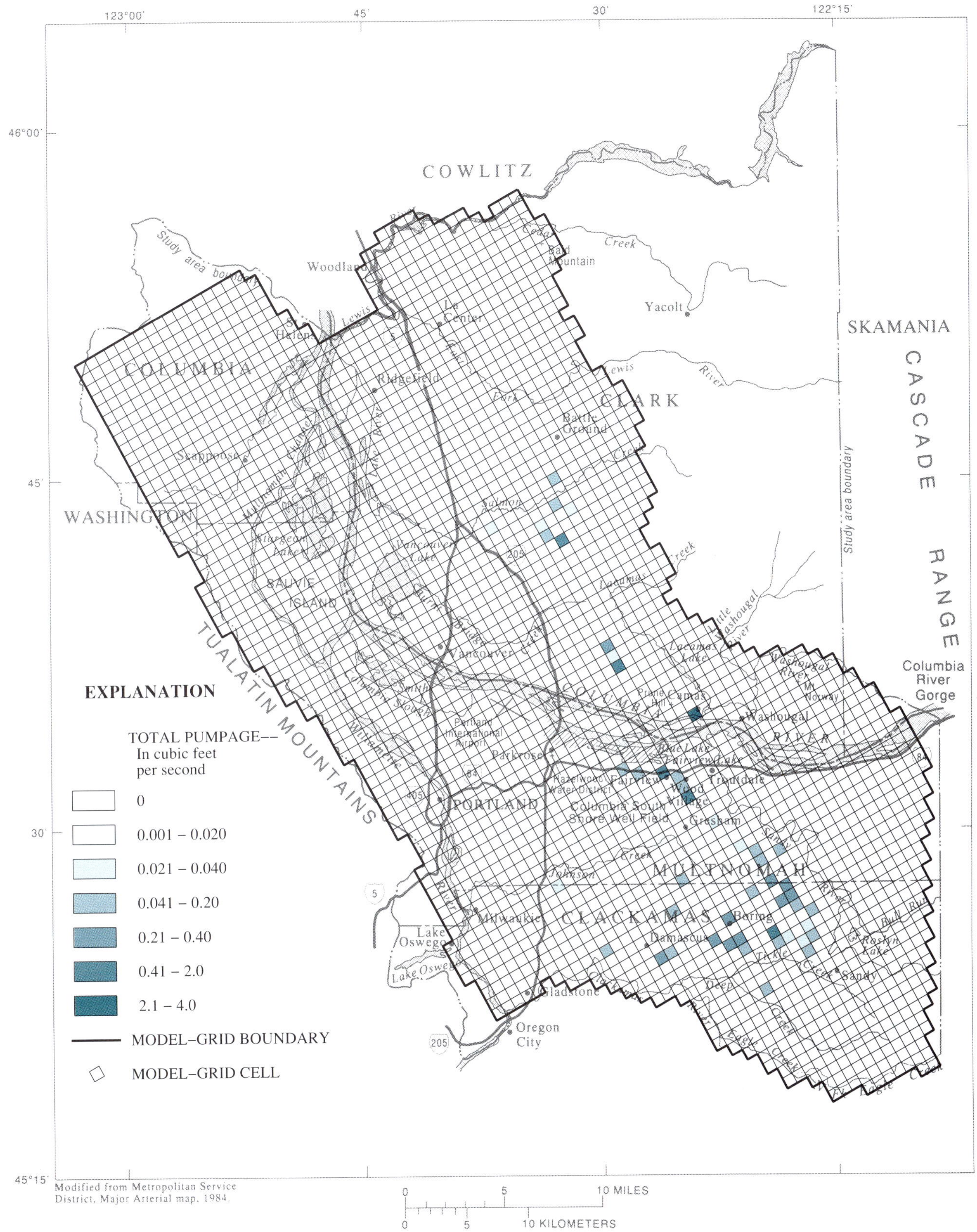

Figure B–3. Estimated rates and distribution of ground-water discharge by wells in the Troutdale sandstone aquifer, 1988.

APPENDIX C

Selected streamflow and leakage characteristics at sites in the Portland Basin, 1987–88

Appendix C

Selected streamflow and leakage characteristics at sites in the Portland Basin, 1987–88

[(A) measured low-flow discharge, September 1988; (B) estimated mean annual base flow; (C) simulated mean base flow, April 1987–March 1988; (D) measured leakage, September 1988; (E) estimated mean annual leakage; and (F) simulated mean leakage, April 1987–March 1988 at selected sites in the Portland Basin. All values are in units of cubic feet per second (ft^3/s). Segment and reach numbers were used to identify the stream in the ground-water flow model. The measurement site numbers are shown on plate 1]

Measurement location	Number		Site	Discharge			Leakage		
	Segment	Reach	ID NO.	(A)	(B)	(C)	(D)	(E)	(F)
Burnt Bridge Creek at 112th Ave.	170	7	1	1.39	5.56	0.148	-0.22	-0.88	0.148
Burnt Bridge Creek at Burbon Rd.	170	11	2	2.82	11.28	.000	-.16	-.64	.000
Burnt Bridge Creek at Evergreen St.	170	17	3	3.67	14.68	.000	-.22	-.88	.000
Burnt Bridge Creek at St. Johns Blvd.	170	19	4	3.62	14.48	.000	.01	.04	.000
Burnt Bridge Creek at Leverich Park	170	23	5	4.55	18.20	.000	-.45	-1.80	.000
Cold Canyon Creek at Hazel Dell Dr. by Burnt Bridge Creek	171	3	6	.45	1.80	.000	--	--	.000
Burnt Bridge Creek	172	1	7	4.61	18.44	.000	.08	.32	5
Cedar Creek Tributary	82	5	23	.53	2.12	.495	--	--	-.023
Pup Creek at Spurrel Rd. near Cedar Creek	84	1	24	1.11	4.44	4.400	--	--	-.337
Cedar Creek at Grist Mill	85	4	27	13.60	54.40	48.641	-.85	-3.40	-.983
Cedar Creek Tributary	86	2	29	--	--	.057	--	--	.057
Cedar Creek Tributary	88	2	30	--	--	.041	--	--	.041
Cedar Creek at confluence	89	2	31	12.70	50.80	56.261	.03	.12	-.019
East Fork Lewis River Tributary	92	3	41	.10	.40	.000	--	--	.000
East Fork Lewis River at Lewisville Park	94	9	42	43.70	174.80	189.087	.02	.08	-.965
East Fork Lewis River at Daybreak Park	94	18	43	47.80	191.20	211.734	-.52	-2.08	-2.238
Mill Creek near East Fork Lewis River	96	5	44	.15	.60	.211	--	--	-.057
East Fork Lewis River Tributary	95	6	45	.52	2.08	1.423	--	--	-.037
East Fork Lewis River above Mason Creek	97	10	48	58.10	232.40	240.816	-.64	-2.56	-1.267
Mason Creek at confluence near East Fork Lewis River	98	19	51	1.45	5.80	11.012	--	--	-.022
East Fork Lewis River Tributary	100	9	52	.14	.56	.000	--	--	.000
East Fork Lewis River Tributary	102	5	53	--	--	.000	--	--	.000
East Fork Lewis River above Lockwood Creek	103	2	54	65.00	260.00	262.567	-.23	-.92	-1.088
Lockwood Creek near East Fork Lewis River	104	14	55	1.07	4.28	3.120	--	--	-2.575
Brezee Creek near East Fork Lewis River	106	7	56	.68	2.72	1.308	--	--	-.341
East Fork Lewis River Tributary	108	3	57	.10	.40	.000	--	--	.000
McCormick Creek near East Fork Lewis River	110	8	58	.16	.64	.000	--	--	-.557
Jenny Creek near East Fork Lewis River	112	6	59	.29	1.16	.000	--	--	-.037
Jenny Creek Tributary near East Fork Lewis River	112	7	60	--	--	.037	--	--	.002
Fifth Plain Creek NE 212th Ave.	176	1	61	.09	.36	.000	-.05	-.20	.000
Fifth Plain Creek at NE Davis Rd.	176	6	62	.26	1.04	.248	-.06	-.24	-.789
Shanghai Creek at 222nd Ave. near Fifth Plain Creek	177	3	63	.56	2.24	2.110	--	--	-.319
Shanghai Creek at NE 212 Ave. near Fifth Plain Creek	177	4	64	.60	2.40	2.429	-.03	-.12	-.671
Shanghai Creek at Fifth Plain Creek	177	8	65	.54	2.16	4.320	--	--	-.595
Fifth Plain Creek at Ward Rd.	178	2	66	.76	3.04	6.408	.00	.00	-.076
China Ditch at Ward Rd. near Fifth Plain Creek	175	6	67	.15	.60	2.597	--	--	-.116
Fifth Plain Creek at Hwy 500	179	4	68	3.10	12.40	10.267	-.43	-1.72	-.279
Gee Creek at Park	127	5	69	.12	.48	.293	-.03	-.12	-.140
Gee Creek at NW Royle Rd.	127	7	70	.43	1.72	.421	-.19	-.76	-.111
Gee Creek Tributary	128	4	71	.10	.40	.000	--	--	.000
Gee Creek at Hwy 501	129	1	72	--	--	1.689	--	--	.061
Gee Creek at Abrams Park	129	3	73	.78	3.12	1.643	-.07	-.28	-.044

Appendix C

Measurement location	Number		Site ID NO.	Discharge			Leakage		
	Segment	Reach		(A)	(B)	(C)	(D)	(E)	(F)
Matney Creek at NE 68th St. near Lacamas Creek	181	2	75	1.06	4.24	0.000	--	--	-0.615
Lacamas Creek at NE 217th Ave.	182	2	76	6.32	25.28	16.732	-0.03	-0.12	-.130
Lacamas Creek at Hwy 500	182	6	77	6.54	26.16	19.402	-.07	-.28	-1.726
Lacamas Creek at Lacamas Park	183	7	79	12.30	49.20	36.606	-.68	-2.72	-.909
Little Washougal River at SE Blair Rd.	190	3	80	8.49	33.96	34.755	--	--	-2.581
Little Washougal River at Blair (Gage Site)	190	6	81	7.98	31.92	42.310	.13	.52	-.528
Salmon Creek at 182nd Ave.	138	8	84	4.04	16.16	18.650	-.03	-.12	-1.308
Salmon Creek at 167th Ave.	138	9	85	4.27	17.08	19.958	-.22	-.88	-2.263
Morgan Creek at NE 167th Ave. near Salmon Creek	137	2	86	1.64	6.56	4.513	--	--	-.309
Salmon Creek at Hwy 503	139	9	87	7.20	28.80	37.111	-.10	-.40	-.820
Weaver Creek near Hwy 503 near Salmon Creek	140	11	88	1.43	5.72	.192	--	--	-.320
Salmon Creek Tributary at end of 112th Ave.	142	6	89	.52	2.08	.000	--	--	-.061
Salmon Creek at 112th Ave.	143	1	90	9.13	36.52	39.999	.02	.08	-.696
Curtin Creek at 139th St. near Salmon Creek	144	11	93	3.43	13.72	1.342	--	--	-.748
Salmon Creek at 72nd Ave.	145	1	94	13.20	52.80	48.417	.00	.00	-.182
Mill Creek at Salmon Creek Rd. near Salmon Creek	148	15	95	.64	2.56	5.990	--	--	-.073
Salmon Creek at I-205 Overpass	149	5	96	17.80	71.20	60.089	-.33	-1.32	-2.002
Salmon Creek Tributary at 119th St.	150	4	97	.80	3.20	.000	--	--	.000
Salmon Creek at Hwy 99	151	3	98	19.50	78.00	57.858	-.22	-.88	3.026
Salmon Creek Tributary 200 feet east of Interstate-5	152	7	99	.38	1.52	.000	--	--	.000
Salmon Creek below Klineline Pond	153	2	101	19.60	78.40	60.028	.38	1.52	-1.625
Cougar Canyon Creek at 119th St. near Salmon Creek	154	6	103	.77	3.08	.000	--	--	.000
Salmon Creek above 36th Ave.	155	3	104	23.40	93.60	74.896	-.75	-3.00	-.684
Whipple Creek at Union Rd.	158	5	105	--	--	.000	--	--	.000
Whipple Creek at 11th Ave.	158	8	106	.42	1.68	.000	--	--	.000
Packard Creek at 179th St. near Whipple Creek	159	6	108	.16	.64	.000	--	--	.000
Whipple Creek at 179th St.	160	1	109	2.65	10.60	.000	-.04	-.16	.000
Whipple Creek Tributary at end of 189th St.	161	4	110	.18	.72	.000	--	--	-.231
Beaver Creek at Cochran Rd.	15	12	401	.04	.16	.000	--	--	.000
Kelly Creek at Mt. Hood Community College near Beaver Creek	14	11	402	.64	2.56	.000	--	--	.000
Beaver Creek at Troutdale Rd.	16	3	403	1.44	5.76	.000	-.57	-2.28	.000
Beaver Creek at Columbia Rd.–Troutdale	16	6	404	.91	3.64	.344	.07	.28	-.026
Mt. Scott Creek at Mt. Scott Creek Rd.	51	4	409	--	--	.000	--	--	.000
Mt. Scott Creek at Sunnyside	51	5	410	.02	.08	.000	-.01	-.04	.000
Mt. Scott Creek at SE 97th	51	7	411	.04	.16	.000	-.01	-.04	-.011
Mt. Scott Creek behind COSTCO	51	9	412	.15	.60	.000	-.04	-.16	-.776
Mt. Scott Creek at Rusk Rd.	51	12	415	2.05	8.20	2.275	-.91	-3.64	-1.019
Mt. Scott Creek near Kellogg Creek	51	13	416	1.91	7.64	3.294	.09	.36	-1.508
Rock Creek at Troge Rd.	40	8	421	.05	.20	.361	--	--	-.301
Rock Creek at Sunnyside Rd.	40	12	422	.20	.80	1.390	-.08	-.32	-.800
Rock Creek at Hwy 224	40	15	423	.58	2.32	3.094	-.18	-.72	-2.701

Appendix C

Measurement location	Number Segment	Number Reach	Site ID NO.	Discharge (A)	Discharge (B)	Discharge (C)	Leakage (D)	Leakage (E)	Leakage (F)
Richardson Creek at Royer Rd.	38	3	428	0.02	0.08	0.000	--	--	0.000
Richardson Creek at Hwy 224	38	8	429	.25	1.00	1.703	-0.07	-0.28	-.726
Kellogg Creek at Webster Rd.	52	2	434	--	--	.000	--	--	.000
Kellogg Creek at Theisen Rd.	52	4	435	.13	.52	.009	-.07	-.28	-.015
Kellogg Creek at Rusk Rd.	52	7	436	1.89	7.56	.444	-1.40	-5.60	-.085
Kellogg Creek at Oatfield Rd.	53	3	437	3.70	14.80	6.937	-.73	-2.92	-.041
Tributary to Kellogg Creek at McLoughlin Blvd.	54	2	439	.15	.60	.000	--	--	.000
Tickle Creek at Hwy 211	25	10	444	.05	.20	.576	--	--	-.063
Tickle Creek at 362nd	25	14	446	.40	1.60	1.451	-.12	-.48	-.742
Tickle Creek at Colorado Rd.	25	18	447	1.18	4.72	5.276	-.12	-.48	-.625
Tickle Creek Tributary at Colorado Rd.	26	6	448	.15	.60	3.486	--	--	-.825
Tickle Creek at Deep Creek Rd.	27	4	450	2.39	9.56	13.224	-.40	-1.60	-2.758
Deep Creek at Crane Rd.	22	14	455	.39	1.56	13.561	--	--	-1.614
Deep Creek Tributary near Hwy 211	23	11	457	.13	.52	3.601	--	--	.10
Deep Creek at Hwy 211	24	1	458	1.02	4.08	25.049	-.03	-.12	-.103
Deep Creek at Hoist Rd.	24	8	459	1.22	4.88	30.340	-.04	-.16	-.523
Deep Creek at Aemisegger Rd.	28	2	460	4.07	16.28	49.866	-1.50	-6.00	-.593
Deep Creek Tributary at Aemisegger Rd.	29	4	461	.05	.20	1.083	--	--	-.195
North Fork Deep Creek at confluence near Deep Creek	33	17	462	.78	3.12	9.475	--	--	-.081
Noyer Creek at Deep Creek Rd. near Deep Creek	35	7	463	.03	.12	.257	--	--	-2.591
Deep Creek at Hwy 224	36	1	464	5.28	21.12	67.512	-.02	-.08	-.567
Johnson Creek at Palmblad Rd.	47	6	469	.08	.32	3.001	--	--	-.537
Johnson Creek at Walters Rd.	47	10	470	.42	1.68	4.120	-.08	-.32	-.192
Johnson Creek at SE 190th Ave.	47	14	471	.47	1.88	4.933	-.02	-.08	-.159
Johnson Creek at Gaging Station near Foster Rd.	47	17	472	.64	2.56	7.199	-.05	-.20	-.593
Johnson Creek at SE 112th & Brookside	47	21	473	.74	2.96	9.600	-.03	-.12	-2.604
Johnson Creek at 82nd Ave.	47	26	474	.92	3.68	8.712	-.01	-.04	-.145
Johnson Creek at SE 45th and Johnson Creek Blvd.	47	31	475	1.35	5.40	2.558	-.09	-.36	-1.147
Johnson Creek at Tidemar Johnson Park	47	32	478	3.45	13.80	3.705	-.67	-2.68	.811
Johnson Creek above Crystal Springs Creek	47	34	479	2.91	11.64	2.190	.30	1.20	1.792
Crystal Springs Creek at Sherrett St. by Johnson Creek	48	6	480	10.50	42.00	.000	--	--	-.708
Johnson Creek at SE River Rd.	49	4	482	17.70	70.80	1.142	-.84	-3.36	-.343
Sandy River near Sandy	2	10	483	219.00	876.00	904.238	--	--	-7.481
Cedar Creek at Fish Hatchery near Sandy River	1	13	484	3.55	14.20	34.006	--	--	-2.209
Bull Run River above Sandy River at Dodge Park	6	7	485	160.00	640.00	755.737	--	--	-.392
Sandy River at Gaging Station near Bull Run, Oregon	7	1	486	370.00	1,480.00	1,730.455	.68	2.72	-1.437
Walker Creek at confluence near Sandy River	8	6	487	3.03	12.12	.956	--	--	.014
Sandy River at Oxbow Park	9	10	493	294.00	1,176.00	1,769.988	1.98	7.92	-2.890
Trout Creek at Gordon Creek Rd. near Sandy River	10	14	494	4.35	17.40	.000	--	--	-.000
Gordon Creek at Gordon Creek Rd. near Sandy River	12	14	495	11.70	46.80	53.724	--	--	-.076
Sandy River at Stark St.	13	18	496	317.00	1,268.00	1,866.029	-.56	-2.24	-2.002
Sandy River at Troutdale	13	23	497	318.00	1,272.00	1,881.473	-.21	-.84	-3.907

☆ U.S. GOVERNMENT PRINTING OFFICE: 1997-547-442

Hannsjörg Schröder • Rob A. I. de Vos
Stefan Huggenberger • Lennart Müller-Thomsen
Annemieke Rozemuller • Farman Hedayat
Natasha Moser

The Human Brainstem

Anatomy and Pathology

Springer

Hannsjörg Schröder
Porto, Portugal

Stefan Huggenberger
Institute of Anatomy and Clinical Morphology
Witten/Herdecke University
Witten, Germany

Annemieke Rozemuller
Department of Pathology
Amsterdam Universitair Medische
Centra, location VUmc
Amsterdam, The Netherlands

Natasha Moser
Heilig Geist-Hospital
Dept. of Neurology
Köln, Germany

Rob A. I. de Vos
Alkmaar, The Netherlands

Lennart Müller-Thomsen
Institute of Anatomy and Clinical Morphology
Witten/Herdecke University
Witten, Germany

Farman Hedayat
Rückenzentrum Grafenberg
Neurochirurgische Praxis
Düsseldorf, Germany

ISBN 978-3-030-89979-0 ISBN 978-3-030-89980-6 (eBook)
https://doi.org/10.1007/978-3-030-89980-6

Cover illustration: Darrow red-stained horizontal section through the human brainstem at the level of the upper medulla oblongata. LabPON Twente.

This Springer imprint is published by the registered company Springer Nature Switzerland AG
The registered company address is: Gewerbestrasse 11, 6330 Cham, Switzerland

Preface

The idea for this book goes back to the scientific collaboration of Rob de Vos and Hannsjörg Schröder—supported by Farman Hedayat—and various seminars on the human brainstem and its disorders. The goal was to present a book that synoptically shows the normal and pathological anatomy of the human brainstem, together with the peripheral targets of the cranial nerves and their living anatomy, supplemented by a histological MRT and topography atlas of the human brainstem. To the best of our knowledge, this is the first book which jointly deals with all these various aspects.

Our first sincere thanks go to Dr. Ina Stoeck, Springer Heidelberg, who cared for this project in the same excellent way she did for our previous book on the mouse neuroanatomy. She did the utmost to meet the ideas of the authors and showed never-ending patience for the problems that are inherent to a book with several authors and their often challenging ideas.

In the end, we are convinced that she cared for an outstanding synthesis which hopefully will be of value to the potential readers of this book. In our opinion, their spectrum will include neuroanatomists, neuropathologists, neuroradiologists, neurologists, neurosurgeons, and ENT specialists.

From a technical point of view, this book would never have been possible without the relentless assistance of Sophia Lara Schröder, M.Sc., and Elena Schmidt, M.Sc., who cared for the organization of the book as to proofreading, copyright permissions, checking photographs and references, and a lot of things that were essential for the generation of this book. These extra tasks could only be realized by the financial support of the Human Brainstem Foundation, Enschede, The Netherlands.

Many illustrations in this book originate from the inventories of the authors. However, it would not be complete without the kind contribution of colleagues who provided additional material from their special field of expertise.

We would like to express our sincere gratitude to those who granted us this courtesy (in alphabetical order):

Dr. M. Bugiani
Department of Pathology, Amsterdam Universitair Medische Centra, Amsterdam, The Netherlands
R. Charles
Medneo Diagnostikzentrum, Düsseldorf, Germany
3T-MRT of the brainstem
B. Dawidowski
Zentrum Anatomie, Universitätsklinikum Köln, Köln, Germany
R. Hamoen, MSc—Directeur—Bestuurder
Laboratorium Pathologie Oost-Nederland Twente, Hengelo, The Netherlands
– Contributions labeled in the book as: **LabPON Twente**

Prof. G. Feigl
Institute of Anatomy and Clinical Morphology, Witten/Herdecke University, Witten, Germany
Prof. em. Jürgen Koebke†
Anatomisches Institut II, Universitätsklinikum Köln, Köln, Germany
- Provided the plastinated sections of the head to the Sammlung des Zentrums Anatomie der Universität zu Köln

Prof. G. Reiss
Institute of Anatomy and Clinical Morphology, Witten/Herdecke University, Witten, Germany
- Contributions labeled in the book as **Collection A. Rozemuller** have been provided by the following colleagues

Prof. I. Huitinga
Director of the Netherlands Brain Bank, Netherlands Institute of Neuroscience, Amsterdam, The Netherlands
Prof. M. J. van de Vijver
Department of Pathology, Amsterdam Universitair Medische Centra, Amsterdam, The Netherlands
Prof. J. van Diest
Department of Pathology, Universitair Medisch Centrum Utrecht, Utrecht, The Netherlands

Prof. em. M.-M. Ruchoux
Faculté de Médecine de Lille, Hôpital Roger Salengro, CHRU Lille, Lille, France
Prof. U. Rüb†
Dr. Senckenbergischen Anatomie, Goethe-Universität Frankfurt am Main, Frankfurt am Main, Germany
Critical advice as to human brainstem anatomy
Prof. Y. Saito
Brain Bank for Aging Research, Tokyo Metropolitan Institute of Gerontology, Tokyo, Japan
Prof. M. Scaal
Zentrum Anatomie, Universitätsklinikum Köln, Köln, Germany
- Contributions labeled in the book as: **Sammlung des Zentrums Anatomie der Universität zu Köln**

S. L. Schröder, M.Sc.
Department of Biology, Köln, Germany
Prof. em. E. Stennert
Universitätsklinikum Köln, Köln, Germany
Dr. C. Stückle
Privatpraxis für offene MRT, Europaplatz 11 44269 Dortmund, Germany
Prof. D.R. Thal
Department of imaging and Pathology, KU Leuven, Leuven, Belgium
Prof. Dr. D. Uhlenbrock
MVZ Prof. Dr. Uhlenbrock GmbH, Dortmund, Germany
Thin layer MRT of the brainstem
Dr. T. Voigt
Institute of Anatomy and Clinical Morphology, Witten/Herdecke University, Witten, Germany

For technical assistance, we are greatly indebted to (in alphabetical order):
Laboratorium Pathologie Oost-Nederland Twente LabPON
R. Rieksen for outstanding photographic work
L. Stokkink for excellent technical assistance

Zentrum Anatomie Unversität zu Köln
I. Koch for excellent photographic work
P. Lück for excellent secretarial assistance
H.-P. Notermanns for the excellent technical execution of the plastinates of the human brain and the human head
L. E. Huggenberger, J. Ribbers, and H.-J. Stoffels for outstanding graphic work

Porto 2022

Hannsjörg Schröder, Prof. em.,
Department of Anatomy
University of Köln
Germany

Material and Methods

A protocol-like documentation of methods is beyond the scope of this book. Short descriptions of the specimens and techniques used will be provided with appropriate references.

Processing of the Tissue for Histology and Immunohistochemistry

Following removal at autopsy, brains from neurologically healthy and diseased individuals and spinal cords were fixed in buffered 4% aqueous solution of formaldehyde for at least three weeks.

Samples from one hemisphere were taken according to a standard protocol from several neocortical and allocortical regions, the cerebellum, and the brainstem. The samples were embedded in paraplast, cut at 8–10 μm, and stained with hematoxylin-eosin (H & E stain), Congo red, modified Bielschowsky silver impregnation, Gallyas silver-iodide method, Elastica van Gieson, and Klüver-Barrera.

For immunohistochemistry, sections were processed following standard procedures using polyclonal antibodies to ubiquitin, beta amyloid A4, a monoclonal antibody to hyperphosphorylated tau protein (AT8), and another one to α-synuclein. When required antibodies to CR3/43, calbindin and ataxin were employed (see De Vos et al. 1995).

Samples from the other hemisphere, i.e., tissue blocks from neo- and allocortex, striatum and pallidum, thalamus, hypothalamus, cerebellar cortex, and brainstem were embedded in polyethylene-glycol (PEG 1000, Merck, Darmstadt, Germany), cut at 100 μm, and stained with Darrow red and Campbell Switzer (see below).

- **Visualization of neuronal perikarya**
  Aldehyde fuchsin/Darrow red staining (see atlas part Darrow red). The staining results in the blue appearance of lipofuscin granula while Nissl substance will stain red (for details, see Braak and Braak 1991; Braak et al. 2000).
- **Demonstration of fiber tracts**
  Campbell silver impregnation.
  The impregnation renders myelinated fibers dark-red, melanin-containing cells appear black (cf. e.g. locus caeruleus [LC], see Chap. 17 and atlas part Campbell 15). It was also being used for detection of amyloid deposits, Lewy bodies, and Lewy neurites as well as neuromelanin.

Processing of the Tissue for Plastination

Plastination is the impregnation of biological specimens using special embedding material. Human brains and one head were obtained from the body donor system of the Institute of Anatomy in Köln and fixed in formalin. Subsequently, with the exception of the head specimen, they were cut in 1.5 mm thick coronal and horizontal slices. The slices were put in a bath of acetone to remove water and soluble lipids from the tissue. Acetone then replaced the water inside the cells. The following plastination was performed using BIODUR tm P35 (for details, see https://bodyworlds.com/plastination/plastination-technique/).

One head of a body donor was plastinated in total and subsequently cut into sections of approx. 5 mm with a precision saw.

MRT Imaging of the Brainstem

One of us (FH) served as a proband to perform a horizontal brainstem series using a 3T magnetic resonance tomography (T2 sequence) as well as for obtaining a thin-layer MRT series (3D CIS sequence) of the brainstem. The sequences chosen for the individual images are indicated in the atlas.

By courtesy Dr. Christoph Stückle (Privatpraxis für offene MRT, Dortmund, Germany) we were able to include a single example of a MRI scan of a fluid fixed human specimen demonstrating the image quality using routine scanning methods.

Standard books and inventories used throughout the generation of the manuscript which for reasons of clearness have not been everywhere cited in the text are

- Faller A (1978) Die Fachwörter der Anatomie, Histologie und Embryologie. 29. Aufl. JF Bergmann München
- Dauber W (2005) Feneis‘ Bildlexikon der Anatomie. Thieme
- Hausman L (1951) Atlases of the spinal cord and brainstem and the forebrain. 1st edition Charles C. Thomas
- Mai JK, Majtanik M (2017) Human brain in standard MNI space. Academic Press
- Nieuwenhuys R, Voogd J et al. (2008) The Human Central Nervous System. 4th edn. Springer
- Parent A (1996) Carpenter's Human Neuroanatomy (2006). 9th edn. Williams & Wilkins
- PROMETHEUS Kopf, Hals und Neuroanatomie Michael Schünke, Erik Schulte, Udo Schumacher. Thieme
- Schaltenbrand G, Wahren W (eds) (1977) Atlas for Stereotaxy of the Human Brain. Thieme
- Sobotta (2017) Atlas der Anatomie Band 3 Kopf, Hals und Neuroanatomie. 24. Auflage. Paulsen, F (Autor); Waschke, J (Autor) Elsevier
- Ten Donkelaar HJ, Kachlik D et al. (2018) An Illustrated Terminologia Neuroanatomica. A Concise Encyclopedia of Human Neuroanatomy. Springer
- Tillmann BN (2005) Atlas der Anatomie des Menschen. Springer
- Von Hagens G, Whalley A et al. (1990) Schnittanatomie des menschlichen Gehirns. Ein photographischer Atlas plastinierter Serienschnitte. Steinkopff Verlag

References

Braak H, Braak E (1991) Demonstration of amyloid deposits and neurofibrillary changes in whole brain sections. Brain Pathol 1:213–216

Braak H, Rüb U et al. (2000) Parkinson's disease: affection of brain stem nuclei controlling premotor and motor neurons of the somatomotor system. Acta Neuropathol 99:489–495

De Vos RA, Jansen EN et al. (1995) "Lewy body disease": clinico-pathological correlations in 18 consecutive cases of Parkinson's disease with and without dementia. Clin Neurol Neurosurg 97:13–22

https://bodyworlds.com/plastination/plastination-technique/

Contents

About the Authors

Rob A. I. de Vos, MD worked for many years as clinical pathologist and neuropathologist in the Laboratorium Pathologie Oost Nederland (LabPON) Twente in Enschede (now in Hengelo), one of the largest pathology institutes of the Netherlands. His main scientific interests are neurodegenerative diseases, with Parkinsonism and related disorders, dementias, and spinocerebellar ataxias as main topics.

Farman Hedayat, MD is a senior consultant in neurosurgery and a certified spine surgeon. He wrote his dissertation at the Department II of Anatomy in Cologne, Germany, on the Distribution of Opioid Receptors in the Human Brainstem. He has been working during his medical education as a tutor in Anatomy and Neuroanatomy courses for many years.

Stefan Huggenberger, PhD is zoologist and lecturer at the Institute of Anatomy and Clinical Morphology at the University of Witten/Herdecke, Germany. His research deals with the comparative anatomy of the senses of vertebrates, especially whales and dolphins. His expertise is reflected in numerous scientific publications and several reference books.

Natasha Moser, Dr. rer. medic. is a graduate biologist and a former lecturer at the Department II of Anatomy at the University of Köln, Germany. She has been teaching macroscopic anatomy and neuroanatomy courses for many years for human and dental medicine students. Since 2008, she has been teaching neuroanatomy to prospective physiotherapists and recently to prospective orthoptists. Her scientific interest focuses on neurodegeneration (Alzheimer's disease, Parkinson's disease) and nicotine receptors. Since 2022 she is active in the neurological diagnostic facilities of the Heilig Geist-Hospital in Köln, Germany.

Lennart Müller-Thomsen, PhD is biologist and lecturer at the Institute of Anatomy and Clinical Morphology at the University of Witten/Herdecke, Germany. His research deals with neurodegenerative diseases of the brain. The focus of interest are the electrophysiological and morphological neuronal changes due to hyperphosphorylated tau protein and pretangles in tauopathies.

Annemieke Rozemuller, MD is professor of neuropathology in Amsterdam UMC with extensive expertise in neurodegeneration. As neuropathologist for the Netherlands Brain Bank and, until 2022, for the prion lab in UMC Utrecht, she does the diagnostics of both rapidly progressive and slowly progressive dementias, movement disorders, and psychiatric disorders.

Hannsjörg Schröder, MD was professor at the Department II of Anatomy of the University of Köln, Germany, from 1992 to 2019. His teaching focused on the clinical neuroanatomy of humans and the comparative neuroanatomy of rodentia for students of medicine and neuroscience. He founded the BSc/MSc program Neuroscience in 2003. His research focuses on topics of neurodegeneration in man and mouse models using molecular histochemical models.
E-mail contact: akb10@uni-koeln.de

Part I

Neuroanatomy and Neuropathology of the Human Brainstem

General Considerations

1

Contents

Abstract

This chapter deals with the basic neuroanatomical facts necessary to understand the human brainstem. It demonstrates the three-dimensional terminology used in neuroanatomy and neuroimaging. Subsequently, the most important macroscopical and histological features are presented, including the cranial nerves, the bony surroundings and the ventricular system, including examples for the occlusion of the fourth ventricle and the mesencephalic aqueduct. Furthermore, the important issue of brainstem and brain vessels as well as vascular disorders is dealt with. The chapter closes with synoptical data on the major fiber tracts (with an example for a brainstem tumor) and explanations on the anatomical terminology used throughout the following chapters.

H. Schröder et al., *The Human Brainstem*, https://doi.org/10.1007/978-3-030-89980-6_1

Fig. 1.1 Anatomical terms of location and direction. These denominations refer to three main axes: (1) Longitudinal or vertical axis of the body (shown here and in Fig. 1.2B), (2) the sagittal axis (Fig. 1.2A), and (3) the horizontal axis (Fig. 1.2C)

The following terms of direction are used: (caudal [caudalis] <> cranial [caudalis/cranialis], medial [medialis] <> lateral [lateralis]). Related to the body (and the brain) structures near—or directed to—the head (cranium) are designated as cranial, those near—or directed to—the feet are called caudal. Caudal is a zoological term that alludes to the tail most mammalian animals dispose of. Again, related to the longitudinal axis of the body are the terms medial <> lateral [medialis <> lateralis] where medial means toward the midline (medium) and lateral (latus) to the side.

Finally, related to the extremities, proximal [proximalis] means toward/near the trunk while distal [distalis] points to the acra. The terms ventral [ventralis] (anterior) /dorsal [dorsalis] (posterior) are not included in the graphic. Ventral (from Latin venter = belly) means toward the belly and dorsal (from Latin dorsum = back) means to the back.

Albrecht Dürer (1471–1528): Adam und Eva (1504) Metropolitan Museum of Art, New York, USA. Public domain

1.1 Anatomical terms of direction and location/planes and axes of the human body

The majority of anatomical terms of direction, location, and the planes and axes of the human body are derived from Latin denominations (see legend Fig. 1.1).

1.1.1 Terms of direction and location

Using the famous etching by Albrecht Dürer, the main terms are explained here below.

1.1.2 Planes and axes of the human body

The following figures (Figs. 1.2 and 1.3) illustrate the most important terms used for planes and axes of the human body and brain.

1.2 The human brainstem

The brainstem is the anatomical interface between the caudally located spinal cord, the cranially located thalamus, and the telencephalon with the hypothalamus, the amygdala, corpus striatum, hippocampus, and neocortex. The descending and ascending fiber tracts (see Tables 1.3 and 1.4) run through the brainstem, which means that in every section through the brainstem you will find at least one of these bundles.

Traditionally, the brainstem is thought to be composed in caudocranial order of the

- Medulla oblongata (Medulla [Latin] = marrow, pulp/ oblongatus [Latin] = extended) [Myelencephalon].
- Pons (Pons [Latin] = bridge, here symbolizing the connection of both halves of the cerebellum).
- Mesencephalon (midbrain) (μέσο- [meso-, Greek] = in between; ὁ εγκεφαλός [ho enkephalos, Greek] = brain).

For a critical account of this classification, see Watson et al. (2019), Box 1.1 and paragraph 1.4 here below.

In the adult human brain, this subdivision can best be appreciated combining a midsagittal (Figs. 1.4 and 1.5) and a ventral view (Fig. 1.6) of the brain.

Figure 1.5 shows a detail of the midsagittal specimen in Fig. 1.4. The brainstem (medulla oblongata, pons, and mesencephalon) and the cerebellum are contained in the posterior cranial fossa [Fossa cranii posterior] (see Fig. 1.35). The border between the medulla oblongata and the pons is indicated by the medullopontine sulcus [Sulcus bulbopontinus] (white arrow) which is very distinct in the human brain (see Fig. 1.4). At its caudal end, the medulla oblongata merges with the spinal cord (Fig. 1.5 ①).

The medulla oblongata (MO) has a rostral, open part in the region of the fourth ventricle where it forms the floor of the fourth ventricle (rhomboid fossa).

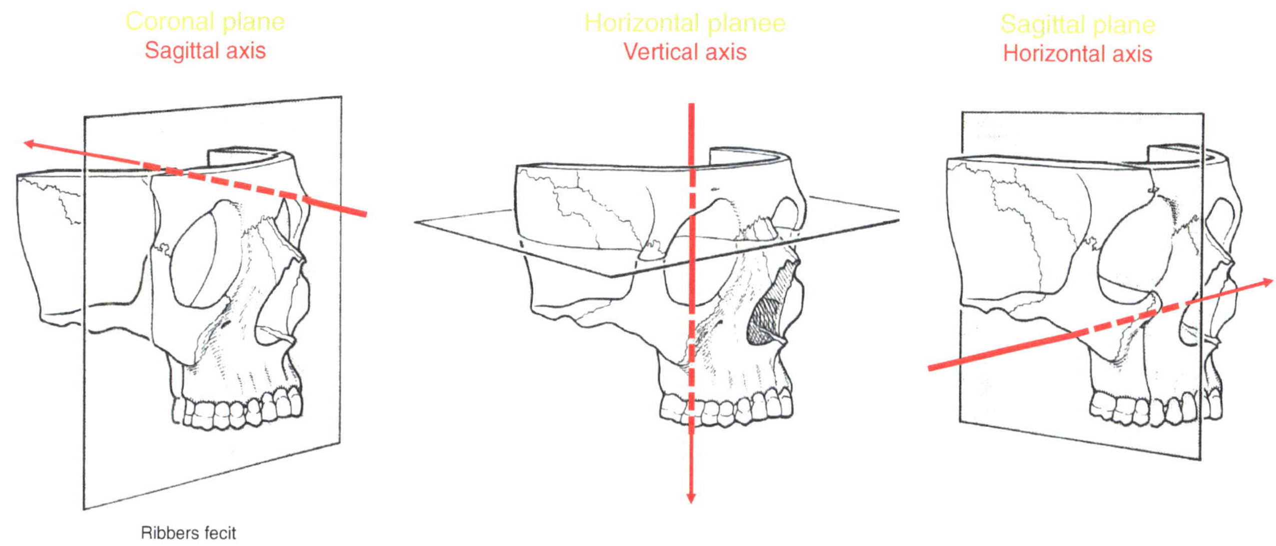

Fig. 1.2 Anatomical terms for the planes and axes of the human body. With the exception of coronal and sagittal the terms are self-explanatory. Coronal is related to the coronal suture of the skull (see **below** Fig. 1.3). Sagittal alludes to the Latin word sagitta = arrow. The sagittal axis "perforates" the skull frontally like an arrow. Each of the three planes has a corresponding axis. These terms are of great importance for the description of anatomical and imaging findings

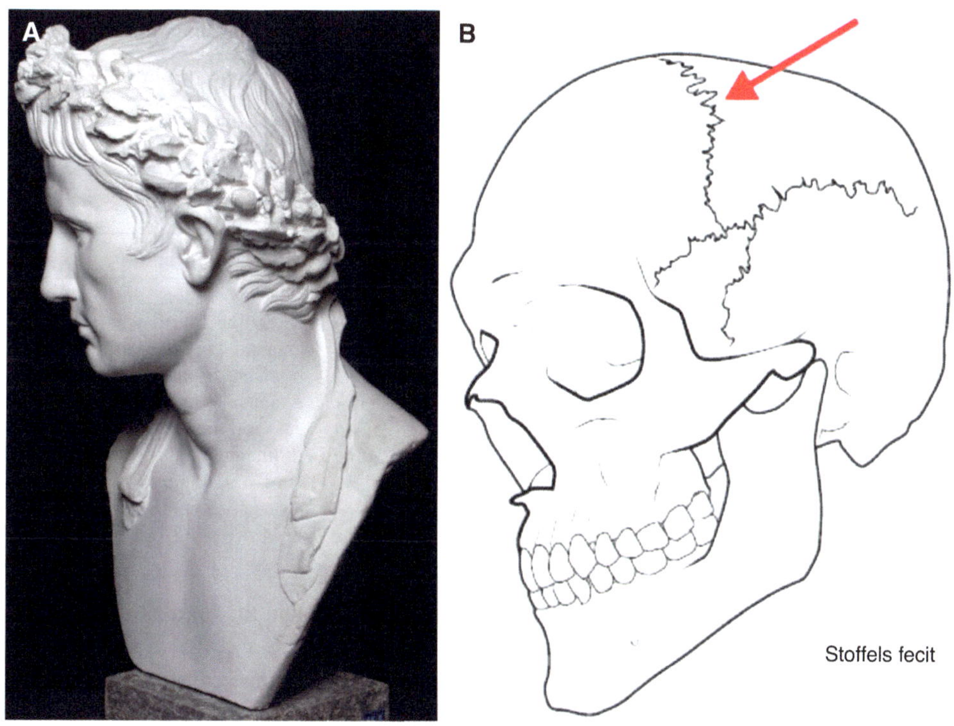

Fig. 1.3 The origin of the term coronal (not coronary alluding to the heart) has its origins in the Corona civica of the ancient Romans. It means civic crown made of interwoven oak leaves and was regarded the second highest military decoration a citizen could aspire to (**A**). (**B**) The coronal suture [Sutura coronalis] (red arrow) between frontal and parietal bone displays an orientation similar to the Corona civica. © Archäologisches Institut der Unversität Göttingen, Germany. Photograph Stephan Eckardt with permission

In this region, the roof of the fourth ventricle is formed by the superior and inferior medullary velum [Velum medullare superius and inferius]. At the caudal end of the open MO, the fourth ventricle continues into the central canal, located inside the MO as in the spinal cord. This part is therefore called the closed part. The border between both parts is the obex (see Fig. 1.9).

The pons shows a larger ventrodorsal diameter as compared with the medulla oblongata and the mesencephalon (see Box 1.1).

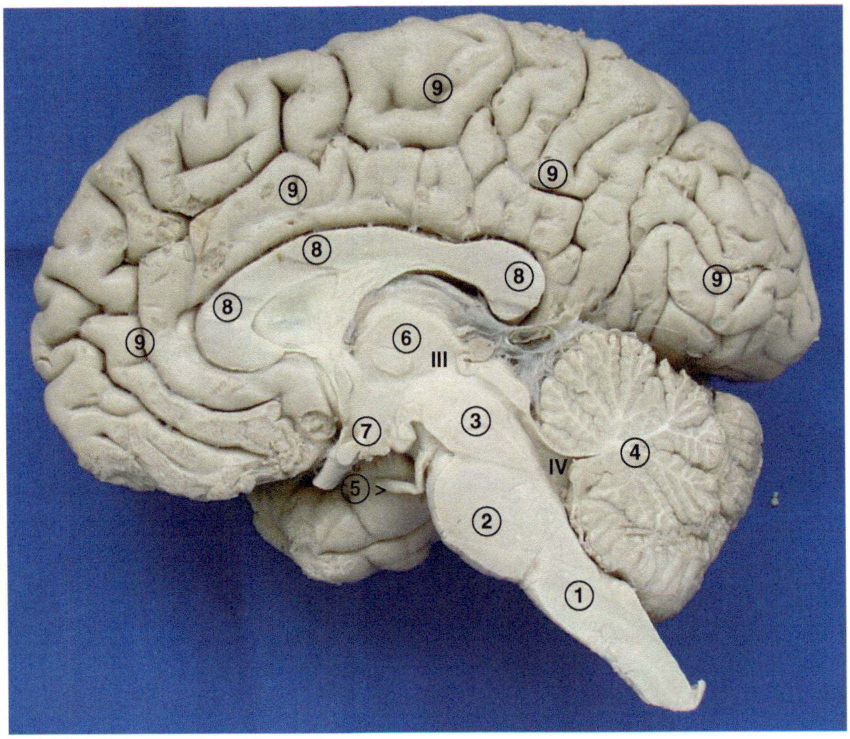

①	**Medulla oblongata**	⑥	**Thalamus**
②	**Pons**	⑦	**Hypothalamus**
③	**Mesencephalon**	⑧	**Corpus callosum**
④	**Cerebellum**	⑨	**Telencephalon / Neocortex**
⑤	**Oculomotor nerve CNIII**	**III**	**3rd ventricle**
		IV	**4th ventricle**

Fig. 1.4 Midsagittal view of the human brain. Formalin-fixed specimen. The largest part of the human brain is the telencephalic neocortex ⑨. In the midsagittal view, it is almost completely separated from the subcortical regions by the corpus callosum ⑧. Caudal to the posterior parts of the latter the third ventricle (III) is visible. The paired lateral ventricles are not visible here because they are lateral to the cutting plane. The third ventricle, the mesencephalic aqueduct (see Fig. 1.5), and the fourth ventricle (IV) are midline structures. The aqueduct connects the third with the fourth ventricle. In the lateral parts (not visible here), the fourth ventricle is connected to the leptomeningeal (subarachnoid) space via the lateral apertures (see Fig. 3.9A). Since the brain has been removed from its original position in the skull, the topographical correlation with the leptomeningeal space (see Figs. 1.7 and 1.8) and the surrounding bones (Fig. 1.7) is not visible here. You have to imagine that the whole of the brain as shown here—and the spinal cord—are kind of embedded in a liquid cushion, the leptomeningeal space filled with cerebrospinal fluid. Sammlung des Zentrums Anatomie der Universität zu Köln

At the dorsal surface of the pons, the cerebellum (see Fig. 1.5 ②) is located.

The midbrain shows at its ventral surface an indentation—exit region of the oculomotor nerve (CNIII) (see Fig. 1.4)—the so-called interpeduncular fossa [Fossa interpeduncularis], a depression between the cerebral peduncles [Pedunculi cerebri] (see Fig. 1.5).

A clear-cut macroscopical border between thalamus [Thalamus] and mesencephalon and the latter and the hypothalamus is difficult to draw.

A ventral view of the human brainstem with the cranial nerves is displayed in Fig. 1.6A. The olfactory nerve (CNI) and the optic nerve (CNII) are belonging to the telencephalon and the diencephalon, respectively. Derivatives of the brainstem neuromeres are the oculomotor nerve (CNIII, m1/m2) supplying the majority of eye muscles, the trochlear (CNIV, isthmus, superior obliquus muscle), and the abducens nerve (CNVI, see Sect. 9.2.1.1, lateral rectus muscle).

The trigeminal nerve (CNV, see Sects.11.3.1 and 12.2.3.2) is the first of the branchial arch nerves (see Sect. 2.8) providing most of the sensory supply of the head region as well as the innervation of the masticatory muscles. The second branchial nerve is the facial nerve (CNVII, see Sect. 8.2.1) providing innervation of the mimic muscles, the lacrimal,

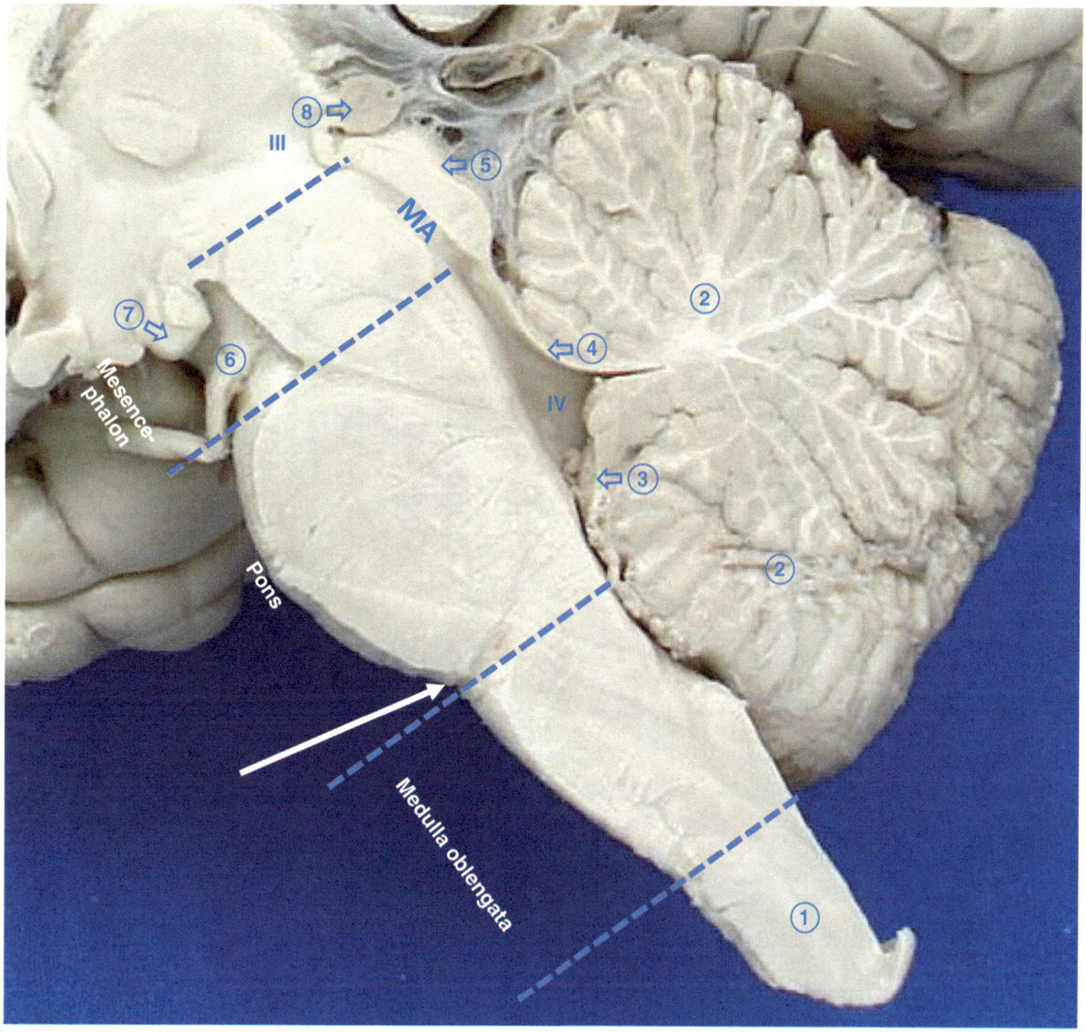

①	Spinal cord	⑦	Mammillary body
②	Cerebellum	⑧	Pineal body
③	Inferior medullary velum	III	3rd ventricle
④	Superior medullary velum	IV	4th ventricle
⑤	Quadrigeminal plate	MA	Mesencephalic aqueduct
⑥	Interpeduncular fossa		

Fig. 1.5 Midsagittal view of the human brain. Formalin-fixed specimen. At higher magnification (for survey, see Fig. 1.4), the parts of the brainstem in the region of the mesencephalic aqueduct (MA) and the fourth ventricle (IV) can be seen in detail. This cerebrospinal fluid (CSF) system is an important landmark for orientation in macroanatomic specimens and in neuroimaging but also in histological sections (see further below). The arrow indicates the border between pons and medulla oblongata, the medullopontine sulcus. Sammlung des Zentrums Anatomie der Universität zu Köln

submandibular and sublingual glands and the gustatory structures of the tongue. CNVII exits the brainstem together with the vestibulocochlear nerve (CNVIII, see Sects. 5.3.1.1.3 and 8.1.1.3) providing afferent innervation of the organs of hearing and equilibrium, in the so-called cerebellopontine angle. The third branchial nerve is the glossopharyngeal nerve (CNIX, see Sect. 5.4.2.1.3) which mainly provides motor and sensory input to the pharyngeal region and innervation of the parotid gland. It is followed by the fourth branchial nerve, the vagus nerve (CNX, see Sect. 7.3.2.1.2) which provides the

A

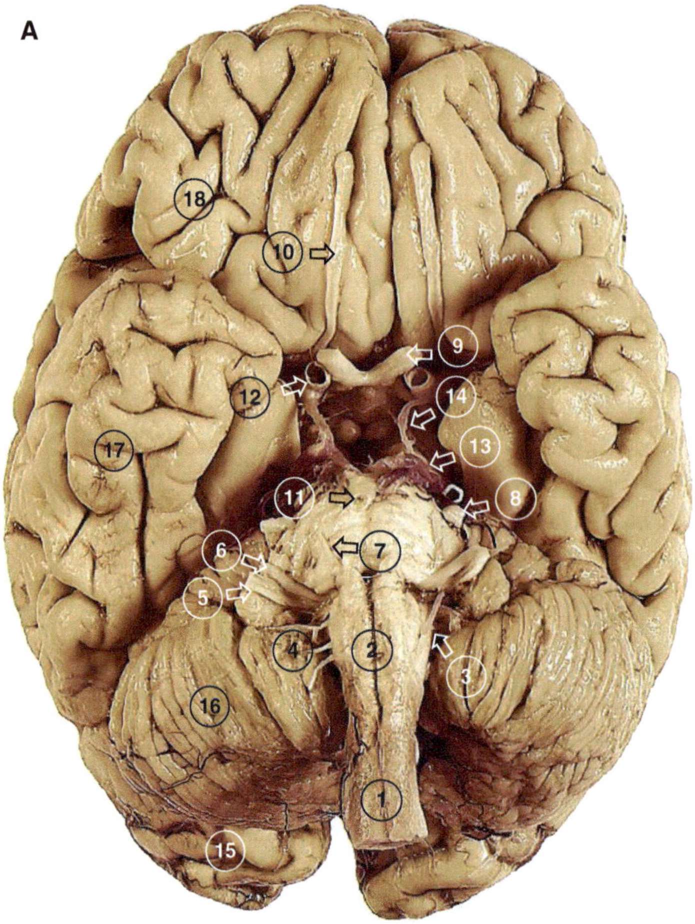

①	Spinal cord	⑩	Olfactory nerve CNI
②	Medulla oblongata	⑪	Basilar artery
③	Hypoglossal nerve CNXII	⑫	Internal carotid artery
④	Caudal cranial nerves CNXI, X, IX	⑬	Posterior cerebral artery
⑤	Vestibulocochlear nerve CNVIII	⑭	Posterior communicating artery
⑥	Facial nerve CNVII	⑮	Occipital lobe
⑦	Abducens nerve CNVI	⑯	Cerebellum
⑧	Trigeminal nerve CNV	⑰	Temporal lobe
⑨	Optic nerve CNII	⑱	Frontal lobe

Fig. 1.6 (**A**) Ventral view of the human brain. For details, see text. LabPON Twente. A dorsal view counterpart of Fig. 1.6A is provided in Fig. 1.6B.

B

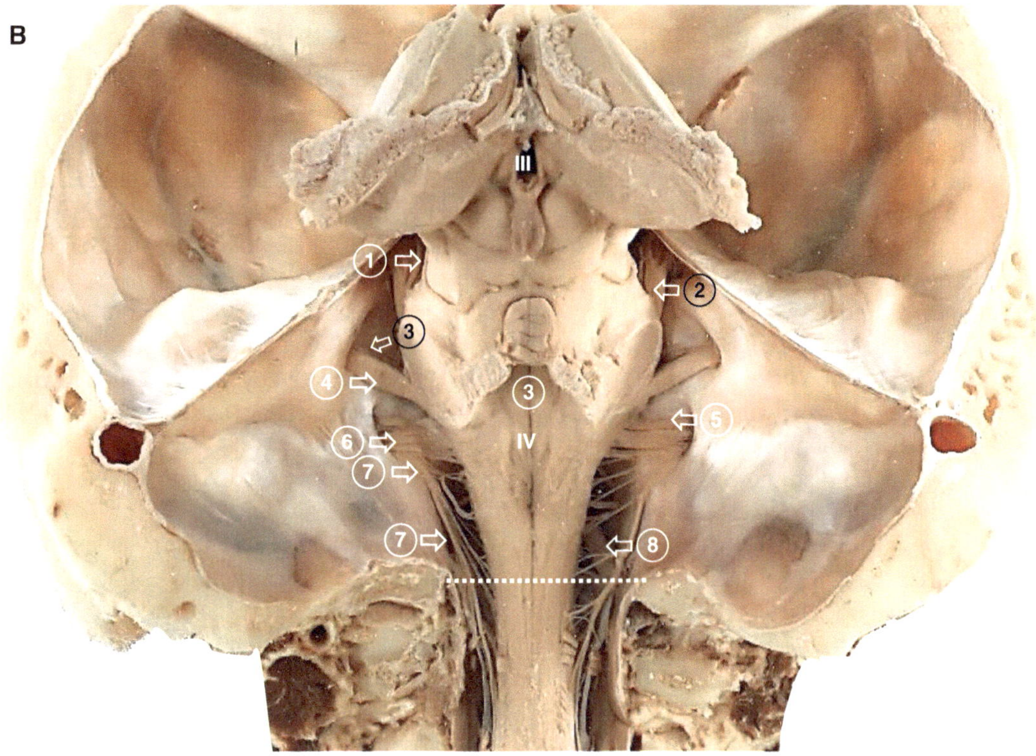

①	Trochlear nerve CNIV	⑦	Accessory nerve CNXI
②	Trigeminal nerve CNV	⑧	Hypoglossal nerve CNXII
③	Facial Nerve CNVII	---	Approximate level of the Foramen magnum
④	Vestibulocochlear nerve CNVIII	III	3rd ventricle
⑤	Glossopharyngeal nerve CNIX	IV	4th ventricle
⑥	Vagus nerve CNX		

Fig. 1.6 (continued) (**B**) Dorsal view of the human brain within the skull. The cerebral peduncles are cut to remove the cerebellum, the diencephalon is still in place. For details, see text. Sammlung des Zentrums Anatomie der Universität zu Köln

innervation of almost all internal organs, which means that CNX is the only cranial nerve with supply territories in the head region as well as in the periphery of the body. The eleventh cranial nerve, the accessory nerve (CNXI, see Sect. 3.3.1.2) innervates the sternocleidomastoid and the trapezius muscles of the neck region. The 12th cranial nerve, the hypoglossal nerve (CNXII, see Sect. 4.3.1.1) provides the motor innervation of the tongue muscles.

Box 1.1 Traditional anatomical classification of the brainstem and genoarchitecture (neuromeres)

The following description of the brainstem structures starting with rhombomere 11 (neuromere 11 of the rhombencephalon, see Sect. 2.4) is based on their classification in the BrainNavigator mouse brain atlas (Paxinos and Watson 2010). The basis of this classification is the developmental attribution of nuclei to the different neuromeres (cf. e.g., Puelles 2009, see Sect. 2.4) based on the differential expression of transcription factors (see Watson et al. 2019).

For each neuromere—starting with the caudal most brainstem neuromere, the rhombomere 11 (r11)—we will provide a synoptical list showing the nuclear groups of the individual rhombomeres. Some structures are derivatives of several rhombomeres. They will be usually described and explained the first time they show up in the lists and the text. This book deals with the most important CNS nuclei and structures.

The traditional anatomical conception of the pons as the voluminous formation on the ventral surface of the human brainstem (see Figs. 1.4, 1.5, and 1.6) has been challenged by Watson et al. (2019) based on the rhombomere model and the developmental expression of certain key transcription factors.

The cardinal objection of Watson and coworkers points to the fact that in many mammals the basilar pontine nuclei (r3/4) and the reticulotegmental nucleus (r3/4) aggregate at the ventral surface of these rhombomeres. Thus, the pontine bulge in those species is restricted to the area of these two rhombomeres. Neurons destined to become the pontine nuclei in the rhombic lip (see "Development" Chap. 2) develop in r6 and r7 then migrating to their final location in r3 and r4 (Watson et al. 2019).

Due to the extensive development of the pontine bulge in the human brain, large parts from the isthmus to r2 and the rhombomeres containing the abducens (r5) and facial nucleus (r6) as well as the superior olivary complex (r5) are covered.

The developmentally correct solution suggested by Watson and coworkers would be to confine the use of the term pons for all mammals to the basilar pontine formation of rhombomeres r3/r4. A prepontine brainstem would embrace the neuromeres isthmus, r1 and r2. The postpontine brainstem would include rhombomeres 5, 6. The "medulla oblongata" would encompass r7 through r11. These regions appear subdivided in transversal rhombomeric/neuromeric domains. The traditional "mesencephalon" on genoarchitectonic grounds should be included into the forebrain.

In consideration of further changes introduced the subdivision of the hindbrain into pons and medulla oblongata is suggested to be abandoned. Instead the hindbrain should subsume rhombomers r0 through r11 where the isthmus is considered its first rhombomere (r0) (Watson et al. 2019).

As summarized by Watson and coworkers, the following recommendations should be regarded "....

1. Abandon the subdivision of the hindbrain into 'pons' and 'medulla.'
2. Restrict the use of the term 'pons' to refer to the nuclei and fiber bundles of the basilar pontine formation.
3. Recognize the isthmus (rhombomere 0) as the first segment of the hindbrain.
4. Recognize that the cerebellum is a derivative of the rostral prepontine hindbrain.
5. Recognize that the posterior commissure and associated nuclei, the nucleus of Darkschewitsch, the interstitial nucleus of Cajal, and the rostral part of the red nucleus belong to the caudal diencephalon and not to the midbrain.
6. Consider the evidence for including the midbrain in the forebrain on genoarchitectural grounds, which would have the effect of making the old term 'brainstem' synonymous with the hindbrain.
7. Adopt a modern functional and segmental nomenclature for the classification of the monoamine cell groups of the brainstem...."

Realistically estimating the time course of this endeavor, the authors continue to write:

".... we realize that in order to make embryological and physiological rhombomere-related scientific progress accessible to clinical topographic analysis of pathology and surgery within the conventional 'pons' region ...... it may take decades to extinguish its indiscriminative use as a regional descriptor for the whole rostral hindbrain."

Wherever reasonable we will get back to the above cited recommendations in the individual sections of this book.

1.2.1 The cerebrospinal fluid system and the meninges

The CSF system of the brainstem consists of midline cavities. The 3rd ventricle (III), also midline, separates both halves of the thalamus and the hypothalamus and connects to the mesencephalic aqueduct (MA) [Aqueductus mesencephali] (Cerebral aqueduct, Aqueduct of Sylvius) (see Box 1.2).

Box 1.2 FrancIscus Sylvius

Franciscus Sylvius, latinized from Franz de le Boë (1614–1672), is reckoned to be the founder of natural science-based medicine and of clinical chemistry. Some sources have it that Sylvius, to cure heartburn and digestive trouble, applied a mixture of alcohol, juniper (Juniperus) berries and herbs that he took in market as "Genever" (Dutch version of Juniperus). In the United Kingdom where Genever was not only used as a remedy the name gin was coined.

Notabene: The name Franciscus Sylvius is related as eponym to the Fissura Sylvii/Sylvian fissure, today Sulcus lateralis/lateral sulcus between the temporal lobe vs. the frontal and parietal lobes. The term Aqueductus Sylvii/Sylvian aqueduct for the mesencephalic aqueduct (see Fig. 1.5)—a structure already recognized in Galen's time—probably refers to the French physician Jacobus Sylvius (1478–1555), Sylvius being the latinized version of Dubois (see Clarke and O'Malley 1996).

The CSF is a fluid secreted by the so-called choroid plexus [Plexus choroideus] of the lateral (I + II), third (III), and fourth (IV) ventricles. It circulates to the fourth ventricle where it leaves the inner CSF compartment (ventricles) to reach the subarachnoid (leptomeningeal) space (Fig. 1.7 ⑤). The latter surrounds the complete CNS. The term subarachnoid space [Spatium subarachnoideum] (Figs. 1.7 and 1.8) for the space deep to the outer layer of the leptomeninx, containing the arachnoid trabeculae, is not correct since the spatium is bounded internally by the outer layer of the pia mater. The most appropriate designation is therefore leptomeningeal space [Spatium leptomeningeum] (cf. Terminologia neuroanatomica FIPAT, http://fipat.library.dal.ca/TNA/).

The brain is protected by the osseous skull ①–③ followed by the meninges ④–⑥ (ἡ μῆνιγξ, μήνιγγ- he meninx, mening- = linings of the brain, meninges) (Figs. 1.7 and 1.8).

The meninges can be differentiated into the pachymeninx (παχύς, pachys = thick), the cranial dura mater [Dura mater cranialis] (Fig. 1.7 ④), and the leptomeninx (λεπτός leptos = thin) with the cranial arachnoid mater [Arachnoidea mater cranialis]) (ἡ ἀράχνη he arachne = spider, τό ἀράχνιον, to arachnion = spider web) and the cranial pia mater [Pia mater cranialis] ⑥ (Latin pia = tender).

Note that by contrast to the spinal cord/vertebral column there is no epidural space inside the skull.

A dreaded cerebrovascular incident is the so-called subarachnoid hemorrhage (see also Sect. 1.2.1.1), a bleeding into the CSF-filled leptomenigeal space. The most common cause is a local dilatation of a cerebral vessel, the so-called aneurysm. Since the meninges are very well equipped with trigeminal sensory endings (see Sect. 12.2.3.2), patients usually complain about acute unbearable headache accompanied by neurological symptoms. The bleeding into the leptomeningeal space may cause contraction of cerebral vessels and thus massively impair cerebral circulation. Even today, the prognosis unfortunately is rather poor.

The mesencephalic (or formerly cerebral) aqueduct [Aqueductus mesencephali] (see Box 1.2) is one of the tightest parts of the internal CSF system. It is the link between the third ventricle cranially and the fourth ventricle caudally. As the name says, the aqueduct is located inside the mesencephalon. The tectum (see Fig. 1.12) of the mesencephalon, the so-called quadrigeminal plate [Lamina quadrigemina] (Fig. 1.5 ⑤) forms the roof of the aqueduct, the mesencephalic tegmentum (see below Fig. 1.12) is its floor. This "bottleneck" of CSF circulation is a predilection site for narrowing or closure of the ventricular system (see Sect. 1.2.1.1.2). Under normal conditions, the CSF flows caudally into the fourth ventricle (IV) which in midsagittal sections displays a triangular shape. Toward the cerebellum the fourth ventricle is sealed by the superior and inferior medullary vela [Velum medullare superius Fig. 1.5 ④/inferius ③]. The floor of the fourth ventricle is the rhomboid fossa [Fossa rhomboidea] (see Fig. 1.9). The fourth ventricle displays three apertures via which the CSF in the ventricular system flows into the leptomeningeal space where it is eventually resorbed.

At the site of the foramen magnum (see Figs. 1.35 and 1.36), the medulla oblongata merges with the cervical spinal

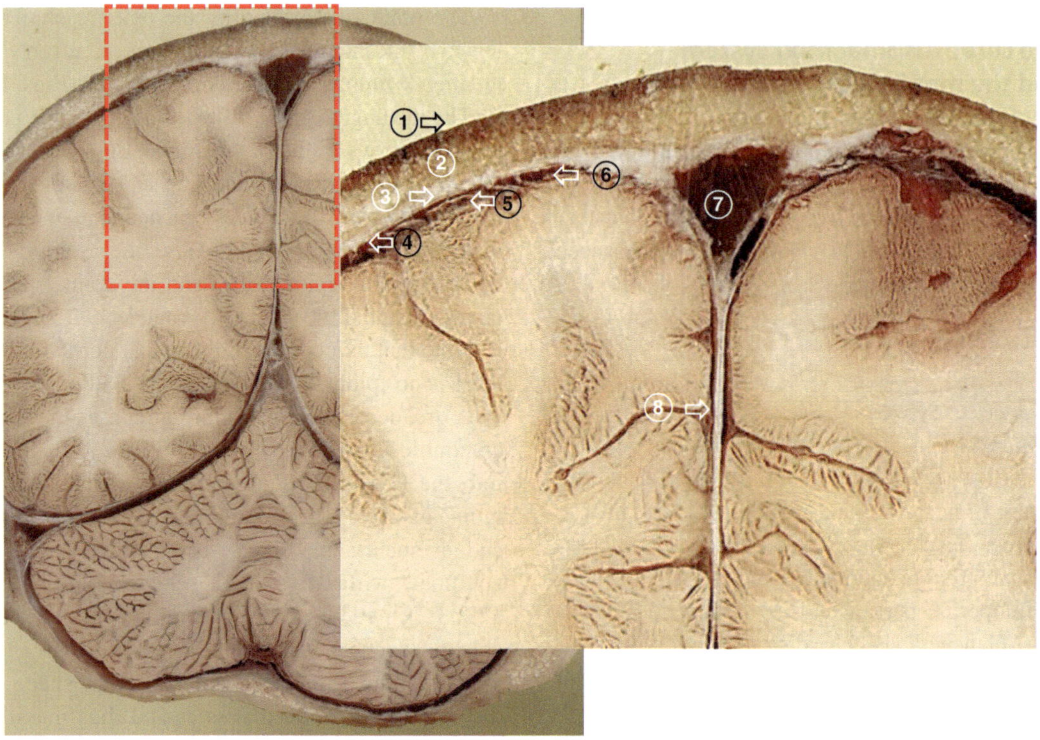

①	**Lamina externa**	⑤	**Leptomeningeal space with arachnoid trabeculae**
②	**Diploe**	⑥	**Pial surface of the brain**
③	**Lamina interna**	⑦	**Superior sagittal sinus**
④	**Cranial dura mater + Cranial Arachnoid**	⑧	**Falx cerebri**

Fig. 1.7 Coronal section through the human skull with the occipital lobes and the cerebellum. The inset shows the bony and connective tissue coverings of the brain (for details, see text). The falx cerebri is a dura duplication separating both telencephalic hemispheres. At its dorsal end, the dural folds open to form the superior sagittal sinus one of the dural venous sinuses [Sinus durae matris] (for details, see Fig. 3.11). Sammlung des Zentrums Anatomie der Universität zu Köln

cord and the fourth ventricle continues—starting already in the medulla oblongata—into the narrow cylindric central canal [Canalis centralis].

A closer look into the fourth ventricle and on the surrounding structures is possible only after the overlying cerebellum and medullary vela (see Fig. 1.5) have been removed. This situation is illustrated in Fig. 1.9.

Figure 1.10 shows a detailed photograph of the specimen shown in Fig. 1.9 allowing for a closer view of the protuberances of the floor of the fourth ventricle. All these structures are separated from their contralateral counterparts by the midline posterior median sulcus [Sulcus medianus posterior] (④ in Fig. 1.9).

Starting cranially, the first hill is the so-called facial colliculus [Colliculus facialis] (collis, dim. colliculus Latin for hill) due to the (internal) genu (knee) of the facial nerve—innervating among other targets the mimic muscles and the gustatory receptors of the tongue (see Sect. 8.2.1)—which at this site has an arcuate course around the nucleus of abducens nerve (see Sect. 9.2.1.1), an eye muscle nerve. The next prominence is the hypoglossal trigone [Trigonum nervi hypoglossi] caused by the nucleus of hypoglossal nerve (CNXII, see Sect. 4.3.1.1) which here comes remarkably close to the floor of the fourth ventricle. The CNXII innervates the tongue muscles.

The vagal trigone [Trigonum nervi vagi] is due to the close apposition of the posterior nucleus of the vagus nerve [Nucleus posterior nervi vagi] (see Sect. 7.3.2.1)—innervating the smooth muscles of the majority of internal organs—to the ventricular floor.

The vestibular area [Area vestibularis] is the macroscopic equivalent of the vestibular nuclei (equilibrium system, see Sect. 5.3.1) located in the floor of the fourth ventricle.

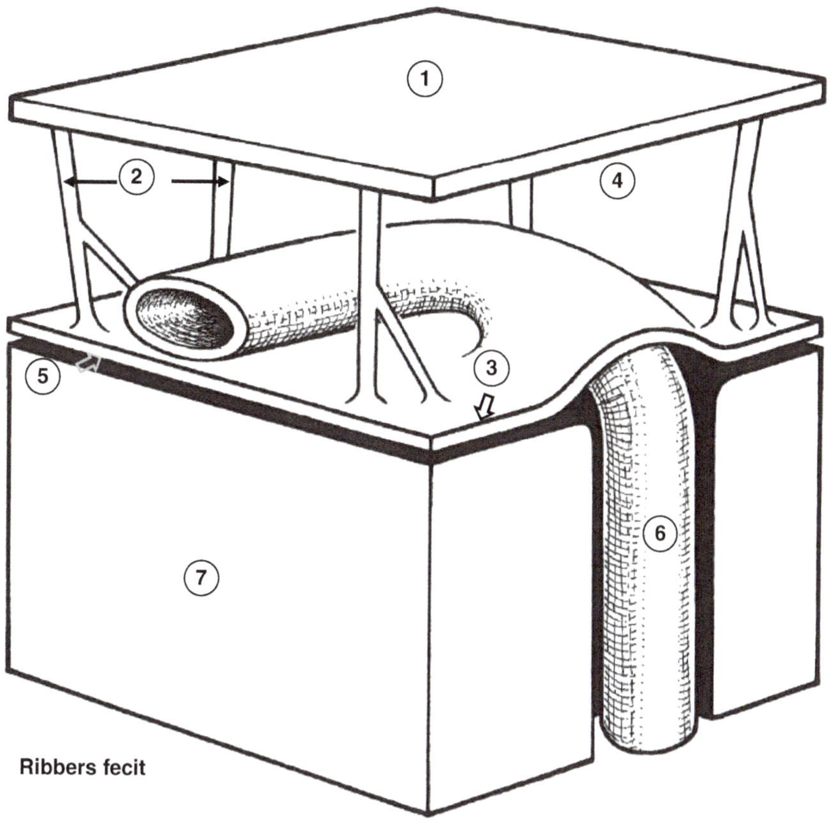

①	Cranial arachnoid	⑤	Subpial space
②	Arachnoid trabeculae	⑥	Pial vessel
③	Cranial pia mater	⑦	Brain
④	Leptomeningeal space (CSF)		

Fig. 1.8 Schematic illustration of the leptomeninx. In vivo, the cranial arachnoid ① would be covered by the dura mater (see Fig. 1.7). Small processes of the arachnoid, arachnoid trabeculae ② merge with the cranial pia mater ③. Between both, the CSF-filled leptomeningeal space ④ is located. The pia is separated from the brain ⑦ by the subpial space ⑤. The large cerebral vessels ⑥ are running in the leptomeningeal space, then dive below the pia mater to enter the brain tissue ⑦

Finally, the horizontally running medullary striae of the fourth ventricle [Striae medullares ventriculi quarti] indicate the presence of slender fiber fascicles of extending transversely below the ependymal floor of the ventricle from the median sulcus to enter the inferior cerebellar peduncle. They arise from the arcuate nuclei (Sect. 4.1.1.4) on the ventral surface of the medullary pyramid.

1.2.1.1 Pathology of the ventricular system (fourth ventricle and mesencephalic aqueduct)

There is a number of potential causes for the occlusion of the ventricular system of the brainstem comprising the general spectrum of neuropathological lesions like developmental disorders, hemorrhages, tumors, and inflammation. The closure of the ventricular system eventually leads to the stop of CSF circulation with a dilatation of the upstream located parts of the cerebrospinal fluid system.

As examples we will mention here one occlusion of the fourth ventricle caused by a colloid cyst and we will discuss the occlusion of the cerebral aqueduct.

1.2.1.1.1 Colloid cyst

A variety of benign cysts occur at several sites in and around the CNS. Hirano and Hirano (2004) classified CNS cysts into two categories. The first group includes cysts derived from the CNS, for example, arachnoid and ependymal cysts. The second group consists of intracranial or intraspinal lesions that are derived from non-CNS tissues such as endoderm or ectoderm.

Arachnoid cysts can occur all along the craniospinal axes and have a preponderance of location for the middle cranial

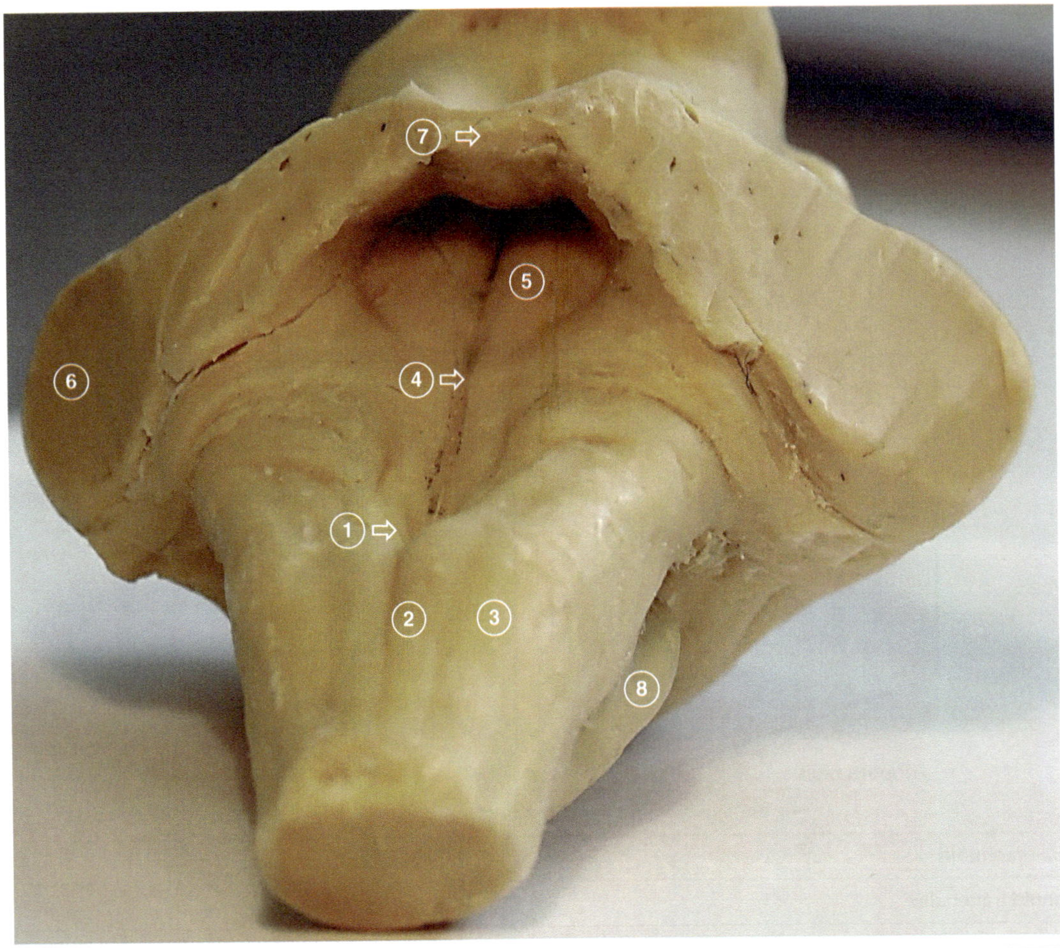

①	**Obex**	⑤	**Facial colliculus**
②	**Gracile tubercle**	⑥	**Medial cerebellar peduncle (cut surface)**
③	**Cuneate tubercle**	⑦	**Superior medullary velum (cut surface)**
④	**Posterior median sulcus (midline)**	⑧	**Inferior olive**

Fig. 1.9 Dorsocaudal view of the rhomboid fossa [Fossa rhomboidea], floor of the fourth ventricle. Siliconized specimen. In the living brain, the fourth ventricle is covered by the superior and inferior medullary vela (see Fig. 1.5) and posterior to these by the cerebellum. The latter and the vela have been removed here to enable the view into the ventricle. Its floor is not plane but has various protuberances (for details Fig. 1.10). At low magnification, the most prominent is the facial colliculus ⑤. These little hills are due to peculiarities of structures located below the floor of the ventricle. The tearing edge of the inferior medullary velum is located along the rostral edge of the gracile tubercle ② at either side of the obex. The gracile ② and cuneate tubercles ③ (r11) are located outside of the ventricle on the dorsal surface of the medulla oblongata. At its lateral surface, the olive (inferior olive [Oliva] r10, see Sect. 4.1.1.3) ⑧ is clearly visible. The medulla oblongata can be distinguished from the spinal cord—of similar shape and diameter—by inferior olive and pyramids (see Fig. 1.6A). Sammlung des Zentrums Anatomie der Universität zu Köln

fossa (Hayashi 2016). The cerebellopontine angle (see Fig. 5.4) represents the second most common location and accounts for about 11% of all arachnoid cysts. Cerebral convexity arachnoid cyst, suprasellar arachnoid cyst, intrasellar arachnoid cyst, quadrigeminal cistern arachnoid cyst, intraventricular arachnoid cyst, and posterior fossa arachnoid cyst are less frequently seen. Ependymal (neuroectodermal) cysts can be found deep within the parenchyma of the brain and are lined by a single layer of ependymal cells (Hirano and Hirano 1991). Dermoid, epidermoid cysts, and Rathke's cleft cysts are known to arise from ectodermal tissue, and belong, as well as enterogeneous cysts and endodermal cysts, to the second group.

Colloid cysts are located mostly in the third ventricle (see Fig. 1.5) where they represent 15% of all tumors here. They could cause acute hydrocephalus (see here below), but the primary presenting complaint is a specific type of headache, described as intermittent, severe, and intense, and of short

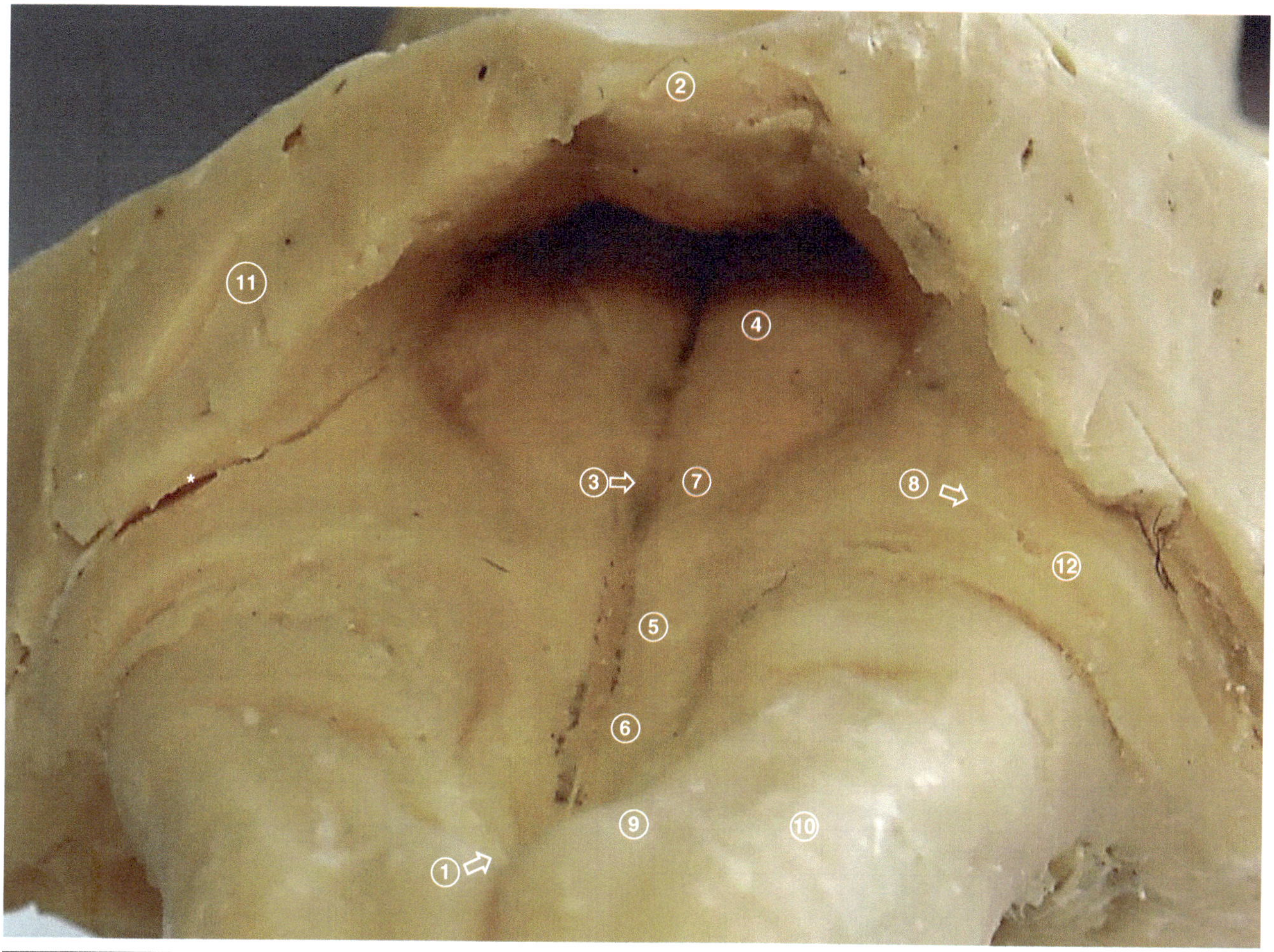

①	**Obex**
②	**Superior medullary velum**
③	**Posterior median sulcus**
④	**Facial colliculus**
⑤	**Hypoglossal trigone**
⑥	**Vagal trigone**
⑦	**Vestibular area**
⑧	**Medullary striae of 4th ventricle**
⑨	**Gracile tubercle**
⑩	**Cuneate tubercle**
⑪	**Superior cerebellar peduncle**
⑫	**Region of the lateral aperture**
*	**Artifact**

Fig. 1.10 Magnified view of the rhomboid fossa as shown in Fig. 1.3. For details, see text. Sammlung des Zentrums Anatomie der Universität zu Köln

duration and usually located frontally (Spears 2004). These cysts can be a cause of hydrocephalus and sudden death has been reported to occur in approximately 10% of patients (Young and Silberstein 1997). Papilloedema was the most frequent sign in a study of 36 cases (Nitta and Symon 1985). Colloid cysts have rarely been reported in the fourth ventricle (Jan et al. 1989; Wang et al. 2012; Hasegawa et al. 2015) (Fig. 1.11).

Colloid cysts are most often small, 2–50 mm in diameter and are often found incidentally at autopsy in the midline of the brain near the foramen of Monro. There are no histological differences between incidental colloid cysts found at autopsy and those with symptoms requiring surgery (Hirano and Hirano 2004). Both computed tomography (CT) and MRI can be used in radiologic diagnosis of colloid cysts. On CT, most colloid cysts are seen as hyperdense when compared with brain parenchyma.

The cysts have a characteristic gelatinous aspect, contain colloid-like material, and are ususally lined by a single layer of columnar cells, some of which are ciliated or contain mucin (Fig. 1.11D). Immunohistochemical stains show immunoreactivity for cytokeratins and epithelial membrane

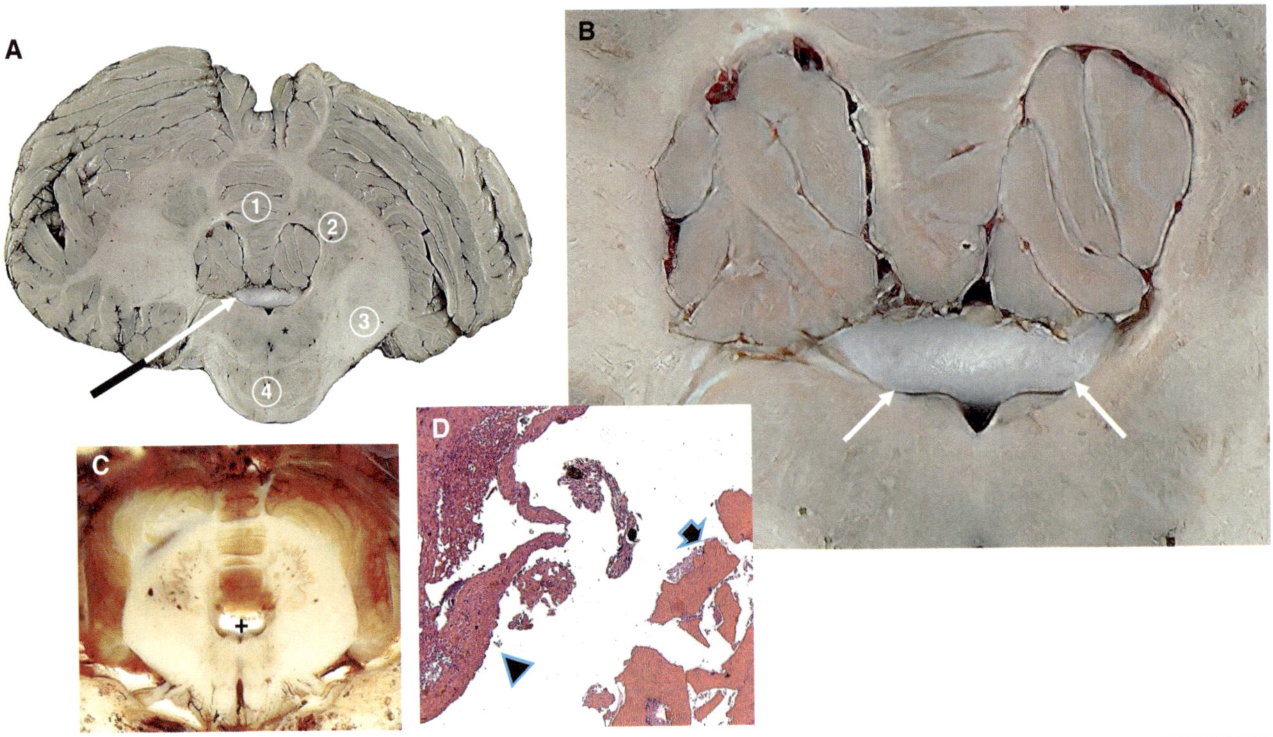

①	Vermis	④	Pyramidal (corticospinal) tract
②	Dentate ncl.	+	4th ventricle
③	Inferior cerebellar peduncle		

Fig. 1.11 (**A**) A grayish, sharply bordered, oval cystic structure is seen inside the fourth ventricle in A (arrow), detail in (**B**). (**C**) Normal fourth ventricle (IV). In (**D**) microscopical picture showing the cyst containing eosinophilic gelatinous material (thick arrow), lined by a single layer of columnar cells (triangle). H & E stain. LabPON Twente, (**C**) Sammlung des Zentrums Anatomie der Universität zu Köln

antigen (EMA) and are negative for GFAP (astrocyte marker), suggesting that colloid cysts are of endodermal origin (Hirano and Hirano 2004; Mackenzie and Gilbert 1991).

There have been several reported instances of a familial association of colloid cysts in the third ventricle (Aggerwal et al. 1999; Weisbrod et al. 2018). Although the mode of inheritance is unkown, it has been suggested that a genetic component is involved. The increasing number of reports of the familial incidence of colloid cysts suggest the possibility of an autosomal dominant inheritance pattern (Aggerwal et al. 1999). Most of the cysts can be removed successfully by various neurosurgical approaches, for example, transcortical or transcallosal operations or stereotactic or endoscopic aspiration (Spears 2004; Hemesniemi and Leivo 1996). Recurrences are rare.

1.2.1.1.2 Aqueductal stenosis and hydrocephalus

The narrowing or closure of the mesencephalic aqueduct is called aqueductal stenosis. This condition may be caused by a tumor in the immediate vicinity of the midbrain, for example, originating from the pineal gland (see Fig. 1.5 ⑧)—belonging to the epithalamus—or a meningioma originating from the neighboring meninges (Corbett and Haines 2018). Clogging of the aqueduct may also be due to cellular debris seen after bleeding into the ventricles (intraventricular hemorrhage), fungal or bacterial infections or by proliferation of the ependyma following viral infections of the CNS, in particular mumps (Corbett and Haines 2018). The aqueductal stenosis leads to enlargement of the third ventricle (pressure on thalamus, hypothalamus) and the lateral ventricles known under the name hydrocephalus (from Greek τό ὕδωρ, to hydor = water and ἡ κεφαλή he kephale = head). Another form of hydrocephalus is of congenital origin: hydrocephalus with stenosis of the aqueduct of Sylvius (HSAS) is a historical term used to describe a phenotype now considered to be part of the X-linked L1 clinical spectrum (L1 syndrome). HSAS is characterized by severe hydrocephalus mostly with prenatal onset, signs of intracranial hypertension, adducted thumbs, spasticity, and severe intellectual deficit. HSAS represents the severe end of the spectrum and is associated with poor prognosis.

https://www.orpha.net/consor/cgi-bin/OC_Exp.php?lng=en&Expert=2182.

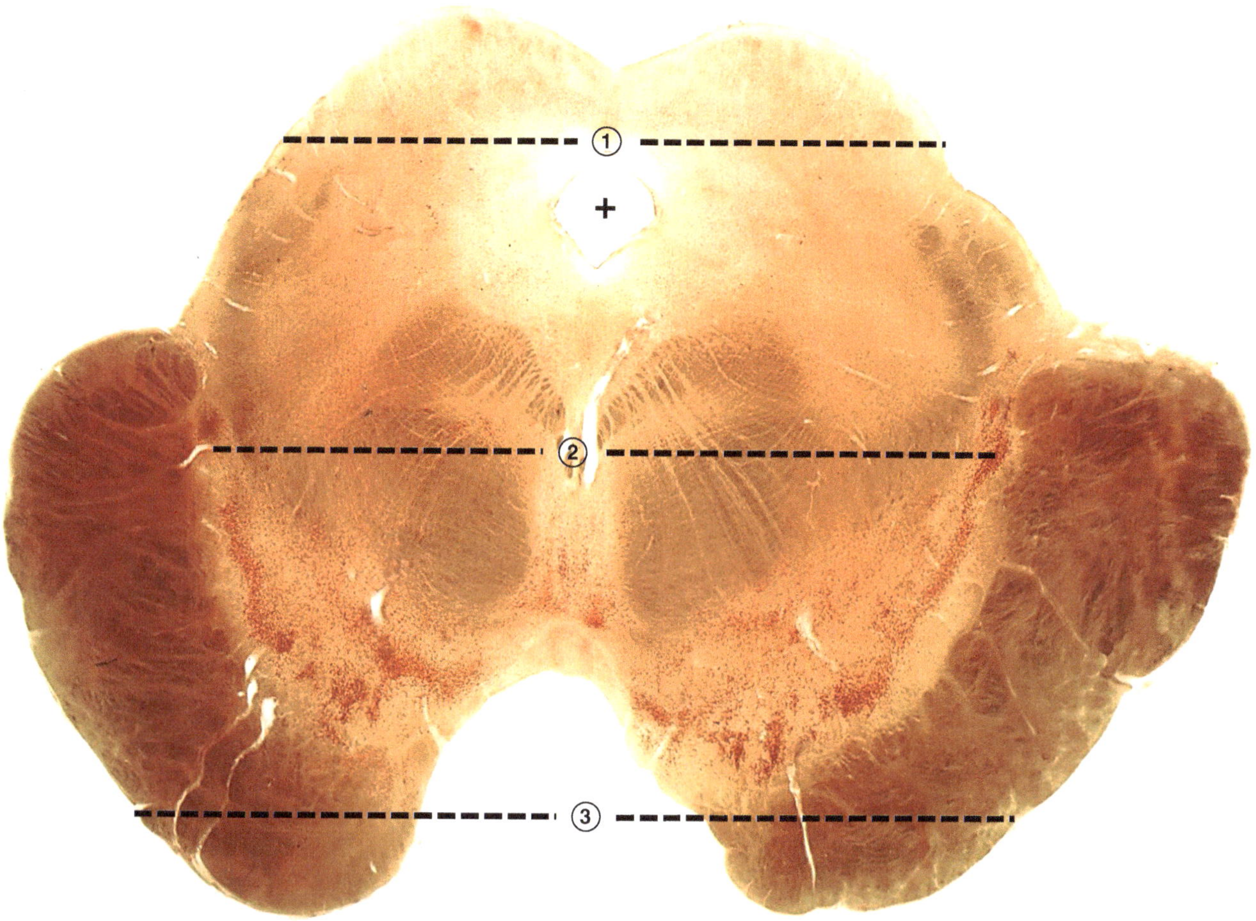

①	Tectum	③	Basis
②	Tegmentum	+	Mesencephalic aqueduct

Fig. 1.12 Horizontal section through the human mesencephalon. Darrow red staining. The inner CSF system is represented by the mesencephalic aqueduct (see Fig. 1.5). For details see text. LabPON Twente

1.2.2 Internal subdivision of the brainstem

In ventrodorsal direction the traditional subdivision of the adult brainstem is as follows (see Fig. 1.12):

- Basis (ἡ βάσις, -εως [he basis, baseos, Greek] = underlay, groundwork)
- Tegmentum (Tegmentum [Latin] = hood, cap)
- Tectum (Tectum [Latin] = roof)

The tectum is the dorsal most part of the brainstem (here shown paradigmatically for the mesencephalon). It is forming the roof (tectum) of the cerebral aqueduct (mesencephalon, colliculi/quadrigeminal plate) and of the fourth ventricle (pons, medulla oblongata) where it consists of the superior and inferior medullary velum [Velum medullare superius/inferius] (see Fig. 1.5).

The tegmentum is that part of the brainstem between tectum and basis located ventral to the mesencephalic/cerebral aqueduct and the fourth ventricle. The brainstem tegmentum contains the cranial nerve and other brainstem nuclei, the reticular formation of the brainstem, and a variety of ascending and descending tracts.

The basis is the ventral most part of the brainstem. Over most of its extension the pyramids are a morphological landmark of the basis. In the example shown in Fig. 1.12 the tectum is the quadrigeminal plate (see Fig. 1.5), the main structures of the tegmentum are the red nucleus (see Sect. 16.5.1) and the substantia nigra (Sect. 16.6.1). The basis consists of the cerebral peduncles with important fiber tracts (see Tables 1.3 and 1.4) descending from the cerebral cortex and ascending to the thalamus and the cerebral cortex.

The horizontal composition of the brainstem as explained above has its root in the neural tube (see Sect. 2.1). During development, the cylindrical neural tube in the region of the prospective spinal cord and the brainstem can be differentiated into four different plates (see Fig. 2.5): As in the spinal cord, as a rule of thumb, the prospective motor centers are located ventrally in the brainstem while the dorsal parts house the prospective sensory centers. Again, motor and sensory qualities can be subdivided into somatic and visceral portions. The brainstem, however, unlike the neural tube is not an almost ideal cylinder since the fourth ventricle is larger than the central canal and extends laterally. Therefore, the nuclei located ventrally during development eventually gain a position close to the midline, i.e. inner most the somatomotor nuclei (GSE, for abbreviations see Table 1.1), followed laterally by visceromotor nuclei (GVE/SVE), then viscerosensory (GVA/SVA) and somatosensory (GSA/SSA) nuclei.

The situation is illustrated paradigmatically by a coronal histological section through the medulla oblongata indicating selected main nuclei and their qualities (Fig. 1.13).

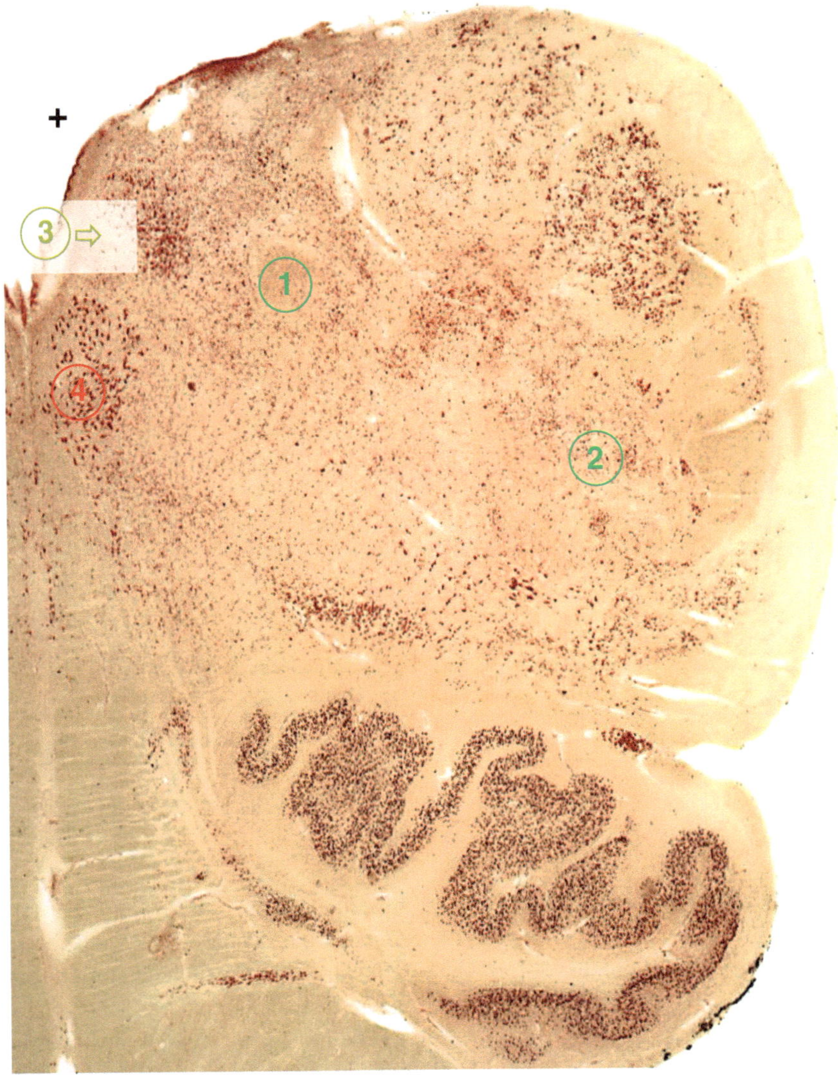

①	**Solitary tract and nucleus** - Viscerosensory	③	**Posterior ncl. of vagus nerve** - Visceromotor
②	**Spinal ncl. of trigeminal nerve** - Somatosensory	④	**Ncl. of hypoglossal nerve** - Somatomotor
		+	**4th Ventricle**

Fig. 1.13 Schematic demonstration of the functional topography of the human brainstem using a histological section. Darrow red staining. While in the further caudal parts—here level of the inferior olive—the anatomy still resembles that of the spinal cord (central canal, dorsal of it sensory domains, ventral motor domains), here, the canal has opened into the fourth ventricle. The orientation of nuclear domains has changed from ventrodorsal (motor > sensory) to mediolateral. The motor nuclei (hypoglossal, vagus) are located medial most in the ventricular floor, sensory nuclei dislocated laterally and the solitary nucleus and the spinal nucleus of trigeminal nerve are following. In the spinal cord, the visceral domains are in between the somatomotor and somatosensory domains showing here an corresponding mediolateral sequence. This general scheme gets increasingly blurred the farther we move cranially, mainly due to the fact that the specific structures of the head (sensory organs, foregut, mimic muscles, glands etc.) are represented in the brainstem and the simple organization gets more complicated (see survey in Table 1.1). In general, however, even with Watson's objections in mind (Box 1.1), the mesencephalon, displays sensory structures, for example, the colliculi in the tectum (dorsolateral), motor domains like the substantia nigra near the basis. LabPON Twente

1.3 Cranial nerves

Brainstem nuclei endowed with one or several of the above mentioned qualities give rise to efferent or afferent cranial nerves, most of them anatomical counterparts of the spinal nerves. As to the sensory cranial nerves this means that—with the exception of the mesencephalic nucleus of trigeminal nerve (see Sect. 13.2.6.1)—the perikarya of the first-order neurons (first-order neurons conduct impulses from peripheral receptors to the CNS) are located outside the CNS in ganglia analogous to the spinal ganglia [Ganglia spinalia] (formerly dorsal root ganglia). The cranial nerve nuclei as well as the other brainstem nuclei will be dealt with individually for the different neuromeres. A survey list of the 12 (CNI-XII) cranial nerves with a short description of their function is provided in Fig. 1.14 and Table 1.1.

1.4 Cerebral arteries and vascular disorders

1.4.1 Normal anatomy of cerebral vessels

1.4.1.1 Circle of Willis and arterial supply of the brainstem

Two pairs of large arteries—originating from the aortic arch and it branches, respectively—supply the blood to the brain, approximately 70% through the internal carotid arteries and 30% through the vertebral arteries (Kalaria et al. 2015) (Figs. 1.15A, B and 1.19).

The systems anastomose via the anterior and posterior communicating arteries and form the circle of Willis (see Figs. 1.16, 1.17, 1.18, and 1.19). The circle of Willis (see Box 1.3) demonstrates considerable morphologic variation among relatively healthy individuals. Segments are fre-

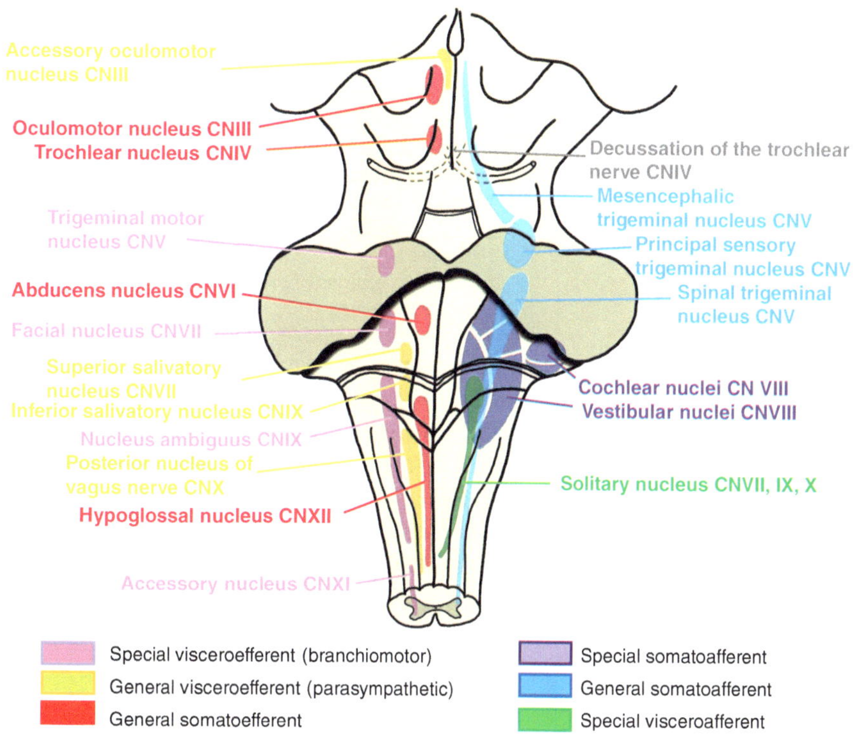

Fig. 1.14 Schematic representation of the cranial nerve nuclei in the human brainstem in dorsal view. The topographical situation of the exits and entries of the cranial nerves is displayed in Fig. 1.6A, B. Details concerning the course of cranial nerves will be provided in the individual sections below. Modified after Huggenberger et al. 2019, Fig. 15.5 with permission

Table 1.1 The cranial nerves (cf. Fig. 1.14)

Cranial nerve	Name	Qualities	Function
I	Olfactory[a]	SSA	Perception of smell
II	Optic[a]	SSA	Perception of visual stimuli
III	Oculomotor	GSE GVE	Extraocular eye muscles Intraocular eye muscles
IV	Trochlear	GSE	Extraocular eye muscle
V	Trigeminal	GSA SVE	Facial skin, mucosae, meninges Masticatory muscles
VI	Abducens	GSE	Extraocular eye muscle
VII	Facial	GVE SVE SVA	Lacrimal gland, minor salivatory glands Mimic muscles Gustation
VIII	Vestibulocochlear	SSA	Sense of equilibrium, Perception of acoustic stimuli
IX	Glossopharyngeal	GVE SVE SVA	Parotid gland Pharynx muscles Gustation
X	Vagus	GVE SVE GVA SVA	Inner organs Pharynx/Larynx muscles Inner organs Gustation
XI	Accessory[b]	GSE	Trapezius and sternocleidomastoid muscles
XII	Hypoglossal	GSE	Tongue muscles

GSA General somatic afferent (somatosensory), *GSE* General somatic efferent (somatomotor), *GVA* General visceral afferent (viscerosensory), *GVE* General visceral efferent (visceromotor), *SSA* Special sensory afferent (as to sensory organs of the head), *SVA* Special visceroafferent (sensory afferents from branchial arches), *SVE* Special visceroefferent (motor efferents to the branchial arches)

[a] CN I and II are peripheral parts of the telencephalon (I) and the diencephalon (II), not "real" cranial nerves

[b] CN XI: neurons of origin are located in the cervical spinal cord

quently hypoplastic or absent and the role of the morphology of the circle of Willis in patients with carotid artery disease is the subject of discussion in the literature (Krabbe-Hartkamp et al. 1998; van Seeters et al. 2015).

The name carotis is derived from Greek ἡ καρωτίς = he karotis, from καρόειν = karoein = to anesthetize since it was already known to the anatomist Galenos (130–200 BCE) that compression of the carotid arteries results in drowsiness.

The absence of an ipsilateral posterior communicating artery has been reported to be related to the risk of watershed infarction in patients with carotid artery occlusion (Schomer et al. 1994). Others conclude that the risk is associated with the absence of a functional anterior communicating artery (Miralles et al. 1995). An anastomosing pathway exists between the internal and external carotid arteries via the ophthalmic arteries. Abnormal vascular networks are believed to form collateral vessels, secondary to

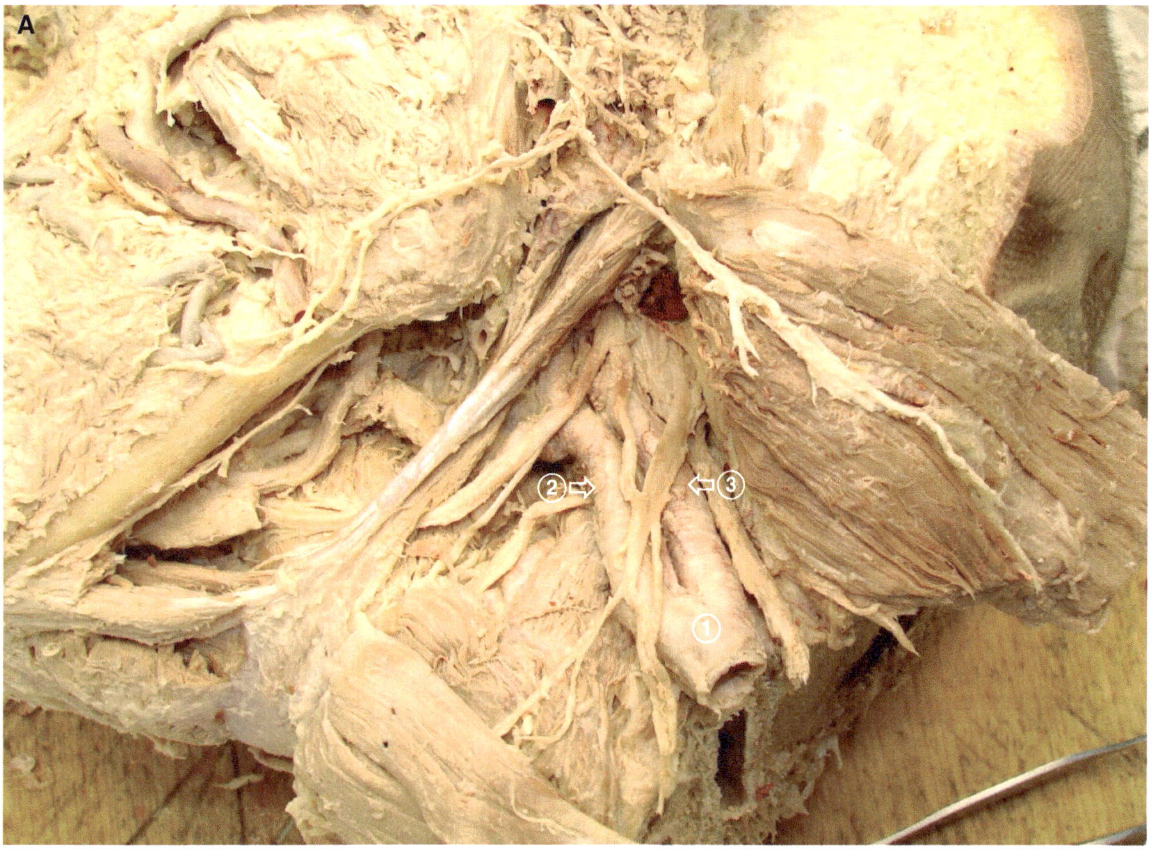

①	**Common carotid artery (CCA)**	③	**Internal carotid artery (ICA)**
②	**External carotid artery (ECA)**		

Fig. 1.15 (**A**) Lateral view onto the region of the carotid bifurcation of the common carotid artery ① into the external carotid artery ② (arterial supply to the face region and the mucosae of mouth and nose) and the internal carotid artery ③ (arterial supply to the eye and the anterior part of the circle of Willis (see here below). Müller-Thomsen fecit. Sammlung des Zentrums Anatomie der Universität zu Köln

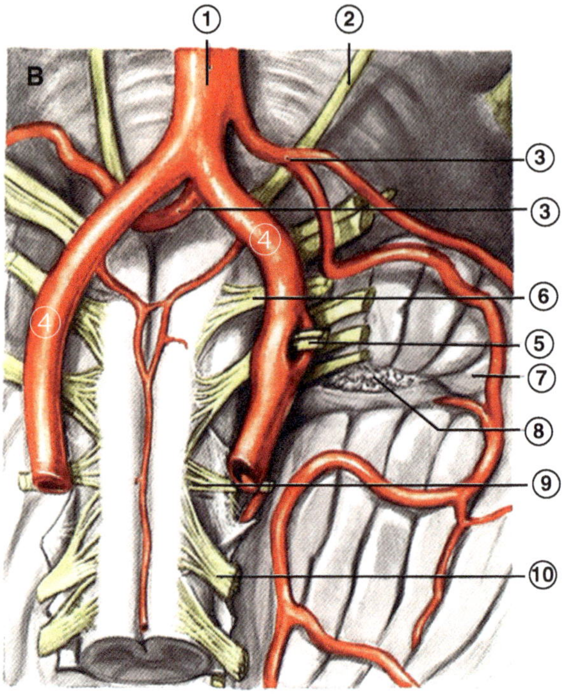

①	**Basilar artery**	⑥	**Superior root bundle of CNXII**
②	**Abducens nerve CNVI**	⑦	**Flocculus**
③	**Posterior inferior cerebellar artery The right PICA originates from the contralateral vertebral artery (variation)**	⑧	**Lateral aperture of the 4th ventricle**
④	**Vertebral artery**	⑨	**Root C_1**
⑤	**Inferior root bundle of CNXII passing through an ‚insula' in the vertebral artery (rare variation)**	⑩	**Root C_2**

Fig. 1.15 (continued) (**B**) Dorsal view onto the course of the vertebral artery [A. vertebralis]. The cerebellum and the brainstem have been removed and the arteries have been injected with a red dye. The vertebral artery is a branch of the subclavian artery [A. subclavia], then ascends posterior to the internal carotid artery. It then enters the transverse foramina of the cervical vertebrae, then leaves the vertebral column, forms a dorsally directed arch and then enters the dura. It gives off the posterior inferior cerebellar artery (PICA) ⑤ and eventually joins the contralateral vessel at the level of the bulbopontine sulcus (see Fig. 1.5) to form the basilar artery (see Figs. 1.17A, B). From von Lanz, Wachsmuth et al. 2003, Fig. 162 with permission

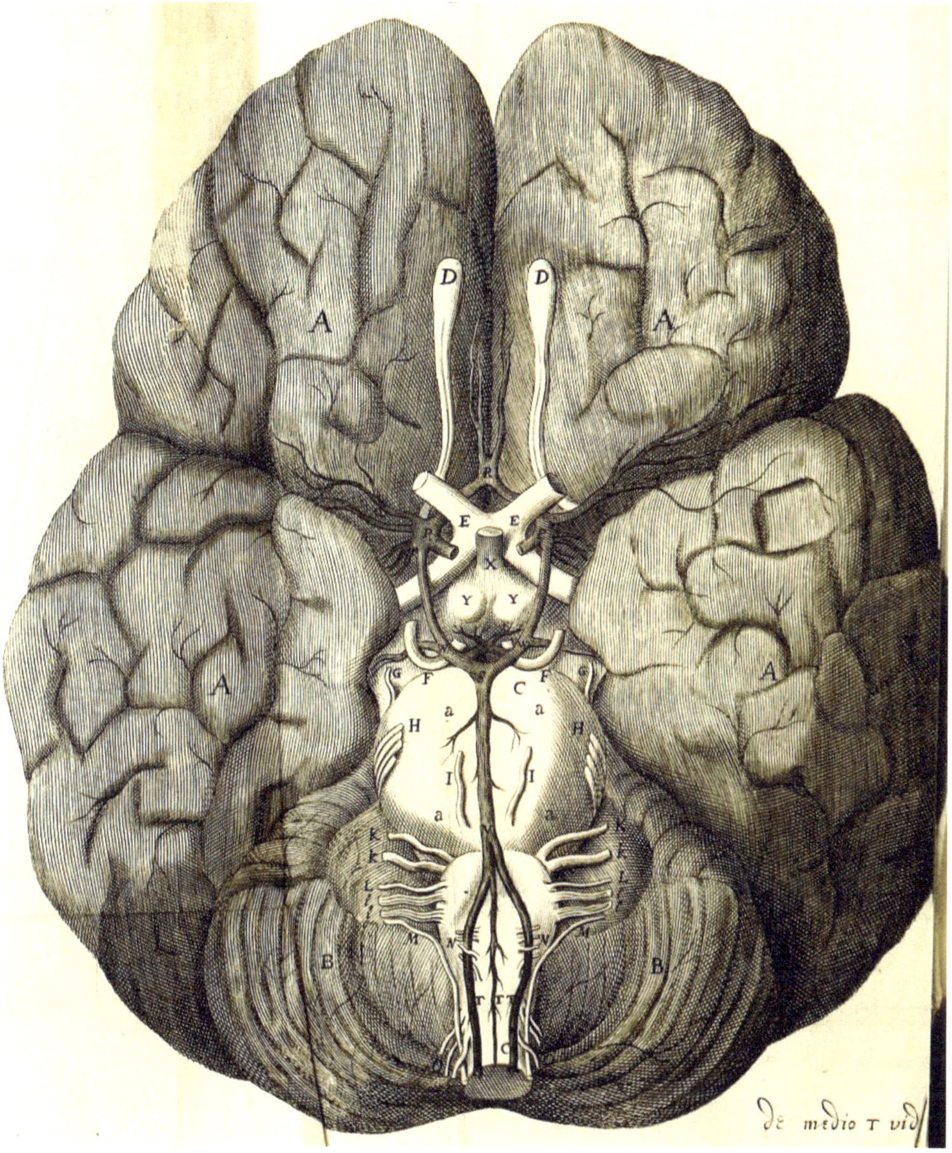

Fig. 1.16 Figure 1 (Ia, figura prima) from Willis's *Cerebri anatome* (1664). Ventral view of the human brain with the basal arteries (circle of Willis) and the cranial nerves. Compare with Figs. 1.6, 1.17, 1.18, and 1.19 and Table 1.2. Drawing attributed to Christopher Wren Sponsor: Open Knowledge Commons and Harvard Medical School. https://archive.org/stream/cerebrianatomecu00will#page/n71/mode/2up

Table 1.2 Explanatory survey of the labels in Figs. 1.16, 1.17 and 1.18

	Vessels Willis[1]			
[1]Translation of Willis' Latin text see behind see below		Latin name	English name	
P. P.	Arteriae carotidis truncus abscissus, ubi	A. carotis interna	Internal carotid artery	ICA
	in ramum anteriorem &	A. cerebri anterior	Anterior cerebral artery	ACA
	posteriorem dividitur	A. communicans posterior	Posterior communicating artery	PCoA
Q. Q.	Ejus ramus inter duos cerebri lobos incedens	A. cerebri media	Middle cerebral artery	MCA
R	Carotidum rami anteriores uniti abscedunt,	A. communicans anterior	Anterior communicating artery	ACoA
	in cerebri fissuram pergentes	A. cerebri anterior	Anterior cerebral artery	ACA
S	Carotidum rami posteriores uniti	A. cerebri posterior	Posterior cerebral artery	PCA
	& trunco vertebrali occurentes	A. basilaris	Basilar artery	BA
T. T. T.	Arteriae vertebrales, & earum tres rami ascendentes	A. vertebralis	Vertebral artery	VA
V	Vertebralium rami in eundem truncum coalescentes	A. vertebralis A. basilaris	Vertebral artery Basilar artery	VA BA
S - V	–		Course of BA	
W. W.	Locus designatur, ubi Arteriae vertebrales & carotides uniuntur,	A. cerebri posterior	Posterior cerebral artery	PCA
	& utrimque ramus ad plexum choroeiden (sic!) ascendit	A. communicans posterior	Posterior communicating artery	PCoA

	Cranial nerves Willis	Figs. 1.17 and 1.18		FIPAT
D. D.	Nervi olfactorii, sive Par primum	CNI	N. olfactorius	Olfactory nerve
E. E.	Nervi optici, sive Par secundum	CNII	N. opticus	Optic nerve
F. F.	Nervi oculorum motorii, sive Par tertium	CNIII	N. oculomotorius	Oculomotor nerve
G. G.	Nervi oculorum pathetici, sive Par quartum	CNIV	N. trochlearis	Trochlear nerve
H. H.	Nervorum Par quintum	CNV	N. trigeminus	Trigeminal nerve
I. I.	Nervorum Par sextum	CNVI	N. abducens	Abducens nerve
K. K.	Nervi auditorii, & eorum utrimque bini processus,	CNVII	N. facialis	Facial nerve
k.k.	Par septimum	CNVIII	N. vestibulocochlearis	Vestibulocochlear nerve
L.L.	Par vagum, sive octavum, pluribus fibris constans	L CNIX	N. glossopharyngeus	Glossopharyngeal nerve
l.l.l. & c.		l CNX	N. vagus	Vagus nerve
M. M.	Nervus spinalis, ad originem Paris vagi à loginquo accedens	CNXI	N. accessorius	Accessory nerve
N. N.	Par nonum, pluribus etiam fibris constans, (quae deorsum tendentes, in E in eundem truncum coalescunt) qui paulo supra processum occipitis emergit	(CNXII)	N. hypoglossus	Hypoglossal nerve

Translation		
	Vessels	Englisch translation Modern English terms in *italics*
P. P.	Arteriae carotidis truncus abscissus, ubi	Division of the trunk of the carotid artery (*Internal carotid artery ICA*) where it divides into
	in ramum anteriorem &	an anterior branch (*Anterior cerebral artery ACA*) and
	Posteriorem dividitur	a posterior branch (*Middle and posterior communicating artery MCA / PCoA*)
Q. Q.	Ejus ramus inter duos cerebri lobos incedens	Where its branch (Middle cerebral artery MCA) enters between the two cerebral lobes (i.e., *temporal and frontal lobes*)
R	Carotidum rami anteriores uniti abscedunt, in cerebri fissuram pergentes	Where the anterior carotid branches (i.e., *anterior cerebral arteries ACA*) divide and continue into the Fissura cerebri
S	Carotidum rami posteriores uniti & trunco vertebrali occurentes	Where the posterior united carotid branches (i.e., *posterior communicating arteries PCoA*) meet with the vertebral trunk (i.e., *the basilar artery*)
T. T. T.	Arteriae vertebrales, & earum tres rami ascendentes	*Vertebral arteries* (VA) and their three ascending branches (i.e., the two *vertebral arteries* proper, in their middle the unpaired *anterior spinal artery*)

Table 1.2 (continued)

Translation			
	Vessels	Englisch translation	Modern English terms in *italics*
V	Vertebralium rami in eundem truncum coalescentes	Where the vertebral branches come together into the same trunk (i.e., *the basilar artery*)	
S - V	—	Course of the basilar artery from the exit of the posterior cerebral arteries to the fusion of the vertebral arteries (no anatomical term available)	
W. W.	Locus designatur, ubi Arteriae vertebrales & carotides uniuntur, & utrimque ramus ad plexum choroeiden (sic!) ascendit	Site where the vertebral arteries unite with the carotid ones and from where bilaterally a branch to the choroid plexus ascends	
	Cranial nerves		
D. D.	Nervi olfactorii, sive Par primum	Olfactory nerves, or first couple (of cranial nerves) *CNI*	
E. E.	Nervi optici, sive Par secundum	Optic nerves, or second couple *CNII*	
F. F.	Nervi oculorum motorii, sive Par tertium	Oculomotor nerves, or third couple *CNIII*	
G. G.	Nervi oculorum pathetici, sive Par quartum	Pathetic nerves[a], or fourth couple *CNIV*	
H. H.	Nervorum Par quintum	Couple of fifth nerves *CNV*	
I. I.	Nervorum Par sextum	Couple of sixth nerves *CNVI*	
K. K. k.k.	Nervi auditorii, & eorum utrimque bini processus, Par septimum	Auditory nerves (*CNVIII*) and their two bilateral processes, seventh couple (*CNVII*)	
L.L. l.l.l. & c.	Par vagum, sive octavum, pluribus fibris constans	Couple of vagus, or eighth couple (*?*), consisting of fibers which are many times stronger (*?*)	
M. M.	Nervus spinalis, ad originem Paris vagi à loginquo accedens	Spinal nerve (*CNXI*), approaching the origin of the vagus nerves from a distance	
N. N.	Par nonum, pluribus etiam fibris constans, (quae deorsum tendentes, in E in eundem truncum coalescunt) qui paulo supra processum occipitis emergit	Ninth couple (*CNXII !*), also consisting of many times stronger fibers (extending caudally, coming together in the same trunk in E (?)) and which a little bit above the Processus occipitis (Occipital condyle, Condylus occipitalis) emerges	

[a] Since the muscle moves the eye inwards and down during the baroque era the trochlear nerve was called N. patheticus (Nerf pathétique)

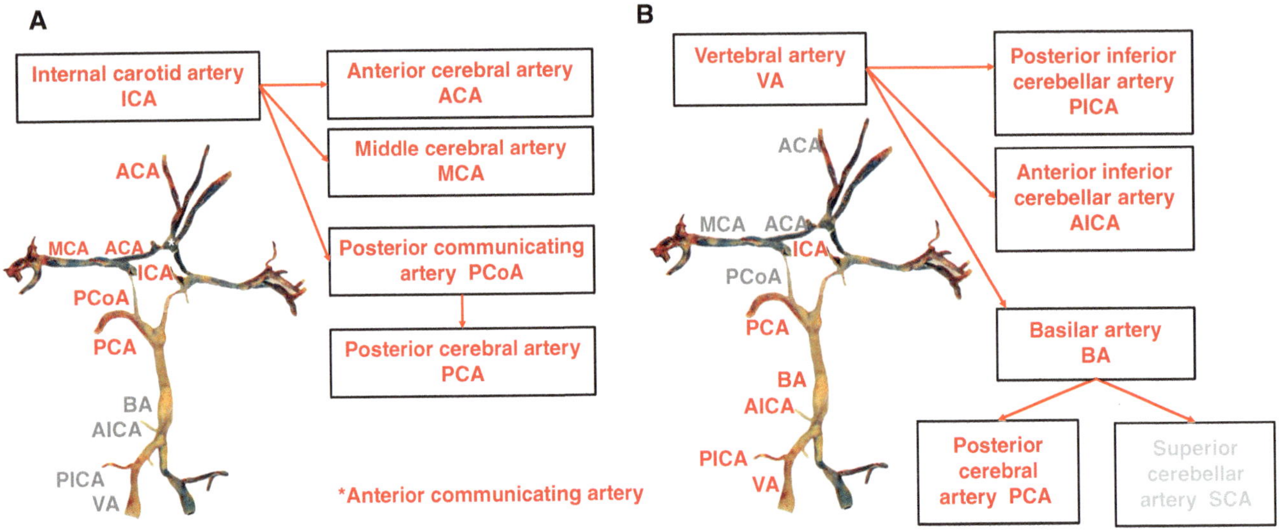

Fig. 1.17 (**A**) Anterior part of the human circle of Willis. Müller-Thomsen fecit. Sammlung des Zentrums Anatomie der Universität zu Köln. (**B**) Posterior part of the human circle of Willis. Müller-Thomsen fecit. Compare with Fig. 1.17A and Fig. 1.18A, B. Sammlung des Zentrums Anatomie der Universität zu Köln

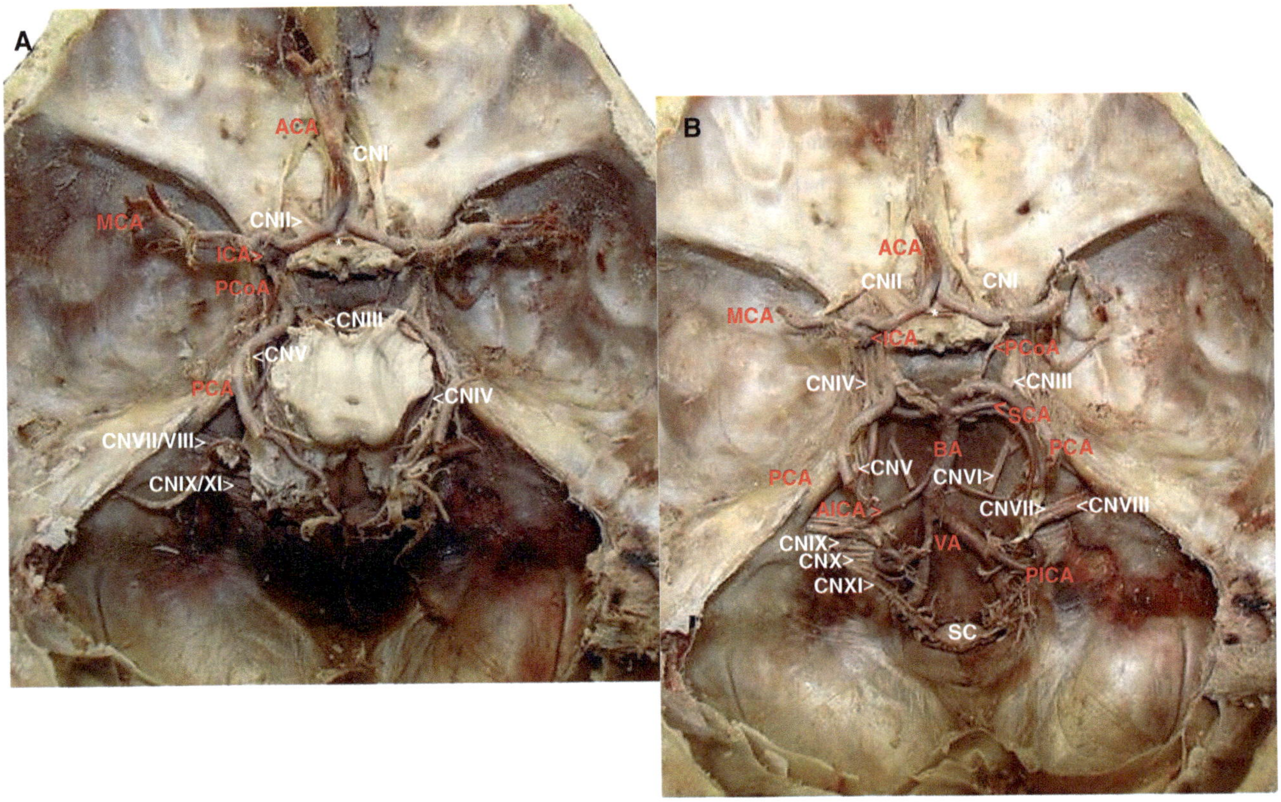

Fig. 1.18 (**A**) Dorsal view of the human skull base. Formalin-fixed specimen. Forebrain and cerebellum removed, brainstem sectioned horizontally at the mesencephalic level. Müller-Thomsen fecit. Sammlung des Zentrums Anatomie der Universität zu Köln. (**B**) Brainstem removed enabling the view on the complete circle of Willis and the cranial nerves I -XI. A doubled superior cerebellar artery (SCA) has been found in 14–22% of studied cases as it is present here on the right-hand side (Lang 1991). Müller-Thomsen fecit. Sammlung des Zentrums Anatomie der Universität zu Köln. *ICA* Internal carotid artery (for Latin names see Table 1.2), *ACA* Anterior cerebral artery * Anterior communicating artery, *MCA* Middle cerebral artery, *PCoA* Posterior communicating artery, *PCA* Posterior cerebral artery, *CNI* Olfactory nerve (for Latin names, see Table 1.2), *CNII* Optic nerve, *CNIII* Oculomotor nerve, *CNIV* Trochlear nerve, *CNV* Trigeminal nerve, *CNVI* Abducens nerve, *CNVII* Facial nerve, *CNVIII* Vestibulocochlear nerve, *CNIX* Glossopharyngeal nerve, *CNX* Vagus nerve, *CNXI* Accessory nerve, SC Spinal cord, *BA* Basilar artery, *SCA* Superior cerebellar artery, *AICA* Anterior inferior cerebellar artery, *VA* Vertebral artery, *PICA* Posterior inferior cerebellar artery

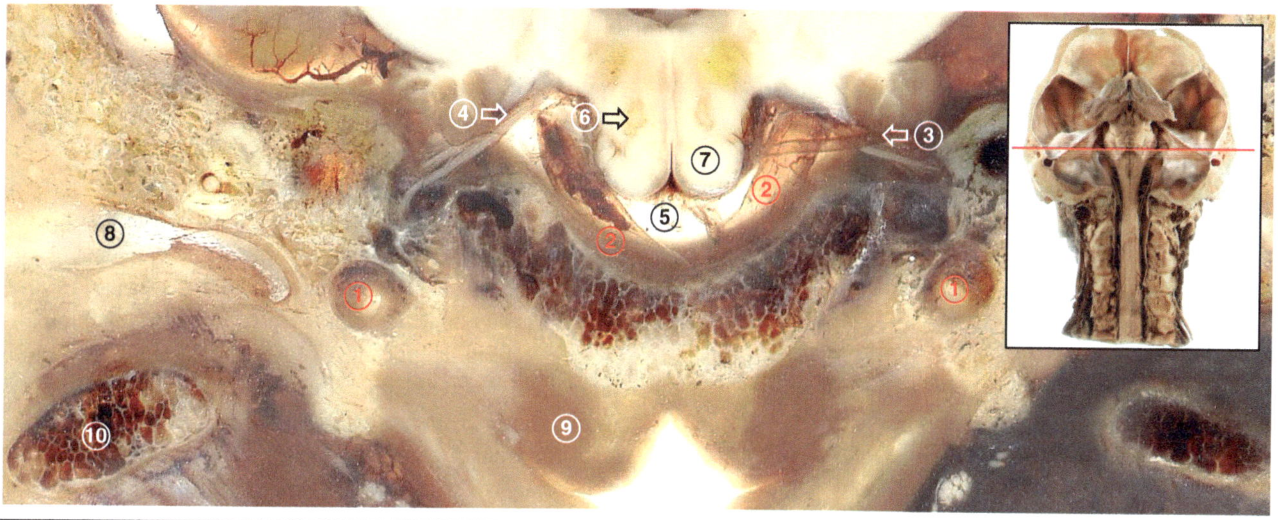

①	Internal carotid arteries in the skull base	⑥	Inferior olivary complex
②	Vertebral arteries (region of merging)	⑦	Pyramidal (corticospinal) tract
③	Labyrinthine artery (branch of the basilar artery)	⑧	External auditory meatus
④	Vestibulocochlear nerve CNVIII	⑨	Longus capitis muscle
⑤	Prepontine cistern	⑩	Mandible / Condylar process

Fig. 1.19 Horizontal section through the head at the level of the cerebellopontine angle. Plastinated section. The vertebral ② and internal carotid arteries ① are located close together. The former lie ventral of the medulla oblongata and are seen at approximately the level where they unite to form the basilar artery (basilaris = belonging to the base, latinized form of French "basilaire"). The internal carotid arteries are seen in the skull base (see Sect. 1.5) in close proximity to the external acoustic meatus and the inner ear. Sammlung des Zentrums Anatomie der Universität zu Köln

obstructive lesions around the terminal portions of the internal carotid arteries, often without leading to infarction (the so-called Moyamoya vessels) (Ikeda and Hosoda 1992; Haltia et al. 1982).

Box 1.3 Thomas Willis (1621–1675) (for details on the life of Willis see Arráez-Aybar et al. 2015)

Thomas Willis was born on January 27, 1621 in Great Bedwyn, Wiltshire, UK, about 70 miles northwest of London (see Arráez-Aybar et al. 2015).

Willis originally had a clerical vocation that led him to learn Latin. Although he disliked the inflexibility of theology, his great command of Latin was very useful in his later medical writings. He studied medicine at the University of Oxford. He later became a fellow of a select club of eminent experimentalists and a Fellow of the Royal Society. As a physician, Willis did not base his practice on aphorisms but on a "bench to bedside" approach to medicine, while studying neuroanatomy as a basis for the comprehension of neurological pathology (Arráez-Aybar et al. 2015).

He developed innovative anatomical methods for the preservation and dissection of the brain, injection of colored substances and illustrations of his findings. In his famous book *Cerebri anatome* (Fig. 1.16), Willis recognized the cerebral cortex as the substrate of cognition. He explained for the first time the pathological and functional meaning of the brain's circular arterial anastomosis, which is named after him and coined the term "Neurology" (Arráez-Aybar et al. 2015). He died of tuberculosis and pneumonia on November 11, 1675. His remains rest next to those of his wife in the north transept of Westminster Abbey, London (Arráez-Aybar et al. 2015).

Detailed data on the smaller arterial branches of the brainstem can be found in the books by Duvernoy (1999), Lang (1991) and Parent (1996) and for the inner ear under Sect. 5.3.1.1.5 "Arteries of the inner ear."

Please note that among all terminological problems related to the brainstem that of the arteries is one of the most complicated since all three publications mentioned do not use the FIPAT terminology. While FIPAT does not provide any pictures, the vessels in Duvernoy's ink injection histology pictures and in Parent's figure are not labeled. We did our best to provide a paradigmatic view of the three levels shown in the latter indicating the FIPAT terms (see Fig. 1.20). Nevertheless, this is an issue that urgently needs a revision by experts in cerebral vasculature.

1.4.1.2 Morphology and histology of cerebral vessels

The extracranial parts of the carotid and vertebral arteries have the same structure as all large arteries. The intracranial arteries have a different structure. The endothelial cells are joined by tight junctions, they have no fenestrations and their metabolism has specific features related to barrier and transcellular transport functions (see Fig. 1.21A, B).

The muscle layer of intracranial arteries is thinner compared to that of extracranial arteries and the external elastic lamina is lacking. The adventitia is very thin (Kalaria et al. 2015) (Fig. 1.21A, B).

The common carotid artery is an elastic artery and the internal carotid artery (intracranial artery) becomes muscular at a varying distance (0.5–2 cm) from the bifurcation (Janzen 2004). Histologically, transition between elastic and muscular-type arteries is marked by a gradual rather than abrupt change in architecture. Elastic lamellar units predominate in the proximal regions whereas these components are dissociated and reduced more distally (Fig. 1.21A and B).

In the transitional zone (TZ), a different ratio of elastin and collagen versus vascular smooth muscle cells is seen. Preponderance of atherosclerosis at branching points potentially corresponding to transitional zones has been well documented (Samuel 1956; Hori et al. 2008; Masuoka et al. 2010) (see Fig. 1.23).

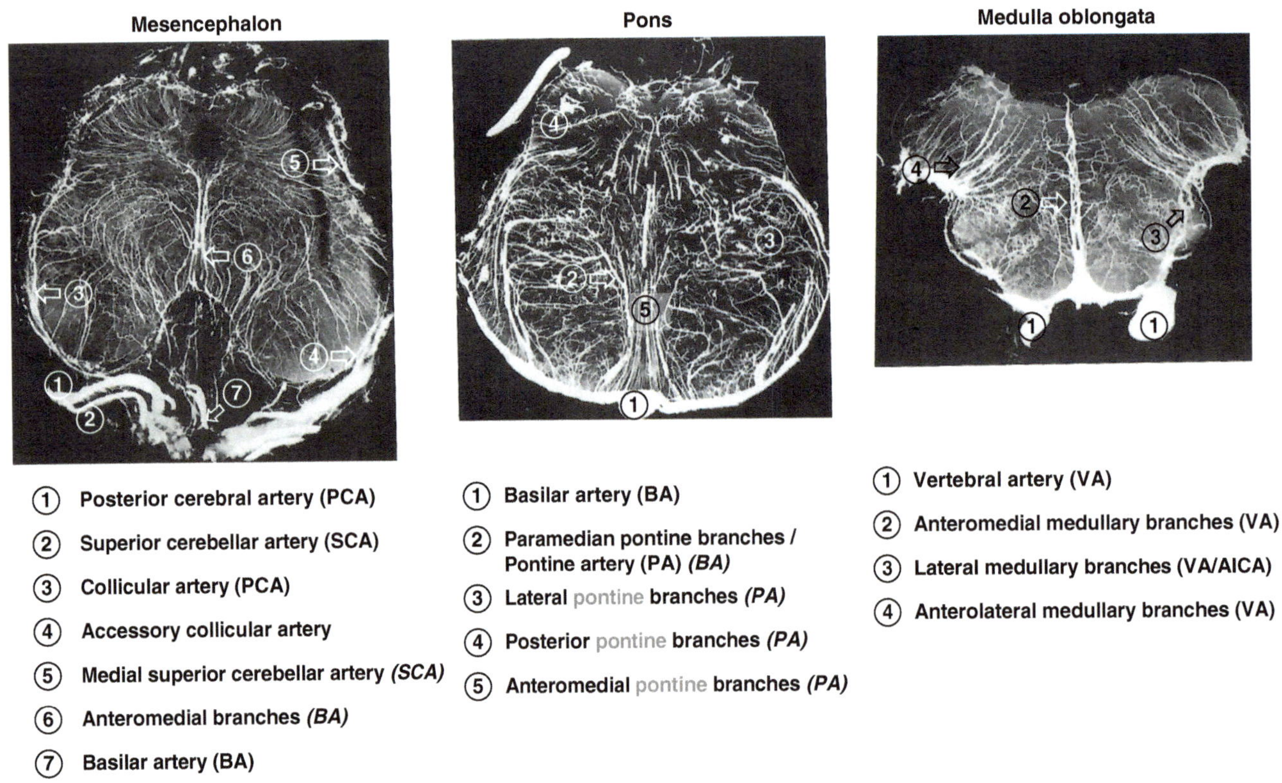

Fig. 1.20 Vascular architecture of paradigmatic levels of the human brainstem. Postmortem X-ray pictures of horizontal brainstem sections after injection of contrast medium (see Thron 1988 for technical details and X-ray anatomy of the human spinal cord). The original illustration from Parent (1996) does not provide labels. This is an attempt to indicate the most important vessels using the FIPAT terminology based on the books of Duvernoy and Lang. The abbreviations in italics indicate the vessel(s) of origin. "Pontine" in gray is—unfortunately—not shown in FIPAT Ch.1. Compare with Fig. 1.25 (Wallenberg syndrome) where the medullary lesion was caused by unilateral occlusion of the vertebral artery. On the microscopic level, obviously, mainly the anteromedial medullary branches were affected. From Parent 1996 Fig. 4.19 with permission

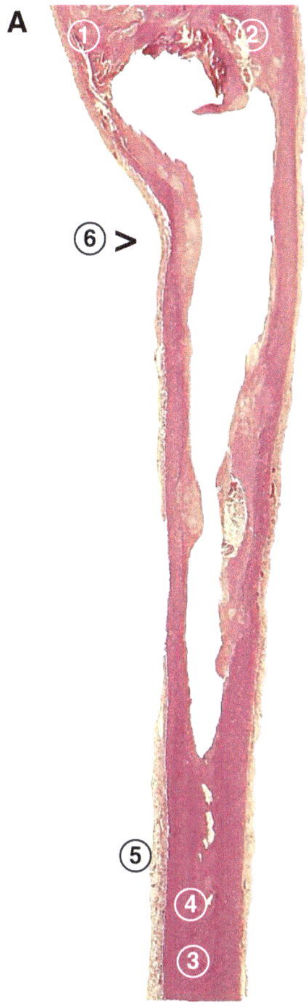

①	Internal carotid artery ICA	④	Tunica media
②	External carotid artery ECA	⑤	Tunica adventitia
③	Common carotid artery CCA	⑥	Transitional zone

Fig. 1.21 (**A**) Longitudinally cut transitional zone ⑥ of the carotid bifurcation. This transitional zone varies in length between 5 mm and 15 mm. Compare with Fig. 1.21B. LabPON Twente

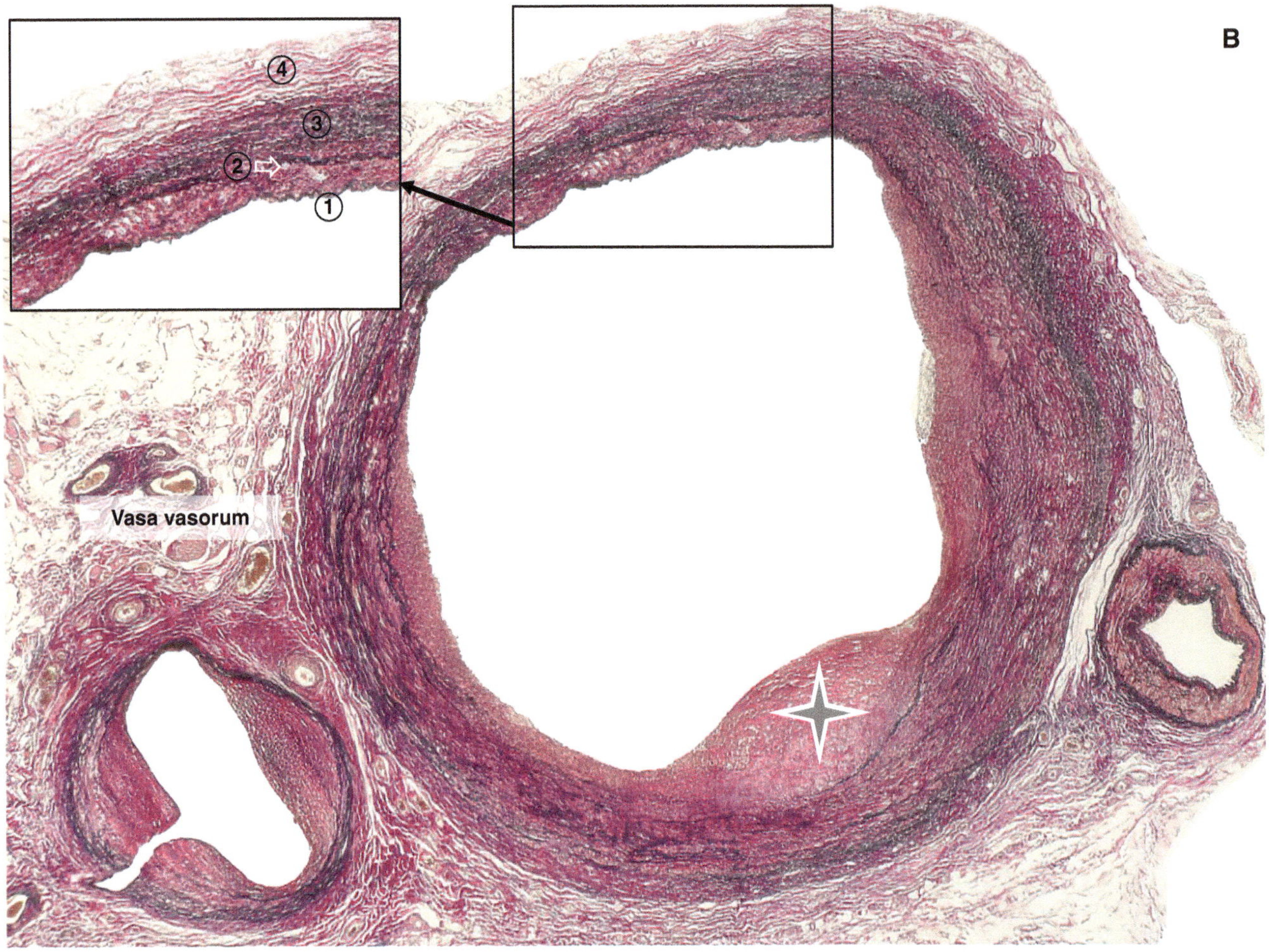

①	Tunica intima	③	Tunica media
②	Internal elastic lamina	④	Tunica adventitia

Fig. 1.21 (continued) (**B**) Cross-sectioned external carotid artery (ECA). Elastica von Gieson stain. Example of the external carotid artery, just beyond the bifurcation, with Vasa vasorum (Artery-supplying small vessels). The wall structure is shown in the inset. The Tunica intima borders the arterial wall vs. the vessel lumen with endothelial cells. The Tunica intima is followed by the internal elastic lamina which encircles the Tunica media (smooth muscle cells and elastic fibers) together with the external elastic lamina (in this example not sharply delineated). The vessel is anchored in the surrounding tissue by the Tunica adventitia, mainly consisting of collagen fibers. Intracranial arteries, like the internal carotid artery, lack the external elastic lamina and the Vasa vasorum and their media is smaller than in extracranial vessels with a decreased density of elastic fibers. Note that this is not a section of a completely healthy external carotid artery. At 17:00, the intima is massively thickened by a fibrous plaque (star). The border to the media, however, can clearly be identified by the internal elastic lamina. LabPON Twente

1.4.2 Pathological anatomy of cerebral vessels

In this book, distributed to the respective chapters, we will deal with the following cerebrovascular entities:

1. Vascular diseases
   - Atherosclerosis
   - Arterio- and arteriolosclerosis
   - Amyloid angiopathy
2. Consequences of vascular diseases: Stroke

   Disorders listed under 1. and 2. here above will be dealt with in the following to lay the base for the understanding of the other vascular disorders:
3. Vascular hamartomas
   - Capillary telangiectasias
   - Cavernous angiomas
   - Arteriovenous malformations
   - Venous malformations

   These entities are described in Chap. 5 under Sect. 5.3.1.1.6.
4. Vascular dementia

   This disorder is dealt with in Chap. 10 under Sect. 10.1.1.3.

1.4.2.1 Atherosclerosis

The most common vascular disease is atherosclerosis. It can affect both intracranial and extracranial large arteries, but regional variations exist. The carotid arteries are one of the

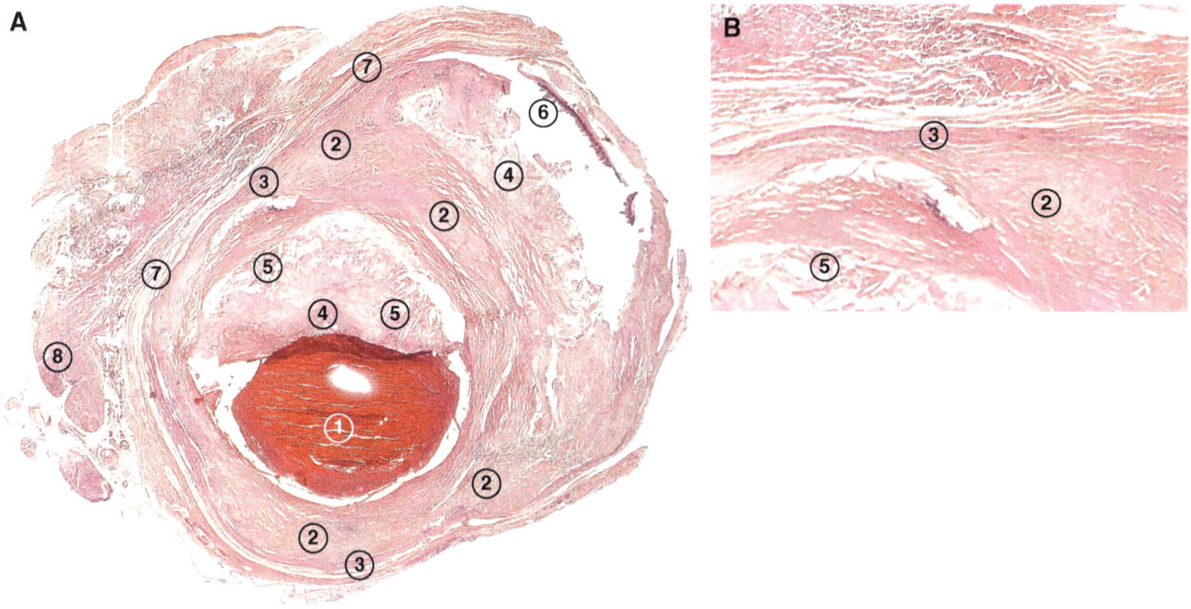

①	Thrombotic occlusion	⑤	- cholesterol clefts
②	Collageneous fibrosis	⑥	- calcification
③	Loss of smooth muscle cells	⑦	Tunica adventitia
④	Necrotic core with (⑤ ⑥):	⑧	Nerve cell complexes

Fig. 1.22 Thrombotic occlusion in a complicated plaque in the internal carotid artery. Overview (**A**) and Detail (**B**). H & E stain. Severe atherosclerosis in the ICA: Loss of smooth muscle cells, collagenous fibrosis, necrotic core with cholesterol clefts, calcification, and superimposed thrombus formation. Note the nerve cell complexes in the surrounding connective tissue. LabPON Twente

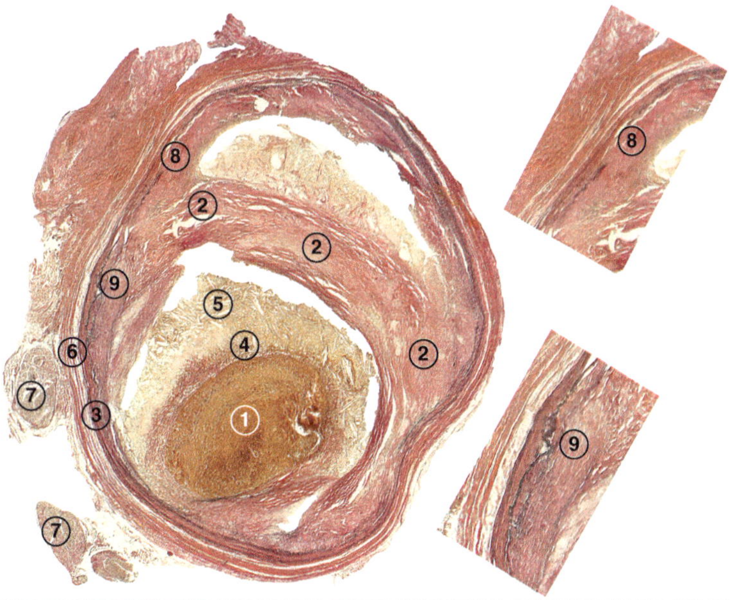

①	Thrombotic occlusion	⑥	Tunica adventitia
②	Collageneous fibrosis	⑦	Nerve cell complexes
③	Loss of smooth muscle cells	⑧	Patchy loss of elastic fibres
④	Necrotic core with:	⑨	Elastic fragmentation
⑤	- cholesterol clefts		

Fig. 1.23 Thrombotic occlusion in a complicated plaque in the internal carotid artery. Overview and Details. Cross-section. Elastica von Gieson stain. Complicated plaque rupture. Elastic fragmentation and patchy loss of elastic fibers. LabPON Twente

most often affected extracranial sites and the progression is similar to that of the aorta and coronary arteries (Kalaria et al. 2015) (Figs. 1.22 and 1.23).

Intracranial atherosclerotic disease is the process of atherosclerosis that affects intracranial large arteries and is thought to be responsible for at least 10% of ischemic stroke worldwide (Battistella and Elkind 2014). Among the etiologies of ischemic stroke, carotid atherosclerosis plays a key role as one third of the leading cause (Adams et al. 1993; Noh and Kang 2019).

Well-known major risk factors for atherosclerosis are age, family history, diabetes mellitus, tobacco use, hypertension, high blood-fat levels, and obesity. The severity of atherosclerosis can vary in different vessels. The degree of aortic or coronary artery atherosclerosis does not always predict cerebral atherosclerosis. Some studies however, showed that carotid artery bifurcation lesions are associated with the development of coronary artery lesions (Kallikazaros et al. 1999; Polak et al. 2013).

Large intracranial vessels including the basilar arteries, vertebral arteries, and the proximal ends of the middle cerebral arteries often show atheromatous disease, especially in African and Asian people whereas in European people the internal carotid artery is most often involved (Kalaria et al. 2015). An important feature is disruption of the endothelial barrier through injury to endothelial cells followed by an accumulation of lipids in the arterial intima. The initial visible lesions are the fatty streaks, formed by clusters of foam cells. In most cases, the fatty streaks do develop into more advanced lesions, as atherosclerotic plaques (atheroma), over a long period of time. The plaques are called fibrous plaques if the collagenous connective tissue component predominates over the lipids (Kalaria et al. 2015). Inflammatory cells enter the plaques from the circulation. The cascade of inflammatory reactions involves monocytes, macrophages, T lymphocytes, and vascular smooth muscle cells (Chen et al. 2008; Kalaria et al. 2015). The smooth muscle cells are responsible for the production of extracellular matrix components of the fibrous cap. Thinning of the cap leads to plaque instability, mediated by inflammatory cells, which are an important finding with plaque rupture (Carr et al. 1997). Vascular and inflammatory cells produce metalloproteinases which can weaken the fibrous cap. Thin fibrous caps become unstable and, consequently, ulceration and thrombosis can occur.

Complicated plaques always show a disrupted endothelial lining which activates coagulation and thrombus forms over the plaque, causing narrowing of the lumen and predisposing to embolism (Figs. 1.22 and 1.23).

Recent research has shown that substenotic intracranial plaques may cause cerebral infarcts just as substenotic coronary plaques cause myocardial infarction (Battistella and Elkind 2014). Compared with extracranial vessels of a similar size, the adventitia and media of the intracranial arteries are thinner and their internal elastica lamina is fenestrated differently and thicker (Mohr et al. 1998; Chen et al. 2008). Luminar stenosis, percentage of lipid area, and the presence of intraplaque neovasculature may play a key role in ischemic stroke (Chen et al. 2008).

1.4.2.2 Pathology of extracranial and intracranial carotid arteries

Stenosis due to severe atherosclerosis at the origin of the internal carotid artery (see Figs. 1.15A, 1.16, and 1.17B) is an important cause of ischemic stroke (see Sect. 1.4.2.7). Rupture of the atherosclerotic lesion can induce an intraluminal thrombus that results in artery-to-artery embolism to the brain or result in an occlusion of the internal carotid artery (ICA) (Adams 2010) (see Figs. 1.22 and 1.23).

Occlusion of the ICA can lead to a stroke or death but can also be completely asymptomatic, especially in patients with an adequate collateral flow.

A stroke caused by ICA occlusion can present with clinical features similar to those due to any other etiology (Thanvi and Robinson 2007). An episode of transient monocular blindness (Amaurosis fugax), however, is highly suggestive of ICA disease proximal to the ophthalmic branch.

Symptomatic ICA occlusion increases the risk of future cerebrovascular events (Thanvi and Robinson 2007). Surgical intervention to lower the risk of stroke among people with severe symptomatic ICA stenosis include endarterectomy (CEA) and carotid angioplasty and stenting (CAS) (Adams 2010). MR vessel wall imaging research investigating the relationship of atherosclerosis between intracranial and extracranial carotid arteries suggests that extracranial carotid plaque burden might be an independent indicator for the severity of intracranial atherosclerosis (Xu et al. 2018).

The internal and external carotid arteries originate from the common carotid artery and a bifurcation forming an angle between the two arteries (see Fig. 1.15A). The angle leads to hemodynamic stress. The clinical significance of the angle and tortuosity has been reported (Adams 2010; Noh and Kang 2019) which could be related to atherosclerosis. An association of kinking and coiling with internal carotid artery dissection has also been described (Saba et al. 2015). Although the internal carotid angle could be a possible risk factor for ischemic stroke, results for the clinical significance are not consistent. Infarcts resulting from occlusion of the internal carotid artery (at the bifurcation of the common carotid artery) resemble in size and distribution to those caused by thrombosis of the middle cerebral artery.

The histologic tissue composition at transitional zones, associated with bifurcations and branching, may play an important role in the development of "wear and tear" arterial wall damage predisposing these segments to atherosclerosis (Janzen 2004). The so-called complicated plaques show cholesterol clefts, macrophages, and often old hemorrhages, in older lesions, with a significant narrowing of the vessel lumen. Ulceration and overlying thrombosis can be present. The histologic features are best demonstrated with stains that differentiate between elastica, smooth muscle, and fibrous tissue (e.g., Elastica von Gieson, Masson's trichrome stain) (Figs. 1.24A–C).

Unilateral infarction of the dorsolateral medulla oblongata is caused by an arteriosclerotic-thrombotic obstruction of the intracranial vertebral artery or posterior inferior cerebellar artery (Wallenberg syndrome) (see Figs.1.25 and 1.26). In general, it can be said that occlusion of large arteries frequently results from embolization of atherosclerotic debris originating from the common or internal carotid arteries or from a cardiac source and in only a small number is caused by plaque ulceration and thrombosis.

1.4.2.3 Wallenberg syndrome

The earliest known description of lateral medullary infarction was given in 1810 by Gaspard Vieusseux of Geneva at the Medical and Chirurgical Society of London. The symptoms, including the accurate location of the lesion, were described by Adolf Wallenberg in 1895 (Pearce 2000). He described a

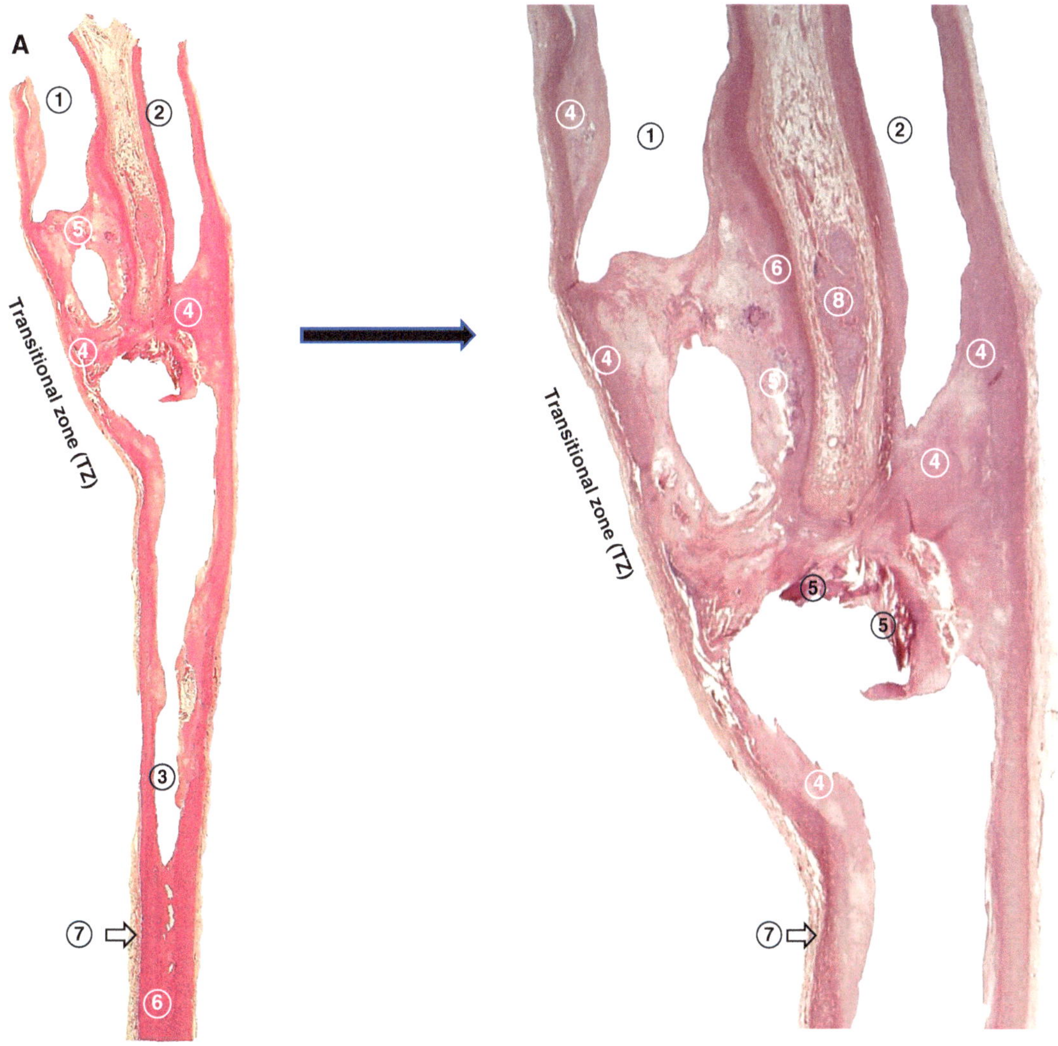

①	**Internal carotid artery ICA**	⑤	**Necrotic core with calcification**
②	**External carotid artery ECA**	⑥	**Tunica media**
③	**Common carotid artery CCA**	⑦	**Tunica adventitia**
④	**Collagenous fibrosis**	⑧	**Nerve cell complexes**

Fig. 1.24 (**A**) Longitudinal section of the Sinus (bulbus) portion of the internal carotid artery (ICA) Overview (left hand-picture) including the common carotid artery and the TZ and detail of the TZ (→ right hand-picture). H & E stain. LabPON Twente. (**B**) Elastica von Gieson stain. Overview (**B**) and detailed view of sites X and Y. LabPON Twente. (**C**) ICA at the transitional zone, cross-section, showing a complicated plaque (left). Elastica von Gieson stain. In the picture on the right cross-section through the (muscular) internal carotid artery, distal to the transition zone with fibrosis, elastic fragmentation (arrow), and thromboembolus. Elastica stain. LabPON Twente

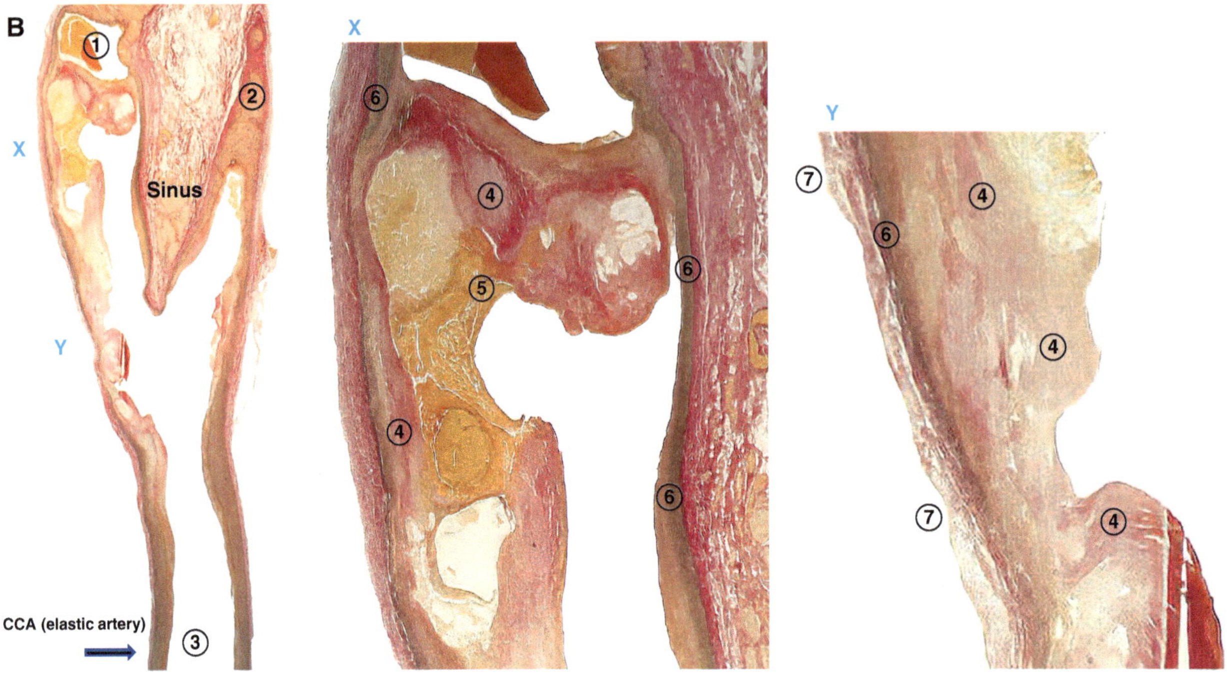

①	Internal carotid artery ICA	⑤	Necrotic core with calcification
②	External carotid artery ECA	⑥	Tunica media
③	Common carotid artery CCA	⑦	Tunica adventitia
④	Collagenous fibrosis		

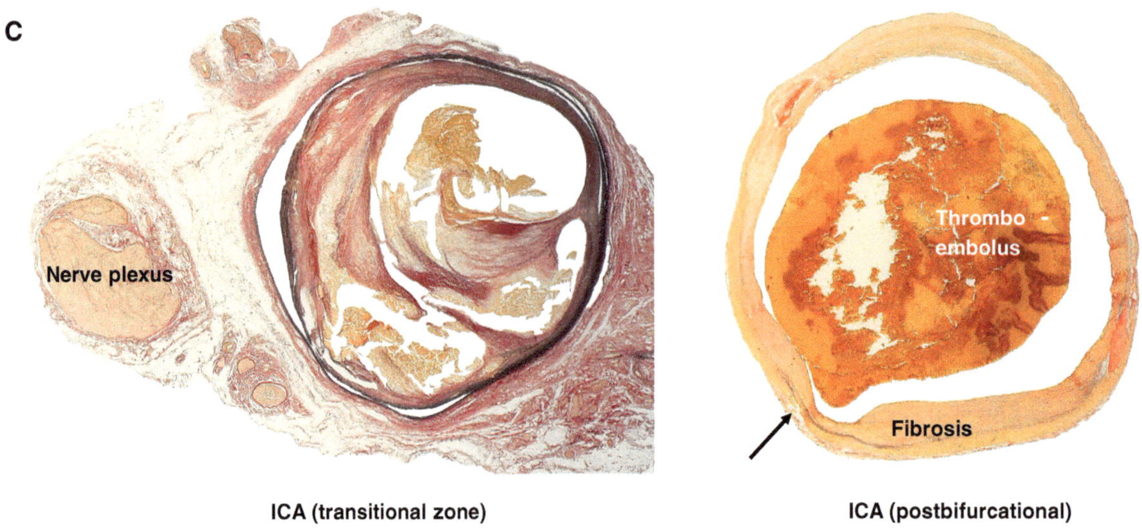

Fig. 1.24 (continued)

syndrome characterized by neurological symptoms produced by unilateral infarction of the dorsolateral medulla oblongata and caused by an arteriosclerotic-thrombotic obstruction of the intracranial vertebral artery or by the posterior inferior cerebellar artery (Wallenberg 1895; Wilkins and Brody 1970) (see Table 5.1 and Fig. 5.25). Occasionally, an occlusion of the basilar artery or distal extracranial vertebral artery was responsible and spontaneous or exercise-induced dissection of the vertebral arteries has also been described (Rodriguez et al. 2020). Vertebral occlusion is the most common cause of this syndrome (Wilkins and Brody 1970; Sacco et al. 1993; Kitis et al. 2004) (Figs. 1.25 and 1.26).

The syndrome can also be caused by tumors, metastases, or encephalitis in the region of the dorsolateral medulla

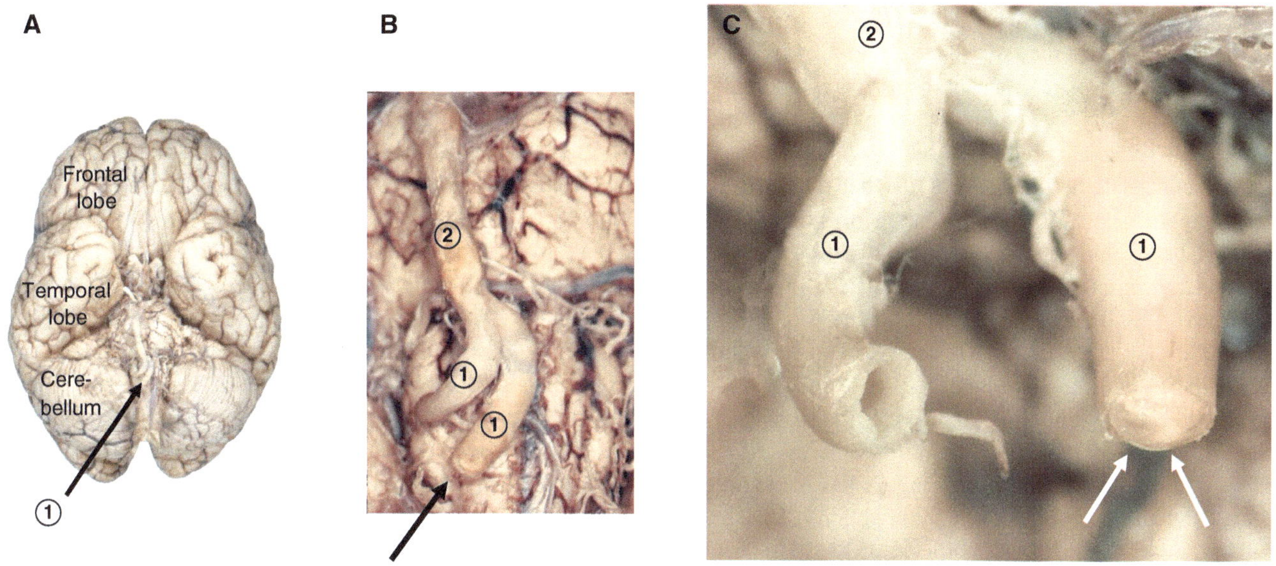

①	Vertebral artery	②	Basilar artery

Fig. 1.25 Brain from a patient presenting with a lateral medullary syndrome (Wallenberg syndrome) (see Fig. 1.26). Severe atherosclerosis of the circle of Willis and total occlusion of (**A**) the left vertebral artery (①, black arrow), detail in (**B**) (black arrow) and (**C**) (white arrows). LabPON Twente

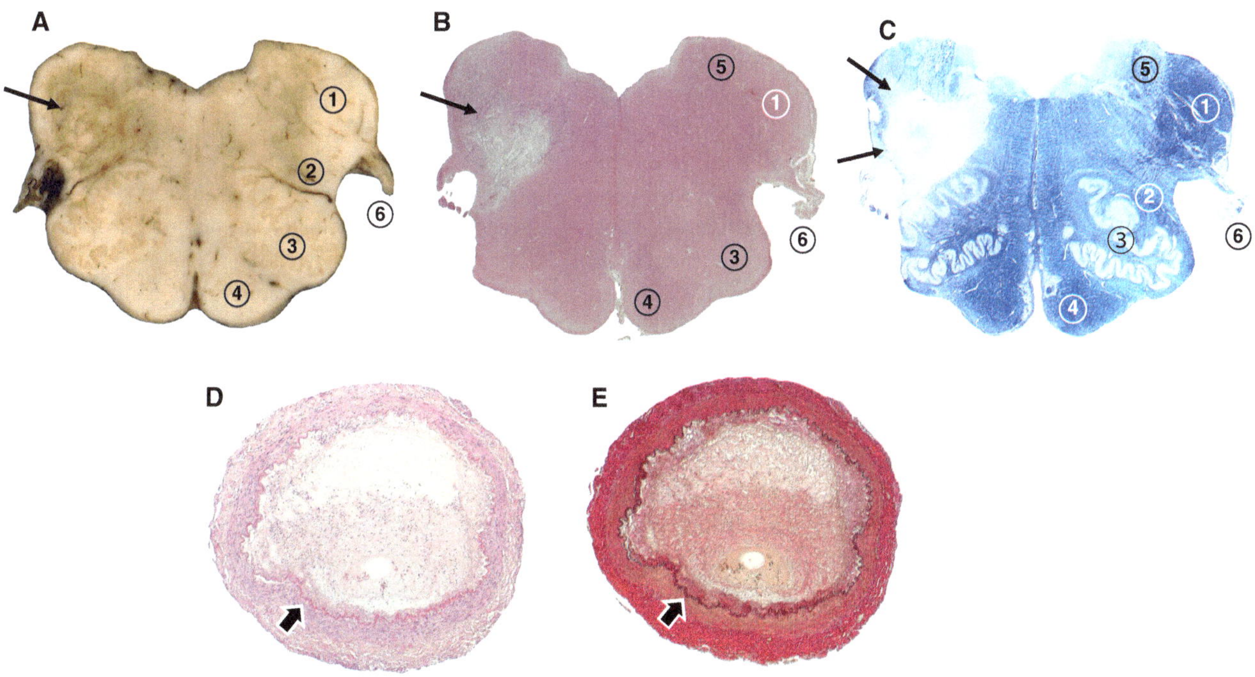

①	Inferior cerebellar peduncle	④	Pyramidal tract
②	Anterolateral tract	⑤	Vestibulospinal tract
③	Inferior olivary complex	⑥	Glossopharyngeal nerve CNIX

Fig. 1.26 (**A**) Horizontal section of the medulla oblongata, showing a recent infarction, dorsolateral to the inferior olivary complex ③ (see Sect. 4.1.1.3) (**B**, **C**) Microscopical sections of the infarct zone. The infarct was caused by a total occlusion of the left vertebral artery (**D** and **E**). Note the lack of an external elastic lamina, a typical feature for intracranial arteries (arrows in D and E = internal elastic lamina), H & E stain (**B** and **D**), Klüver-Barrera stain for myelin (**C**), Elastica von Gieson (**E**). LabPON Twente

oblongata (Quast and Liebegott 1975). Clinical features of Wallenberg's syndrome (see also Sect. 5.4.2.1.6) caused by neoplastic diseases are often characterized by gradual development and steady progression of symptoms. The syndrome is clinically characterized by Horner's syndrome, ipsilateral ataxia, contralateral paralysis of the body, and ipsilateral facial hypoalgesia. The manifestation of symptoms, however, is broad and includes numbness, dysphagia, vertigo, nausea-emesis, hoarseness, hiccups, facial pain, swallowing dysfunction, and visual disturbance (Shetty et al. 2012). The most common acute cause of a central Horner syndrome is Wallenberg syndrome (see Sect. 5.4.2.1.6, Table 5.1) (Sacco et al. 1993; Pellegrini et al. 2020), and it is present in up to three quarters of Wallenberg syndrome cases (Kim et al. 1994). Magnetic resonance imaging (MRI) is the preferred diagnostic method for brainstem infarctions (Kistler et al. 1984; Kitis et al. 2004; Shetty et al. 2012). Diffusion-weighted imaging (DWI) is a valuable technique in the examination of patients presenting signs and symptoms comparable to Wallenberg's syndrome (Kitis et al. 2004).

A number of other vascular brainstem syndromes have been described. For overview see Cuoco et al. 2016.

1.4.2.4 Arterio- and arteriolosclerosis

Arteriolosclerosis (sometimes described as lipohyalinosis) involves small parenchymal arteries of 300–400 μm or less in diameter which show thickening of their walls by concentric smooth muscle hyperplasia and/or collagen deposition, leading to narrowing of the lumen (Figs. 1.27 and 1.28).

Rarely, macrophages are seen within the hyalinized material (Vinters et al. 2018). The frequency of arteriolosclerosis rises with age and several factors, such as hypertension, abnormal endothelial permeability, and basement membrane components may play a role in its pathogenesis (Lammie et al. 1997; Vinters et al. 2018). The definition of small vessels is arbitrarily defined and the term small vessel disease is sometimes used as a synonym for arteriosclerosis, lipohyalinosis, and arteriolosclerosis (see Figs. 1.27 and 1.28). Lacunar infarcts have been linked to arteriolar disease caused by hypertension (Vinters et al. 2018; Bailey et al. 2012) but white matter disease and vascular dementia also tend to be associated with microvascular pathology. Although mixed vascular and neurodegenerative pathologies are common in older people, microvascular pathology on itself can be the cause of cognitive impairment (Kalaria et al. 2012). It can lead to severe secondary demyelination of the white matter, especially frontally (in the past called Binswanger's disease) (Caplan 1995).

1.4.2.5 Amyloid angiopathy

Cerebral amyloid angiopathy (CAA) is a generic morphological term describing deposition of amyloid in the walls of arteries, arterioles and, less often, capillaries and veins of the grey matter of the central nervous system (Revesz et al. 2002). It is an important cause of cerebral hemorrhage in subcortical white matter (lobar bleedings) and in the cortex affected by CAA, and of ischemic lesions and dementia.

Because of the characteristic binding of Congo red to amyloid-laden vessels, the condition was originally described as congophilic angiopathy. Among the different amyloidogenic proteins, seven are associated with CAA. In sporadic CAA, the amyloid fibrils are composed of Aß, whereas in the hereditary CAAs all seven proteins are involved. The most common types of CAA are those associated with deposition of Aß, a cleavage product of amyloid-ß precursor protein (APP), encoded on chromosome 21. Sporadic CAA is commonly associated with Alzheimer's disease (AD) but can be found in brains of individuals who are cognitively intact and may also be an incidental finding at autopsy (Vinters and Gilbert 1983). The involvement of different brain areas in CAA follows a hierarchical sequence similar to that of Alzheimer-related senile plaques (see Sect. 13.2.2.1.4).

The deposition of Aß within capillaries distinguishes two types of CAA. One with capillary Aß-deposition is characterized by a strong association with the APOE-epsilon 4 allele and by its frequent occurrence in Alzheimer's disease whereas the other one lacking capillary Aß-deposits is not associated with APOE-epsilon 4, the latter more often associated with microinfarcts and larger hematomas (Thal et al. 2008). AD patients show more widespread CAA, exhibiting lesions in the cerebellum, amygdala, and sometimes the basal ganglia, whereas in most non-demented patients, CAA is restricted to cortical areas (Thal et al. 2003). Capillary CAA leads to capillary occlusion in the human brain (Thal et al. 2008).

Using immunohistochemistry, microscopically Aß deposits are visible among smooth muscle cells within the media of parenchymal arterioles. Later, degeneration of the smooth muscle cell layer starts and can lead to fibrinoid necrosis and/or microaneurysms. Vessels severely affected by CAA can rupture and cause hemorrhages or infarction (Mandybur 1986; Vonsattel et al. 1991).

Whereas atherosclerosis and arteriolosclerosis are associated with subcortical infarcts, cortical microinfarcts are linked to CAA (Arvanitakis et al. 2017; Haglund et al. 2006; Soontornniyomkij et al. 2010). The severity of vascular amyloid deposition as well as CAA-related intracerebral hemorrhage has been linked to a variation in the CR1 gene (Biffi et al. 2012). APOE-epsilon 2 is also associated with CAA-related cerebral hemorrhage (Nicoll et al. 1997).

1.4.2.6 Hereditary CAA

In familial AD, CAA is caused by mutations of the amyloid-ß precursor protein (AßPP), presenilin-1, or presenilin-2 genes (Revesz et al. 2002). The first mutation described in the AßPP gene was found in the autosomal condition heredi-

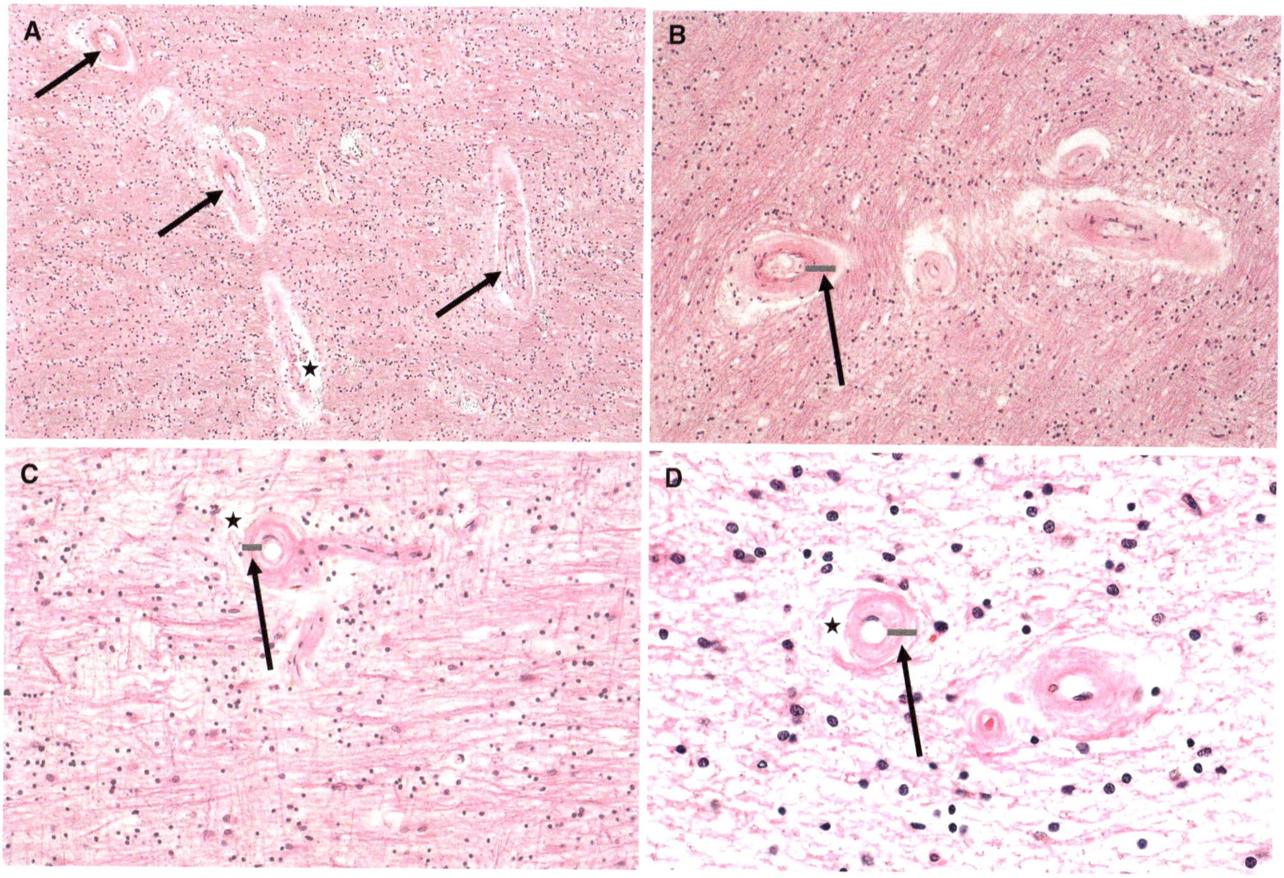

Fig. 1.27 Arteriolosclerosis (**A–D**), thickening of the vessel walls (arrows, white label) by concentric smooth muscle hyperplasia and collagen deposition, leading to narrowing of the lumen (sometimes called lipohyalinosis), perivascular lacunae (stars), and pallor of the surrounding white matter, marked in (**C**) and (**D**). H & E stain. LabPON Twente

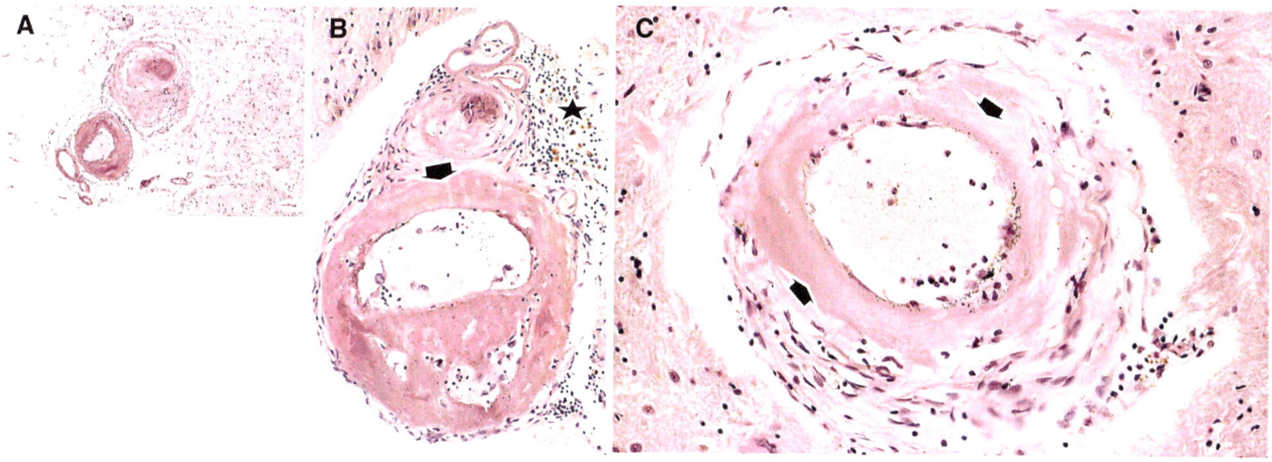

Fig. 1.28 Arteriosclerosis (**A–C**) with severe hyalinization of the vessel wall (thick arrows), surrounding mononuclear infiltrate (star). H & E stain. LabPON Twente

tary cerebral hemorrhage with amyloidosis-Dutch type (HCHWA-D) which causes severe CAA with hemorrhagic strokes of mid-life onset and dementia (Levy et al. 1990; Wang et al. 1997; Maat-Schieman et al. 2005).

The autosomal dominant disorder cystatin C-related familial CAA, HCHWA-I, in families living in western Iceland, is characterized by severe amyloid deposition within small arteries and arterioles of leptomeninges, cerebral cortex, basal ganglia, brainstem, and cerebellum (Jensson et al. 1989). Other rare conditions are amyloidoses of the transthyretin gene (TTR) (Benson 1996), gelsolin-related amyloidosis caused by G654A or G654T point mutations of the gelsolin gene (Kiuru 1998), prion protein-related cerebral amyloid angiopathy (Jansen et al. 2010), and BR12 Gene-related cerebral amyloid angiopathies (Familial British dementia and familial Danish dementia) caused by mutations on the B12 gene (Rostagno et al. 2010; Vidal et al. 1999).

1.4.2.6.1 Other hereditary angiopathies

A small number of familial cerebral microangiopathies caused by mutations in single genes has been described which present with cognitive impairment progressing to dementia (Vinters et al. 2018; Yamamoto et al. 2011). The most common genetically determined form of stroke and vascular dementia in adults is cerebral autosomal dominant arteriopathy with subcortical infarcts and leukoencephalopathy (CADASIL) (Kalimo et al. 2002). The recessive form (CARASIL) and the Cathepsin A-related angiopathy with subcortical infarcts and leucoencephalopathy (CARASAL) are rare disorders (Bugiani et al. 2016) (see Sect. 3.2.3.5).

1.4.2.7 Consequences of vascular diseases: stroke

– Ischemic infarct (lacunar infarct, infarct)
– Hemorrhagic infarct (parenchymal brain hemorrhage).

Diseases affecting the cerebral blood vessels can cause two basic types of sequelae: ischemic damage, resulting from obstruction of the blood vessels, and hemorrhage, produced by rupture of the vessel wall. Stroke is the second leading cause of death, after ischemic heart disease, responsible for 11% of the world's total deaths (WHO 2020 https://www.who.int/news-room/fact-sheets/detail/the-top-10-causes-of-death). Acute stroke is defined as the acute onset of focal neurological symptoms in a vascular territory as a result of underlying cerebrovascular disease (Tadi and Lui 2018). Most strokes are ischemic, caused by large-artery atherosclerosis, cardioembolism, or small vessel diseases (Yew and Cheng 2009; Adams et al. 1993) (Fig. 1.29).

The remaining 13% of strokes are hemorrhagic in intracerebral or subarachnoid locations, caused by bursting of a blood vessel, sometimes because of ischemia (hemorrhagic infarct). The symptoms must continue for more than 24 h for

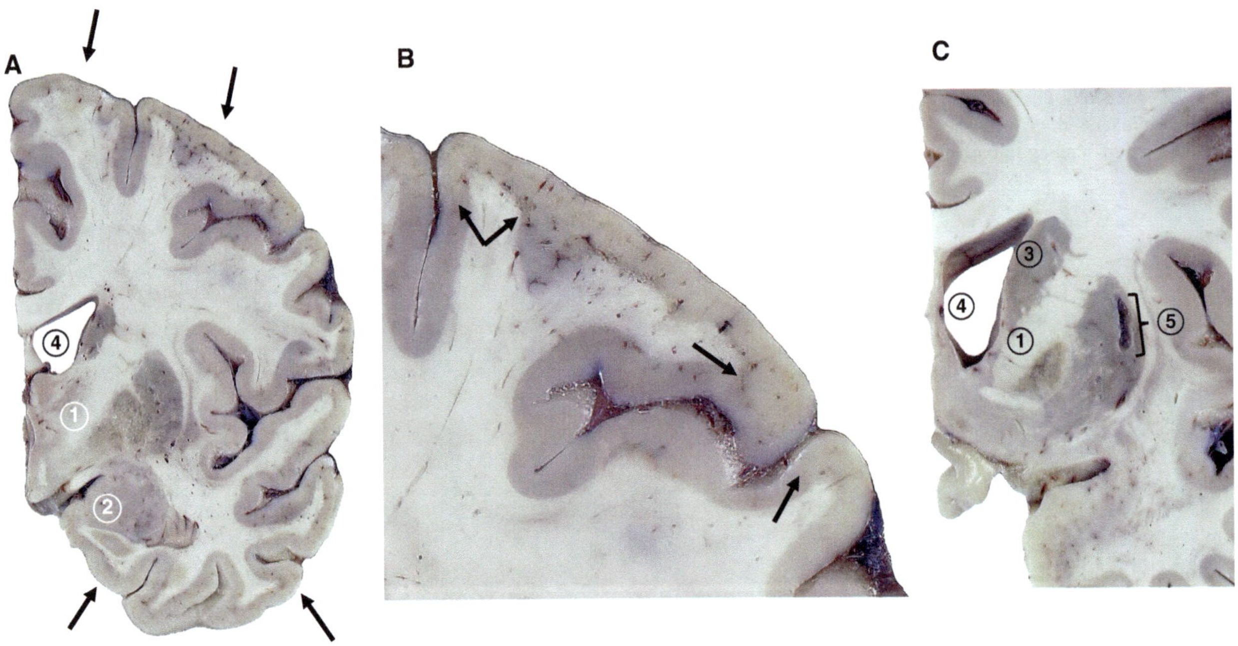

①	Internal capsule	④	Lateral ventricle
②	Amygdala	⑤	Lacunar infarct in the putamen
③	Caudate nucleus		

Fig. 1.29 Coronal section of the right hemisphere, large recent territory infarct middle cerebral artery (MCA), discoloration/infarction of cortex and pallidum, patchy blurring of the gray/white matter junction (**A** and **B**, long arrows). In (**C**) detail, showing a lacunar infarct in the putamen ⑤. LabPON Twente

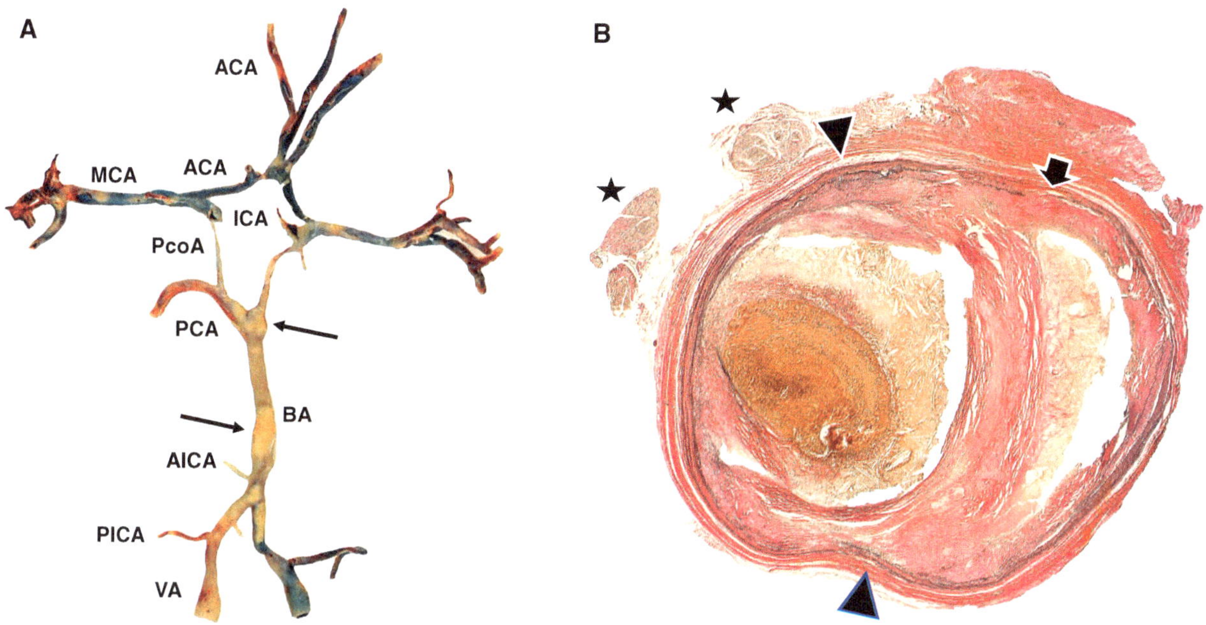

Fig. 1.30 Atherosclerosis of the circle of Willis and its major branches (**A**). Multiple patchy yellow discolorations (arrows). For abbreviations see Fig. 1.17. In (**B**), cross-section of an artery with a complicated plaque, collagenous fibrosis, elastic fragmentation (triangle), and loss of elastic fibers (short thick arrow) and superimposed thrombus formation. Nerve cell plexus in the surrounding connective tissue (stars). LabPON Twente

a diagnosis of stroke, which is usually associated with permanent damage of the brain. A TIA (transient ischemic attack) is a temporary period of symptoms similar to those of a stroke, usually lasting a few minutes and not causing permanent damage to the brain. The cause of TIA in most cases is small emboli from an extracranial source, either cardiac or an atherosclerotic plaque in the carotid or vertebrobasilar artery (Kalaria et al. 2012, 2015) (Fig. 1.30).

Two etiologic classifications commonly used are the TOAST classification (Adams et al. 1993) and the ASCOD classification (Amarenco et al. 2013). Both are based on the etiology of ischemic stroke, but the ASCOD classification differs because it gives suitable secondary prevention measures based on the disease linked to stroke.

Different from TOAST, ASCOD gives a proper indication of the patient's present causative factor (similar to TOAST) and other factors that can possibly lead to further recurrences (Patel et al. 2019). Common risk factors include hypertension, smoking, diabetes mellitus, hypercholesterolemia, physical inactivity, obesity, genetic risk factors, and atrial fibrillation. Cerebral emboli commonly originate from the heart, especially in patients with preexistent heart arrhythmias, valvular disease, atrial and ventricular defect, and chronic rheumatic heart disease. Carotid artery stenosis (Fig. 1.23) increases the risk of a completed stroke, especially if it is associated with a previous TIA (Kalaria et al. 2015).

Reliable distinguishing between intracerebral hemorrhage and ischemic stroke can only be done through neuroimaging.

Hypercoagulability can be the cause of strokes, such as in the antiphospholipid antibody syndrome (APS, formely called Sneddon's syndrome). APS, which was described in 1965, is characterized by the combination of livedo reticularis—a reddish blue mottling of the skin especially of the extremities—and cerebrovascular disease in the absence of a recognized collagen vascular disease or infection (Tietjen and Levine 1995). Venous and arterial thromboses are present in the presence of antiphospholipid antibodies. The principal pathologic findings are multiple small cortical infarcts associated with occlusion of medium-sized arteries and prominent focal smooth muscle hyperplasia of smaller arterial vessels (Hilton and Footitt 2003) (Fig. 1.31).

Subarachnoid hemorrhage presents differently from intracerebral hemorrhage. The most common symptom is acute, severe headache. The neurological symptoms depend on the site of the bleeding.

Cortical or cerebellar lesions and brainstem or subcortical hemispheric infarcts greater than 1.5 cm in diameter on CT or MRI are considered to be of potential large-artery atherosclerotic origin (Adams et al. 1993) (Fig. 1.32).

Lacunar infarcts are the most common type of infarcts (Boiten et al. 1993). This type of infarct is the most commonly identified cause of stroke. The major risk factors for lacunar infarcts are hypertension and diabetes. Lacunar infarcts with a diameter of 10–15 mm are thought to be caused by the occlusion of the perforating small arteries/arterioles with outer diameters of 100–200 μm (Fig. 1.29C)

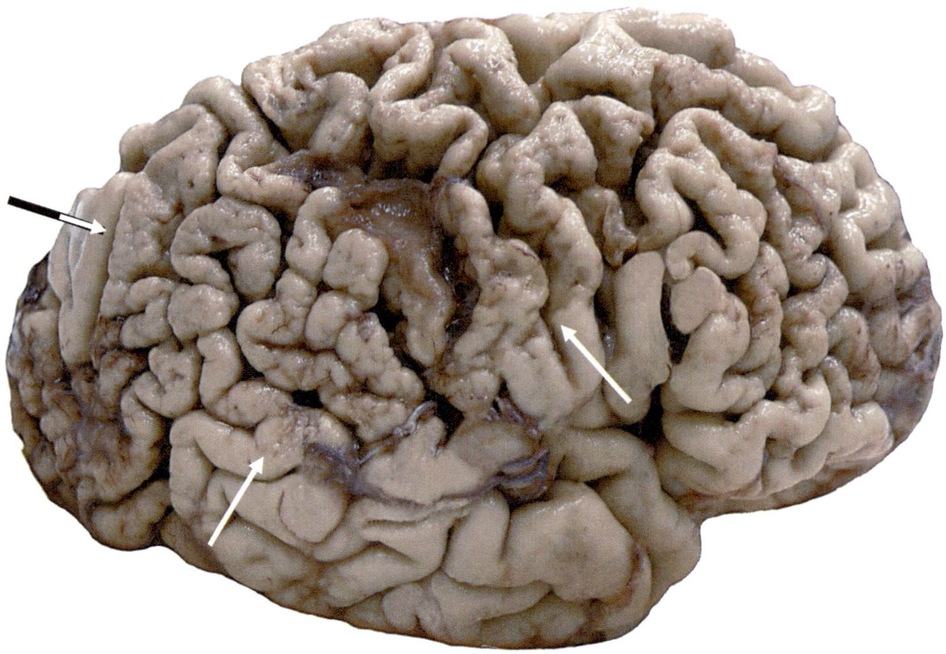

Fig. 1.31 Brain from a patient with APS syndrome, viewed from lateral (R), showing a characteristic granular surface of the cortex, caused by multiple cortical infarcts (arrows). Collection A. Rozemuller

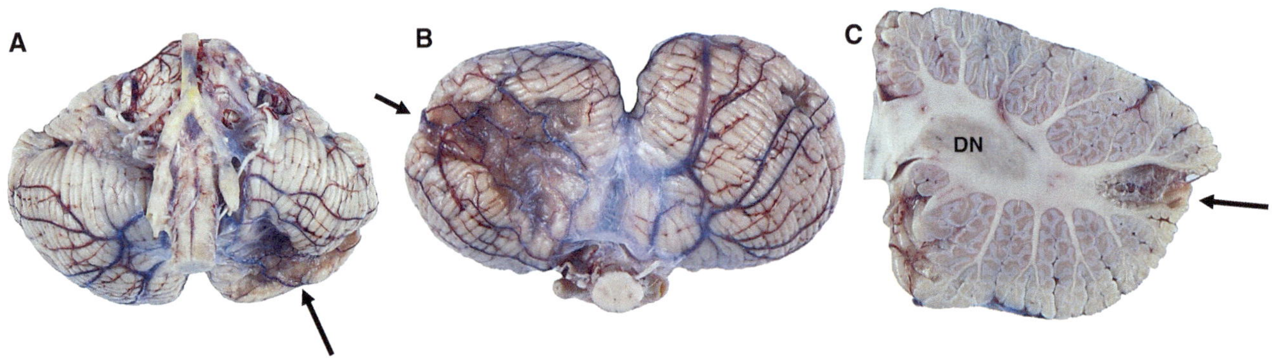

Fig. 1.32 Ventral (**A**) and dorsal (**B**) views of an old infarct (arrow) in the left cerebellar hemisphere. Note severe atherosclerosis of the vertebral/basilar arteries (yellow patches). Cross-section of cerebellar hemisphere (**C**) showing old cystic infarct (arrow). *DN* dentate nucleus. LabPON Twente

(Kalaria et al. 2012, 2015). Multiple, often asymptomatic, lacunar infarcts reflect arteriolosclerosis. However, single symptomatic lacunar infarcts probably result from microemboli or microatheromatosis.

The extracranial arteries and the heart have been implicated as potential sources of emboli in 10–20% of lacunar infarcts. Lacunar infarcts occur in the basal ganglia, internal capsule, thalamus, and pons (Fig. 1.33).

The macroscopic appearance in the deep grey matter shows an irregular shaped aspect. Microscopically, old lesions were described as small pale cavities with irregular borders, often trabeculated with connective tissue and accompanied by gliosis, macrophages, and fibrous astrocytes (Fig. 1.33B, C, E).

Acute lesions had macrophages, liquefaction necrosis, and little gliosis. The surrounding parenchyma showed edema, spongiosis as well as reactive astrocytes (Fig. 1.33E) (Bailey et al. 2012; Mancardi et al. 1988). Dilated perivascular spaces with a round regular border by rarefied and gliotic nervous tissue were often found close to lacunar lesions. This appearance was common in normal aging brains and in those with vascular dementia (see Sect. 10.1.1.3). Microscopically, small vessel changes are often seen in the perforating arteries and arterioles. Miller Fisher (1991) described a characteristic vascular pathology in the penetrating small size arteries and arterioles and called it segmental arterial disorganization, fibrinoid degeneration, and lipohya-

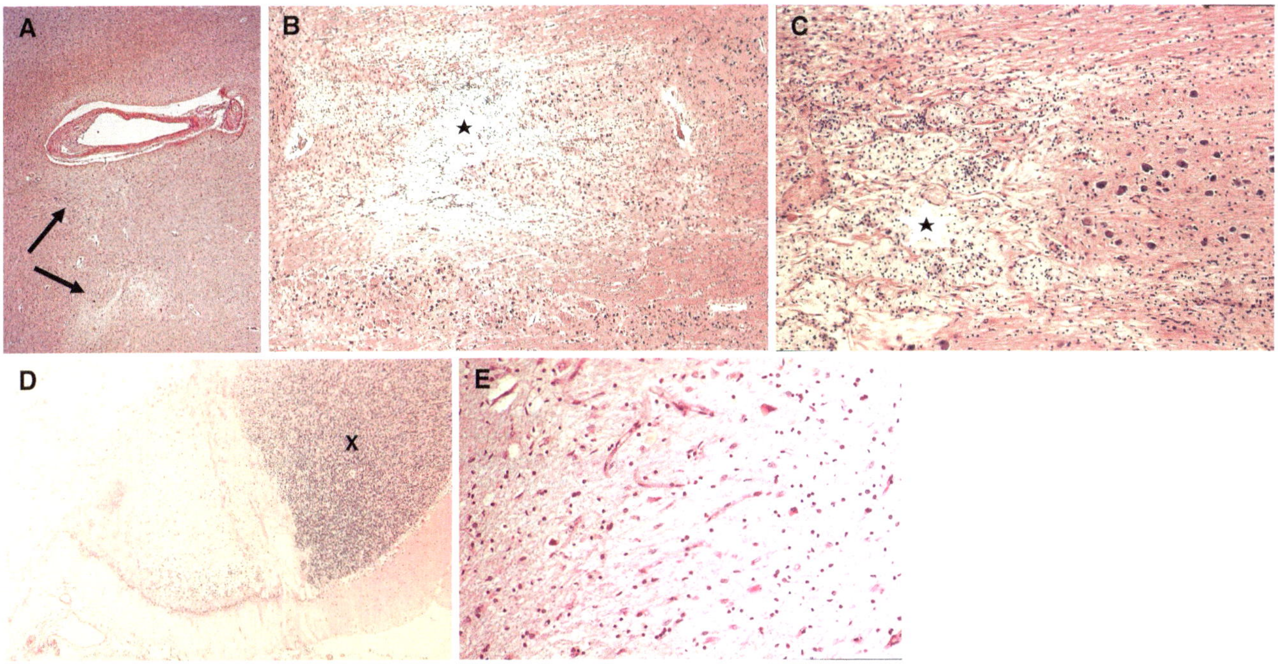

Fig. 1.33 Microscopical sections of the pons with old lacunar micro-infarcts (arrows) in (**A**) and arteriosclerosis. In (**B**, **C**), high power view showing many macrophages (star) and in (**E**) capillary proliferation, neovascularisation, and astrogliosis. In (**D**) old cerebellar infarct, preexistent cerebellar tissue (X). LabPON Twente

linosis. Fibrinoid necrosis was also called lipohyalinosis although the term is variably applied (Miller Fisher 1991).

Two major vascular pathologies underlie brain damage in patients with disease of small size penetrating arteries and arterioles: Thickening of the media by fibrinoid deposition and hypertrophy of smooth muscle and obstruction of the origins of penetrating arteries by intimal plaques (Caplan 2015). Hyaline thickening of 40–150 µm diameter arteriolar vessels is called arteriolosclerosis, although there is overlap with the definition of lipohyalinosis of Miller Fisher.

In the first 12 h after an arterial blockage, no significant macroscopic changes can be seen. There is cytotoxic edema. This early infarction can be visualized by diffusion-weighted MRI. Microscopically, the earliest and most characteristic hallmark of an infarction are the eosinophilic neurons with pycnotic nuclei, visible after about 4 h. Thereafter, neutrophil leucocyte infiltration begins and after 2 days macrophages begin to appear. Around 1 week later, there is an increased proliferation of astrocytes around the core, vessels show hyperplasia, and neovascularization commences (Kalaria et al. 2015). Phagocytosis of the tissue debris, visible as foam cells, may be apparent for months to years in large infarcts. The proportion of infarcts that is hemorrhagic varies from 18 to 48%, the majority being embolic. Two generally accepted mechanisms by which an infarct becomes hemorrhagic are:

1. Reperfusion of necrotic, leaky blood vessels.
2. Occlusion of venous drainage (Kalaria et al. 2012, 2015). Subarachnoid hemorrhage (SAH) decribes an acute extravasation of blood into the space between the arachnoid membrane and pia mater (see Fig. 1.8). Spontaneous SAH occurs when a weakened vessel in the subarachnoid space ruptures. Known causes are berry aneurysms (the most common cause), infective aneurysms, fusiform aneurysms, arteriovenous malformations (see Sect. 5.3.1.1.6), and cerebral amyloid angiopathy (see Sect. 1.4.2.5).

Parenchymal brain hemorrhage (PBH) encompasses cerebral, cerebellar, and brainstem parenchymal hemorrhages. PBH commonly extends to ventricles, leptomeninges, and basal cisterns, producing secondary intraventricular hemorrhage and subarachnoidal hemorrhage (Ellison et al. 2004a, b). Hypertension is a major risk factor for PBH and counts for about 40–50% of PBHs. Hypertensive PBHs most commonly originate in the putamen, thalamus, cerebellum, or pons.

Macroscopically, hematomas present as a soft red mass distinct form brain parenchyma (Fig. 1.34).

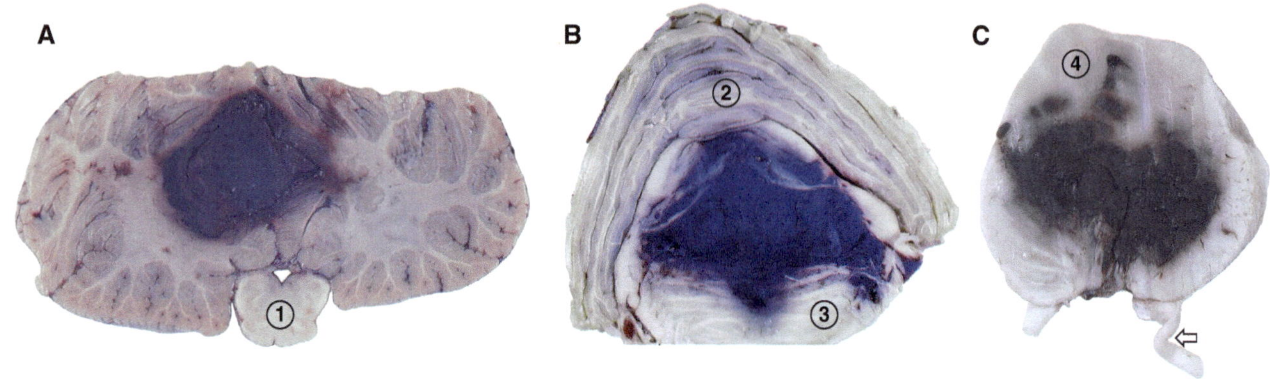

①	Medulla oblongata	③	Pons
②	Cerebellum	④	Mesencephalon with oculomotor nerve ⇨

Fig. 1.34 Hematomas of the posterior cranial fossa. (**A**) Cerebellar hematoma, (**B**) Pontine hematoma, and (**C**) Mesencephalic hematoma. See "Parenchymal brain hemorrhage" here above. LabPON Twente

PBH may extend into the ventricular system. Brain swelling and herniation is often seen. Microscopy reveals thickening of walls of arteries and arterioles ("hyaline arteriosclerosis").

1.5 The bony surroundings of the brainstem

The following Figs. 1.35 and 1.36 will present the skull base in a dorsal and a ventral view. They are meant to show the most important openings of the skull base mainly with respect to the exiting and entering cranial nerves.

Figure 1.35 shows a skull preparation after removal of the calvarium, the meninges and the central nervous system. Thus, you have a view on the whole of the skull base. The posterior cranial fossa [Fossa cranii posterior] in which the brainstem and the cerebellum are located in vivo (see Figs. 1.4 and 1.5) is delineated by the red line. With the exception of the most rostral part, this line also demarcates the line of anchorage of the cerebellar tentorium [Tentorium cerebelli], a sheet of dural tissue that separates the posterior and the middle cranial fossae (MCF) (for details, see Box 3.5).

When considering the cranial nerves, it is important to note that the third (oculomotor), the fourth (trochlear), the fifth (trigeminal), and the sixth (abducens) cranial nerve nuclei send their axons in the homonymous nerves outside the posterior fossa where they leave or enter the skull via several openings that are located in the adjoining middle cranial fossa.

The majority of cranial nerves passes the foramina shown in Fig. 1.35 and then exits at the ventral surface of the skull or into the orbita (see Fig. 1.36).

Only the vestibulocochlear nerve (CNVIII) reaches its final targets, the cochlea (organ of acoustic perception, see Sect. 8.1.1.2) and the labyrinth (organ of equilibrium, see Sect. 5.3.1) in the petrous bone.

As described above, the oculomotor nerves oculomotor, trochlear, abducens (CNIII, IV, VI) exit the cranial cavity via the superior orbital fissure (Fig. 1.35 ⑨) [Fissura orbitalis superior]. Inside the orbita they get in contact with the eye muscles (see Table 9.1) and thus do not appear at the outside of the skull.

Certain branches of the trigeminal nerve (CNV) appear at the ventral surface after passing through the skull base.

The first main trigeminal branch, the ophthalmic nerve (V_1) [N. ophthalmicus]—mainly providing sensory supply to the orbita—enters the latter via superior orbital fissure and immediately splits off into a number of branches (see Sect. 12.2.3.2.1).

The maxillary nerve (Fig. 1.35 ⑧) [N. maxillaris]—second main trigeminal branch—leaves the cranial cavity via the Foramen rotundum. Its numerous branches reach different areas of supply via, for example, the greater palatine foramen, (Fig. 1.36 ⑨) the greater palatine nerve supplying the hard (bony) palate [Palatum durum].

The mandibular nerve—third main branch of CNV (V_3)—is running from the intracranial to the extracranial opening of the Foramen ovale (Figs. 1.35 and 1.36 ⑦), reaches the infratemporal fossa [Fossa infratemporalis] from where it provides numerous branches for the sensory innervation of the mandibular region (r2) as well as motor rami for the masticatory muscles (see Sect. 11.3.1.1.2). The meningeal branch of the mandibular nerve enters again the cranial vault via the foramen spinosum (Figs. 1.35 and 1.36 ⑥) to run to the meninges.

The petrotympanic fissure [Fissura petrotympanica] (Fig. 1.36 ⑧) Glaseri (Johann Heinrich Glaser, 1629–1675, Swiss anatomist, surgeon, and botanist) is traversed by the Chorda tympani, a gustatory and secretomotor branch of the facial nerve.

The main bulk of the facial nerve (CNVII) (Sect. 8.2.1.3)—entering the skull base at the internal acoustic opening (see Fig. 1.35 ④) and traversing it via the facial

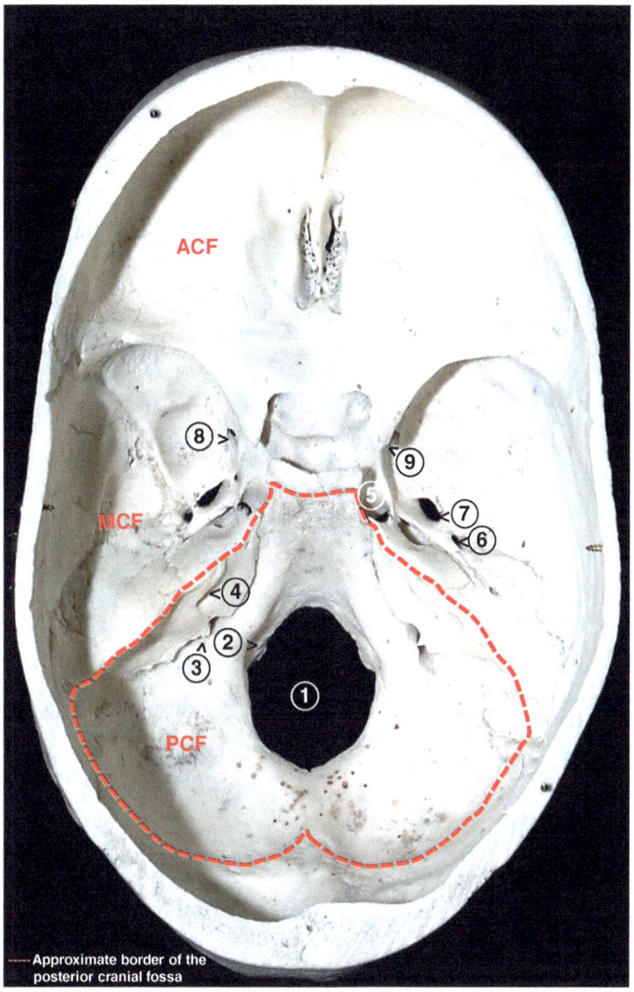

①	**Foramen magnum** **- Medulla oblongata** **- Spinal root of accessory nerve**	⑤	**Internal opening of carotid canal** **- Internal carotid plexus**
②	**Hypoglossal canal** **- Hypoglossal nerve**	⑥	**Foramen spinosum** **- Meningeal branch of mandibular nerve**
③	**Jugular foramen** **- Glossopharygeal nerve** **- Vagus nerve** **- Cranial roots of accessory nerve**	⑦	**Foramen ovale** **- Mandibular nerve**
④	**Internal acoustic opening** **- Facial nerve** **- Vestibulocochlear nerve**	⑧	**Foramen rotundum** **- Maxillary nerve**
		⑨	**Superior orbital fissure** **- Ophthalmic nerve: lacrimal, frontal, nasociliary** **- Oculomotor, trochlear, abducens nerves**

Fig. 1.35 Dorsal view of the skull base. The different openings (foramina, fissures) through which the cranial nerves or their branches enter/exit the skull base are labeled. Note that not all cranial nerves seen here reappear at the external surface of the skull. For details, see text. *PCF* Posterior cranial fossa, *MCF* Medial posterior fossa [Fossa cranii media], *ACF* Anterior cranial fossa [Fossa cranii anterior]. Sammlung des Zentrums Anatomie der Universität zu Köln

canal [Canalis nervi facialis]—reaches the skull base exiting at the stylomastoid foramen [Foramen stylomastoideum] (Fig. 1.36 ④). From there the nerve turns anteriorly to reach the mimic muscles (see Sect. 8.2.1.6.1). The foramen is easily found by first identifying the prominent styloid process [Processus styloideus] (see Fig. 1.36, arrow).

The jugular foramen (Figs. 1.35 and 1.36 ③), visible inside the skull as well as at the external surface of the skull

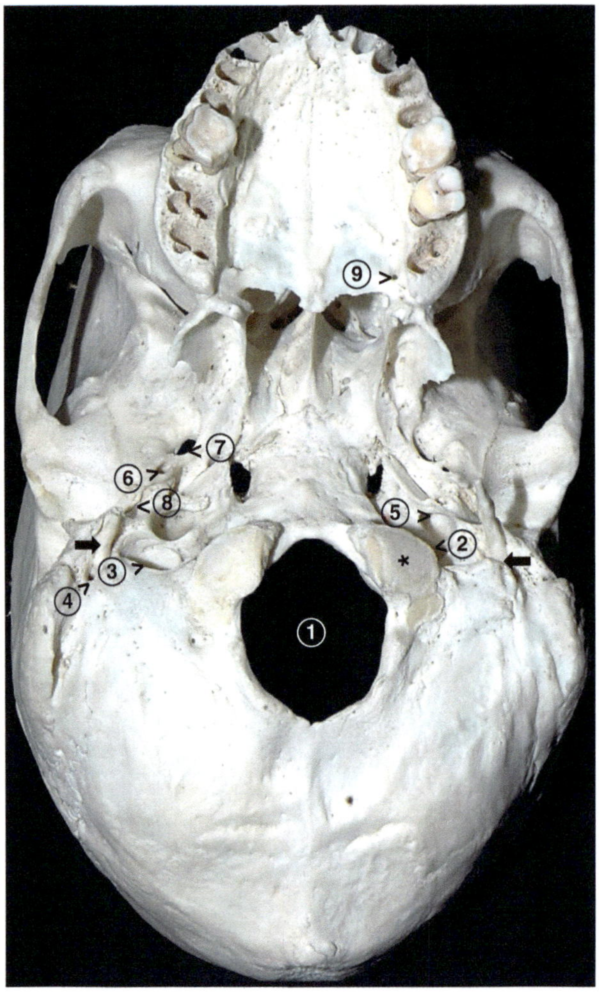

①	**Foramen magnum** **- Medulla oblongata** **- Spinal root of accessory nerve**	⑥	**Foramen spinosum** **- Meningeal branch of mandibular nerve**
②	**Hypoglossal canal** **- Hypoglossal nerve**	⑦	**Foramen ovale** **- Mandibular nerve**
③	**Jugular foramen** **- Glossopharygeal nerve** **- Vagus nerve** **- Cranial roots of accessory nerve**	⑧	**Petrotympanic fissure (Glaseri)** **- Chorda tympani**
④	**Stylomastoid foramen** **- Facial nerve ➡ Styloid process**	⑨	**Greater palatine foramen** **- Greater palatine nerve**
⑤	**External opening of carotid canal** **- External carotid plexus**	*	**Occipital condyle**

Fig. 1.36 Ventral view of the skull base. The different openings (foramina) through which the cranial nerves or their branches pass the skull base to their exits at the ventral surface are labeled. For details, see text. Sammlung des Zentrums Anatomie der Universität zu Köln

base, releases the glossopharyngeal (CNIX), the vagus (CNX), and the accessory nerve (CNXI, see Sect. 3.3.1.2). The glossopharyngeal nerve is involved in motor innervation of the pharynx as well as in chemo- and baroreceptive information to the brainstem.

The jugular foramen (Figs. 1.35 and 1.36 ③), visible inside the skull as well at the external surface of the skull base, releases the glossopharyngeal (CNIX) (see Sect. 5.4.2.1.3), the vagus (CNX) (see Sect. 7.3.2.1.2), and the accessory nerve (CNXI, see Sect. 3.3.1.2). The glossopharyngeal nerve (see Sect. 5.4.2.1.3) is involved in motor innervation of the pharynx as well as in chemo- and baroreceptive information to the brainstem.

The vagus nerve has the most extended area of innervation of all cranial nerves, not only providing motor innervation to the larynx and the upper airways but also to all thoracic (heart, lungs) and abdominal organs (stomach, small intestine, colon) down to the left flexure of the colon. In the same area, it collects chemoceptive information.

The accessory nerve (see Sect. 3.3.1.2) with a contribution from the cervical spinal cord entering the skull via the foramen magnum (Figs. 1.35 and 1.36 ①) is running to the dorsal part of the neck region where it innervates the sternocleidomastoid [M. sternocleidomastoideus] and the trapezius [M. trapezius] muscles.

The hypoglossal nerve (CNXII) (see Sect. 4.3.1.1) leaves the hypoglossal canal (Figs. 1.35 and 1.36 ②) running then in anterior and medial direction to the tongue muscles (see Sect. 4.3.1.1.2).

The foramen magnum (Figs. 1.35 and 1.36 ①) harbors the medulla oblongata and the spinal roots of the accessory nerve (see Sect. 3.3.1.2) that enter the cranial cavity.

1.6 Major fiber tracts of the human brainstem

In the following (Tables 1.3 and 1.4), we will provide survey information on the most important tracts that are located in the brainstem. Most of them have their targets and/or origins in the brainstem. They will be dealt with in detail in the corresponding chapters. The table includes—with some exceptions—those important brainstem tracts that can be identified in routine fiber tract stainings (see Fig. 3.14 A–G).

Please note that FIPAT Ch.1 as an endnote (# 34) to the anterolateral tract states that "... A tract may be defined as a projection (a set of fibres with one main source and one main site of termination) which manifests itself as a fibre concen-

Table 1.3 Major descending fiber tracts of the human brainstem (in alphabetical order) (in italics terminology deviant from FIPAT)

Name	Origin	Target	Function	Reference
Central tegmental tract Fig. 3.14 C–F	Red ncl., parvocellular part Ipsilateral	Inferior olivary complex	Part of the dentato-rubro-olivary pathway[a]	Nathan and Smith (1982)[b]
Corticobulbar/nuclear system[g] Fig. 3.14 C–F *[Bulbar corticonuclear fibers]*	Cortical motor fields contralateral	Nuclei of CNV[c], VII, IX, X, XII, not VI	Control of brainstem motor nuclei	ten Donkelaar (2011) (reviewing the findings of Kuypers[b])
Corticospinal tract Fig. 3.14 A–F Pontine tracts	Primary motor cortex Layer V ipsilateral[d]	Spinal interneurons/Spinal α-motoneurons	Voluntary movements	Nathan et al. (1990)[b], Nieuwenhuys et al. (1988), ten Donkelaar (2011)
Lateral vestibulospinal tract[h] Fig. 3.14 A–G	Lateral vestibular ncl. Ipsilateral	Total length of spinal cord (anterolateral funiculus)	Excitation of paravertebral/proximal limb extensors ("anti-gravity")	Nathan et al. (1996)[b], Jang et al. (2018), Mihailoff and Haines (2018), ten Donkelaar (2011)
Medial vestibulospinal tract[h] Fig. 3.14 A–G	Medial/spinal vestibular ncll. Ipsi-/contralateral	Cervical/upper thoracic spinal cord (anterior funiculus)	Control of neck musculature	Nathan et al. (1996)[b], Jang et al. (2018)[e] Mihailoff and Haines (2018)[f], ten Donkelaar (2011)
Medial tegmental tract[h]	Ncl. of Darkschewitsch	Inferior olive	?	ten Donkelaar (2011)
Rubrospinal tract[i] Fig. 3.14 A–G	Red nucleus, magnocellular part Contralateral	Cervical spinal cord	Non-voluntary movements	Nathan and Smith (1982)[b], ten Donkelaar (2011)
Tectospinal tract[h] Fig. 3.14 A–G	Superior colliculi Contralateral	Cervical spinal cord	Motor reaction to auditory stimuli	Reynolds and Al Khalili (2019)

[a] For details, see red nucleus, Chap. 16
[b] Silver impregnation (Nauta, Marchi aut similia) in neuropathological specimens with various lesions
[c] Corticobulbar input into the trigeminal motor nucleus and the principal sensory trigeminal nucleus
[d] 80% of fibers cross at the border between the spinal cord and the medulla oblongata in the pyramidal decussation
[e] Diffusion tensor imaging
[f] Claim bilateral course
[g] Together with corticospinal tract
[h] Inside medial longitudinal fasciculus
[i] Inside central tegmental tract

Table 1.4 Major ascending fiber tracts of the human brainstem (in alphabetical order) (in italics terminology deviant from FIPAT)

Name	Origin	Target	Function	Reference
Spinothalamic tract Fig. 3.14 A, B, D–F	Nociceptors in the periphery of the body – contralateral[a]	Ventral posterolateral ncl. of thalamus (VPL)	Protopathic sensory function body	Nieuwenhuys et al. (1988)
Cuneate tract Fig. 3.14 A see medial lemniscus	Peripheral mechanoreceptors – ipsilateral[b]	Cuneate ncl.	Epicritic sensory function body	Nieuwenhuys et al. (1988)
Gracile tract Fig. 3.14 A see medial lemniscus	Peripheral mechanoreceptors – ipsilateral[c]	Gracile ncl.	Epicritic sensory function body	Nieuwenhuys et al. (1988)
Dorsal trigeminothalamic tract *[posterior trigemi - nothalamic tract]*	Principal sensory ncl. – ipsilateral	Ventral posteromedial ncl. of thalamus (VPM)	Epicritic sensory function face	ten Donkelaar (2011)
Gustatory fibers[i] Fig. 3.14 C–F	Ncl. of solitary tract – ipsilateral	Parvocellular part of VPL	Taste[d]	ten Donkelaar (2011)
Lateral lemniscus Fig. 3.14 A–G	Cochlear nuclei[e] Superior olive[e]	Inferior colliculi	Ascending auditory system	Nieuwenhuys et al. (1988)
Lateral trigeminothalamic tract[j]	Nociceptors face – contralateral	Spinal trigeminal ncl. Caudal part	Protopathic sensory function face	ten Donkelaar (2011), FIPAT Ch1 48
Medial lemniscus Fig. 3.14A–G	Cuneate/gracile ncll. – contralateral	VPL	Epicritic sensory function body	Nieuwenhuys et al. (1988), ten Donkelaar (2011)
Trigeminal lemniscus[k] *[anterior trigeminothalamic tract]*	Spinal trigeminal ncll. Principal sensory ncl.[f] – contralateral	VPM	Epicritic and protopathic sensory function face	ten Donkelaar (2011), FIPAT Ch1 48
Medial longitudinal fasciculus[h] Fig. 3.14 A–G	Contains different tracts	Coordination of vestibular system	Eye and neck muscles	Nieuwenhuys et al. (1988)
Mesencephalic trigeminal tract	Mesencephalic trigeminal ncl.	Thalamus	Proprioceptors face muscles	ten Donkelaar (2011), Usunoff et al. (1997)
Spinocerebellar tracts	Spinal cord	Cerebellum	Proprioception	Nieuwenhuys et al. (1988)
Vestibulothalamic tract[l] Fig. 3.14 A–G	Vestibular nuclei – ipsilateral	Lateral thalamic ncll.	Sense of equilibrium	Jang et al. (2018)[g]

[a] Fibers cross in the spinal cord after entry of the primary-order neurons
[b] Afferents from the upper extremities, neck region
[c] Afferents from the lower extremities, trunk
[d] Do not confuse taste and smell. Smell / olfactory signals are conveyed via the olfactory nerve
[e] Various crossing and non-crossing courses
[f] Mainly ventral principal trigeminal nucleus
[g] Diffusion tensor imaging
[h] The medial longitudinal fasciculus is composed of different tracts (see details in the tables 1.3 and 1.4 itself)
[i] Inside central tegmental tract (see "Descending tracts")
[j] Joins the spinothalamic (anterolateral) tract
[k] Inside medial lemniscus
[l] Inside medial longitudinal fasciculus

tration over at least part of its course (Nieuwenhuys et al. 1988). For fibre systems with a more diffuse organization, the term Fibrae is advocated…."

The corticospinal tract [Tractus corticospinalis] (cst) is described here in more detail since it is the only tract visible on all atlas plates. It is running through the whole length of the brainstem while its origin and targets are cranial/caudal of the brainstem (Nathan et al. 1990; ten Donkelaar 2011). It originates in the primary motor cortex, enters the internal capsule and forms the pyramidal tract of the cerebral peduncle. From there, the cst is located in the ventral parts of the brainstem. In mesencephalon and pons, the cst is covered by the pontine nuclei (see Watson et al. 2019) until the tract reaches the ventral surface of the medulla oblongata. At the border between the latter and the cervical spinal cord 80% of the fibers cross to the contralateral side, forming the lateral corticospinal tract in the anterolateral funiculus. The remaining 20% form the ventral cst in the anterior funiculus, crossing to the contralateral side in its further caudal course (for details, see ten Donkelaar 2011).

1.7 Tumors of the Brainstem

In general, brain tumors can develop in any given site of the CNS. Paradigmatically, we will here present a tumor located in the fiber-rich pontine basis. The meninges form the most common tumor site of primary brain tumors and CNS tumors. Whereas 21% of tumors are located within the frontal, temporal, parietal, and occipital lobes, brainstem tumors account for 1.6% of all tumors (Dolecek et al. 2012) (Fig. 1.37).

The most frequent is the meningioma followed by glioblastoma. The broad category gliomas represent approximately 30% of all tumors, 4.2% of which are located in the brainstem. The most common childhood brain tumor are the low-grade gliomas, representing over 30% of all primary brain tumors in pediatric patients (Dolecek et al. 2012). Most commonly seen in children are pilocytic astrocytoma and diffuse (fibrillary) astrocytoma, but oligodendroglioma, ganglioglioma, pilomyxoid astrocytoma, and pleomorphic astrocytoma are also seen.

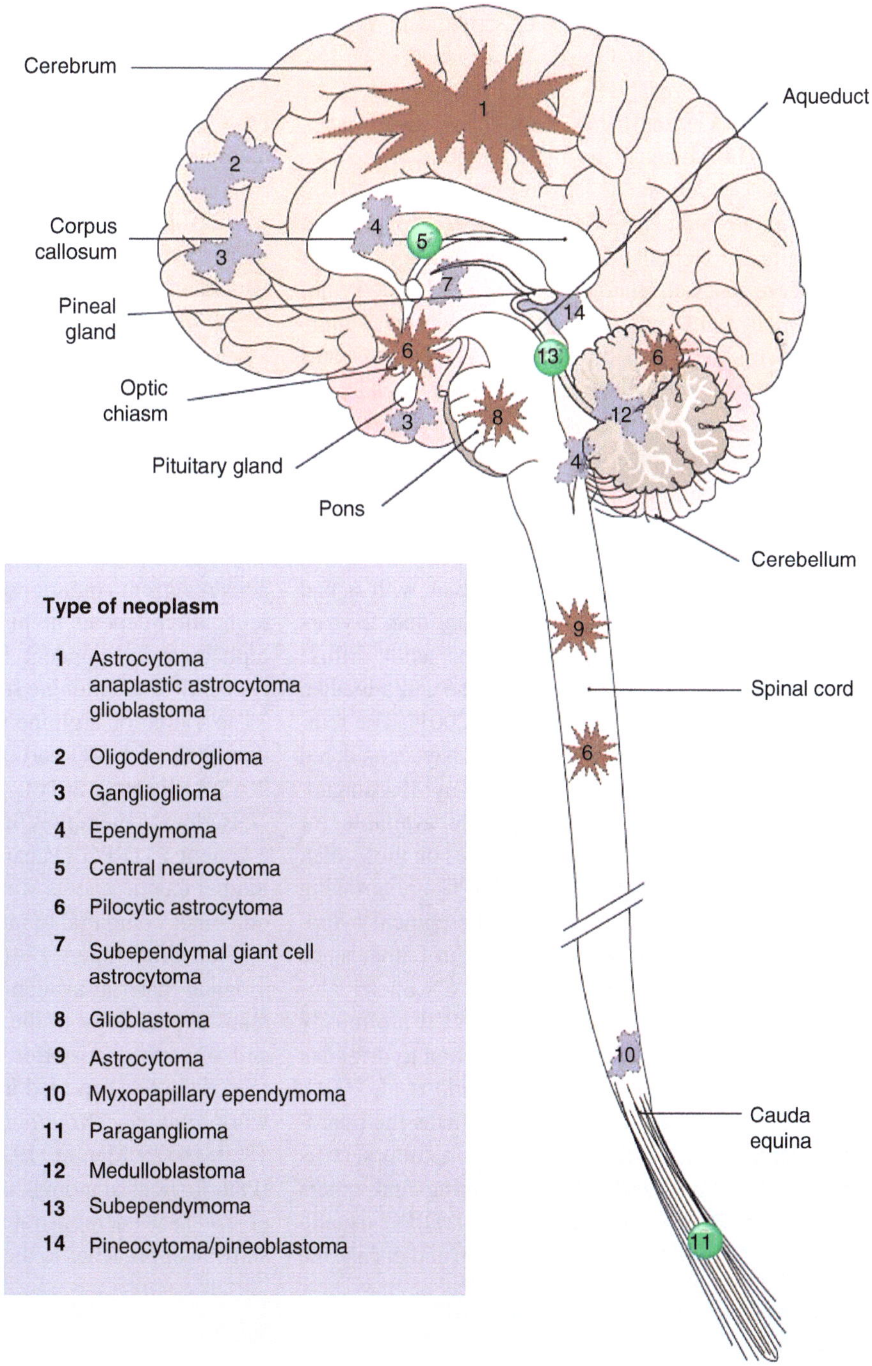

Fig. 1.37 Common locations of various neuroepithelial neoplasms. The clinical presentations of CNS neoplasms depend largely on their site and nature. Primary neoplasms account for ~2% of all cancers and ~20% of all cancers in children under 15 years of age. From Ellison et al. 2004a, b, Fig. 34.1 with permission

In the pediatric group, brainstem tumors represent 10–20% of all central nervous system tumors (Walker et al. 1999; Vanan and Eisenstat 2015; Tanrikulu and Özek 2020). Brainstem tumors include tumors that arise from the midbrain, pons, medulla oblongata, and upper cervical cord and their peak incidence is between 7 and 9 years of age (Berger et al. 1983). Brainstem tumors are associated with distinctive clinical presentations. If the symptoms have started recently and the neurological examination reveals involvement of multiple cranial nerve nuclei, a diffuse growth pattern is likely. Symptoms with insidious onset and single cranial nerve involvement point to focal brainstem tumors (Tanrikulu and Özek 2020).

Brainstem gliomas (BSG) can be classified according to their configuration. Focal and cystic tumors tend to be low grade and diffuse tumors (75% of all brainstem gliomas) tend to be high grade and behave agressively (Roth et al. 2020). BSG may be located in the midbrain, pons, or medulla with some overlap depending on the extent of the tumor. Clinical symptoms that have been described are elevated intracranial pressure, cranial nerve deficits such as facial or abducens nerve palsy in pontine tumors, long-tract symptoms, cerebellar symptoms (especially in pontine tumors involving the middle cerebellar peduncle) and symptoms such as respiratory abnormalities (Tanrikulu and Özek 2020).

The most common brainstem tumor of childhood is the diffuse intrinsic pontine gliomas subtype (DIPG). DIPG, H3 K27M-mutant is a highly malignant glial tumor, almost equally distributed between males and females, with a peak incidence between 4 and 7 years of age at diagnosis. DIPG remains among the most challenging to treat with a bad prognosis, where less than 10% survive more than 2 years after diagnosis (Hargrave et al. 2006). The adult diffuse brainstem gliomas have a more benign course and a median survival of up to 7.3 years (Guillamo et al. 2001). The reintroduction of biopsies, which were previously considered dangerous and unnecessary in the context of an MRI diagnosis, and autopsies have made DIPG tissue available for molecular and genomic studies and the studies on molecular, genetic and epigenetic pecularities of DIPG are growing exponentially (Noureldine et al. 2020). The clinical symptoms often progress over a short period, with bulbar signs and symptoms, particularly those related to CN nuclei situated at the level of the pons. CNVI and CNVII are mostly involved and DIPG seems to have a predilection to originate close to the region of CNVI nuclei (Guillamo et al. 2001). Diffuse expansion within the basis pontis affects the ponto-cerebellar and long tract fibers, leading to symptoms such as loss of balance, ataxia, difficulty in walking and others (Noureldine et al. 2020; Sandri et al. 2006). DIPG usually expands anteriorly into the prepontine cistern and encase the basilar artery to some degree. Upwards expansion toward the midbrain can potentially lead to obstructive hydrocephalus from aqueductal stenosis secondary to mass effect (Noureldine et al. 2020). MRI imaging is the modality of choice for the evaluation of lesions involving the brainstem. It allows for evaluation of signal characteristics, lesion location, intrinsic features, and overall extent (Poretti et al. 2012). On structural MRI, DIPG is often presenting as an expansile, infiltrative, and often asymmetric mass occupying more than 50% of the pons (Hargrave et al. 2006). Long debates have taken place about the role of diagnostic biopsies but in a large study of 130 pediatric DIPGs the authors concluded that stereotactic biopsy can be considered as a safe procedure in well-treated neurosurgical teams (Puget et al. 2015).

In pathological terms, DIPG is classified under the newly defined entity, diffuse midline glioma, H3 K27M-mutant. DIPG is a WHO grade IV CNS tumor with significant tumor heterogeneity in terms of histological phenotype, ranging from grade I to grade IV and the relatively high frequency of grade I-like regions may lead to false-negative biopsy results underscoring the importance of the choice of the biopsy site (Buckowicz et al. 2014; Bugiani et al. 2017). Interestingly, higher grade features are mostly limited to the pons, whereas lower grade features are found in the pons as well as in the nearby infiltrated structures, implying spatial histological heterogeneity (Bugiani et al. 2017).

Although there is no significant difference in overall survival based on histology, the age of diagnosis increases with higher tumor grade (Buckowicz et al. 2014). It seems that the majority of pediatric midline and high-grade gliomas undergoes recurrent somatic mutations in genes encoding for the replication-dependent histone H3 isoform, H3.1, and the replication-independent H3.3 isoform, mostly resulting in lysine 27 to methionine substitution (H3.3 K27M) or glycine 34 to valine or arginine substitution (H3.3 G34v/R). More than 90% of DIPG harbor H3 K27 mutations (Gielen et al. 2013; Lulla et al. 2016).

Radiation therapy is the only form of treatment that offers a benefit in DIPG (Vanan and Eisenstat 2015). So far, chemotherapeutic agents were not able to show better survival outcomes compared to radiotherapy alone.

Jean Cruveilhier (1791–1874) studied the extensive pathological material available for him during his post as chairman of pathology at the University of Paris. He described and illustrated posterior fossa pathology including that of one of the first reported infiltrating tumors of the pons in his work *Anatomie Pathologique du Corps Humain* (Cruveilhier 1829–1842; Flamm 1973; Kobets and Goodrich 2020) (Pathological anatomy of the human body). Robert Bright (1789–1858) demonstrated the detailed anatomical relationships of approaches to the brainstem in the second volume of

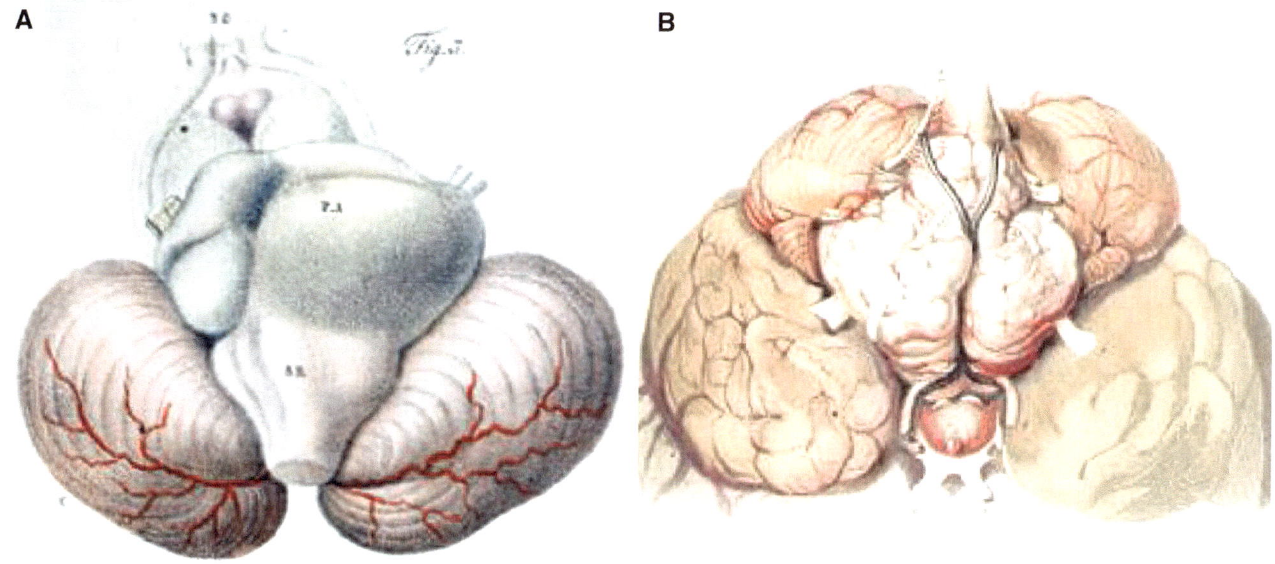

Fig. 1.38 Jean Cruveilhier's depiction of one of the first reported infiltrating tumors of the pons drawn in his work *Anatomie Pathologique du Corps Humain* in (**A**). A case of a pontine glioma demonstrated from a surface view in the work of Richard Bright in (**B**). From Jallo et al. 2020, from the personal collection of the author (Goodrich JT)

his work *Reports of Medical Cases, selected with a view of illustrating the symptoms structures cure of ideases* (Bright 1827–1831; Kobets and Goodrich 2020). He nicely demonstrated the surface view of a pontine glioma (Fig. 1.38).

Macroscopically, an often symmetric enlargement with an irregular, sometimes nodular suface is seen. On cut surface a firm and white aspect is seen, with a more or less diffuse enlargement and absence of a discrete tumor mass. The medulla is often spared (Figs. 1.39 and 1.40).

Microscopically, in nearly all cases heterogeneity is found. A high degree of cellular density alternates with low density areas. Combination of diffuse infiltrating glial cells, high mitotic activity, microvascular proliferation, and/or necrosis is characteristic (Fig. 1.41).

Proliferative activity is prominent, containing typical and atypical mitoses. Often a variable positivity is seen in a GFAP stain (Fig. 1.42). Vimentin and S-100 positivity is often present. The pediatric tumors have a set of genetic alterations that is different from adults, such as mutations in the chromatin remodeling gene H3F3A (Fig. 1.42).

As stated earlier, H3F3A mutations can lead to an amino acid substitution either at K27 or at G34. K27-mutated tumors occur in the young children, whereas the G34-mutated tumors occur in teenagers and young adults (Schwartzentruber et al. 2012; Sturm et al. 2012). Sturm et al. also showed that each H3F3A mutation defines an epigenetic subgroup of glioblastoma with a distinct global methylation pattern. While K27-mutated tumors are predominantly seen in midline locations such as thalamus, pons, and spinal cord, tumors from all other subgroups almost exclusively arose in the cerebral hemispheres (Fig. 1.43).

1.8 Terminology used

The Latin terminology used in this book has mainly been chosen according to the Terminologia Neuroanatomica and the Terminologia Anatomica (International Federation of Associations of Anatomists (FIPAT terminology:http://www.ifaa.net/committees/anatomical-terminology-fipat/) Terminologia anatomica as laid down on the website © 2020—Anatonomina (2020) http://terminologia-anatomica.org/en/Terms/View?sitemapItemId=173&termId=2635.

The FIPAT terminology is widely recommended also by Watson et al. 2019 (see here below, second paragraph).

The FIPAT terminology is subdivided into three chapters:

- Chapter/Ch. 1: Central nervous system [Systema nervosum centrale].
- Chapter/Ch. 2: Peripheral nervous system [System nervosum periphericum] (including the cranial nerves).
- Chapter/Ch. 3: Sense organs [Organa sensum].

There is also an embryological chapter of FIPAT for the organogenesis of the nervous system (TE2, PART IV).

The Latin terms provided by FIPAT are given in brackets []. In those cases where no Latin term was available, we have usually included our own translation of the corresponding English term [underlined]. The remarks on terminology hold true for the whole book.

For FIPAT terms differing from the ones used by Paxinos in his BrainNavigator, we have first mentioned the English Paxinos term then the English FIPAT term, for example.

Ambiguus nucleus/Ncl. ambiguus r9 [Nucleus ambiguus].

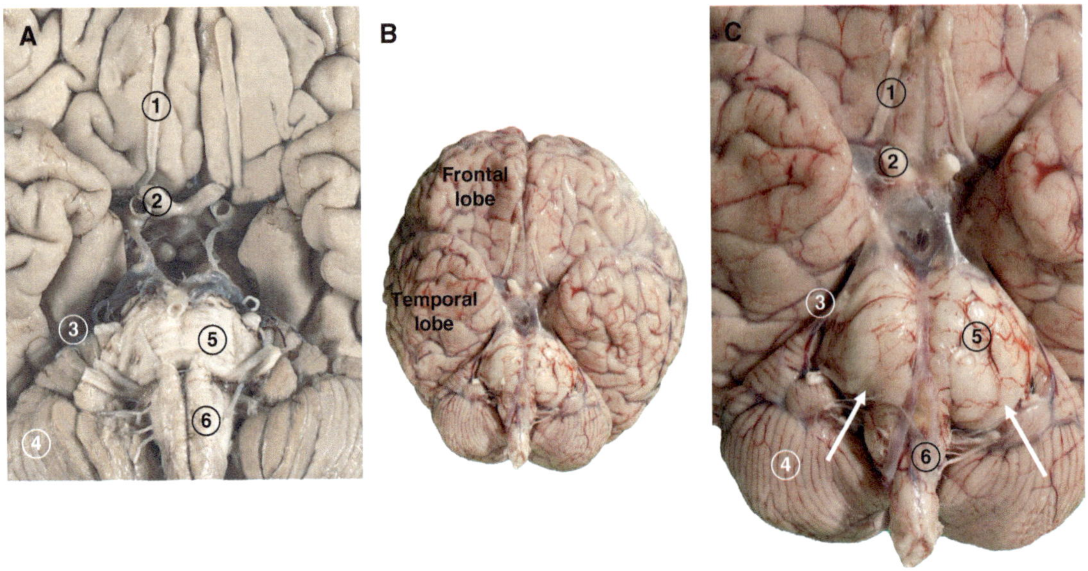

①	Olfactory nerve CNI	④	Cerebellum
②	Optic nerve CNII	⑤	Pons
③	Trigeminal nerve CNV	⑥	Medulla oblongata

Fig. 1.39 Ventral view of the human brain from a patient with DIPG (**B**, **C**). (**C**) Diffuse enlargement of the pons (arrow), extending to the midbrain and medulla oblongata (overview B and detail C, unfixed specimen), compared to a control case in (**A**). Collection A. Rozemuller and LabPON Twente

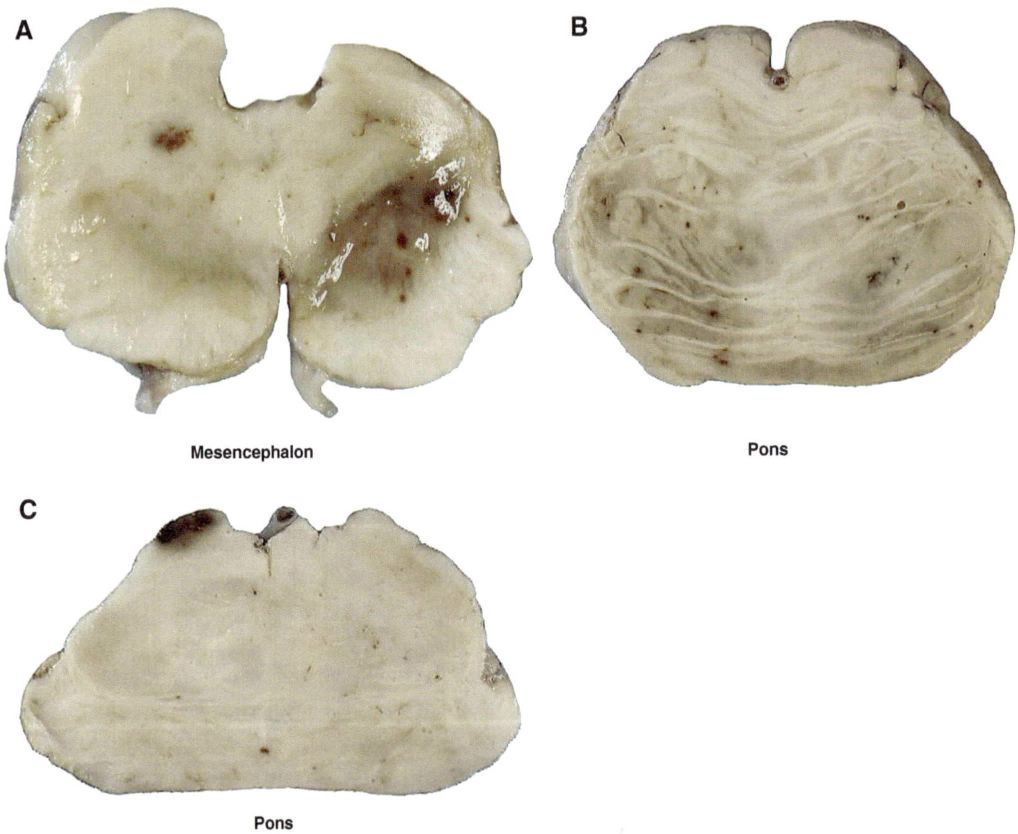

Fig. 1.40 Same patient as in Fig. 1.39. Midbrain (**A**) and pons (**B** and **C**) showing diffuse tumor growth, mucoid gray neoplastic tissue with foci of necrosis and hemorrhage, disrupting the normal structures. Compare with normal findings in Fig. 3.14 D–F. Collection A. Rozemuller

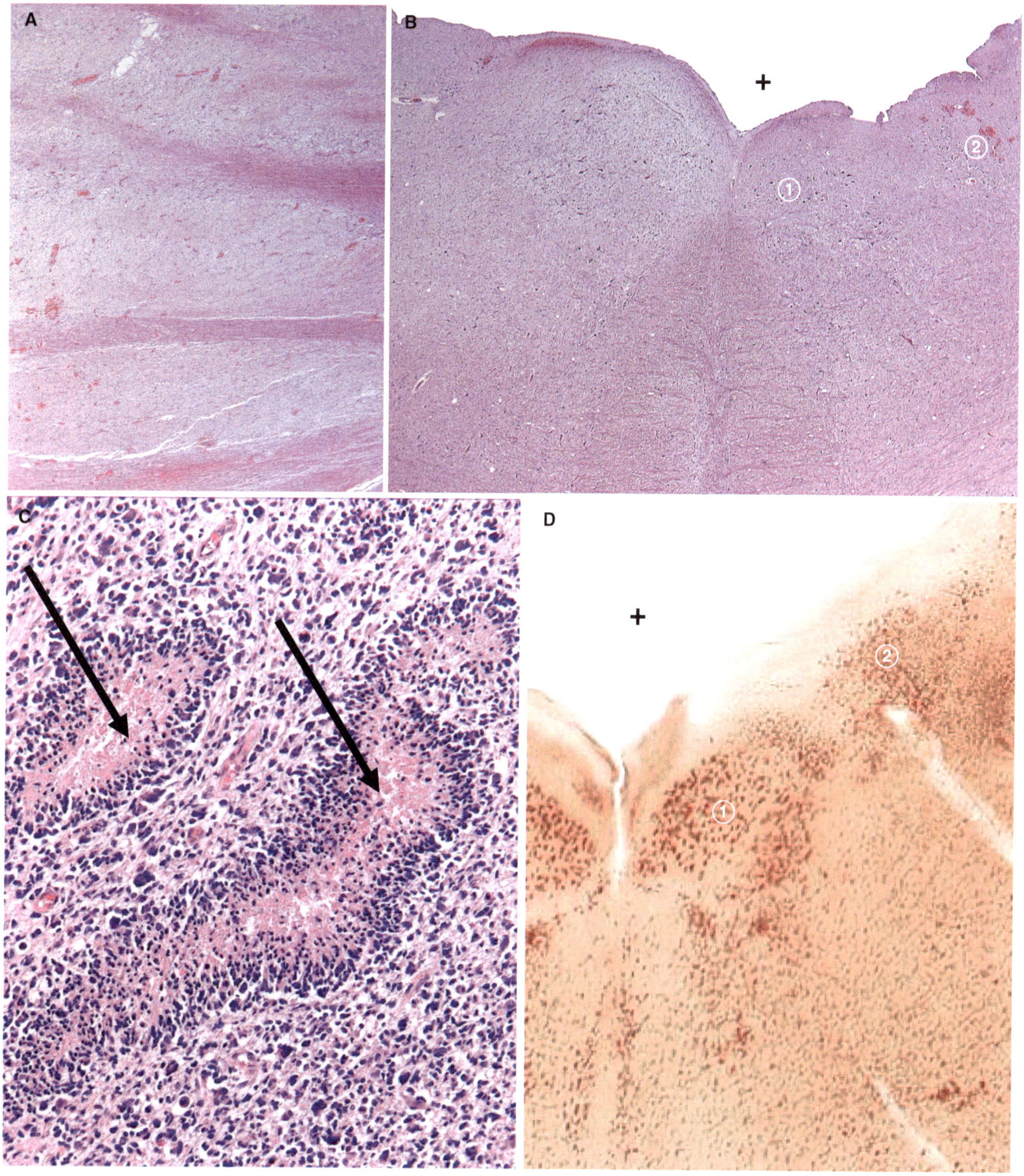

①	Ncl. of hypoglossal nerve CNXII	②	Posterior ncl. of vagus nerve CNX

Fig. 1.41 Microscopical picture of the pons (**A**) showing diffuse infiltration of tumor cells with extension into the medulla oblongata in (**B**), infiltrating in the preexistent structures of CNX and CNXII. H & E stain. Compare with Darrow red stain in (**D**) (control case). DIPGs often show a significant tumor heterogeneity in terms of histological phenotype, including areas with WHO grade II-IV histology. In (**C**) palisading of pleomorphic tumor cells around necrosis, typical for glioblastoma. Arrow points to necrosis. + = fourth ventricle. Collection A. Rozemuller, (**D**) LabPON Twente

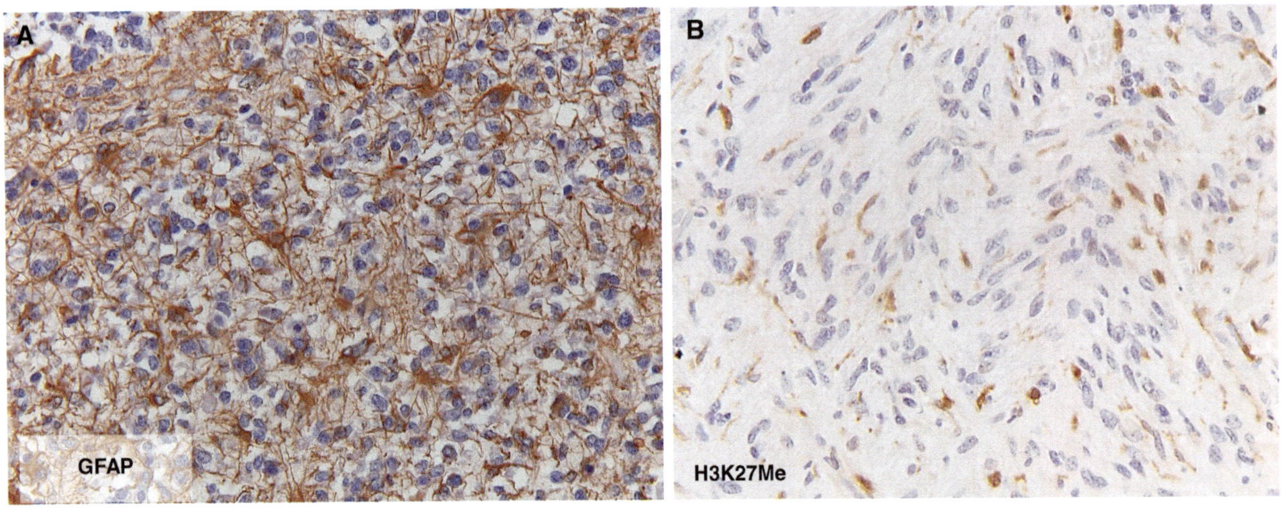

Fig. 1.42 DIPG tumor cells showing astrocytic differentiation, confirmed with a GFAP immunostain (**A**). In (**B**), microscopical picture of a DIPG, H3K27M wild type, stained for H3K27Me showing methylation loss in part of the tumor, which is the cause of tumor formation. Courtesy Dr. M. Bugiani

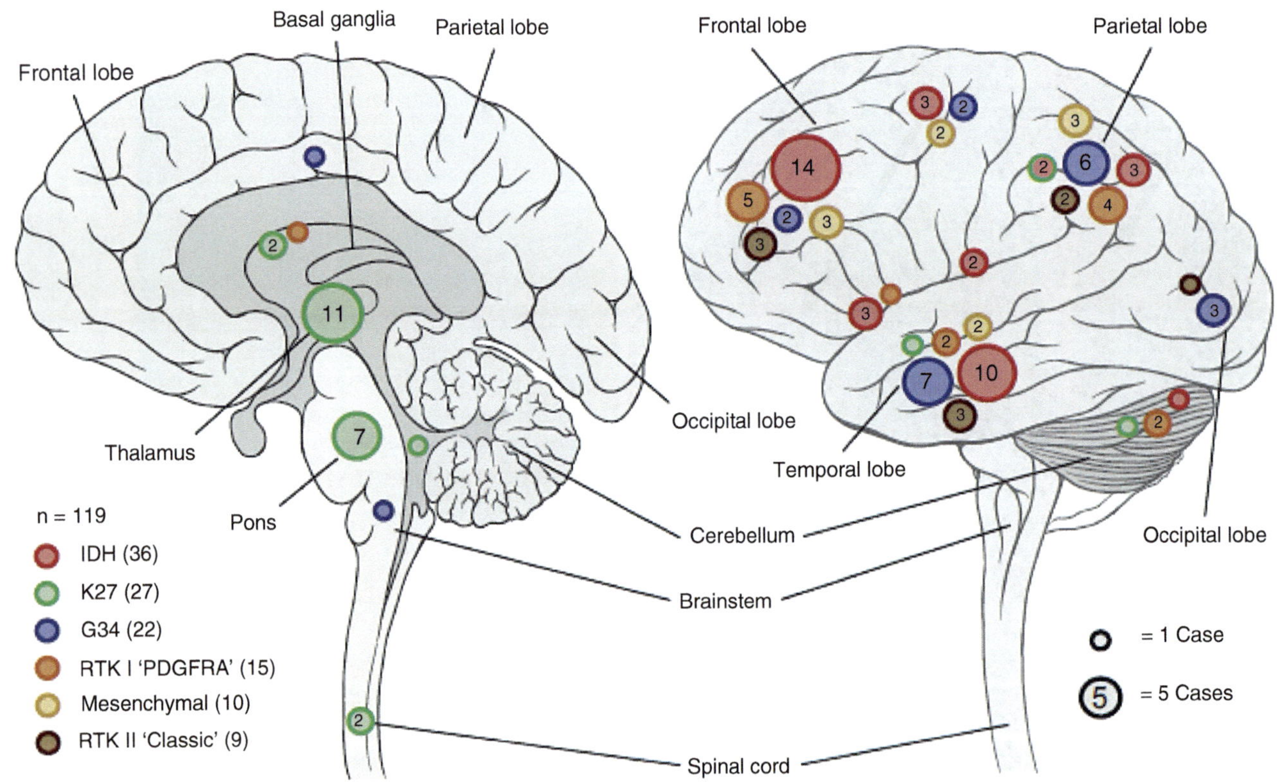

Fig. 1.43 Epigenetic subgroups of glioblastoma correlate with distinct clinical characteristics. Location of 119 GBMs in the human CNS grouped by methylation clusters. The number of cases in each group is indicated with the circles. Circles without numbers represent single cases. Different colors indicate methylate cluster affiliation. Tumors in the midline locations are depicted in the sagittal view (left panel), tumors occurring in the cerebral and cerebellar hemispheres are depicted in the exterior view (right panel). From Sturm D et al. 2012, Fig. 3 with permission

References

Adams HP (2010) Management of carotid artery stenosis: endarteriectomy or stenting? Mayo Clin Proc 85:1071–1072

Adams HP, Bendixen BH et al (1993) Classification of subtype of acute ischemic stroke. Definitions for use in a multicenter clinical trial. TOAST. Trial of org 10172 in acute stroke treatment. Stroke 24:35–41

Aggerwal A, Corbett A et al (1999) Familial colloid cyst of the third ventricle. J Clin Neurosci 6:520–522

Amarenco P, Bogousslavski J et al (2013) The ASCOD phenotyping of ischemic stroke (updated ASCO phenotyping). Cerebrovasc Dis 36:1–5

Arráez-Aybar L-E, Pedro Navia-Álvarez P et al (2015) Thomas Willis, a pioneer in translational research in anatomy (on the 350th anniversary of *Cerebri anatome*). J Anat 226:289–300

Arvanitakis Z, Capuano AW et al (2017) The relationship of cerebral vessel pathology to brain microinfarcts. Brain Pathol 27:77–85

Bailey EL, Smith C et al (2012) Pathology of lacunar ischemic stroke in humans. A systematic review. Brain Pathol 22:583–591

Battistella V, Elkind M (2014) Intracranial atherosclerotic disease. Eur J Neurol 21:956–962

Benson MD (1996) Leptomeningeal amyloid and variant transthyretins. Am J Pathol 148:351–354

Berger MS, Edwards MS et al (1983) Pediatric brains stem tumors: radiographic, pathological, and clinical correlations. Neurosurgery 12:298–302

Biffi A, Shulman JM et al (2012) Genetic variation at CR1 increases risk of cerebral amyloid angiopathy. Neurology 78:334–341

Boiten J, Lodder J et al (1993) Two clinically distinct lacunar infarct entities? A hypothesis. Stroke 24:652–656

Bright R (1827–1831) Reports of medical cases, selected with a view of illustrating the symptoms and cure of diseases by a reference to morbid anatomy. Richard Taylor for Longman et al, London

Buckowicz P, Bartels U et al (2014) Histopathological spectrum of paediatric diffuse intrinsic pontine gliomas: diagnostic and therapeutic implications. Acta Neuropathol 128:573–581

Bugiani M, Kevelam S et al (2016) Cathepsin A-related arteriopathy with strokes and leukoencephalopathy (CARASAL). Neurology 87:1777–1786

Bugiani M, van Zanten SEV et al (2017) Deceptive morphologic and epigenetic heterogeneity in diffuse intrinsic pontine glioma. Oncotarget 8:60447

Caplan LR (1995) Binswanger's disease—revisited. Neurology 45:626–633

Caplan LR (2015) Lacunar infarction and small vessel disease: pathology and pathophysiology. J Stroke 17:2–6

Carr SC, Farb A et al (1997) Activated inflammatory cells are associated with plaque rupture in carotid artery stenosis. Surgery 122:757–763

Chen XY, Wong KS et al (2008) Middle cerebral artery atherosclerosis: histological comparison between plaques associated with and not associated with infarct in a postmortem study. Cerebrovasc Dis 25:74–80

Clarke E, O'Malley CD (1996) The human brain and spinal cord, 2nd edn. Norman Publishing, San Francisco

Corbett JJ, Haines DE (2018) The ventricles, choroid plexus, and cerebrospinal fluid. In: Fundamental neuroscience for basic and clinical applications, 5th edn

Cruveilhier J (1829–1842) Anatomie pathologique du corps humain ou description des diverses altérations morbides dont le corps humain est susceptible avec figures lithographiées et coloriées. J-B. Baillière, Paris

Cuoco JA, Hitscherich K et al (2016) Brainstem vascular syndromes: a prectical guide for medical students. Edorium J Neurol 3:4–16

Dolecek TA, Propp JM et al (2012) Central brain tumor registry of the United States 2012. CBTRUS statistical report: primary brain and central nervous system tumors diagnosed in the United States in 2005–2009. Neuro Oncol 14(sup5):v1–v49

Duvernoy HM (1999) Human brain stem vessels: including the pineal gland and information on brain stem infarction. Springer

Ellison D, Love S et al. (eds) (2004a) Chapter 3. Neuropathology. A reference text of CNS pathology. Vascular disorders. 2nd edn. Mosby

Ellison D, Love S et al. (eds) (2004b) Chapter 34. Neuropathology. A reference text of CNS pathology. Neoplasms. Classification and general concept of CNS neoplasms. 2nd edn. Mosby

Flamm ES (1973) The neurology of Jean Cruveilhier. Med Hist 17:343–353

Gielen GH, Gessi M et al (2013) H3F3A K27M mutation in pediatric CNS tumors: a marker for diffuse high-grade astrocytomas. Am J Clin Pathol 139:345–349

Guillamo J-S, Monjour A et al (2001) Brainstem gliomas in adults: prognostic factors and classification. Brain 124:2528–2539

Haglund M, Passant U et al (2006) Cerebral amyloid angiopathy and cortical microinfarcts as putative substrates of vascular dementia. Int J Geriatr Psychiatry 21:681–687

Haltia M, Livanainen M et al (1982) Spontaneous occlusion of the circle of Willis (Moyamoya syndrome). Clin Neuropathol 1:11–22

Hargrave D, Barteks U et al (2006) Diffuse brainstem glioma in children: critical review of clinical trials. Lancet Oncol 7:241–248

Hasegawa M, Nouri M et al (2015) Neuroepithelial cyst of the fourth ventricle. Childs Nerv Syst 31:155–159

Hayashi (2016) Arachnoid cysts: epidemiology, treatment and clinical outcomes. In: Nguyen K (ed) Neuroscience research progress. Nova Science Publishers, New York, pp 1–35

Hemesniemi J, Leivo S (1996) Management outcome in thrid ventricle colloid cysts in a defiend population: a series of 40 patients treated mainly by transcallosal microsurgery. Surg Neurol 45:2–14

Hilton DA, Footitt D (2003) Neuropathologic findings in Sneddon's syndrome. Neurology 60:1181–1182

Hirano A, Hirano M (1991) Benign cysts in the central nervous system: light and electron microscopic observations of cyst walls. In: Ikuta F (ed) Neuropathology in brain research. Excerpta Medica, Amsterdam, pp 173–185

Hirano A, Hirano M (2004) Benign cysts in the central nervous system: neuropathological observations of the cyst walls. Neuropathology 24:1–7

Hori E, Hayashi N et al (2008) A development of atheromatous plaque is restricted by characteristic arterial. Wall structure at the carotid bifurcation. Surg Neurol 69:586–591

Huggenberger S, Moser N et al. (2019) Neuroanatomie des Menschen. Springer

Ikeda E, Hosoda Y (1992) Spontaneous occlusion of the cricle of Willis (cerebrovascular Moyamoya disease) with special reference to its clinicopathological identity. Brain and Development 14:251–253

Jallo GI, Noureldine MAH, Shimony M (eds) (2020). Brainstem tumors: diagnosis and management. Springer Nature Switzerland AG

Jan M, Zeze VB et al (1989) Cyst of fourth ventricle: diagnostic problems and pathogenic considerations. Neurosurgery 24:939–942

Jang SH, Kwon JW et al (2018) Three dimensional identification of medial ad lateral vestibulospinal tract in the human brain: a diffusion tensor imaging study. Front Hum Neurosci 12:229

Jansen C, Parchi P et al (2010) Prion protein amyloidosis with divergent phenotype associated with two novel nonsense mutations in PRNP. Acta Neuropathol 119:189–197

Janzen J (2004) The microscopic transitional. zone between elastic and muscular arteries. Arch Mal Coeur Vaiss 97:909–914
Jensson O, Palsdottir A et al (1989) The saga of cystatin C gene mutation causing amyloid angiopathy and brain hemorrhage—clinical genetics in Iceland. Clin Genet 36:368–377
Kalaria RN, Perry RH et al (2012) Atheromatous disease in small intracerebral vessels, microinfarcts and dementia. Neuropathol Appl Neurobiol 38:505–508
Kalaria R, Ferrer I et al. (2015) Vascular disease, hypoxia, and related conditions. In: Love S, Perry, A, Ironside J, Budka H (eds) Greenfield's neuropathology, 9th edn. pp 59–210
Kalimo H, Ruchoux M-M et al (2002) CADASIL: a common form of hereditary arteriopathy causing brain infarcts and dementia. Brain Pathol 12:371–384
Kallikazaros I, Tsioufis C et al (1999) Carotid artery disease as a marker for the presence of severe coronary artery disease in patients evaluated for chest pain. Stroke 30:1002–1007
Kim JS, Lee JH et al (1994) Spectrum of lateral medullary syndrome. Correlation between clinical findings and magnetic resonance imaging in 33 subjects. Stroke 25:1405–1410
Kistler JP, Buonanno FS et al (1984) Vertebral-basilar posterior cerebral territory stroke—delineation by proton nuclear magnetic resonance imaging. Stroke 15:417–426
Kitis O, Calli C et al (2004) Wallenberg's lateral medullary syndrome: diffusion-weighted imaging findings. Acta Radiol 45:78–84
Kiuru S (1998) Gelsolin-related familial amyloidosis, Finnish type (FAF), and its variants found worldwide. Amyloid 5:55–66
Kobets AJ, Goodrich JT (2020) History of brainstem glioma surgery. In: Jallo GI, Noureldine MAH, Shimony N (eds). Brainstem tumors: diagnosis and management. Springer Nature Switzerland
Krabbe-Hartkamp MJ, van der Grond J et al (1998) Cicrcle of Willis: morphological. variation on three-dimensional time-of-flight MR angiograms. Radiology 207:103–111
Lammie GA, Brannan F et al (1997) Non-hypertensive cerebral small-vessel disease. An autopsy study. Stroke 28:2222–2229
Lang J (1991) Clinical anatomy of the posterior cranial fossa and its foramina. Thieme, New York
Levy E, Carman MD et al (1990) Mutation of the Alzheimer's disease amyloid gene in hereditary cerebral hemorrhage, Dutch type. Science 248:1124–1126
Lulla R, Saratsis A et al (2016) Mutations in chromatin machinery and pediatric high-grade glioma. Sci Adv 2:1501354
Maat-Schieman M, Roos R et al (2005) Hereditary cerebral hemorrhage with amyloidosis-Dutch type. Neuropathology 25:288–297
Mackenzie IRA, Gilbert JJ (1991) Cysts of the neuraxis of endodermal origin. J Neurol Neurosurg Psychiatry 54:572–575
Mancardi GL, Romagnoli P et al (1988) Lacunae and cribriform cavities of the brain. Eur Neurol 28:11–17
Mandybur TI (1986) Cerebral amyloid angiopathy: the vascular pathology and complications. J Neuropathol Exp Neurol 45:79–79
Masuoka T, Hayashi N et al (2010) Distribution of internal elastic lamina and external elastic lamina in the internal carotid artery: possible relationship with atherosclerosis. Neurol Med Chir (Tokyo) 50:179–182
Mihailoff GA, Haines DE (2018) Motor system I. In: Fundamental neuroscience for basic and clinical applications
Miller Fisher C (1991) Lacunar infarct. A review. Cerebrovasc Dis 1:311–320
Miralles M, Dolz JL et al (1995) The role of the circle of Willis in carotid occlusion: assessment with phase contrast MR angiography and transcranial duplex. Eur J Vasc Endovasc Surg 10:424–430
Mohr JP, Lazar RM et al (1998) Middle cerebral artery disease. In: Barnett HJM, Mohr JP, Stein BM, Yatsu FM (eds) Stroke: pathophysiology, diagnosis and management, 3rd edn. Churchill Livingstone, Philadelphia, pp 427–479
Nathan PW, Smith MC (1982) The rubrospinal and central tegmental tracts in man. Brain 105:223–269
Nathan PW, Smith MC et al (1990) The corticospinal tracts in man. Brain 113:303–324
Nathan PW, Smith MC et al (1996) Vestibulospinal, reticulospinal and descending propriospinal nerve fibres in man. Brain 119:1809–1833
Nicoll JAR, Burnett C et al (1997) High frequency of apolipoprotein E epsilon 2 allele in hemorrhage due to cerebral amyloid angiopathy. Ann Neurol 41:716–721
Nieuwenhuys R, Voogd J et al. (1988) The human central nervous system. 3rd revised edn. Springer
Nitta M, Symon L (1985) Colloid cysts of the third ventricle. A review of 36 cases. Acta Neurochir 76:99–104
Noh S-M, Kang HG (2019) Clinical significance of the internal carotid artery angle in ischemic stroke. Sci Rep 9:1–5
Noureldine MAH, Shimony M et al. (2020) Diffuse midline glioma—diffuse intrinsic potine glioma. In: Jallo GI, Noureldine MAH, Shimony M (eds). Brainstem tumors: diagnosis and management. Springer Nature Switzerland
Parent A (1996) Carpenter's human neuroanatomy, 9th edn, Williams & Wilkins, Philadelphia
Patel AR, Patel AR et al (2019) The underlying stroke etiology: a comparison of two classifications in a rural setup. Cureus 11:e5157
Paxinos G, Watson C (2010) BrainNavigator: interactive atlas and 3D brain software for research, structure analysis, and education. Elsevier, Amsterdam
Pearce JMS (2000) Wallenberg's syndrome. Historical note. J Neurol Neurosurg Psychiatry 68:570
Pellegrini F, Interlandi E et al (2020) Corneal involvement in Wallenberg syndrome: case report and literature review. Neuroophthalmology 44:54–58
Polak JF, Tracy R et al (2013) Carotid artery plaque and progression of coronary artery calcium: the multi-ethnic study of atherosclerosis. J Am Soc Echocardiogr 26:548–555
Poretti A, Meoded A et al (2012) Neuroimaging of pediatric posterior fossa tumors including review of the literature. J Magn Reson Imaging 35:32–47
Puelles L (2009) Forebrain development: prosomere model. In: Lemke G (ed) Developmental neurobiology. Academic, London, pp 315–319
Puget S, Beccaria K et al (2015) Biopsy in a series of 130 pediatric diffuse intrinsic pontine gliomas. Childs Nerv Syst 31:1773–1780
Quast M, Liebegott G (1975) Zur Pathogenese des Wallenberg-Syndroms. Beitr Pathol 54(3):308–323
Revesz T, Holton JL et al (2002) Sporadic and familial cerebral amyloid angiopathies. Brain Pathol 12:343–357
Reynolds N, Al Khalili Y (2019) Neuroanatomy, tectospinal tract. StatPearls, Treasure Island
Rodriguez R, Santiago Rodriguez P et al (2020) Wallenberg syndrome: exercise induced vertebral artery dissection. J Am Coll Cardiol 75(11_Supplement_1):2590
Rostagno A, Holton JL et al (2010) Cerebral amyloidosis: amyloid subunits, mutants and phenotypes. Cell Mol Life Sci 67:581–600
Roth J, Kozyrev D et al. (2020) Low-grade pediatric brainstem gliomas. In: Jallo GI, Noureldine MAH, Shimony M (eds). Brainstem tumors: diagnosis and management. Springer Nature Switzerland AG
Saba L, Argiolas GM et al (2015) Association between internal carotid artery dissection and arterial tortuosity. Neuroradiology 57:149–153
Sacco RL, Freddo L et al (1993) Wallenberg's lateral medullary syndrome. Clinical-magnetic resonance imaging correlations. Arch Neurol 50:609–614
Samuel KC (1956) Atherosclerosis and occlusion of the internal carotid artery. J Pathol Bacteriol 71:391–402

Sandri A, Sardi N et al (2006) Diffuse and focal brain stem tumors in childhood: prognostic factors and surgical outcome. Childs Nerv Syst 22:1127–1135

Schomer DF, Marks MP et al (1994) The anatomy of the posterior communicating artery as a risk factor for ischemic cerebral infarction. N Engl J Med 330:1565–1570

Schwartzentruber J, Korshunov A et al (2012) Driver mutations in histone H3.3 and chromatin remodelling genes in paediatric glioblastoma. Nature 482:226–231

Shetty S, Anusha R et al (2012) Wallenberg's syndrome. J Neurosci Rural Practice 3:100–102

Soontornniyomkij V, Lynch MD et al (2010) Cerebral microinfarcts associated with severe cerebral β-amyloid angiopathy. Brain Pathol 20:459–467

Spears RC (2004) Colloid cyst headache. Curr Pain Headache Rep 8:297–300

Sturm D, Witt H et al (2012) Hotspot mutations in H3F3A and IDH1 define distinct epigenetic and biological subgroups of glioblastoma. Cancer Cell 22:425–437

Tadi P, Lui F (2018) Acute stroke. StatPearls (Internet)

Tanrikulu B, Özek MM (2020) Clinical presentation and assessment for brainstem tumors. In: Jallo GI, Noureldine MAH, Shimony M (eds). Brainstem tumors: diagnosis and management. Springer Nature Switzerland

ten Donkelaar HJ (2011) Clinical neuroanatomy: brain circuitry and its disorders. Springer

Thal DR, Ghebremedhin E et al (2003) Vascular pathology in Alzheimer's disease: correlation of cerebral amyloid angiopathy and arteriosclerosis/lipohyalinosis with cognitive decline. J Neuropathol Exp Neurol 62:1287–1301

Thal DR, Griffin WST et al (2008) Cerebral amyloid angiopathy and its relationship to Alzheimer's disease. Acta Neuropathol 115:599–609

Thanvi B, Robinson T (2007) Complete occlusion of extracranial internal carotid artery: clinical features, pathophysiology, diagnosis and management. Postgrad Med J 83:95–99

Thron AK (1988) Vascular anatomy of the spinal cord. Neuroradiological Investigations and Clinical Syndromes. Springer Wien New York

Tietjen GE, Levine SR (1995) Sneddon's syndrome and Divry-Van Bogaert syndrome. In: Bogousslavsky J, Caplan L (eds) Stroke syndromes. Cambridge University Press, Cambridge, pp 422–429

Usunoff KG, Marani E et al (1997) The trigeminal system in man. Adv Anat Embryol Cell Biol 136:1–126

Van Seeters T, Hendrikse J et al (2015) Completeness of the circle of Willis and risk of ischemic stroke in patients without cerebrovascular diease. Neuroradiology 57:1247–1251

Vanan MI, Eisenstat DD (2015) DIPG in children—what can we learn from the past? Front Oncol 5:237

Vidal R, Fangione B et al (1999) A stop-codon mutation in the BRI gene associated with familial British dementia. Nature 399:776–781

Vinters HV, Gilbert JJ (1983) Cerebral amyloid angiopathy: incidence and complications in the aging brain. II. The distribution of amyloid vascular changes. Stroke 14:924–928

Vinters HV, Zarow C et al (2018) Review: vascular dementia: clinicopathologic and genetic considerations. Neuropathol Appl Neurobiol 44:247–266

Von Lanz T, Wachsmuth W et al. (2003) Praktische Anatomie: Kopf: Teil B, Gehirn- und Augenschädel. Springer Berlin Heidelberg

Vonsattel JP, Myers RH et al (1991) Cerebral amyloid angiopathy without and with cerebral hemorrhages: a comparative histological study. Ann Neurol 30:637–649

Walker DA, Punt JA et al (1999) Clinical management of brain stem glioma. Arch Dis Child 80:558–564

Wallenberg A (1895) Acute Bulbäraffection. Archiv f Psychiatrie 27:504–520

Wang ZZ, Jensson O et al (1997) Microvascular degeneration in hereditary cystatin C amyloid angiopathy of the brain. APMIS 105:41–47

Wang Z, Yan HM et al (2012) A colloid cyst in the fourth ventricle complicated with aseptic meningitis: a case report. Clin Neurol Neurosurg 114:1095–1098

Watson C, Bartholomaeus C et al (2019) Time for radical changes in brain stem nomenclature: applying the lessons from developmental gene patterns. Front Neuroanat 13:10

Weisbrod LJ, Smith KA et al (2018) Familial colloid cyst of the third ventricle. Interdiscip Neurosurg 14:186–187

Wilkins RH, Brody IA (1970) Wallenberg's syndrome. Arch Neurol 22:379

Xu Y, Yuan C et al (2018) Association of severity between carotid and intracranial artery atherosclerosis. Ann Clin Transl Neurol 5:843–849

Yamamoto Y, Craggs L et al (2011) Review: molecular genetics and pathology of hereditary small vessel diseases of the brain. Neuropathol Appl Neurobiol 37:94–11

Yew KS, Cheng E (2009) Acute stroke diagnosis. Am Fam Physician 80:33–40

Young WB, Silberstein SD (1997) Paroxysmal headache caused by colloid cysts of the third ventricle: case report and review of the literature. Headache 37:15–20

Web Links

Anatonomina (2020) http://terminologia-anatomica.org/en/Terms

https://archive.org/stream/cerebrianatomecu00will#page/n71/mode/2up

http://fipat.library.dal.ca/TNA

http://www.ifaa.net/committees/anatomical-terminology-fipat/

https://www.neurosurgicalatlas.com/neuroanatomy/posterior-view-of-occipito-atlantal-region

https://www.orpha.net/consor/cgi-bin/OC_Exp.php?lng=en&Expert=2182

https://www.who.int/news-room/fact-sheets/detail/the-top-10-causes-of-death

General Brain Development

2

Contents

Abstract

Basic knowledge about the prenatal development of the human brain in general, the brainstem and the cranial nerve-innervated structures of head and neck is a prerequisite for the understanding of the adult findings and a number of developmental diseases.

After the 20th day after fertilization, the neural tube begins to develop which is the first trace of the central nervous system (CNS) in the embryo. The neural tube expands in the head region to three initial cerebral vesicles. These are in rostrocaudal direction the prosencephalic, mesencephalic, and rhombencephalic vesicles. The prosencephalic vesicle develops into the telencephalon and diencephalon.

Just to mention a relatively frequent disorder at this early stage of development are closure disturbances of the neural tube. The best-known of these is probably Spina bifida at the spinal level, but closure disorders can exist also at more cranial levels of the central nervous system.

The brainstem is formed by the mesencephalic vesicle, which develops into the mesencephalon, and the rhombencephalic vesicle, which becomes the pons, the cerebellum (including its peduncles), and the medulla oblongata.

During development, the brainstem is highly segmentally organized in the so-called neuromeres, and the cranial nerves are related to these brainstem segments. These neuromeres—called rhombomeres in most of the pro-

H. Schröder et al., *The Human Brainstem*, https://doi.org/10.1007/978-3-030-89980-6_2

spective brainstem—are the basis of arrangement of the different chapters of this book.

Functionally, these cranial nerves can be divided simplified into motor nerves and sensory nerves although most nerves of the brainstem unite both qualities. The examination of the cranial nerves during physical examination and the wide spectrum of disorders involving the cranial nerves, for example, trigeminal neuralgia, paralysis of the facial nerve, or tumors of the vestibular component of the vestibulocochlear nerve are frequent disorders relevant for neurology, neurosurgery, ENT, and neuroradiology.

A peculiarity during the development of the head and neck region is the fact that the region below the future mouth is organized in five pharyngeal arches, and each arch is associated with a specific cranial nerve, which arises from the adjacent hindbrain. This unique arrangement is the basis of the clinical examination of the cranial nerves where disturbance of the periphery like sensory impairment or motor restrictions allows to draw conclusion on the state of a given cranial nerve as the basis for further, for example, imaging diagnostics.

2.1 General brain development

The first embryonic trace of the nervous system is the neural plate which is a dorsal thickening of the ectoderm beginning to form around the 16th day but becomes clearly visible around the 18th to 19th day after fertilization (embryonic day). Already in the following days, the lateral edges (neural folds) of the neural plate rise and form the so-called neural groove. The neural folds rise dorsally and then close in the midline. This closure starts in the head region and migrates gradually in caudal direction to form the neural tube (neuraxis) (Fig. 2.1). Defects in the closure of the neural tube are the cause of anencephaly or spina bifida, a congenital malformation that can manifest in various clinical forms. A failure of the neural tube to close throughout its entire length results in a condition called rachischisis.

The two open ends of the neural tube, the neuropori, close between the 25th (Neuroporus anterior) and the 27th to 28th day of development (Neuroporus posterior) resulting in a simple cone-shaped caudal end. This cone is going to be the adult medullary conus, the caudal end of the adult spinal cord.

Along the lateral edge of the neural fold, clusters of cells (all of ectodermal origins) group to form the neural crest [Crista neuralis]. When the neural fold closes to form the neural tube, the cells of the neural crest are placed initially as an intermediate layer between the surface ectoderm and the neural tube. Subsequently, the elements of the neural crest are arranged dorsolaterally on both sides of the neural tube.

The neural tube develops so that there is a basal plate for motor function and an alar plate for sensory functions—a construction principle preserved in the adult spinal cord (see Fig. 2.5).

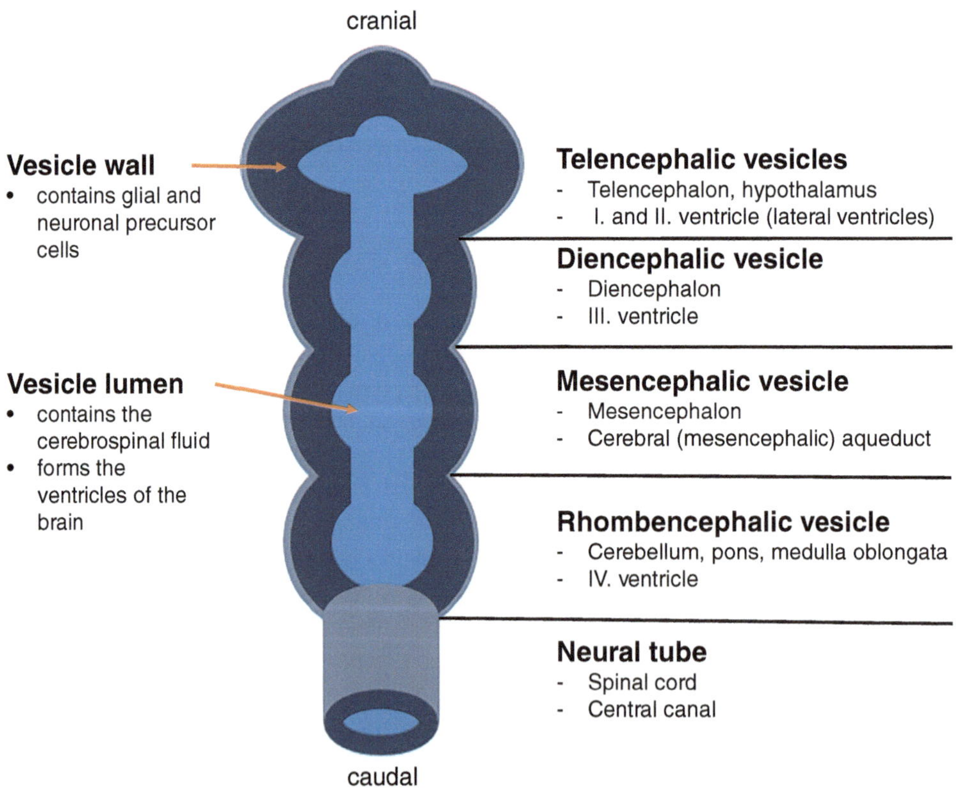

Fig. 2.1 In principle, the development of the central nervous system (CNS) in mammals starts with the formation of the neural tube, a cylindric structure with a central lumen filled with cerebrospinal fluid (CSF). The cylindrical structure of the neural tube is still visible in the adult spinal cord (see Fig. 3.5). At the cranial end, however, the neural tube widens into several vesicles. The walls of these vesicles give rise to the neuro- and glioblasts—as do the developing spinal cord walls—that make up the prospective parts of the brain surrounding the individual vesicles. S. Huggenberger fecit

As development progresses, the cells that constitute the wall of the neural tube begin to divide. The dividing cells are densely packed in the neural tube close to the inner fluid-filled cavity (future ventricles and central canal) and initially form the neuroepithelial cells of the ventricular zone (a pseudostratified epithelium) with complex connections to the inner limiting membrane. This condition also persists during the following developmental stages. The ventricular zone (germinal, primitive ependymal, or matrix layer) corresponds to the apical region of the radial glial cells and contains the nucleated parts of the columnar cells and rounded cells undergoing mitosis (Fig. 2.2).

The zone of cellular proliferation is located close to the inner limiting membrane of the neural tube so that a sparsely packed cell layer (marginal zone; Fig. 2.2B) is formed with the outer lining membrane as the external enclosure of the neural tube. This outer zone initially consists of the external (basal) cytoplasmic processes of radial glia cells. It is soon invaded by tracts of axonal processes that grow from neuroblasts developing in the mantle zone, together with varieties of non-neuronal cells (glial cells and, later, vascular endothelium and perivascular mesenchyme; Standring 2016).

From the inner ventricular zone, successive cell bodies migrate distally along their radial extensions and form an intermediate layer (mantle zone) by pushing the marginal zone further outwards. The mantle zone, or intermediate zone, contains thus the migrant cells from the divisions occurring in the ventricular zone.

2.1.1 Brain vesicles and their derivatives

After closure of the neural tube, various parts, especially in the head region, modify. Thus, one can recognize different expansions in this region from fourth week up to sixth week. The proliferation of the most rostrally located parts of the neural tube is not uniform and forms three cerebral vesicles. These are in the rostrocaudal direction prosencephalic, mesencephalic, and rhombencephalic vesicles. These vesicles then form further subdivisions (Figs. 2.1, 2.3, and 2.4).

1. The prosencephalic vesicle develops into the following:
   (a) The telencephalon from which the cerebral cortex [Cortex cerebri], the basal ganglia, and the two lateral ventricles develop. The cerebral cortex can be divided into the phylogenetic older allocortex, which is the dorsal archicortex and the ventral palaeocortex, and the isocortex (neocortex) which evolved in between the latter two.
   (b) The diencephalon from which the thalamus, hypothalamus, and other structures, all related to the third ventricle, develop. Derivatives of the diencephalon precursor are also the retina, the neurohypophysis, and the pineal gland [Corpus pineale, Epiphysis cerebri]. Modern studies on the pattern of gene expression during brain development show that the hypothalamus belongs to the telencephalon (see below). In this book, however, we stick to the classical view mentioned above for easier comparability with the still existing literature.
2. The mesencephalic vesicle develops into the mesencephalon (with a narrow internal cavity, the mesencephalic aqueduct [Aqueductus mesencephali]) separated from the subsequent vesicle by a constriction, isthmus rhombencephali (Fig. 2.3A, B). It develops into the cerebral peduncles, the tegmentum, and the tectum.
3. The rhombencephalic vesicle becomes the metencephalic and myelencephalic vesicles, separated by the pontine flexure (Fig. 2.3E). These develop into the following divisions:
   (a) the isthmus rhombencephali which give rise to the superior medullary velum and the superior cerebellar peduncles
   (b) the pons
   (c) the cerebellum with the middle cerebellar peduncle, both (b and c) from the metencephalic vesicle
   (d) the medulla oblongata and the inferior cerebellar peduncle of the myelencephalic vesicle

These four parts (a–d) are spatially related to the fourth ventricle.

Caudal of the myelencephalic vesicle, the neural tube is not subject to any relative volumetric changes of importance during development and maintains a similar diameter in all parts without the formation of differentiated vesicles. Here, the spinal cord develops.

Due to the contact of different disciplines, such as human anatomy, comparative anatomy, and human and veterinary clinical medicine, a definition of common English terms seems to be useful:

Spinal cord—Medulla spinalis

Hindbrain—Medulla oblongata + pons + cerebellum (derivatives of the myelencephalon [Medulla oblongata] + metencephalon [pons and cerebellum] both are derived from the rhombencephalon)

Midbrain—tectum + tegmentum (both derivatives of the mesencephalon). From a descriptive point of view, the crura cerebri are the rostral-most part of the midbrain.

Brainstem—hindbrain + midbrain

Forebrain—diencephalon + telencephalon (both derivatives of the prosencephalon)

The entire brain develops segmentally (metamerically). This was shown by studies of the expression patterns of regulatory genes during brain development (Puelles et al. 2013; Puelles and Rubenstein 2015), enabling to define a large

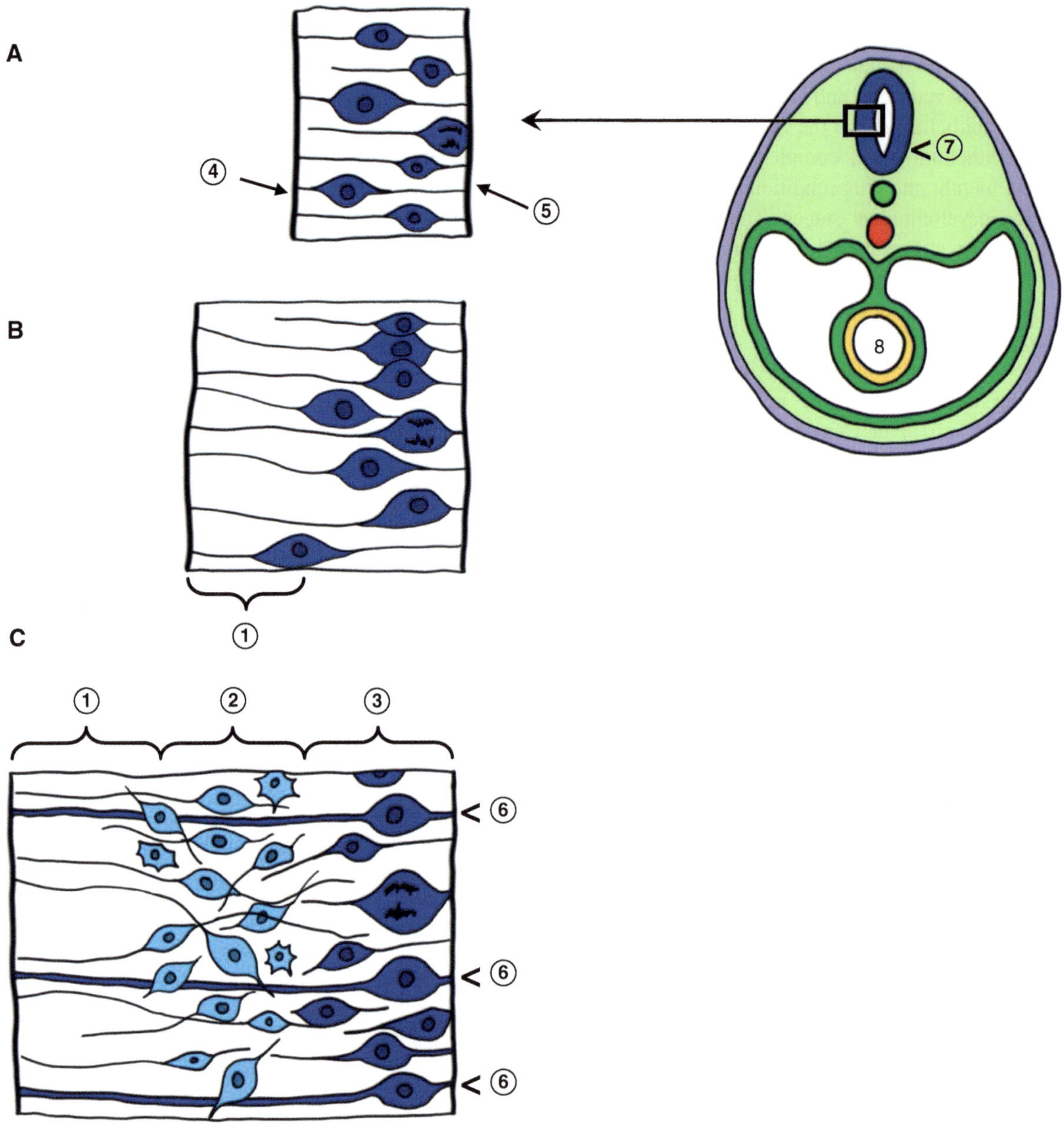

①	**Marginal zone**	⑤	**Inner limiting membrane**
②	**Intermediate zone / layer**	⑥	**Radial glial cells**
③	**Ventricular zone**	⑦	**Neural tube**
④	**Outer limiting membrane**	⑧	**Foregut**

Fig. 2.2 Cellular proliferation within the early neural tube. The earliest layer of the neural tube is the ventricular zone with neuroepithelial cells (**A**). Three layers are delineated in the embryological spinal cord and brainstem (**C**). The inset on the right shows a schematic transverse cut of an embryo with the dorsally closed neural tube ⑦. S. Huggenberger fecit

number of neuromeres (transversal brain segments). These genes provide information for the position along the longitudinal axis. According to these results, the rhombencephalon is divided into 12 rhombomeres (isthmus and r1 to r11) and the mesencephalon into two mesomeres (m1 and m2; see below). The prosencephalon is divided into at least three caudal prosomeres for the diencephalon (t1 to t3: the later thalamus) and an additional rostral portion. There is also evidence

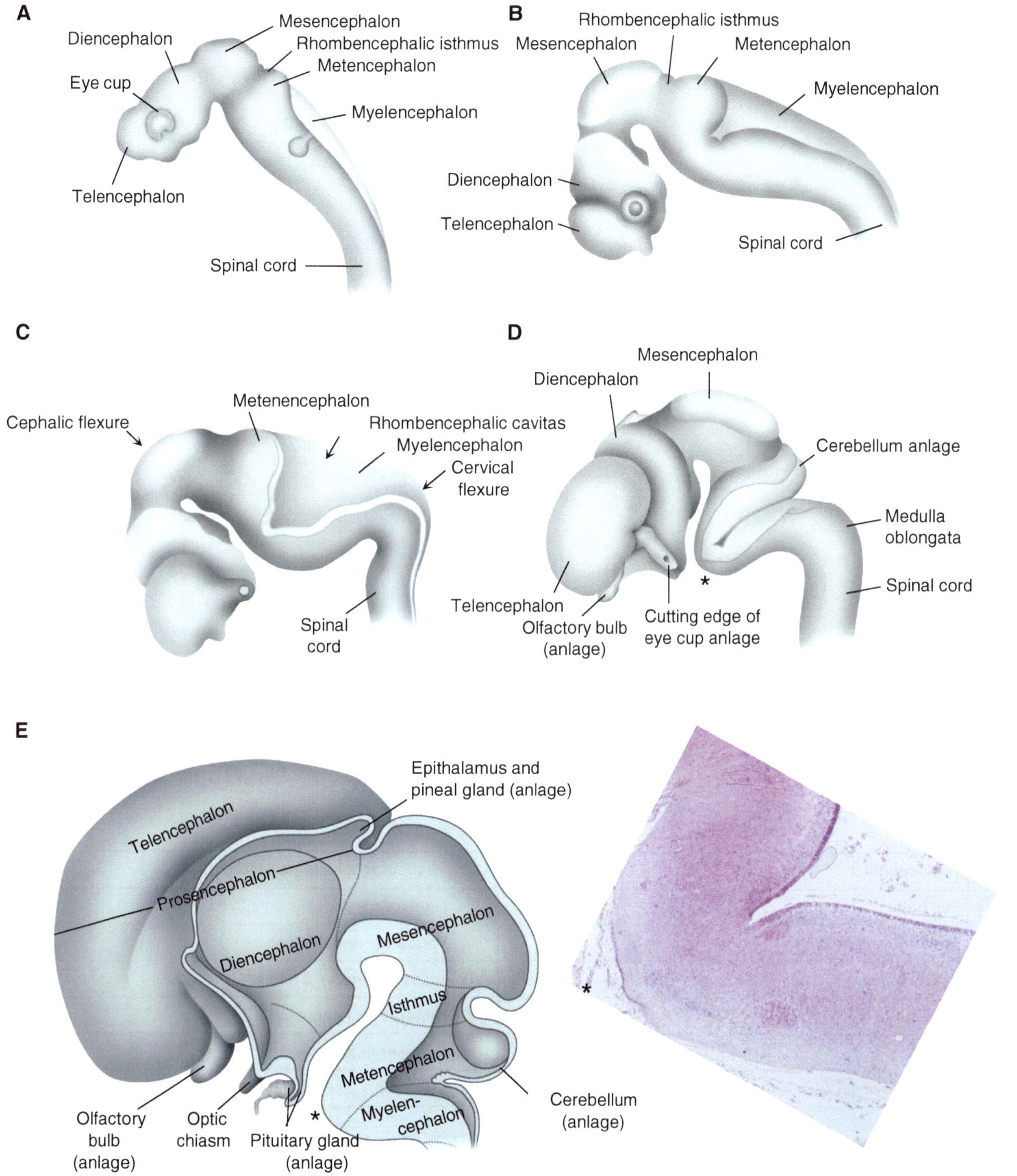

Fig. 2.3 Developmental stages of the neural tube (**A**, third week; **B**, fourth week; **C**, fifth week; **D**, eighth week). As a consequence of the craniocaudal folding of the brain tube, two dorsal flexures develop: the mesencephalic (cephalic) flexure and, further caudal, a cervical flexure between the myelencephalon and the spinal cord [Medulla spinalis]. In between, a ventral flexure develops, the pontine flexure (* in **E** and in the inset). Brain vesicles of a human fetus of 3 months (**E**). The inset shows a histological detail (H & E stain) of the pontine flexure (*) of a human fetus of 12 weeks. Modified after Huggenberger et al. 2019, Figs. 2.1 and 2.2 with permission

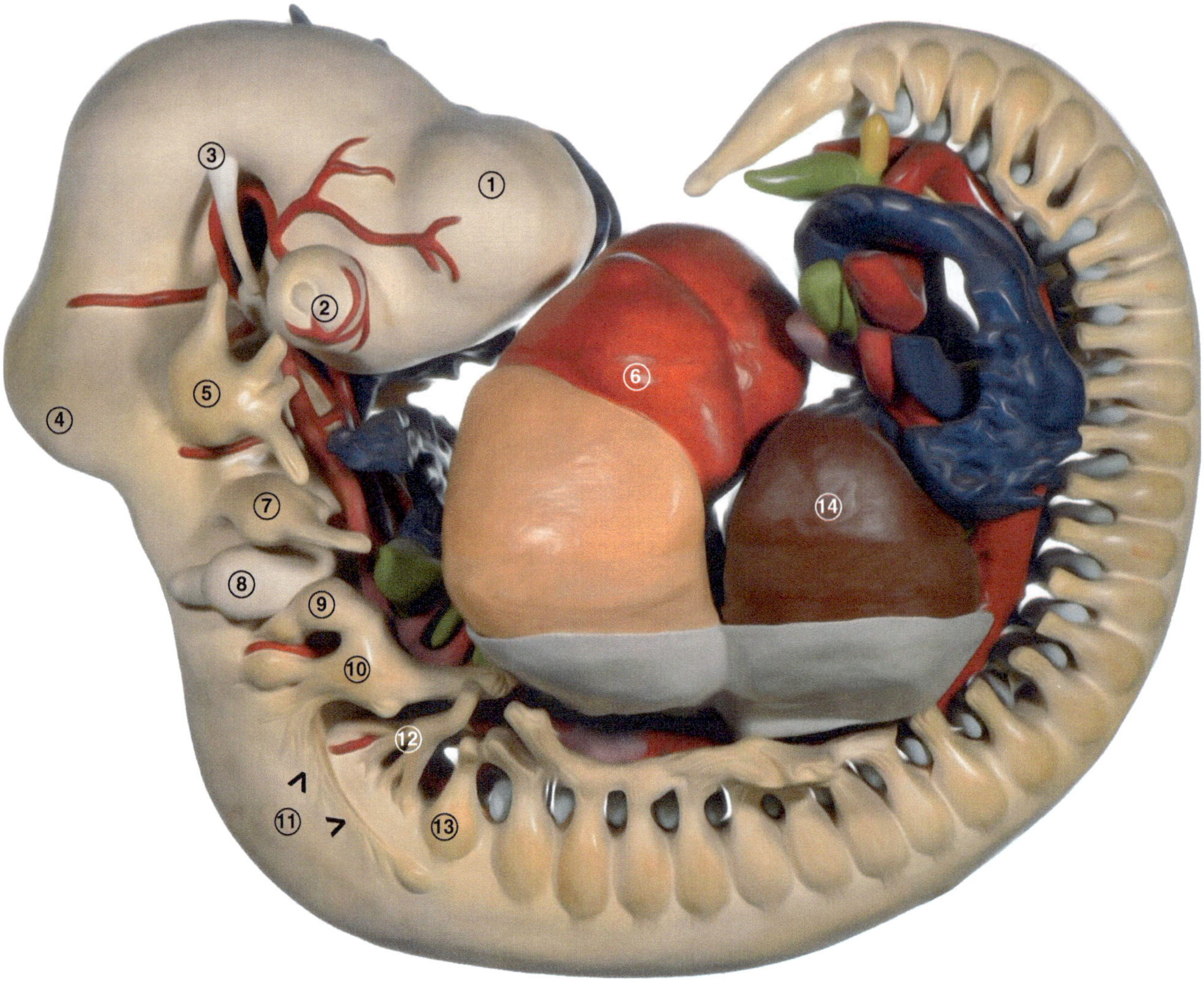

①	Telencephalon	⑧	Vestibulocochlear nerve CNVIII
②	Eye bulb	⑨	Glossopharyngeal nerve CNIX
③	Oculomotor nerve CNIII	⑩	Vagus nerve CNX
④	Cervical flexure	⑪	Accessory nerve CNXI
⑤	Trigeminal ganglion CNV	⑫	Hypoglossal nerve CNXII
⑥	Heart	⑬	First cervical nerve
⑦	Facial nerve CNVII	⑭	Liver

Fig. 2.4 Right side view of a human embryo (26 days, 5 mm, Homo Nürnberger 28.07.1950, after a 100x model by Rosenbauer, 1954). Sammlung des Zentrums Anatomie der Universität zu Köln

that this rostral part consists of two other prosomeres and forms the secondary prosencephalon. The location within this secondary prosencephalon determines the development of the telencephalon (dorsal) and the hypothalamus (ventral). These more recent developmental results show that the hypothalamus (ht1 and ht2) is not a part of the diencephalon although it is termed purely descriptively—as in this book—as part of it.

2.1.2 General organization of the CNS during development

After closure of the neuropori, a more complex organization develops from the stratum neuroepitheliale in the wall of the neural tube. Cellular layers become visible: an inner layer (ventricular zone), a subventricular zone, an intermediate layer (mantle zone), the cortical plate, and a cell-poor outer layer (marginal zone). In the ventricular zone, the neuroepithelial cells differentiate into radial glia cells, which become the precursors of most cells (neurons and glial cells) in the central nervous system (Fig. 2.2). The first neuronal cells of the ventricular layer migrate into the marginal layer where they form the Cajal-Retzius cells (see Box 2.1). They release Reelin as chemical messengers into the extracellular matrix. Reelin then causes the formation of radial extensions in the radial glia cells, which in the further course of development leads the new cells to their final localization.

Box 2.1 Ramón y Cajal and Magnus Gustaf Retzius

Santiago Ramón y Cajal (1852–1934) was a Spanish neuroscientist and pathologist specialized in the histology of the central nervous system (see also Box 4.2). He received the Nobel Prize in Physiology or Medicine in 1906 together with Camillo Golgi. Magnus Gustaf Retzius (1842–1919) was a Swedish physician and anatomist who dedicated a large part of his life to researching the histology of the sense organs and nervous system.

https://en.wikipedia.org/wiki/Santiago_Ram%C3%B3n_y_Cajal

https://en.wikipedia.org/wiki/Gustaf_Retzius

At the end of this developmental process, the ventricular zone is reduced to a single-celled layer of ependymal cells lining the central canal of the spinal cord and the interior of the cerebral ventricles. The mantle zone, originating from migrated cells of the ventricular zone, gives rise to the substantia grisea (gray matter) in adults. The axons of the developing neurons of the mantle zone may group together in tracts [Tractus] and together with the majority of the glial cells in the marginal zone form the substantia alba (white substance).

The cerebral and cerebellar cortices are an exception to this general developmental process in so far as the neuroblasts migrate to the outside. As a result, in the cerebrum and cerebellum, the gray substance lies as a "mantle" (cortex) on the brain's surface.

As can be seen from the subdivision of the cerebral vesicles, it is the most rostrally located parts of the neural tube that undergo the greatest volumetric growth during development. Thus, these rostral parts characterize the external appearance of the brain of the adult. This rapid development of its volume is particularly evident in the cerebral cortex, which begins to rotate around an imaginary transverse axis through the insula region. This proliferation is responsible for the rotation and thus for the shape of the adult human brain characterized externally by the hypertrophy of the endbrain with its massive occurrence of cortical convolutions (gyri) and furrows (sulci).

There are no volumetric differences along the spinal cord during development. It initially extends parallel to the development of the spinal column. In the third month of development, the spinal cord occupies the entire length of the spinal canal and the spinal nerves leave the CNS via the intervertebral foramen [Foramen intervertebrale] very close to their exit from the spinal cord. Nevertheless, the growth of the vertebrae and the dura mater is faster than that of the spinal cord. At birth, the spinal cord is at the level of the third lumbar vertebra, in adults finally at the level of the second lumbar vertebra. This is a perceived ascensus of the spinal cord. As a result of the extended length growth of the spinal column, the spinal nerves get more inclined caudally and cover larger distances within the spinal canal before exiting via the corresponding Foramen intervertebrale.

The roots of the spinal nerves, which run caudally within the vertebral canal below the caudal end of the spinal cord in direction of their foramina, form the cauda equina (so named because it resembles a ponytail). The dura mater has an attachment at the coccygeal level. Here, one finds the Filum terminale, a remnant of the embryonic spinal cord.

2.2 The prenatal fate of the brainstem

The caudal part of the cranial portion of the neural tube gives rise to the brainstem. The brainstem is a derivative of the mesencephalon and the rhombencephalon (see Fig. 2.3). The cervical flexure (see Fig. 2.4) demarcates the embryonic brainstem, i.e., the myelencephalon, from the spinal cord. In the adult brain, this junction approximates the level of the superior rootlet of the first cervical nerve, which is located roughly at the foramen magnum (see Figs. 1.35 and 1.36). By the fifth to sixth gestational week, the pontine flexure, located in the future pontine region, divides the hindbrain into caudal (myelencephalon) and rostral (metencephalon) parts. The myelencephalon becomes the medulla oblongata while the pons and the cerebellum are derived from the metencephalon. The vesicle lumen of the myelencephalon becomes the inferior part of the fourth ventricle and the central canal in the medulla oblongata, respectively.

2.2.1 Myelencephalon

The caudal part of the myelencephalon resembles the spinal cord developmentally and structurally. Here, the neural canal of the neural tube forms the small central canal. Unlike those of the spinal cord, neuroblasts from the alar plates in the myelencephalon migrate into the marginal zone and form isolated areas of gray matter: the gracile nuclei (see Sect. 3.2.3.4) medially and the cuneate nuclei laterally (see Sect. 3.2.3.3, Moore et al. 2020). In the adult brain, these nuclei are associated with correspondingly named tracts (see Sect.

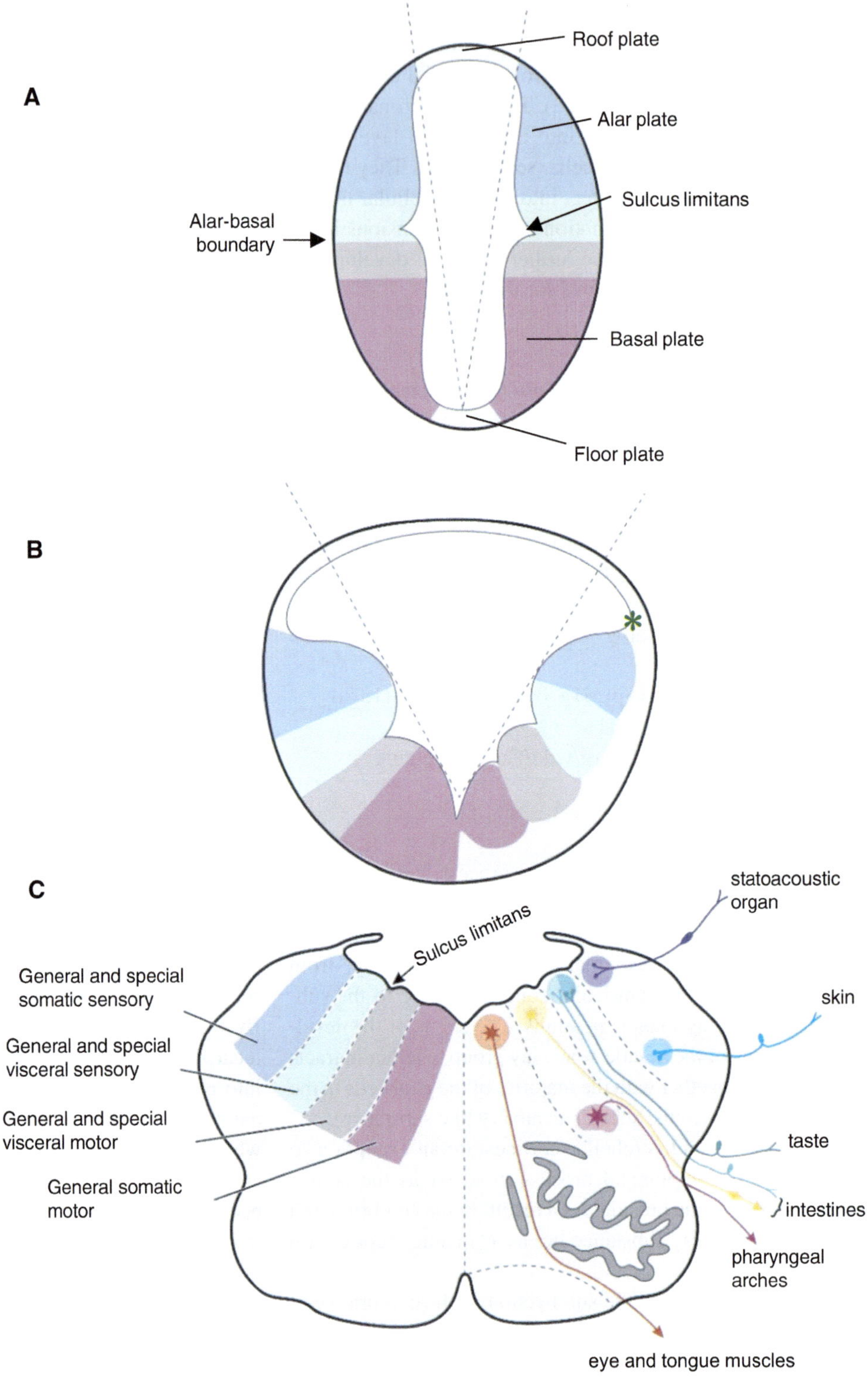

Fig. 2.5 Schematic cross-section through the brainstem during ontogenesis. (**A**) Brain tube with the basal and alar plates, divided by the sulcus limitans. (**B**) Expansion of the central canal and formation of the fourth ventricle with unfolding of the nuclear columns * area of roof lip (rhombic lip). (**C**) Scheme of the adult brainstem at level of the inferior olive. Left: the different nuclear columns and their physiological quality. Right: the origin of the components (qualities) of the cranial nerves and their respective periphery. Modified after Cozzi et al. 2017 Fig. 6.5 with permission

3.2.3) that enter the medulla from the dorsal spinal cord. However, thick fiber bundles sprouting from the primordial motor cortex in direction of the spinal cord form the pyramids (see Table 1.3, corticospinal—pyramidal tract).

The rostral part of the myelencephalon opens dorsally so that it becomes wide and rather flat, especially opposite the pontine flexure. The pontine flexure causes the lateral walls of the medulla to divert laterally (resembling the opening of a book; Fig. 2.5).

As a result, its roof plate is stretched and greatly thinned. The cavity of this part of the myelencephalon (part of the future fourth ventricle, see Fig. 1.9) becomes rhomboidal

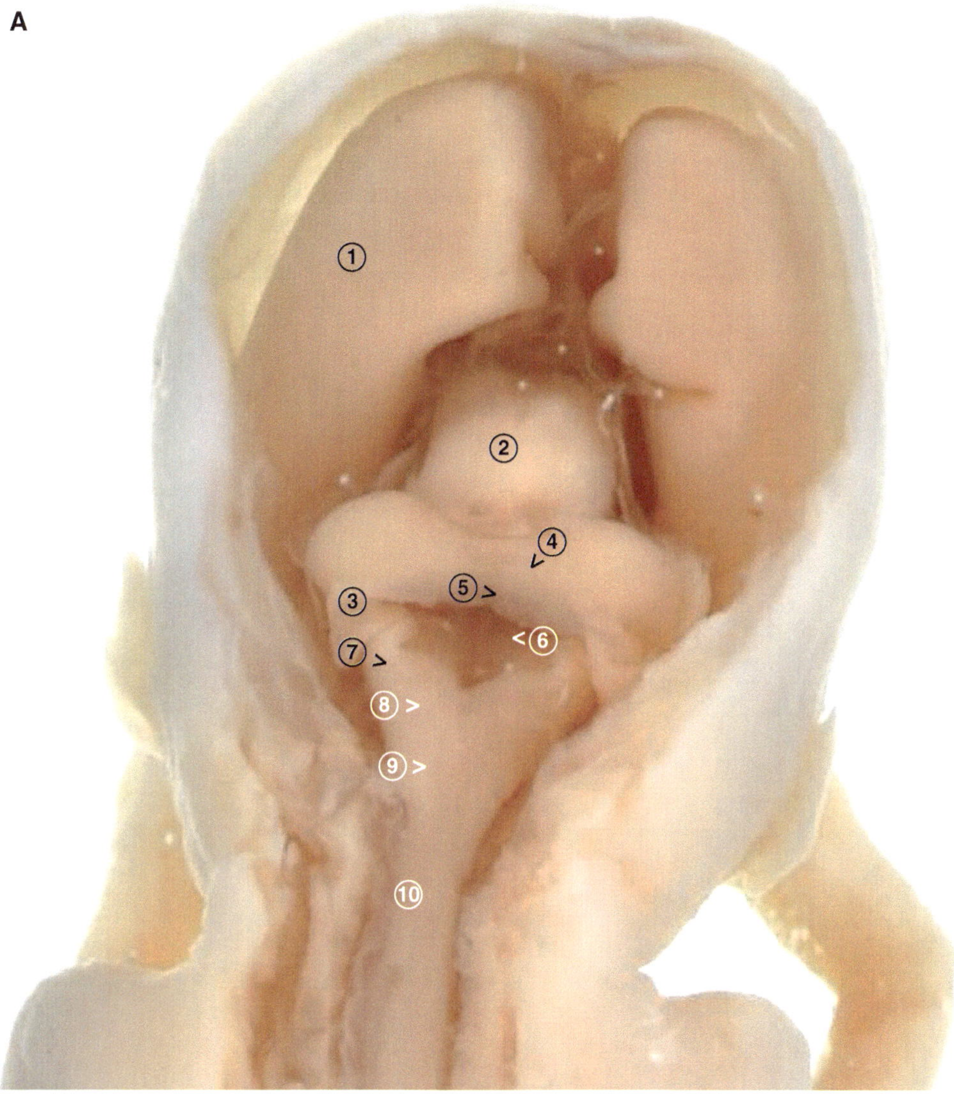

①	Cerebral cortex	⑥	Rhomboid fossa
②	Mesencephalic tectum	⑦	Rhombic lip
③	Cerebellar hemisphere	⑧	Cuneate tubercle
④	Fissura prima	⑨	Gracile tubercle
⑤	Vermis	⑩	Spinal cord

Fig. 2.6 Dorsal view of CNS of a human fetus of approximately 12 weeks (**A**) and 24 weeks (**B**). The distance between the dura and the surface of the occipital lobe/telencephalon is due to shrinkage induced by fixation. Sammlung des Zentrums Anatomie der Universität zu Köln. Photographs by MedizinFotoKöln, Cologne, Germany

(rhomboidal fossa; Fig. 2.6). Due to the lateral movement of the medulla's walls, the alar plates get a position lateral to the basal plates (Fig. 2.5). Therefore, the motor nuclei usually develop medial to the sensory nuclei and not ventral as is the case in the spinal cord (see Fig. 3.5).

Within the basal plates of the medulla, like in the spinal cord, efferent (motor) neurons develop. These neurons form nuclei and organize into three cell columns on each side (Fig. 2.5). From medial to lateral, the columns are as follows:

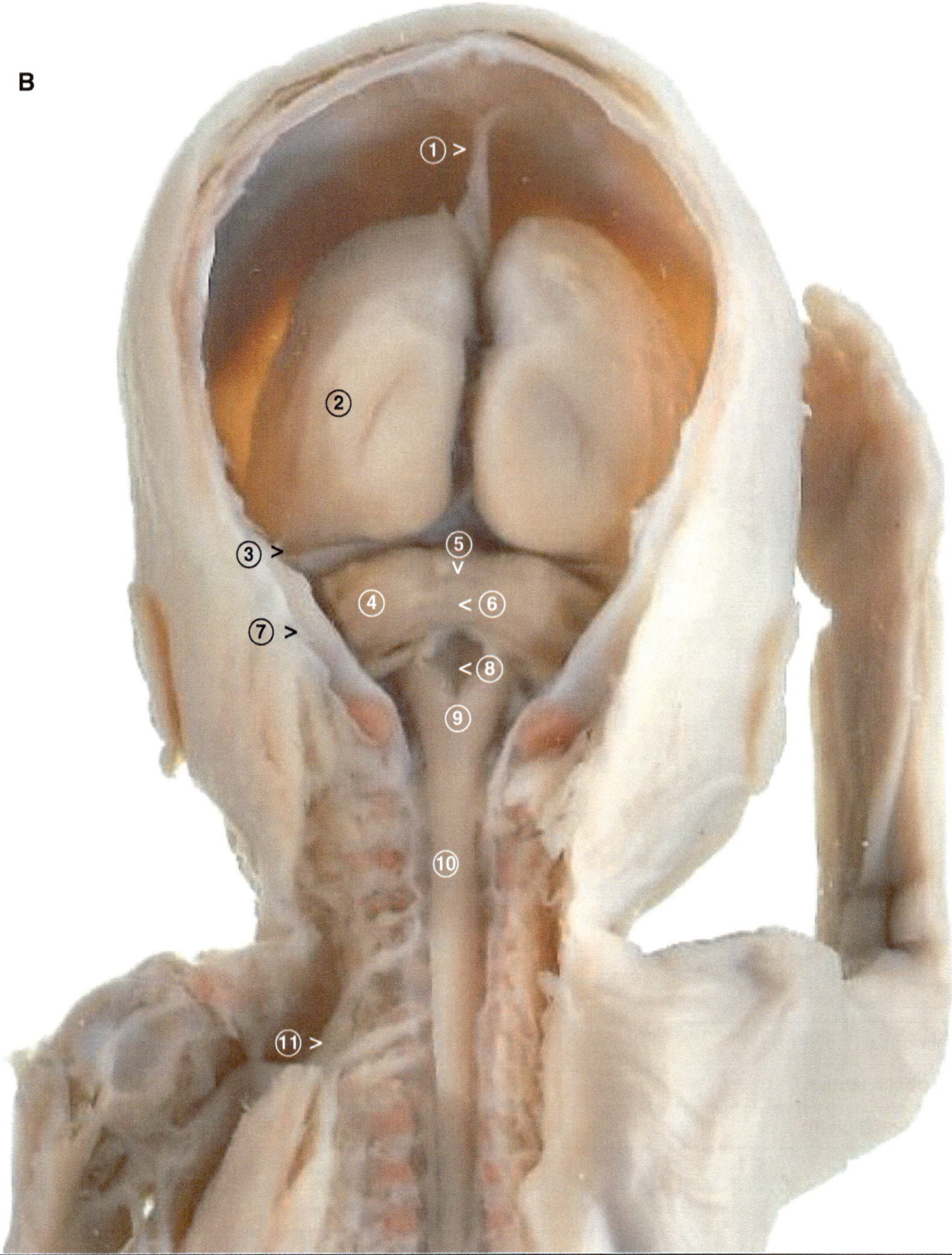

①	**Falx cerebri**	⑥	**Fissura secunda**
②	**Occipital lobe**	⑦	**Dura mater**
③	**Cerebellar tentorium**	⑧	**Rhomboid fossa**
④	**Cerebellar hemisphere**	⑨	**Rhombencephalon**
⑤	**Fissura prima**	⑩	**Spinal cord**
		⑪	**Brachial plexus**

Fig. 2.6 (continued)

- general somatic efferent represented by motor neurons of the hypoglossal nerve
- special visceral efferent represented by neurons innervating muscles derived from the pharyngeal arches
- general visceral efferent represented by motor neurons of the vagus and glossopharyngeal nerves

The alar plates of the medulla give rise to neurons that receive sensory information. They are arranged in four columns from medial to lateral, which harbor neurons getting

- general visceral afferents from the viscera
- special visceral afferents of taste fibers
- general somatic afferents of the head
- special somatic afferents from the ear (see also Table 1.1 and Figs. 1.13 and 1.14)

All cells of the cranial nerve nuclei are born by the end of the embryonic period. Shortly after their formation, the nuclei lie ordered next to each other. After that, there are major changes in the positional relationship of these nuclei due to the ingrowth of fibers, the development of the reticular formation (see Sect. 3.3.2), and the immigration of new neurons from the rhombic lip (see Fig. 2.5B). An example of this cellular migration is the displacement of the facial nucleus (see Sect. 8.2.1). At the beginning of the second month, it lies close to the median sulcus and slightly dorsal to the abducent nerve. The efferent fibers have already reached the surface of the medulla oblongata. The facial nucleus is then shifted further ventrally, the abducens nucleus (see Sect. 9.2.1.1) dorsally. As a result, the facial fibers are bent dorsally around the abducens nucleus and form the inner knee of the facial nerve—genu nervi facialis (see Figs. 8.13 and 8.15).

With the formation of the cranial nerve nuclei, the proliferation activity of the ventricular zone of the fourth ventricle is completed. All other neurons, that are still formed for the remaining structures of the rhombencephalon, originate mainly from the subventricular zone at the interface of the alar plate and the roof plate, the rhombic lip (Fig. 2.6A). The rhombic lip forms the dorsal part of the wing plate at the transition to the cover plate over the entire extent of myelencephalon and metencephalon. It develops a strong proliferative activity during several months. All post-mitotic cells of the rhombic lip migrate to other areas during this period. The migration takes place either in the marginal zone (superficial route) or in the intermediate zone (deep route). The neurons, which take the superficial route, form the pontobulbar body from which the complex of the inferior olive (see Sect. 4.1.1.3), the arcuate (see Sect. 4.1.1.4), and the nuclei at the base of the pons emerge (Wingate 2001; Moore et al. 2020). The neurons, which migrate in the intermediate zone, form the accessory olivary nuclei and medial parts of the inferior olive. After the sixth month, the rhombic lip disappears completely (Benninghoff and Drenckhahn 2004).

In the rostral part of the rhombencephalon, there is the isthmus rhombencephali and the nucleus of trochlear nerve (see Sect. 14.5.2.1). According to recent findings, the isthmus is the rostral-most part of the rhombencephalon (see below).

2.2.2 Metencephalon

The metencephalon forms the pons and cerebellum. The ventricle of the metencephalon forms the superior part of the fourth ventricle (Fig. 2.3E).

2.2.2.1 Pons

Similar to the rostral part of the myelencephalon, the pontine flexure causes divergence of the lateral walls of the pons so that the gray matter constructs the floor of the fourth ventricle. This way the three efferent qualities described above (see Fig. 2.5) come to the same relative position in the rostral myelencephalon: visceral and somatic afferent nuclei develop dorsolaterally of the efferent nuclei giving rise to the pontine nuclei, cochlear and vestibular nuclei, and the sensory nuclei of the trigeminal nerve (see Sect. 12.2.3).

2.2.2.2 Cerebellum

The dorsal parts of the alar plates, however, constitute the future cerebellum. At the anterolateral edges of the roof of the metencephalon, the rhomboid lips (Fig. 2.6A), which are thickened inwards and outwards to the same extent, become visible. From here, the cerebellar swellings project first into the fourth ventricle. By a progressive enlargement, these swellings fuse in the mid-sagittal plane dorsal of the fourth ventricle and later overlap the pons and medulla to represent the uniform structure of the cerebellum. Through the growth of the cerebellum and through further sinking of the pontine flexure, a continuous, transverse cerebellar plate is formed. The vermis develops from its narrow middle part, the hemispheres develop from the lateral parts (Fig. 2.6A). During this developmental phase, there is the so-called extra- and intraventricular cerebellum. The extraventricular cerebellum grows outwards, i.e., dorsally and laterally, the intraventricular part grows into the lumen of the fourth ventricle. However, the extraventricular part becomes larger than the intraventricular part (cerebellar eversion). Toward the end of the third month, the largest part of the cerebellum is extraventricular (Fig. 2.6A).

The hemispheres develop very rapidly, while the middle sector remains behind at first, but catches up from the third month onwards. From the 12th week onwards, transverse fissures develop. They first become visible on the vermis and

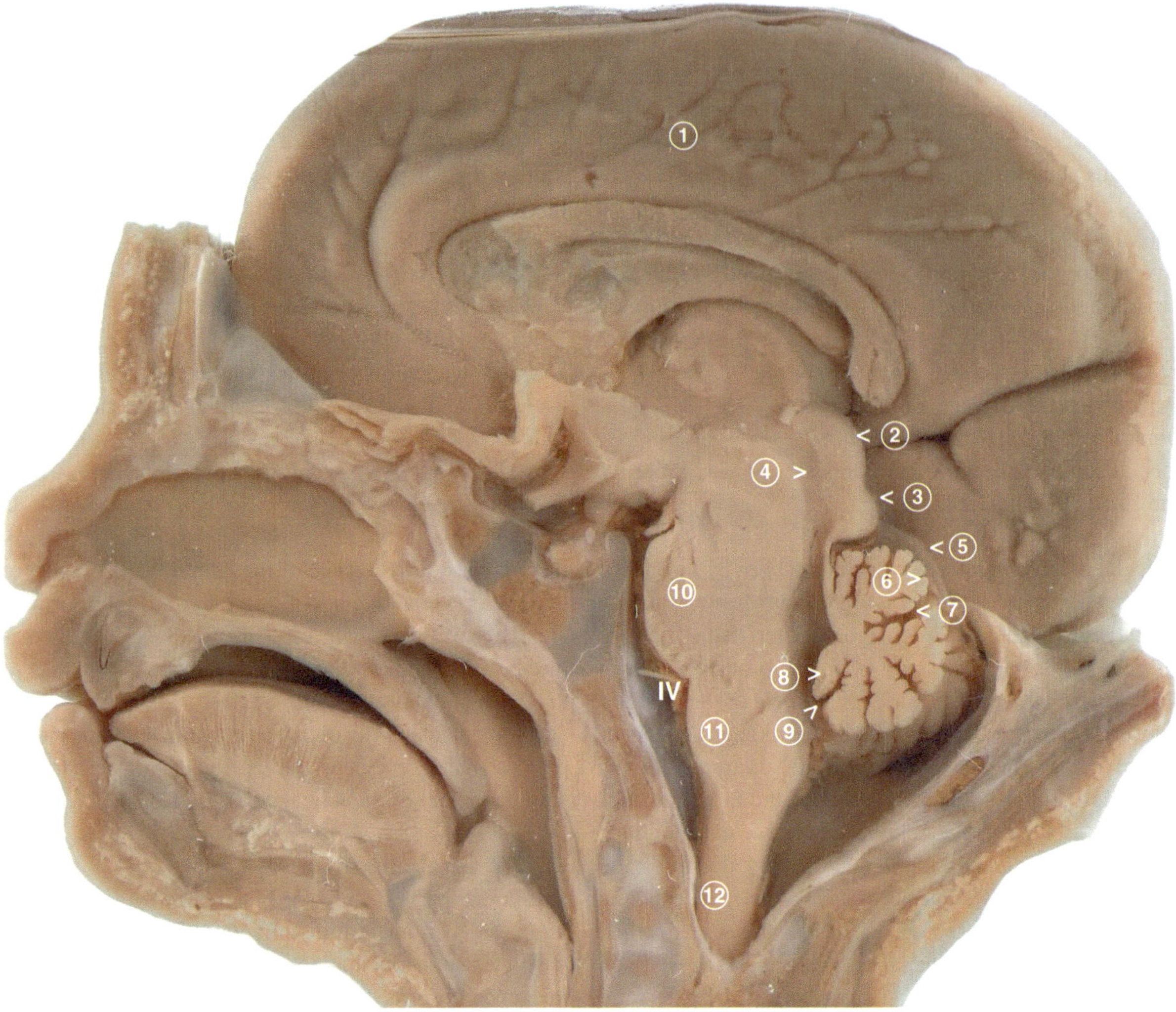

①	Cerebral cortex	⑦	Fissura prima
②	Superior colliculus	⑧	Nodule
③	Inferior colliculus	⑨	Fissura secunda
④	Mesencephalic aqueduct	⑩	Pons
⑤	Cerebellar tentorium	⑪	Medulla oblongata
⑥	Foliae vermis	⑫	Spinal cord
		IV	4th ventricle

Fig. 2.7 Left view of a mid-sagittal section of the CNS of a human fetus, fourth to fifth month. Sammlung des Zentrums Anatomie der Universität zu Köln. Photograph by MedizinFotoKöln, Cologne, Germany

then spread to the hemispheres (see Figs. 1.4, 1.5, and 1.6A). The first to develop is the posterolateral fissure, which separates the Flocculonodular lobe from the rest of the cerebellum. This is followed by the primary fissure [Fissura prima] (Fig. 2.6), which divides the cerebellum into an anterior and posterior lobe, and the secondary fissure [Fissura secunda], which separates the uvula from the pyramid (Fig. 2.7). Afterwards, numerous other fissures appear of which only the horizontal fissure should be emphasized because it is particularly well visible. After the fissures, the numerous foliae form (Fig. 2.7). The fourth ventricle ventral of the cerebellum is covered by the velum medullare superius and inferius (see Fig. 1.5).

In the early stages of cerebellar development, the paired box transcription factor Pax6 seems to play a crucial role (Moore et al. 2020). At the interface between the dorsal alar plate and the roof plate, the embryonic rhombic lip develops (cf. Figs. 2.5 and 2.6A) which is thought to give rise of the cerebellar granule cells—by far the most numerous neurons in the cerebellar cortex. More specifically, according to Watson et al. (2019) the cerebellar cortex is an outgrowth of the dorsal most alar plate of the caudal isthmus and the first rhombomere r1. It is therefore part of the prepontine hindbrain, contradicting the old assumption that it forms a developmental unit with the pons. The vermis of the cerebellum is derived mainly from the rhombic lip of the isthmic alar plate, and the hemisphere of the cerebellum is mainly derived from the rhombic lip of the r1 alar plate. The Purkinje cells and the (deep) cerebellar nuclei, however, arise from the ventricular zone of the dorsal metencephalic alar plate (ten Donkelaar 2011).

The structure of the cerebellar cortex reflects its phylogenetic development as follows:

- The phylogenetically oldest part is the archicerebellum (flocculonodular lobe). It has mainly connections with the vestibular nuclei.
- The paleocerebellum (vermis and anterior lobe), of more recent development in mammals, is associated with sensory data from the limbs. Deficiencies of the archicerebellum, which gets mainly afferents from the vestibular system (see Sect. 5.3.1), and the paleocerebellum, which gets proprioceptive information, result in ataxic disorders.
- The phylogenetically newest part is the neocerebellum (posterior lobe), which mainly controls limb movements. This part additionally gets afferents from the motor cortex and the "extrapyramidal system" so that patients with posterior lobe failures show intentional tremor and rigor.

Nerve fibers connecting the cerebral and cerebellar cortices with the spinal cord pass through the marginal layer of the ventral region of the metencephalon. This region of the brainstem is the pons (Latin for "bridge") because of the robust band of nerve fibers that crosses the median plane and forms a bulky ridge on its anterior and lateral aspects. This is a prominent feature characteristic for humans (see Fig. 1.5).

2.3 Mesencephalon

In the midbrain (mesencephalon), the neural canal narrows and becomes the cerebral (mesencephalic) aqueduct (Fig. 2.7), the channel that connects the third and fourth ventricles (see Fig. 1.5). Neuroblasts migrate from the alar plates of the midbrain into the mesencephalic tectum (Figs. 2.6A and 1.12) dorsal to the aqueduct and aggregate to form four large groups of neurons, the paired superior and inferior colliculi (Fig. 2.7). Neuroblasts from the basal plates may give rise to groups of neurons in the tegmentum of the midbrain (red nuclei, nuclei of third and fourth cranial nerves, and reticular nuclei; see below) (see Fig. 1.12). The substantia nigra (see Sect. 16.6.1) may also differentiate from the basal plate (Puelles et al. 2019). Fibers growing from the forebrain (including the diencephalon and cerebral hemispheres) form the crura cerebri (cerebral peduncles; sing. crus cerebri) (see Figs. 1.9, 1.10, and 1.11) anteriorly. The peduncles become progressively more prominent as more descending fiber groups (corticopontine, corticobulbar, and corticospinal) pass through the developing midbrain on their way to the brainstem and spinal cord (Standring 2016).

2.4 Segmental development of the brainstem

The midbrain is composed of two embryonic neuromeres known as the mesomeres and the hindbrain is composed of 12 rhombomeres (isthmus and r1 to r11; Puelles et al. 2013; see above). The borders of the brain's constituents are defined by gene expression patterns and their boundaries, respectively. The boundary between diencephalon and midbrain is marked by the caudal boundary of Pax6 expression and the boundary between midbrain and hindbrain is defined by the expression patterns of the homeobox genes Otx2 and Gbx2 (Puelles et al. 2013). The midbrain proper can be subdivided into a thick rostral mesomere m1 and a thin caudal mesomere m2 (Fig. 2.8). The latter, characterized by the expression of Otx2 and Pax2, separates its future colliculus inferior from the isthmus.

According to Puelles et al. (2013), the developing hindbrain can be divided into three parts, a rostral part (isthmus and r1), a central part (r2 to r6), and a caudal part (r7 to r11). The rostral and caudal parts are not overtly segmented but by their genes. Puelles and colleagues (Puelles et al. 2013, 2019; Watson et al. 2019) expand the classical segmentation of eight rhombomeres (see, e.g., ten Donkelaar 2011, Standring 2016) where the rostral and caudal parts cf. Puelles et al. (2013) are one rhombomere each.

According to Watson et al. (2019), the cerebellum is a derivative of the rhombomere r1 organized by expressions of various genes (Fgf8, Gbx2) of the isthmus. The pons, in turn, is a derivative of the following rhombomeres r3 and r4. This means that, in the human, the pons hides much of the rostral "prepontine" hindbrain (isthmus and r1, r2) and the retropontine hindbrain of r5 and r6 (Watson et al. 2019). Starting from r3/r4 the crossed pontocerebellar fibers of the middle cerebellar peduncle stretch via r2 into r1 to enter the cerebellum. Following this developmental path, the fibers encircle the trigeminal root. The trapezoid body and neighboring trapezoid and superior olivary nuclei (see Sect. 8.1.1.5) as well as the nucleus abducens are embryologically retropontine and asso-

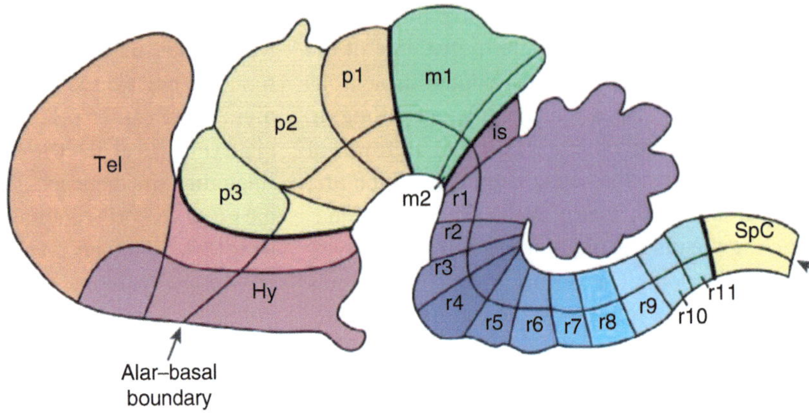

Fig. 2.8 Diagram of lateral view of the developing mouse brain. The telencephalon (Tel) and hypothalamus (Hy) form the forebrain. The prosomeres p1 to p3 develop into the later diencephalon, the mesomeres m1 and m2 form the midbrain and the 12 rhombomeres the hindbrain (isthmus (is) and rhombomeres r1 to r11). *SpC* spinal cord. Modified after Puelles et al. 2013, Fig. 1 with permission

ciated with r5. The pyramidal decussation lies in r11 (Puelles et al. 2013) and in the cranial medulla spinalis, respectively, as mentioned in recent papers (Watson et al. 2019).

The segmental developmental patterns described above are further altered by cellular migrations. The interpeduncular nucleus (see Sect. 12.3.3) is formed by migratory cells from diverse alar and basal areas at the isthmic and r1 midline. Pontine neurons migrate from the caudal rhombic lip (r6, r7) to the ventral surface of r3 and r4 to form the pontine nuclei (Puelles et al. 2013). Cells of the caudal rhombic lip (r8 to r11) migrate to the lateral reticular nucleus (see Sect. 4.1.1.1), external cuneate nucleus (see Sect. 5.2.1.1), and into the inferior olive (see Sect. 4.1.1.3).

On the other hand, some structures classically (as in this book) included into the midbrain are embryologically of diencephalic origin: the nucleus of Darkschewitsch (see Sect. 16.4.1), the interstitial nucleus of Cajal (see Sect. 16.4.2), the parvocellular red nucleus (see Sect. 16.5.1), the pre-Edinger-Westphal nucleus (see Sect. 16.4.3), the subcommissural organ (see Box 5.1), the posterior commissure and its related nuclei, and the medial terminal nucleus of the accessory optic tract (Watson et al. 2019). Moreover, the classical posterior pretectal nucleus is part of the rostral midbrain (m1) while only the caudal most parts of the substantia nigra and of the ventral tegmental area are of midbrain origin.

2.5 Development of the cranial nerves of the brainstem

During development, the brainstem is organized highly segmentally and so are the cranial nerves (see Table 1.1). However, functionally, they can be divided and simplified into motor nerves and sensible nerves although most nerves of the brainstem sum both qualities.

2.6 Nuclei of motor cranial nerves

The nucleus of the oculomotor nerve (CNIII, see Sect. 15.3.1) emerges from the somatomotor column of the mesencephalon. Within the oculomotorius nerve, the parasympathetic fibers for the pupillary reflex run to the ciliary ganglion. They originate in the visceral motor column in the Edinger-Westphal nucleus (see Sect. 15.3.1.2). According to recent findings, the nucleus of the trochlear nerve (see Sect. 14.5.2) does not belong to the mesencephalon but originates from the base plate of the isthmus (Watson et al. 2019). The exit point of the trochlear nerve (CNIV) is shifted to the dorsal side of the brainstem because fiber bundles run from its nucleus in the isthmus dorsally. The boundary between mesencephalon and rhombencephalon (isthmus) must therefore be placed between the nucleus of the oculomotor nerve and the nucleus of trochlear nerve.

The nucleus of abducens nerve (CNVI, see Sect. 9.2.1.1) is located in rhombomere r5 (cf. Fig. 2.8). Due to the enlarged pons of the human brain, the abducens nerve has a long course within the brainstem to exit caudally of the pons.

The branchial motor column for the muscles of the pharyngeal arches can be divided into the motor nucleus of the mandibular nerve (V_3, masticatory muscles, see Sect. 11.3.1) and the nucleus of the facial nerve (CNVII; mimic muscles, see Sect. 8.2.1). Since the mammalian facial motor nucleus migrates from the medial part of rhombomere r4 to its ventral (retropontine) position in r6, the facial nerve fibers—following this migratory course—run around the nucleus of the abducens nerve, forming its inner knee, before they exit the brainstem (see Figs. 8.13 and 8.15).

The branchial motor areas of the glossopharyngeal nerve (CNIX), vagus nerve (CNX) and accessory nerve (CNXI) for the striated pharyngeal muscles unite with the visceral motor areas of these nerves for the smooth pharyngeal muscles to form the nucleus ambiguus (see Sect. 5.4.2.1) in the rhombomeres r7 to r10. In front of those lies the motor nucleus of the facial nerve and, more dorsally, the salivatory nuclei (see Sect. 7.3.1.1). The fibers for the hypoglossal nerve (CNXII) (see Sect. 4.3.1.1) originate from the ventral somatomotor column.

2.7 Nuclei of sensory cranial nerves

The long sensory nucleus of the trigeminal nerve (CNV), extending from the mesencephalon to the myelencephalon, lies in the somatosensory column of the alar plate (see Sects. 3.2.2 and 12.2.3).

The nuclei from the branchiosensory column for the chorda tympani (taste fibers whose pseudounipolar ganglion cells are located in the ganglion geniculi of the facial nerve), for the glossopharyngeal and the vagus nerves unite with the parasympathetic viscerosensory nuclei of the glossopharyngeal and vagus nerves to form the solitary nucleus (see Sect. 3.2.4).

2.8 Cranial nerves of the pharyngeal arches

A peculiarity during the development of the head and neck region is the fact, that here a lot of structures, such as the musculature of mastication and facial expression, are not derived from somites (see Box 3.14) but from pharyngeal arches. These pharyngeal arches have their evolutionary origin in the gill apparatus of fishes, which is why these pharyngeal arches are also called branchial arches. The original function of the gills was lost in terrestrial vertebrates but the tissues of the arches function as anlages of species-specific structures of the visceral skeleton and the pharynx. Thus, they develop, for example, to the upper and lower jaws, the middle ear, the hyoid apparatus, and the larynx.

One can distinguish five pharyngeal arches during development and each arch is associated with a specific cranial nerve (Fig. 2.9), which arise from the adjacent hindbrain (Table 2.1). The fifth and sixth pharyngeal arches are often described as one structure since the fifth pharyngeal arch is often only rudimentary or not formed at all. Initially, the motor roots extend from the basal plate of the hindbrain to innervate the striated muscle of the arches and convey special visceral efferent axons to innervate pharyngeal musculature. Sensory nerves extend peripherally from cranial ganglia that are formed partly from neural crest and partly from cells that delaminate from epipharyngeal placodes (Standring 2016). They form general and special somatic afferent nerves. Accordingly, the nerves of the pharyngeal arches are mixed nerves. Additionally, each nerve of the first three arches has a sensory branch that innervates the arch rostral to it. This nerve is called the pretrematic branch because it extends rostral to the cleft (ancient Greek: τό τρῆμα—to trêma, "hole") between the two arches. Therefore, the nerves to the mandibular arch include the mandibular division of the trigeminal nerve (see Sect. 12.2.3.2.3) and, additionally, the chorda tympani (see Sect. 3.2.4.1.2) and greater petrosal nerve (see Box 8.5) which are branches of the facial nerve. The maxillary branch of the trigeminal (see Sect. 12.2.3.2.3) and the tympanic branch of the glossopharyngeal (see Sect. 5.4.2.1.3) are also considered to be pretrematic nerves. The ophthalmic branch of the trigeminal nerve (see Sect. 12.2.3.2.3), which supplies the frontonasal area, is not an arch nerve because, in fishes, its ganglion is separate from the first arch nerve ganglion (Standring 2016). The vagus nerve (see Sect. 7.3.2.1.2) supplies the fourth and sixth arches (Fig. 2.9). The recurrent laryngeal branch initially loops under the sixth arch artery on both sides. However, since the right aortic arch lateral to the right pulmonary artery degenerates during early development, the recurrent laryngeal nerve stretches under the fourth arch-derived subclavian artery while, on the left hand side, the nerve loops under the aortic arch at the ductus arteriosus.

2.9 Development of the meninges

The meninges develop from cells of the neural crest and mesenchyme in embryos between 20 and 35 days. The cells migrate to surround the neural tube and form the primordial meninges. Mesenchymal cells of the external layer thicken to form the dura mater (Waxman 2017). An internal cell layer around the neural tube forms the layer of the leptomeninx that later condenses into the pia mater and the arachnoid mater. Fluid-filled spaces appear within the leptomeninges that coalesce to form the subarachnoid / leptomeningeal space. The origin of the pia mater and arachnoid from a single leptomeningeal layer is indicated in the adult by arachnoid trabeculae, which are numerous, delicate strands of connective tissue that pass between the pia and arachnoid (Moore et al. 2020). Cerebrospinal fluid (CSF) begins to form during the fifth week.

2.10 Rhombencephalic choroid plexus and cerebrospinal fluid (CSF)

The thin rhombencephalic roof of the fourth ventricle is covered externally by pia mater. This vascular membrane, together with the ependymal roof, forms the tela choroidea, the sheet of pia covering the lower part of the fourth ventricle. The pontine flexure separating the metencephalon and the myelencephalon results in a transverse fold in the rhomb-

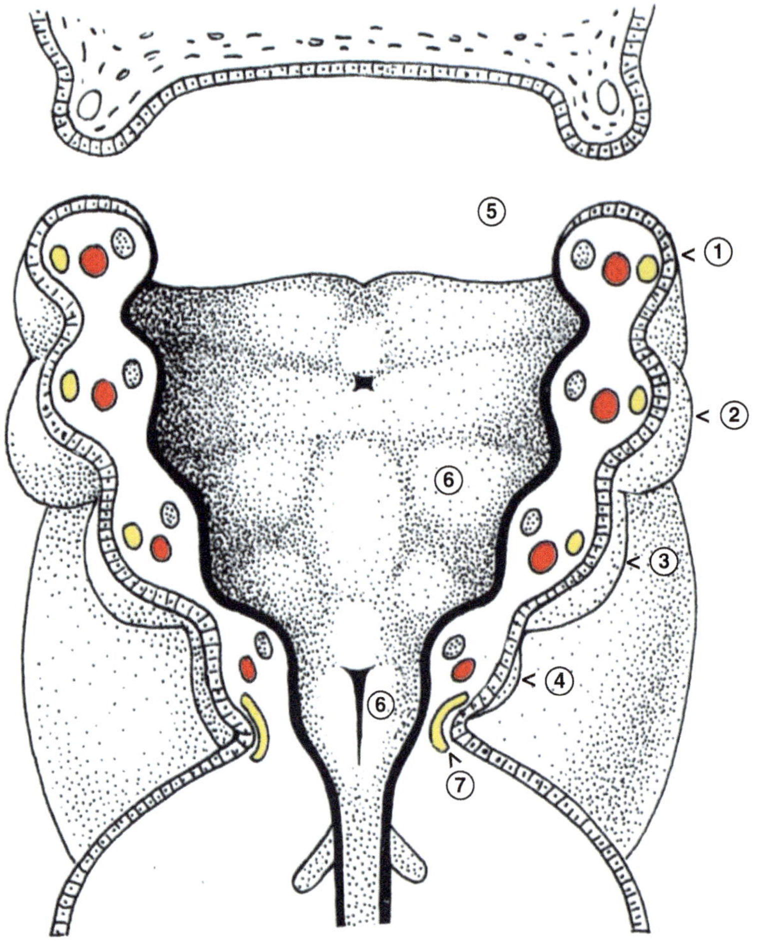

①	Mandibular pharyngeal arch	⑤	Stomodeum (mouth)
②	Hyoid pharyngeal arch	⑥	Pharynx
③	Third pharyngeal arch	⑦	Vagus nerve CNX
④	Fourth pharyngeal arch		

Fig. 2.9 Drawing of a horizontal section of a human embryo through the pharynx in dorsal view. Note that this embryo resembles approximately Carnegie stage 13 (28–32 days) when the sixth pharyngeal arch is not developed yet. The nerves are marked in yellow, the arch arteries in red, compare Table 2.1. Ribbers fecit

encephalic roof, the plica choroidea. This plica, together with choroidal arteries of the pia, invaginates in the fourth ventricle forming the choroid plexus (Larsen 2003; Garel et al. 2011). The choroid plexuses, which additionally develop in the third and lateral ventricles, secrete most of the CSF. Various signaling morphogens are found in CSF and the choroid plexus that are necessary for brain development (Moore et al. 2020).

The thin roof of the fourth ventricle evaginates in three locations. These outpouchings rupture to form openings, one median and lateral apertures (foramen of Magendie and foramina of Luschka, see Fig. 3.9, aperturae medialis et laterales ventriculi quarti, respectively). Here, the CSF enters the subarachnoid space from the fourth ventricle.

The epithelial lining of the choroid plexus is derived from neuroepithelium, whereas the stroma develops from mesenchymal cells. The CSF is thought to be mainly absorbed into the venous system through the arachnoid villi, which are protrusions of arachnoid mater into the dural venous sinuses. The arachnoid villi consist of a thin cellular layer derived from the epithelium of the arachnoid and the endothelium of the sinus.

Table 2.1 Structures and derivatives of pharyngeal arches and their cranial nerves (modified after Standring 2016)

Arch number and name	Nerve	Derivatives (examples)
1. Mandibular arch	Trigeminal nerve (CNV), mandibular division (V_3)	Incus, malleus, tensor tympani muscle, mastication muscles, mylohyoid muscle, anterior belly of digastric muscle, tensor veli palatini muscle
2. Hyoid arch	Facial nerve (CNVII)	Stapes, stapedius muscle, stapedial artery, styloid process of temporal bone, facial muscles
3. Third arch	Glossopharyngeal nerve (CNIX)	Main part of hyoid bone, stylopharyngeus muscle, common carotid artery
4. Fourth arch	Vagus nerve (CNX), pharyngeal branch	Thyroid, corniculate, cuneiform cartilages, pharyngeal and extrinsic laryngeal muscles, levator veli palatini muscle, proximal part of subclavian artery on the right, arch of aorta from left common carotid to subclavian arteries
5. Sixth arch	Vagus nerve (CNX), recurrent laryngeal branch	Arytenoid cartilages, intrinsic laryngeal muscles

2.11 The early postnatal fate of the brainstem

There are only few studies so far describing the early normal postnatal development of the brainstem which seems to be in contrast to the scientific focus on the development of the forebrain (see, e.g., Gao et al. 2017). Regarding the early postnatal development of the brainstem, the myelinization is of special interest. In contrast to most areas of the telencephalon and diencephalon, the brainstem shows an early myelination pattern at birth or shortly after birth (Brody et al. 1987).

The superior, middle, and inferior cerebellar peduncles are among those white matter tracts that appear early during brain development. Using diffusion tensor imaging (DTI), the cerebellar peduncles as well as the corticospinal tract could be traced in the fetal brain not before the 14th week. All other major brainstem fibers could be reproducibly traced by DTI in fetal brains from the 37th week (Ouyang et al. 2015).

Among the brainstem structures that are not myelinated at birth are cerebellar white matter, pontocerebellar fibers, the middle cerebellar peduncle, the brachium of the inferior colliculus, the corticospinal tracts, and the crura cerebri. However, in most fiber tracts, the degree of myelination becomes higher during the first 4–7 months of life, which can be seen also in MRI studies for the crura cerebri. However, some fiber tracts, such as the solitary and the central tegmental tracts, do not fully myelinate until 2 years of age (Brody et al. 1987) or later. The crura cerebri, the red nucleus, and the substantia nigra show a mature MRI signal starting several years after birth (Martin et al. 1991).

References

Benninghoff A, Drenckhahn D (2004) Anatomie—Makroskopische Anatomie, Histologie, Embryologie, Zellbiologie, 16. Aufl. Elsevier Urban & Fischer, München

Brody BA, Kinney HC, Kloman AS, Gilles FH (1987) Sequence of central nervous system myelination in human infancy. I. An autopsy study of myelination. J Neuropathol Exp Neurol 46(3):283–301

Cozzi B, Huggenberger S, Oelschläger HHA (2017) The anatomy of dolphins—insights into body structure and function. San Diego: Elsevier/Academic Press. 438pp

Gao W, Lin W, Grewen K (2017) Functional connectivity of the infant human brain: plastic and modifiable. Neuroscientist 23(2):169–184

Garel C, Fallet-Bianco C, Guibaud L (2011) The fetal cerebellum: development and common malformations. J Child Neurol 26(12):1483–1492

Huggenberger S, Moser N, Schröder H, Cozzi B, Granato A, Merighi A (2019) Neuroanatomie des Menschen. Springer, Heidelberg. isbn:978-3-662-56460-8

Larsen W (2003) Brain and cranial nerves development. In: Larsen W (ed) Human embryology. Bruxelles, De Boeck & Larcier, pp 419–464

Martin E, Krassnitzer S, Kaelin P, Boesch C (1991) MR imaging of the brainstem: normal postnatal development. Neuroradiology 33:391–395

Moore KL, Persaud TVN, Torchia MG (2020) The developing human—clinical oriented embryology. Elsevier Health Sciences, Edinburgh

Ouyang A, Jeon T, Sunkin SM, Pletikos M, Sedmak G, Sestan N, Lein ES, Huang H (2015) Spatial mapping of structural and connectional imaging data for the developing human brain with diffusion tensor imaging. Methods 73:27–37

Puelles L, Rubenstein LR (2015) A new scenario of hypothalamic organization: rationale of new hypotheses introduced in the updated prosomeric model. Front Neuroanat 19(9):27

Puelles L, Harrison M, Paxinos G, Watson C (2013) A developmental ontology for the mammalian brain using the prosomere model. Trends Neurosci 36(10):570–578

Puelles L, Tvrdik P, Martínez-de-la-Torre M (2019) The postmigratory alar topography of visceral cranial nerve efferents challenges the classical model of hindbrain columns. Anat Rec (Hoboken) 302:485–504

Standring S (2016) Gray's anatomy—the anatomical basis of clinical practice. 41st edn. Elsevier Health Sciences

ten Donkelaar HJ (2011) Clinical neuroanatomy—brain circuitry and its disorders. Springer, Heidelberg

Watson C, Bartholomaeus C, Puelles L (2019) Time for radical changes in brain stem nomenclature—applying the lessons from developmental gene patterns. Front Neuroanat 13:10

Waxman SG (2017) Clinical neuroanatomy, 28th edn. McGraw-Hill Education, New York

Wingate RJT (2001) The rhombic lip and early cerebellar development. Curr Opin Neurobiol 11:82–88

Rhombomere 11 r11

3

Contents

H. Schröder et al., *The Human Brainstem*, https://doi.org/10.1007/978-3-030-89980-6_3

Abstract

Starting with this chapter, the description of the neuromeres (in caudocranial order rhombomeres 11-1, isthmus, m1/m2) follows. Rhombomere 11 is the neuromere that directly connects to the spinal cord. This is in particular obvious with regard to the protopathic spinal trigeminal nucleus and tract as a continuation of the sensory dorsal horn of the spinal cord. The epicritic sensory dorsal column nuclei provide the second-order neurons the axons of which traverse the whole brainstem up to the thalamus in the medial lemniscus. In addition to the trigeminal, the fifth cranial nerve (CNV), the vagus (with the solitary nuclei), the tenth cranial nerve (CNX) and the accessory nerve, the eleventh cranial nerve (CNXI) are represented at this level. The monoaminergic nuclei which are distributed throughout the brainstem are present here as the A2 noradrenaline and C1 adrenaline cell groups. Among the reticular nuclei at this level, the rostral ventral respiratory group is of utmost importance for the regulation of breathing.

With regard to brainstem pathology, this chapter will deal with multiple system atrophy (MSA), an adult-onset, sporadic, progressive, neurodegenerative disorder affecting both men and women. The main features include autonomic failure, parkinsonism, cerebellar ataxia, and pyramidal signs in any combination. MSA has been included here because of the affection of the rostral ventral respiratory group.

3.1 Rhombic lip r11 (Rhombic: ὁ ῥόμβος [ho rhombos] Greek = rhomboid)

The rhombic lip comprises the interface between dorsal neuroepithelium and the roof plate during development (for details, see Sects. 2.2.1, 2.3, and 2.4).

3.1.1 Precerebellar r11

Precerebellar nuclei r11 (Pre- prae- Latin = prior to, here upstream to the cerebellum) [Nuclei precerebellares]

The term precerebellar nuclei refers to a mixed group of hindbrain and spinal nuclei which all project to the cerebellum. In developmental terms, neurons of the major hindbrain precerebellar nuclei are derived from the rhombic lip (see Sects. 2.2.1, 2.3, and 2.4). (For details, see Rodriguez and Dymecki 2000).

3.1.1.1 Intercalated nucleus of the medulla/ Intercalated nucleus r11

[Ncl. intercalatus] (Intercalatus [Latin] = lying in between)

The intercalated nucleus (belonging to the perihypoglossal nuclei, see Box 3.1) is bordered medially by the hypoglossal nucleus (see Sect. 4.3.1.1) and laterally by the posterior nucleus of the vagus nerve (see Sects. 3.2.4 and 7.3.2.1) (see Fig. 3.1).

Box 3.1 Perihypoglossal nuclei

The perihypoglossal nuclei are a group of nuclei in the lower medulla oblongata close to the hypoglossal nucleus (see Sect. 4.3.1.1) involved in the control of eye movements (Kheradmand et al. 2016), comprising the prepositus hypoglossi nucleus (see Sect. 6.2.1.5), the nucleus of Roller (subhypoglossal nucleus) (see Sect. 6.2.1.6), and the intercalated nucleus (see Sect. 3.1.1.1).

Direct anatomical evidence on the connectivity and function of these nuclei is not available for humans. Tracing studies in the macaque monkey (Brodal and Brodal 1983) have shown that these nuclei belong to the precerebellar nuclei, i.e., a group of brainstem nuclei projecting to the cerebellum (see Sect. 3.1.1).

From clinical studies it is known that the perihypoglossal nuclei are involved in gaze holding with lesions of the prepositus nucleus mainly evoking downbeat gaze-evoked nystagmus (see Box 3.2) and lesions of the intercalated nucleus and the nucleus of Roller causing upbeat gaze-evoked nystagmus (Kheradmand et al. 2016).

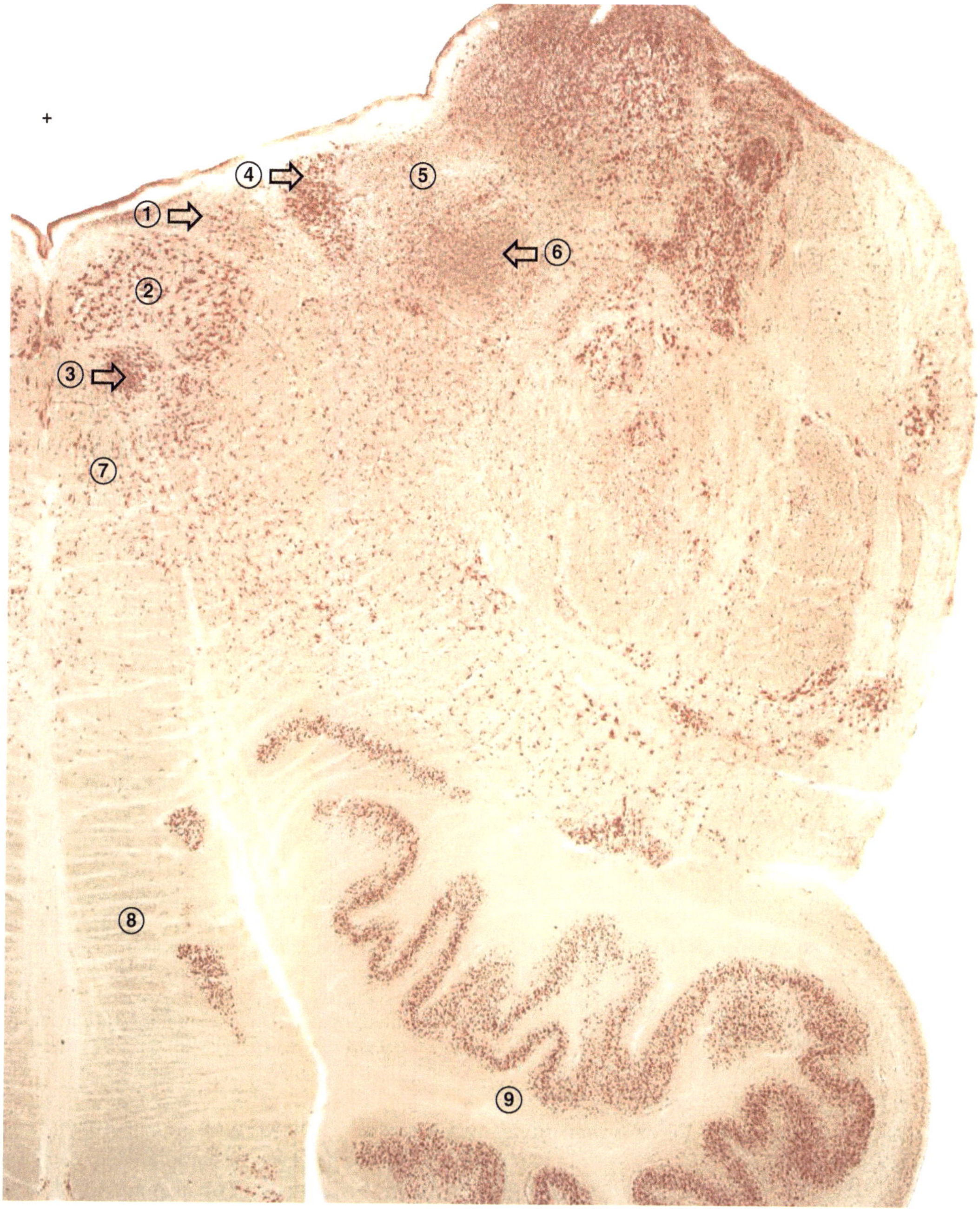

①	**Intercalated ncl.**	⑥	**Solitary tract**
②	**Ncl. of hypoglossal nerve CNXII**	⑦	**Medial longitudinal fasciculus**
③	**Subhypoglossal ncl. (Roller)**	⑧	**Medial lemniscus**
④	**Posterior ncl. of vagus nerve CNX**	⑨	**Inferior olivary complex**
⑤	**Ncll. of solitary tract**	+	**4th ventricle**

Fig. 3.1 Horizontal section in Darrow red staining through the medulla oblongata (open part of fourth ventricle, cranial of the central canal, see Fig. 3.13A) with two perihypoglossal nuclei (Box 3.1). The intercalated nucleus ① contains medium-sized neurons and is bordered medially by the hypoglossal nucleus ② (see Sect. 4.3.1.1) and dorsolaterally by the posterior nucleus of vagus nerve ④ (see Sect. 7.3.2.1). See also atlas part Darrow red 8, 10A. LabPON Twente

Box 3.2 Gaze-evoked nystagmus

Gaze-holding mechanisms mediated by brainstem nuclei and the cerebellum keep the eyes steady when the head is not in motion. Abnormal spontaneous eye movements, such as nystagmus (ὁ νυσταγμός, ho nystagmos Greek = drowsiness), are threats to steady fixation and clear vision (Shaikh 2019). In gaze-evoked nystagmus, alternating drifts and corrective movements and their gaze position are the characteristic clinical phenomena seen (Shaikh 2019).

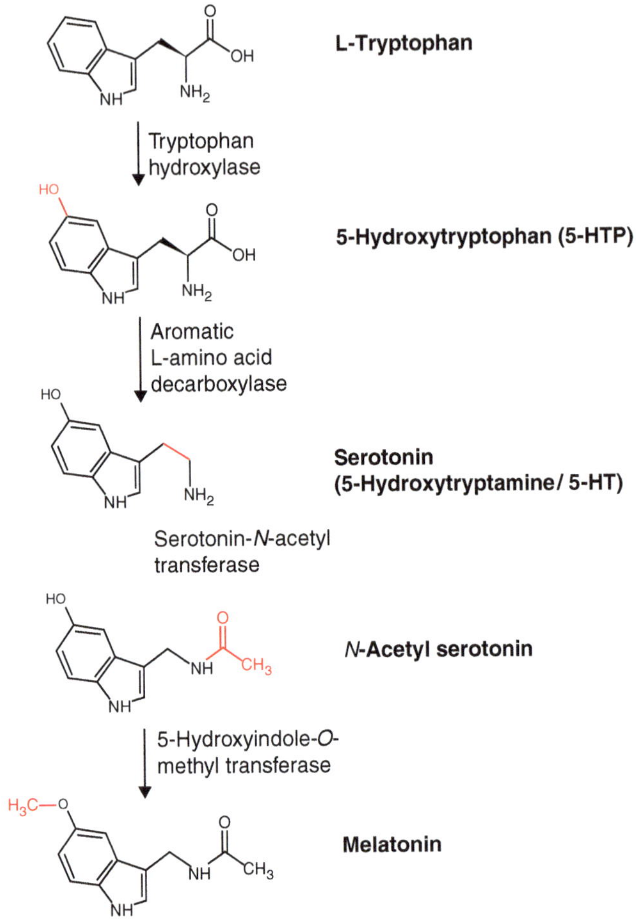

Fig. 3.2 Schematic representation of serotonin biosynthesis. The essential amino acid L-tryptophan is converted by the enzyme tryptophan hydroxylase to 5-hydroxytryptophan. The aromatic amino acid decarboxylase catalyzes the decarboxylation of 5-hydroxytryptophan to 5-hydroxytryptamine or serotonin. Two more conversion steps result in the synthesis of melatonin**,** the hormone released into the circulation in the pineal gland of the epithalamus. Sophia Schröder fecit

3.2 Alar r11

3.2.1 Monoamine nuclei r11

Monoamines are amines (organic derivatives of ammonia, formed by decarboxylation of amino acids) containing one amino group. They play an important role as neurotransmitters like serotonin (5-HT) (see Fig. 3.2), the catecholamines dopamine (DA), noradrenaline (NA) (norepinephrine), and adrenaline (A) (epinephrine) (see Fig. 3.3).

In experimental animals, the histochemical evaluation of the serotonergic raphe nuclei started with the introduction of the formaldehyde vapor technique. This rather simply applicable technique is based on the property of formaldehyde to form fluorescent derivatives of biogenic amines. Based on the different colors of the fluorescent compounds, serotonin with its yellow-orange color can be clearly differentiated from the green fluorescence emitted by the catecholamine group (DA, NA, A). Using this approach, Dahlström and Fuxe (1964) mapped the monoaminergic cell groups of the brainstem.

According to the putative transmitters contained, rat brain monoamine nuclei have been classified as follows (Dahlström and Fuxe 1964; Hökfelt et al. 1984):

A	Noradrenaline (NA)	A1–A12 (caudal to rostral)
B	Serotonin (5-HT)	B1–B9 (caudal to rostral)
C	Adrenaline (A)	C1–C4 (caudal to rostral)

This classification must not be confused with the abbreviations for monoamines (see above) or the abbreviations C1–C8 for cervical spinal.

Today these classical histochemical techniques are supplemented by immunohistochemical techniques which allow for the visualization of serotonin itself or of the synthesizing enzyme tryptophan hydroxylase (see Fig. 3.2). A general marker for catecholaminergic neurons (DA, NA, A) is the synthesizing enzyme tyrosine hydroxylase (TH) (see Fig. 3.3). The presence of noradrenaline can be made probable by the visualization of dopamine-β-hydroxylase (DBH) and that of adrenaline by phenylethanolamine-N-methyltransferase (PNMT).

Anatomical classification of brainstem monoaminergic nuclei

The details of the different monoaminergic brainstem nuclei will be dealt with in the individual rhombomere chapters. Here, only a broad overview of the anatomy and functional implications will be given.

Anatomical classification of brainstem catecholaminergic nuclei

A comprehensive visualization is possible by the use of antibodies against tyrosine hydroxylase (see Fig. 3.3) as

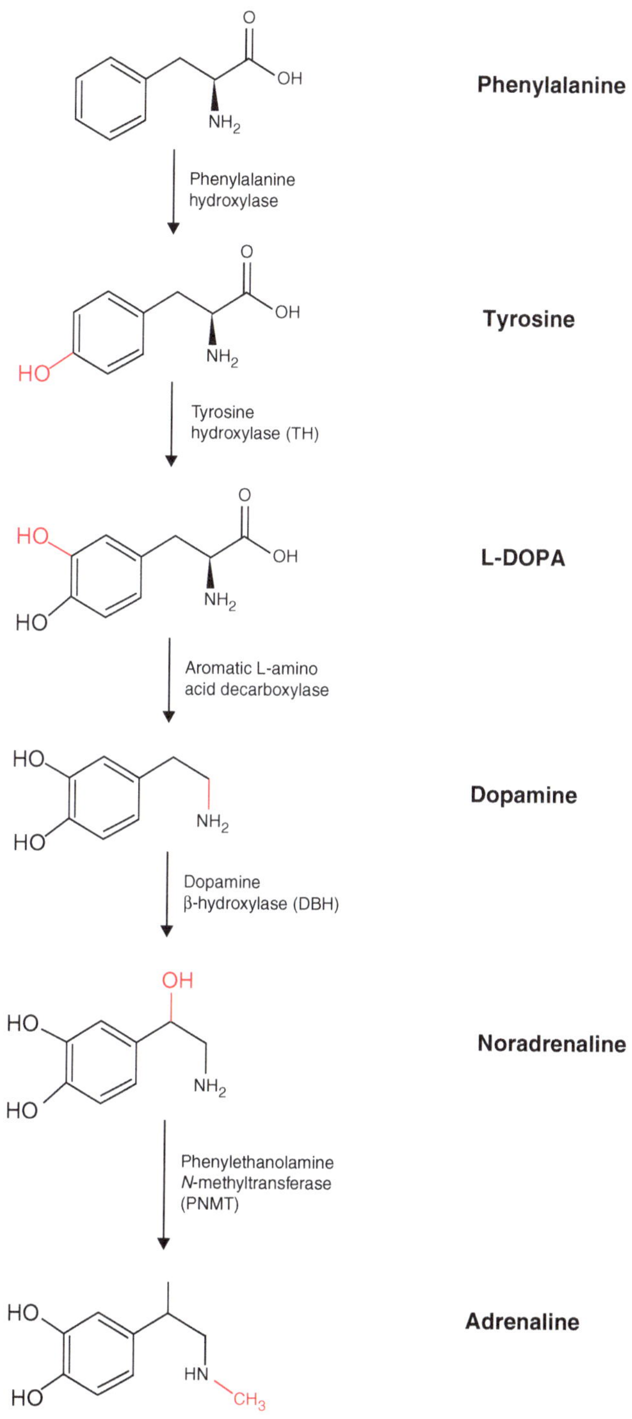

Fig. 3.3 Schematic representation of catecholamine biosynthesis. The pathway starts with the conversion of the essential amino acid phenylalanine to tyrosine by the enzyme phenylalanine hydroxylase. Tyrosine is then converted to L-DOPA (L-3,4-dihydroxyphenylalanine) by tyrosine hydroxylase (TH), an important marker peptide for the immunohistochemical demonstration of catecholaminergic structures. The conversion of L-DOPA to the widespread transmitter dopamine (DA) is catalyzed by the aromatic L-amino acid decarboxylase. Dopamine is the precursor compound for the synthesis of noradrenaline (or norepinephrine) by dopamine β-hydroxylase (DBH) and eventually of adrenaline (or epinephrine) by phenylethanolamine-N-methyltransferase (PNMT). Besides TH (dopamine, L-DOPA), DBH (noradrenaline), and PNMT (adrenaline) can be used as immunocytochemical markers (antibodies) for the differential visualization of catecholamine expression in tissue. Sophia Schröder fecit

immunohistochemical markers (VanderHorst and Ulfhake 2006).

Dopamine:	Dopaminergic cell groups A8, A9, A10, A11
Noradrenaline	Noradrenergic cell groups A1, A2, A4, A5, A6, A7, LC
Adrenaline	Adrenergic cell groups C1, C2

Central nervous catecholamines are involved in a number of behavioral and physiological processes by interacting with a plethora of different postsynaptic targets (see individual cell groups).

The FIPAT inventory Chap. 1 (https://fipat.library.dal.ca/wp-content/uploads/2017/02/FIPAT-TNA-Front-Matter.pdf) differentiates between dopaminergic, (see synthesis in Fig. 3.3), noradrenergic and adrenergic cell groups under the rubrum "neuromodulatory nuclei." It is not clear whether this terminology (see below) is advantageous as compared to the rather simple original classification (see above).

- Adrenergic cells in medulla oblongata (C1, C2)
- Noradrenergic cells in medulla oblongata (A1, A2)
- Caerulean nucleus [Ncl. caeruleus] (A6)
- Noradrenergic cells of superior cerebellar peduncle (A5)
- Noradrenergic cells in caudolateral pons (A4)
- Noradrenergic cells in nucleus of lateral lemniscus (A7)
- Dopaminergic cells in retrorubral area (A8)
- Dopaminergic cells in compact part of substantia nigra (A9)
- Dopaminergic cells in ventral tegmental area (A10)
- Dopaminergic cells of periaqueductal grey (A11)

The dopaminergic cell group with the foremost clinical relevance is A9 (Substantia nigra) in the pathogenesis of Parkinson's disease (see Sect. 16.6.1.5).

Additional catecholaminergic cells (A12, C3, C4) are present in the more rostral parts of the mammalian brain.

3.2.1.1 A2 noradrenaline cells r11

This noradrenergic cell group is contained in the solitary nuclei (see Sect. 3.2.4 further below). From animal studies it is known that this cell group has direct axonal projections to several hypothalamic nuclei like the paraventricular and supraoptic nuclei which contain vasopressin and oxytocin as well as corticotrophin-releasing factor (CRF) (Blessing and Benarroch 2012).

An A1 as well as an A3 noradrenaline cell group is not present in the human brain (Szabadi 2012).

http://braininfo.rprc.washington.edu/centraldirectory.aspx?ID=1799

In primates and rodents, the term A1 cell group refers to a group of cells in the vicinity of the lateral reticular nucleus of the medullary reticular formation that label for noradrenaline.

http://braininfo.rprc.washington.edu/centraldirectory.aspx?ID=1864

3.2.1.2 C1 adrenaline cells r11

Using an antibody to phenylethanolamine-N-methyl-transferase (PNMT) (see Fig. 3.3), Kitahama et al. (1985) could demonstrate immunoreactive neurons forming a nucleus just ventrolateral to the area postrema and in the dorsal part of the nuclei of the solitary tract (NTS) (see below Sect. 3.2.4).

Larger adrenergic neurons were also present in and around the lateral reticular nucleus (see Sect. 4.1.1.1) and in relationship with the posterior nucleus of vagus nerve (see Sect. 7.3.2.1) in regions equivalent to the rat C1 and C2 groups (Kitahama et al. 1985).

In mice (VanderHorst and Ulfhake 2006), the C1 cells (TH-immunoreactivity) are located in the rostral part of a ventrolateral longitudinal column (from the caudal pole of the facial nucleus to the spinal C3 segment). C1 is the caudalmost adrenergic cell group in the rodent brainstem.

3.2.1.3 Anatomical classification of brainstem serotonergic nuclei

Serotonin is the main transmitter in almost all neurons of the so-called raphe system (raphe nuclei [Nuclei raphes] ἡ ῥάφη, τῆς ῥαφῆς he raphe genitive: tes raphes Greek = seam). The raphe nuclei are a group of several serotonergic—exception raphe interpositus nucleus—brainstem nuclei from the rostral hindbrain to the junction of the brainstem with the spinal cord. They are located close to the midline (seam!) and are composed of rather small neurons which render their unequivocal identification in routine histological specimens difficult. A comprehensive visualization is possible by the use of antibodies against serotonin or tryptophan hydroxylase (see Fig. 3.2) as immunohistochemical markers (see Hornung 2003).

Central nervous serotonin is involved in a number of behavioral and physiological processes by interacting with a plethora of different postsynaptic 5HT-receptor subtypes (Zarindast et al. 2014). Serotonin-(5-hydroxytryptamine, 5HT) containing neurons are the major constituents of the raphe nuclei which also contain heterogeneous populations of neurons with distinct morphologies, projections, and neurochemical characteristics (Hornung 2003).

As with other nuclear groups of the brain, the terminology of the raphe nuclei unfortunately is not unequivocal. We start here with the terminology provided by Hornung (2003) because he and his group visualized the human raphe nuclei based on their 5-HT content by means of immunocytochemistry.

Two groups of serotonergic neurons can be differentiated, a rostral and a caudal one. The first is located in the mesencephalon and the rostral pons (projections to the forebrain) and the second one extends from the caudal pons to the caudal medulla oblongata (projections to the caudal brainstem and the spinal cord, Hornung 2003).

The rostral group accounts for 85% of all serotonergic neurons in the human brain (Hornung 2003) and comprises:

- The caudal linear nucleus (m1/m2, r0 see Sect. 14.5.3.1)
- The dorsal raphe nucleus (Isthmus/r1, see Sect. 14.5.3.2)
- The median raphe nucleus (r1–r3, most likely the central raphe nucleus acc. to Rüb et al. 2000, (Rüb, personal communication) not listed in FIPAT Ch. 1, see Sect. 11.3.2.1, see also Ohm et al. 1989 classification here below).

The caudal group (approx. 15% of 5-HT neurons) consists of:

- The raphe magnus nucleus (r6/r7, see Sect. 7.3.3.2)
- The raphe pallidus nucleus (r6–r10, see Sect. 4.3.3.2)
- The raphe obscurus nucleus (r6–r10, see Sect. 4.3.3.1)

A classification of the human oral raphe system based on pigment-Nissl preparations (see Braak 1976) has been presented by Ohm et al. 1989. They differentiate between:

- Nucleus raphes* dorsalis
- Nucleus raphes linearis
- Nucleus raphes centralis *raphes = genitive of raphe, see above

The FIPAT inventory Ch. 1 differentiates between the raphe nuclei [Nuclei raphes] of the medulla oblongata, of the pons and those of the mesencephalon. Here, the following nuclei are listed for the medulla oblongata:

- Raphe pallidus nucleus [Nucleus raphes pallidus] Serotonergic cell group B1
- Raphe obscurus nucleus [Nucleus raphes obscurus] Serotonergic cell group B2
- Raphe magnus nucleus [Nucleus raphes magnus] Serotonergic cell group B3

For the pons, the following nuclei are listed:

- Raphe magnus nucleus [Nucleus raphes magnus] Serotonergic cell group B3
- Pontine raphe nucleus [Nucleus raphes pontis] Serotonergic cell group B5
- Median raphe nucleus [Nucleus raphes medianus] Serotonergic cell group B6
- Dorsal raphe nucleus [Nucleus raphes dorsalis] Serotonergic cell group B7
- Serotonergic cells of vestibular area [Cellulae serotonergicae areae vestibularis]
- Serotonergic cell group B4

For the mesencephalon, the following nuclei are listed:

- Dorsal raphe nucleus [Nucleus raphes dorsalis] Serotonergic cell group B7
- Linear raphe nucleus [Nucleus raphes linearis] Serotonergic cell group B8

Notabene: The raphe interpositus nucleus (see Sect. 9.2.4.1, not listed in FIPAT Ch. 1) is not a serotonergic nucleus.

3.2.1.3.1 Clinical implications of serotonin

The major biomedical relevance of serotonin (5-HT) is for major depression. There is now a plethora of data from animal experiments as well as clinical studies and neuroimaging showing that the serotonergic raphe nuclei play a role in major depression. The prevalence of major depression (the percentage of a population that is affected with a particular disease at a given time, here lifetime) is about 20% in US citizens (Hasin et al. 2018).

Signs and symptoms of major depression acc. to the NIMH (National Institute of Mental Health; http://www.nimh.nih.gov/health/publications/depression/what-are-the-signs-and-symptoms-of-depression.shtml) are:

- Persistent sad, anxious, or "empty" feelings
- Feelings of hopelessness or pessimism
- Feelings of guilt, worthlessness, or helplessness
- Irritability, restlessness
- Loss of interest in activities or hobbies once pleasurable, including sex
- Fatigue and decreased energy
- Difficulty concentrating, remembering details, and making decisions
- Insomnia, early-morning wakefulness, or excessive sleeping
- Overeating or appetite loss
- Thoughts of suicide, suicide attempts
- Aches or pains, headaches, cramps, or digestive problems that do not ease even with treatment

Since decreased concentrations of serotonin have been identified in patients suffering from major depression, drugs have been developed that selectively inhibit the reuptake of serotonin from the synaptic cleft thus prolonging the time for released serotonin to interact with serotonin receptors (selective serotonin reuptake inhibitors [SSRI]) (cf. Wenthur et al. 2014). SSRIs are the most commonly prescribed antidepressants. They can ease symptoms of moderate to severe depression and are relatively safe.

3.2.2 Trigeminal sensory nuclei r11

The sensory root of the trigeminal nerve (cranial nerve V, the fifth cranial nerve, CNV) transfers protopathic (from Middle Greek = *prōtopathēs* affected first) (responsive only to strong, rather crude stimuli) and epicritic sensory information (from Middle Greek *epikritikos* determinative) (accurate discrimination between small degrees of sensation) from the face region to the trigeminal sensory nuclei (see also Sect. 3.2.3 Dorsal column nuclei).

The term "trigeminus/trigeminal" (Latin = triplet) is related to the three main branches of the trigeminal nerve in the head region the names reflecting the area of supply of the individual branches:

Ophthalmic nerve, V_1 [N. ophthalmicus] from Greek ὁ ὀφθαλμός, ho ophthalmos = eye

Maxillary nerve, V_2 [N. maxillaris] from Latin maxilla = upper jaw

Mandibular nerve, V_3 [N. mandibularis] from Latin mandibula = lower jaw

There are four different trigeminal nuclei in caudocranial order:

1. Spinal trigeminal nucleus/Spinal nucleus of trigeminal nerve (subdivided in caudal, interpolar, and oral part) [Ncl. spinalis n. trigemini] (see Sect. 3.2.2.2)
2. Principal sensory trigeminal nucleus/Principal sensory nucleus of trigeminal nerve [Ncl. principalis n. trigemini] (see Sect. 12.2.3.2, see Fig. 12.1)
3. Motor trigeminal nucleus/Motor nucleus of trigeminal nerve [Ncl. motorius n. trigemini] (see Sect. 11.3.1.1, Fig. 11.2)
4. Mesencephalic trigeminal nucleus/Mesencephalic nucleus of trigeminal nerve [Ncl. mesencephalicus n. trigemini] (see Sect. 13.2.6.1, see Fig.13.31)

As you can see in Fig. 3.4, in addition to the three sensory nuclei, the trigeminal nerve disposes of a motor root supplied by axons from the trigeminal motor nucleus (see Sect. 11.3.1.1. Fig. 11.2) innervating the masticatory (chewing) muscles (see Sect. 11.3.1.1.2). The spinal trigeminal nucleus is responsible for nociceptive (protopathic) signal transmission. The principal sensory nucleus receives epicritic sensory signals.

The mesencephalic trigeminal nucleus (see Sect. 13.2.6.1) is the receiver of proprioceptive muscle spindle input from the masticatory and mimic muscles (see Sects. 8.2.1.6.1 and 11.3.1.1.2).

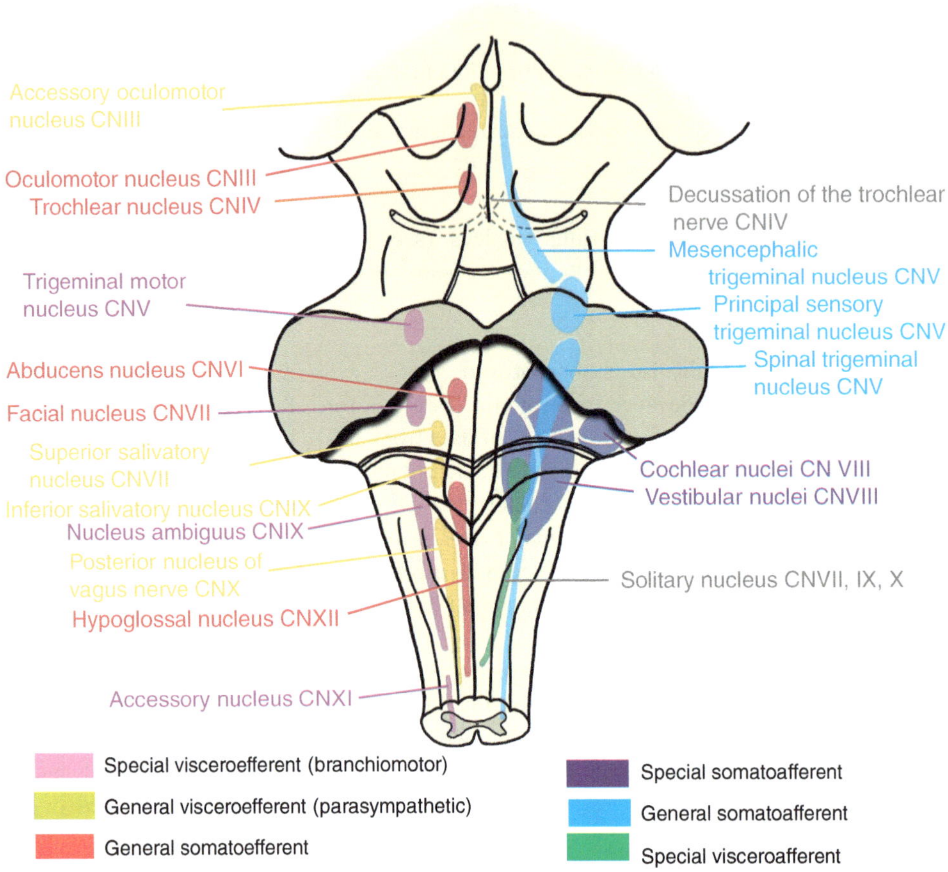

Fig. 3.4 Approximate position of the sensory trigeminal nuclei inside the human brainstem. The caudal most nucleus of the trigeminal system is the spinal nucleus of trigeminal nerve which stretches from the lower medulla oblongata to the midpontine level. It is followed cranial by the principal nucleus of trigeminal nerve and cranial most by the mesencephalic nucleus of trigeminal nerve. Please note that due to space limitation not all terms correspond to FIPAT nomenclature. Modified after Huggenberger et al. 2019, Fig. 15.5 with permission

3.2.2.1 Supratrigeminal nucleus r11

Detailed information on this nucleus are provided in Chap. 12 under Sect. 12.2.3.1.

3.2.2.2 Spinal trigeminal nucleus/Spinal nucleus of trigeminal nerve r11

(subdivided in caudal, interpolar, and oral part) [Ncl. spinalis n. trigemini]

3.2.2.2.1 Location and fine structure of the spinal nucleus of trigeminal nerve

The spinal trigeminal nucleus is the most extended component of the three trigeminal sensory nuclei (see Figs. 3.4 and 3.6). It starts at the border with the spinal cord and extends up to the pons where it is followed cranially by the principal sensory trigeminal nucleus (see Fig. 3.4). It is located within the lateral part of the brainstem, bordered by the spinal tract of trigeminal nerve [Tractus spinalis n. trigemini] which in turn is separated from the lateral surface by the inferior cerebellar peduncle [Pedunculus cerebellaris inferior].

The spinal nucleus of trigeminal nerve can be divided into three different parts in caudo-cranial order:

- Caudal subnucleus [Subnucleus caudalis]
- Interpolar subnucleus [Subnucleus interpolaris]
- Oral subnucleus [Subnucleus oralis]

The **caudal subnucleus** is a rostral prolongation of the gray matter of the posterior horn [Cornu posterius] of the cervical spinal cord (see Fig. 3.5).

The cytoarchitecture is similar to that of the apex of the posterior horn of the spinal cord, i.e.,

- Lamina zonalis (FIPAT: Trigeminal lamina I [Lamina trigeminalis I]
- Lamina gelatinosa (FIPAT: Trigeminal lamina II [Lamina trigeminalis II]
- Lamina magnocellularis (FIPAT: Trigeminal laminae III-IV [Laminae trigeminales III-IV]
- For details, see Olszewski and Baxter (1982).

Lamina I is the outer most, lamina III the inner most of the layers. The perikarya differ in shape from rather large and medium-sized ones in laminae I and III-IV to rather small ones in the subnucleus gelatinosus (Olszewski and Baxter 1982).

The **interpolar subnucleus** is caudally continuous with the caudal one, extends along the inferior olive (see Sect. 4.1.1.3), and is cranially continuous with the oral subnucleus. The cells of the interpolar subnucleus are a uniform population of small or medium-sized, oval fusiform, or triangular perikarya (for details, see Olszewski and Baxter 1982).

The **oral subnucleus** extends from the border with the interpolar one to the caudal limits of the principal sensory

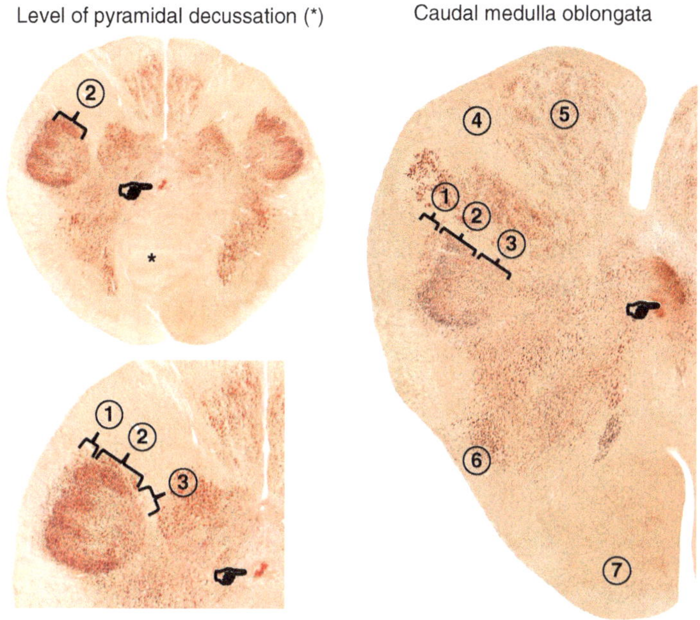

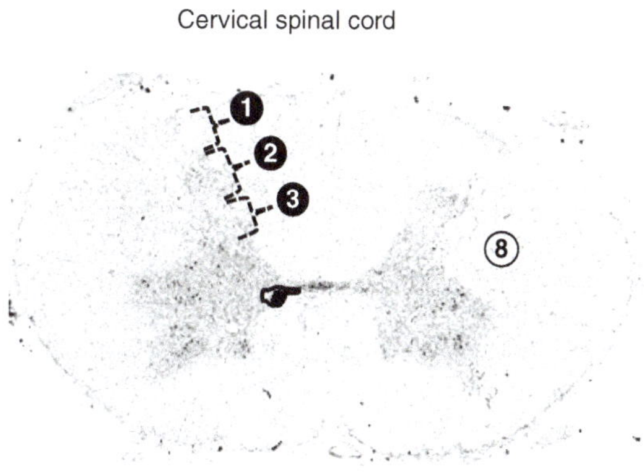

❶	**Spinal lamina I Zona marginalis**
❷	**Spinal lamina II Substantia gelatinosa**
❸	**Spinal lamina III-IV Ncl. proprius**
☛	**Central canal**

①	**Trigeminal lamina I Lamina zonalis**	⑤	**Gracile ncl.**
②	**Trigeminal lamina II Lamina gelatinosa**	⑥	**Retroambiguus ncl.**
③	**Trigeminal lamina III-IVLamina magnocellularis**	⑦	**Pyramidal tract**
④	**Cuneate ncl.**	⑧	**Lateral corticospinal tract**

Fig. 3.5 Horizontal section through the cranial cervical spinal cord and caudal medulla oblongata of man. Darrow red staining. In the dorsal/posterior horn [Cornu posterius], the three superficial laminae are visible. In native specimens, the Substantia gelatinosa appears kind of transparent because it does not contain myelinated fibers. The laminae I–III continue into the spinal nucleus of trigeminal nerve (see Fig. 3.6A). * = pyramidal decussation [Decussatio pyramidum]. LabPON Twente/ Sammlung des Zentrums Anatomie der Universität zu Köln

nucleus of trigeminal nerve [Nucleus principalis nervi trigemini] (see Sect. 12.2.3.2). It displays small perikarya which partly are difficult to distinguish from parvocellular nuclei in the surroundings like the nucleus of the solitary tract (NTS) and the parvocellular reticular nucleus (see Sect. 6.5.5.2) (for details, see Olszewski and Baxter 1982).

The spinal tract of trigeminal nerve (Fig. 3.6B ⑨) is the caudal most part of the trigeminal system. It is the continuation of the Substantia gelatinosa (it appears glassy in unfixed state) ⑩ [Spinal lamina II] and stretches from the medulla oblongata up to the pontine region where it is continued by the principal sensory ncl. of trigeminal nerve ⑦. The first-order peripheral part of the spinal and principal sensory nuclei reaches the brainstem from the periphery (sensory root ⑤) via the trigeminal ganglion ③ where the pseudounipolar perikarya of the neurons are located.

Their centripetal fibers form the spinal tract of trigeminal nerve ⑨ and end in the spinal nucleus of trigeminal nerve ⑧. Second-order ascending fibers from the caudal part of the latter (Lateral trigeminothalamic tract [Tractus trigeminothalamicus lateralis]) ⑯, join the anterolateral tract (protopathic sensory function) which crosses in the decussation of the spinothalamic tract* [Decussatio tractus spinothalamici] ⑱ to the contralateral side and reaches the ventral posteromedial ncl. of the thalamus ⑬ [Ncl. ventralis posteromedialis] (see FIPAT Ch.1 endnote 48).

The second-order epicritic fibers, originating in the principal sensory nucleus of trigeminal nerve ⑦ form the anterior trigeminothalamic tract ⑭ [Tractus trigeminothalamicus anterior]—together with fibers from the spinal trigeminal nucleus ⑧—cross to the contralateral side ⑰ (Decussation of the anterior trigeminothalamic tract* [Decussatio tractus trigeminothalamici anterioris]), join the medial lemniscus (see FIPAT Ch.1 endnote 48) and end—as the protopathic fibers—in the ventral posteromedial ncl. of the thalamus ① (see FIPAT Chap. 1 endnote 48). Sensory input from the periphery of the body ends in the ventral posterolateral ncl. of the thalamus.

*both terms are non-canonical and are not listed in FIPAT Ch.1.

3.2.2.2.2 Course of the trigeminal nerve/Protopathic part

The majority of peripheral receptive structures of the trigeminal nerve is depicted in Fig. 3.7 (see also Sect. 12.2.3.2). These kind of receptors are also present in the peripheral parts of the skin (for review, see Chouchkov 1978, Abraira and Ginty 2013). Skin is innervated by complex combinations of low- and high-threshold mechanoreceptors, each with unique physiological profiles and response properties elicited by distinct tactile stimuli (Abraira and Ginty 2013). The terminology used in Fig. 3.7 (Abraira and Ginty 2013) is mostly non-canonical but widely used. In the following text, the English FIPAT terms for the receptive structures are given in (brackets). We have refrained from citing the FIPAT Latin terms since their use would be a rare exception.

- Merkel cells (Sensory epithelial cells) are sensors for sustained touch applied onto the skin (indentation) (Friedrich Sigmund Merkel, 1845–1919, German anatomist and histopathologist) in the basal layer of epidermis in glabrous skin and around Guard hair follicles in hairy skin (see also FIPAT Chap. 2, endnote 152).
- The Ruffini corpuscles (Bulbous corpuscles) are sensors for the degree of skin stretching located in the dermis of glabrous skin (Angelo Ruffini,1864–1929, Italian histologist and embryologist) (see also FIPAT Chap. 2, endnote 154).

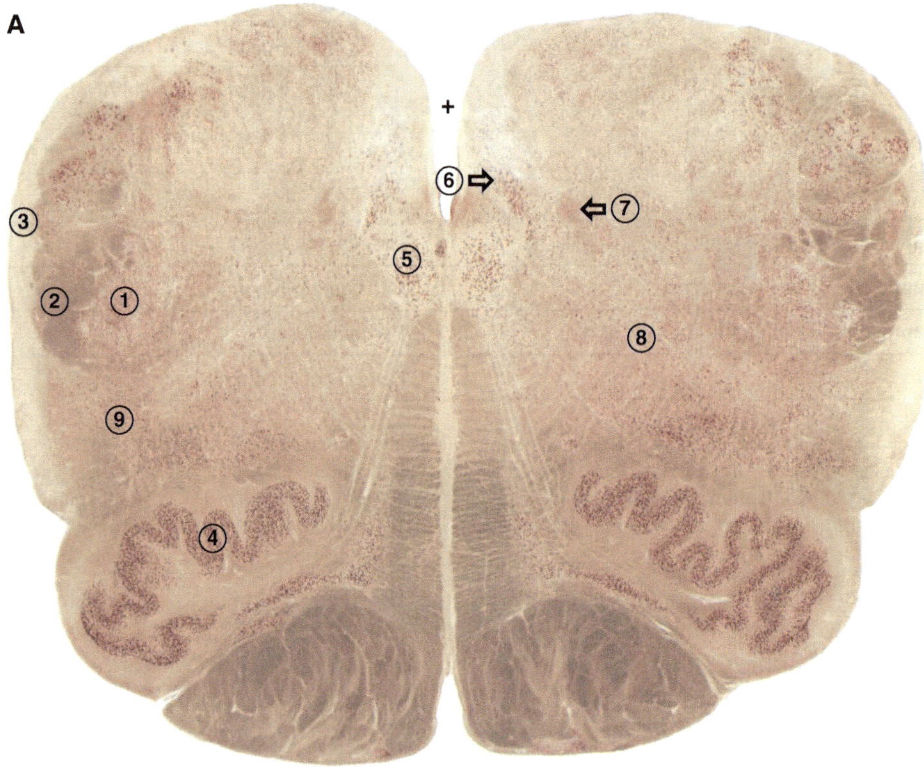

①	Spinal ncl. of trigeminal nerve CNV	⑥	Posterior ncl. of vagus nerve CNX
②	Spinal tract of trigeminal nerve	⑦	Solitary tract
③	Inferior cerebellar peduncle	⑧	Intermediate reticular zone
④	Inferior olivary complex	⑨	Lateral reticular ncl.
⑤	Ncl. of hypoglossal nerve CNXII	+	4th ventricle

Fig. 3.6 (**A**) Horizontal section through the human brainstem at the level of the medial accessory nucleus (see also Fig. 4.3) of the inferior olive. Darrow red staining. The spinal trigeminal complex ① ② is located in the middle of the medullary tegmentum. The massive fiber bundle of the trigeminal tract ② (compare with Fig. 3.6B) lies lateral of the corresponding nucleus ①. See atlas part Darrow red 4. LabPON Twente. (**B**) Graphic visualization of the course of the different tracts of the trigeminal system. As for the periphery of the body (compare with Fig. 3.13), there is a protopathic (Spinal ncl. and tract of trigeminal nerve) [Ncl. spinalis n. trigemini] and an epicritic sensory component (Principal sensory ncl. of trigeminal nerve) [Ncl. principalis N. trigemini]. In addition, proprioceptive signals from the face region are represented in the mesencephalic ncl. ⑥ and tract of the trigeminal nerve [Ncl./Tractus mesencephalicus n. trigemini] (see Sect. 13.2.6.1) (for details, see text, cranial nerves are labeled by the appropriate Roman numbers). Modified acc. to Nieuwenhuys et al. 2008, Fig. 16.7 with permission

B

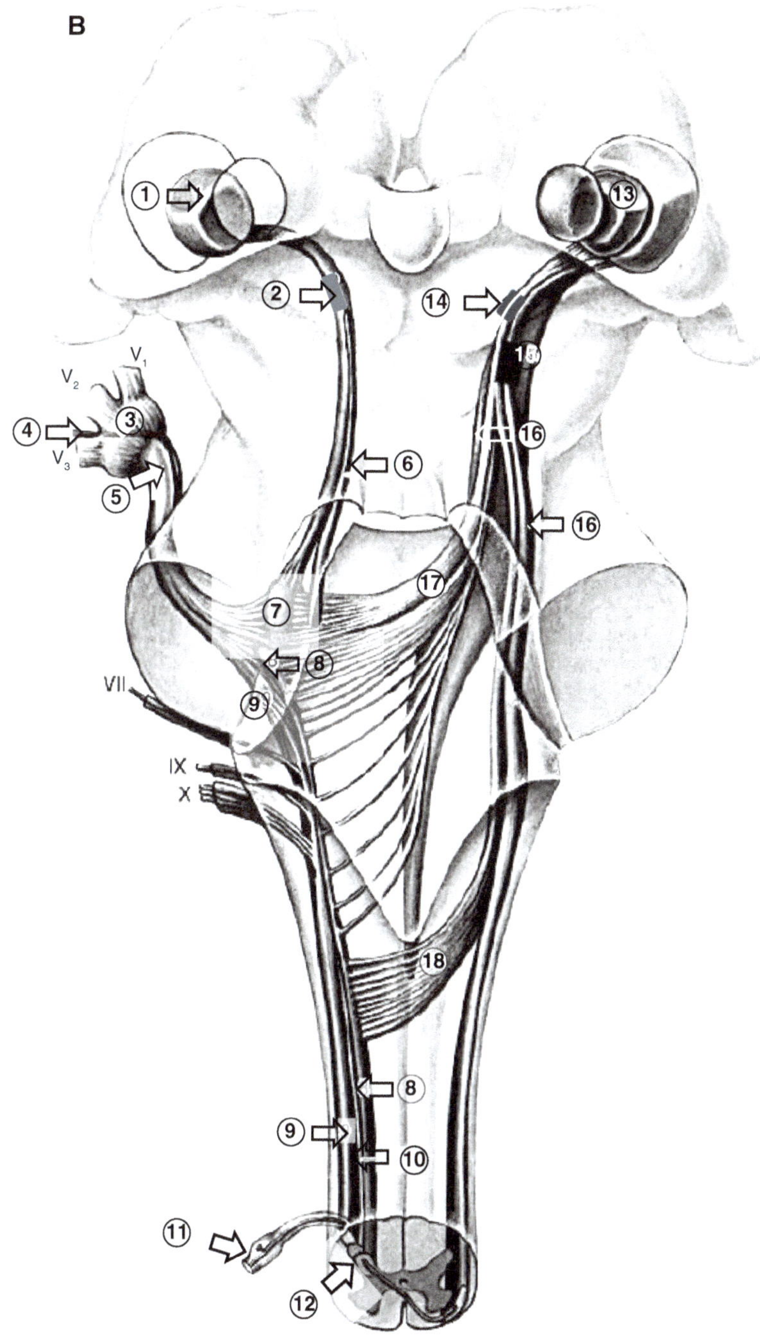

①	Ventral posteromedial ncl.	⑩	Substantia gelatinosa
②	Medial lemniscus	⑪	Posterior root
③	Trigeminal ganglion	⑫	Ncl. proprius of the dorsal horn
④	Motor root	⑬	Ventral posteromedial ncl.
⑤	Sensory root	⑭	Anterior trigeminothalamic tract
⑥	Mesencephalic ncl. of trigeminal nerve	⑮	Medial lemniscus
⑦	Principal sensory ncl. of trigeminal nerve	⑯	Lateral spinothalamic tract
⑧	Spinal ncl. of trigeminal nerve	⑰	Decussation of anterior trigeminothalamic tract
⑨	Spinal tract of trigeminal nerve	⑱	Decussation of the spinothalamic tract

Fig. 3.6 (continued)

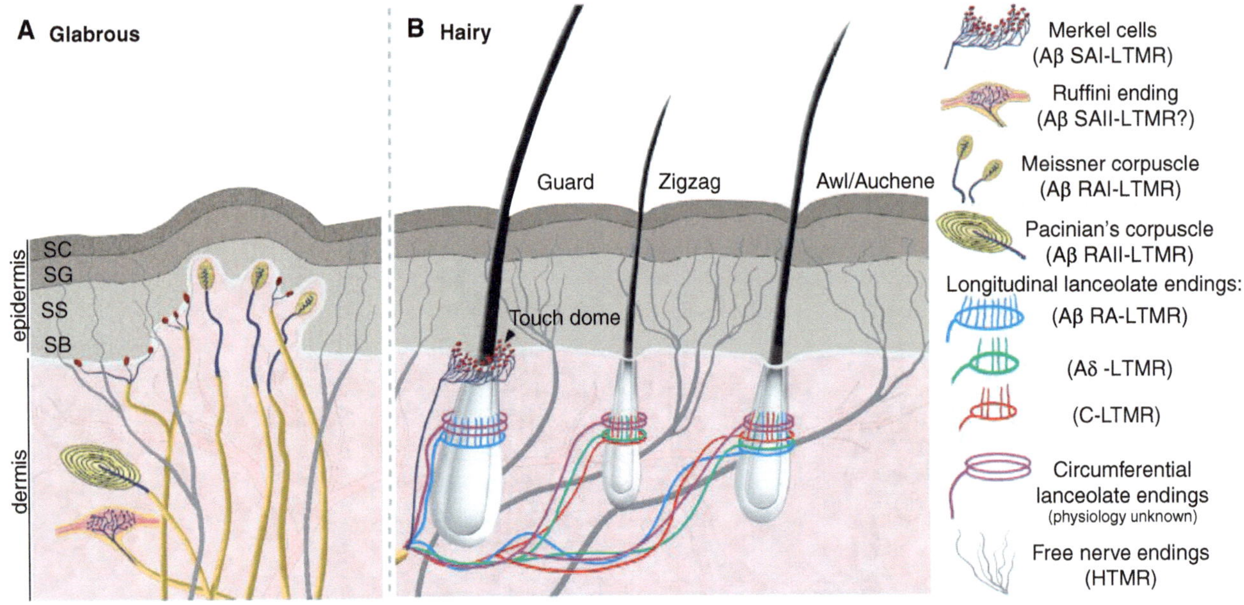

Fig. 3.7 Tactile receptors of the skin. For explanation, see text. In brackets below the individual receptors, the associated fiber types are indicated. *SC* Stratum corneum, *SG* Stratum granulosum, *SS* Stratum spinosum, *SB* Stratum basale, *SA* slowly adapting, *RA* rapidly adapting, *LTMR* low-threshold mechanoreceptor, *HTMR* high-threshold mechanoreceptor. From Abraira and Ginty 2013, Fig. 1 with permission

- Meissner's corpuscles (Tactile corpuscles), located in dermal papillae, record the speed with which a stimulus is indenting the skin (skin movement) in glabrous skin dermis and hair follicle deflection, located there as longitudinal lanceolate endings at the Guard/Awl-Auchene hair follicles (Georg Meissner, 1829–1905, German anatomist and physiologist) (see also FIPAT Chap. 2, endnote 155).
- The (Vater-)Pacinian corpuscles (Lamellar corpuscles) are sensors for vibration (epicritic signal in the classical sense) located in the deep dermis of glabrous skin (Filippo Pacini, 1812–1883, Italian anatomist) (see also FIPAT Chap. 2, endnote 156).
- At the Awl/Auchene / Zigzag hair follicles again (see Meissner's corpuscles) the longitudinal lanceolate endings are present which are stimulated by hair follicle deflection (see also FIPAT Chap. 2, endnote 151).
- The free nerve endings located in epidermis and dermis are sensitive to thermal, chemical, and noxious mechanical stimuli (pain) (protopathic signals in the classical sense) (see also FIPAT Chap. 2, endnote 150).
- The root hair plexus (Hair follicle endings) record the movement of hairs (see also FIPAT chap. 1, endnote 151).
- Krause's end bulbs (Tactile corpuscles) record the sensations of touch and cold (Johann Friedrich Wilhelm Krause, 1833–1910, German anatomist) (see also FIPAT Chap. 1, endnote 155).

For the physiology of touch receptors in the skin, see Abraira and Ginty (2013) and Owens and Lumpkins (2014). In addition to these skin receptors, there are proprioceptors of the skeletal muscles and tendons which record the state of strain.

The sensory receptors of the protopathic portion of the trigeminal nerve (free nerve endings) are located in the skin, the mucous membranes (oral and nasal cavity, nasal sinuses), the teeth and the dura mater of the head/cranium. Via a great number of small peripheral branches, the protopathic component eventually is collected in the ophthalmic nerve [N. ophthalmicus] (orbita), the maxillary nerve [N. maxillaris] (region of the upper jaw), and the mandibular nerve [N. mandibularis] (region of the lower jaw) (see Sect. 3.2.2 Trigeminal sensory nuclei).

An account of the most important peripheral branches is given under Sect. 12.2.3.2.1.

The peripheral processes (dendrites) of the trigeminal nerve branches are connected with their perikarya in the trigeminal ganglia (see Fig. 3.8).

The afferent sensory trigeminal fibers arise from their cell bodies in the trigeminal ganglion Fig. 3.8 ① [Ggl. trigeminale, semilunare] (Ganglion Gasseri; First description by Johann Lorenz Gasser, Austrian anatomist, 1723–1765). This ganglion is located in a duplication of the dura mater, called Cavum trigeminale (trigeminal cave) (cavum Latin = cave). The Ligamentum petrolinguale or Ligamentum sphenopetrosum inferius (Lang 1979) largely forms the bottom of the cavum. The Porus trigeminus (trigeminal opening) is a dural opening communicating the cavum with the infratentorial space (see below Box 3.3) (Lang 2001). Dupont et al. (2019) report about an ossification of the dura above the Porus trigeminus mentioning that ossification of

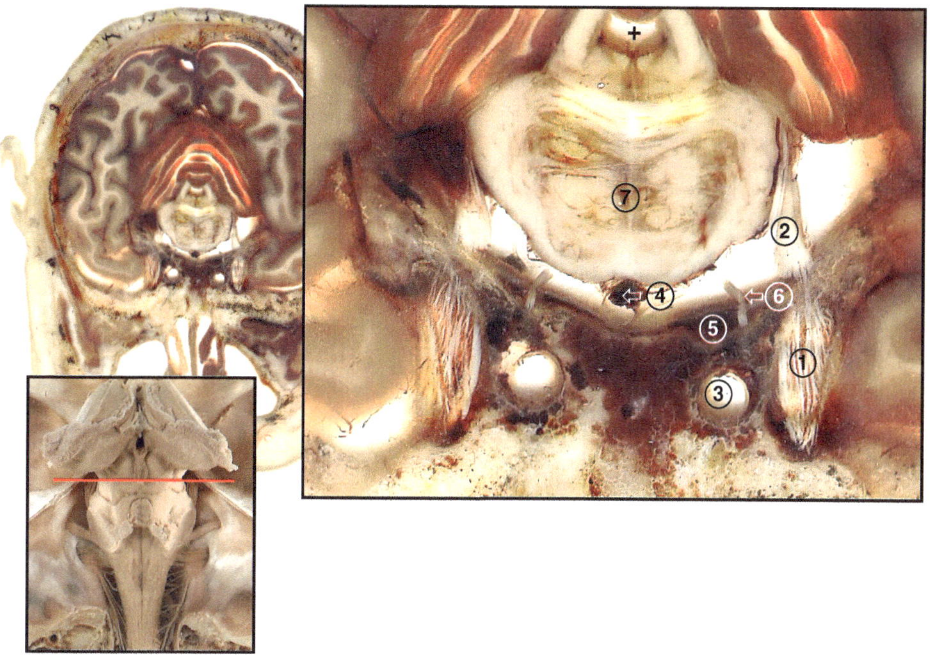

①	**Trigeminal ganglion**	⑤	**Cavernous sinus**
②	**Trigeminal nerve CNV**	⑥	**Abducens nerve CNVI**
③	**Internal carotid artery inside ⑤**	⑦	**Pons**
④	**Basilar artery**	+	**4th ventricle**

Fig. 3.8 Horizontal section (plastinated specimen) through the human brain at the level of the entry (sensory root, see Sect. 12.2.3.2.1)/exit (motor root, see Sect. 11.3.1.1.3) of the trigeminal nerve ②. The perikarya of the primary sensory neurons are located in the trigeminal ganglion ①. The central processes of the primary sensory trigeminal neurons enter the brainstem after the trigeminal nerve has passed through the prepontine cisterns (Box 3.3). Inside the brainstem, the fibers form a descending nociceptive bundle to reach the spinal trigeminal nucleus and an ascending sensory bundle targeting the principal sensory and the mesencephalic nuclei (see Sect. 3.2.2) of trigeminal nerve. Sammlung des Zentrums Anatomie der Universität zu Köln

parts of the dura is a common finding. The centripetal fibers of the ganglion cells enter the lateral pons and course in a dorsomedial direction.

As with the dorsal roots of spinal nerves, the primary nociceptive neuron is located in a ganglion outside the central nervous system, the trigeminal ganglion. The central processes of the ganglion cells run in the subarachnoid/leptomeningeal space of the prepontine cistern (see Box 3.3), enter the pons and form (1) a descending bundle, the trigeminal spinal tract, which contacts the spinal trigeminal nucleus, and (2) an ascending bundle reaching the principal sensory nucleus (see Sect. 12.2.3.2) and the mesencephalic nucleus of the trigeminal nerve (Miller et al. 2005).

Signals are conveyed to the thalamus (ventral posteromedial nucleus) from the spinal trigeminal nucleus in the anterior trigeminothalamic tract and from the caudal part of the spinal trigeminal nucleus via the lateral trigeminothalamic tract [Tractus trigeminothalamicus lateralis] which joins the spinothalamic fibers of the anterolateral tract [Tractus anterolateralis] (formerly spinal lemniscus) in the brainstem.

Box 3.3 Cisterns of the brainstem (Infratentorial cisterns [Cisternae infratentoriae])

Cisterns are enlargements of the leptomeningeal space containing cerebrospinal fluid (CSF). The infratentorial cisterns are located in the posterior cranial fossa, i.e., caudal of the tentorium cerebelli. The CSF produced in the choroid plexus of the cerebral ventricles is flowing from the fourth ventricle via two lateral and one median aperture into the subarachnoid (old term)/ leptomeningeal space. The brain is kind of swimming in the CSF which serves as a cushion. The cranial nerves on their way to / away from the brainstem have to pass the leptomeningeal space on their way in the posterior or middle cranial fossa and to their targets in the head/neck region. The brainstem cisterns belong to the group of infratentorial cisterns (Tentorium see Box 3.5). The most important of these cisterns related to the cranial nerves (and large brainstem vessels) are listed here (cf. Standring 2016). In most cases, the names indicate the location of the cisterns (Most important arteries shown in red, veins in blue):

1. Interpeduncular cistern [Cisterna interpeduncularis] in the region of the interpeduncular fossa (mesencephalon): Passage of the oculomotor nerve (CNIII) (see Fig. 15.7).
   Basilar artery (see Fig. 12.3 ⑤), Superior cerebellar [A. superior cerebelli] and anterior inferior cerebellar [A. inferior anterior cerebelli] arteries (see Figs. 1.15 and 1.17).
2. Prepontine cistern [Cisterna prepontina]: Passage of trigeminal nerve (CNV) (see Fig. 12.3), abducens nerve (CNVI) (see Fig. 9.9 A), facial nerve (CNVII) (see Fig. 8.17 ①), vestibulocochlear nerve (CNVIII) (see Fig. 8.17 ③).
   Basilar artery [A. basilaris] (see Fig. 12.3 ⑤), origins of the anterior inferior cerebellar [A. inferior anterior cerebelli] and superior cerebellar [A. superior cerebelli] arteries (see Figs. 1.15 and 1.17). Labyrinthine artery [A. labyrinthi] (see Sect. 5.3.1.1.5, Fig. 5.8).
3. Cerebellomedullary cistern [Cisterna cerebellomedullaris] (Cisterna magna):
   Bridges the arachnoid interspace between the medulla oblongata and the inferior surface of the cerebellum. Connection between fourth ventricle and the leptomeningeal space via the median aperture [Aperture mediana] Magendie. Passage of glossopharyngeal (CNIX, see Fig. 1.6 B), vagus (CNX) (see Figs. 1.6 B and Fig. 5.15), accessory (CNXI) (see Fig. 1.6 B), and hypoglossal (CNXII) (see Fig. 1.6 A) nerves.
   Vertebral arteries [A. vertebralis] (see Fig. 1.15 B) origins of the posterior inferior cerebellar arteries [A. inferior posterior cerebelli] (see Figs. 1.15 and 1.17).
4. Cerebellopontine cistern [Cisterna cerebellopontina]: Connection between the fourth ventricle and the leptomeningeal space via the lateral aperture [Apertura lateralis] (Luschka) (Fig. 3.9A, B). Transversed by the trigeminal (CNV), facial (CNVII) and the vestibulocochlear (CNVIII) nerves (see Fig. 1.6 B ②–④), the anterior inferior cerebellar arteries [A. inferior anterior cerebelli] (see Figs. 1.15 and 1.17) and the superior petrosal veins [Vv. petrosae superiores].
5. Quadrigeminal cistern [Cisterna quadrigeminalis] (superior cistern, vein of Galen cistern): posterior of the brainstem and third ventricle. Occupies the space between the splenium of the corpus callosum and the superior cerebellar surface. Basal vein [Vena basalis] (Rosenthal), P3 segment of the posterior cerebral arteries [A. cerebri posterior].
6. Cisterna ambiens [Cisterna ambiens]: Extension of the quadrigeminal cistern [Cisterna quadrigeminalis]. Located lateral of the cerebral peduncles [Pedunculus cerebri] / Passage of the trochlear nerve (CNIV) (see Fig. 14.5). Contains the superior cerebellar artery [A. superior cerebelli] and the P2 segment of the posterior cerebral artery [A. cerebri posterior] (i.e., distal of the posterior communicating artery) (Fig. 1.17 A), basal vein [Vena basalis] (Rosenthal).

This particular anatomical situation shown in Fig. 3.9 enables the positioning of flat electrodes on the surface of the cochlear nuclei as a therapy option in case of neurofibromatosis type 2 (NF2) (Mahboubi et al. 2020). NF2 is a rare autosomal dominant syndrome. Bilateral vestibular schwannomas—benign tumors originating from the myelin producing Schwann cells—are the hallmark of the disease occurring in 90 to 95% of the patients (Mahboubi et al. 2020). This will eventually lead to hearing loss. Depending on the individual case history an auditory brainstem implant (ABI) can directly stimulate the cochlear nucleus, bypassing the cochlea or cochlear nerve (Mahboubi et al. 2020). The ABI electrode is inserted into the lateral recess of the fourth ventricle via the foramen Luschka and placed in contact with the cochlear nucleus, mainly the ventral cochlear nucleus, the main relay for cochlear nerve input (Mahboubi et al. 2020) (see also Sect. 8.1.1).

Box 3.4 Georg (Georges) Schaltenbrand (1897–1979)
The publication of the stereotaxic atlas of the human brain is undoubtedly a merit of Schaltenbrand—director of the Department of Neurology at the University of Würzburg from 1937 to 1969—who on the other hand was responsible for one of the most infamous series of experiments on patients without consent (Martin et al. 2020). This happened in the framework of his studies on the pathogenesis of multiple sclerosis (see Sect. 9.2.1.3). Convinced of a viral etiology, he injected the CSF (see Sect. 1.2.1) of allegedly infected monkeys and multiple sclerosis patients into severely handicapped patients without their consent. He was never trialed for his activities.

Of neurosurgical importance are various intercisternal membranes [Membranae intercisternales]. For the interpeduncular cistern, Liliequist (1959) described an intercisternal membrane, the basal intercisternal membrane/Liliequists membrane [Membrana intercisternalis basalis] (see Lang 1991) separating the interpeduncular cistern from the more rostral chiasmatic cistern. For details on the microanatomy and surgical anatomy of the cisterns, see Lang (1991).

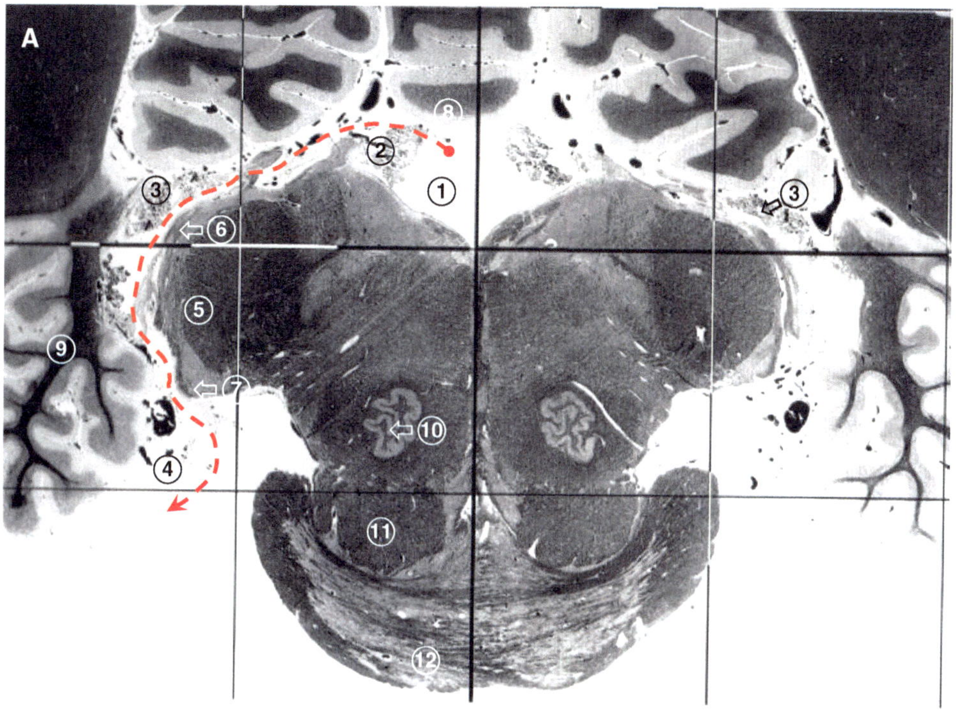

①	4th ventricle	⑦	Anterior cochlear ncl.
②	Choroid plexus in the 4th ventricle	⑧	Nodule
③	Choroid plexus in the lateral aperture	⑨	Cerebellar hemisphere
④	Choroid plexus in the leptomeningeal space	⑩	Inferior olivary complex
⑤	Inferior cerebellar peduncle	⑪	Pyramidal tract
⑥	Posterior cochlear ncl.	⑫	Basis of the pons

Fig. 3.9 (**A**) Horizontal section through the human brainstem at the level of the lateral aperture / lateral recess. Myelin stain. The course of the CSF from the fourth ventricle ① to the leptomeningeal space ④ is indicated by the red line. The roof and the lateral wall of the lateral aperture is formed by the cerebellum ⑧⑨, the floor and medial wall by the posterior (dorsal) ⑥ and anterior (ventral) ⑦ cochlear nucleus and the underlying inferior cerebellar peduncle ⑤. (Modified after G. Schaltenbrand (Box 3.4) and W. Wahren (1977), page 62 with permission, © Thieme Gruppe). (**B**) Dorsal oblique view onto the lateral margin of the fourth ventricle. The midline structure is the median sulcus ⑦, flanked by the facial colliculus (F) (see Figs. 1.9 and 1.10). The upper lateral margin of the ventricle is formed by the cerebellar peduncles ①. At their foot the auditory tubercle ③ is located. The entrance into the lateral aperture (see Fig. 3.9 A), the lateral recess ⑤, lies ventrolateral of the auditory tubercle. See also Fig. 1.10. From Quester and Schröder 1999, Fig. 5 with permission

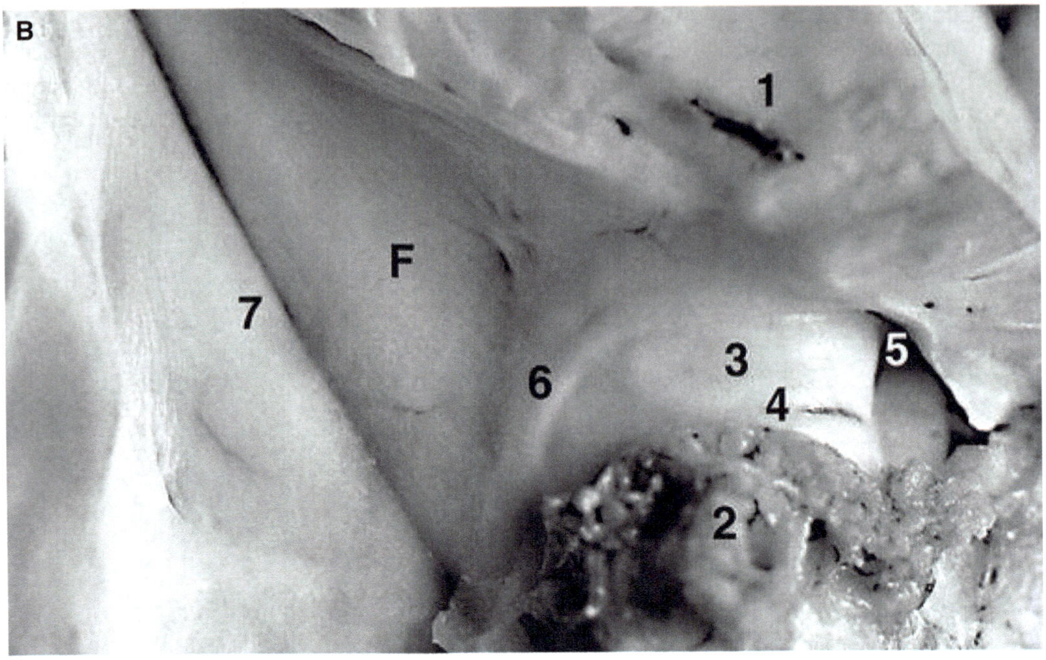

①	**Cerebellar peduncle**	⑤	**Lateral recess**
②	**Choroid plexus in the 4th ventricle**	⑥	**Course of CNIX and CNX**
③	**Auditory tubercle**	⑦	**Median sulcus**
④	**Vessel at caudal margin of auditory tubercle**	F	**Facial colliculus**

Fig. 3.9 (continued)

Box 3.5 Tentorium cerebelli [Tentorium cerebelli] (Tentorium Latin = tent]

The tentorium is a duplication of the dura mater which in fashion spreads tent-like above the superior cerebellar surface. It separates the posterior cranial fossa [Fossa cranii posterior] with the cerebellum (see Fig. 3.10 ③) from the middle cranial fossa [Fossa cranii media] with the telencephalon (see Fig. 3.10 ④). It has free and fixed margins. The latter are attached to the superior borders of the petrous part of the temporal bone, known as the posterior clinoid process via anterior and posterior petroclinoid folds and along the grooves of the transverse sinuses on the occipital bone posteriorly (Rai et al. 2018) (Fig. 3.11 ⑩). The free margin lies at the anterior edge forming a U-shape gap, the tentorial notch or incisura tentoria (Incisure of the tentorium). This gap, bordered by the free margins, contains the midbrain (anterior) and the superior cerebellar vermis or the splenium of the corpus callosum (posterior).

At its rostral margin, the tentorium continues in dural plicae originating at the apex of the petrosal bone, anchoring the tentorium at the clinoid process. Inside the plicae the abducens nerve (see Sect. 9.2.1.1) is running towards the cavernous sinus (see Fig. 3.11 ②) in the canal of Dorello (Primo Dorello, 1872–1963, Italian anatomist). Between the free margin of the tentorium, the plicae and the posterior margin of the Sella turcica opens the tentorial notch with the mesencephalon (see Rai et al. 2018).

The transverse sinus [Sinus transversus] (Fig. 3.11 ⑩) originates in the midline confluence of sinuses [Confluens sinuum] ⑪ of the superior sagittal sinus [Sinus sagitalis superior] and the straight sinus [Sinus rectus] at the superior surface of the cerebellar tentorium [Tentorium cerebelli] (see Fig. 3.10). It continues into the sigmoid sinus [Sinus sigmoideus] ⑨ which runs to the jugular bulb [Bulbus venae jugularis] ⑧ and continues into the internal jugular vein [V. jugularis interna] at the jugular foramen [Foramen jugulare]. The sigmoid sinus is connected with the superior petrosal sinus [Sinus petrosus superior] ⑦ at the apex of the petrosal

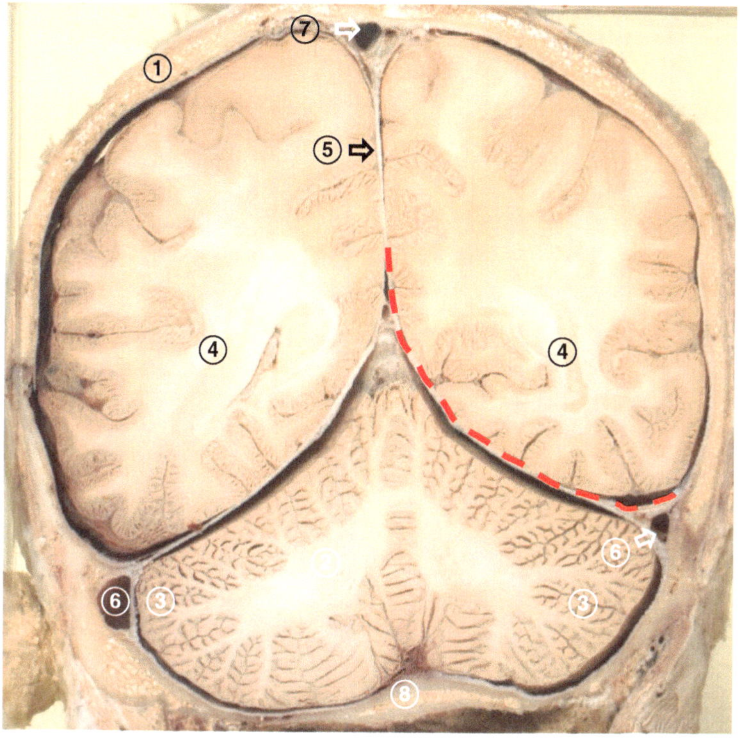

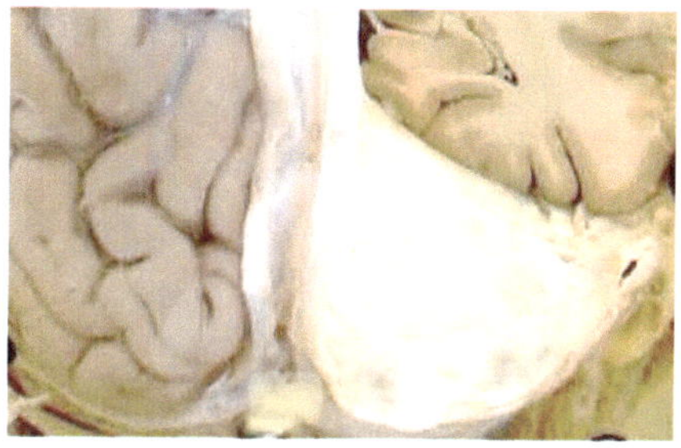

---	Outline of the Tentorium cerebelli (right side of the brain)	⑤	Falx cerebri
②	Cerebellum (white matter)	⑥	Transverse sinus
③	Cerebellar cortex	⑦	Superior sagittal sinus
④	Occipital lobe of telencephalon	⑧	Skull base (Posterior cranial fossa)

Fig. 3.10 Coronal section through the human skull at the level of the middle third of the cerebellum. The tentorium cerebelli separates the cerebellum (② ③) from the occipital lobes ④ of the telencephalon. The tentorium is continuous with the walls of the transverse sinus ⑥ [Sinus transversus] (Intradural venous vessel) laterally on both sides (see Fig. 3.11). In direction to the convexity of the brain, the tentorium merges with the Falx cerebri ⑤ which on its dorsal surface contains the superior sagittal sinus ⑦ [Sinus sagittalis superior]. The superior sagittal sinus is part of the intradural veins which drain into the dural sinus system of the posterior and middle fossa (see Fig. 3.11) and eventually into the internal jugular vein [Vena jugularis]. The inset—dorsal view of the brain, the right-hand occipital pole has been removed—shows a view of the mother of pearl color tentorium. Sammlung des Zentrums Anatomie der Universität zu Köln. Photograph by MedizinFotoKöln, Cologne, Germany

bone and the inferior petrosal sinus [Sinus petrosus inferior] ⑥ at the base of the petrosal bone. The petrosal sinus connects with the posterior intercavernous sinus ④. The basilar plexus [Plexus basilaris] ⑤ on the clivus is connected with the inferior petrosal sinus. At both sides of the sella, the cavernous sinus ② [Sinus cavernosus] is located (see Box 9.5).

Both parts of the cavernous sinus are connected via the anterior ③ and the posterior (inferior) intercavernous sinuses [Sinus intercavernosus anterior/posterior]. The rostral most part of the dural venous sinuses [Sinus durae matris] is the sphenoparietal sinus [Sinus sphenoparietalis] ① which runs along the greater wing (of the sphenoideal bone) [Ala major] to the parietal bone. The marginal sinus [Sinus marginalis] ⑫ runs around the posterior part of the Foramen magnum and connects to the jugular bulb.

3.2.2.2.3 Living anatomy and clinical implications

As with other regions of the human body, the function of the nociceptive innervation clinically can be tested by simple pin-prick application, a small puncture made as if by a pin. This can be achieved, for example. by the broken ends of a Q-tip. This allows for the alternate testing of sharp (pain) vs. dull (touch). The eyes of the patient should be shut. The patient is asked to describe the sensations and about any differences in sensation on one side compared to the other side. This should be performed in the whole area of sensory supply of the trigeminal nerve and separately for the three main trigeminal branches (ophthalmic, maxillary, mandibular nerve). For anatomical details, see Sect. 12.2.3.2.3.

Epicritic sensory function (see here below Sect. 3.2.3 "Dorsal column nuclei" and Sect. 12.2.3.2) is mainly checked

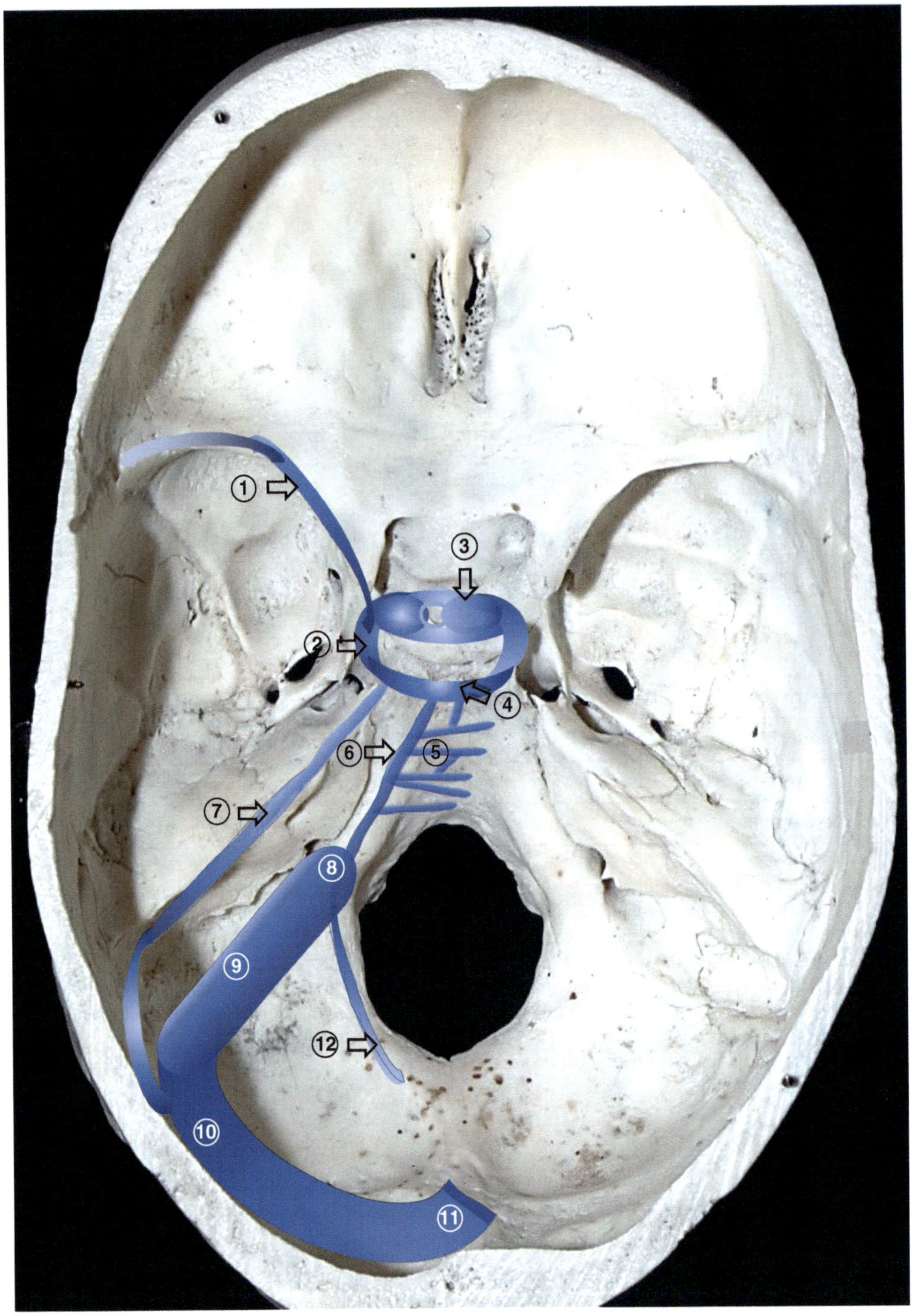

①	Sphenoparietal sinus	⑦	Superior petrosal sinus
②	Cavernous sinus	⑧	Jugular bulb
③	Anterior intercavernous sinus	⑨	Sigmoid sinus
④	Posterior intercavernous sinus	⑩	Transverse sinus
⑤	Basilar plexus	⑪	Confluence of sinuses
⑥	Inferior petrosal sinus	⑫	Marginal sinus

Fig. 3.11 Semi-schematic representation of the dural venous sinus of the skull base. For details, see text. Sammlung des Zentrums Anatomie der Universität zu Köln. H. Schröder fecit

for by use of a tuning fork (normally 128 Hz) placed on predefined bony prominences as well as on the skin. The resulting vibration signal is perceived mostly by Pacinian and Meissner corpuscles (see Fig. 3.7) and transferred as a neural signal to the central nervous system (Klingner and Witte 2018).

A rather frequent clinical entity related to nociception by the trigeminal nerve is trigeminal neuralgia (also called tic douloureux). Patients complain about extreme, sporadic, sudden burning, or shock-like facial pain that lasts from a few seconds to as long as 2 min per episode (typical form). The "atypical" form is characterized by constant aching, burning, stabbing pain of somewhat lower intensity than the typical form. The intensity of pain can be physically and mentally incapacitating.

https://www.ninds.nih.gov/Disorders/Patient-Caregiver-Education/Fact-Sheets/Trigeminal-Neuralgia-Fact-Sheet

Before the advent of microsurgical techniques and the basic microsurgical anatomy those forms that could not be treated by analgesic drugs were treated by electrocoagulation of the trigeminal ganglion (Brisman 1989) by radiofrequency lesion. For this procedure, a needle probe is inserted in the lower parts of the cheek and advanced to the foramen ovale under X-ray control. The foramen ovale (see Figs. 1.35 and 1.36) is the exit foramen of the mandibular nerve (third branch of the trigeminal nerve, V_3) in close vicinity to the trigeminal ganglion (see Fig. 3.8). The electrode is then introduced into the ganglion and a thermal lesion is produced. This approach was far from a causal therapy and patients often had to cope with problems related to bilateral chewing (Menzel et al. 1975). Furthermore, it has a high risk of facial numbness.

It turned out, however, that in many cases of trigeminal neuralgia the trigeminal nerve was compressed by a small arterial vessel (Jannetta 1967). The microvascular decompression (MVD) according to the technique of Jannetta has become a routine procedure for the treatment of trigeminal neuralgia and hemifacial spasm (facial nerve). It provides a high rate of cure, and it is safe in general. Side effects and permanent neurological deficits are rare (Broggi et al. 2012; Chang and Ho 2018). Chang and Ho (2018) summarized more recent approaches for the treatment of trigeminal neuralgia.

In terms of topographical anatomy, the approach to the trigeminal nerve is a very interesting issue which is illustrated in a survey manner in Fig. 3.12 (see legend for explanation). In addition, there is a number of videos on this procedure available on the Internet (e.g., https://www.youtube.com/watch?v=mXHbBPX6lm8).

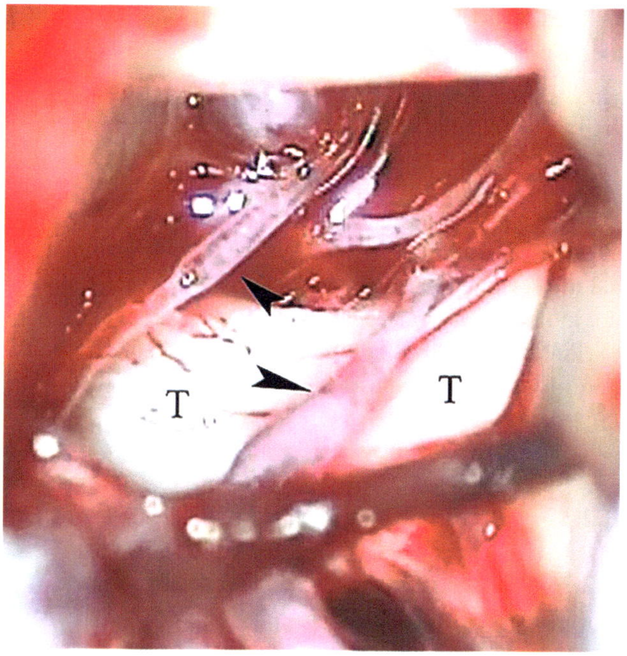

Fig. 3.12 Intraoperative dorsal view into the posterior cranial fossa at the exit point of the trigeminal nerve (T). The arrowheads point to two sites, where vessels are in close contact with the trigeminal nerve (vascular compression) as a most likely cause of trigeminal pain. The treatment consists of the interposition of small pieces of Teflon between the nerve and the compressing vessels. This approach can also be performed using an endoscope with the advantage of less intraoperative compression of the cerebellum (Farid and ElKheshin 2020). To reach the location shown, the so-called lateral suboccipital or retro-sigmoid approach (for details, see Gulec et al. 2019) to the cranial nerves is chosen. The patient is in a sitting position, the head fixed in a Crutchfield cervical traction tong, deflected away from the symptomatic side. A vertical skin incision is made approx. two finger-breadths behind the ear. After the incision of the skin, the occipital bone is exposed by the overlying muscles being pushed aside. Using a cranial rongeur, kind of forceps to remove skull bone, a round portion of the bone (ca. 30 mm in diameter) is removed. Now the cerebellum, still covered by the dura mater, can be seen. The dura is then opened and the cerebellum is carefully moved away from the brainstem and fixed with a flexible spatula to a retractor system. Now the view onto the brainstem with the trigeminal and facial nerves is unobscured. The next steps of the procedure are performed using a surgical microscope. After microsurgical dissection of the arachnoid, facial and trigeminal nerves can be seen in their intracisternal course together with the surrounding vessels as shown here above. The compressing artery—mostly the superior cerebellar artery (SCA) (see Fig. 1.18 B)—is then carefully separated from the nerve which often displays on its surface a discrete indentation at the site of compression. Originally a small piece from a patient's own skeletal muscle, nowadays, some sponge-like material like teflon, is interposed between the artery and the nerve to reduce the impact of the vessel on the nerve. Subsequently, the dura is closed, the removed bone is put back into place and fixed with a number of sutures to the surrounding skull. Eventually, the skin is closed using a cutaneous suture. Farid and ElKheshin 2020, Fig. 1 with permission

3.2.2.3 Paratrigeminal nucleus r11 [Ncl. paratrigeminalis] (not listed in FIPAT)

The term paratrigeminal nucleus refers to a group of cells located lateral to the spinal trigeminal tract (see above) of the medulla and medial to the spinocerebellar tracts at the level of the spinal trigeminal nucleus (see above) present in different mammalian species like man, macaque, rat, and mouse. The cells are partially embedded in the adjacent tracts; their distribution varies by species.

The paratrigeminal nucleus may represent an extension of the lateral cervical nucleus (see Sect. 3.3.2.4) into the medulla.

http://braininfo.rprc.washington.edu/centraldirectory.aspx?ID=761

In mice, the different parts of the paratrigeminal nucleus have been shown to belong—with a number of other brainstem nuclei—to the interstitial system of the trigeminal spinal tract. They project bilaterally to the magnocellular—but not to the parvocellular—red nucleus [Ncl. ruber] (see Sect. c) (Pinto et al. 2007). This projection may mediate modulation of the facial muscles by pain and other sensory information.

3.2.3 Dorsal column nuclei r11

3.2.3.1 Location and connectivity of the dorsal column nuclei

The term dorsal column refers to the fiber tracts (cuneate and gracile fasciculi) contained in the dorsal funiculi of the spinal cord.

The cuneate and gracile fasciculi [Fasciculi cuneatus (Burdach) et gracilis (Goll)] convey epicritic sensory signals from the periphery of the body to the CNS via the central processes of epicritic first-order neurons of which the perikaryon is located in the spinal ganglion (formerly dorsal root ganglion, pl. ganglia) [Ggl. spinale]. In the lower medulla oblongata, both tracts contact the homonymous nuclei (Fig. 3.13A, B) in the dorsolateral parts of the medulla. They give rise to the second-order axons of the epicritic pathway which form the medial lemniscus [Lemniscus medialis], running directly dorsal of the pyramids—cross the midline and end in the ventral posterolateral nucleus of the thalamus [Ncl. ventralis posterolateralis, VPL) (see Fig. 3.13B).

The medial lemniscus is formed upstream to the decussation of the second-order epicritic cuneate and gracile nuclei (see Fig. 3.13A). It first shows a ventrodorsal orientation in two fiber columns directly lateral of the midline (see Fig. 3.14A, B). In the pons, they change their position gradually into a horizontal one during which the tracts are difficult to identify (Fig. 3.14C). In the cranial parts of the pons, their new position is clearly visible (Fig. 3.14D). At the transition to the mesencephalon, the medial lemnisci gradually change from the vertical position to a location lateral of the red nucleus (Fig. 3.14E, F). From there they run laterally to the thalamus.

3.2.3.2 Peripheral receptive structures of the dorsal column nuclei

The receptors that transmit signals to the first-order neurons of the dorsal column are the vibration receptors (Pacinian corpuscles) and the proprioceptors (Muscle spindles, tendon organs) (see Fig. 3.7).

By contrast to man, the rodent dorsal funiculi do not only contain the cuneate and gracile fasciculi but in its ventral most part the corticospinal tract (for details in mice, see Schröder et al. 2020).

In man, the (main) lateral corticospinal tract is located in the dorsolateral region of the ventrolateral funiculus.

Fig. 3.13 (**A**) Horizontal section through the human caudal medulla oblongata (Closed part of the medulla oblongata with the central canal (+). Compare with Fig. 3.14A and Darrow red 2, 2A). Darrow red staining. Dorsally, the medially located gracile fasciculus (from the lower part of the body) and the homonymous nucleus are seen. Lateral of it the cuneate fasciculus (upper part of the body) and nucleus are located (epicritic sensory system). The protopathic sensory system is mainly represented by the spinal tract and nucleus of trigeminal nerve (nociception of the face region) and the lateral spinothalamic tract—lateral of the lateral reticular nucleus—transmitting nociceptive, thermoceptive, and coarse mechanoceptive signals to supraspinal thalamic levels (see Fig. 3.13B). LabPON Twente. (**B**) Graphic visualization of the epicritic and protopathic sensory tracts ascending to and/or through the human brainstem. (1) The epicritic sensory first-order axons from the periphery of the body enter the spinal cord via the posterior root ⑧ and ascend uncrossed to the first nuclei of the system, the cuneate and gracile nuclei ④ ⑤. The second-order axons (Internal arcuate fibers [Fibrae arcuatae internae] cross in the decussation of the medial lemniscus [Decussatio lemnisci medialis] ③ to the contralateral side and ascend in the medial lemniscus ② to the ventral posterolateral nucleus of the thalamus [Ncl. ventralis posterolateralis, VPL] ①. (2) The protopathic first-order axons ⑦ enter the spinal cord as the epicritic fibers do but directly make synaptic contacts with the second-order neurons in the spinal cord. Their axons cross to the contralateral side and ascend as the spinothalamic tract/anterolateral system to the intralaminar and posterior ventrolateral nuclei of the thalamus ②. *Nieuwenhuys et al. (2008) have defined several parts of the reticular formation depending on their location. For the medial reticular formation, see Box 4.5. Modified after Nieuwenhuys et al. 2008, Figs. 16.1 and 16.3 with permission

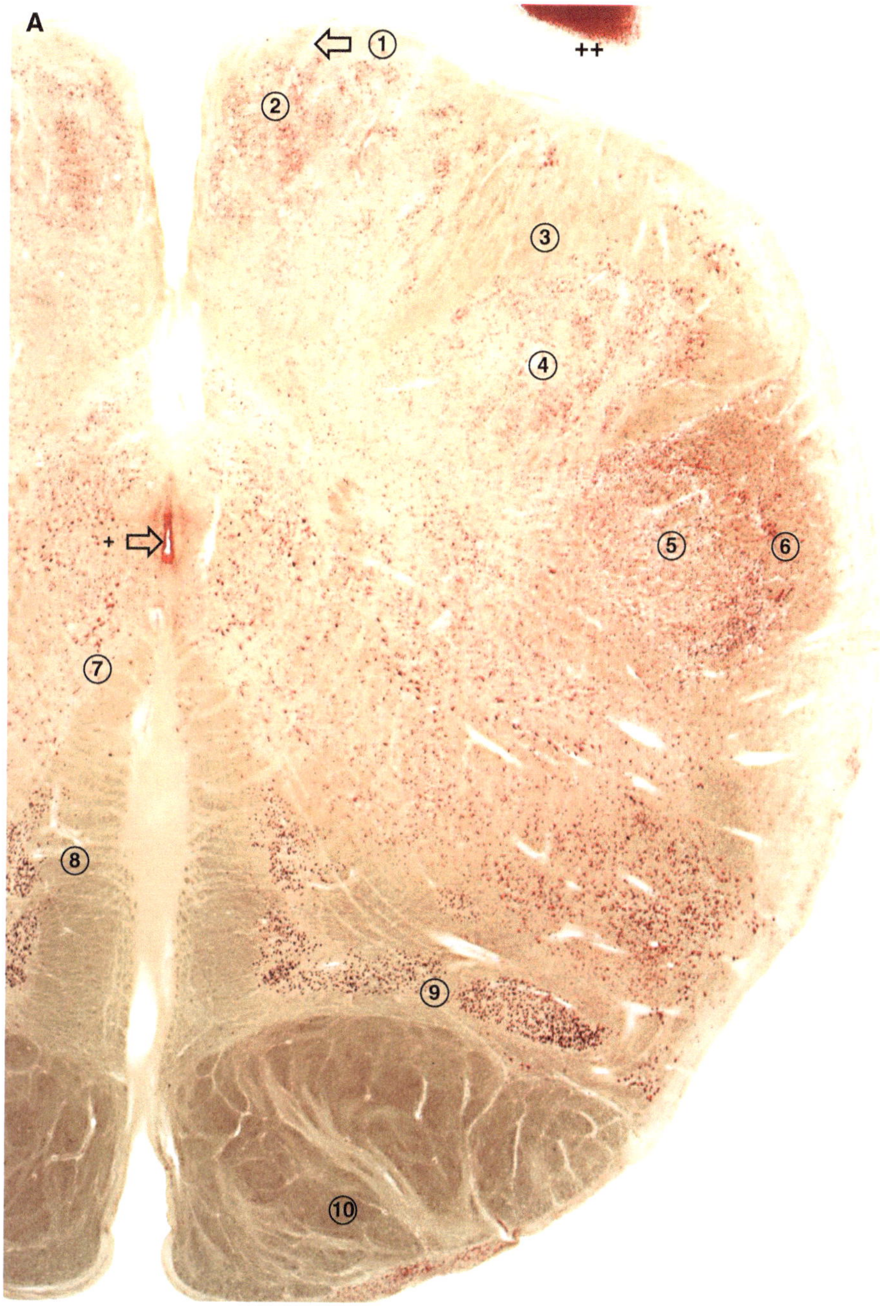

①	Gracile fasciculus	⑦	Medial longitudinal fasciculus
②	Gracile ncl.	⑧	Medial lemniscus
③	Cuneate fasciculus	⑨	Inferior olivary complex
④	Cuneate ncl.	⑩	Pyramidal (corticospinal) tract
⑤	Spinal ncl. of trigeminal nerve	+	Central canal
⑥	Spinal tract of trigeminal nerve	++	Cerebellum

B

(1) Epicritic sensory system
- Dorsal column tracts
- Medial lemniscus

(2) Protopathic sensory system
- Spinothalamic tract (Anterolateral system)

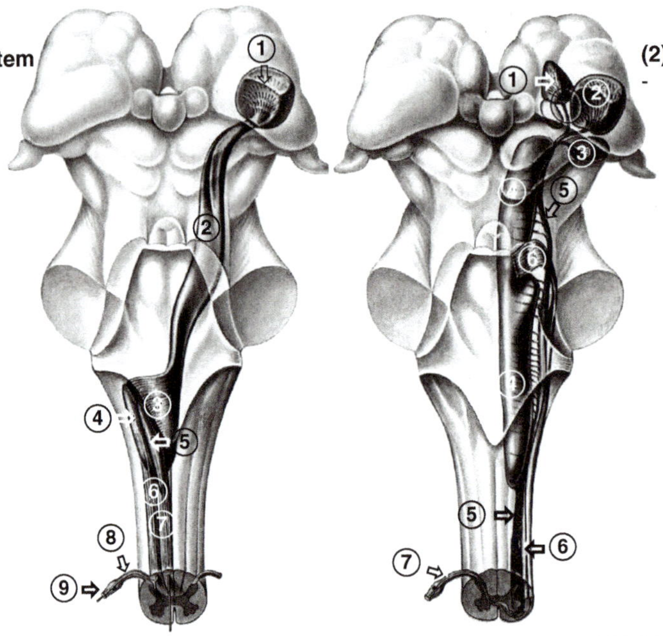

①	**Ventral posterolateral ncl.**
②	**Medial lemniscus**
③	**Internal arcuate fibers / Decussation of the medial lemnisci**
④	**Cuneate ncl.**
⑤	**Gracile ncl.**
⑥	**Cuneate fasciculus**
⑦	**Gracile fasciculus**
⑧	**Posterior root**
⑨	**Spinal ganglion**

①	**Intralaminar ncll.**
②	**Posterior ventrolateral ncl.**
③	**Medial geniculate body**
④	**Medial reticular formation***
⑤	**Spinothalamic tract**
⑥	**Anterior spinocerebellar tract**
⑦	**Posterior root**

Fig. 3.13 (continued)

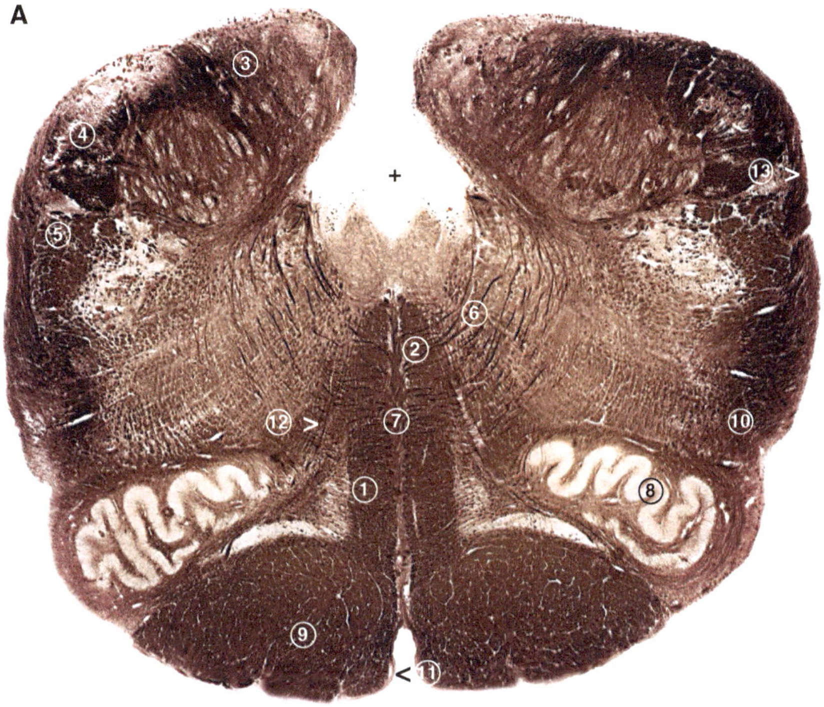

①	Medial lemniscus	⑧	Inferior olivary complex
②	Medial longitudinal fasciculus	⑨	Pyramidal (corticospinal) tract
③	Gracile fasciculus	⑩	Region of the spinothalamic fibers / Anterolateral tract
④	Cuneate fasciculus	⑪	Anterior external arcuate fibers
⑤	Spinal trigeminal tract	⑫	Hypoglossal nerve CNXII
⑥	Internal arcuate fibers	⑬	Inferior cerebellar peduncle
⑦	Decussation of medial lemnisci	+	4th ventricle

Fig. 3.14 (**A**) Horizontal section through the human caudal medulla oblongata at the level of the dorsal column nuclei. Campbell fiber staining. At the dorsolateral surface of the brainstem, you see the heavily stained fibers of the dorsal column tracts ③ ④ which here contact the second-order neurons of the epicritic pathways in the cuneate and gracile nuclei (see Fig. 3.13A). The cuneate and gracile nuclei axons form the internal arcuate fibers ⑥ which run to the midline and cross to the contralateral side ⑦ and get concentrated in the medial lemniscus ①. Do not confuse internal with external arcuate fibers (⑪, see Sect. 4.1.1.4) which run from the arcuate nucleus (see Sect. 4.1.1.4) to the inferior olive ⑧. The protopathic pathways are less well delimitable: the spinothalamic fibers ⑩ are located at the lateral surface between the spinal trigeminal complex ⑤ and the inferior olive ⑧. The vertical orientation of the medial lemnisci ① close to the midline is sustained up to the pons (see Fig. 3.14B, C). See also atlas part Campbell 1.

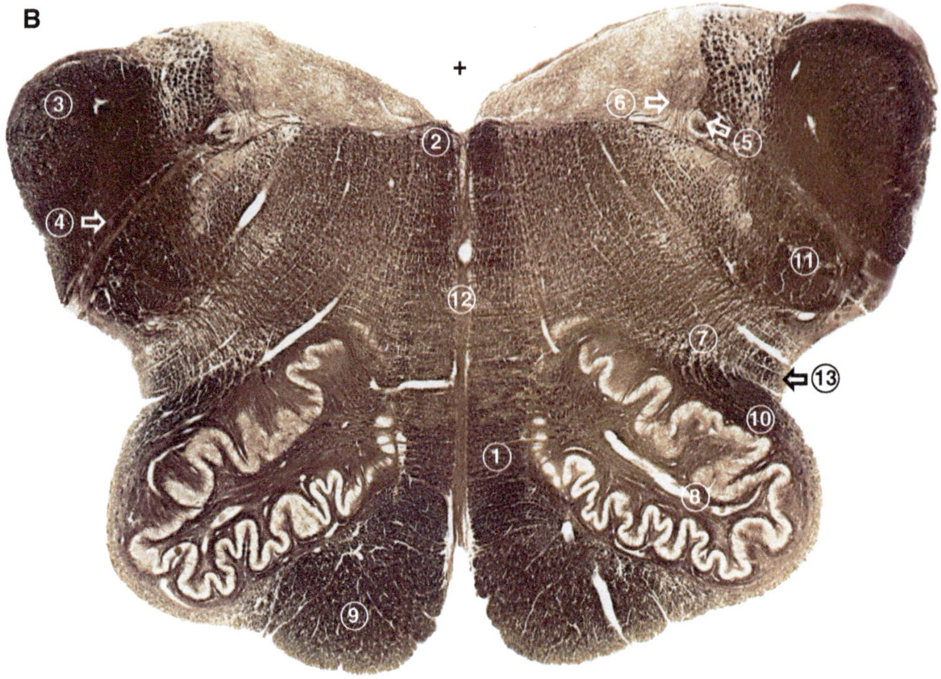

①	**Medial lemniscus**	⑧	**Inferior olivary complex**
②	**Medial longitudinal fasciculus**	⑨	**Pyramidal tract**
③	**Inferior cerebellar peduncle**	⑩	**Central tegmental tract**
④	**Glossopharyngeal nerve CNIX**	⑪	**Spinal trigeminal tract**
⑤	**Solitary tract**	⑫	**Decussation of medial lemnisci**
⑥	**Vestibular nerve CNVIII**	⑬	**Region of the spinothalamic fibers / Anterolateral tract**
⑦	**Olivocerebellar tract**	+	**4th ventricle**

Fig. 3.14 (**B**) Horizontal section through the human caudal medulla oblongata at the level of the glossopharyngeal nerve ④. Campbell fiber staining. The cuneate and gracile fasciculi are no longer present (see Fig. 3.14A). At their place lateral most of the massive inferior cerebellar peduncle ③ is seen followed medially by the vestibular area with the vestibular nerve ⑥. The medial lemniscus ① is in a position comparable to Fig. 3.14A, between inferior olivary complex ⑧ lateral, corticospinal tract ⑨ (pyramids) anterior (ventral) and medial longitudinal fasciculus ② (MLF) posterior (dorsal). As before, the spinothalamic fibers ⑬ are located dorsal to the inferior olive ⑧. See also atlas part Campbell 4, 5 Darrow red 12.

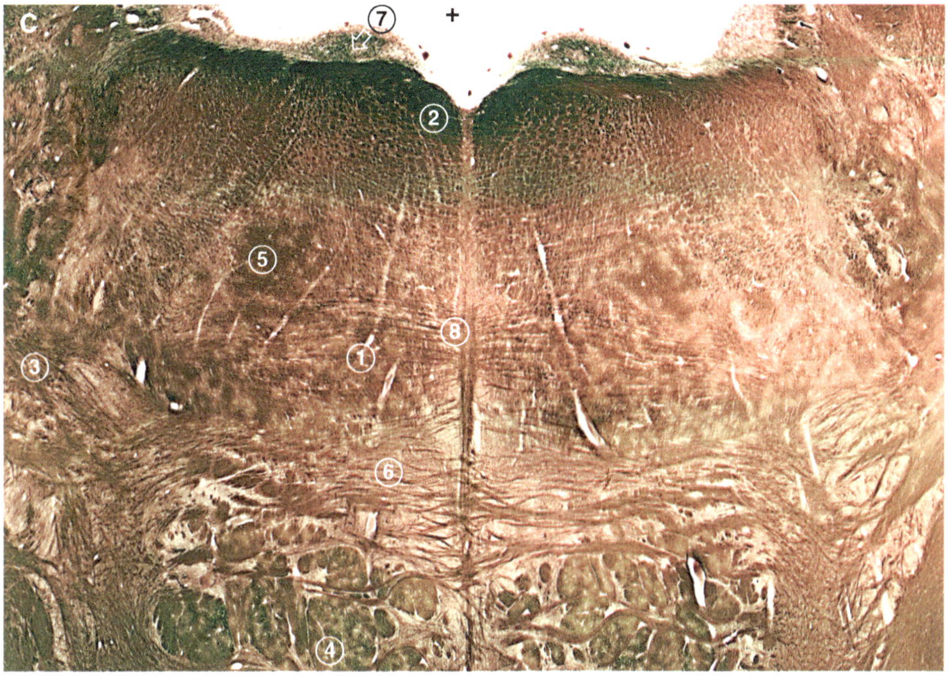

①	Medial lemniscus	⑥	Pontocerebellar fibers
②	Medial longitudinal fasciculus	⑦	Posterior longitudinal fasciculus
③	Anterolateral tract	⑧	Trapezoid body
④	Pyramidal tract	+	4th ventricle
⑤	Central tegmental tract		

Fig. 3.14 (**C**) Horizontal section through the human pons at the level of the ventral sensory trigeminal decussation ⑧. Campbell fiber staining. The medial lemniscus ① no longer maintains its vertical position and is difficult to identify at the midpontine level before it takes over a horizontal position further rostrally (see Fig. 3.14D). See also atlas part Campbell 11A.

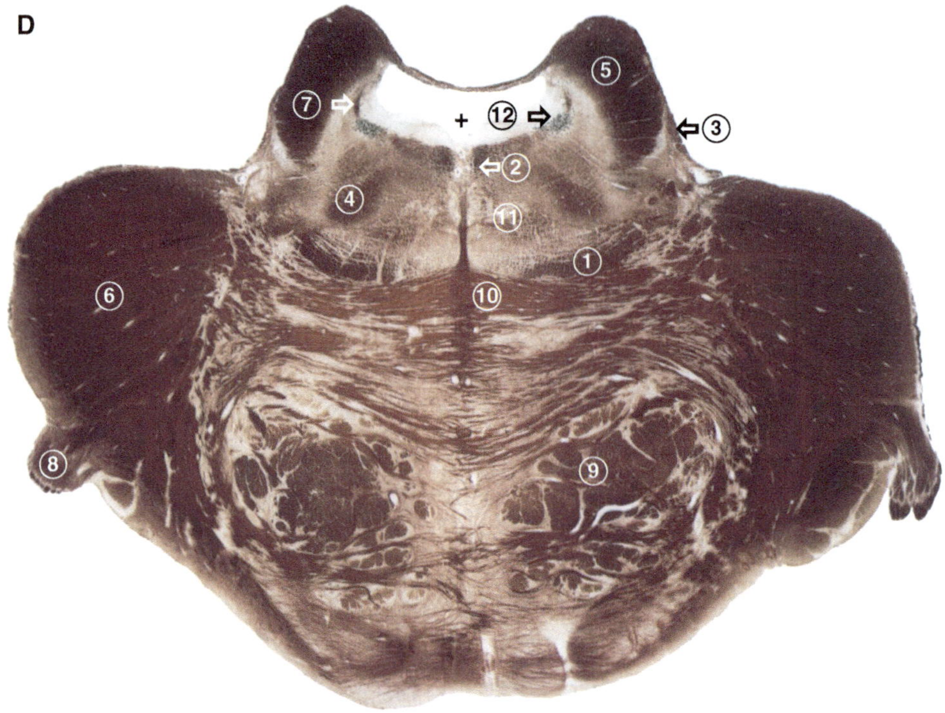

①	Medial lemniscus	⑦	Mesencephalic tract of trigeminal nerve
②	Medial longitudinal fasciculus	⑧	Trigeminal nerve CNV
③	Lateral lemniscus	⑨	Corticospinal tract
④	Central tegmental tract	⑩	Transverse pontine (pontocerebellar) fibers
⑤	Superior cerebellar peduncle	⑪	Tectospinal tract
⑥	Middle cerebellar peduncle	⑫	Locus caeruleus
		+	4th ventricle

Fig. 3.14 (**D**) Horizontal section through the rostral pons at the level of the exit of the trigeminal nerve ⑧ and the locus caeruleus ⑫. Campbell fiber staining. The picture is dominated by the massive fiber bundles of the superior ⑤ and middle cerebellar peduncles ⑥. Medially to the latter, the medial lemnisci ① are located horizontally (compare with Figs. 3.14A–C). The lateral lemniscus ③ lies between superior and medial cerebellar peduncles. Do not confuse medial and lateral lemnisci. The latter are the ascending auditory system from the cochlear nuclei to the inferior colliculi.

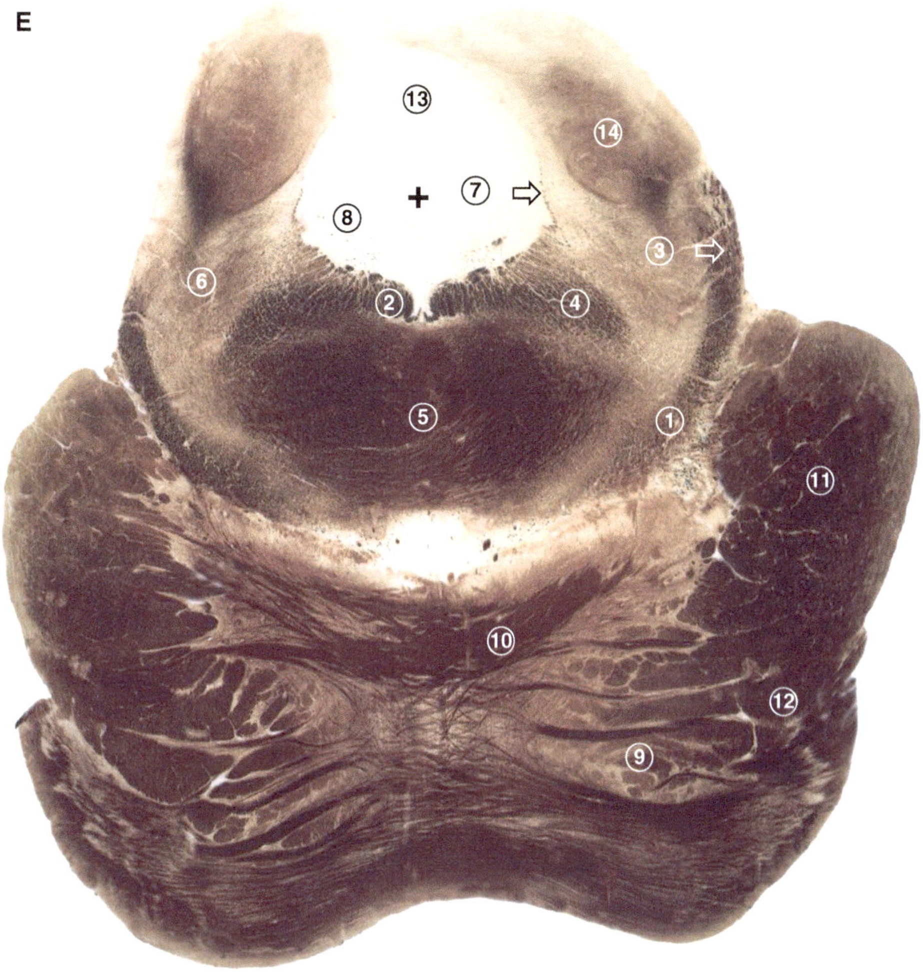

①	Medial lemniscus	⑧	Posterior longitudinal fasciculus
②	Medial longitudinal fasciculus	⑨	Pyramidal (corticospinal) tract
③	Lateral lemniscus	⑩	Transverse pontine fibers
④	Central tegmental tract	⑪	Parietotemporopontine fibers
⑤	Decussation of superior cerebellar peduncles	⑫	Frontopontine fibers
⑥	Anterolateral tract	⑬	Periaqueductal gray substance
⑦	Mesencephalic tract of trigeminal nerve	⑭	Inferior colliculus
		+	Mesencephalic aqueduct

Fig. 3.14 (E) Horizontal section through the brainstem (transition pons to midbrain), level of the decussation of the superior cerebellar peduncles ⑤ and the inferior colliculi ⑭. Campbell fiber staining. Medial ① and the lateral lemnisci ③ now form an arcuate structure between parieto-temporopontine fibers ⑪ and inferior colliculi ⑭. The fibers of the anterolateral system ⑥ are located ventral (anterior) of the inferior colliculi.

F

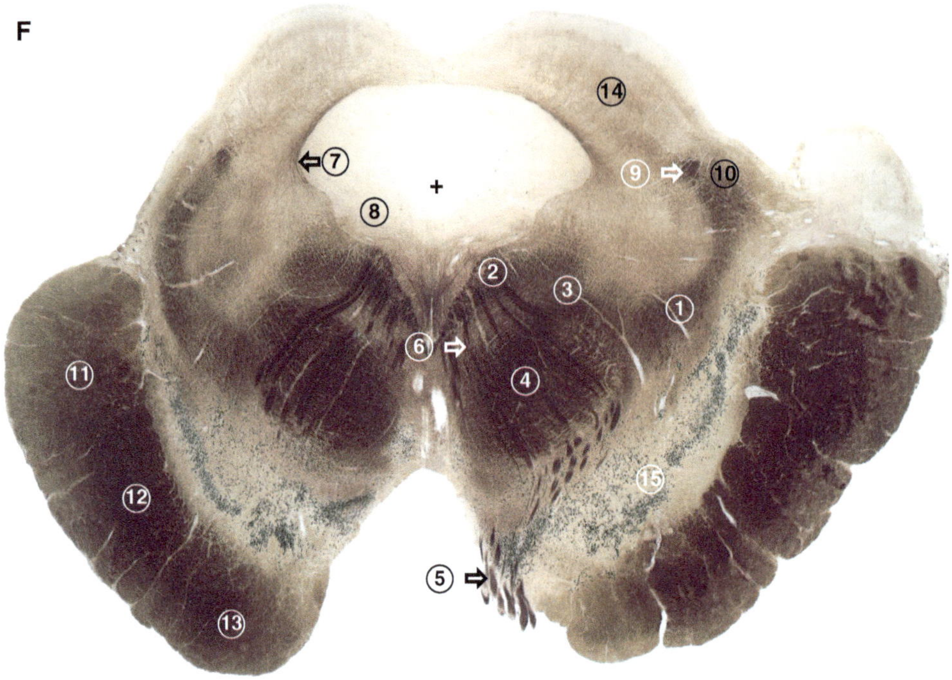

①	Medial lemniscus	⑨	Anterolateral tract
②	Medial longitudinal fasciculus	⑩	Brachium of inferior colliculus
③	Central tegmental tract	⑪	Parietotemporopontine tract
④	Superior cerebellar peduncle (Red nucleus)	⑫	Corticospinal tract
⑤	Oculomotor nerve CNIII	⑬	Frontopontine tract
⑥	CNIII inside red ncl.	⑭	Superior colliculus
⑦	Mesencephalic tract CNV	⑮	Substantia nigra
⑧	Posterior longitudinal fasciculus	+	Mesencephalic aqueduct

Fig. 3.14 (**F**) Horizontal section through the human mesencephalon at the level of the intracerebral course of the oculomotor nerve ⑤. Campbell fiber staining. The medial lemniscus ① is still in a position as in Fig. 3.14E, but the lateral lemniscus has merged into the inferior colliculi (see Fig. 3.14E) which here have been replaced by the superior colliculi ⑭. The fibers of the anterolateral system (anterolateral tract) ⑨ are located dorsomedial of the medial lemniscus at the base of the superior colliculi. See also atlas part Darrow red 40, 40A.

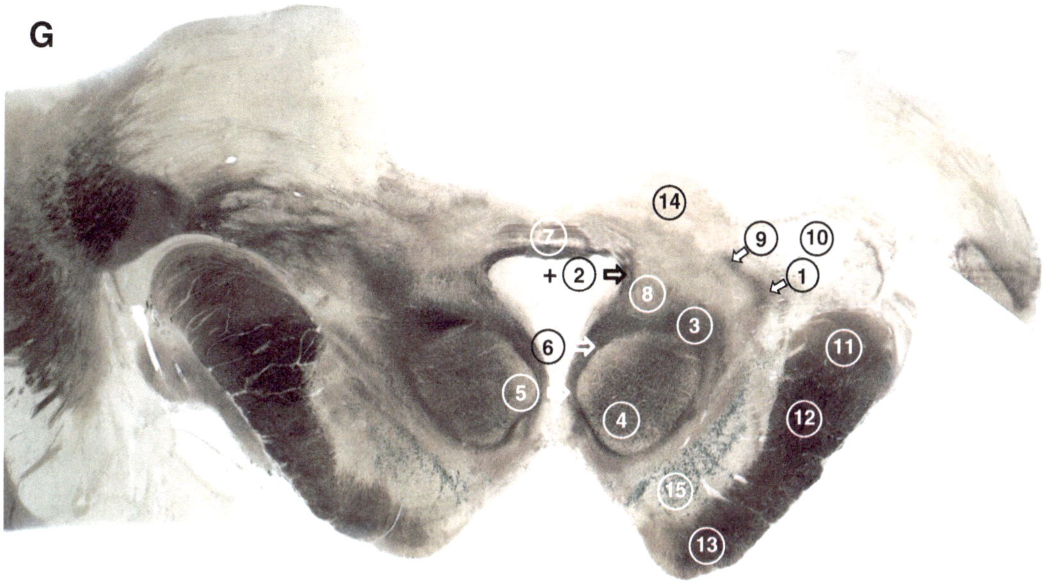

①	Medial lemniscus	⑨	Anterolateral tract
②	Dorsal longitudinal fasciculus	⑩	Medial geniculate body
③	Central tegmental tract	⑪	Parietotemporopontine tract
④	Superior cerebellar peduncle (Red nucleus)	⑫	Pyramidal tract
⑤	Habenulointerpeduncular tract	⑬	Frontopontine tract
⑥	Medial longitudinal fasciculus	⑭	Superior colliculus
⑦	Posterior commissure	⑮	Substantia nigra
⑧	Posterior trigeminothalamic tract	+	Mesencephalic aqueduct

Fig. 3.14 (**G**) Horizontal section through the human brainstem at the level of the transition between the rostral mesencephalon and the thalamus ⑦ ⑩. Campbell fiber staining. The medial lemniscus ① is now seen as a hook-shaped structure dorsolateral of the red nucleus/superior cerebellar peduncle ④ and the posterior trigeminothalamic tract ⑧. The spinothalamic fibers ⑨ sit at the mediodorsal end of this hook. Both fiber bundles from here enter the thalamus (see Box 3.6). See also atlas part Darrow red 41

3.2.3.3 Cuneate nucleus r11 [Ncl. cuneatus]

See Sect. 3.2.3.4

3.2.3.4 Gracile nucleus r11 [Ncl. gracilis]

(cuneatus = wedge-shaped, gracile = slender)

The cuneate and gracile fasciculi [Fasciculi cuneatus/gracilis] target the nuclei of the same name in the medulla oblongata. Both nuclei are located at the dorsal surface of the medulla oblongata (see Fig. 3.13A). The axons of the epicritic second-order neurons originating in either nucleus cross to the contralateral side in the decussation of the medial lemnisci [Decussatio lemniscorum medialium] and eventually reach the thalamus (see Figs. 3.13B and Fig. 3.14A–F).

However, based on findings in monkeys, a new subdivision in four subnuclei is also used in the FIPAT Ch. 1 terminology (see Florence et al. 1989). The term decussation / decussatio is derived from Latin decim = X = ten. The X is thought as a symbol for a crossing.

As to the human cuneate nucleus, Del Fiacco Quartu et al. (2014) have shown the existence of subregions, characterized by neurochemical and histological features, which may be committed to the neurotransmission of protopathic transmission.

Olszewski and Baxter (1982) have described the following histological features:

- Cuneate nucleus: Round or oval plump cells with short dendrites, excentric nuclei.
- Gracile nucleus: Wide morphological variety, the majority of cells are medium-sized, plump or elongated, multipolar with the nucleus in a central position (for details, see Olszewski and Baxter 1982).

The course of the medial lemniscus (epicritic signals from the dorsal columns) together with that of the spinothalamic fibers (protopathic signals from the anterolateral columns) is shown at selected levels of the brainstem in the figures above (Fig. 3.14A–F) (All figures: LabPON Twente).

Box 3.6 Tracts of the human brainstem (see also Tables 1.3 and 1.4)
For reasons inherent to experimental work, the direct visualization of fiber tracts in humans is impossible. Some data can be gained by very laborious approaches like the postmortem investigation of degenerating terminals in cases where the neurological deficits were well documented (see Sect. 4.3.1.1.4 "Cerebrocortical control of brainstem motor nuclei," see Box 3.8). Thus, projection and connectivity pattern of most human brainstem structures have to be correlated with animal findings. Due to the minute size of most brainstem structures, diffusion tensor imaging (DTI) (Box 3.7) will only fulfill anatomical requirements in special selected cases. This will remain a problem as long as the spatial resolution of in vivo tracing is limited.

Box 3.7 Diffusion tensor imaging (DTI)
DTI is a variant of diffusion-weighted magnetic resonance imaging (DW-MRI). In contrast to DW-MRI, DTI not only visualizes the movement of water molecules but also their direction by calculating a tensor enabling to monitor the H_2O movement in 3D. The H_2O movement is more pronounced in parallel to the fiber orientations rather than diagonally. It is used to investigate fiber tracts of the brain in vivo.

Box 3.8 The history of tract identification in the human brain
A pioneer in the correlation of clinical findings and deficits with the postmortem assessment of the pathologic and topographical anatomy was the Austrian neurologist Ludwig Türck. Already in 1835, he wrote (first the German original text followed by the English translation):

"*Die Experimente, welche am lebenden Thiere angestellt wurden, um die Leitungsvorgänge im Rückenmarke zu ermitteln, können beim Menschen zum Theil durch die klinische Beobachtung solcher Fälle ersetzt werden, in welchen Segmente des Rückenmarkes pathologisch-anatomische Veränderungen eingegangen sind.*

Hierbei tritt an die Stelle der partiellen künstlichen Trennungen am Thiere die Erkrankung einzelner Stellen des menschlichen Rückenmarkes; in beiden Fällen hat man zu untersuchen, wie sich Sensibilität und Motilität in den unterhalb gelegenen Theilen verhalten.

Those experiments that have been performed on living animals in order to elucidate potential ways of signal transmission in the spinal cord can partly be substituted in man by the clinical observation in cases in which segments of the spinal cord were subject of pathologic-anatomical changes.

Instead of the partial artificial transection in animals, in man the lesions of individual parts of the spinal cord have to be assessed. In either case, it is mandatory to examine how sensory and motor performance in the distal parts behave."

A classic example using this approach by an involuntary experiment of nature is the investigation of dorsal column degeneration in Tabes dorsalis. This condition belongs to the spectrum of symptoms observable in the stadium of syphilis with affection of the central nervous system. Cases of Tabes dorsalis were quite frequent in the nineteenth century due to the lack of an efficient therapy which is available today.

The main clinical signs of Tabes (decay, from Latin *tabere* = to decay) dorsalis (referring to the affection of the dorsal columns, see here below) are a locomotor ataxia, impaired proprioception, and decreased deep tendon reflexes. All of these can be attributed to a degeneration of the dorsal columns:

– Impaired proprioception: The peripheral sensors of the epicritic system are proprioceptors of muscles and tendons.
– Monosynaptic or deep tendon reflexes in their afferent part rely on intact proprioceptors. When the afferent input is strongly reduced, this is reflected in a reduced motor response.
– Locomotor ataxia (Patients are moving like passengers on a ship in huge seas to improve balance): In Tabes dorsalis, the proprioceptive input which humans are particularly dependent on for proper locomotion is lacking. Such patients are visually dependent when walking (they look down) and perform poorly in the dark. They walk slowly, have a postural instability—a tendency to be unstable when standing—and frequently fall to the ground. Nowadays locomotor ataxia may, for example, rather be due to chemotherapy or cobalamin deficiency (Vitamin B12 deficiency) (Viswanathan and Sudarsky 2012).

In patients who suffer from Tabes dorsalis, the neuropathological postmortem examination shows a partial or complete demyelination of the dorsal column(s) (for details, see Legend Fig. 3.15).

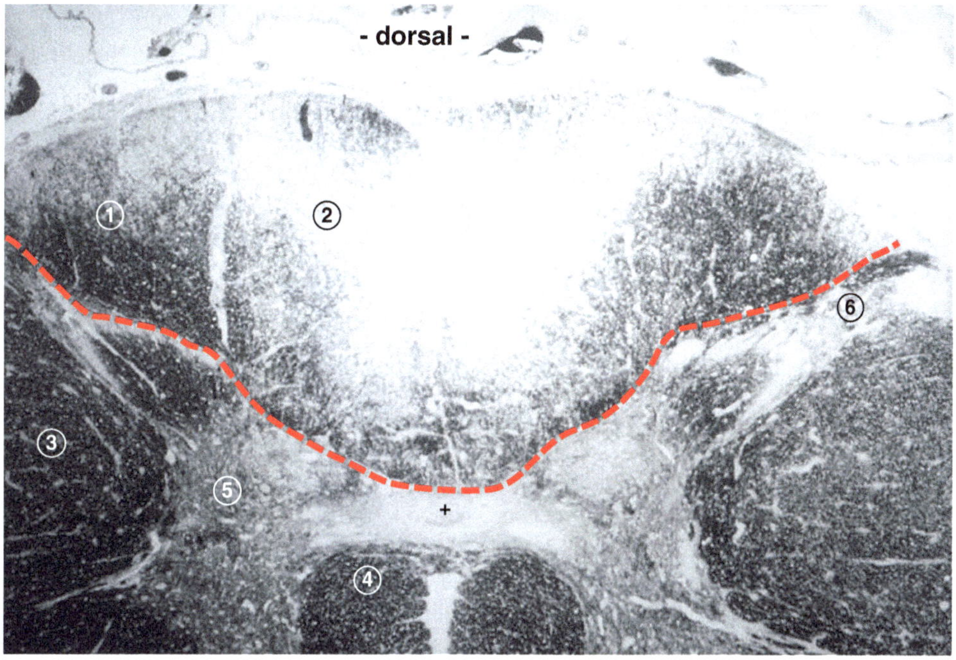

---	Approximate ventral border of the dorsal column	④	Anterior spinal fasciculus
①	Cuneate fasciculus	⑤	Spinal gray matter
②	Gracile fasciculus	⑥	Dorsal horn / dorsal root entry zone
③	Ventrolateral spinal fasciculus	+	Spinal central canal

Fig. 3.15 Horizontal section through the human spinal cord (dorsal part, rostral segment). Tabes dorsalis. Myelin stain. The dorsal most you see the dorsal column with its fasciculi ①②. The gracile nucleus contains the epicritic sensory fibers from the lower part of the body; those from the upper part are attached during the caudo-rostral course of the tracts. The ventral neighbor of the dorsal column is the spinal gray matter ⑤, in particular the dorsal horn ⑥ which separates the dorsal column from the more ventral parts of the white matter ③. The myelin stain is supposed to label all parts of the CNS with myelinated fibers in high density (the macroscopically white matter) while those regions which harbor more perikarya like here the grey matter ⑤ are less intensely stained. Lack of staining—in this specimen clearly visible in the gracile fasciculus ②—points to loss of myelinated fibers. This finding fits in with the locomotor ataxia given the fact that the gracile fasciculus collects the proprioceptive afferents from the lower half of the body. From the Centers for Disease Control and Prevention (CDC), Susan Lindsley, Public Health Image Library (PHIL), identification number #2345. Public domain

3.2.3.5 CADASIL, CARASIL and CARASAL

Cadasil (cerebral autosomal dominant arteriopathy with subcortical infarcts and leucoencephalopathy) was first described by van Bogaert in 1955 as Binswanger's disease with a rapid course in two sisters (van Bogaert 1955). The disease is dealt with at this position here since in Fig. 3.17A you can see a cystic infarction of the medial lemniscus (compare with Fig. 3.14A–G here above). It is a hereditary disease, characterized by recurrent subcortical ischemic strokes, starting in early or midadulthood and leading in a number of patients to dementia (Tournier-Lasserve et al. 1993). CADASIL is an angiopathy caused by mutation in the Notch 3 gene, which encodes NOTCH3 receptor expressed in vascular smooth muscle cells (VSMC). Notch3 extracellular domain (N3ECD) accumulates in arterial walls followed by VSMC degeneration, fibrosis and stenosis of arterioles, predominantly in cerebral white matter (Tikka et al. 2014). Linkage studies assigned the disease locus to chromosome 19q12 (Tournier-Lasserve et al. 1993; Joutel et al. 1996). Since then, CADASIL has been reported in more than 500 families worldwide, but its overall prevalence is unknown (Chabriat et al. 2020). In 1960, CARASIL (also known as Maeda syndrome), autosomally inherited, was described by Nemoto similarly as a form of Binswanger syndrome and then by

Maeda as multifocal softenings in the cerebral white matter caused by cerebral angiitis (Maeda et al. 1976; Tikka et al. 2014). The gene defect of CARASIL was mapped to a 2.4-Mb region on chromosome 10q (Hara et al. 2009). CARASAL (cathepsin A-related arteriopathy with strokes and leukoencephalopathy) is a very rare adult-onset small vessel disease in which endothelin-1 seems to play a role in the pathogenesis of the leukoencephalopathy (Bugiani et al. 2016; Haffner and Vinters 2016).

CADASIL is the most common small vessel disease that has been reported to account for 2% of cases of lacunar stroke with leucoaraiosis in patients younger than 65 years and for 11% of cases in those younger than 50 years (Dong et al. 2003). The clinical presentation varies between and within families but the disease is mainly characterized by migraine (its frequency is five times higher than in the general population) with aura, subcortical ischemic events (TIA's and ischemic strokes), mood disturbances, apathy, and cognitive impairment (the second most frequent clinical manifestation of CADASIL) (Chabriat et al. 2020; Dichgans et al. 1998). MRI changes, such as subcortical infarcts and leucoencephalopathy, precede the onset of symptoms by 10–15 years (Chabriat et al. 1998, 2020; Dichgans et al. 1998, 1999). There is currently no treatment for CADASIL (Chabriat et al. 2020; Bersano et al. 2017).

Macroscopic examination of the brain may show a more or less uniform pattern of changes. Multiple small infarcts, particularly in the basal ganglia, rarefaction of the periventricular white matter, thalamus, mesencephalon, and pons. Often a variable degree of cortical white matter atrophy and a diffuse myelin pallor is seen (Sourander and Wålinder 1977) (Fig. 3.16).

Widespread neuronal apoptosis is seen in the cortex, especially in the presence of a subcortical ischemic lesion (Viswanathan et al. 2006). Furthermore, microscopic findings consist of focally intensive pallor of myelin staining throughout the white matter, preserving the U-fibers. Multiple infarcts are present in the internal and external capsules and in the deep white matter. The corpus callosum could be involved. The lesions are nearly symmetrical in both hemispheres. In the mesencephalon and pons, the longitudinal tracts, like the medial lemniscus (Fig. 3.17), show pallor of myelin and infarcts have been described (Sourander and Wålinder 1977; Ruchoux and Maurage 1997) (Figs. 3.17, 3.18, and 3.19).

The vascular changes were most prominent in small muscular arteries affecting mainly the small penetrating cerebral and leptomeningeal arteries, characterized by a concentric thickening of the wall leading to narrowing of the vascular lumen, a largely normal endothelium and a smudgy hyaline degeneration within the intima (PAS- positive) (Fig. 3.18B).

The granular osmiophilic material (GOM) (Figs. 3.19 and 3.20), located within the media and extending into the adventitia, is the pathological hallmark of CADASIL. It is not only found in the brain, but also in the peripheral vascular tree (Ruchoux and Maurage 1997; Chabriat et al. 2020). GOM consists of electron-dense, extracellular, granular depositis, ranging in size from very small, barely detectable deposits to large deposits (0.2–0.8 mm). They are located close to the smooth muscle cells (Ruchoux and Maurage 1997) (Figs. 3.19 and 3.20). GOMs commonly have a mushroom shape with the foot closely apposed to the VSMC-plasma membrane and the head projecting to the extracellular membrane (ECM). Alternatively, GOMs are irregularly rounded bodies, usually located in indentations of VSMCs (Tikka et al. 2014). Immunostaining with a NOTCH3 monoclonal antibody is highly sensitive and specific (Chabriat et al. 2020; Lesnik Oberstein et al. 2003) for GOM. This kind of arteriopathy is also present in other organs, such as spleen, liver, kidneys, muscle, and skin (Ruchoux and Maurage 1997).

The important message is that a skin biopsy can be used as a diagnostic tool for a strictly neurological disease, despite an immunohistochemical test, that is now available. Later on, it became an important research tool and information obtained by the knowledge of CADASIL provided new tools for research and clues to understand the consequences of vascular impairments in dementia (Ruchoux et al. 2002, 2006). Because vascular smooth muscle cells secrete a powerful endothelial permeability factor (VEGF), their destruction could lead to a decrease in vascular permeability. The degeneration of VSMCS may rather be triggered by the disruption of their normal anchorage, based on the important role of adhesion for cell survival (Ruchoux et al. 2003).

CARASAL (cathepsin A-related arteriopathy with strokes and leukoencephalopathy) is a hereditary adult-onset cerebral vessel disease in which endothelin-1 may have a role, next to the vascular abnormalities (Bugiani et al. 2016) (Fig. 3.21).

Fig. 3.16 CADASIL. Coronal section through the cerebral hemispheres at frontal level (**A** and **B**). Grayish-brown periventricular white matter (small arrows) and involvement of the genu of the corpus callosum (long arrrows). In (**C**) coronal section at the level of the amygdala. Necrotic defect in the pallidum (long arrow), lacunar state in the putamen (triangle). In (**D**) histological section at the same level as in (**C**). H & E Stain. LabPON Twente

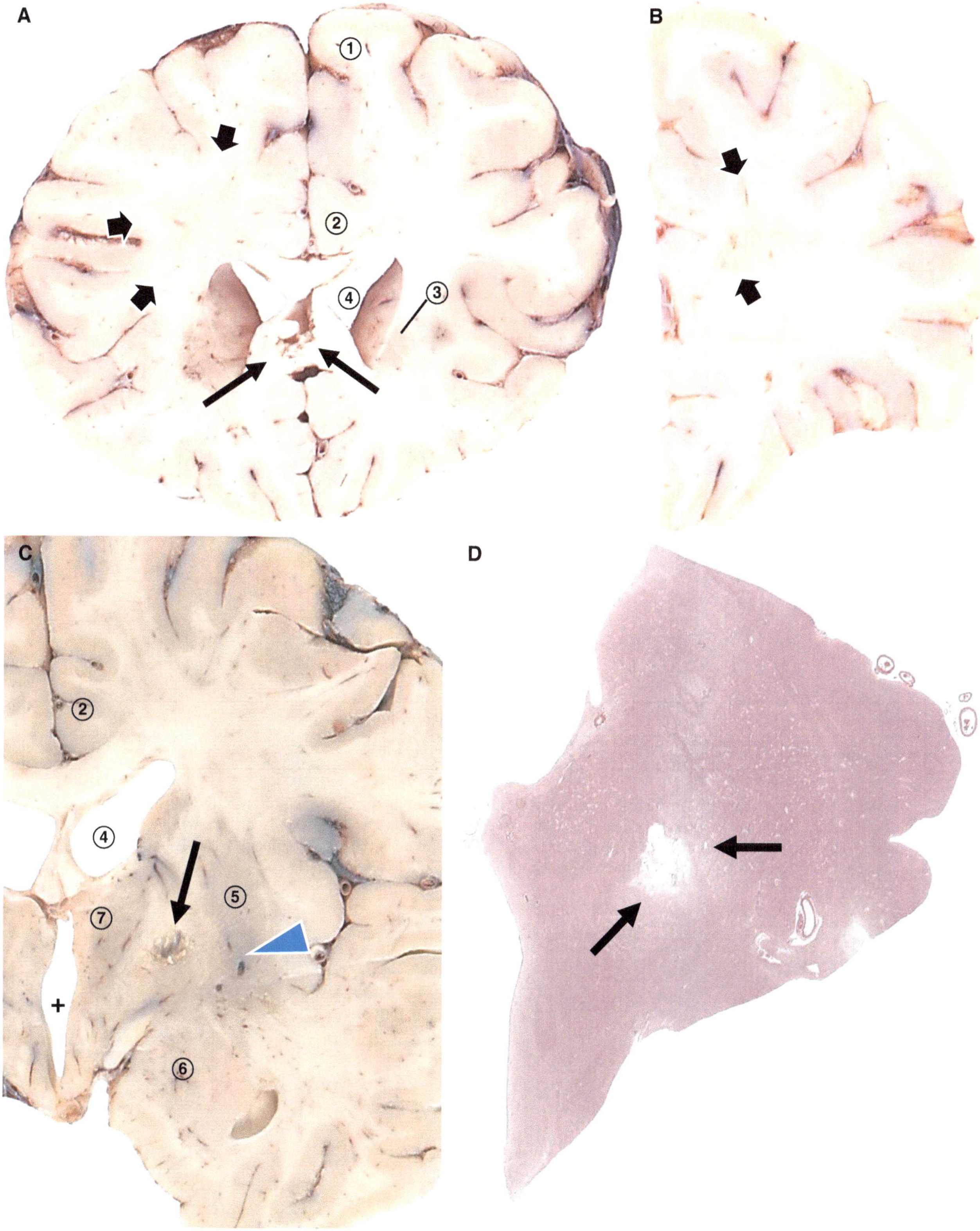

①	**Superior frontal gyrus**	⑤	**Putamen**
②	**Cingulate gyrus**	⑥	**Amygdala**
③	**Caudate ncl.**	⑦	**Thalamus**
④	**Lateral ventricle**	+	**3rd ventricle**

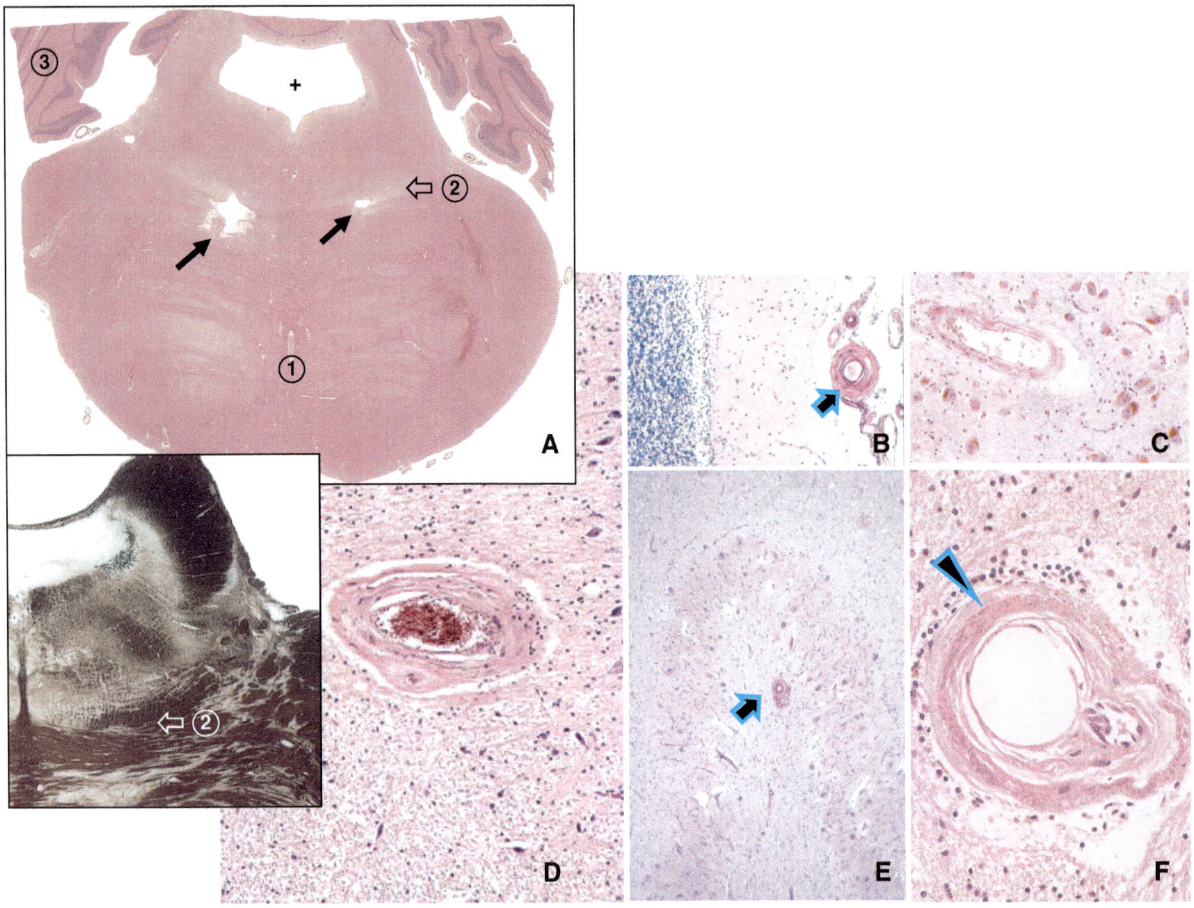

①	Pons	③	Cerebellum
②	Medial lemniscus	+	4th ventricle

Fig. 3.17 CADASIL. (**A**) Cystic infarctions in the pons (arrows) in the medial parts of the medial lemniscus in a control case. H & E stain. Compare with the inset, Campbell fiber staining. In (**B**) pons artery with concentric lamellar fibrosis and narrowing of the lumen. Vessel wall changes in the cerebellar cortex (**C**), the substantia nigra (**D** and **F**), inferior olivary nucleus (**E**) (short arrow). Typical smudgy hyaline degeneration in (**D**) and (**F**) (triangle). H & E stain. (see Fig. 3.14D, see Sect. 3.2.2.2.2). LabPON Twente

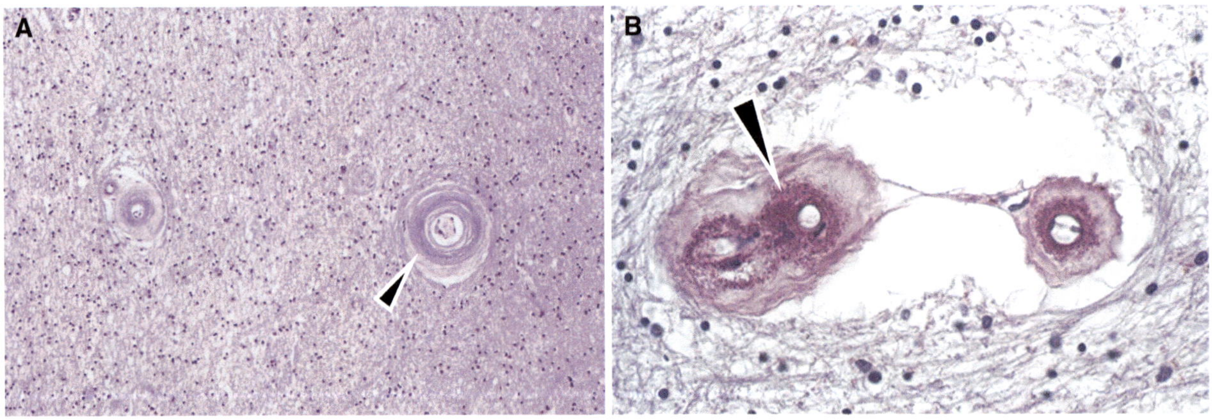

Fig. 3.18 CADASIL. Vessel changes in the cortical white matter showing the typical smudgy hyaline degeneration of the intima. Strong periodic acid-Schiff reaction (PAS stain) in the thickened intima and media in (**B**) (triangle).There is an underlying arteriopathy, i.e., widely spread occlusive vascular changes characterized by a degeneration of vascular smooth muscle cells (VSMCs) and the abnormal accumulation of electron-dense granular material called GOM and Notch 3 protein (Figs. 3.19 and 3.20). LabPON Twente

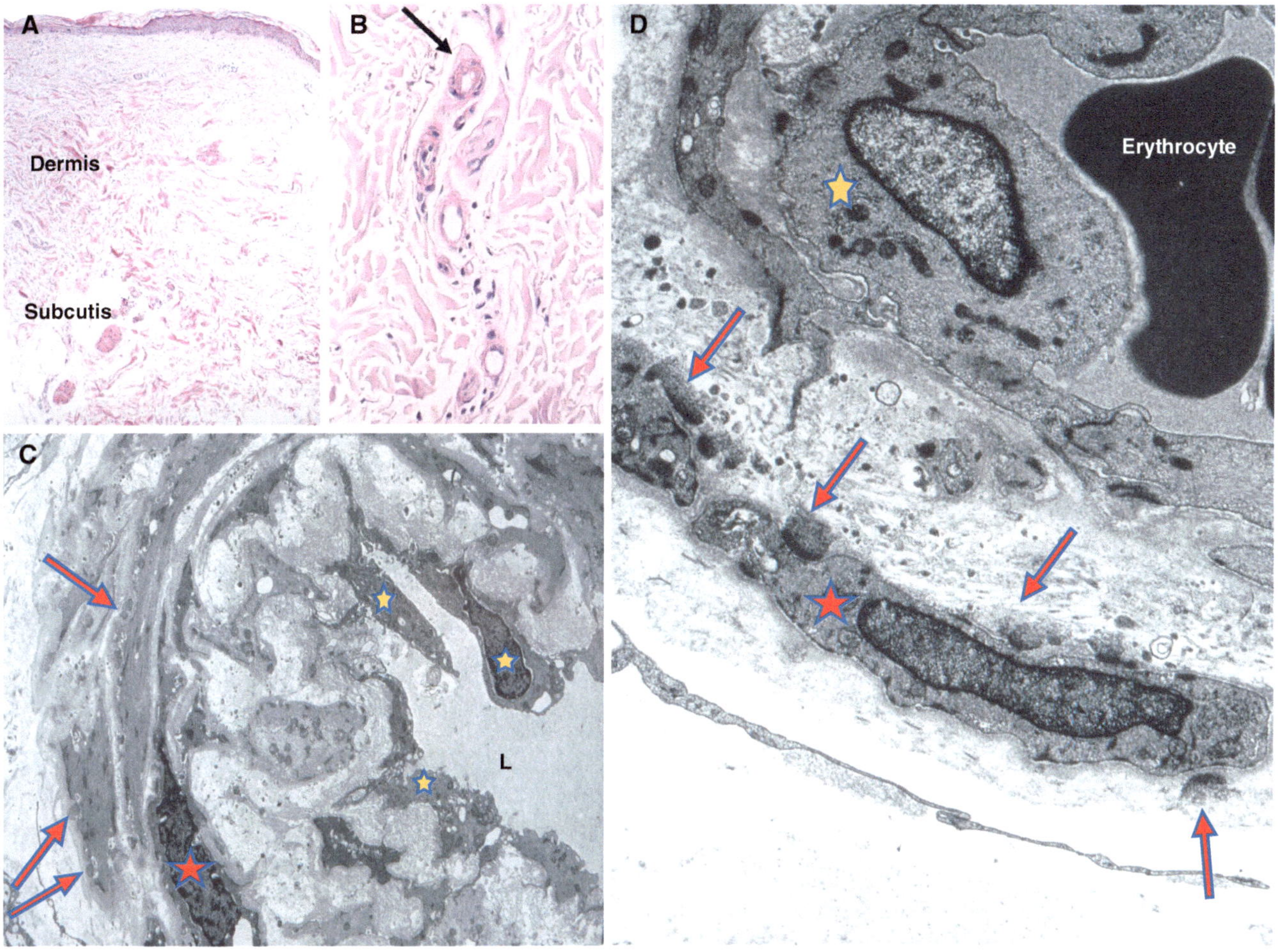

Fig. 3.19 (**A**, **B**) Skin biopsy, arrow points to an arteriole in a H & E stain. (**C**, **D**) Skin arterioles, electron microscopical picture (**C**, detail in **D**) from a patient with CADASIL, showing numerous sites with granular osmiophilic material (GOM), GOM deposits (red arrows). GOMs are characteristic for CADASIL. Endothelial cells = yellow stars; VSMC = red stars; L = lumen. Courtesy Prof. M.-M. Ruchoux

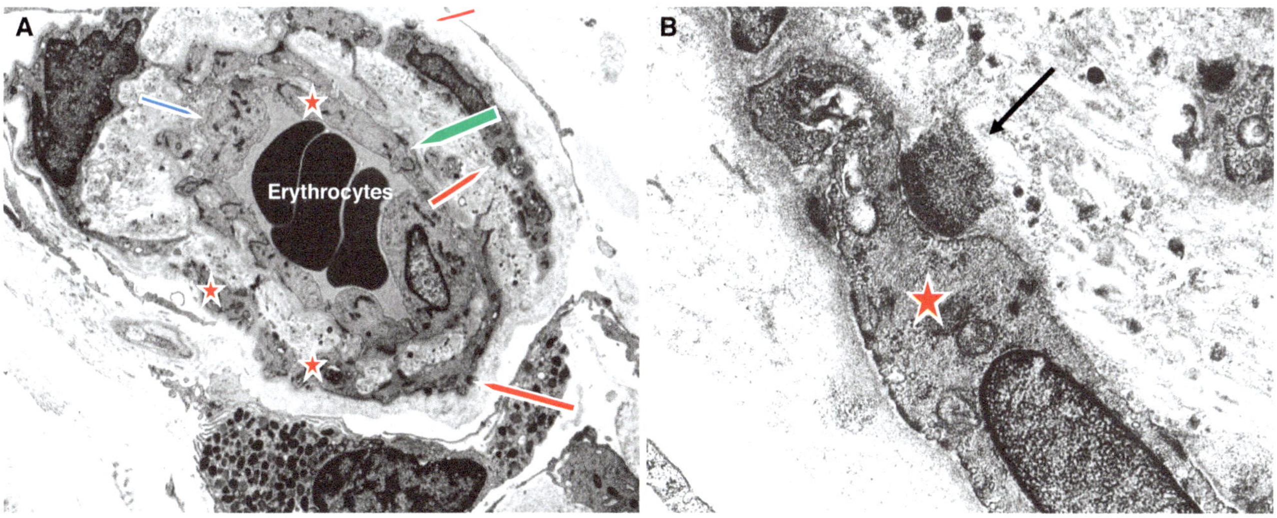

Fig. 3.20 CADASIL. (**A**) Electron microsopy. Arteriole showing characteristic smooth muscle cell fragmentation (green arrow) and GOMs (red arrows). VSMC = red stars; elastic lamina = blue arrow. (**B**) Detail illustration. Typical GOM in a smooth muscle cell notch (arrow). Smooth muscle cell = red star. Courtesy Prof. M.-M. Ruchoux

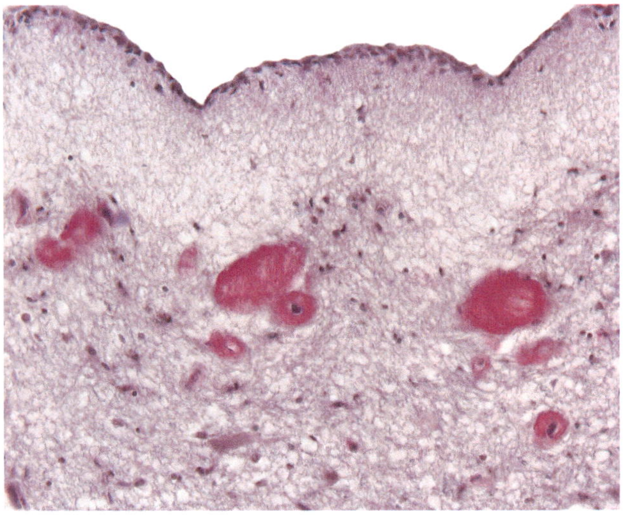

Fig. 3.21 Frontal white matter with subependymal, thick-walled, small vessels, typical for CARASAL. Elastica von Gieson stain. Courtesy Dr. M. Bugiani

3.2.4 Solitary nuclei r11

3.2.4.1 Solitary nucleus / Ncll. of solitary tract r11 [Ncll. tractus solitarii] (NTS)

(*solitarius* Latin = person living alone, hermit)

3.2.4.1.1 Location and fine structure of the NTS

The NTS (Fig. 3.22) comprises a number of subnuclei (Box 3.9) which are arranged along the caudo-cranial axis of this complex that reaches from the lower medulla oblongata up to rhombomere 6 (Chap. 8). The neurons of the NTS are arranged around the solitary tract [Tractus solitarius]. In routinely stained sections as in our atlas figures, the solitary tract appears as a round, clear, and homogeneous structure (see Fig. 3.23). As in rodents, the NTS can be subdivided into ten subnuclei (see Box 3.9).

Fig. 3.22 Approximate position of the nucleus of the solitary tract (NTS) inside the human brainstem. The NTS extends from the lower medulla oblongata—where the nuclei of both sides are connected via the commissural subnucleus in the midline—to approximately the level of the inferior salivatory nucleus, assuming a more lateral position and increasing in volume in rostral direction. Please note that due to space limitation not all terms correspond to FIPAT nomenclature. From Huggenberger et al. 2019, Fig. 15.5 with permission

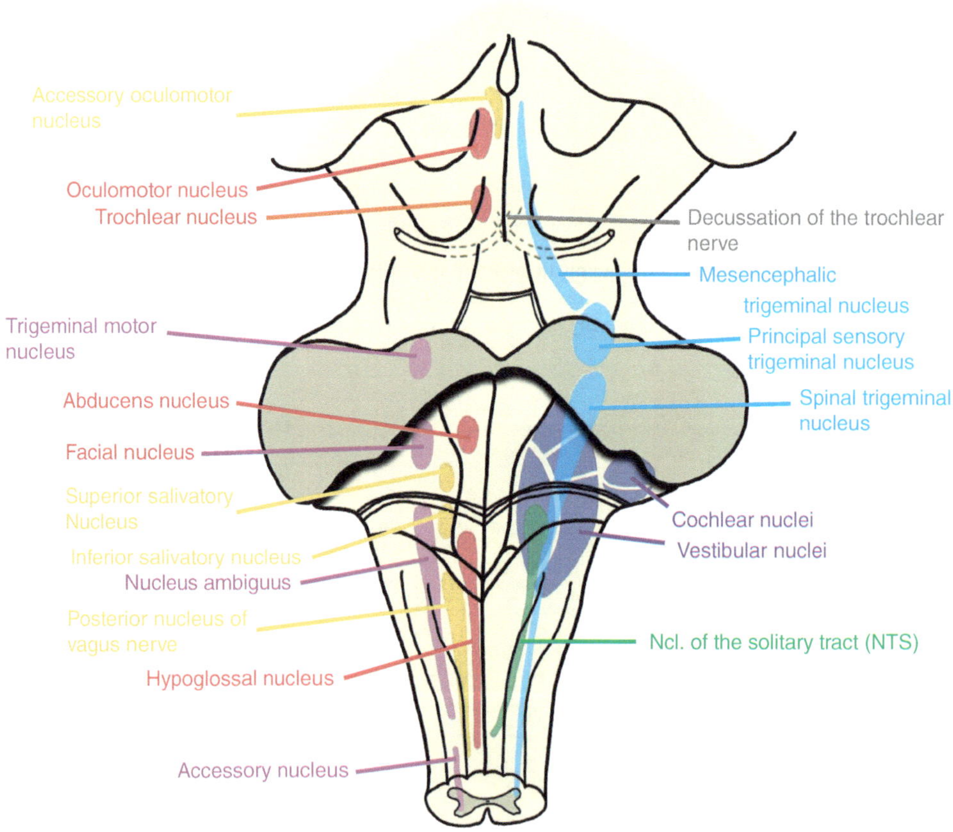

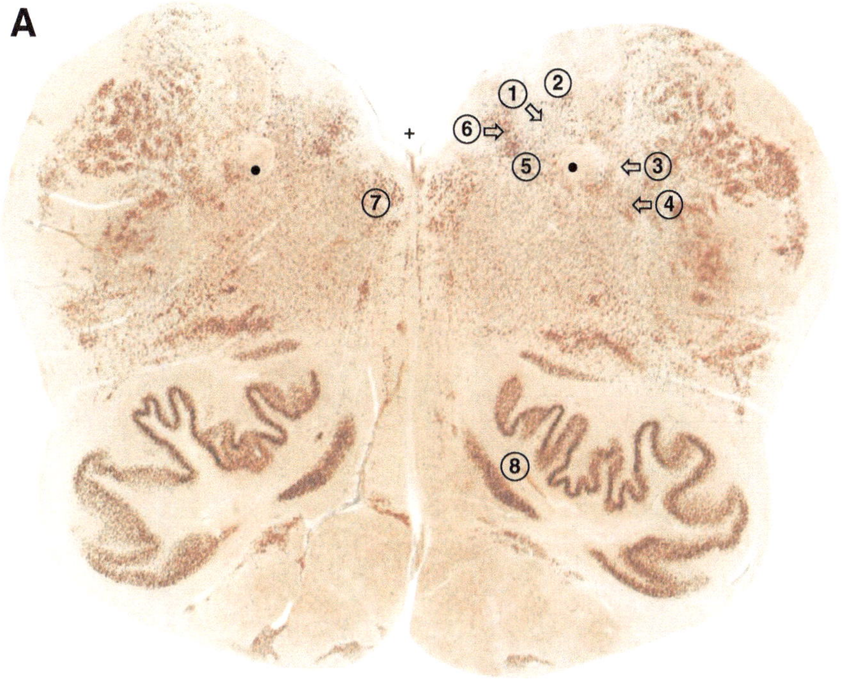

•	Solitary tract
①	Medial solitary ncl.
②	Gelatinous solitary ncl.
③	Posterolateral solitary ncl.
④	Anterolateral solitary ncl.
⑤	Intermediate solitary ncl.
⑥	Posterior ncl. of vagus nerve
⑦	Ncl. of hypoglossal nerve
⑧	Inferior olivary complex
+	4th ventricle

Fig. 3.23 (**A**) Paradigmatic representation of the solitary tract and the NTS at the level that approximates Fig. 1E of McRitchie and Törk (1993). Darrow red staining. Compare with Fig. 3.23B. The region of the NTS is bordered dorsomedially in direction to the central canal/floor of the fourth ventricle by the posterior nucleus of the vagus nerve (⑥, see Sect. 7.3.2.1), medially by the hypoglossal nucleus (⑦, see Sect. 4.3.1.1) and laterally by the cuneate nucleus. See also atlas Darrow red 6, 10A LabPON Twente. (**B**) Graphic representation of the structures involved in the signalization of gustation. The primary-order neurons in the geniculate ganglion (see Fig. 3.25B) receive information from the gustatory receptors in the tongue (see Fig. 3.24A). Their central processes reach the brainstem together with the facial nerve (VII; see Fig. 8.17) and then assemble to form the thick solitary tract ⑪. The latter contacts the nuclei of the solitary tract ⑧ (NTS). Without further switching, the axons of the NTS run ② to the parvocellular part of the ventral posteromedial ncl. of the thalamus ①. The oval nucleus ⑥ is the cranial tip of the NTS. Modified after Nieuwenhuys et al. 2008, Fig. 16.9 with permission

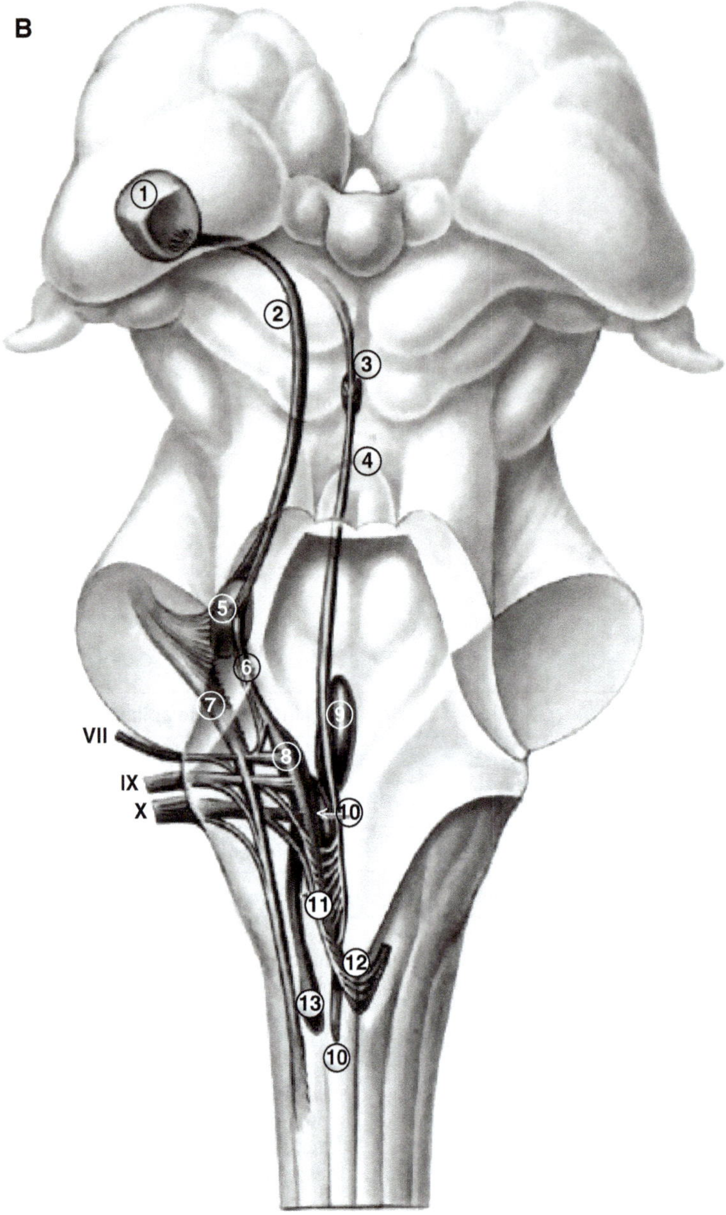

①	Ventral posteromedial ncl., parvocellular part	⑧	Ncll. of solitary tract
②	Posterior trigeminothalamic tract + uncrossed ascending gustatory fibers	⑨	Prepositus hypoglossi ncl.
③	Dorsal tegmental ncl.	⑩	Posterior ncl. of vagus nerve
④	Posterior longitudinal fasciculus	⑪	Solitary tract
⑤	Principal sensory ncl. of trigeminal nerve	⑫	Obex
⑥	Oval ncl.	⑬	Ncl. ambiguus
⑦	Spinal tract of trigeminal nerve		

Fig. 3.23 (continued)

Box 3.9 Subnuclei of the NTS

The human NTS comprises ten different subnuclei: the paracommissural, commissural, gelatinous, medial, ventral, ventrolateral, dorsal, dorsolateral, intermediate, and interstitial subnuclei. The terminology is using the solitary tract as the reference structure. The term "commissural" refers to the commissure of the solitary nuclei at their caudal end. Gelatinosus is the Latin word for congealed substance, jelly.

The FIPAT Terminology (https://fipat.library.dal.ca/wp-content/uploads/2017/02/FIPAT-TNA-Front-Matter.pdf) has the NTS nuclei (Nuclei of solitary tract [Nuclei tractus solitarii]) as follows:

- Commissural solitary nucleus [Nucleus solitarius commissuralis]
- Gelatinous solitary nucleus [Nucleus solitarius gelatinosus]
- Intermediate solitary nucleus [Nucleus solitarius intermedius]
- Interstitial solitary nucleus [Nucleus solitarius interstitialis]
- Medial solitary nucleus [Nucleus solitarius medialis]
- Paracommissural solitary nucleus [Nucleus solitarius paracommissuralis]
- Posterior solitary nucleus [Nucleus solitarius posterior]
- Posterolateral solitary nucleus [Nucleus solitarius posterolateralis]
- Anterior solitary nucleus [Nucleus solitarius anterior]
- Anterolateral solitary nucleus [Nucleus solitarius anterolateralis]

The so-called oval nucleus is the cranial tip of the NTS.

The historical designation—in italics as used by McRitchie and Törk 1993—differs from the FIPAT terminology as follows:

- Posterior/*Dorsal* solitary nucleus [Nucleus solitarius posterior/*dorsalis*]
- Anterior/*Ventral* solitary nucleus [Nucleus solitarius anterior/*ventralis*]
- Anterolateral/*Ventrolateral* solitary nucleus [Nucleus solitarius anterolateralis/*ventrolateralis*]

The human subnuclei could be identified by McRitchie and Törk (1993) based on their acetylcholinesterase (AChE) (see Box 3.10) reactivity and cytoarchitecture (see Box 3.11). The authors claim the delineation of the subnuclei also to be feasible on cresyl-violet stained specimens. We experienced the identification of the solitary subnuclei as described by McRitchie and Törk (1993) in our material (Darrow red staining) as rather challenging (see Fig. 3.23A). We had to refrain from the detection of smaller subnuclei as, for example, the interstitial solitary nucleus. The unequivocal identification of human NTS subnuclei should rely on AChE histochemistry. Cheng et al. (2006) have studied the human fetal appearance of the NTS.

Box 3.10 Acetylcholinesterase (AChE)

AChE is the enzyme that breaks down acetylcholine (ACh) (see Sect. 3.3.1) in the synaptic cleft. A great part of chemical warfare is based on AChE-inhibitors (AChEI). They inhibit ACh breakdown and prolong ACh interaction with nicotinic receptors at the neuromuscular junction. Eventually, this results in uncoordinated contractions of the diaphragm (see Sect. 3.3.2.2 "rostral ventral respiratory group r11" here below, see Fig. 3.23) with subsequent lethal respiratory insufficiency. AChE-inhibitors have recently come into the limelight due to the probable use of NOVICHOK—Russian Новичок: "newcomer," a new type of drug irreversibly inhibiting AChE—for onslaughts on Russian dissidents in 2018 and 2020 (for details, see Nepovimova and Kuca 2018).

Box 3.11 Cytoarchitecture

Cytoarchitecture (cytoarchitectonics) means the characterization of regions of the brain based on the microscopic appearance and distribution of neurons (form, size, staining intensity, morphology of the nucleus, and chromatin distribution) often performed using the cresyl-violet/Nissl stain (Nissl 1894) (Franz Nissl, German psychiatrist, and neuropathologist, 1860–1919). The classical example is the work of Korbinian Brodmann (Brodmann 1909) (German neuroanatomist and psychiatrist, 1868–1918) who provided us with a cytoarchitectonic subdivision of the cerebral cortex not only of man but also of other vertebrates (Brodmann 1909). Another prominent example is the work of Olszewski and Baxter on the human brainstem often cited in the following (Olszewski and Baxter 1982). While the intracranial CNS is the main field of cytoarchitectonic studies, the principle can be applied to all other parts of the CNS. A prominent example for a subcortical classification is the cytoarchitectonic subdivision of the gray matter of the spinal cord by the Swedish neuroscientist Bror Rexed (1952, 1954) (1914–2002).

Based on data from the rat, Hamilton and Norgren (1984) concluded that the NTS appears to have two major functional divisions, an anterior-lateral oral-gustatory half, and a posterior-medial visceral afferent half (see Sect. 3.2.4.1.2). To the best of our knowledge no human-specific data are available on this issue. This may be related to the fact that the current resolution of fMRI is not high enough to come to any serious conclusions about this topic.

For a detailed description of the NTS and its subnuclei in mice, see Ganchrow et al. (2013).

3.2.4.1.2 Afferent and efferent connections of the NTS and their course

General considerations

With regard to connectivity, we have to differentiate between

- Gustatory afferents
- Viscerosensory afferents

Their course to the NTS is provided by a number of branches of different cranial nerves (facial nerve CNVII, see Sect. 8.2.1, glossopharyngeal nerve CNIX, see Sect. 5.4.2.1.3, and the vagus nerve CNX, see Sect. 7.3.2.1).

Gustatory afferents (from Latin *gustatio, gustare* = to taste)

This in themselves show a kind of topic arrangement in that the signals from the anterior part of the tongue (Fig. 3.24) are mainly associated with the facial nerve by interposition of the Chorda tympani (see Fig. 3.25). Smaller areas of taste perception are connected with the NTS via the glossopharyngeal (tongue base) and the vagus nerves (epiglottis). The facial nerve taste afferents are the functionally and clinically most important ones.

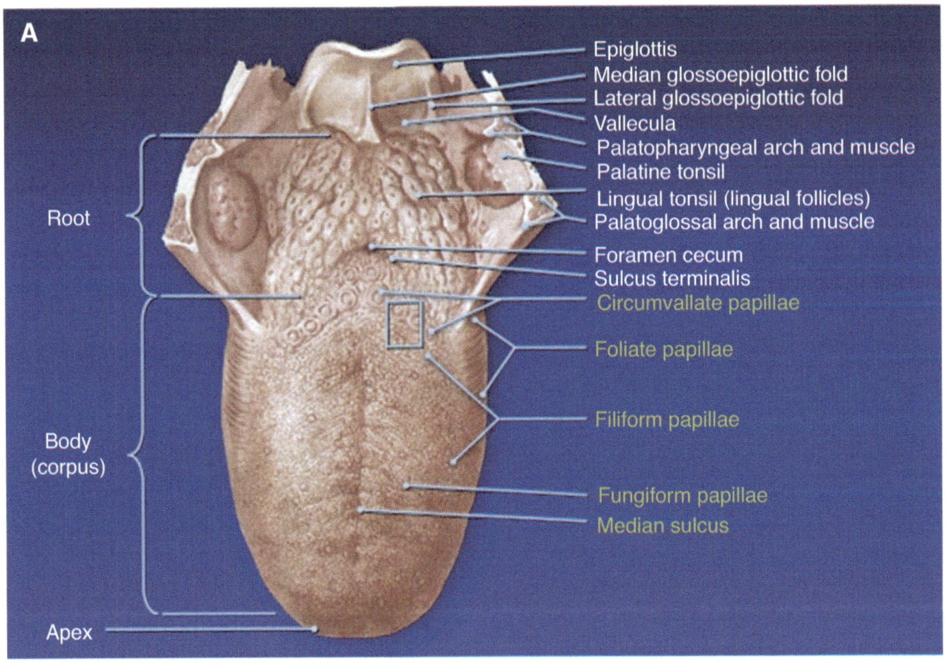

Fig. 3.24 (**A**) Microscopic anatomic and ultrastructural view of the papillae on the human tongue. Overview of the dorsum of the tongue, the tip (apex) is down, transition to pharynx and larynx top. Three types of papillae are indicated in the drawing. The (circum)vallate papillae (see Fig. 3.24B) are located in front of the Sulcus terminalis. Ultrastructurally, they consist of an oval center surrounded by a ditch the walls of which contain the taste buds. The filiform papillae—as the fungiform papillae—are located on the more anterior parts of the tongue. The filiform papillae show a thread-like appearance, fungiform ones look like mushrooms. From Mangold et al. 2016, Fig. 1 with permission. (**B**) Histological appearance of the circumvallate papillae (see also Fig. 3.24A) of the human tongue. Sudan stain. For details, see text. Sammlung des Zentrums Anatomie der Universität zu Köln

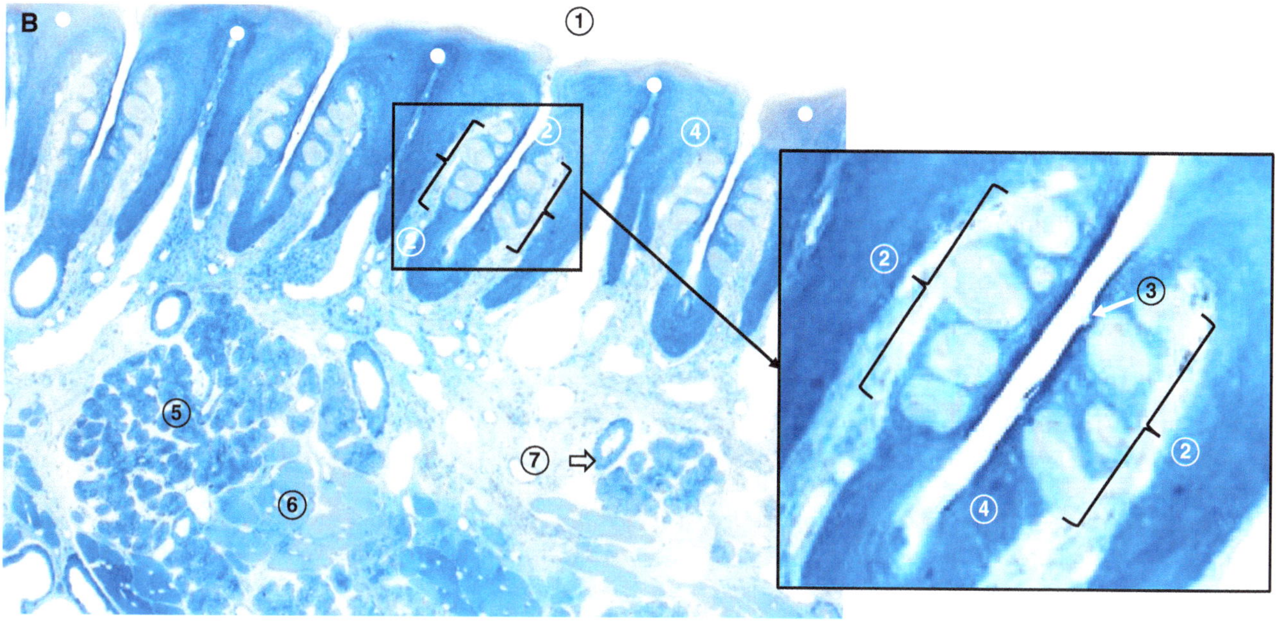

○	Papillae (circum)vallatae	④	Gustatory epithelium
①	Oral cavity	⑤	Serous glands
②	Taste buds with taste cells	⑥	Striated muscle
③	Taste pore	⑦	Arteriole

Fig. 3.24 (continued)

Viscerosensory afferents

These are the domain of the vagus nerve providing information from mucosal mechanoreceptors, chemoreceptors, and tension receptors (esophagus, stomach, and proximal small intestine). They also provide information from sensory endings in the liver and pancreas (Breit and Kupferberg 2018), from the lungs, bronchi, the larynx and from baro- and chemoreceptors (large vessels) to the nodose ganglion [Ganglion nodosum] (see here below). Nociceptive signals from the inner organs are transferred via peripheral sympathetic fibers to the dorsal horn of the spinal cord, not directly to the brainstem. They ascend in the anterolateral, spinothalamic system, and can be identified on their way from the medulla oblongata to the thalamus (see Fig. 3.14C, Anterolateral tract).

In addition, the vagus nerve provides visceromotor innervation (see Sect. 7.3.2.1, posterior nucleus of the vagus nerve) of the pharynx (together with CNIX, see there Sect. 5.4.2.1.3), respiratory system, the heart, and all parts of the gastrointestinal system down to the so-called Cannon's point or ring (in the German literature Cannon-Böhm Punkt = point). This region in the last third of the transverse colon [Colon transversum] near the left colic or splenic flexure [Flexura coli sinistra/splenica] is named after the US-American physiologist Walter B. Cannon (1871–1945) and the German radiologist Gottfried Böhm (1880–1952). Unfortunately, an unequivocal location in anatomical specimens is not possible. Distally of that point the pelvic nerves from the sacral parasympathetic system innervate the colon and the other inner organs of the pelvis.

Gustation

The most peripheral part of the gustatory system are the taste receptors with their taste buds of the different papillae. The taste buds (see below) are sensory end organs associated with four morphologically different papillae [Papillae linguales] (Latin *papilla* = nipple). The papillae are processes or foldings of the mucosa (see Fig. 3.24A) of the oral epithelium.

While the papillae are microscopic-anatomic structures identifiable by the use of a magnifying len, the different taste buds are visible only on the histological level.

- Fungiform papillae [Papillae linguales]
- Filiform papillae [Papillae filiformes]
- Circumvallate papillae [Papillae (circum)vallatae]
- Foliate papillae [Papillae foliatae]

Microanatomically and histologically, the papillae show the following characteristics:

- Fungiform papillae [Papillae fungiformes]: As the name implies (Fungus [Latin] = mushroom), the morphology of these epithelial folds resembles mushrooms. They are located as isolated structures in particular at the tip [Apex linguae] and the margins [Margo linguae] of the tongue.
- Filiform papillae [Papillae filiformes]: Filis in Latin means string and these epithelial processes look like threads.
- (Circum)vallate papillae [Papillae vallatae]: Circumvallation means entrenchment in Latin, and this type of papillae with a central round body is surrounded by a ditch, the wall of which contains the taste buds. They are arranged in a V-form shape just in front of the terminal sulcus (see Fig. 3.24A) of the tongue.
- Foliate papillae [Papillae foliatae]: Folia in Latin means leaf and these leaf-like structures are arranged parallel to each other as mucosal folds equipped with taste buds.

Each type of papilla contains the so-called taste buds (not to confuse with the papillae themselves) which are clusters of columnar sensory cells embedded in the stratified epithelium of the papillae in tongue, palate, and epiglottis (see Roper and Chaudhari 2017). As mainly known from animal studies, there are three different types of cells in a taste bud, arranged three dimensionally resembling a garlic bulb (for details, see (Roper and Chaudhari 2017).

Taste buds react to more than one tastant since they contain gustatory cells of different specificities (Roper and Chaudhari 2017).

Molecularly, taste perception—data mostly gained in animals—is provided by the action of different G-coupled

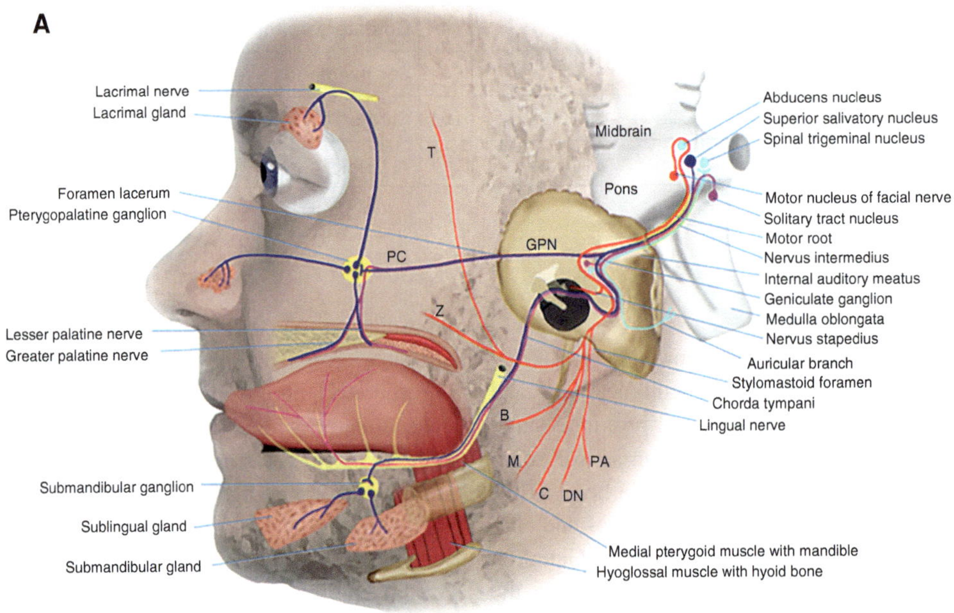

Fig. 3.25 (**A**) Schematic illustration of the motor and sensory supply zone of the facial nerve (for details, see Sect. 8.2.1.2). For a detailed description of the course of gustatory afferents, see legends Fig. 3.25B. Leander Huggenberger fecit. *GPN* Greater petrosal nerve, Branches of facial nerve, *PA* posterior auricular, *DN* digastric, *C* cervical, *M* mandibular, *B* buccal, *Z* zygomatic, *T* Temporal, *PC* Nerve of pterygoid canal. (**B**) In addition to Fig. 3.25A, the anatomical and topographic relationships of the Chorda tympani are depicted here in more detail. The chorda tympani (chorda [Greek] ἡ χορδή, he chorde = string and τό τύμπανον, to tympanon = drum, here the middle ear) runs centripetally from the lingual taste receptors (see Fig. 3.24). It then joins the lingual nerve (branch of mandibular nerve V_3, see Fig. 12.16 ④), leaves the latter at the skull base, and enters the anterior canaliculus of chorda tympani ① in the petrotympanic fissure (Glaseri, Johann Heinrich Glaser, 1629–1675, professor of anatomy at the University of Basel) (see Fig. 3.26A ①). It reaches the tympanic cavity (Fig. 3.26B ⑧) via the tympanic aperture of (anterior) canaliculus of chorda tympani [Apertura tympanica canaliculi chordae tympani]. There, it runs freely between the malleus and the incus (see Fig. 8.6)—this section not visible with current imaging techniques—and leaves the tympanon via the (posterior) canaliculus of chorda tympani ② and joins the facial nerve inside the facial canal (Fig. 3.26B ⑩) above the stylomastoid foramen (Fig. 3.26A ③). The perikarya of the gustatory neurons of the chorda tympani are located in the geniculate ganglion. The central gustatory processes leave the skull through the internal acoustic meatus and reach together with the facial nerve (Fig. 3.26B ⑨) the cerebellopontine angle via the prepontine cistern. The gustatory fibers run medially in the brainstem to form the solitary tract, which eventually reaches the nuclei of the solitary tract (see Fig. 3.23B) and is not canonical.

(see http://terminologia-anatomica.org/en/Terms/Occurence/557, https://www.imaios.com/en/e-Anatomy/Anatomical-Parts/Canaliculus-for-chorda-tympani; https://radiopaedia.org/articles/chorda-tympani). From Tillmann 2005 Atlas der Anatomie, Fig. 2.148 with permission

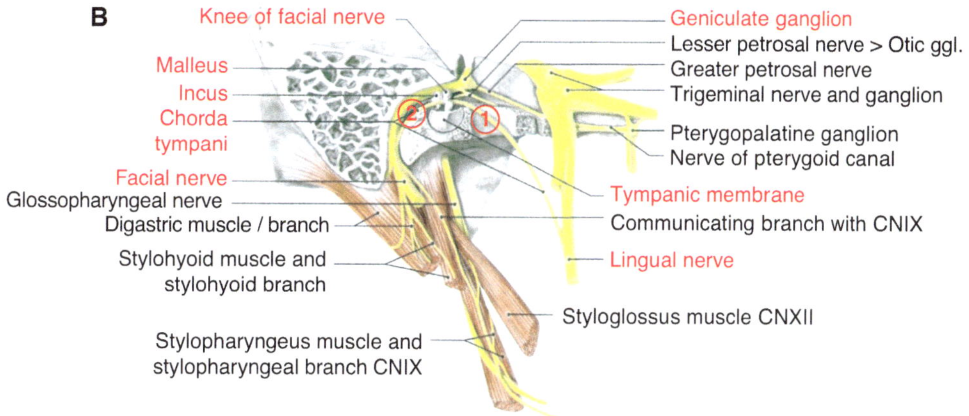

①	Chorda tympani in the anterior canaliculus of chorda tympani	②	Chorda tympani in the posterior canaliculus of chorda tympani

Fig. 3.25 (continued)

receptors (Roper and Chaudhari 2017; Matsumoto 2013), specific for individual tastes. A heterooligomer of T1R2 (T1R = taste receptor) and T1R3 (aka Tas1r2 and Tas1r3, respectively) functions as a sweet receptor (Matsumoto 2013). Umami is perceived by a heterooligomer of T1R1 (aka Tas1r1) and T1R3 and bitter by T2Rs (aka Tas2r). Umami (Japanese = delicious) is a gustatory sensation elicited by monosodium glutamate, enriched in the alga *Laminaria japonica* used in Japanese cuisine. Cells expressing PKD2LI and Car4 are sour- and carbonation-sensing cells (Matsumoto 2013). The idea of a strict mapping of individual tastant reception in defined areas of the tongue has been abandoned.

The taste receptors are in contact with the peripheral processes of pseudounipolar neurons belonging to the facial (see Sect. 8.2.1.1), glossopharyngeal (see Sect. 5.4.2.1.3), and vagus nerves (see Sect. 7.3.2.1).

As described above, the anterior two thirds of the human tongue are the gustatory supply zone of the facial nerve (CNVII, see also Sect. 8.2.1). The pseudounipolar first-order neurons of the facial gustatory fibers are located in the geniculate ganglion [Ggl. geniculi] (see Fig. 3.25A, B) in the facial canal (see Fig. 8.17).

Do not confuse the canaliculi of chorda tympani with the tympanic canaliculus (Fig. 3.26A ②), entry of the tympanic nerve of CNIX (see Sect. 5.4.2.1.3) into the tympanic cavity.

The rather complicated course of the chorda tympani is described in detail in the legend of Fig. 3.25B. A detailed account of the clinical anatomy of the chorda tympani has been provided by McManus et al. (2011).

Inside the facial nerve gustatory—and vegetative, secretomotor efferent fibers (see Sect. 8.2.1)—are running in the so-called intermediate nerve [N. intermedius].

In addition to the facial nerve which contains the gustatory afferents mainly from the anterior part of the tongue (see Fig. 3.24A), the glossopharyngeal nerve (see Sect. 5.4.2.1.3) and the vagus nerve (see Sect. 7.3.2.1) are also involved in gustatory perception. Fibers from the caudal third of the tongue reach the NTS via the inferior/petrous ganglion [Ggl. inferius/petrosum] of the glossopharyngeal nerve (CNIX see Sect. 5.4.2.1.3) [N. glossopharyngeus].

Gustatory fibers from the epiglottis (see Fig. 3.24A) reach the NTS via the superior laryngeal nerve [N. laryngeus superior], branch of the vagus nerve, and the nodose ganglion.

Afferent ascending connections of the NTS

Gustatory information

Although still presently not completely elucidated, the further course of the gustatory fibers from the NTS in humans is thought to ascend directly to the thalamus, i.e., the parvocellular part of the ventral posteromedial nucleus [Ncl. ventralis posteromedialis, pars parvocellularis], missing out the parabrachial nuclei (see Sect. 13.2.3) [Ncll. parabrachiales] which in rodents are supposed to be included in the gustatory pathway (see here below). From there the primary cortical location in the insular region (Viscerosensory cerebral cortex) is reached (for details, see Iannilli and Gudziol 2019). The question of how and where the solitaryfugal fibers cross is still subject of discussion.

Mascioli et al. (2015) could show that stimulation of each hemitongue evoked a bilateral activation of the anterior insula-frontal operculum, ascribable to the gustatory component of the stimulation, and a bilateral activation of the inferior part of the postcentral gyrus, ascribable to the tactile component of the stimulation (see Trigeminal nerve, Sect. 12.2.3.2).

Beckstead et al. (1980), using electrophysiologically guided injections of tritiated amino acids at five different anterior-posterior levels of the monkey NTS, concluded that projections from the rostral, primarily gustatory part of the

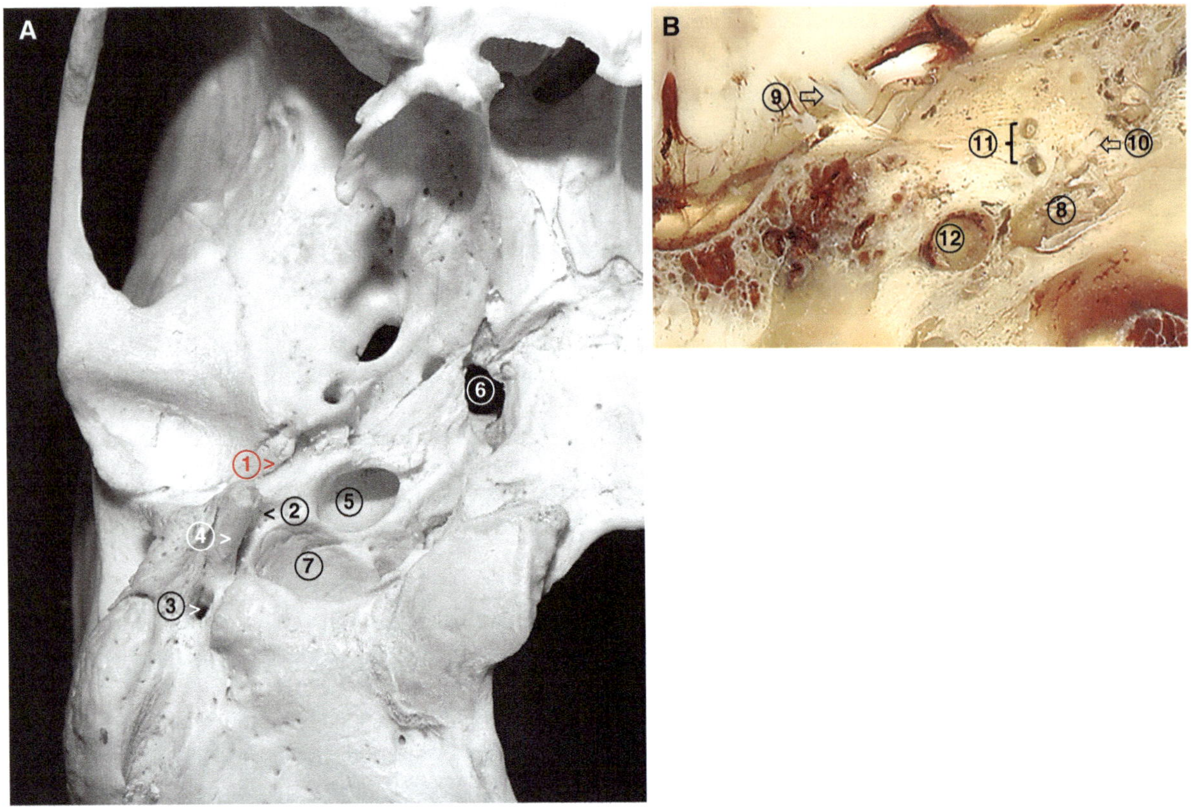

①	Petrotympanic fissure	⑦	Jugular fossa
②	Tympanic canaliculus	⑧	Tympanic cavity
③	Stylomastoid foramen	⑨	Facial nerve in cerebellopontine angle
④	Stylomastoid process	⑩	Facial nerve in facial canal
⑤	Carotid canal	⑪	Anterior vestibular canal
⑥	Foramen lacerum	⑫	Internal carotid artery

Fig. 3.26 Anatomical hallmarks of the course of the chorda tympani and the facial nerve in the region of the skull base. (**A**) Human skull, ventral view. (**B**) Section through the human head at the level of the cerebellopontine angle. Plastinated specimen. Sammlung des Zentrums Anatomie der Universität zu Köln

NTS ascend in the ipsilateral central tegmental tract, bypass the PB, and terminate within the caudal half of the parvocellular ventral posteromedial nucleus (VPMpc) of the thalamus.

The traditional view of the central gustatory pathway in mice (Tokita et al. 2009, 2010) is a connection from the NTS to the parabrachial nuclei [Ncll. parabrachiales] (this step lacking in humans) and from there to the thalamus (Ganchrow et al. 2013). A combined molecular and WGA (wheat germ agglutinin) tracing approach in the mouse, however, has shed some doubt on the parabrachial nucleus being part of the gustatory pathways (Matsumoto 2013). Eventually, the thalamocortical fibers reach the mouse gustatory cortex where the basic tastes are represented in a gustotopic map (Chen et al. 2011). In mice, the solitary nucleus gives rise to numerous cerebellopetal fibers (Fu et al. 2011). Mouse solitary subnuclei have been shown to be involved in regulation of peristalsis in the oropharynx and the esophagus (Sang and Goyal 2000).

Viscerosensory information from inner organs

When dealing with viscerosensory function in terms of anatomical organization, it is important to differentiate between nociception (pain) and other qualities like mechanoreception, chemoreception, and reception of tension (see below).

1. Visceral pain is subject of the spinal afferent signaling (pelvic and sympathetic spinal afferents), as evidenced by studies on patients with transection of the spinal cord or loss of spinal afferent innervation due to nerve ablation or trauma (for details, see Bulmer and Roza 2018). Information is transferred to the supraspinal levels via the

spinothalamic system (see Fig. 3.13B). Three different classes of nociceptors can be differentiated in the gastrointestinal system due to their electrophysiological properties.

2. Another population of sensory receptors based on their electrophysiological properties most likely are not nociceptors but thought to relay physiological information, for example, about the filling of the gut, (Bulmer and Roza 2018) to the CNS via the vagus nerve (see gastrointestinal tract here below).

   In addition, afferents from baroreceptors and chemoreceptors are transferred to the NTS via vagal afferents (see baro- and chemosensitivity).

The different peripheral branches of the vagus nerve are dealt with under Sect. 7.3.2.1: Posterior nucleus of the vagus nerve."

Gastrointestinal tract

Although the gastrointestinal tract is equipped with an intrinsic, enteric nervous system (for relevance in Parkinson's disease, see Fig. 16.21 C), CNS control of gastrointestinal functions is provided in efferent and afferent direction by the vagus nerve (Browning et al. 2017). The visceromotor fibers for the innervation of the gastrointestinal system (but also for larynx, pharynx, respiratory system, heart) originate in the posterior (dorsal motor) nucleus of vagus nerve (see Sect. 7.3.2.1). The viscerosensory afferents from all these organs are conveyed to the NTS via different branches of the vagus nerve (see Sect. 7.3.2.1.3).

The abdominal vagal afferents, include mucosal mechanoreceptors, chemoreceptors, and tension receptors in the esophagus, stomach, and proximal small intestine, and sensory endings in the liver and pancreas (Breit and Kupferberg 2018). The sensory afferent cell bodies are located in the nodose ganglia and send information to the nucleus tractus solitarii (NTS) (Breit and Kupferberg 2018) (for vagal efferents, see Sect. 7.3.2.1.3). The nodose ganglia also receive sensory input from the respiratory organs and the heart (Hirota et al. 2016).

Based on animal studies, Mei (1983) could confirm that numerous vagal receptors exist in all layers of the small intestine as simple free nerve endings, mainly connected to nonmyelinated fibers. By means of electrophysiology, it could be shown that mechanical, thermal, chemical, or osmotic events in the intestine induce action potentials in one or several kinds of receptors. In a broader view, intestinal receptors are involved in the regulation of gastrointestinal motility, homeostasis, and alimentary behavior (Mei 1983).

In mice, Bai et al. (2019) could show that vagal sensory neurons are involved in the control of feeding. Interestingly, they could show that—counterintuitively—stimulation of mechanoreceptors in the intestine plays a crucial role in food intake regulation, not mucosal afferents or other cell types. The mechanoreceptors durably inhibit the hunger-promoting AgRP (Agouti-related protein) neurons in the hypothalamus.

Although vagal afferents are not nociceptive, vagal afferent endings are sensitive to many of the same noxious chemical mediators as spinal afferent nociceptors (for details, see Bulmer and Roza 2018). Consequently, the simultaneous activation of vagal afferents by painful stimuli provides a mechanistic explanation for some of the marked autonomic responses that accompany visceral pain. In particular, the activation of gastrointestinal vagal afferents during visceral pain will produce sensations of nausea and trigger vomiting due to their termination within the Area postrema (see Sect. 5.1.1), the vomiting center within the human brainstem (for details, see Bulmer and Roza 2018). For obvious reasons, there is no experimental proof for the human brain of vagal terminals in the Area postrema.

Respiratory organs

Larynx

Viscerosensory information from the larynx is transferred to the NTS via the superior and inferior laryngeal nerves [Nn. laryngei superior / inferior]. The perikarya of these first-order sensory fibers are located in the nodose ganglion [Ganglion nodosum]—from Latin *nodus* = knot, here *nodosus* = having one or more swellings—also known as inferior ganglion [Ganglion inferius]. The central processes of the ganglionic neurons end in the NTS. Anatomical information on the nodose ganglion is mainly from animal studies, in particular from dogs. The ganglion is located 1–2 cm below the jugular foramen (Job and Branstetter IV 2017) and of small size in man (25 mm length, 5 mm width) (Hirota et al. 2016). The histological appearance is characterized mainly by pseudounipolar and fusiform bipolar neurons which based on their ultrastructural appearance can be classified into three groups (for details, see Hirota et al. 2016). The ganglion cells are equipped with a number of peptides, nNOS (neuronal nitric oxide synthase) and TH (see Fig. 3.3) (Hirota et al. 2016). The presence of vanilloid receptor subtype 1 (VR1) points to nociceptive function. Furthermore, the presence of acid-sensing ion channels was shown (for details, see Hirota et al. 2016).

For the mouse, Kupari et al. (2019) have provided a molecular identification of neuronal types in the vagal ganglion complex. Prdm12+ jugular neurons share features with spinal somatosensory neurons while Phox2b+ neurons of the nodose ganglion are viscerosensory, molecularly versatile, and highly specialized. The properties of the different nodose neurons are consistent with chemo-, baro-, stretch, tension, and volume sensors (Kupari et al. 2019).

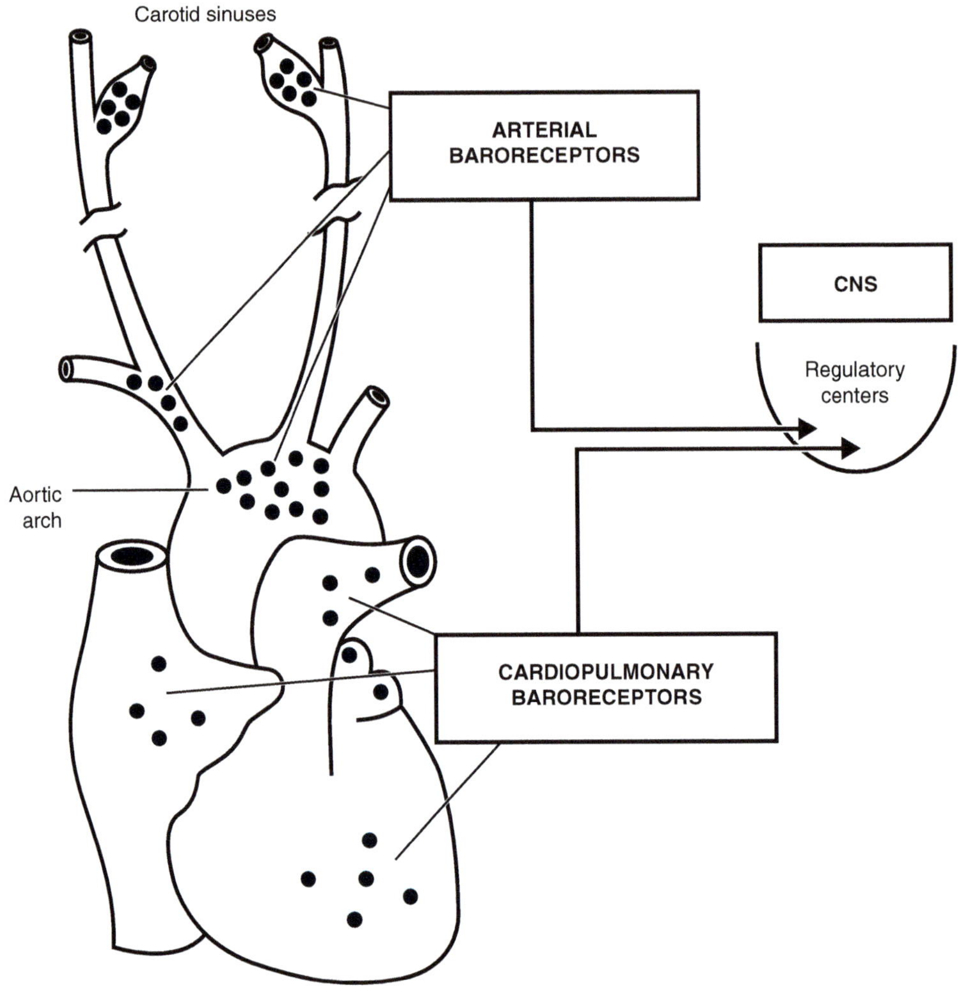

Fig. 3.27 Schematic drawing of baroreceptors and cardiopulmonary receptors in humans. Arterial baroreceptors are located in the large arterial thoracic and cervical vessels (Aortic arch [Arcus aortae], carotid arteries and carotid bifurcation [Bifurcatio carotidis] (see Fig. 1.15 A), carotid sinus [Sinus caroticus]). The cardiopulmonary receptors are present in the cardiac atria [Atrium/atria cordis] and ventricles [Ventriculus/ventriculi cordis] as well as in the pulmonary artery [A. pulmonalis]. From Seravalle et al. 2019, Fig. 1 with permission

Baro- and chemosensitivity

Information about blood pressure and the respiration-related O_2 and CO_2 pressures are transferred to the brain from peripheral receptors via afferent fibers of the vagus nerve.

Arterial and cardiopulmonary baroreceptors are mechanoreceptors that are located close to the bifurcation of the carotid sinus and in the aortic arch (see Fig. 3.27) (Seravalle et al. 2019).

Toorop et al. (2013) used antibodies to vesicular glutamate transporter 2 (VGLUT2) and protein gene product 9.5 (PGP9.5) to visualize baroreceptors in the region of the human carotid bifurcation (see Fig. 3.27). Immunoreactivity for both markers was present in the adventitia of the carotid arteries with highest nerve density in the medial wall of the first centimeters of the internal carotid artery (Toorop et al. 2013). The clinical implication of their findings lies in the removal of the carotid adventitia as treatment for carotid sinus syndrome (CSS, see here below) indicating that stripping of the proximal portion of the internal carotid artery should be sufficient (cf. Trout II et al. 1979). The Tunica adventitia (see Fig. 1.21 A, B) is the outermost layer of blood and lymphatic vessels—made of connective tissue—which contains small vessels and nerves, in this case also the afferent nerves of the baroreceptors.

CSS may lead to carotid sinus syncope (loss of consciousness) elicited by stimulation of the hypersensitive carotid sinus. This kind of syncopal attack must be differentiated from other, more common causes like cardiac disease, vasovagal syncope (see Box 3.12), postural hypotension, and cerebrovascular insufficiency (Trout II et al. 1979).

Box 3.12 Vasovagal syncope and postural hypotension

Vasovagal syncope (short lasting loss of consciousness) is due to a common autonomic reflex involving the cardiovascular system. It is associated with bradycardia (cardioinhibitory response) and/or hypotension (vasodepressor response), likely mediated by parasympathetic activation and sympathetic inhibition (Gopinathannair et al. 2018) (for details, see da Cunha et al. 2019, Gopinathannair et al. 2018, van Lieshout et al. 1991). Postural (orthostatic) hypotension is an excessive fall of blood pressure which may lead to faintness, light-headedness, dizziness, confusion, or blurred vision as far as falls, syncope, or even generalized seizures.

https://www.msdmanuals.com/professional/cardiovascular-disorders/symptoms-of-cardiovascular-disorders/orthostatic-hypotension

Via mechanosensitive ion channels the mechanical state of the vessel and cardiac walls is reported to the NTS by the generation of action potentials. Under normal conditions, an increase in blood pressure is accompanied by deactivation of the carotid and aortic baroreceptors (Seravalle et al. 2019) which eventually leads to increased parasympathetic activity and among changes in other parameters to a decrease in blood pressure and heart rate. On the other hand, a blood pressure decrease is followed by an increased sympathetic outflow followed by vasoconstriction and an increase in heart rate, evoking a blood pressure increase. When the blood pressure (BP) changes persist for approximately 24 h (in contrast to the rather quick changes related to the baroreflex), the baroreceptors will reset to the new BP levels.

https://www.datasci.com/solutions/cardiovascular/baroreceptor-sensitivity-(brs)

Arterial chemoreceptors are located bilaterally at the bifurcation of the common carotid artery [A. carotis communis] (see Fig. 3.27 above) (Prabhakhar and Joyner 2015). In general, only few structural data on these receptors in the human inner organs are available. Chemoreceptor tissue is composed of two major cell types: the type I (also called glomus) cells and type II cells. A substantial body of evidence suggests that type I cells are the initial sites of sensory transduction and they work in concert with the nearby afferent nerve ending as a "sensory unit," whereas the type II cells are supporting cells resembling glial cells of the nervous system (see Kumar and Prabhakar 2012, for references).

Sensory innervation to the carotid body [Glomus caroticum] is provided by a branch of the glossopharyngeal nerve (see Sect. 5.4.2.1.3) called the "carotid sinus nerve (CSN)" [N. sinus carotici].

Hayashi et al. (1995) described the ultrastructure of coronary chemoreceptors in man and dog. These receptors are supposed to cause a hypertensive reflex. The chemoreceptors were composed of the so-called chief cells, sustentacular cells, Schwann cells (Box 3.13), nerve fibers—forming complex junctions—blood vessels and connective tissue (Hayashi et al. 1995). A glomoid cluster was made of three to four chief cells and a Schwann cell. The chief cells are filled with a plethora of osmiophilic dense granules. Schwann cells are the cells that provide the myelin sheath of neuronal axons in the peripheral central nervous system (Salzer 2015), as the oligodendrocytes do it in the CNS.

Box 3.13 Theodor Schwann

Theodor Schwann was born in 1810 in Neuss. He studied medicine in Bonn, Würzburg, and Berlin where he was a doctoral student of the famous physiologist Johannes Müller. In 1838, he became Full Professor of Anatomy at the University of Leuven/Louvain, Belgium. In 1839, he published the book "Mikroskopische Untersuchungen über die Übereinstimmung in der Struktur und dem Wachstum der Tiere und Pflanzen" (Microscopical researches into the accordance in the structure and growth of animals and plants) together with the botanist Matthias Schleiden. They defined the individual cells as the modules organisms are made of. In 1845, he was awarded the Copley medal of the Royal Society. In 1848, he became full professor at the University of Liège/Luik in Belgium where he worked till his retirement in 1880. He coined the term "metabolism." Theodor Schwann can be considered the founder of histology and cell biology. He died in 1882 in Cologne/Köln, Germany.

http://dibb.de/theodor-schwann.php

Postmortem tract tracing in human brains has revealed projections from the NTS to the regions of the intermediate reticular zone (see Sect. 3.3.2.1) [Zona reticularis intermedia], the dorsal raphe nucleus (see Sect. 13.3.4.2) [Ncl. raphes dorsalis], the gigantocellular reticular nucleus [Ncl. gigantocellularis], and the spinal trigeminal nucleus (see Sect. 3.2.2.2) [Ncl. spinalis n. trigemini]. In addition, there were terminals in the NTS itself, the dorsal/posterior nucleus of vagus nerve [Ncl. dorsalis/posterior n. vagi] (see Sect. 7.3.2.1) (formerly also dorsal motor nucleus of the vagus nerve, DMV)—supposedly for the triggering of vasovagal reflexes—and to the reticular formation [Formatio reticularis] (Ruggiero et al. 2000).

In human postmortem midgestational fetuses, Zec and Kinney (2003) have shown connections of the NTS to all its future subnuclei by DiI tracing. These subnuclei were ipsilateral of the application site of the DiI crystal, in the caudal raphe, arcuate nucleus (see Sect. 4.1.1.4), the posterior (dorsal motor) nucleus of the vagus nerve (see Sect. 7.3.2.1), the nucleus ambiguus (see Sect. 5.4.2.1) the ventral respiratory group (see Sect. 3.3.2.2), and the rostral and caudal ventrolateral medulla.

3.2.4.1.3 Living anatomy and clinical implications

The function of the gustatory system can be checked for in probands or patients by applying tastants solutions of the five basic tastes sweet (sugar), bitter (chinin), umami (umami paste), salty (sodium chloride), and sour (citric acid). Diluted solutions of the five substances can be applied separately on both halves of the tongue.

Please note that in the medical sense only the above-mentioned tastes are received by the taste organs of the tongue and transferred via the facial (CNVII), glossopharyngeal (CNIX), and vagus (CNX) nerves to the NTS.

By contrast, smell with a much broader spectrum of different sensations contributes far more to the personal perception of taste. Olfaction, however, is received via the olfactory epithelium of the nose and transferred to the CNS by the olfactory nerve (CNI).

Taste dysfunction may be caused by upper respiratory and middle ear infections, head and neck radiation, certain drugs, head injury, certain surgical procedures (middle ear, teeth), or poor oral hygiene and dental problems. Just very recently, loss of taste was reported as one symptom of the mild-to-moderate forms of the new coronavirus disease (COVID-19) (Lechien et al. 2020).

3.2.4.2 Parasolitary nucleus r11 [Ncl. parasolitarius]

The term parasolitary nucleus refers to a cell group located in the medulla just lateral to the solitary nucleus. It is found in the human, the macaque, the rat, and the mouse. Functionally, it is one of seven nuclear groups that comprise postcerebellar and precerebellar nuclei of the subcortical motor system.

http://braininfo.rprc.washington.edu/centraldirectory.aspx?ID=7

3.2.5 Alar tegmentum r11

3.2.5.1 Matrix region of the medulla r11 [Regio matricis medullae]

For this region, no human-specific data are available nor is it listed in FIPAT.

In mice, the matrix region of the medulla is an area between the cuneate nucleus and the spinal trigeminal nucleus. It has a weak efferent bilateral connection with the cerebellum (Fu et al. 2011).

3.3 Basal r11

3.3.1 Branchial motor nuclei r11 (branchiae Latin = gills)

The term branchial refers to the developmental appearance of a series of cartilaginous arches—no real gills!—that develop in the walls of the oral cavity and pharynx of vertebrate embryos (see Fig. 2.7). Branchial motor nuclei are also referred to as special visceral efferent (SVE) (see Table 1.1) since they provide motor supply to skeletal muscles derived from the branchial arches not from somites (see Box 3.14) (GSE; General somatic efferent). The terminals of branchial motor neurons release acetylcholine as their transmitter.

Box 3.14 Somites

At that time during embryonic development when the neural folds begin to form (see Chap. 2), the paraxial mesoderm separates into blocks of cells called somites which are extremely important in organizing the segmental pattern of vertebrate embryos. The somites determine the migration paths of neural crest cells and spinal nerve axons. Somites give rise to the cells that form the vertebrae and ribs, the dermis of the dorsal skin, the skeletal muscles of the back, and the skeletal muscles of the body wall and limbs (Gilbert 2000).

Acetylcholine (ACh) is one of the best characterized neurotransmitters of the peripheral and the central nervous system. The first investigations date back to the pioneering work of Otto Loewi (see Box 3.15) who simultaneously confirmed the chemical nature of neuronal transmission in vertebrates. In 1921, Loewi, using frog hearts for his studies, was able to show that a chemical substance was responsible for the effects, for example, on heart rate. He coined the name *Vagusstoff* for this compound. Later, Loewi and Navratil (1926) could verify the chemical equivalent of the *Vagusstoff* to be ACh. It was for these investigations that Loewi received the Nobel Prize in 1936, jointly with Sir Henry Dale. This and other discoveries in the fields of chemistry, physics, and pharmacology have since then led to a complete renewal of the concepts of the parasympathetic and the sympathetic nervous system and are the basis for studying the central cholinergic system.

Box 3.15 Otto Loewi and the Nazis

When the Germans invaded Austria in 1938, Loewi, being a professor at the University of Graz, Austria, was forced to leave his homeland but only after he had been compelled to instruct the Swedish bank in Stockholm to transfer the Nobel Prize money to a prescribed Nazi-controlled bank.

www.nobelprize.org/nobel_prizes/medicine/laureates/1936/loewi-bio.html

Table 3.1 Cholinergic cell groups of the mammalian brain [Ncll. cholinergici]

Ch cell group	Anatomical location	Projection to
Ch1	Cholinergic cells of medial septal nucleus	Hippocampal formation
Ch2	Cholinergic cells of vertical limb of diagonal band	Hippocampal formation
Ch3	Cholinergic cells of horizontal limb of diagonal band	Olfactory bulb
Ch4	Cholinergic cells of basal nucleus (Meynert)	Cerebral cortex, amygdala
Ch5 (Sect. 13.1.1.1.1)	**Pedunculopontine tegmental nucleus** [Ncl. tegmentalis pedunculopontinus]	Thalamus
Ch6 (Sect. 13.1.1.2)	**Laterodorsal tegmental nucleus** [Ncl. tegmentalis laterodorsalis]	Thalamus
Ch7	Cholinergic cells of epithalamus/Medial habenular nucleus [Ncl. habenularis medialis]	Interpeduncular nucleus (Sect. 14.5.4.4)
Ch8 (Isthmus)	**Parabigeminal nucleus** [Ncl. parabigeminalis]	Superior colliculus (see Sect. 15.1.3)

Brainstem cholinergic nuclei in bold

In analogy to the monoaminergic cell groups of the brain (see Sect. 3.2.1), Mesulam et al. (1983) provided a classification of cholinergic cell groups in addition to the cholinergic cranial nerve nuclei. Table 3.1 lists these groups (Ch1–Ch8 in rostrocaudal order) as described for the rat CNS correlated with the conventional neuroanatomical terminology. Ch groups of the brainstem are given in bold.

3.3.1.1 Retroambiguus nucleus r11 [Ncl. retroambiguus]

The term retroambiguus nucleus refers to a group of neurons in the medulla at the caudal end of the nucleus ambiguus with which it is continuous (see Fig. 3.5 ⑥ and Sect. 5.4.2.1). It receives input from respiration-related nuclei and from the periaqueductal grey (ten Donkelaar et al. 2018). It projects to the motoneurons of the nucleus ambiguus, the spinal motoneurons for the respiratory, abdominal, and pelvic floor muscles (ten Donkelaar et al. 2018). From studies in mammals, it is known that in addition to respiration-related function the retroambiguus nucleus is involved in vocalization and vomiting (for details, see Vanderhorst 2005, see also Sect. 5.1.1 Area postrema).

In mice, the retroambiguus nucleus comprises a group of neurons in the lateral medulla oblongata (cf. Vanderhorst 2005) located inside the ventral third of the intermediate reticular nucleus. Tracing studies in mice [C57BL/6, CD-1] have shown the retroambiguus nucleus mainly to project to motoneurons of the thoracic, and upper lumbar spinal cord that supply abdominal wall [Musculi abdominis] and cremaster muscles [M. cremaster] (Muscle of the spermatic cord, partially raises the testicle) (Vanderhorst 2005).

3.3.1.2 Accessory nerve nucleus/Ncl. of accessory nerve r11 [Ncl. n. accessorii]

The accessory nerve is the 11th cranial nerve. Historically, it was thought to dispose of a cranial root and a spinal root (see Fig. 3.28).

The cranial root, joining the vagus nerve, provides innervation of palatal, pharyngeal, and laryngeal muscles. The spinal root originates from a small cell column from C1 through C5 (C6) and gives rise to axons that innervate the sternocleidomastoid [M. sternocleidomastoideus] and the trapezius muscle [M. trapezius]. While up to now the nerves from both roots were classified as special visceral efferent (SVE, visceromotor see Table 1.1), recently claims have been made that the spinal root innervates muscles derived from somites (see Box 3.14 here above) and therefore has to be general somatic efferent (GSE, somatomotor).

Based on the investigations of Benninger and McNeil (2010), there is reasonable evidence to regroup the two roots as follows: The cranial root is a branch of the vagus nerve called laryngopalatopharyngeal nerve [N. laryngopalatopharyngeus] (not listed in FIPAT).

The spinal root—a nerve with both cranial and spinal characteristics—has been proposed to be designated as a kind of transitional nerve since the muscles it innervates are mesoderm-derived striatal ones with connective tissue and osseous attachments of neural crest origin (see Sect. 2.1, Benninger and McNeil 2010). The neural crest is a cell group developing from the wall of the neural tube but splitting off from the CNS. Important examples for neural crest derivatives are the Schwann cells, meninges, sympathetic nerve ganglia, parts of the adrenal gland, melanocytes of the skin and odontoblasts (anlage of teeth). The fibers of the laryngopalatopharyngeal nerve leave the cranial cavity with the vagus nerve via the jugular foramen (see Fig. 3.28).

The fibers of the transitional nerve enter the cranial cavity via the foramen magnum and exit from the skull together with the vagus and the glossopharyngeal nerve. They consti-

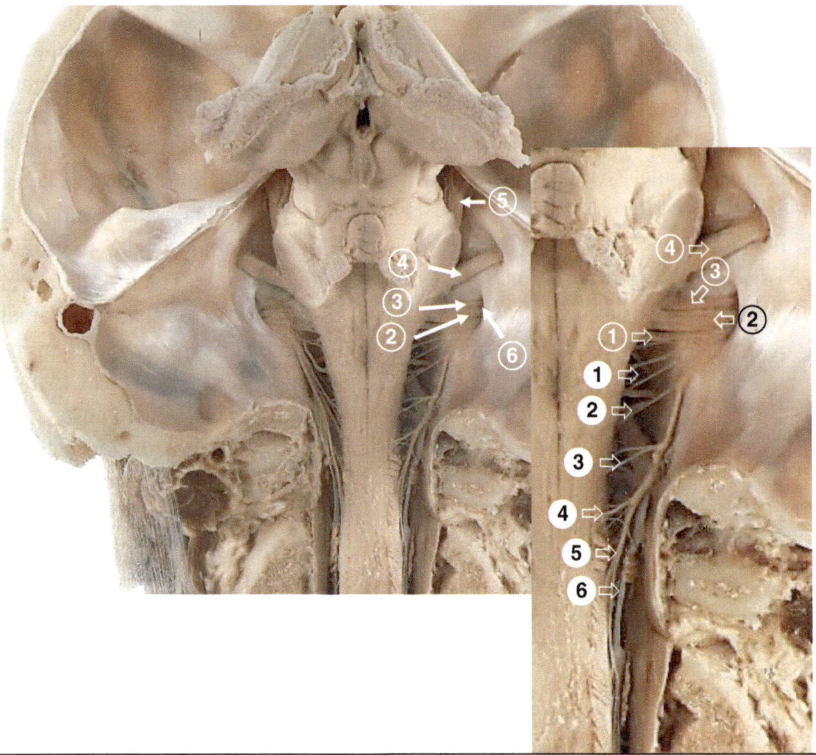

①	**Cranial root / laryngo-palatopharyngeal nerve**	③	**Glossopharyngeal nerve CNIX**
❶	**Spinal root(s) /**	④	**Vestibulocochlear nerve CNVIII**
❻	**cervical roots C1 – C6**	⑤	**Trigeminal nerve CNV**
②	**Vagus nerve CNX**	⑥	**Jugular foramen**

Fig. 3.28 Dorsal view of the human brainstem. Formalin-fixed specimen. The posterior parts of the skull bone as well as the spinous process and laminae have been removed to enable the view onto the brainstem and the upper cervical spinal cord. ① labels the cranial root of the accessory nerve (for details, see text here below). ❶–❻ indicate the six cervical spinal roots which form the spinal root of CNXI. Sammlung des Zentrums Anatomie der Universität zu Köln

tute the external branch [Ramus externus] with two muscular branches, one for the sternocleidomastoid and the other for the trapezius muscle (Fig. 3.29A, B).

1. Sternocleidomastoid muscle [M. sternocleidomastoideus]: The muscle originates from the sternum (Greek τό στέρνον, to sternon) and the clavicula (Greek ἡ κλείς, he kleis, kleidos = key, clavicula) and inserts at the mastoid process (Greek ὁ μαστός, ho mastos = breast, μαστοειδής, mastoeides = resembling a breast) [Processus mastoideus] and the superior nuchal line [Linea nuchalis superior]. Unilateral contraction causes a bend of the head to the same side and a rotation of the face/head to the contralateral side. Bilateral contraction leads to lifting of the head.
2. Trapezius muscle [M. trapezius] (Greek ἡ τράπεζα, he trapeza = flat plane, desk): The muscle consists in craniocaudal order of a descending, a transverse, and an ascending part. The descending part originates from the occipital bone and inserts at the clavicula. The two more inferior located parts originate mainly from the spinous processes [Processus spinosi] of the vertebral column and insert mainly at the scapula. The latter pull the scapula toward the vertebral column. The descending part counteracts a downward drag (see Fig. 3.29), rotates, and adducts the scapula. When the scapula is fixed, it rotates the head.

3.3.2 Reticular nuclei r11 [Ncll. reticulares]

The reticular nuclei are the neuronal agglomerations of the so-called reticular formation (RF) [Formatio reticularis]. The RF is a morphologically ill-defined region stretching through the whole brainstem. The term "reticular" from rete [Latin] = net was chosen because of the netlike arrangement of neurons in the reticular formation.

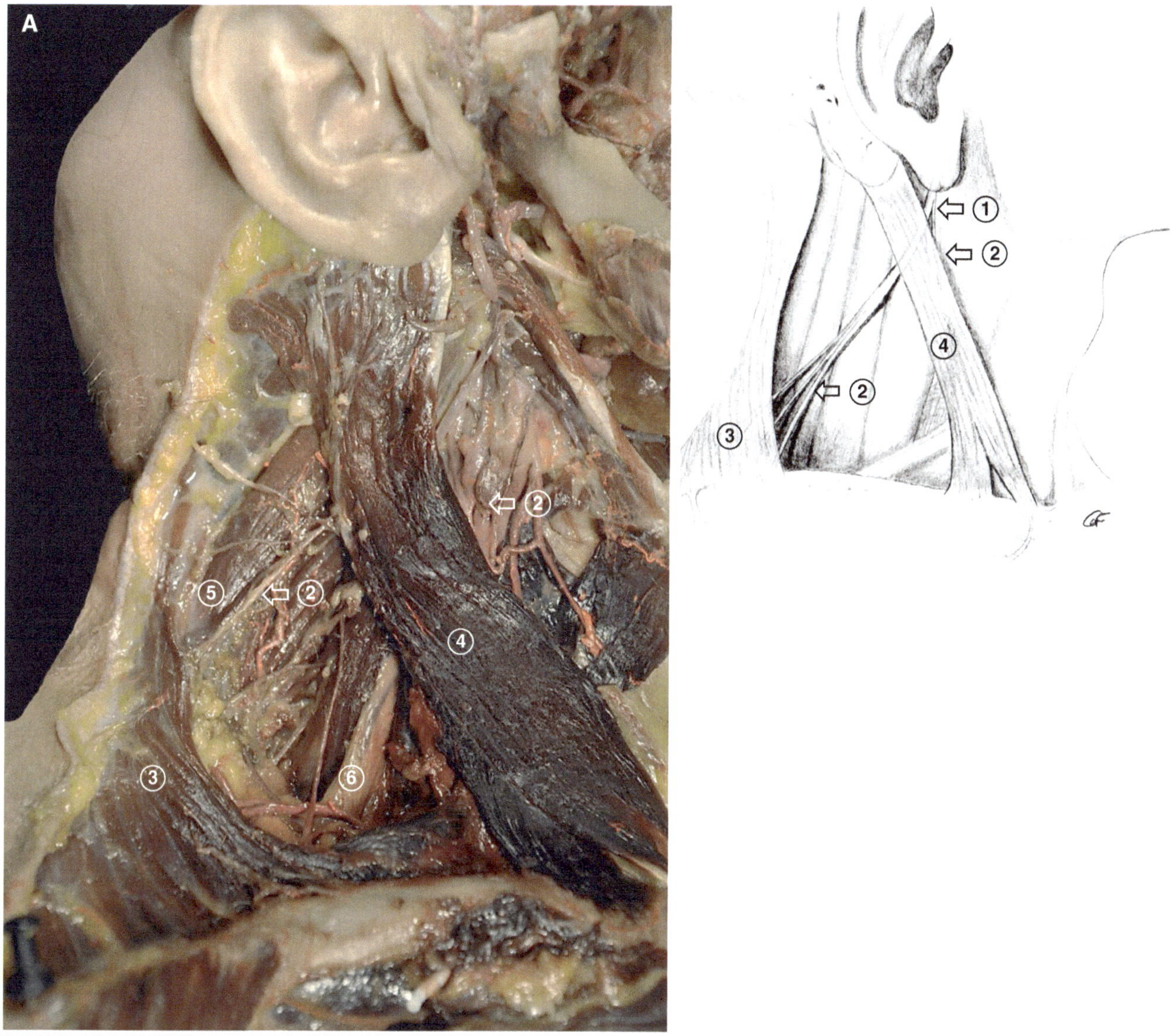

①	External branch of CNXI	④	Sternocleidomastoid muscle
②	Muscular branches	⑤	Levator scapulae muscle
③	Trapezius muscle	⑥	Plexus brachialis

Fig. 3.29 (**A**) Accessory nerve and targets: extracranial course of the accessory nerve. After leaving the skull via the jugular foramen (see Fig. 3.28 ⑥), the trunk of the accessory nerve [Truncus nervi accessorii] is running laterally and splits off in the internal branch [R. internus] (fibers of laryngopalatal nerve, see above) and the external branch [R. externus] ① with muscular branches [Rr. musculares] ② for the sternocleidomastoid [Sternocleidomastoid branch/R. sternocleidomastoideus] and the trapezius [Trapezius branch/Ramus trapezius] ③ muscles. The nerve crosses the jugular vein laterally, then runs on the ventral surface of the levator scapulae muscle (guiding muscle) ⑤, provides branches to the sternocleidomastoid muscle and eventually reaches further caudally the trapezius muscle. Lesions of the accessory nerve may, for example, occur when lymph nodes of the lateral neck region are removed surgically. Courtesy Prof. Georg Feigl, Private Universität Witten/Herdecke GmbH. (**B**) Clinical examination of the sternocleidomastoid and trapezoid muscles. Trapezius muscle (left): The patient is asked to lift the shoulders without and against the resistance of the investigator's hands. Sternocleidomastoid muscle (right): The patient is asked to turn the head in either direction against the resistance of the investigator. This maneuver has to be performed with caution in particular in older subjects since it may lead to the mobilization of atherosclerotic plaques in the common carotid artery increasing the danger of ischemic cerebral infarction (see Fig. 1.23)

B

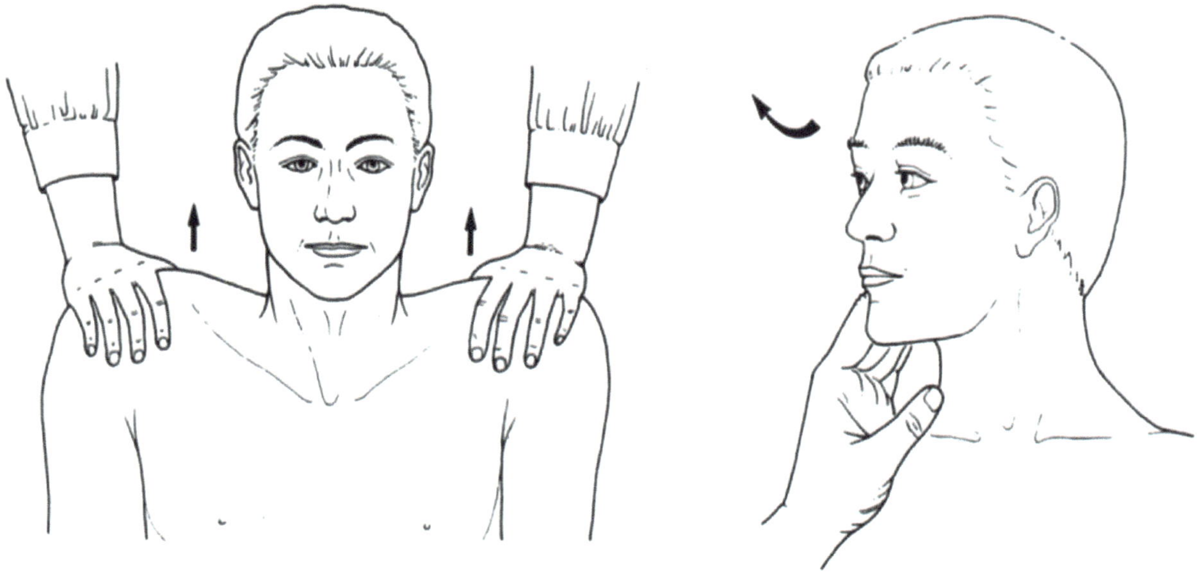

Ribbers fecit

Fig. 3.29 (continued)

However, there is no commonly accepted classification of cell groups belonging to the reticular formation which renders dealing with the individual nuclei confusing.

Nuclei which without any doubt do **not** belong to the reticular formation but often included are the lateral reticular nucleus (see Sect. 4.1.1.1), the reticulotegmental nucleus of the pons (see Sect. 10.1.1.1) (see also Watson et al. 2019), the raphe nuclei (Chaps. 4 through 14), and the locus caeruleus (see Sect. 13.2.2.1) (Fu et al. 2011).

Nieuwenhuys et al. (2008) have defined several parts of the reticular formation depending on their location:

- The medial reticular formation as defined by Nieuwenhuys (1998) and Nieuwenhuys et al. (2008) consists of six overlapping structures (see Box 4.5).
- The lateral reticular formation—confined to the rhombencephalon—as defined by Nieuwenhuys (1998), Nieuwenhuys et al. (2008) also consists of six overlapping structures:
  - The parvocellular reticular area or nucleus [Ncl. reticularis parvocellularis]
  - The ventrolateral superficial reticular area [Ncl. reticularis ventrolateralis superficialis]
  - The lateral pontine tegmentum [Tegmentum pontinum laterale]
  - The noradrenergic cell group A1–A7 [Cellulae noradrenergicae]
  - The adrenergic cell groups C1 and C2 [Cellulae adrenergicae]
  - The cholinergic cell groups Ch5 and Ch6 [Cellulae cholinergicae]

These structures contain the following nuclei:

- The parvocellular reticular area or nucleus
  - Ncl. dorsalis reticularis not listed in FIPAT Ch.1
  - Parvocellular reticular ncl. (see Sect. 6.5.5.2) FIPAT Ch.1 1065 [Ncl. reticularis parvocellularis]
- The ventrolateral superficial reticular area
  - Lateral paragigantocellular ncl. (see Sect. 7.3.5.6) FIPAT Ch.1 1064 [Ncl. paragigantocellularis lateralis]
  - Retroambiguus ncl. (Sect. 3.3.1.1) FIPAT Ch.1 1084 under limbic nuclei [Ncl. retroambiguus]
  - A1-A5*
    - A1 not present in humans
    - A2 (see Sect. 3.2.1.1) FIPAT Ch.1 1082 Noradrenergic cells in medulla oblongata
    - A3 not present in humans
    - A4 FIPAT Ch.1 1272 Noradrenergic cells of superior cerebellar peduncle
    - A5 (see Sect. 8.1.3.1) FIPAT Ch.1 1273 Noradrenergic cells in caudolateral pons
    - C1 (see Sect. 3.2.1.2) FIPAT Ch.1 1080 Adrenergic cells in medulla oblongata (+ C2)
- The lateral pontine tegmentum
  - Parabrachial ncll. (see Sect. 13.2.3) [Ncll. parabrachiales]
  - Kölliker Fuse ncl. (see Sect. 13.2.4) [Ncl. subparabrachialis]
  - Locus caeruleus/Caeruleus ncl. (see Sect. 13.2.2.1)

[Locus caeruleus]
Subcaeruleus ncl. (see Sect. 12.2.4.2)
[Ncl. subcaeruleus]
A7 (see Sect. 12.2.4.1) FIPAT Ch.1 1274 Noradrenergic cells in Ncl. of lateral lemniscus*
Ch5 (see Sect. 13.1.1.1.1) Ch.1 1277 Pedunculopontine nucleus
[Ncl. pedunculopontinus]

- The noradrenergic cell group A1–A7*
  A1–A5 see here above
  A6 (Caeruleus ncl.) FIPAT Ch.1 1270
  A7 (see Sect. 12.2.4.1) FIPAT Ch.1 1274 Noradrenergic cells in Ncl. of lateral lemniscus
- The adrenergic cell groups* C1 (see Sect. 3.2.1.2) and C2 (see Sect. 3.2.1.2)
- The cholinergic cell groups Ch5 and Ch6
  Ch5 (see Sect. 13.1.1.1.1) FIPAT Ch.1 1277 Pedunculopontine nucleus
  [Ncl. pedunculopontinus]
  Ch6 (see Sect. 13.1.1.2) FIPAT Ch.1 1276 Laterodorsal tegmental nucleus
  [Ncl. tegmentalis laterodorsalis]
  *FIPAT displays Latin names for the noradrenergic and adrenergic cell groups. Since their use, however, is absolutely uncommon, they are not listed here.

According to Fu et al. (2011), based on mouse data, the hindbrain reticular formation consists of three zones:

- A medial gigantocellular zone
  Ventral medullary reticular nucleus [Ncl. reticularis medullaris ventralis]
  Gigantocellular reticular nucleus [Ncl. gigantocellularis]
  Paramedian reticular nucleus [Ncl. reticularis paramedianus]
  Caudal part of the pontine reticular nucleus/Caudal pontine recticular
  nucleus [Ncl. reticularis pontinus caudalis]
  Oral part of the pontine reticular nucleus/Oral pontine recticular nucleus
  [Ncl. reticularis pontinus oralis]
- An intermediate zone
  Intermediate reticular nucleus/zone [Zona reticularis intermedia]
- A lateral parvocellular zone
  Dorsal medullary reticular nucleus [Ncl. reticularis medullaris dorsalis]
  Parvocellular reticular nucleus [Ncl. reticularis parvocellularis]

A number of the reticular nuclei can be considered as precerebellar nuclei like the pontine reticular nucleus, the gigantocellular reticular nucleus, and the ventral medullary reticular nucleus (lateral zone): In mice, in the intermediate zone and in the paramedian reticular nucleus (lateral zone) a few cells project to the cerebellum. This holds also true for the parvocellular nucleus of the lateral reticular zone while no cerebellopetal fibers were seen in the dorsal medullary reticular nucleus (for details, see Fu et al. 2011).

3.3.2.1 Intermediate reticular nucleus/zone r11 [Zona reticularis intermedia]

The intermediate reticular nucleus or zone (see Fig. 3.6A ⑧) was first described in the rat as the region between gigantocellular and parvocellular reticular nuclei containing a variety of different neuron types and displaying slightly more AChE-reactivity (see Box 3.10) than the neighboring nuclei (Paxinos and Watson 1986, see also Watson et al. 2019).

In terms of development, it is the junctional zone between the alar and basal plates (see Table 2.1, see Fig. 2.5) in the medulla oblongata and the retropontine tegmentum (FIPAT Ch.1 endnote 44).

In man, the intermediate reticular nucleus can be visualized by TH-immunoreactivity (see Fig. 3.3, catecholamine synthesis). It includes the A1 cell group (see Sect. 3.2.1.) and extends dorsally to reach the A2/C2 groups (see Sect. 3.2.1.1) (see also Kitahama et al. 1985).

In a wider sense, it contains the motoneurons of the ambiguus (see Sect. 5.4.2.1), the retrofacial and facial nuclei (Sect. 8.2.1) as well as the retroambiguus nucleus (see Sect. 3.3.1.1). Furthermore, it includes the Bötzinger and pre-Bötzinger complex (see Sect. 3.3.2.2), and the rostral and ventral respiratory and vasomotor groups (FIPAT Ch.1 endnote 44).

In mice, it belongs to the minor precerebellar nuclei bilaterally projecting to the cerebellum (Fu et al. 2011).

3.3.2.2 Rostral ventral respiratory group r11 [Centrum respiratorium ventrale rostrale] Respiration / Respiratory networks

The term respiration has at least two different denotations (see Box 3.16).

Although of vital importance to man and animals, the underlying mechanisms of breathing control are still not

Box 3.16 Respiration
Macroscopically, respiration is the process of inhaling (inspiration) and exhaling (expiration) by means of the respiratory muscles the most important of which is the diaphragm (From τό διάφραγμα Greek = to diaphragma). The diaphragm is innervated by the phrenic nerve [N. phrenicus] (Latin adjective from Greek φρένες, phrenes [pl.] = emotion, possibly related to the influence of emotions on the respiratory activity). The

phrenic nerve has its main contribution from cervical spinal motor roots C4, not from the brainstem.

Biochemically, respiration is the oxidative process occurring within living cells by which the chemical energy of organic molecules is released in a series of metabolic steps involving the consumption of oxygen and the liberation of carbon dioxide and water.

completely understood (for review, see Smith et al. 2009). In general, innate rhythmic movements like inspiration/expiration are most likely produced by central pattern generators, specialized CNS neuronal circuits (see Smith et al. 2009 and further below). These are capable of generating rhythmic activity and motor output without rhythmic inputs from other central circuits and sensorimotor feedback (Smith et al. 2009).

The core structures responsible for the control of inspiration and expiration are (Fig. 3.30):

The Bötzinger complex/Medullary expiratory center/Bötzinger group of expiratory neurons (see Box 3.17) [Centrum expiratorium medullae] (r7)

The Pre-Bötzinger complex/Medullary expiratory center (Pre-Bötzinger cells, generating respiratory rhythm) [Centrum generans motuum respiratorium] (r6/r8)

The rostral ventral respiratory group (RVRG) [Centrum respiratorium ventrale rostrale] (r11)

While the human pre-Bötzinger complex has been described in detail by Schwarzacher et al. (2011) (see below), this does not hold true for the other components of the ventral respiratory group (VRG). Furthermore, most of the experimental and functional data are derived from animal studies. In this vein, recent experimental data on the developmental origin of the two main rhythm generators in the respiratory circuit show the pre-Bötzinger complex to develop from r6 and the retrotrapezoid nucleus (RTN) (see further below) from r4/r5 (van der Heijden and Zoghbi 2020).

From animal studies it is known that the "respiratory column" in the brainstem commences rostrally with the RTN and the so-called parafacial respiratory group, followed caudally by the paragigantocellular reticular nucleus (see Sect. 5.4.4.4), the Bötzinger complex, and the pre-Bötzinger complex, subsequently the RVRG and the caudal VRG.

The VRG contains Bötzinger complex (expiratory) neurons (Box 3.17), pre-Bötzinger complex (inspiratory) neurons, rostral ventral respiratory group (predominantly inspiratory) neurons, and caudal ventral respiratory group (predominantly expiratory) neurons (Horner and Malhotra 2016). Although these three structures originate from three different rhombomeres, they will be dealt here together with for functional reasons.

Box 3.17 Bötzinger complex

The term Bötzinger complex refers to a group of cells ventral to the nucleus ambiguus in the medulla of vertebrates. The term is the only geographical one in neuroanatomy since Bötzinger is not the eponym for a scientist by this name but the place name of a wine-growing village (Bötzingen) in Southern Germany (Kaiserstuhl/Baden). At a symposium at the University of Heidelberg on central nervous breathing control, the nucleus now known as Bötzinger was "baptized" with Bötzinger Pinot noir (Spätburgunder Weißherbst/Blanc de noirs) (Dr. H. Wicht/Badische Zeitung 23.10.2008).

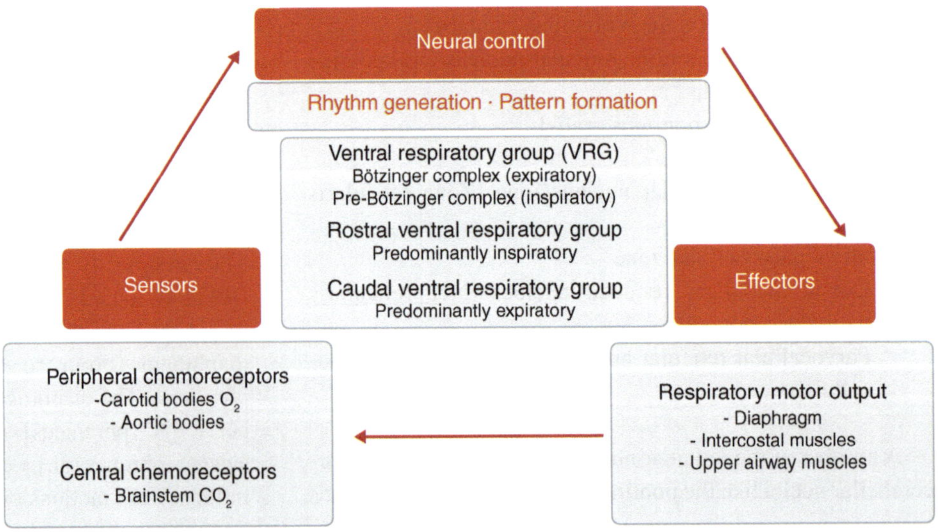

Fig. 3.30 Schematic representation of the mammalian respiratory control system. For explanation, see text

Chemosensory and mechanosensory information originates from peripheral (carotid and aortic bodies, see NTS above) and central receptors (chemosensitive areas throughout the brainstem). The carotid bodies located at the bifurcation of the external and internal carotid arteries are responsible for the majority of the hypoxemic ventilatory responses in mammals (see NTS) (Johnson and de Morais 2006). Chemosensitive areas found throughout the brainstem are the primary receptors mediating the hypercapnic ventilatory response. Hypercapnia means an elevated amount of CO_2 in the blood ($pCO_2 > 45$ mmHg).

The superordinate brainstem center for respiratory control is the retrotrapezoid nucleus (RTN, not listed in FIPAT Ch.1 and Paxinos and Watson 2009) located rostral to the ventral respiratory column, ventral to the facial nucleus (see Sect. 8.2.1), and caudal to the superior olive (see Sect. 8.1.1.5). Based on animal studies, there is growing evidence that the principal central respiratory chemoreceptors are located within the RTN and that RTN neurons are directly sensitive to [H^+] (Guyenet et al. 2016). Stimulation of the RTN activates premotor neurons in the VRG, which then activate cranial and spinal motor neurons to modulate respiratory rhythm and pattern of breathing.

Output from motor neurons reaches the respiratory effector muscles (e.g., diaphragm, intercostal muscles) producing a breath. Information from the effectors feeds back to the sensors to modulate breathing.

Studies on rats have shown that the acute silencing of neurons within the parafacial region, including RTN and rostral C1 circuit—which is not present in humans—leads to a dramatic reduction of exercise capacity, the maximum amount of physical exertion that a subject can sustain, in a treadmill paradigm (Korsak et al. 2018).

The absence of RTN neurons probably underlies the sleep apnea and lack of chemoreflex that characterize congenital central hypoventilation syndrome (Guyenet et al. 2016).

Inspiratory motor output is generated by excitatory neurons in the pre-Bötzinger complex (Fig. 3.31) and the RVRG.

Inhibitory neurons in the pre-Bötzinger complex provide inspiratory inhibition.

Expiratory inhibition is mediated by inhibitory neurons in the Bötzinger complex.

These nuclei are located in craniocaudal order in the brainstem. Please note that the pre-Bötzinger complex against expectation (pre/cranial) is located caudally to the Bötzinger complex. The RVRG is an important part of the ventral respiratory column near the ventral surface of the brainstem. The cell group is located ventrally of the nucleus ambiguus (Sect. 3.3.1.1).

As described by Schwarzacher et al. (2011), the complex is characterized by an aggregation of loosely scattered, small, and lipofuscin-rich neurons. These authors have studied the human pre-Bötzinger complex in neurologically normal subjects and in neurodegenerative brainstem diseases. They could show that in patients suffering from multiple system atrophy (see Sect. 3.3.2.3) (with central respiratory deficits but without swallowing problems) neurons of the pre-Bötzinger complex were reduced while pharyngeal motoneurons of the nucleus ambiguus (see Fig. 3.31) were not affected. On the other hand, in spinocerebellar ataxia 3 (SCA3) patients (see Sect. 7.2.1.3.3) (without central respiratory deficits but with dysphagia) the pre-Bötzinger neurons were preserved but the motoneurons of the nucleus ambiguus (see Sect. 3.3.1.1) were diminished.

3.3.2.3 Multiple system atrophy (MSA) (progressive autonomic failure with multiple system atrophy)

Multiple system atrophy is an adult-onset, sporadic, progressive, neurodegenerative disorder affecting both men and women. The main features include autonomic failure, parkinsonism, cerebellar ataxia, and pyramidal signs in any combination (Wenning et al. 2004). Parkinsonian features predominate in 80% of patients and cerebellar ataxia in 20% of patients. Autonomic dysfunction, which includes prominent urogenital dysfunction and orthostatic hypotension, is common to both variants and reflects the degeneration of central and peripheral autonomic pathways (Ozawa 2007; Wenning et al. 2008). MSA appears to affect males and females in equal numbers. The peak onset of MSA is between 55 and 60 years of age. 50% of individuals are wheelchair-bound within 5–6 years of the onset of motor symptoms. The range, severity, and distribution of symptoms vary greatly among affected individuals.

Historically, patients with MSA were classified under three clinical categories: The Shy-Dräger syndrome, striatonigral degeneration, and olivopontocerebellar atrophy. The term Shy-Dräger syndrome was often applied to patients with early prominent autonomic failure.

Patients presenting with a predominantly parkinsonian syndrome were often referred to as cases of striatonigral degeneration (SND), now referred to as MSA-P.

The term "sporadic olivopontocerebellar atrophy (OPCA)" was used for those patients with cerebellar syndrome, now referred to as MSA-C.

Since most patients suffer from some degree of autonomic failure, the term Shy-Dräger syndrome is no longer used (Shy and Dräger 1960; Gilman et al. 1998). Differential diagnosis of MSA-P and PD may be very difficult in the early stages owing to several overlapping features such as resting tremor or asymmetric akinesia and rigidity. Levodopa-induced improvement of parkinsonism may be seen in 30% of patients with parkinsonism. 90% of these patients are unresponsive to levodopa in the long term (Wenning et al. 2004). Dysautonomia is characteristic of both MSA motor presentations, primarily comprising urogenital and ortho-

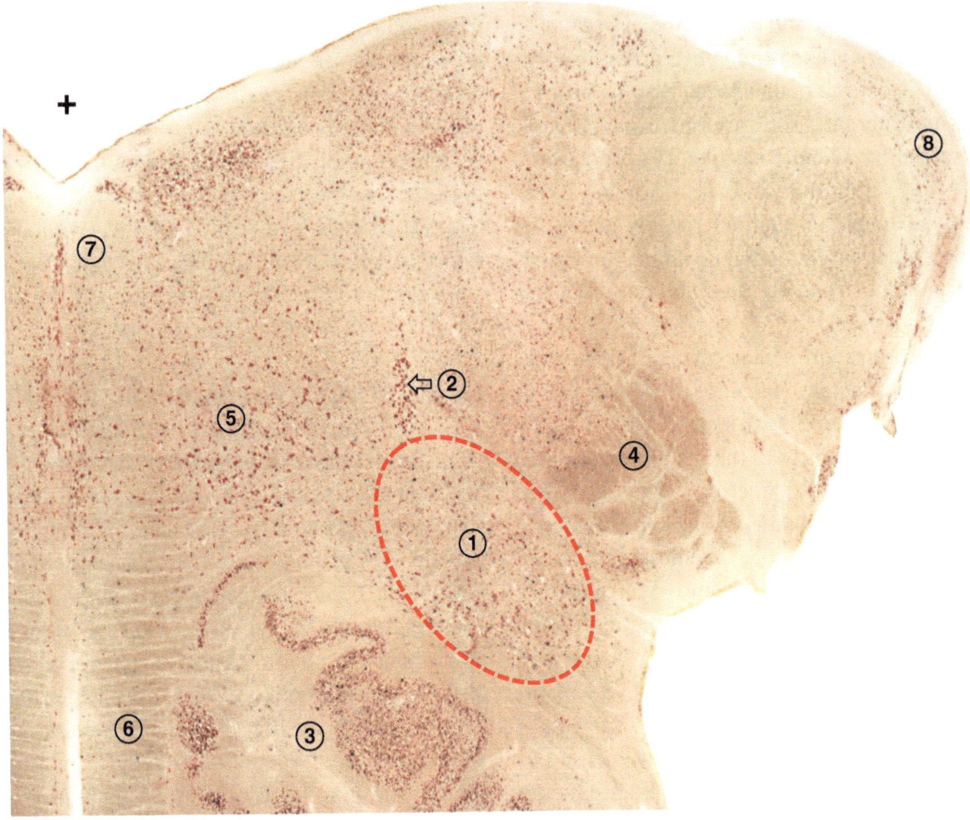

①	**Region of the pre-Bötzinger complex**	⑥	**Medial lemniscus**
②	**Ncl. ambiguus CNIX, CNX**	⑦	**Medial longitudinal fasciculus**
③	**Inferior olivary complex**	⑧	**Posterior cochlear ncl. CNVIII**
④	**Spinal trigeminal tract CNV**	+	**4th ventricle**
⑤	**Gigantocellular reticular ncl.**		

Fig. 3.31 Horizontal section through the human medulla oblongata at the level of the cochlear nuclei ⑧. Darrow red staining. The area of the pre-Bötzinger complex is delineated by the red line following the description by Schwarzacher et al. (2011). It is located between the nucleus ambiguus ② (see Sect. 3.3.1.1), the spinal trigeminal tract ④ (see Fig. 3.5B) and the inferior olivary complex ③. See also atlas part Darrow red 14, 14 A. LabPON Twente

static dysfunction. Disorders of micturition in MSA are caused by changes in the complex peripheral and central innervation of the bladder. Degeneration of Onuf's nucleus—a small group of neurons located in the ventral horns of the sacral spinal cord—is related to incontinence. Symptomatic orthostatic hypotension is present in 68% of clinically diagnosed patients (Wenning et al. 2004).

Cognitive dysfunction is almost never a dominant symptom though it does occur (Hovestad 2007). Patients with MSA-C, especially, have shown impaired executive memory and visuospatial functions as well as deficits in verbal fluency and attention. It is suggested that cognitive dysfunction result from disruption of the corticostriatal and pontocerebellar circuits (Stankovic et al. 2014; Lee et al. 2016). So far, no pathological differences were identified between MSA cases with cognitive dysfunction and cases with normal cognition (Asi et al. 2014). Evidence favoring a role of genetic predisposition in MSA is not strong at present (Stankovic et al. 2019). It is generally considered to be a sporadic disease though there are occasional reports of familial occurrence (Itoh et al. 2014).

The macroscopical appearance of the brain in MSA ranges from normal to focal atrophy of the affected brain regions, which is generally related to the clinical pattern of the disease. On cut surface, the putamen shows variable atrophy, being most pronounced in patients with MSA-P. There is often a greyish-green discoloration of the atrophic putamen (Fig. 3.32). The substantia nigra and the locus caeruleus

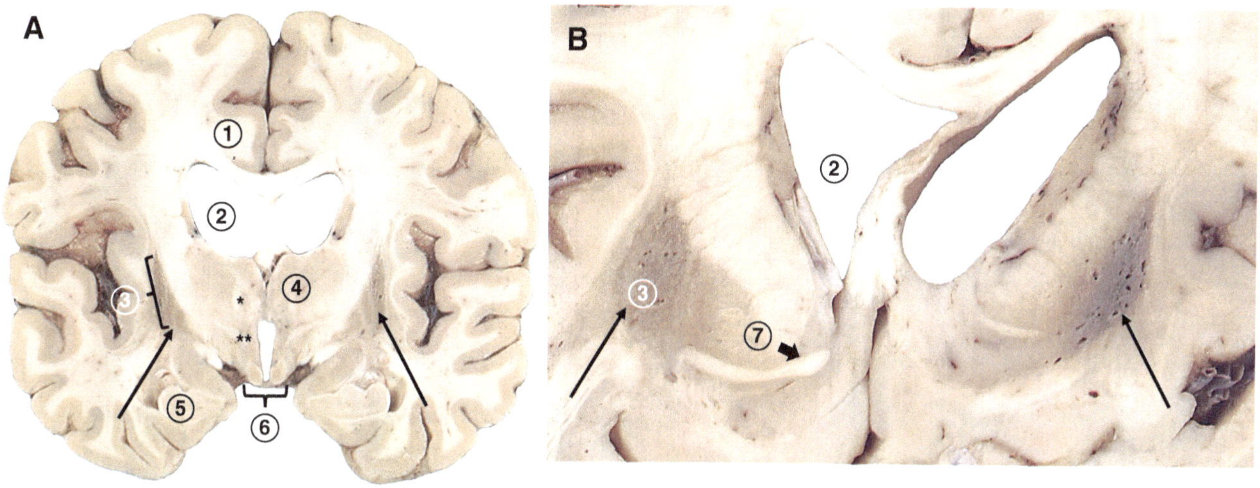

①	Cingulate gyrus	⑥	Mammillary bodies
②	Lateral ventricle	⑦	Anterior commissure
③	Putamen	*	Mammillotegmental tract
④	Thalamus	**	Mammillothalamic tract
⑤	Hippocampus		

Fig. 3.32 Coronal section of a brain from an MSA-P patient (**A** and **B**). Atrophy, dark discoloration, and cribriform appearance of the putamen ③ is especially well visible in (**B**) (arrows). Symmetrical ventricular dilatation. LabPON Twente

are pale as the result of a loss of pigment. Atrophy of the cerebellum, the cerebellar peduncles, pons and medulla oblongata is most marked in MSA-C (Figs. 3.33 and 3.34).

In 1989, Papp et al. described glial cytoplasmic inclusions in the central nervous system of 11 patients with various combinations of striatonigral degeneration, olivopontocerebellar atrophy, and Shy-Dräger syndrome (Papp et al. 1989). Glial cytoplasmatic inclusions (GCIs) are argyrophilic (Gallyas-positive), and their shape varies from triangular, sickle, half-moon, oval or conical, to being occasionally flame-shaped. They are of variable size, often completely occupying the cytoplasm to displace the nucleus eccentrically. Structures rich in GCIs include the primary motor and higher motor areas of the cerebral cortex and their subjacent white matter, pyramidal, extrapyramidal and corticocerebellar systems, the supraspinal autonomic systems, and their targets. Also the dorsolateral part of the putamen (Lantos 1998), the dorsolateral smaller part of the caudate nucleus, the pons, the globus pallidus, the internal and external capsules, the reticular formation, the middle cerebellar peduncles, and the cerebellar white matter show many GCIs. The GCIs are positive for α-synuclein, ubiquitin, tubulin, and αB-crystallin. α-synuclein and ubiquitin can be found in association with a number of other proteins, such as the microtubule-associated protein tau (in a phosphorylation state that differs from that seen in Alzheimer's and related tauopathies) and prion disease-linked 14-3-3 (Wenning et al. 2008). Degenerating neurites are also strongly immunopositive for α-synuclein. Furthermore, α-synuclein immunoreactivity can be seen in neuronal cytoplasmic inclusions in the Substantia nigra (see below Figs. 3.35 and 3.36).

Ultrastructural studies have shown GCIs to be present in all three types of oligodendrocytes: in satellite cells adjacent to neurons in the grey matter, in the interfascicular variant in the white matter, and in the perivascular oligodendrocytes of the affected areas (Gilman et al. 1998; Matsuo et al. 1998).

In the cerebral cortex, there is neuronal loss in the primary and supplementary motor cortex, with an accompanying increase in glial cells. Severe oligodendrocytic alterations with GCIs are seen in the fiber tracts of the motor system (corticopontine, corticobulbar, corticospinal. pontocerebellar, olivocerebellar, reticulocerebellar, spino-reticular, spinoolivary, spinocerebellar tracts, for anatomical survey see Tables 1.3 and 1.4) (Braak et al. 2003).

In a study of MSA-C patients, GCIs were associated with myelinated axons and the severity of GCIs correlated with demyelination (Brettschneider et al. 2017). The white matter shows demyelination, especially in affected areas (Matsuo et al. 1998).

Neuronal loss/gliosis is severe in the fifth layer of the motor cortex, lateral horn of the spinal cord, striatum (especially dorsolateral zone of the caudal putamen), and substantia nigra (see Sect. 16.6.1). Affected neurons bearing NCIs were seen in all of the precerebellar nuclei (inferior olivary complex (see

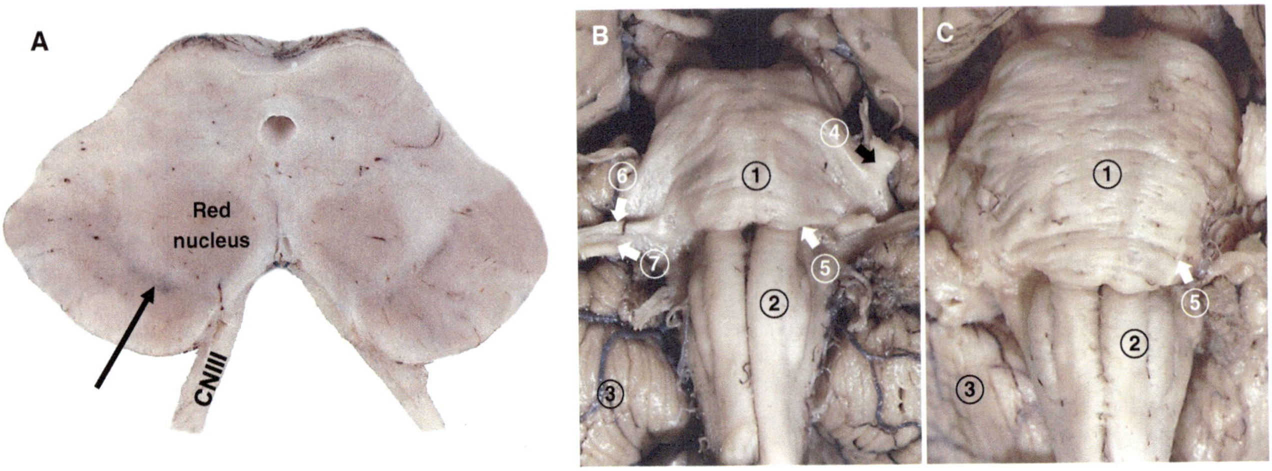

①	Pons
②	Medulla oblongata
③	Cerebellum
④	Trigeminal nerve CNV
⑤	Abducens nerve CNVI
⑥	Facial nerve CNVII
⑦	Vestibulocochlear nerve CNVIII

Fig. 3.33 (**A**) Mesencephalon in an MSA-P patient showing pallor (depigmentation) of the substantia nigra of the mesencephalon (arrow). (**B**) Ventral view of the brainstem from an MSA-C patient showing marked atrophy of the basal pons as compared to a control case (**C**). LabPON Twente

Sect. 4.1.1.3), pontine nuclei (see Sect. 10.1.1.2), lateral reticular nucleus (Sect. 4.1.1.1), external cuneate nucleus (see Sect. 5.2.1.1), subventricular nucleus, nucleus of Roller (see Sect. 6.2.1.6), conterminal nucleus, arcuate nucleus (see Sect. 4.1.1.4), pontine grey (see Sect. 13.2.7), and pontobulbar body with the lateral reticular nucleus and the accessory subnuclei of the inferior olive sustaining the most severe neuronal damage (Braak et al. 2003) and cerebellar Purkinje cells. It is less severe in the dentate nucleus, nucleus ambiguus, vestibular nuclei, dorsal motor nucleus of the vagal nerve, intermediolateral column of the spinal cord, Onuf's nucleus, and anterior horn cells. The pre-Bötzinger complex (see Sect. 3.3.2.2) in the ventrolateral medulla oblongata, highlighted by showing expression of high levels of neurokinin 1 receptor and somatostatin, is of functional relevance for central respiratory control (Schwarzacher et al. 2011) and is heavily damaged in MSA. Involvement of the respiratory network could contribute to the occurrence of breathing problems in MSA (Nogués and Benarroch 2008). Excessive iron accumulation has been demonstrated, particularly in the basal ganglia. The cell types that contain iron pigment include predominantly oligodendroglia and microglia, but neurons and astrocytes are also affected.

GCIs have also been found in progressive supranuclear palsy (PSP) (see Sect. 4.1.1.3.3), corticobasal degeneration (CBD) (Sect. 16.6.1.9), Alzheimer's disease (AD) (see Sect. 13.2.2.1.4), Pick's disease (see Sect. 4.3.1.1.5), and frontotemporal lobar degeneration FTLDP-17 (see Sect. 4.3.1.1.5) where they contain tau inclusions. Furthermore, they were seen in spinocerebellar ataxia type 1 (SCA1), and a subgroup of hereditary familial olivopontocerebellar atrophy (spinocerebellar atrophy, SCA2) (see Sect. 7.2.1.3.2), containing ubiquitin but not tau, α-synuclein, or polyglutamine (Yokoyama and Ishiyama 2014). Some FTLD-TDP forms show TDP inclusions in glial cells (Lee et al. 2017) and GCI with FUS (functionally related protein fused in sarcoma) were found in FTLD-FUS (Munoz et al. 2009). In addition, glial inclusions containing α-synuclein have been described in Parkinson's disease (PD) (see Sect. 16.6.1.5.) cases (Braak et al. 2007).

Widespread GCIs are the defining morphological feature of MSA and are thought to play a central role in the pathogenesis of the disease (Trojanowski et al. 2007; Ahmed et al. 2012).

Pathology in MSA is not only found in the central nervous system, but also in the peripheral nervous system, i.e., demyelination in the sural nerve, and abnormal EMG in muscles (Kanda et al. 1996; Pramstaller et al. 1995). Given the clinical, pathological, and biochemical overlap of Lewy body disease (see Sect. 16.6.1.6) and MSA, it seems reasonable to suggest that they define a disease spectrum ("synucleinopathies").

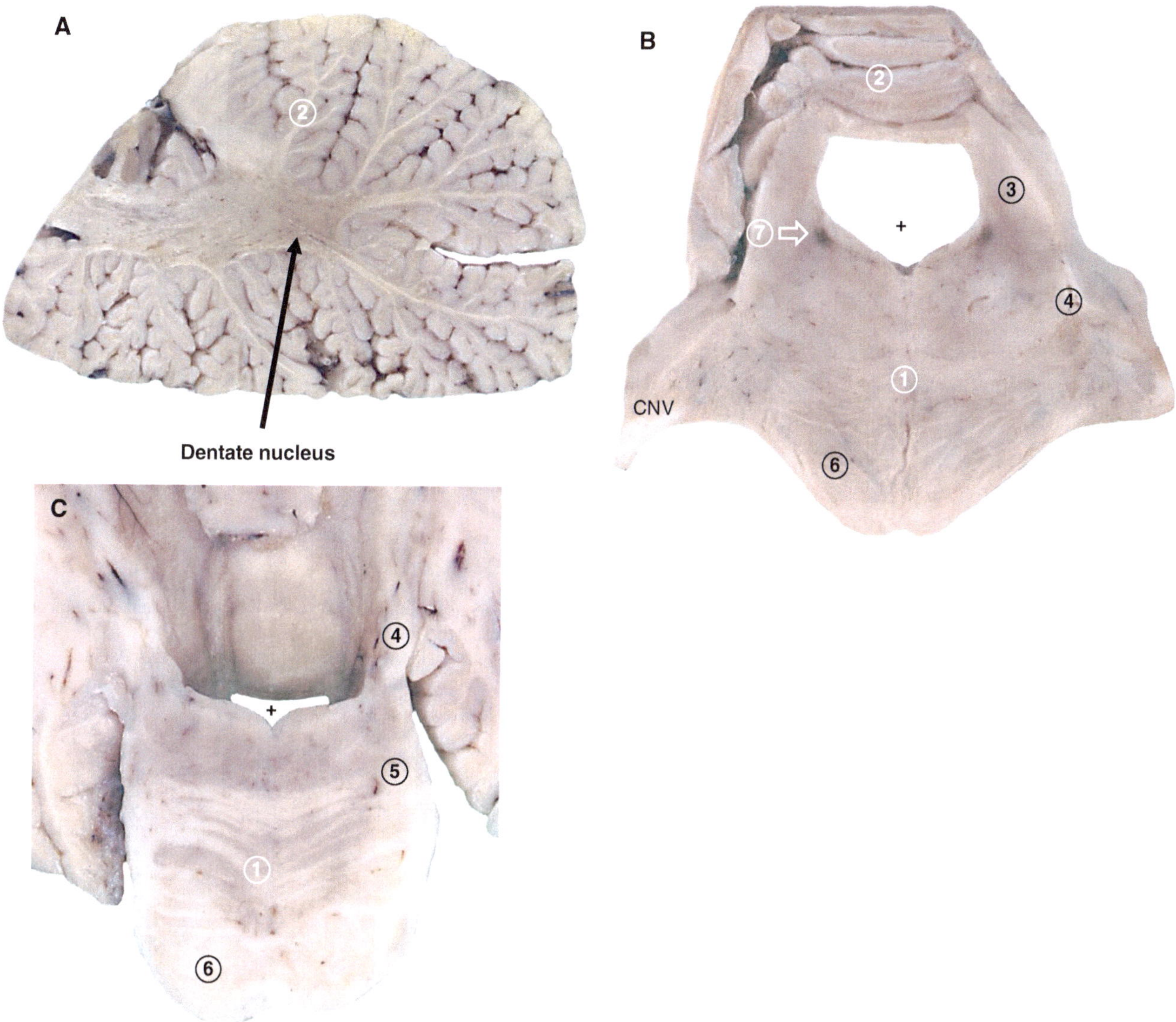

①	Pons	⑤	Inferior cerebellar peduncle
②	Cerebellum	⑥	Corticospinal tract
③	Superior cerebellar peduncle	⑦	Locus caeruleus
④	Middle cerebellar peduncle	+	4th ventricle

Fig. 3.34 Brain of an MSA-C patient. In the cerebellum (**A**) slight cortical atrophy and discoloration of the (reduced) white matter. In (**B**) and (**C**) marked atrophy of the cerebellar peduncles and pontine base. LabPON Twente

3.3.2.4 Central cervical nucleus of the spinal cord/Central cervical nucleus r11 [Ncl. cervicalis centralis]

The term central cervical nucleus refers to a cell column that occupies lamina VII in the upper four cervical segments of the spinal cord. It is located lateral to the intermediomedial cell column.

In the mouse, it extends into the medulla. In the functional model of central nervous system organization, it is classified as part of the subcortical somatosensory system. This nucleus receives primary afferents from neck muscles and the vestibular apparatus (cf. Watson et al. 2008).

http://braininfo.org/centraldirectory.aspx?ID=1648

3.3.2.5 Medullary reticular nucleus r11 [Ncl. reticularis medullaris]

The term medullary reticular nucleus refers to the portion of the medullary reticular formation located caudal to the

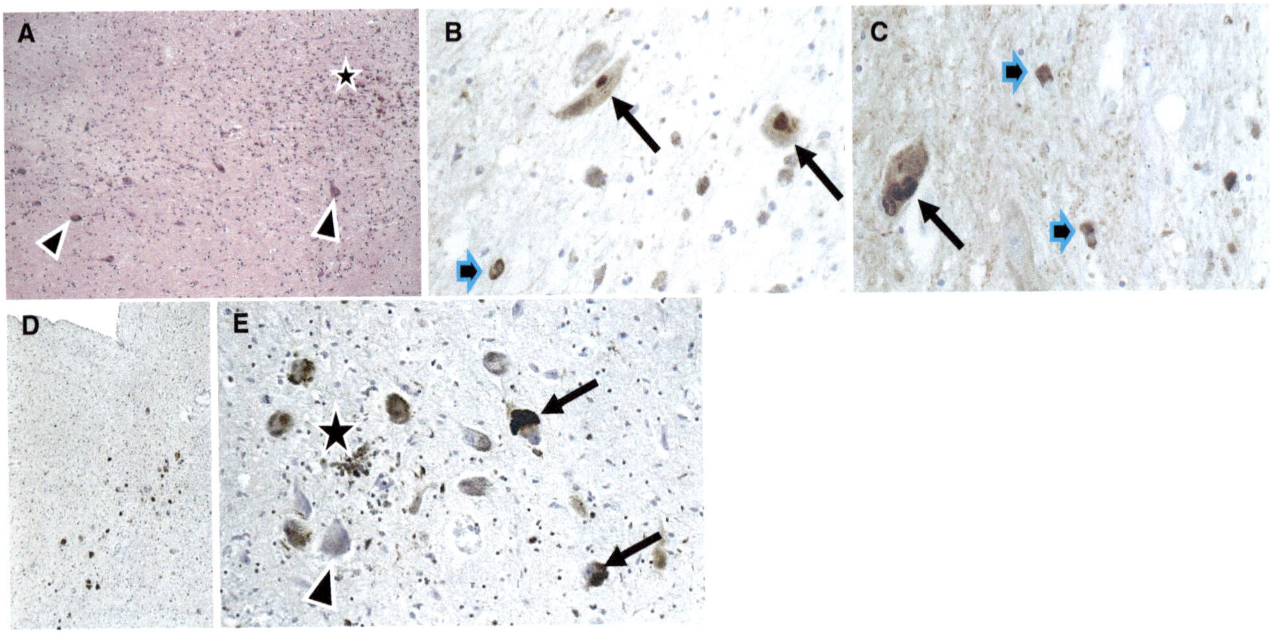

Fig. 3.35 Severe neuronal loss in the substantia nigra is a common feature in MSA-P patients (**A**, H & E stain). Neuronal cytoplasmic inclusions (NCIs, long arrow) and glial cytoplasmic inclusions GCIs, the hallmark of MSA (short thick arrow) in the substantia nigra (**B** and **C**), and locus caeruleus (**D** and **E**) (α-synuclein immunostain). Extraneuronal pigment, indicative of neuronal loss (star). Triangle points to neuromelanin-containing neurons. LabPON Twente

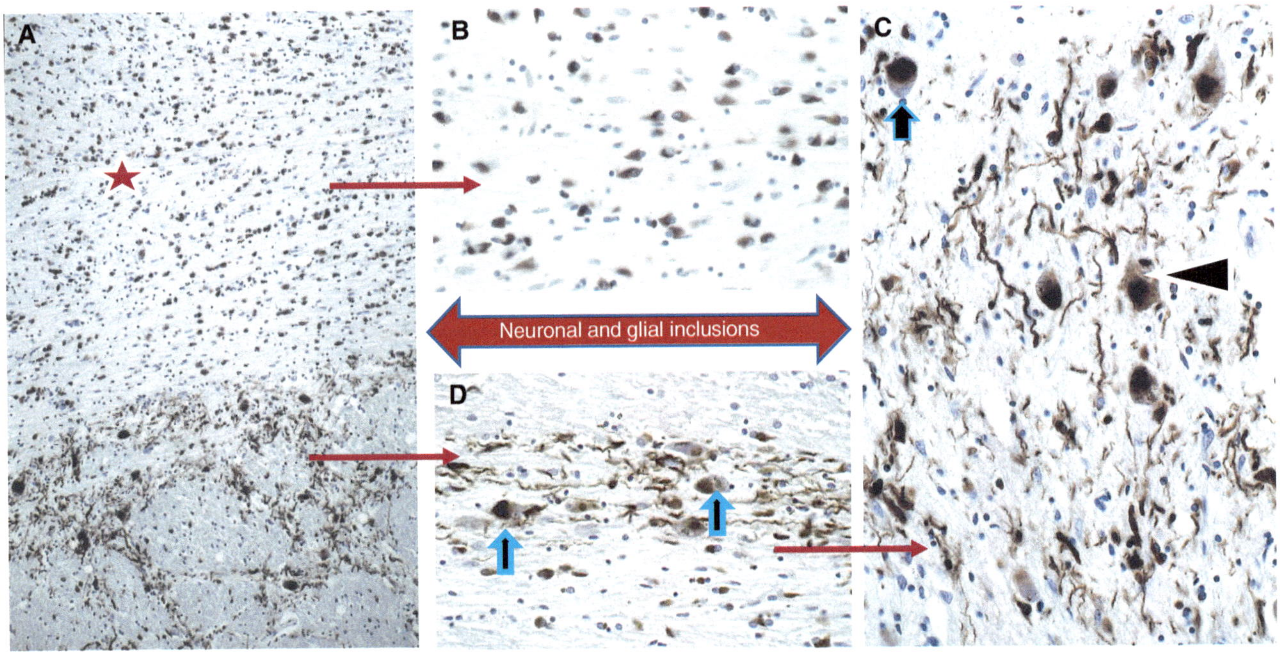

Fig. 3.36 Microscopical sections of the pons (**A-C**) in MSA showing both GCIs (star), NCIs (thick arrows), neuronal nuclear inclusions (NNIs) (triangle) and immuno-positive neuropil threads in an α-synuclein immunostain. LabPON Twente

gigantocellular nucleus and the inferior olivary complex of human, macaque, rat, and mouse. It is generally accepted that it has two parts, the dorsal part of the medullary reticular nucleus and the ventral part of the medullary reticular nucleus. Functionally, the medullary reticular nucleus belongs to the reticular formation of the subcortical motor system.

http://braininfo.org/centraldirectory.aspx?ID=1789

In mice, the ventral part of the medullary reticular nucleus is a minor precerebellar nucleus with bilateral projections to the cerebellum (Fu et al. 2011).

References

Abraira VE, Ginty DD (2013) The sensory neurons of touch. Neuron 79:618–639

Ahmed Z, Asi YT et al (2012) Review: the neuropathology, pathophysiology and genetics of multiple system atrophy. Neuropathol Appl Neurobiol 38:4–24

Asi JT, Ling H et al (2014) Neuropathological features of multiple system atrophy with cognitive impairment. Mov Disord 29:884–888

Bai L, Mesgarzadeh S et al (2019) Genetic identification of vagal sensory neurons that control feeding. Cell 179:1129–1143

Beckstead RM, Morse JR et al (1980) The nucleus of the solitary tract in the monkey: projections to the thalamus and brain stem nuclei. J Comp Neurol 190:259–282

Benninger B, McNeil J (2010) Transitional nerve: a new and original classification of a peripheral nerve supported by the nature of the accessory nerve (CN XI). Neurol Res Int 2010:476018

Bersano A, Bedini G et al (2017) CADASIL: treatment and mangement options. Curr Treat Options Neurol 19:31

Blessing WW, Benarroch EE (2012) Lower brainstem regulation of visceral, cardiovascular, and respiratory function. In: Mai JK, Paxinos G (eds.) The human nervous system. Elsevier, pp 1058–1073

Braak H (1976) On the striate area of the human isocortex. A golgi- and pigmentarchitectonic study. J Comp Neurol 166:341–364

Braak H, Rüb U et al (2003) Involvement of precerebellar nuclei in multiple system atrophy. Neuropathol Appl Neurobiol 29:60–76

Braak H, Sastre M et al (2007) Development of alphasynuclein immunoreactive astrocytes in the forebrain parallels stages of intraneuronal pathology in sporadic Parkinson's disease. Acta Neuropathol 114:231–241

Breit S, Kupferberg A (2018) Vagus nerve as modulator of the brain–gut axis in psychiatric and inflammatory disorders. Front Psychiatry 9:44

Brettschneider J, Irwin DJ et al (2017) Progression of alpha-synuclein pathology in multiple system atrophy of the cerebellar type. Neuropathol Appl Neurobiol 43:315–329

Brisman R (1989) Treatment of trigeminal neuralgia by radiofrequency electrocoagulation. In: Brisman R (ed) Neurosurgical and medical management of pain: trigeminal neuralgia, chronic pain, and cancer pain, Topics in neurosurgery, vol 3. Springer, Boston

Brodal A, Brodal P (1983) Observations on the projection from the perihypoglossal nuclei onto the cerebellum in the macaque monkey. Arch Ital Biol 121:151–166

Brodmann K (1909) Vergleichende Lokalisationslehre der Großhirnrinde in ihren Prinzipien dargestellt auf Grund des Zellenbaues. Verlag von Johann Ambrosius Barth, Leipzig

Broggi G, Broggi M et al (2012) Surgical technique for trigeminal microvascular decompression. Acta Neurochir 154:1089–1095

Browning KN, Verheijden S et al (2017) The vagus nerve in appetite regulation, mood and intestinal inflammation. Gastroenterology 152:730–744

Bugiani M, Kevelam SH et al (2016) Cathepsin A-related arteriopathy with strokes and leukoencephalopathy (CARASAL). Neurology 87:1–10

Bulmer DC, Roza C (2018) Visceral pain. In: Wood JN (ed) The Oxford handbook of the neurobiology of pain

Chabriat H, Levy C et al (1998) Pattern of MRI lesions in CADASIL. Neurology 51:452–457

Chabriat H, Joutel A et al (2020) Cadasil: yesterday, today, tomorrow. Eur J Neurol 27:1588–1595

Chang SD, Ho A (eds.) (2018) Effective techniques for managing trigeminal neuralgia. IGI Global

Chen X, Gabitto M et al (2011) A gustotopic map of taste qualities in the mammalian brain. Science 333:1262–1126

Cheng G, Zhu H et al (2006) Development of the human nucleus of the solitary tract: a cyto- and chemoarchitectural study. Auton Neurosci 128:76–95

Chouchkov C (1978) Introduction. In: Cutaneous receptors. Advances in anatomy, embryology and cell biology/Ergebnisse der Anatomie und Entwicklungsgeschichte/revues d'anatomie et de morphologie expérimentale, vol 54/5. Springer, Berlin

da Cunha GJL, Rocha BML et al (2019) A systematic review on recurrent cardioinhibitory vasovagal syncope: does pacing therapy break the fall? Pacing Clin Electrophysiol 42:1400–1407

Dahlström A, Fuxe K (1964) Evidence for the existence of monoamine-containing neurons in the central nervous system. I. Demonstration of monoamines in the cell bodies of brain stem neurons. Acta Physiol Scand Suppl 2323:1–55

Del Fiacco Quartu M et al (2014) The human cuneate nucleus contains discrete subregions whose neurochemical features match those of the relay nuclei for nociceptive information. Brain Struct Funct 219:2083–2101

Dichgans M, Mayer M et al (1998) The phenotypic spectrum of CADASIL: clinical findings in 102 cases. Ann Neurol 44:731–739

Dichgans M, Filippi M et al (1999) Quantitative MRI in CADASIL: correlation with disability and cognitive performance. Neurology 52:1361–1367

Dong Y, Hassan A et al (2003) Yield of screening for CADASIL mutations in lacunar stroke and leukoaraiosis. Stroke 34:203–205

Dupont G, Altafulla J et al (2019) Ossification of the roof of the porus trigeminus with duplicated abducens nerve. Anat Cell Biol 52:211–213

Farid AM, ElKheshin SE (2020) Endoscope assisted microvascular decompression: are the advantages still worthy towards more? Egypt J Neurosurg 35:1

Florence SL, Wall JT et al (1989) Somatotopic organization of inputs from the hand to the spinal gray and cuneate nucleus of monkeys. J Comp Neurol 286:48–70

Fu Y, Tvrdik P et al (2011) Precerebellar cell groups in the hindbrain of the mouse defined by retrograde tracing and correlated with cumulative Wnt1-Cre genetic labeling. Cerebellum 10:570–584

Ganchrow D, Ganchrow JR et al (2013) The nucleus of the solitary tract in the C57BL/6J mouse: subnuclear parcellation, chorda tympani nerve projections and brainstem connections. J Comp Neurol 522:1565–1596

Gilbert SF (2000) Developmental biology, 6th edn. Sinauer Associates, Sunderland

Gilman S, Low PA et al (1998) Consensus statement on the diagnosis of multiple system atrophy. J Auton Nerv Syst 74:189–192

Gopinathannair R, Salgado BC et al (2018) Pacing for vasovagal syncope. Arrhythm Electrophysiol Rev 7:95–102

Gulec S, Spedicato F et al. (2019) lateral suboccipital approach (retrosigmoid). In: July J, Wahjoepramono EJ. Neurovascular surgery. Surgical approaches for neurovascular diseases. Springer, pp 43–48

Guyenet PG, Bayliss DA et al (2016) Proton detection and breathing regulation by the retrotrapezoid nucleus. J Physiol 594:1529–1551

Haffner C, Vinters HV (2016) CADASIL, CARASIL, CARASAL. The linguistic subtleties of cerebral small vessel disease. Neurology 87:1752–1753

Hamilton RB, Norgren R (1984) Central projections of gustatory nerves in the rat. J Comp Neurol 222:560–577

Hara K, Shiga A et al (2009) Association of HTRA1 mutations and familial ischemic cerebral small-vessel disease. N Engl J Med 360:1729–1739

Hasin DS, Sarvet AL et al (2018) Epidemiology of adult DSM-5 major depressive disorder and its specifiers in the United States. JAMA Psychiatry 75:336–346

Hayashi T, James TN et al (1995) Ultrastructure and immunohistochemistry of the coronary chemoreceptor in human and canine hearts. Am Heart J 129:946–959

Hirota R, Okano H et al (2016) Nodose Ganglion. In: Neuroanatomy and neurophysiology of the larynx. Springer, pp 73–82

Hökfelt T, Johansson O et al (1984) Chemical anatomy of the brain. Science 225:1326–1334

Horner RL, Malhotra A (2016) Control of breathing and upper airways during sleep. In: Murray and Nadel's textbook of respiratory medicine, 6th edn., vol. 2. Elsevier, pp 1511–1526

Hornung J-P (2003) The human raphe nuclei and the serotonergic system. J Chem Neuroanat 26:331–343

Hovestad A (2007) Multiple system atrophy. In: van Laar T, Berendse HW (eds) Parkinsonism and related disorders. VU University Press, Wellington, p 251

Huggenberger S, Moser N et al (2019) Neuroanatomie des Menschen. Springer

Iannilli E, Gudziol V (2019) Gustatory pathway in humans: a review of models of taste perception and their potential lateralization. J Neurosci Res 97:230–240

Itoh K, Kasai T et al (2014) Definite familial multiple system atrophy with unknown genetics. Neuropathology 34:309–313

Jannetta PJ (1967) Arterial compression of the trigeminal nerve at the pons in patients with trigeminal neuralgia. J Neurosurg 26:159–162

Job J, Branstetter IV BF (2017) Imaging of the posterior skull base. In: Aygun N (ed) Skull base imaging. Radiologic clinics of North America, vol. 55, 1, pp 103–122

Johnson RA, de Morais A (2006) Respiratory acid-base disorders in fluid, electrolyte, and acid-base disorders in small animal practice, 3rd edn

Joutel A, Corpechot C et al (1996) Notch3 mutations in CADASIL, a hereditary adult-onset condition causing stroke and dementia. Nature 383:707–710

Kanda T, Tsukagoshi H et al (1996) Changes of unmyelinated nerve fibres in sural nerve in amyotrophic lateral sclerosis, Parkinson's disease and multiple system atrophy. Acta Neuropathol 91:145–154

Kheradmand A, Colpak AI et al. (2016) Eye movements in vestibular disorders. In: Furman JM, Lempert T (eds.) Handbook of clinical neurology, vol. 137 (3rd series). Elsevier

Kitahama K, Pearson J et al (1985) Adrenergic neurons in human brain demonstrated by immunohistochemistry with antibodies to phenylethanolamine-N- methyltransferase (PNMT): discovery of a new group in the nucleus Tractus Solitarius. Neurosci Lett 53:303–308

Klingner CM, Witte OW (2018) The parietal lobe. In: Vallar G, Coslett HB (eds.), Handbook of clinical neurology, 1st edn. Elsevier

Korsak A, Sheikhbahaei S et al (2018) The role of Parafacial neurons in the control of breathing during exercise. Sci Rep 8:400

Kumar P, Prabhakar NR (2012) Peripheral chemoreceptors: function and plasticity of the carotid body. Compr Physiol 2:141–219

Kupari J, Häring M et al (2019) An atlas of vagal sensory neurons and their molecular specialization. Cell Rep 27:2508–2523

Lang J (1991) Clinical anatomy of the posterior cranial fossa and its foramina. Thieme, Stuttgart

Lang J (1979) Praktische Anatomie: Ein Lehr- und Hilfsbuch der anatomischen Grundlagen ärztlichen Handelns/begr. von T von Lanz, W Wachsmuth. Fortgef u hrsg von J Lang, W Wachsmuth. Springer, Berlin Heidelberg New York. NE: Lang, Johannes (Hrsg); Lanz, Titus von (Begr) Teil 1: Bd 1 Kopf. Teil B Gehirn- und Augenschädel/von J Lang. In Zsarb mit K-A Bushe, W Buschmann u D Linnert

Lang J (2001) Skull Base and related structures, atlas of clinical anatomy, 2nd edn. Schattauer, Stuttgart

Lantos PL (1998) The definition of multiple system atrophy: a review of recent developments. J Neuropathol Exp Neurol 57:1099–1111

Lechien JR, Chiesa-Estomba CM et al (2020) Olfactory and gustatory dysfunctions as a clinical presentation of mild-to-moderate forms of the coronavirus disease (COVID-19): a multicenter European study. Eur Arch Otorhinolaryngol 277(8):2251–2261

Lee MJ, Shin JH et al (2016) Cognitive impairments associated with morphological changes in cortical and subcortical structures in multiple system atrophy of the cerebellar type. Eur J Neurol 23:92–100

Lee EB, Porta S et al (2017) Expansion of the classification of FTLD-TDP: distinct pathology associated with rapidly progressive frontotemporal degeneration. Acta Neuropathol 134:65–78

Lesnik Oberstein SAJ, van Duinen SG et al (2003) Evaluation of diagnostic NOTCH3 immunostaining in CADASIL. Acta Neuropathol 106:107–111

Liliequist B (1959) The subarachnoid cisterns. An anatomic and roentgenologic study. Acta Radiol Suppl 185:1–108

Loewi O (1921) Über humorale Übertragbarkeit der Herznervenwirkung. Pflügers Arch 189:239–242

Loewi O, Navratil E (1926) Über den Mechanismus der Vaguswirkung von Physostigmin und Ergotamin. Klin Wochenschr 5:894–895

Maeda S, Nakayama H et al (1976) Familial unusual encephalopathy of Binswanger's type without hypertension. Folia Psychiatr Neurol Jpn 30:165–177

Mahboubi H, Slattery WH et al (2020) Options and strategies for hearing restoration in pediatric neurofibromatosis type 2. Childs Nerv Syst 36(10):2481–2487. Epub ahead of print

Mangold AR, Torgerson RR et al (2016) Diseases of the tongue. Clin Dermatol 34:458–469

Martin M, Fangerau H et al (2020) Georg Schaltenbrand (1897–1979) und seine "entgrenzte Forschung " zur Multiplen Sklerose. (Georg Schaltenbrand (1897–1979) and his research without moral boundaries on multiple sclerosis). Nervenarzt 91:43–52

Mascioli G, Berlucchi G et al (2015) Functional MRI cortical activations from unilateral tactile-taste stimulations of the tongue. Physiol Behav 151:221–229

Matsumoto I (2013) Gustatory neural pathways revealed by genetic tracing from taste receptor cells. Biosci Biotechnol Biochem 77:1359–1362

Matsuo A, Akiguchi I et al (1998) Myelin degeneration in multiple system atrophy detected by ubiquitin antibodies. Am J Pathol 153:735–744

McManus LJ, Dawes PJD et al (2011) Clinical anatomy of the chorda tympani: a systematic review. J Laryngol Otol 125:1101–1108

McRitchie DA, Törk I (1993) The internal organization of the human solitary nucleus. Brain Res Bull 31:171–193

Mei N (1983) Recent studies on intestinal vagal afferent innervation. Functional implications. J Auton Nerv Syst 9:199–206

Menzel J, Piotrowski W et al (1975) Long-term results of Gasserian ganglion electrocoagulation. J Neurosurg 42:140–143

Mesulam MM, Mufson EJ et al (1983) Central cholinergic pathways in the rat: an overview based on an alternative nomenclature (Ch1-Ch6). Neuroscience 10:1185–1201

Miller NR, Walsh BF et al (eds.) (2005) Walsh and Hoyt's clinical. Neuroophthalmology, vol. 1

Munoz D, Neuman M et al (2009) FUS pathology in basophilic inclusion body disease. Acta Neuropathol 118:617–627

Nepovimova E, Kuca K (2018) Chemical warfare agent NOVICHOK—mini-review of available data. Food Chem Toxicol 121:345–350

Nieuwenhuys R (1998) Structure and organisation of fibre systems. In: Nieuwenhuys R, ten Donkelaar HJ, Nicholson C (eds.) The central nervous system of vertebrates. 4th edn. Springer, pp 113–157

Nieuwenhuys R, Voogd J, van Huijzen C (2008) The human central nervous system. 4th edn. Springer

Nissl F (1894) Ueber eine neue Untersuchungsmethode des Centralorgans speziell zur Feststellung der Localisation der Nervenzellen. Zentralbl Nervenheilkd Psychiatr 17:337–344

Nogués MA, Benarroch E (2008) Abnormalities of respiratory control and the respiratory motorunit. Neurologist 14:273–288

Ohm TG, Heilmann R et al (1989) The human oral raphe system. Architectonics and neuronal types in pigment-Nissl preparations. Anat Embryol 180:37–43

Olszewski J, Baxter D (1982) Cytoarchitecture of the human brain stem, 2nd edn. Karger, Basel

Owens DM, Lumpkins EA (2014) Diversification and specialization of touch receptors in skin. Cold Spring Harb Perspect Med 4(6):a013656

Ozawa T (2007) Morphological substrate of autonomic failure and neurohormonal dysfunction in multiple system atrophy: impact on determining phenotype spectrum. Acta Neuropathol 114:201–211

Papp MI, Kahn JE et al (1989) Glial cytoplasmic inclusions in the CNS of patients with multiple system atrophy (striatonigral degeneration, olivopontocerebellar atrophy and Shy-Drager syndrome). J Neurol Sci 94:79–100

Paxinos G, Watson C (1986) The rat brain in stereotaxic coordinates. 2nd edn

Paxinos G, Watson C (2009) Interactive atlas and 3D brain software for research, structure analysis, and education (BrainNavigator). Elsevier, Amsterdam

Pinto ML, Olyntho-Tokunaga HH et al (2007) The interstitial system of the trigeminal spinal tract projects to the red nucleus in mice. Somatosens Mot Res 24:221–225

Prabhakhar NR, Joyner MJ (2015) Tasting arterial blood: what do the carotid chemoreceptors sense? Front Physiol 5:524

Pramstaller PP, Wenning GK et al (1995) Nerve conduction studies, skeletal muscle EMG, and sphincter EMG in multiple system atrophy. J Neurol Neurosurg Psychiatry 58:618–621

Quester R, Schröder R (1999) Topographic anatomy of the cochlear nuclear region at the floor of the fourth ventricle in humans. J Neurosurg 91:466–476

Rai R, Iwanaga J et al (2018) The tentorium cerebelli: a comprehensive review including its anatomy, embryology, and surgical techniques. Cureus 10:e3079

Rexed B (1952) The cytoarchitectonic organization of the spinal cord in the cat. J Comp Neurol 96:414–495

Rexed B (1954) A cytoarchitectonic atlas of the spinal cord in the cat. J Comp Neurol 100:297–379

Rodriguez CI, Dymecki SM (2000) Origin of the precerebellar system. Neuron 27:475–486

Roper SD, Chaudhari N (2017) Taste buds: cells, signals and synapses. Nat Rev Neurosci 18:485–497

Rüb U, Del Tredici K et al (2000) The evolution of Alzheimer's disease-related cytoskeletal pathology in the human raphe nuclei. Neuropathol Appl Neurobiol 26:553–567

Ruchoux MM, Maurage CA (1997) CADASIL: cerebral autosomal dominant arteriopathy with subcortical infarcts and leukoencephalopathy. J Neuropath Exp Neurol 56:947–964

Ruchoux MM, Brulin P et al (2002) Lessons from CADASIL. Ann N Y Acad Sci 977:224–231

Ruchoux MM, Domenga V et al (2003) Transgenic mice expressing mutant MOTCH3 develop vascular alteration characteristic of cerebral autosomal dominant arteriopathy with subcortical infarcts and leukoencephalopathy. Am J Pathol 162:329–342

Ruchoux MM, Brulin P et al (2006) Skin biopsy value and leukoaraiosis. Ann N Y Acad Sci 903:285–292

Ruggiero DA, Underwood MD, Mann JJ, Anwar M, Arango V (2000) The human nucleus of the solitary tract: visceral pathways revealed with an "in vitro" postmortem tracing method. J Auton Nerv Syst 79:181–190

Salzer JL (2015) Schwann cell myelination. Cold Spring Harb Perspect Biol 7:a020529

Sang Q, Goyal RK (2000) Lower esophageal sphincter relaxation and activation of medullary neurons by subdiaphragmatic vagal stimulation in the mouse. Gastroenterology 119:1600–1609

Schaltenbrand G, Wahren W (1977) Atlas for stereotaxy of the human brain, 2nd edn. Thieme, New York

Schröder H, Moser N et al (2020) Neuroanatomy of the mouse. An introduction. Springer

Schwarzacher SW, Rüb U et al (2011) Neuroanatomical characteristics of the human pre-Bötzinger complex and its involvement in neurodegenerative brainstem diseases. Brain 134:24–35

Seravalle G, Dell'Oro R et al (2019) Baroreflex activation therapy systems: current status and future prospects. Expert Rev Med Devices 16:1025–1033

Shaikh AG (2019) Human gaze holding and its disorders. Ann Otol Neurotol 2:33–40

Shy GM, Dräger GA (1960) A neurological syndrome associated with orthostatic hypotension: a clinicopathologic study. Arch Neurol 2:511–527

Smith JC, Abdala APL et al (2009) Structural and functional architecture of respiratory networks in the mammalian brainstem. Philos Trans R Soc Lond Ser B Biol Sci 364:2577–2587

Sourander P, Wålinder J (1977) Hereditary multi-infarct dementia. Morphological and clinical studies of a new disease. Acta Neuropathol 39(3):247–254

Standring S (2016) Gray's anatomy—the anatomical basis of clinical practice. 41st edn. Elsevier Health Sciences

Stankovic I, Krismer F et al. (2014) Cognitive impairment in multiple system atrophy: a position statement by the Neuropsychology Task Force of the MDS Multiple System Atrophy (MODIMSA) study group. Mov Disord 29:857–867

Stankovic I, Quinn N et al (2019) A critique of the second consensus criteria for multiple system atrophy. Mov Disord 34:975–984

Szabadi E (2012) Modulation of physiological reflexes by pain: role of the locus coeruleus. Front Integr Neurosci 6:94

ten Donkelaar HJ, Kachlik D et al (2018) An illustrated terminologia neuroanatomica. A concise encyclopedia of human neuroanatomy. Springer

Tikka S, Baumann M et al (2014) CADASIL and CARASIL. Brain Pathol 24:525–544

Tillmann BN (2005) Atlas der Anatomie. Springer 2005

Tokita K, Inoue T et al (2009) Afferent connections of the parabrachial nucleus in C57BL/6J mice. Neuroscience 161:475–488

Tokita K, Inoue T et al (2010) Subnuclear organization of parabrachial efferents to the thalamus, amygdala and lateral hypothalamus in C57BL/6J mice: a quantitative retrograde double labeling study. Neuroscience 171:351–365

Toorop RJ, Ousrout R et al (2013) Carotid baroreceptors are mainly localized in the medial portions of the proximal internal carotid artery. Ann Anat 193:248–252

Tournier-Lasserve E, Juotel A et al (1993) Cerebral autosomal dominant arteriopathy with subcortical infarcts and leucoencephalopathy maps to chromosome 19q12. Nat Genet 3:256–259

Trojanowski JQ, Revesz T, for the Neuropathology Working Group on MSA (2007) Proposed neuropathological criteria for the post mortem diagnosis of multiple system atrophy. Neuropathol Appl Neurobiol 33:615–620

Trout HH II, Brown LL et al (1979) Carotid sinus syndrome: treatment by carotid sinus denervation. Ann Surg 189:575–580

Türck L (1835) Beobachtungen über das Leitungsvermögen des menschlichen Rückenmarks. Aus dem Maihefte des Jahrganges 1835 der Sitzungsbericht der mathem.-naturw. Classe der kais. Akademie der Wissenschaften [Bd. XVI, S. 329] besonders abgedruckt, Wien

Van Bogaert L (1955) Encephalopathie sous-corticale progressive (Binswanger) à évolution rapide chez deux soeurs. Med Hellen 24:961–972 van der Heijden ME, Zoghbi HY (2020) Development of the brainstem respiratory circuit. Wiley Interdiscip Rev Dev Biol 9(3):e366

van Lieshout JJ, Wieling W et al (1991) The vasovagal response. Clin Sci (Lond) 81(5):575–586

Vanderhorst VG (2005) Nucleus retroambiguus-spinal pathway in the mouse: localization, gender differences, and effects of estrogen treatment. J Comp Neurol 488:180–200

VanderHorst VG, Ulfhake B (2006) The organization of the brainstem and spinal cord of the mouse: relationships between monoaminergic, cholinergic, and spinal projection systems. J Chem Neuroanat 31:2–36

Viswanathan A, Sudarsky L (2012) Balance and gait problems in the elderly. Ataxic disorders. In: Aminoff MJ, Boller F, Swaab DF (eds.) Handbook of clinical neurology, vol. 103. Elsevier, pp 633–634

Viswanathan A, Gray F et al (2006) Cortical neuronal apoptosis in CADASIL. Stroke 37:2690–2695

Watson C, Paxinos G, Kayalioglu G (eds) (2008) The spinal cord, 1st edn. Academic, New York

Watson C, Bartholomaeus C et al (2019) Time for radical changes in brain stem nomenclature—applying the lessons from developmental gene patterns. Front Neuroanat 13:10

Wenning GK, Colosimo C et al (2004) Multiple system atrophy. Lancet Neurol 3:93–103

Wenning GK, Stefanova N et al (2008) Multiple system atrophy: a primary oligodendrogliopathy. Ann Neurol 64:239–246

Wenthur CJ, Bennett MR et al (2014) Classics in chemical neuroscience: fluoxetine (Prozac). ACS Chem Neurosci 5:14–23

Yokoyama T, Ishiyama M (2014) Novel neuronal cytoplasmic inclusions in a patient carrying SCA8 expansion mutation. Neuropathology 34:27–31

Zarindast M, Nasehi M et al (2014) A mini review of serotonin and its receptors. Int J Med Rev 1:39–45

Zec N, Kinney HC (2003) Anatomic relationships of the human nucleus of the solitary tract in the medulla oblongata. Auton Neurosci 105:131–144

Web Links

http://braininfo.rprc.washington.edu/centraldirectory.aspx?ID=7

http://braininfo.rprc.washington.edu/centraldirectory.aspx?ID=761

http://braininfo.org/centraldirectory.aspx?ID=1648

http://braininfo.org/centraldirectory.aspx?ID=1789

http://braininfo.rprc.washington.edu/centraldirectory.aspx?ID=1799

http://braininfo.rprc.washington.edu/centraldirectory.aspx?ID=1864

https://www.datasci.com/solutions/cardiovascular/baroreceptor-sensitivity-(brs)

http://dibb.de/theodor-schwann.php

https://es.m.wikipedia.org/wiki/Archivo:Tabes_Dorsalis.jpg

https://fipat.library.dal.ca/wp- content/uploads/2017/02/FIPAT-TNA-Front-Matter.pdf

https://www.imaios.com/en/e-Anatomy/Anatomical-Parts/Canaliculus-for-chorda-tympani

https://www.msdmanuals.com/professional/cardiovascular-disorders/symptoms-of-cardiovascular-disorders/orthostatic-hypotension

http://www.nimh.nih.gov/health/publications/depression/what-are-the-signs-and-symptoms-of-depression.shtml

http://www.katieroorda.com/medical/microvascular-decompression-trigeminal-neuralgia-2/

www.nobelprize.org/nobel_prizes/medicine/laureates/1936/loewi-bio.html

https://radiopaedia.org/articles/chorda-tympani

https://rarediseases.org/rare-diseases/multiple-system-atrophy/

http://terminologia-anatomica.org/en/Terms/Occurence/557

https://www.ninds.nih.gov/Disorders/Patient-Caregiver-Education/Fact-Sheets/Trigeminal-Neuralgia-Fact-Sheet

https://www.youtube.com/watch?v=mXHbBPX6lm8

Rhombomere 10 r10

4

Contents

H. Schröder et al., *The Human Brainstem*, https://doi.org/10.1007/978-3-030-89980-6_4

Abstract

Ascending cranially in the brainstem, this chapter deals with some more precerebellar nuclei—projecting to the cerebellum—like the lateral reticular nucleus (not a nucleus of the reticular formation which Sect. 4.3.4 deals with in a general manner together with the gigantocellular nucleus), the large inferior olivary complex, the arcuate nucleus, and the linear nucleus of the medulla.

The group of cranial nerves is represented by the hypoglossal nucleus which innervates the muscles of the tongue. After the general introduction into the raphe nuclei in Chap. 3, here starts the description of the individual nuclei with the caudal most ones of this group, the raphe obscurus, and pallidus nuclei.

From a neuropathological point of view, Chap. 4 deals with progressive supranuclear palsy (PSP), frontotemporal lobar degeneration (FTLD), and amyotrophic lateral sclerosis (ALS).

4.1 Rhombic lip r10

Section 2.2.1 deals in more detail with the rhombic lip.

4.1.1 Precerebellar nuclei r10

4.1.1.1 Lateral reticular nucleus r10 [Ncl. reticularis lateralis]

The lateral reticular nucleus (LRN) is a precerebellar nucleus located in the ventrolateral medulla oblongata near its surface (see Fig. 4.1 ①) Despite the designation "reticular," this nucleus does not belong to the reticular formation. It is also present in macaques, rats, and mice. In man, it contains a large number of serotonergic neurons (for details, see Hornung 2003, 2012).

http://braininfo.rprc.washington.edu/centraldirectory.aspx?ID=727

As shown in a pathoanatomical study by Mehler (1962, 1966 reported by ten Donkelaar 2011) of the brainstem after a bilateral cordotomy (surgical lesion of the nociception conducting anterolateral tracts of the spinal cord), the LRN receives sensory spinal input which is in accordance with findings in cats (see ten Donkelaar 2011). The human LRN receives cerebrocortical input as shown in silver impregnation studies by Kuypers (1958) and Schoen (1969) (both reported by ten Donkelaar 2011).

The LRN displays massive pathological changes in spinocerebellar ataxia type 3 (see Sect. 7.2.1.3.3, see Fig. 7.21A) depriving the cerebellum from an important precerebellar input.

4.1.1.2 Intercalated nucleus of the medulla

Detailed information on this nucleus is available under Sect. 3.1.1.1.

4.1.1.3 Inferior olive (IO)/Oliva, inferior olivary complex r10 [Complexus olivaris inferior]

4.1.1.3.1 Location of the inferior olive

The IO (Figs. 4.2 and 4.3) is the longitudinally most wide-stretched precerebellar nucleus, appearing at the ventrolateral surface of the medulla oblongata as an olive-like protrusion. It is located in the oral 1.5 cm of the medulla oblongata, dorsal to the pyramids, and lateral to the medial lemniscus (see Figs. 4.2 and 4.3) (Olszewski and Baxter 1982). The inferior olive extends as a folded, cell-rich band from the level of the hypoglossal nucleus almost to that of the facial nucleus.

In man, the IO is composed of the following nuclei (Fig. 4.3, for details see Olszewski and Baxter 1982).

- Principal olivary nucleus[Ncl. olivaris principalis]
  - Posterior lamella [Lamella posterior] ①
  - Anterior lamella [Lamella anterior] ③
  - Lateral lamella [Lamella lateralis] ②

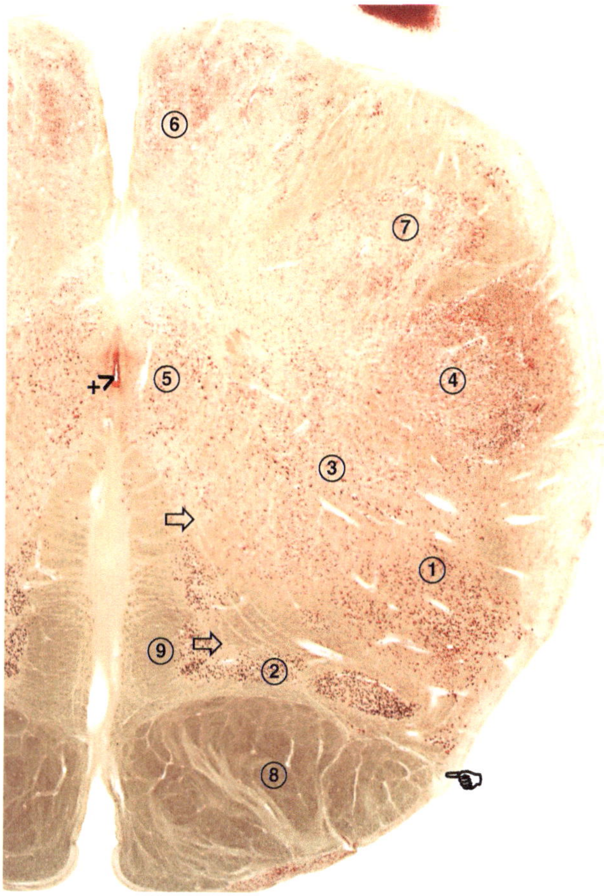

①	Lateral reticular ncl.	⇨	Course of hypoglossal nerve ☞ = Exit zone
②	Inferior olivary complex	⑥	Gracile ncl.
③	Intermediate reticular zone	⑦	Cuneate ncl.
④	Spinal ncl. of trigeminal nerve CNV	⑧	Pyramidal (corticospinal) tract
⑤	Ncl. of hypoglossal nerve CNXII	⑨	Medial lemniscus
		+	Central canal

Fig. 4.1 Horizontal section through the human medulla at the level of the dorsal column nuclei. Darrow red staining. The lateral reticular nucleus ① is located beneath the ventrolateral surface of the medulla oblongata. Its neighbors are the inferior olive ② (see Sect. 4.1.1.3), the intermediate reticular zone ③ (see Sect. 3.3.2.1) and the spinal nucleus ④ and tract of trigeminal nerve (see Sect. 3.2.2.2). See also atlas part Darrow red 2 and Campbell 1. LabPON Twente

- Hilum of inferior olivary nucleus [Hilum nuclei olivaris inferioris] ④
- Posterior accessory olivary nucleus [Ncl. olivaris accessorius posterior] ⑤
- Medial accessory olivary nucleus [Ncl. olivaris accessorius medialis] ⑥

The cells of the principal and accessory nuclei are morphologically similar, medium- sized, plump shaped and multipolar, darkly stained, containing lipofuscin (see Box 4.4) already early in life. Nuclei are often eccentric (Olszewski and Baxter 1982).

The neurochemical organization of the human inferior olive with emphasis on calcium-binding proteins, the nitric oxide synthesizing enzyme (nNOS) and non-phosphorylated neurofilament protein (NPNFP) has been described by Baizer et al. (2011). Based on these markers, in 2014, Baizer has published a paper on the comparative anatomy of the inferior olive.

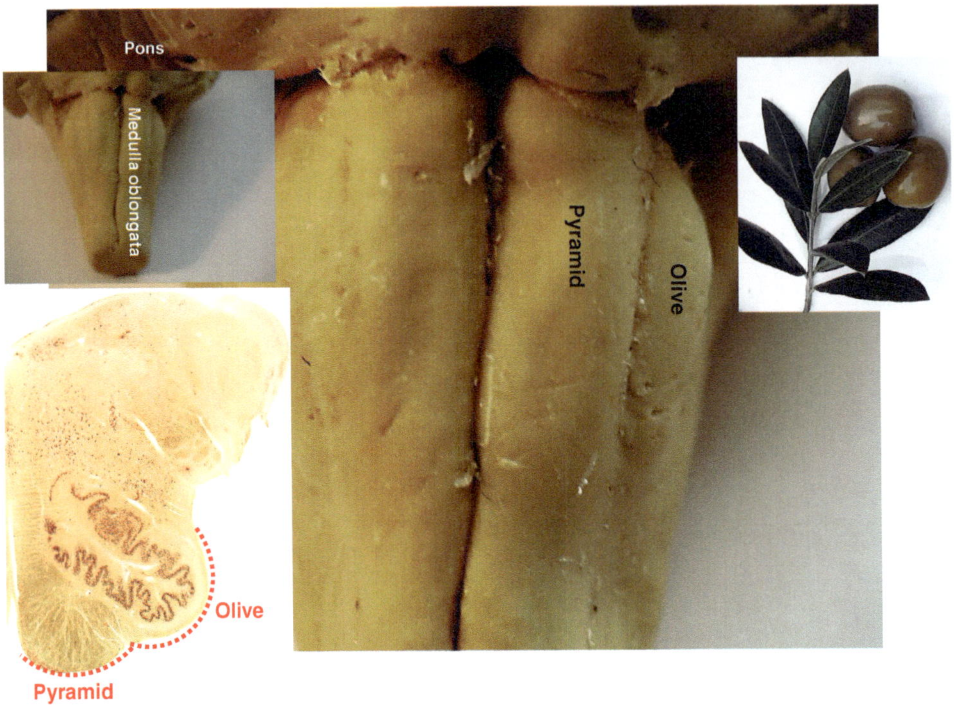

Fig. 4.2 Ventral view of the human brainstem. Siliconized specimen. Leptomeninges removed. Directly lateral of the anterior median fissure [Fissura mediana anterior] the pyramis containing the corticospinal tract [Tractus corticospinalis] appears accompanied laterally by another protrusion, the inferior olive (compare with Fig. 4.3)—reminiscent to the fruit of the olive tree (*Olea europea*)—which is due to the prominence of the inferior olivary complex (IO) at the surface of the medulla oblongata. Sammlung des Zentrums Anatomie der Universität zu Köln and LabPON Twente (Darrow red staining). For a review of data on the mouse inferior olive—which is morphologically different from the human one—see Schröder et al. (2020)

4.1.1.3.2 Connectivity of the inferior olive

The most important connections of the human inferior olive have been demonstrated by silver impregnation of lesioned brains postmortem (see Holmes and Stewart 1908; Voogd 2004; ten Donkelaar 2011).

The afferent connections to the human inferior olive are mainly provided by two tegmental tracts in mammals including man (Voogd 2004, ten Donkelaar 2011).

1. The central tegmental tract [Tractus tegmentalis centralis]—particularly well developed in humans has its origin in the parvocellular part of the red nucleus [Ncl. ruber] (see Sects. 15.3.4 and 16.6.2). As its name says, it is running toward the inferior olive in the tegmentum (see Fig. 1.12). Regarding its course in detail, see Fig. 3.14 A–F. Furthermore, afferents from the lateral vestibular nucleus are running via the central tegmental tract to the IO (ten Donkelaar 2011).
2. The medial tegmental tract [Tractus tegmentalis medialis] originates around the Fasciculus retroflexus (now habenulo-interpeduncular tract [Tractus habenulo-interpeduncularis]) (see Fig. 16.2), including the elliptic nucleus of the prerubral tegmentum (Ncl. of Darkschewitsch, see Sect. 16.4.1). It subsequently descends in the ventral part of the medial longitudinal fasciculus [Fasciculus longitudinalis medialis] and terminates in the medial accessory olive and posterior accessory olive (Voogd 2004) (see Fig. 4.3). The course of the medial longitudinal fasciculus can be traced in Fig. 3.14 A–F.

In addition, the IO receives input from the superior colliculi (see Sect. 15.1.3), the nucleus of the optic tract and from the spinal cord (ten Donkelaar 2011).

In terms of efferents, the IO provides the climbing fiber input to the Purkinje cells of the contralateral half of the cerebellum via the contralateral inferior cerebellar peduncle [Pedunculus cerebellaris inferior] (see Box 14.1). They have been called climbing fibers since they wrap around the dendrites of the Purkinje cells like vines around a tree trunk.

All other precerebellar nuclei reach the cerebellum using the mossy fiber system. The branches of the mossy fibers have numerous swellings, called rosettes, synapsing on cerebellar granule cell dendrites. This arrangement appeared to early neuroanatomists as a kind of moss. Already in 1908, Holmes and Steward described a diagram—remaining the only one of this sort—of the olivocerebellar projections in

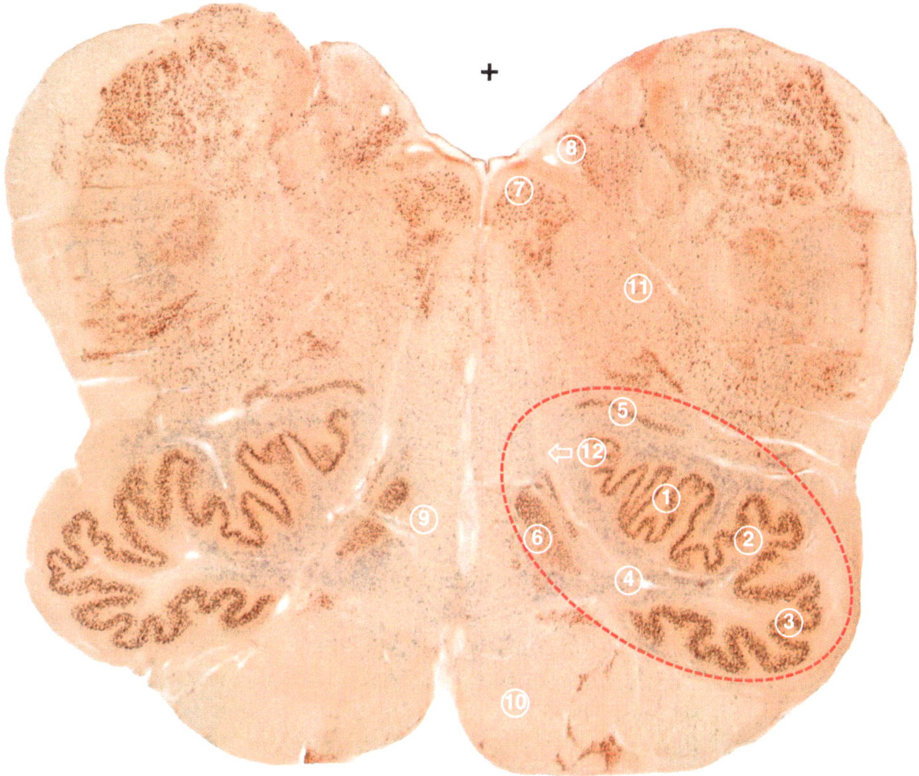

---	Inferior olivary complex, approximate circumference	⑦	Ncl. of hypoglossal nerve CNXII
①	Principal olivary ncl., Posterior lamella	⑧	Posterior ncl. of vagus nerve CNX
②	Principal olivary ncl., Lateral lamella	⑨	Medial lemniscus
③	Principal olivary ncl., Anterior lamella	⑩	Pyramidal (corticospinal) tract
④	Hilum of inferior olivary complex	⑪	Intermediate reticular zone
⑤	Posterior accessory olivary ncl.	⑫	Hypoglossal nerve CNXII
⑥	Medial accessory olivary ncl.	+	4th ventricle

Fig. 4.3 Horizontal histological section through the human brainstem at the level of the posterior nucleus of vagus nerve ⑧. Darrow red staining. The inferior olivary complex - do not confuse the inferior olivary complex with the superior olivary complex (see Sect. 8.1.1.5), a part of the auditory system—has as its ventral neighbor the pyramis (corticospinal tract) ⑩ and is dorsally bordered by the reticular formation of the medulla oblongata. It comprises the inferior olivary complex proper, subdivided in three lamellae ①–③, and two accessory nuclei ⑤, ⑥. Three-dimensionally, the IO can be conceived as a kind of folded bag which is open medially (hilum of IO [Hilum nuclei olivaris inferioris] ④). It happens that on histological sections like that shown in Fig. 4.13 the cerebellar dentate nucleus is confused with the IO because of the similar histological appearance. Note that only the IO reaches the outer surface of the brainstem. See also atlas part Darrow red 7. LabPON Twente

humans. They could show that the different subdivisions of the inferior olive project in a topical fashion to different lobules of the cerebellum (see Voogd 2004).

The human inferior olive receives a homogeneous noradrenergic innervation (Powers et al. 1990).

As in man, in mice, inferior olivary climbing fibers carry sensorimotor signals that trigger motor learning by controlling cerebellar Purkinje cell synaptic plasticity and discharge (Chaumont et al. 2013). Purkinje cells phasically control the discharge of their own IO afferents and thus might participate in the regulation of cerebellar motor learning discharge (Chaumont et al. 2013). The contact pattern of mouse climbing fibers with cerebellar interneurons has been studied by Galliano et al. (2013).

The cerebellar Purkinje cells contact the (deep) cerebellar nuclei, the output structure of the cerebellum, including an inhibitory GABAergic projection to the IO. This circuit suggests a closed loop in the olivo-cerebello-cortical network.

4.1.1.3.3 Aging and pathology of the inferior olive/Progressive supranuclear palsy (PSP)

Olszewski and Baxter (1954) already described a loss of IO neurons in older individuals. Using silver impregnation and immunohistochemistry, Baizer et al. (2018) could show that in their whole sample (age range 25–71 years) silver-labeled intraneuronal granules were present but use of antibodies to abnormal tau did not yield any results. There are several brainstem pathologies affecting the IO like Multiple system atrophy (MSA) (see Sect. 3.3.2.3) or Progressive supranuclear palsy (PSP) (see Sect. 4.1.1.3.3). Main clinical symptoms found in these diseases are related to the loss of cerebellopetal connections from the medulla oblongata, here the IO.

Progressive supranuclear palsy (PSP)

In 1963, Richardson, Olszewski, and Steele described a small series of patients who presented a distinctive clinical picture characterized by defects of ocular gaze, particularly of downward gaze. Downgaze palsy is, however, not required for diagnosing PSP by NINDS/SPSP criteria (Litvan et al. 1996). Furthermore, both downgaze and upgaze palsy can occur in PSP patients (Chen et al. 2010). Although some aspects of all forms of eye movements are affected in PSP, the predominant defects concern vertical saccades (both up and down), impaired vergence, and inability to modulate the linear vestibulo-ocular reflex appropriately for viewing distance (Chen et al. 2010). Two major, contiguous midbrain structures are important for the generation of torsional and vertical eye movements: the rostral interstitial nucleus of the medial longitudinal fasciculus (riMLF) and the interstitial nucleus of Cajal (iC) (Helmchen et al. 1998). Both the riMLF (early in the course of PSP) and the iC are affected in PSP (Chen et al. 2010). For the interstitial nucleus of Cajal (see Sect. 16.4.2, see Box 4.2), disturbance of gaze is a cardinal clinical feature of PSP that can be of great help in establishing the diagnosis.

Furthermore, spasticity of facial musculature (see Sect. 8.2.1.6.1) with dysarthria, pseudobulbar palsy, dystonic rigidity of neck—with head retraction—and upper trunk, cognitive, and behavioral changes have been described (Steele et al. 1964). Richardson, Olszewski, and Steele coined the term progressive supranuclear palsy in referring to the supranuclear ophthalmoplegia (concerning vertical eye movements). It remains the most distinctive diagnostic feature not to confuse with internuclear ophthalmoplegia (see Box 9.3) due to medial longitudinal fasciculus lesions (see Box 9.2).

PSP usually presents as an akinetic, rigid parkinsonian syndrome, nonresponsive to L-DOPA, associated with supranuclear gaze palsy, pseudobulbar palsy, axial dystonia, and frontal lobe-like symptomatology. It is a disorder of tau protein aggregation and its clinical spectrum is much wider than originally described with a phenotype resembling Parkinson's disease (PD) (see Sect. 16.6.1.5) accounting for a third of cases. Many patients with PSP are initially thought to have PD or multiple system atrophy (MSA) (see Sect. 3.3.2.3). Fewer than 50% of patients with pathologically diagnosed PSP will have received a diagnosis of PSP at presentation, and 20% will have had a different diagnosis at the time of death. The pathological events and processes leading to the accumulation of phosphorylated tau protein in the brain are best considered as dynamic processes that can develop at different rates, leading to different clinical phenomena (Williams and Lees 2009). PSP is almost entirely sporadic although nonmendelian genetic risk factors exist (Golbe 2014). Inheritance of H1 haplotype and H1/H1 genotype is a recognized risk factor for PSP (Höglinger et al. 2011; Kovacs 2015) and the increased risk of PSP that is associated with the H1 haplotype is driven by 4–5 specific H1 subhaplotypes (Heckman et al. 2019). There is an indication that PET imaging with a tau radiotracer could be of help in diagnosing and differentiating patients with suspected PSP (Brendel et al. 2020). Tau fibrils in PSP contain only a 4-repeat microtubule-binding isoform having unique biochemical properties that differ from those in other neurodegenerative disorders.

Apparent macroscopical abnormalities are not always visible and if so, they are restricted to diencephalon, brainstem, and cerebellum. The midbrain may be shrunken (superior colliculi, see Sect. 15.1.3, tegmentum, and periaqueductal gray see Sect. 13.2.7.1), the SN (see Sect. 16.6.1) can be pale, and the red nucleus (see Sect. 15.3.4) discolored (Lantos 1994). The pontine tegmentum is occasionally atrophied and the locus caeruleus (LC) (see Sect. 13.2.2.1) less pigmented. In the cerebellum, the superior cerebellar peduncles and dentate nuclei are sometimes shrunken, while the hilus appears discolored (Figs. 4.4 and 4.5). The inner segment of the globus pallidus and the subthalamic nucleus may be atrophied and discolored. There may be mild dilatation of the third and fourth ventricles and/or aqueduct.

Light microscopy in PSP cases shows neuronal loss and gliosis in the SN (see Sect. 16.6.1), globus pallidus, Ncl. subthalamicus, cerebellar dentate, and red nucleus (see Sect. 15.3.4). Microscopical abnormalities are related to tau changes, i.e., NFT with neuronal loss, neuropil threads, and glial tau changes (coiled bodies, thorn-shaped, and tufted astrocytes) (Dickson 1999; Ikeda et al. 1995; Komori 1999; Yamada et al. 1993). Changes are not confined to the areas mentioned above, but can extend also to neocortex and subcortical white matter, striatum, internal capsule, substantia innominata, thalamus (especially ventral anterior and lateral nuclei, Dickson 1999), superior colliculus (see Sect. 15.1.3), PAG (see Sect. 13.2.7.1), CNIII nuclei (see Sect. 15.3.1.1), interstitial nucleus of Cajal (see Sect. 16.4.2), rostral

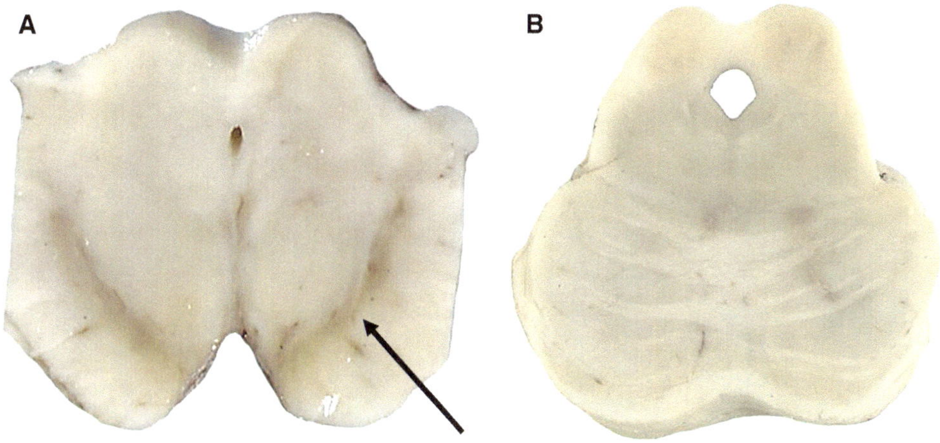

Fig. 4.4 PSP. Pale substantia nigra in the mesencephalon (arrow) in (**A**). Atrophy of the rostral pons (tegmentum) in (**B**). Collection A. Rozemuller

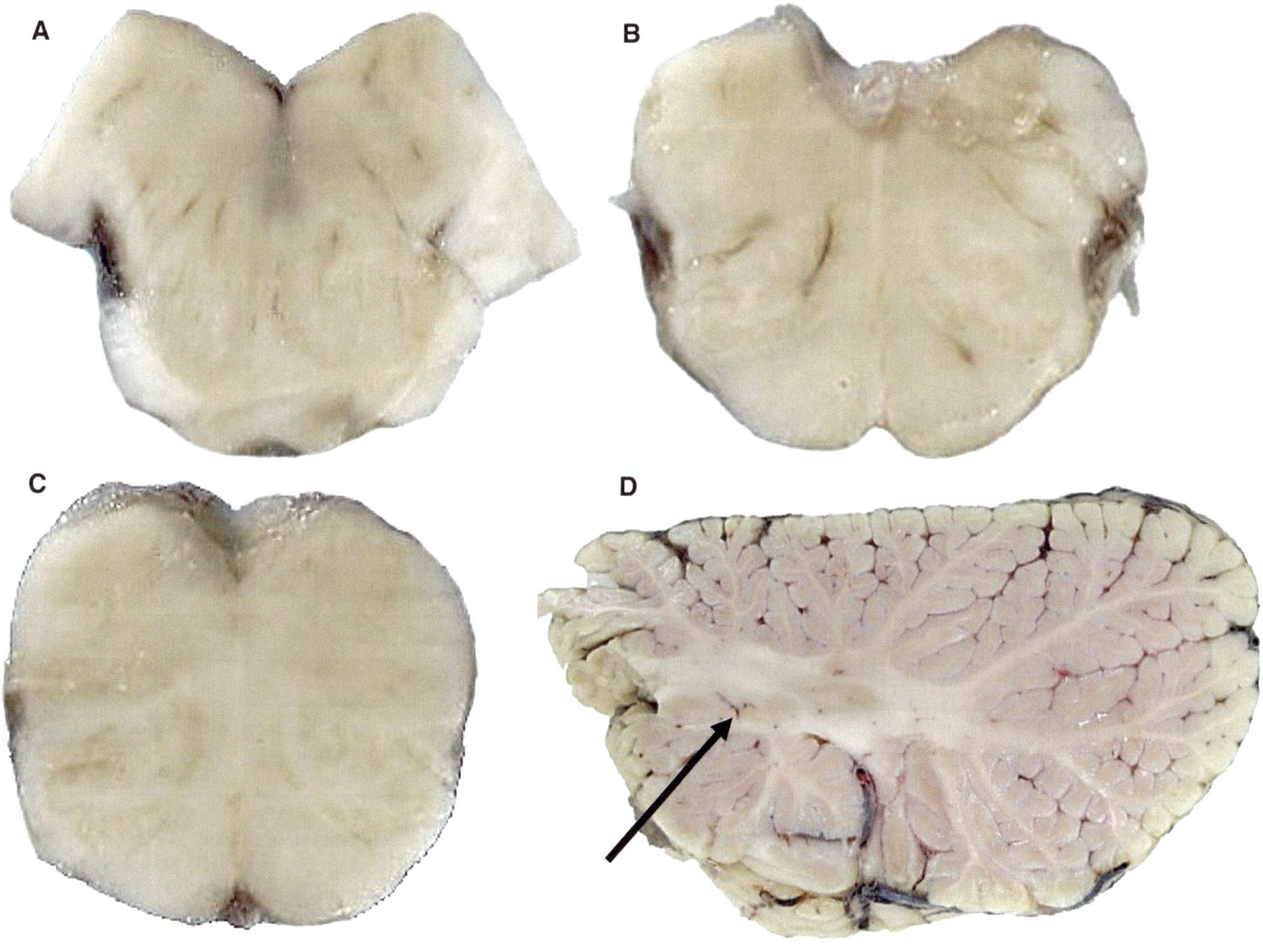

Fig. 4.5 PSP. Atrophy of the pons (tegmentum) and medulla oblongata in **A**–**C**. In **D** shrinkage of the dentate nucleus (arrow). Collection A. Rozemuller

interstitial nucleus of the medial longitudinal fasciculus (see Sect. 16.4.5), locus caeruleus (see Sect. 13.2.2.1), intermediate reticular zone (see Sect. 3.3.2.1), gigantocellular reticular (see Sect. 4.3.4.2), parabachial nucleus (see Sect. 13.2.3), and raphe nuclei (Rüb et al. 2002), amygdala, hippocampus, entorhinal cortex, nuclei of CNIV (see Sect. 14.5.2.1), VIII, X (see Sect. 7.3.2.1), XII (see Sect. 4.3.1.1), pontine and medullary tegmentum, IO (see Sect. 4.1.1.3), and spinal cord (Kikuchi et al. 1999) (Figs. 4.6 and 4.7).

The severity of the neuronal skeletal pathology in the parabrachial nuclei and IRZ (see Sect. 3.3.2.1) is comparable to that seen in the final stages of AD (see Sect. 13.2.2.1.4). In

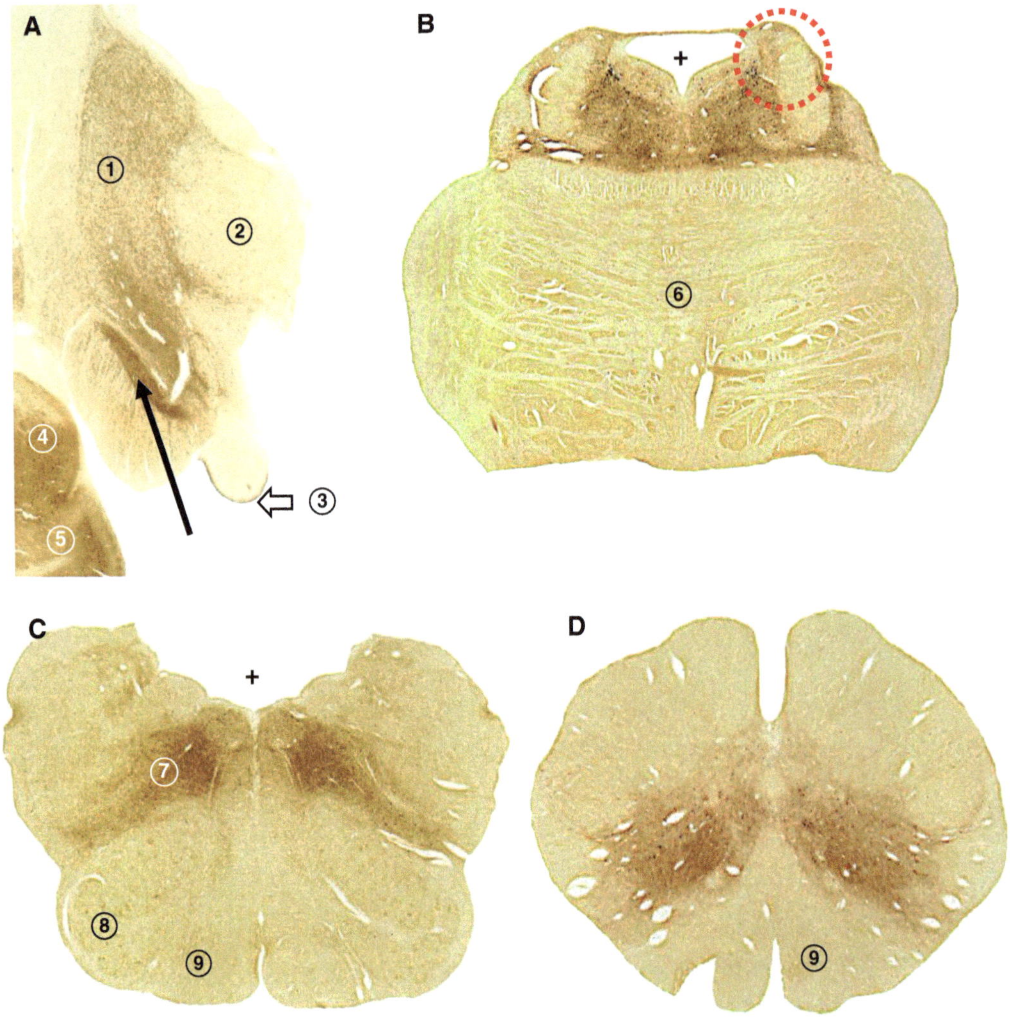

①	Ventral anterior thalamic ncl.	⑦	Intermediate reticular zone
②	Mediodorsal thalamic ncl.	⑧	Inferior olivary complex
③	Mammillary bodies	⑨	Corticospinal tract
④	Amygdala	+	4th ventricle
⑤	Entorhinal cortex	◌	Region of the parabrachial ncll. on either side of the superior cerebellar peduncle
⑥	Pons		

Fig. 4.6 (**A**) Coronal section of the thalamic region in a PSP patient (PEG 100 μm, AT8-immunostain). Tau-positivity in the ventral anterior nucleus and the subthalamic nucleus (arrow), as well as in the allocortex. (**B–D**) Horizontal sections through pons (**B**), medulla oblongata (**C**), and spinal cord (**D**). AT-8 immunoreactivity is seen in the pontine tegmentum including the parabrachial nuclei (**B**) ○ (see Sect. 13.2.3), the intermediate reticular zone (IRZ), slightly the IO (**C**), and the ventral horn of the spinal cord (**D**). LabPON Twente

PSP, it is, however, predominantly of the globose-type, accompanied by a dense network of diseased cell processes. The glial pathology in these nuclei does not occur in AD (Rüb et al. 2002). In contrast to AD, neurofibrillary tangles, neuronal loss, gliosis, and neuropil threads in PSP are primarily seen in the motor and premotor cortex including the white matter in these areas (Verny et al. 1996) (Fig. 4.8).

Neurofibrillary tangles are fibrillar intracytoplasmic inclusions in the cell bodies and proximal dendrites of affected neurons, whereas neuropil threads and neurites are predominantly swollen, filament-containing dendrites and distal axons and terminals, respectively (Savonenko et al. 2015). In the cortex, glial tau is more prominent than tangles (Dickson 1999). Two tangle subtypes can be distinguished:

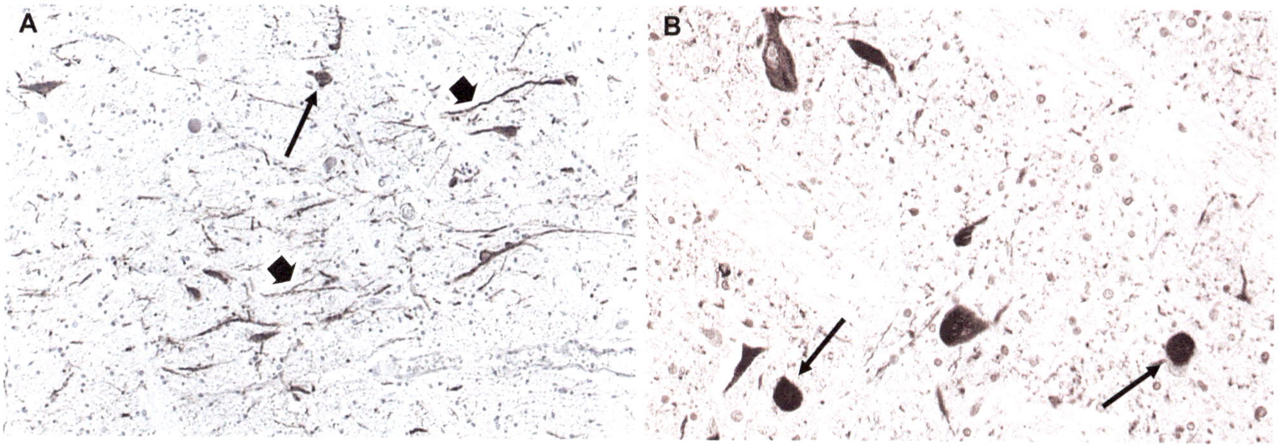

Fig. 4.7 PSP. Intermediate reticular zone of the medulla oblongata, showing neuropil threads (short thick arrows) and neurofibrillary tangles (NFTs). Typical globose-type NFT (long small arrows). AT8-immunostain. LabPON Twente

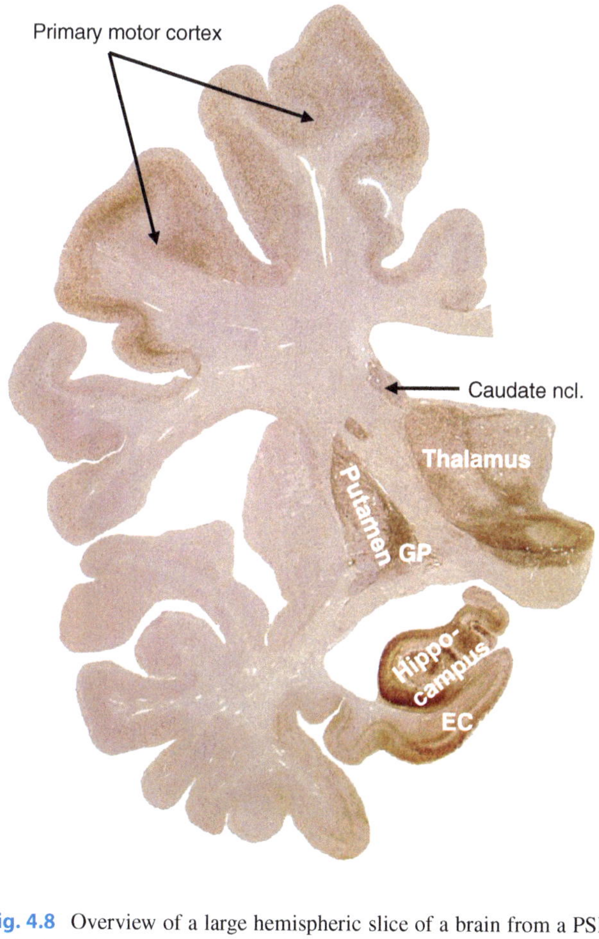

Fig. 4.8 Overview of a large hemispheric slice of a brain from a PSP patient showing a typical distribution pattern in an AT8 immunostain. In contrast to Alzheimer's disease, tau immuno positivity is especially seen in the primary motor cortex, furthermore in the thalamus, putamen and (especially) pallidum, caudate nucleus, and allocortex. *EC* entorhinal cortex, *GP* Globus pallidus. LabPON Twente

globose type and flame-shaped (AD-like) form, the first being far more common (Fig. 4.9).

Ultrastructurally, the neurofibrillary tangles (NFT) are composed mostly of bundles of straight filaments of indeterminate length with a diameter of 15 nm (Flament et al. 1991). Immunocytochemically, neurofibrillary tangles are visualized with antibodies against 4-repeat tau and hyperphosphorylated tau (AT8), but not or much less with antibodies against the 3-repeat form.

Immunostaining with antibodies to ubiquitin or ubiquitin-related molecule P62 gives a negative or weakly positive reaction. Neuropil threads in PSP, demonstrated by the Gallyas stain and tau antibodies, are in part oligodendrocytic (Dickson 1999) in contrast to those in AD. They often run along axons extending across gray and white matter. On Gallyas staining, the aggregation pattern, orientation, and appearance of these threads are different from those seen in AD. The density of neuropil threads is highest at the cortico-medullary junction and is overall less pronounced than it is in corticobasal degeneration (see Sect. 16.6.1.9) (Feany and Dickson 1996). Their distribution overlaps that of coiled bodies. In generally used terminology of neuropathologists, coiled body means oligodendroglial inclusion, appearing as bundles of filaments winding themselves around a translucent oligodendrocytic nucleus and extend in the proximal parts of oligocendrocytic cell processes.

Neuropil threads are especially prominent in the pre- and postcentral gyrus and also present in the internal capsule, pencil fibers in the globus pallidus and midbrain tegmentum (Fig. 4.10).

In the pontine base, coiled bodies and threads are rare in PSP, in contrast to the consistent presence of neurofibrillary tangles in this region. Thorn-shaped astrocytes (with a gemistocytic profile and often eccentric nuclei) show positive

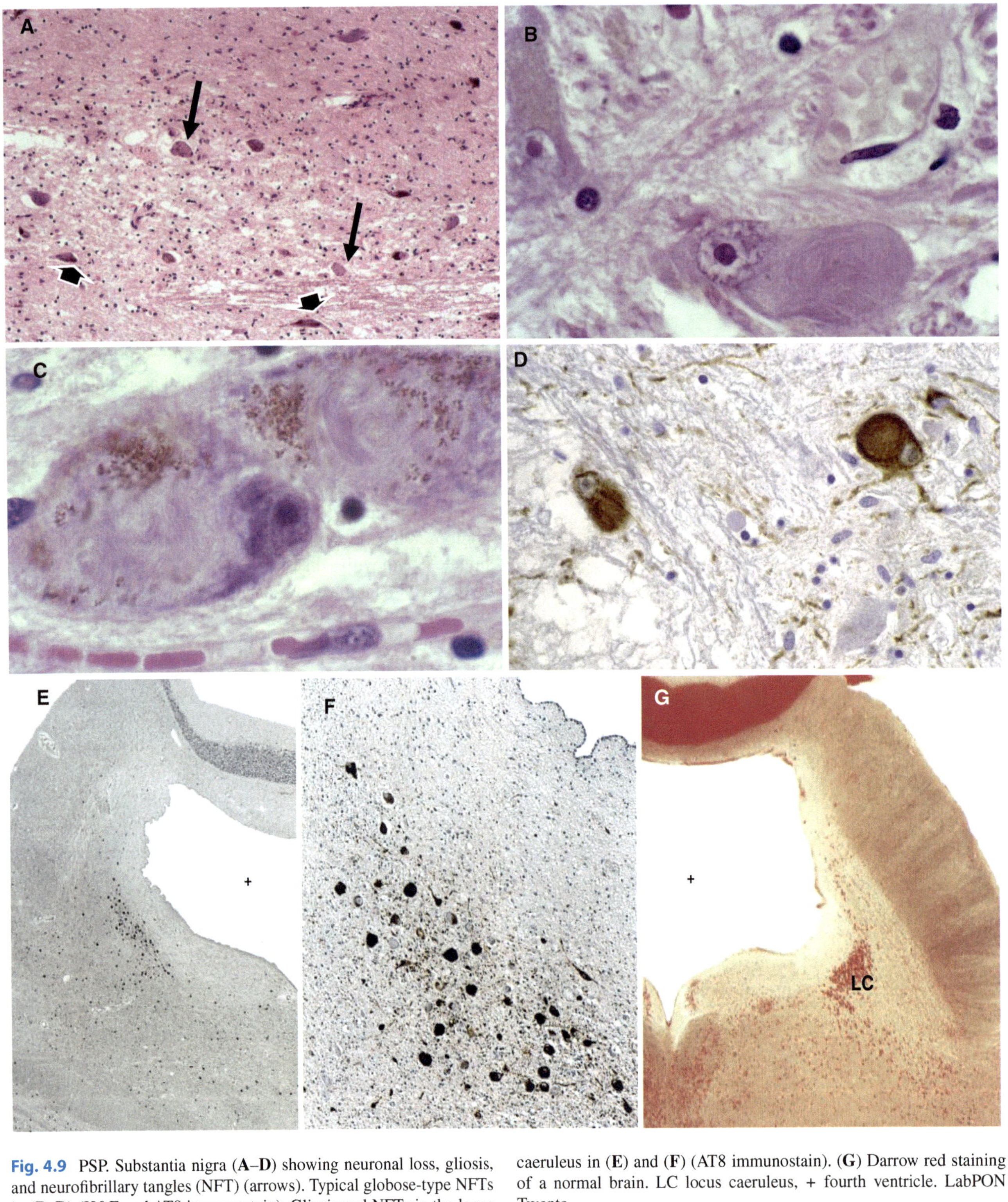

Fig. 4.9 PSP. Substantia nigra (**A**–**D**) showing neuronal loss, gliosis, and neurofibrillary tangles (NFT) (arrows). Typical globose-type NFTs in (**B**–**D**) (H&E and AT8 immunostain). Gliosis and NFTs in the locus caeruleus in (**E**) and (**F**) (AT8 immunostain). (**G**) Darrow red staining of a normal brain. LC locus caeruleus, + fourth ventricle. LabPON Twente

Gallyas and tau staining. They are seen in subpial and subependymal areas. Glial fibrillary tangles (tuft-shaped astrocytes), a characteristic feature of PSP, are immunoreactive against antibodies to GFAP and CD44 (Dickson 2007) (Fig. 4.10).

Tau-immunoreactivity is concentrated in the cytoplasmic center. On Gallyas staining, tufts are observed as conglomerated, fine or thick processes in a concentric arrangement. They are tree-shaped, branching without collaterals. They

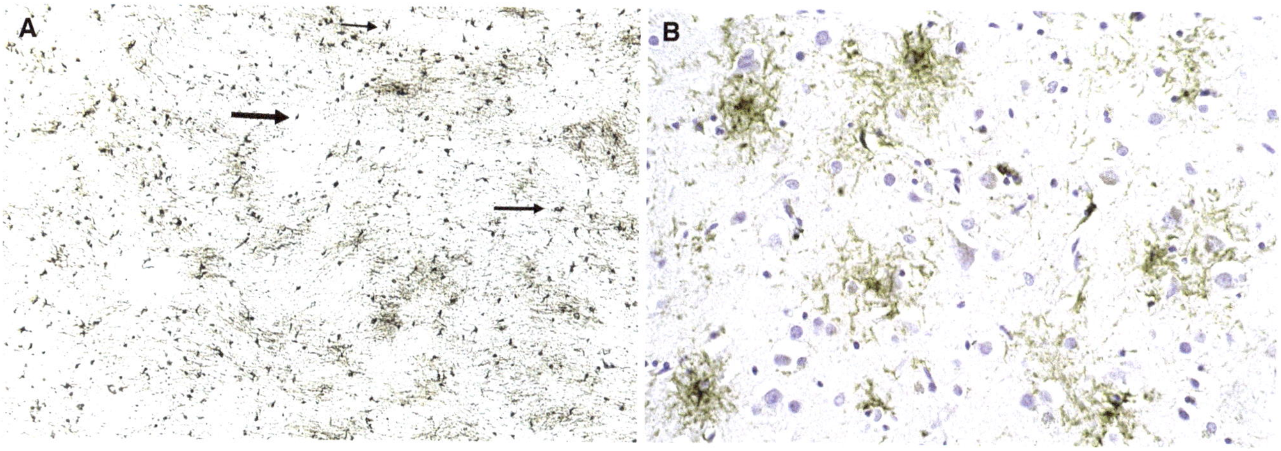

Fig. 4.10 Argyrophilic threads and coiled bodies (arrows) in the internal capsule (**A**) and tufted astrocytes, highly characteristic for PSP, in the premotor cortex (**B**). AT8 immunostain. LabPON Twente

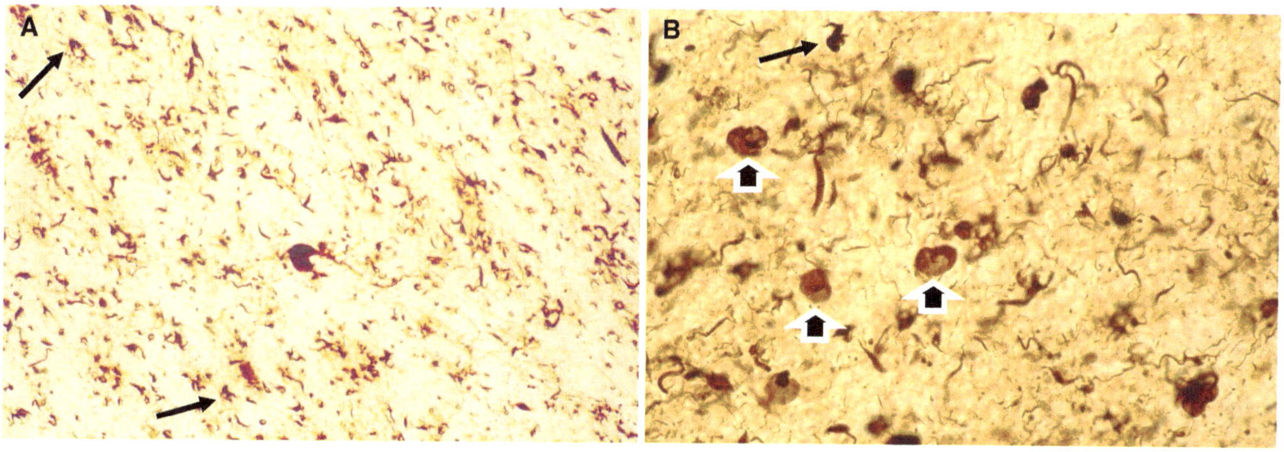

Fig. 4.11 PSP. Globus pallidus (**A**) and subthalamic nucleus (**B**) in an AT8 immunostain (PEG, 100 μm) showing multiple threads and NFTs, globose-type (short thick arrows) and thorn-shaped (long arrrows) astrocytes. LabPON Twente

are restricted to the gray matter and are located especially in the precentral and premotor gyri, putamen, medial thalamus, subthalamic nucleus, and brainstem tegmentum. Tuft-shaped astrocytes can have double nuclei. Tau-positive coiled bodies are most consistently found in PSP. They contain enlarged nuclei with thin coil-like, thick comma-like, and spine-like structures with frequent branching (Fig. 4.11).

Fine, branching coiled bodies (see here above) are characteristic of PSP, while thick comma-like ones are often observed in corticobasal degeneration (CBD, see Sect. 16.6.1.9). They are limited to the precentral cortex, internal capsule, pencil fibers in the lenticular nuclei, midbrain, and tegmentum (Figs. 4.10 and 4.11). Their presence in the cerebellar white matter is relatively specific to PSP. Tau-positive coiled bodies are characteristic, but can also be found in diseases such as CBD, Pick's disease (see Sect. 4.3.1.1.5), dementia with argyrophilic grains (DAG) (see Sect. 16.6.1.7), and also in aging (primary age-related tauopathy, PART, Crary et al. 2014). In PD and DLB, α-synuclein-positive coiled bodies have also been described (Seidel et al. 2015). Western blotting of tau from patients with PSP showed a doublet of 64 and 69 kDa consistent with 4-repeat tau forms (Kovacs 2015). Tau pathologies of PSP principally contain 4-repeat (4R)-tau but rare cases show 3R-tau-positive astrocytes (Taniguchi et al. 2020).

Hypertrophy of the inferior olivary nucleus, caused by lesions in the dentate nucleus, and of the central tegmental tract of the pons tract (see above Sect. 4.1.1.3.2, see Fig. 3.14 B–G) is occasionally seen (Fig. 4.12).

Furthermore, granulovacuolar degeneration (for recent data on granulovacuolar degeneration see Puladi et al. 2021) can be present in the substantia nigra (see Sect. 16.6.1), red nucleus (see Sect. 15.3.4), locus caeruleus (see Sect. 13.2.2.1), basis pontis, and dentate nucleus. Typical "grumose degeneration" may develop around dentate neurons, and shows eosinophilic granular structures, some of which stained by antibodies against phosphorylated neurofilaments. This represents a type of degeneration associated with clusters of degenerating

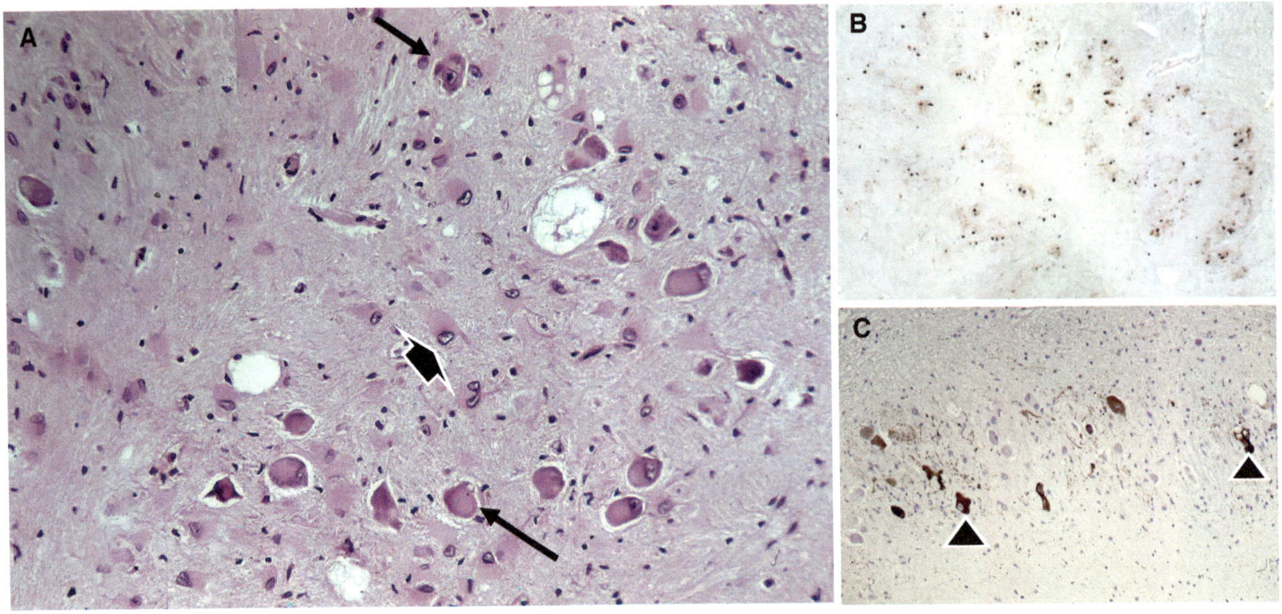

Fig. 4.12 PSP. Typical case of olivary hypertrophy (**A**). Mild neuronal loss and gliosis. Enlarged neurons, occasionally with vacuolated cytoplasm (long arrows), H&E stain. In (**B**) and (**C**) AT8 immunostain, overview and detail. Monstruous immunoreactive neurons (short thick arrows). LabPON Twente

presynaptic terminals around dentate neurons (Ishizawa et al. 2000). Ballooned (achromatic) neurons in the entorhinal/transentorhinal area/neocortex appear less often (MacKenzie and Hudson 1995).

PSP is a heterogenous disorder (Gearing et al. 1994). In addition to typical and atypical PSP forms, PSP may also occur in combination with other neurodegenerative disorders, such as dementia with Lewy bodies (DLB) (see Sect. 16.6.1.6), multiple system atrophy (MSA) (see Sect. 3.3.2.3), Alzheimer's disease (AD) (see Sect. 13.2.2.1.4), and with vascular changes. Criteria and grading systems for PSP have been proposed (Litvan et al. 1996). When a 12-tiered grading system is used based on coiled body and thread lesions in the substantia nigra, caudate and dentate nucleus, tau pathology is severest in the clinical subtype called Richardson's syndrome, when compared to PSP-parkinsonism type and the pure akinesia with gait freezing type (Williams et al. 2007).

4.1.1.4 Arcuate nucleus r10 [Ncl. arcuatus] (ARC) (arcuatus Latin = bent)

The ARC in humans is located at the ventral surface of the lower brainstem. It extends from the caudal border of the pons to the caudal pole of the inferior olivary complex (see Fig. 4.13) (Mikhail and Ahmed 1975; Paradiso et al. 2018). Paradiso et al. (2018) have described the histological appearance of the nucleus. Its neurons appear loosely arranged, medium-sized, oval, polygonal or elongated with eccentric large nuclei and long dendrites, frequently well discernible. Occasionally, small groups of neurons are embedded among the fascicles of the pyramids.

Observations on children, who died from sudden infant death syndrome (SIDS), showed hypoplasia of the arcuate nucleus. Accordingly, the hypothesis was developed that this nucleus is vitally important with regard to breathing. To test for the situation in adults, recently, Paradiso et al. (2018) have studied morphometrically the arcuate nuclei of 25 adults (34–89 years of age) who died from various cases. Surprisingly, in 56% of the cases studied, a hypodevelopment of the arcuate nucleus was found. This casts some doubt on the suggestion that the arcuate nucleus is indeed involved in respiratory regulation.

In the study of Filiano and Kinney (1992), only two of 41 SIDS cases showed a deficiency of volume of the ARC. This could represent biologic variability without pathophysiologic significance and not be related to the death of the children.

The arcuate nucleus projects via the anterior (ventral) arcuate fibers (see Fig. 3.14 A), the medullary striae of the fourth ventricle (see Fig. 4.15 ⑧) and the inferior cerebellar peduncle to the cerebellum (ten Donkelaar 2011).

In mice, the arcuate nucleus can be regarded as a subgroup of the rostral part of the inferior olivary complex (Fu and Watson 2012).

Do not confuse the r10 arcuate nucleus with the hypothalamic arcuate nucleus.

4.1.1.5 Linear nucleus of the medulla r10 [Ncl. medullaris linearis]

This nucleus is supposed to be a rostral extension of the lateral reticular nucleus (see here above Sect. 4.1.1.1) as already described by Ramón y Cajal (see Watson et al. 2019) in man,

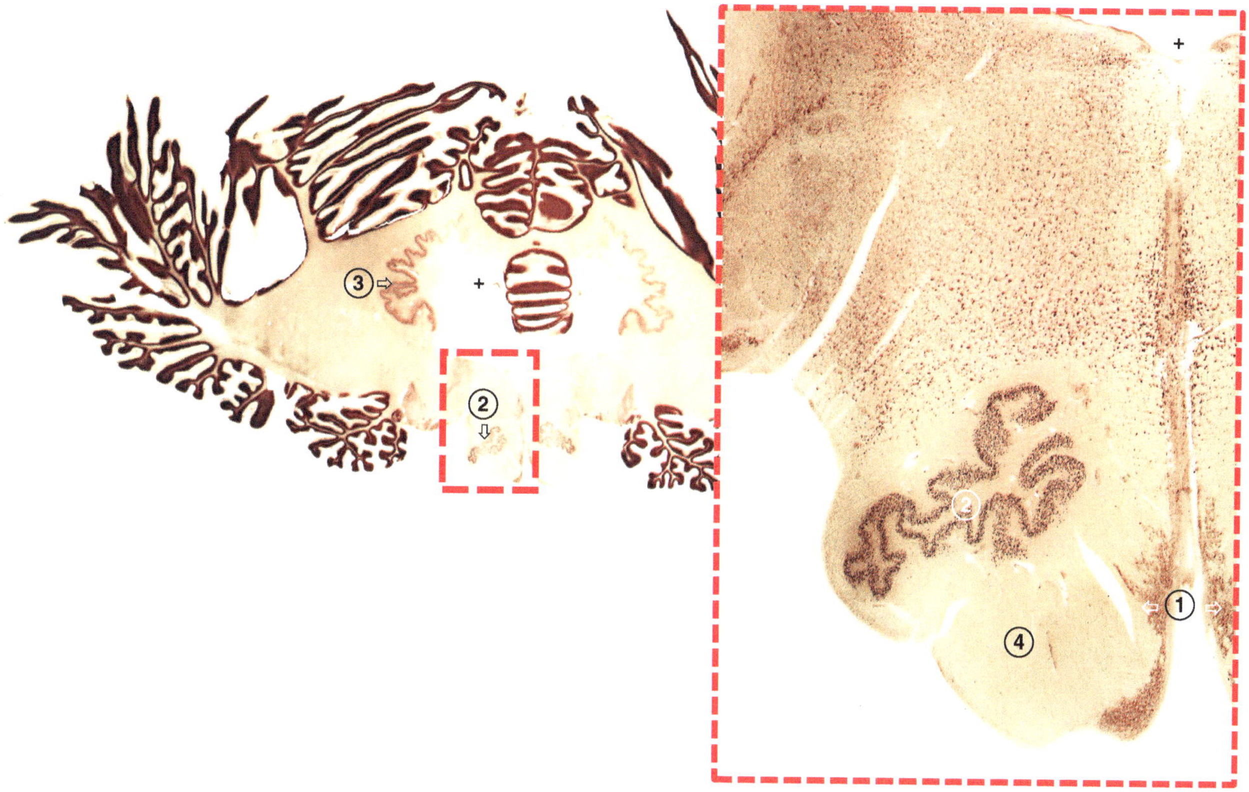

①	Arcuate ncl.	④	Corticospinal tract
②	Inferior olivary complex	+	4th ventricle
③	Dentate ncl. (cerebellum)		

Fig. 4.13 Horizontal section through the human brainstem at the level of the inferior olivary complex ② of the upper medulla oblongata. Darrow red staining. The arcuate nucleus ① is located bilaterally between the ventral surface of the medulla and the corticospinal tract. Note that unexperienced observers may confuse the dentate nucleus of the cerebellum ③ with the inferior olivary complex ②. LabPON Twente

macaque, rat, and mouse. In mice, it projects bilaterally to the cerebellum (Fu et al. 2011).

http://braininfo.rprc.washington.edu/centraldirectory.aspx?ID=1187

4.2 Alar r10

4.2.1 Monoamine nuclei r10

4.2.1.1 A2 noradrenaline cells

For detailed information on the A2 cells, see Sect. 3.2.1.1.

4.2.1.2 C1 adrenaline cells

For detailed information on the C1 cell group, see Sect. 3.2.1.2.

4.2.2 Trigeminal sensory nuclei r10

4.2.2.1 Spinal trigeminal nucleus

For detailed information on the spinal trigeminal nucleus, see Sect. 3.2.2.2.

4.2.3 Dorsal column nuclei r10

4.2.3.1 Cuneate nucleus

For detailed information on the cuneate nucleus, see Sect. 3.2.3.3.

4.2.3.2 Gracile nucleus

For detailed information on the gracile nucleus, see Sect. 3.2.3.4.

4.2.4 Solitary nuclei r10

4.2.4.1 Solitary nucleus

For detailed information on the solitary nuclei, see Sect. 3.2.4.

4.2.5 Alar tegmentum r10

4.2.5.1 Matrix region of the medulla

For detailed information on the matrix region of the medulla, see Sect. 3.2.5.1.

4.3 Basal r10

4.3.1 Somatic motor nuclei r10

Somatic motor nuclei (General somatic efferent; GSE) (see Table 1.1) contain neurons giving rise to axons innervating striated muscles of somite (see Box 3.14) origin. Typical examples are the ventral horn neurons of the spinal cord. As other somatomotor neurons (e.g., hypoglossal nucleus—here below) or the oculomotor nuclei (see Sect. 15.3.1), they are large with dark and distinct Nissl staining and a distinct nucleolus (see Box 4.1). They can be visualized specifically by choline acetyltranferase (ChAT) or vesicular acetylcholine transporter (VAChT) immunostaining. ChAT, catalyzing the formation of acetylcholine from acetyl-CoA and choline, is the synthesizing enzyme of acetylcholine, the transmitter of neuromuscular junctions and cholinergic/acetylcholine-releasing neurons (see Sect. 3.3.1).

Box 4.1 Nucleolus

The nucleolus is a clearly visible round substructure within the cell. Ramón y Cajal (see Box 4.2) could show that a prominent nucleolus is a characteristic of (large) neurons. In smaller neurons and glial cells, the chromatin is not arranged in this way.

The nucleolus is the nuclear subdomain that assembles ribosomal subunits in eukaryotic cells. The nucleolar organizer regions of chromosomes, which contain the genes for pre-ribosomal ribonucleic acid (pre-rRNA), serve as the foundation for nucleolar structure (Olson and Dundr 2015).

Box 4.2 Ramón y Cajal

Santiago Ramón y Cajal was born on May 1, 1852, at Petilla de Aragón, Spain. As a boy, he was apprenticed first to a barber and then to a cobbler. He himself wished to be an artist—his gift for draughtsmanship is evident in his published works. His father, a Professor of Applied Anatomy at the University of Saragossa, persuaded him to study medicine. In 1873, he took his Licentiate in Medicine at Saragossa and served as an army doctor. In 1877, he obtained the degree of Doctor of Medicine at Madrid and in 1883, he was appointed Professor of Descriptive and General Anatomy at Valencia. In 1887, he was appointed Professor of Histology and Pathological Anatomy at Barcelona and in 1892, he was appointed to the same Chair at Madrid.

Using the silver nitrate staining developed by Camillo Golgi (1844–1926, Professor of Histology, Universities of Siena and Torino, Italy and of General Pathology at the University of Pavia, Italy) led Ramón y Cajal to the groundbreaking finding that each neuron is an independent entity and that synapses transfer nerve impulses from one neuron to another.

To cite only some of his published studies:

Les nouvelles idées sur la fine anatomie des centres nerveux (New ideas on the fine anatomy of the nerve centres), 1894; *Textura del sistema nervioso del hombre y de los vertebrados* (Textbook on the nervous system of man and the vertebrates), 1897–1899; *Die Retina der Wirbelthiere* (The retina of vertebrates), 1894.

In 1906, he shared the Nobel Prize "in recognition of their work on the structure of the nervous system" with Camillo Golgi (see here above).

Ramón y Cajal died on October 17, 1934, in Madrid.

From: Nobel Lectures, Physiology or Medicine 1901–1921, Elsevier Publishing Company, Amsterdam, 1967.

https://www.nobelprize.org/prizes/medicine/1906/cajal/biographical/

4.3.1.1 Hypoglossal nucleus/Ncl. of hypoglossal nerve r10 [Ncl. n. hypoglossi]

4.3.1.1.1 Location and structure of the hypoglossal nucleus

The hypoglossal nucleus (CNXII) is a purely somatomotor nucleus (GSE, see Table 1.1), providing innervation to the muscles of the tongue (ἡ γλῶσσα, he glossa, Greek = tongue, ὑπό-, hypo- = beneath). The nucleus is located in the lower medulla oblongata, ventral to the central canal and—further cranially—in the floor of the fourth ventricle (Figs. 4.14, 4.15, and 4.16).

The histological sections in Fig. 4.16 show the position of the hypoglossal nucleus in the wall of the central canal and the fourth ventricle, respectively. Lying close to the midline, its dorsal neighbor is the posterior ncl. of vagus nerve ④

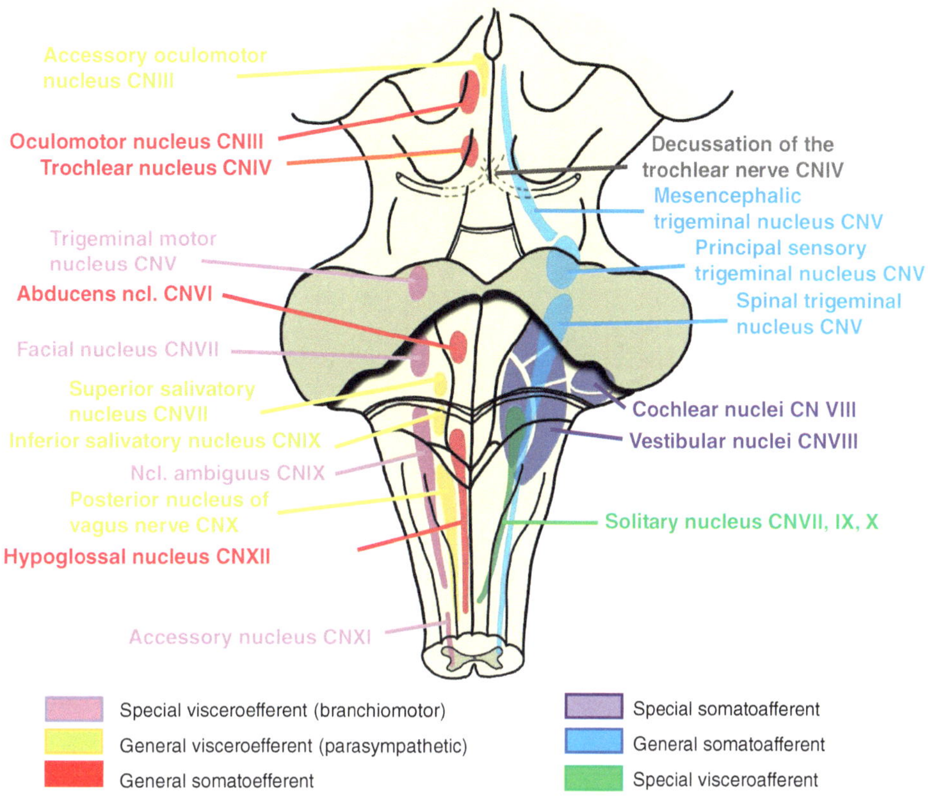

Fig. 4.14 The nucleus of hypoglossal nerve is the caudal most in the row of general somatic efferent (GSE, somatomotor) nuclei (shown in red) in the human brainstem. From Huggenberger et al. 2019, Fig. 15.5 with permission

(see Sect. 7.3.2.1), ventrally it borders the medial longitudinal fasciculus (see Box 9.2 and Fig. 3.14 A–F) and the lateral aspect borders the reticular formation. The hypoglossal neurons are of the typical somatomotor phenotype (large multipolar neurons with a high density of Nissl substance, ChAT-immunopositive).

4.3.1.1.2 Target muscles of the hypoglossal nerve

The extrinsic muscles of the tongue (origin outside the tongue) are the genioglossus (from the chin, Figs. 4.17GG and 4.21 ⑧), the hyoglossus (from the hyoid bone, see Figs. 4.17HG and 4.21 ④) and the styloglossus muscles (from the styloid process, see Fig. 1.36 ④) (see Figs. 4.17SG and 4.21 ⑦). In addition, there is the palatoglossus muscle (from the palate), innervated by the glossopharyngeus nerve (CNIX, see Sect. 5.4.2.1.3).

The geniohyoid part of the hypoglossal nucleus is a small group of neurons lateral of the hypoglossal nucleus which provides the innervation of the geniohyoid muscle [M. geniohyoideus] extending from the mandibula to the hyoid body [Os hyoideum].

The intrinsic muscles of the tongue are the vertical, transverse, superior and inferior longitudinal muscles (see Figs. 4.17 and 4.21).

Saigusa et al. (2006) (see Fig. 4.21) have shown that fibers of the trigeminal motor root (see Sect. 11.3.1) join the lingual nerve and reach the tongue where they branch in the superior and inferior longitudinal muscles. The authors claim that probably these two muscles are not innervated by the hypoglossal nerve.

Touré and Vacher (2006) have studied a large number of human fetal specimens for the arrangement of tongue muscles.

> **Box 4.3 Visible human project**
>
> The Visible Human project is based on the body donation of one man—having been on death row and executed—and a woman, whose bodies were scanned (CT, MRI) after death, deep-frozen, and horizontally sectioned. Thus, a complete imaging and anatomical series of the human body was created. The Visible Man data set was publicly released in 1994 and the Visible Woman in 1995. The U.S. National Library of Medicine provides the data sets to serve as (1) a reference for the study of human anatomy, (2) public-domain data for testing medical imaging algorithms, and (3) a test bed and model for the construction of network-accessible image libraries. (https://www.nlm.nih.gov/research/visible/visible_human.html)

4.3.1.1.3 Course of the hypoglossal nerve

The axons of the hypoglossal neurons run in ventrolateral direction between the medial lemniscus [Lemniscus medialis] and the inferior olivary complex (see Figs. 4.1 and 4.18).

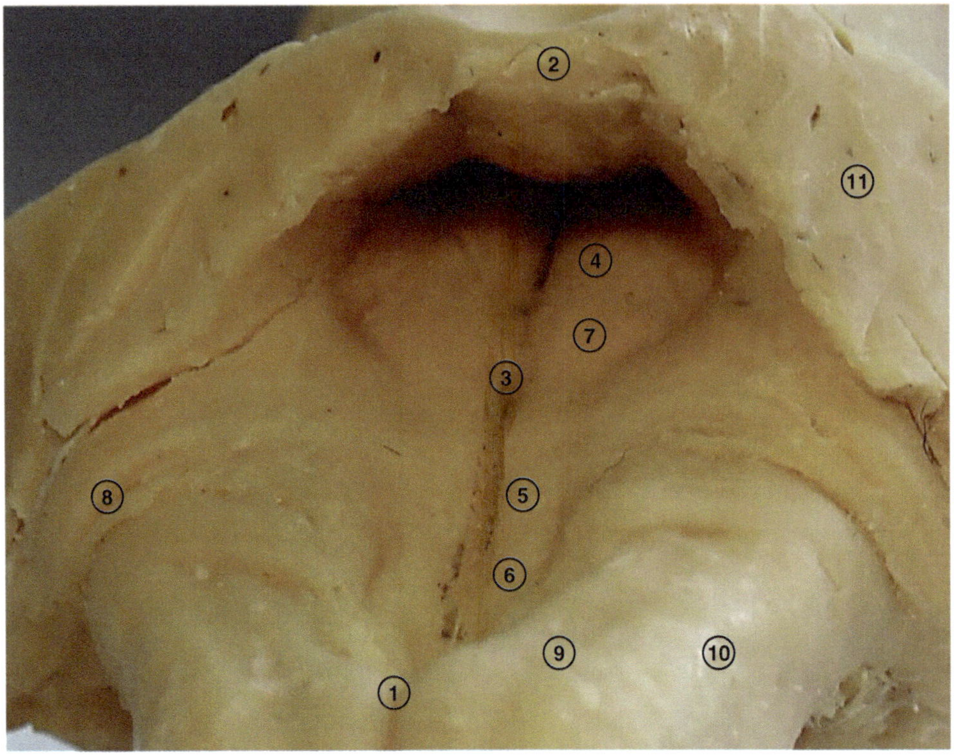

①	Obex	⑦	Vestibular area
②	Superior medullary velum	⑧	Medullary striae of 4th ventricle
③	Posterior median sulcus	⑨	Gracile tubercle
④	Facial colliculus	⑩	Cuneate tubercle
⑤	Hypoglossal trigone	⑪	Medial cerebellar peduncle
⑥	Vagal trigone		

Fig. 4.15 Siliconized sample of the dissected human brainstem. The cerebellum (attached to the cerebellar peduncles) and the inferior medullary velum have been removed to open the view into the region of the fourth ventricle/rhomboid fossa. In the floor of the ventricle, several trigona and colliculi are visible which are the equivalents of the underlying cranial nerve nuclei or cranial nerves. The medullary striae of fourth ventricle ⑧ [Striae medullares ventriculi quarti] arise from the arcuate nuclei (see Sect. 4.1.1.4), extend transversely below the ependymal floor of the fourth ventricle from the posterior median sulcus ③, and enter the inferior cerebellar peduncle (see Box 14.1). Sammlung des Zentrums Anatomie der Universität zu Köln

They reach the preolivary groove [Sulcus preolivaris] and leave the brainstem between the inferior olive (see Fig. 4.1) and the pyramis.

The other caudal cranial nerves (CNIX, CNX, CXI, exception CNXII) leave the brainstem dorsal to the inferior olive in the retroolivary groove [Sulcus retroolivaris].

Inside the leptomeningeal space (Lateral cerebellomedullary cistern, see Box 3.3), the hypoglossal nerve turns laterally (see Fig. 4.19) and inferior to enter the hypoglossal canal [Canalis n. hypoglossi]. The latter is located in the occipital bone on the lateral walls of the foramen magnum (see Fig. 4.20).

After exiting the hypoglossal canal, C1 and C2 cervical plexus branches join the hypoglossal nerve. In its further course, the nerve is running inferior and reaches the angle of the mandible crossing internal and external carotid arteries (see Fig. 4.19B). Subsequently, the nerve is running anteriorly between the stylohyoid and hyglossus muscles to reach the lateral side of the tongue. It gives off several branches for the innervation of the extrinsic and intrinsic muscles of the tongue (see Fig. 4.21).

https://geekymedics.com/the-hypoglossal-nerve-cn-xii/#

4.3.1.1.4 Cerebrocortical control of brainstem motor nuclei

The anatomical knowledge about the cerebrocortical control of brainstem motor nuclei we dispose of is based on investigations by Kuypers in the 1950s on the brains of patients who died after an anatomically defined stroke. Upon autopsy, the brains were dissected and prepared for the so-called Gygax technique (see Giolli 1965). Histological sections of the lesioned area in the cerebral cortex and of the brainstem were studied for the presence of degenerating axons. This allows for conclusions about the in vivo targets of fiber tracts.

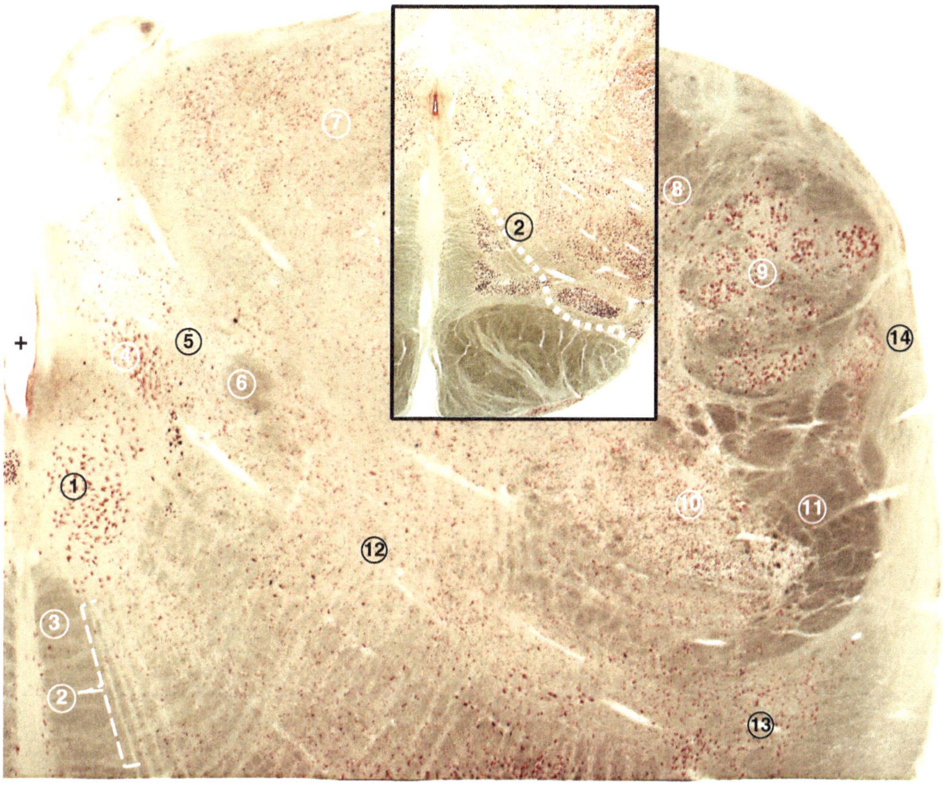

①	Ncl. of hypoglossal nerve CNXII	⑨	Accessory cuneate ncl.
②	Course of hypoglossal nerve	⑩	Spinal ncl. of trigeminal nerve CNV
③	Medial longitudinal fasciculus	⑪	Spinal tract of trigeminal nerve
④	Posterior ncl. of vagus nerve CNX	⑫	Intermediate reticular zone
⑤	Solitary ncll.	⑬	Lateral reticular ncl.
⑥	Solitary tract	⑭	Inferior cerebellar peduncle
⑦	Gracile ncl. and tract	+	4th ventricle
⑧	Cuneate ncl. and tract		

Fig. 4.16 Horizontal section through the human medulla oblongata at the level of the dorsal column nuclei ⑦, ⑧. Darrow red staining. The histological picture shows the easily recognizable hypoglossal nucleus ①, the spinal trigeminal nucleus ⑩, and the medial longitudinal fasciculus ③. The intracerebral course of the hypoglossal nerve can easily be followed ②. The nerve exits the anterior median sulcus between the inferior olive and the pyramid (see Fig. 4.2). See also atlas part Darrow red 4 and Campbell 1 and 2. LabPON Twente

The thorough investigation of the specimens (here cited for a unilateral subtotal obstruction of the middle cerebral artery (acc. to ten Donkelaar 2011) revealed that degenerating corticobulbar (corticonuclear) fibers are distributed to both halves of the brainstem, the hypoglossal nuclei (particularly contralaterally), the ambiguus nucleus (see Sect. 5.4.2.1), the spinal trigeminal nucleus (see Sect. 3.2.2.2), the facial nucleus (see Sect. 8.2.1; in particular contralaterally), and the motor trigeminal nucleus (see Sect. 11.3.1). The abducens nucleus (see Sect. 9.2.1.1) was free of degenerating terminals.

Living anatomy

A simple clinical test for the integrity of tongue motor innervation requires to ask the patient for the protrusion of the tongue (see Fig. 4.22). Under normal conditions, the tongue leaves the mouth in the midline (A). In case of a unilateral lesion of tongue innervation, the intact half will shift the tongue to the lesioned side (B).

4.3.1.1.5 Frontotemporal lobar degeneration

Frontotemporal dementia (FTD) is a heterogenous disorder with speech abnormalities (primary progressive aphasia, PPA), behavioral disorders, extrapyramidal signs, and memory disorders in various combinations. Clinically, the PPA can be either a nonfluent, agrammatic variant, or a semantic variant. Extrapyramidal symptoms can occur later in the disease course. The pathological substrate of FTD, frontotemporal lobar degeneration (FTLD), is heterogeneous as well

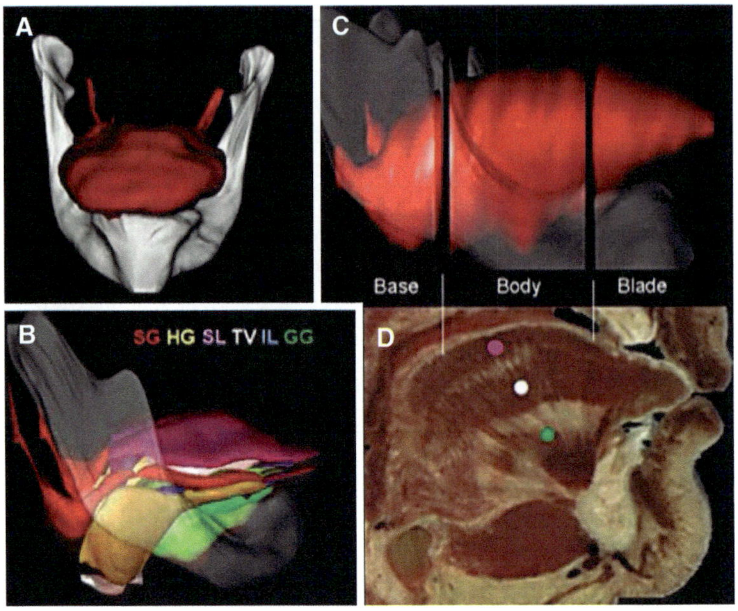

SG	Styloglossus muscle	TV	Transverse / vertical muscle
HG	Hyoglossus muscle	IL	Inferior longitudinal muscle
SL	Superior longitudinal muscle	GG	Genioglossus muscle

Fig. 4.17 Three-dimensional model of the human tongue based on data from "Visible. Human Project" (see Box 4.3). (**A**) Anterior view of the 3D model of the human tongue and mandible. (**B**) Lateral view of 3D model with the surface removed. For the identification of individual muscles, see color code and right-hand side table. (**C**) Lateral view of the 3D model with boundaries shown between base, body, and blade. (**D**) A sagittal view of the Visible Human showing the approximate borders between base, body, and blade of the tongue as well as three of the tongue muscles displayed in **B**. From Sanders and Mu 2013, Fig. 2 with permission

and affecting frontal and temporal lobes, basal ganglia, and hippocampus as well as other regions such as parietal cortex and brainstem.

Depending on the abnormal protein accumulating there are three subtypes of FTLD:

1. FTLD-TDP (transactive response DNA-binding protein 43 KDa),
2. FTLD-Tau
3. FTLD-FET (family of proteins such as Fused in Sarcoma (FUS) protein, Ewing sarcoma protein and TAF15 protein).

Aggregated (misfolded) proteins can often be stained using antibodies against the chaperone proteins of the ubiquitin-proteasome system such as ubiquitin and P62. Rare subtypes with unknown inclusions that do stain with these chaperone proteins are called FTLD-U (ubiquitin).

Frontotemporal lobar degeneration FTLD-TDP

This is the most frequent cause of FTLD and 20–50% of the cases are familial forms (Pottier et al. 2016). Five different histopathological subtypes (A-E) exist depending on the distribution and morphology in the neocortex of TDP-positive neuronal cytoplasmic inclusions (NCI) and TDP threads, the presence of TDP-positive neuronal intranuclear inclusions (NII) and the presence of TDP inclusions in oligodendrocytes (Lee et al. 2017) (Fig. 4.23).

This classification was based on the harmonized criteria (MacKenzie et al. 2011) and correlates well with clinical and genetic findings. The subcortical distribution in FTLD-TDP types fits well in these different subtypes A–E (Mackenzie and Neumann 2020). All TDP subtypes have different biochemical fingerprints and conformations and appear to be different "strains" that can propagate in a prion like manner (Aguzzi et al. 2007).

FTLD-TDP Type A

Part of the cases with a type A pathology have mutations in the progranuline gene (GRN) (chromosome 17) (Baker et al. 2006; Cruts et al. 2006). The mutations lead to an autosomal dominant disorder with especially progressive aphasia. Sporadic mutations occur as well. Macroscopically, symmetric, globally small brains are seen with slightly more atrophy frontally and temporally. In the H & E stain, there is some spongiosis of the superficial cortical second and superficial third layer. The pattern of cortical ubiquitin inclusions was first described in 2006 (Mackenzie et al. 2006) and later,

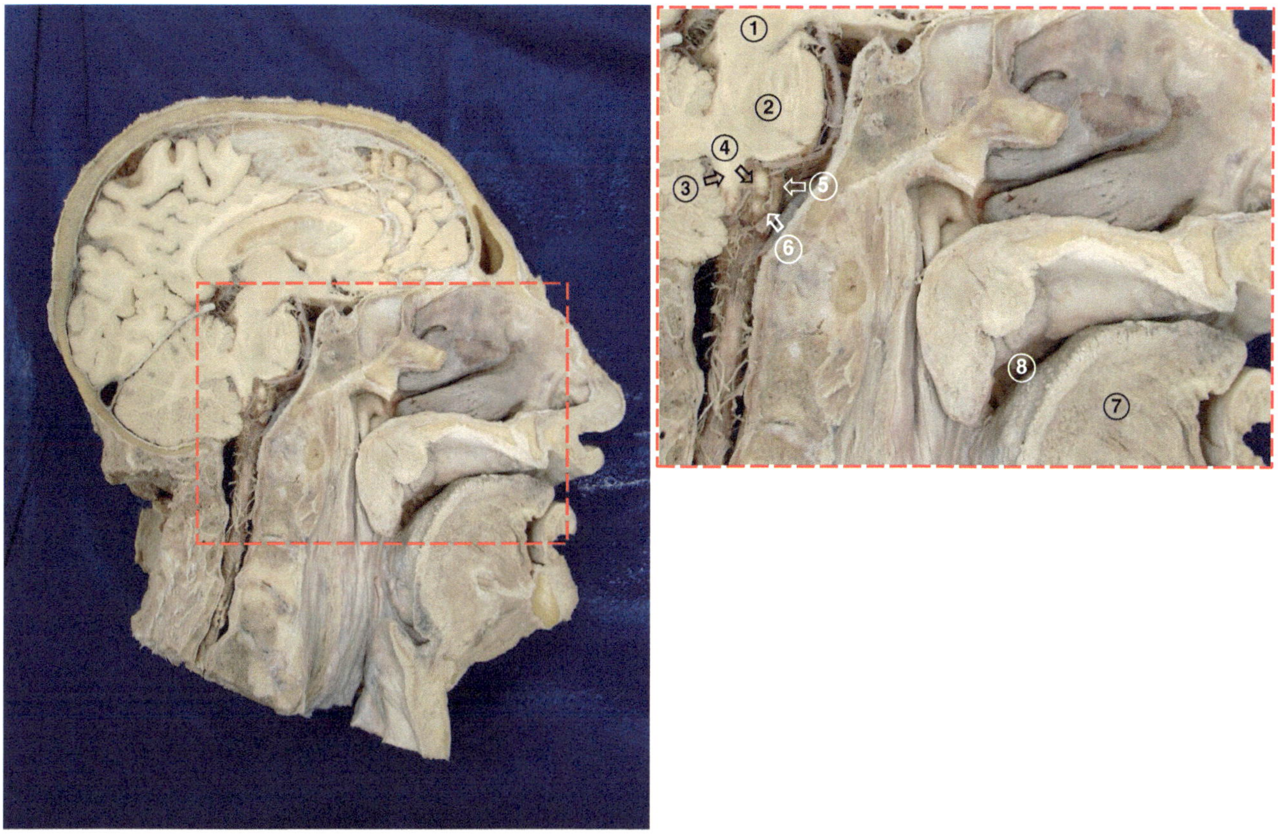

①	**Mesencephalon**	⑤	**Pyramis**
②	**Pons**	⑥	**Exit zone of hypoglossal nerve into the lateral cerebellomedullary cistern**
③	**Medulla oblongata**	⑦	**Tongue**
④	**Oliva**	⑧	**Oral cavity**

Fig. 4.18 Midsagittal section through the human head. Formalin-fixed specimen. At higher power you can identify the brainstem with mesencephalon ①, pons ②, and medulla oblongata ③. At the caudal part of the pontine bulging, the arrow ④ points to the inferior olive (dorsal landmark of CNXII exit, see Fig. 4.1) and ⑤ to the pyramis (ventral landmark). Between both, CNXII exits the preolivary groove [Sulcus preolivaris], enters the leptomeningeal space, and runs to its foramen, the canal of hypoglossal nerve [Canalis nervi hypoglossi] (see Fig. 4.20 ①). Photograph by MedizinFotoKöln, Cologne, Germany. Sammlung des Zentrums Anatomie der Universität zu Köln

accumulation of TDP (see Sect. 4.3.1.1.5 (1)) in these aggregates was found (Cairns et al. 2007). The histopathology of patients with GRN mutations shows a type A with lentiform NCIs, some compact NCIs, short thick TDP threads, and subcortical threads, especially in the superficial cortical layers. The other part of type A has C9orf72 mutations (see below).

FTLD-TDP Type B

The most frequent genetic disorder causing FTLD-TDP was found as a hexanucleotide repeat expansion (GGGGCC) on gene C9orf72 (DeJesus-Hernandez et al. 2011; Renton et al. 2011). The number of expansions is at least above 30 and can vary to thousands of expansions. At least 10–20% of the patients also develop motor neuron symptoms, as in amyotrophic lateral sclerosis. These two disorders have overlapping pathogenesis, forming a spectrum. In amyotrophic lateral sclerosis (see Sect. 4.3.1.1.6), at least 5% is caused by C9orf9 mutations. Macroscopically, atrophy is most pronounced temporally, occasionally with "knife-edge" gyri. In general, motor, sensorimotor, and posterior cerebral cortices are largely unaffected and cerebellum and brainstem appear macroscopically normal (Fig. 4.24).

Upon coronal section, the atrophy within the frontal, anterior temporal, anterior parietal, anterior cingulate, and anterior insular regions is obvious. The amygdala is frequently atrophied, as is the caudate nucleus. The medial temporal lobe and hippocampus can be atrophic as well. The lateral ventricles are always enlarged. In some instances, the SN (see Sect. 16.6.1) is pale, whereas the LC (see Sect. 13.2.2.1) is pigmented. The hypoglossal nucleus (see Sect. 4.3.1.1) is affected with neuronal loss.

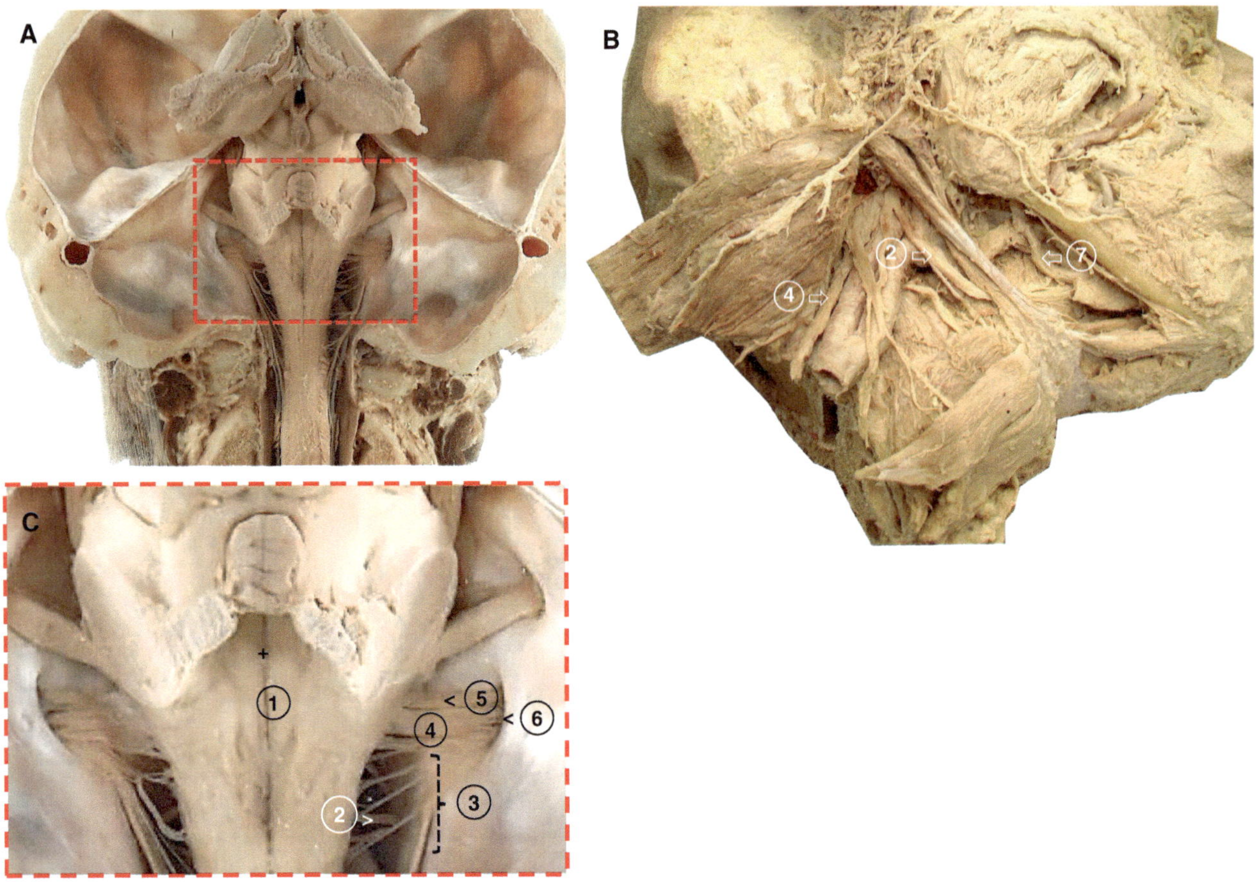

①	**Hypoglossal trigone**	⑤	**Glossopharyngeal nerve CNIX**
②	**Hypoglossal nerve CNXII**	⑥	**Jugular foramen**
③	**Accessory nerve CNXI Spinal root**	⑦	**Lingual nerve V_3**
④	**Vagus nerve CNX**	+	**4th ventricle**

Fig. 4.19 (**A**) Dorsal view of the human brainstem. Formalin-fixed specimen. Posterior parts of the skull removed. At higher magnification, the fourth ventricle with the hypoglossal trigone ① is clearly visible. Ventral of the spinal roots of the accessory nerve ③ the hypoglossal nerve is visible traversing the lateral cerebellomedullary cistern to the canal of hypoglossal nerve (see Fig. 4.20 ①). (**B**) Müller-Thomsen fecit. At the lateral side of the neck, the hypoglossal nerve becomes visible below the external acoustic meatus, contacts the Ansa cervicalis, and runs parallel to CNIX and CNX. It then crosses the internal external carotid arteries and reaches the tongue from caudal in an arcuate course. Photographs by MedizinFotoKöln, Cologne, Germany. Sammlung des Zentrums Anatomie der Universität zu Köln

The ventral roots in the spinal cord can be atrophic. TDP inclusions were found by Neumann et al. (2006) and Arai et al. (2006). Next to familial cases also sporadic cases with type B pathology are found. Microscopically, H & E stain shows spongiosis of the superficial layers (mostly layer II) next to gliosis. This is especially seen in the temporal lobe, less in the frontal lobe. Occasionally, swollen chromatolytic neurons are seen, especially in layers V and VI (immunoreactive for αB-crystallin). Astroglial reaction is slight, the loss of myelin and axons is often inconspicuous. The caudate is gliotic and the hippocampus can show sclerosis with loss of neurons in CA1. Immunohistochemically, type A or type B can be found or a combination of both. In FTLD-type B diffuse granular NCIs and threads are found in all cortical layers.

In addition, subcortical glial inclusions can be found in the white matter. CNXII also shows NCIs with TDP, correlating with the motor neuron symptoms (Fig. 4.25).

Because of RAN (Repeat-associated non-AUG) translation, dipeptides form and aggregate as small cytoplasmic inclusions in neurons in the cerebellar granular layer as typical for C90rf72 mutations.

FTLD-TDP Type C

FTD patients with this type of pathology show clinically especially semantic disorders. These cases are often spo-

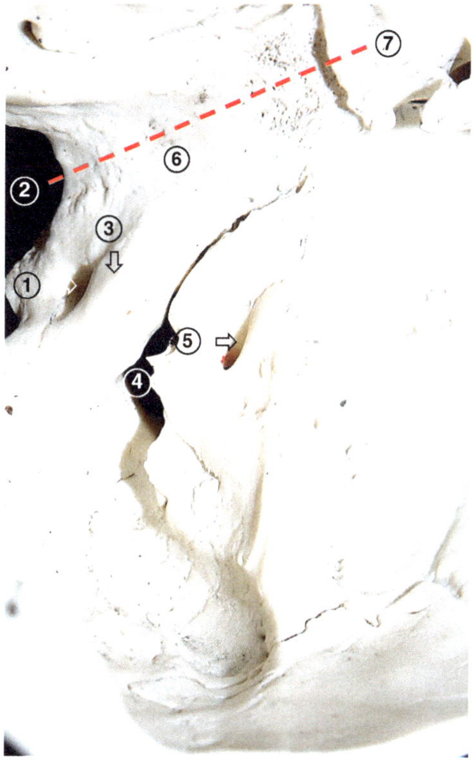

①	Canal of hypoglossal nerve CNXII	⑤	Internal acoustic porus / meatus
②	Foramen magnum	⑥	Clivus
③	Jugular tubercle	⑦	Sella turcica
④	Jugular foramen	---	Midline

Fig. 4.20 Dorsal view onto the human skull base. As can be expected for the caudal most cranial nerve (CNXII), it enters the skull base via the caudal most of paired foramina, the canal of hypoglossal nerve ①. Sammlung des Zentrums Anatomie der Universität zu Köln

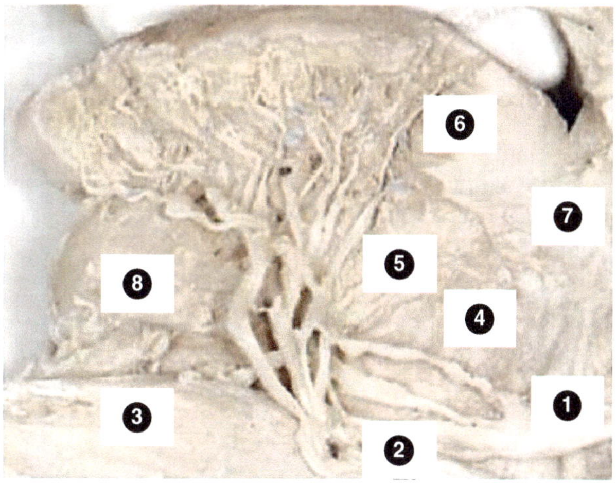

①	Hypoglossal nerve CNXII	⑤	Inferior longitudinal muscle
②	Lingual nerve CNV V_3	⑥	Superior longitudinal muscle
③	Anterior belly of digastric muscle	⑦	Styloglossus muscle
④	Hyoglossal muscle	⑧	Genioglossus muscle

Fig. 4.21 Lateral view of the dissected human tongue. The superficial structures have been removed to show the course of the hypoglossal and lingual nerves in vicinity to the tongue as well as the tongue muscles (cf. with Fig. 4.17). Modified after Saigusa et al. 2006, Fig. 4 with permission

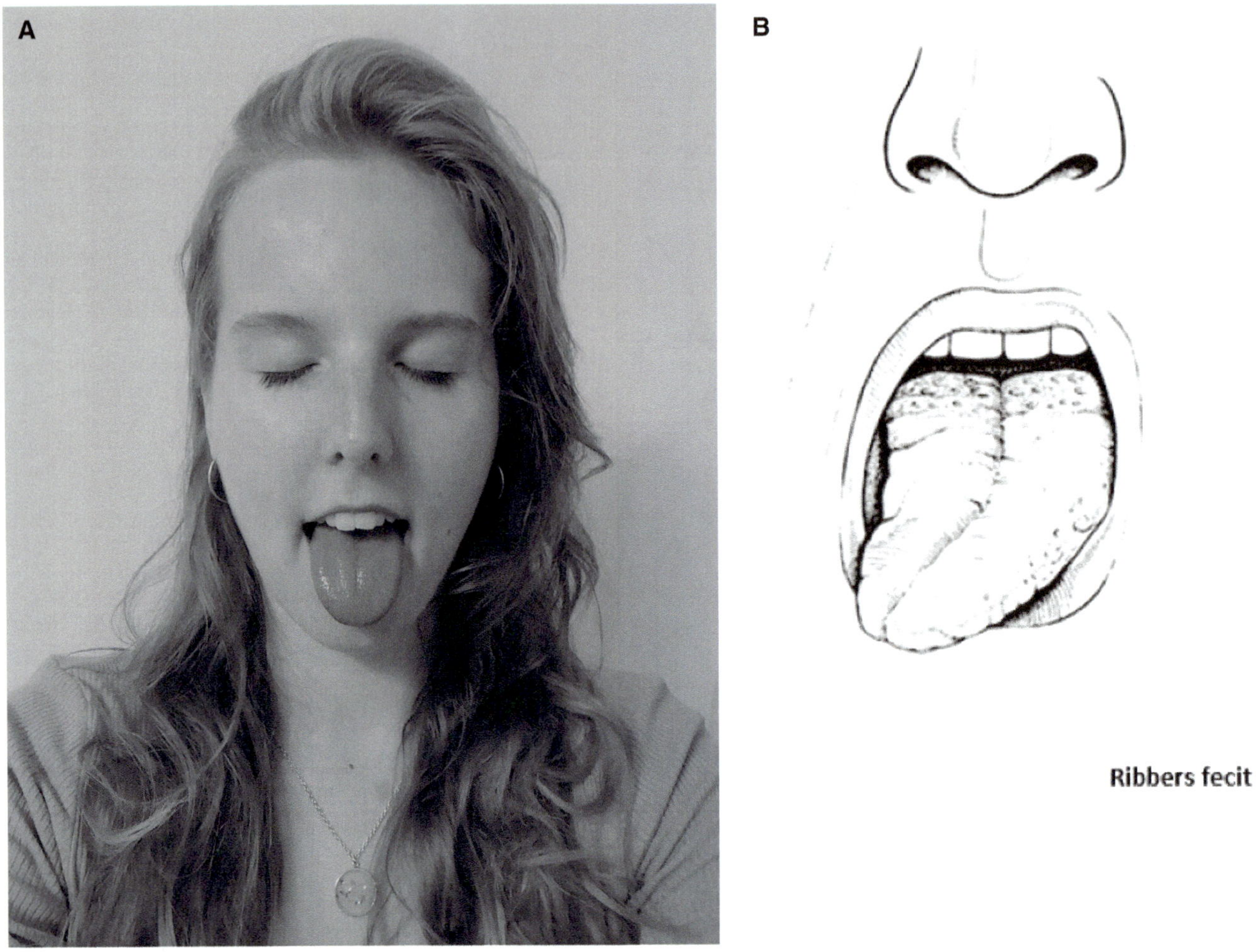

Fig. 4.22 Voluntary protrusion of the tongue. (**A**) Normal finding. (**B**) Typical finding in a right-sided paralysis of the tongue. The intact muscles of the left half of the tongue shift the tongue to the paralyzed right side. Courtesy Sophia Schröder

radic, the pathogenesis is unknown. Cases with right predominant pathology may show loss of empathy and behavioral disorders. Extrapyramidal movement disorders are infrequent. Macroscopically, moderate frontotemporal atrophy can be found. Microscopically, these brains show long, tortuous and thick TDP threads in the superficial layers and few inclusions. The dentate gyrus shows small, sharply edged, rounded NCIs in the TDP stain as well as. Caudate also has these typical neuronal inclusions. Involvement of the brainstem has not been described.

FTLD-TDP Type D

The rare cases with this type of FTLD pathology, showing only NII with TDP in the cortex without threads, have mutations in the VCP molecule (Holm et al. 2007). This disorder is associated with myopathy and Paget disease of the bone. In this type, the basal ganglia, nucleus basalis, amygdala, thalamus, and midbrain can show threads and NCIs, but the rest of the brainstem is spared.

FTLD-TDP Type E

Recently, a rapidly progressive behavioral variant of FTD with a course of 2–3 years until death has been described with granulofilamentous NCIs in all cortical layers and involvement of oligodendroglia (Lee et al. 2017). Histopathology is related to type B.

Frontotemporal lobar degeneration FTLD-Tau

Several subtypes of FTLD-Tau exist, including hereditary tau mutations, the sporadic Pick's disease, and the FTD with histopathologically progressive supranuclear paralysis (see Sect. 4.1.1.3.3) or corticobasal degeneration (see Sect. 16.6.1.9).

Pick's disease

The term Pick's disease is now restricted to the sporadic, behavioral variant FTD cases that show numerous Pick bodies in the histology. Macroscopically, lobar or circumscribed cerebral atrophy affecting the frontal and/or anterior temporal lobes is seen which may extend to the parietal lobe (Figs. 4.26 and 4.27).

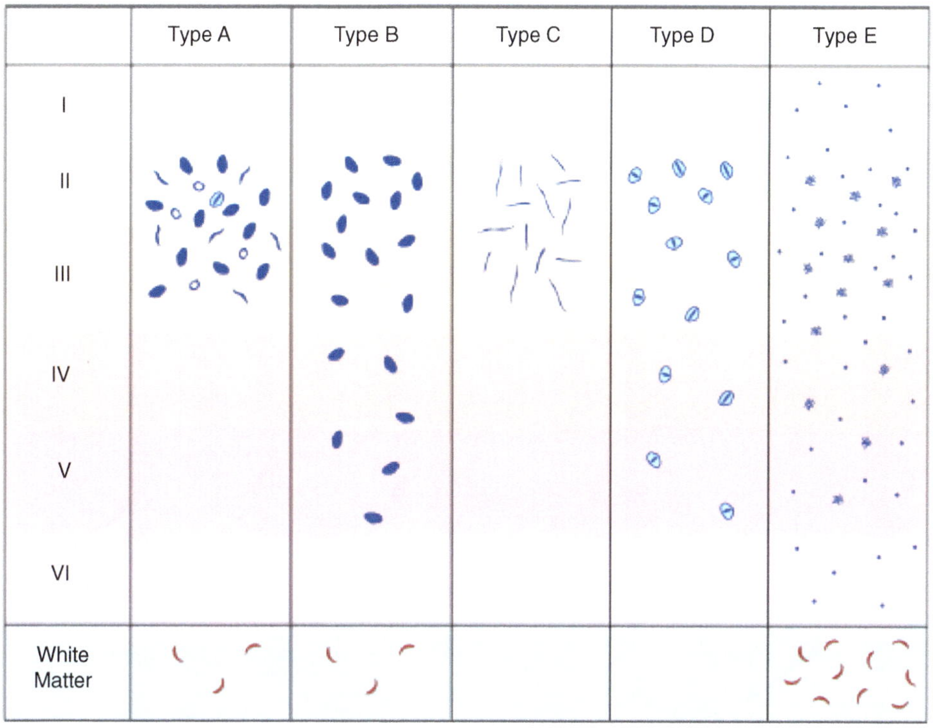

	Type A	Type B	Type C	Type D	Type E
Cortical Pathology	▪ NCI's including ▪ ring inclusions ▪ Short DN's ▪ /- Lentiform NII's ▪ /- Oligo inclusions ▪ Superficial	▪ NCI's ▪ Few DN's ▪ Short DN's ▪ /- Oligo inclusions ▪ /- Superficial and deep	▪ Long DN's ▪ Few NCI's ▪ Superficial	▪ Lentiform NII's ▪ Few NCI's ▪ Superficial and deep	▪ GFNI's ▪ Grains ▪ Cuvilinear oligodendroglial inclusions ▪ Superficial and deep
Common Phenotype	bvFTD naPPA	bvFTD +/- MND	svPPA bvFTD	IBMPFD-ALS	bvFTD
Genetic Associations	*GRN* mutations	*C9orf72* mutations	None	*VCP* mutations	Uncertain

Fig. 4.23 Illustration of TDP-43 aggregate morphology that define Types A, B, C, D, and E. Below each FTLD-TDP type is a summary of the pathological features, clinical phenotypes, and genetic features of each subtype. From Lee et al. 2017, Fig. 6 with permission

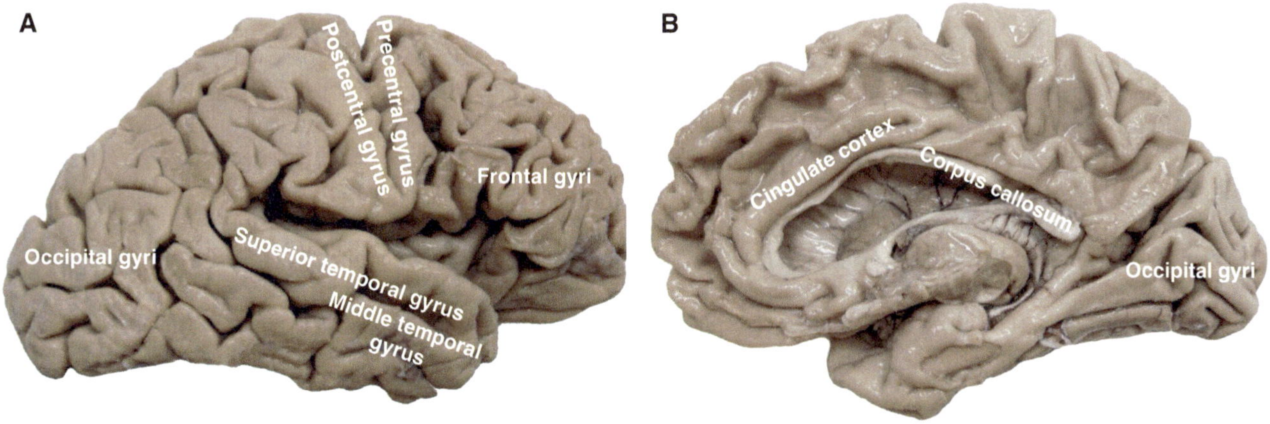

Fig. 4.24 Lateral (**A**) and (**B**) midsagittal view of a brain from a patient with FTD-TDP/ALS due to a C9orf72 mutation. Mixed type A and B. Frontotemporal atrophy. The motor and sensorimotor cortices are relatively spared. Collection A Rozemuller

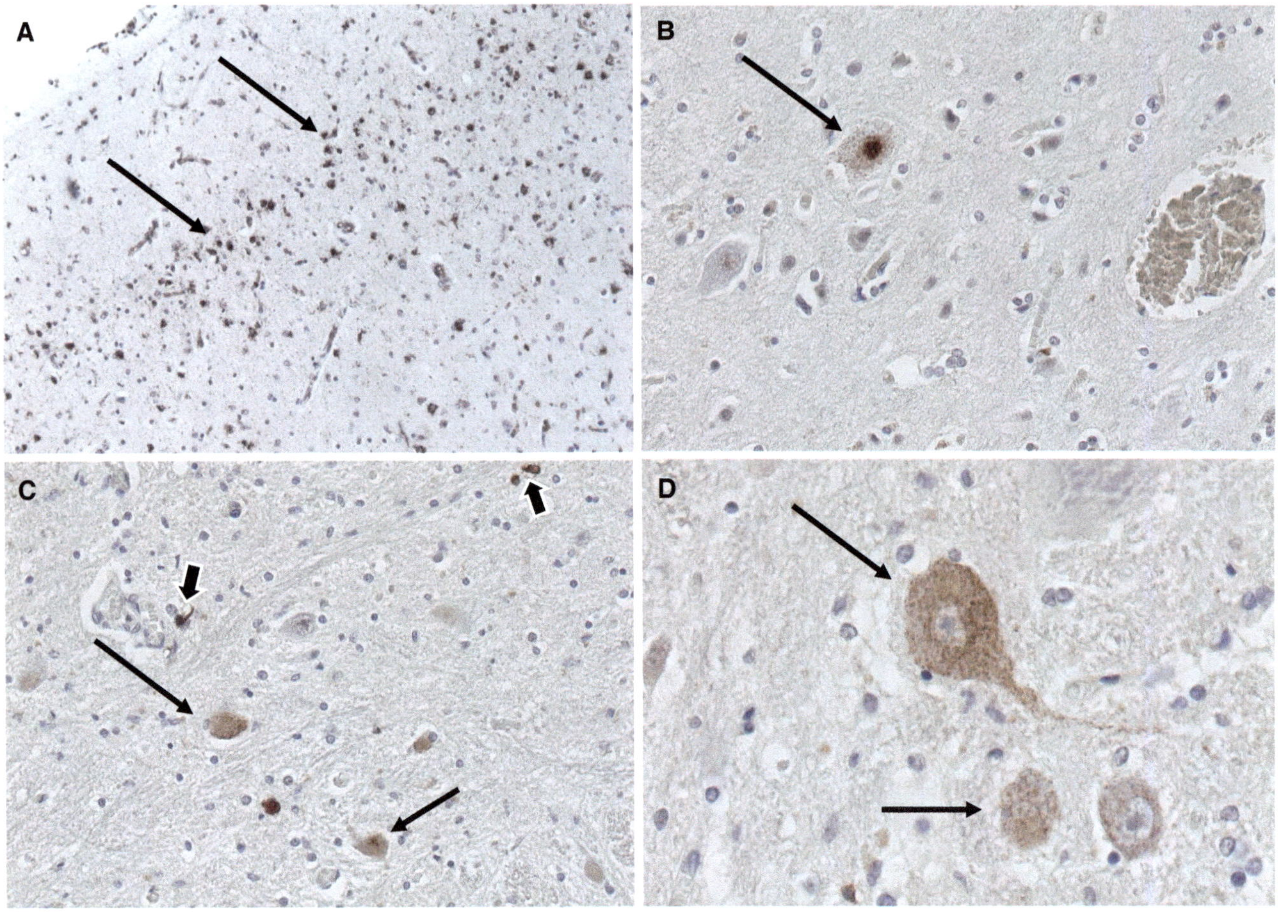

Fig. 4.25 Frontal cortex (same patient as in Fig. 4.24), TDP-positive NCIs (arrows) in all cortical layers, more pronounced in the second layer. In (**B**) NCI in the motor cortex, in (**C**) CNXII with loss of neurons and NCIs (long arrows) and a few coiled bodies (short thick arrows), in (**D**) NCIs in preserved neurons of the anterior horn of the cervical spinal cord. pTDP-43 immunostain. Collection A Rozemuller

Marked gyral atrophy after long duration of the disease is sometimes referred to as "knife edge" atrophy or "walnut brain." Areas spared include the posterior part of the superior temporal gyrus (Figs. 4.26B and 4.27C) and pre- and post-central gyri. The atrophy may be asymmetrical with the dominant hemisphere (usually the left side) being more affected. Rarely only temporal atrophy is seen. On sectioning, the cortical ribbon is thinner than usual and the gray-white junction is indistinct. The white matter can be atrophied, rubbery, and granular. The ventricles, especially the frontal and temporal horns of the lateral ventricles, are dilated (Fig. 4.27C). Severity of atrophy results in a brain weight often less than 1000 g. In areas with severe pathology, cytoarchitectural features of the cortex become obscured with loss of large pyramidal neurons, collapse of the parenchyma, spongiosis, and astrocytic gliosis (Fig. 4.28).

Gliosis is often marked in the upper cortical layers and at the gray-white matter junction. Ballooned neurons are present in the middle and lower cortical layers. Cerebrocortical neurons contain round to irregularly shaped inclusions (Pick bodies), intensely argyrophilic with silver impregnation. The nucleus lies remarkably eccentric (Fig. 4.28C). Anti-tau (3-repeat tau) and anti-neurofilament immunostains strongly label Pick bodies and Pick neurites. Characteristically, Pick bodies contain 3-repeat tau isoforms and are negative for 4-repeat tau antibodies. Cell loss is greater in the frontal and temporal lobes than in the parietal lobe (Braak and Braak 1998). Pick bodies are widespread in different subcortical and brainstem nuclei as the locus caeruleus (Fig. 4.29E, F). The hippocampus and amygdala are usually severely affected, with many Pick bodies in CA1, subiculum, and dentate gyrus (Fig. 4.29).

Pick bodies can be found in the olfactory nucleus as well. In regions where atrophy is greatest, the subjacent white matter displays a loss of myelinated fibers, accompanied by gliosis and axonal degeneration. Western blotting shows two major bands (55 and 66 kDa) (Delacourte et al. 1998) consistent with 3-repeat tau. These isoforms also contain exon 3. Ubiquitin has been identified as another constituent of Pick bodies, but frequently the staining is weak and not found in all Pick bodies. Pick bodies are negative for α-synuclein, which differentiates them from Lewy bodies (LBs) (see Sect.

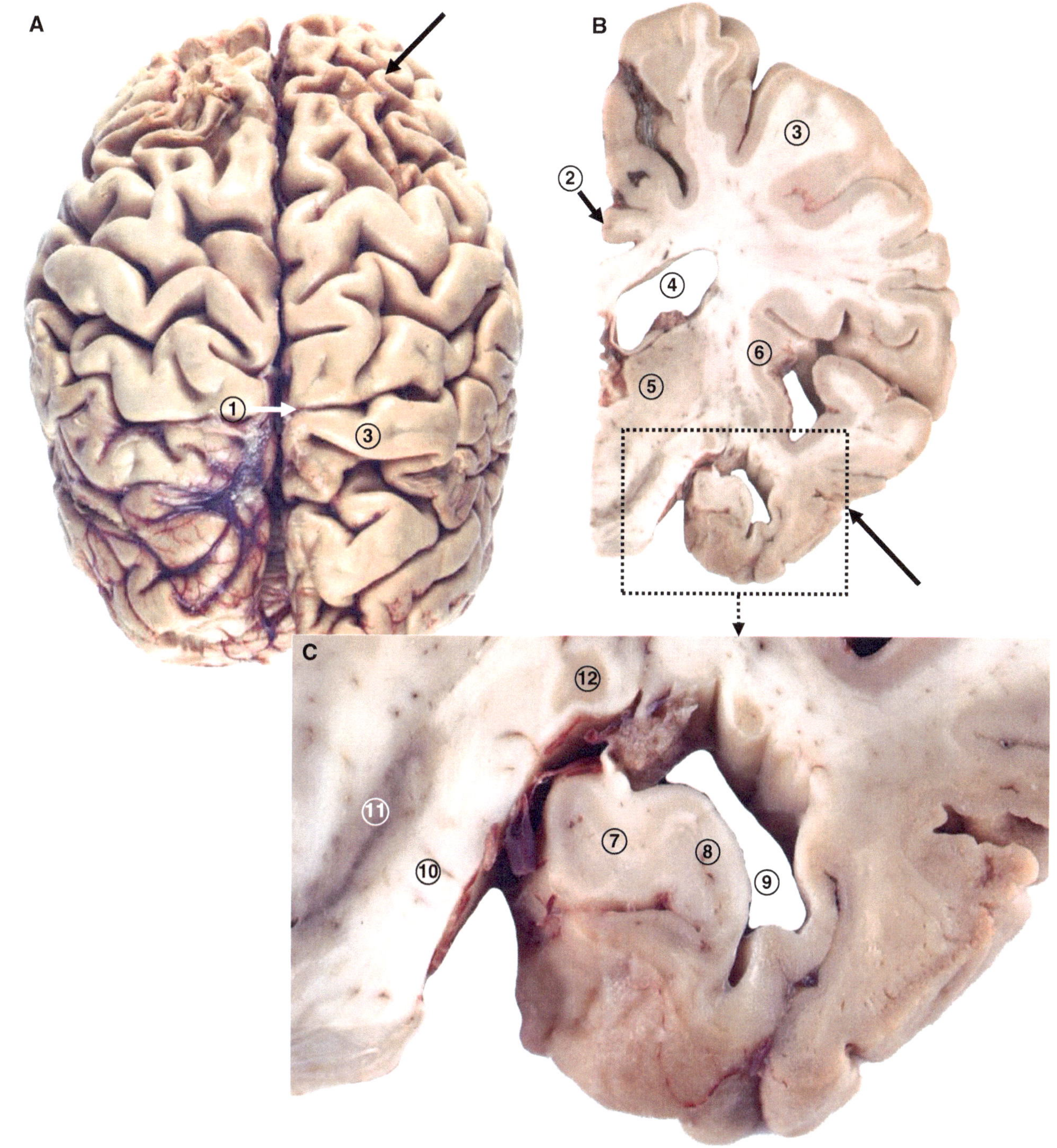

①	Central sulcus	⑦	Hippocampus
②	Cingulate gyrus	⑧	CA1 region
③	Postcentral gyrus	⑨	Temporal horn of lateral ventricle
④	Lateral ventricle	⑩	Cerebral peduncle
⑤	Thalamus	⑪	Substantia nigra
⑥	Insula	⑫	Lateral geniculate body

Fig. 4.26 Brain from a patient with Pick's disease (FTLD-Tau) seen from above, showing frontal atrophy (**A**, black arrow). In (**B**) coronal section showing severe frontotemporal atrophy (**C** detail, arrow). LabPON Twente

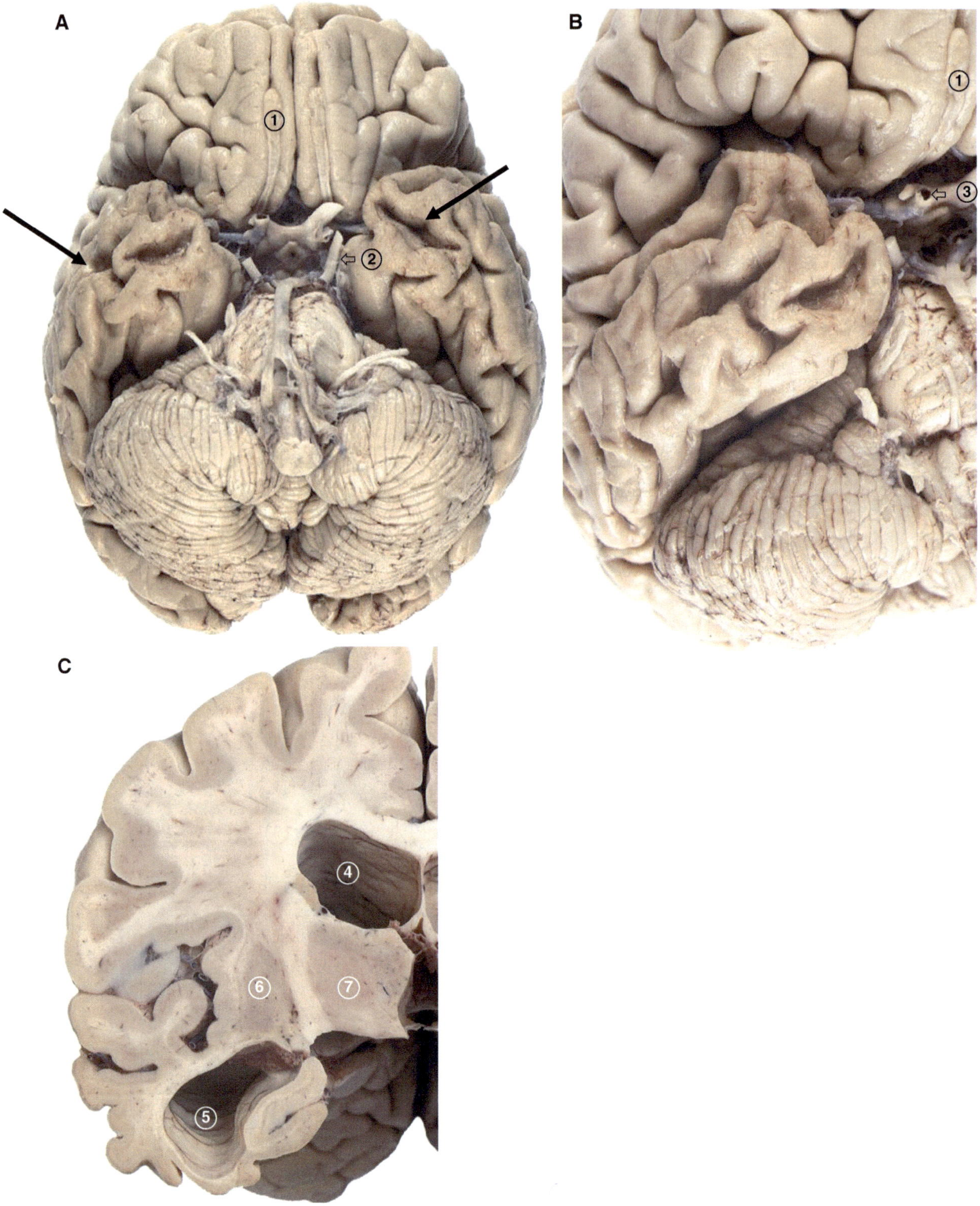

①	**Olfactory bulb CNI**	⑤	**Temporal horn of lateral ventricle**
②	**Oculomotor nerve CNIII**	⑥	**Putamen**
③	**Internal carotid artery**	⑦	**Thalamus**
④	**Lateral ventricle**		

Fig. 4.27 (**A**) Ventral view, (**B**) detail and (**C**) coronal section at the thalamic level of a brain from a patient with Pick's disease. Temporal atrophy (arrows), the posterior part of the superior temporal gyrus is spared. Dilated ventricular system. LabPON Twente

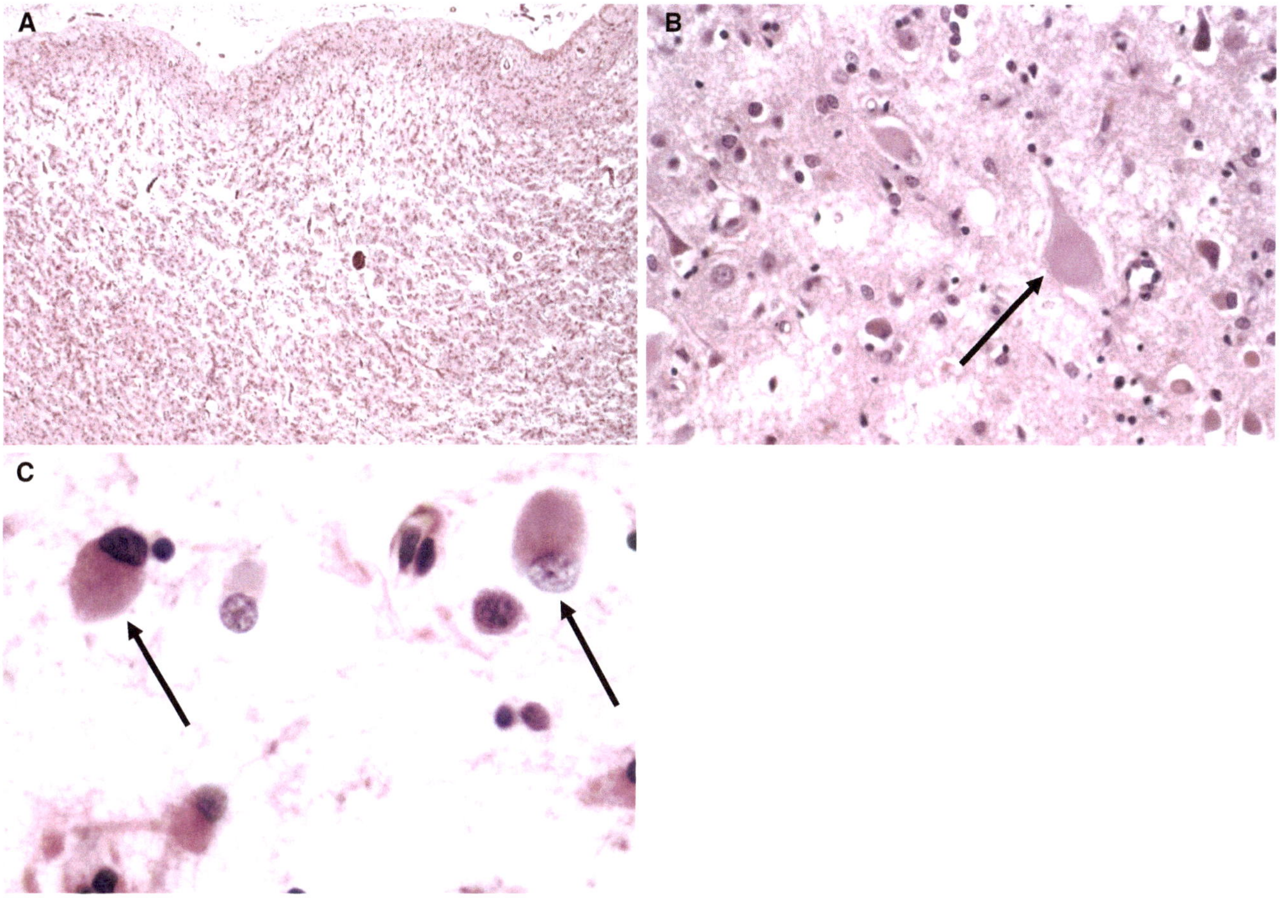

Fig. 4.28 Pick's disease. In (**A**) frontal cortex with collapse of the parenchyma and spongiosis, in (**B**) and (**C**) swollen neurons and Pick bodies (arrows). H & E stain. LabPON Twente

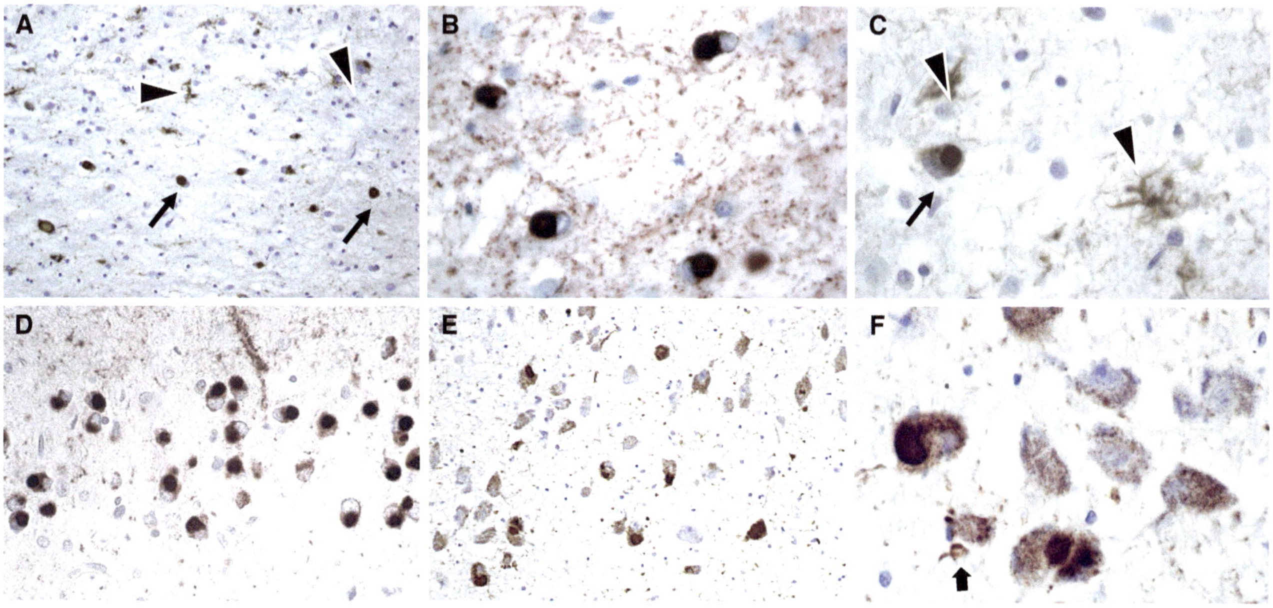

Fig. 4.29 Pick's disease. Tau-positive Pick bodies and astrocytes in the frontal cortex (**A**). In (**B**) tau-positive Pick bodies and threads, in (**C**) Pick body surrounded by "ramified" astrocytes. In (**D**) Pick bodies in the granular cells of the dentate gyrus. Locus caeruleus in **E** and **F** (detail), many Pick bodies, some threads, and a coiled body. Long arrows point to Pick bodies, triangles to astrocytes, small arrow to coiled body. AT8 immunostain. LabPON Twente

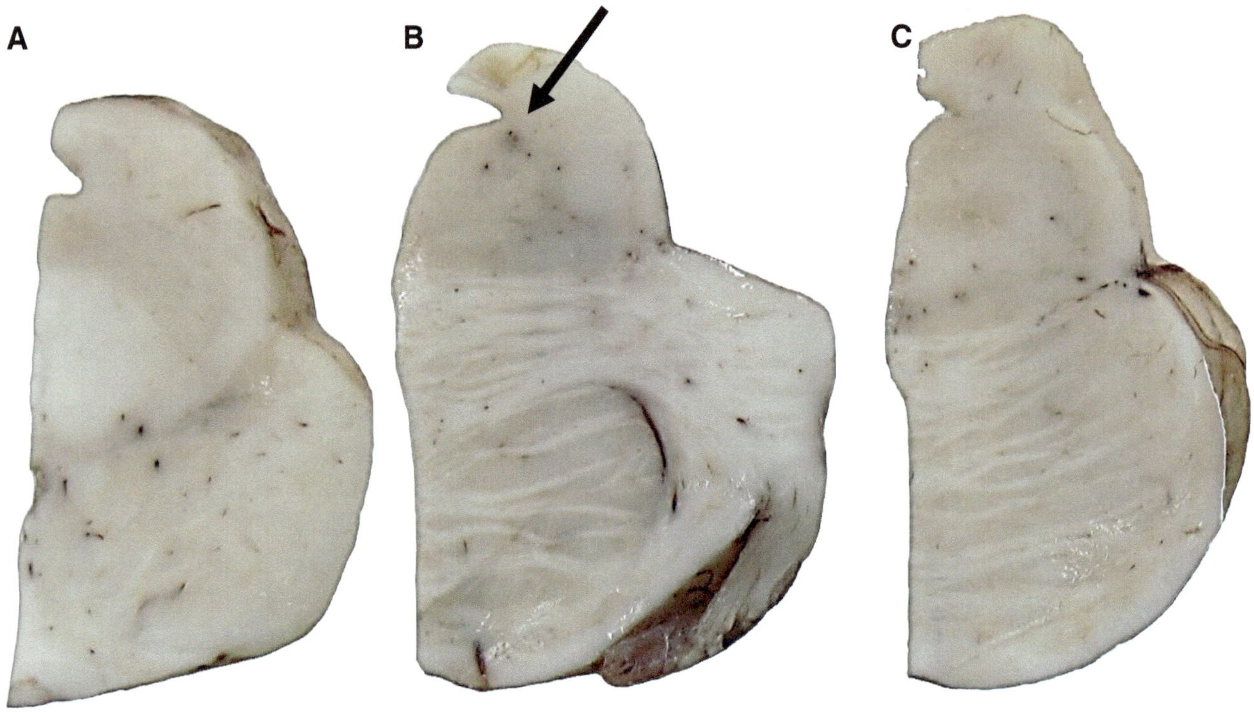

Fig. 4.30 Horizontal sections of the pons. Patient with a gene mutation in the FTLD-MAPT gene. Atrophy of the pons, pale locus caeruleus (arrow). Collection A Rozemuller

16.6.1.5). Ultrastructurally, they are composed of randomly arranged straight filaments, ranging from 14 to 16 nm in diameter and admixed with a variable number of twisted filaments with a diameter ranging from 22 to 24 nm and a half period of 120–160 nm. This contrasts with the twisted paired helical filaments of AD, which are 22 nm wide with a half period of 60–80 nm. Ballooned neurons are best demonstrated with antibodies that recognize phosphorylated neurofilament epitopes. They also contain stress proteins (αB-crystallin positive). They are composed of granulofilamentous material. Tau-positive astrocytes have been described in the cerebral cortex and in the white matter (Braak and Braak 1998; Komori 1999).

Most of them are 3-repeat tau-positive, and a small amount is stained for 4-repeat tau. In the cerebral cortex, ramified tau-positive astrocytes are present with a few thick processes and eccentric nuclei. In the white matter, they show dense tau-positivity and have a sharp margin. The distribution of the ramified astrocytes is limited to the affected cortex, such as the frontal, temporal, and insular cortices, and the hippocampus (Fig. 4.29). Coiled bodies are far less numerous than in PSP and CBD and are limited to the affected areas (Komori 1999). Neuropil threads in Pick's disease can better be visualized using 3-repeat tau antibodies and are rare in Gallyas silver staining that visualizes especially 4-repeat tau forms.

FTLD-Tau in MAPT mutations

Primary tauopathies caused by mutations in the tau gene (chromosome 17) can be either aggregates of 3-repeat tau, 4-repeat tau or both, depending on the type of mutations. Macroscopically, the degree of atrophy depends on the type of the tau mutation and the duration of the disease but can be severe. Atrophy of the hemispheres can be mild in intermediate stages. Frontal and temporal atrophy, especially of part of the anterior temporal lobes, is prominently seen in later stages. On cut surface, the substantia nigra and locus caeruleus are pale (Fig. 4.30). The caudate nucleus, putamen, globus pallidus, amygdala, hippocampus, and ventral hypothalamus are variably involved. A loss of frontotemporal white matter is seen, with enlargement of the ventricles. The parietal and occipital lobes are less frequently affected. The midbrain, pons, and cerebellar cortex can be atrophic (Fig. 4.30).

The microscopic findings vary substantially depending on the type of the mutation (Dickson 2003; Bird et al. 1999; Yen et al. 1999). In general, tau can accumulate in neurons and glial cells in the cortex, basal nuclei, and brainstem nuclei, and in glial cells in the white matter (Fig. 4.31).

The deposits can be found in the perikaryon or in the extensions. The neuronal inclusions may resemble AD tangles (as in the P310L mutation), globose tangles, or Pick-like inclusions (Fig. 4.31). Glial tau inclusions resemble coiled

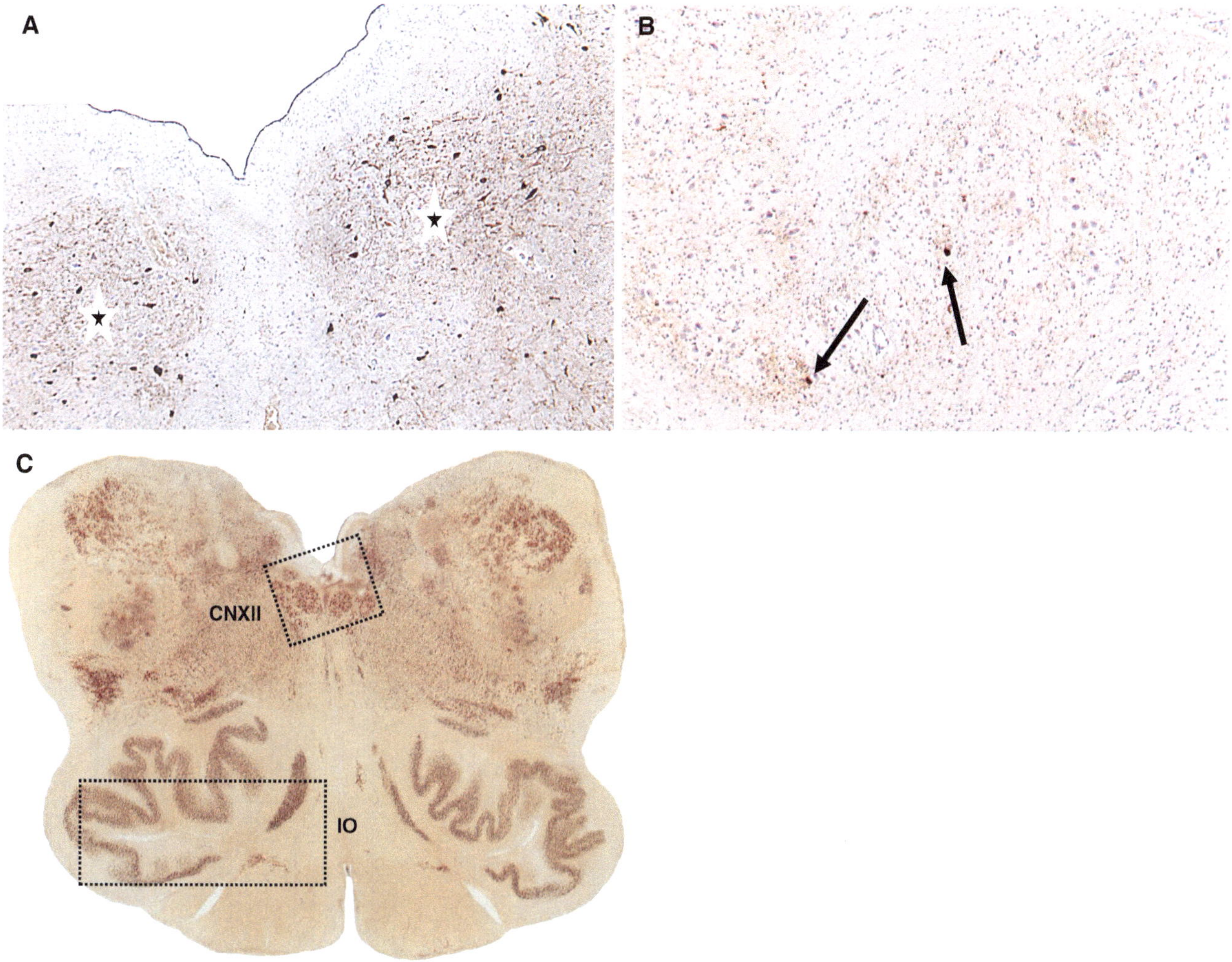

Fig. 4.31 Medulla oblongata (same patient as in Fig. 4.30). (**A**) Tau-positive NCIs and threads in the CNXII nucleus and reticular formation. In (**B**) inferior olivary complex (IO), some globose-like tangles (arrows) and tau-positive threads. AT8 immunostain. (**C**) Darrow red staining. Collection A Rozemuller and LabPON Twente

bodies, tuft-shaped astrocytes, or astrocytic plaques. The mutations in exon 10 can lead to more 4-repeat tau in neurons and glial cells. A deletion of exon 10, as seen in the deltaK 280 mutation, shows many Pick-like inclusions of 3-repeat tau (Van Swieten et al. 2007). Some mutations in exons 9, 11, 12, and 13 also lead to Pick-like inclusions. They can be found in frontotemporal cortex, caudate nucleus, pallidum, hippocampus (CA1, subiculum, dentate gyrus), and substantia nigra. In addition to Pick-like inclusions and tangles, neuronal tau staining can be diffuse (as in the P310L mutation), ring-like, granular, rounded, oval, and dot-like. Ballooned neurons can be visualized with αB-crystallin. In areas with severe neuronal loss and gliosis, the architecture cannot be recognized, and spongiosis is present. Western blot patterns for tau varies from Pick-like doublets to AD-like triplets and PSP-like doublets depending on the type of mutation (amount of 3- /4-repeat tau).

FTLD-FET

In 2009, the protein FUS (Fused in Sarcoma) was found in inclusions in a family with ALS (Kwiatkowski Jr et al. 2009; Vance et al. 2009). Later, a new, rare, sporadic FTLD was found with FUS inclusions, clinically characterized by severe behavioral disorder in young adults. Ventricular dilatation and atrophy of the caudate in these cases is striking. Brainstem atrophy and a pale substantia nigra can be found (Fig. 4.32).

Neuronal intracytoplasmic inclusions (NCIs) with the protein FUS are found, spread in cortex and in the brainstem, such as locus caeruleus (Fig. 4.33).

Fused in sarcoma protein (FUS) is a member of a family of RNA-binding proteins, including Ewing Sarcoma (ES) protein and TAF 15 protein. All three proteins can be found aggregated in neurons in sporadic FTLD-FUS though in different cell regions (Neumann et al. 2011). Prion-like proper-

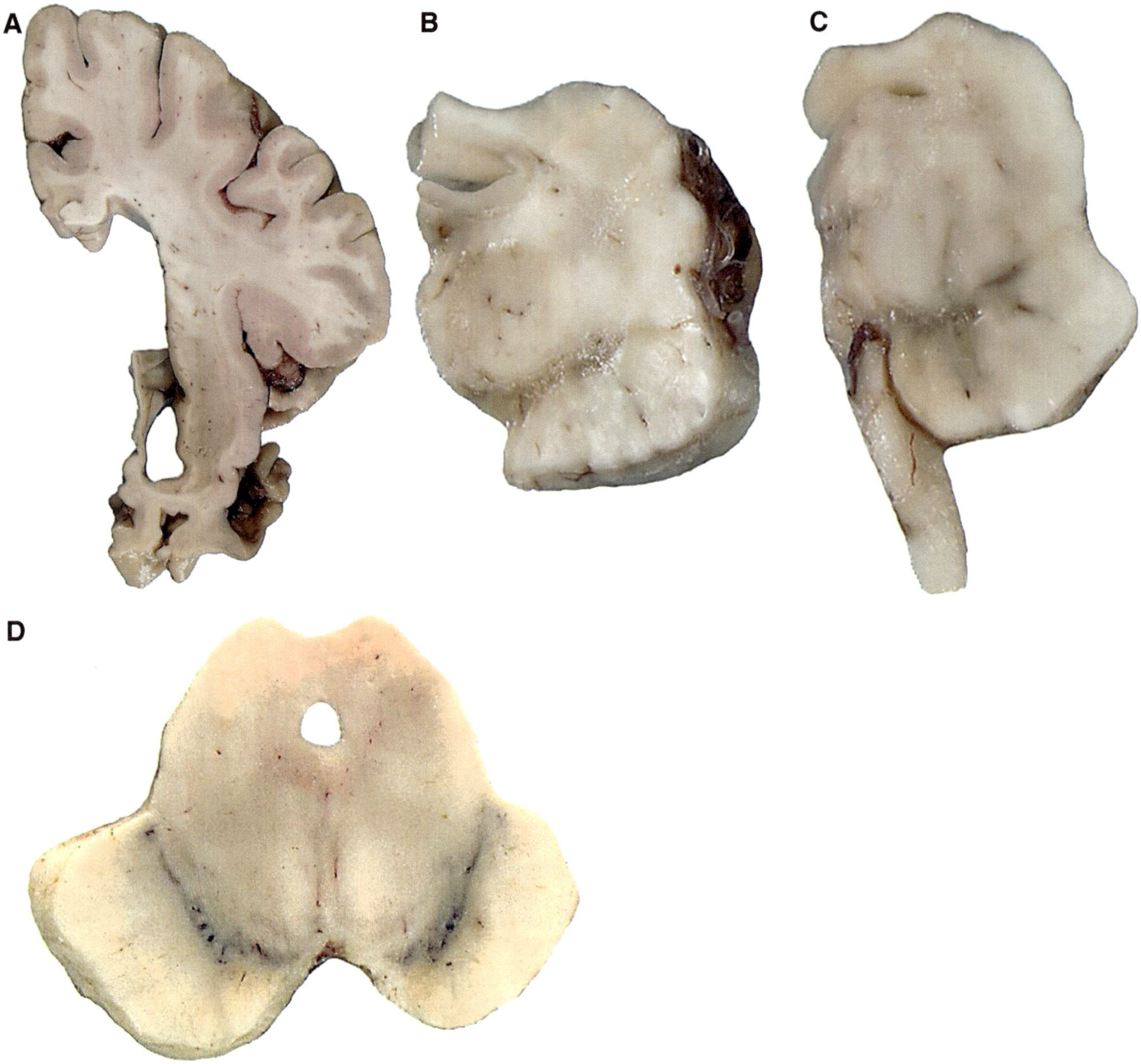

Fig. 4.32 (**A**) Frontal coronal section of the right hemisphere of a patient with FTLD-FUS, severe atrophy of the corpus callosum and a dilated ventricular system, in (**B**) and (**C**) pale substantia nigra. Compare with normal substantia nigra in (**D**). Collection A Rozemuller and LabPON Twente

ties of FUS seems to lead to this aggregation in line with other proteinopathies.

4.3.1.1.6 Amyotrophic lateral sclerosis (ALS)

The term "motor neuron disease" (MND) is generally accepted as a synonym for amyotrophic lateral sclerosis (ALS) and related diseases. In the USA, the disease is also known as Lou Gehrig's disease, named after a Henry Louis Gehrig (1903–1941), a famous professional baseball first baseman who suffered and died from ALS.

ALS is the most frequent cause of motor neuron disease, affecting upper and/or lower motor neurons, respectively, located in the motor gyrus and in the nucleus of CNXII (bulbar ALS) (see Sect. 4.3.1.1) and anterior horn of the spinal cord (see Fig. 3.5) (Ince et al. 1998; Foster and Salajegheh 2019). In about 20% of cases, ALS begins with bulbar symptoms (dysarthria and dysphagia), progressing to neurogenic wasting of the tongue, and palatal paralysis, often termed progressive bulbar palsy (Ince et al. 2015). The incidence rates are about 2 per 100,000 (Chio et al. 2013). Several diseases can be associated with motor neuron loss which means that secondary causes of motor neuron disease must be excluded. The findings of a characteristic inclusion body in anterior horn cells, i.e., ubiquitin-filament inclusions, have facilitated the precise diagnosis (Lowe et al. 1988).

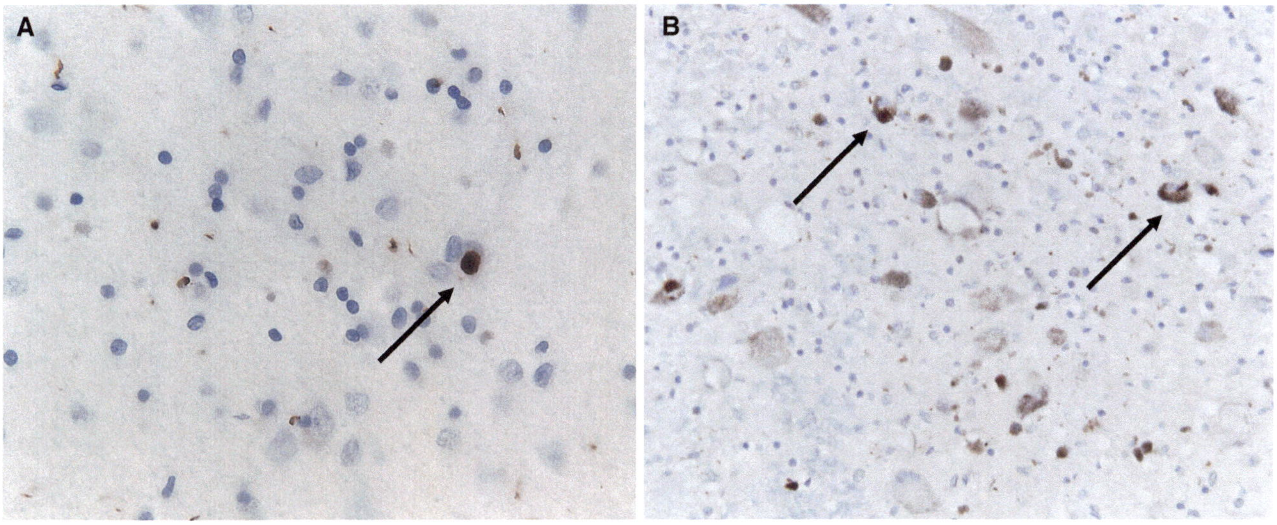

Fig. 4.33 (A) NCI (arrow) in the hippocampus. (B) Numerous NCIs (arrows) in the locus caeruleus. Anti-FUS protein immunostain. Collection A Rozemuller

Sporadic and most familial forms of ALS/MND are similar in terms of molecular pathology (Ince et al. 2015). The clinical standard for the diagnosis of ALS is the revised El Escorial World Federation of Neurology criteria, also known as the Airlie House criteria (Brooks et al. 2000). On average, it starts in the seventh decade although it can start already in the twenties, especially in familial ALS. ALS affects men and women evenly. Patients can feel weakness in one limb (80%) and have for instance difficulties with writing and raising the arm. Fasciculations, muscle atrophy or stiffness can be noted. Pain can also be present. In 20% of the patients, the disease starts with bulbar signs. Cognitive impairment is found in about 60% of the patients, while 30% have behavioral impairment. 15% met criteria for FTD (Ringholz et al. 2005; Beeldman et al. 2016). There is an overlap with frontotemporal lobar degeneration (FTLD)-TDP and FTLD-FUS (see Sect. 4.3.1.1.5). In the so-called ALS plus cases, next to FTD, also parkinsonism, supranuclear gaze paresis, autonomic insufficiency, and/or sensory loss can be found.

The duration of ALS is about 3–5 years. The diagnosis until now is based on history and physical examination. MRI and laboratory test can exclude other diagnoses. The pathogenesis is unknown but recently the role of dysregulated microRNA has been stressed (Rinchetti et al. 2018). Three genetic loci, for ALS/MND, TARDBP, FUS, and C9orf72 have provided major advances in understanding the pathology and biology of ALS/MND (motor neuron disease). The TARDBP gene encodes for the protein TDP-43, which is concerned with RNA metabolism, including regulation of RNA splicing and other mechanisms and TDP-43 has been shown to be the underlying proteinopathy associated with ubiquitylated inclusions in neurons and glia in sporadic ALS/MND and FTD (Ince et al. 2015). FUS encodes a protein that is involved in RNA metabolism (Strong and Volkening 2011). At least 10% is hereditary with an autosomal dominant trait and the most frequent gene involved is C9orf72 (about 80%). In about 10% of the familial cases, Cu2+/Zn2+ superoxide dismutase (SOD)-1 mutations are found; in about 5% mutations in TARD-DP-43 and in 5% mutations in fused in sarcoma (FUS) gene (Foster and Salajegheh 2019) (see Sect. 4.3.1.1.5).

The disease cannot be cured and symptomatic treatment is given.

Macroscopically, the brain is unremarkable although atrophy of the precentral gyrus can be seen in some cases. Frontal and/or temporal lobe atrophy may be present in cases with associated dementia. The spinal cord may be atrophic with shrunken gray anterior nerve roots when compared with the posterior sensory roots. The most important histological findings are the loss of motor neurons and astrocytosis in the anterior horns of the spinal cord, brainstem motor nuclei, and motor cortex. In the brainstem, the tracts from the motor cortex are degenerated. This is found in cerebral peduncles, pontine base, medullary pyramids, and the lateral corticospinal tracts of the spinal cord. At autopsy, the region containing the lateral corticospinal tract at the dorsolateral area of the spinal cord is gliotic and hardened. Microscopically, the CNXII nucleus is mostly affected with loss of neurons, ballooned neurons, and in most cases cytoplasmic staining for TDP (Fig. 4.34).

Not only the CNXII nucleus, but also the nucleus ambiguus (see Sect. 5.4.2.1), motor nuclei of the trigeminal nerve and facial nerve are involved. The nuclei of the CNIII, CNIV, and CNVI cranial nerves appear normal (Schröder and Reske-Nielsen 1984) although neuronal cytoplasmic inclusions have been observed (Okamoto et al. 1993). Bunina

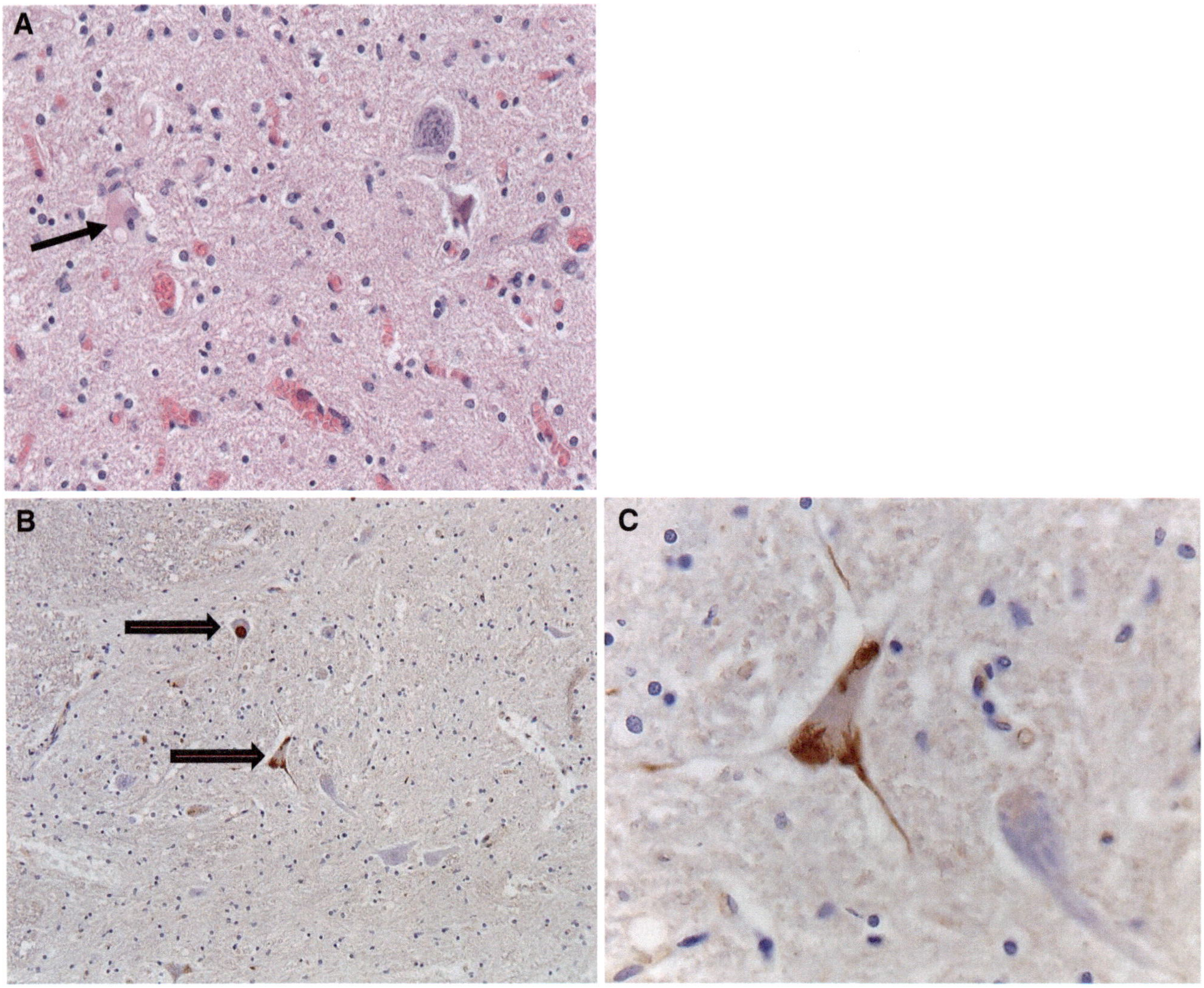

Fig. 4.34 (**A**) Ballooned neuron in the motor nucleus of CNXII (H & E stain). In (**B**) and (**C**, detail) hypoglossal nucleus stained with pTDP antibodies showing cytoplasmic TDP positivity. Collection A Rozemuller

bodies, small eosinophilic inclusions, occur in at least 85% of cases of ALS/MND and almost always ubiquitin-immunoreactive skein and/or spherical inclusions can be detected (Ince et al. 1998; Piao et al. 2003). Inclusion bodies in motor neurons, containing TDP-43, can be detected by immunohistochemical staining for ubiquitin and p62 and are said to be specific for sporadic ALS and some familial types. They can be found in both upper and lower motor neurons (Lowe et al. 1988; Lowe 1994; Neumann et al. 2006). In 5% nuclear and cytoplasmic staining for FET proteins (including Fused in Sarcoma, Ewing Sarcoma protein and TAF-15 protein) is found. In ALS plus cases, the substantia nigra, locus caeruleus, and the medulla can also be involved. In ALS, the white matter of the spinal cord shows myelin loss from the corticospinal tracts associated with astrocytic gliosis and accumulation of microglia (Ince et al. 2003). Myelin loss from the spinocerebellar tracts and posterior columns can be seen in up to 50% of sporadic cases and in familial cases. The myelin loss can be demonstrated by Luxol fast blue and the Marchi method for degraded myelin, but it is best demonstrated by using immunohistochemistry for the microglial lysosomal marker CD68 (Ince et al. 2003).

4.3.2 Branchial motor nuclei r10

4.3.2.1 Retroambiguus nucleus

Detailed information on this nucleus is available under Sect. 3.3.1.1.

4.3.3 Raphe nuclei r10 [Nuclei raphes] (ἡ ῥαφή, he raphe, Greek = seam)

For a general survey of the raphe nuclei, see Sect. 3.2.1.3 "Anatomical classification of brainstem serotonergic nuclei."

The raphe nuclei are a group of several serotonergic—exception raphe interpositus nucleus (see Sect. 9.2.4.1)—brainstem nuclei extending from the rostral hindbrain to the junction of the brainstem with the spinal cord. They are located close to the midline ("seam") and are composed of rather small neurons which render their unequivocal identification in routine histological specimens difficult.

Two groups of serotonergic neurons can be differentiated, a rostral and a caudal one. The first is located in the mesencephalon and the rostral pons (projections to the forebrain) and the second one extends from the caudal pons to the caudal medulla oblongata (projections to the caudal brainstem and the spinal cord (Hornung 2003, for details see Sect. 3.2.1.3).

A comprehensive visualization is possible using serotonin antibodies as immunohistochemical markers (cf. Morcinek et al. 2013).

Central nervous serotonin is involved in several behavioral and physiological processes by interacting with at least 14 different postsynaptic receptor subtypes. Serotonin-(5-hydroxytryptamine, 5HT) containing neurons are present in the raphe nuclei intermingled with non-serotonergic neurons (cf. Deneris 2011).

In mice, upon tracer injection into the cerebellum, labeled cells were found in the dorsal raphe, the median raphe, the paramedian raphe, the raphe magnus, the pontine raphe, the raphe interpositus, the raphe pallidus, and the raphe obscurus nuclei. No labeled cells were found in the caudal linear nucleus of the raphe (Fu et al. 2011).

4.3.3.1 Raphe obscurus nucleus r10 [Ncl. raphes obscurus; B1 group/Cellulae serotonergicae B1]

The raphe obscurus nucleus is described here below together with the raphe pallidus nucleus.

4.3.3.2 Raphe pallidus nucleus r10 [Ncl. raphes pallidus; B2 group/Cellulae serotonergicae B2]

These two nuclei are the caudal most serotonergic cell groups in the mammalian brain. They extend longitudinally from the region just cranial of the pyramidal decussation [Decussatio pyramidum] to the level of the posterior (dorsal) nucleus of vagus nerve (see Sect. 7.3.2.1). The raphe pallidus (pallidus Latin = pale) is located ventrally of the raphe obscurus nucleus (obscurus Latin = dark) (Fig. 4.35). The designations are not really helpful in identifying and distinguishing both nuclei.

The raphe obscurus nucleus neurons are located near the midline in the dorsal half of the medulla oblongata along its entire extent (Fig. 4.35 ①) (Hornung 2003, 2012, see also for data on peptide expression).

The smallest group of serotonergic neurons among the raphe nuclei is contained in the raphe pallidus nucleus (Fig. 4.35 ②). Its rostrocaudal extension corresponds to the inferior olive (Hornung 2003).

The location of the raphe pallidus nucleus by Olszewski and Baxter (1954) has been mistaken for that of the raphe magnus nucleus (for details, see Hornung 2012).

As known from animal studies, the efferent projections of the raphe obscurus and raphe pallidus nuclei project to the spinal and to visceral and somatic motor nuclei of the lateral reticular formation (for details, see Hornung 2003, Šimić et al. 2017).

Afferent fibers reach the rostral raphe pallidus nucleus from the hypothalamus, the periaqueductal gray (see Sect. 13.2.7.1), the amygdala, the bed nucleus of the stria terminalis and the medullary reticular formation (for details, see Hornung 2003). In addition, raphe pallidus and raphe obscurus receive input from visceral sensory afferents.

Most of the raphe obscurus neurons are large multipolar cells, larger than those of the raphe magnus nucleus, which can be used as a criterion to delineate both nuclei. The perikarya and the proximal dendrites are filled with lipofuscin granula (see Box 4.4). Besides the large neurons, medium-sized cells with tightly packed lipofuscin granula and small neurons with only few lipofuscin granula are found.

Box 4.4 Lipofuscin

Lipofuscin consists of highly oxidized cross-linked macromolecules. It can neither be degraded nor cleared by exocytosis (Moreno-García et al. 2018). Therefore, it accumulates within the lysosomes and cytoplasm of long-live post-mitotic cells like neurons. Lipofuscin is autofluorescent with a maximum of emission spectra around 578 nm for excitation at 364 nm (Moreno-García et al. 2018).

Lipofuscin may interfere with immunofluorescence signals used to mark antibodies for immunohistology (see Pyon et al. 2019).

Comparative studies in primates and other mammals give rise to the conclusion that descending serotonergic fibers to the spinal cord originate in the raphe obscurus nucleus and the raphe magnus nucleus (see Sect. 7.3.3.2).

In mice, because of their close topographical relationship, both nuclei are sometimes difficult to separate (VanderHorst and Ulfhake 2006). Both nuclei project to the spinal cord with a preferential innervation of the ventral horns (VanderHorst and Ulfhake 2006).

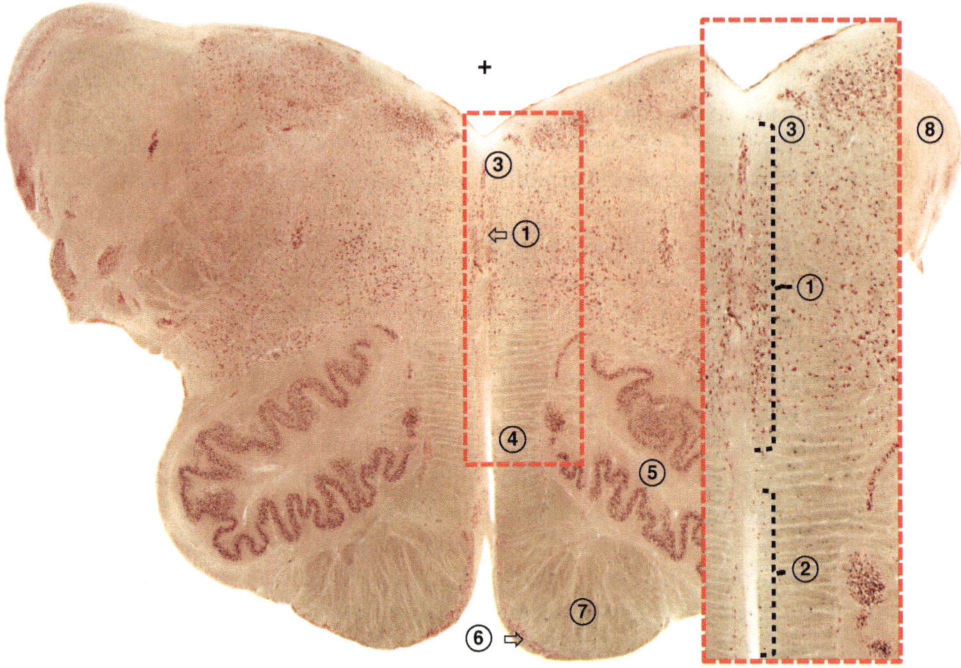

①	Raphe obscurus ncl.	⑤	Inferior olivary complex
②	Raphe pallidus ncl.	⑥	Arcuate ncl.
③	Medial longitudinal fasciculus	⑦	Corticospinal tract / Pyramis
④	Medial lemniscus	⑧	Posterior cochlear ncl. CNVIII
		+	4th ventricle

Fig. 4.35 Horizonal section of the medulla oblongata at the level of the posterior cochlear nucleus. Darrow red staining. The raphe obscurus nucleus ① is located in the tegmentum at either side of the midline, the pallidus nucleus ② bilaterally at the level of the ventral inferior olive ⑤. It comprises only a very small number of neurons. See also atlas part Darrow red 14, 14 A. LabPON Twente

4.3.3.3 Clinical implications of the raphe nuclei

As shown in a detailed study by Rüb et al. (2000), the raphe nuclei show stage-dependent pathology in Alzheimer's disease. Implications depend on the location and connectivity of the individual raphe nuclei (see there).

For the raphe pallidus nucleus, Šimić et al. (2017) claim that its lesion should result in a loss of pain-controlling input from the periaqueductal gray (see Sect. 13.2.7.1), to the spinal cord and in changes of excitability of cranial and spinal motoneurons. The same authors postulate for lesions of the raphe obscurus nucleus loss of regulation of the sympathetic system, in particular regarding the cardiovascular system and changes in the excitability of cranial and spinal motoneurons.

4.3.4 Reticular nuclei r10

4.3.4.1 Intermediate reticular nucleus

For detailed information on this nucleus, see Sect. 3.3.2.1.

4.3.4.2 Gigantocellular reticular nucleus r10 / Gigantocellular nucleus [Ncl. gigantocellularis]

The human gigantocellular reticular nucleus is located lateral of the medial longitudinal fasciculus (Fig. 4.36) and flanked laterally by the intermediate reticular zone (r11). The gigantocellular nucleus is part of the medial reticular formation (see Box 4.5).

Box 4.5 The medial reticular formation

The medial reticular formation, as defined by Nieuwenhuys et al. 1988, consists of six overlapping structures:

1. The ventral nucleus [Ncl. ventralis]

   This nucleus—caudal medulla oblongata—corresponds to the ventral subnucleus of the central nucleus of the medulla oblongata acc. to Olszewski and Baxter.

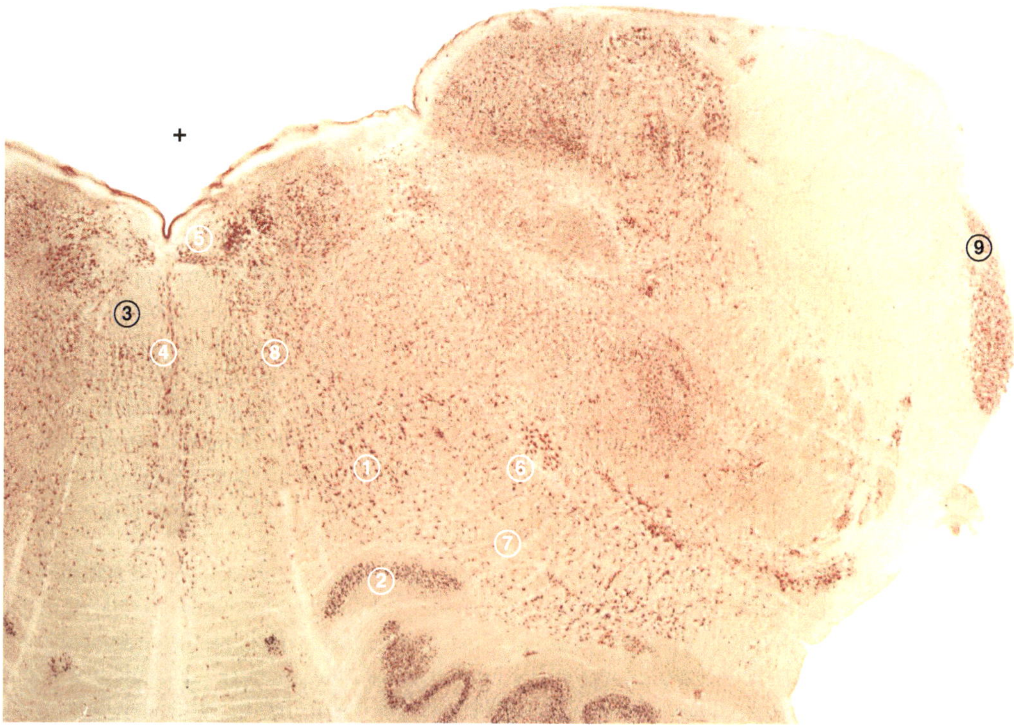

①	Gigantocellular reticular ncl.	⑥	Ncl. ambiguus
②	Posterior accessory olivary ncl.	⑦	Lateral paragigantocellular ncl.
③	Medial longitudinal fasciculus	⑧	Posterior paragigantocellular reticular ncl.
④	Raphe obscurus ncl.	⑨	Posterior cochlear ncl.
⑤	Ncl. of hypoglossal nerve	+	4th ventricle

Fig. 4.36 Horizontal section through the human medulla oblongata at the level of the gigantocellular reticular nucleus ①. Darrow red staining. As the name says, the gigantocellular reticular nucleus is composed mainly of large neurons. It is located dorsal of the posterior accessory olivary nucleus ②. Its dorsal neighbor is the posterior paragigantocellular reticular nucleus ⑧ (see Sect. 5.4.4.4). The prominent structure at its lateral side is the ncl. ambiguus ⑥. See also atlas part Darrow red 13. LabPON Twente

2. The gigantocellular nucleus (rhombomere 5–rhombomere 10) [Ncl. gigantocellularis]

   The gigantocellular nucleus lies in the rostral medulla oblongata and is characterized by large neuronal perikarya (for details, see Sect. 4.3.4.2.1 here below)
3. The caudal pontine reticular nucleus (see Sect. 7.3.5.3 rhombomere 7) [Ncl. reticularis pontis caudalis]
4. The oral pontine reticular nucleus (see Sect. 13.4.1 rhombomere 1, Sect. 10.3.5.4 rhombomere 4) [Ncl. reticularis pontis oralis]

   As a kind of prolongation of the gigantocellular nucleus, these nuclei appear at the level of the motor trigeminal nucleus (see Sect. 11.3.1).

   According to ten Donkelaar (2011), parts of these nuclei correspond to the horizontal gaze center, also known as the pontine paramedial reticular formation (PPRF). The PPRF includes most elements of the brainstem burst generator involved in the generation of horizontal saccades.

   In mice, the oral and caudal part of the pontine reticular nucleus gives rise to few ipsilateral and contralateral cerebellopetal fibers (Fu et al. 2011).
5. The (mesencephalic) cuneiform nucleus (isthmus)

   This nucleus (see Sect. 14.4.1.1) with mainly small, densely packed cells together with the pedunculopontine nucleus (Sect. 13.1.1.1.1) and the mesencephalic subcuneiform nucleus (see here below 6.) are the main areas of the mesencephalic locomotor region.

6. The mesencephalic sub(pre)cuneiform nucleus (isthmus; Sects. 14.4.1.1 and 15.4.4) is less densely packed and lies ventrolateral of the cuneiform nucleus (see also FIPAT Ch 1 endnote 63).

The details concerning the medial reticular nuclei 3 through 6 will be dealt with in the individual chapters (for cross references, see here above).

4.3.4.2.1 Location and cytoarchitecture of the human gigantocellular reticular nucleus

This nucleus exends from midolivary levels into the caudal pons and displays giant cells in the dorsal and large cells in its ventral parts (ten Donkelaar 2011). The ventral part is also known as the pars alpha of the gigantocellular nucleus (Meesen and Olszewski 1949). At the level of the motor trigeminal nucleus (see Sect. 11.3.1.1), the gigantocellular nucleus is replaced cranially by the caudal and oral pontine reticular nuclei (see Sects. 7.3.5.3 and 10.3.5.4).

The gigantocellular nucleus is located mainly in the medulla oblongata while its oral pole is extending into the lower pons, located ventrally to the caudal pontine reticular nucleus [Ncl. reticularis pontis caudalis] (see Sect. 7.3.5.3). The nucleus is bordered caudally by the central reticular nucleus (formerly central nucleus of medulla oblongata, part of the medial reticular formation, see Box 4.5 here above), [Ncl. reticularis centralis], and medially by the raphe nuclei. Its ventral neighbors are the medial lemniscus (rostrally) and the inferior olive (caudally). Ventrolaterally, it is bordered by the lateral paragigantocellular nucleus [Ncl. paragigantocellularis lateralis] and dorsolaterally by the parvocellular reticular nucleus [Ncl. reticularis parvocellularis].

Note that, although both last mentioned nuclei are reticular nuclei, the FIPAT terminology does not provide the adjunct "reticular" to the lateral gigantocellular nucleus.

4.3.4.2.2 Connectivity of the human gigantocellular reticular nucleus

Connectivity data of the human gigantocellular nucleus are scarce but ten Donkelaar (2011) summarizes an interesting case report by Mehler (1962, 1966), which is important in multiple aspects for brainstem tracts and their targets.

A patient underwent a bilateral cordotomy—surgical transection of the spinothalamic system (anterolateral fasciculus)—in case of otherwise untractable pain, which he unfortunately survived for only 12 days. During the postoperative period, he was free of pain and a sensory loss to approximately cervical level C6. After his death, the whole brainstem and the diencephalon were examined neuropathologically, using the Nauta-Gygax (Giolli 1965) silver technique which reveals degenerating fiber tracts.

As expected, the spinal anterolateral fasciculus with the spinothalamic tracts (see Fig. 3.14 B) showed massive degeneration. On their way craniad, the degenerating fibers took a more dorsolateral position as the anterolateral tract (see Fig. 3.14 A–G).

The pontine portion of the gigantocellular nucleus harbors the so-called alpha part (pars alpha [Pars alpha]) (Meesen and Olszewski 1949), which forms a kind of cap over the raphe magnus nucleus and is bordered laterally by the central tegmental tract (see above Sect. 4.1.1.3.2, see Fig. 3.14 B–G). The alpha part of the nucleus, together with the raphe magnus nucleus, forms the serotonergic B3 cell group of the brainstem. The gigantocellular nucleus has also been shown to contain cholinergic neurons (Mizukawa et al. 1986).

The first medullary targets were the lateral reticular nucleus (see Sect. 4.1.1.1) and the region medial of it. The fibers then coursed through this nucleus and terminated in the gigantocellular nucleus (and paramedian reticular nuclei, see Sect. 5.4.3.1). The input into the gigantocellular nucleus together with terminals to the raphe pallidus (see Sect. 4.3.3.2) and to facial subnuclei (see Sect. 8.2.1) was seen at the level of the facial nucleus. The fibers passed the mesencephalon and eventually reached the ventral posterolateral nucleus (posterior ventrolateral) of thalamus (see Fig. 3.14 A–G).

Postmortem tract tracing in human brains has revealed that the gigantocellular nucleus receives input from the solitary nucleus (see Sect. 3.2.4) (Ruggiero et al. 2000).

In the rat, tracing studies have revealed connections to the diencephalon, in particular, the intralaminar and midline nuclei of the thalamus. Furthermore, efferent fibers reach Forel's field, the zona incerta, the hypothalamus, and the mammillary nuclei complex, parts of the facial nucleus, the locus caeruleus (LC), nucleus subcoeruleus, the mesencephalic and pontine raphe nuclei (see above), the red nucleus (RN), and other reticular nuclei. Descending fibers are projecting diffusely to all levels of the spinal cord including autonomic motoneurons (Martin et al. 1990).

Tracing studies in mice have shown the origin of descending (spinal cord) and ascending fibers (midbrain) from cell clusters in the gigantocellular nucleus (Martin et al. 2011). The gigantocellular nucleus projects bilaterally to the cerebellum (Fu et al. 2011).

Based on the connectivity data, the gigantocellular nucleus has been supposed to be involved in autonomic, nociceptive function, and/or arousal. Furthermore, investigations in the cat have revealed the dorsal part of the gigantocellular nucleus to play a role in the voluntary motor system (axial movements, posture, orientation), and the ventral part to participate in the emotional motor system (global effect on overall level of activity of motoneurons, preparing animals to act on emotional cues) (Martin et al. 2011). In macaque monkeys, input into the gigantocellular nucleus from the corticobulbar tract (see Sect. 4.3.1.1.3) (Fregosi et al. 2018) has been shown.

http://braininfo.rprc.washington.edu/centraldirectory.aspx?ID=727

4.3.4.2.3 Pathology of the human gigantocellular reticular nucleus

The gigantocellular reticular nucleus is afflicted in the classical neurodegenerative diseases, Alzheimer's (AD) (see Sect. 13.2.2.1.4) and Parkinson's disease (PD) (see Sect. 16.6.1.5). In later stages of AD, neurofibrillary changes are seen in the nucleus.

In PD, Lewy bodies (LB) and Lewy neurites can be found in the reticular formation and in particular in the gigantocellular nucleus. The LBs of the brainstem show a classical morphology comprising an eosinophilic hyaline core and a pale staining peripheral halo (see Sect. 16.6.1.5, see Fig. 16.14).

4.3.4.3 Rostral ventral respiratory group

For detailed information on this nucleus, see Sect. 3.3.2.2.

4.3.4.4 Central cervical nucleus of the spinal cord

For detailed information on this nucleus, see Sect. 3.3.2.4.

References

Aguzzi A, Heikenwalder M et al (2007) Insights into prion strains and neurotoxicity. Nat Rev Mol Cell Biol 8:552–561

Arai T, Hasegawa M et al (2006) TDP-43 is a component of ubiquitin-positive tau-negative inclusions in frontotemporal lobar degeneration and amyotrophic lateral sclerosis. Biochem Biophys Res Commun 351:602–611

Baizer JS (2014) Unique features of the human brainstem and cerebellum. Front Hum Neurosci 8:202

Baizer JS, Sherwood CC et al (2011) Neurochemical and structural organization of the principal nucleus of the inferior olive in the human. Anat Rec (Hoboken) 294:1198–1216

Baizer JS, Wong KM et al (2018) Individual variability in the structural properties of neurons in the human inferior olive. Brain Struct Funct 223:1667–1681

Baker M, Mackenzie IR et al (2006) Mutations in progranulin cause tau-negative frontotemporal dementia linked to chromosome 17. Nature 442:916–919

Beeldman E, Raaphorst J et al (2016) The cognitive profile of ALS: a systematic review and meta-analysis update. J Neurol Neurosurg Psychiatry 87:611–619

Bird TD, Nochlin D et al (1999) A clinical pathological comparison of three families with frontotemporal dementia and identical mutations in the tau gene (P301L). Brain 122:741–756

Braak H, Braak E (1998) Involvement of precerebellar nuclei in Pick's disease. Exp Neurol 153:351–365

Brendel M, Barthel H et al (2020) Assessment of 18F-PI-r2620 as a biomarker in progressive supranuclear palsy. JAMA Neurol 77:1408–1419

Brooks BR, Miller RG et al (2000) El Escorial revisited: revised criteria for the diagnosis of amyotrophic lateral sclerosis. Amyotroph Lateral Scler Other Motor Neuron Disord 1:293–299

Cairns NJ, Neumann M et al (2007) TDP-43 in familial and sporadic frontotemporal lobar degeneration with ubiquitin inclusions. Am J Pathol 171:227–240

Chaumont J, Guyon N et al (2013) Clusters of cerebellar Purkinje cells control their afferent climbing fiber discharge. Proc Natl Acad Sci U S A 110:16223–16228

Chen AL, Riley DE et al (2010) The disturbance of gaze in progressive supranuclear palsy: implications for pathogenesis. Front Neurol 1:1–19

Chio A, Logroscino G et al (2013) Global epidemiology of amyotrophic lateral sclerosis: a systematic review of the published literature. Neuroepidemiology 41:118–130

Crary JF, Trojanowski JQ et al (2014) Primary age-related tauopathy (PART): a common pathology associated with human aging. Acta Neuropathol 128:755–766

Cruts M, Gijselinck I et al (2006) Null mutations in progranulin cause ubiquitin-positive frontotemporal dementia linked to chromosome 17q21. Nature 442:920–924

DeJesus-Hernandez M, Mackenzie IR et al (2011) Expanded GGGGCC hexanucleotide repeat in noncoding. Region of C9ORF72 causes chromosome 9p-linked FTD and ALS. Neuron 72:245–256

Delacourte A, Sergeant N et al (1998) Vulnerable neuronal subsets in Alzheimer's and Pick's disease are distinguished by their tau isoform distribution and phosphorylation. Ann Neurol 43:193–204

Deneris ES (2011) Molecular genetics of mouse serotonin neurons across the lifespan. Neuroscience 197:17–27

Dickson DW (1999) Neuropathologic differentiation of progressive supranuclear palsy and corticobasal degeneration. J Neurol 246(Suppl2):S6–S15

Dickson DW (2003) Neurodegeneration: the molecular pathology of dementia and movement disorders. ISN Neuropath Press, Basel

Dickson DW (2007) Progressive supranuclear palsy: pathology and genetics. Brain Pathol 17:74–82

Feany MB, Dickson DW (1996) Neurodegenerative disorders with extensive tau pathology: a comparative study and review. Ann Neurol 40:139–148

Filiano JJ, Kinney HC (1992) Arcuate nucleus hypoplasia in the sudden infant death syndrome. J Neuropathol Exp Neurol 51:394–403

Flament S, Dealcourte A et al (1991) Abnormal tau proteins in progressive supranuclear palsy. Similarities and differences with the neurofibrillary degeneration of the Alzheimer type. Acta Neuropathol 81:591–596

Foster LA, Salajegheh MK (2019) Motor neuron disease: pathophysiology, diagnosis, and management. Am J Med 132:32–37

Fregosi M, Contestabile A et al (2018) Changes of motor corticobulbar projections following different lesion types affecting the central nervous system in adult macaque monkeys. Eur J Neurosci 48:2050–2070

Fu YH, Watson C (2012) The arcuate nucleus of the C57BL/6J mouse hindbrain is a displaced part of the inferior olive. Brain Behav Evol 79:191–204

Fu Y, Tvrdik P et al (2011) Precerebellar cell groups in the hindbrain of the mouse defined by retrograde tracing and correlated with cumulative Wnt1-Cre genetic labeling. Cerebellum 10:570–584

Galliano E, Baratella M et al (2013) Anatomical investigation of potential contacts between climbing fibers and cerebellar Golgi cells in the mouse. Front Neural Circuits 7:59

Gearing M, Olson DA et al (1994) Progressive supranuclear palsy: neuropathologic and clinical heterogeneity. Neurology 44:1015–1024

Giolli R (1965) A note on the chemical mechanism of the Nauta-Gygax technique. J Histochem Cytochem 13:206–210

Golbe LI (2014) Progressive supranuclear palsy. Semin Neurol 34:151–159

Heckman MG, Brennan RR et al (2019) Association of MAPT subhaplotypes with risk of progressive supranuclear palsy and severity of tau pathology. JAMA Neurol 7:710–717
Helmchen C, Rambold H et al (1998) Deficits in vertical and torsional eye movements after uni- and bilateral muscimol inactivation of the interstitial nucleus of Cajal of the alert monkey. Exp Brain Res 119:436–452
Höglinger GU, Melhem NM et al (2011) Identification of common variants influencing risk of the tauopathy progressive supranuclear palsy. Nat Genet 43:699–705
Holm IE, Englund E et al (2007) A reassessment of the neuropathology of frontotemporal dementia linked to chromosome 3. J Neuropathol Exp Neurol 66:884–891
Holmes G, Stewart TG (1908) On the connections of the inferior olives with the cerebellum in man. Brain 31:125–137
Hornung J-P (2003) The human raphe nuclei and the serotonergic system. J Chem Neuroanat 26:331–343
Hornung J-P (2012) Raphe nuclei. In: Mai JK, Paxinos G (eds). The human nervous system, 3rd edn. Elsevier. pp 401–424
Huggenberger S, Moser N et al (2019) Neuroanatomie des Menschen. Springer
Ikeda K, Akiyama H et al (1995) Thorn-shaped astrocytes: possibly secondarily induced tau-positive glial fibrillary tangles. Acta Neuropathol 90:620–625
Ince PG, Lowe J et al (1998) Amyotrophic lateral sclerosis: current issues in classification, pathogenesis and molecular pathology. Neuropathol Appl Neurobiol 24:104–117
Ince P, Evans J et al (2003) Corticospinal tract degeneration in the progressive muscular atrophy variant of ALS. Neurology 60:1252–1258
Ince P, Higlhey JR et al (2015) Chapter 14. Motor neuron disorders. In: Greenfield's neuropathology, 9th edn. CRC Press, Boca Raton
Ishizawa K, Lin WL et al (2000) A qualitative and quantitative study of grumose degeneration in progressive supranuclear palsy. J Neuropathol Exp Neurol 59:513–524
Kikuchi H, Doh-ura K et al (1999) Preferential neurodegeneration in the cervical spinal cord of progressive supranuclear palsy. Acta Neuropathol 97:577–584
Komori T (1999) Tau-positive glial inclusions in progressive supranuclear palsy, corticobasal degeneration and Pick's disease. Brain Pathol 9:663–679
Kovacs GG (2015) Neuropathology of tauopathies: principles and practice. Neuropathol Appl Neurobiol 41:3–23
Kuypers HGJM (1958) Corticobulbar connections to the pons and lower brain stem in man. Brain 81:364–388
Kwiatkowski TJ Jr, Bosco DA et al (2009) Mutations in the FUS/TLS gene on chromosome 16 cause familial amyotrophic lateral sclerosis. Science 323:1205–1208
Lantos PL (1994) The neuropathology of supranuclear palsy. J Neural Transm Suppl 42:137–152
Lee EB, Porta S et al (2017) Expansion of the classification of FTLD-TDP: distinct pathology associated with rapidly progressive frontotemporal degeneration. Acta Neuropathol 134:65–78
Litvan I, Hauw JJ et al (1996) Validity and reliability of the preliminary NINDS neuropathologic criteria for progressive supranuclear palsy and related disorders. J Neuropathol Exp Neurol 55:97–105
Lowe J (1994) New pathological findings in amyotrophic lateral sclerosis. J Neurol Sci 124(Suppl):38–51
Lowe J, Lennox G et al (1988) A filamentous inclusion body within anterior horn neurones in motor neurone disease defined by immunocytochemical localisation of ubiquitin. Neurosci Lett 94:203–210
Mackenzie IRA, Hudson LP (1995) Achromatic neurons in the cortex of progressive supranuclear palsy. Acta Neuropathol 90:615–619
Mackenzie IR, Neumann M (2020) Subcortical TDP-43 pathology patterns validate cortical FTLD-TDP subtypes and demonstrate unique aspects of C9orf72 mutation cases. Acta Neuropathol 139:83–98
Mackenzie IR, Baker M et al (2006) The neuropathology of frontotemporal lobar degeneration caused by mutations in the progranulin gene. Brain 129(Pt 11):3081–3090
Mackenzie IR, Neumann M et al (2011) A harmonized classification system for FTLD-TDP pathology. Acta Neuropathol 122:111–113
Martin GF, Holstege G et al (1990) Reticular formation of the pons and medulla. In: Paxinos G (ed) The human nervous system. Academic, San Diego, pp 203–220
Martin EM, Devidze N et al (2011) Molecular and neuroanatomical characterization of single neurons in the mouse medullary gigantocellular reticular nucleus. J Comp Neurol 519:2574–2593
Meesen H, Olszewski J (1949) A cytoarchitectonic atlas of the rhombencephalon of the rabbit. Karger, Basel
Mehler WR (1962) The anatomy of the so-called "pain tract" in man: an analysis of the course and distribution of the ascending fibres of the fasciculus anterolateralis. In: French JD, Porter RW (eds) Basic research in paraplegia. Thomas, Springfield, pp 26–55
Mehler WR (1966) The posterior thalamic region in man. Confin Neurol 27:18–29
Mikhail Y, Ahmed YY (1975) Outline of the arcuate nucleus in the human medulla oblongata. Acta Anat (Basel) 92:285–291
Mizukawa K, McGeer PL et al (1986) The cholinergic system of the human hindbrain studied by choline acetyltransferase immunohistochemistry and acetylcholinesterase histochemistry. Brain Res 379:39–55
Morcinek K, Köhler C et al (2013) Pattern of tau hyperphosphorylation and neurotransmitter markers in the brainstem of senescent tau filament forming transgenic mice. Brain Res 1497:73–84
Moreno-García A, Kun A et al (2018) An overview of the role of lipofuscin in age-related neurodegeneration. Front Neurosci 12:464
Neumann M, Sampathu DM et al (2006) Ubiquitinated TDP-43 in frontotemporal lobar degeneration and amyotrophic lateral sclerosis. Science 314:130–133
Neumann M, Bentmann E et al (2011) FET proteins TAF15 and EWS are selective markers that distinguish FTLD with FUS pathology from amyotrophic lateral sclerosis with FUS mutations. Brain 134:2595–2609
Nieuwenhuys R, Voogd J et al (1988) The human central nervous system, 3rd edn. Springer, Heidelberg
Okamoto K, Hirai S et al (1993) Oculomotor nuclear pathology in amyotrophic lateral sclerosis. Acta Neuropathol 85:458–462
Olson MOJ, Dundr M (2015) Nucleolus: structure and function. In: eLS. Wiley, Chichester
Olszewski J, Baxter D (1954) Cytoarchitecture of the human brain stem. Karger, Basel
Olszewski J, Baxter D (1982) Cytoarchitecture of the human brain stem, 2nd edn. Karger, Basel
Paradiso B, Ferrero S et al (2018) Variability of the medullary arcuate nucleus in humans. Brain Behav 8:e01133
Piao Y-S, Wakabayashi K et al (2003) Neuropathology with clinical correlations of sporadic amyotrophic lateral sclerosis: 102 autopsy cases examined between 1962 and 2000. Brain Pathol 13:10–22
Pottier C, Ravenscroft TA et al (2016) Genetics of FTLD: overview and what else we can expect from genetic studies. J Neurochem 138(Suppl):32–53
Powers RE, O'Connor DT et al (1990) Noradrenergic innervation of human inferior olivary complex. Brain Res 523:151–155
Puladi B, Dinekov M et al (2021) The relation between tau pathology and granulovacuolar degeneration of neurons. Neurobiol Dis 147:105–138
Pyon WS, Gray DT et al (2019) An alternative to dye-based approaches to remove background autofluorescence from primate brain tissue. Front Neuroanat 13:73
Renton AE, Majounie E et al (2011) A Hexanucleotide repeat expansion in 9ORF72 is the cause of chromosome 9p21-linked ALS-FTD. Neuron 72:257–268

Richardson JC, Steele J et al (1963) Supranuclear ophthalmoplegia, pseudobulbar paralysis, nuchal dystonia and dementia. A clinical report on eight cases of "heterogenous system degeneration". Trans Am Neurol Assoc 88:25–29

Rinchetti P, Rizzuti M et al (2018) MicroRNA metabolism and dysregulation in amyotrophic lateral sclerosis. Mol Neurobiol 55:2617–2630

Ringholz GM, Appel SH et al (2005) Prevalence and patterns of cognitive impairment in sporadic ALS. Neurology 65:586–590

Rüb U, Del Tredici K et al (2000) The evolution of Alzheimer's disease-related cytoskeletal pathology in the human raphe nuclei. Neuropathol Appl Neurobiol 26:553–567

Rüb U, del Tredici K et al (2002) Progressive supranuclear palsy: neuronal and glial cytoskeletal pathology in the higher order processing autonomic nuclei of the lower brainstem. Neuropathol Appl Neurobiol 28:12–22

Ruggiero DA, Underwood MD et al (2000) The human nucleus of the solitary tract: visceral pathways revealed with an "in vitro" postmortem tracing method. J Auton Nerv Syst 79:181–190

Saigusa H, Tanuma K et al (2006) Nerve fiber analysis for the lingual nerve of the human adult subjects. Surg Radiol Anat 28:59–65

Sanders I, Mu L (2013) A three-dimensional atlas of human tongue muscles. Anat Rec (Hoboken) 296:1102–1114

Savonenko AV, Melnikova T et al (2015) Biological basis of neurological and psychiatric disorders. In: Zigmond MJ, Rowland LP, Coyle JT (eds.) Neurobiology of brain disorders. pp 321–338

Schoen JHR (1969) The corticofugal projection on the brain stem and spinal cord in man. Psychiatr Neurol Neurochir 72:121–128

Schröder HD, Reske-Nielsen E (1984) Preservation of the nucleus X-pelvic floor motosystem in amyotrophic lateral sclerosis. Clin Neuropathol 3:210–216

Schröder H, Moser N et al (2020) Neuroanatomy of the mouse. Springer

Seidel K, Mahlke J et al (2015) The brainstem pathologies of Parkinson's disease and dementia with Lewy bodies. Brain Pathol 25:121–135

Šimić G, Babić Leko M et al (2017) Monoaminergic neuropathology in Alzheimer's disease. Prog Neurobiol 151:101–138

Steele JC, Richardson JC et al (1964) Progressive supranuclear palsy: a heterogeneous degeneration involving the brain stem, basal ganglia and cerebellum with vertical gaze and pseudobulbar palsy, nuchal dystonia and dementia. Arch Neurol 10:333–359

Strong MJ, Volkening K (2011) TDP-43 and FUS/TLS: sending a complex message about messenger RNA in amyotrophic lateral sclerosis? FEBS J 278:3569–3577

Taniguchi D, Takanashi M et al (2020) Astrocytic 3-repeat tau pathologies in progressive supranuclear palsy. J Neuropathol Exp Neurol 79:1015–1018

ten Donkelaar HJ (2011) Clinical neuroanatomy: brain circuitry and its disorders. Springer

Touré G, Vacher C (2006) Anatomic study of tongue architecture based on fetal histological sections. Surg Radiol Anat 28:547–552

Van Swieten JC, Bronner IF et al (2007) The DeltaK280 mutation in MAP tau favors exon 10 skipping in vivo. J Neuropathol Exp Neurol 66:17–25

Vance C, Rogelj B et al (2009) Mutations in FUS, an RNA processing protein, cause familial amyotrophic lateral sclerosis type 6. Science 323:1208–1211

VanderHorst VG, Ulfhake B (2006) The organization of the brainstem and spinal cord of the mouse: relationships between monoaminergic, cholinergic, and spinal projection systems. J Chem Neuroanat 31:2–36

Verny M, Duyckaerts C et al (1996) The significance of cortical pathology in progressive supranuclear palsy. Clinico-pathological data in 10 cases. Brain 119:1123–1136

Voogd J (2004) Cerebellum and precerebellar nuclei. In: Paxinos G, Mai JK (eds) The human nervous system, 2nd edn. Elsevier, Amsterdam, pp 321–392

Watson C, Bartholomaeus C et al (2019) Time for radical changes in brain stem nomenclature: applying the lessons from developmental gene patterns. Front Neuroanat 13:10

Williams DR, Lees AJ (2009) Progressive supranuclear palsy: clinicopathological concepts and diagnostic challenges. Lancet Neurol 8:270–279

Williams DR, Holton JL et al (2007) Pathological tau burden and distribution distinguishes progressive supranuclear plasy-parkinsonism from Richardson's syndrome. Brain 130:1566–1576

Yamada T, Calne DB et al (1993) Further observations on tau-positive glia in the brains with progressive supranuclear palsy. Acta Neuropathol 85:308–315

Yen SH, Hutton M et al (1999) Fibrillogenesis of tau: insights from tau missense mutations in FTDP-17. Brain Pathol 9:695–705

Web Links

https://geekymedics.com/the-hypoglossal-nerve-cn-xii/#
http://braininfo.rprc.washington.edu/centraldirectory.aspx?ID=727
http://braininfo.rprc.washington.edu/centraldirectory.aspx?ID=1187
https://www.nlm.nih.gov/research/visible/visible_human.html
https://www.nobelprize.org/prizes/medicine/1906/cajal/biographical/

Rhombomere 9 r9

5

Contents

H. Schröder et al., *The Human Brainstem*, https://doi.org/10.1007/978-3-030-89980-6_5

Abstract

Cranial nerve nuclei appearing here, in rhombomere 9, for the first time are the spinal (inferior) vestibular nucleus, belonging to a group of nuclei involved in the transmission of signals from the organ of equilibrium in the inner ear, as well as the nucleus ambiguus, providing branchiomotor input to the glossopharyngeal nerve (CNIX). The Area postrema is a roof plate derivative, a chemosensitive, vomiting center. The rhombic lip gives rise to the accessory cuneate nucleus at this level.

In addition to the raphe nuclei described in Chap. 4, here, the paramedian reticular nucleus appears which is affected in spinocerebellar ataxia (SCA) 2 and 3 (see Chap. 7). In the group of the reticular nuclei, the already described nuclei are complemented by the dorsal paragigantocellular nucleus.

In terms of pathology, this chapter deals with vascular hamartomas like cavernous angiomas of which one is shown paradigmatically located in the spinal vestibular nucleus (Sect. 5.3.1.1).

5.1 Roof plate r9

5.1.1 Area postrema r9 [Area postrema] (postremus, -a, -um [Latin] = backmost)

The Area postrema (Fig. 5.1) belongs to the group of circumventricular organs, CNS structures located near the ventricles lacking the blood–brain barrier. Their neurons are in direct contact with compounds in the blood (cf. McKinley et al. 2003; see Box 5.1).

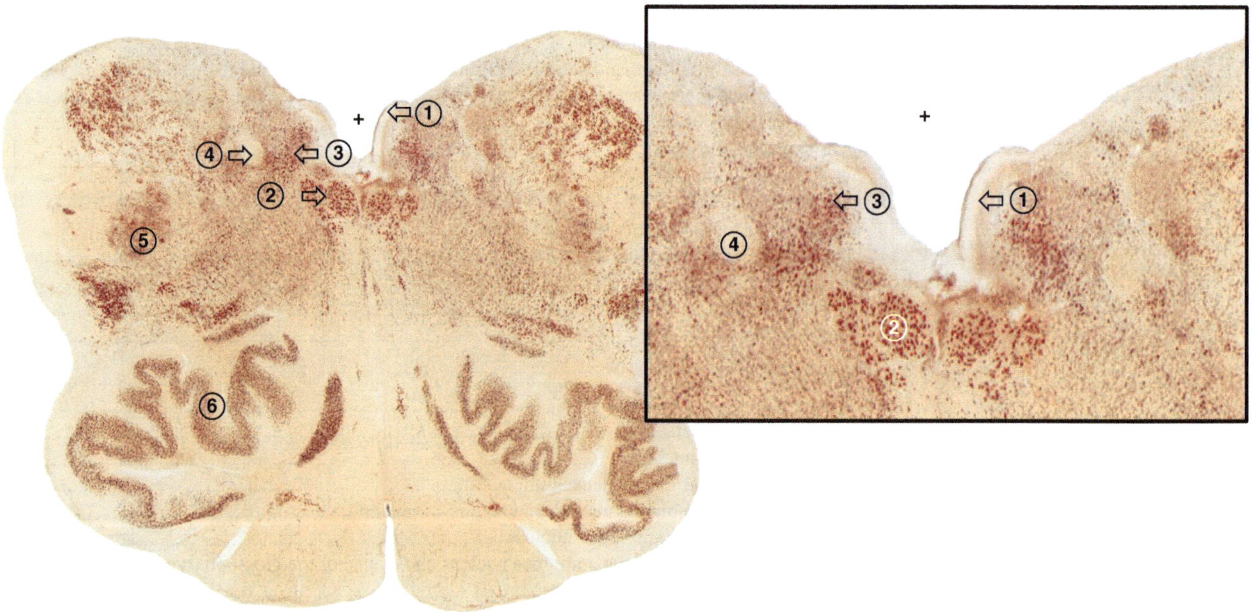

①	Area postrema	⑤	Spinal ncl. of trigeminal nerve
②	Ncl. of hypoglossal nerve	⑥	Inferior olivary complex
③	Posterior ncl. of vagus nerve	+	4th ventricle
④	Solitary tract		

Fig. 5.1 Horizontal section through the human medulla oblongata at the level of the paired Area postrema (AP) ①. The AP is located directly under the ventricular ependyma, neighbored ventrolaterally by the posterior nucleus of vagus nerve ③ (see Sect. 7.3.2.1) and the solitary tract ④ (see Sect. 3.2.4) ("dorsal vagal complex," for details, see text). LabPON Twente

The human Area postrema (AP) is a small, highly vascularized (Lindstrom and Brizzee 1962; Duvernoy et al. 1972) region, which caudally starts at the region of the obex (see Figs. 1.9 and 1.10) as a midline structure. Further cranially, where the central canal continues into the fourth ventricle (Fig. 5.1), it becomes V-shaped with one limb at either side in the bottom of the fourth ventricle.

An elegant way to visualize the human AP is via ink injection which visualizes the AP clearly due to its high content of vessels (Duvernoy et al. 1972). A role for the Area postrema as chemoreceptor trigger zone for vomiting has been suggested for decades (Miller and Leslie 1994). It belongs to the group of circumventricular organs (CVO) (Box 5.1), ultrastructurally characterized by the presence of fenestrated capillaries. Functionally, this means a lack of the blood–brain barrier (BBB) and that therefore certain molecules, among them most likely emetic toxins, can reach the AP. The term dorsal vagal complex has been coined for the triad of AP, solitary nucleus (see Sect. 3.2.4) and posterior nucleus of vagus nerve (see Sect. 7.3.2.1), all of which—known from animal studies—receive vagal afferent fibers from the inner organs (for details, see Sect. 3.2.4). AP also receives visceral afferent input from the glossopharyngeal (see Sect. 5.4.2.1.3) and the vagus nerves (see Sect. 7.3.2.1) (Low 2016) and is targeted by several hypothalamic nuclei. Efferent projections of the AP target the NTS (see Sect. 3.2.4), the ventral lateral medulla, and the parabrachial nucleus (see Sect. 13.2.3). Consistent with its role as a sensory organ, the AP has receptors for several neuropeptides, including glucagon-like peptide-I and amylin (Low 2016), and contains chemosensory neurons including osmoreceptors. The AP has also been suggested to be involved in the control of blood pressure. AP lesions in rats blunt the rise in blood pressure induced by angiotensin II (Low 2016).

Box 5.1 Circumventricular organs (CVO)

Max Lewandowsky (1900) (German neurologist, 1876–1918) coined the term "Blut-Hirn-Schranke" (blood–brain barrier, BBB) based on his observation that central nervous—but not peripheral—injections of sodium ferrocyanide caused seizures in dogs. Edwin Goldmann (1913) (German surgeon, 1862–1913), upon intravenous injection of trypan blue in rabbits, found the CNS unstained while the majority of inner organs stained blue. The only exceptions were those parts of the CNS counting among what today we call CVOs. Injection of dye into the cerebral ventricles, by contrast, resulted in a blue staining of brain and spinal cord, while the extracerebral organs lacked coloration.

Today, it is known that the most important structural and functional difference between peripheral and central blood vessels is the lack of fenestrated capillaries in most of the CNS. This type of capillaries displays small gaps (windows, Latin fenestra) in the endothelial wall. This is the structural basis for the passage of rather large molecules from vessel lumen into tissue which is not possible in the CNS.

The term CVO comprises the following structures (in caudocranial order): Area postrema (brainstem), pineal gland (diencephalon), Organum vasculosum laminae terminalis (OVLT) (hypothalamus), subfornical organ (diencephalon), median eminence (hypothalamus), pituitary gland (intermediate/neural lobe; hypothalamus) and subcommissural organ (telencephalon). In all other parts of the mammalian brain, the free passage of molecules is hindered by the BBB which is complemented on the abluminal site by astrocytic processes. In the BBB-equipped regions blood-borne substances can enter the brain only by use of special transporters. Glucose, for example, the foremost energy carrier for brain metabolism (molecular mass ~ 180 g/mol), cannot cross the BBB but enters the brain via the glucose transporter (Mergenthaler et al. 2013).

The BBB is of great importance for the delivery of drugs to the brain. It is common knowledge that ethanol readily reaches the brain due to its lipophilic nature. Most therapeutically used drugs, however, are not able to enter the brain unless their physicochemical properties are adapted. Dopamine (DA), theoretically optimal to substitute its loss in Parkinson's disease (see Sect. 16.6.1.5), as a charged molecule cannot pass the BBB. The lipophilic, non-charged L-DOPA— converted in the Substantia nigra neurons into DA (see Fig. 3.3)—enables the substitution of DA in the brain.

A more anecdotal but anatomically interesting piece of evidence for the vomiting-related function of the human AP is a report by Lindstrom and Brizzee (1962) on the neurosurgical treatment of otherwise intractable vomiting. The AP was identified intraoperatively as a discrete bulge of the ventricular floor rostral to the gracile tubercle (see Fig. 1.10) and a thermal lesion applied to ablate the AP. This led to a relief from vomiting in all five treated patients and a postoperatively raised threshold of apomorphine reactions—a dopamine receptor agonist acting on the AP chemoreceptor trigger zone and used as a central emetic in the therapy of drug overdose.

Today, the AP is an important landmark for neuroendoscopy of the fourth ventricle (Longatti et al. 2015). By i.v. administration of fluorescein sodium, then using an appropriately equipped endoscope (fluorescent mode), the AP can be selectively visualized: The lacking BBB (see Box 5.1) lets the fluorescein pass into the CSF of the fourth ventricle. Endoscopically, the AP is visible as two coupled leaves on the floor of the fourth ventricle, diverging from the central canal rostrally. The leaves normally appear short and thick but show interindividual varieties (Longatti et al. 2015).

In rodents, the AP is a small midline structure of inverted triangular shape (coronal sections) located ventral to the cerebellum in the caudal parts of the fourth ventricle at the transition zone to the central canal of the spinal cord (Glattfelder et al. 2008). Most mammals display a paired AP, but rodents have a midline AP terminating in rostral direction in two limbs (see Rohrschneider et al. 1972). The mouse AP projects to the parabrachial nuclei (see Sect. 13.2.3) (Tokita et al. 2009). By contrast to man, in all rodents tested so far, including rats and mice, classical drugs, which elicit vomiting by stimulating the AP like apomorphine (see here above), did induce neither retching nor vomiting (Horn et al. 2013). This is in line with previous observations showing that rodents do not vomit. The inability to vomit is important for the efficacy of "blood thinning drugs" ("rat poison") used to eradicate rodents.

5.2 Rhombic lip r9

5.2.1 Precerebellar nuclei r9

5.2.1.1 Accessory/external cuneate nucleus [Ncl. cuneatus accessorius/externus] (cuneatus Latin = wedge-shaped)

The human accessory cuneate nucleus (Fig. 5.2) is located between the cuneate nucleus and the dorsal surface of the medulla oblongata. It has rather large neurons.

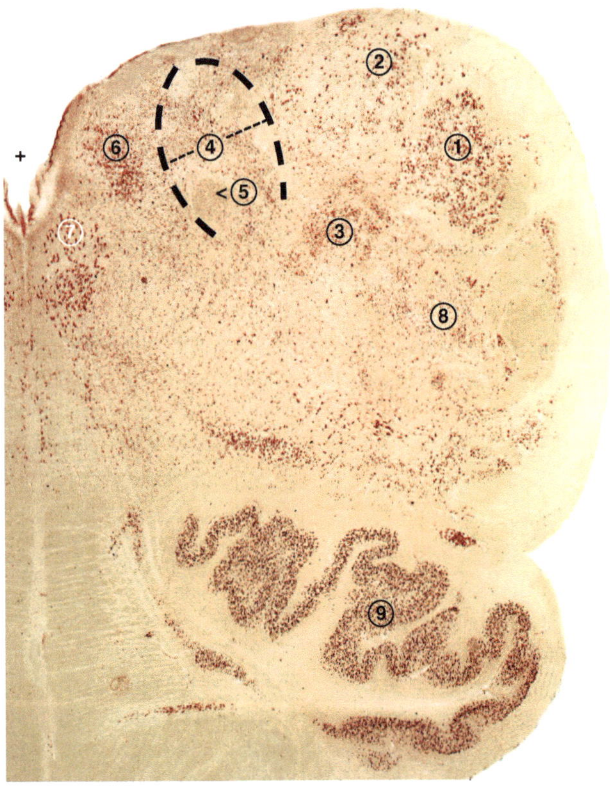

①	Accessory cuneate ncl.	⑥	Posterior ncl. of vagus nerve
②	Cuneate ncl.	⑦	Ncl. of hypoglossal nerve
③	Parvocellular reticular ncl.	⑧	Spinal ncl. of trigeminal nerve
④	Ncll. of solitary tract	⑨	Inferior olivary complex
⑤	Solitary tract	+	4th ventricle

Fig. 5.2 Horizontal section through the human medulla oblongata at the level of the external (accessory) cuneate nucleus ①. See also atlas part Darrow red 5, 5 A. LabPON Twente

The accessory cuneate nucleus is the origin of the cuneocerebellar tract which, together with fibers from the cuneate nucleus (Sect. 3.2.3.3), conveys mainly proprioceptive signals from the spinal cord to the cerebellum, as known from animal experiments (ten Donkelaar 2011).

In mice, by contrast to man, there is a rather large area bulging at the dorsal surface of the medulla oblongata lateral of the posterior nucleus of vagus nerve. Fu et al. (2011) were able to show bilateral staining of neurons in the mouse external cuneate nucleus after tracer injection (HRP = horseradish peroxidase) into the cerebellum.

Do not confuse the accessory cuneate nucleus with the cuneate nucleus (see Sect. 3.2.3.3) or the cuneiform nucleus (see Sect. 14.4.1.1).

5.2.1.2 Lateral reticular nucleus

For details on the lateral reticular nucleus, see Sect. 4.1.1.1.

5.2.1.3 Nucleus of Roller

For details on the Ncl. of Roller, see Sect. 6.2.1.6.

5.2.1.4 Intercalated nucleus of the medulla

For details about the intercalated nucleus of the medulla, see under Sect. 3.1.1.1.

5.3 Alar r9

5.3.1 Vestibular nuclei r9 [Nuclei vestibulares] (Vestibulum Latin = chamber, cavity)

The vestibular nuclei are housing the perikarya of the secondary neurons of the vestibular system which subserves the equilibrium sense. The mammalian brain displays four different vestibular nuclei (in caudocranial direction with their eponyms).

Spinal/inferior vestibular nucleus [Ncl. vestibularis inferior] (Roller) (see Fig. 5.7 ①) Christian Friedrich Wilhelm Roller (1802–1878) Psychiatrist, Achern, Germany. Please note that the subhypoglossal nucleus (see Sect. 6.2.1.6) also bears the eponym Roller.

Medial vestibular nucleus [Ncl. vestibularis medialis] (Schwalbe) (see Fig. 5.7 ②) Gustav Albert Schwalbe (1844–1916), German anatomist, University of Strasbourg, France.

Lateral vestibular nucleus [Ncl. vestibularis lateralis] (Deiters) Otto Friedrich Karl Deiters (1834–1863) German neuroanatomist, Bonn, Germany (see Sect. 10.2.2.1).

Superior vestibular nucleus [Ncl. vestibularis superior] (Bechterew) Vladimir Mikhailovich Bekhterev (Bechterew; 1857–1927), Russian neurologist, University of Petrograd/St. Petersburg, Russia (see Sect. 11.2.2.1, see Box 16.1 for the eponym problem).

The peripheral receptive sense organ of all four vestibular nuclei is the labyrinth in the inner ear. The mythical creature Minotaur was kept by King Minos of Crete in a complicated building by the name of labyrinth (Greek Λαβύρινθος, Labyrinthos). Figuratively, labyrinth means a structure in which to orientate oneself is complicated like the inner ear. The labyrinth is a complex membranous structure in which the organs of the inner ear (organ of equilibrium/vestibular organ; cochlea) are contained (Fig. 5.3).

As shown in detail in mice, the labyrinth develops from a cystic anlage in the skull, about day 11 post conception (Morsli et al. 1998). From this geometrically simple auditory vesicle a complex membranous structure develops. The vesicle (Membranous labyrinth [Labyrinthus membranaceus]) is filled with a fluid, the so-called endolymph (Endolymphatic space [Spatium endolymphaticum]). After the membranous labyrinth has formed, bone tissue develops around it (Bony labyrinth [Labyrinthus osseus]). Between bone and membranous labyrinth, a fluid-filled (perilymph) space [Spatium perilymphaticum] develops (Fig. 5.4). At the ventral end of the anlage, the cochlea (sensory organ for auditory perception) develops. At the dorsal end, it differentiates into five different structures containing the sensory organs for equilibrium: the saccule [Sacculus] (little bag, purse, Latin saccus = sack, bag) and Utricle [Utriculus] (small skin or leather bottle) and the three semicircular ducts [Ductus semicirculares] (Figs. 5.3 and 5.4).

The saccule, the utricle, and the three semicircular ducts (lateral, posterior, anterior) (Fig. 5.3) contain sensory receptors, the vestibular hair cells (VHC). In the saccule and utricle, they are located in the so-called macula region. Saccular and utricular receptors detect linear accelerations, including the constant pull of gravity. In the semicircular canals, the receptors or crests [cristae], which detect angular accelerations, are situated in the ampullae. Each of the vestibular receptors is afferently innervated by fibers of the vestibular nerve and receives also efferent innervation as shown for the cat (Gacek 1984). The ribbon synapses of VHC type I are organized as calices (calix = cup, goblet) (see Box 5.2). The perikarya of the afferent neurons are located in the vestibular ganglion [Ganglion vestibulare]. The central processes of the vestibular nerve [Nervus vestibularis] join the cochlear nerve [Nervus cochlearis] to form the vestibulocochlear nerve [Nervus vestibulocochlearis] (CNVIII) (see Figs. 5.4, 5.5, 5.6, and 5.7).

The light microscopic and ultrastructural appearance of human vestibular hair cells, obtained at the occasion of translabyrinthine surgery (see Box 5.3), has been studied by Taylor et al. (2015). Interestingly, they could show that in the

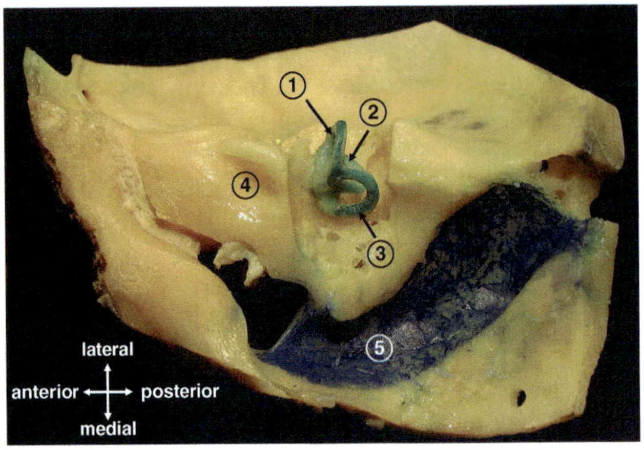

①	**Anterior semicircular duct**	④	**Internal acoustic meatus**
②	**Lateral semicircular duct**	⑤	**Sigmoid sinus**
③	**Posterior semicircular duct**		

Fig. 5.3 Dorsal view onto the dissected human labyrinth. The three semicircular ducts are shown in green. The adjacent sigmoid sinus (see Fig. 3.11) is stained blue. The three semicircular ducts contain the semicircular canals (see Fig. 5.5). The ducts describe three perpendicular arches. Each of them corresponds to two third of an arc of circle. The anterior semicircular duct stands up straight vertically, approximately perpendicular to the longitudinal axis of the petrous bone. The lateral semicircular duct is oriented horizontally and lateral most of all ducts. The posterior semicircular duct is oriented approximately vertical in a plane which is running parallel to the longitudinal axis of the petrous bone. Courtesy Prof. Eberhard Stennert, Köln

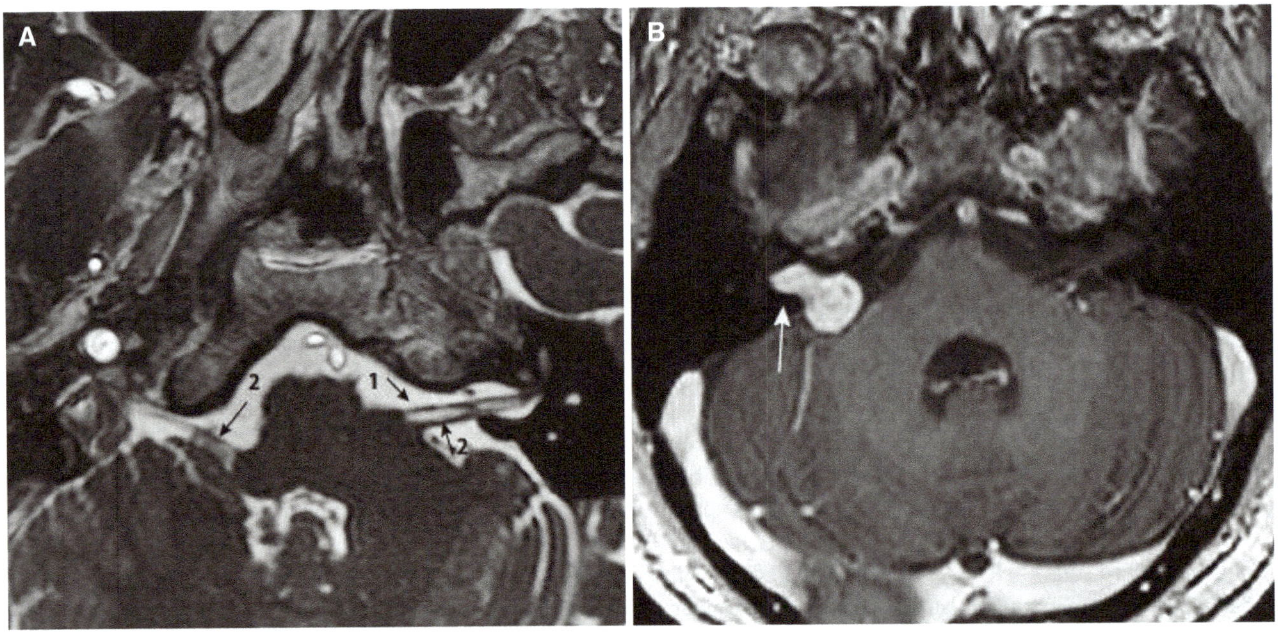

Fig. 5.4 (**A**) Normal findings of the cerebellopontine angle. (**B**) T_1-weighted horizontal MRT. Right cerebellopontine angle tumor. A hyperdense mass extends from the angle between cerebellum and pons in direction to the inner acoustic porus/meatus (arrow). (**A**) In comparison, a T_2-weighted horizontal MRT of a normal human brainstem. CNVII (1, approximate location) and CNVIII (2) are running through the CSF-filled prepontine cistern (see Box 3.3) from the cerebellopontine angle to the inner acoustic porus and meatus. Compare with Fig. 5.8. In a situation as shown in (**B**) the nerves may be compressed by the tumor. MRT does not allow for an unequivocal diagnosis of the tumor biology. The preferential neoplasm in the cerebellopontine angle is the schwannoma, a nerve sheath tumor. Although biologically benign, depending on the extension of the tumor, it may compress cranial nerves up to CNV and down to CNIX/X (see Fig. 5.8). From Huggenberger et al. 2019, Fig. 14.13 with permission

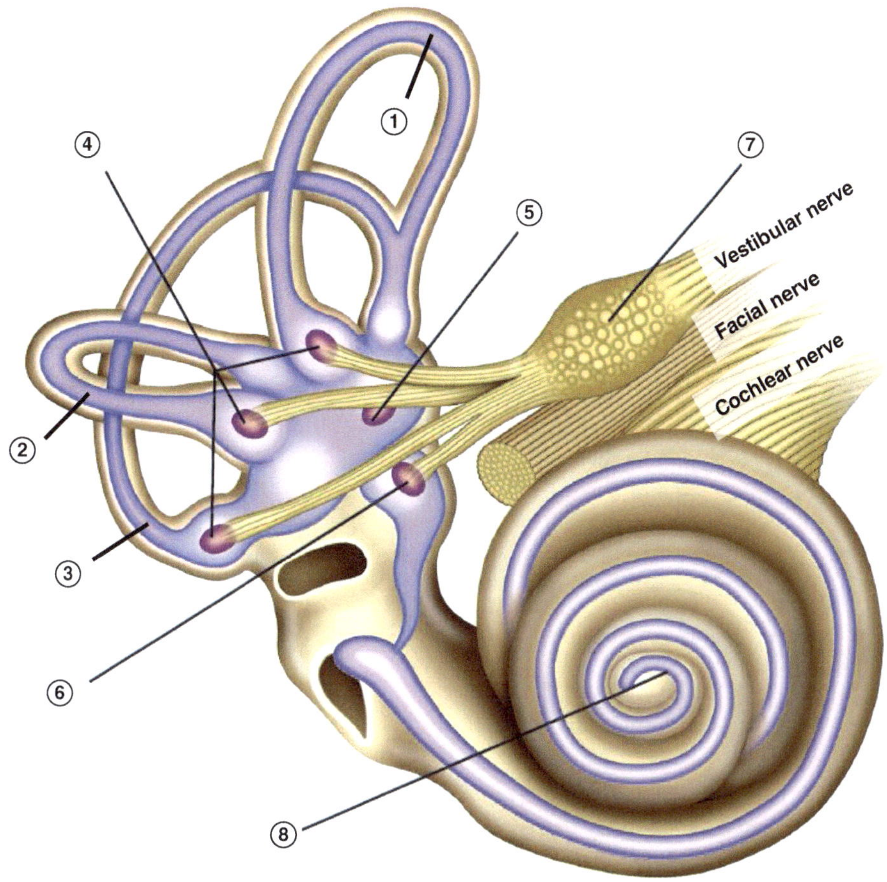

[purple]	Endolymphatic space /duct	④	Ampullar crests of the semicircular ducts
[yellow]	Perilymphatic space / canal	⑤	Macula of utricle
①	Anterior semicircular duct / canal	⑥	Macula of saccule
②	Lateral semicircular duct / canal	⑦	Vestibular ganglion
③	Posterior semicircular duct / canal	⑧	Cochlear duct [purple]

Fig. 5.5 Schematic illustration of the human labyrinth. Purple: the endolymph compartment of the labyrinth, located inside the perilymph compartment (beige), which in vivo would be followed in equal shape by the bony labyrinth. The sensory apparatus for hearing (cochlea) and that for the sense of equilibrium (semicircular ducts, saccule, utricle) are located in the endolymph space. Inside of these five sensory regions in a transmitter zone, mechanical changes elicited in the ducts, utricle and saccule are transferred via hair cells to nerve terminals of the vestibular nerve. The perikarya form the vestibular ganglion ⑦. Like spinal nerves, the central process forms the hair cell (receptor)-contacting nerve terminals, and a peripheral process, forming together with the cochlear nerve the vestibulocochlear nerve. Both run through the prepontine cistern (see Box 3.3) to the cerebellopontine angle, where they reach the brainstem, then split off to reach their respective nuclei. For topographical aspects, see Fig. 5.8. From Huggenberger et al. 2019, Fig. 15.13 with permission

oldest patients there was a significant loss of hair cells. On the other hand, indicating cell regeneration, they found immature hair bundles in older patients, a phenomenon observed in mice after stimulation of developmental pathways.

The vestibulocerebellum integrates information from the vestibular organ with visual and proprioceptive inputs, which generates an estimate of the position and movements of the head in space (Beraneck et al. 2012).

As concluded from studies in several animal species, there is also an efferent innervation of the vestibular hair cells (see Gacek 1984). The most likely candidate for a nucleus of origin is the caudal parvocellular reticular nucleus (see Sect. 6.5.5.2).

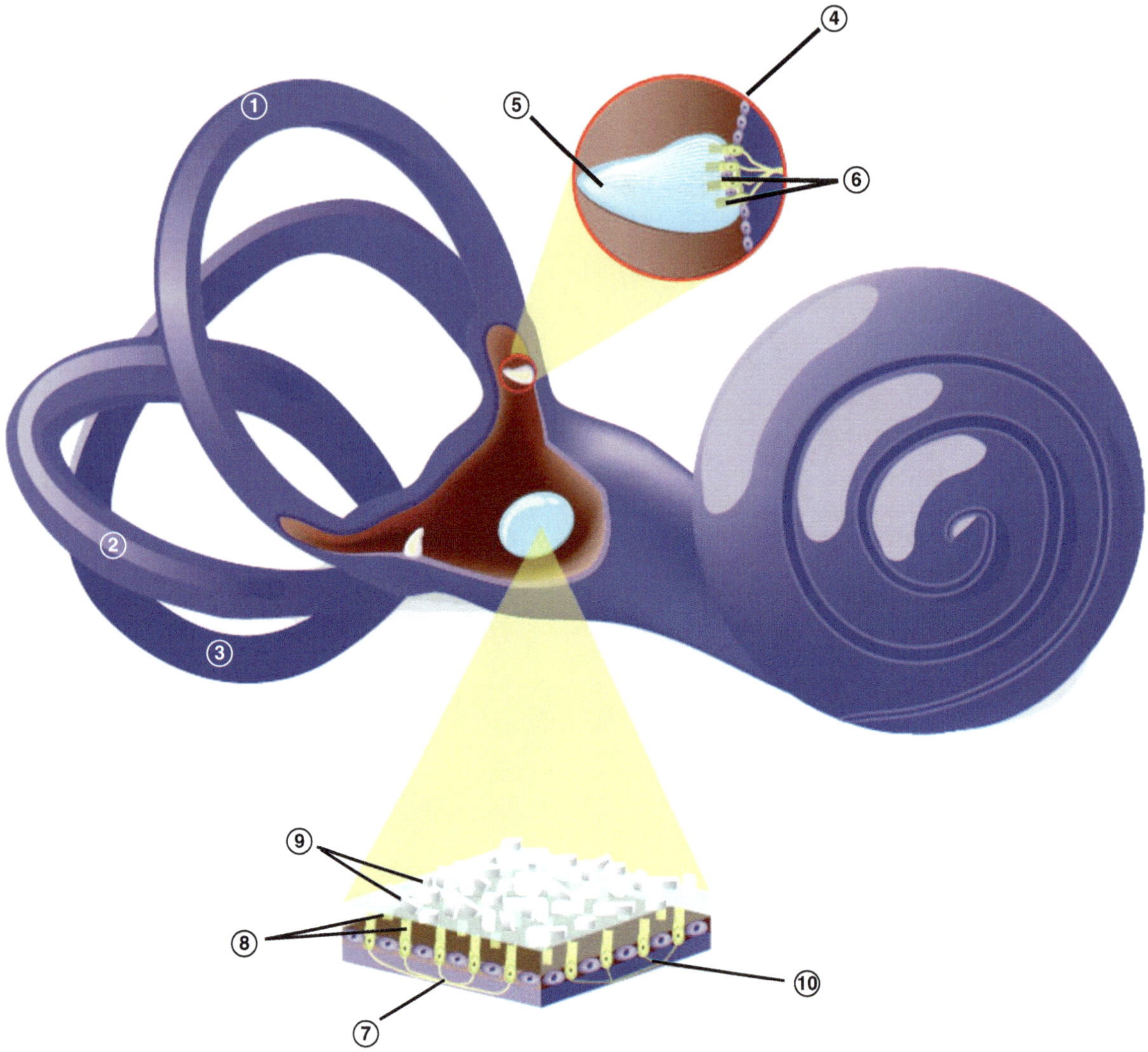

	Endolymphatic space /duct	⑥	Hair cells
①	Anterior semicircular duct	⑦	Macula of utricle
②	Lateral semicircular duct	⑧	Hair cells
③	Posterior semicircular duct	⑨	Otoliths / Statoconia
④	Ampullar crest of anterior semicircular duct	⑩	Nerve terminals
⑤	Cupula		

Fig. 5.6 Schematic illustration of the endolymph portion of the human membranous labyrinth. To the left you see the three semicircular ducts. The sense organ—contained in each of the three ducts—of the anterior duct is shown at higher magnification. The so-called ampullary crest [Crista ampullaris] contains nerve endings of the vestibular nerve which surround the sensory hair cells. Their apical processes are embedded in a kind of gelatinous substance, the cupula. The cupula spans almost the whole diameter of the duct. From Huggenberger et al. 2019, Fig. 15.14 with permission

Kong et al. (1998) have shown ChAT-immunoreactive terminals in the human vestibular apparatus which provides evidence for a cholinergic nature of efferent vestibular innervation in humans.

Box 5.2 Ribbon synapse

Sensory-transduction mechanisms, in which a synapse does not respond to action potentials but to a graded receptor potential, are characteristic for all sensory cell types having synaptic active zones equipped with a presynaptic electron-dense structure known as ribbon tethering synaptic vesicles. Also common to all ribbon synapses is the fact that they are glutamatergic, wherein glutamate is released at high and sustained rates (Safieddine and El-Amraoui 2012). Synaptic ribbons are also present in auditory hair cells (see Sect. 8.1.1.2), retinal photoreceptors and bipolar cells as well as in pinealocytes (pineal gland).

Box 5.3 Translabyrinthine surgery

An historically older approach to the cerebellopontine region—as compared with the transoccipital approach described under Sect. 3.2.2—is the translabyrinthine surgery. It is advised in case of large tumors extending toward the midline or anterior of the pons. Its big disadvantage is the sacrifice of hearing since the labyrinthine sensory organs including the cochlea (see Fig. 5.5 ⑧, see Sect. 8.1.1.2) are destroyed (for details, see Nickele et al. 2012). The transoccipital approach (see Fig. 3.12) would probably be chosen for the tumor shown above in Fig. 5.4.

In terms of function, in the semicircular ducts (see Figs. 5.3, 5.5, and 5.6), angular accelerations deflect the cupula (Fig. 5.6) and thereby the belonging hair cells. The hair cells signal the acceleration to the CNS. The activation of individual ampullar hair cells depends on the direction of acceleration in regard to the individual semicircular ducts (Curthoys 2019).

In the utricle and saccule, a second kind of sensory apparatus is present, the macula. Here—as with the crest organs—the floor of utricle and saccule is covered with nerve terminals of the vestibular nerve contacting sensory hair cells. By contrast to the static labyrinth, there is no cupola but the surface of the receptor cells is covered by an accumulation of little calcium carbonate crystals, the so-called otoliths, statoliths, or statoconia (from ancient Greek ὠτο- oto to οὖς ous = ear and ὁ λίθος lithos = stone and στατός statos = standing and ἡ κονία konia = dust).

Linear acceleration in one direction displaces the dense crystals of the otoliths of the maculae and the belonging hair cells, tuned to that direction, are deflected and activated (for details, see Curthoys 2019).

5.3.1.1 Spinal/inferior vestibular nucleus [Ncl. vestibularis inferior]

5.3.1.1.1 Location of the spinal vestibular nucleus

The inferior/spinal vestibular nucleus is the most caudal of the vestibular nuclei (Fig. 5.7 ①). It is located between the inferior cerebellar peduncle laterally and the medial vestibular nucleus medially. It accompanies the medial vestibular nucleus (see Fig. 5.7 ②) laterally for most of the extension of the latter and takes an extension of approximately 5 mm (Augustine 2017). Further rostrally, for most of its extension, it lies medially to the inferior cerebellar peduncle. The nucleus ends at the level of the caudal pole of the facial nucleus. At this location, it is continuous with the caudal pole of the lateral vestibular nucleus (see Sect. 10.2.2.1). The spinal vestibular nucleus can be distinguished by its lower neuronal density (Fig. 5.7A) and the internuclear presence of fibers from the medial vestibular nucleus (see Fig. 5.7B ③). The vertical fibers in the spinal vestibular nucleus belong to descending portions of primary vestibular, vestibulospinal and cerebellovestibular fibers (Augustine 2017). The vestibular nuclei, due to their location in the floor of the fourth ventricle, form a certain surface feature the so-called vestibular area (see Fig. 1.10 ⑦) in the floor of the rhomboid fossa.

In Fig. 5.10 (see Sect. 5.3.1.1.6 "Vascular hamartomas"), you will find the location of a brainstem cavernous angioma exactly inside the spinal vestibular nucleus.

A

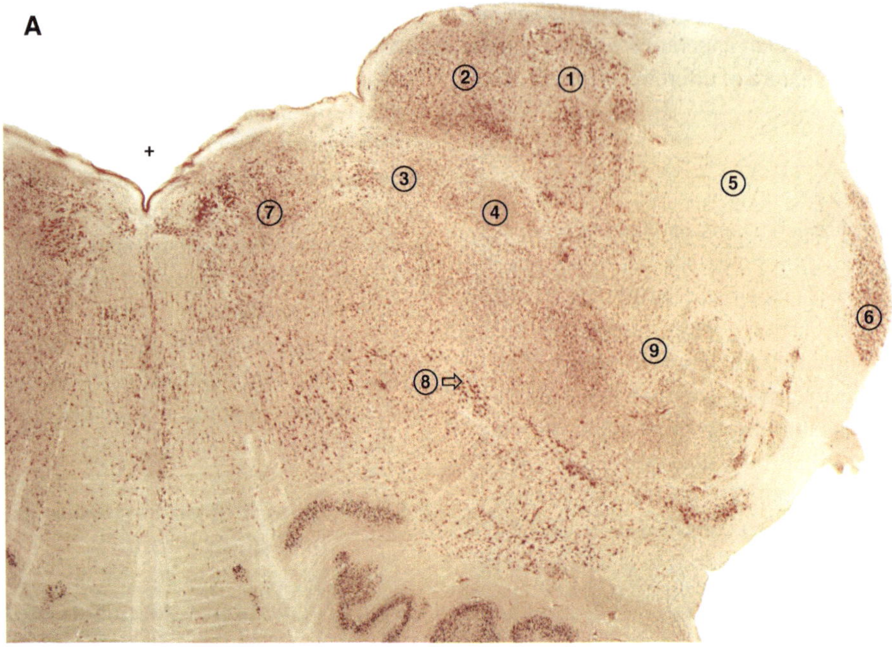

①	Inferior vestibular ncl. CNVIII	⑥	Posterior cochlear ncl. CNVIII
②	Medial vestibular ncl. CNVIII	⑦	Prepositus hypoglossi ncl.
③	Solitary ncll.	⑧	Ncl. ambiguus CNIX
④	Solitary tract	⑨	Spinal ncl. of trigeminal nerve CNV
⑤	Inferior cerebellar peduncle	+	4th ventricle

B

①	Inferior vestibular ncl. CNVIII	⑥	Medial lemniscus
②	Medial vestibular ncl. CNVIII	⑦	Glossopharyngeal nerve CNIX
③	Vestibular nerve CNVIII	⑧	Olivocerebellar tract
④	Inferior cerebellar peduncle	⑨	Spinal tract of trigeminal nerve
⑤	Medial longitudinal fasciculus	+	4th ventricle

Fig. 5.7 Horizontal section through the human medulla oblongata at the level of the anterior cochlear nucleus. (**A**) Darrow red stain. See also atlas part Darrow red 13 (**B**) Campbell fiber stain. See also atlas part Campbell 5. For details, see text. LabPON Twente

5.3.1.1.2 Target and connectivity of the spinal vestibular nucleus

The peripheral input to the spinal vestibular nucleus is from the labyrinth (see Figs. 5.3, 5.5, and 5.6) in the inner ear (petrosal part of the temporal bone) via the vestibulocochlear nerve (see Figs. 5.5 and 5.7). The bulk of knowledge on the connectivity of the human vestibular nuclei is derived from animal studies. In general, vestibular nuclei of either side are strongly interconnected by commissural projections in all vertebrates (Horn 2020).

Commissural pathways allow the comparison of activity within the vestibular nuclei on both sides. In monkeys, two massive commissural systems are present: one system in the medulla, originating from peripheral neurons in medial and dorsal/inferior vestibular nuclei. The other system travels dorsally through the cerebellum and arises mainly from the peripheral parvocellular superior vestibular nucleus (see Sect. 11.2.2.1) and ventral Y-group (Horn 2020).

The medial vestibular nucleus gives rise to the largest homotopic commissural projection system, whereas the most impressive non-corresponding projections to the medial vestibular nucleus arise from the superior vestibular nucleus (see Sect. 11.2.2.1) and the ventral Y-group. No commissural connections derive from the lateral vestibular nucleus (see Sect. 10.2.2.1, for review see Goldberg et al. 2012). It has been suggested that the peripheral neurons of the vestibular nuclei are interconnected by intrinsic neurons, which project into the magnocellular central regions representing the major output zone of the vestibular nuclei giving rise to vestibulo-ocular and vestibulospinal projections.

Ascending vestibular tracts include the vestibulomesencephalic tracts (Büttner-Ennever and Gerrits 2004): The medial vestibulomesencephalic tract from the vestibular nuclei to the oculomotor nuclei (see Sect. 15.3.1) in the medial longitudinal fasciculus (see Fig. 3.14 A–G), the lateral vestibulomesencephalic tract from the lateral vestibular nucleus (see Sect. 10.2.2.1) to the oculomotor nuclei, running just lateral of the medial longitudinal fasciculus and a ventral vestibulomesencephalic tract from the Y-group and the superior vestibular nucleus (see Sect. 11.2.2.1), which crosses in the ventral tegmentum (in or below the superior cerebellar peduncle, see Box 14.1).

A human ipsilateral vestibulothalamic tract adjacent to the medial lemniscus (see Fig. 3.14 A–G) has been shown by Zwergal et al. 2008.

5.3.1.1.3 Course of vestibulocochlear nerve

At their abluminal end, the vestibular hair cells (see Figs. 5.5 and 5.6) form afferent synapses with the peripheral processes of the vestibular ganglion cells. The pseudounipolar perikarya of these first-order sensory neurons give rise to central processes, forming the vestibular nerve [N. vestibularis] (see Fig. 5.5), which unites with the cochlear nerve [N. cochlearis] to form the vestibulocochlear nerve, exiting from the internal acoustic meatus/opening [Meatus/Porus acusticus internus]. It passes the cerebellopontine cistern and enters the brainstem in the cerebellopontine angle in close vicinity to the facial nerve (see Sect. 8.2.1) (Fig. 5.8). From there, the central processes are running to the four vestibular nuclei (see Sect. 5.3.1 above). The different contributions from the periphery and from both ganglia have a precise somatotopic arrangement in the vestibular nerve.

In mice, the spinal vestibular nucleus is the caudal most of the vestibular nuclei, located between the external cuneate nucleus laterally and the medial vestibular nucleus medially. Retrograde tracing from mouse cervical and lumbar spinal cord showed labeled neurons in the spinal and the lateral vestibular nucleus (Liang et al. 2014).

5.3.1.1.4 Living anatomy, clinical and pathological implications of the vestibular system

As to diagnostics of vestibular disorders, a reasonably complete account is way beyond the scope of this book. We will survey here some bedside tests which have been found useful to ascertain labyrinth problems in the framework of occupational medicine (Zamysłowska-Szmytke et al. 2015). Three tests simple to perform turned out to be suited:

1. Head impulse (thrust) test (HIT)
2. Dynamic visual acuity test (DVA)
3. Unterberger (stepping) test

1. The HIT allows for a clinical evaluation of the vestibulo-ocular reflex (see Box 5.4). The basis of the test is thrusting the head—30° bended forward with low amplitude and high acceleration to one side resulting in the functional elimination of the contralateral labyrinth.
2. The DVA test is performed based on the ophthalmological Snellen charts (Fig. 5.9) when the patient holds the head still and when moving the head.
3. In the Unterberger test, the patient is asked to perform stationary stepping for 1 min with their eyes closed. Under normal conditions, the patient does not perform an additional rotational movement, which occurs, however, directed to the lesioned side when a vestibular lesion is present. Take precautions for the case the patient would start to stumble during the test.

Box 5.4 Vestibulo-ocular reflex

Rotations of the head are detected by the vestibular apparatus of the membranous labyrinth. The vestibular nuclei are connected with the oculomotor nuclei (CN III, IV, VI) via the medial longitudinal fasciculus (see Fig. 3.14 A–G). The rotation-elicited signal triggers an inhibitory signal to the extraocular muscles on one side and an excitatory signal to the contralateral muscles. This results in compensatory eye movements (Sekirnjak and du Lac 2006).

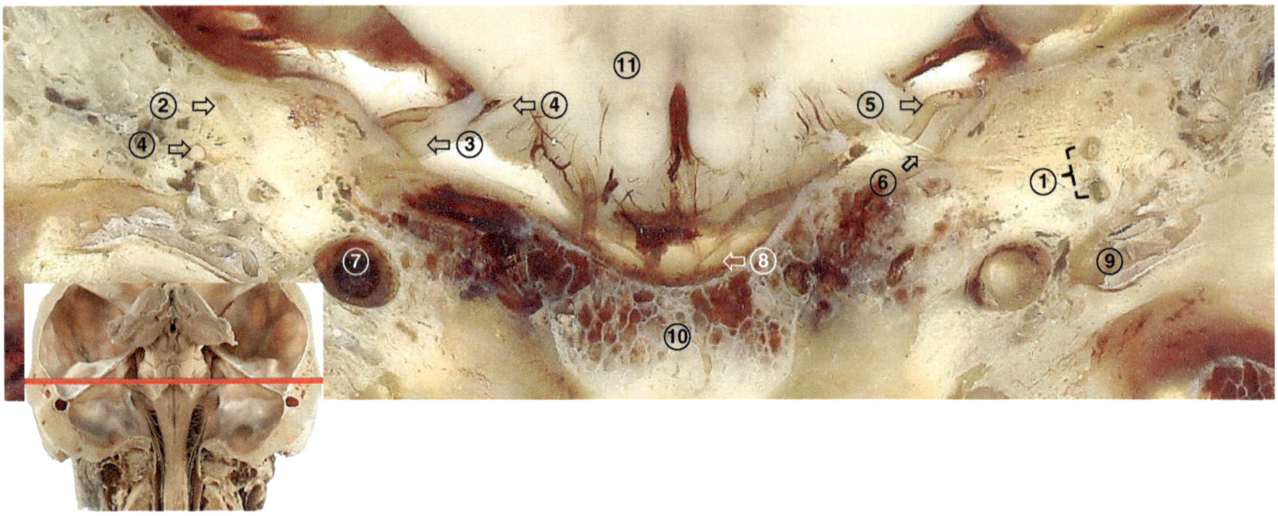

①	**Anterior semicircular canal**	⑥	**Labyrinthine artery**
②	**Posterior semicircular canal**	⑦	**Internal carotid artery**
③	**Vestibulocochlear nerve CNVIII in prepontine cistern**	⑧	**Vertebral artery**
④	**Facial nerve CNVII**	⑨	**Tympanic cavity**
⑤	**Inferior anterior cerebellar artery**	⑩	**Clivus**
		⑪	**Pons**

Fig. 5.8 Horizontal section through the human brainstem at the level of the cerebellopontine angle (see Fig. 5.4) showing parts of the labyrinthine semicircular canal/duct system ① ② and of the vestibulocochlear nerve ③. The labyrinthine artery ⑥, branch of the anterior inferior cerebellar artery ⑤ (branching cranial of the level shown here), supplies the inner ear (see Sect. 5.3.1). Note the direct neighborhood of the labyrinth to the internal carotid artery ⑦, tympanic cavity ⑨, and facial nerve (see Sect. 8.2.1) in the facial canal ④. Inset: Red line shows approximate level of the section. See also atlas part Darrow red 17, 19 and Campbell 8. Sammlung des Zentrums Anatomie der Universität zu Köln

The spinal vestibular nucleus—as well as the other vestibular nuclei (see Sect. 5.3.1, see Fig. 7.22)—is affected in the polyglutamine spinocerebellar ataxias (SCA, see Sect. 7.2.1.3) (SCA1, 2, 3, 6, and 7) (Rüb et al. 2013).

5.3.1.1.5 Arteries of the inner ear

The labyrinthine artery [A. labyrinthi], branch of the anterior inferior cerebellar artery (AICA) [A. cerebelli inferior anterior] (see Fig. 5.8), splits off in an anterior vestibular artery [A. vestibuli anterior] and the common cochlear artery [A. cochlearis communis] (Haidara et al. 2015). The anterior vestibular artery provides arterial supply to the utricle, the superior part of the saccules and the anterior and horizontal semicircular canals (Haidara et al. 2015). The cochlear artery splits again into the common cochlear artery [A. cochlearis communis]—supplies the apical parts of the cochlea—and the vestibulo-cochlear artery [A. vestibulocochlearis]. The latter gives off a posterior vestibular branch [Ramus vestibularis posterior]—inferior part of saccule, posterior semicircular canal—and the cochlear branch [Ramus cochlearis] for the basal parts of the cochlea (Haidara et al. 2015).

The infarction of the labyrinthine artery leads to sudden vertigo and deafness (Haidara et al. 2015).

5.3.1.1.6 Vascular hamartomas

- Capillary teleangiectases
- Arteriovenous malformations
- Cavernous angiomas
- Venous malformations

The group of vascular hamartomas consists of capillary teleangiectases, cavernous angiomas, arteriovenous malformations (AVMs), and venous malformations.

This classification, published by McCormick in 1966, proposed that nearly all vascular malformations involving the CNS could be classified into one of the four categories. This system is based upon macroscopic gross and microscopic pathological features. The overall incidence of cerebrovascular malformations ranges from 0.1 to 4.6% in different series (McCormick 1966; Russell and Rubinstein 1998; Challa et al. 1995).

Venous malformations are most common, followed by capillary teleangiectasia (0.8%), AVMs (0.5%) and cavern-

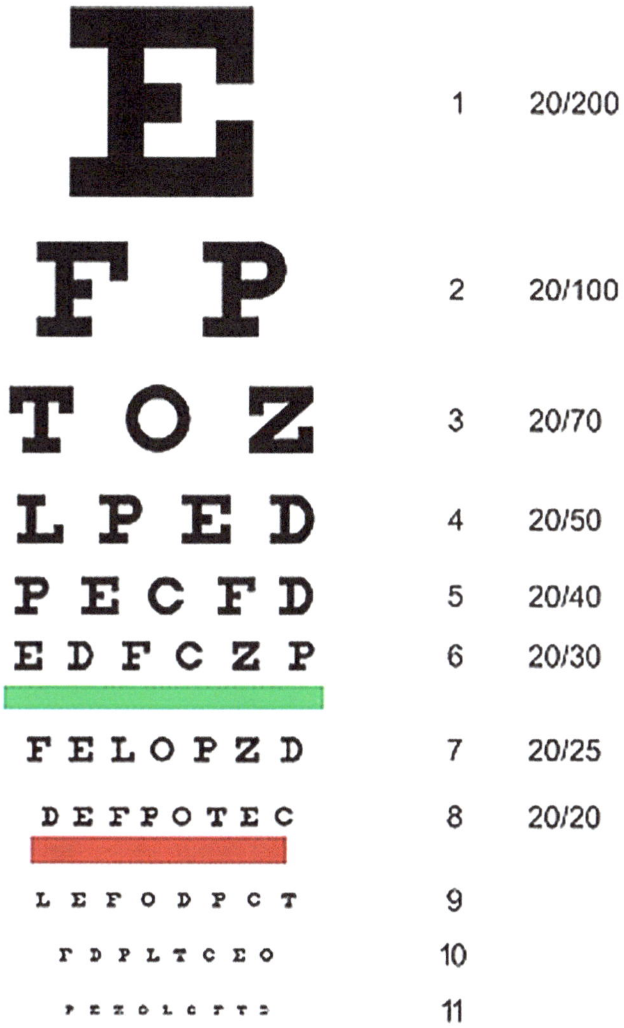

Fig. 5.9 Snellen chart developed in 1862 by the Dutch ophthalmologist Herman Snellen (1834–1908). The Snellen fraction represents the visual acuity in the form of a fraction (e.g., 6/6, 6/24, and 6/60 if measured in meters). Here, the numerator is the testing distance (6 m in a clinician's office) and the denominator is the viewing distance at which the height of the letters subtends 5 min of arc (Bailey and Jackson 2016). Graphics from Lakshminarayanan 2016, Fig. 1 with permission

ous malformations (0.3%) (Zabramski et al. 1999). It is important to realize that combined forms exist so that some lesions simply do not fit into one of the categories. Furthermore, all types are regarded as developmental malformations and not true neoplasms (McCormick 1966).

Capillary teleangiectases

Capillary teleangiectases are seldom of clinical importance and are usually chance findings at autopsy. Spontaneous hemorrhage is a rare complication (Rigamonti et al. 1991), but in the brainstem it can be rapidly fatal. Although mostly solitary, capillary teleangiectases are occasionally multiple. The pons is the most frequent site (Russell and Rubinstein 1989, 1998).

Cavernous angiomas

Cavernous angiomas or Cavernous malformations are more important than the capillary teleangiectases. They occur throughout the brain and spinal cord as well as in the meninges but are most commonly found in the subcortical white matter, external capsule, and pons (Zabramski et al. 1999). Implication of the cerebellum is rare. The incidence of cavernous angiomas is difficult to estimate because of the confusion that may exist between these lesions and other forms of cerebrovascular malformation. These lesions have been variably referred to as "cavernous angioma," "cavernomas," and "capillary hemangiomas," which makes the terminology very complex and confusing. Most lesions are observed in patients in the third to fifth decades (Garner et al. 1991). They occur in sporadic and familial forms (Gangemi et al. 1990; Giombini and Morello 1978; Zabramski et al. 1994). Solitary examples of cavernous angiomas are about three times as common as multiple ones. Multiple cavernous angiomas are more commonly associated with the familial form of the disease (Zabramski et al. 1994) (Figs. 5.10, 5.11, and 5.12).

In the familal form, a strong history of seizures is often seen (Zabramski et al. 1999). The most common symptoms in affected patients are epileptical seizures, hydrocephalus, and spontaneous hemorrhage (see Sect. 1.2.1.1.2) (Chanda and Nanda 2002; Louis and Marsh 2016), which may be recurrent or, occasionally, fatal. However, the risk of bleeding is lower than that of arteriovenous malformations (Chanda and Nanda 2002). In the brainstem such recurrent bleeding may produce cranial nerve palsies (Fritschi et al. 1994; Giombini and Morello 1978). They vary considerably in size, but most examples are of smaller size and range from 1 to 2 cm in diameter. Macroscopically, the lesions are well defined, have a lobulated appearance with a dark red or purple color, and have been described as having a "mulberry-like"appearance (Zabramski et al. 1999) (Fig. 5.11).

Microscopically, a complex of markedly dilated vascular channels with little or no intervening parenchym is seen. The walls are lined by a single layer of endothelium surrounded by a thin layer of fibrous tissue. The vascular channels closely resemble dilated capillaries (Fig. 5.12B). Recurrent episodes of thrombosis and hemorrhage lead to depostion of iron and other breakdown products of blood surrounding the malformation. No abnormal large arteries or veins are seen in the neighborhood of cavernous angiomas. Multiplicity varies from two independent foci to as many as 42 (Russell and Rubinstein 1989). They may be inherited as an autosomal dominant disorder known as familial cerebral cavernoma

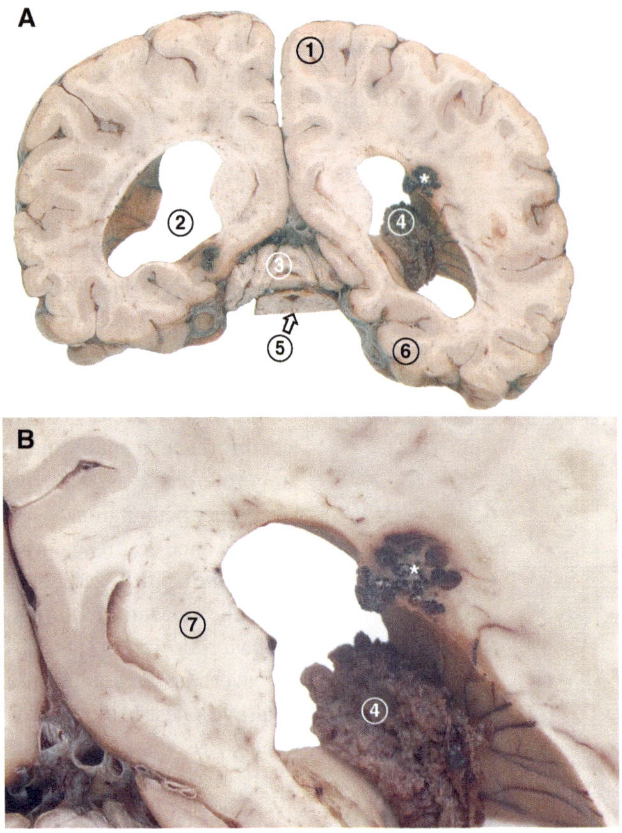

①	**Superior parietal lobule**	⑤	**Pons**
②	**Lateral ventricle**	⑥	**Fusiform gyrus**
③	**Cerebellum**	⑦	**Splenium of corpus callosum**
④	**Choroid plexus**		

Fig. 5.10 Patient with multiple familial cavernous angiomas. (**A**) Angioma (*) in the right parietal lobe, extruding into the lateral ventricle. Detail in (**B**), typical, lobulated "mulberry-like" appearance, dark bluish-red color. LabPON Twente

(FCC) (Labauge et al. 1998). With the introduction of the MRI, it has become apparent that familial cavernous angiomas or malformations are more common (Mori et al. 1996).

Arteriovenous malformations (AVMs)

AVMs are the most common clinically reported CNS malformations because of their propensity for hemorrhage. The incidence varies from 0.5 to 1% of the population. AVMs typically occur in a sporadic fashion although familial inheritance have been reported. They have been found in association with genetic disorders causing the Osler-Weber-Rendu and Wyburn-Mason syndromes (Martin and Vinters 1995; Putman et al. 1996). A well-known association exists of single or multiple intracranial berry aneurysms with cerebral arteriovenous malformations (Wilkins 1985). These malformations demonstrate fistulous communications between arteries and veins. They are second in importance only to aneurysms as a cause of spontaneous subarachnoid hemorrhage and are also a significant cause of intracerebral hemorrhage (Zabramski et al. 1999). They are found throughout the CNS and may also involve the meningeal coverings and the subarachnoidal space (Fig. 5.13).

The area supplied by the middle cerebral artery is most often involved by AVMs. The cerebellum as well as the spinal cord are rarely affected. They can become large to produce a tortuous vascular mass. Clinically, they usually present either with hemorrhage or with epilepsy. Radiographic features are typical and rarely confused with other lesions. MRI-scans of typical unruptured arteriovenous malformations show a honeycomb shape of signal void with or without areas of increased signals due to thrombosed vessels. MRIs are better in demonstrating the evolving stages of hemorrhage and thrombosis that may be encountered in AVMs (Russel and Rubinstein 1998). Macroscopically, the tangled vessels are covered by a thickened arachnoid membrane (see Fig. 5.12). The size vary greatly from several millimeters to

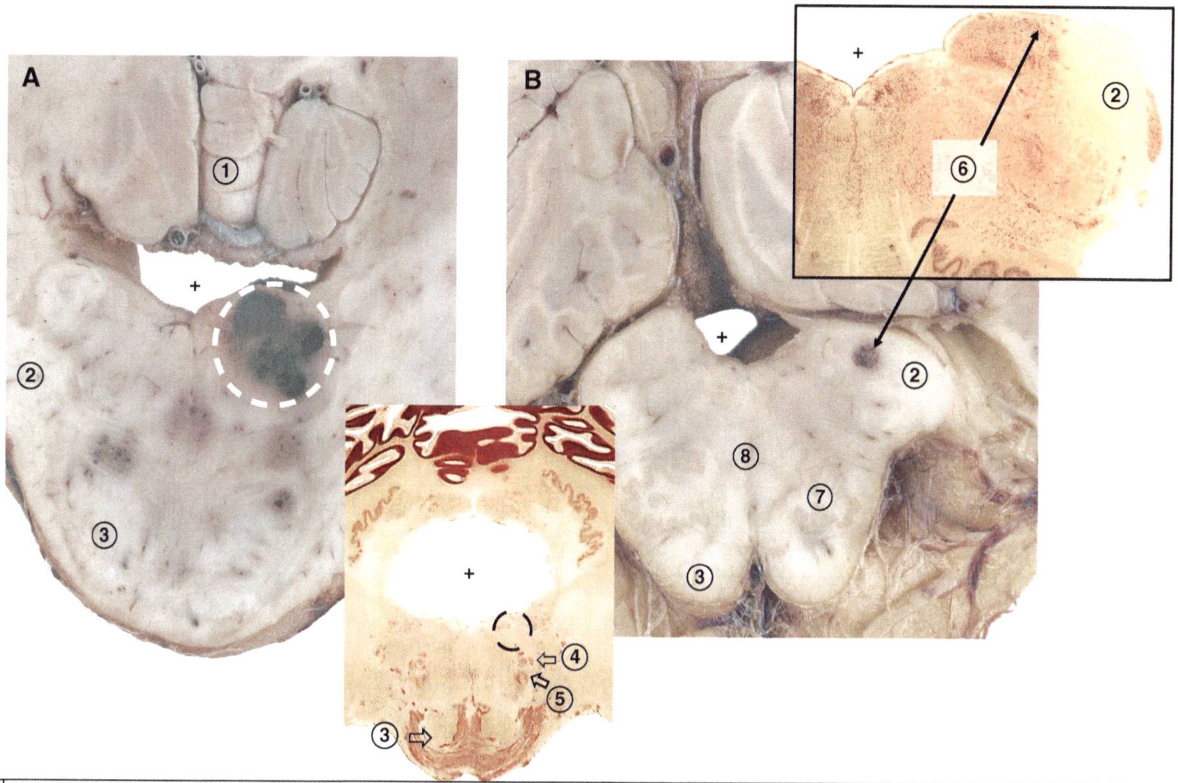

①	Vermis	⑥	Inferior / spinal vestibular ncl.
②	Inferior cerebellar peduncle	⑦	Inferior olivary ncl.
③	Pyramidal (Corticospinal) tract	⑧	Medial lemniscus
④	Motor ncl. of facial nerve	+	4th ventricle
⑤	Superior olivary complex		

Fig. 5.11 In (**A**) (pons) and (**B**) (medulla oblongata), cavernous angiomas in the tegmentum. (**A**) Besides several hemorrhages in the pontine basis a large part of the pontine tegmentum—dorsal of the facial nucleus and the superior olivar nucleus—is infested with the angioma. (**B**) A circumscript lesion is located in the inferior vestibular nucleus (Compare with Fig. 5.7). LabPON Twente

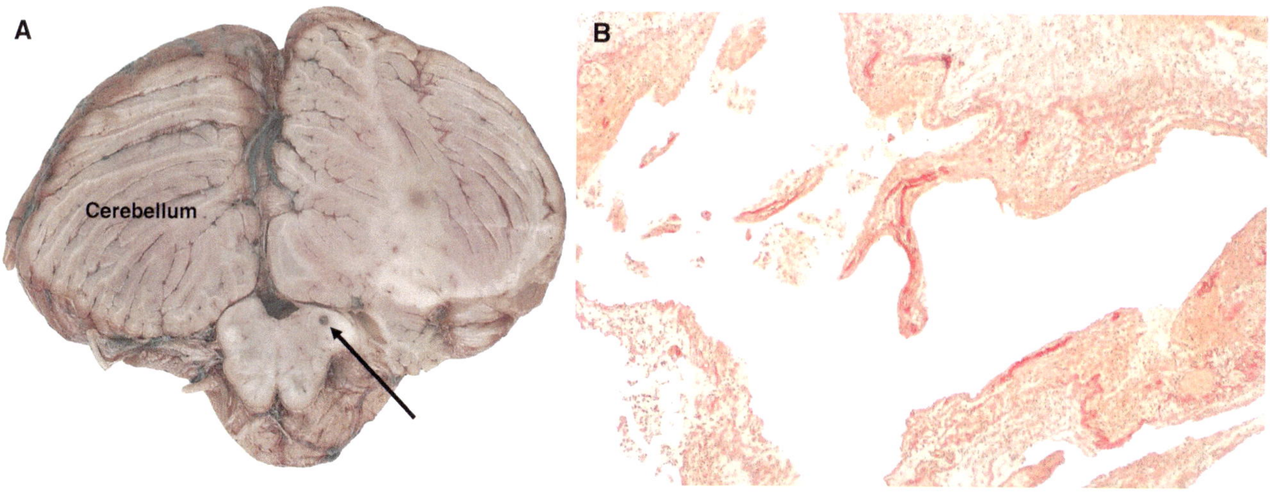

Fig. 5.12 The same patient as in Fig. 5.11 with cavernous angioma in the medulla oblongata (**A**, arrow). In (**B**) microscopical picture, irregular vascular channels lined by endothelium and thin hyalinized walls. Von Gieson stain. LabPON Twente

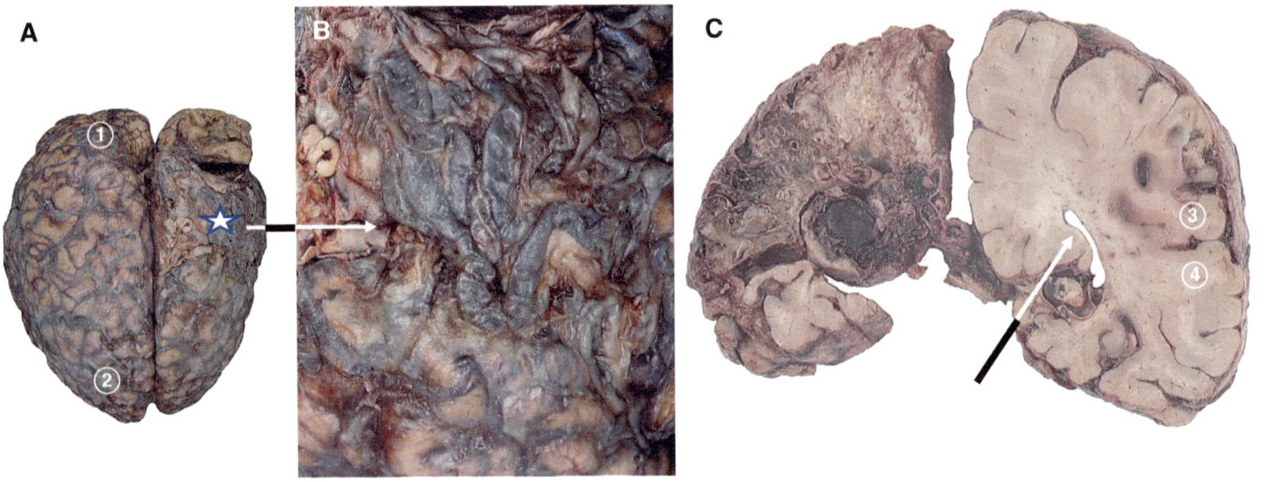

①	Occipital lobe	③	Superior temporal gyrus
②	Frontal lobe	④	Inferior temporal gyrus

Fig. 5.13 (**A**) Brain seen from above. Typical arteriovenous malformation (AVM). Tortuous, vascular mass on the surface of the left hemisphere (**A** and **B**, detail). In (**C**) coronal section showing malformed vessels, penetrating the brain, extending from the subarachnoidal space to the ventricle. Subjacent convolutions show atrophy due to the pressure, accompanied by secondary changes. Narrowing of the right lateral ventricle (arrow). LabPON Twente

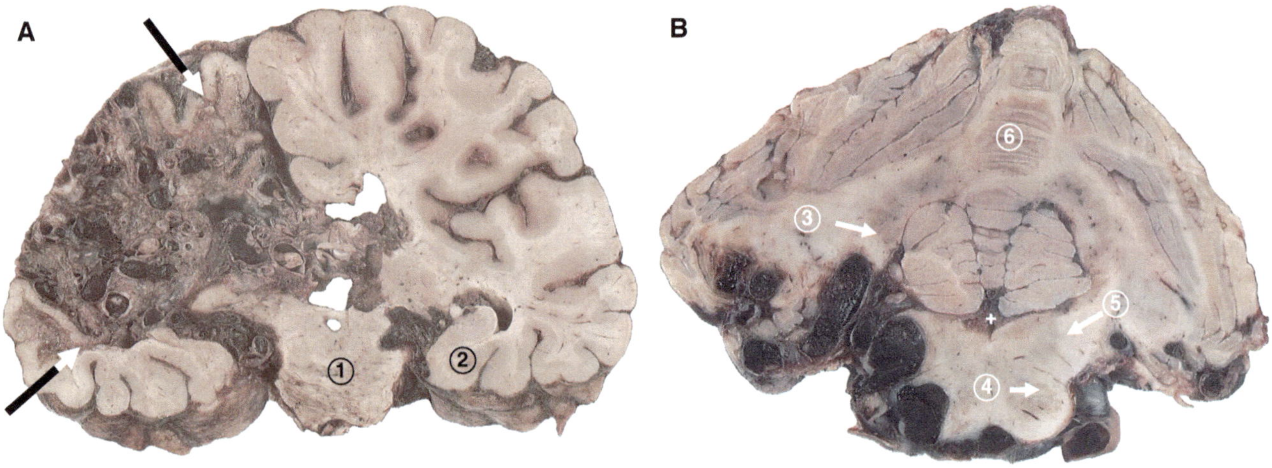

①	Pons	⑤	Inferior cerebellar peduncle
②	Hippocampus	⑥	Vermis
③	Dentate ncl.	+	4th ventricle
④	Inferior olivary ncl.		

Fig. 5.14 (**A**) Coronal section at the level of the parietal lobe. AVM extending from the thickened subarachnoidal space to the ventricle, multiple malformed vascular channels of varying thickness and secondary changes. Pressure atrophy of the subjacent gyri (arrows) and pons and (**B**) involvement of the cerebellum and medulla oblongata. LabPON Twente

centimeters. On cut surface AVMs are typically pyramidal-shaped lesions, base parallel to the meninges, apex pointing to the ventricles or deep brain (Zabramski et al. 1999) (Figs. 5.13 and 5.14). The subjacent convolutions show a variable degree of atrophy from pressure and secondary degenerative changes induced by the vessels.

The wall of the ventricle may be reached and can explain the spontaneous ventricular hemorrhage that may occasionally be fatal (Russell and Rubinstein 1998). Microscopically, the structures of the vessels are abnormal. The vessels have increased diameters with varying degrees of ectasia and the arterial walls are thickened by the proliferation of fibroblasts, connective tissue, and smooth muscle cells (Martin and Vinters 1995). Thinning of the arterial walls may lead to the formation of aneurysms. Thickening of veins are also seen and enlarged "arterialized veins" may show proliferation of fibroblasts as well as increased cellularity. They are however, not truly arteries because they do not contain an organized elastica. The distinction between arteries and veins may be difficult in some places.

Venous malformations

Venous malformations seem to be more frequent in the spinal cord than in the brain. Relatively few cases have been described (Zabramski et al. 1999; Wilkins 1985). The area of the middle cerebral artery is mainly affected and most of the lesions are found in the frontal lobes. They are only rarely associated with hemorrhage. They are composed of abnormally enlarged veins separated by neural parenchyma that appears normal. The vessels are arranged in a radial pattern extending out from a dilated central venous trunk (Zabramski et al. 1999). Angiographically, this arrangement produces the typical Caput medusae, or umbrella appearance, seen in the normal venous phase (Rengachary and Kaylan-Raman 1996; Rigamonti et al. 1990). Histologically, the vessels, which lack smooth muscle and elastic tissue, resemble normal venous structures, with a thin layer of endothelium and a wall of fibrous tissue composed mainly of collagen.

Mixed forms of cerebral vascular malformations have been described in the literature, but their frequency is unknown.

With respect to the genetics, disease-causing mutations were identified in cerebral cavernous malformations (CCM) in three genes, KRIT1, MGC4607, and PDCD10 (Felbor et al. 2006). These genes are thought to play a role in angiogenesis. Mutations causing hereditary CCMs exhibit autosomal dominant transmission (Zhu et al. 2014). Genetic counseling of patients with a CCM is strongly advisable for patients with a positive family history and for seemingly sporadic cases with multiple lesions (Felbor et al. 2006).

5.3.2 Monoamine nuclei r9

5.3.2.1 A2 noradrenaline cells (NA2)

For details on these cells, see Sect. 3.2.1.1

5.3.2.2 C1 adrenaline cells (C1)

For details on these nuclei, see Sect. 3.2.1.2.

5.3.3 Trigeminal sensory nuclei r9

5.3.3.1 Spinal trigeminal nucleus

For details on the spinal trigeminal nucleus, see Sect. 3.2.2.

5.3.4 Dorsal column nuclei r9

5.3.4.1 Cuneate nucleus

For details on the cuneate nucleus, see Sect. 3.2.3.3.

5.3.5 Solitary nuclei r9

For details on the solitary nuclei, see Sect. 3.2.4.

5.4 Basal r9

5.4.1 Somatic motor nuclei r9

5.4.1.1 Hypoglossal nucleus

For details on the hypoglossal nucleus, see 4.3.1.1.

5.4.2 Branchial motor nuclei r9

5.4.2.1 Nucleus ambiguus r9 [Nucleus ambiguus]

5.4.2.1.1 Location of the ambiguus nucleus (ambiguus = wavering, uncertain)

The nucleus ambiguus (the branchiomotor nucleus of the third branchial arch) (see Sect. 2.8, see Fig. 2.9) is located approximately in the dorsoventral half of the medulla oblongata (Fig. 5.7A ⑧).

Longitudinally, it represents a column caudal from the inferior olive to the caudal pole of the facial nucleus (Olszewski and Baxter 1982). Neighbors of the nucleus in Fig. 5.6, in clockwise direction, starting at 12 h, are: the reticular formation, the spinal trigeminal nucleus (see Sect. 3.2.2.2), the intermediate reticular zone (see Sect. 3.3.2.1), and the gigantocellular reticular nucleus (see Sect. 4.3.4.2).

Cytoarchitectonally, the nucleus is mostly composed of large, multipolar neurons (Olszewski and Baxter 1982) with many Nissl granules, the typical appearance of a motor nucleus (compare, for example, the hypoglossal nucleus, see Sect. 4.3.1.1, or the trigeminal motor nucleus (see Sect. 11.3.1). From animal degeneration studies, there is some evidence for a somatotopic arrangement in the nucleus ambiguus.

5.4.2.1.2 Target organs of the ambiguus nucleus

The branchiomotor neurons of the nucleus provide innervation for the pharyngeal, in particular, the pharyngeal constrictor muscles (see Fig. 5.20), targeted by the glossopharyngeal nerve (CNIX, see Sect. 5.4.2.1.3), partly together with the vagus nerve (see Sect. 7.3.2.1.).

5.4.2.1.3 Efferent connectivity of the ambiguus nucleus and course of the glossopharyngeal nerve

The axons of the ambiguus neurons join the glossopharyngeal nerve (CNIX) [N. glossopharyngeus], the nerve of the third branchial arch (ἡ γλῶσσα, he glossa, Greek = tongue; ἡ φάρυγξ, φαρυγγ- Greek he pharynx, pharyng- = throat, windpipe), (see Sect. 2.8, see Fig. 2.9). Besides the branchiomotor fibers, this nerve has sensory fibers for the pharyngeal mucosa, the pharyngeal tonsils, and gustatory fibers for the posterior one third of the tongue (see also Sect. 3.2.4). Furthermore, the glossopharyngeal nerve leads parasympathetic preganglionic fibers from the inferior salivatory nucleus (see Sect. 7.3.1.1) to the otic ganglion (innervation of the parotid gland, see Fig. 5.18).

In the following, we will describe the course of the glossopharyngeal nerve with its different qualities. The nerve leaves the medulla oblongata in the retroolivary groove [Sulcus retroolivaris] and the cranial cavity via the jugular foramen (Figs. 5.15, 5.16, 5.17).

The small superior ganglion [Ggl. superius] (sensory) is located above or inside the foramen and the larger inferior ganglion [Ggl. inferius, Ggl. nodosum] (viscerosensory)

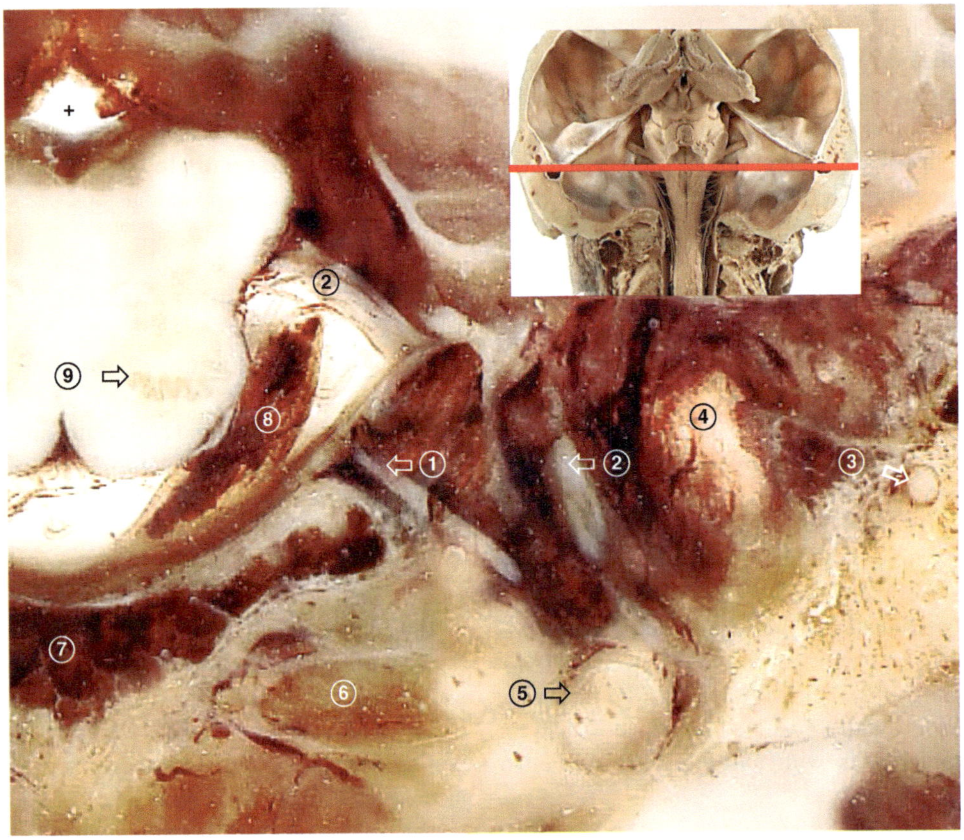

①	Glossopharyngeal nerve CNIX	⑥	Longus capitis muscle
②	Vagus nerve CNX	⑦	Clivus
③	Facial nerve CNVII	⑧	Vertebral artery
④	Jugular bulb	⑨	Inferior olivary complex
⑤	Internal carotid artery	+	4th ventricle

Fig. 5.15 Horizontal section through the human medulla oblongata at the level of the exit of the glossopharyngeal nerve ①. Plastinated specimen. The nerve leaves the medulla oblongata between the inferior olive ⑨—exit not visible here—and the vagus nerve ②, and together with the latter is running through the lateral cerebellomedullary cistern into the jugular foramen (see Figs. 5.16 and 5.17). It leaves the foramen into the jugular fossa (see Fig. 5.16 ②) where it is directly neighboring the internal carotid artery ⑤. See also atlas part Darrow red 13. Sammlung des Zentrums Anatomie der Universität zu Köln

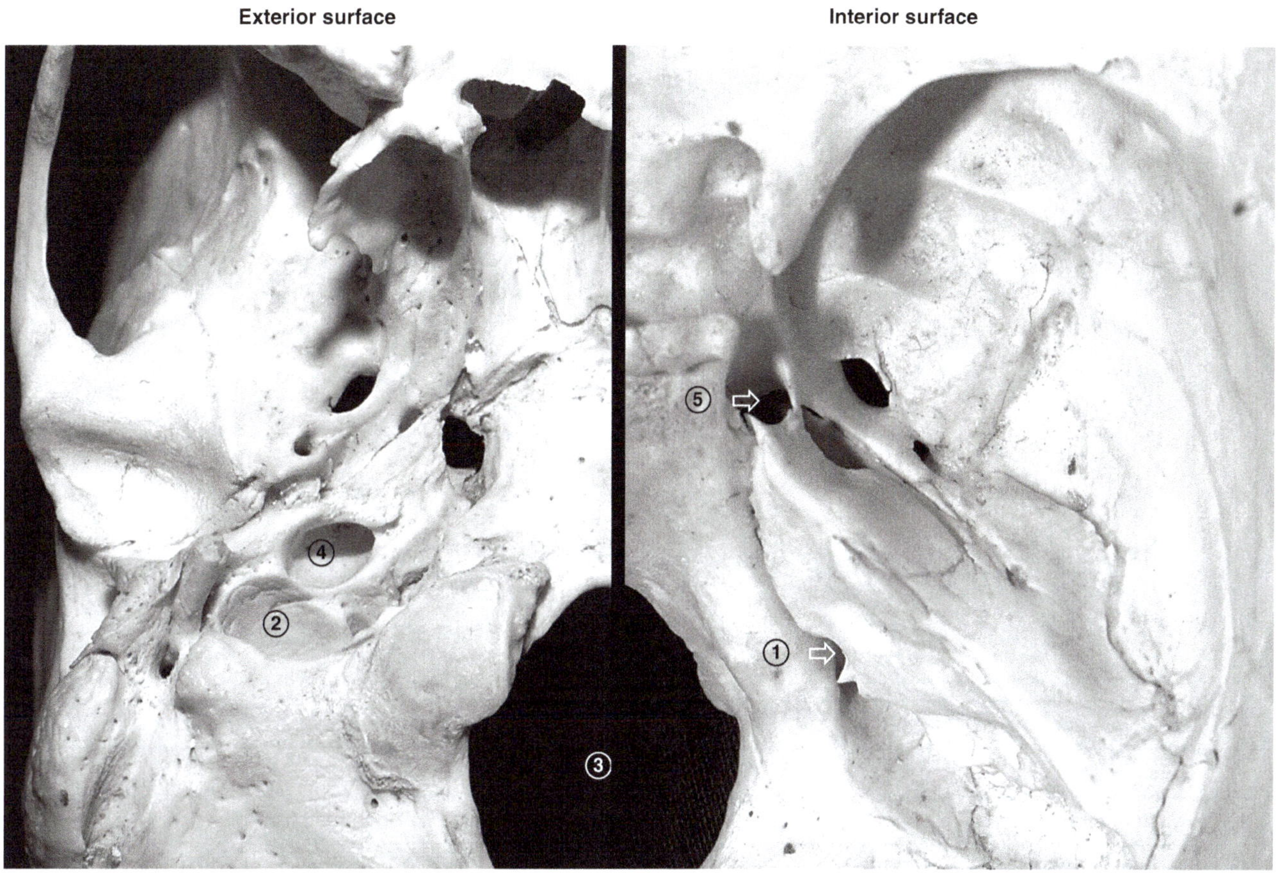

①	**Jugular foramen**	④	**External opening of carotid canal**
②	**Jugular fossa**	⑤	**Internal opening of carotid canal**
③	**Foramen magnum**		

Fig. 5.16 Passage of the glossopharyngeal nerve. The nerve leaves the leptomeningeal space into the jugular foramen ① and exits the skull base into the jugular fossa ②. Sammlung des Zentrums Anatomie der Universität zu Köln

below the foramen. The first branch of the glossopharyngeal nerve (CNIX) is the tympanic nerve [N. tympanicus], which reaches the inferior ganglion and is running between the carotid canal and the foramen lacerum through the tympanic canaliculus [Canaliculus tympanicus] into the tympanic cavity. It contains preganglionic parasympathetic fibers from the inferior salivatory nucleus (see Sect. 7.3.1.1) for the parotid gland [Glandula parotidea] (see Fig. 5.18) and somatoafferent fibers from the tympanic cavity and the auditory tube/ Eustachian tube [Tuba auditiva Eustachii] (see below).

Together with sympathetic fibers from internal carotid plexus, the nerve forms the tympanic plexus, which releases the lesser petrosal nerve [N. petrosus minor] with its preganglionic parasympathetic fibers (Arslan 1960). Entering the medial cranial fossa via the hiatus for minor petrosal nerve [Hiatus canalis n. petrosi minoris], they leave the fossa via the Fissura sphenopetrosa and enter the otic ganglion [Ganglion oticum]. The postganglionic parasympathetic fibers join the auriculotemporal nerve (V_3, r2) [N. auriculotemporalis] to innervate the parotid gland (Fig. 5.18).

5.4.2.1.4 The auditory tube

The auditory or pharyngotympanic tube also known as Eustachian tube after the description by the Italian anatomist Bartolomeo Eustachi (Eustachius), (✝ 1574) is a partially bony and partially cartilaginous connection between the superior pharynx and the tympanic cavity which provides the ventilation of the middle ear (Okada et al. 2018).

There are three pharyngeal muscles working on the tube:

– The tensor veli palatini [M. tensor veli palatini].
– The levator veli palatini [M. levator veli palatini].
– The salpingopharyngeus [M. salpingopharyngeus].

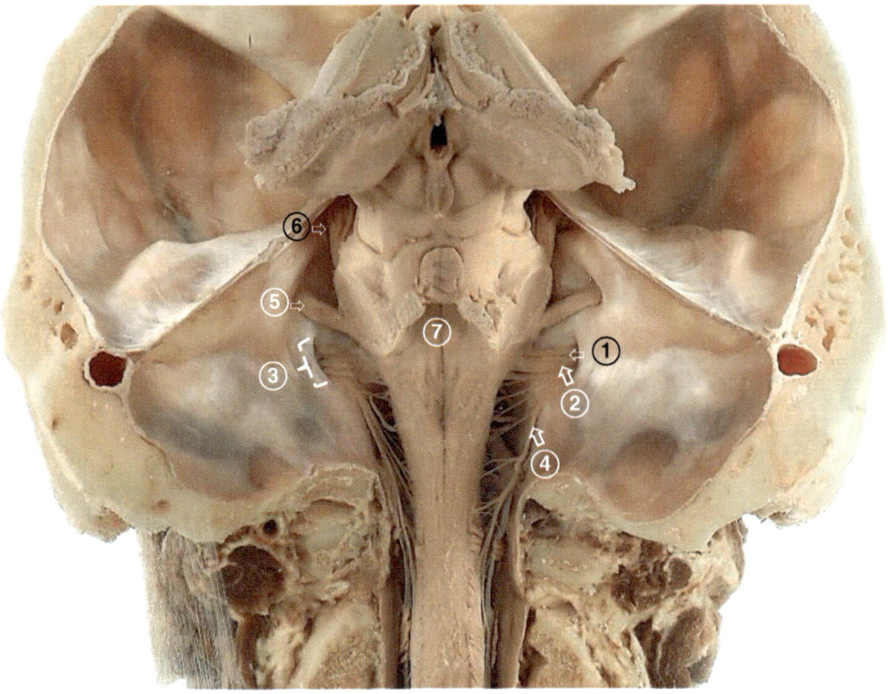

①	**Glossopharyngeal nerve CNIX**	⑤	**Vestibulocochlear nerve CNVIII**
②	**Vagus nerve CNX**	⑥	**Trigeminal nerve CNV**
③	**Jugular foramen**	⑦	**4th ventricle / Rhomboid fossa**
④	**Spinal root of accessory nerve CNXI**		

Fig. 5.17 Dorsal view of the human brainstem. Upper vertebral column, posterior parts of occipital bone, calvarium, meninges, telencephalon, and cerebellum removed. CNIX ① and CNX ② run through the lateral cerebellomedullary cistern from their exit to the jugular foramen ③ (compare with Figs. 5.15 and 5.16). Sammlung des Zentrums Anatomie der Universität zu Köln

The action of the first two muscles arises from their names. The velum palatinum (sail/canvas of the palate) is synonymous with the soft palate [Palatum molle] (see Fig. 5.19 ②). It is made up of connective tissue rich in fibers and moves according to the actions exerted by the connected muscles. The tensor (from Latin tendere = to tense) tenses the soft palate while the levator (from Latin levare = to lift) lifts it.

The salpingopharyngeus (Greek ἡ σάλπιγξ, –ιγγος-, he salpinx, salpingos = trumpet), interdigitating with the palatopharyngeus, opens the orifice of the Eustachian tube.

The topographical situation of the glossopharyngeal, the vagus, and spinal accessory nerve is shown in Figs. 5.17 and 5.18.

As recently reinvestigated in a detailed study (Okada et al. 2018), there is evidence that all three pharyngeal muscles are involved in the opening of the auditory tube (Fig. 5.19)—the walls normally are collapsed—which can be performed voluntarily during swallowing (see Box 5.5) by talking, yawning, and chewing. Whoever experienced a rapid descent in an airplane while suffering from a cold with occlusion of the auditory tube knows how important these techniques are to eventually accomplish a pressure equalization between pharynx (outer world) and tympanic cavity. This painful discomfort turns into a potentially life-threatening condition in case of rapid decompression/depressurization in an airplane. In addition to more serious problems like hypoxia, the gas expansion in the middle ear (in particular when the tube is closed) may lead to severe pain (innervation of the middle ear by the glossopharyngeal nerve: tympanic nerve, plexus) and in the worst case to rupture of the eardrum which easily disables the pilot. This is a serious issue in flight medicine.

https://www.skybrary.aero/index.php/Middle_Ear_and_Sinus_Problems

5.4.2.1.5 Pharyngeal muscles

Subsequently, the motor branches, i.e., the pharyngeal branches [Rr. pharyngei] and the stylopharyngeal branches [Rr. stylopharyngei], leave the CNIX.

The CNIX pharyngeal branches—together with those from the vagus nerve (CNX)—form the pharyngeal (motor) plexus [Plexus pharyngeus] for the muscles derived from branchial arches 3 and 4. As to innervation of pharyngeal muscles, the differentiation between CNIX and CNX is not unequivocal but rather provided by either nerve.

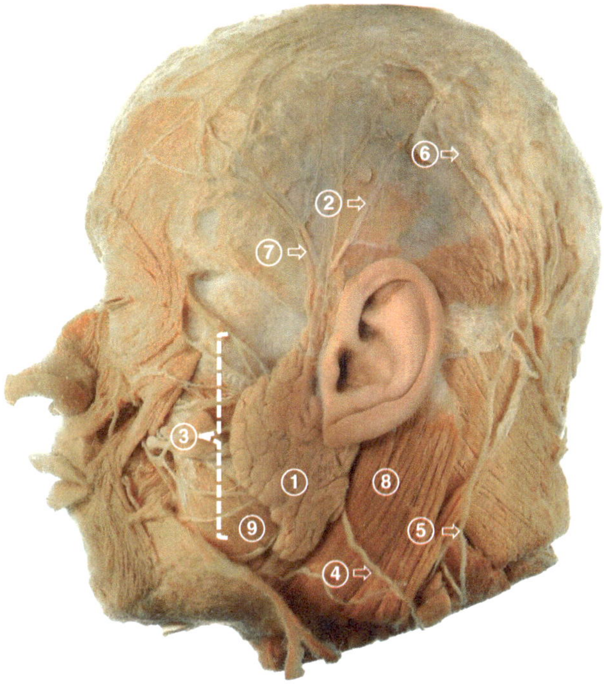

①	Parotid gland	⑥	Greater occipital nerve
②	Auriculotemporal nerve V_3	⑦	Ascending temporal artery, frontal branch
③	Peripheral branches of the facial nerve CNVII	⑧	Sternocleidomastoid muscle
④	Great auricular nerve	⑨	Masseter muscle
⑤	Lesser occipital nerve		

Fig. 5.18 Lateral view of the human head. Formalin-fixed specimen. Skin, subcutaneous fat and muscle fascia removed. The parotid gland ① [Glandula parotidea] (Greek παρα para = next to, τό οὖς, ωτός, to ous, otos = ear) is located directly anterior of the external ear. At its anterior edge, a number of nerve branches all belonging to the facial nerve [N. facialis] (CNVII) (see Sect. 8.2.1)—are running into the periphery of the face ③. The main trunk of CNVII reaches the face beneath the parotid. The nerve crossing the sternocleidomastoid muscle [M. sternocleidomastoideus] at the posterior edge of the parotid gland is the great auricular nerve [N. auricularis magnus] ④. Together with the greater ⑥ and lesser ⑤ occipital nerve [Nn. occipitales major et minor], they provide somatosensory innervation of the posterior head regions (see Fig. 12.2). All these nerves originate in the cervical spinal cord while the face region is innervated by the trigeminal nerve (see Sect. 12.2.3.2). Photograph by MedizinFotoKöln, Cologne, Germany. Sammlung des Zentrums Anatomie der Universität zu Köln

In craniocaudal direction, there are three pharyngeal constrictor muscles (constringere [Latin] = to constrict, to narrow)

- Superior constrictor [M. constrictor pharyngis superior]
- Middle constrictor [M. constrictor pharyngis medius]
- Inferior constrictor [M. constrictor pharyngis inferior]

which form the external circular muscle layer (Fig. 5.20). In addition, there is a stylopharyngeus muscle [M. stylopharyngeus] (From Greek ὁ στῦλος ho stylos = pillar), here referring to the styloid process of the temporal bone and the pharynx (= throat) (see Fig. 5.20). The stylopharyngeus originates from the styloid process (Fig. 5.21) and inserts mainly into the posterior thyroid cartilage of the larynx (see Fig. 7.44). Contraction of the muscle results in elevation of pharynx and larynx (Fig. 5.19). Together with the other longitudinal muscles (innervation by CNX), this movement is executed during swallowing (Bui and Das 2019) (see Box 5.5). The stylopharyngeus is the only of the latter with third arch origin and is exclusively innervated by the glossopharyngeal nerve.

Contraction of the superior constrictor muscle constricts the upper portion of the pharynx. The middle constrictor muscle—constricting the middle portion of the pharynx—originates at the hyoid bone (Fig. 5.20) and stylohyoid liga-

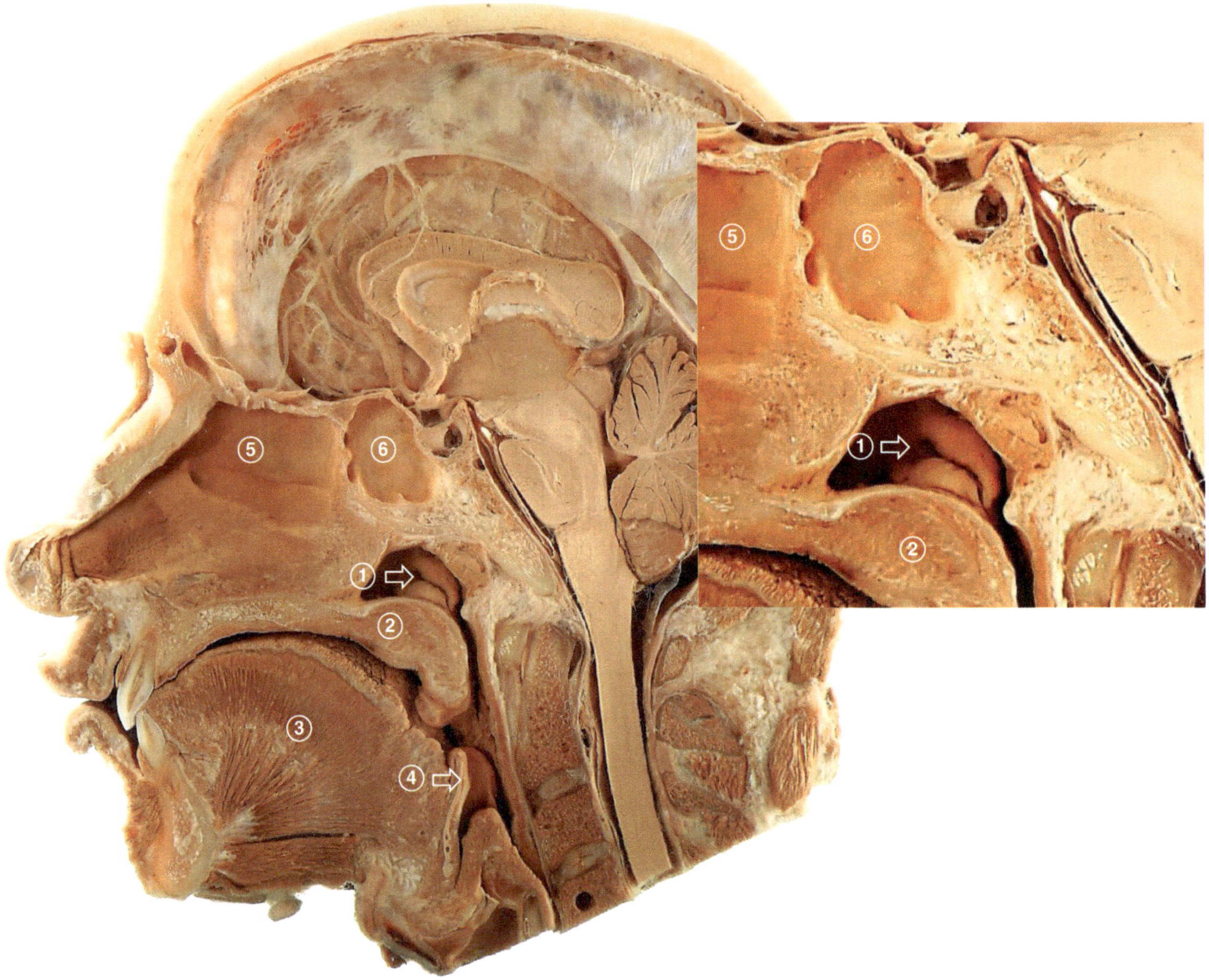

①	**Ostium of the auricular tube**	④	**Epiglottis**
②	**Soft palate**	⑤	**Nasal septum**
③	**Tongue**	⑥	**Sphenoid sinus**

Fig. 5.19 Midsagittally sectioned human head. The inset shows parts of the naso- and oropharynx, separated by the soft palate. Above the latter, a small ovoid fossa is visible. This is the entrance into the auditory tube connecting the nasopharynx with the middle ear (tympanic cavity). For the medical relevance of this structure, see text. Photographs by MedizinFotoKöln, Cologne, Germany. Sammlung des Zentrums Anatomie der Universität zu Köln

ment and inserts into the pharyngeal raphe (Fig. 5.20) as well as blending in with fibers of the superior and inferior constrictors (Bui and Das 2019).

The inferior pharyngeal constrictor muscle further divides into two smaller muscles, which originate from the larynx (Fig. 5.20) and insert into the pharyngeal raphe and the esophagus. They constrict the lower portion of the pharynx.

The constrictor muscles work together and contract to push food after swallowing to propel it from the oral cavity into the esophagus (Bui and Das 2019) (see Box 5.5).

The pharyngeal inner longitudinal layer consists of the palatopharyngeus [M. palatopharyngeus], the salpingopharyngeus [M. salpingopharyngeus], and stylopharyngeus [M. stylopharyngeus] muscles (Fig. 5.21).

The salpingopharyngeus muscle originates from the inferior auditory tube and inserts into the palatopharyngeus muscle. This muscle contracts to elevate the pharynx as well as to open the auditory or pharyngotympanic tube [Tuba auditiva/formerly also auditoria] (connection between pharynx and tympanic cavity/middle ear, see Fig. 5.19) when swallowing (see Box 5.5). In a wider sense, the levator veli palatini (see Fig. 5.21 ⑤) and the tensor veli palatini (see Fig. 5.21 ⑦) can be subsumed among the pharyngeal muscles. Note that the levator veli palatini is the

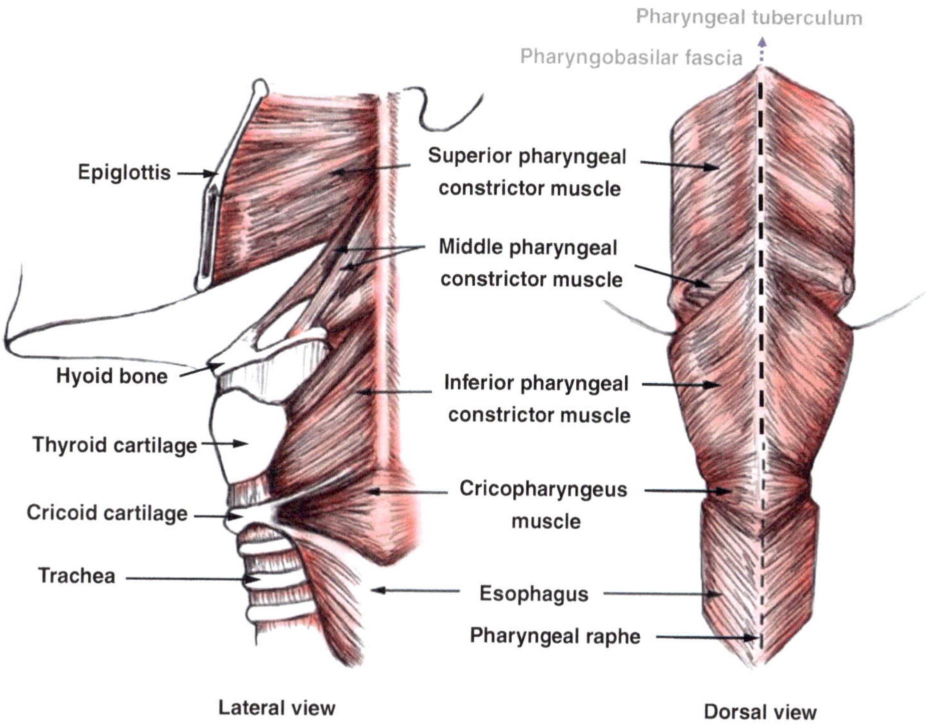

Fig. 5.20 The pharyngeal constrictor muscles. The individual muscles are arranged like roof tiles, nested on top of each other. The superior constrictor muscle of either side joins in the pharyngobasilar fascia, which runs to the pharyngeal tubercle (see Fig. 5.21 ⑨ ⑩). This way, together with its middle and inferior counterparts, a muscular tube—the pharynx—is formed. The constrictor muscles of either side meet in the pharyngeal raphe [Raphe pharyngis]. Modified after Morrison and Postma 2013, Fig. 59.1 with permission

only one of these muscles with a horizontal course. All the others run in vertical direction.

Of special clinical-anatomical importance is the act of swallowing, which under pathological conditions may lead to aspiration of food or liquids into the airways with subsequent development of pneumonia or even suffocation (see Box 5.5).

Box 5.5 Swallowing

The act of swallowing is integral to the process of eating and drinking and thereby for the proper energy input into the human body (Sasegbon and Hamdy 2017). It is a highly complex process requiring sensory input and motor output to the mouth, pharynx, larynx, and esophagus. It is therefore also a classical example of the interaction of different cranial nerves. Food has to be snapped and advanced into the oral cavity. There, by chewing and appropriate tongue movements, a bolus is formed. This is then transported into and down the pharynx—while simultaneously, the respiratory pathways are closed—and from the pharynx into the esophagus, eventually into the stomach. These events have first a voluntary and then an involuntary component and also receive modulatory input from higher CNS centers (for details, see Sasegbon and Hamdy 2017).

Among the three classical phases of swallowing the oral one is voluntary involving the lips (see Sect. 7.3.2.1), the teeth and masticatory muscles (CNV, see Sect. 11.3.1.1.2, note the typing error in table1 of Sasegbon and Hamdy 2017, the masticatory muscles are innervated by CNV not CNVII) and the tongue (CNXII, see Sect. 4.3.1.1). With the aid of saliva secreted from the salivatory glands under the control of CNIX (parotid gland, see Fig. 5.18) and CNV and CNVII (submandibular and sublingual glands, see Box 12.1) the food gains a more liquid consistency (Sasegbon and Hamdy 2017). The food bolus is then positioned in the anterior portion of the mouth between the tongue and the soft palate forming a temporary posterior wall. By attaching the tongue tip to the hard palate and by anterior-posterior movements the tongue propulses the bolus—located now at the superior tongue surface—to the pharynx. The following steps are increasingly automatic and kind of involuntary (Sasegbon and Hamdy 2017).

As to the muscles involved in addition to the muscles already mentioned, part of the mimic muscles, in particular the orbicularis oris muscle (see Sect. 8.2.1.6.1), prevent the spilling of food or liquids from the mouth. Furthermore, the muscles of the soft palate (tensor veli palatini, palatoglossus, palatopharyngeus, levator veli palatini, and M. uvulae (see Fig. 5.21) elevate the soft palate during swallowing and by sealing the nasopharynx they prevent the reflux of food or fluids into the nasal cavity (see Fig. 5.22) (Sasegbon and Hamdy 2017).

In terms of function, the so-called upper esophageal sphincter has been suggested which is kept close during non-swallowing. After the initiation of swallowing, the sphincter relaxes (for details, see Sasegbon and Hamdy 2017) via activation of the vagus nerve. Another important step in swallowing is the change of the upright vertical resting position of the epiglottis (see Fig. 5.19 ④) into a horizontal one, as well as the adduction of the vocal cords (see Fig. 5.45), to prevent aspiration of food or liquids by several laryngeal muscles innervated by the vagus nerve (for details, see Malone and Arya 2020).

At the brainstem level, afferent sensory signals are transferred by CNV (oral mucosa) as well as by CNVII, CNIX, and CNX for the pharynx (Sasegbon and Hamdy 2017). From animal studies it is known that the sensory input converges onto the nuclei of the solitary tract (NTS) (see Sect. 3.2.4), which can induce swallowing. Via interneurons the ncl. ambiguus (see here above) is activated together with CNV, CNVII, CNX, and CNXII (see above) as the efferent limbs of the swallowing process (Sasegbon and Hamdy 2017). In terms of function, the term central pattern generator (CPG) has been coined as a bilaterally present and interconnected unit for the medullary nuclei network involved. In this vein, two brainstem reflexes merit mention, the (1) swallowing reflex and (2) the gag reflex, the testing of which can be easily integrated into the neurological examination. The swallowing reflex can be triggered by touching the palatal arches (Fig. 5.23), the posterior pharyngeal wall with afferents via the glossopharyngeal and vagus nerves and efferents as described here above.

The gag reflex usually can be elicited by touching the uvula (Fig. 5.23) and/or by pressing down the root of the tongue. This leads to nausea and retching, in the worst case to vomiting. It may lack in case of lesions of the glossopharyngeal nerve.

Given the numerous neural and muscular components involved in swallowing it does not come as a surprise that a plethora of disorders on the several levels may cause dysphagia (see Table 5.1 and Sasegbon and Hamdy 2017 for details). Spinocerebellar Ataxia Type 3 (SCA3) (see Sect. 7.2.1.3.3) affects a number of swallowing-related nuclei (see Table 7.1 for details) (Rüb et al. 2002).

5.4.2.1.6 Living anatomy and clinical implications of the glossopharyngeal nerve

The clinical signs elicited by lesions of the glossopharyngeal and/or vagus nerves are scarce. A site easy to approach during physical examination is the palatine arch (see Fig. 5.23). Under normal conditions, together with the midline uvula, they form kind of a symmetric curtain which demarcates the oral cavity from the pharynx. Between both arches the palatine tonsils are located (see below sensory tonsillar branches of the glossopharyngeal nerve). In case of unilateral lesions of CNIX/CNX, the whole "curtain" is dragged to the healthy side, accompanied with a corresponding dislocation of the uvula (see Fig. 5.23).

In its further course, the carotid sinus nerve [N. sinus carotici] (Toorop et al. 2013) from the baroreceptors of the carotid sinus [Sinus caroticus] and the carotid glomus [Glomus caroticum] (Sect. 3.2.4.1.2 "Arterial chemoreceptors") join the CNIX. Then the tonsillar branches [Rami tonsillares] from the palatine tonsils and their surroundings reach the nerve. Gustatory fibers from the posterior one third of the tongue (Lingual branches/Rr. linguales—do not confuse with the lingual nerve, branch of the mandibular (trigeminal) nerve) join the glossopharyngeal nerve.

In mice, the nucleus ambiguus has been described to innervate the pharyngeal and laryngeal muscles. Furthermore, this nucleus contains preganglionic parasympathetic neurons targeting cardiac ganglia in the mouse cardiac atria (cf. Li et al. 2010). The latter may act as a local integration center of central and local inputs on cardiac function (cf. Li et al. 2010).

A didactically excellent example of vascular brainstem syndromes in this context is Wallenberg's (lateral medullary) syndrome (see also Sect. 1.4.2.3) because it includes several of the structures dealt with in Chap. 5:

- Ncl. ambiguus
- Spinal vestibular nucleus
- Ncl. of solitary tract/Solitary nucleus (see Sect. 3.2.4)
- Spinal trigeminal nucleus/Spinal trigeminal tract (see Sect. 3.2.2.2)
- Spinothalamic tract (see Fig. 3.13B)
- Posterior ncl. of vagus nerve (see Sect. 7.3.2.1)

The vessel responsible for the development of Wallenberg's syndrome is the vertebral artery (see Fig. 1.25), respectively, the posterior inferior cerebellar artery (PICA) (see 1–17). Its supply territory is the lateral segment of the medulla oblongata posterior / dorsal of the inferior olive (Lui et al. 2020). The earliest known description of lateral medullary infarction was given in 1810 by Gaspard Vieusseux of Geneva at the Medical and Chirurgical Society of London (Pearce 2000). However, Wallenberg's case report in 1895 amplified the clinical signs with accurate localization of the

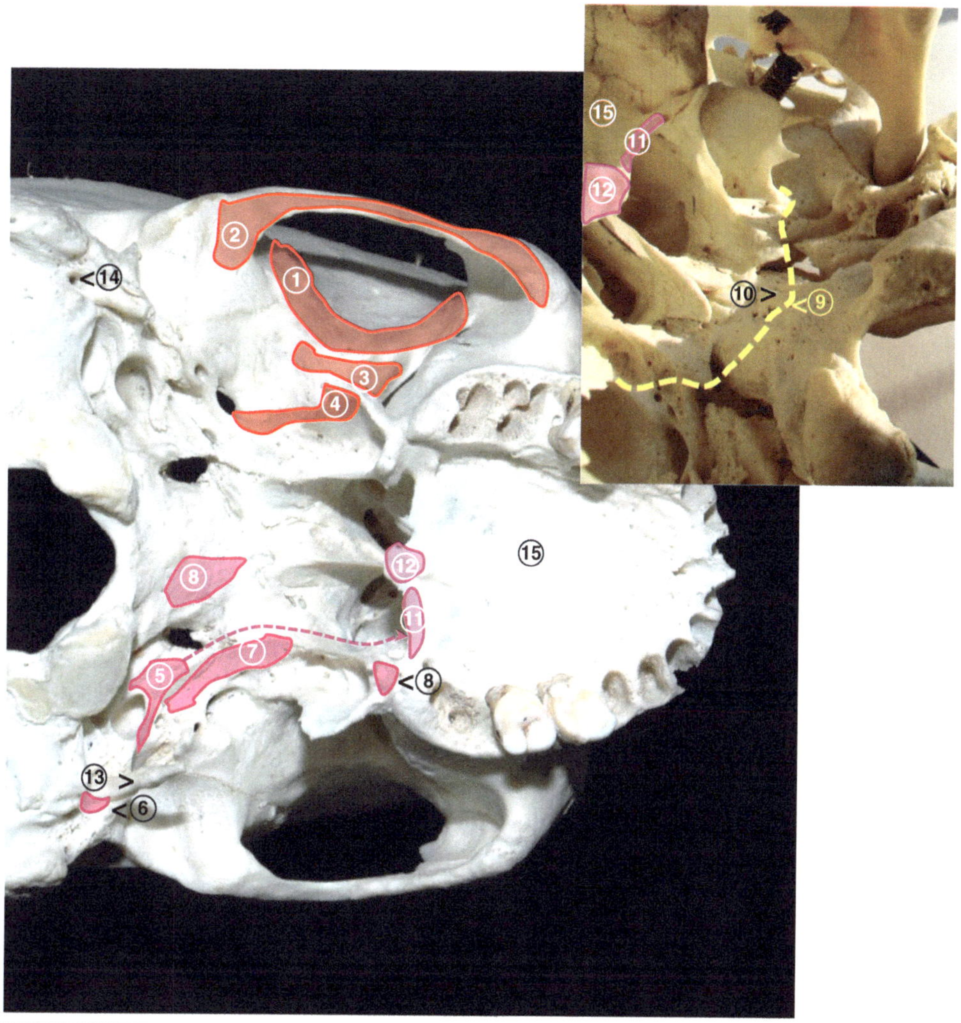

	Masticatory muscles V_3
①	Temporal muscle
②	Masseter muscle
③	Lateral pterygoid muscle
④	Medial pterygoid muscle
	Pharyngeal muscles IX, X
⑤	Levator veli palatini muscle with approximate course ---
⑥	Stylopharyngeus muscle
⑦	Tensor veli palatini muscle
⑧	Superior pharyngeal constrictor muscle

⑨	Pharyngobasilar fascia With approximate course ---
⑩	Pharyngeal tubercle
⑪	Palatopharyngeus muscle
⑫	Uvula muscle
⑬	Styloid process
⑭	Stylomastoid foramen
⑮	Hard palate

Fig. 5.21 Bony origins of pharynx muscles at the skull base (caudal view). Note that there is a muscle-free pharyngobasilar fascia ⑨ between the outer surface of the skull base and the upper margin of the superior constrictor muscle (see Fig. 5.20). The superior constrictor muscle ⑧ originates from the skull base, the mandible, and the pterygomandibular ligament/raphe [Raphe pterygomandibularis]. Further pharyngeal muscles are the stylopharyngeus ⑥ from the styloid process ⑬ to the pharynx and the palatopharyngeus ⑪ from the hard palate [Palatum durum] ⑮ to the pharynx. The palatopharyngeus muscle originates from the posterior hard palate, the palatine aponeurosis, and inserts into the thyroid cartilage [Cartilago thyroidea] of the larynx. When the palatopharyngeus muscle contracts, it elevates the pharynx. Sammlung des Zentrums Anatomie der Universität zu Köln

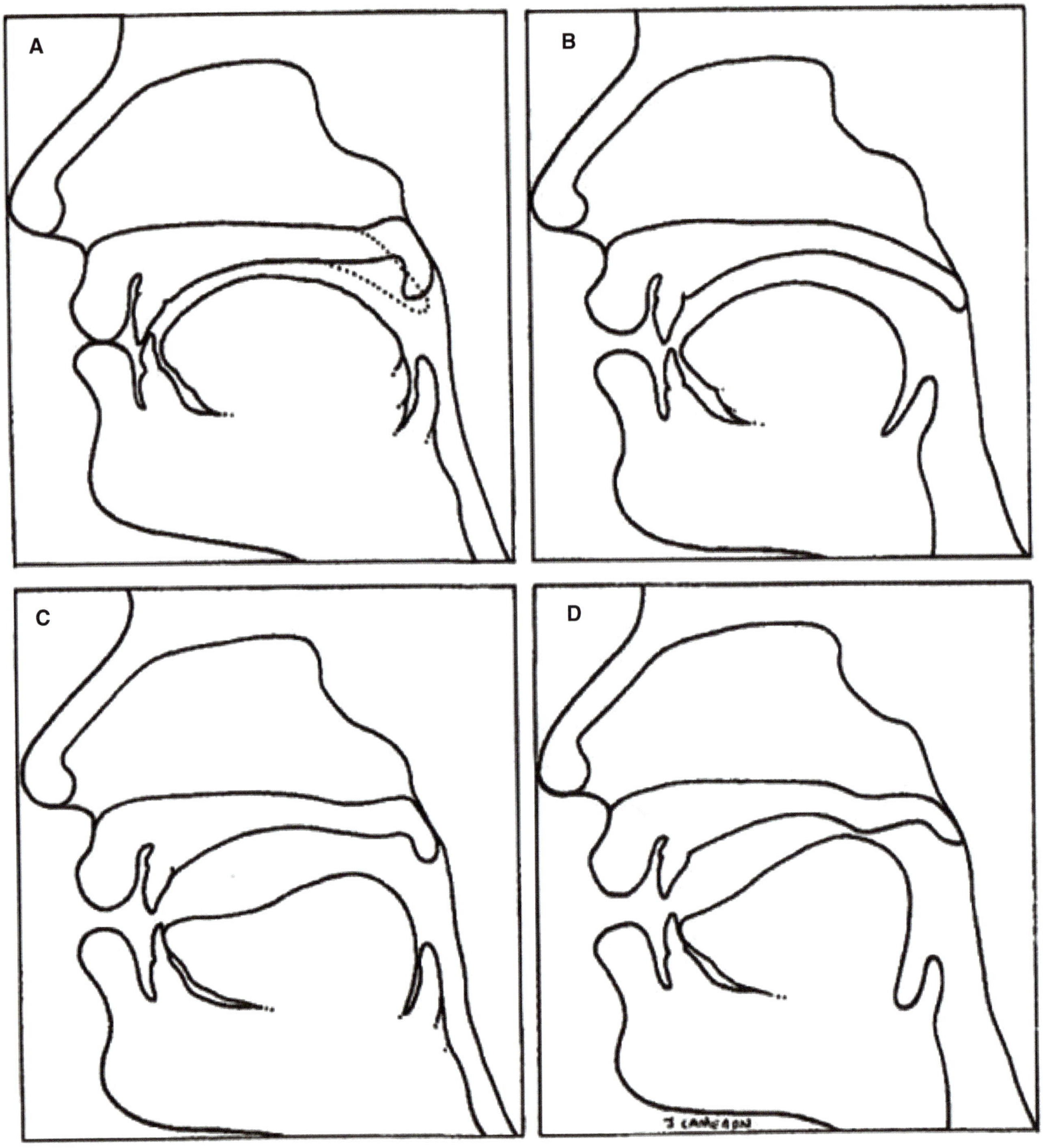

Fig. 5.22 Illustration of the mouth during swallowing and vocalization. (**A**) Shows schematically how the tensor and levator veli palatini muscles together with a part of the superior constrictor muscle close off the nasopharynx from the mouth during swallowing. The soft palate (see Figs. 5.19 and 5.23) is the lid which seals the passage from mouth to nasopharynx. In addition, the soft palate plays an important role in forming sounds like "i" (**B**), "ah" (**C**), or "u" (**D**). From Aiello and Dean 2002, Fig. 13.2 with permission

lesion, which he later proved by postmortem investigations. Wallenberg (see Box 5.6) published four papers on the syndrome (for details see Pearce 2000).

The typical symptoms of Wallenberg's syndrome and their structural equivalents are summarized in Table 5.1. The numbers in red refer to the microphotography shown in

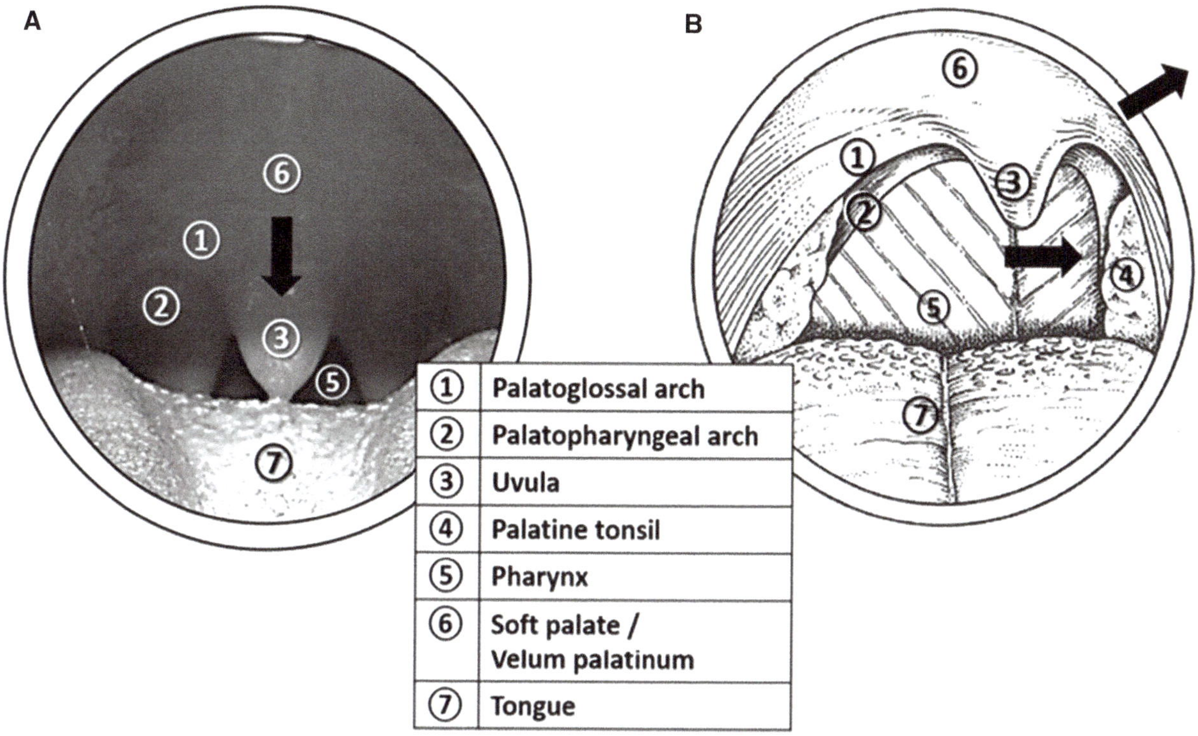

Fig. 5.23 View of the oral cavity with tongue, palatal arches, soft palate, and tonsils under normal conditions (**A**) and in case of glossopharyngeal/vagal lesions (**B**). (**A**) shows the view of the oral cavity in vivo, the inspection of which is compulsory part of any physical examination. Under normal circumstances (**A**), the uvula ③ is located in the midline indicated by the midline groove of tongue [Sulcus medianus linguae]. In (**B**), the typical finding in case of a right-sided paralysis of CNIX and/or CNX is shown. The right palatoglossal muscle (covered by the mucosa of the palatoglossal arch ①) and the right palatopharyngeal muscle (hidden under the mucosa of the palatopharyngeal arch ②) are unable to contract (compare with Fig. 5.21). This invests their left-hand counterparts with an overbalance, and the velum palatinum ⑥ with the uvula is dragged to the left side (→). Note that there is no simple clinical test to differentiate between lesions of CNIX vs. CNX. Ribbers et Müller-Thomsen fecerunt

Table 5.1 Localization of symptoms/Wallenberg syndrome

Ipsilateral symptoms	Structure (s. Fig. 5.24)
Dysphonia, dysarthria, and dysphagia Ipsilateral loss of gag reflex (see Box 5.5)	⑥ Posterior ncl. of vagus nerve / ⑤ Nucleus ambiguus
Impaired taste sensation	② Ncll. of solitary tract / Solitary tract
Vertigo with nystagmus	① Inferior (spinal) vestibular ncl.
Ipsilateral ataxia/tendency to fall to the ipsilateral side	③ Inferior cerebellar peduncle
Pain and numbness with impaired facial sensation	④ Spinal trigeminal ncl. / Spinal trigeminal tract
Contralateral symptoms	
Impaired pain + temperature sensation in arms and legs	◌ Spinothalamic tract

Fig. 5.23). The histological sections displayed in Fig. 5.24 show in (A) the normal anatomical conditions of the lateral medulla oblongata while in (B) the lesion zone of a Wallenberg case is shown.

Box 5.6 Adolph Wallenberg and the Nazis

Adolph Wallenberg, the son of the physician Samuel Wallenberg, was born in 1862 in Stargard (near Danzig, then in Western-Prussia). His grandfather was a Rabbi. Wallenberg studied medicine in Heidelberg and Leipzig and wrote a thesis on poliomyelitis. He returned to Danzig where he was appointed chief physician to the city hospital in 1907. He received the Erb medal for his "merits in the field of anatomy, physiology, and pathology of the nervous system" in 1929. He published 47 papers before retiring, but in 1938, he had to finally end his career under intolerable pressure from the Nazis. Eventually, he and his wife fled through Holland to Oxford, where he worked with LeGros Clark. In 1943, he emigrated to the USA. Wallenberg died from ischemic heart disease in 1949 (Pearce 2000).

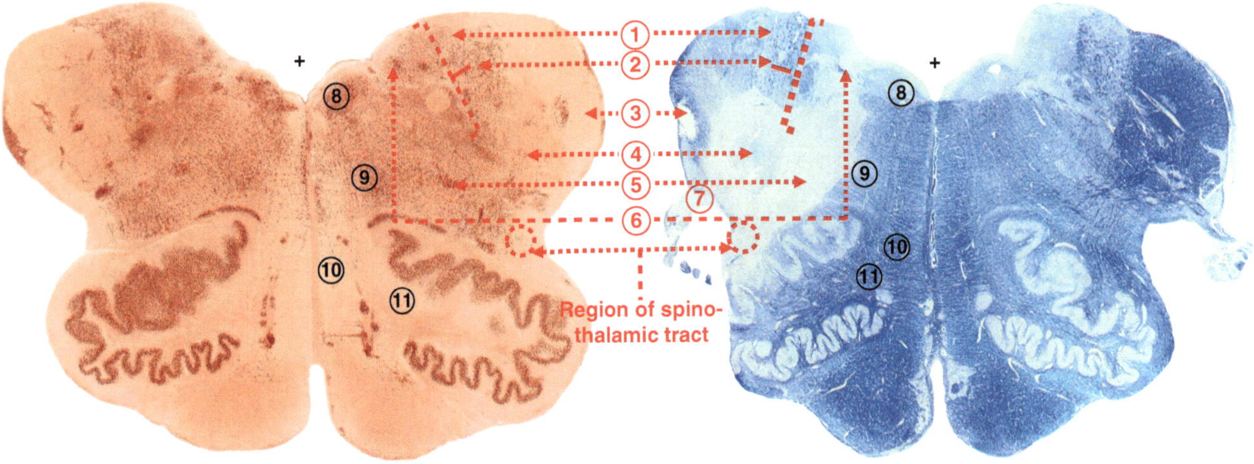

①	Inferior vestibular ncl.	⑦	Vagus nerve
②	Solitary ncll. + tract	⑧	Ncl. of hypoglossal nerve
③	Inferior cerebellar peduncle	⑨	Gigantocellular reticular ncl.
④	Spinal trigeminal ncl. + tract	⑩	Medial lemniscus
⑤	Nucleus ambiguus	⑪	Inferior olivary complex
⑥	Posterior ncl. of vagus nerve	+	4th ventricle

Fig. 5.24 Horizontal section through the medulla oblongata at the level of the exit of the vagus nerve. (**A**) Normal anatomy. Darrow red stain. Figures in red indicate those structures that are lesioned in the Wallenberg syndrome. (**B**) Klüver-Barrera fiber stain of the medulla oblongata at the same level as (**A**). Compare with Table 5.1

5.4.3 Raphe nuclei r9

5.4.3.1 Paramedian reticular nucleus [Ncl. reticularis paramedianus]

FIPAT Ch1 (see Sect. 1.8) has a posterior/dorsal paramedian nucleus [Ncl. paramedianus posterior/dorsalis] (Fig. 5.25), formerly Ncl. funiculi anterioris or Ncl. of Jacobsohn. The corresponding footnote mentions that this nucleus possibly is a precerebellar nucleus (Olszewski and Baxter 1982; Baizer et al. 2007).

The paramedian reticular nucleus (PRN) is small, located in the floor of the fourth ventricle near to the raphe obscurus nucleus and laterally flanked by the prepositus nucleus (Box 3.1).

Extensive retro- and anterograde tracing in the cat (Pan et al. 1992) suggests that the paramedian reticular nucleus projects to the pressor area of the posterior nucleus of vagus nerve and the nucleus of solitary tract as well as to the intermedial lateral cell column (IML) of the spinal cord. This constellation may allow the nucleus to exert its inhibitory functions on the sympathetic pressor actions. In addition, PRN may suppress the vagal bradycardic action through its reciprocal connections with the mesencephalic cardioinhibitory mechanism (CIM), the gigantocellular reticular nucleus and the ncl. ambiguus (Pan et al. 1992). This nucleus is affected in the spinocerebellar ataxias (see Sect. 7.2.1.3) SCA2 and SCA3 (Rüb et al. 2013).

In mice, a few labeled cells were found by Fu et al. (2011) in the intermediate reticular zone and the paramedian reticular nucleus after tracer injection into the cerebellum. VanderHorst and Ulfhake (2006) report no data on potential 5HT-immunoreactivity.

Do not confuse this nucleus with the bilateral paramedian pontine reticular formation.

5.4.3.2 Raphe obscurus nucleus

For details on this nucleus, see Sect. 4.3.3.1.

5.4.3.3 Raphe pallidus nucleus

For details on this nucleus, see Sect. 4.3.3.2.

5.4.4 Reticular nuclei r9

5.4.4.1 Intermediate reticular nucleus

For details on the intermediate reticular nucleus/zone, see Sect. 3.3.2.1.

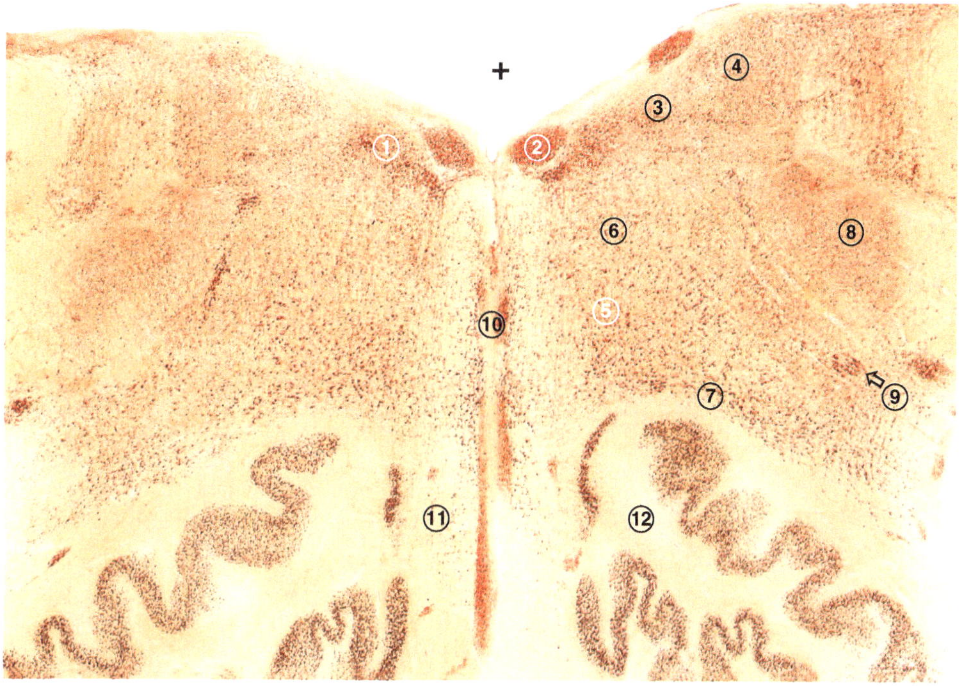

①	Prepositus hypoglossi ncl.	⑧	Parvocellular reticular ncl.
②	Posterior paramedian ncl.	⑨	Motor ncl. of facial nerve
③	Interpositus ncl.	⑩	Raphe obscurus ncl.
④	Medial vestibular ncl.	⑪	Medial lemniscus
⑤	Gigantocellular reticular ncl.	⑫	Inferior olivary complex
⑥	Posterior paragigantocellular reticular ncl.	+	4th ventricle
⑦	Lateral paragigantocellular ncl.		

Fig. 5.25 Horizontal section of the human medulla oblongata at the level of the posterior paramedian nucleus ③. See also atlas part Darrow red 15, 15 A. For details, see text. LabPON Twente

5.4.4.2 Parvocellular reticular nucleus

For details on the parvocellular reticular nucleus, see Sect. 6.5.5.2.

5.4.4.3 Gigantocellular reticular nucleus

For details on the gigantocellular reticular nucleus, see Sect. 4.3.4.2.

5.4.4.4 Dorsal paragigantocellular nucleus/ Posterior paragigantocellular reticular nucleus [Ncl. paragigantocellularis posterior]

5.4.4.4.1 Location of the dorsal paragigantocellular nucleus

The dorsal (posterior) paragigantocellular nucleus (DPN) (Fig. 5.25 ⑥ ⑦) is—as the name implies—the dorsal neighbor of the gigantocellular nucleus (Fig. 5.25 ⑤). Its longitudinal extension is from the oral to middle third of the inferior olive to the caudal pole of the facial nucleus. It is of rectangular shape, occupying the dorsomedial medullary tegmentum and displays mainly small to medium-sized neurons (Olszewski and Baxter 1982). Its neighbors are, in addition to the gigantocellular nucleus (see Sect. 4.3.4.2), the prepositus hypoglossi nucleus (dorsal, see Sect. 6.2.1.5), the parvocellular ncl. (lateral, see Sect. 6.5.5.2), and the medial longitudinal fasciculus (medial, see Fig. 3.14 A–G).

Some histochemical data on the human DPN are available. In human infants, Rikard-Bell et al. (1990) detected sparse substance P-like immunoreactivity—predominantly labeling terminals—similar to that in the gigantocellular nucleus.

Calcitonin-binding sites have been demonstrated by Sexton et al. (1994) in the gigantocellular, the dorsal and lateral paragigantocellular nuclei. In the human fetal brain, somatostatin receptors were shown in the DPN, the gigantocellular, the parabrachial, and the prepositus hypoglossi nucleus (Carpentier et al. 1996).

Studies in rats have shown that neurons of the dorsal paragigantocellular nucleus are involved in sleep regulation. These neurons project to the locus caeruleus and are active after REM (rapid eye movements) sleep and during recovery from REM sleep deprivation (Dergacheva et al. 2010). REM sleep is a sleep phase characterized by rapid eye movements (REM) with changes in autonomic and motor activity and dreaming.

5.4.4.5 Rostral ventral respiratory group

For details about this group, see Sect. 3.3.2.2.

5.4.4.6 Central cervical nucleus of the spinal cord

For details on this nucleus, see Sect. 3.3.2.4.

References

Aiello L, Dean C (2002) Chapter 13. An Introduction in human evolutionary anatomy. In: The anatomy of the vocal tract. Elsevier, pp 232–243

Arslan M (1960) The innervation of the middle ear. Proc R Soc Med 53:1068–1074

Augustine JR (2017) Human neuroanatomy, 2nd edn. Wiley-Blackwell, Hoboken

Bailey IL, Jackson AJ (2016) Changes in the clinical measurement of visual acuity. J Phys Conf Ser 772:012046

Baizer JS, Baker JF et al (2007) Neurochemical organization of the nucleus paramedianus dorsalis in human. Brain Res 1176:45–52

Beraneck M, Bojados M et al (2012) Ontogeny of mouse vestibulo-ocular reflex following genetic or environmental alteration of gravity sensing. PLoS One 7:e40414

Bui T, Das JM (2019) Anatomy, head and neck, pharyngeal muscles last StatPearls. In: StatPearls [Internet]. StatPearls, Treasure Island

Büttner-Ennever J, Gerrits NM (2004) Vestibular system. In: The human nervous system. Elsevier, pp 1212–1240

Carpentier V, Vaudry H et al (1996) Anatomical distribution of somatostatin receptors in the brainstem of the human fetus. Neuroscience 73:865–879

Challa VR, Moody DM et al (1995) Vascular malformations of the central nervous system. J Neuropathol Exp Neurol 54:609–621

Chanda A, Nanda A (2002) Multiple cavernomas of brain presenting with simultaneous hemorrhage in two lesions: a case report. Surg Neurol 57:340–345

Curthoys IS (2019) An overview of the peripheral vestibular system. In: Reference module in neuroscience and biobehavioral psychology

Dergacheva O, Wang X et al (2010) The lateral paragigantocellular nucleus modulates parasympathetic cardiac neurons: a mechanism for rapid eye movement sleep-dependent changes in heart rate. J Neurophysiol 104:685–694

Duvernoy H, Koritké JG et al (1972) Sur la vascularisation de l'area postrema et de la face posterieure du bulbe chez l'homme. Z Anat Entwickl Gesch 138:41–66

Felbor U, Sure U et al (2006) Genetics of cerebral cavernous angioma. Zentralbl Neurochir 67:110–116

Fritschi JA, Reulen HJ et al (1994) Cavernous malformations of the brainstem. A review of 139 cases. Acta Neurochir 130:35–46

Fu Y, Tvrdik P et al (2011) Precerebellar cell groups in the hindbrain of the mouse defined by retrograde tracing and correlated with cumulative Wnt1-Cre genetic labeling. Cerebellum 10:570–584

Gacek RR (1984) Efferent innervation of the labyrinth. Am J Otolaryngol 5:206–224

Gangemi M, Maiuri F et al (1990) Familial cerebral cavernous angiomas. Neurol Res 12:131–136

Garner TB, Del Curling O et al (1991) The natural history of intracranial venous angiomas. J Neurosurg 75:715–722

Giombini S, Morello G (1978) Cavernous angiomas of the brain. Account of fourteen personal cases and review of the literature. Acta Neurochir 40:61–82

Glattfelder KJ, Ng LL et al (2008) Area postrema. Nat prec

Goldberg JM, Wilson VJ et al (2012) The vestibular system; A sixth sense. Oxford University Press, New York

Goldmann EE (1913) Vitalfärbung am Zentralnervensystem: Beitrag zur Physio-Pathologie des Plexus choroideus und der Hirnhäute. Ausgabe 1 von Abh. der Königl. Preuss. Akad. der Wiss. 1913. Phys.-Math. Classe

Haidara A, Peltier J et al (2015) Microsurgical anatomy of the labyrinthine artery and clinical relevance. Turk Neurosurg 25:539–543

Horn AKE (2020) Vestibulo-vestibular commissural fibers. Neuroanatomy of central vestibular connections. In: Reference module in neuroscience and biobehavioral psychology, 2020

Horn CC, Kimball BA et al (2013) Why can't rodentia vomit? A comparative behavioral, anatomical, and physiological study. PLoS One 8:e60537

Huggenberger S, Moser N et al (2019) Neuroanatomie des Menschen. Springer

Kong W-J, Hussl B et al (1998) Ultrastructural localization of ChAT-like immunoreactivity in the human vestibular periphery. Heart Res 119:96–103

Labauge P, Laberge S et al (1998) Hereditary cerebral cavernous angiomas: clinical and genetic features in 57 French families. Lancet 352:1892–1897

Lakshminarayanan V (2016) Visual acuity. In: Chen J, Cranton W, Fihn M (eds) Handbook of visual display technology. Springer, Cham

Lewandowsky M (1900) Zur Lehre von der Cerebrospinalflüssigkeit. Zeitschrift für klinische Medicin 40:480–494

Li L, Huang C et al (2010) Structural remodelling of aortic arch and nucleus ambiguus (NA) projections to cardiac ganglia in a transgenic mouse model of type I diabetes. J Comp Neurol 518:2771–2793

Liang H, Bácskai T et al (2014) Projections from the lateral vestibular nucleus to the spinal cord in the mouse. Brain Struct Funct 219:805–815

Lindstrom PA, Brizzee KR (1962) Relief of intractable vomiting from surgical lesions. J Neurosurg 19:228–236

Longatti P, Porzionato A et al (2015) The human area postrema: clear-cut silhouette and variations shown in vivo. J Neurosurg 122:989–995

Louis N, Marsh R (2016) Simultaneous and sequential hemorrhage of multiple cerebral cavernous malformations: a case report. J Med Case Rep 10:36

Low MJ (2016) Area postrema. In: Williams textbook of endocrinology, 13th edn

Lui F, Tadi P et al (2020) Wallenberg syndrome. StatPearls

Malone JC, Arya NR (2020) Anatomy, head and neck, swallowing. In: StatPearls [Internet]. StatPearls, Treasure Island

Martin NA, Vinters HV (1995) Arteriovenous malformations. In: Carter LP, Spetzler RF, Hamilton MG (eds) Neurovascular surgery. Mc Graw-Hill, New York, pp 875–903

McCormick WF (1966) The pathology of vascular ("arteriovenous") malformations. J Neurosurg 24:807–816

McKinley MJ, McAllen RM et al (2003) The sensory circumventricular organs of the mammalian brain. Adv Anat Embryol Cell Biol 172:1–122

Mergenthaler P, Lindauer U et al (2013) Sugar for the brain: the role of glucose in physiological and pathological brain function. Trends Neurosci 36:587–597

Miller AD, Leslie RA (1994) The area postrema and vomiting. Front Neuroendocrinol 15:301–320

Mori T, Fujimoto M et al (1996) Familial presumed cerebral cavernous angiomas diagnosed by MRI: three generations. Neuroradiology 38:641–645

Morrison MP, Postma GN (2013) Surgical treatment of Zenker's diverticulum. In: Morrison MP (ed) Principles of deglutition. Springer, pp 847–858

Morsli H, Choo D et al (1998) Development of the mouse inner ear and origin of its sensory organs. J Neurosci 18:3327–3335

Nickele CM, Akture E et al (2012) A stepwise illustration of the translabyrinthine approach to a large cystic vestibular schwannoma. Neurosurg Focus 33:E11

Okada R, Muro S et al (2018) The extended bundle of the tensor veli palatini: Anatomic consideration of the dilating mechanism of the Eustachian tube. Auris Nasus Larynx 45:265–272

Olszewski J, Baxter D (1982) Cytoarchitecture of the human brain stem, 2nd edn. Karger, Basel

Pan CM, Wang SD et al (1992) Afferent and efferent connections of the paramedian reticular nucleus in the brain stem of cats. Chin J Physiol 35:181–196

Pearce JM (2000) Wallenberg's syndrome. J Neurol Neurosurg Psychiatry 68:570

Putman CM, Chaloupka JC et al (1996) Exceptional multiplicity of cerebral arteriovenous malformations associated with hereditary hemorrhagic telangiectasia (Osler-Weber-Rendu syndrome). AJNR Am J Neuroradiol 17:1733–1742

Rengachary SS, Kaylan-Raman UP (1996) Teleangiectasias and venous malformations. In: Wilkins RH, Rengachary SS (eds) Neurosurgery. McGraw-Hill, New York, pp 2509–2514

Rigamonti D, Spetzler RF et al (1990) Cerebral venous malformations. J Neurosurg 73:560–564

Rigamonti D, Johnson PC et al (1991) Cavernous malformations and capillary teangiectasia: a spectrum within a single pathological entity. Neurosurgery 28:60–64

Rikard-Bell GC, Törk I et al (1990) Distribution of substance P-like immunoreactive fibres and terminals in the medulla oblongata of the human infant. Neuroscience 34:133–148

Rohrschneider I, Schink I et al (1972) Der Feinbau der Area postrema der Maus. Z Zellforsch 123:251–276

Rüb U, de Vos RAI et al (2002) Spinocerebellar ataxia type 3 (Machado-Joseph disease): severe destruction of the lateral reticular nucleus. Brain 125:1–10

Rüb U, Schöls L et al (2013) Clinical features, neurogenetics and neuropathology of the polyglutamine spinocerebellar ataxias type 1, 2, 3, 6 and 7. Prog Neurobiol 104:38–66

Russell DS, Rubinstein LJ (1989) Pathology of tumors of the nervous system. Williams & Wilkins, Baltimore, pp 727–765

Russell DS, Rubinstein LJ (1998) Pathology of tumors of the nervous system, 6th edn. Arnold, London, pp 257–294

Safieddine S, El-Amraoui A (2012) The auditory hair cell ribbon synapse: from assembly to function. Annu Rev Neurosci 35:509–528

Sasegbon A, Hamdy S (2017) The anatomy and physiology of normal and abnormal swallowing in oropharyngeal dysphagia. Neurogastroenterol Motil 29:e13100

Sekirnjak C, du Lac S (2006) Physiological and anatomical properties of mouse medial vestibular nucleus neurons projecting to the oculomotor nucleus. J Neurophysiol 95:3012–3023

Sexton PM, Paxinos G et al (1994) In vitro autoradiographic localization of calcitonin binding sites in human medulla oblongata. J Comp Neurol 341:449–463

Taylor RR, Jaggera DJ et al (2015) Characterizing human vestibular sensory epithelia for experimental studies: new hair bundles on old tissue and implications for therapeutic interventions in ageing. Neurobiol Aging 36:2068–2084

ten Donkelaar HJ (2011) Clinical neuroanatomy: brain circuitry and its disorders. Springer

Tokita K, Inoue T et al (2009) Afferent connections of the parabrachial nucleus in C57BL/6J mice. Neuroscience 161:475–488

Toorop RJ, Ousrout R et al (2013) Carotid baroreceptors are mainly localized in the medial portions of the proximal internal carotid artery. Ann Anat 193:248–252

VanderHorst VG, Ulfhake B (2006) The organization of the brainstem and spinal cord of the mouse: relationships between monoaminergic, cholinergic, and spinal projection systems. J Chem Neuroanat 31:2–36

Wilkins RH (1985) Natural history of intracranial malformations: a review. Neurosurgery 16:421–430

Zabramski JM, Wascher TM et al (1994) The natural history of familial cavernous malformations: results of an ongoing study. J Neurosurg 80:422–432

Zabramski JM, Henn JS et al (1999) Pathology of cerebral vascular malformations. Neurosurg Clin N Am 10:395–410

Zamysłowska-Szmytke E, Szostek-Rogula S et al (2015) Bedside examination for vestibular screening in occupational medicine. Int J Occup Med Environ Health 28:379–387

Zhu H, Guo Y et al (2014) Familial cerebral cavernous angiomas: clinical and genetic features in a Chines family with a frame-shift mutation in the CCM1 gene (krit1). J Mol Neurosci 54:790–795

Zwergal A, Büttner-Ennever JA et al (2008) An ipsilateral vestibulothalamic tract adjacent to the medial lemniscus in humans. Brain 131:2928–2935

Web Link

https://www.skybrary.aero/index.php/Middle_Ear_and_Sinus_Problems

Rhombomere 8 r8

6

Contents

H. Schröder et al., *The Human Brainstem*, https://doi.org/10.1007/978-3-030-89980-6_6

Abstract

As to rhombic lip derivatives, this chapter deals with the prepositus and subhypoglossal nuclei which together with the intercalated nucleus form the group of perihypoglossal nuclei (see Chap. 3). They play an important role in gaze holding.

Dorsomedial spinal trigeminal and intermedius nucleus have been shown for other mammals but are most likely not present in humans. This also holds true for the caudoventral reticular nucleus.

The parvocellular reticular nucleus is the most likely candidate nucleus of origin for the efferent innervation of the vestibular hair cells (see Chap. 5). Related to this nucleus, the concept of the lateral reticular formation is described and discussed in this chapter.

6.1 Roof plate r8

6.1.1 Area postrema r8

For details on the Area postrema, see Sect. 5.1.1.

6.2 Rhombic lip

6.2.1 Precerebellar nuclei r8

6.2.1.1 Linear nucleus of the medulla

For details on the linear nucleus of the medulla, see Sect. 4.1.1.5.

6.2.1.2 Inferior olive

For details on the inferior olive, see Sect. 4.1.1.3.

6.2.1.3 External cuneate nucleus

For details on the external cuneate nucleus, see Sect. 5.2.1.1.

6.2.1.4 Lateral reticular nucleus

For details on lateral reticular nucleus, see Sect. 4.1.1.1.

6.2.1.5 Prepositus hypoglossi nucleus [Ncl. prepositus hypoglossi]

Details about the prepositus hypoglossi nucleus are described below together with the nucleus of Roller (Sect. 6.2.1.6) (Fig. 6.1).

6.2.1.6 Nucleus of Roller (Subhypoglossal ncl.) [Ncl. subhypoglossalis]

Christian F. W. Roller (1802–1878) Psychiatrist, Achern, Germany.

The human prepositus nucleus and the nucleus of Roller (Fig. 6.2) belong to the group of perihypoglossal nuclei (Box 3.1), together with the intercalated nucleus (see Sect. 3.1.1.1). Do not confuse the nucleus of Roller, i.e., the subhypoglossal nucleus, with the inferior/spinal vestibular nucleus (see Sect. 5.3.1.1) which also bears the eponym Roller.

Direct anatomical evidence on the connectivity and function of these nuclei is not available for humans. Lesioning of the medial vestibular nucleus (see Fig. 6.1 ④, Sect. 5.3.1) and the prepositus nucleus in monkeys resulted in massive difficulties with fixation and slow eye movements (see ten Donkelaar 2011) (see also Box 9.6 "Basic eye movements").

From studies in cats and monkeys, it is known that the prepositus hypoglossi nucleus projects to the abducens nucleus (see Sect. 9.2.1.1), the oculomotor nucleus, (see Sect. 15.3.1.1), the vestibulocerebellum, the pontomedullary reticular formation (see Box 6.1), and the superior colliculi (see Sect. 15.1.3) (for details, see ten Donkelaar 2011).

Büttner-Ennever and Horn (1997) have claimed that the prepositus nucleus is, just as the interstitial nucleus of Cajal (see Sect. 16.4.2), involved in the integrator function for horizontal eye movements and receives afferents from all oculomotor structures.

Rüb et al. (2008a) have stated "… *Recent pathoanatomical studies using Nissl-pigment stained thick serial tissue sections revealed that the premotor oculomotor brainstem network is, in fact, severely affected in terminal SCA3* (spinocerebellar ataxia type 3) *patients (*Rüb et al. 2008b*)* (see Sect. 7.2.1.3.3 for SCA3). *The oculomotor* (see Sect. 15.3.1.1), *trochlear* (see Sect. 14.5.2.2)*, and abducens nuclei* (see Sect. 9.2.1.1)*, the reticulotegmental nucleus of the pons* (see Sect. 10.1.1.1)*, the area of the excitatory burst neurons for horizontal saccades,*

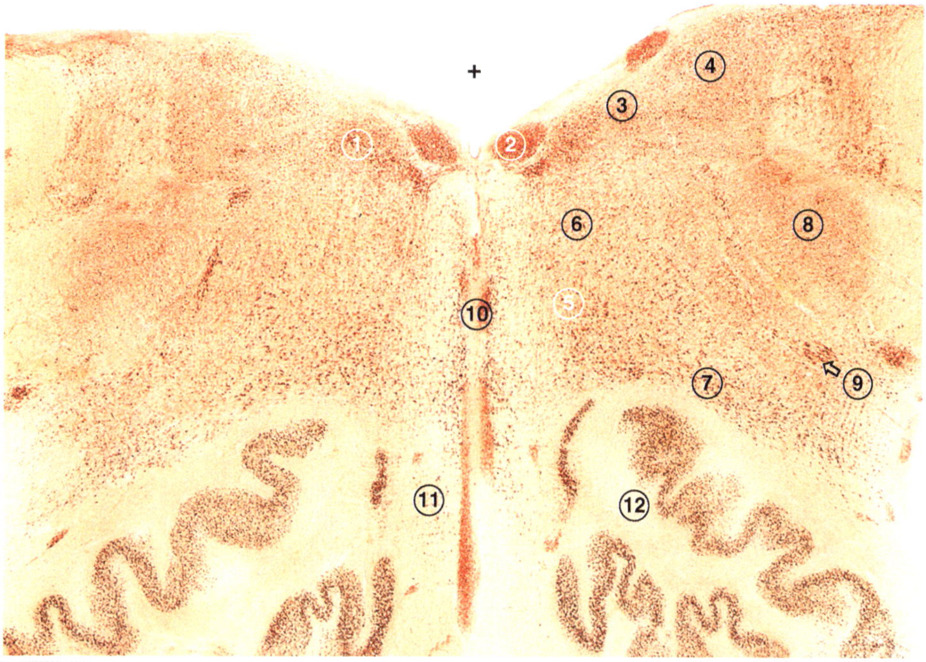

①	Prepositus hypoglossi ncl.	⑥	Posterior paragigantocellular reticular ncl.	⑪	Medial lemniscus
②	Posterior paramedian ncl.	⑦	Lateral paragigantocellular ncl.	⑫	Inferior olivary complex
③	Interpositus ncl.	⑧	Parvocellular reticular ncl.	+	4th ventricle
④	Medial vestibular ncl.	⑨	Motor ncl. of facial nerve CNVII		
⑤	Gigantocellular reticular ncl.	⑩	Raphe obscurus ncl.		

Fig. 6.1 Horizontal section through the medulla oblongata at the level of the posterior paramedian nucleus. The prepositus hypoglossi nucleus ① lies lateral of the posterior paramedian nucleus ② in the floor of the fourth ventricle. See also atlas part Darrow red 15, 15A and Campbell 5, 6A. LabPON Twente

the raphe interpositus nucleus (see Sect. 9.2.4.1), *the superior, medial and lateral vestibular nuclei* (see Sects. 7.2.1.1, 10.2.2.1, and 11.2.2.1 respectively) *as well as the prepositus hypoglossi nucleus* (*see* Sect. 6.2.1.5) *are among the consistent targets of the neurodegenerative process in SCA3* (Rüb et al. 2008b). *In view of the functional role of these brainstem nuclei, these findings explained several hitherto poorly understood SCA3 oculomotor symptoms (*i.e., *saccadic smooth pursuits, initially hypermetric and later hypometric saccades, impairment of the vestibuloocular reflex* (see Box 6.2) *and optokinetic nystagmus, gradual slowing and sometimes loss of vertical and horizontal saccades, and horizontal gaze-evoked nystagmus)"* (see Box 9.6 Basic eye movements).

Box 6.1 Pontomedullary reticular formation

The pontomedullary reticular formation (PMRF) is a collective term for the medullary gigantocellular reticular nucleus (see Sect. 4.3.4.2), the ventral reticular nucleus, the magno- and parvocellular (see Sect. 6.5.5.2) reticular nuclei, a paramedian reticular nuclear group (see Sect. 5.4.3.1) plus the pontine reticular caudal (see Sect. 7.3.5.3) and oral (Sect. 10.3.5.4, see Sect. 7.3.5.3) nuclei. Together with the reticulospinal tract, it provides a major descending system for the control of movement and a critical hub for sensorimotor integration (MacKinnon 2018). It allows the CNS to appropriately couple voluntary actions with posture and locomotion. Lesions to this structure or to the reticulospinal pathways, in both cats and primates, lead to an inability to maintain an upright posture during normal everyday activities (Lawrence and Kuypers 1968a, b).

The PMRF receives input from the motor and premotor cortex, the cerebellum, the vestibular nuclei (see Chaps. 5, 7, 10, 11) and, via the spinoreticular tract from peripheral proprioceptive and exteroceptive afferents (for details, see MacKinnon 2018, Miller et al. 2017). In addition, it is under indirect control of the basal ganglia via the mesencephalic locomotor region, comprising the cuneiform nucleus (see Sect. 14.4.1.1) and the pedunculopontine nuclei (see Sect. 13.1.1.1.1).

The output of the PMRF is mediated via the lateral (medullary) and the medial (pontine) reticulospinal tracts (MacKinnon 2018) to spinal interneurons (further to spinal motoneurons) and to gamma motoneurons.

Projections from the medullary reticulospinal formation run bilaterally just lateral to the medial longitudinal fasciculus (see Fig. 3.14A–G), then enter the spinal cord and either join with the pontine fibers that compose the medial reticulospinal tract or travel via the ventrolateral funiculus in the lateral reticulospinal tract (Parent 1996).

In mice, the perihypoglossal nuclei enclose the rostral part of the hypoglossal nucleus (Fu et al. 2011). They are involved in eye movement coordination (Kolkman et al. 2011). Upon intracerebellar tracer injection, labeled neurons were detected bilaterally in all three perihypoglossal nuclei (Fu et al. 2011). The majority of labeled perihypoglossal neurons are located in the nucleus of Roller and the caudal portion of the prepositus nucleus. Together with the medial vestibular nucleus (see Sect. 7.2.1.1), the latter is involved in the vestibulo-ocular reflex (see Box 6.2).

Box 6.2 Vestibulo-ocular reflex

Rotations of the head are detected by the vestibular apparatus of the membranous labyrinth (see Sect. 5.3.1). The vestibular nuclei are connected with the oculomotor nuclei (CNIII, IV, VI) (see Sects. 9.2.1.1, 14.5.2.1, and 15.3.1.1) via the medial longitudinal fasciculus (see Fig. 3.14A–G, Box 9.2, Medial longitudinal fasciculus).

The rotation-elicited signal triggers an inhibitory signal to the extraocular muscles on one side and an excitatory signal to the contralateral muscles. This results in compensatory eye movements (shown by Sekirnjak and du Lac 2006 for the mouse).

Clinical studies have shown the perihypoglossal nuclei to be involved in gaze holding. Lesions of the prepositus nucleus mainly evoke downbeat gaze-evoked nystagmus

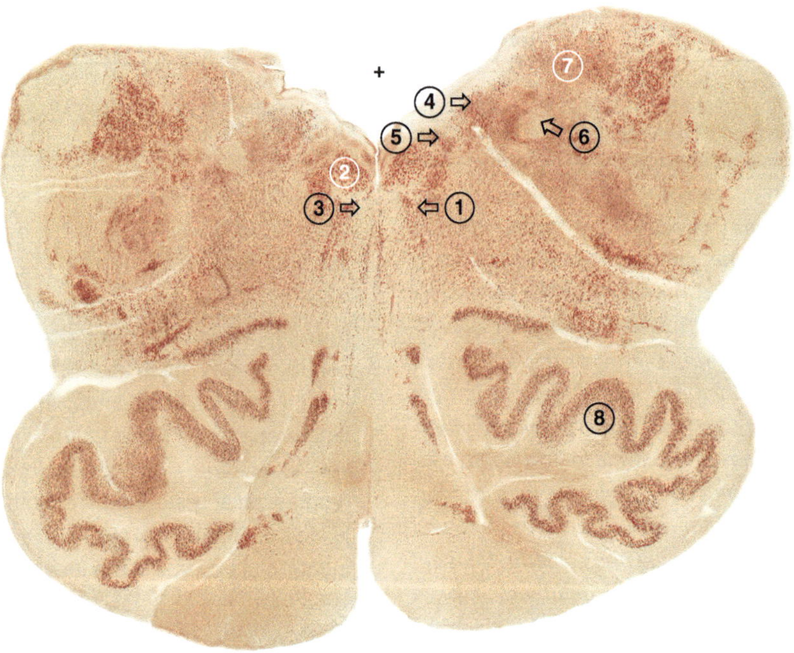

①	Subhypoglossal ncl. Roller	⑥	Solitary tract
②	Hypoglossal ncl.	⑦	Medial vestibular ncl.
③	Medial longitudinal fasciculus	⑧	Inferior olivary complex
④	Posterior ncl. of vagus nerve	+	4th ventricle
⑤	Intercalated ncl.		

Fig. 6.2 Horizontal section through the medulla oblongata at the level of the subhypoglossal nucleus of Roller ① (see Box 3.1). See also atlas part Darrow red 10, 10A and Campbell 2, 3. LabPON Twente

(see Box 6.3) and lesions of the intercalated nucleus and the Roller nucleus cause upbeat gaze-evoked nystagmus (Kheradmand et al. 2016).

In SCA2 and SCA3, patient's widespread involvement of these nuclei such as vestibular, Roller, intercalate and prepositus hypoglossi nuclei is present (see also Sect. 7.2.1.3).

Box 6.3 Gaze-evoked nystagmus
Gaze-holding mechanisms mediated by brainstem nuclei and the cerebellum keep the eyes steady when the head is not in motion. Abnormal spontaneous eye movements such as nystagmus are threats to steady fixation and clear vision (Shaikh 2019). In gaze-evoked nystagmus, alternating drifts and corrective movements and their gaze position are the characteristic clinical phenomena (Shaikh 2019; compare Box 3.2). Lesions of the interstitial nucleus of Cajal (see Sect. 16.4.2) lead to vertical gaze-evoked nystagmus and lesions of the medial vestibular nucleus (Sect. 7.2.1.1) and the prepositus hypoglossi nucleus lead to horizontal gaze-evoked nystagmus (acc.to Rüb et al. 2008a).

6.3 Alar r8

6.3.1 Vestibular nuclei r8

6.3.1.1 Spinal vestibular nucleus

For details on the spinal vestibular nucleus, see Sect. 5.3.1.1.

6.3.2 Monoamine nuclei r8

6.3.2.1 A2 noradrenaline cells (NA2)

For details on these cells, see Sect. 3.2.1.1.

6.3.2.2 C1 adrenaline cells

For details on these cells, see Sect. 3.2.1.2.

6.3.3 Trigeminal sensory nuclei r8

6.3.3.1 Spinal trigeminal nucleus

For details on the spinal trigeminal nucleus, see Sect. 3.2.2.2

6.3.3.2 Dorsomedial spinal trigeminal nucleus / Nucleus of trigeminal nerve r8 [Ncl. spinalis n. trigemini dorsomedialis]

In mice, the dorsomedial spinal trigeminal nucleus is located at the dorsomedial angle of the spinal trigeminal nucleus and extends from the caudal end of the parvocellular reticular nucleus to the level of the facial nucleus (Chap. 8 r6). This nucleus projects to the parabrachial nuclei (see Sect. 13.2.3) (Tokita et al. 2009).

It has not been established for the human brain.

http://braininfo.rprc.washington.edu/whatspecies.aspx?questID=1103

6.3.4 Dorsal column nuclei r8

6.3.4.1 Cuneate nucleus

For details on the cuneate nucleus, see Sect. 3.2.3.3.

6.3.5 Solitary nuclei r8

6.3.5.1 Solitary nucleus

For details on the solitary nuclei, see Sect. 3.2.4.

6.4 Alar tegmental r8

6.4.1 Matrix region of the medulla

For details on the matrix region of the medulla, see Sect. 3.2.5.1.

6.4.2 Intermedius nucleus of the medulla r8 [Ncl. intermedius medullae]

This nucleus has not been established for the human brain.

http://braininfo.rprc.washington.edu/whatspecies.aspx?questID=1066

In mice, it is a small perihypoglossal nucleus lateral of the hypoglossal nucleus at the level of the Area postrema. From studies in other mammals, it is known to receive afferent signals from the neck muscles and descending input from the vestibular nuclei, the gustatory portion of the solitary nucleus (NTS) and cortical areas involved in movements of the tongue (Edwards et al. 2009). Efferents of the nucleus reach the NTS and the hypoglossal nucleus. The connectivity pattern suggests a possible role in the coordination of tongue movements and autonomic responses to changes in posture.

6.5 Basal r8

6.5.1 Somatic motor r8

6.5.1.1 Hypoglossal nucleus

For details on the hypoglossal nucleus, see Sect. 4.3.1.1.

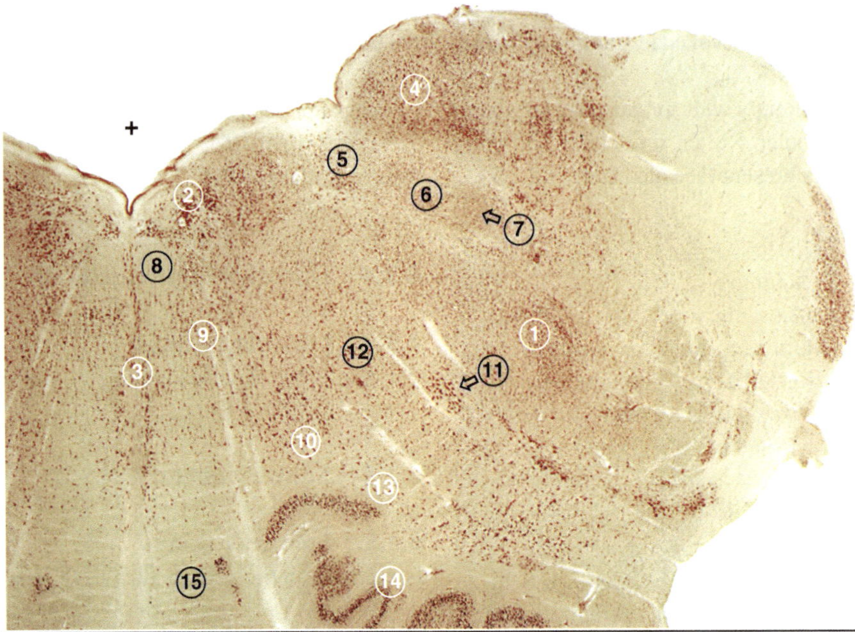

①	Parvocellular reticular ncl.	⑨	Posterior paragigantocellular reticular ncl.
②	Prepositus hypoglossi ncl.	⑩	Gigantocellular reticular ncl.
③	Raphe obscurus ncl.	⑪	Ncl. ambiguus
④	Medial vestibular ncl.	⑫	Intermediate reticular zone
⑤	Posterior ncl. of vagus nerve	⑬	Lateral paragigantocellular ncl.
⑥	Solitary ncll.	⑭	Inferior olivary complex
⑦	Solitary tract	⑮	Medial lemniscus
⑧	Medial longitudinal fasciculus	+	4th ventricle

Fig. 6.3 Horizontal section through the human brainstem at the level of the medial vestibular nucleus ④. The parvocellular reticular nucleus ① is located lateral of the ambiguus nucleus ⑪. Its neurons appear small compared to those of the gigantocellular reticular nucleus ⑩. See also atlas part Darrow red 13. LabPON Twente

6.5.2 Branchial motor r8

6.5.2.1 Ncl. ambiguus

For details on the ambiguus nucleus, see Sect. 5.4.2.1.

6.5.3 Raphe nuclei r8

6.5.3.1 Paramedian reticular nucleus

For details on the paramedian reticular nucleus, see Sect. 5.4.3.1.

6.5.3.2 Raphe obscurus nucleus

For details on this nucleus, see Sect. 4.3.3.1.

6.5.3.3 Raphe pallidus nucleus

For details on this nucleus, see Sect. 4.3.3.2.

6.5.4 Basal tegmentum r8

6.5.4.1 Intercalated nucleus of the medulla

For details on the intercalated nucleus of the medulla, see Sect. 3.1.1.1.

6.5.5 Reticular nuclei r8

6.5.5.1 Intermediate reticular nucleus

For details on the intermediate reticular nucleus, see Sect. 3.3.2.1.

6.5.5.2 Parvocellular reticular nucleus [Ncl. reticularis parvocellularis]

The parvocellular reticular nucleus (Fig. 6.3 ⑬) (parvus Latin = small, alluding to the small size of its neurons compared to the large ones in the neighboring gigantocellular reticular nucleus, Fig. 6.3 ⑩) is part of the lateral reticular formation (see Box 6.4). The official term is parvocellular, sometimes you find parvicellular (Olszewski and Baxter 1982).

Box 6.4 The lateral reticular formation (see also Sect. 3.3.2)

The lateral reticular formation as defined by Nieuwenhuys et al. (1988) consists of six overlapping structures:

1. The parvocellular nucleus or area (see Sect. 6.5.5.2).
2. The ventromedial superficial area (see below).
3. The lateral pontine tegmentum (see below).
4. The noradrenergic cell groups A1–A7 (A1—not present in human brain, A2 see Sect. 3.2.1.1, A3—not present in human brain, A4 noradrenergic cells in caudolateral pons, A5 see Sect. 8.1.3.1, A6 see Sect. 13.2.2.1 “locus coeruleus,” A7 see Sect. 11.2.3.1).
5. The adrenergic cell groups C1 and C2 (see Sect. 3.2.1.2).
6. The cholinergic cell groups Ch5 (see Sect. 13.1.1.1) and Ch6 (see Sect. 13.1.1.2).

Holstege and Kuypers (1977) have suggested, based on studies in the cat, that the majority of the rhombencephalic reticular formation is part of the lateral reticular zone.

Ad 2. The ventromedial superficial area [Area superficialis ventromedialis]

This area extends from the level of the facial nucleus (see Sect. 8.2.1) to the spinomedullary junction and contains in its rostral part the lateral paragigantocellular nucleus (see Sect. 7.3.5.6) and the retroambiguus nucleus in its caudal part (ten Donkelaar 2011). It is involved in cardiovascular and respiratory control. Furthermore, the noradrenergic cell groups A1 (not present in human brain) and A5 (Sect. 8.1.3.1), as well as the adrenergic cell group C1 (Sect. 3.2.1.2), are located in the ventromedial superficial area.

Ad 3. The lateral pontine tegmentum [Tegmentum pontis laterale]

The lateral pontine tegmentum forms the enlarged rostral end of the lateral reticular zone (ten Donkelaar 2011) and contains the parabrachial nuclei (see Sect. 13.2.3) and the Kölliker-Fuse nucleus (see Sect. 13.2.4). In addition, the locus coeruleus (see Sect. 13.2.2.1, A6), the subcaeruleus area (see Sect. 12.2.4.2) and the cell group A7 (see Sect. 11.2.3.1) are located within the lateral pontine tegmentum (ten Donkelaar 2011). The area is mainly involved in respiratory, cardiovascular, and gastrointestinal regulation. Finally, the principal sensory trigeminal nucleus (see Sect. 12.2.3.2) is located in the lateral pontine tegmentum.

In addition to the lateral, there is a medial reticular formation comprising the ventral, gigantocellular, caudal, and oral pontine nuclei as well as the mesencephalic cuneiform and subcuneiform nuclei (ten Donkelaar 2011) (for details, see Box 4.5).

The details concerning the nuclei mentioned under 2 through 6 will be dealt with in the corresponding chapters.

As concluded from studies in several animal species, the caudal parvocellular reticular nucleus is the most likely candidate nucleus of origin for the efferent innervation of the vestibular hair cells (see Gacek 1984, see Sect. 5.3.1). Kong et al. (1998) have shown ChAT-immunoreactive terminals in the human vestibular apparatus, which provides evidence for a cholinergic nature of efferent vestibular innervation in humans.

Interestingly, in the rat, the vestibular efferent neurons inside the parvocellular reticular neurons have been shown to be ChAT-immunoreactive (Ohno et al. 1991).

Mizukawa et al. (1986) have mapped the human brainstem using AChE-histochemistry (Box 3.10) and ChAT-immunohistochemistry (for technical aspects, see e.g., Ohno et al. 1991). Among many other structures, they have seen ChAT-positive cells in a region they had coined Formatio reticularis centralis (FRC). This term is not canonical and a comparison of the map in their Fig. 1 (an immunostained microphotography of their FRC was not provided) with plates XVI/XVII of Olszewski and Baxter’s atlas (1982) suggests that the FRC of Mizukawa et al. (1986) is at least partly identical with the parvocellular reticular nucleus. That means that similar to the rat there are cholinergic neurons in the human parvocellular reticular nucleus which could belong to the cholinergic vestibular terminals seen by Kong et al. (1998).

Tracing studies in mice have revealed a few labeled cells in the parvocellular reticular nucleus and its alpha part after HRP (horseradish peroxidase) injection into the cerebellum (see Fu et al. 2011). Han et al. (2017) have shown that the parvocellular reticular formation is the target of efferents of the central nucleus of the amygdala (CeA), which controls the cervical-mandibular muscular system. The co-activation of the CeA with the parvocellular nucleus—periaqueductal gray system induces robust hunting behavior in mice (Han et al. 2017).

In rats, a neuronal connection between the premotor neurons of the dorsal parvocellular nucleus and hypoglossal motoneurons has been shown (Zhang and Luo 2003), as well as with the trigeminal motor nucleus (Fay and Norgren 1997). This aligns the parvocellular reticular nucleus/formation with the premotor system of masticatory motoneurons (ten Donkelaar 2011).

6.5.5.3 Gigantocellular reticular nucleus

For details on the gigantocellular reticular nucleus, see Sect. 4.3.4.2.

6.5.5.4 Lateral paragigantocellular nucleus

For details on the lateral paragigantocellular nucleus, see Sect. 7.3.5.6.

6.5.5.5 Dorsal paragigantocellular nucleus

For details on the dorsal paragigantocellular nucleus, see Sect. 5.4.4.4.

6.5.5.6 Caudoventral reticular nucleus [Ncl. reticularis caudalis ventrolateralis]

The term caudolateral reticular nucleus refers to a group of cells in the medullary reticular formation of the mouse. The neurons have been reported to contain tyrosine hydroxylase (cf. Tanii et al. 2000). It has not been established for the human brain. http://braininfo.rprc.washington.edu/whatspecies.aspx?questID=1201.

6.5.5.7 Pre-Bötzinger complex

For details on the Pre-Bötzinger complex, see Sect. 3.3.2.2.

6.5.5.8 Rostral ventral respiratory group

For details on the rostral ventral respiratory group, see Sect. 3.3.2.2.

References

Büttner-Ennever JA, Horn AK (1997) Anatomical substrates of oculomotor control. Curr Opin Neurobiol 7:872–879

Edwards IJ, Deuchars SA, Deuchars J (2009) The intermedius nucleus of the medulla: a potential site for the integration of cervical information and the generation of autonomic responses. J Chem Neuroanat 38:166–175

Fay RA, Norgren R (1997) Identification of rat brainstem multisynaptic connections to the oral motor nuclei using pseudorabies virus. I. Masticatory muscle motor system. Brain Res Brain Res Rev 25:255–275

Fu Y, Tvrdik P et al (2011) Precerebellar cell groups in the hindbrain of the mouse defined by retrograde tracing and correlated with cumulative Wnt1-cre genetic labeling. Cerebellum 10:570–584

Gacek RR (1984) Efferent innervation of the labyrinth. Am J Otolaryngol 5:206–224

Han W, Tellez LA et al (2017) Integrated control of predatory hunting by the central nucleus of the amygdala. Cell 168:311–324

Holstege G, Kuypers HGJM (1977) Propriobulbar fibre connections to the trigeminal, facial and hypoglossal motor nuclei. I. An anterograde degeneration study in the cat. Brain 100:239–264

Kheradmand A, Colpak AI et al. (2016) Eye movements in vestibular disorders. In: Furman JM, Lempert T (eds.) Handbook of clinical neurology, vol. 137 (3rd series). Elsevier

Kolkman KE, Moghadam SH et al (2011) Intrinsic physiology of identified neurons in the prepositus hypoglossi and the medial vestibular nuclei. J Vestib Res 21:33–47

Kong W-J, Hussl B et al (1998) Ultrastructural localization of ChAT-like immunoreactivity in the human vestibular periphery. Heart Res 119:96–103

Lawrence DG, Kuypers HGJM (1968a) The functional organization of the motor system in the monkey. I. The effects of bilateral pyramidal lesions. Brain 91:1–15

Lawrence DG, Kuypers HGJM (1968b) The functional organization of the motor system in the monkey. II. The effects of lesions of the descending brain-stem pathways. Brain 91:16–36

MacKinnon CD (2018) Pontomedullary reticular formation and reticulospinal tract. In: Handbook of clinical neurology

Miller DM, DeMayo WM et al. (2017) Neurons in the pontomedullary reticular formation receive converging inputs from the hindlimb and labyrinth. Exp Brain Res 235:1195–1207

Mizukawa K, McGeer PL et al (1986) The cholinergic system of the human hindbrain studied by choline acetyltransferase immunohistochemistry and acetylcholinesterase histochemistry. Brain Res 379:39–55

Nieuwenhuys R, Voogd J et al. (1988) The human central nervous system. 3rd edn. Springer

Ohno K, Takeda N et al (1991) Coexistence of acetylcholine and calcitonin-gene-related peptide in the vestibular efferent neurons of the rat. Brain Res 566:103–107

Olszewski J, Baxter D (1982) Cytoarchitecture of the human brain stem. Karger, Basel

Parent A (1996) Carpenter's human neuroanatomy, 9th edn. Williams and Wilkins, Philadelphia

Rüb U, Jen JC, Braak H, Deller T (2008a) Functional neuroanatomy of the human premotor oculomotor brainstem nuclei: insights from postmortem and advanced in vivo imaging studies. Exp Brain Res 187:167–180

Rüb U, Brunt ER et al (2008b) New insights into the pathoanatomy of spinocerebellar ataxia type 3 (Machado-Joseph disease). Curr Opin Neurol 21:111–116

Sekirnjak C, du Lac S (2006) Physiological and anatomical properties of mouse medial vestibular nucleus neurons projecting to the oculomotor nucleus. J Neurophysiol 95:3012–3023

Shaikh AG (2019) Human gaze holding and its disorders. Ann Otol Neurotol 2:33–40

Tanii H, Zang X-P et al (2000) Involvement of GABA neurons in allylnitrile-induced dyskinesia. Brain Res 887:454–459

ten Donkelaar HJ (2011) Clinical neuroanatomy: brain circuitry and its disorders. Springer

Tokita K, Inoue T et al (2009) Afferent connections of the parabrachial nucleus in C57BL/6J mice. Neuroscience 161:475–488

Zhang J, Luo P (2003) Ultrastructural features of synapse from dorsal parvocellular reticular formation neurons to hypoglossal motoneurons of the rat. Brain Res 963:262–273

Web Links

http://braininfo.rprc.washington.edu/whatspecies.aspx?questID=1201
http://braininfo.rprc.washington.edu/whatspecies.aspx?questID=1103
http://braininfo.rprc.washington.edu/whatspecies.aspx?questID=1066

Rhombomere 7 r7

7

Contents

H. Schröder et al., *The Human Brainstem*, https://doi.org/10.1007/978-3-030-89980-6_7

Abstract

Rhombomere 7 is the origin of another vestibular nucleus—previously the spinal/inferior vestibular nucleus (see Chap. 5)—the medial vestibular nucleus. Among the branchial motor nuclei, the inferior salivatory nucleus shows up. The innervation of the parotid gland will be explained here. Another important cranial nerve is the vagus nerve—the viscerosensory part of which had been dealt with in Chap. 3—with its motor nucleus. The course of this nerve will be shown here to the plethora of inner organs to which it provides the parasympathetic innervation. The last of the caudal raphe nuclei—previously the raphe pallidus and obscurus nuclei (see Chap. 4)—the raphe magnus nucleus, the largest of the caudal raphe nuclei, is described. The pontine reticular nucleus is involved in the generation of horizontal saccades.

Among the reticular nuclei listed here, the human one is the lateral paragigantocellular nucleus, located in a region that regulates homeostatic functions, e.g., blood pressure and cardiovascular reflexes.

The pathological units dealt with here are Huntington's disease and the wide spectrum of the spinocerebellar ataxias (SCA) which despite their traditional name affect not only the spinal cord and the cerebellum but quite a number of brainstem nuclei.

7.1 Rhombic lip r7

7.1.1 Precerebellar nuclei r7

7.1.1.1 Linear nucleus of the medulla

For detailed information on this nucleus, see Sect. 4.1.1.5.

7.1.1.2 Inferior olive

For detailed information on the inferior olive, see Sect. 4.1.1.3.

7.1.1.3 External cuneate nucleus

For detailed information on this nucleus, see Sect. 5.2.1.1.

7.1.1.4 Lateral reticular nucleus

For detailed information on this nucleus, see Sect. 4.1.1.1.

7.1.1.5 Prepositus nucleus

For detailed information on this nucleus, see Sect. 6.2.1.5.

7.1.1.6 Nucleus of Roller

For detailed information on this nucleus, see Sect. 6.2.1.6.

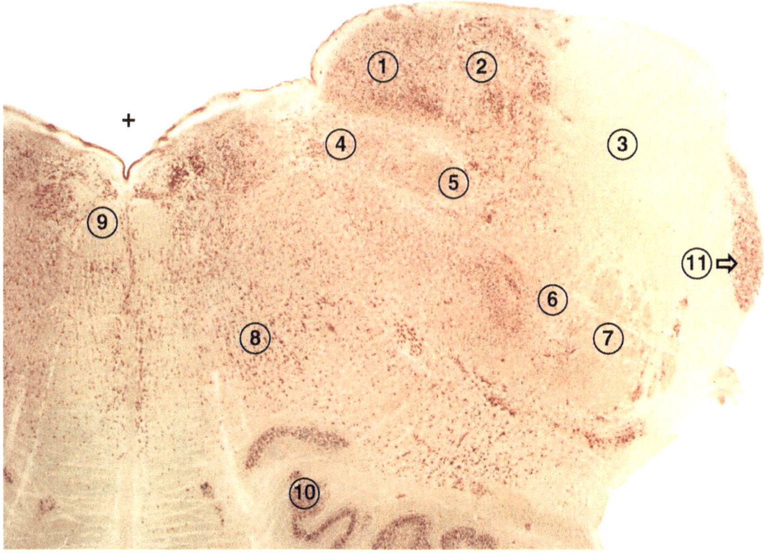

①	Medial vestibular ncl. CNVIII	⑦	Spinal trigeminal tract CNV
②	Inferior vestibular ncl. CNVIII	⑧	Gigantocellular reticular ncl.
③	Inferior cerebellar peduncle	⑨	Medial longitudinal fasciculus
④	Solitary ncl.	⑩	Inferior olivary complex
⑤	Solitary tract	⑪	Cochlear nuclei CNVIII
⑥	Spinal trigeminal ncl. CNV	+	4th ventricle

Fig. 7.1 Horizontal section through the medulla oblongata (open part of the fourth ventricle in the medulla oblongata) at the level of the peri-hypoglossal nuclei (see Box 3.1). Darrow red staining. The medial vestibular nucleus ① lies in the lateral wall of the fourth ventricle (Area vestibularis, see Fig. 1.10). Its ventral neighbors are the solitary nucleus ④ and tract ⑤. Laterally, it is bordered by the inferior cerebellar peduncle ③. See atlas part Darrow red 13. LabPON Twente

7.2 Alar r7

7.2.1 Vestibular nuclei r7

7.2.1.1 Medial vestibular nucleus [Ncl. vestibularis medialis]

The medial vestibular nucleus—(Fig. 7.1) consisting of rather small neurons—is located for most of its dorsal surface in the floor/wall of the fourth ventricle.

7.2.1.1.1 Location and histology of the medial vestibular nucleus

The medial vestibular nucleus (MVN) [Ncl. vestibularis medialis] extends over a distance of about 9 mm from the caudal third of the pons—caudal tip of the nucleus of abducens nerve (see Sect. 9.2.1.1)—to the upper medulla oblongata, about 1 mm caudal to the rostral tip of the nucleus of hypoglossal nerve (Sect. 4.3.1.1) (Augustine 2017). It is located beneath the widest part of the floor of the fourth ventricle (see Fig. 7.1 and Fig. 1.10, vestibular area) with densely packed neurons displaying bundles of presumed vestibulocerebellar fibers (Augustine 2017). In the more caudal parts, it is located between the inferior/spinal vestibular nucleus (see Sect. 5.3.1.1) and the solitary nucleus/tract (Fig. 7.1) (see Sect. 3.2.4).

7.2.1.1.2 Afferent and efferent connectivity of the medial vestibular nucleus

As described in detail under Sect. 5.3.1, the MVN, like the other vestibular nuclei, receives afferent signals from the vestibular organ in the inner ear (see Sect. 5.3.1, Figs. 5.3–5.6).

For connectivity data, see Sect. 5.3.1.1.2.

7.2.1.1.3 Functional considerations and disorders

In monkeys, lesions of the MVN—and the adjacent prepositus hypoglossi nucleus (see Box 3.1)—have been reported to result in severe impairment of fixation and slow eye movements (Cannon and Robinson 1987) (see Box 9.6). In cases of Wallenberg syndrome (see Table 5.1) with involvement of the MVN, disturbances of the vestibulo-ocular reflex (see Box 6.2) and nystagmus (see Box 6.3) occur (cf. ten Donkelaar 2011) (see also Box 9.6 "Basic eye movements").

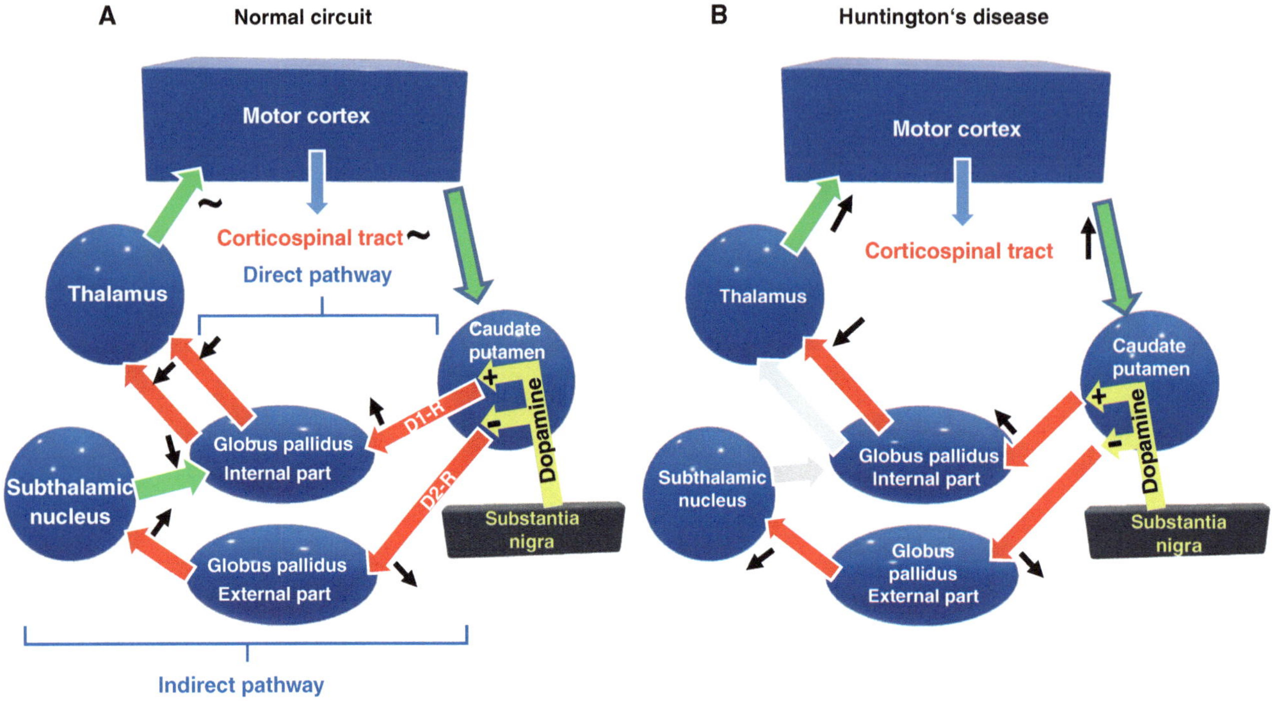

Fig. 7.2 The normal course of transmission in this network is described in detail in Fig. 16.6. Under normal conditions, the combined effects of the direct and the indirect pathway would result in a balanced excitation (~) of the motor cortex represented clinically by a "normal" motor behavior as an adjusted velocity of movements (*PD* hypoactivity, hypokinesia, *HD* hyperactivity, hyperkinesia), a normal posture as well as accompanying movements. The hypokinesia in PD can be explained by the decreased dopaminergic modulation of the caudate putamen (see Fig. 16.6). A crucial lesion site in PD is the substantia nigra (see Sect. 16.6.1.5). First supposed for unilateral chorea but increasingly shown to be important in HD and corroborated by HD animal models, the motor changes seen in HD could be also due to malfunctioning of the subthalamic nucleus. As shown for HD in the above scheme, the loss of input from the subthalamic nucleus (a glutamatergic stimulation of part of the internal globus pallidus thalamopetal GABAergic fibers) will impair inhibition in the cortico-thalamo-cortical loop resulting in the shift of motor cortex input to excitation or less inhibition. Of course, the pathogenetic mechanisms playing a role in the generation of the clinical symptoms are much more complicated but the subthalamic nucleus may play here a hitherto underestimated role. H. Schröder fecit

In mice, the MVN stretches from the level of the linear nucleus of the medulla to the level of the locus caeruleus. A detailed classification of neurons in the MVN of the mouse has been provided by Bagnall et al. (2007) and by Kolkman et al. (2011). Together with the MVN, the prepositus nucleus (see Sect. 6.2.1.5) is involved in the vestibulo-ocular reflex (see Box 6.2).

7.2.1.2 The vestibular nuclei and Huntington's disease

It is common medical knowledge that putamen and the caudate nucleus are the main battlefields of this devastating disease. What is usually less known is the fact that Huntington's disease (HD) also affects the brainstem—see here below—and the subthalamic nucleus (see Fig. 7.9) [Ncl. subthalamicus]. As kind of the clinical antipole of Parkinson's disease (PD) (see Sect. 16.6.1.5) with an uncontrollable motor activity, it may be interesting to first have a look on the connectivity scheme of the Substantia nigra section of this book (see Fig. 16.6) and Fig. 7.2 here.

7.2.1.2.1 Clinics and genetics of Huntington's disease

HD is a rare autosomal dominantly inherited, progressive, and ultimately fatal neuropsychiatric disorder clinically characterized by involuntary movements (particularly chorea), psychiatric symptoms including psychosis, depression, and obsessive-compulsive disorder together with progressive cognitive impairment and dementia (Box 7.1).

Box 7.1 George Huntington

George Huntington (*1850 East Hampton, NY, USA, † 1916 New York City, NY, USA) provided the most comprehensive paper on chorea although he was not the first to describe the disease and its symptoms. As Wagner et al. (2008) in their contribution on Huntington's disease state: "… *George Huntington* *was only 8 years old when he first observed patients with hereditary chorea; riding with his father (a physician) on his professional rounds around East Hampton, Long Island. At the age of 21, George Huntington summarized the cardinal features of the disease that now bears his name in a report entitled "On Chorea,"* which appeared in the Medical and Surgical Reporter of Philadelphia on April 13, 1872 (see Fig. 7.3). *This concise treatise outlined the distinctive choreiform movement disorder, the distinct hereditary nature of the chorea, and the frequent association of the chorea with psychiatric disease ("a tendency to insanity and suicide") ..."*

The estimated prevalence is 4–8/100.000 (Harper 1992). Like other members of the group of polyglutamine diseases (e.g., specific forms of spinocerebellar ataxias, see Sect. 7.2.1.3), it belongs to the polyglutamine or CAG-repeat diseases, and is caused by an expanded trinucleotide repeat in a gene on chromosome 4p16.3 (Rawlins et al. 2016). The human HD gene (also called IT15) contains meiotically unstable sequences of CAG-repeats in its exon 1.

In healthy individuals, the length of the CAG-repeat sequences commonly ranges from 6 to 35 CAG triplets. Sequences of 35 or more CAG-repeats are definitely in the pathological range and provoke HD disease symptoms. 36 to 40 CAG-repeat triplets are associated with an incomplete clinical penetrance of HD and 41 or more CAG-repeats with a fully developed clinical picture (Rüb et al. 2016). The pathologically elongated CAG-repeat sequences are translated into abnormally long polyglutamine stretches in the mutated protein. These stretches are held to be responsible for the tendency of this protein to lose its normal solubility in nerve cells and to aggregate into intra-neuronal inclusions in different components of vulnerable brain regions (Rüb et al. 2016, Gil and Rgeo 2008). Currently, there are no treatment options available for HD patients and treatment is based on symptomatic approaches to ameliorate disease symptoms.

7.2.1.2.2 Macroscopical findings in Huntington's disease

The macroscopic findings depend on the stage of the disease. Early cases may show little or no macroscopic changes. Advanced cases show cerebral atrophy with a reduction in brain weight (often 900–1000 g). Advanced cases show bilateral symmetrical atrophy of the caudate nucleus, putamen, and globus pallidus (Figs. 7.4 and 7.5).

In late stages, atrophy of the cerebral cortex and white matter is evident, greater in frontal, parietal, and occipital regions than in the temporal lobe (Hedreen and Roos 2003). The Vonsattel grading system for striatal degeneration is based on the macroscopic appearance of striatal atrophy and the visible microscopic extent of the striatal neuronal loss and degree of reactive astrogliosis (Vonsattel et al. 1985, Vonsattel and DiFiglia 1998). This grading system distinguishes among five grades (grade 0–4) of striatal involvement in HD.

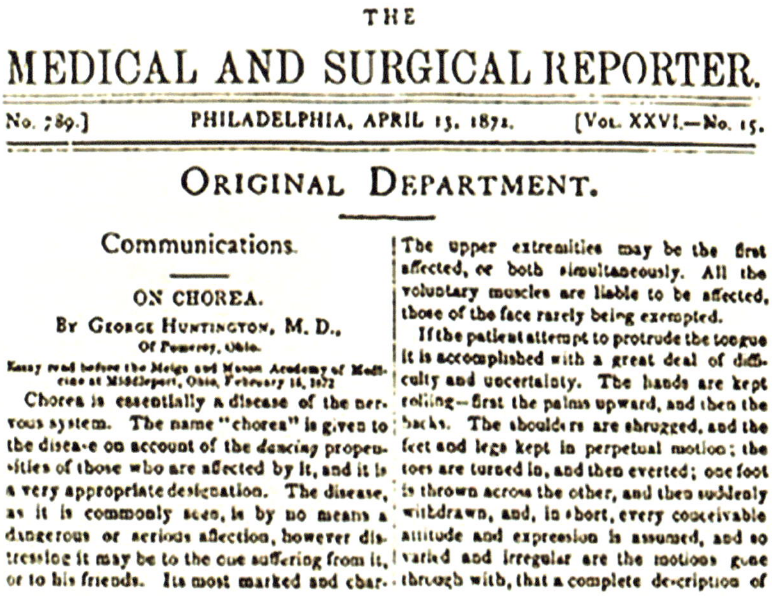

THE

MEDICAL AND SURGICAL REPORTER.

No. 789.] PHILADELPHIA, APRIL 13, 1872. [Vol. XXVI.—No. 15.

ORIGINAL DEPARTMENT.

Communications.

ON CHOREA.

By George Huntington, M. D.,
Of Pomeroy, Ohio.

Essay read before the Meigs and Mason Academy of Medicine at Middleport, Ohio, February 15, 1872

Chorea is essentially a disease of the nervous system. The name "chorea" is given to the disease on account of the *dancing* propensities of those who are affected by it, and it is a very appropriate designation. The disease, as it is commonly seen, is by no means a dangerous or serious affection, however distressing it may be to the one suffering from it, or to his friends. Its most marked and char-

The upper extremities may be the first affected, or both simultaneously. All the voluntary muscles are liable to be affected, those of the face rarely being exempted.

If the patient attempt to protrude the tongue it is accomplished with a great deal of difficulty and uncertainty. The hands are kept rolling—first the palms upward, and then the backs. The shoulders are shrugged, and the feet and legs kept in perpetual motion; the toes are turned in, and then everted; one foot is thrown across the other, and then suddenly withdrawn, and, in short, every conceivable attitude and expression is assumed, and so varied and irregular are the motions gone through with, that a complete description of

Fig. 7.3 Portrait of George Huntington and the first columns of his famous paper "On Chorea" from 1872. From Bates 2005, Fig. 1 with permission

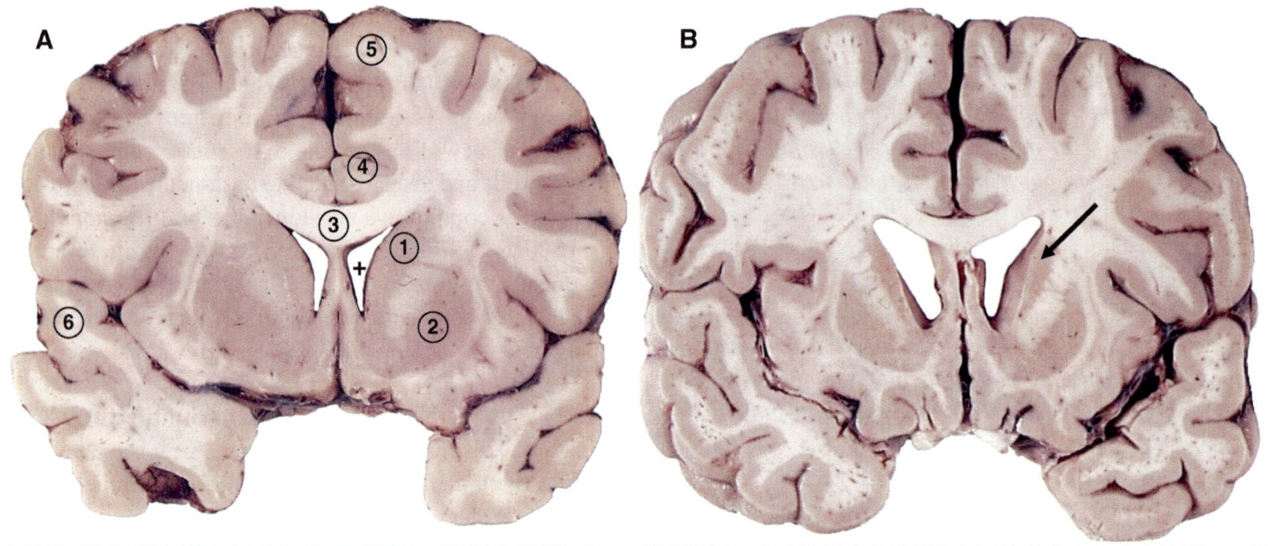

①	Caudate ncl.	④	Cingulate gyrus
②	Putamen	⑤	Superior frontal gyrus
③	Corpus callosum	⑥	Superior temporal gyrus
		+	Lateral ventricle

Fig. 7.4 Coronal section, frontal level, of a brain from a HD patient (**B**) compared to that of a control (**A**). Characteristic atrophy of the caudate nucleus (arrow), mild gyral atrophy. LabPON Twente

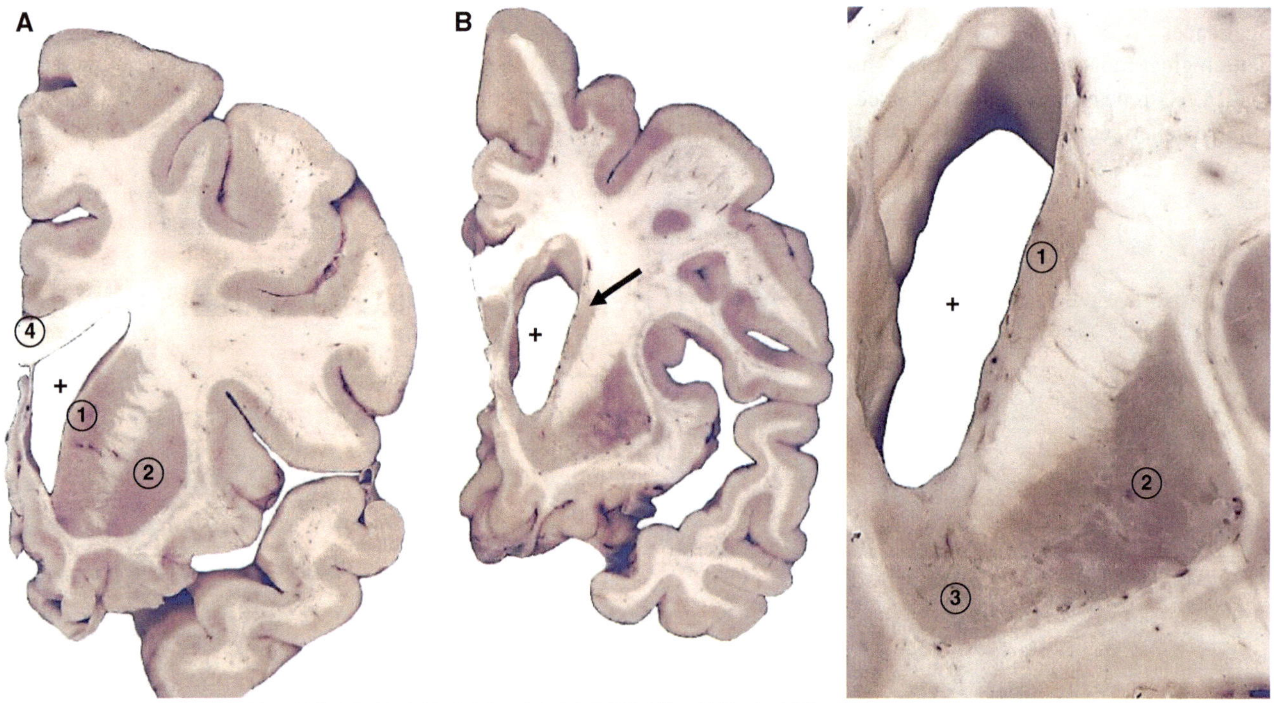

①	Caudate ncl.	④	Corpus callosum
②	Putamen	+	Lateral ventricle
③	Accumbens nucleus		

Fig. 7.5 Coronal section through the right basal forebrain of a control individual (**A**). (**B**) Frontal section through the right basal forebrain of a HD patient, atrophic caudate nucleus and dilated frontal horn of the lateral ventricle. Right hand photograph: detailed documentation of (**B**). Note the shrinkage of the diseased brain (**B**) in comparison to (**A**). LabPON Twente

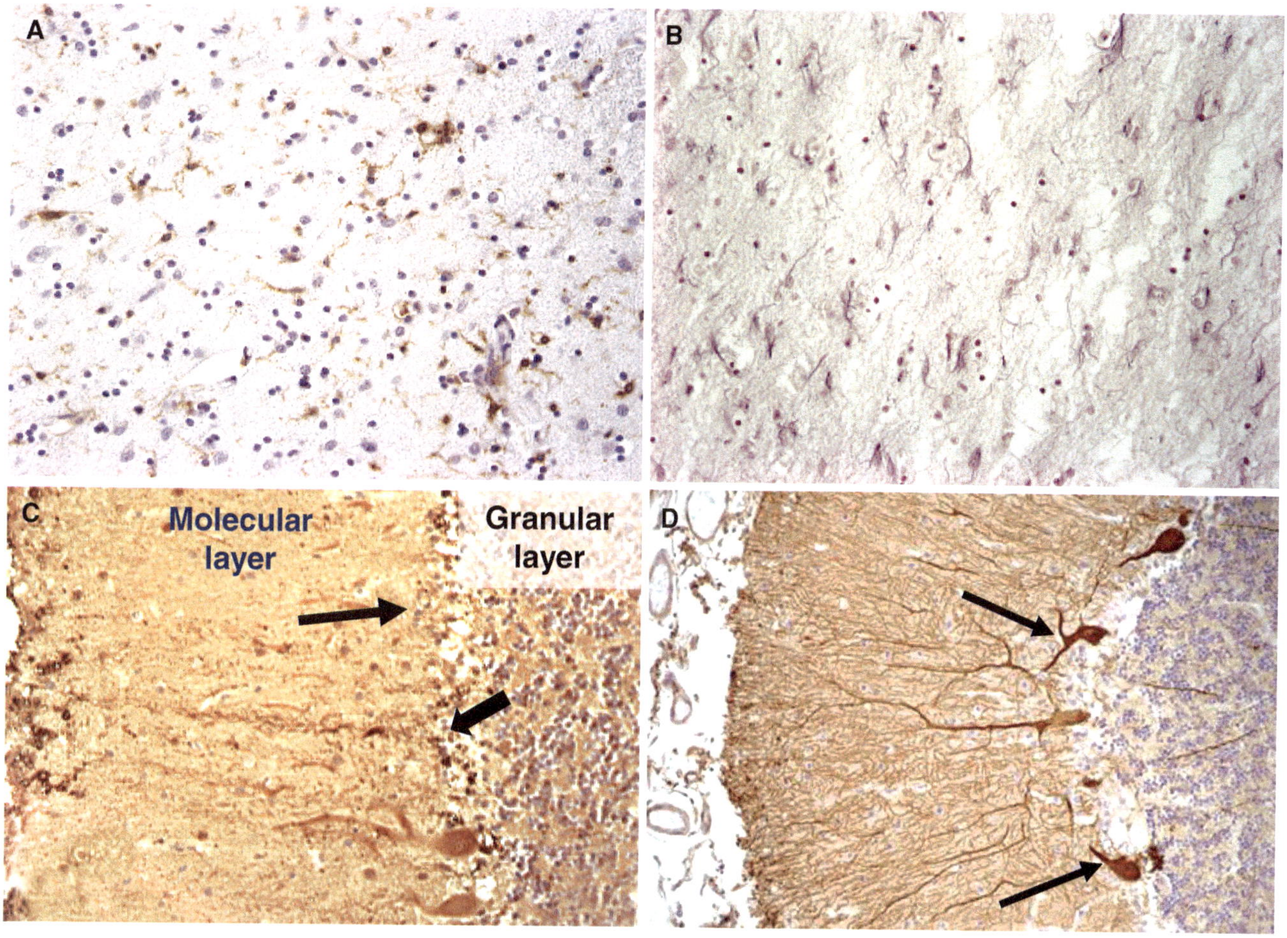

Fig. 7.6 Huntington's disease. Marked astrocytic gliosis in the striatum, in a GFAP immunostain (**A**) and a PTAH-glia stain (**B**). Cerebellar atrophy with loss of Purkinje cells (thin arrows), slight proliferation of Bergmann astrocytes (thick arrow) at the junction of the granular and molecular layers in (**C**) (calbindin immunostain) compared to a control case (**D**). Arrows in (**D**) point to normal Purkinje cells. LabPON Twente

7.2.1.2.3 Microscopical findings in Huntington's disease

Recent studies in HD have shown a layer-specific neuronal loss in various areas of the cerebral allocortex and neocortex (Rüb et al. 2016). This neuronal loss, with associated astrocytosis and microgliosis, is now regarded as the most important morphological correlate of the psychiatric symptoms, neuropsychological defects, and cognitive decline. The neostriatal neuronal loss is mainly caused by the extensive loss of GABA-ergic medium spiny stellate projection neurons, which belong to the main targets of striatal input and provide the main efferent output of the striatum. These highly vulnerable neurons account for 95% of all neurons occurring in the neostriatum (Rüb et al. 2016). There is often a widespread neuronal loss in the cerebellar cortex and in the cerebellar nuclei (Figs. 7.6 and 7.7).

The Purkinje cell layer is affected at circumscribed predilection sites. The four cerebellar nuclei (fastigial, globose, emboliform, and dentate nuclei) consistently undergo neurodegeneration in HD, with the fastigial nucleus being the most affected. The reticular part of the substantia nigra (see Sect. 16.6.1) is commonly affected (Fig. 7.7).

All polyQ diseases display formation of aggregates containing the mutant disease protein in neuronal cells. Different types of inclusions can be detected in brains from polyQ patients, such as neuronal nuclear inclusion bodies (NNI), that are small, rounded, condensed, and located within the neuronal nucleus resulting in diffuse nuclear staining. Small- to medium-sized granules are found in the perinuclear cytoplasm and, finally, aggregates can be found in axons (Seidel et al. 2010, Rüb et al. 2016). All these aggregate types occur in both HD and SCA3 patients (see Sect. 7.2.1.3.3) and can be demonstrated by immunohistochemistry using antibodies against Huntingtin (in Huntington disease, especially against the amino-terminal region), ubiquitin, or expanded polyglutamine tracts, such as the monoclonal antibody IC2 (Herndon et al. 2009) and p62 (Fig. 7.8).

Abnormal neurites in the cortex, medial temporal lobe, and striatum, with similar immunohistochemical properties, are more common than nuclear inclusions. Neuritic aggregates appear to be age-related (Revesz et al. 2015).

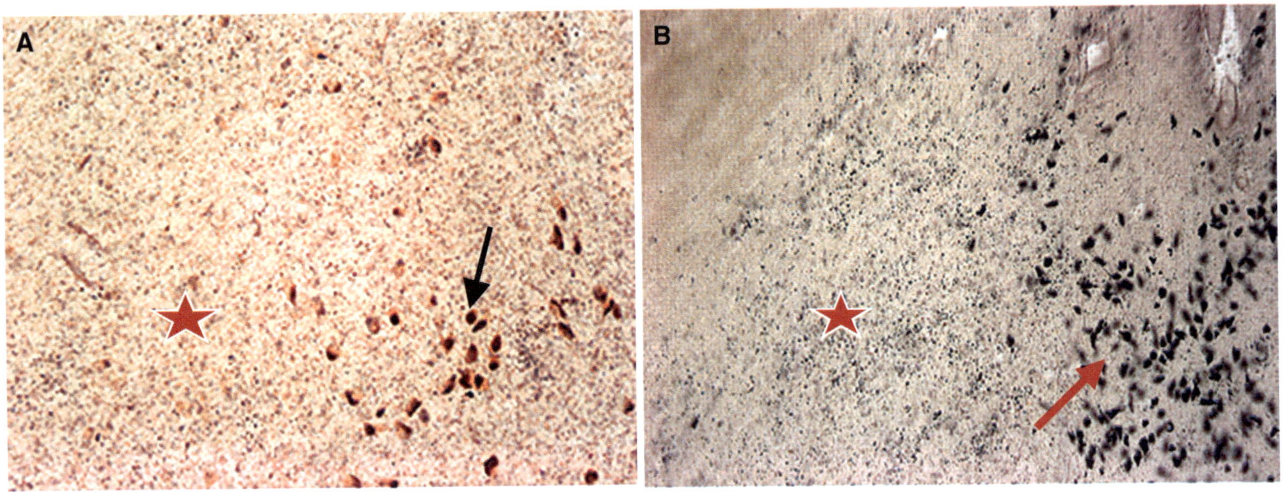

Fig. 7.7 Huntington's disease. Substantia nigra with severe loss of neurons (red stars) in the reticular part and pigment incontinence (**A**, PEG 100 µm, Darrow red stain and **B**, PEG 100 µm, Campbell stain). Arrows point to preexistent neurons. LabPON Twente

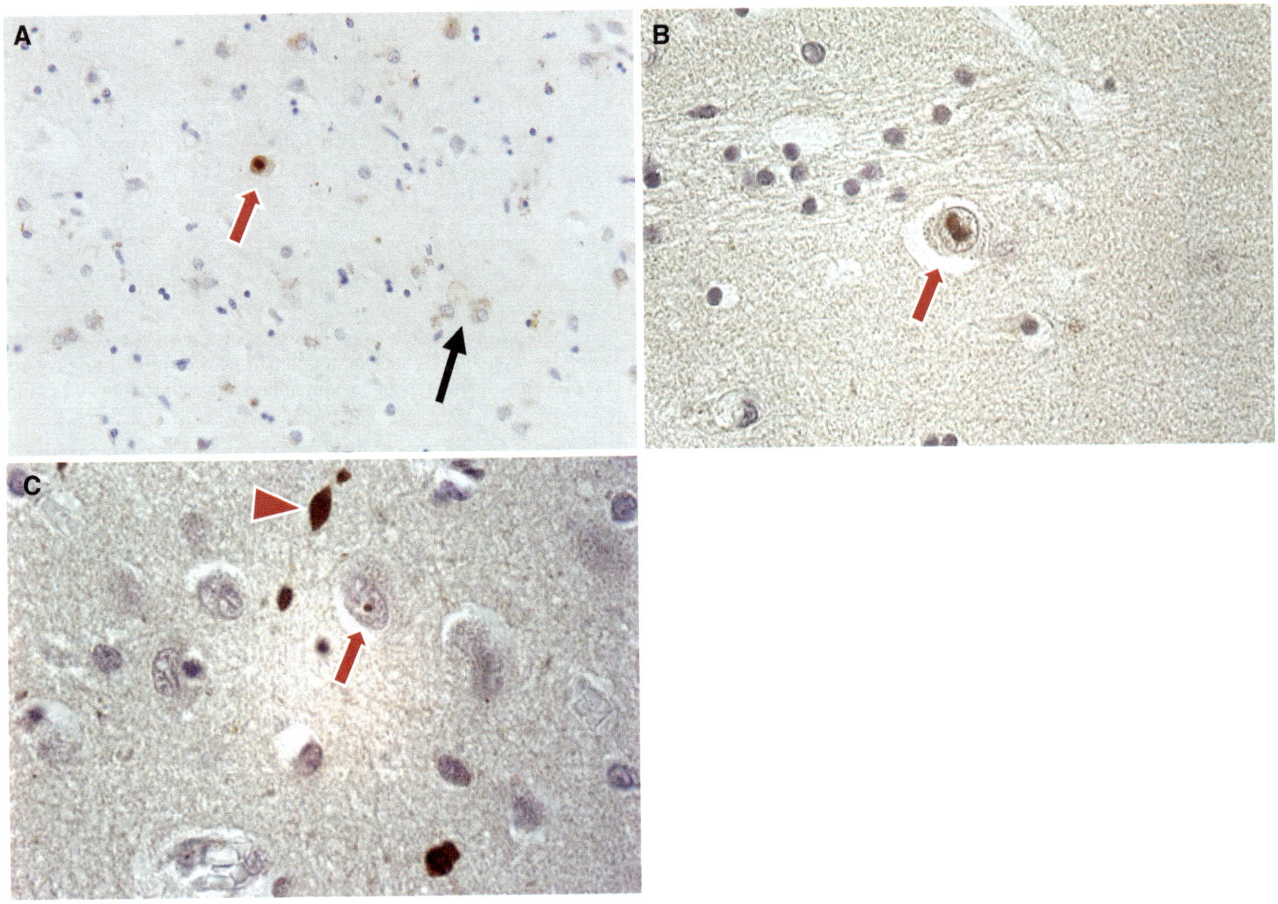

Fig. 7.8 Huntington's disease. Neuronal nuclear inclusions (NNIs) in the striatum in 1C2 immunostain (**A**) and ubiquitin immunostain (**B**) and NNIs (red arrows) and neurites (red arrowhead) in ubiquitin immunostain (**C**). Black arrow in (**A**) points to astrogliosis. LabPON Twente

Recent data show that astrocyte dysfunction is an important contributor to onset and progression of some HD symptoms (Khakh et al. 2017). This has culminated in the current concept of HD as a multisystem neurodegenerative disorder emphasized by a brain distribution pattern which, apart from the striatum, includes neo- and allocortex, thalamus, pallidum, brainstem, and cerebellum (Rüb et al. 2016).

7.2.1.2.4 The human brainstem in Huntington's disease

Despite its unremarkable macroscopic appearance, the brainstem of HD patients consistently shows widespread neuronal loss upon light-microscopic examination. Not only the reticulate part of the substantia nigra (see Sect. 16.6.1) is involved, but also the precerebellar pontine nuclei (see Sect. 10.1.1.2) and inferior olive (see Sect. 4.1.1.3) and the oculomotor reticulotegmental nucleus of the pons (see Sect. 10.1.1.1) (see Box 10.1) (Rüb et al. 2016). The pontine nuclei and the inferior olive (see Sect. 4.1.1.3) are intimately connected with the cerebellum and are therefore assigned to the so-called precerebellar nuclei, which act as a gateway for all afferent input destined for the cerebellum. HD patients also suffer from disease symptoms such as oculomotor dysfunction. It is well known that the reticulotegmental nucleus of the pons (see Sect. 10.1.1.1) is involved in the generation of smooth pursuit eye movements (see Box 9.6). Furthermore, neuronal loss can be observed in the superior olivary complex (auditory system; see Sect. 8.1.1.5) as well as in the lateral and medial vestibular nuclei (see Sect. 10.2.2.1) (Rüb et al. 2016) (Fig. 7.9).

The lateral vestibular nucleus (see Sect. 10.2.2.1) subserves the optokinetic nystagmus (aka "railroad nystagmus") and the medial vestibular nucleus (see Sect. 7.2.1.1 here above) the vestibulo-ocular reflex (see Box 6.2). Their degeneration probably plays a role in the pathogenesis of impairments of these kinds of eye movements (Rüb et al. 2016). P62—used as immunostaining for the assessment of polyQ aggregation (Fig. 7.10)—and ubiquitin-immunoreactive neuronal intranuclear inclusions as well as p62-immunoreactive axonal neuronal inclusions can be seen in brainstem nuclei and brainstem fiber tracts (Rüb et al. 2016) (Figs. 7.10 and 7.11).

The destructive process leads to a disease-specific distribution pattern of brain degeneration, sharing many neuropathological characteristics with the spinocerebellar types SCA1, SCA2, and SCA3 (Rüb et al. 2016) (see Sect. 7.2.1.3 here below).

7.2.1.3 Spinocerebellar ataxias (SCA)

Spinocerebellar ataxias (SCAs) can be divided into two major categories on the basis of their genetic mutations: (1) SCAs caused by microsatellite expansions and (2) SCAs caused by point mutations. The first group can be further divided into those caused by polyglutamine (polyQ)-coding CAG-repeat expansions and those caused by non-protein-coding repeats (Ashizawa et al. 2018). Here, we will focus on SCAs caused by CAG-repeat expansion diseases as much knowledge has been obtained in recent years through extensive neuropathological research.

Polyglutamine or CAG-repeat diseases are diseases in which the expansion of a poly-CAG-repeat sequence within a gene coding for a disease-specific protein causes expansion of a polyQ sequence resulting in aggregation of the disease protein. This is considered to be a major contributing factor to the severe degeneration present in these diseases. Spinocerebellar ataxias (SCAs) belong to the CAG-repeat disorders and are characterized by progressive ataxia due to degeneration of the cerebellum and its afferent and efferent pathways. This family of neurological disorders comprises SCAs, Huntington's disease (see Sect. 7.2.1.2), spinobulbar muscular atrophy as well as dentatorubro-pallidoluysian atrophy (Paulson 2000, Rüb et al. 2003). The prefix "SCA" with an associated number (reflecting the order of genetic discovery) is assigned to dominantly inherited ataxias when their genetic loci are defined (Ashizawa et al. 2018). SCAs comprise more than 40 autosomal dominant neurodegenerative disorders that present principally with progressive ataxia. Although clinical characteristics can help to guide physicians toward the correct diagnosis, the diagnosis of SCAs depends on genetic testing and the testing technologies have improved, especially in repeat-expansion disorders. (Ashizawa et al. 2018). All SCAs display classic cerebellar signs, and many display disabling noncerebellar features, most commonly, brainstem dysfunction (Paulson 2009). Eye movement abnormalities in SCA (see Box 9.6) are common, reflecting cerebellar and brainstem degeneration. Conventional MRI in patients presenting with progressive ataxia demonstrates three main patterns of microscopic damage, namely spinal, olivopontocerebellar, and cortical cerebellar atrophy (Mascalchi 2008).

With a few exceptions (SCA6), SCAs are progressive, fatal diseases. No drugs for SCAs have been approved by FDA (Food and Drug Administration) or European Medicines Agency (EMA) although some promising candidate drugs are being developed (Ashizawa et al. 2018). In general, studies support the efficacy of physical therapy.

Cerebellar ataxia is a result of degeneration in every part of efferent or afferent connections with the cerebellum. The various forms of SCA are distinguishable by their specific morphological appearance of the cerebellar, extrapyramidal, and/or oculomotor systems (Iwabuchi et al. 1999). To inspect these areas, sagittal slices of both vermis and hemispheres are necessary. In addition, the (deep) cerebellar nuclei, brainstem, and spinal cord must be meticulously studied. The use of unconventionally thick serial brain sections has allowed the recognition of subtle alterations and can provide more

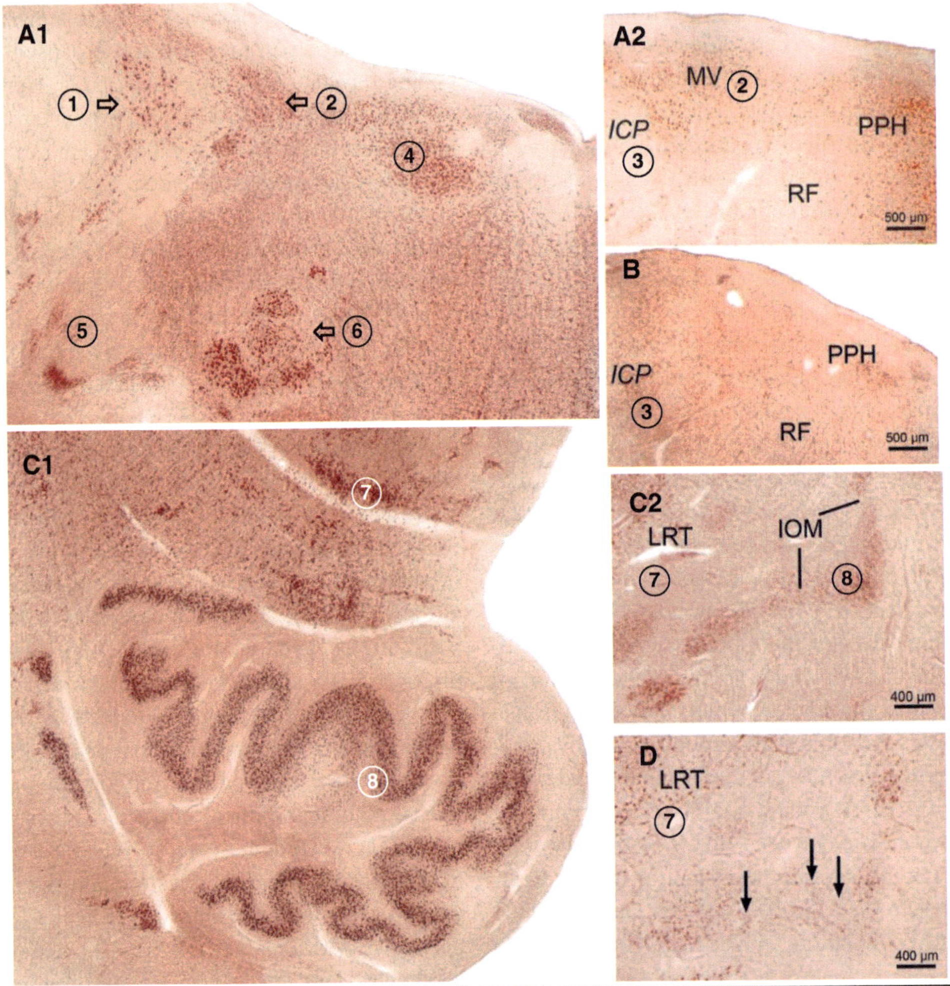

①	**Lateral vestibular ncl.**	⑤	**Superior olivary complex**
②	**Medial vestibular ncl. MV**	⑥	**Facial ncl. CNVII**
③	**Inferior cerebellar peduncle ICP**	⑦	**Lateral reticular ncl. LRT**
④	**Abducens ncl. CNVI**	⑧	**Inferior olive IOM**

Fig. 7.9 Horizontal sections of the brainstem showing vestibular nuclei in a control (A1, arrow and A2), compared to the severe neuronal loss of the medial vestibular nucleus of a Huntington's disease patient (B). Compare with Fig. 7.1. (C1) and (C2) show the inferior olive (see Sect. 4.1.1.3) and the lateral reticular nucleus (LRT) (see Sect. 4.1.1.1) in a control, compared to the neuronal loss in both nuclei of a Huntington's disease patient (D). Arrows point to reaming extraneuronal lipofuscin (see Box 4.4). *PPH* prepositus hypoglossi nucleus (see Sect. 6.2.1.5), *RF* reticular formation. From Rüb et al. 2016, Fig. 10 with permission

precise information with respect to the extent and severity of the pathological process underlying neurodegenerative diseases (Rüb et al. 2008a).

7.2.1.3.1 Spinocerebellar ataxia type 1 (SCA 1)

SCA1 corresponds to olivopontocerebellar atrophy (OPCA 1) or the progressive and dominantly inherited ataxia (Eadie 1975a, b). Like most other SCAs, SCA1 begins as a gait disorder, evolving to sincere limb ataxia with dysarthria and leaving most patients wheelchair-bound within 15–20 years (Paulson 2009).

SCA1 represents the first molecular, genetically characterized, autosomal dominantly inherited cerebellar ataxia. The gene that causes SCA1 has been localized to 6p22-p23

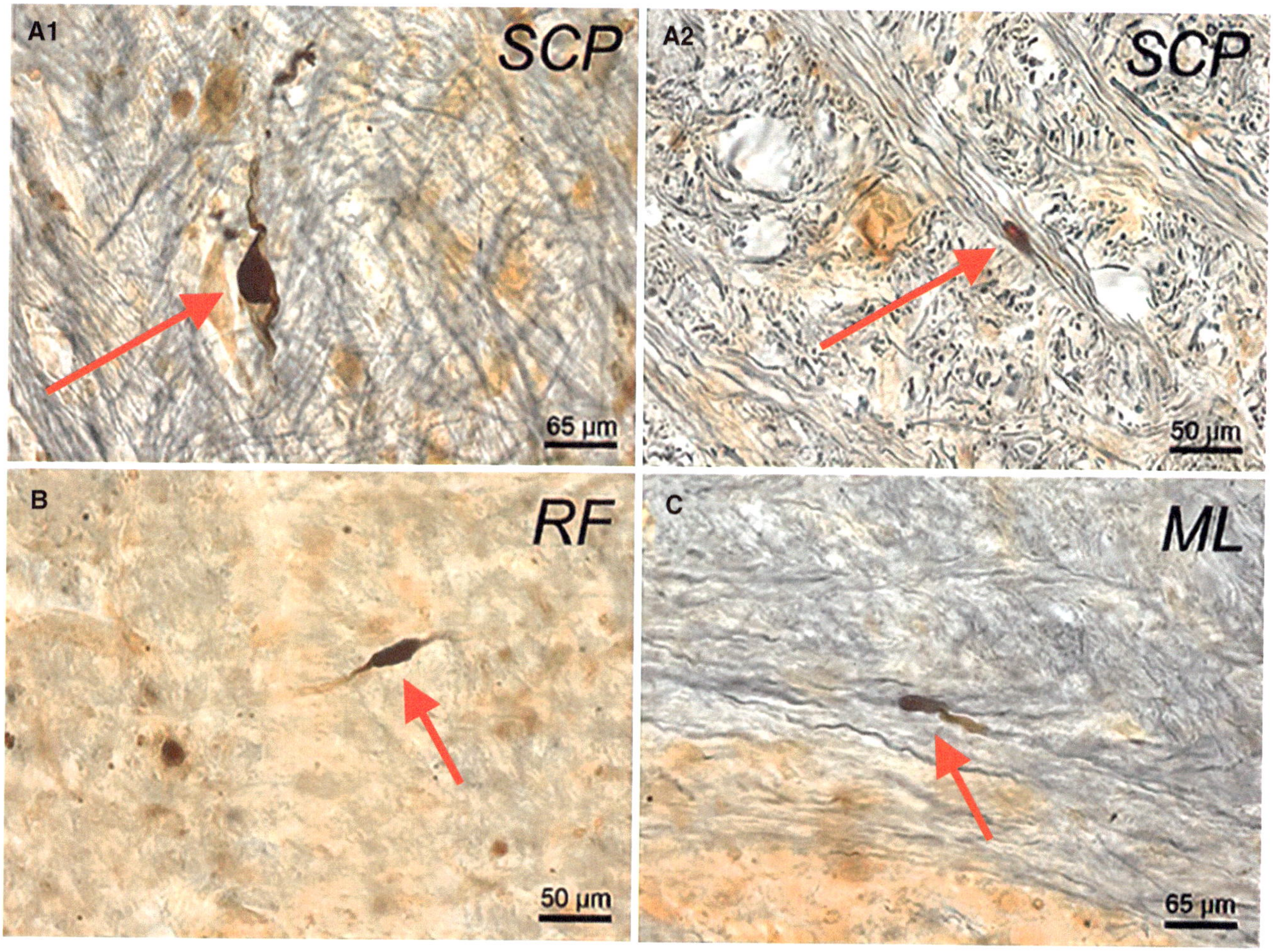

Fig. 7.10 P62-immunoreactive axonal neuronal inclusions (red arrows) in brainstem fiber tracts in Huntington's disease. Superior cerebellar peduncles (SCP) in (**A1**) and (**A2**), reticular formation (RF) in (**B**) and medial lemniscus (ML) in (**C**). From Rüb et al. 2016, Fig. 2 with permission

and the mutation being responsible for SCA1 has been determined to be an expansion of a CAG trinucleotide repeat (Orr et al. 1993, Banfi et al. 1994). Rüb et al. (2012) were the first to investigate serial tissue sections through the complete brains of clinically diagnosed and genetically confirmed SCA1 patients. The wild-type form of ataxin-1 is widely expressed in nerve cells of healthy human brains and its mutated form tends to aggregate into insoluble intranuclear neuronal inclusion bodies (NI) in the brains of affected SCA1 patients (Robitaille et al. 1995).

Macroscopically, the brain weight is reduced, and atrophy is clearly visible in the pons, cerebellum, inferior olives (see Sect. 4.1.1.3), and spinal cord (Fig. 7.12) (Robitaille et al. 1997).

Frontotemporal and parietal cerebral cortices can be atrophic, too. Widespread neuronal loss in the motor cortex (considerable loss of layer V giant Betz pyramidal cells), basal forebrain, thalamus, pallidum, cerebellum, midbrain (substantia nigra, see Sect. 16.6.1; red nucleus, see Sect. 15.3.4) and brainstem auditory, oculomotor, somatosensory, precerebellar, and cranial nerve nuclei is seen (Rüb et al. 2012). In some patients, there is pallor of the substantia nigra. In the cerebellum, loss of Purkinje cells (especially in vermis) and of neurons in the dentate nucleus (with eosinophilic degeneration) is prominent. The pontine (see Sect. 10.1.1.2) and inferior olivary nuclei show neuronal loss, resulting in atrophy of the superior, medial, and inferior cerebellar peduncles (see Box 14.1). The reticulotegmental nucleus (see Sect. 10.1.1.1) is heavily damaged as well (Fig. 7.12B, C).

The nuclei of the third (CNIII, see Sect. 15.3.1), tenth (CNX, see Sect. 7.3.2.1), and twelfth (CXII, see Sect. 4.3.1.1) cranial nerves show a loss of neurons and gliosis. Myelin pallor and/or atrophy is present in the cerebellar white matter, cerebellar peduncles, cranial nerves, as well as

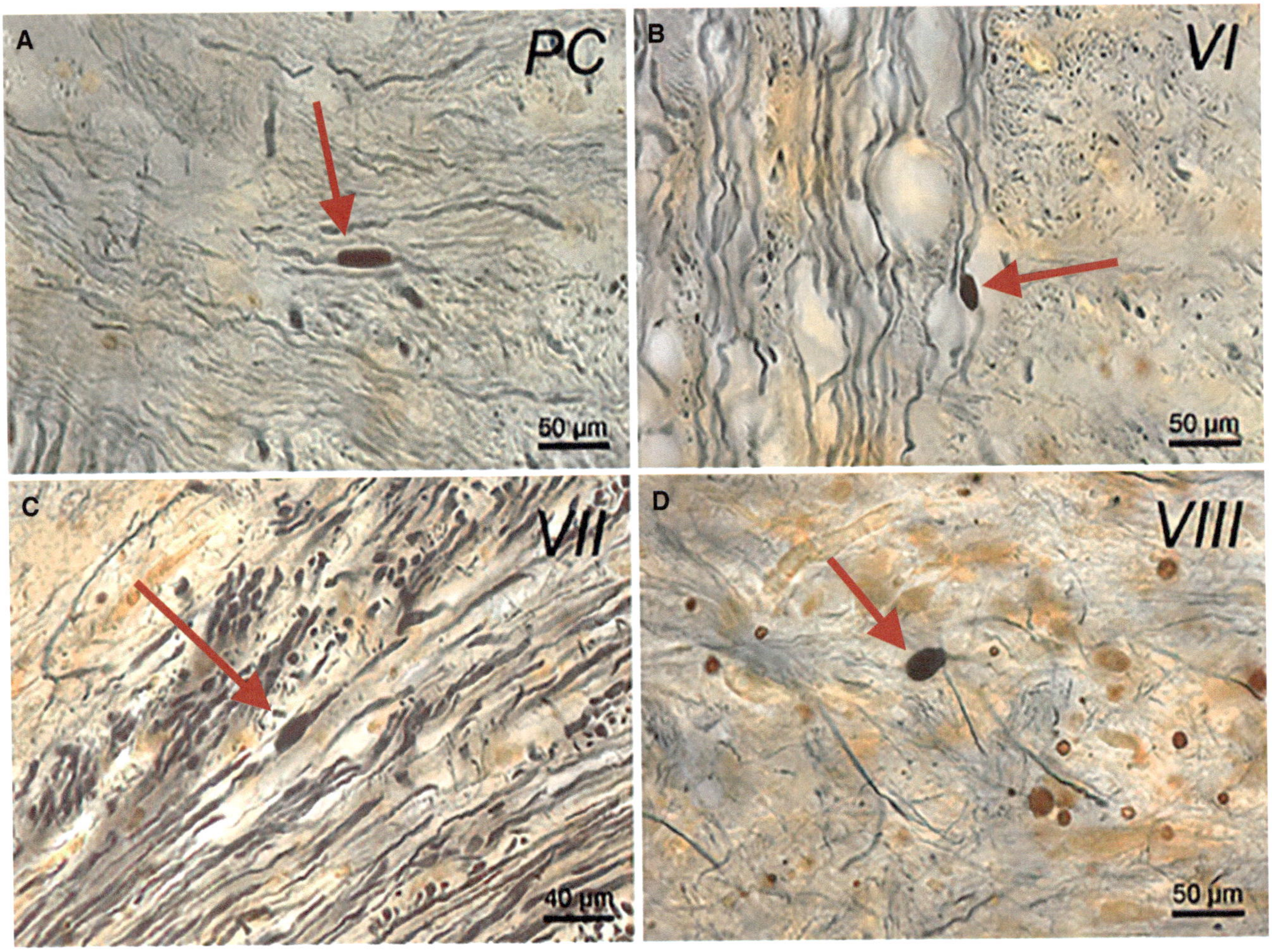

Fig. 7.11 Huntington's disease. P62-immunoreactive axonal neuronal inclusions (red arrows) in pontocerebellar fibers (PC) in (**A**), abducens nerve (VI) in (**B**), facial nerve (VII) in (**C**), and vestibulocochlear nerve (VIII) in (**D**). From Rüb et al. 2016 Fig. 2 with permission

in the somatosensory auditory and precerebellar fiber tracts (Fig. 7.13).

The vestibular nuclei can be involved as well (see Sect. 7.2.1). In end-stage disease, loss of motor neurons in spinal cord and of neurons in dorsal nuclei of Clarke [formerly Nucleus dorsalis, FIPAT: Posterior thoracic ncl./Ncl. thoracicus posterior (in the dorsal horn of the spinal cord) is seen. In the posterior columns, the gracile nucleus (see Sect. 3.2.3.4) is most heavily involved. Intranuclear inclusions mimicking nucleoli can be found using antibodies against protein chaperones, such as ubiquitin (Robitaille et al. 1997) or P62. Antibodies against polyglutamine stretches (IC2) show abnormal nuclear staining as well (Duyckaerts et al. 1999).

7.2.1.3.2 Spinocerebellar ataxia type 2 (SCA2)

SCA2 has a highly variable phenotype, most often characterized by ataxia, dysarthria, slow saccades (see Box 9.6), and peripheral neuropathy. Emotional incontinence, sleep disturbances, attentional deficits, impairments of memory, frontal-executive, and visuospatial dysfunctions can be present as well. Dysfunction of the somatosensory system is among the clinical signs that characterize a variety of polyglutamine or CAG-repeat diseases, not only SCA1 and SCA2, but also SCA3, SCA6, and dentatorubro-pallidoluysian atrophy (DRPLA) (Rüb et al. 2003). The unstable and expanded CAG repeats reside in the codon region on chromosome 12q23-24.1 and encode the protein ataxin-2 of unknown function, widely expressed in neural and non-neural tissue.

The brain weight is strongly reduced (usually less than 1000 g), especially of the brainstem and cerebellum (Robitaille et al. 1997). There is atrophy of the frontotemporal and occipital lobes, the cerebellum and brainstem.

Microscopically, there is a loss of neurons in pallidum, thalamus, and precerebellar nuclei and degeneration of the middle and inferior cerebellar peduncles, in addition to the severe loss in the inferior olivary nuclei (Rüb et al. 2005b).

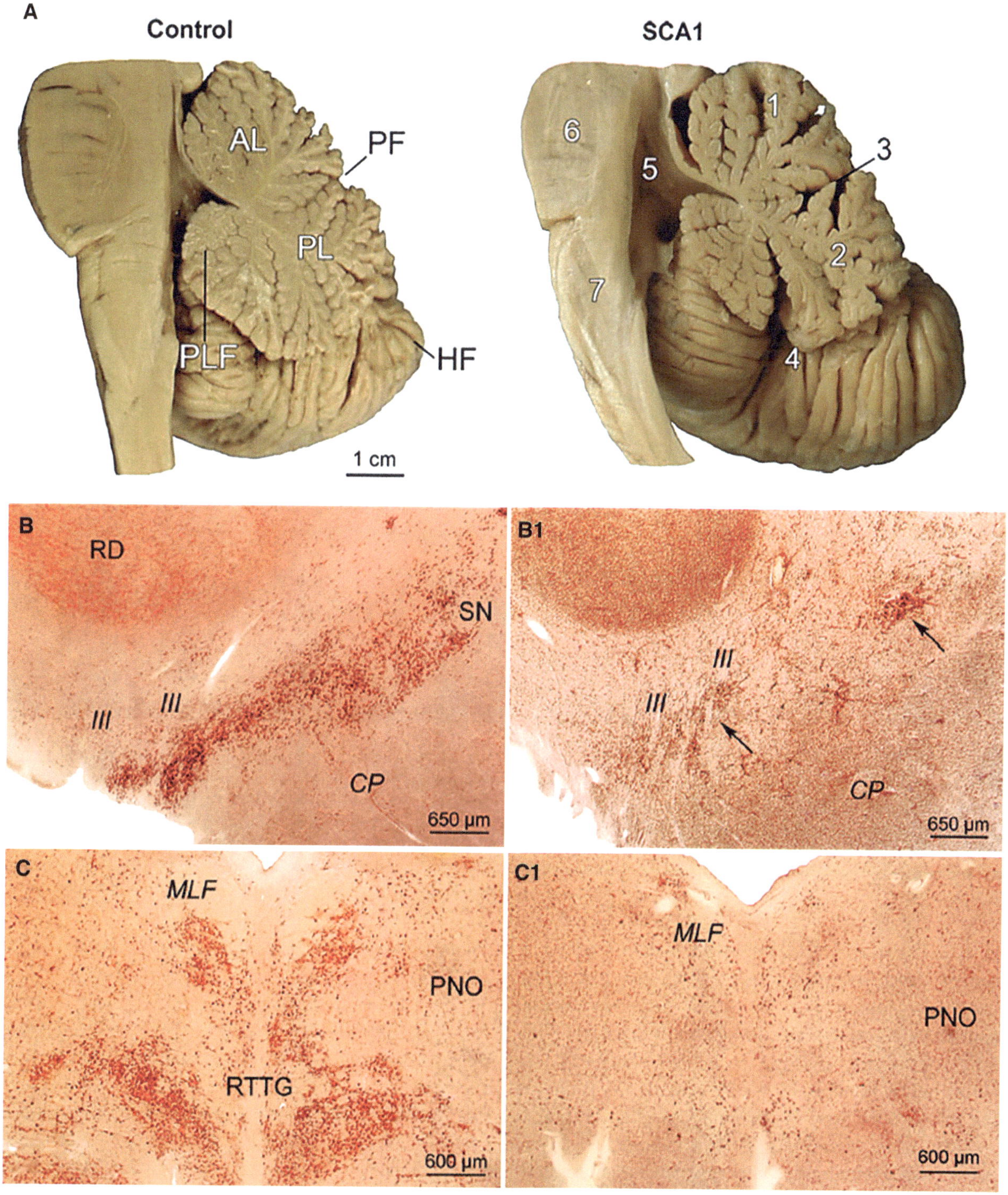

Fig. 7.12 (**A**) Neurodegeneration in cerebellum and brainstem of a patient with SCA1. a: medial aspect of the left cerebellum and brainstem of a control case. b: left cerebellum and brainstem of a SCA1 patient with atrophic anterior lobe (1), atrophic posterior lobe (2), widened posterior fissure (3), enlarged fourth ventricle (5), flattened pons (6), and atrophic medulla oblongata (7). *AL* anterior lobe, *HF* horizontal fissure, *PF* posterior fissure, *PL* posterior lobe, *PLF* posterolateral fissure. (**B**) Horizontal section through the midbrain of a control individual shows the rostral substantia nigra (SN). In (**B1**), the rostral SN of an SCA1 patient is markedly degenerated. Note the severe neuronal loss in the red nucleus (RD) (see Sect. 15.3.4). (**C**) Horizontal section through the rostral pons of a control individual with the reticulotegmental nucleus of the pons (RTTG, nucleus of Bechterew) (see Sect. 10.1.1.1, see Box 16.1). Severe damage to the RTTG of a patient with SCA1 in (**C1**). PEG Darrow red stain, 100 µm. *CP* cerebral peduncle, *MLF* medial longitudinal fasciculus, *PNO* oral pontine reticular formation, *RD* red nucleus (see Sect. 15.3.4), III rootlets of CNIII. For comparison, see atlas part Darrow red 40A (**B**) and 26, 27 (**C**). (**A**) from Seidel et al. 2012, Fig. 1 with permission (**B**, **B1**, **C**, **C1**) from Rüb et al. 2012, Fig. 5 with permission

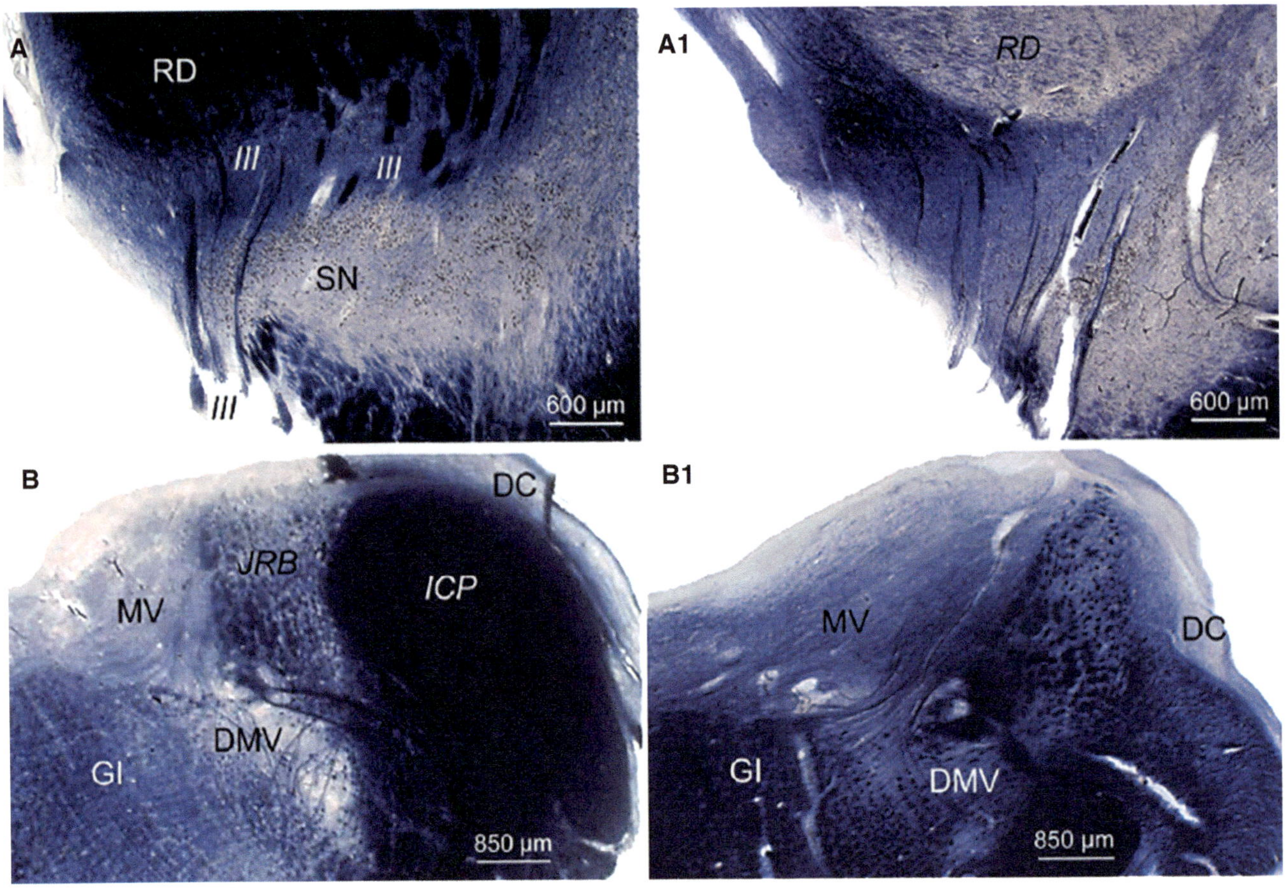

Fig. 7.13 White matter involvement in SCA1. PEG 100 µm sections, modified Heidenhain stain for myelin. (**A**) Horizontal section through the rostral midbrain of a control case depicting the rootlets of the oculomotor nerve (III) (see Sect. 15.3.1) leaving the midbrain at the interpeduncular fossa. In (**A1**) atrophic rootlets of the III of an SCA1 patient shows additional myelin pallor. Note the loss of neuromelanin-containing nerve cells (see Box 13.5) in the substantia nigra (see Sect. 16.6.1). (**B**) Horizontal section of the rostral medulla oblongata of a control case. Myelin pallor and atrophy of the inferior cerebellar peduncle (ICP) of an SCA1 patient in (**B1**). *DC* dorsal cochlear nucleus, *DMV* dorsal motor vagus nucleus, *GI* gigantocellular reticular nucleus (see Fig. 7.1 ⑧), *JRB* juxtarestiform body, *MV* medial vestibular nucleus (see Fig. 7.1 ①), *RD* red nucleus (see Sect. 15.3.4), *SN* Substantia nigra (see Sect. 16.6.1). From Rüb et al. 2012, Fig. 6 with permission

There is a consistent moderate to severe neuronal loss in all cholinergic basal forebrain nuclei. In the cerebellum, the Purkinje cells are particularly affected. The extrapyramidal system shows degeneration of the substantia nigra (see Sect. 16.6.1), subthalamic and red nuclei (see Sect. 15.3.4). Degeneration of the dorsal columns (gracile fasciculus) is prominent. The majority of the interconnecting brainstem fiber tracts (dorsal spinocerebellar tract, cuneate and gracile fascicles, medial lemniscus, spinal trigeminal tract, trigeminal nerve, and mesencephalic trigeminal tract (see Table 1.4 and Fig. 3.13B) show atrophy accompanied by demyelinization (see Fig. 7.15). Neuronal loss occurs in nearly all relay stations of the somatosensory system such as cuneate, external cuneate and gracile nuclei, spinal, principal, and mesencephalic trigeminal nuclei, and ventral posterior lateral and medial nuclei of the thalamus (Rüb et al. 2003) (Figs. 7.14 and 7.15).

These findings explain the patient's impaired sense of vibration, position, (see Fig. 3.15) and temperature. Degenerative changes in the ingestion-related brainstem nuclei are present in most patients (Rüb et al. 2006). Neuronal loss combined with astrogliosis has been described in the trigeminal, facial, and hypoglossal nuclei. The nucleus ambiguus and the solitary and reticular nuclei can also be involved. Damage to the reticulotegmental nucleus of the pons (see Sect. 10.1.1.1), an important component of the oculomotor circuits, crucial for the performance of horizontal smooth pursuits (see Box 9.6) and the accuracy of horizontal saccades have been described in SCA2 patients, as well as SCA1 and SCA3 patients (Rüb et al. 2004).

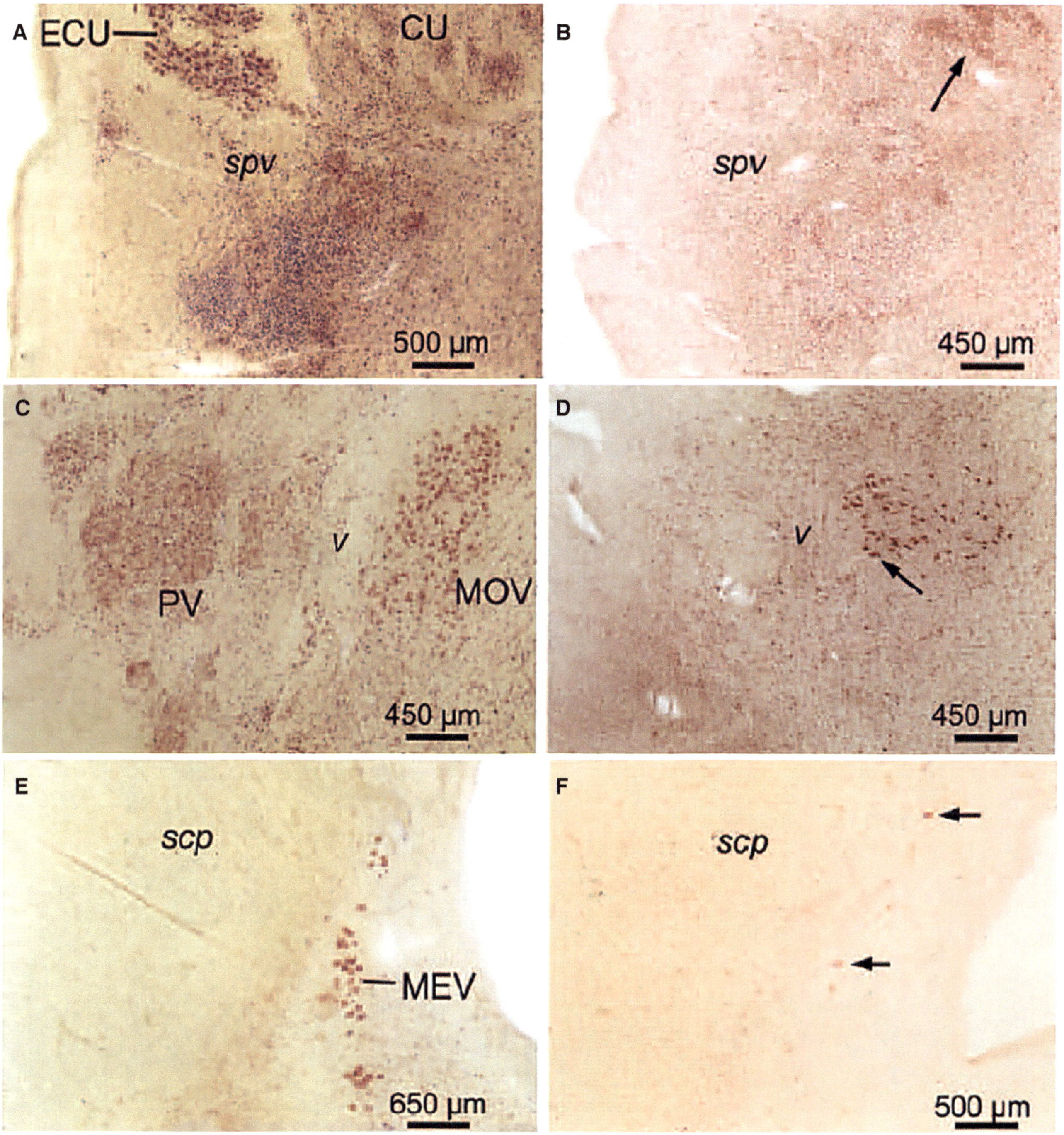

Fig. 7.14 SCA2. Somatosensory nuclei of the trigeminal nerve. (**A**, **C**, **E**) Control cases. In (**B**) marked neuronal loss in the spinal trigeminal nucleus (spv) (see Sect. 3.2.2.2), in (**D**) severely damaged principal trigeminal nucleus (PV, see Sect. 12.2.3.2) and in (**F**) marked involvement of the mesencephalic trigeminal nucleus (MEV) (see Sect. 13.2.6.1) of an SCA2 patient is shown. Note the loss of nerve cells in the external (accessory) cuneate nucleus (ECU) (see Sect. 5.2.1.1), the damage to the cuneate nucleus (CU) (see Sect. 3.2.3.3) and the additional loss in the motor trigeminal nucleus (MOV) (see Sect. 11.3.1.1). Arrows point to remaining nerve cells in the CU, MOV, and MEV. PEG. Darrow red stain, 100 μm slice thickness. *scp* superior cerebellar peduncle, *spv* spinal trigeminal tract. For comparison, see atlas part Darrow red 5A (**A**, **B**), 23 (**C**, **D**). 25 (**E**, **F**). From Rüb et al. 2003, Fig. 5 with permission

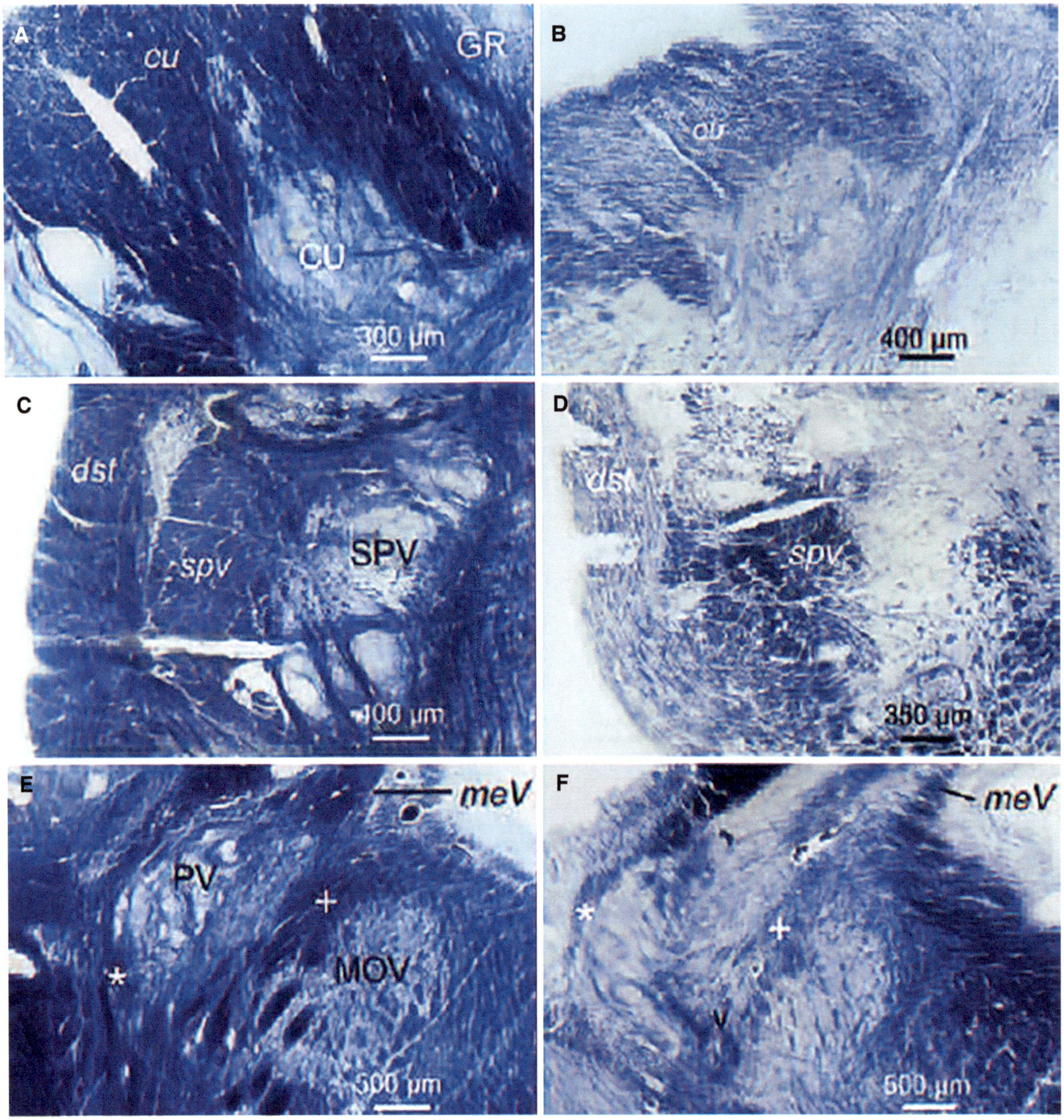

Fig. 7.15 Fiber tracts of the somatosensory system of a control case (**A**, **C**, **E**) in comparison to those of a SCA2 case (**B**, **D**, **F**). Atrophy and loss of myelin in the cuneate fascicle (cu) in (**B**), as well as in the spinal trigeminal tract (spv) (see Sect. 3.2.2.2) and in the dorsal spinocerebellar tract (dst) in (**D**). Marked atrophy and demyelinization of the sensory (*) and motor portions (+) of the trigeminal nerve as well as moderate demyelinization of the mesencephalic trigeminal tract (meV) (see Sect. 13.2.6.1) in (**F**). *CU* cuneate ncl., *GR* gracile ncl., *MOV* trigeminal motor ncl., *PV* principal sensory trigeminal ncl., *SPV* spinal trigeminal ncl. From Rüb et al. 2003, Fig. 7 with permission

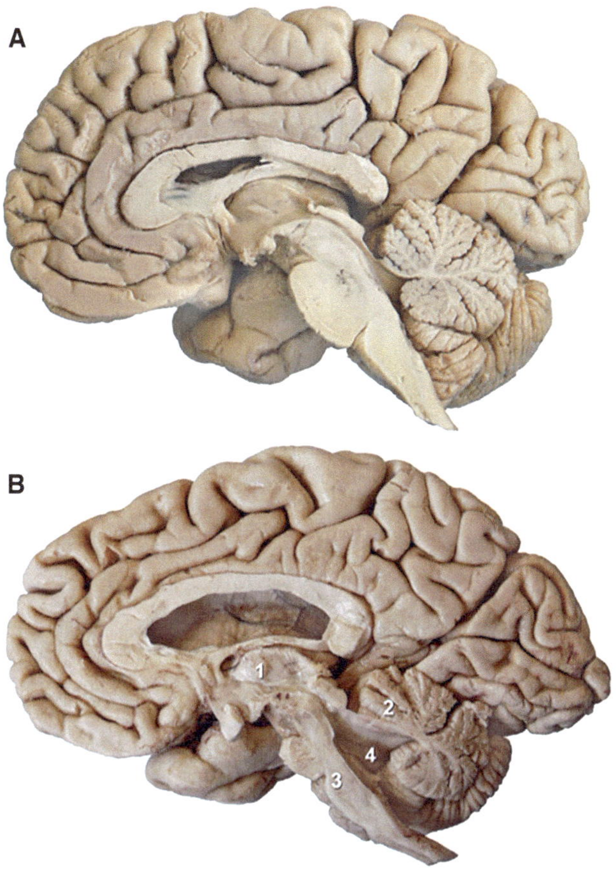

Fig. 7.16 Medial aspect of the right cerebral hemisphere, cerebellum, and brainstem of a representative control individual (**A**) compared to that of a (**B**) SCA3 patient (64 CAG-repeats in the mutated SCA3 allele). Note the atrophic thalamus (1), cerebellum (2), midbrain, pons, and medulla oblongata (3), as well as the widenened fourth ventricle (4). From Rüb et al. 2008a, Fig. 1 with permission

7.2.1.3.3 Spinocerebellar ataxia type 3 (SCA3)

SCA 3, also known as Machado-Joseph's disease, is the most frequent but also the most heterogeneous form of the SCAs. It accounts for about 30–35% of the SCA patients. The disease starts as a progressive ataxia accompanied by lid retraction and infrequent blinking (creating "staring eyes"), ophthalmoparesis, impaired speech, and swallowing (see Box 5.5). The phenotype varies widely depending on repeat length: early-onset rigidity and dystonia for the largest expansions, adult-onset ataxia for the medium-sized expansions, and late-onset ataxia accompanied by neuropathy for the smallest expansions (Paulson 2009). The gene responsible for SCA3 is located on chromosome 14q32.1, is comprised of 12–40 CAG triplets in healthy individuals and expanded to approximately 56–84 CAG triplets in SCA3 patients (Kawaguchi et al. 1994). The product of this gene, ataxin-3, is involved in ubiquitin-dependent protein surveillance (Burnett and Pittman 2005). It may support "shuttling"of polyubiquitinylated proteins to the proteasome and tends to aggregate into neuronal intranuclear inclusions (Paulson 2007, Schmidt et al. 2002, Rüb et al. 2008a).

The weight of the brain of nearly all SCA3 patients with a disease duration of more than 15 years is considerably reduced (Iwabuchi et al. 1999). Macroscopically, the most characteristic features are atrophy of the cerebellum, medial cerebellar peduncle, bulge of the inferior olives (see Sect. 4.1.1.3), spinal cord, pontine base, and a variable atrophy of the outgoing cranial nerves II to XII. The substantia nigra (see Sect. 16.6.1) is depigmented (Iwabuchi et al. 1999, Robitaille et al. 1997) (Figs. 7.16, 7.17, 7.18, 7.19, and 7.20).

The major microscopical abnormalities are found in the spinocerebellar system and fastigial and dentate nuclei ("eosinophilic grumose" degeneration) representing degenerating axon terminals of Purkinje cells around neurons of the cerebellar nuclei that can be visualized nicely using silver impregnation or synaptophysin immunostaining, in the cerebellar Purkinje cell layer, and in the pontine nuclei (see Sect. 10.1.1.2) (Fig. 7.21).

Within the basal ganglia-thalamocortical loop (see Fig. 7.2) the globus pallidus (especially the internal segment), thalamus (ventral anterior nucleus), subthalamic nucleus, substantia nigra (see Sect. 16.6.1), and red nucleus (see Sect. 15.3.4) are involved (see Fig. 16.6). In the neocortex, most of terminal SCA3 patients show severe depletion of the giant pyramidal Betz cells in the primary motor cortex (Rüb et al. 2008a). Degeneration of the visual system may occur in the thalamus (lateral and inferior nuclei) and the lateral geniculate body and damage to the auditory system may be seen at the level of the inferior colliculus (see Sect. 15.1.4), nuclei of the lateral lemniscus, superior olivary complex (see Fig. 7.22, here below), and cochlear nuclei (see Sect. 8.1.1). In the somatosensory system, the thalamus, all trigeminal nuclei, the cuneate, external cuneate (see Sect. 5.2.1.1), and gracile nuclei (see Sect. 3.2.3.4) as well as Clarke's column in the spinal cord may undergo neurodegeneration (Fig. 7.22).

In the lower brainstem, in additon to the nuclei of the vestibular system, the cranial nerve nuclei and select premotor nuclei of the oculomotor system show degenerative changes (Fig. 7.22). Degenerative changes also have been described in the ingestion-related nuclei, such as the trigeminal, facial,

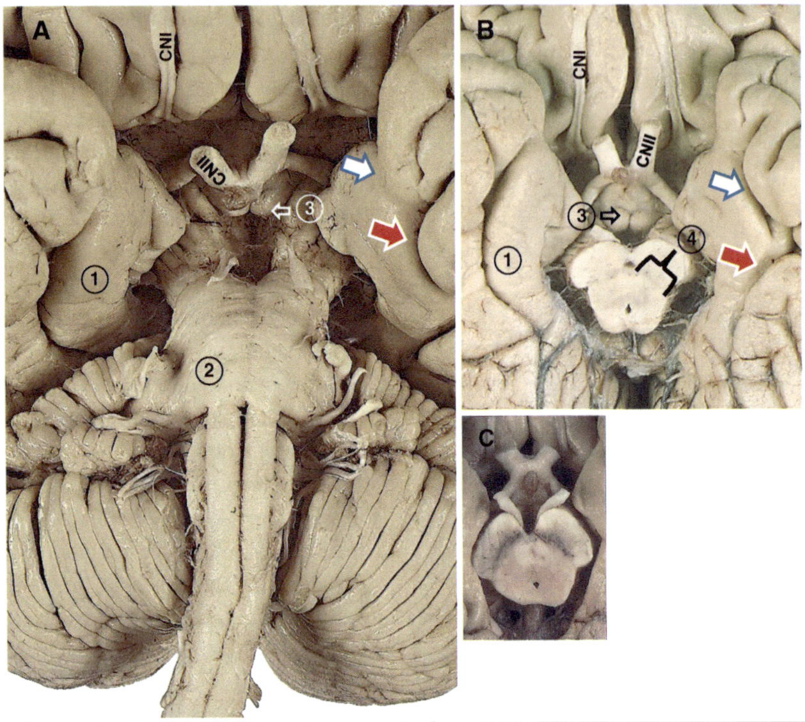

①	Entorhinal region	③	Mammillary bodies
②	Pons	④	Substantia nigra
red arrow	Collateral sulcus	white arrow	Rhinal sulcus

Fig. 7.17 (**A**) Ventral view of the brain of an SCA3 patient showing a remarkable atrophy (arrows) of the cerebellum, pons, and medulla oblongata. In (**B**) ventral view of the brain of an SCA3 patient after removal of the brainstem (higher magnification), showing atrophy of the mesencephalon. Note the pallor of the substantia nigra (black arrow) (compare with Fig. 16.3, see 16.6.1). Rhinal sulcus (white small arrow), collateral sulcus (red small arrow). LabPON Twente

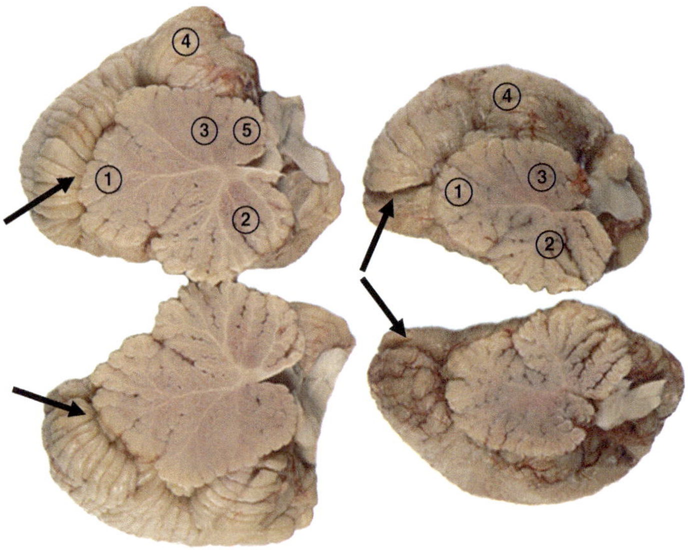

①	Posterior cerebellar lobe	④	Tonsil
②	Anterior cerebellar lobe	⑤	Nodule
③	Flocculonodular cerebellar lobe		

Fig. 7.18 Paramedian aspect (and vermis) of the cerebellum of a control case (left) compared to that of an SCA3 patient (right). Remarkable depth of the horizontal fissure (arrows), atrophy of the anterior (A), posterior (P), and flocculonodular (F) lobes, as well as of the tonsil (T). From Rüb et al. 2003, Fig. 1 with permission

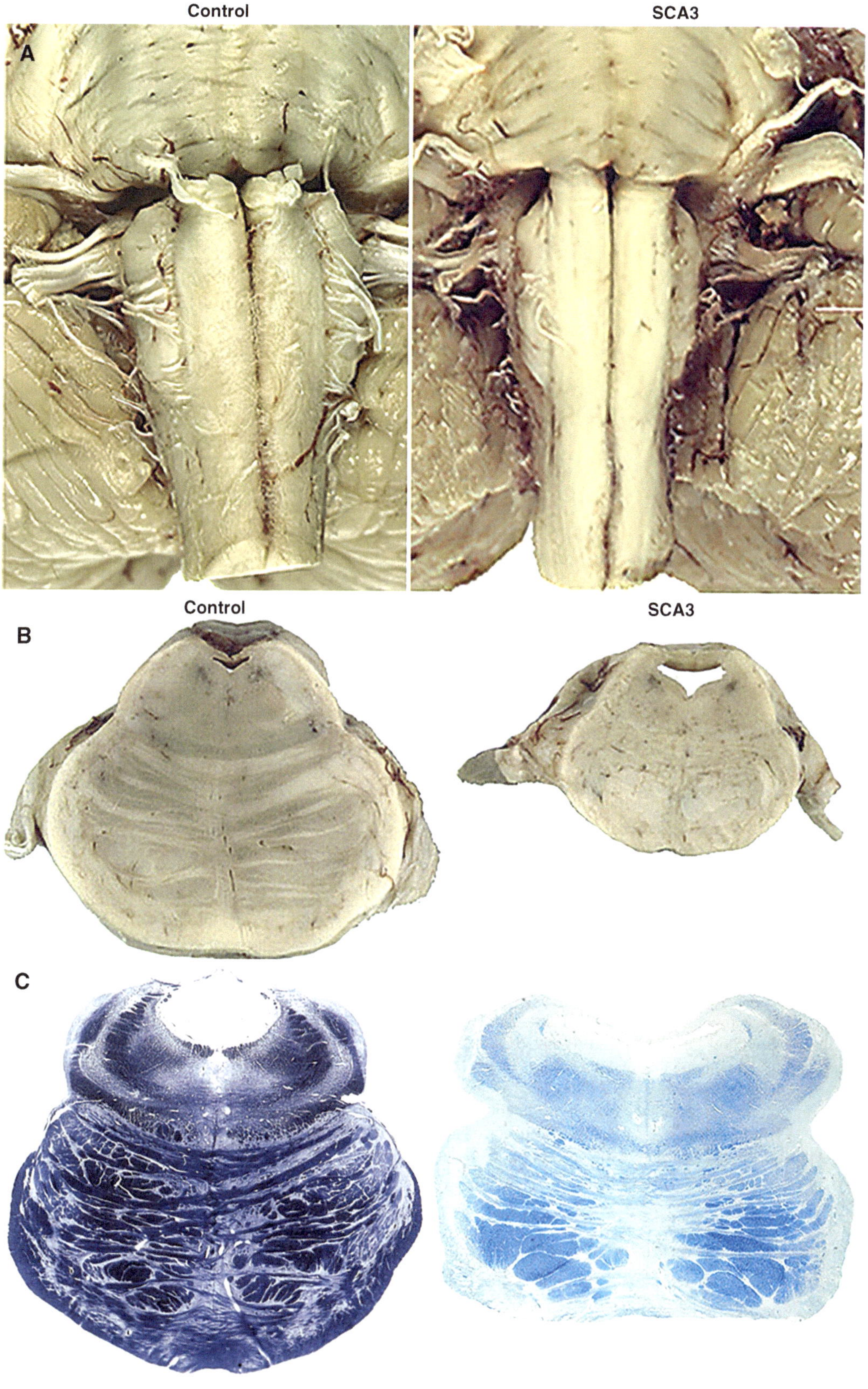

Fig. 7.19 (**A**) Ventral view of pons and medulla oblongata. Severe atrophy of the pons and medulla oblongata in an SCA3 patient compared to a control case. (**B**) Horizontal sections of the pons. Severe atrophy of the pons in the same patient compared to a control case. (**C**) Microscopical sections of the atrophic pons in the SCA3 patient showing a severe loss of myelin (Klüver-Barreira stain for myelin) compared to the rostral pons of a control case (modified Heidenhain stain for myelin). LabPON Twente

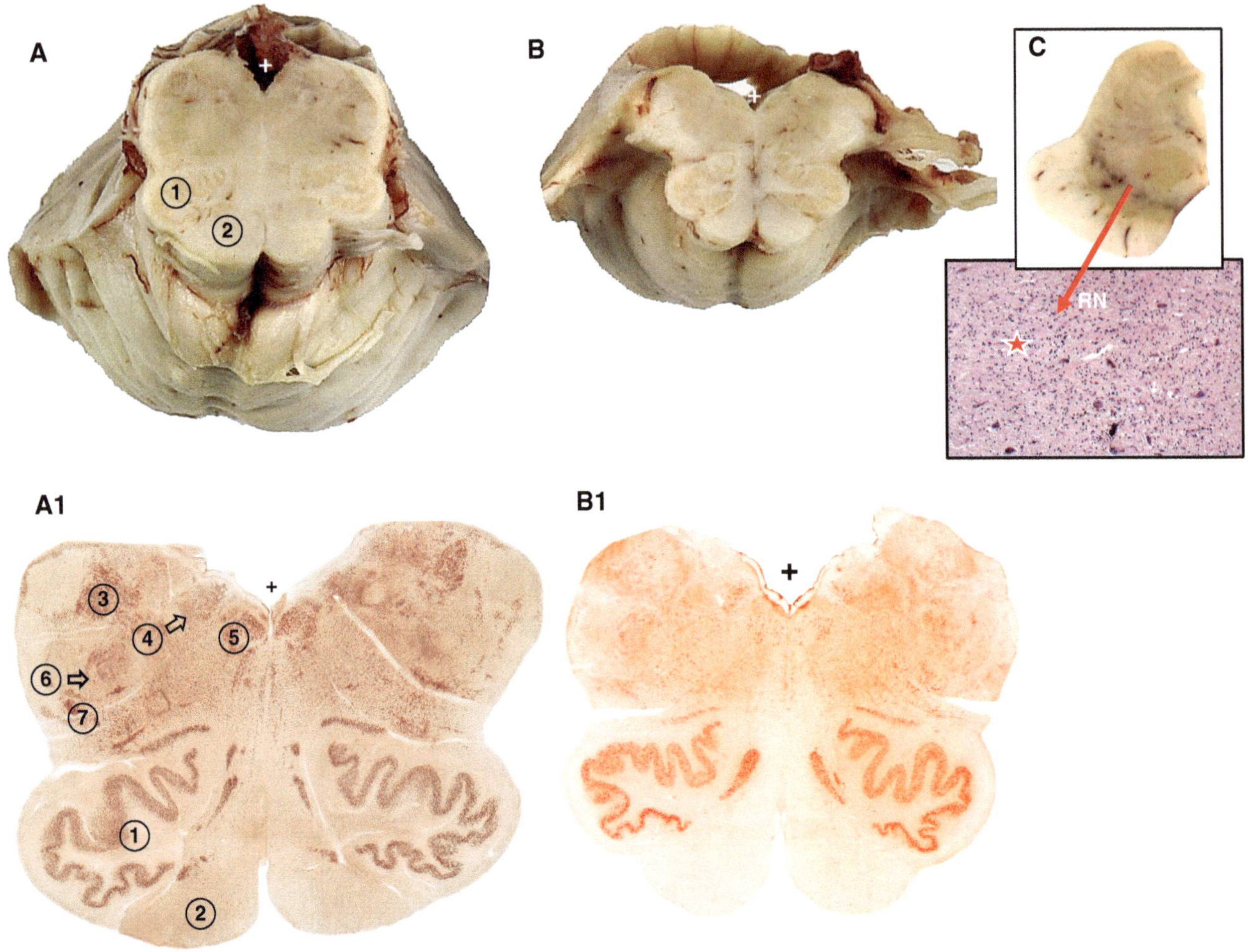

①	Inferior olive	⑤	Ncl. of CNXII
②	Corticospinal tract	⑥	Spinal trigeminal ncl. CNV
③	Accessory cuneate ncl.	⑦	Lateral reticular ncl.
④	Posterior ncl. of CNX	+	4th ventricle

Fig. 7.20 (**A**) Horizontal section of the medulla oblongata on level of the inferior olive in a (**A**) control case compared to that of a SCA3 patient (**B**). Note the severe atrophy of the medulla oblongata and the atrophic roots of the cranial nerves (see also Fig. 7.29A). Microscopical sections (PEG, 100 μm, Darrow red stain) show the medulla oblongata of a control case in (**A1**) compared to the medulla oblongata of an SCA3 patient in (**B1**), e.g., severe destruction of the external cuneate nucleus (see Sect. 5.2.1.1), the cranial nerve nuclei CNX (see Sect. 7.3.2.1) and CNXII (see Sect. 4.3.1.1), the spinal trigeminal nucleus (see Sect. 3.2.2.2), and the lateral reticular nucleus (see Sect. 4.1.1.1). A mild neuronal loss is seen in the inferior olive (see Sect. 4.1.1.3). See also atlas part Darrow red 10. The horizontal section of the atrophic mesencephalon (C) shows pallor of the substantia nigra (see Sect. 16.6.1). In the accompanying microscopical section, the substantia nigra shows severe neuronal loss and gliosis (star). RN = red nucleus. LabPON Twente

hypoglossal, ambiguus, dorsal motor vagal, solitary, reticular nuclei, in addition to the precerebellar nuclei (Rüb et al. 2008a, b) (Figs. 7.20, 7.22, and 7.23).

In the motor nuclei of the aforementioned cranial nerves, there is a loss of motor neurons (Figs. 7.20 and 7.22). Neuronal loss is associated with a marked astrogliosis (Fig. 7.22). The brain regions spared by the neurodegenerative process of SCA3, along with regions of the cerebral neocortex and the striatum, include components of the limbic,

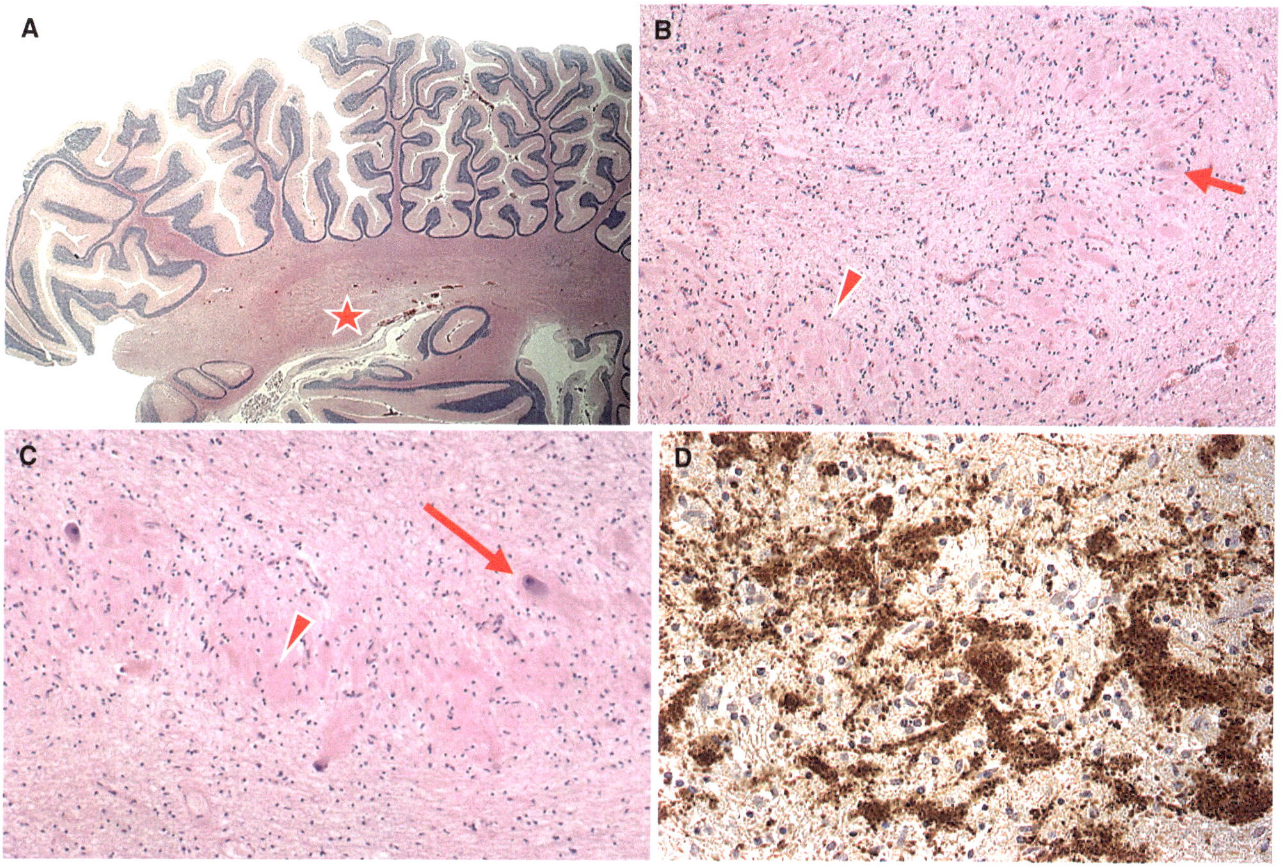

Fig. 7.21 SCA3. (**A**) Microscopical section of the cerebellar cortex and dentate nucleus (star). Detail (**B** and **C**) of the dentate nucleus in a SCA3 patient, showing eosinophilic granular structures (triangles) representing degenerating axon terminals of Purkinje cells around neurons of the dentate nucleus, some of which are swollen (red arrow). H & E stain. The granular structures are labeled in a synaptophysin immunostain (**D**). LabPON Twente

cholinergic, and serotonergic systems, as well as the hippocampus, amygdala, and hypothalamus (Rüb et al. 2008a).

Outside the brain, neurons in the anterior horn and in the nuclei of Clarke of the spinal cord and in the dorsal (spinal) ganglia are affected (see Fig. 7.25).

The intactness of the major components of the limbic and basal forebrain cholinergic systems may account for the phenomenon that dementia is often absent in the majority of patients (Manuff et al. 1996). In cases where cognitive symptoms were present, most often Alzheimer's disease (see Sect. 13.2.2.1.4)-related cytoskeletal pathology has been found. The affection of the white matter is less severe in SCA3 and is confined to the cerebellum, brainstem, and spinal cord.

Widespread ataxin-3 and p62-positive axonal aggregates in fiber tracts known to undergo neurodegeneration in SCA3 have been described, indicating involvement of neuronal protein quality control mechanisms (Seidel et al. 2010). The axonal aggregates vary in regards to morphology (Fig. 7.24).

A typical morphological feature represents the occurrence of neuronal intranuclear ataxin-3 immunopositive inclusions, not only in degenerated but also in spared central nervous components of SCA3 patients (Rüb et al. 2008a). The intranuclear inclusions can be visualized using antibodies against ataxin-3 or the chaperone proteins ubiquitin and P62, especially in the pontine neurons (Paulson et al. 1997, Rüb et al. 2008a) (Fig. 7.25).

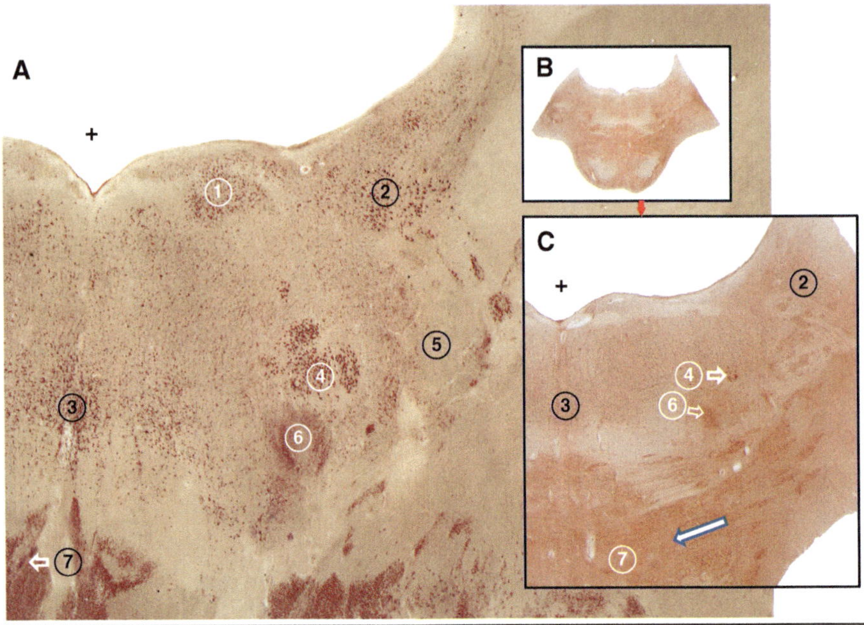

①	Abducens ncl. CNVI	⑤	Spinal trigeminal tract CNV
②	Vestibular ncll. CNVIII	⑥	Superior olivary complex
③	Raphe magnus ncl.	⑦	Pontine ncll.
④	Facial ncl. CNVII	+	4th ventricle

Fig. 7.22 Microscopical section (PEG, 100 µm, Darrow red stain) of the pontomedullary junction in a control case (**A**) and an SCA3 patient, overview (**B**) and detail (**C**), showing a severe loss of neurons in the vestibular nuclei ② (see Sect. 5.3.1), cranial nerve nuclei CNVI ① (see Sect. 9.2.1.1) and CNVII ④ (see Sect. 8.2.1) and raphe nuclei ③ (see Sect. 4.3.3), accompanied by severe astrogliosis in the pons base (white arrow). Some preserved neurons in the superior olivary complex ⑥ (see Sect. 8.1.1.5). See also atlas part Darrow red 19, 19A. LabPON Twente

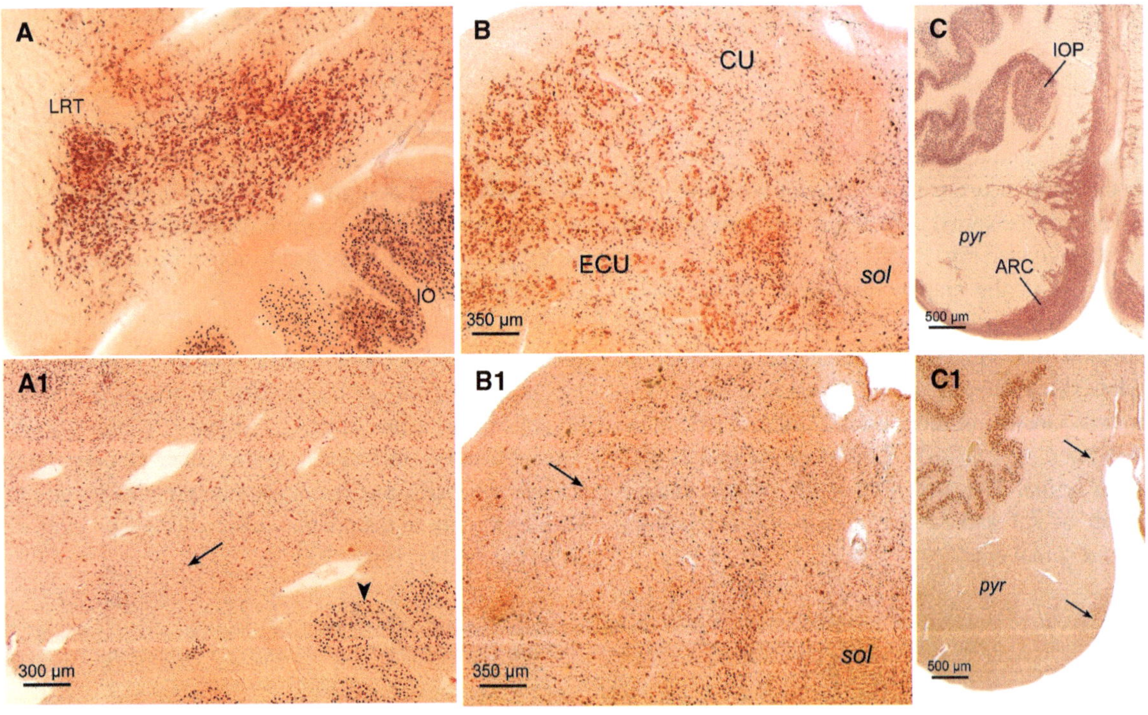

Fig. 7.23 The lateral reticular nucleus (LRT) (see Sect. 4.1.1.1), the external (accessory) cuneate nucleus (ECU) (see Sect. 5.2.1.1), and the arcuate nucleus (ARC) (see Sect. 4.1.1.4) in a control case (**A–C**), compared to the severe damage of the LRT and marked affection of the principal nucleus of the inferior olive (IOP) (see Sect. 4.1.1.3) in (**A1**), severe degeneration of the ECU in (**B1**), and obvious neuronal loss in the ARC in (**C1**) in a patient with SCA3. Arrows point to remaining nerve cells in LRT, ECU, and ARC, arrowhead to surviving IO nerve cells. *CU* cuneate ncl., *Pyr* pyramidal tract, *sol* solitary tract. From Rüb et al. 2005b, Figs. 6 and 7 with permission

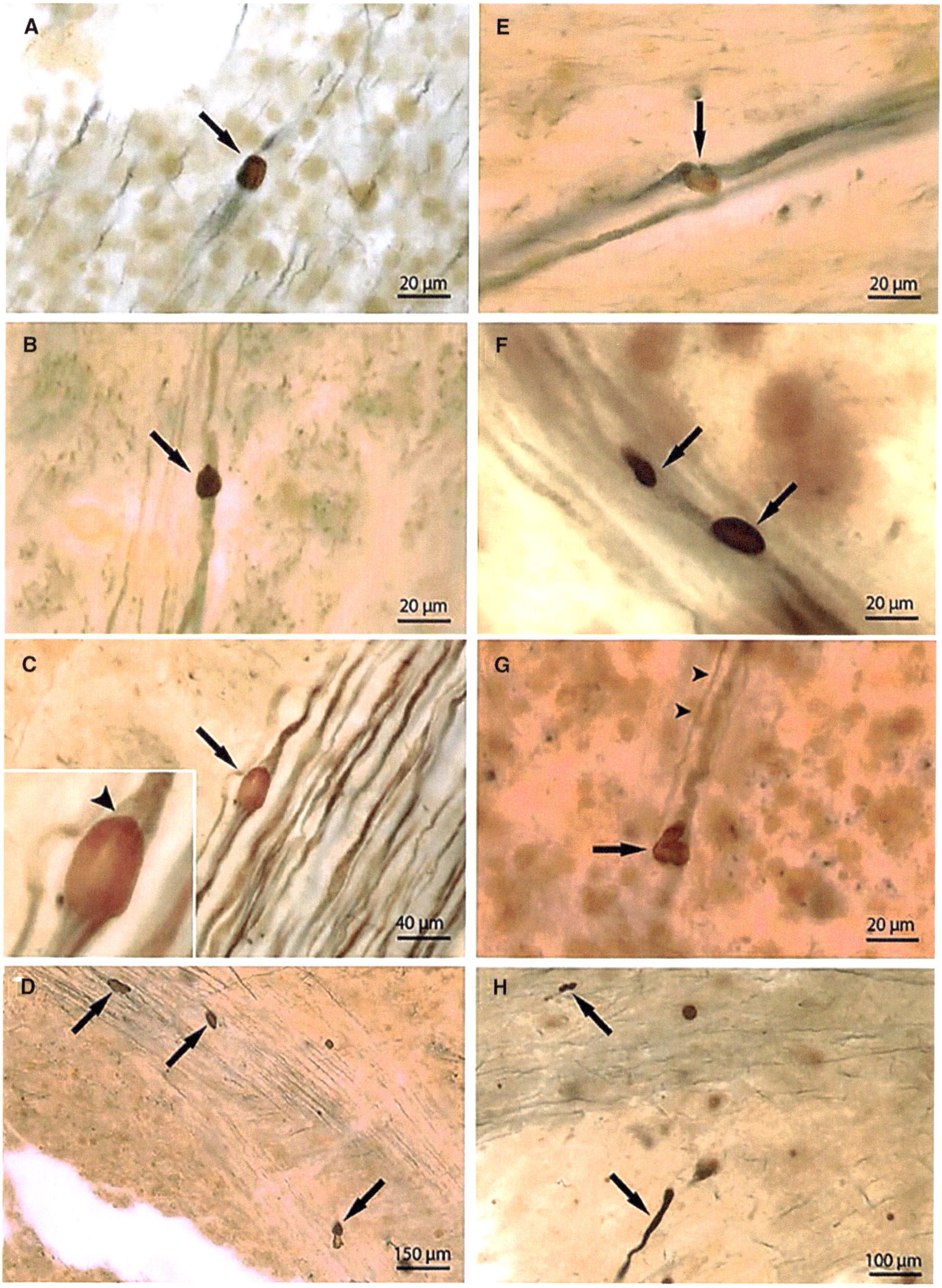

Fig. 7.24 Axonal aggregegates in the pyramidal tract (**A**) (see Sect. 1.6), oculomotor nerve (**B**) (see Sect. 15.3.1), facial nerve (**C**, **D**) (see Sect. 8.2.1.3), hypoglossal nerve (**E**–**G**) (Sect. 4.3.1.1) and nigrostriatal tract (**H**) (see Sect. 16.6.1.3) in SCA3 patients. Axonal aggregates vary in morphology between ovoid (**A**–**D**), elongated (**H**), or irregular (**E**, **G**). Size is also variable. Note the pallor of the tau staining, indicating axonal dilatation (arrowhead, **C**), and the comparatively large axonal diameter compared to the unaffected neighboring axons (arrowheads, **G**). Double immunostaining with anti-ataxin-3, brown, and the axonal marker AT270, blue-gray. PEG, 100 µm sections. From Seidel et al. 2010, Fig. 1 with permission

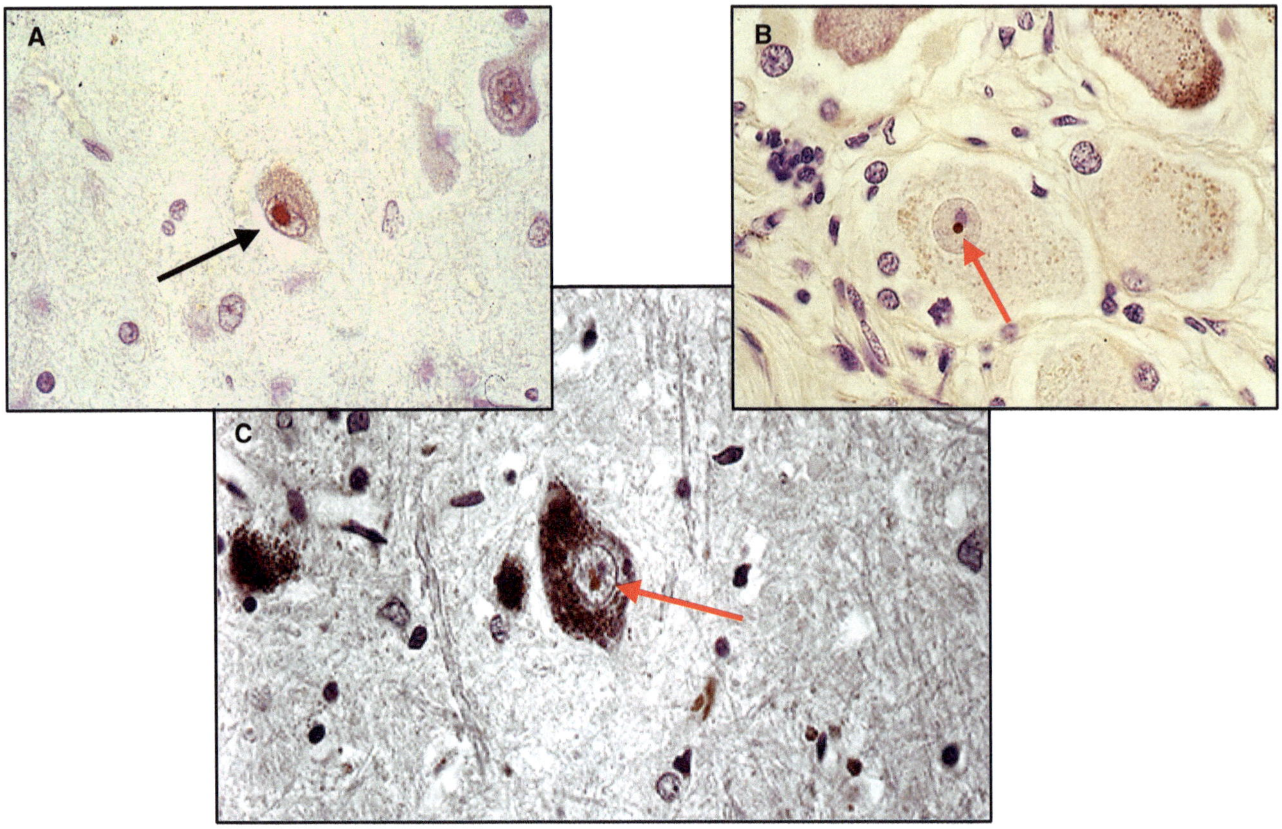

Fig. 7.25 Intranuclear inclusions (arrows) in the pons (**A**), dorsal ganglion (**B**), and substantia nigra (**C**) (see Sect. 16.6.1). Ubiquitin immunostain in **A** and **B**, anti-ataxin-3 immunostain in (**C**). Their association with the underlying pathological process and their precise role in the pathognenesis of SCA3, however, is still enigmatic. In summary, the pathological process of SCA3 affects a widespread distribution pattern of lesions which can account for a variety of symptoms. LabPON Twente

7.2.1.3.4 Spinocerebellar ataxia type 6 (SCA6)

SCA6 is the prototype of a pure cerebellar ataxia (Gomez et al. 1997, Yang et al. 2000). SCA6 is an unusual type of polyglutamine disease in which the affected gene is located on chromosome 19p13 and encodes for the α-subunit of a calcium channel (CACNA1A) that is highly expressed in Purkinje cells (Zhuchenko et al. 1997). This is the same gene in which other non-repeat mutations cause episodic ataxia type 2 and familial hemiplegic migraine. Those conditions are categorized as channelopathies and there is evidence to suggest that the expanded polyglutamine seen in SCA6 may also result in channel dysfunction. The expanded mutant alleles in SCA6 (21–27 repeats) are remarkably smaller than the expanded alleles observed in any of the other neurodegenerative diseases (36–121 repeats) and are well within the normal range of polyglutamine tracts seen at the other loci in many unaffected individuals (Zhuchenko et al. 1997). SCA6 represents a "milder" disease, manifesting as a "pure" cerebellar ataxia, accompanied by dysarthria and nystagmus. The onset of the disease commonly starts after the age of 50 years and has a slow progression (Paulson 2009). MRI brain scans show pure cerebellar atrophy. Compared with other SCAs, noncerebellar symptoms occur much less frequently. Clinical features include dysmetria, dysarthria, dysphagia, ataxia, and abnormal eye movements, ranging from difficulty fixating on moving objects to diplopia and an abnormal vestibulo-ocular reflex (see Box 6.2) (Schöls et al. 1998, Paulson 2009, Casamento-Moran et al. 2015). Higher cognitive function has been considered generally preserved. The disease progresses slowly compared to other SCAs and usually is compatible with a normal lifespan. It is a common form of ataxia in some populations, such as northern Europe (Germany) and Japan. Currently, there is no treatment. Successful therapeutic strategies must target a valid pathological mechanism (Solodkin and Gomez 2012). Given the fact that the affected gene is highly expressed in Purkinje cells, it is not surprising that a severe loss of Purkinje cells is seen in all cases, most pronounced in the vermis. Therefore, atrophy in the cerebellum is most conspicious and can be visualized optimally in sagittal sections. Changes in the vermis are most severe in the anterior part. The pons is frequently normal (Figs. 7.26 and 7.27).

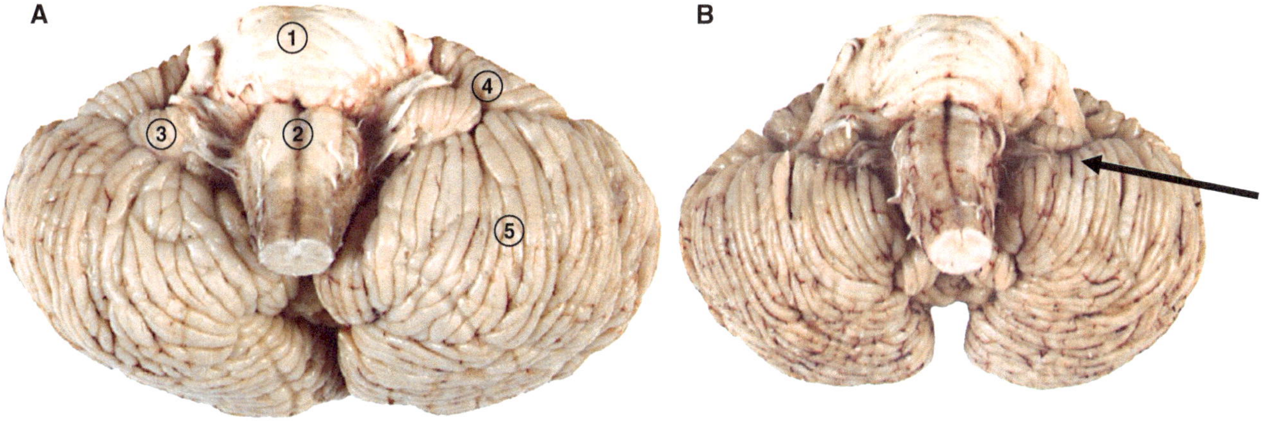

①	**Pons**	④	**Anterior cerebellar lobe**
②	**Medulla oblongata**	⑤	**Posterior cerebellar lobe**
③	**Flocculonodular cerebellar lobe**		

Fig. 7.26 (**A**) Cerebellum, pons, and medulla oblongata of a control case, compared to that of a SCA6 patient (**B**). Atrophy of the cerebellar folia, widening of the sulci in (**B**). Note the atrophic flocculonodular lobe (arrow). Mild atrophy of the pons. LabPON Twente

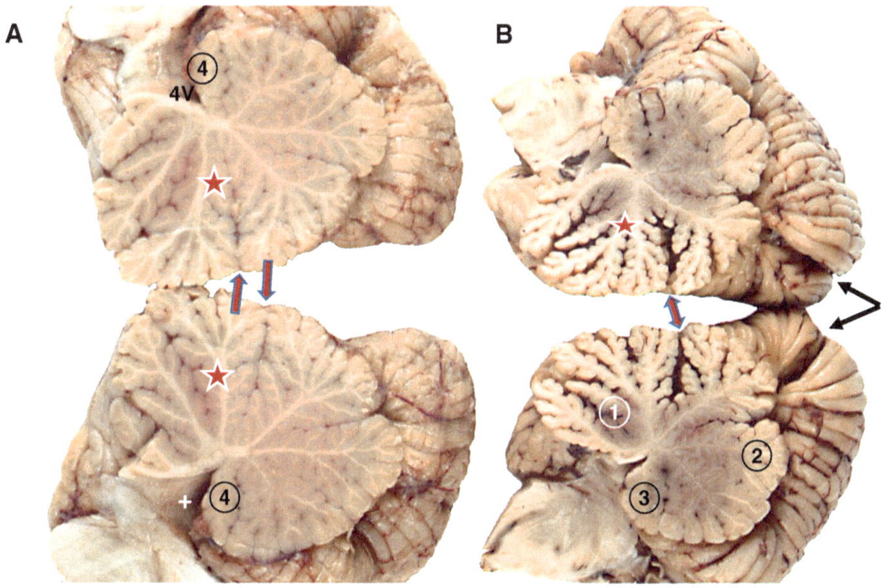

①	**Anterior cerebellar lobe**	→	**Horizontal fissure**
②	**Posterior cerebellar lobe**	⇒	**Primary fissure**
③	**Flocculonodular cerebellar lobe**	+	**4th ventricle**
④	**Nodule**		

Fig. 7.27 (**A**) Midsagittal section of the left (top) and right (bottom) cerebellum of a control case, compared to that of an SCA6 patient in (**B**). Remarkable atrophy of the vermis in **B**, especially of the anterior lobe (red star). Note the conspicuous widening of the horizontal fissure (black arrows). LabPON Twente

Microscopically, the marked loss of the Purkinje cells can be easily demonstrated using antibodies against calbindin, a calcium-binding protein. The dendritic arborization is almost completely lost. Swollen axons (torpedos) can be found (Yang et al. 2000). A proliferation of Bergmann glial cells is often present (Figs. 7.28 and 7.29).

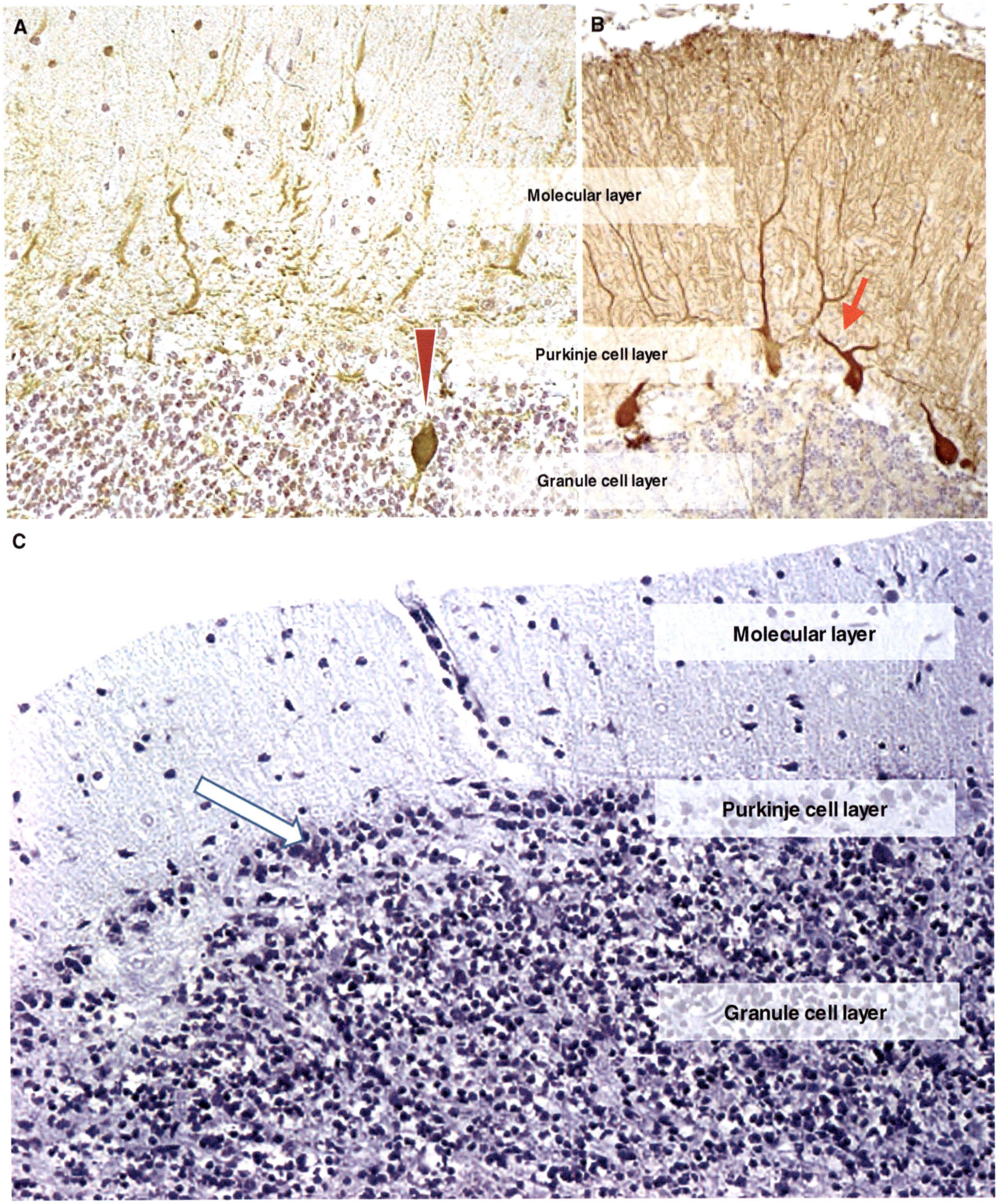

Fig. 7.28 Loss of Purkinje cells in atrophic cerebellum (**A** and **C**), in a calbindin immunostain (**A**, **B**) and in a Nissl stain (**C**) in a SCA6 patient, compared to the cerebellum of a control case (**B**). Torpedo (swelling of Purkinje cell axon) in **A** (red triangle). White arrow points to Bergmann glia proliferation. Black arrow points to Purkinje cell. LabPON Twente

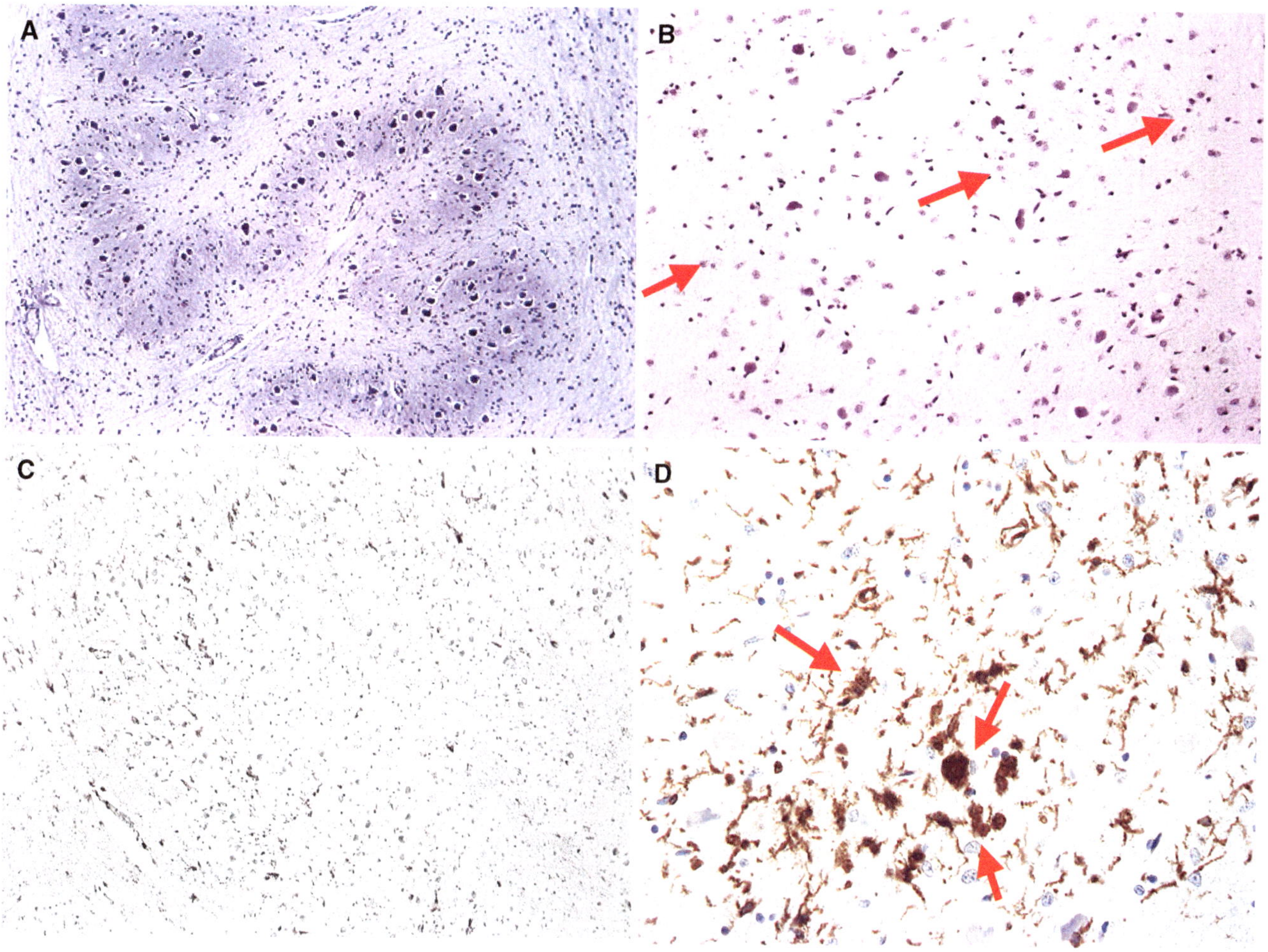

Fig. 7.29 SCA6. Microscopical sections of the inferior olive nucleus (**A** overview and **B** detail) (see Sect. 4.1.1.3), showing mild loss of neurons and microglia proliferation (arrows). Nissl stain (**A** and **B**). CR3/43 immunostain (**C**, overview) shows activated microglial cells (arrows in **D**). SCA6 patient. LabPON Twente

The dentate nucleus is usually normal. The inferior olives (see Sect. 4.1.1.3) show a variable loss of neurons as well as gliosis, probably because of transsynaptic degeneration. A proliferation of microglial cells can be seen (Fig. 7.29).

Degenerative changes have been described in some brainstem nuclei, especially in ingestion-related nuclei (see Table 7.1), including the mesencephalic trigeminal, the spinal trigeminal, facial, solitary, and reticular nuclei (Rüb et al. 2006) (Fig. 7.30).

Malfunctions during all phases of the ingestive process possibly contribute to dysphagia in SCA6 patients, as well as in patients with SCA2, SCA3, and SCA7 (Rüb et al. 2006). Microglia activation in areas showing neuronal loss can be visualized using microglia markers (CD68, CR3/43) (Fig. 7.29). Mild to moderate neuronal loss with gliosis in the inferior olive nucleus (see Sect. 4.1.1.3) may be detected, especially when disease's duration is long (Yang et al. 2000, Solodkin and Gomez 2012). Intranuclear neuronal inclu-

Table 7.1 Ingestion-related structures of the human brainstem (in caudo-cranial order) involved in spinocerebellar ataxia type 3 pathology such as neuronal loss and astrocytosis (modif. acc to Rüb et al. 2006)

Name	*r*	Function
Ambiguus ncl. see Sect. 5.4.2.1	9	Innervation of pharynx muscles
Intermediate reticular zone see Sect. 3.3.2.1	11	Premotor swallowing neurons
Hypoglossal ncl. see Sect. 4.3.1.1	10	Innervation of tongue muscles
Spinal trigeminal ncl. see Sect. 3.2.2.2	11	Nociceptive afferents of the head region
Posterior vagus ncl. see Sect. 7.3.2.1	7	Innervation of thoracic and abdominal organs down to Cannon's point (at distal one third of the transverse colon)
Solitary nucleus see Sect. 3.2.4	11	Gustatory and enteroceptive afferents
Subhypoglossal nucleus see Sect. 6.2.1.6	8	Eye movement
Parvocellular reticular ncl. see Sect. 6.5.5.2	8	Premotor swallowing neurons
Facial ncl. see Sect. 8.2.1	6	Innervation of mimic muscles
Motor trigeminal ncl. see Sect. 11.3.1.1	3	Innervation of masticatory muscles
Principal sensory trigeminal ncl. see Sect. 12.2.3.2	2	Epicritic sensory afferents of the head region, including the mucosae
Mesencephalic trigeminal ncl. see Sect. 13.2.6.1	1	Muscle spindle input from the masticatory muscles

sions have not been observed. The mutated protein is found in the cytoplasm of Purkinje cells (Ishikawa et al. 1999).

7.2.1.3.5 Spinocerebellar ataxia type 7 (SCA7)

SCA 7 is distinguished from other SCAs by the nearly universal presence of retinal degeneration (Garden and La Spada 2008). It is the only SCA in which patients tend to go blind and careful ophthalmoscopic examination is necessary in any patient presenting with progressive ataxia (Paulson 2009). It is a rare form of ataxia and accounts for 1–11% of genetically diagnosed SCAs (Rüb et al. 2005a). The SCA7 gene has been mapped on chromosome 3p12-p21.1 and encodes for the diseased protein designated ataxin-7. In healthy individuals, it contains repeat sequences of 4–35 CAG triplets. The normal form of ataxin-7 is found predominantly neuronal within the CNS and retina, and in a variety of peripheral tissues. It is localized to the cytoplasmic compartment, whereas the mutated form of ataxin-7 is restricted to neuronal nuclei, where it tends to aggregate into nuclear inclusions (NI) (Rüb et al. 2005a). Almost all SCA7 patients have more than 36 CAG-repeats on one ataxin-7 allele (Garden and La Spada 2008). CAG-repeat expansions of approximately 60 and more are associated with early onset and rapid progression of SCA7. Patients with less than 60 expanded CAG-repeats in the mutated allele suffer from late SCA7 onset, first having ataxic symptoms, which may be followed by a progressive visual impairment (Rüb et al.

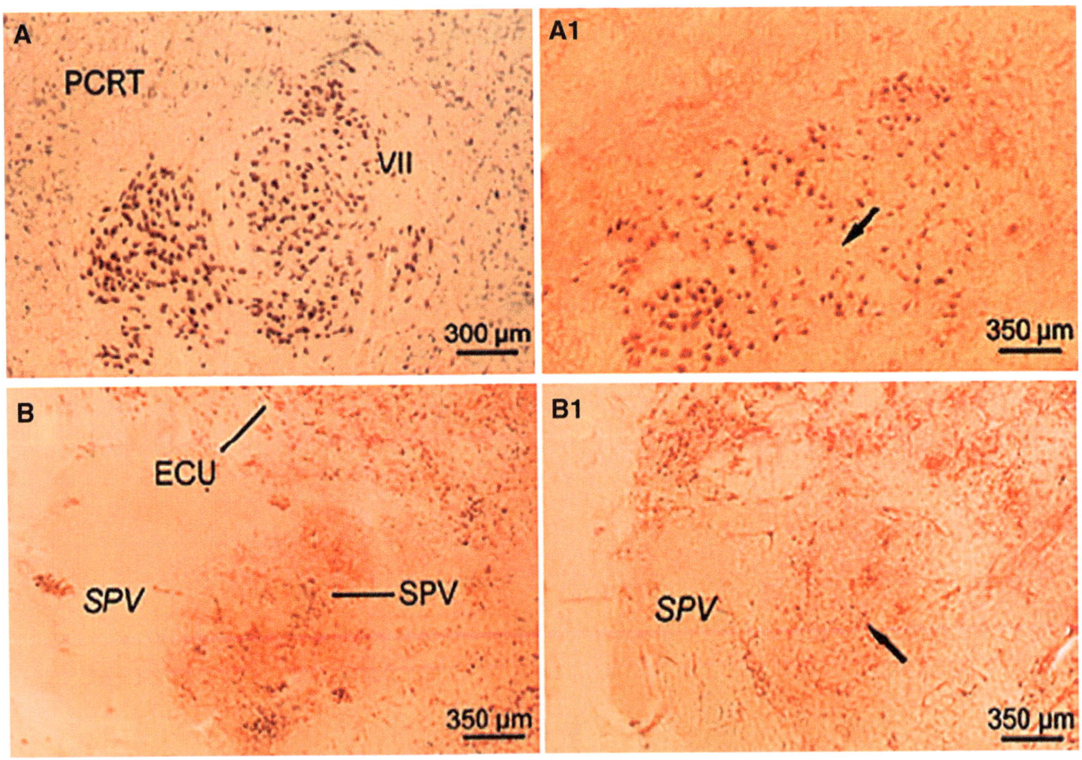

Fig. 7.30 (**A**) Nucleus of CNVII (VII; see Fig. 7.22, here above) and the parvocellular reticular nucleus (PCRT) (see Sect. 6.5.5.2) in a control case, compared to the marked damage to VII and PCRT in a SCA6 patient (**A1**). (**B**) Spinal trigeminal nucleus (SPV) (see Sect. 3.2.2.2) in a control case, compared to the marked involvement of the SPV in an SCA6 patient (**B1**). Arrow indicates surviving neurons. *ECU* external cuneate nucleus. From Rüb et al. 2006, Figs. 2 and 5 with permission

2005a, Martin et al. 1999). MRI of SCA7 patients typically reveals marked atrophy of the cerebellum and pons. Moreover, MRI-based volumetric analysis demonstrated that SCA7 patients have significantly more pontine atrophy than patients with other SCAs. The volume loss may develop in the pons prior to the development of cerebellar atrophy, suggesting that the primary site of disease onset may be in brainstem structures rather than in the cerebellar folia (Bang et al. 2004, Garden and La Spada 2008).

The brainweight is often reduced in SCA7. Atrophy of the cerebellar hemispheres and the vermis is seen, as well as reduction of the cerebellar white matter and, furthermore, pons atrophy with loss of pontocerebellar fibers and reduction of the medial cerebellar peduncle (Fig. 7.31).

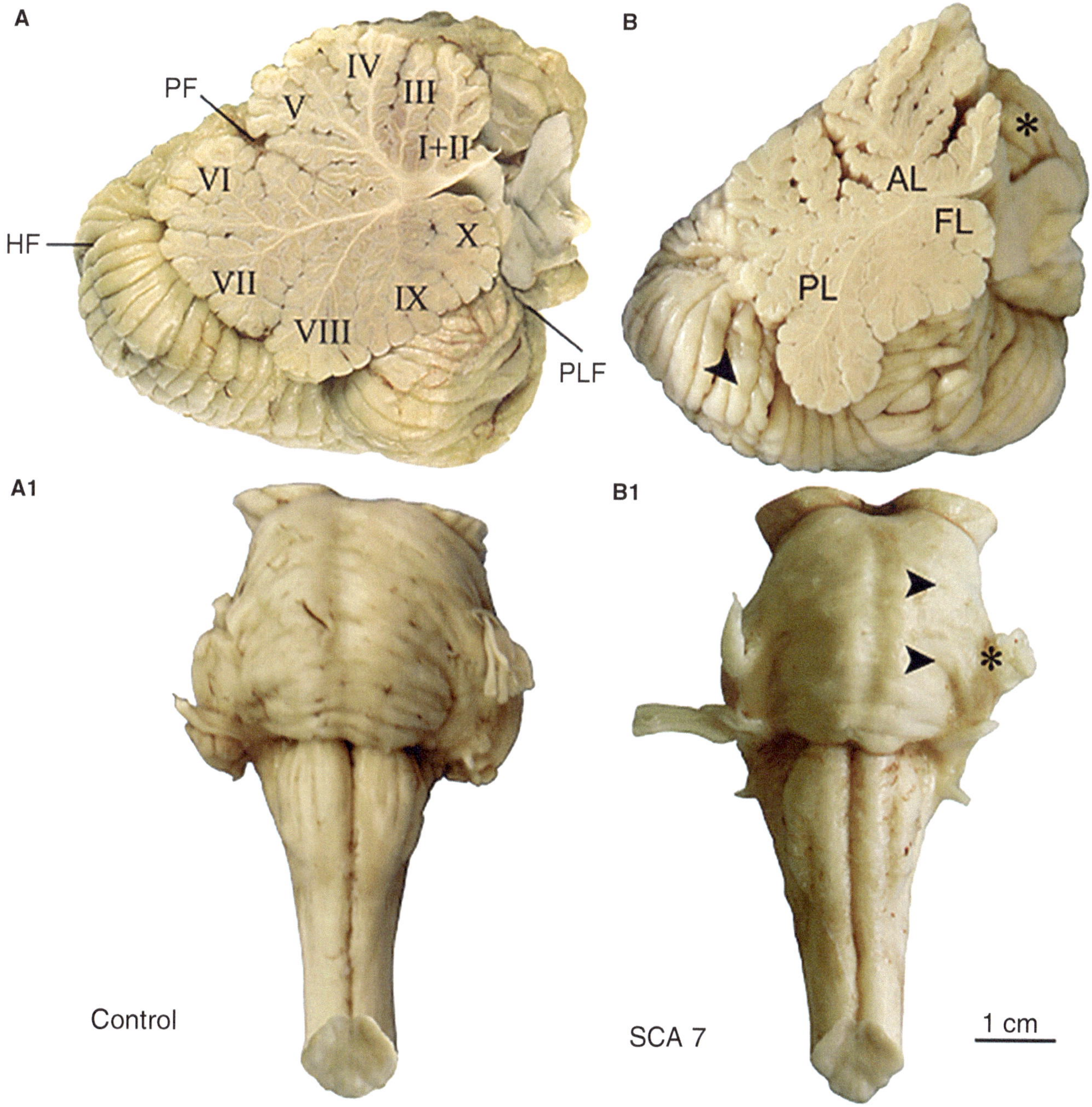

Fig. 7.31 Cerebellum, pons, and medulla oblongata in a control case (**A** and **A1**), compared to a SCA7 patient (**B** and **B1**). In (**B**) remarkable atrophy of the vermis, anterior and posterior lobe and the tonsil (arrow). In (**B1**) atrophy of the pons with loss of pontocerebellar fibers (arrowheads) and reduction of the medial cerebellar peduncle (asterisk). Cerebellar lobules are marked by their according roman numbers (lobules I, II, III = lingula and central lobule; lobules IV, V = culmen; lobule VI = declive; lobule VII = tuber and folium; lobule VIII = pyramis; lobule IX = uvula; lobule X = nodule). *AL* anterior lobe, *PL* posterior lobe, *FL* flocculonodular lobe, *HF* horizontal fissure, *PF* primary fissure, *PLF* posterolateral fissure. From Rüb et al. 2005a, Fig. 1 with permission

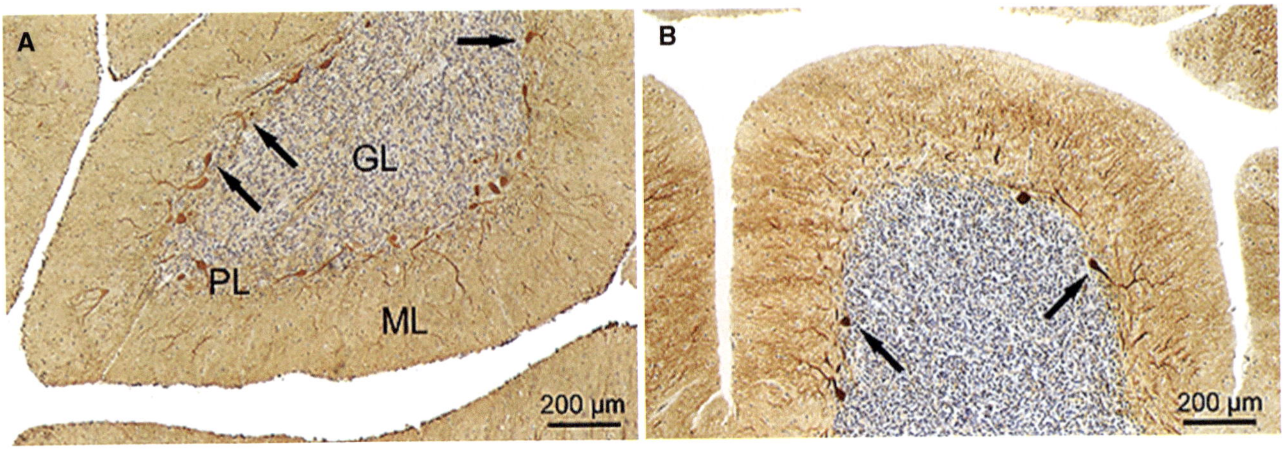

Fig. 7.32 (**A**) Microscopical section through the cortex of the cerebellar declive of a control case, showing the molecular (ML), Purkinje (PL), and granular (GL) cell layers. Arrows point to calbindin-immunopositive Purkinje cells. In (**B**) loss of Purkinje cell bodies and dendrites in the cortex of the cerebellar declive of an SCA7 patient. Arrows point to surviving calbindin-immunopositive Purkinje cells. From Rüb et al. 2005a, Fig. 4 with permission

Microscopical findings are atrophy of the cerebellar molecular and granular layers and diffuse loss of Purkinje cells (Fig. 7.32). Furthermore, neuronal loss and eosinophilic granular degeneration of the cerebellar dentate nucleus, neuronal loss in the substantia nigra (see Sect. 16.6.1), red nucleus (see Sect. 15.3.4), internal and external pallidal segments, thalamus, and cerebral cortex, myelin pallor of the dorsal columns, shrunken spinal cord motoneurons, and destruction of the Clarke column in the thoracic part of the spinal cord are observed (Iwabuchi et al. 1999, Rüb et al. 2005b). Widespread degeneration of brainstem nuclei involved in the ingestive process has also been found (Rüb et al. 2006). The brain pathology is more or less identical to the brain lesion pattern of terminal SCA2 and SCA3 patients (see Sects. 7.2.1.3.2 and 7.2.1.3.3) (Rüb et al. 2008a, b). An earlier SCA7 onset and a longer CAG-repeat expansion tends to be associated with the occurrence of visual impairments and the affection of the visual pathway (Rüb et al. 2008b, Fig. 7.33).

Among the degenerative changes in the retina, changes of the photoreceptor cell layer, the outer nuclear layer, and the ganglion cell layer have been described in the macular area, as well as discontinuity of the retinal pigment epithelium (Gouw et al. 1994, Michalik et al. 2004, Rüb et al. 2008b). Ataxin-7-immunopositive intranuclear inclusions (NIS) were seen in degenerated brain and retinal regions (Figs. 7.34 and 7.35).

In contrast to the other polyQ disorders, such as Huntington's disease, SCA1, SCA2, and SCA3, the distribution of inclusions is not restricted to areas of severe neuronal loss (Holmberg et al. 1998, Rüb et al. 2005a).

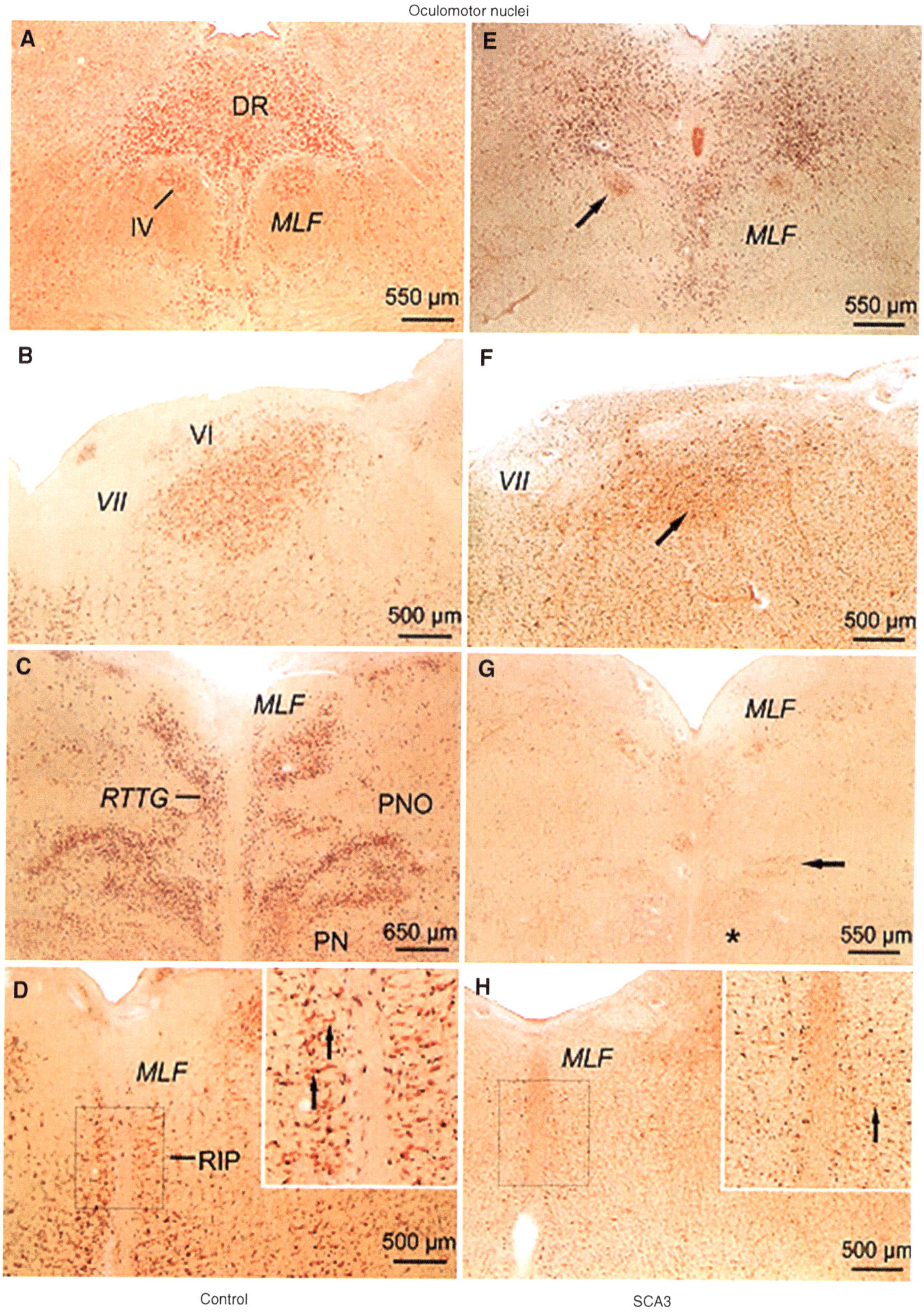

Fig. 7.33 (**A–D**) Horizontal sections of oculomotor cranial nerve nuclei and premotor oculomotor nuclei through the caudal midbrain within the trochlear nucleus (**A**) (see Sect. 14.5.2.1), through the pontomedullary junction with the abducens nucleus (see Sect. 9.2.1.1) (**B**), through the rostral pons showing the reticulotegmental nucleus (see Sect. 10.1.1.1) (RTTG) (**C**), and through the pontomedullary junction with the nucleus raphe interpositus (RIP) (**D**) of a control case. SCA7 patient with degeneration of the trochlear nucleus (see Sect. 14.5.2.1) in (**E**), marked neuronal loss in the abducens nucleus in (**F**) (see Sect. 9.2.1.1), severely impaired RTTG and marked degeneration of the PN (see Sect. 10.1.1.2) (asterisk) in (**G**), and degeneration in the RIP in (**H**) of an SCA7 patient. Framed areas in (**D**) and (**H**) are depicted at higher magnification at top right. Arrows in (**D–H**) point to remaining nerve cells. PEG 100 μm setions, Darrow red stain. *DR* dorsal raphe nucleus, supratrochlear part, *PN* pontine nuclei (see Sect. 10.1.1.2), *PNO* pontine reticular formation, oral nucleus, *MLF* medial longitudinal fascicle, *RIP* raphe interpositus nucleus, *RTTG* reticulotegmental nucleus (Bechterew), *IV* trochlear nucleus, *VI* abducens nucleus, *VII* facial nerve. See also atlas part Darrow red 36, 36A for the ncl. of the trochlear nerve A, E IV. From Rüb et al. 2005a, Fig. 3 with permission

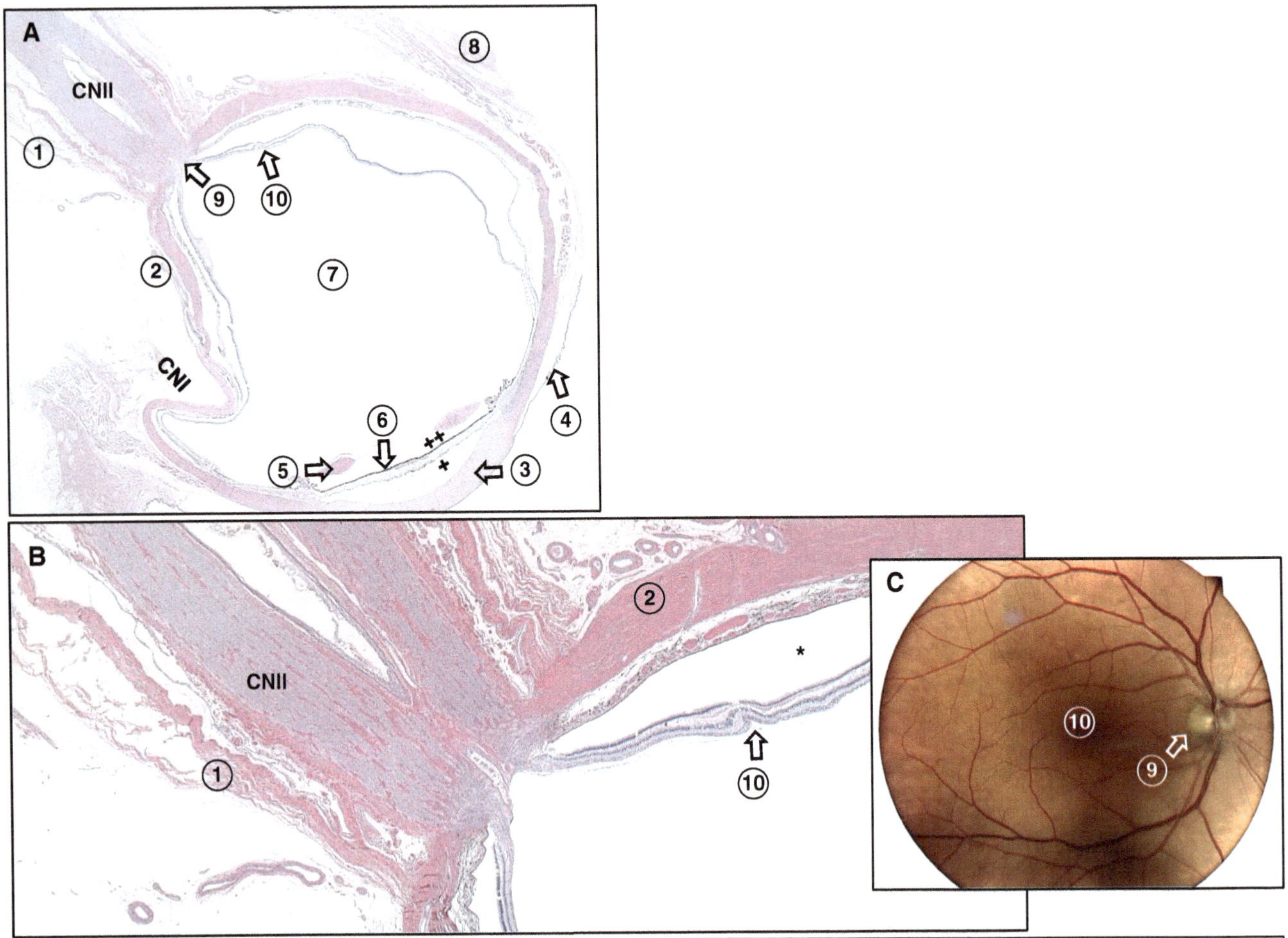

①	Dura mater	⑦	Vitreous body
②	Sclera	⑧	External eye muscles
③	Cornea	⑨	Optic disc
④	Conjunctiva	⑩	Macula / Central fovea region
⑤	Ciliary body	+	Anterior chamber of eyeball
⑥	Iris	++	Posterior chamber of eyeball

Fig. 7.34 SCA7 patient with visual impairments. Sagittal sections through the eye including the initial part of the optic nerve, overview (**A**) and detail (**B**). In (**B**) retinal cell layer, including the macular area (see Fig. 7.35) and fovea center. (**C**) Fundoscopic/ophthalmoscopic view of the retina. (**A**, **B**) LabPON Twente. (**C**) Courtesy Optica Boavista, Porto, Portugal / H. Schröder

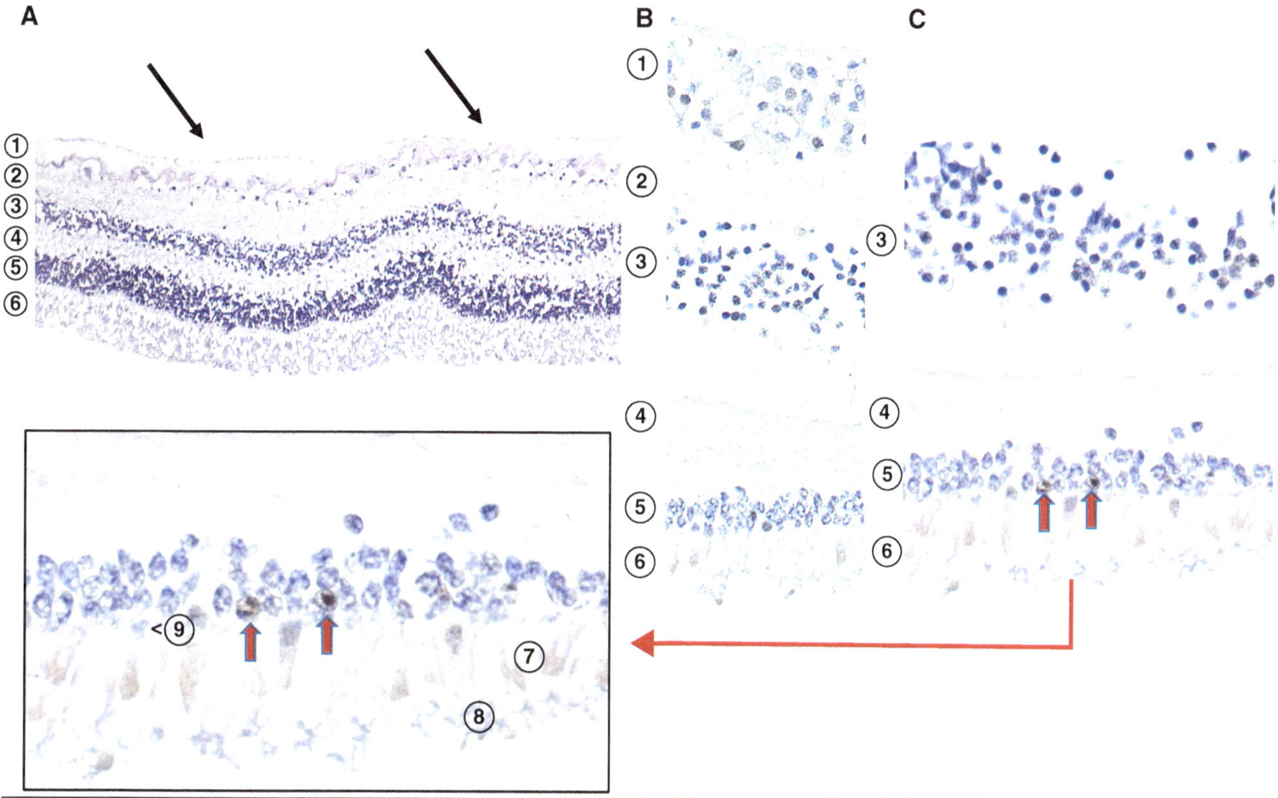

①	Ganglionic layer of retina	⑥	Layer of inner and outer segments of retina
②	Inner plexiform layer of retina	⑦	Outer segment
③	Inner nuclear layer of retina	⑧	Inner segment
④	Outer plexiform layer of retina	⑨	Outer limiting layer of retina
⑤	Outer nuclear layer of retina		

Fig. 7.35 Retinal findings in an SCA7 patient with visual impairments. (**A**) Extensive degenerative changes in the macular area, especially in the ganglion cell layer in (**A**) (black arrows). In (**B**) peripheral retina, showing the different cell layers and in (**C**) ataxin-7 immunopositive neuronal inclusion bodies (NIs) in the outer nuclear cell layer (red arrows). In this case, NIs were mainly present in the outer nuclear layer. The outer nuclear layer ⑤, containing rods and cones, is subdivided into the photoreceptor outer segment (os) and inner segment (is) layers ⑥, distal to the external limiting membrane (ELM). LabPON Twente

7.2.1.3.6 Spinocerebellar ataxia 17 (SCA17)

SCA17 has a variable phenotype including late-onset ataxia, psychiatric symptoms such as dementia, psychosis and paranoia, oculomotor dysfunction, and extrapyramidal symptoms. Especially, chorea and diverse forms of dystonia are frequent in SCA17, but are rare in other types of SCA (Rolfs et al. 2003). The disease onset is childhood or adulthood. It is caused by the expansion of polyglutamine tracts in the TATA-binding protein (TBP), a general transcription initiation factor. The expansion in TBP bears 47–55 repeats, whereas the normal repeat number ranges from 29 to 42 (Nakamura et al. 2001). MRI studies display volume loss of the cerebrum, cerebellum, and the brainstem (Rolfs et al. 2003, Seidel et al. 2012). Neuropathology reveals neuronal loss in the cingulate and parahippocampal gyri, striatum, ventral thalamic nuclei, cerebellar Purkinje cell layer and dentate nucleus, substantia nigra (see Sect. 16.6.1), and inferior olive (see Sect. 4.1.1.3). Intranuclear neuronal inclusion bodies, stained with 1C2 and anti-TBP-antibodies, are more widely distributed througout the brain gray matter than in other SCAs (Rolfs et al. 2003).

In summary, the autosomal dominant cerebellar ataxias represent a heterogeneous group of neurodegenerative diseases and the current classification is based on the underlying genetic defects. Extensive neuropathological studies, especially those using unconventionally thick serial brain sections, have allowed the recognition of the differences, subtle alterations, and similarities and will provide more precise information with respect to the extent and severity of the pathological process underlying this group of neurodegen-

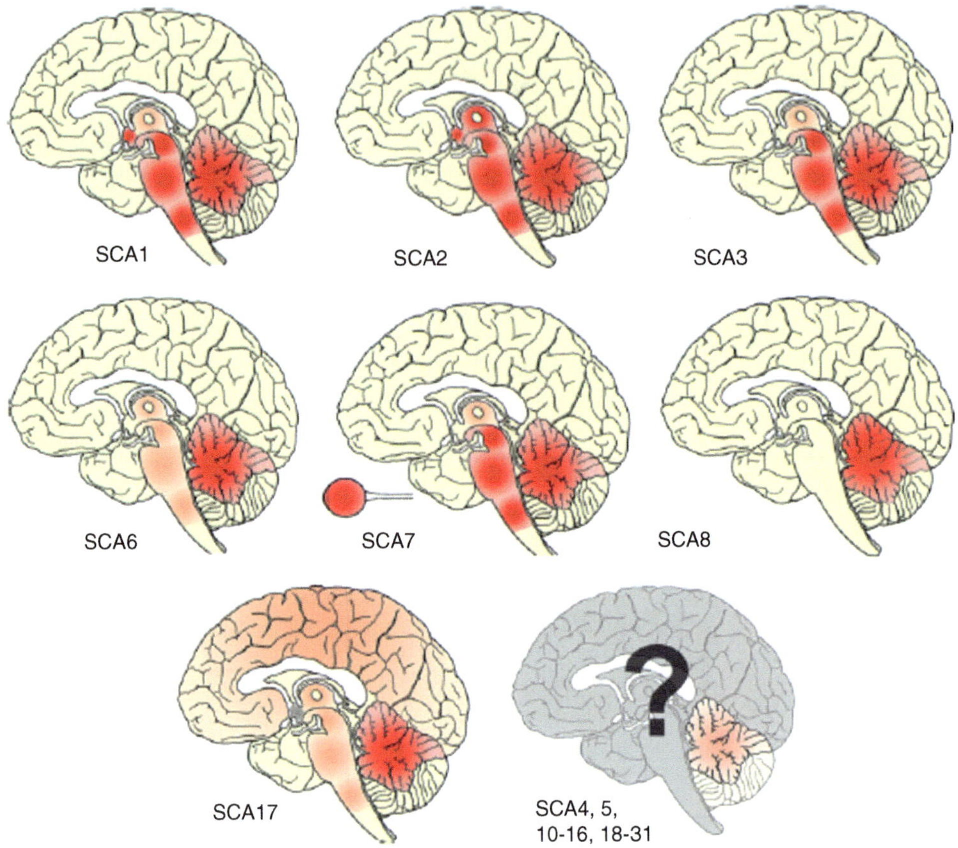

Fig. 7.36 Distribution of neuronal loss in the brain of patients with SCA1, SCA2, SCA3, SCA6, SCA7, SCA8, substitute , behind SCA17 by. SCA17. Diagrammatic representation of midsagittal sections through the cerebrum, cerebellum, and brainstem of the human brain. The degree of neuronal loss in the cerebral cortex, basal forebrain, thalamus, cerebellum, and brainstem is indicated by red (severe) or light red (marked) color. From Seidel et al. 2012, Fig. 2 with permission

erative diseases (Rüb et al. 2008a). As a result of the neuropathological data that has become available, the clinical symptoms of SCA1, SCA2, SCA3, SCA6, and SCA7 are now better understood, symptomatic treatment strategies for some symptoms (i.e., ataxia, dysphagia) are improved, and a pathoanatomical basis is defined for the validation of experimental animal models and further biochemical, genetic, or morphological studies (Seidel et al. 2012) (Fig. 7.36).

The disease proteins are known to misfold and aggregate in affected neurons, but the role of these aggregates in the pathological process remains unclear so far. Their aggregation into neuronal intranuclear inclusions (NIS) is an established feature of SCA1, SCA3, SCA7, and SCA17, while in SCA 2 and SCA6 primarily cytoplasmic aggregations are reported (Seidel et al. 2012). But the fact that they do occur in affected as well as in non-affected brain regions questions the possible toxic nature of these aggregates and argues more for the idea that they serve as reliable morphological markers of the polyglutamine ataxias (Rüb et al. 2006).

There is not much information available for the non-polyglutamine SCAs. Most patients with these rare entities exhibit cerebellar degeneration, usually detected by means of MRI. Neuropathological data are scarce and are needed in the future for better understanding of symptoms and underlying mechanisms (Seidel et al. 2012).

7.2.1.4 Spinal vestibular nucleus

For detailed information on this nucleus, see Sect. 5.3.1.1.

7.2.2 Monoamine nuclei r7

7.2.2.1 A2 noradrenaline cells

For detailed information on these nuclei, see Sect. 3.2.1.1

7.2.2.2 C1 adrenaline cells (C1)

For detailed information on these nuclei, see Sect. 3.2.1.2.

7.2.3 Trigeminal sensory nuclei r7

7.2.3.1 Spinal trigeminal nucleus

For detailed information on this nucleus, see Sect. 3.2.2.2.

7.2.3.2 Dorsomedial spinal trigeminal nucleus

For detailed information on this nucleus. see Sect. 6.3.3.2.

7.2.4 Solitary nuclei r7

7.2.4.1 Solitary nucleus

For detailed information on this nucleus, see Sect. 3.2.4.

7.2.5 Alar tegmentum r7

7.2.5.1 Matrix region of the medulla

For detailed information on this region, see Sect. 3.2.5.1.

7.3 Basal r7

7.3.1 Branchial motor nuclei r7

7.3.1.1 Inferior salivatory nucleus r7 [Ncl. salivatorius inferior]

7.3.1.1.1 Location and histology of the salivatory nuclei

The parasympathetic secretomotor innervation of the salivatory glands (see Box 12.1 and here below) originates from a longitudinal column of neurons in the medulla, called the salivatory nucleus, which can be differentiated into the inferior and the superior salivatory nucleus. As stated in FIPAT TNA Chap. 1, endnote 47, based on Blessing (2004); however, both nuclei represent a distributed set of neurons that do not condense in nuclei as such. That renders their identification in routine histological specimens difficult. The unequivocal detection requires the use of ChAT-immunohistochemistry (see e.g., Huang et al. 1993).

7.3.1.1.2 Efferent connectivity of the salivatory nuclei

The superior salivatory nucleus provides the preganglionic fibers for the innervation of the lacrimal gland (see Fig. 12.6, see Sect. 12.2.3.2.2), submandibular, and sublingual glands (for details, see Box 12.1).

The inferior salivatory nucleus provides the preganglionic parasympathetic neurons for the innervation of the parotid gland (Fig. 7.37). The ganglion targeted is the otic ganglion.

7.3.1.1.3 Course of the parasympathetic fibers to the parotid gland

In detail, the preganglionic parasympathetic fibers join the glossopharyngeal nerve (CNIX) (see also Sect. 5.4.2.1.3). They accompany the first branch of CNIX, the tympanic nerve, which enters the tympanic cavity (for details, see Sect. 5.4.2.1.3). The tympanic nerve forms the tympanic plexus, fibers of which form the lesser petrosal nerve (see Box 8.5) which enters the otic ganglion [Ganglion oticum] located closely below the oval foramen, medial of the mandibular nerve (see Sect. 12.2.3.2.3). The postganglionic parasympathetic fibers join the auriculotemporal nerve (see Sect. 12.2.3.2.3, "Mandibular nerve," see Fig. 7.37). The latter has communicating branches with the facial nerve (see Sect. 8.2.1) via which the secretomotor fibers reach the parotid gland.

The innervation pattern of the parotid gland differs among mammals. In the mouse, parotid gland innervation from the otic ganglion is likely but the preganglionic input has not been shown unequivocally (CNVII, CNIX) (Inoue 1984).

In rats, the preganglionic fibers for the innervation of the parotid to the otic ganglion are provided by the facial nerve. This is in contrast to the human situation where the facial nerve traverses the parotid gland, but the innervation of the gland is provided by the glossopharyngeal nerve.

Both, the parasympathetic and sympathetic nerves innervate salivary glands, with parasympathetic stimulation leading to a serous salivary product and ion secretion, and sympathetic stimulation leading to increases in the secretion of proteins (for details, see Ghannam and Singh 2020, Norberg et al. 1968).

7.3.1.1.4 Living anatomy and clinical aspects of the parotid gland

The excretory parotid duct [Ductus parotideus] traverses the anterior edge of the masseter and opens into the oral cavity at the level of the second superior molar tooth. It is visible by mirror-based self-inspection. By gentle massage of the parotid gland, the flowing of saliva into the oral cavity can be sensed and/or seen.

Mumps (Parotitis epidemica) is a virus-induced inflammation of the parotid gland. The swelling of the gland is usually painful due the stretching of superficial sensory trigeminal branches (see Sect. 12.2.3.2.3).

7.3.1.2 Ncl. ambiguus, compact part

Detailed information on the ambiguus nucleus is provided under Sect. 5.4.2.1.

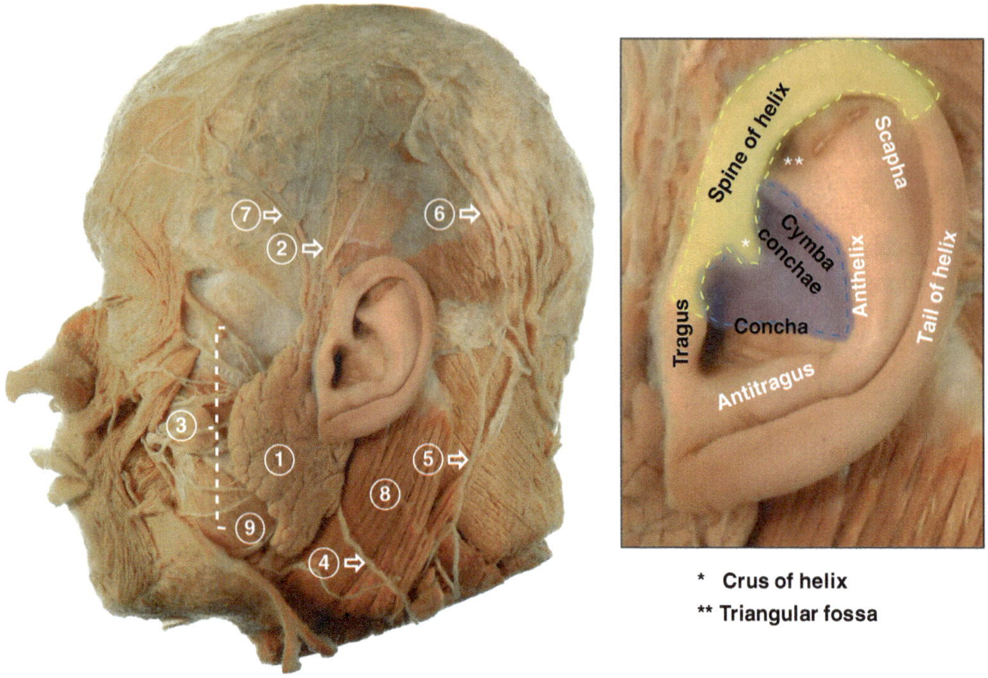

①	Parotid gland	⑥	Greater occipital nerve
②	Auriculotemporal nerve V_3	⑦	Ascending temporal artery, frontal branch
③	Peripheral branches of the facial nerve CNVII	⑧	Sternocleidomastoid muscle
④	Greater auricular nerve	⑨	Masseter muscle
⑤	Lesser occipital nerve		

Fig. 7.37 Lateral view of the human face region. Skin removed. Formalin-fixed specimen. The large parotid gland ①—largest of the human salivatory glands (for the small salivatory glands see Sect. 8.2.1.6.2, Figs. 12.14B and 12.15)—is located directly under the skin, anterior of the auricle, bordering posteriorly the sternocleidomastoid muscle ⑧ (see Sect. 3.3.1.2), and anteriorly it reaches the masseter muscle ⑨ (see Fig. 11.3). At the anterior edge of the gland, the peripheral branches of the facial nerve (③ see Sect. 8.2.1.4) become visible after traversing the parotid gland. They do not, however, innervate the gland. Inset: Anatomy of the external ear. The largest part of the external ear's somatosensory innervation is provided by the greater auricular nerve (C2/C3) ④, region uncolored in the inset. The blue region (Cymba conchae) is innervated by the auricular branch of the vagus nerve, the yellow part (mainly spine of helix) by the auriculotemporal nerve ②. Some words to the very special terminology for the external ear: concha from Greek ἡ κόγχη, he konche conch, shell, crus, cruris Latin = flank, cymba from Greek ἡ κύμβη, he kymbe = barge (trough), helix ἡ ἕλιξ -ικος, he helix, –ikos = turn, convolution, scapha from Greek ἡ σκάφη, he skaphe = trough (also scaphoid fossa), tragus from Greek ὁ τράγος, ho tragos = he-goat. The tragus grows hairs in older subjects which remembered the old anatomists a he-goat's beard (barbula tragi). Photograph by MedizinFotoKöln, Cologne, Germany. Sammlung des Zentrums Anatomie der Universität zu Köln

7.3.2 Visceral motor nuclei r7

7.3.2.1 Dorsal motor nucleus of vagus/posterior nucleus of vagus nerve r7 [Ncl. posterior n. vagi]

The posterior nucleus of vagus nerve (Fig. 7.38)—formerly dorsal motor nucleus of vagus—provides the general visceral efferent (visceromotor, see Table 1.1) contribution to the vagus nerve (CNX).

7.3.2.1.1 Location and histology of the posterior nucleus of vagus nerve

In accordance with the principal arrangement of the basic qualities of cranial nerves (see Chaps. 1 and 2, see Table 1.1, Fig. 1.13, Fig. 2.4), the visceromotor qualities of the nucleus of vagus nerve are located lateral of the somatomotor hypoglossal nucleus and medial of the special somatosensory solitary nucleus. The aminergic A2 (A3 is absent in the human brainstem) group is located in the posterior nucleus of the vagus nerve and the nucleus of solitary tract.

Huang et al. 1993 mapped the human posterior nucleus of vagus nerve by means of ChAT-immunohistochemistry in nine subnuclei. They also described a granular substance P-like immunoreactivity. Only two of the subnuclei also contained substance P-like immunopositive neurons.

In mice, the wedge-shaped or triangular nucleus is composed of medium-sized, ChAT-immunoreactive neurons

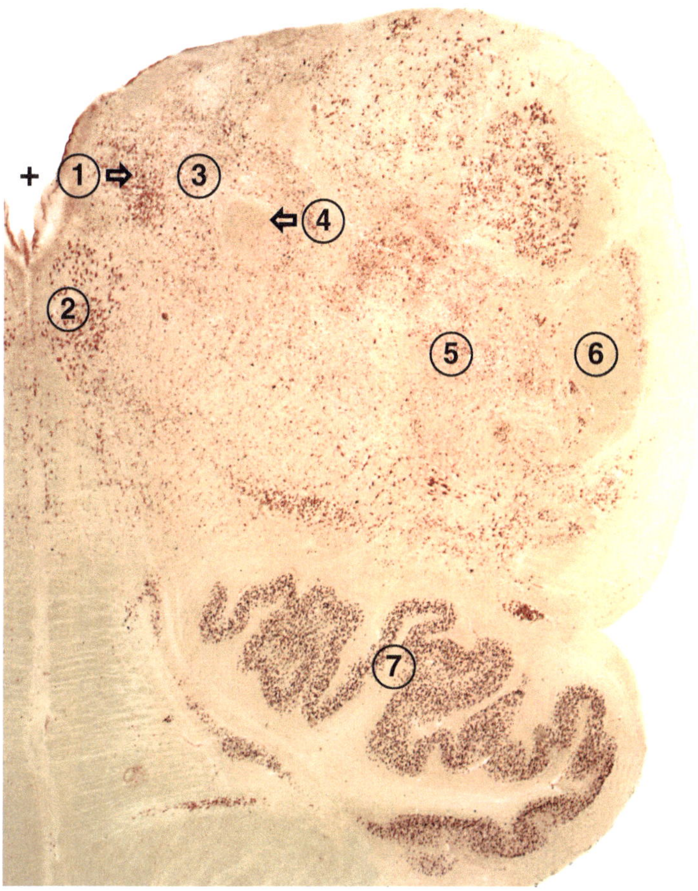

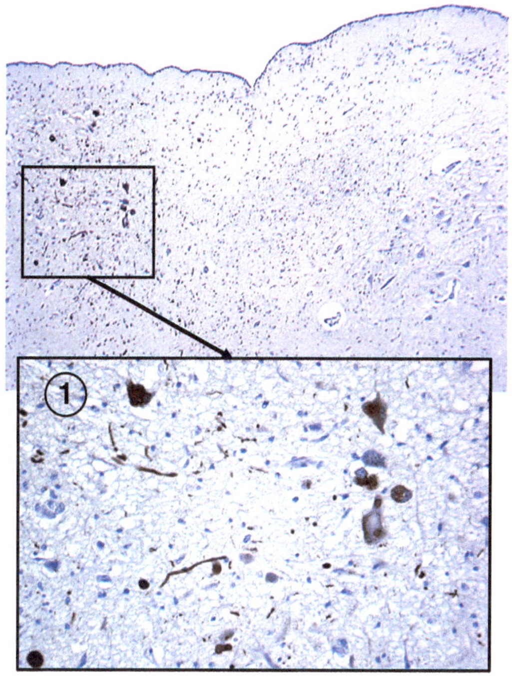

①	Posterior ncl. of vagus nerve	⑤	Spinal trigeminal ncl.
②	Hypoglossal ncl.	⑥	Spinal trigeminal tract
③	Solitary ncll.	⑦	Inferior olivary complex
④	Solitary tract	+	4th ventricle

Fig. 7.38 Horizontal section through the middle human medulla oblongata. Darrow red stain. The posterior nucleus of the vagus nerve ① with its relatively large neuronal perikarya is easily identifiable dorsolateral of the hypoglossal nucleus ② and medial of the solitary nucleus ③. See also atlas part Darrow red 5A. The right-hand insets show the affection of the posterior nucleus of vagus nerve in neurodegeneration with iron accumulation (NBIA) (for details, see Sect. 16.6.1.8) showing α-synuclein immunoreactivity of CNX neurons. LabPON Twente

comparable to the situation in man. Again, as in humans, the neurons of the posterior nucleus of vagus nerve give rise to cholinergic axons that innervate via interposed autonomic ganglia all the inner organs of the thoracic cavity and almost all organs of the abdominal cavity. In addition, the vagus nerve innervates the pharyngeal constrictor, cricothyroid, and levator veli palatini muscles as well as the intrinsic laryngeal muscles (Kaufman and Bard 1999).

7.3.2.1.2 Course of vagus nerve

The vagus nerve is the nerve of the fourth and fifth branchial arch (see Sect. 2.8, see Fig. 2.9). A special feature of the vagus nerve among the other cranial nerves, in particular the branchial nerves, is the fact that beyond the muscular and mucosal structures of the corresponding arches it also provides the extracranial parasympathetic innervation of the human body down to Cannon's point near the left colic or splenic flexure. Figure 7.39 provides a schematic overview of the main branches of the vagus nerve.

The vagus nerve leaves the posterior nucleus at its ventral side and runs ventrolaterally through the medulla oblongata (Fig. 7.40 ①) to its exit site ventral of the inferior cerebellar peduncle (Fig. 7.40 ⑤). It exits from the brainstem close to the glossopharyngeal nerve dorsal of the retroolivary groove of the medulla oblongata (Fig. 7.41) and enters the lateral cerebellomedullary cistern (Fig. 7.42A, B).

As the glossopharyngeal nerve (see Sect. 5.4.2.1.3), the vagus nerve has a superior ganglion [Ggl. superius] inside

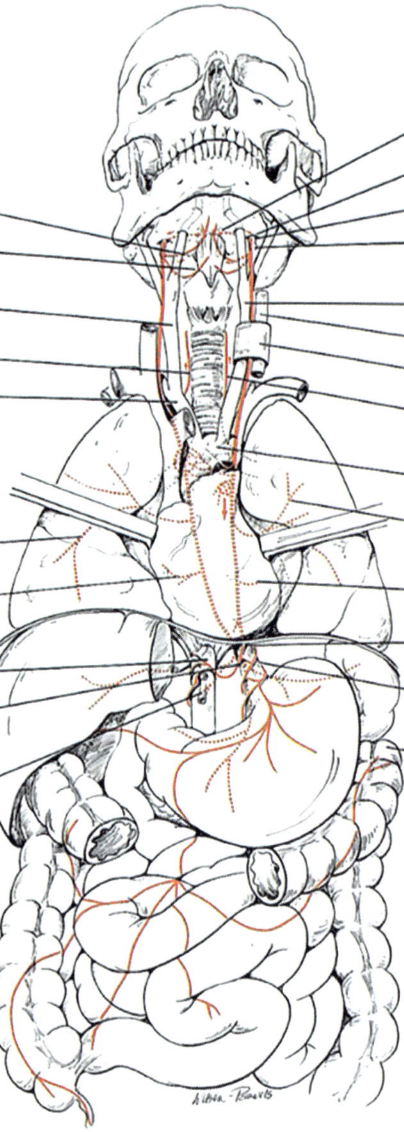

Fig. 7.39 Schematic survey of the main branches of the vagus nerve (orange). It becomes clear that in addition to the branchial arch-derived organs (pharynx, larynx) the vagus nerve innervates the heart, the bronchial system, and the lungs, the esophagus, the stomach, the liver, and large parts of the small and large intestine (see also Tables 7.1 and 7.2). From Wilson-Pauwels et al. 1988. Cranial Nerves: Anatomy and Clinical Comments. Toronto: BC Decker

the jugular foramen (see Fig. 7.42B ①). This ganglion gives off a recurrent branch for the sensory supply of the dura in the posterior cranial fossa (region of the transverse sinus and the occipital region, see Fig. 3.11), the meningeal branch [Ramus meningeus].

The next branch is the auricular one [auris Latin = ear, Ramus auricularis] which enters the mastoid canaliculus [Canaliculus mastoideus] (Fig. 7.42B ⑥) and reaches via the tympanomastoid fissure [Fissura tympanomastoidea] (mastoid from Greek ὁ μαστός, ho mastos = breast, μαστοειδής, mastoeides = resembling the papilla of the breast) (Fig. 7.42B ⑦) the posterior surface of the pinna (auricle) [Auricula] (see Fig. 7.37) and the region below the external acoustic meatus [Meatus acusticus externus].

Next in the course of the CNX lies the larger inferior ganglion [Ggl. inferius] and the communicating branch with glossopharyngeal nerve [R. communicans cum nervo glossopharyngeo] (not listed in FIPAT Ch. 2), a communication between the auricular branch and CNIX.

The next branch is the pharyngeal branch [R. pharyngeus] giving off branches into the pharyngeal plexus [Plexus pharyngeus] which is formed together with CNIX and sympathetic input for the innervation of the pharyngeal constrictors (see Sect. 5.4.2.1.5).

The superior laryngeal nerve [N. laryngeus superior] (see Fig. 7.43 ⑤) originates from the inferior ganglion, descends medial of the internal carotid artery ④, and splits off in an external and an internal branch. The following branches sup-

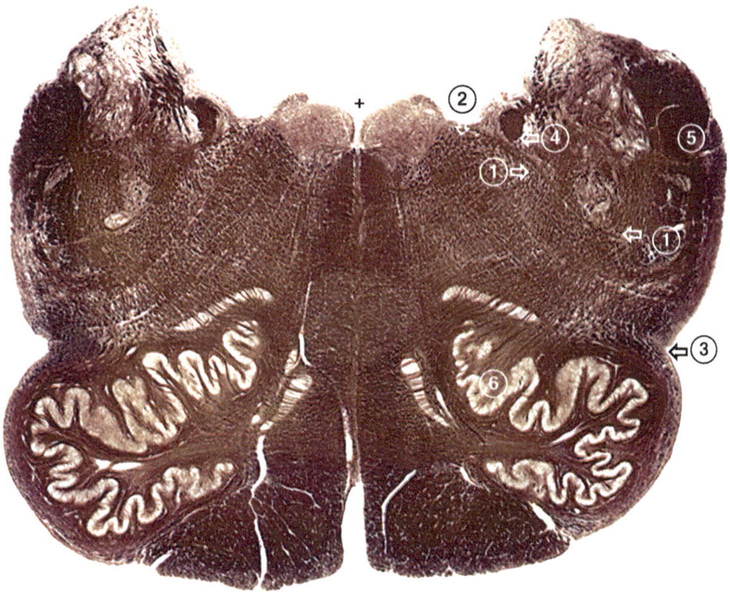

①	Intracerebral course of CNX	⑤	Inferior cerebellar peduncle
②	Region of the posterior ncl. of vagus nerve CNX	⑥	Inferior olivary complex
③	Retroolivary groove	+	4th ventricle
④	Solitary tract		

Fig. 7.40 Horizontal section at the level of the dorsal column nuclei. Campbell staining. The fibers of the vagus nerve are running from its posterior nucleus ② in a ventrolateral arcuate course to the surface of the medulla oblongata. See also atlas part Campbell 3. LabPON Twente

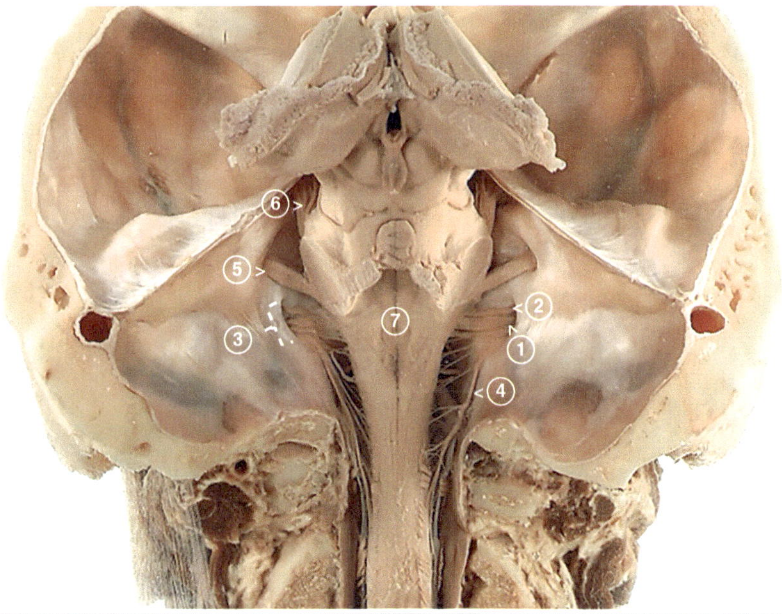

①	Vagus nerve CNX	⑤	Vestibulocochlear nerve CNVIII
②	Glossopharyngeal nerve CNIX	⑥	Trigeminal nerve CNV
③	Jugular foramen	⑦	4th ventricle / rhomboid fossa
④	Spinal root of accessory nerve CNXI		

Fig. 7.41 Dorsal view of the human head/neck region. Skull and vertebral column partly removed to enable the view on the diencephalon, brainstem, and cervical spinal cord. The photograph of the posterior cranial fossa shows the region of the jugular foramen ③ with the glossopharyngeal ② and vagus ① nerves running from the medulla oblongata to the foramen. The spinal root of the accessory nerve ④ (Sect. 3.3.1.2) is running craniad into the skull and leaves it via the jugular foramen with CNX and CNXI. Sammlung des Zentrums Anatomie der Universität zu Köln

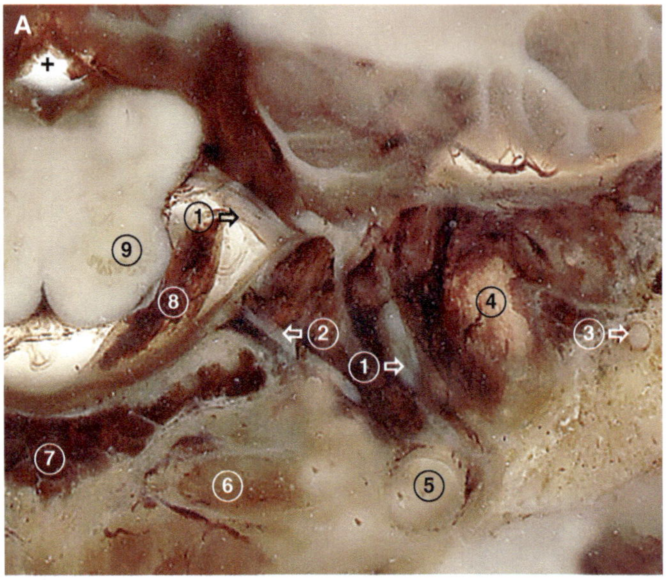

①	Vagus nerve CNX	⑥	Longus capitis muscle
②	Glossopharyngeal nerve CNIX	⑦	Clivus
③	Facial nerve CNVII	⑧	Vertebral artery
④	Jugular bulb	⑨	Inferior olivary complex
⑤	Internal carotid artery	+	4th ventricle

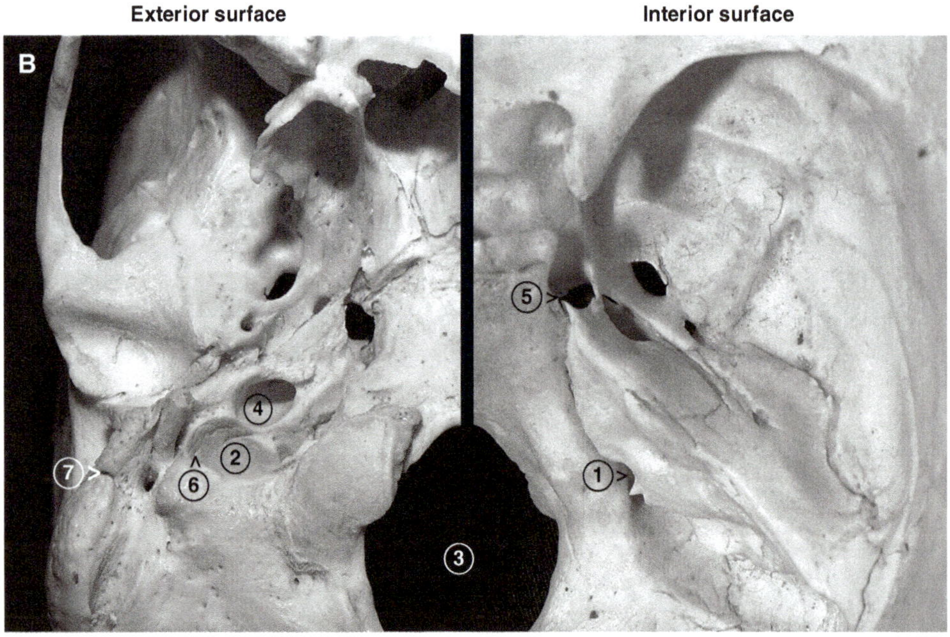

①	Jugular foramen	⑤	Internal opening of carotid canal
②	Jugular fossa	⑥	Mastoid canaliculus
③	Foramen magnum	⑦	Tympanomastoid fissure
④	External opening of carotid canal		

Fig. 7.42 (**A**) Horizontal section through the human medulla oblongata at the exit level of the vagus and the glossopharyngeal nerves. Plastinated specimen. Note that the vagus ① runs toward the internal carotid artery ⑤ at the dorsal edge of which it will appear in the neck region (see Fig. 7.43). (**B**) Exterior and interior surfaces of the human skull. Relevant bony structures for the vagus nerve are the jugular foramen ① and fossa ② (containing the trunk of the vagus nerve and ganglia) as well as the mastoid canaliculus ⑥ and the tympanomastoid fissure ⑦ (auricular branch) (For details, see text). Sammlung des Zentrums Anatomie der Universität zu Köln

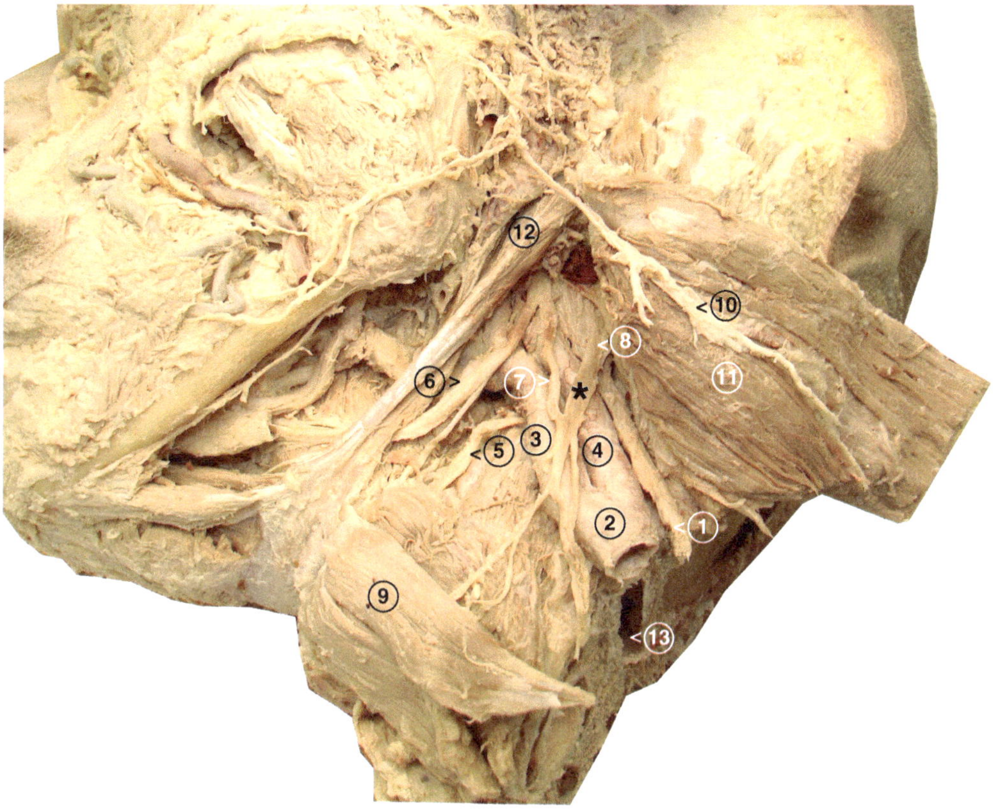

①	Vagus nerve CNX	⑧	Ansa cervicalis, inferior root
②	Common carotid artery	⑨	Omohyoid muscle
③	External carotid artery	⑩	Ramus colli of facial nerve (dislocated)
④	Internal carotid artery	⑪	Sternocleidomastoid muscle
⑤	Superior laryngeal nerve	⑫	Venter posterior of digastric muscle
⑥	Hypoglossal nerve	⑬	Trachea
⑦	Ansa cervicalis, superior root	*	Ansa cervicalis

Fig. 7.43 Lateral view of the human neck region. The most easily recognizable landmark is the sternocleidomastoid muscle ⑪ (innervated by the accessory nerve CNXI, see Sect. 3.3.1.2). Note that the nerve ⑩ lying here on top of the sternocleidomastoid muscle is not the CNXI but the ramus colli of CNVII. At the anterior edge of the muscle, the course of the vagus nerve ① below the exit from the jugular foramen on its way to the larynx and the superior thoracic aperture (thoracic inlet) [Apertura thoracis superior] is seen. The nerve is accompanied anteriorly by the internal ④ and the common ② carotid arteries, crossed on its ventral surface by the inferior root of the ansa cervicalis ⑧. The superior laryngeal nerve ⑤ as branch of CNX is running parallel to CNXII ⑥. Photograph by MedizinFotoKöln, Cologne, Germany. Sammlung des Zentrums Anatomie der Universität zu Köln. Müller-Thomsen fecit

ply the larynx (see Fig. 7.46) (Ancient Greek ἡ λάρυγξ, –υγγος he larynx, laryngos = gullet, throat).

The first one provides branches to the inferior constrictor muscle (see Fig. 5.20) and eventually reaches the cricothyroid muscle at the ventral surface of the larynx.

The internal branch together with the superior laryngeal artery perforates the thyrohyoid membrane and is then located below the mucosa of the piriform recess [Recessus piriformis]. It provides the sensory innervation of the mucosa of the epiglottic vallecula [Vallecula epiglottica], the epiglottis, and the larynx down to approximately the vocal folds [Plica vocalis] (compare with Fig. 7.46).

Finally, a Ramus communicans cum nervo laryngeo recurrente (not listed in FIPAT Ch. 2) provides the connection with the inferior laryngeal nerve [N. laryngeus inferior] (see here further below and Fig. 7.44B).

The next two branches, inferior and superior cervical cardiac [Rr. cardiaci cervicales superiores et inferiores] contribute to the cardiac plexus [Plexus cardiacus].

The inferior part of the larynx below the vocal cord gets sensory supply from the recurrent laryngeal nerve [N.

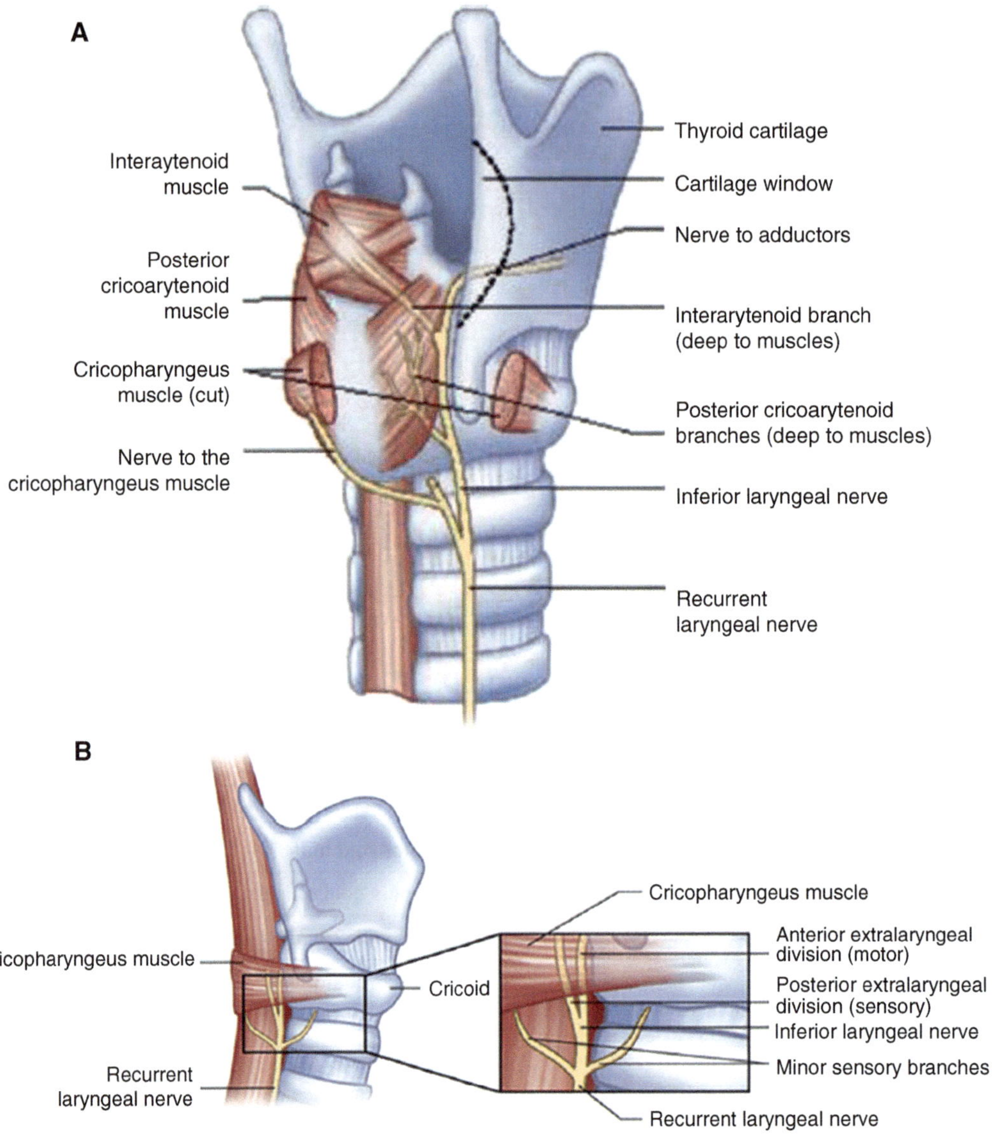

Fig. 7.44 (**A**) Oblique posterior view on the larynx and the cranial part of the trachea. The larynx is a particularly important supply territory of the vagus nerve and its branches. The superior laryngeal nerve (not shown here) provides motor innervation of the cricothyroid muscle (only muscle innervated by this nerve and only external muscle, not shown here) and sensory innervation of the mucosa above the rima glottidis with the vocal folds (see Fig. 7.46). As a rule of thumb, the inferior/recurrent laryngeal nerve innervates all internal laryngeal muscles and the mucosa below the vocal folds. For details, see text. (**B**) Detailed lateral view on the larynx and the cranial part of the trachea. The inset shows the communicating zone between the recurrent laryngeal nerve and the inferior laryngeal nerve. For details, see text. From Feehs et al. 2020 Figs. 30.1 and 30.2 with permission

laryngeus recurrens] (recurrere Latin = to run back], which also innervates all laryngeal muscles with the exception of the cricothyroid muscle. It has a particular course in so far as it runs around the subclavian artery on the right side and around the aortic arch on the left side (see Sect. 2.8). During development, the recurrent laryngeal branch initially loops under the sixth arch artery on both sides. However, since the right aortic arch lateral to the right pulmonary artery degenerates during early development, the recurrent laryngeal nerve stretches under the fourth arch-derived subclavian artery while, on the left-hand side, the nerve loops under the aortic arch at the ductus arteriosus.

The nerves of either side then run cranially between the trachea and esophagus close to the thyroid gland [Gld. thyroidea] to the larynx and perforate the inferior constrictor muscle. The connection with the superior laryngeal nerve [N. laryngeus superior] is provided via the communicating branch with recurrent laryngeal nerve (Galen's anastomosis) (not listed in FIPAT Ch.2) [R. communicans laryngei recurrentis (Galen)] (see Box 7.2).

In its descending course, the vagus nerve gives off branches to the trachea (tracheal branches) [Rr. tracheales], the esophagus (oesophageal branches) [Rr. oesophagei], and the inferior pharyngeal constrictor (pharyngeal branches) [Rr. pharyngei].

Box 7.2 Claudius Galenus (Galen)

Greek Γαληνός = Galenos (129 BCE Pergamos – 200 Rome), most influential physician of ancient Rome. He gained most of his anatomical knowledge by the dissection and vivisection of animals. He was the personal physician (ἀρχίατρος = archiatros) of the Roman emperor Marcus Aurelius (Pasipoularides 2014).

Galen introduced the concept of the Central Nervous System by proposing that the spinal cord is an extension of the brain. Among other findings on the CNS, he described seven cranial nerves which in his classification include the optic (I), oculomotor (II), trigeminal (sensory root) (III), trigeminal (motor root) (IV), facial-vestibulocochlear complex (V), glossopharyngeal-vagus-accessory complex (VI), and hypoglossal (VII) nerves (Ng et al. 2019).

Note that the numbering does not correspond to the currently used one (for details on the historical aspects of cranial nerve terminology, see Ng et al. 2019 and Fig. 7.45). Galen considered each nerve to represent a pathway for the "animal spirit" that distributed itself to different parts of the body, where hard nerves were motor and soft nerves were sensory.

His ideas were widely distributed in the Christian Middle Ages and beyond. During the Renaissance, anatomists like Vesalius (see Box 7.3) corrected several of his findings and expanded the anatomical knowledge on the human body based on the autopsies performed.

The anatomical work of Galen is accessible in a Greek-Italian version by Garofalo (1991).

Box 7.3 Andreas Vesalius (Vesal)

Andreas Vesalius (Andries Wytinck van Wesel) was born on December 31, 1514, in Sablon near Brussels, today Belgium. Interestingly, the house in which he was raised was close to Gallow's hill which allowed him to approach the corpses of executed criminals (Cambiaghi 2017).

He matriculated at the University of Leuven/Louvain, today Belgium. After completing his studies in 1533, he moved to the University of Paris. He completed his studies of medicine and his dissertation in Leuven in 1537 and was authorized to perform public dissections (Cambiaghi 2017). In 1537, he moved to the University of Padua/Padova, Italy, to start his career as an academic teacher and researcher. By contrast to tradition, he based his lectures on direct observation of the corpses and based on this he published his first atlas book "Tabulae anatomicae sex" (six anatomical tables) in 1538 and, in 1543, his groundbreaking masterpiece "De Humani Corporis Fabrica" (On the fabric of the human body) (1543) (Cambiaghi 2017). This atlas was adorned by more than 200 illustrations designed in the Venetian workshop of the outstanding Renaissance artist Titian.

This atlas dealt in great detail with the human brain, including the cranial nerves (see Fig. 7.45). He provided a description on how to perform the inspection of the brain. We know about the lectures he performed—among other topics—on this subject from the notes of a German student by the name of Baldasar Heseler (Eriksson 1959).

After having served as a physician at the court of emperor Charles V, he followed Charles' son, Philip II, King of Spain at his court in Madrid. He died in 1564 at the return from a pilgrimage to the Holy Land at the Greek Island of Zakynthos (Cambiaghi 2017).

The distribution of branches is followed by the thoracic cardiac branches [Rr. cardiaci thoracici] at the superior thoracic aperture (thoracic inlet) [Apertura thoracis superior], the bronchial branches [Rr. bronchiales] to the hilum of lung [Hilum pulmonis] (Entry of main bronchi and vessels into the lung parenchyma) and the pulmonary plexus [Plexus pulmonalis] innervating the bronchi, pulmonary vessels, and the visceral pleura.

Before the formation of the large vagal trunks [Trunci vagales] by the vagus nerves of either side, a plexus is formed around the esophagus.

The main branches of the vagal trunks are as follows:

Anterior vagal trunk [Truncus vagalis anterior]

- Anterior gastric branches [Rr. gastrici anteriores] (front face of the stomach).
- Anterior nerve of lesser curvature [N. anterior curvaturae minoris] (lesser curvature).
- Hepatic branches [Rr. hepatici] (liver).
- Pyloric branch [R. pyloricus].

Posterior vagal trunk [Truncus vagalis posterior]

- Posterior gastric branches [Rr. gastrici posteriores] (posterior face of the stomach).
- Posterior nerve of lesser curvature [N. posterior curvaturae minoris] (lesser curvature).
- Coeliac branches [Rr. coeliaci] (Plexus coeliacus).
- Renal branches [Rr. renales] (Kidneys).

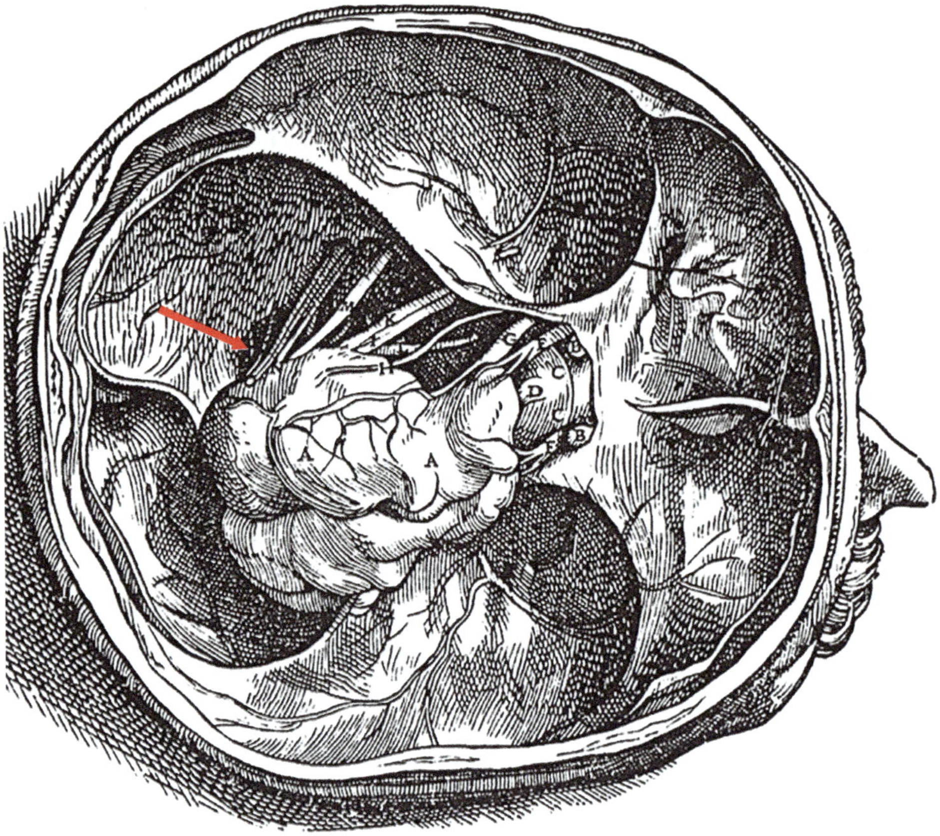

A	Brainstem	H	Trochlear nerve CNIV
B	Optic nerve	I	Trigeminal nerve CNV
C	Pituitary gland	K	Trigeminal nerve CNV
D	Infundibulum	L	Abducens nerve CNVI
E	Internal carotid artery	M	Facial-acoustic complex CNVII, CNVIII
F	Internal carotid artery	N	Glossopharyngeus-vagus-accessory nerve CNIX, X, XI
G	Oculomotor nerve CNIII	O	Hypoglossal nerve CNXII -Arrow inserted by authors -

Fig. 7.45 Figure 14 of Chap. VII from Vesal's "De Humani Corporis Fabrica' (1543). Identification of the cranial nerves according to the annotations of Saunders and O'Malley. Please note that the modern classification of cranial nerves is based on the work "Vom Hirn und Rückenmark" (On brain and spinal cord) in 1788 by the German physician and anatomist Samuel Thomas von Soemmerring (1755–1830) although (see Box 7.2) the existence of these nerves was already known in ancient Rome. Thus, it is not surprising that Vesal's classification differs from the modern one (acc. to de CM Saunders and O'Malley (1950): *CNIII* second pair of Vesal, *CNIV* third pair of Vesal, *CNV* third and fourth pairs of Vesal, *CNVI* minor root of fifth pair, *CNVII/VIII* fifth pair, *CNIX/X/XI* sixth pair, *CNXII* seventh pair. Compare with Fig. 1.16. From Sanders and O'Malley 1950, Plate 72:1 [VII Fig. 14])

For the afferent branches of the vagus nerve which represent approx. 75% of vagal nerve fibers to the solitary nucleus, see Sect. 3.2.4.1.

7.3.2.1.3 Afferent and efferent connectivity of the posterior nucleus of vagus nerve

The posterior nucleus of vagus nerve is part of the extended Area postrema (AP see Sect. 5.1.1), i.e., signals entering the solitary nucleus from the inner and taste organs are conveyed to the nucleus. This allows for a direct motor reaction to the sensory input. In this vein, the term dorsal vagal complex has been coined for the triad of AP, nucleus of solitary tract (see Sect. 3.2.4), and posterior nucleus of vagus nerve all of which—known from animal studies—receive vagal afferent fibers from the inner organs (see Sect. 3.2.4.1.2).

Anatomical evidence for a reciprocal connection between the posterior vagal nucleus and the solitary nucleus has been provided by Zec and Kinney (2003) using DiI tracing in the human fetal brain.

Central afferents—as studied in animals—originate in the amygdala, the paraventricular nucleus of the hypothalamus, the dorsomedial and posterior hypothalamus, the midbrain periaqueductal gray (see Sect. 15.2.1), the parabrachial nuclei (Sect. 13.2.3), and various parts of the reticular formation (ten Donkelaar 2011). In the cat, the posterior paramedian reticular nucleus, which is involved in cardiovascular control, has been shown to project to the posterior nucleus of vagus nerve and solitary nucleus (for details, see Sect. 5.4.3.1).

7.3.2.1.4 Living anatomy, potential therapeutic and pathological aspects

A clinical examination likewise important for neurology, neurosurgery, and otolaryngology is the so-called laryngoscopy. It is easy to perform, only a laryngeal mirror is needed. To avoid fogging during examination, the mirror—not the backside, this might lead to thermal lesions of the pharynx—is heated. The patient opens its mouth as far as possible. Note that in some patients,for example, touching the uvula may lead to retching and vomiting (vagal sensory afferents to the brainstem).

The mirror is inserted as shown in Fig. 7.46 and the normal finding that is expected to see is shown in this figure.

The rima glottidis between the vocal folds [Plicae vocales] should be located in the midline. Moving laterally, the Plica vocalis is followed by the vestibular fold [Plica vestibularis], separated by the laryngeal ventricle [Ventriculus laryngis Morgagni].

Unilateral paralysis of the recurrent laryngeal nerve would lead to a shift of the rima glottidis to the non-paralyzed side. Bilateral paralysis with a central but very narrow rima is a life-threatening condition because of the resulting deficit in respiratory air.

The two tubercula hide (compare with Fig. 7.46) below the laryngeal mucosa.

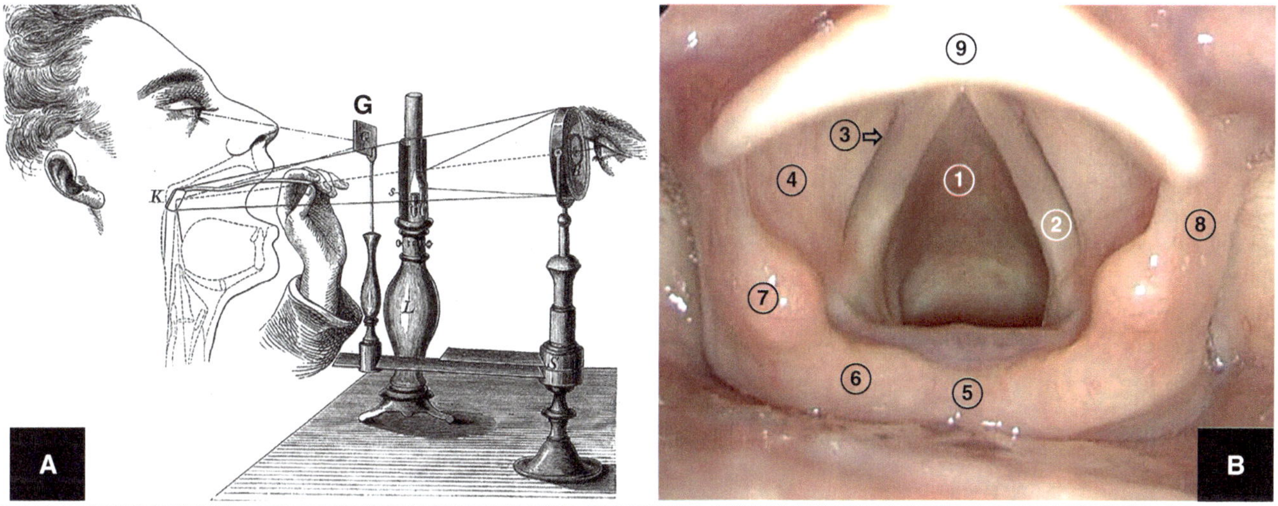

①	Rima glottidis	⑥	Corniculate tubercle
②	Vocal fold	⑦	Cuneiform tubercle
③	Laryngeal ventricle (Morgagni)	⑧	Ary-epiglottic fold
④	Vestibular fold	⑨	Epiglottis
⑤	Interarytenoid part		

Fig. 7.46 (**A**) Contemporal illustration of the principles of indirect laryngoscopy acc. to Czermak (Lübbers and Lübbers 2021). The Austrian physician Johann Nepomuk Czermak (1828–1873) is one of the founders of indirect laryngoscopy based on the invention of the laryngeal mirror by the Spanish opera singer Manuel Patricio Rodríguez García (1805–1806). The examiner (right hand side) looks through the central hole of a mirror [S, right hand side] which reflects light from a light source [L] to the laryngeal mirror [K]. This way the investigator can see an enlarged picture of the larynx as shown in (**B**). This illustration is peculiar in so far as it shows kind of an ‚auto-laryngoscopy." When you observe the picture carefully, you will see that the patient/proband holds the mirror [K]—which usually is held by the examiner—and a second mirror [G] is fixed in such a way that the patient can see the picture of the larynx reflected from mirror [K]. The photograph in (**B**) was obtained by direct laryngoscopy, i.e. by use of an endoscope instead of the instruments shown in (**A**)—displaying the normal situation during breathing with a widely opened rima. In case of phonation, the vocal folds will closely narrow. For Giovanni Battista Morgagni, see Box 7.4. The typical picture of a unilateral cord paralysis shows an incomplete closure of the rima. The rendering of the anatomical situation is comparable between both techniques, but the direct laryngoscopy allows for the photographic documentation of possible pathological changes like, e.g., vocal cord (fold) paralysis due an injury to the recurrent laryngeal nerve(s) (see Fig. 7.44). Unilateral paralysis of the recurrent laryngeal nerve would lead to a shift of the rima glottidis to the non-paralyzed side. Bilateral paralysis with a central but very narrow rima is a life-threatening condition because of the resulting deficit in respiratory air. The most frequent cause of recurrent lesions is an injury during thyroid gland surgery. From (**A**) Lübbers and Lübbers 2021, Fig. 2 and (**B**) Fleischer 2020, Figs. 3a and b with permission

Box 7.4 Giovanni Battista Morgagni

Giovanni Battista Morgagni (1682–1771) was an Italian physician who laid down the foundations of modern pathological anatomy in his famous work (1761) "De sedibus et causis morborum per anatomen indagatis libri quinque" (Five books about the location and causes of diseases investigated by Anatomy) by meticulously describing the autopsy findings and correlating them with the course of the diseases.

The larynx is an especially important supply territory of the vagus nerve and its branches. The superior laryngeal nerve provides motor innervation of the cricothyroid muscle (only muscle innervated by this nerve).

Concluding the functional anatomy of the vagus nerve in Table 7.2, you will find an overview about the parasympathetic (vagal) and sympathetic innervation of major inner organs. The innervation of the pelvis is not listed here since the parasympathetic part is not provided by the vagus nerve but by the spinal parasympathetic system.

It is known for quite some time that the transcutaneous stimulation of the vagus nerve, in particular of the auricular branch (see Sect. 7.3.2.1.2), which provides somatosensory innervation of certain parts of the external ear—in particular the cymba conchae (see Fig. 7.37) can be used for the treatment of drug-resistant epilepsy and depression. The selection of the correct stimulation point is not trivial since the greater part of the external ear is innervated by the auriculotemporal nerve (V_3) (see Fig. 7.37, Sect. 12.2.3.2.3) and the greater auricular nerve (C2/C3 spinal nerves) (see Fig. 7.37). A critical review of this issue has been provided by Butt et al. (2020).

Since weight loss is a known common side effect of CNX stimulation in patients with implanted devices, the procedure was evaluated related to potential treatment of obesity: Alicart et al. (2021) studied fMRI signals in 21 healthy participants presented with 72 food pictures and asked to rate how much they liked that food. Before scanning a 1 h sham (Left scapha, scaphoid fossa, see Fig. 7.37) or verum (left cymba conchae) stimulation was applied (Alicart et al. 2021). Significant activations were found mainly in left brainstem, thalamus, temporal pole, amygdala, insula, hippocampus, and supplementary motor area for interaction between ratings (high vs low) and session (verum vs. sham) (Alicart et al. 2021). According to the authors, this shows the effect of CNX stimulation on food image processing as a prerequisite for a therapeutic application to be evaluated in long-term studies.

Interestingly, the authors refrained from defining a region of interest (ROI) as to the presumable site of the nucleus of solitary tract (NTS) which may be wise given the resolution problems of small brainstem structures (see e.g., Gerlach et al. 2019).

In the midbrain ROI of Alicart and coworkers, there is a strong signal lateral of the red nucleus (see Sect. 15.3.4) on the right side. From an anatomical point the best bet for

Table 7.2 Regulation of inner organs by the autonomous nervous system

Organ	Parasympathetic innervation	Effect	Effects of sympathetic innervation
Heart	Cardiac branches of CNX	Heart rate ↓ Contraction ↓ Conduction ↓ Negative chronotropic Negative inotropic Negative dromotropic	Heart rate ↑ Contraction ↑ Conduction ↑ Positive chronotropic Positive inotropic Positive dromotropic
Arterial vessels	Organ-specific branches of CNX	Vasodilatation	Vasoconstriction
Trachea	Tracheal branches	Constriction	Dilation
Bronchi	Bronchial branches of CNX Pulmonary plexus	Bronchoconstriction	Bronchodilation
Esophagus	Oesophageal branches CNX	Contraction ↑[a] Relaxation ↑[a]	Small modulatory effect [a]
Stomach	Gastric branches CNX	Acid secretion ↑	Moderate effect on motility [b]
Small intestine		Motility ↑	Motility ↓
Pylorus	Pyloric branches CNX		
Liver	Hepatic branches CNX	Regulation of gluconeogenesis	Glycogenolysis ↑
Kidneys	Renal branches	? [d]	Renin secretion rate ↑ Urinary Na^+ excretion ↓ Renal blood flow ↓ [c]

[a] Depending on different neuron populations in the posterior nucleus of vagus nerve (see Mashimo and Goyal 2006)

[b] Browning and Travagli (2014)

[c] See Kopp (2011)

[d] Little evidence exists for parasympathetic innervation of the kidney (see Kirkpatrick et al. 2021)

the site of the signal is the central tegmental tract (see Fig. 3.14G ③).

Beckstead et al. (1980) upon electrophysiologically guided injections of tritiated amino acids into the Ncl. of the solitary tract (NTS) of monkeys could show that projections from the rostral, primarily gustatory part of the NTS ascend in the ipsilateral central tegmental tract bypass the parabrachial nuclei and terminate within the caudal half of VPMpc of the thalamus.

In Pårkinson's disease, the posterior nucleus of vagus nerve is one of the brainstem structures infested with Lewy neurites (LN) (for details Sect. 16.6.1.5) (for NBIA, see Fig. 7.38 and Sect. 16.6.1.8). LNs consist of intraneuritic Lewy body material which is invisible by conventional hematoxylin-eosin staining but is highlighted by ubiquitin and α-synuclein stainings, specifically seen in the nucleus basalis of Meynert, amygdala, reticular formation, and the posterior nucleus of vagus nerve. They are also described in the substantia nigra (see Sects. 16.6.1 and 16.6.1.5), pedunculopontine nucleus (see Sect. 13.1.1.1.1), raphe nuclei (see Sect. 3.2.1.3), and neocortex. Pathological changes in the posterior nucleus of the vagus nerve have been described also in progressive supranuclear palsy (PSP) (see Sect. 4.1.1.3.3) and multiple system atrophy (MSA) (see Sect. 3.3.2.3).

Another interesting finding with regard to neurodegeneration is the correlation of apolipoprotein D (Apo D) expression with age in the human brainstem, also in the posterior nucleus of vagus nerve (Navarro et al. 2013). Apo D is a marker which gets upregulated upon peripheral and/or central nervous lesions. The authors measured Apo D expression in the posterior nucleus of vagus nerve, the vestibular nuclei, the inferior olive (see Sect. 4.1.1.3), the hypoglossal nucleus (see Sect. 4.3.1.1), facial nucleus (see Sect. 8.2.1), oculomotor nucleus (see Sect. 15.3.1), nucleus of the solitary tract (see Sect. 3.2.4), and Roller's nucleus (see Sect. 6.2.1.6). The conclusion of the authors is that a strong expression of Apo D throughout life (32–96 years) may contribute to resistance against age-related degeneration and/or neurodegeneration.

7.3.3 Raphe nuclei r7

7.3.3.1 Paramedian reticular nucleus

For detailed information on this nucleus, see Sect. 5.4.3.1.

7.3.3.2 Raphe magnus nucleus [Ncl. raphes magnus] serotonergic cell group B3

7.3.3.2.1 Location and histology of the Raphe magnus nucleus

The raphe magnus nucleus (RMg) (Fig. 7.47) is situated in the border region of the medulla oblongata with the pontine tegmentum at the level of the facial nucleus.

The serotonergic neurons are located adjacent to the midline in continuity with those located in the gigantocellular reticular nucleus (see Sect. 4.3.4.2) (Hornung 2003, 2012). The raphe magnus is the largest nucleus of the caudal raphe group. The serotonergic neurons have various morphologies ranging from bipolar to multipolar. Besides serotonin, neuropeptide-synthesizing neurons are present.

7.3.3.2.2 Efferent and afferent connectivity of the Raphe magnus nucleus

The RMg is the rostral most nucleus of the caudal group of serotonergic nuclei and, as the raphe pallidus and obscurus nucleus, projects to the spinal cord and to a number of structures in- and outside the brainstem (for details, see Šimić et al. 2017). Inside the brainstem, it targets the sensory nuclei, cranial motor nuclei, the locus caeruleus, the superior colliculus, the pretectum, parabrachial nuclei, A7 and A5 cell groups, the solitary nucleus and, in the spinal cord, the laminae I and II, the intermediolateral nuclei.

According to Šimić et al. (2017), the raphe magnus nucleus receives afferents from the spinal trigeminal nucleus (see Sect. 3.2.2.2), the gracile and cuneate nuclei (Sect. 3.2.3), the periaqueductal gray (see Sect. 13.2.7.1), the locus caeruleus (see Sect. 13.2.2.1), the cuneiform nucleus (see Sect. 14.4.1.1), the gigantocellular reticular nucleus (see Sect. 4.3.4.2), the inferior and superior colliculi (see Sect. 15.1.3), pretectal nuclei (see Sect. 16.4), from the hypothalamus, the parafascicular ncl. of the thalamus, the preoptic area, the lateral habenular nucleus, the central ncl. of the amygdala, the zona incerta, the bed ncl. of the stria terminalis (BNST), and the prefrontal cortex.

For lesions of the nucleus, one would expect the lack of endogenous analgesia which relies on the inhibition of nociceptive transmission in the superficial laminae of the caudal spinal nucleus of the trigeminal nerve and the spinal dorsal horn and a reduced motor response to nociceptive stimuli.

In mice, the raphe magnus nucleus is a midline nucleus located at the level of the superior olivary complex and the facial nucleus. The raphe magnus as well as the raphe pallidus and obscurus nuclei (see above) and the lateral paragigantocellular nucleus (r7 below) contain 5HT-immunoreactive neurons that project bilaterally to the spinal cord (cf. VanderHorst and Ulfhake 2006).

7.3.3.2.3 Pathological aspects of the Raphe magnus nucleus

In a systematic study on the affection of raphe nuclei in Alzheimer's disease (see Sect. 13.2.2.1.4), Rüb et al. (2000) could show that the caudal nuclei show mild neurofibrillary changes late in the course of the disease. By contrast, the changes in the oral raphe nuclei (see Sect. 3.2.1.3) are massive and progress in close connection with the pathology in the cortical target sites (Rüb et al. 2000).

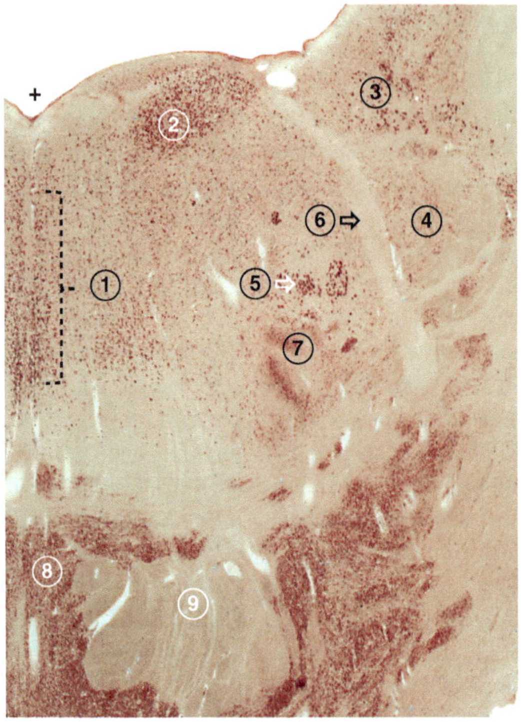

①	Raphe interpositus ncl.	⑥	Facial nerve CNVII
②	Abducens ncl. CNVI	⑦	Superior olivary complex
③	Vestibular ncll. CNVIII	⑧	Pontine nuclei
④	Spinal trigeminal ncl. CNV	⑨	Corticospinal tract
⑤	Facial ncl. CNVII	+	4th ventricle

Fig. 7.47 Horizontal section through the human brainstem at the level of abducens ② and facial ⑤ nuclei (see Sect. 9.2.1.1). Darrow red staining. As all raphe nuclei (for survey, see Sect. 3.2.1.3), the raphe interpositus nucleus ①—not that big as the name suggests—is located at either side of the midline in the tegmentum. See also atlas part Darrow red 20. LabPON Twente

7.3.3.3 Raphe obscurus nucleus

For detailed information on this nucleus is provided, see Sect. 4.3.3.1.

7.3.3.4 Raphe pallidus nucleus

For detailed information on this nucleus, see Sect. 4.3.3.2

7.3.4 Basal tegmentum r7

7.3.4.1 Dorsomedial tegmental area r7 [area tegmentalis dorsomedialis]

The term dorsomedial tegmental (DMTg) area refers to a group of cells in the pontine tegmentum of the mouse. In rats, Colussi-Mas et al. (2007) could show a massive Fos-expression in the DMTg upon amphetamine application. The DMTg projects to the ventral tegmental area (VTA; see Sect. 15.5.1). Amphetamine is a member of the phenethylamine family, which includes a range of substances that may be stimulants or hallucinogens with addiction potential. The nucleus is absent in humans.

7.3.5 Reticular nuclei r7

7.3.5.1 Intermediate reticular nucleus

For detailed information on this nucleus, see Sect. 3.3.2.1.

7.3.5.2 Parvocellular reticular nucleus

For detailed information on this nucleus, see Sect. 6.5.5.2.

7.3.5.3 Pontine reticular nucleus, caudal part/caudal pontine reticular nucleus r7 [Ncl. reticularis pontis caudalis]

The term caudal pontine reticular nucleus refers to a major cellular component of the reticular formation (classical) in the pontine reticular formation (Figs. 9.1 and 9.18).

Identified by Nissl stain, it is variously described as continuous across the midline or bounded medially by the raphe nuclei (classical) and the reticulotegmental nucleus (see Sect. 10.1.1.1). It is bounded dorsomedially by the medial longitudinal fasciculus of the pons (see Fig. 3.14A–F), ventromedially by the trapezoid body (see Sect. 8.1.1.5) and ventrolaterally by the superior olivary complex (see Sect. 8.1.1.5). The remainder of its boundaries is defined by several smaller reticular and cranial nerve nuclei. Its rostral neighbor is the rostral pontine reticular nucleus, the caudal one the central medullary reticular group. It is found in the human, the macaque, in rat and mouse.

In the rat, the majority of noradrenergic neurons of the A5 cell group are located in this nucleus. In mice, the oral and caudal part of the pontine reticular nucleus gives rise to few ipsilateral and contralateral cerebellopetal fibers (Fu et al. 2011).

http://braininfo.rprc.washington.edu/centraldirectory.aspx?ID=566

7.3.5.4 Rostroventrolateral reticular nucleus [Ncl. reticularis rostralis ventrolateralis]

The rostroventrolateral reticular nucleus comprises a group of cells in the caudal medullary reticular formation of the mouse. It is not present in man.

The rostral ventrolateral medulla (RVLM) is traditionally known for regulating respiration and the autonomic nervous system. In its medial portion, it contains many ChAT-immunoreactive neurons. Stornetta et al. (2013) could show that in mice these cholinergic neurons do not belong to the autonomous nervous or the motor system but have connections with dorsal column nuclei, the medullary trigeminal complex, the cochlear nuclei, the superior olivary complex, and lamina 3 of the spinal cord. Most likely these cholinergic neurons regulate sensory afferent information selectively and presumably have little to do with respiration or circulatory control.

http://braininfo.rprc.washington.edu/whatspecies.aspx?questID=1079

7.3.5.5 Gigantocellular reticular nucleus

For details on the gigantocellular reticular nucleus see Sect. 4.3.4.2.

7.3.5.6 Lateral paragigantocellular nucleus [Ncl. paragigantocellularis lateralis]

The lateral paragigantocellular nucleus is located anterolateral of the gigantocellular nucleus (see Sect. 4.3.4.2) and directly adjacent to it (see Fig. 6.1 ⑦). It is known as a sympathoexcitatory area involved in the control of blood pressure (Sirieix et al. 2012).

The mouse lateral paragigantocellular nucleus is located laterally of the inferior olive (see Sect. 4.1.1.3) and contains serotonin-immunoreactive neurons (VanderHorst and Ulfhake 2006). Combined neuroanatomical and neurophysiological studies in rats have shown that the lateral paragigantocellular nucleus is involved in the regulation of rapid eye movement sleep (Sirieix et al. 2012). In mice, 5HT-immunoreactive (ir) cells of the lateral paragigantocellular nucleus project bilaterally to the spinal cord (VanderHorst and Ulfhake 2006).

7.3.5.7 Dorsal paragigantocellular nucleus

For details on this nucleus, see Sect. 5.4.4.4.

7.3.5.8 Bötzinger complex

For details on the Bötzinger complex, see Sect. 3.3.2.2.

7.3.5.9 Rostral ventral respiratory group

For details on the Rostral ventral respiratory group, see Sect. 3.3.2.2.

References

Alicart H, Heldmann M et al (2021) Modulation of visual processing of food by transcutaneous vagus nerve stimulation (tVNS). Brain Imaging Behav 15(4):1886–1897. Epub ahead of print

Ashizawa T, Öz G et al (2018) Spinocerebellar ataxias: prospects and challenges for therapy development. Nat Rev Neurol 14:590–605

Augustine JR (2017) Human neuroanatomy. 2nd ed. Wiley

Bagnall MW, Stevens RJ et al (2007) Transgenic mouse lines subdivide medial vestibular nucleus into discrete neurochemically distinct populations. J Neurosci 27:2318–2330

Banfi S, Servadio A et al (1994) Identification and characterization of the gene causing type 1 spinocerebellar ataxia. Nat Genet 7:513–520

Bang OY, Lee PH et al (2004) Pontine atrophy precedes cerebellar degeneration in spinocerebellar ataxia 7: MRI-based volumetric analysis. J Neurol Neurosurg Psychiatry 75:1452–1456

Bates GP (2005) History of genetic disease: the molecular genetics of Huntington disease—a history. Nat Rev Genet 6:766–773

Beckstead RM, Morse JR et al (1980) The nucleus of the solitary tract in the monkey: projections to the thalamus and brain stem nuclei. J Comp Neurol 190:259–282

Blessing WB (2004) Lower brain stem regulation of visceral, cardiovascular, and respiratory function. In: Paxinosm G, Mai G (eds) The human nervous system, 2nd edn. Elsevier, Amsterdam, pp 465–478

Browning KN, Travagli AT (2014) Central nervous system control of gastrointestinal motility and secretion and modulation of gastrointestinal functions. Compr Physiol 4:1339–1368

Burnett BG, Pittman RN (2005) The polyglutamine neurodegenerative protein ataxin 3 regulates aggresome formation. Proc Natl Acad Sci U S A 102(12):4330–4335

Butt MF, Albusoda A et al (2020) The anatomical basis for transcutaneous auricular vagus nerve stimulation. J Anat 236:588–611

Casamento-Moran A, Chen Y-T et al (2015) Force dysmetria in spinocerebellar ataxia type 6 correlates with functional capacity. Front Hum Neurosci 9:184

Cambiaghi M (2017) Andreas Vesalius (1514-1564). J Neurol 264:1828–1830

Cannon SC, Robinson DA (1987) Loss of the neural integrator of the oculomotor system from brain stem lesions in monkey. J Neurophysiol 57:1383–1409

Colussi-Mas J, Geisle S et al (2007) Activation of afferents to the ventral tegmental area in response to acute amphetamine: a double labeling study. Eur J Neurosci 26:1011–1025

Duyckaerts C, Dürr A et al (1999) Nuclear inclusions in spinocerebellar ataxia type 1. Acta Neuropathol 97:201–207
Eadie MJ (1975a) Olivo-ponto-cerebellar atrophy (Déjérine-Thomas type). In: Vinken PJ, Bruyn GW (eds) Handbook of clinical neurology. North-Holland, Amsterdam, pp 415–431
Eadie MJ (1975b) Olivo-ponto-cerebellar atrophy (Menzel type). In: Vinken PJ, Bruyn GW (eds) Handbook of clinical neurology. North-Holland, Amsterdam, pp 433–449
Eriksson R (1959) Andreas Vesalius first public anatomy at Bologna 1540. An eyewitness report by Baldasar Heseler, Medicinae Scolaris
Feehs KR, Thomas RW et al (2020) Laryngeal reinnervation. In: Weissbrod PA, Francis DO (eds) Neurologic and neurodegenerative diseases of the larynx. Springer, pp 355–363
Fleischer S (2020) Tipps und Tricks für die Laryngoskopie. HNO Nachrichten 50:30–37
Fu Y, Tvrdik P et al (2011) Precerebellar cell groups in the hindbrain of the mouse defined by retrograde tracing and correlated with cumulative Wnt1-Cre genetic labeling. Cerebellum 10:570–584
Garden GA, La Spada AR (2008) Molecular pathogenesis and cellular pathology of spinocerebellar ataxia type 7 neurodegeneration. Cerebellum 7:138–149
Garofalo I (1991) Galeno Procedimenti anatomici. vol. I–III. ΑΝΑΤΟΜΙΚΑΙ ΕΓΧΕΙΡΗΣΕΙΣ. Biblioteca Universale Rizzoli, Milano
Gerlach DA, Manuel J et al (2019) Novel approach to elucidate human Baroreflex regulation at the brainstem level: pharmacological testing during fMRI. Front Neurosci 13:193
Ghannam MG, Singh P (2020) Anatomy, head and neck, salivary glands. StatPearls Publishing, Treasure Island
Gil JM, Rgeo AC (2008) Mechanisms of neurodegeneration in Huntington's disease. Eur J Neurosci 27:2803–2820
Gomez CM, Thompson RM et al (1997) Spinocerebellar type: gaze-evoked and vertical nystagmus, Purkinje cell degeneration, and variable age of onset. Ann Neurol 42:933–950
Gouw LG, Digre KB et al (1994) Autosomal dominant cerebellar ataxia with retinal degeneration: clinical, neuropathologic, and genetic analysis of a large kindred. Neurology 44:1441–1447
Harper PS (1992) The epidemiology of Huntington's disease. Hum Genet 89:365–376
Hedreen JC, Roos RAC (2003) Huntington's disease. In: Dickson DW (ed) Neurodegeneration: The molecular pathology of dementia and movement disorders. vol 5, pp 229–242
Herndon ES, Hladik CL et al (2009) Neuroanatomic profile of polyglutamine immunoreactivity in Huntington disease brains. J Neuropathol Exp Neurol 68(3):250–261
Holmberg M, Duyckaerts C et al (1998) Spinocerebellar ataxia type 7 (SCA7): a neurodegenerative disorder with neuronal intranuclear inclusions. Hum Mol Genet 7:913–918
Hornung J-P (2003) The human raphe nuclei and the serotonergic system. J Chem Neuroanat 26:331–343
Hornung J-P (2012) Raphe nuclei. In: Mai JK, Paxinos G (eds). The human nervous system, 3rd edn. Elsevier, pp 401–424
Huang X-F, Törk I et al (1993) Dorsal motor nucleus of the vagus nerve: a cyto- und chemoarchtectonic study in the human. J Comp Neurol 8:158–182
Inoue M (1984) Structure and innervation of mouse parotid gland. J Juzen Med Soc 93:534–549
Ishikawa K, Fujigasaki H et al (1999) Abundant expression and cytoplasmic aggregations of alpha1A voltage-dependent calcium channel protein associated with neurodegeneration in spinocerebellar ataxia type 6. Hum Mol Genet 8:1285–1293
Iwabuchi K, Tsuchiya K et al (1999) Autosomal dominant spinocerebellar degenerations. Clinical, pathological, and genetic correlations. Rev Neurol (Paris) 155:255–270
Kawaguchi Y, Okamoto T et al (1994) CAG expansion in a novel gene for Machado-Joseph disease at chromosome14q32.1. Nat Genet 8:221–228
Kaufman MH, Bard JBL (1999) The anatomical basis of mouse development. Academic, San Diego
Khakh BS, Beaumont V et al (2017) Unraveliong and exploiting astrocyte dysfunction in Huntington's disease. Trends Neurosci 40:422–437
Kirkpatrick JJ, Foutz S et al (2021) Anatomy, abdomen and pelvis, kidney nerves. StatPearls Publishing, Treasure Island
Kolkman KE, Moghadam SH et al (2011) Intrinsic physiology of identified neurons in the prepositus hypoglossi and the medial vestibular nuclei. J Vestib Res 21:33–47
Kopp UC (2011) Neuroanatomy. Chapter 2 of 'neural control of renal function'. Morgan & Claypool Life Sciences
Lübbers W, Lübbers C (2021) Eine "Selbstbetrachtung" der HNO-Ärzte. HNO Nachrichten 51:138–139
Mashimo H, Goyal RK (2006) Physiology of esophageal motility. GI Motility online
Manuff P, Tyler P et al (1996) Cognitive decline in Machado-Joseph disease. Ann Neurol 40:421–427
Martin J, van Regemorter N et al (1999) Spinocerebellar ataxia type 7 (SCA7)—correlations between phenotype and genotype in one large Belgian family. J Neurol Sci 168:37–46
Mascalchi M (2008) Spinocerebellar ataxias. Neurol Sci 29:S311–S313
Michalik A, Martin J-J et al (2004) Spinocerebellar ataxia type 7 associated with pigmentary retinal dystrophy. Eur J Hum Genet 12:2–15
Nakamura K, Jeong S-Y et al (2001) SCA17, a novel autososomal dominant cerebellar ataxia caused by an expanded polyglutamine in TATA-binding protein. Hum Mol Genet 10(14):1441–1448
Navarro A, Méndez E et al (2013) Lifelong expression of apolipoprotein D in the human brainstem: correlation with reduced age-related neurodegeneration. PLoS One 8(10):e77852
Ng ALC, Rosenfeld JV et al (2019) Cranial nerve nomenclature: historical vignette. World Neurosurg 128:299–307
Norberg K-A, Hökfelt T et al (1968) The autonomic innervation of human submandibular and parotid glands. J Neurovisc Relat 31:280–290
Orr H, Chung M-Y et al (1993) Expansion of an unstable trinucleotide CAG repeat in spinocerebellar ataxia type 1. Nat Genet 4:221–226
Pasipoularides A (2014) Galen, father of systematic medicine. An essay on the evolution of modern medicine and cardiology. Int J Cardiol 172:47–58
Paulson HL, Perez MK et al (1997) Intranuclear inclusions of expanded polyglutamine protein in spinocerebellar ataxia type 3. Neuron 19:333–344
Paulson HL (2000) Toward an understanding of polyglutamine neurodegeneration. Brain Pathol 19:759–782
Paulson HL (2007) Dominantly inherited ataxias: lessons learned form Machado-Joseph disease/spinocerebellar ataxia type 3. Semin Neurol 27:133–142
Paulson H (2009) The spinocerebellar ataxias. J Neuroophthalmol 29:227–237
Rawlins MD, Wexler NS et al (2016) The prevalence of Huntington's disease. Neuroepidemiology 46:144–153
Revesz T, Clark HB et al (2015) Extrapyramidal disorders of movement. In: Love S, Perry A, Ironside J, Budka H (eds) Greenfield's neuropathology, 9th edn., p 740
Robitaille Y, Schut L et al (1995) Structural and immunocytochemical features of olivopontocerebellar atrophy caused by the spinocerebellar ataxia type 1 (SCA-1) mutation define a unique phenotype. Acta Neuropathol 90:572–581
Robitaille Y, Lopes-Cendes I et al (1997) The neuropathology of CAG repeat diseases: review and update of genetic and molecular features. Brain Pathol 7:901–926

Rolfs A, Koeppen AH et al (2003) Clinical features and neuropathology of autosomal dominant spinocerebellar ataxia (SCA17). Ann Neurol 54:367–375

Rüb U, Del Tredici K et al (2000) The evolution of Alzheimer's disease-related cytoskeletal pathology in the human raphe nuclei. Neuropathol Appl Neurobiol 26:553–567

Rüb U, Schultz C et al (2003) Anatomically based guidelines for systematic investigation of the central somatosensory system and their application to a spinocerebellar ataxia type 2 (SCA2) patient. Neuropathol Appl Neurobiol 29:418–433

Rüb U, Burk K et al (2004) Damage to the reticulotegmental nucleus of the pons in spinocerebellar ataxia type 1, 2, and 3. Neurology 63:1258–1263

Rüb U, Brunt ER et al (2005a) Spinocerebellar ataxia type 7 (SCA7): first report of a systematic neuropathological study of the brain of a patient with a very short expanded CAG-repeat. Brain Pathol 15:287–295

Rüb U, Gierga K et al (2005b) Spinocerebellar ataxias type 2 and 3: degeneration of the precerebellar nuclei isolates the three phylogenetically defined regions of the cerebellum. J Neural Transm 112:1523–1545

Rüb U, Brunt ER et al (2006) Degeneration of ingestion-related brainstem nuclei in spinocerebellar ataxia type 2, 3, 6, and 7. Neuropathol Appl Neurobiol 32:635–649

Rüb U, Brunt ER et al (2008a) New insights into pathoanatomy of spinocerebellar ataxia type 3 (Machado-Joseph disease). Curr Opin Neurol 21:111–116

Rüb U, Brunt ER et al (2008b) Spinocerebellar ataxia type 7 (SCA7): widespread brain damage in an adult-onset patient with progressive visual impairments in comparison with an adult-onset without visual impairments. Neuropathol Appl Neurobiol 34:155–168

Rüb U, Burk K et al (2012) Spinocerebellar ataxia type 1 (SCA1): new pathoanatomical and clinico-pathological insights. Neuropathol Appl Neurobiol 38:665–680

Rüb U, Seidel K et al (2016) Huntington's disease (HD): the neuropathology of a multisystem neurodegenerative disorder of the human brain. Brain Pathol 26:726–740

Saunders JB, O'Malley CD (1950) The illustrations from the works of Andreas Vesalius of Brussels. Dover Publications, New York

Schmidt T, Lindenberg KS et al (2002) Protein surveillance machinery in brains with spinocerebellar ataxia type 3: redistribution and differential recruitment of 26S proteasome subunits and chaperones to neuronal intranuclear inclusions. Ann Neurol 51:302–310

Schöls L, Krüger R et al (1998) Spinocerebellar ataxia type 6: genotype and phenotype in German kindreds. J Neurol Neurosurg Psychiatry 64:67–73

Seidel K, den Dunnen WF et al (2010) Axonal inclusions in spinocerebellar ataxia type 3. Acta Neuropathol 120:449–460

Seidel K, Siswanto S et al (2012) Brain pathology of spinocerebellar ataxias. Acta Neuropathol 124:1–21

Solodkin A, Gomez CM (2012) Spinocerebellar ataxia type 6. Handb Clin Neurol 103:461–473

Šimić G, Babić Leko M et al (2017) Monoaminergic neuropathology in Alzheimer's disease. Prog Neurobiol 151:101–138

Sirieix C, Gervasoni D et al (2012) Role of the lateral paragigantocellular nucleus in the network of paradoxical (REM) sleep: an electrophysiological and anatomical study in the rat. PLoS One 7(1):e28724

Stornetta RL, Macon CJ et al (2013) Cholinergic neurons in the mouse rostral ventrolateral medulla target sensory afferent areas. Brain Struct Funct 218:455–475

ten Donkelaar HJ (2011) Clinical neuroanatomy: brain circuitry and its disorders. Springer

VanderHorst VG, Ulfhake B (2006) The organization of the brainstem and spinal cord of the mouse: relationships between monoaminergic, cholinergic, and spinal projection systems. J Chem Neuroanat 31:2–36

Vesalius A (1543) De humani corporis fabrica (Andreae Vesalii Bruxellensis, Scholae medicorum Patavinae professoris, de Humani corporis fabrica. Libri septem) Joannes Oporinus Basileae (Basel)

Vonsattel JP, Myers RH et al (1985) Neuropathological classification of Huntington's disease. J Neuropathol Exp Neurol 44:559–577

Vonsattel JP, DiFiglia M (1998) Huntington's disease. J Neuropathol Exp Neurol 57:369–384

Wagner A, Menalled L et al. (2008) Chapter 6. Huntington's disease. In: McArthur RA, Borsini F (eds.) Animal and translational models for CNS drug discovery. Academic, pp 207–266

Wilson-Pauwels L, Akesson EJ et al (1988) Cranial nerves: anatomy and clinical comments. BC Decker, Toronto

Yang Q, Hashizume Y et al (2000) Morphological Purkinje cell changes in spinocerebellar ataxia type 6. Acta Neuropathol 100:371–376

Zec N, Kinney HC (2003) Anatomic relationships of the human nucleus of the solitary tract in the medulla oblongata. Auton Neurosci 105:131–144

Zhuchenko O, Bailey J et al (1997) Autosomal dominant cerebellar ataxia (SCA6) associated with small polyglutamine expansions in the α1A-voltage-dependent calcium channel. Nat Genet 15:62–69

Rhombomere 6 r6

8

Contents

H. Schröder et al., *The Human Brainstem*, https://doi.org/10.1007/978-3-030-89980-6_8

Abstract

In this chapter, an important sensory system is described: the auditory system, with the cochlear part of CNVIII, including and anticipating the further steps of the ascending auditory system. Starting with the dorsal and ventral (assigned to Chap. 9) cochlear nucleus here, auditory-related structures listed in the following chapters are the lateral lemniscus nuclei (Chaps. 10, 12, 13) and the inferior colliculi (Chap. 15).

The series of noradrenergic nuclei is continued here with the A5 cell group.

A cranial nerve of utmost practical and clinical importance is the facial motor nucleus (CNVII), which, via its motor root, innervates the mimic muscles, uniquely distinct in humans and important for social interaction. Via its parasympathetic part, this nerve regulates the signal transmission of the minor salivatory nuclei as well as the secretion of the lacrimal gland. The most important afferent part is located in the chorda tympani which conveys the signals from the gustatory receptors in the anterior two thirds of the tongue (see also Chap. 3).

8.1 Alar r6

8.1.1 Cochlear/auditory nuclei r6 [Ncll. acustici] Posterior (dorsal) and anterior (ventral) cochlear nuclei

8.1.1.1 General considerations and location

The human brainstem contains two cochlear nuclei (see Fig. 8.1A, B), a dorsal (posterior) (PCN) [Ncl. cochlearis posterior] [formerly Ncl. cochlearis dorsalis] and a ventral (anterior) one (ACN) [Ncl. cochlearis anterior] [formerly Ncl. cochlearis ventralis].

The dorsal cochlear nucleus forms the acoustic tubercle [Tuberculum acusticum] on the posterior surface of the middle cerebellar peduncle and is continuous medially with the area of the vestibular nuclei in the rhomboid fossa (Fig. 8.2).

The ventral cochlear nucleus lies between the cochlear and vestibular fibers of the vestibulocochlear nerve (CNVIII). The ventral cochlear nucleus is derived from the alar plate (see Fig. 2.5) of rhombomere 5 but is described here together with its dorsal counterpart.

There are approximately 25,000 axons in the human cochlear nerve (Standring 2016), and they project on to a much larger number of neurons in the cochlear nucleus. As a result, the number of fibers in the following auditory pathway, i.e., the lateral lemniscus (see below), greatly exceeds that in the cochlear nerve.

The cochlear nuclei are the first brainstem representation of the auditory system (audire Latin = to hear). The cochlear nuclei receive input from the cochlea (see Sect. 8.1.1.2) via the cochlear nerve (cn) which is the auditory portion of the VIIIth cranial nerve, the vestibulocochlear nerve (CNVIII, for the vestibular part see Sect. 5.3.1).

The common auditory brainstem pathway is the lateral lemniscus (ll) [Lemniscus lateralis] (see Fig. 8.3) (not to confuse with the medial lemniscus, see Fig. 3.14 A–G) which connects the cochlear nuclei directly via the dorsal acoustic stria (said in Fig. 8.1) with the inferior colliculi (IC) [Colliculus inferior] (see Sect. 15.1.4).

An alternative route runs as intermediate or ventral acoustic striae [Striae cochleares intermedia et anterior] to the inferior colliculi. However, this alternative route is modulated by at least one synapse in the superior olivary complex (SO) [Complexus olivaris superior] (see Sect. 8.1.1.5) or in nuclei within the trapezoid body (ct) [Corpus trapezoideum] and the lateral lemniscus.

From the IC the auditory fibers reach their thalamic relay station, the medial geniculate body (MGB) [Corpus geniculatum mediale] (see Fig. 8.10 ①, not to confuse with the lateral geniculate nucleus for the visual system, see Fig. 8.10 ②) via the brachium of the inferior colliculus (bci) [Brachium colliculi inferioris]. Eventually, the auditory pathways reach the Brodmann area 41 (BA 41), located on the temporal plane (pt) [Planum termporale] and the transverse temporal gyrus (gtt) [Gyrus temporalis transversus], via the acoustic radiation (ar) [Radiatio acustica].

8.1.1.2 Sensory organ of the auditory system

The cochlear nucleus receives its information from the organ of Corti in the cochlea of the inner ear. The Latin word cochlea means snail, pointing to the helical wound structure of the cochlea. By the organ of Corti (see Box 8.1), sound waves, perceived via the outer and the middle ear (see Fig. 8.6), are transformed from mechanical energy to electrical signals in the receptor cells, the hair cells. The inner hair cells (IHC) of the cochlea (Fig. 8.4 **"Inner Ear"**) make synaptic contacts with peripheral processes of the cochlear

A

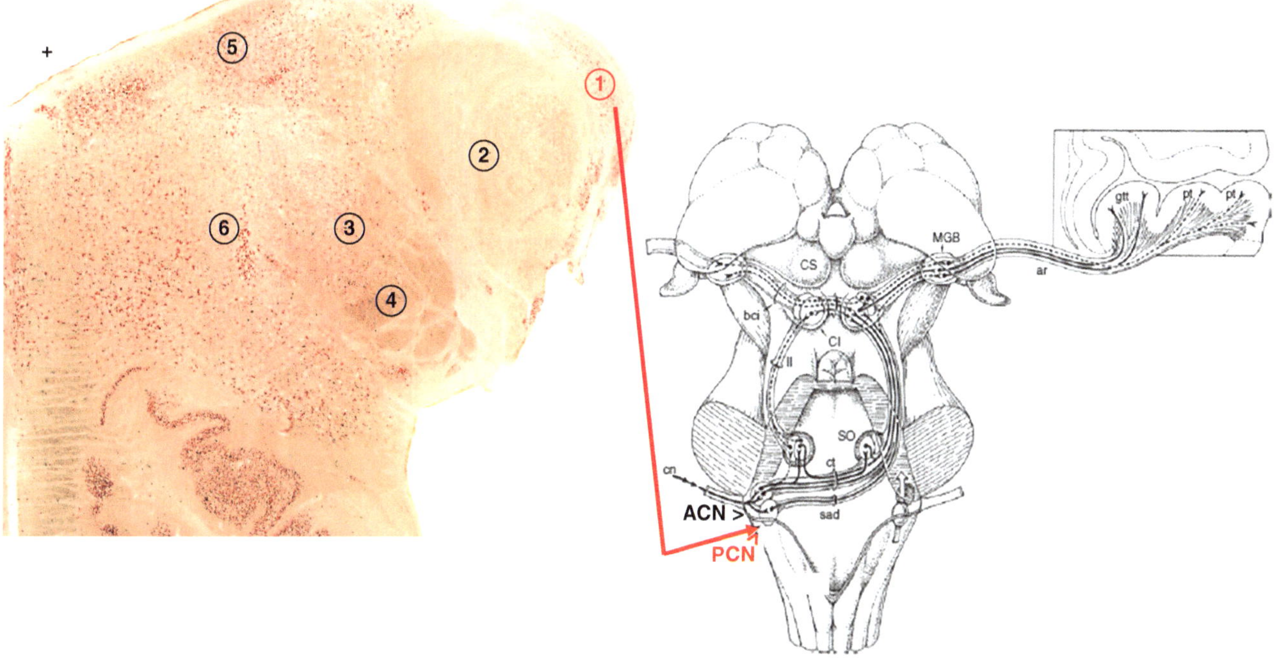

①	Posterior (dorsal) cochlear nucleus (PCN)	⑤	Medial vestibular nucleus
②	Inferior cerebellar peduncle	⑥	Ncl. ambiguus
③	Spinal trigeminal nucleus	+	4th ventricle
④	Spinal trigeminal tract		

Fig. 8.1 (**A**) Horizontal section through the human medulla oblongata at the level of the posterior (dorsal) cochlear nucleus (PCN). Darrow red staining. The PCN ① is located at the dorsolateral "shoulder" of the medulla oblongata in the region of the lateral aperture (see Fig. 3.9) directly under the ependymal surface. Its medial neighbor is the inferior cerebellar peduncle ②. (see also Fig. 8.3). *ar* acoustic radiation, *bci* brachium of colliculus inferior, *CI* colliculus inferior, *cn* cochlear nerve, *CS* colliculus superior, *ct* corpus trapezoideum, *gtt* gyrus temporalis transversus (Heschl's gyrus) of temporal lobe, *ll* lateral lemniscus, *MGB* medial geniculate body, *pt* planum temporale, *sad* stria acoustica dorsalis, *SO* superior olivary complex. See also atlas part Darrow red 14, 14A. LabPON Twente; graphic from ten Donkelaar (2011), Fig. 7.8 with permission. (**B**) Horizontal section through the human medulla oblongata at the level of the cranial portion of the anterior (ventral) cochlear nucleus (ACN) ①. Darrow red staining. Abbreviations: see Fig. 8.1A. The dorsal (posterior) cochlear nucleus first appears at the level of the lateral aperture, followed in cranial direction by the ventral (anterior) cochlear nucleus (see Terr and Edgerton 1985, see Fig. 3.9). In the further cranial course, the lateral aperture "disappears" and both nuclei are direct neighbors of the cerebellum. Then the dorsal cochlear is no longer visible while the ventral nucleus can be followed up to the cerebellopontine angle. This is the situation shown here. See also atlas part Darrow red 17, 17 A. LabPON Twente; graphic from ten Donkelaar 2011, Fig. 7.8 with permission

B

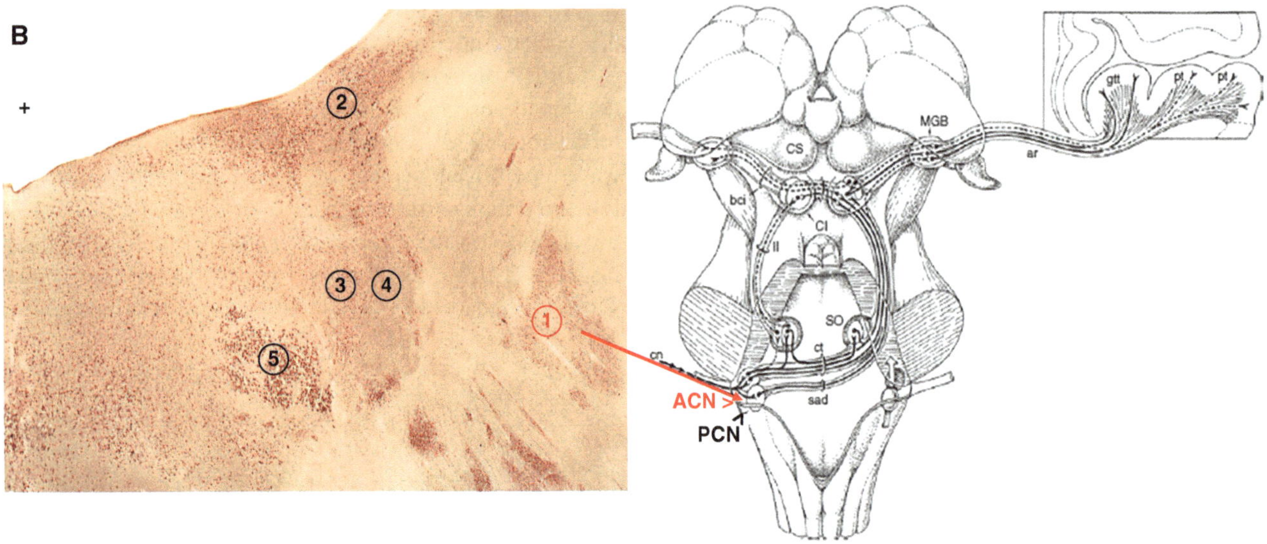

①	Anterior (ventral) cochlear nucleus (ACN)	④	Spinal trigeminal tract
②	Vestibular nuclei	⑤	Ncl. of facial nerve CNVII
③	Spinal trigeminal nucleus	+	4th ventricle

Fig. 8.1 (continued)

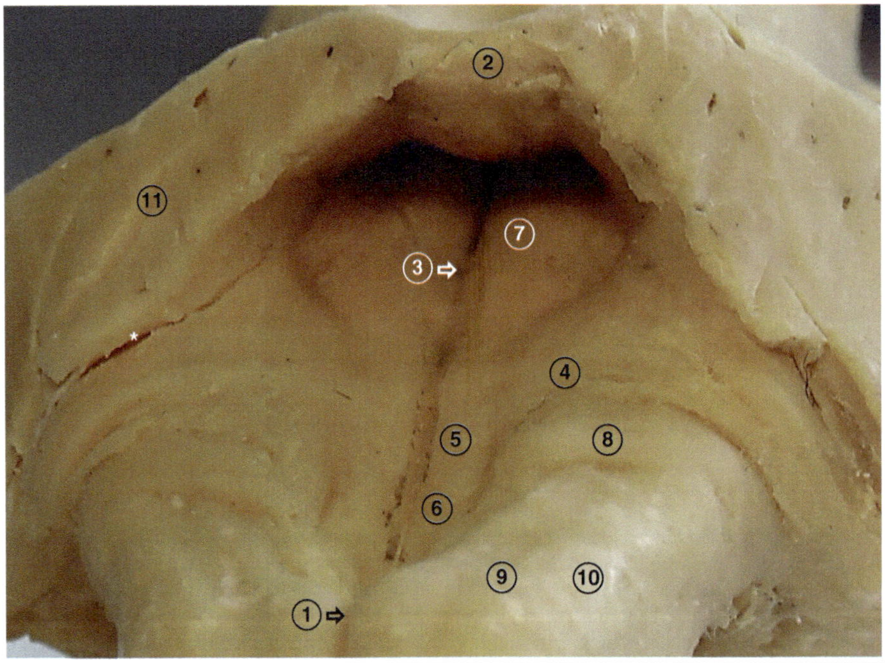

①	Obex	⑦	Facial colliculus
②	Superior medullary velum / Cut surface	⑧	Medullary striae of fourth ventricle
③	Posterior median sulcus	⑨	Gracile tubercle
④	Vestibular area	⑩	Cuneate tubercle
⑤	Hypoglossal trigone	⑪	Medial cerebellar peduncle / Cut surface
⑥	Vagal trigone	*	Artifact

Fig. 8.2 Dorsal view into the rhomboid fossa. Cerebellum and posterior medullary velum (see Fig. 1.5) removed. Siliconized specimen. The medullary striae ⑧, with the posterior (dorsal) cochlear nucleus (see Fig. 8.1A ①) as their lateral point of origin are an indirect indicator for the auditory region. They run to the posterior median sulcus, then join the lateral lemniscus (see Fig. 8.1A ll). Sammlung des Zentrums Anatomie der Universität zu Köln

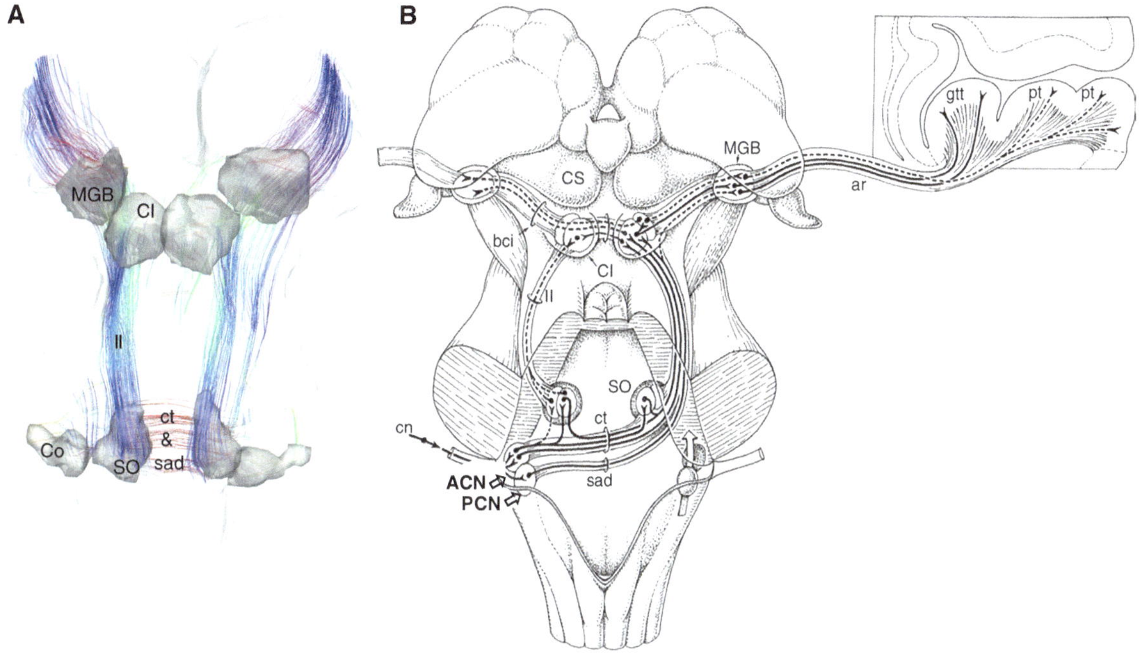

Fig. 8.3 (**A**) In vivo tractography of the subcortical auditory pathway from 7T diffusion-weighted MRI (left). Colors represent the local orientation at each specific point along the streamline: blue is inferior-superior, green is anterior-posterior and red is left-right. (**B**) Schematic overview of the human auditory system in dorsal view. For details, see text. *ar* acoustic radiation, *bci* brachium of colliculus inferior, *CI* colliculus inferior, *cn* cochlear nerve, *Co* cochlear nucleus (PCN, ACN, dorsal and ventral Co), *CS* colliculus superior, *ct* corpus trapezoideum, *gtt* gyrus temporalis transversus (Heschl's gyrus) of temporal lobe, *ll* lateral lemniscus, *MGB* medial geniculate body, *pt.* planum temporale, *sad* stria acoustica dorsalis, *SO* superior olivary complex. (**A**) Modified after Sitek et al. 2019, Fig. 6 with permission. (**B**) From ten Donkelaar 2011, Fig. 7.8 with permission

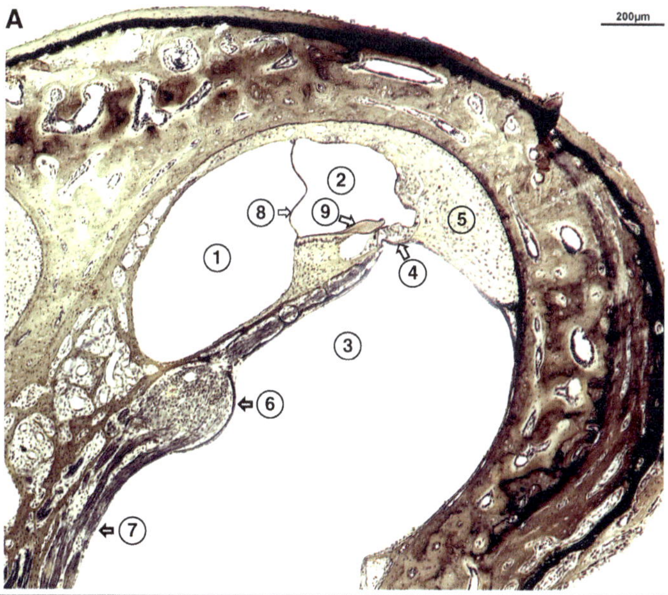

①	Scala vestibuli	⑥	Spiral ganglion
②	Scala media	⑦	Cochlear nerve fibers
③	Scala tympani	⑧	Reissner's membrane
④	Organ of Corti	⑨	Tectorial membrane
⑤	Spiral ligament		

Fig. 8.4 Photomicrographs of the cochlear canal of the cat (iron hematoxylin stain). (**A**) Overview with organ of Corti (**B**, detail) and the spiral ganglion (**C**, detail). Note that in (**A**) the curved course of Reissner's membrane ⑧ is an artifact. Normally, the membrane is running straight from one site of anchorage to the other. Courtesy Dr. T. Voigt, Witten/Herdecke University, Germany

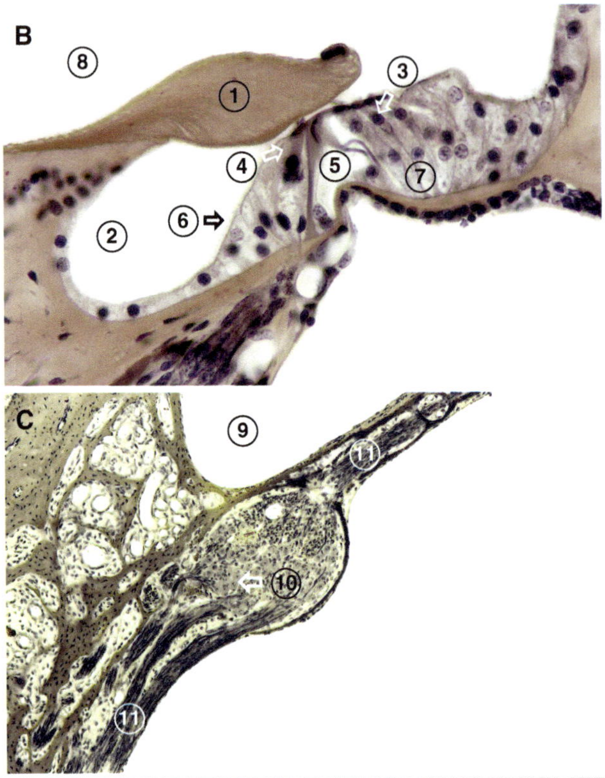

①	Tectorial membrane	⑦	Deiters cells
②	Internal spiral tunnel	⑧	Scala media
③	Outer hair cells	⑨	Scala vestibuli
④	Inner hair cell	⑩	Spiral ganglion cells
⑤	Inner (Corti-) tunnel	⑪	Cochlea nerve fibers
⑥	Inner border cells		

Fig. 8.4 (continued)

nerve so that the signals of the inner hair cells are transmitted to the cochlear nucleus and then to higher auditory centers (see below). Afferent fibers of type I (vast majority in the cochlear nerve) contact the inner hair cells by ribbon synapses. Ribbon synapses use transduction mechanisms of a graded receptor potential (in contrast to synapses that generate action potentials). Ribbon synapses are found at glutamatergic sensory cell types having synaptic active zones equipped with a presynaptic electron-dense structure, known as ribbon, to tether synaptic vesicles (Schröder et al. 2020). Efferent fibers from the lateral superior olivary nucleus (r5, see Fig. 8.8A) form axodendritic synapses with the type I afferents but little is known about their function (Guinan Jr. 2018).

Box 8.1 Alfonso Corti

Alfonso Giacomo Gaspare Corti was born 1822 near Pavia (Italy). He studied at the Universities of Pavia and Vienna. With the outbreak of the 1848 Revolution, he left Vienna and moved 1850 to Würzburg (Germany) to join the laboratory of Albert Kölliker. There, Corti started to study the mammalian auditory system and 1851 he published his famous paper "Recherches sur l'organe de l'ouïe des mammiferes" (Corti 1851). In the same year, after the death of his father, he moved back to Italy. In 1876, Alfonso Corti died in Corvino San Quirico.

https://en.wikipedia.org/wiki/Alfonso_Giacomo_Gaspare_Corti

Hair cells are named for microscopic hair-like extensions, called stereocilia (Fig. 8.5), projecting from their tops in bundles. These "hair bundles" convert sound vibrations into electrical signals. Damage of hair cells, for example, by sound of very high intensity, results in hearing loss since mammalian hair cells cannot regenerate on their own.

Next to the inner hair cells, there are three rows of outer hair cells (OHC) in the cochlea which are known to amplify the acoustic signal. Outer hair cells are reached by type II afferent fibers (approximately 5% in the cochlear nerve). The perikarya of the cochlear nerve afferents (types I and II) are located in the spiral ganglion (see Fig. 8.4C).

Outer hair cells have strong efferent innervation, in addition to the afferents, coming from the superior olivary complex (see Sect. 8.1.1.5). These olivocochlear efferents are part of brainstem-cochlea reflex pathways that allow stimulus-related control of the cochlea and provide a way for the central nervous system to affect hearing at the most

①	**Reissner's membrane**	④	**Outer hair cells**
②	**Tectorial membrane**	⑤	**Organ of Corti**
③	**Inner hair cells**	⑥	**Inner border cells**

Fig. 8.5 Scanning electron micrographs of the organ of Corti of the guinea pig. (**A**) Oblique view into the scala media, vestibular (Reissner's) membrane ruptured (white scale bar 100 μm). (**B**) View on the apical surface of the organ of Corti illustrating the hair bundles of the inner and outer hair cells. (**C**) Similar view as (**B**) (white scale bar 10 μm). Note the gaps within the rows of V-shaped hair bundles of the outer hair cells as a likely result of an acoustic trauma. Courtesy Prof. G. Reiss, Witten/Herdecke University, Germany

peripheral neural level (Guinan Jr. 2018). The outer hair cells are thus involved in modulating sound amplification.

Upstream of the cochlea with the hair cells (internal ear) are the structures of the external and middle ear. The external ear comprises the auricle (see Fig. 8.6, see also Fig. 7.37) with the external acoustic opening, the joining external acoustic meatus [Meatus acusticus externus] and, as a septum to the air-filled middle ear (tympanic cavity), the tympanic membrane [Membrana tympanica].

On its internal side, this membrane is connected with the first of the auditory ossicles, the malleus. The malleus articulates with the incus and the incus with the stapes. The plate of the stapes is fixed at the fenestra ovalis which separates the fluid-filled vestibulum from the middle ear (see Fig. 8.6).

8.1.1.3 Course of the cochlear nerve

The axons of the ganglion cells in the spiral (cochlear) ganglion [Ganglion cochleare] constitute—together with some efferents (see above)—the cochlear nerve [N. cochlearis] which joins with the vestibular nerve [N. vestibularis] to form the vestibulocochlear nerve [N. vestibulocochlearis]. This cranial nerve enters the brainstem at the cerebellopontine angle (Fig. 8.7).

Then, auditory fibers of the vestibulocochlear nerve partially encircle the restiform body [Corpus restiforme] of the inferior cerebellar peduncle laterally ending in the cochlear nuclei (See Fig. 8.1). Cochlear nerve fibers bifurcate on entering the brainstem to terminate in both, dorsal and ventral cochlear nuclei.

8.1.1.4 Histology of the cochlear nuclei

Marked topographical order has been demonstrated in cochlear nerve terminals within the nucleus. Different parts of the spiral ganglion and differing stimulating sound frequencies are related to neurons that are tonotopically arranged in the ventral cochlear nucleus (ACN).

The ACN contains several neuronal types with distinct dendritic field characteristics leading to the description of

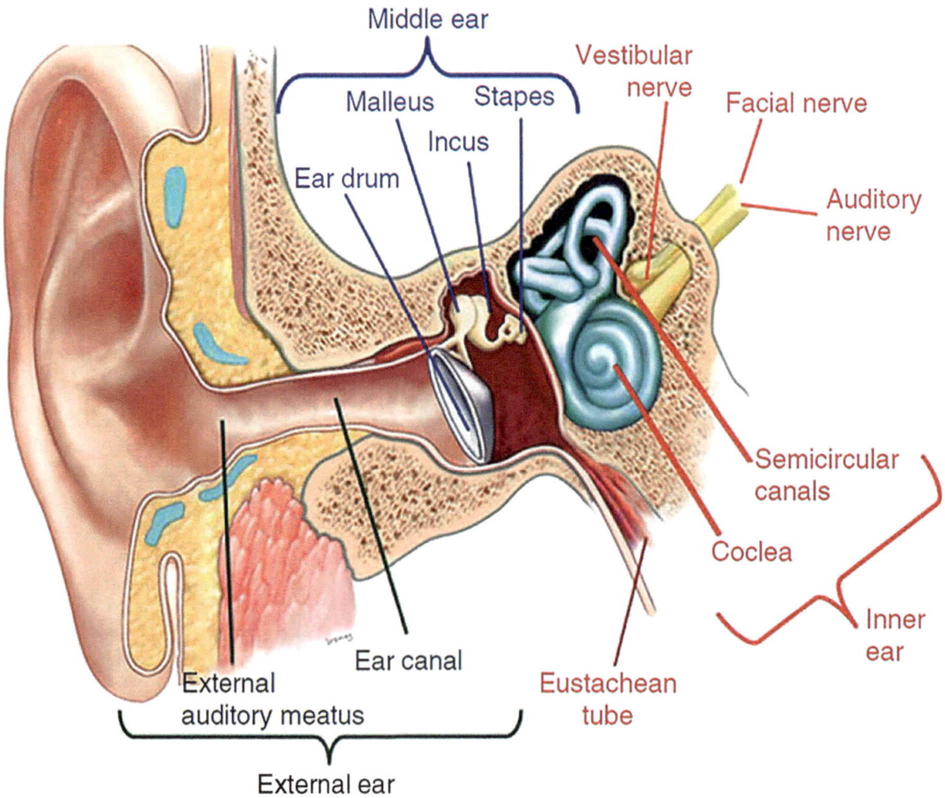

Fig. 8.6 Schematic view of the human ear. For the anatomy of the auricle or pinna, see Fig. 7.37. The external ear, in addition, comprises the external auditory meatus and the ear canal [Porus/Meatus acusticus externus] and reaches to the tympanic membrane [Membrana tympani] which belongs the middle ear [Auris media]. The latter houses the auditory ossicles [Ossicula auditus] of the middle ear together with the tympanic cavity [Cavitas tympani] and the auditory (Eustachean) tube [Tuba auditiva] (see 5.4.2.1.4 and Fig. 5.19). The ossicle chain is fixed to the tympanic membrane by the malleus, articulating with incus and stapes. The base of the stapes joins to the vestibular window [Fenestra vestibuli/ovalis], thereby establishing a mechanical connection to the perilymph-filled vestibulum of the inner ear [Auris interna] with cochlea and organ of equilibrium mainly the semicircular canals, see Figs. 5.4 and 5.5. The hair cells (see Figs. 8.4 and 8.5) of the inner ear are afferently connected with the auditory (cochlear) nerve and the vestibular nerve which together form the vestibulocochlear nerve. The cochlear window [Fenestra cochlearis/rotunda] (not shown here) is an opening at the end of the scala tympani, sealed by a membrane. Malleus (medieval Latin) = hammer, incus (Latin) = anvil; stapes (medieval Latin) = stirrup. From Avelar 2013, Fig. 14.1 with permission

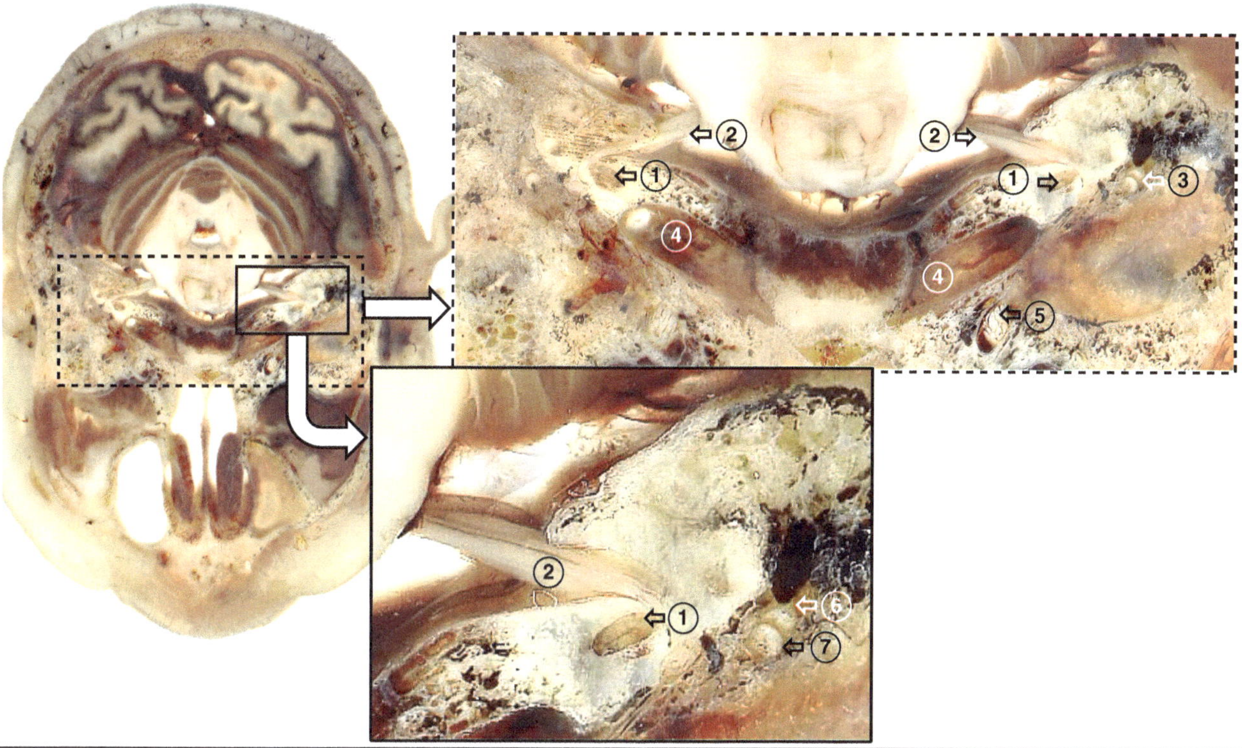

①	**Cochlea**	⑤	**Trigeminal nerve CNV**
②	**Vestibulocochlear nerve CNVIII**	⑥	**Incus**
③	**Tympanic cavity**	⑦	**Malleus**
④	**Internal carotid artery**		

Fig. 8.7 Horizontal plastinated section through the human head at the level of the cerebellopontine angle where the vestibulocochlear (②, CNVIII) and the facial nerve (CNVII) are entering/leaving the brainstem. They run in the internal acoustic meatus via the internal acoustic opening. The cochlear nerve is running to the cochlea ①. A dorsal view into the opened tympanic cavity shows the incus ⑥ and the malleus ⑦ (middle ear, see Fig. 8.6). (see also Sect. 5.3.1.1.5, Arteries of the inner ear). Sammlung des Zentrums Anatomie der Universität zu Köln

different neuron types. Due to its main neuron types, the ACN can be further divided into an anterior (AVCN) [Ncl. cochlearis anterior, pars anterior] and a posterior part (PVCN) [Ncl. cochlearis anterior, pars posterior].

The **AVCN** contains two main types of neurons known as spherical and globular bushy cells. They both receive direct inputs from the auditory nerve. Each spherical bushy cell receives inputs from one to four individual auditory nerve fibers via giant synaptic terminals called end-bulbs of Held (calyces of Held) (see Box 8.2). In contrast to the spherical bushy cells, the globular bushy cells in the AVCN receive their inputs from up to 40 auditory nerve fibers (Pickles 2015).

Box 8.2 Calyx of Held/Hans Held

The Calyx of Held is a particularly large synapse, described as the largest in the brain (Morest 1968), typical for some nuclei in the mammalian auditory pathway. Here, the large presynaptic part wraps around the postsynaptic perikaryon. It was named after Hans Held's (1893) "Faserkörbe" and is based on the fact that this presynaptic part of the synapse resembles a goblet (calyx) around the postsynaptic cell. This type of presynaptic endings provides decreased jitters of synaptic transmissions and is thus crucial for exact timing of neuronal signals. Hans Held, born in 1866, was a medical doctor who studied in Rostock and Leipzig (Germany). In 1899, he became a professor at the Institute of Anatomy at the University of Leipzig. Due to the exchange of academic teachers by the Nazi regime, he became professor emeritus in 1934 and died 1942 in Leipzig.

https://de.wikipedia.org/wiki/Hans_Held_(Mediziner)

Only a minor fraction of the cochlear neurons receives final terminals from the nerve. Terminals are limited to the anterocaudal region of the ventral nucleus where the neurons are probably mostly local interneurons (Standring 2016).

The AVCN releases fibers to the superior olivary complex via the ventral acoustic stria [Stria cochlearis anterior]. The spherical bushy cells project to the medial superior olivary nuclei of both sides and to the ipsilateral lateral superior olivary nucleus. In mammals, the globular bushy cells are known to project mainly to the contralateral medial nucleus of the trapezoid body and from there to the contralateral lateral superior olivary nucleus (see Fig. 8.3).

The **PVCN** is characterized by a large number of octopus cells. Octopus cells receive a very large number (>60) of auditory nerve afferents. These neurons are thought to be specialized for extracting the temporal fluctuations in complex broadband stimuli such as speech sounds (Pickles 2015).

Most PVCN fibers go into the intermediate acoustic stria [Stria cochlearis intermedia]. This small bundle of axons passes dorsally from the PVCN, superficial to descending trigeminal spinal fibers, cerebellar fibers in the restiform body, and axons of the dorsal cochlear nucleus. This stria swerves ventromedially across the midline, ventral to the medial longitudinal fasciculus.

The octopus cells project mainly to the ventral nucleus of the contralateral lateral lemniscus. Other neurons run to neurons surrounding the lateral superior olivary nucleus (periolivary nuclei). The PVCN also has two types of stellate (multipolar) cells. Next to local excitatory connections, they connect inhibitory to the inferior colliculi of both sides and to the ventral nucleus of the contralateral lateral lemniscus (type I) and to the contralateral cochlear nucleus (type II; Cant and Benson 2003).

The dorsal cochlear nucleus (**PCN**) is generally continuous with the ventral nucleus (see legend Fig. 8.1B, see Terr and Edgerton 1985), from which it is separated only by a thin stratum of nerve cell bodies and fibers.

Giant cells predominate within the PCN and their dendritic fields are aligned with the incoming auditory fibers (Standring 2016). Next to this, the PCN contains a multiplicity of cell types and has many interneurons (Pickles 2015). Mainly fusiform and giant cells of the PCN project to the inferior colliculi of both sides; giant cells also to the contralateral cochlear nucleus (Cant and Benson 2003). Their fibers curve dorsomedially round the restiform body toward the midline as the dorsal acoustic striae [Stria cochlearis posterior], ventral to the striae medullares.

Despite the relatively small PCN, the trapezoid body deriving from it, is large (see Fig. 8.8). This is in contrast to the relatively large ventral cochlear nucleus, of which the acoustic striae appear small.

8.1.1.5 Ascending auditory pathways

The next acoustic relay station within the brainstem following the cochlear nucleus is the **superior olivary complex** (SO) [Complexus olivaris superior] (Fig. 8.8A, B) which is situated in the tegmentum of the caudal pons, lateral in the reticular formation at the level of the pontomedullary junction. Accordingly, it is located between the facial nerve root laterally and the abducens nerve root medially (see Fig. 8.8; Amunts et al. 2012). In mammals, this complex includes several named nuclei and nameless smaller groups. The main nuclei are the medial and lateral superior olivary nuclei (Fig. 8.8A).

In humans, the medial superior olivary nucleus (**MSO**) [Ncl. olivaris superior medialis] is large and compact (see Fig. 8.8A, B). It has a flat, disk-like structure extending rostrocaudally for nearly 6 mm (Kulesza Jr 2007) and receives direct inputs from the AVCNs of both sides. There are approximately 15,500 neurons within the MSO and these neurons are mainly bipolar so that their contralateral inputs tend to contact the medial part of their adjacent dendritic arbor, while ipsilateral inputs tend to contact the lateral parts of the dendritic arbors.

The lateral superior olivary nucleus (**LSO**) [Ncl. olivaris superior lateralis] (Fig. 8.8A) has a more complex shape with medial and lateral segments. In humans, the nucleus is relatively small in comparison to the MSO. The LSO contains fusiform, round, and stellate neurons without a preferred orientation and dendrites confined to the body of the nucleus. Neuron counts of the human LSO show a total of approximately 5600 cells (Kulesza Jr 2007) which is nearly similar in absolute numbers to the nucleus in rodents and insectivores (Moore 2000; Amunts et al. 2012). In general, the mammalian LSO receives two major inputs. It receives a direct excitatory input from the ipsilateral AVCN and inhibitory input arriving from neurons in the nucleus of the trapezoid body (see below). The nucleus of the trapezoid body gets its input from the contralateral AVCN. This combination of excitatory and inhibitory input has been seen as the basic mechanism of LSO function so that the LSO has a selectivity based on sound intensity cues (Moore 2000; Pickles 2015).

Other nuclei within the superior olivary complex and the nucleus of the trapezoid body are regarded as relay stations in the mammalian ascending auditory projection. Interestingly, these intricate connections have not been definitively established in humans. According to the nuclei mentioned in the FIPAT Ch. 1 (2017) list, there is a nucleus of the trapezoid body [Ncl. corporis trapezoidei] as well as medial and lateral periolivary nuclei [Ncll. periolivares, nuclei mediales et laterales]. However, there is an ongoing debate about the existence and terminology of periolivary and trapezoid neuron groups due to, for example, the lack of calyxes of Held at perikarya in the potential nucleus of the trapezoid body (Olszewski and Baxter 1982; Moore et al. 1999; Moore 2000; Kulesza Jr 2008; Paxinos et al. 2019).

Interestingly, Schmitt et al. (2010) found perineural nets (PNNs) consisting of chondroitin sulfate proteoglycans that surround the perikarya of several neurons mainly in periolivary nuclei as well as in the ventral cochlear nucleus. Since PNNs may function to promote synaptic stability and prevent plasticity, the authors speculate that the according circuits may be important for human communication (Schmidt et al. 2010).

Neurons of the periolivary nuclei give rise to the **olivocochlear tract**. Moreover, some internuclear connections have been described (Standring 2016). Nevertheless, one of the so-called periolivary nuclei was constantly described in all mammals including humans: the **superior paraolivary nucleus**. It is situated dorsomedially of the MSO and is composed of large fusiform neurons and multipolar neurons with oval or fusiform somata, which are elongated rostrocaudally, and issue dendrites in the parasagittal plane (Kulesza Jr 2007). This nucleus gets its input mainly from contralateral PVCN (Magnusson and Gómez-Álvarez 2019).

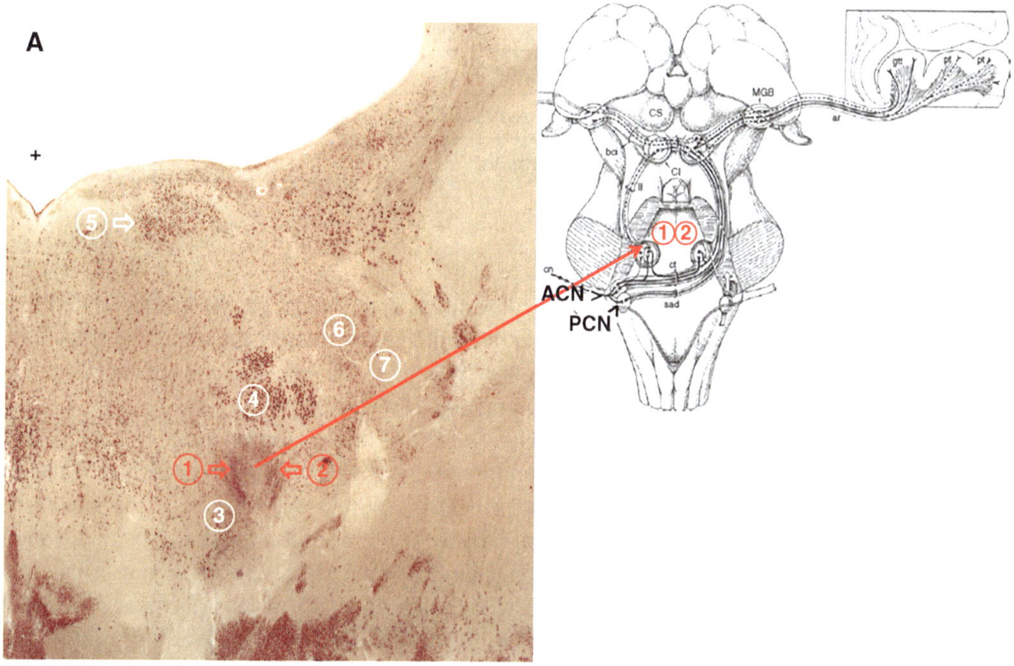

①	Medial superior olivary ncl.	⑤	Ncl. of abducens nerve CNVI
②	Lateral superior olivary ncl.	⑥	Spinal trigeminal ncl. CNV
③	Ncl. of trapezoid body	⑦	Spinal trigeminal tract
④	Motor ncl. of facial nerve CNVII	+	4th ventricle

Fig. 8.8 (**A**) Horizontal section through the human pons. Level of the rostral abducens nucleus. Darrow red staining. Just ventral to the facial nucleus ④ the characteristic horseshoe-shaped superior olivary complex (SO) can be seen. The main constituents of the SO here are the medial ① and lateral superior olivary nuclei ② as well as the nucleus of the trapezoid body ③. For additional histological views of the SO, see Fig. 8.12. See also atlas part Darrow red 19, 19 A. Graphic from ten Donkelaar 2011, Fig. 7.8 with permission. (**B**) Horizontal section through the human pons at the level of the genu of the facial nerve. Left hand side: Campbell fiber staining. Right hand side: Darrow red staining. The massive fiber bundle of the trapezoid body ① crossing the midline at the level of the superior olivary complex ② can clearly be seen in the Campbell fiber staining. See also atlas part Campbell 10A. LabPON Twente; graphic from ten Donkelaar 2011, Fig. 7.8 with permission

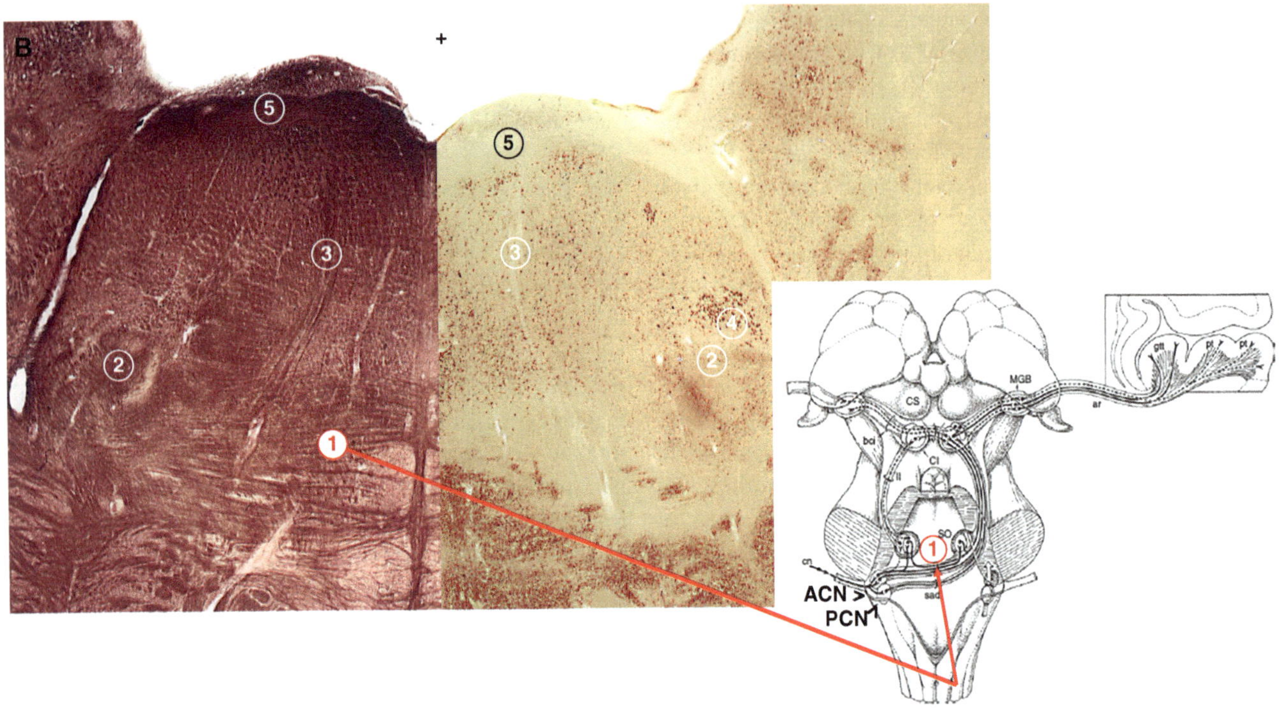

1	Trapezoid body	4	Facial ncl.
2	Superior olivary complex	5	Facial nerve
3	Abducens nerve	+	4th ventricle

Fig. 8.8 (continued)

Functions of the superior olivary complex

The **MSO** is thought to contribute to sound localization by comparing the timing of the stimuli at the two ears. For this, comparison of interaural time differences, including phase differences, is used. The accurate preservation of timing is necessary from both ipsilateral and contralateral inner ear via the cochlear nuclei to the medial and lateral dendrites of MSO neurons (Moore 2000). The MSO is innervated most of its length by axons from low-frequency regions (1.2–5 kHz) of the cochlear nuclei and its output reaches, accordingly, low-frequency region of the central nucleus of the inferior colliculus. This low-frequency bias may account for the fact that the MSO is the largest component of the human SO. Because of the relatively large head size of humans, for example, in relation to rodents, the range of low frequencies broadens that can be utilized as interaural phase difference cues and thus may contribute to the size prominence of the human MSO (Moore 2000).

In the **LSO**, intensities of the stimuli at the two ears are compared. Sounds coming from one side of the head will be more intense in the ipsilateral ear. Therefore, the excitatory/inhibitory interactions mean that neurons in the LSO preferentially respond to sounds on the same side of the head. Since the thresholds and tunings for the ipsilateral and contralateral effects are approximately the same, the cells of the LSO should be able to lateralize sounds in a frequency-dependent manner (Pickles 2015).

The route to the auditory midbrain

All auditory fibers reach the inferior colliculus via the lateral lemniscus (**LL**). Within this fiber tract, the nucleus of the lateral lemniscus is embedded which can be divided into a dorsal nucleus (**DNLL**) [Ncl. dorsalis lemnisci lateralis] (together with an intermediate nucleus; Paxinos et al. 2019) and a ventral nucleus (**VNLL**) [Ncl. ventralis lemnisci lateralis]. Both LL nuclei consist of small groups of 18,000–24,000 neurons that lie among the fibers of the lateral lemniscus. Axonal input comes mainly from the ipsilateral SO and the contralateral cochlear nucleus (for anatomical situation, see atlas part of the book).

DNLL and VNLL efferents enter the midbrain along the lateral lemniscus and terminate in the inferior colliculi (IC). Some efferent axons, however, travel in the cochlear nerves to the hair cells of the organ of Corti. These fibers partake in the olivocochlear tract (bundle of Rasmussen). Though few in number, they are involved in hearing, perhaps by modulating sensory transduction through reflexes via cochlear nuclei. The neurons of origin are located at the hilus and along the lateral border of the lateral superior olivary nucleus and lateral edge of the ventral trapezoid nucleus (Standring 2016).

The DNLL contains small soma-sized neurons which are oriented in parallel to the fibers. Additionally, large neurons with triangular or multipolar soma shapes are scattered within. The DNLL sends GABAergic projections to the contralateral central nucleus of the IC where axons form layers. The structure of the VNLL is dominated by Kv3.1b-positive axonal terminals which are arranged in parallel bands (Amunts et al. 2012).

All auditory fibers merge in the inferior colliculus (**IC**) [Colliculus inferior] either ipsilaterally or contralaterally. The IC is a prominent external feature along the posterior aspect of the mammalian midbrain (Fig. 8.9, see Fig. 15.5).

Based on cytoarchitectural features and projection patterns, the IC can be divided into three subregions: the central nucleus [Ncl. centralis], the dorsal cortex [Ncl. pericentralis] mediodorsally of the central nucleus, and the external cortex [Ncl. externus] laterodorsally of the central nucleus. The human central nucleus of the IC includes approximately 420,000 neurons mainly with dendrites arranged in a stellate pattern with individual dendritic branches that may course parallel or perpendicular to fibrodendritic laminae (Mansour et al. 2019). Thus, the central nucleus has a higher neuron number than the cortices of the IC.

Along the ventrolateral and ventromedial aspect of the central nucleus, the fibers of the lateral lemniscus divide in a Y-shaped manner surrounding the central nucleus and providing its separation from the periaqueductal gray, reticular formation and the external cortex of the IC. The majority of neurons in the external cortex of the IC is smaller than those in the central nucleus and generally has a long axis parallel to the surface of the IC. On the other hand, the dorsal cortex of the IC has a low density of myelinated axons and neurons without clear orientation.

The central IC nucleus receives projections from the contralateral cochlear nucleus, both LSOs, ipsilateral MSO, and the nuclei of the lateral lemniscus. The projections to the

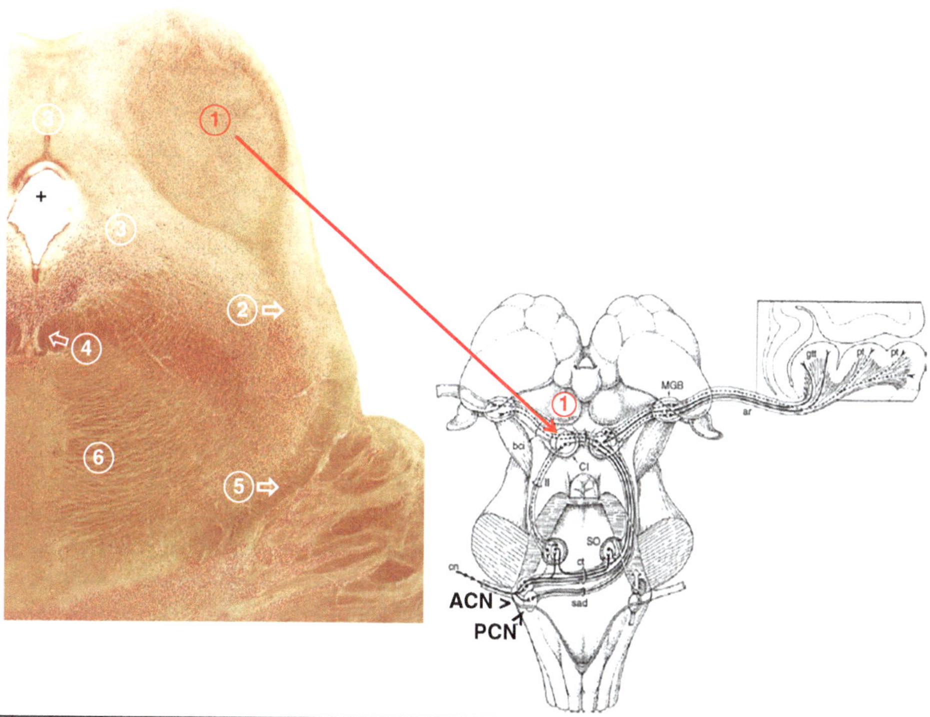

①	**Inferior colliculus**	⑤	**Medial lemniscus**
②	**Lateral lemniscus**	⑥	**Decussation of the superior cerebellar peduncles**
③	**Periaqueductal gray**	+	**Mesencephalic aqueduct**
④	**Medial longitudinal fasciculus**		

Fig. 8.9 Horizontal section through the human mesencephalon at the level of the inferior colliculi ① (collis Latin = hill, colliculus dim. = small hill). Darrow red staining. The lateral lemniscus ② transfers fibers from the cochlear nuclei to the inferior colliculi into which it leads at the collicular ventral circumference. The IC are located directly below the ependymal surface, their medial neighbor is the periaqueductal gray ③. Together with the superior colliculi (see Figs. 8.10 and 15.3) the IC form the quadrigeminal plate at the dorsal surface of the midbrain. See also atlas part Darrow red 34, 34A. LabPON Twente; graphic from ten Donkelaar 2011, Fig. 7.8 with permission

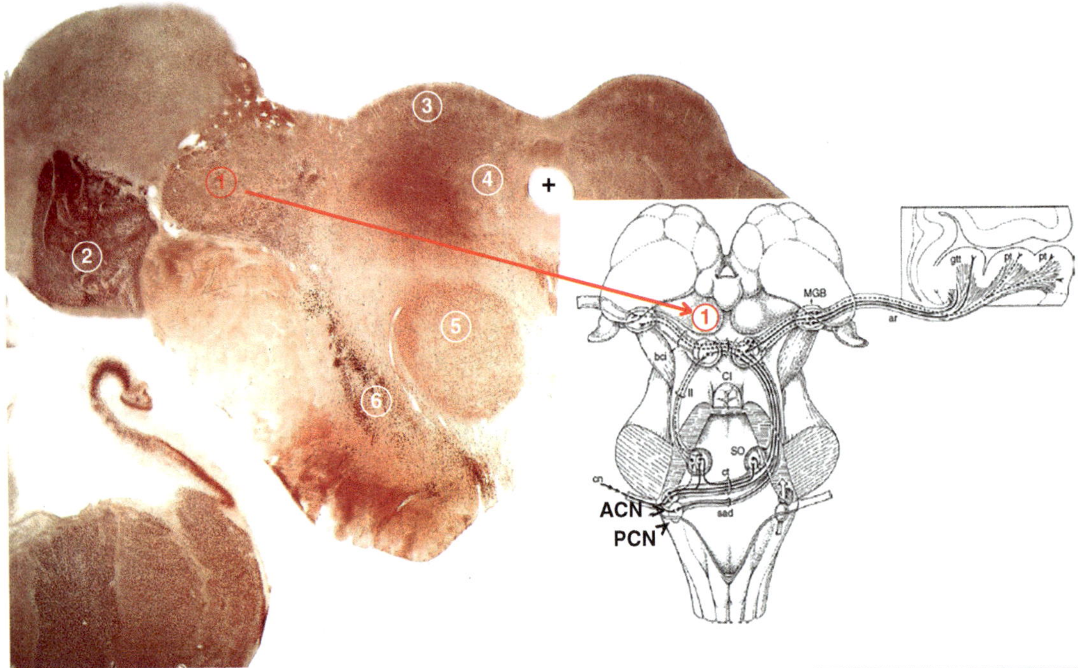

①	Medial geniculate body	⑤	Red ncl.
②	Lateral geniculate body	⑥	Substantia nigra
③	Superior colliculus	+	Mesencephalic aqueduct
④	Periaqueductal gray		

Fig. 8.10 Horizontal section through the human mesencephalon at the level of the superior colliculi ③ and the medial geniculate body (MGB) [Corpus geniculatum mediale] ①, part of the thalamus. Darrow red staining. The brachium of the inferior colliculi (bci) projects to the MGB whose axons run to the primary acoustic cortex. Do not confuse the MGB with the lateral geniculate body (LGB) [Corpus geniculatum laterale] ②, the thalamic endpoint of the visual nerve from the retina. This potential confusion is more of a linguistic problem since—as you can see here clearly—the rather homogenous interior of the MGB is quite different from the lamellate structure of the LGB. See also atlas part Darrow red 41. LabPON Twente; graphic from ten Donkelaar 2011, Fig. 7.8 with permission

external cortex, however, originate in the contralateral PCN and bilaterally from the DNLLs and central IC nuclei (Amunts et al. 2012). Left and right ICs are interconnected via a commissure [Commissura colliculi inferioris] crossing dorsal to the aqueduct. The commissural axons originate in the dorsal nucleus. However, the main efferent tract of the IC is the brachium of the CI [Brachium colliculi inferioris] terminating in the medial geniculate body [Corpus geniculatum mediale] (Fig. 8.10).

8.1.1.6 Descending auditory pathways

The central nucleus of the IC is an important component of the descending auditory pathway and receives centrifugal input from the auditory cortex and the medial geniculate body (Pickles 2015; Mansour et al. 2019). The IC, in turn, sends efferents to the superior olivary complex and the cochlear nucleus which also gets input from the superior olivary complex.

As described above, the olivocochlear system sends fibers from the superior olivary nuclei to the organ of Corti in the cochlea. One component ends on the outer hair cells to reduce the degree of active mechanical amplification of the traveling sound wave. A second component ends on or near the synaptic terminals of the afferent type I fibers on the inner hair cells. It inhibits auditory responses at the neural level. Although animal experiments have shown that the olivocochlear fibers may have an effect in suppressing noise-induced hearing loss, these effects are exceedingly small if existing in the human (Pickles 2015).

8.1.1.7 Living anatomy and clinical implications

With the combination of two straightforward tests using a tuning fork one can distinguish whether a patient may have hearing problems due to conduction hearing loss or due to neural hearing loss. Conduction hearing loss is the result of malfunction of the outer or middle ears while neural, i.e.,

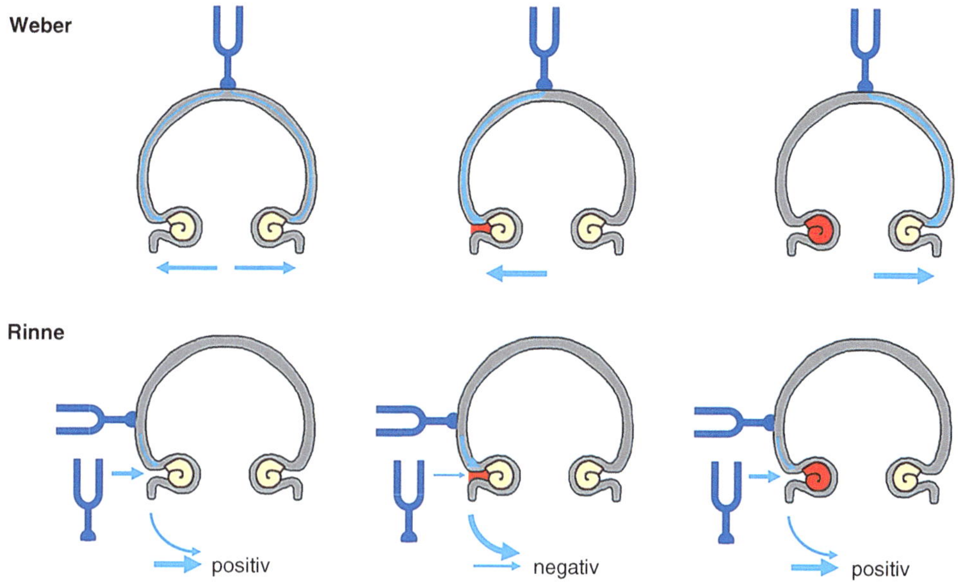

Fig. 8.11 Schematic illustration of the Weber and Rinne tests. First, the Weber test reveals a potential lateralization of bone sound conduction and then the Rinne test whether the lateralization is caused by conductive hearing loss or neural hearing loss. For details, see text below. Modified after Huggenberger et al. 2019, Fig. 14.12 with permission

sensorineural, hearing loss is due to problems in the inner ear or the cochlear nerve.

For the hearing test after Weber, a vibrating tuning fork is placed in the midline on top of the head (bone conduction test). The healthy patient has equal-sided hearing, not lateralized. However, if a patient has conduction hearing loss (Fig. 8.11: red filled meatus), the sound of the tuning fork is louder at the diseased ear. If a patient has neural hearing loss (Fig. 8.11: red filled cochlea), the sound is louder at the healthy ear. To decide now, which one of the two potential problems is the case, the test after Rinne is used. With this test, bone and air sound conductions are compared. A vibrating tuning fork is placed on the mastoid process (Processus mastoideus] behind the ear. The patient should tell when he or she can no longer hear the sound via this bony conduction. Without striking the tuning fork again, the vibrating fork is then held in front of the external auditory canal on the same side (air conduction). The normally hearing patient hears the sound again via the air conduction—"Rinne positive" for this ear. Then the test is repeated for the other ear. However, if the patient with the tuning fork in front of the ear no longer hears the sound, he or she is "Rinne negative" (Rea 2015). This indication speaks for conductive hearing loss. One must keep in mind that the Rinne test can also be positive if the inner ear or cochlear nerve is damaged (neural hearing loss). Thus, both tests together, Weber and Rinne, can locate the side of the hearing impairment.

A noninvasive technique to assess the function of the human auditory brainstem is provided by evoked potentials, i.e., the auditory brainstem response (ABR). ABR is assumed to be generated by synchronized activity of neurons in the ascending auditory pathway. Within the first 10 ms of an ABR recording after a stimulus, there are six typical waves in the voltage trace. It is thought that wave I is generated in the cochlea, while wave II arises from the cochlear nerve. Wave IV may be generated by axons passing directly from CN to the contralateral IC, wave V by axons following the same route but with a delaying synapse within. Thus, ABR results support the idea that two main auditory pathways exist within the brainstem, a synaptic one between the CN and the IC and one with an interposed synapse within the SO (Moore 2000).

The location of the dorsal cochlear nucleus at the lateral periphery of the brainstem (see Fig. 3.9 A) is the basis for experimental and clinical approaches for the surgical treatment of deafness due to sensorineural hearing impairment by an auditory brainstem implant (ABI). Sensorineural hearing impairment may be caused by damage to the cochlea or the auditory nerve due to, for example, neurofibromatosis type 2 (see page 88) or other conditions. The ABI is similar in design and function to a cochlear implant. As such, the ABI is connected to an external microphone and consists of the signal processing unit and an electrode array. This electrode array is arranged on a small paddle which is placed on the external surface of the cochlear nucleus. The main target of the electrodes is the dorsal cochlear nucleus (Mahboubi et al. 2020). Thus, the superficial location of the cochlear nucleus at the lateral recess (cf. Fig. 3.9) is the prerequisite to place the electrodes without damaging the brainstem. However, in contrast to the cochlea, the cochlear nucleus has not a single linear tonotopic organization but different tonotopic subunits. Accordingly, as the ABI electrode array is placed along the surface of the cochlear nucleus, each electrode likely activates a variety of neuron types with different characteristics. Nevertheless, patients with ABI have variable benefit with regard to sound, but speech comprehension is not as good as with cochlear implants and ABI users often experience side effects due to stimulation of nonauditory axons of passage (Vincent 2012; Tarabichi et al. 2019). In patients performing poorly with ABI or having damage to the brainstem region, auditory midbrain implants (AMIs) provide electrical stimulation higher within the auditory pathway. AMI directly stimulates the inferior colliculus. So far, however, patients with AMI have only been able to achieve auditory sensations, but no independent speech understanding (Mahboubi et al. 2020).

Interestingly, children with autism spectrum disorder (ASD) are known to have auditory dysfunctions including deafness, hyperacusis, difficulty listening with background noise, and problems encoding speech sounds. At least some of these dysfunctions like hypersensitivity result from problems in auditory stimuli perception at the brainstem level. Namely, the networks of the SO seem to be affected by developmental disorganization. The MSO of persons with ASD are significantly smaller and have fewer and smaller neurons in comparison to control persons (Mansour and Kulesza Jr 2020; Smith et al. 2019). In addition, the majority of MSO neurons were abnormally oriented showing round to oval somata (instead of bipolar shapes; see above). Moreover, there are clusters of ectopic neurons in the pons of ASD persons which are thought to be displaced neurons belonging to the SO (Smith et al. 2019).

8.1.2 Vestibular r6

8.1.2.1 Medial vestibular nucleus r6

For details on this nucleus, see Sect. 7.2.1.1

8.1.2.2 Spinal vestibular nucleus r6

For details on this nucleus, see Sect. 5.3.1.1

8.1.2.3 Interstitial nucleus of the vestibulocochlear nerve [Ncl. interstitialis n. vestibulocochlearis]

The interstitial nucleus of the vestibular nerve comprises clusters of large nerve cells distributed within the vestibular component of vestibulocochlear nerve fibers in man, macaque and rat, visible in Nissl-stained sections and with stains for acetylcholinesterase. The nucleus is not listed in FIPAT Ch.1 (2017).

http://braininfo.rprc.washington.edu/centraldirectory.aspx?ID=719

8.1.3 Monoamine nuclei r6

8.1.3.1 A5 noradrenaline cells (NA5)

The A5 noradrenergic cell group in humans is located in the lower pons (see also Sect. 11.2.3). From studies in rats and mice (see here below) it is known that their axons project to the spinal cord.

In mice, the noradrenergic neurons of the A5 group—together with the A7 group being part of a cell column—project predominantly contralaterally to the spinal cord (VanderHorst and Ulfhake 2006). While the A5 cell group is involved in cardiovascular reflexes, the A6 group mediates somatic motor reflexes by mainly ipsilateral projections.

8.1.3.2 A2 noradrenaline cells (NA2)

Details on this cell group are provided under Sect. 3.2.1.1

8.1.4 Trigeminal sensory nuclei r6

8.1.4.1 Spinal trigeminal nucleus, interpolar part

Details on this nucleus are provided under Sect. 3.2.2.2.

8.1.5 Solitary nuclei r6

Details on this nucleus are provided under Sect. 3.2.4

8.2 Basal r6

8.2.1 Facial motor r6 (FIPAT: Motor nucleus of facial nerve) [Ncl. nervi facialis] (facies Latin = face)

As a branchial nerve (second branchial arch, see Sect. 2.8), the facial nerve comprises the somatomotor innervation of the mimic muscles (motor ncl. of facial nerve) and the visceromotor, parasympathetic innervation of the lacrimal gland [Gld. lacrimalis] (Fig. 8.16), the mucous glands of the nose (for ultrastructural details on their innervation, see Cauna et al. 1972), and the small salivatory glands (sublingual and submandibular glands [Gld. submandibularis/sublingualis] (superior salivatory nucleus [Ncl. salivatorius superior] see Sect. 7.3.1.1.1).

Human nasal mucosa has currently come into the limelight related to the intranasal delivery of a SARS-CoV-2 vaccine (see for example, Hassan et al. 2020, general review by Gizurarson 2012).

Afferents from the taste receptors of the tongue reach the nuclei of solitary tract by joining the facial nerve (Sect. 3.2.4.1). Finally, a small somatosensory supply in the ear region is included (spinal ncl. of trigeminal nerve).

The motor nucleus of the facial nerve as well as the superior salivatory nucleus (see Sect. 7.3.1.1.1) are located in a midpontine position (see Fig. 8.12). The solitary nuclei lie in the medulla oblongata (see Sect. 3.2.4.1).

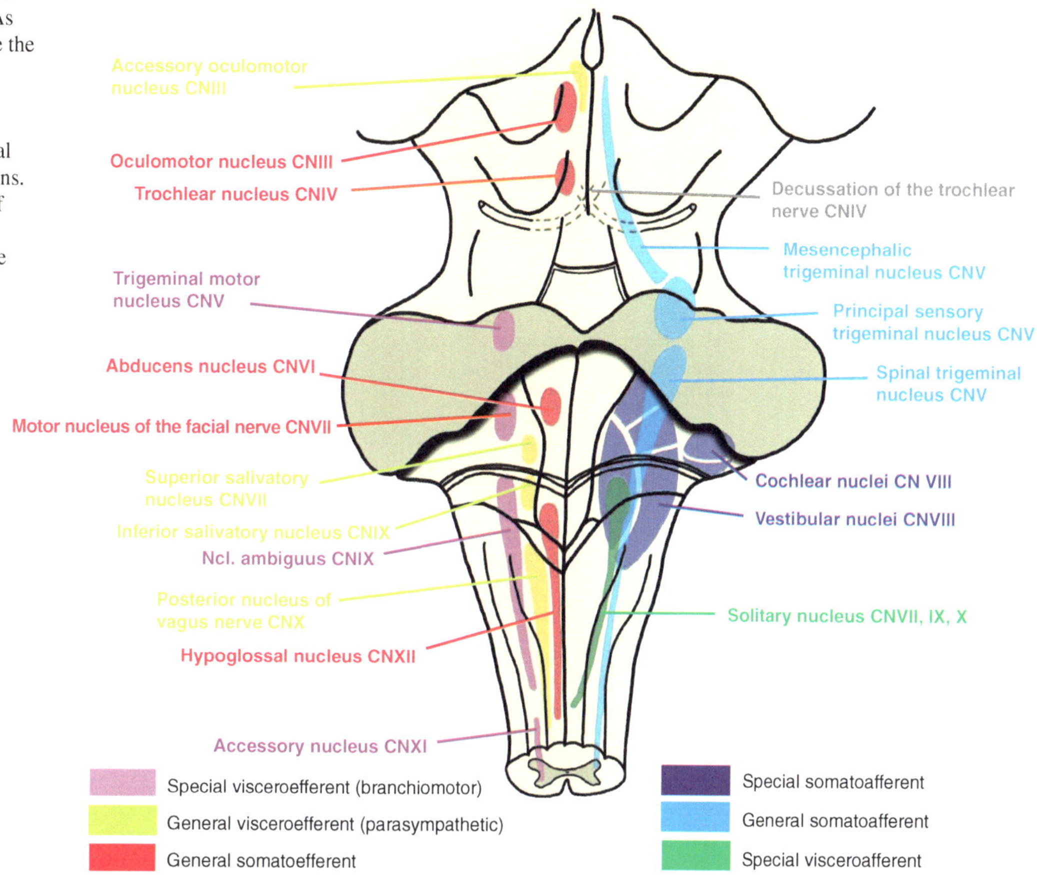

Fig. 8.12 Schematic representation of the human brainstem cranial nuclei. As the second branchial nerve the facial nerve unites special visceroefferent, general visceroefferent, special visceroefferent, and general somatoafferent contributions. Note that due to reasons of space not in all cases the current FIPAT termini have been used. Modified after Huggenberger et al. 2019, Fig. 15.5 with permission

8.2.1.1 Location, histological appearance, and microanatomy of the motor nucleus of facial nerve [Ncl. nervi facialis]

The nucleus is located directly medial of the spinal trigeminal nucleus (see Figs. 8.12 and 8.13) and extends from the transition between medulla oblongata and pons (Fig. 8.13A) to the cerebellopontine angle (see Fig. 8.13D).

Inside the facial nucleus several subnuclei (s. Fig. 8.13) can be differentiated. In caudocranial order the extent of the facial nucleus decreases and not all subnuclei are visible in all sections hitting the nucleus of facial nerve. The largest number of subnuclei is detectable at the caudal most level (see Fig. 8.13A).

FIPAT calls the facial nucleus "motor nucleus of facial nerve" but does not list any subnuclei. The identification of subnuclei shown in this book is based on Olszewski and Baxter (1982) and Arzberger et al. (1994). It is important to note that because of the lack of standardization no internationally accepted terminology for the human facial subnuclei exists which may lead to confusion in particular when it comes to reading historical papers.

As with the spinal and cranial nerve motoneurons, the neurons of the motor nucleus of the facial nerve can unequivocally be identified by using the cholinergic marker choline acetyltransferase (ChAT) (see Fig. 8.14, Arzberger et al. 1994).

A characteristic microanatomical feature of the facial nucleus is the so-called facial colliculus, a bilateral prominence in the floor of the fourth ventricle (rhomboid fossa [Fossa rhomboidea]) (Fig. 8.2). This bulge is due to the turning fibers of the facial nerve (inner knee of the facial nerve) around the nucleus of abducens nerve (see Figs. 8.8, 8.13, and 8.15).

The facial nucleus is located dorsal of the superior olivary complex and medial of the spinal nucleus of trigeminal nerve (Fig. 8.13). The axons of the facial motoneurons are running toward the floor of the fourth ventricle (Fig. 8.13). There, they bend laterally surrounding partially the abducens nucleus (Fig. 8.15).

Subsequently, the fibers are running ventrally to eventually reach their exit zone in the cerebellopontine angle (Fig. 8.17). The peculiar course of the facial nerve inside the brainstem is due to the fact that the facial motor neurons are originally generated in rhombomere 4 (Chap. 10) and then migrate caudally to rhombomere 6. As a result, the CNVII nucleus gains a position caudal to that of the CNVI nucleus, the facial branchiomotor axons ascending to their exit in rhombomere 4 surrounding the abducens nucleus.

A

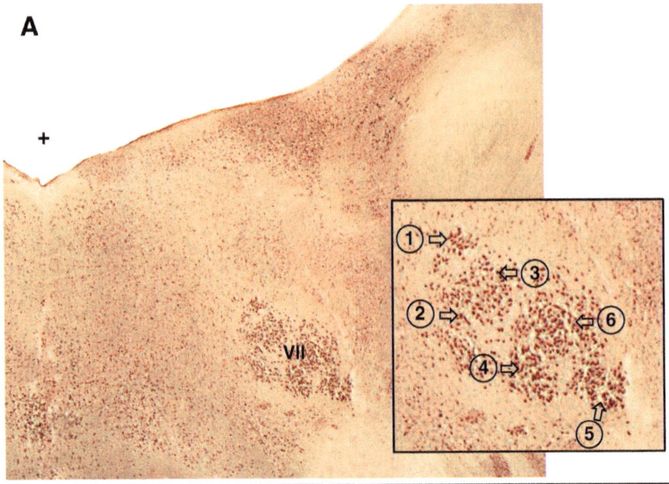

	Motor ncll. of facial nerve CNVII		
1	Dorsal subncl.	5	Ventrolateral subncl.
2	Medial subncl.	6	Ventral subncl.
3	Intermediate subncl.	+	4th ventricle
4	Ventromedial subncl.		

B

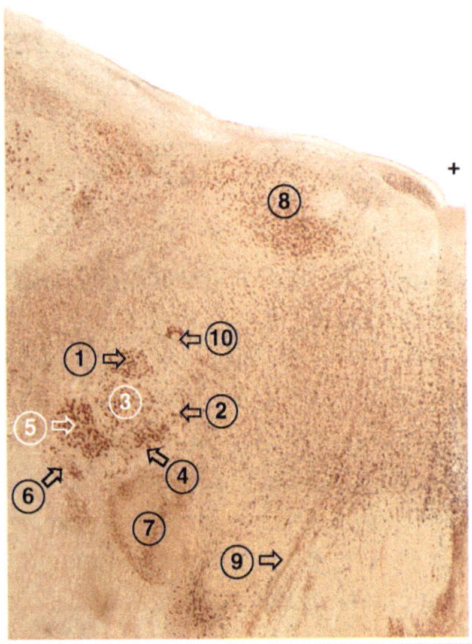

	Motor ncll. of facial nerve CNVII		
1	Dorsal subncl.	7	Superior olivary complex
2	Medial subncl.	8	Abducens ncl. CNVI
3	Intermediate subncl.	9	Abducens nerve CNVI
4	Ventromedial subncl.	10	Retrofacial ncl.
5	Ventral subncl.	+	4th ventricle
6	Ventrolateral subncl.		

Fig. 8.13 Five histological horizontal sections through the human brainstem from the level of the medullopontine transition area (**A**) to the cranial end of the abducens nucleus (**E**) (Darrow red staining). The highest number of subnuclei is present in the caudal part while it decreases in cranial direction. See also atlas part Darrow red 17–21. LabPON Twente. Although it is probable that they reflect a certain somatotopic arrangement, the exact correlation of subnuclei with individual mimic muscles (see Figs. 8.25 and 8.26) has not yet been achieved in humans, in contrast to experimental animals like mice (see Schröder et al. 2020)

C

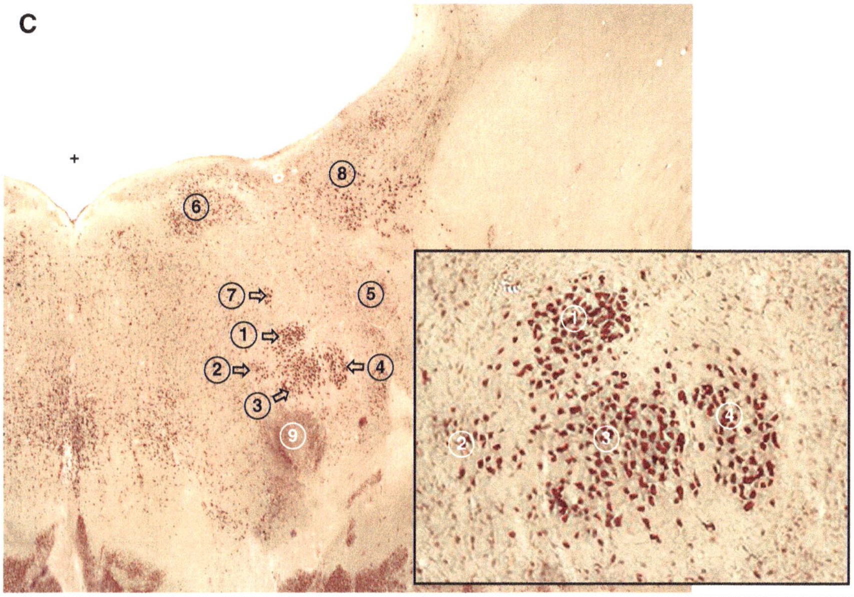

Motor ncll. of facial nerve CNVII			
1	Dorsal subncl.	5	Spinal trigeminal ncl. CNV
2	Medial subncl.	6	Abducens ncl. CNVI
3	Intermediate subncl.	7	Retrofacial ncl.
4	Ventral subncl.	8	Vestibular ncll. CNVIII
		9	Superior olivary complex
		+	4th ventricle

D

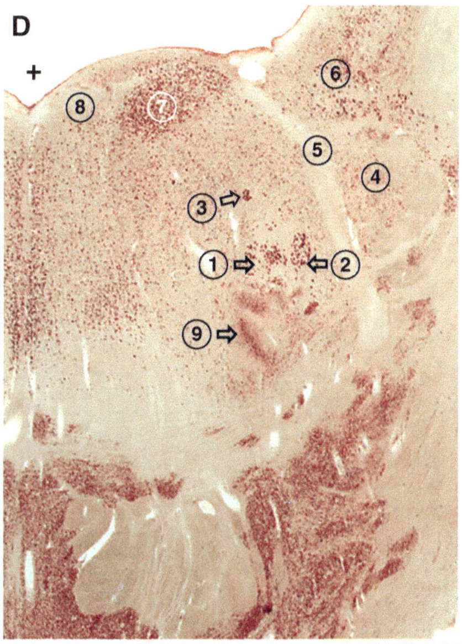

Motor ncll. of facial nerve CNVII			
1	Intermediate subncl.	4	Spinal ncl. of trigeminal nerve CNV
2	Ventral subncl.	5	Facial nerve / descending part
3	Retrofacial ncl.	6	Vestibular ncll. CNVIII
		7	Ncl. of abducens nerve CNVI
		8	Inner knee of facial nerve
		9	Superior olivary complex
		+	4th ventricle

Fig. 8.13 (continued)

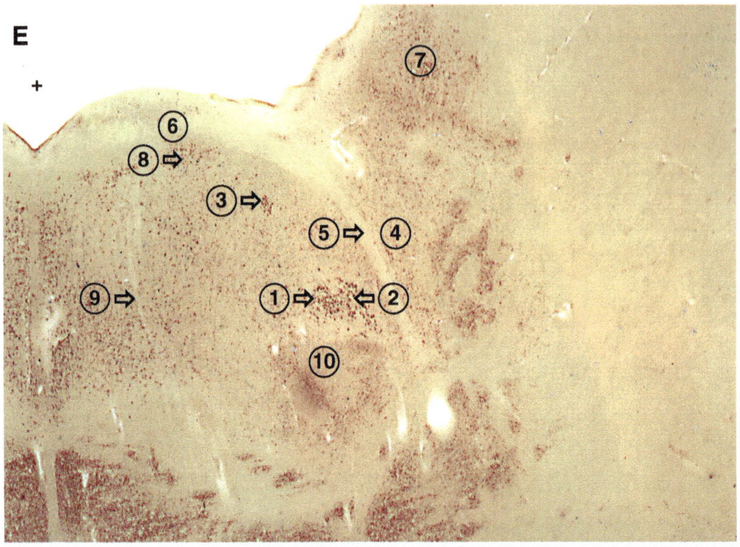

Motor ncll. of facial nerve CNVII			
①	Intermediate subncl.	③	Retrofacial ncl.
②	Ventral subncl.	④	Spinal ncl. of trigeminal nerve CNV
		⑤	Facial nerve / descending part
		⑥	Genu of facial nerve
		⑦	Vestibular ncll. CNVIII
		⑧	Ncl. of abducens nerve CNVI
		⑨	Abducens nerve CNVI
		⑩	Superior olivary complex
		+	4th ventricle

Fig. 8.13 (continued)

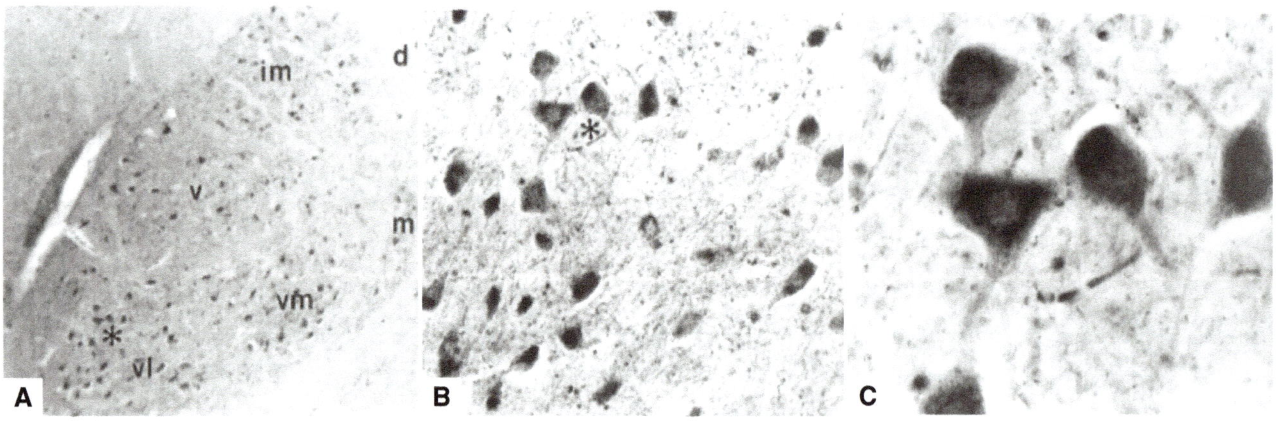

Fig. 8.14 Histological horizontal section through the human brainstem at the level of the facial nucleus. Choline acetyltransferase (ChAT)-immunohistochemistry. The survey microphotograph (**A**; magnification 30×) shows the organization of the human CNVII nucleus (see Olszewski and Baxter 1982) consisting of 6 subnuclei (compare with Fig. 8.13A) with the majority of perikarya being ChAT-immunoreactive. At higher magnification (**B**; 120×), individual immunostained neurons can be identified as mainly large motoneuron-like cells. At high magnification (**C**; 300×), the immunoprecipitate is seen decorating the cytoplasm, sparing the nucleus. *vl* ventrolateral subnucleus, *vm* ventromedial subnucleus, *m* medial subnucleus, *im* intermediate subnucleus, *v* ventral subnucleus, *d* dorsal subnucleus, *indicates the area shown enlarged in (**C**). From Arzberger et al. 1994, Fig. 1a–c with permission

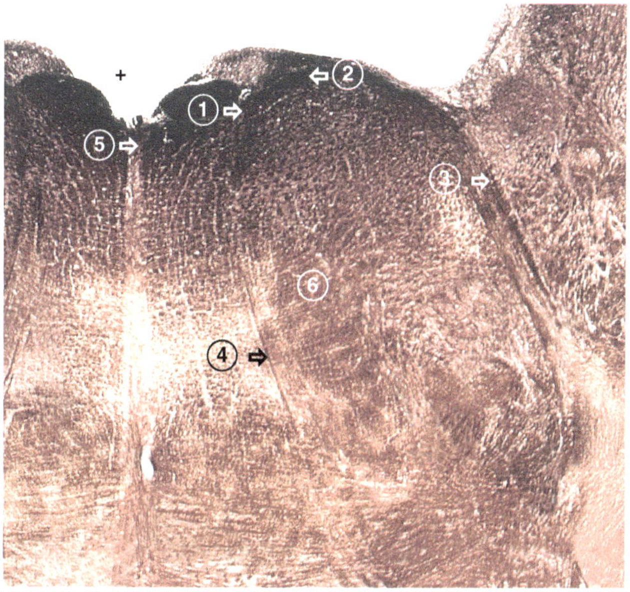

①	Facial nerve CNVII, ascending part	⑤	Medial longitudinal fasciculus
②	Genu of facial nerve CNVII	⑥	Central tegmental tract
③	Facial nerve CNVII, descending part	+	4th ventricle
④	Abducens nerve CNVI		

Fig. 8.15 Horizontal section through the human brainstem at the level of the inner knee of the facial nerve ②. Campbell stain. The ascending portion of CNVII ① is visible lateral of the medial longitudinal fasciculus ⑤. Close to the floor of the fourth ventricle the fiber bundle turns laterally forming the inner knee of CNVII around the nucleus of CNVI (see Fig. 8.13C, D). Subsequently, the nerve turns ventrally on its way to the exit in the cerebellopontine angle (see Fig. 8.17). See also atlas part Campbell 9, 9B. LabPON Twente

The facial nucleus gives rise to the branchiomotor fibers innervating the muscles derived from the second branchial arch. In mice, the facial nucleus is located beneath the ventral surface of the pons stretching from the caudal end of the dorsal cochlear nucleus to the caudal end of the principal sensory trigeminal nucleus. The nucleus is enclosed in a semicircular fashion by the perifacial zone. For the BALB/c mouse, a detailed description was provided by Ashwell (1982). In this mouse strain, the nucleus can be subdivided into six subnuclei and the dorsal accessory nucleus of the facial nerve. The arrangement of the subnuclei is musculotopic or myotopic, i.e., individual subnuclei innervate defined facial muscles. (Ashwell 1982).

8.2.1.2 Schematic representation of the supply area

The facial nerve has a very extended supply area. As you can see in Fig. 8.16, the following qualities are contained in the facial nerve:

Branchiomotor/Special visceroefferent (Ncl. of origin: Motor ncl. of the facial nerve (MN, see also Fig. 8.13): Mimic muscles (Motor root, see legend Fig. 8.16 for the different branches), Stapedius muscle via the stapedius nerve (SN) (see Sect. 8.2.1.4).

The special visceroefferent and special visceroafferent (see below) parts are often designated as the intermediate nerve [N. intermedius].

Special visceroefferent (Ncl. of origin: Superior salivatory nucleus, SSN) Lacrimal (see Fig. 8.20), submandibular and sublingual glands (see Figs. 8.21 and 8.28) as well as mucosal glands of the nasal cavity.

Innervation of the lacrimal gland is provided via the preganglionic parasympathetic fibers to the pterygopalatine ganglion. The postganglionic fibers are running through the inferior orbital fissure and join the communicating branch with the lacrimal gland of the zygomatic nerve, and directly via the lacrimal nerve (see Fig. 8.20).

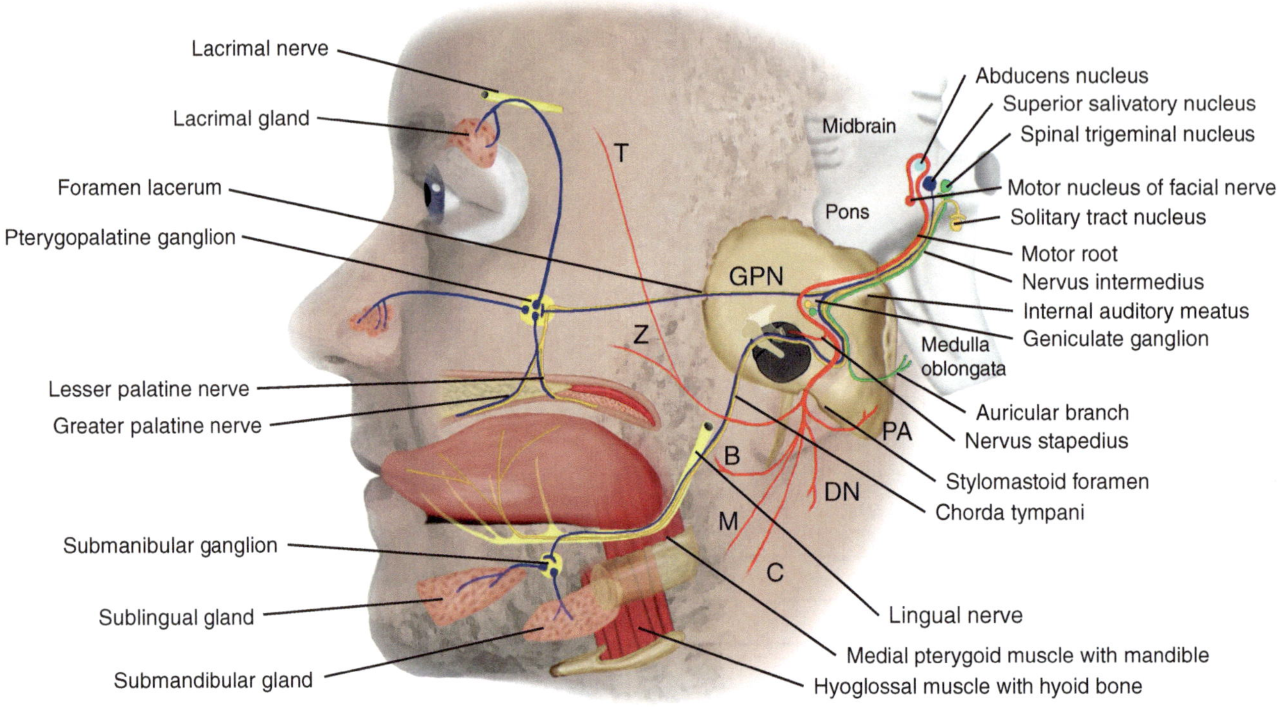

Fig. 8.16 Schematic illustration of the head organs innervated by branches of CNVII. Facial muscles: *B* Buccal branches, *C* Cervical branches (Platysma), *M* Marginal mandibular branches, *PA* Posterior auricular nerve, *T* Temporal branches, *Z* Zygomatic branches. Lacrimal gland: Superior salivatory nucleus > Intermediate nerve > Greater petrosal nerve (GPN) > Nerve of the pterygoid canal (PC) > Pterygopalatine ganglion > Lacrimal nerve (V_1). Gustatory receptors (anterior 2/3 of tongue): Lingual nerve (V_3) > Chorda tympani > Facial nerve > Nuclei of solitary tract. DN, Digastric nerve. Leander Huggenberger fecit Modified acc. to Takezawa et al. 2018, Fig. 2 with permission

Innervation of the submandibular and sublingual glands is realized via the preganglionic fibers in the chorda tympani (CT) via the lingual nerve to the submandibular ganglion and the postganglionic fibers to both glands.

Special visceroafferent (Receiving nucleus: Ncll. of solitary tract): Taste (sweet, salty, bitter, sour, umami) information from the taste receptors of the tongue via the chorda tympani (CT) (see Sect. 3.2.4.1.2).

General visceroafferent (Receiving nucleus: Spinal nucleus of trigeminal tract see Sect. 3.2.2.2), Sensory innervation of parts of external acoustic meatus (auricular branch).

Note that the parotid gland is not innervated by the facial nerve—although the nerve traverses the gland (see Fig. 8.22)—but via the glossopharyngeal nerve (see Sect. 5.4.2.1.3) which relays preganglionic parasympathetic fibers from the inferior salivatory nucleus (see Sect. 7.3.1.1.1) to the otic ganglion. Postganglionic fibers from this ganglion then join the auriculotemporal nerve (see Fig. 8.22, see auriculotemporal nerve V_3, Sect. 12.2.3.2.3 "Mandibular nerve") to innervate the parotid gland (Ghannam and Singh 2020).

8.2.1.3 Anatomy of the intradural, intraosseous, and peripheral course of the facial nerve

The facial nerve exits the brainstem in the cerebellopontine angle (first segment of the peripheral course, see Fig. 8.17), traverses the prepontine cistern (Box 3.3), enters the skull base via the internal acoustic opening and meatus, and leaves it via the stylomastoid foramen (see Fig. 8.18 ①, ②), easy to identify by the prominent styloid process (Fig. 8.18). The

A

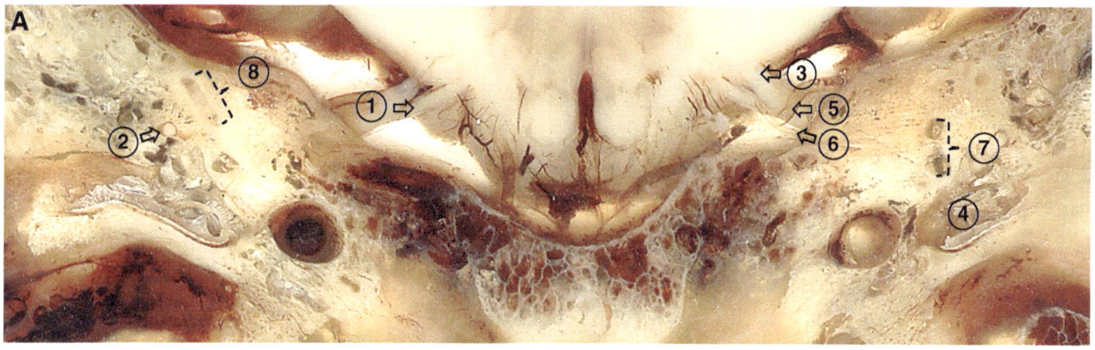

B

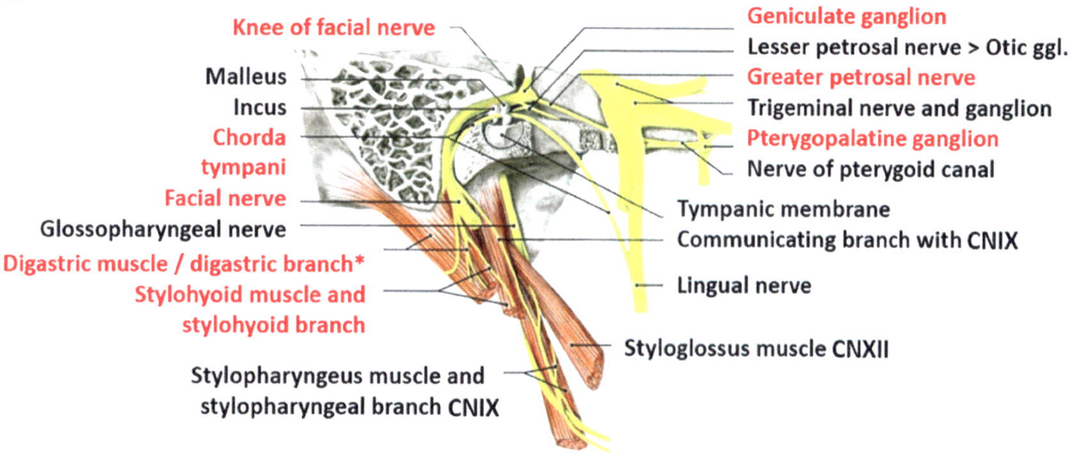

①	Facial nerve CNVII Exit in the cerebellopontine angle	⑤	Anterior inferior cerebellar artery
②	Facial nerve CNVII Course in the skull base	⑥	Labyrinthine artery
③	Vestibulocochlear nerve CNVIII	⑦	Anterior semicircular canal
④	Tympanic cavity	⑧	Posterior semicircular canal

C

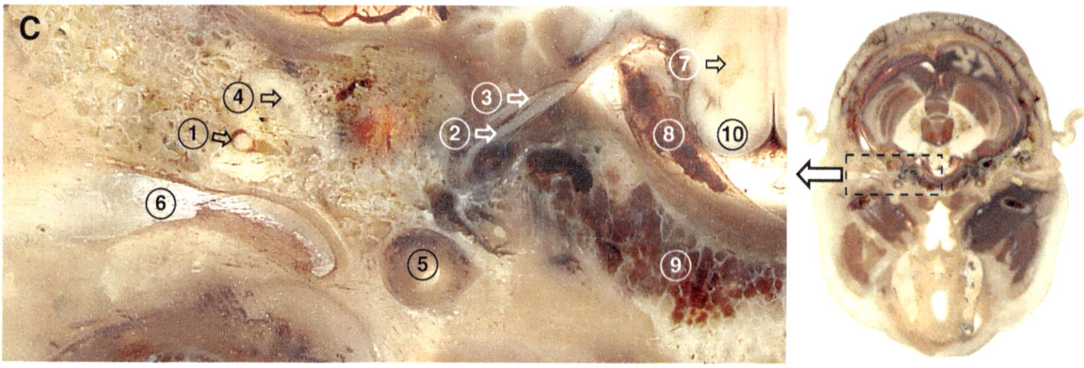

①	Facial nerve CNVII	⑥	External acoustic meatus
②	Cochlear nerve	⑦	Inferior olivary complex
③	Vestibular nerve	⑧	Vertebral artery
④	Lateral semicircular canal	⑨	Clivus
⑤	Internal carotid artery	⑩	Corticospinal tract

Fig. 8.17 Horizontal section through the human brainstem at the cerebellopontine angle (**A**) and the external acoustic meatus (**C**). Plastinated specimen. The facial (CNVII) and vestibulocochlear nerves (CNVIII) exit the brainstem and enter the internal acoustic meatus. (**B**) Drawing of the course of the facial nerve in the skull base. (**A**, **C**) Sammlung des Zentrums Anatomie der Universität zu Köln. (**B**) modified after Tillmann 2005, Fig. 2.149 with permission

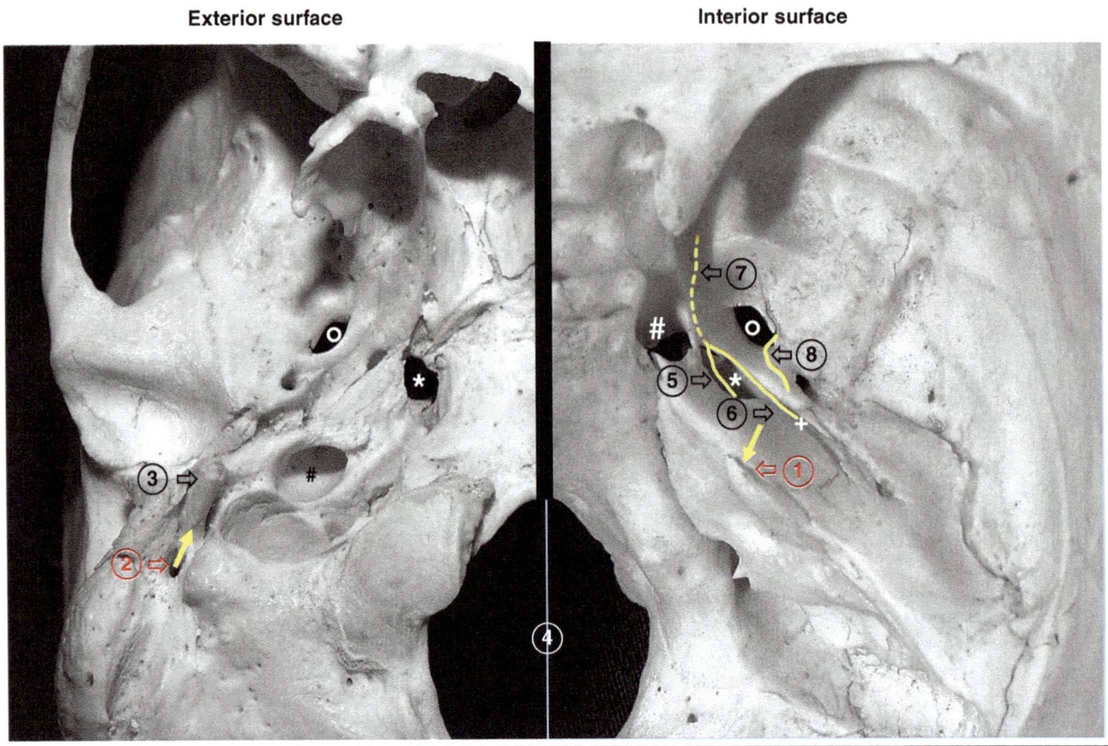

①	Internal acoustic opening CNVII	⑦	Nerve of pterygoid canal
②	Stylomastoid foramen CNVII	⑧	Lesser petrosal nerve = Parasympathetic root of otic ganglion
③	Styloid process	*	Foramen lacerum
④	Foramen magnum	+	Hiatus for greater petrosal nerve
⑤	Deep petrosal nerve = Sympathetic root of pterygopalatine ggl.	#	Carotid canal
⑥	Greater petrosal nerve = Parasympathetic root of pterygopalatine ggl.	o	Foramen ovale

Fig. 8.18 Dorsal and ventral view of the human skull base. The facial nerve, together with the vestibulocochlear nerve, enters the internal acoustic opening (for the course in the skull base, see Fig. 8.17B and text) and exits via the stylomastoid foramen. Sammlung des Zentrums Anatomie der Universität zu Köln

terminology of the extracerebral course of the facial nerve is equivocal (see Box 8.3).

Box 8.3 Anatomical terminology of the facial nerve

Please note that there are no official terms for the different intracerebral sections of the facial nerve, i.e., facial nucleus > knee of facial nerve > exit in the cerebellopontine angle. The additional terms ascending part (facial nucleus > knee of facial nerve) and exiting part (knee of facial nerve > exit in the cerebellopontine angle) (see Fig. 8.17) have been introduced by the authors. Furthermore, we have used the term *inner knee* of the facial nerve to avoid confusion with the *outer knee*, the geniculum (diminutive of genu = knee) at the geniculate ganglion.

The term N. facialis proprius (Facial nerve proper, Nr. 2863) (or N. facialis, Pars extracranialis, see FIPAT Ch. 2 endnote 157) has been introduced in FIPAT Chap. 2. Following Nr. 2863, the peripheral branches of the facial nerve are listed.

Under endnote 157 it reads: …Alternatively, the various parts of the facial nerve may be mentioned (Pars meatica, Pars labyrinthica, Pars pyramidalis, Pars mastoidea, Pars extracranialis)….

Our description here below is based on the paper by Toulgoat et al. (2013), who, however, do not mention a Pars pyramidalis.

When dealing with papers about the extracranial course of the facial nerve, these issues should be kept in mind.

The course of CNVII can be subdivided into a (1) cisternal part (in addition to Box 8.3) (see Fig. 8.17) from the exit in the cerebellopontine angle to the entry into the internal acoustic meatus [Meatus acusticus internus], a (2) meatic part [Pars meatica] for the section in the internal acoustic meatus, a (3) labyrinthine part [Pars labyrinthica], a (4) mastoid part [Pars mastoidea], and an extracranial part [Pars extracranialis].

In the cisternal part, the facial nerve, together with the vestibulocochlear nerve, is traversing the prepontine cistern between the cerebellopontine angle (Fig. 8.17A, C) and the internal acoustic opening/meatus (Fig. 8.18 ①) (meatic part) to the geniculate ganglion [Ggl. geniculi]). The facial nerve gives off fibers of the greater petrosal nerve (see Fig. 8.18 ⑥, see Box 8.3).

The next part is the tympanic or horizontal one (from the geniculate ganglion under the lateral semicircular canal (see Figs. 5.3, 5.5, and 5.6) between vestibule and tympanic membrane (see above Fig. 8.17B, C ⑦). The facial nerve passes the tympanic part without ramifications.

In the mastoid part three branches leave/join the facial nerve: The nerve to stapedius muscle, an auricular branch and the Chorda tympani. The vertical or mastoid part [Pars mastoidea] follows from the ear canal to the stylomastoid foramen [Foramen stylomastoideum], exit of CNVII from the skull base (see Fig. 8.18 ②) eventually, giving off a sensory branch for the auricular region (Toulgoat et al. 2013).

8.2.1.4 Branches of the facial nerve

The first branch given off in the (1) meatic/labyrinthine part is the greater petrosal nerve [N. petrosus major] (for details, see Box 8.5) (see Fig. 8.18). It contains preganglionic parasympathetic fibers from the superior salivatory nucleus.

They run through the foramen lacerum (Figs. 8.18 and 8.19) and form the nerve of the pterygoid canal [N. canalis pterygoidei] (Fig. 8.19) together with the "deep petrosal nerve" (see Box 8.5, see Fig. 8.18 ⑤, Fig. 8.19) to reach the pterygopalatine ganglion [Ggl. pterygopalatinum] (see also

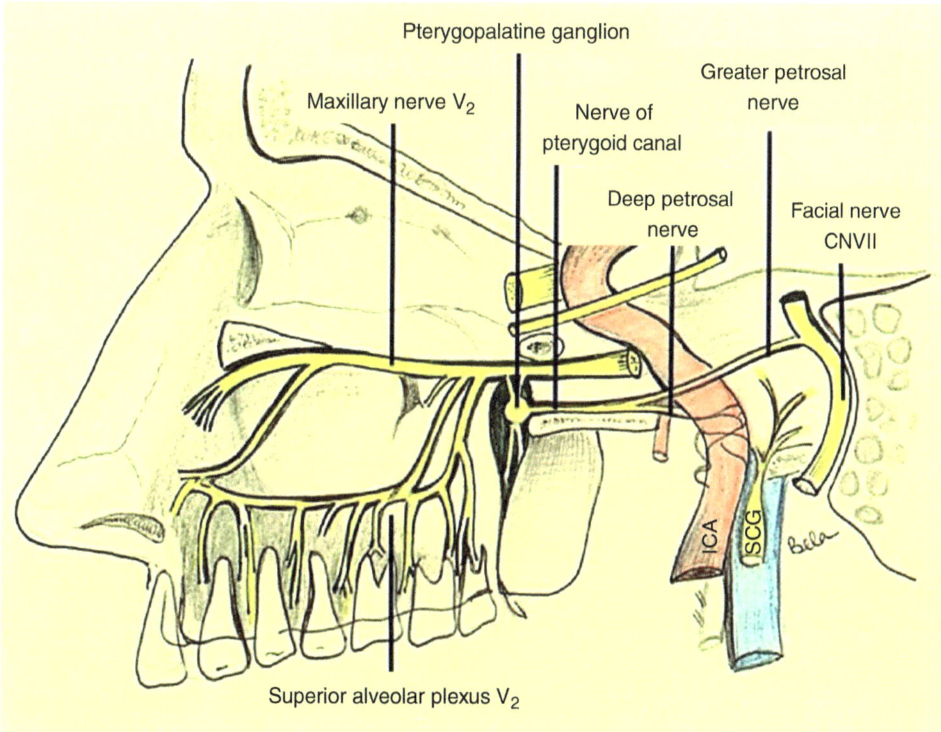

Fig. 8.19 Schematic lateral view of the human skull with the covering soft tissue and bone removed to the point where the facial nerve, the petrosal nerve, the pterygopalatine fossa, and the main branches of the maxillary nerve (V_2) become visible. The illustration allows to trace the course of the greater petrosal and the "deep petrosal" nerves (see Box 8.5 below). The greater petrosal nerve splits off the facial nerve at the external genu of the facial nerve (see Fig. 8.17B), then runs rostrally entering the middle cranial fossa via the hiatus for the greater petrosal nerve (see Fig. 8.18). It leaves the cranial cavity again via the foramen lacerum (see Fig. 8.18✱) medial of the internal carotid artery (ICA). In the pterygoid canal, the deep petrosal nerve (sympathetic root of pterygopalatine ganglion) joins the greater petrosal nerve to form the nerve of pterygoid canal (Vidian nerve). The postganglionic sympathetic fibers of the deep petrosal nerve originate in the superior cervical ganglion (SCG) (see Fig. 8.21, preganglionic cells in the intermediolateral column of the spinal cord) form the internal carotid plexus around the homonymous artery and reach the cranial cavity with the artery via the carotid canal (see Fig. 8.18). In the pterygopalatine ganglion (sphenopalatine ganglion) (see also Fig. 12.16 ①)—located in the pterygopalatine fossa—only the parasympathetic fibers of the nerve of pterygoid canal synapse while the sympathetic ones just cross the ganglion. Beyond the ganglion, the sympathetic fibers continue to run along the lacrimal nerve (V_1) to blood vessels and secretomotor elements in the lacrimal gland, nose, and oral cavity (Goosmann and Dalvin 2018) (see Fig. 8.20). The postganglionic parasympathetic axons join the zygomatic nerve [N. zygomaticus], a branch of the maxillary nerve (V_2), which reaches via its communicating branch to lacrimal nerve [N. lacrimalis]—branch of the ophthalmic nerve (V_1) (see Fig. 8.20)—the lacrimal gland [Gld. lacrimalis]. From Tashi et al. 2016, Fig. 1 with permission

Fig. 12.16 ①) located in the pterygopalatine fossa [Fossa pterygopalatina].

The postganglionic fibers to the lacrimal gland enter the orbit via the inferior orbital fissure, join the zygomaticotemporal nerve and reach the lacrimal gland (Fig. 8.20 ④) directly or via the lacrimal nerve (Fig. 8.20 ③). The labyrinthine segment houses the geniculate ganglion [Ggl. geniculi] (Fig. 8.17B) with the perikarya of the special visceroafferent (taste) and somatosensory fibers (touch, pain) of the posterior auricular nerve [N. auricularis posterior].

In the mastoid part the next branch of the facial nerve is the chorda tympani (Figs. 8.17B and Fig. 8.19) which runs in frontocaudal direction and transverses the tympanic cavity (between malleus and incus) to eventually join the lingual nerve (CNV: V_3) (see Fig. 4.21 ②). The chorda tympani contains visceromotor efferent fibers for the small salivatory glands (submandibular, sublingual, see Figs. 8.21 and 8.28) and afferent gustatory fibers from the anterior two thirds of the tongue (see Sect. 3.2.4.1). Therefore, by joining the lingual nerve, it reaches the tongue region where the chorda

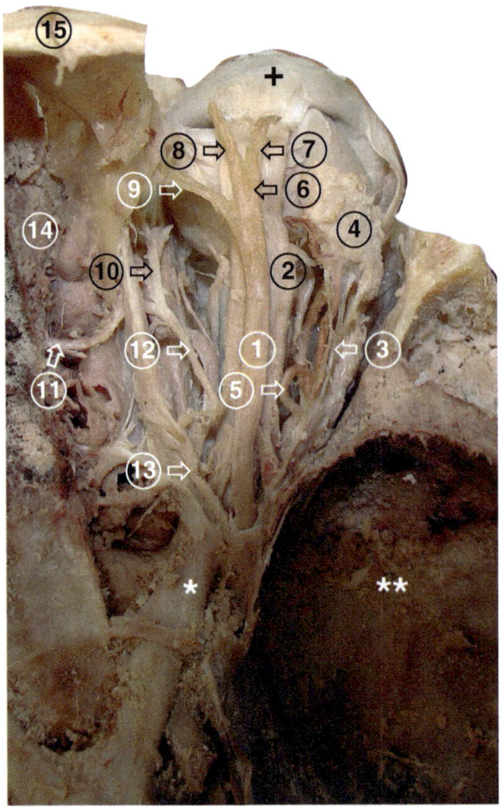

①	**Frontal nerve V_1 lying on the levator palbebrae superioris muscle ②**	⑩	**Infratrochlear nerve V_1**
		⑪	**Anterior ethmoidal nerve V_1**
③	**Lacrimal nerve V_1**	⑫	**Nasociliary nerve V_1**
④	**Lacrimal gland**	⑬	**Trochlear nerve**
⑤	**Lacrimal artery**	⑭	**Nasal cavity**
⑥	**Supraorbital nerve V_1**	⑮	**Frontal bone**
⑦	**Supraorbital nerve, lateral branch**	+	**Optic bulb**
⑧	**Supraorbital nerve, medial branch**	*	**Optic nerve**
⑨	**Supratrochlear nerve V_1**	**	**Middle cranial fossa**

Fig. 8.20 Dorsal view of the human orbita. Roof of orbita and parts of surrounding bones have been removed. Formalin-fixed specimen. The lacrimal nerve reaches the lacrimal gland. A typical feature of peripheral facial paralysis (Bell's) paralysis (see Fig. 8.29) is the dry eye due to the malfunction of the lacrimal gland and the inability to close the eye. This is a serious nursing problem since drying out of the cornea, in the worst case, may lead to blindness. Sammlung des Zentrums Anatomie der Universität zu Köln. Müller-Thomsen fecit

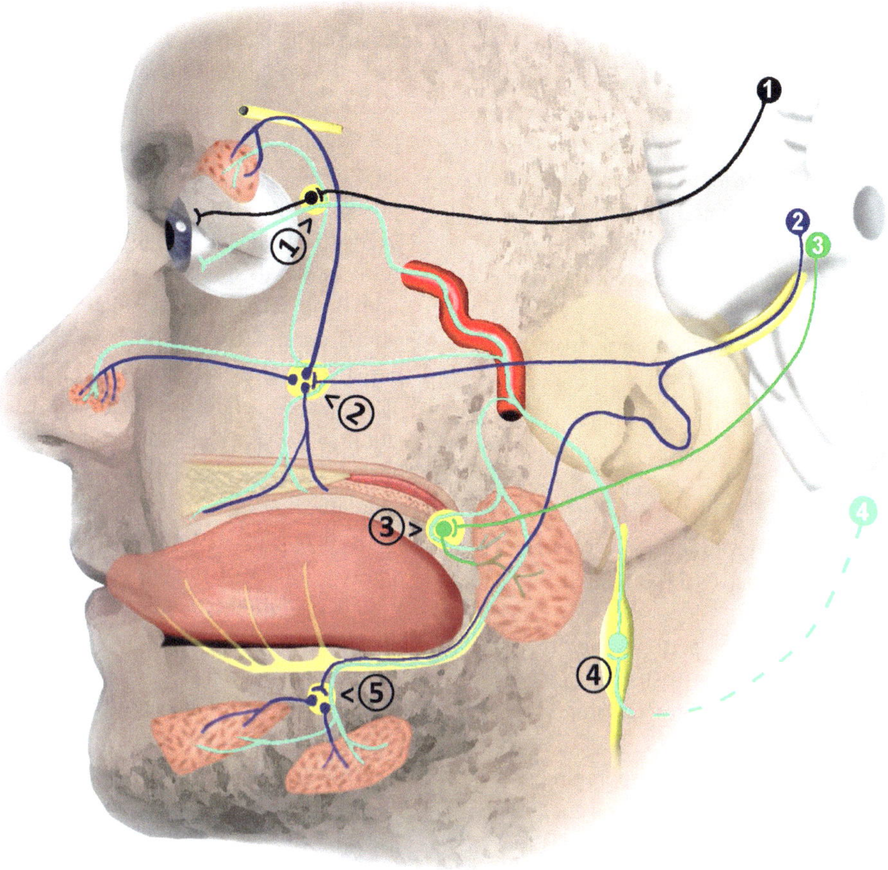

Fig. 8.21 Schematic survey of the parasympathetic and sympathetic innervation of smooth muscles and glands of the head region. (1) Edinger-Westphal nucleus [Ncl. accessorius n. oculomotorii see Sect. 15.3.1.2]. Preganglionic fibers for the ciliary ganglion [Ggl. ciliare, ①]. Postganglionic fibers to the ciliary [M. ciliaris] and the sphincter pupillae [M. sphincter pupillae] of the eye. (2) Superior salivatory nucleus. Preganglionic fibers for the pterygopalatine ganglion [Ggl. pterygopalatinum, ②]. Postganglionic fibers for the lacrimal gland [Gld. lacrimalis]. (2) Superior salivatory nucleus. Preganglionic fibers for the submandibular ganglion [Ganglion submandibulare, ⑤]. Postganglionic fibers for the submandibular [Gld. submandibularis] and sublingual glands [Gld. sublingualis]. (3) Inferior salivatory nucleus. Preganglionic fibers for the otic ganglion [Ggl. oticum]. Postganglionic fibers for the parotid gland [Gld. parotidea, ③]. (4) Intermediolateral nucleus [Ncl. intermediolateralis] of the spinal cord. Preganglionic fibers for the superior cervical ganglion, ④. Postganglionic fibers for the dilator pupillae [M. dilatator pupillae] and the lacrimal gland [Gld. lacrimalis]. Leander Huggenberger fecit Modified acc. to Takezawa et al. 2018, Fig. 2 with permission

tympani afferently innervates the taste receptors and efferently the small salivatory glands.

Furthermore, the facial nerve gives off the nerve to the stapedius muscle [N. stapedius] and an auricular branch. The stapedius muscle—the smallest striated muscle of the human body, ca. 1 mm length—runs from the wall of the tympanic cavity to the stapes (see Fig. 8.6). It is thought to stabilize the stapes. A possible hyperacusis caused by its deficiency in peripheral facial paralysis (see Sect. 8.2.1.6.3) is questionable.

After leaving the skull base via the stylomastoid foramen, the facial nerve gives off branches for the innervation of the digastric (posterior belly) and the stylohyoid muscles (see Box 8.4). The main stem of the facial nerve is running further into the periphery of the face where it branches off in the different motor nerves for the innervation of the facial muscles (see Figs. 8.25 and 8.26).

Box 8.4 Muscles of the mouth floor

The digastric muscle is a muscle with two bellies (hence the name). They originate at the mastoid incisure and reach the digastric fossa (small excavation at the inner surface of the mandibula, anterior region). The bellies are separated by an intermediate tendon which exerts effects on the small cornu of the hyoid bone, elevating the hyoid bone, opening of the mouth. The anterior belly receives innervation from the mylohyoid nerve (CNV), the posterior belly from CNVII.

The stylohyoid (target hyoid bone, moves the hyoid bone upwards and backwards, CNVII), the styloglossus [M. styloglossus] (target tongue, retraction of the tongue, CNXII), and the stylopharyngeus muscle (target pharynx, epiglottis, larynx, CNIX) have as their common denominator the origin at the styloid process (see Fig. 5.21).

They receive, however, their innervation from different cranial nerve branches.

Box 8.5 The petrosal nerves

The traditional anatomical terminology knows three nerves with the designation petrosal [petrosus]:

- The greater petrosal nerve [N. petrosus major]
- The lesser petrosal nerve [N. petrosus minor]
- The deep petrosal nerve [N. petrosus profundus]

The term "petrosal" alludes to the topographic relationship of all three nerves with the petrous bone (Petrous part of temporal bone [Pars petrosa ossis temporalis] (Greek ὁ πέτρος ho petros = stone, rock).

The petrous bone is one of the hardest bones of the human body, hence the name.

Of interest for paleoanthropological purposes is the fact that the petrous bone contains high amounts of well-preserved DNA (Hansen et al. 2017).

The FIPAT terminology has the greater and lesser petrosal nerve but changed the name of the deep petrosal nerve (see below). Even now the terminology is somewhat confusing. This is why we deal with these structures in some detail, also because of their functional importance.

Greater petrosal nerve [N. petrosus major] (Parasympathetic root of pterygopalatine ganglion)

An outdated term for this nerve is the "parasympathetic root of pterygopalatine ganglion" which, however, by contrast to the FIPAT designation, provides some information on the nature of this nerve. Parasympathetic alludes to the origin of these fibers from one of the parasympathetic brainstem nuclei, the superior salivary nucleus [Ncl. salivatorius superior]. The axons of this nerve form the preganglionic part of the innervation of the lacrimal gland and the nasal glands. As parasympathetic ganglia usually are located near to their target organs, the greater petrosal nerve has to cover a certain distance to reach the corresponding pterygopalatine ganglion [Ggl. pterygopalatinum] (see Fig. 8.19 sphenopalatine ganglion) in the pterygopalatine fossa [Fossa pterygopalatina]. There, the preganglionic fibers make synaptic contacts with the postganglionic perikarya. The axons of the latter eventually reach the lacrimal gland (Fig. 8.20)—for anatomical details on the human lacrimal gland see Singh and Basu 2020—and regulate tear secretion (Nturibi and Bordoni 2020).

Very recently, Valstar et al. (2021) reported about a new mucous gland in the region of the entry to the auditory tube which they called the tubarial salivary glands. So far, no data on their innervation are available.

In detail, the secretomotor preganglionic parasympathetic fibers of the superior salivatory nucleus (see Fig. 8.12) join the facial nerve and leave the brainstem in the cerebellopontine angle [Angulus pontocerebellaris].

At the level of the geniculate ganglion [Ganglion geniculi], the secretomotor fibers leave the facial nerve as greater petrosal nerve and enter the cranial cavity via the hiatus of greater petrosal nerve (see Fig. 8.18). It then runs at the anterior surface of the petrous bone and leaves the cranial cavity via the foramen lacerum—in vivo covered with connective tissue—to reach the pterygoid canal [Canalis pterygoideus] (see Fig. 8.18 ⑦) in which it unites with the sympathetic postganglionic fibers of the sympathetic root of pterygopalatine ganglion (see here below, formerly deep petrosal nerve) to form the nerve of pterygoid canal [N. canalis pterygoidei].

Lesser petrosal nerve [N. petrosus minor] (Parasympathetic root of otic ganglion)

As the outdated term in round brackets shows, the lesser petrosal nerve is a parasympathetic (preganglionic) root, too, in this case for the otic ganglion [Ggl. oticum]. Its functional meaning is the innervation of parotid gland and the cheek and lip glands (see Fig. 8.22).

In detail, the preganglionic fibers originate in the inferior salivatory nucleus [Ncl. salivatorius inferior] and first run with the glossopharyngeal nerve (see Sect. 5.4.2.1.3). At the inferior ganglion, they branch together with the tympanic nerve [N. tympanicus] which enters the tympanic cavity between jugular foramen and carotid canal via the tympanic canal. Together with sympathetic fibers of the internal carotid plexus, the nerve forms the tympanic plexus. The latter releases the lesser petrosal nerve with its preganglionic parasympathetic fibers which then enters the medial cranial fossa via hiatus for minor petrosal nerve [Hiatus canalis n. petrosi minoris] (see Fig. 8.18 ⑧). The nerve then runs under the dura at the surface of the petrous part of the temporal bone. It leaves the fossa via the Fissura sphenopetrosa, which is a fissure at the foramen lacerum, and then enters the otic ganglion just below the foramen (Thomas et al. 2020). The ganglion is a flat structure at the medial side of the mandibular nerve (V_2) below the foramen ovale (see Fig. 8.18).

The postganglionic fibers eventually reach the facial nerve via an anastomosis with the auriculotemporal nerve (see Fig. 8.22 ⑨). The secretomotor fibers branch in the parotid gland [Gld. parotis] (see Fig. 8.22 ⑩), the most important human salivary gland.

Secretomotor fibers for the labial [Gld. labiales] and buccal glands [Gld. buccales] reach their targets with the buccal branches of the parotid plexus [Plexus intraparotideus] nerve and the inferior alveolar nerve [N. alveolaris inferior] (V_3) (see Sect. 12.2.3.2.3 "Mandibular nerve V_3").

Surprisingly, only the deep petrosal nerve lost his traditional name but gained a new and functionally better designation which was discarded, however, in case of the greater and lesser petrosal nerves. The current term is "sympathetic root of pterygopalatine ganglion" [Radix sympathica ganglii pterygopalatini]. It is listed in FIPAT under the superordinate term "Perivascular plexuses" [Plexus perivasculares]. In this case, the vessel involved is the internal carotid artery [A. carotis interna] which is equipped with postganglionic sympathetic fibers from the superior cervical ganglion [Ggl. cervicale superius]. These fibers form the internal carotid plexus [Plexus caroticus internus] and get into the interior of the skull via the carotid canal and then join with the greater petrosal nerve (see Fig. 8.18 ⑥). The fibers run through the pterygopalatine ganglion to provide the sympathetic, secreto- and vasomotor innervation of lacrimal gland, the oral and nasal mucosa (Goosmann and Dalvin 2018).

As a rule of thumb, sympathetic ganglia are located far away from their target organs which means that by contrast to parasympathetic fibers the postganglionic sympathetic nerves have to run over relatively long distances to eventually contact their target organs.

There is one more root of the internal carotid plexus which should be mentioned here for reasons of completeness, the sympathetic root of ciliary ganglion [Radix sympathica ganglii ciliaris]. The sympathetic fibers innervate the M. dilatator pupillae (mydriasis = widening of the pupil) (see Fig. 8.21).

The parasympathetic counterpart of the sympathetic root of ciliary ganglion are fibers originating in the Edinger-Westphal nucleus (see Sect. 15.3.1.2). They are preganglionic to the ciliary ganglion and join the oculomotor nerve (see Sect. 15.3.1). With the latter they leave the mesencephalon in the interpeduncular fossa (see Sect. 12.4.1) and reach the orbita via the cavernous sinus and the ciliary ganglion as radix oculomotoria. After having synapsed with postganglionic parasympathetic ganglia, postganglionic fibers form the Nn. ciliares breves, perforating the sclera, reaching the interior of the eye bulb. The parasympathetic fibers innervate the ciliary muscle (accommodation) and the M. sphincter pupillae (miosis) (see also Sect. 15.3.1.2). To summarize this box, Fig. 8.21 provides a simplified scheme of the parasympathetic nerves of the head region.

The CNVII is then running in the facial canal (Fig. 8.17). The first branch it gives off distal of the geniculate ganglion is the greater petrosal nerve with preganglionic parasympathetic fibers to the pterygopalatine ganglion. The postganglionic fibers innervate the lacrimal gland (tear secretion) (see Box 8.5).

The fourth, the mastoid segment stretches further ventrally to the stylomastoid foramen. Above the stylomastoid foramen, the facial nerve gives off preganglionic parasympathetic fibers for the chorda tympani (see Fig. 8.17B) running through the tympanic cavity of the middle ear. It leaves the latter through the anterior canaliculus of the chorda tympani and enters the infratemporal fossa. From there the chorda tympani joins the lingual nerve and reaches the submandibular ganglion on the surface of the hyoglossus muscle (see Figs. 4.17 and 4.21). Postganglionic parasympathetic fibers directly reach the submandibular and sublingual glands (see Fig. 8.21) and their myoepithelial cells (Ghannam and Singh 2020) via the lingual nerve.

Contrastingly, the cell bodies of the sympathetic fibers (Sympathetic root of submandibular ganglion [Radix sympathica ganglii submandibularis]) are found in the superior cervical ganglion, where postganglionic fibers innervate the glands along blood vessels branching off of the carotid plexus (Ghannam and Singh 2020).

The last distal segment shows the distal branches of the facial nerve in the face (see Fig. 8.22). In this region, the nerve is put to risk of getting severed by sharp-edged subjects like knives or glass fragments.

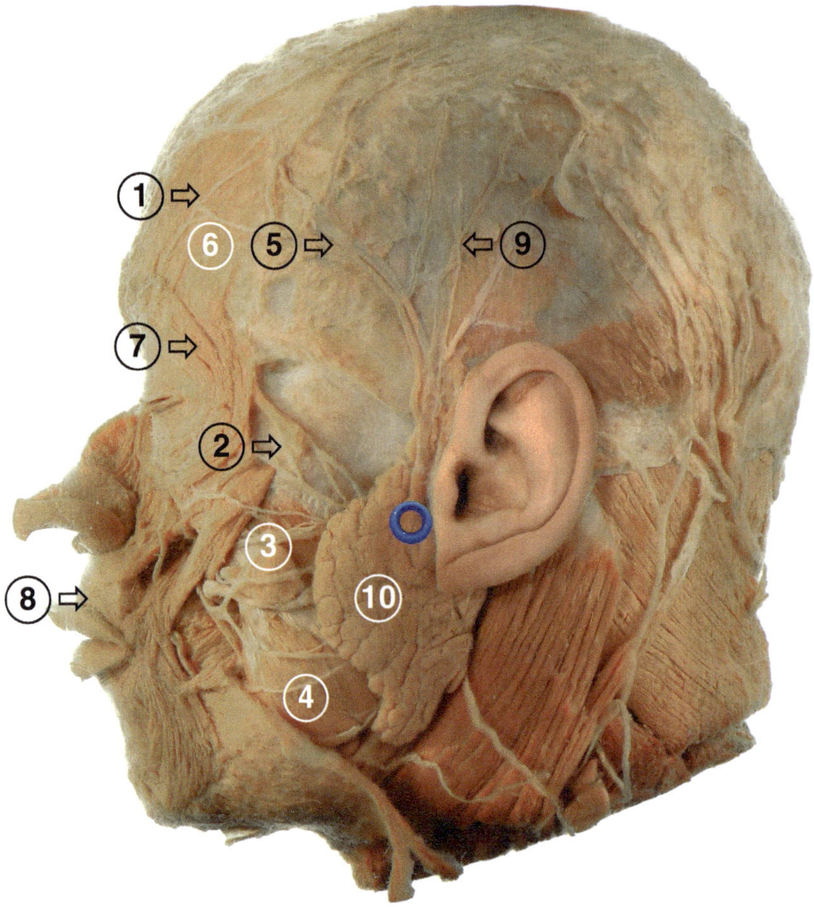

①	Temporal branches	⑦	Orbicularis oculi muscle
②	Zygomatic branches	⑧	Orbicularis oris muscle
③	Buccal branches	⑨	Auriculotemporal nerve V_3
④	Marginal mandibular branches	⑩	Parotid gland
⑤	Ascending temporal artery	O	Projection of the approximate entry of the facial nerve into the parotid gland
⑥	Frontal muscle		

Fig. 8.22 Lateral view of the human face. Skin removed, Formalin-fixed specimen. CNVII runs from the stylomastoid foramen (exit from the skull base) to the face where it traverses the parotid gland (⑩, the gland is innervated by the glossopharyngeal nerve!) and then gives off the temporal branches for the innervation mainly of the (occipito)frontal ⑥ and the orbicularis oculi muscle ⑦, the buccal branches for those mimic muscles (see Fig. 8.25) in the region below the orbicularis oculi muscle and the marginal mandibular branches ④. For individual mimic muscles, see Fig. 8.25. Photographs by MedizinFotoKöln, Cologne, Germany, Sammlung des Zentrums Anatomie der Universität zu Köln

8.2.1.5 Central innervation of the motor nucleus of the facial nerve

The spinal motoneurons and those of the somatomotor and special visceromotor (branchiogenic) cranial nerves are subject to cerebrocortical control from the primary motor cortex. For the spinal motoneurons, this is achieved via the corticospinal (pyramidal) tract and in analogy for the cranial motoneurons the corticonuclear tract. The complex innervation of the motor nucleus of the facial nerve is of crucial importance for clinical neurological diagnostics (see Fig. 8.23 for explanation).

8.2.1.6 Living anatomy and clinical relevance

Due to the plethora of targets in the cranial periphery innervated by the facial nerve (see Fig. 8.16), it is one of the areas which are rather easily accessible for living anatomy and neurological examination.

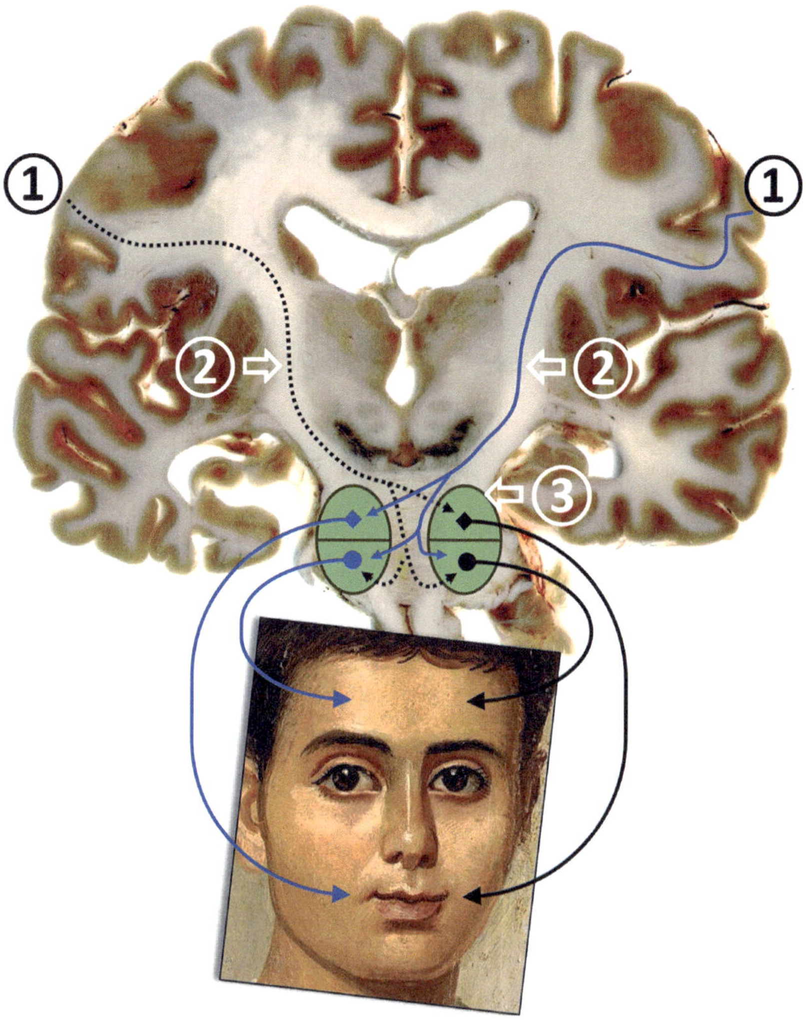

①	Primary motor cortex	●●	Neurons for upper face
②	Internal capsule	◆◆	Neurons for lower face
③	Ncl. of facial nerve		

Fig. 8.23 Schematic illustration of the cerebrocortical input into the motor nucleus of the facial nerve ③. In terms of differential afferent innervation from the primary motor cortices ① via the corticonuclear tract, the facial nucleus can—according to clinical experiences—be subdivided in two subnuclei. One, innervating the upper face, for example, the frontal and orbicularis oculi muscles, receives an ipsi- and contralateral cerebrocortical input. The other, providing innervation to the orbicularis oris muscle and the lower face, gets only a contralateral input. By contrast to peripheral lesions of CNVII (Fig. 8.29) when the whole motor innervation of the face is lacking, a supranuclear lesion of the corticonuclear tract—mainly due to cortical lesions or lesions in the internal capsule ②—resents with intact frontal muscle innervation allowing for furrowing the brows. This type is called central facial palsy, mostly due to a stroke comprising the primary motor cortex and/or the internal capsule and it has to be regarded as a neurological emergency requiring immediate medical treatment. In case of a lesion in the peripheral course of CNVII (③ →periphery, Fig. 8.30) all mimic muscles are paralyzed. Sammlung des Zentrums Anatomie der Universität zu Köln. Anonymous (Egypt) (A.D. 100–150/Roman Period) Portrait of the Boy Eutyches, Metropolitan Museum of Art, New York, with permission

8.2.1.6.1 Mimic muscles

Figure 8.24 shows a clinical aide-memoire for the main CNVII branches (see Fig. 8.22). A graphic representation of mimic muscles is shown in Fig. 8.25. Among higher primates, not to speak of most other mammals, man is the animal with the highest number of mimic muscles and the primate with a really "naked face" being a prerequisite to bring the complex social function of the mimic muscles to bear (see Figs. 8.26 and 8.27). To capture the activity of mimic muscles is an important supplementation to any oral communication.

Compulsory part of any neurological examination is the assessment of the mimic muscles. The minimum program is to ask the patient to produce wrinkles on the forehead

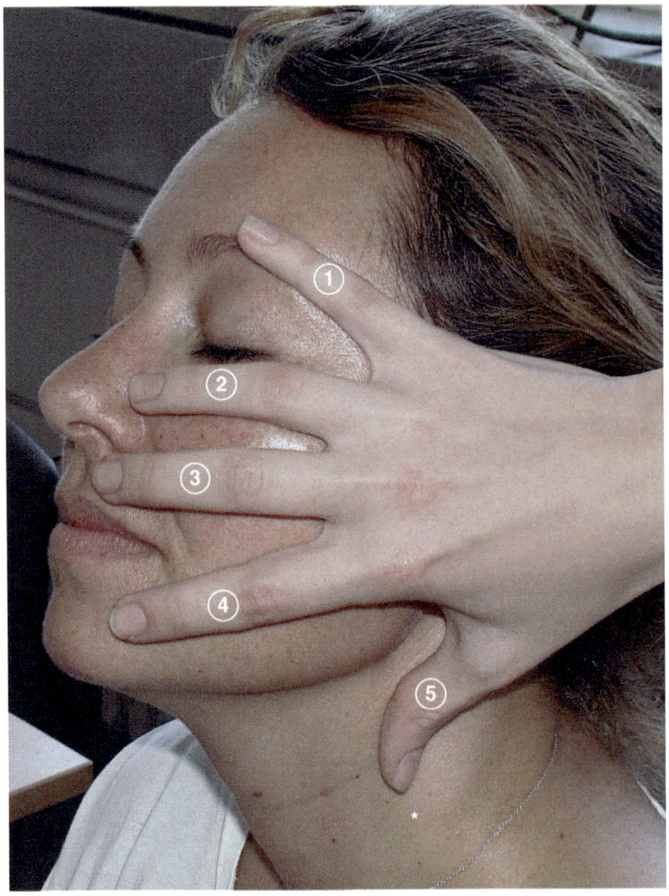

①	**Temporal branches**	④	**Marginal mandibular branch**
②	**Zygomatic branches**	⑤	**Cervical branches**
③	**Buccal branches**		

Fig. 8.24 Aide-mémoire for the five main superficial branches of CNVII. Latin terminology: ① Rr. temporales, ② Rr. zygomatici, ③ Rr. buccales, ④ R. marginalis mandibularis, ⑤ R. cervicalis innervating the platysma (*). Courtesy B. Dawidowski

(Fig. 8.26, occipitofrontalis), to close the eyes (Fig. 8.26, orbicularis oculi), and to pout (Fig. 8.26, Orbicularis oris) (see also Fig. 8.27). In addition to the mimic muscles, the facial nerve also innervates the platysma (see Figs. 8.24 and 8.25). The platysma, broad thin layer of muscle, arises from the fascia covering the upper parts of the pectoralis major and deltoid. Its fibers cross the clavicle and ascend medially in the side of the neck. The anterior fibers interlace across the midline with the fibers of the contralateral ones, below and behind the symphysis menti. Other fibers attach to the lower border of the mandible or attach to the skin and subcutaneous tissue of the lower face. The platysma may assist in depressing the mandible, and via its labial part expresses horror or surprise (see Standring 2005).

Compare this list with the drawing of mimic muscles shown in Fig. 8.25 and with Figs. 8.26 and 8.27.

8.2.1.6.2 Small salivary glands

The sublingual and submandibular glands in the floor of the mouth are visible for inspection upon wide opening of the mouth and bending back of the tongue (see Fig. 8.28). The saliva produced in both glands is secreted into the oral cavity via the sublingual caruncle. Gentle massage of the submandibular gland just medial of the mandibula results in visible secretion of the saliva at the sublingual caruncle.

8.2.1.6.3 Pathology of the facial nerve and muscles

Lesions of the facial nerve in its peripheral course may lead to the so-called peripheral facial paralysis (Bell's palsy) (see Fig. 8.29).

The peripheral facial paralysis (Bell's palsy) (Fig. 8.30) has been named in modern medicine after Sir Charles Bell (1774–1842), a Scottish anatomist, physiologist, and surgeon who described the entity in 1821. However, it was

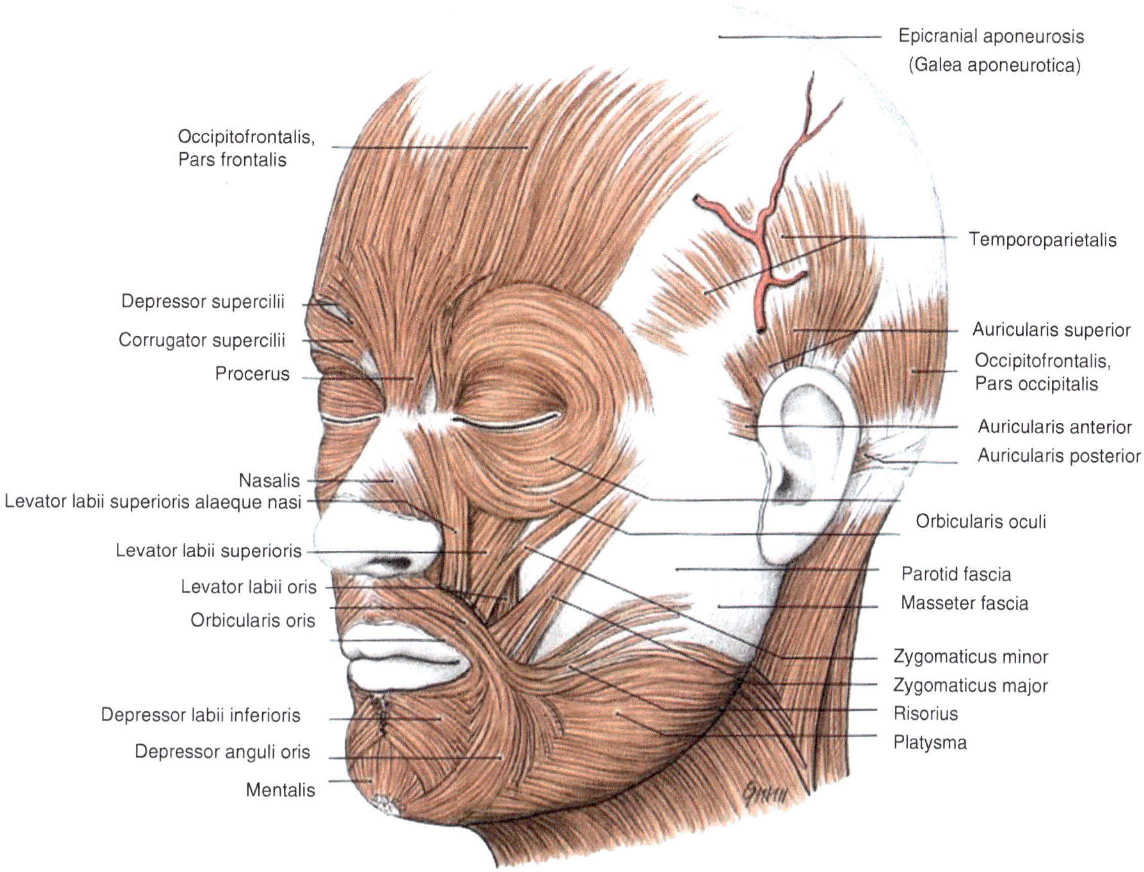

Fig. 8.25 Drawing of the mimic muscles of the human face. From B. Tillmann 2005, Fig. 2.34 with permission

already known in ancient times, and the first comprehensive description was provided by the nineteenth century Persian physician Razi (for details, see Sajadi et al. 2011).

Related to the inability to close the eye in peripheral facial paralysis (see Fig. 8.30), the then observable—physiological—upwards rotation of the eyeball is known as Bell's phenomenon. It is quite interesting, that as early as in the sixteenth century facial paralysis has become the subject of an artistic representation (see Fig. 8.31).

In case of traumatic separation (see Table 8.2) of the facial nerve, special anastomosis surgery has been developed to reconstruct the nerve (Brown et al. 2019).

An example of a tumor of the cerebellopontine angle which can result in a peripheral paralysis of the facial nerve is given in (Fig. 5.4).

Hemifacial spasms are unvoluntary unilateral contractions of the facial muscles. Typically, these symptoms are strictly localized on one side. The beginning of the disease is often involuntary twitching in the area of the orbicularis oculi muscle. During the course of years, the spasms gradually spread to other parts of the affected half of the face (Fig. 8.32). Only in pronounced cases, the platysma may also be affected. In the majority of patients, the symptoms persist even during sleep, and sometimes patients report a "cracking sound" in the ear, which may be caused by contractions of the stapedius muscle (Rosenstengel et al. 2012).

In most cases, the cause of the hemifacial spams is an ectatic or atypical blood vessel that compresses the root of the facial nerve at its exit zone from the brainstem. In this area, the nerve has only a thin arachnoid envelope, the epineurium is missing. Furthermore, this is a transition zone between central (oligodendroglial) and peripheral (Schwann's) myelination. Obviously, these characteristics lead to increased vulnerability. In most cases, the compression is due to the posterior or anterior inferior cerebellar arteries [Aa. cerebelli inferiores posteriores / anteriores]; more rarely, the vertebral artery [A. vertebralis] or in combination of these arteries (Rosenstengel et al. 2012). Accordingly, microsurgical decompression surgery may be beneficial although subcutaneous injections of botulinum toxin are the most popular choice for therapy next to oral pharmacotherapy (Chaudhry et al. 2015).

Important part of the neurological examination is testing facial nerve reflexes. Among the many ones described (see Walker et al. 1990), here, we mention only two because of

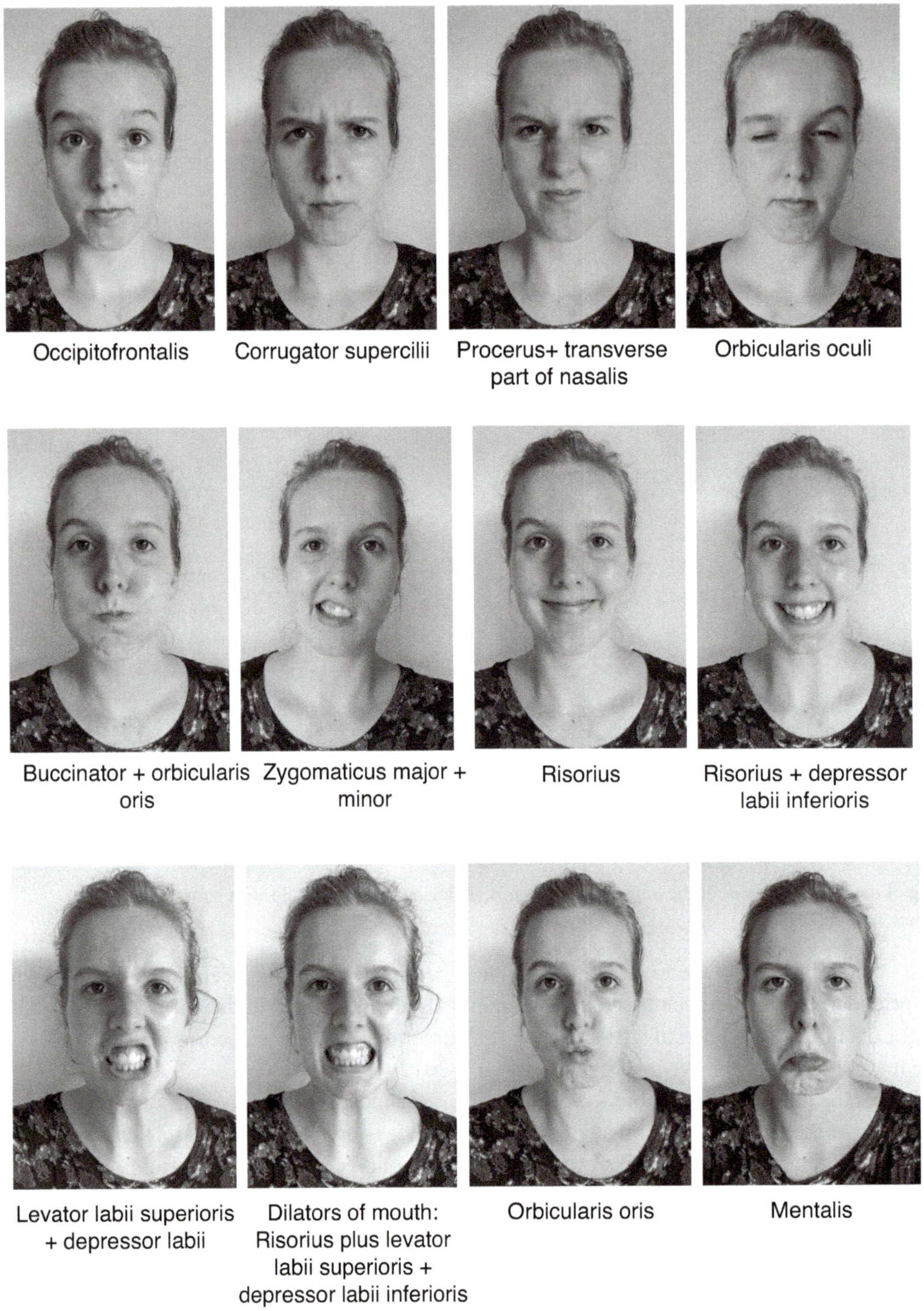

Fig. 8.26 Effects of the most important different mimic muscles (see Table 8.1 for details). Courtesy Sophia Schröder

their feasibility in testing (corneal reflex) or clinical relevance (blink reflex).

To elicit the corneal reflex, the cornea (afferent impulses, CNV) is carefully stimulated by a wisp of cotton which results in bilateral closure of the eyelids (efferent impulses, CNVII, orbicularis oculi muscle) (Walker et al. 1990).

The blink reflex (Pearce 2008) is an adverse-effects reflex similar to the corneal reflex elicited by mechanical stimulation of the cornea and the surroundings of the eye or by acoustic stimulation. For electrophysiological measurement, the supraorbital nerve (see Fig. 12.6, afferent part of the reflex) is stimulated by a surface electrode, and the answer of the orbicularis oculi muscle is registered by a surface electrode above the muscle (Marx 2003).

Interestingly, Seidel et al. (2015) could show the affection of all cranial nerve nuclei with Lewy bodies and Lewy neu-

Table 8.1 Main actions of human mimic muscles (in alphabetical order) (Basic expressions in **bold**)

Muscle	Action	Expression
Buccinator	Compression of cheeks	Pouting
Corrugator supercilii	Tightens eyebrows	Doubt, **sadness**
Depressor anguli oris	Depression of corner of mouth	**Sadness**
Depressor labii inferioris	Depression of lower lip	**Happiness, Joy** **Anger** **Fear**
Depressor supercilii	Depression of eyebrow	Ailing
Levator anguli oris	Elevation of corner of mouth	Disgust
Levator labii superioris alaeque nasi	Elevation of upper lip	Triumph **Anger**
Mentalis	Elevation and protrusion of lower lip	Languidness, indifference
Orbicularis oculi	Eyelid closure	Pouting
Orbicularis oris	Compression and pursing of lips	Languidness
Occipitofrontalis, frontal belly	Lifting eyebrow and upper eyelid, producing horizontal wrinkles of the forehead	**Surprise**
Nasalis	Transverse fibers: Compression of nasal alar cartilages to close the nasal opening. Alar fibers: assisting in opening the nostrils	**Disgust**
Platysma	Depression and wrinkling the skin of lower face and mouth	**Surprise**, Horror
Procerus	Depression of the medial eyebrow angle. Producing a transverse fold across the root of the nose	**Disgust**
Risorius	Drawing of corner of mouth laterally	**Happiness, Joy**
Zygomaticus major	Retraction/elevation of corner of mouth	Contempt
Zygomaticus minor	Retraction and elevation of upper lip	Contempt

rites and of all brainstem fiber tracts with Lewy neurites in dementia with Lewy bodies and Parkinson's disease (see Sect. 16.6.1.5). This, however, does not explain the "normal" results in Parkinson patients when testing the blink reflex,

Comparing groups of patients suffering from dementia with Lewy bodies (DLB, see Sect. 16.6.1.6), multiple system atrophy (see Sect. 3.3.2.3), Alzheimer's disease (Sect. 13.2.2.1.4), Parkinson's disease with or without REM sleep behavior disorder (RBD) (Sect. 16.6.1.5), and progressive supranuclear palsy (Sect. 4.1.1.3.3) without RBD with healthy controls, it turned out that the blink reflexes were significantly delayed only in DLB patients (Bonanni et al. 2007).

8.2.1.6.4 Pathohistology of the facial nerve

Liston and Kleid (1989) in a clinically well-studied patient with Bell's palsy who died during the palsy from a malignant tumor (not the cause) could show on postmortem investigation that the nervous tissue showed a massive infiltration with round inflammatory cells involving the whole length of the nerve from the internal acoustic meatus to the stylomastoid foramen. Partly, the myelin sheaths broke down followed by macrophagic infiltration. Some axons showed narrowing and irregularity of contours.

Basal r6

Details on the following nuclei can be found as indicated:

8.2.2 Visceral motor r6

8.2.2.1 Inferior salivatory nucleus r6

For details on this nucleus, see Sect. 7.3.1.1.

8.2.3 Raphe nuclei r6

8.2.3.1 Paramedian reticular nucleus r6

For details on this nucleus, see Sect. 5.4.3.1.

8.2.3.2 Raphe magnus nucleus r6 (B3 group)

For details on this nucleus, see Sect. 7.3.3.2.

8.2.3.3 Raphe obscurus nucleus r6 (B1 group)

For details on this nucleus, see Sect. 4.3.3.1.

8.2.3.4 Raphe pallidus nucleus r6 (B2 group)

For details on this nucleus, see Sect. 4.3.3.2.

8.3 Basal tegmentum r6

8.3.1 Dorsomedial tegmental area r6

For details, see Sect. 7.3.4.1.

8.3.2 Reticular nuclei r6

8.3.2.1 Intermediate reticular nucleus r6

For details on nucleus, see Sect. 3.3.2.1.

Fig. 8.27 In a non-systematic, artistic way the human possibility to express sentiments via the mimic muscles was already picked up by the famous Dutch painter and engraver Rembrandt Harmenszoon van Rijn (1606–1669)

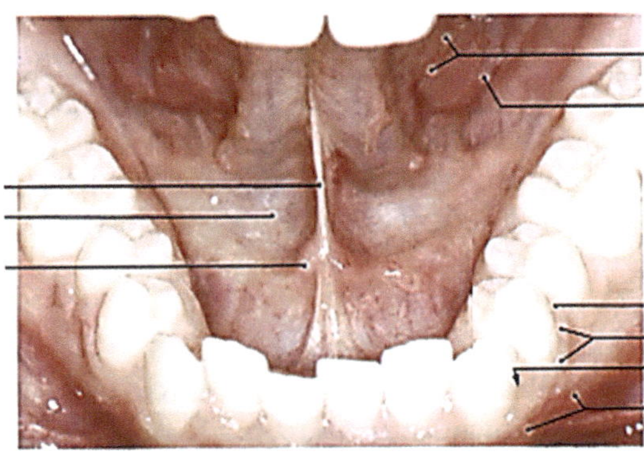

Fig. 8.28 Superior view of the floor of the oral cavity. The dorsal deflection of the tongue allows to see the sublingual gland with the sublingual caruncle (ostium of the sublingual and submandibular duct) and the submandibular gland with the opening of the submandibular gland. The midline is defined by the course of the frenulum of the tongue. From B. Tillmann 2005, Fig. 2.66 with permission

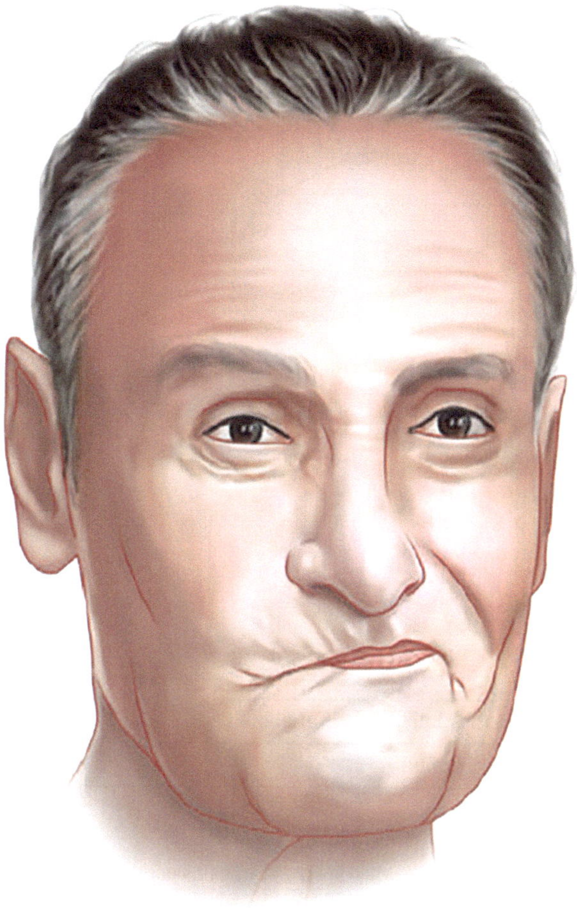

Fig. 8.29 Central facial paralysis on the right side. Note that the mouth is shifted to the left, unlesioned side (Orbicularis oris muscle on the left side intact) and see the less pronounced nasolabial fold on the right side. Frontooccipital and orbicularis oculi muscle are unaffected. In peripheral facial (Bell's) palsy, on the lesioned side the forehead cannot be wrinkled and the eye cannot be closed (cf. with Figs. 8.26 and 8.30). From Jonsson 2021, Fig. 1.1 with permission

8.3.2.2 Parvocellular reticular nucleus r6

For details on this nucleus, see Sect. 6.5.5.2.

8.3.2.3 Rostroventrolateral reticular nucleus r6

For details on this nucleus, see Sect. 7.3.5.4.

8.3.2.4 Gigantocellular reticular nucleus r6

For details on this nucleus, see Sect. 4.3.4.2.

8.3.2.5 Lateral paragigantocellular nucleus r6

For details on this nucleus, see Sect. 7.3.5.6.

8.3.2.6 Dorsal paragigantocellular nucleus

For details on this nucleus, see Sect. 5.4.4.4.

8.3.2.7 Parapyramidal nucleus r6 [Ncl. parapyramidalis]

The term parapyramidal nucleus refers to a small group of cells located lateral to the pyramid of the medulla in the rat and the mouse.

http://braininfo.rprc.washington.edu/centraldirectory.aspx?ID=1771

Human-specific data are not available and this nucleus not listed in FIPAT Ch. 1.

From studies in rats, there is evidence that the parapyramidal neurons project to the intermediolateral cell column of the spinal cord and are involved in vascular regulation and gastric acid secretion (Pelaez et al. 2002; Yang et al. 2000).

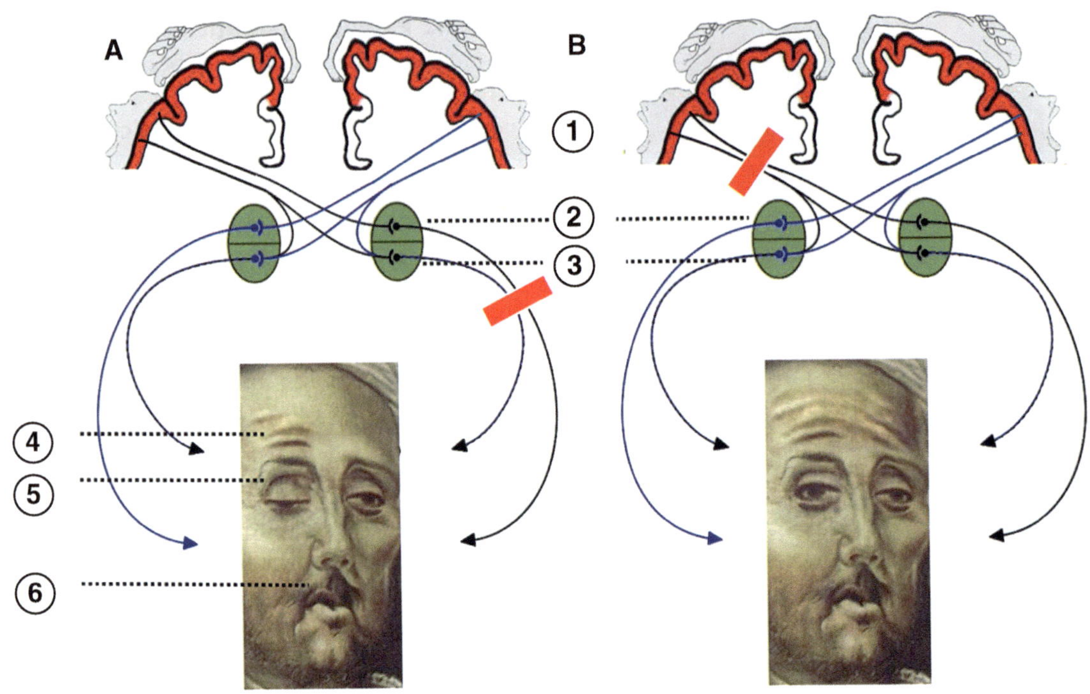

①	**Primary motor cortex**	④	**Occipitofrontalis, pars frontalis**
②	**CNVII ncl. („Frontal branch")**	⑤	**Orbicularis oculi**
③	**CNVII ncl. („Oral branch")**	⑥	**Orbicularis oris**

Fig. 8.30 Comparison between (**A**) peripheral and (**B**) central facial paralysis. A lesion --- distal of the CNVII nucleus hits both, the "frontal branch" (main muscles occipitofrontalis, pars frontalis, and orbicularis oculi) and the "oral branch" (main muscle orbicularis oris). Clinically, on the lesioned side the forehead cannot be wrinkled, the eye cannot be closed. The mouth is distorted to the healthy side, i.e., none of the paradigmatically examined face muscles functions on the lesioned side. In case of a central paralysis (**B**), the clinical picture of the mouth region with the distortion of the mouth is identical to that of the peripheral paralysis. By contrast, the patient can open the eyes and wrinkle the whole forehead. The explanation lies in the central innervation of the CNVII nucleus. Although without known anatomical substrate, the nucleus obviously can be divided into a region supplying the "frontal branch" and another for the "oral branch." Since the "frontal branch" region receives a bilateral innervation, in case of a prenuclear lesion --- the occipitofrontal and orbicularis oculi muscles are still innervated. The "oral branch," however, gets only a contralateral cortical input which explains the orbicularis oris paralysis contralateral to the central lesion. Pictures modified from Fig. 8.31

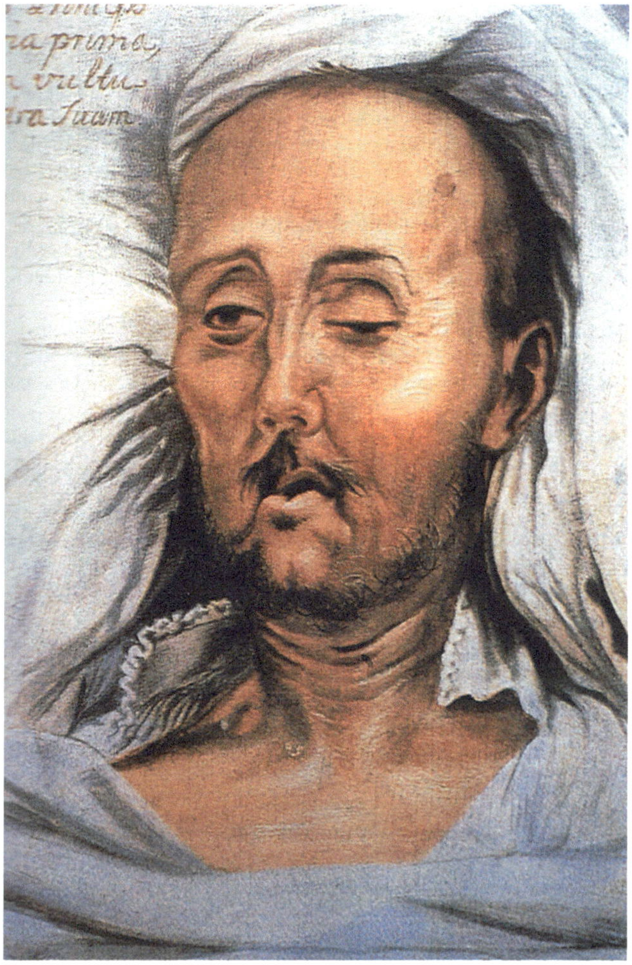

Fig. 8.31 Hans Mielich (1516–1573) Wilhelm IV (1493–1550) Duke of Bavaria, on his deathbed (Wilhelm IV., Herzog von Bayern, auf dem Totenbett). It has been reported that the Duke suffered from a stroke several days before his death. His mouth is drafted to the left, non-affected side. From the history of his case and the post-stroke course one would suppose that he suffered from a central facial paralysis. In this case, one would expect that he still could close his right eye. It is of course pure speculation whether he was not able to do it or the right eye was open voluntarily. The function of the frontal muscle can only be explored in case of cooperative patients. It was also speculated that the duke had a melanoma (left forehead, in the original paint visible) and that he possibly suffered leptomeningeal metastasis with tumor spreading along nerve rootlets. Munich, Germany, Bayerisches Nationalmuseum with permission

Table 8.2 Causes of peripheral facial paralysis (Berghaus and San Nicoló 2015)

Type of palsy	Cause
Idiopathic (Bell's palsy)	Mostly unknown, possibly following a virus infection, often in Diabetes mellitus patients and during pregnancy, ca. 50% of all cases
Traumatic	Fracture of the petrous bone (20% of all cases), face lesions, birth trauma
Inflammatory/ Infectious	Herpes zoster oticus (7% of all cases, see Fig. 12.18), Otitis media, cholesteatoma, and others
Iatrogen	Following surgery in the cerebello-pontine angle (see Fig. 8.30), surgery of the ear and the parotid gland (see Fig. 8.22)
Tumor	Neurinoma, schwannoma of cranial nerves VII and VIII (see Fig. 8.30), meningeoma, glomus tumor and malignant tumor of the parotid gland (see Fig. 8.22)
Syndromal	For example, Möbius syndrome is a rare congenital disorder characterized by complete or partial facial diplegia accompanied by other cranial nerve palsies. MRI shows brainstem hypoplasia with straightening of the fourth ventricle floor, indicating an absence of the facial colliculus (see Fig. 8.2) Pedraza et al. (n.d.)

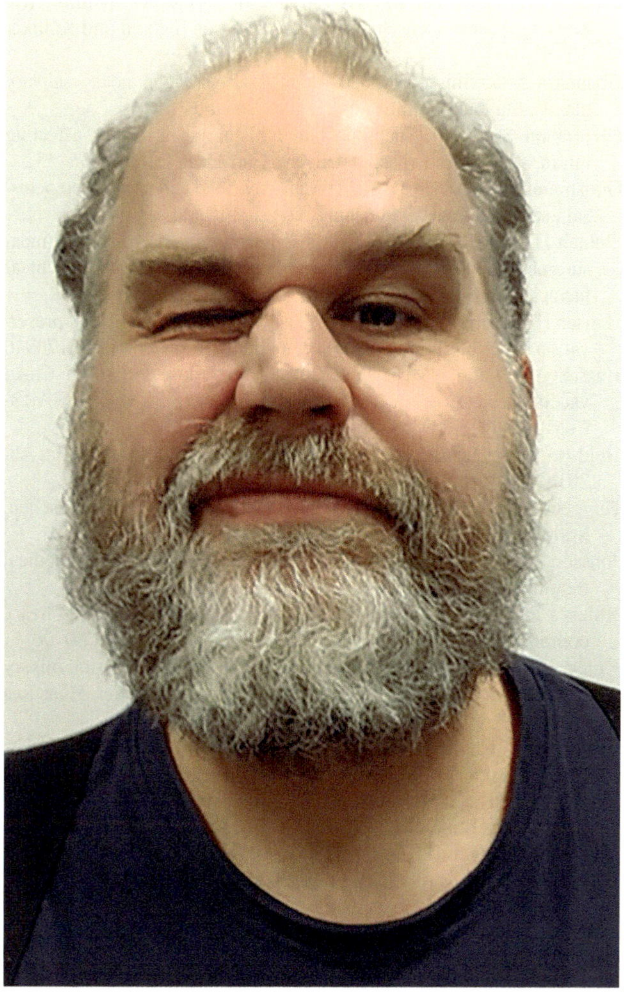

Fig. 8.32 Patient with right hemifacial spasm. Courtesy Dr. S. Huggenberger

References

Amunts K, Morosan P et al (2012) Auditory system. In: Mai JK, Paxinos G (eds) The human nervous system, 3rd edn. Academic, Amsterdam, pp 1270–1300

Arzberger T, Ritter E et al (1994) Human facial nucleus: choline acetyltransferase and calcitonin gene-related peptide. Eur Arch Otorhinolaryngol. S403-6

Ashwell KW (1982) The adult mouse facial nerve nucleus: morphology and musculotopic organization. J Anat 135:531–538

Avelar JM (2013) Surgical anatomy of the ear and neighboring regions. In: Avelar JM (ed) Ear reconstruction. Springer

Berghaus A, San Nicoló M (2015) Fazialisparese—wenn die Mimik erlischt. (Facial palsy—when facial expression disappears). MMW Fortschr Med. 157:42–46

Bonanni L, Anzellotti A et al (2007) Delayed blink reflex in dementia with Lewy bodies. J Neurol Neurosurg Psychiatry 78:1137–1139

Brown S, Isaacson B et al (2019) Facial nerve trauma: clinical evaluation and management strategies. Plast Reconstr Surg 143:1498–1512

Cant NB, Benson CG (2003) Parallel auditory pathways: projection patterns of the different neuronal populations in the dorsal and ventral cochlear nuclei. Brain Res Bull 60:457–474

Cauna N, Cauna D et al (1972) Innervation of human nasal glands. J Neurocytol 1:49–60

Chaudhry N, Srivastava A, Joshi L (2015) Hemifacial spasm: the past, present and future. J Neurol Sci 356:27–31

Corti A (1851) Recherches sur l'organe de l'ouïe des mammiferes. Zeitschrift für wissenschaftliche Zoologie von Siebold und Kölliker 3:1–63

Ghannam MG, Singh P (2020) Anatomy, head and neck, salivary glands. StatPearls Publishing, Treasure Island

Gizurarson S (2012) Anatomical and histological factors affecting intranasal drug and vaccine delivery. Curr Drug Deliv 9:566–582

Goosmann MM, Dalvin M (2018) Anatomy, head and neck, deep petrosal nerve. StatPearls Publishing, Treasure Island

Guinan JJ Jr (2018) Olivocochlear efferents: their action, effects, measurement and uses, and the impact of the new conception of cochlear mechanical responses. Hear Res 362:38–47

Hansen HB, Damgaard PB et al (2017) Comparing ancient DNA preservation in petrous bone and tooth cementum. PLoS One 12:e0170940

Hassan AO, Kafai NM et al (2020) A single-dose intranasal ChAd vaccine protects upper and lower respiratory tracts against SARS-CoV-2. Cell 183(1):169–184

Held H (1893) Die centrale Gehörleitung. Arch Anat Physiol, Anat Abt 201–248

Huggenberger S, Moser N et al. (2019) Neuroanatomie des Menschen. Springer

Jonsson L (2021) Facial paralysis: etiology, diagnosis, and medical treatment. In: Facial palsy. Springer, pp 3–21

Kulesza RJ Jr (2007) Cytoarchitecture of the human superior olivary complex: medial and lateral superior olive. Hear Res 225:80–90

Kulesza RJ Jr (2008) Cytoarchitecture of the human superior olivary complex: nuclei of the trapezoid body and posterior tier. Hear Res 241:52–63

Liston SL, Kleid MS (1989) Histopathology of bell's palsy. Laryngoscope 99:23–26

Magnusson AK, Gómez-Álvarez M (2019) The superior paraolivary nucleus. In: Kandler K (ed) The Oxford handbook of the auditory brainstem. Oxford University Press, Oxford, pp 395–420

Mahboubi H, Slattery WH III et al (2020) Options and strategies for hearing restoration in pediatric neurofibromatosis type 2. Childs Nerv Syst 36(10):2481–2487

Mansour Y, Kulesza RJ Jr (2020) Three dimensional reconstructions of the superior olivary complex from children with autism spectrum disorder. Hear Res 393:107974

Mansour Y, Altaher W et al (2019) Characterization of the human central nucleus of the inferior colliculus. Hear Res 377:234–246

Marx JJ (2003) Die elektrophysiologische Untersuchung des Blinkreflexes/electrophysiological investigation of the blink reflex. Klinische Neurophysiologie 34:8–14

Moore JK (2000) Organization of the human superior olivary complex. Microsc Res Tech 51:403–412

Moore JK, Simmons DD et al (1999) The human olivocochlear system: organization and development. Audiol Neurootol 4:311–325

Morest DK (1968) The collateral system of the medial nucleus of the trapezoid body of the cat, its neuronal architecture and relation to the olivocochlear bundle. Brain Res 9:288–311

Nturibi E, Bordoni B (2020) Anatomy, head and neck, greater petrosal nerve. StatPearls Publishing, Treasure Island

Olszewski J, Baxter D (1982) Cytoarchitecture of the human brain stem. Karger, Basel

Paxinos G, Furlong T, Watson C (2019) Human brainstem: cytoarchitecture, chemoarchitecture, myeloarchitecture. Academic, London

Pearce JMS (2008) Observations on the blink reflex. Eur Neurol 59:221–223

Pedraza S, Gámez J, Rovira A, Zamora A, Grive E, Raguer N, Ruscalleda J (n.d.) Neurology 55(7):1058–1060

Pelaez NM, Schreihofer AM et al (2002) Decompensated hemorrhage activates serotonergic neurons in the subependymal parapyramidal region of the rat medulla. Am J Physiol Regul Integr Comp Physiol 283:R688–R697

Pickles JO (2015) Auditory pathways: anatomy and physiology. In: Celesia GG, Hickok G (eds) The human auditory system: fundamental organization and clinical disorders. Elsevier, Amsterdam, pp 3–25

Rea P (2015) In: Essential clinical anatomy of the nervous system. Elsevier

Rosenstengel C, Matthes M, Baldauf J, Fleck S, Schroeder H (2012) Hemifacial spasm—conservative and surgical treatment options. Dtsch Arztebl Int 109:667–673

Sajadi MM, Sajadi M-R et al (2011) The history of facial palsy and spasm: hippocrates to Razi. Neurology 77:174–178

Schmidt E, Wolski Jr. TP, Kulesza Jr. RJ (2010) Distribution of perineuronal nets in the human superior olivary complex. Hearing Research 265:15–24

Schröder H, Moser N et al (2020) Neuroanatomy of the mouse. Springer

Seidel K, Mahlke J et al (2015) The brainstem pathologies of Parkinson's disease and dementia with Lewy bodies. Brain Pathol 25:121–135

Singh S, Basu S (2020) The human lacrimal gland: historical perspectives, current understanding, and recent advances. Curr Eye Res 45(10):1188–1198

Sitek KR, Gulban OF et al (2019) Mapping the human subcortical auditory system using histology, postmortem MRI and in vivo MRI at 7T. elife 8:e48932

Smith A, Storti S et al (2019) Structural and functional aberrations of the auditory brainstem in autism spectrum disorder. J Am Osteopath Assoc 119:41–50

Standring S (ed) (2005) Gray's anatomy: the anatomical basis of clinical practice, 39th edn. Elsevier, Philadelphia, pp 535–536

Standring S (ed) (2016) Gray's anatomy—the anatomical basis of clinical practice. 41st Ed. Elsevier Health Sciences

Takezawa K, Townsend G et al (2018) The facial nerve: anatomy and associated disorders for oral health professionals. Odontology 106:103–116

Tarabichi O, Kanumuri V et al (2019) Three-dimensional surface reconstruction of the human cochlear nucleus: implications for auditory brain stem implant design. J Neurol Surg Part B Skull Base 81:114–120

Tashi S, Purohit BS et al (2016) The pterygopalatine fossa: imaging anatomy, communications, and pathology revisited. Insights Imaging 7:589–599

ten Donkelaar HJ (2011) Clinical neuroanatomy: brain circuitry and its disorders. Springer
Terr LI, Edgerton BJ (1985) Surface topography of the cochlear nuclei in humans: two- and three-dimensional analysis. Hear Res 17:51–59
Thomas K, Minutello K et al (2020) Neuroanatomy, Cranial nerve 9 (Glossopharyngeal)
Tillmann BN (2005) Atlas der Anatomie des Menschen. Springer, Heidelberg
Toulgoat F, Sarrazin JL et al (2013) Facial nerve: from anatomy to pathology. Diagn Interv Imaging 94:1033–1042
Valstar MH, de Bakker BS et al (2021) The tubarial salivary glands: a potential new organ at risk for radiotherapy. Radiother Oncol 154:292–298
VanderHorst VG, Ulfhake B (2006) The organization of the brainstem and spinal cord of the mouse: relationships between monoaminergic, cholinergic, and spinal projection systems. J Chem Neuroanat 31:2–36
Vincent C (2012) Auditory brainstem implants: how do they work? Anat Rec (Hoboken) 295:1981–1086
Walker HK, Hall WD et al (eds) (1990). Chapter 62, Cranial nerve VII: the facial nerve and taste. Clinical methods: in: the history, physical, and laboratory examinations. 3rd edn. Butterworths, Boston
Yang H, Yuan PQ et al (2000) Activation of the parapyramidal region in the ventral medulla stimulates gastric acid secretion through vagal pathways in rats. Neuroscience 95:773–779

Web Links

https://en.wikipedia.org/wiki/Alfonso_Giacomo_Gaspare_Corti
https://de.wikipedia.org/wiki/Hans_Held_(Mediziner)
http://braininfo.rprc.washington.edu/centraldirectory.aspx?ID=719
http://braininfo.rprc.washington.edu/centraldirectory.aspx?ID=1771
http://slideplayer.com/slide/5693994/18/images/13/Muscles+of+The+Face+(Muscle+of+Facial+Expressions).jpg

Rhombomere 5 r5

9

Contents

H. Schröder et al., *The Human Brainstem*, https://doi.org/10.1007/978-3-030-89980-6_9

Abstract

The structures of the auditory pathways (see Sect. 9.1.6 Nuclei of the superior olivary complex) have been dealt together with the cochlear nuclei and other belonging structures in Chap. 8.

The superior olivary complex—not to confuse with the inferior olivary complex (see Sect. 4.1.1.3)—is an important relay station in the auditory pathways receiving bilateral input from the cochlea which subserves its function in sound localization.

The central gray will be described together with the periaqueductal portions in Chap. 13.

The nucleus of the abducens nerve (CNVI) is the most caudally located oculomotor nucleus. The abducens nerve leaves the brainstem in the bulbopontine sulcus and runs with the trochlear nerve (CNIV, see Sect. 14.5.2.1) and the oculomotor nerve (CNIII, see Sect. 15.3.1.1) through the cavernous sinus to the orbita and the eye muscles. The abducens nerve innervates the lateral rectus muscle, which moves the eyeball outwards.

Multiple sclerosis, a very frequent neurological disease is dealt with here from the neuropathological point of view because of its connection with internuclear ophthalmoplegia.

A neuropathologically important raphe nucleus is the raphe interpositus which is heavily affected in spinocerebellar ataxia type 3 (SCA3) (see Sect. 7.2.1.3.3).

9.1 Alar r5

9.1.1 Cochlear nuclei r5

9.1.1.1 Dorsal cochlear nucleus

For detailed information on the cochlear nuclei, see Sect. 8.1.1.

9.1.1.2 Ventral cochlear nucleus

For detailed information on the cochlear nuclei, see Sect. 8.1.1.

9.1.2 Monoamine nuclei r5

9.1.2.1 A5 noradrenaline cells

For detailed information on these cells, see Sect. 8.1.3.1.

9.1.3 Vestibular nuclei r5

9.1.3.1 Medial vestibular nucleus

For detailed information on this nucleus, see Sect. 7.2.1.1.

9.1.3.2 Spinal vestibular nucleus

For detailed information on this nucleus, see Sect. 5.3.1.1.

9.1.4 Interstitial nucleus of the vestibulocochlear nerve

For detailed information on this nucleus, see Sect. 8.1.2.3.

9.1.5 Trigeminal sensory nuclei r5

9.1.5.1 Spinal trigeminal nucleus

For detailed information on this nucleus, see Sect. 3.2.2.2.

9.1.6 Nuclei of the superior olivary complex r5

9.1.6.1 Dorsal periolivary region (DPO)

For detailed information on this region, see Sect. 8.1.1.5.

9.1.6.2 Lateral superior olive (LSO)

For detailed information, see Sect. 8.1.1.5.

9.1.6.3 Medial superior olive (MSO)

For detailed information, see Sect. 8.1.1.5.

9.1.6.4 Caudal periolivary nucleus (CPO)

For detailed information, see Sect. 8.1.1.5.

9.1.6.5 Superior paraolivary nucleus (SPO)

For detailed information, see Sect. 8.1.1.5.

9.1.6.6 Lateroventral periolivary nucleus (LVPO)

For detailed information, see Sect. 8.1.1.5.

9.1.6.7 Nucleus of the trapezoid body (Tz)

For detailed information, see Sect. 8.1.1.5.

9.1.7 Central gray r5

For detailed information on the central gray, see Sect. 13.2.7.

9.1.8 Supragenual nucleus [Ncl. supragenualis] PMT cell group 3

The supragenual nucleus is a group of neurons embedded in the pontine central gray of the pontine tegmentum (Fig. 9.1). It is located dorsal to the internal genu of the facial nerve (hence the name) under the ependyma of the fourth ventricle. It has been found in the human, the macaque, the rat, and the mouse. http://braininfo.rprc.washington.edu/centraldirectory.aspx?ID=1981

Besides the anatomical data nothing is known about the functional meaning of this nucleus in humans. It has been suggested to be involved in the head direction circuitry based on animal studies (for details, see Biazoli et al. 2006). Vestibular input is known to be crucial for the head direction network and rat data suggest that the supragenual nucleus might be an interface for vestibular input. Vestibular information is sent to the cortex via the anterior/ventral vestibulothalamic tract (see Sect. 5.3.1.1.2) which comprises projections from the vestibular nuclei to the nucleus prepositus (see Sect. 6.2.1.5) and the supragenual nucleus, further on to the anterior dorsal thalamus via the head direction network and via the posterior vestibulothalamic pathway with projections from the vestibular nuclei further on to the ventral posterior lateral nucleus of thalamus (Cullen 2019).

The term PMT alludes to the traditional terminology of the paramedian tract (PMT) neurons [Nuclei tractus paramediani] (s. FIPAT Ch. 1 endnote 46) as a collection of six or more clusters of neurons with a common projection to the monkey floccular region (Langer et al. 1985). In the human

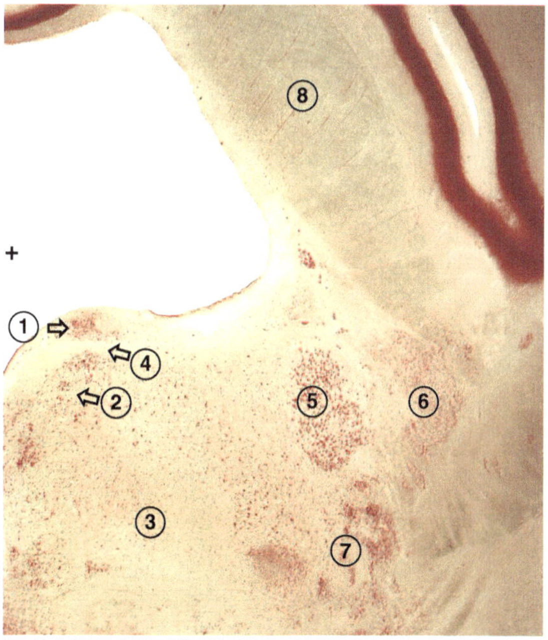

①	Supragenual ncl.	⑥	Principal sensory ncl. of trigeminal nerve
②	Dorsomedial tegmental area	⑦	Superior olivary complex
③	Pontine reticular ncl., caudal part	⑧	Superior cerebellar peduncle
④	Facial nerve / genu	+	4th ventricle
⑤	Motor ncl. of trigeminal nerve		

Fig. 9.1 Horizontal section through the human pons at the level of the supragenual nucleus ①. Darrow red staining. See atlas part Darrow red 23. LabPON Twente

brainstem, they have been numbered PMT1–PMT6. They lie around the paramedian fiber tracts (see Buresch 2005).

PMT cell group 1 is the medullary interfascicular nucleus, group 2 the pararaphe nuclei, group 3 the supragenual ncl., group 4a the rostral cap of the abducens nucleus, 4b neurons of the abducens nucleus projecting to the flocculus, group 5a the intrafascicular nucleus of the preabducent area, groups 5b + 5c the dorsal subnucleus of the nucleus raphes pontis, and group 6 the dorsal midline pontine group (see FIPAT endnotes 46 and 51).

9.2 Basal r5

9.2.1 Somatic motor nuclei r5

9.2.1.1 Abducens nucleus/Nucleus of abducens nerve [Ncl. n. abducentis]

9.2.1.1.1 Location and fine structure of the abducens nucleus

The abducens nucleus is the motor nucleus of CNVI (abducens nerve [N. abducens]) (see Fig. 9.2 ①), located in the pontine tegmentum. The cholinergic neuronal perikarya of this nucleus lie in direct ventromedial vicinity to the genu (inner knee) of facial nerve [Genu n. facialis] (see Sect. 8.2.1, Fig. 9.2 ②) beneath the facial colliculus [Colliculus facialis] (Fig. 9.3 ⑤) (Colliculus Latin = Small hill) in the floor of the fourth ventricle (Olszewski and Baxter 1982; Horn et al. 2018). As the neurons of the principal oculomotor (CNIII) (see Sect. 15.3.1.1) and trochlear nuclei (CNIV, see Sect. 14.5.2.1), those of the abducens nucleus are similar to other somatomotor neurons but in general more plump, smaller, only lightly stained and their Nissl granules are smaller and less regular (Olszewski and Baxter 1982). The nerve cells of the abducens nucleus can be visualized by choline acetyltransferase immunohistochemistry (Horn et al. 2018, Fig. 9.2C, E). Cholinergic fibers contained in CNVI and CNVII (Fig. 9.2C, E) are simultaneously visualized this way. Functionally, the CNVI cholinergic neurons can be differentiated according to the type of muscle fibers they innervate (Box 9.1) into singly (SIF) and multiply innervated (MIF) fiber neurons. Histochemically, they can further be differentiated by use of the marker chondroitin sulfate proteoglycan (CSPG) forming perineuronal nets around SIFs but not around MIFs (see Fig. 9.2).

Besides these "principal" neurons, the abducens nucleus displays the so-called internuclear neurons (INT) whose axons travel within the contralateral medial longitudinal fasciculus (MLF) (Box 9.2) (see Fig. 3.14A–G) to target the medial rectus muscle neurons in the contralateral oculomotor nucleus. The motoneurons as well as the INTs are innervated by the same premotor neurons in the prepositus hypoglossi nucleus (see Sect. 6.2.1.5), the paramedian pontine reticular formation (PPRF) [Formatio reticularis pontis paramediana]—includes most elements of the brainstem burst generator involved in the generation of horizontal saccades—and the medial vestibular nucleus [Ncl. vestibularis medialis] (see Sect. 7.2.1.1). These connections provide the basis for conjugate eye movements (Horn et al. 2018, see Boxes 9.2 and 9.6).

PMT neurons (see here above), first detected in monkeys, belong to small midline-near groups of neurons in the pontine and medullary reticular formation receiving input from premotor neurons of the oculomotor system (see Sect. 16.4) and project to the cerebellum (see Horn et al. 2018).

Fig. 9.2 (**A**) Horizontal histological section through the human brainstem at the level of the inner knee of the facial nerve. Darrow red staining. The CNVII ② originates in the CNVII nucleus ④ a certain distance caudal of the level shown here. It runs medially and cranially, then becomes visible here close to the midline dorsal of the medial longitudinal fasciculus (MLF). In its further course, it surrounds the CNVI nucleus ① (the so-called inner knee of the facial nerve ②, see Sect. 8.2.1.1) and turns to the ventral surface of the brainstem ⑤ where it exits at the cerebellopontine angle (not visible here, see Fig. 8.17). See atlas part Darrow red 20, Campbell 9B. (**B–E**) Horizontal histological sections through the human brainstem approximately at the level of (**A**). The sections have undergone double immunostaining: Choline acetyltransferase (ChAT)-immunoreactive (ir) structures appear black, CSPG (see here above)-ir ones brown (**C**, **E**). Both, SIF- and MIF motoneurons (MN) of CNVI are ChAT-positive appearing as black dots. Quantitative results of the immunostains are shown in the left-hand graphics. SIF-MNs appear as gray, MIF-MNs as red dots, INTs as open circles, PMTs as green dots and arrows in (**C**) and (**E**). (**A**) LabPON Twente. Scale bar = 500μm (applies to **B-E**). (**B-E**) From Horn et al. 2018, Fig. 5 with permission

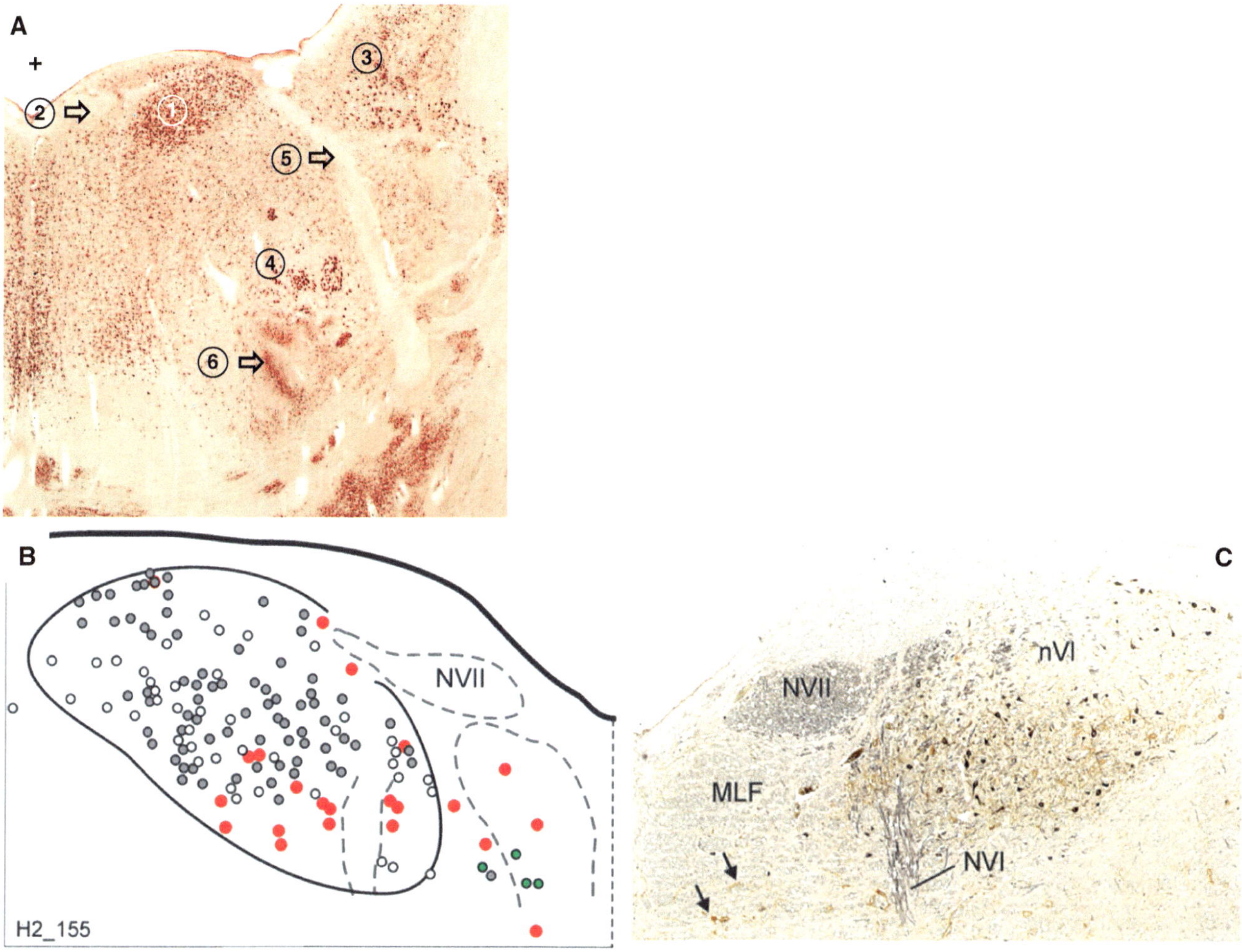

①	Ncl. of abducens nerve	nVI		⑥	Superior olivary complex	
②	Genu of facial nerve	NVII		+	4th ventricle	
③	Vestibular ncll.				Abducens nerve	NVI
④	Facial ncl.				Medial longitudinal fasciculus	MLF
⑤	Facial nerve, descending part					

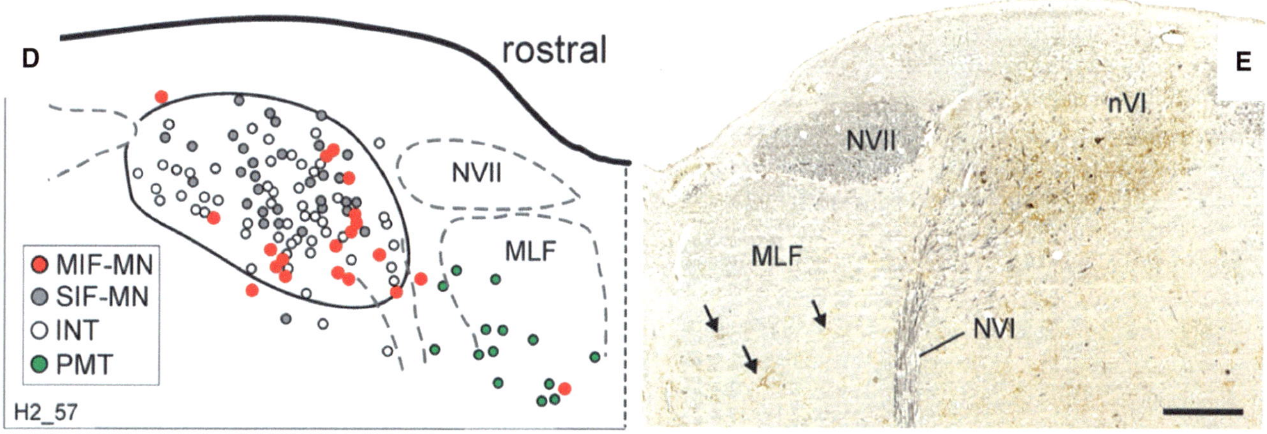

MIF-MN	Multiply innervated fiber neuron	INT	Internuclear neurons
SIF-MN	Singly innervated fiber neuron	PMT	Paramedian tract neurons

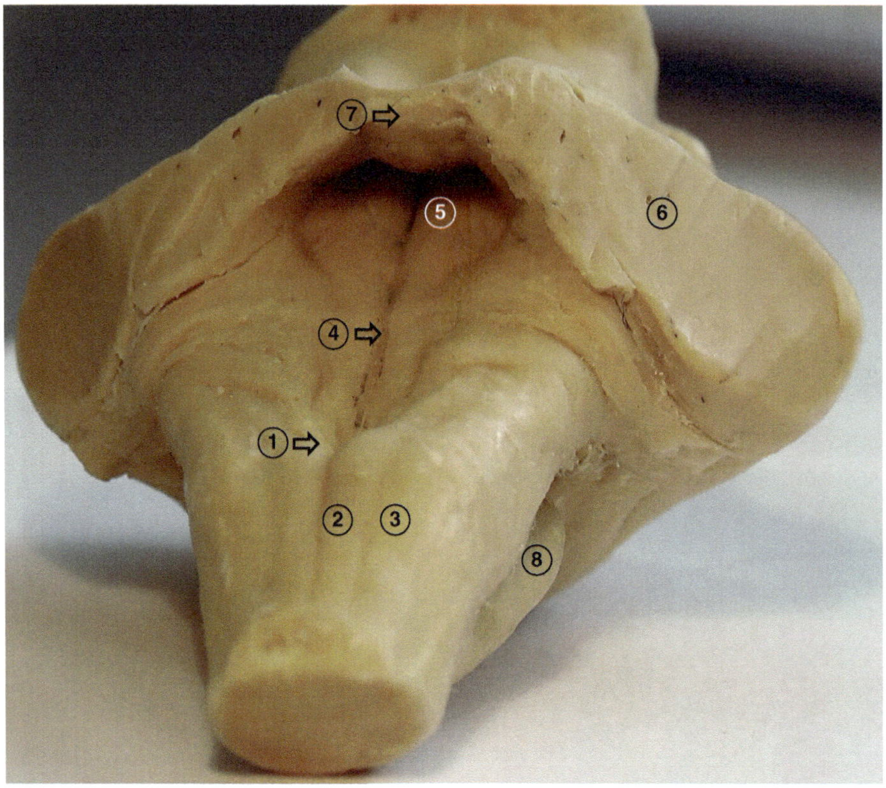

①	**Obex**	⑤	**Facial colliculus**
②	**Gracile tubercle**	⑥	**Medial cerebellar peduncle (cutting surface)**
③	**Cuneate tubercle**	⑦	**Superior medullary velum (cutting surface)**
④	**Posterior median sulcus (midline)**	⑧	**Inferior olive**

Fig. 9.3 Dorsal view of the human rhomboid fossa. Siliconized specimen. Cerebellum removed. At both sides of the posterior median, prominences are visible. In case of the facial colliculus ⑤, the elevation of the ventricle floor is due to the internal knee of the facial nerve (see Fig. 9.2). Directly ventral of the internal knee the abducens nucleus is located (see Fig. 9.2). For inferior olive also, see Fig. 4.2. Sammlung des Zentrums Anatomie der Universität zu Köln

Box 9.1 Eye muscles [Musculi bulbi]

Eye muscles are those muscles that move the eye bulb (external or extraocular eye muscles, [Musculi bulbi] (see Table 9.1 and Fig. 15.12) or regulate the width of the pupil or the conformation of the lens (internal eye muscles, see Sect. 15.3.1.1.2).

The extraocular eye muscles consist of different muscle fiber types (Horn et al. 2018). The distinction feature is the type of innervation by the human extraocular eye muscle neurons. In general, there are two morphologically different nerve terminals, (1) the en plaque endings and (2) the en grappe endings. The en plaque endings (also present on striated limb muscles) are the axon terminals on the so-called switch singly innervated muscle fibers (SIF). They react with a twitch upon electrical stimulation. The name is derived from the ultrastructural appearance of the terminals in a single isolated fashion (cf. Dietert 1965). The en grappe (la grappe, French = grape, here bunch of grapes) endings are the axon terminals on the so-called multiply-innervated muscle fibers (MIF) which display a slow tonic contraction after stimulation (Horn et al. 2018).

Electron microscopically, this ending displays numerous small drops or beads in a linear array (Dietert 1965). Both ending types belong to two different types of motoneurons in extraocular eye muscle nuclei, SIF, and MIF neurons (see Fig. 9.2) which both are cholinergic—visualizable by, for example, choline acetyltransferase—but can be differentiated by certain other markers (see Horn et al. 2018).

Box 9.2 Medial longitudinal fasciculus (MLF) [Fasciculus longitudinalis medialis]

The MLF (see Fig. 3.14) is a rather small fiber bundle stretching from the cervical spinal cord to the mesencephalon. In the second cervical "segment" of the spinal cord, it lies directly lateral of the anterior corticospinal tract. At the level of the pyramidal decussation (see Fig. 3.5), it is located lateral of the pyramids. In the medulla oblongata, it first takes a position lateral of the decussation of the medial lemnisci (see Fig. 3.14A ⑦), then reaches the top of the fiber column of the medial lemnisci. It remains in this position throughout the medulla oblongata, reaching the floor of the fourth ventricle in the pons near the abducens nucleus. At the transition to the mesencephalon, the MLF lies approximately in the middle part of the tegmentum on both sides of the midline (see Fig. 3.14D ②). In the midbrain, it assumes a position directly ventral of the reticular formation, getting in close contact with the nuclei of the trochlear and oculomotor nerve (see Fig. 3.14E ②) to reach a position medial of the red nucleus (Büttner-Ennever et al. 1982) (see Fig. 3.14F ② and G ⑥). It connects the three oculomotor nerve nuclei and also contains the medial vestibulospinal and tectospinal tracts, which innervate neck musculature and upper limbs (Rea 2015). Its descending part originates in the vestibular nuclei (Glover 2004) with input from CNVIII and cerebellum, processing proprioceptive information from head and neck. Descending fibers from the superior colliculus (integration of visual input) (see Sect. 15.1.3) are present in the MLF, which is connected to the pontine reticular formation (extensor muscle tone) (Rea 2015). This explains the suitability of the MLF to integrate eye with head movements and also provides the basis for certain features of eye motility. MLF fibers, that connect the CNVI nucleus to the contralateral CNIII nucleus, for example, are necessary to perform horizontal conjugate lateral gaze (see Box 9.6) (Fenichel 2009). It forms a major part of the optokinetic and vestibulo-ocular reflexes (see Box 5.4). To test the former, a patient follows an object with both smooth and saccadic eye movements (see Box 9.6) which allows following an object out of the field of vision and the eye returning to the position it held viewing the object initially (Rea 2015). The latter is a reflex which results in stabilizing the image which falls onto the retina by moving the eye in the opposite direction to which the head moves (Rea 2015). Fibers originating in the vestibular nucleus are involved in the vestibulo-ocular reflex (Rea 2015) (see Box 5.4). MLF lesions cause internuclear ophthalmoplegia (INO) (Box 9.3) (see also Sect. 10.3.5.4). For details on the human medial longitudinal fasciculus and additional clinical implications, see Frohman et al. (2008).

Box 9.3 Internuclear ophthalmoplegia (INO) (from Greek ὁ ὀφθαλμός = ho ophthalmos = eye and ἡ πληγή = he plege = paralysis)

INO is caused by lesions in the MLF (Box 9.2, see Fig. 3.14). The features of INO are (1) incomplete adduction of one eye on lateral gaze (i.e., due to the "weak" medial rectus muscle, see Table 9.1) and (2) jerk nystagmus (horizontal to-and-fro eye movements, the movement in one direction is faster than in the other) of the contralateral abducting eye (McGee 2018). For example, when the patient is asked to look to the far right (see Figs. 9.5, 9.11, and 9.12) and his left eye is unable to completely adduct and the right eye develops a jerk nystagmus, the patient has a left INO, i.e., a lesion in the left medial longitudinal fasciculus. When the same patient is asked to look to the left, both eyes move normally, but when the patient looks to the right, the left eye fails to adduct, and the contralateral eye develops a jerk nystagmus (McGee 2018). Ninety-seven percent of patients with ***bilateral*** INOs have multiple sclerosis (see Sect. 9.2.1.3), whereas ***unilateral*** INO has many causes, although the most common one is vertebrobasilar cerebrovascular disease (McGee 2018) (see Sect. 1.4).

9.2.1.1.2 Target muscle of the abducens nerve

The abducens nerve innervates the lateral rectus muscle [M. rectus lateralis] (see Fig. 9.4A, B, Box 9.1 "Eye muscles" and Table 9.1 above).

9.2.1.1.3 Course of the abducens nerve

The CNVI fibers leave the nucleus (Fig. 9.6A) at its ventral margin, run through the brainstem (Fig. 9.6B) and exit in the medullopontine sulcus [Sulcus bulbopontinus, from the old term bulbus lat. onion for the medulla oblongata] near the midline (see Fig. 9.6C) just lateral of the corticospinal tract (Somani and Adesina 2019).

After entering the subarachnoid (leptomeningeal) space ventral of the pons, the fibers run rostrally along the surface of the clivus (Fig. 9.7). They pass under the petroclinoid ligament through Dorello's canal (see Box 9.4 and Fig. 9.7) and then enter the cavernous sinus (see Figs. 9.8, 9.9 and Box 9.5) (cf. Somani and Adesina 2019; Ambekar et al. 2012).

Table 9.1 Human external/extraocular eye muscles (see Figs. 9.4 and 15.8). For the internal eye muscles, see Sect. 15.3.1.1.2

Muscle Innervating nerve	Main functions/muscle moves eye
Lateral rectus muscle [M. rectus lateralis] Abducens nerve CNVI	Laterally
Medial rectus muscle [M. rectus medialis] Oculomotor nerve CNIII	Medially
Superior rectus muscle [M. rectus superior] Oculomotor nerve CNIII	Upwards
Inferior rectus muscle [M. rectus inferior] Oculomotor nerve CNIII	Downwards
Inferior oblique muscle [M. obliquus superior] Oculomotor nerve CNIII	Upwards/outward rotation
Superior oblique muscle [M. obliquus superior] Trochlear nerve CNIV	Downwards/inward rotation

Humans lack the retractor bulbi muscle which retracts the bulb into the orbita (see Sect. 9.2.1.2)

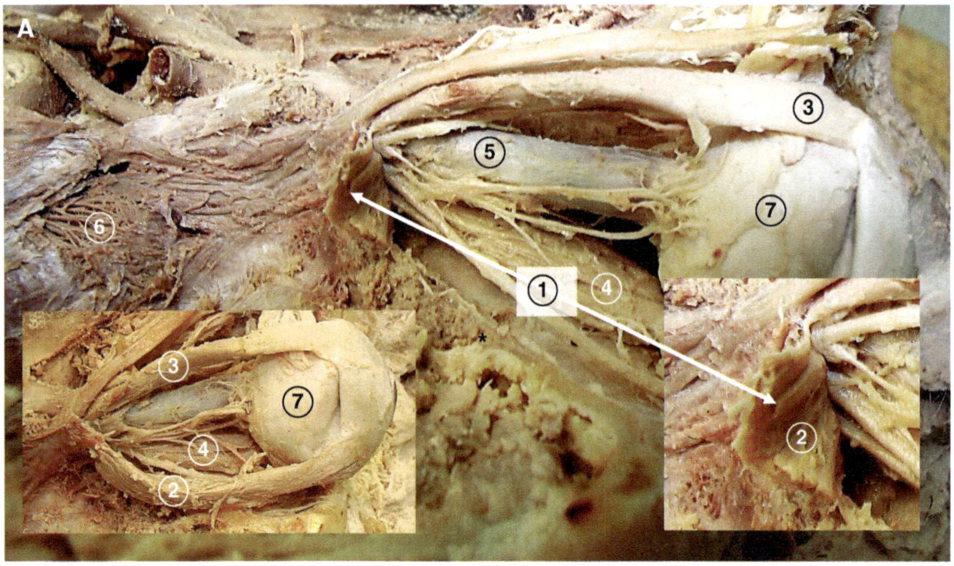

①	**Abducens nerve CNVI upon the cut and laterally bent lateral rectus muscle ②**	④	**Inferior rectus muscle CNIII**
		⑤	**Optic nerve**
②	**Lateral rectus muscle**	⑥	**Trigeminal ganglion**
③	**Superior rectus muscle CNIII**	⑦	**Ocular bulb**

Fig. 9.4 (**A**) Lateral view onto the human orbita. Formalin-fixed dissected specimen. Lateral orbital wall and roof, the levator palpebrae superioris muscle, orbital fat, and the lacrimal gland have been removed. The lateral rectus muscle ② is located directly beneath the lateral orbital wall and moves the eye laterally (abduction, hence the name of CNVI). Müller-Thomsen fecit. Sammlung des Zentrums Anatomie der Universität zu Köln

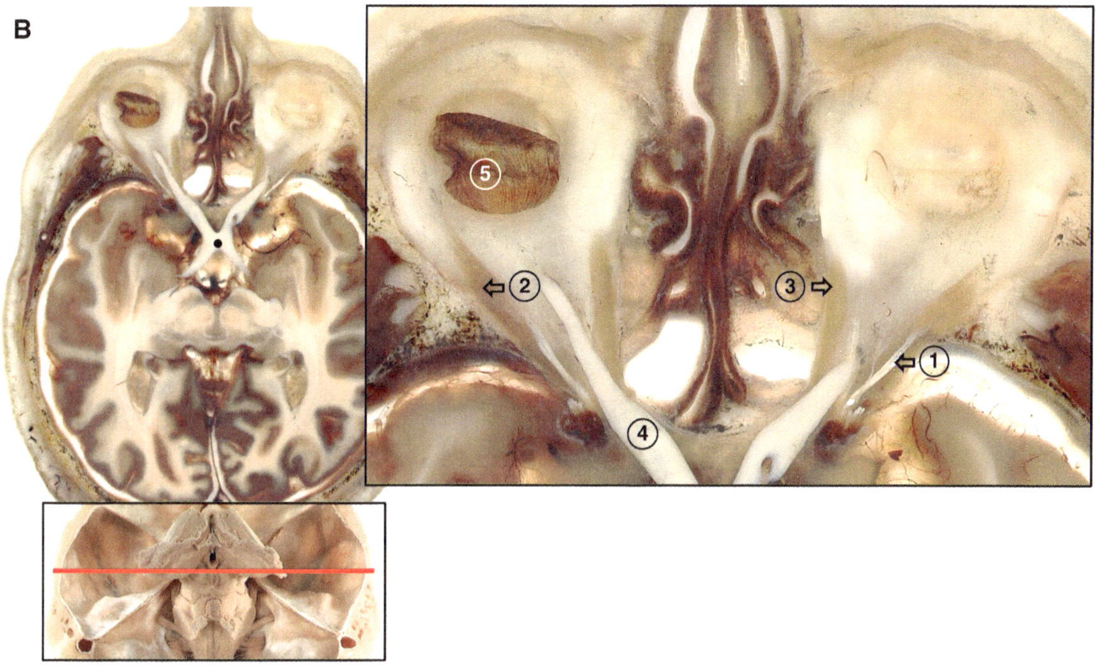

1	Abducens nerve CNVI	4	Optic nerve CNII
2	Lateral rectus muscle CNVI	5	Retina, partly detached
3	Medial rectus muscle CNIII		

Fig. 9.4 (continued) (**B**) Horizontal section through the human head at the level indicated in the lower inset which shows the optic chiasm as a major landmark (— approximate level of section). Lateral to the optic nerve the abducens nerve enters the lateral rectus muscle. Sammlung des Zentrums Anatomie der Universität zu Köln

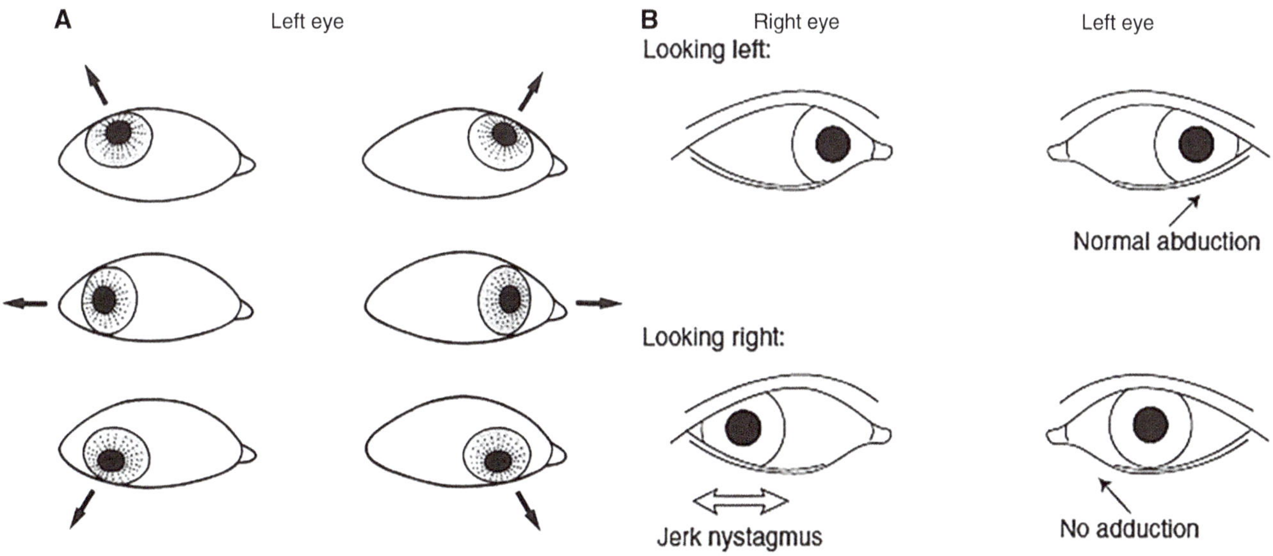

Fig. 9.5 Schematic representation of normal eye motility (**A**) (six cardinal movements) (shown for the left eye). (**B**) Findings in internuclear ophthalmoplegia (INO). The patient is asked to look to the left, the right eye performs an adduction (toward the nose), the left eye an abduction (conjugate eye movements) (see Sect. 10.3.5.4, no diplopia). When the patient is asked to look to the far right and his left eye is unable to completely adduct and the right eye develops a jerk nystagmus, the patient has a left INO, i.e., a lesion in the left medial longitudinal fasciculus (see Box 9.2) (see also Figs. 9.11 and 9.12). A jerk nystagmus is a slow drift of the eyes in one direction, followed by a rapid recovery movement, always described in the direction of the recovery movement. (**A**) Ribbers fecit. (**B**) from McGee 2012, Fig. 57.5 with permission

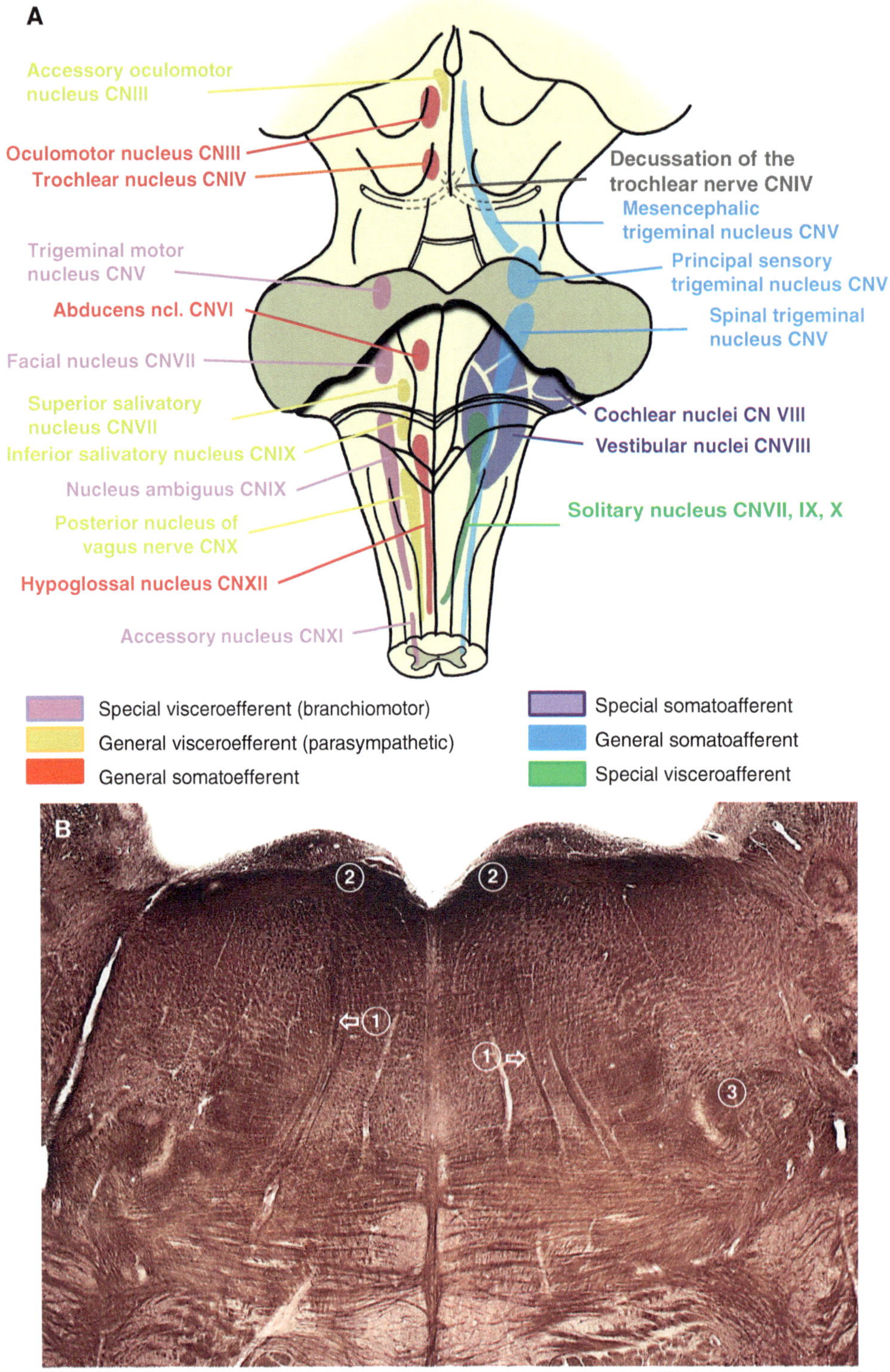

①	Abducens nerve CNVI	③	Superior olivary complex
②	Inner knee of facial nerve CNVII		

Fig. 9.6 (**A**) Schematic view of the human cranial nerve nuclei. The CNVI nucleus belongs to the general somatoefferent nuclei, together with the oculomotor, trochlear, and hypoglossal nuclei shown in red. From Huggenberger et al. 2019, Fig. 15.5 with permission. (**B**) Horizontal section through the human pons at the level of the inner knee of the facial nerve ②. Campbell fiber staining. Slightly ventral of the latter the CNVI nucleus is located (compare with Fig. 9.2). The axon bundles of the nucleus are running in ventral direction where they leave the brainstem in the bulbopontine sulcus (**C**). See atlas part Campbell 9B, 10A. LabPON Twente

C

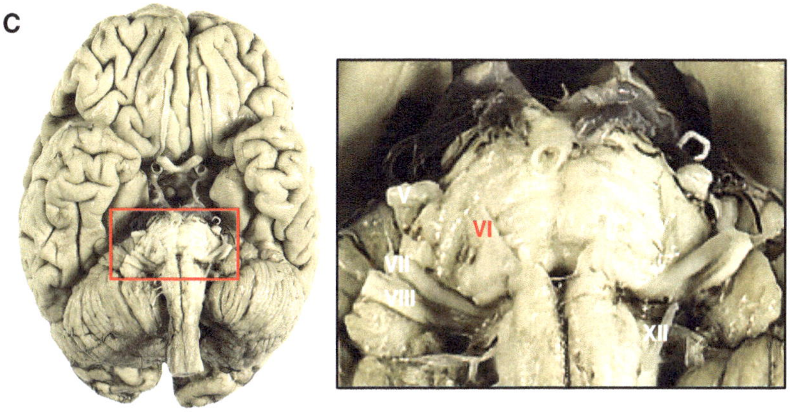

V	Trigeminal nerve	VIII	Vestibulocochlear nerve
VI	Abducens nerve	XII	Hypoglossal nerve
VII	Facial nerve		

Fig. 9.6 (continued) (**C**) Ventral view onto the human brainstem. The photograph shows the exit of the CNVIII from the medullopontine sulcus which separates the caudally located medulla oblongata from the cranially located pons. LabPON Twente

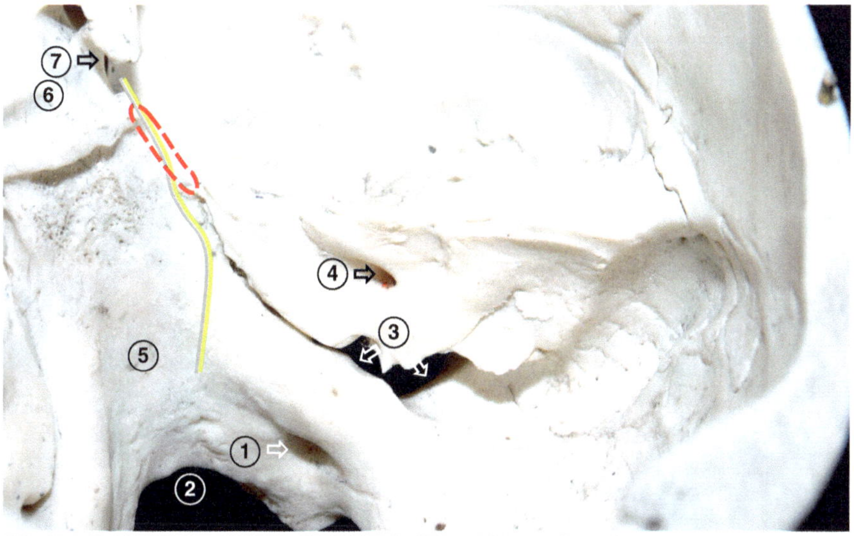

①	Canal of hypoglossal nerve CNXII	⑥	Sella turcica
②	Foramen magnum	⑦	Superior orbital fissure
③	Jugular foramen	(red dashed line)	Location of the petroclinoid ligament
④	Internal acoustic meatus	(yellow line)	Approximate course of the abducens nerve CNVI
⑤	Clivus		

Fig. 9.7 Dorsal view onto the human skull base. Graphic reconstruction of the location of the petroclinoid ligament (---) and the course of the abducens nerve (—). The latter runs in the canal of the abducens nerve (Dorello) under the petroclinoid ligament (Gruber). This is a predilection site for lesions of the abducens nerve resulting in loss of abduction of the eye bulb. From here the abducens nerve reaches the superior orbital fissure and enters the orbita (see Figs. 9.8 and 9.10). Sammlung des Zentrums Anatomie der Universität zu Köln

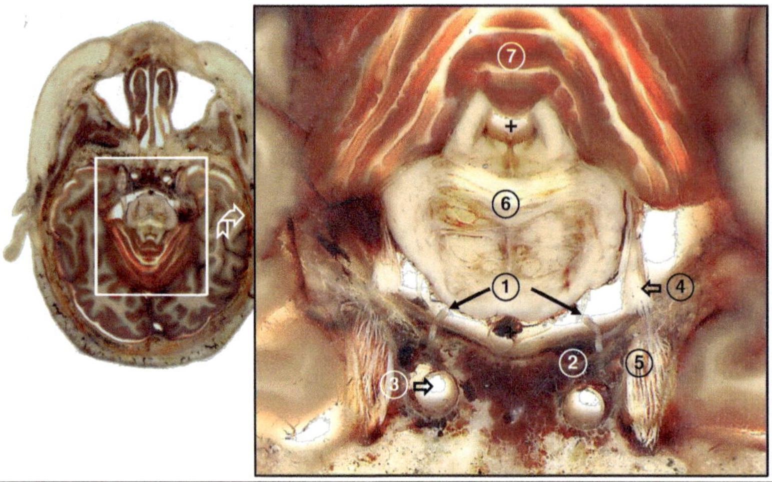

①	Abducens nerve CNVI	⑤	Trigeminal ganglion CNV
②	Cavernous sinus	⑥	Pons
③	Internal carotid artery	⑦	Cerebellum
④	Trigeminal nerve CNV	+	4th ventricle

Fig. 9.8 Horizontal section through the human head at the level of the trigeminal ganglion ⑤. Plastinated specimen. In the picture, detail the abducens nerve ① is visible in the terminal ascending portion before it turns rostrally and enters the cavernous sinus ② (see Box 9.5). The internal carotid artery ③ is running through the cavernous sinus (see Dolenc 1989). Sammlung des Zentrums Anatomie der Universität zu Köln

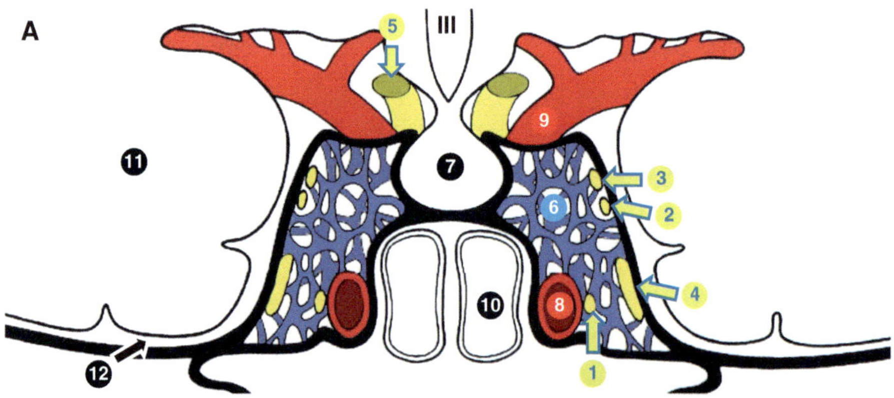

①	Abducens nerve CNVI	⑧	Internal carotid artery, cavernous part
②	Trochlear nerve CNIV	⑨	Middle cerebral artery
③	Oculomotor nerve CNIII	⑩	Sphenoid sinus
④	Ophthalmic nerve CNV_1	⑪	Telencephalon
⑤	Optic nerve	⑫	Leptomeningeal space
⑥	Cavernous sinus / trabeculae	▬	Dura mater
⑦	Pituitary gland	III	3rd ventricle

Fig. 9.9 (**A**) Schematic frontal/coronal section through the human head. The midline is indicated by the unpaired pituitary gland ⑦ and the paired sphenoidal sinus ⑩ [Sinus sphenoidalis]. Lateral of both the trabecular meshes of the cavernous sinus ⑥, in vivo filled with venous, deoxygenated blood. Its walls are formed by the dura mater (-). The abducens ①, trochlear ②, and oculomotor nerves ③ run through the cavernous sinus rostrally to the superior orbital fissure (see Fig. 9.7 ⑦). CNIII and CNIV are located in the lateral wall while CNVI runs freely through the cavernous sinus close to the internal carotid artery ⑧ whose cavernous part is located in the cavernous sinus. Ribbers, H. Schröder, and Huggenberger fecerunt

B

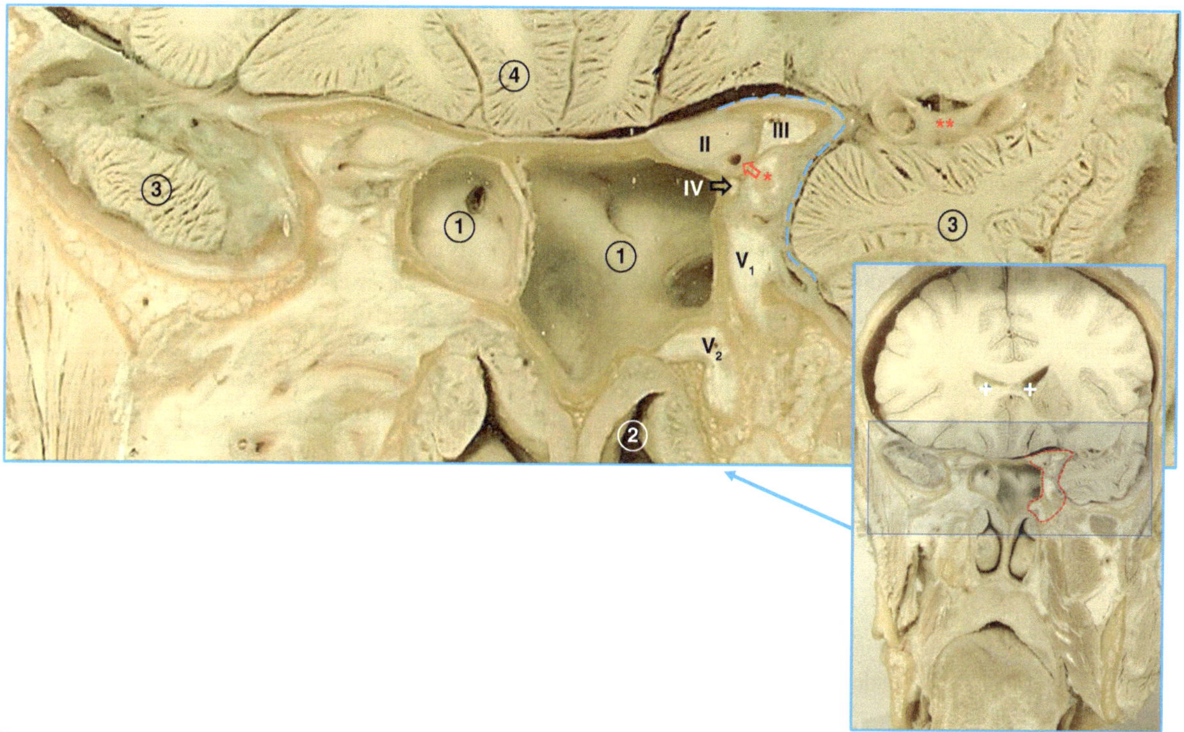

---	Dorsolateral border of cavernous sinus	④	Frontal lobe
①	Sphenoid sinus	*	Ophthalmic artery
②	Nasopharynx	**	Middle cerebral artery in lateral sulcus
③	Temporal lobe	+	Lateral ventricles

Fig. 9.9 (continued) (**B**) Frontal view on frontal/coronal section through the human head at the level of the sphenoidal sinus ① (compare with **A**). Formalin-fixed specimen. Note that the section is oblique so that the bilateral cavernous sinus is only seen on the left in the wall of the sphenoid sinus (see red outline in the survey photograph). The optic nerve II is situated in the roof of the sinus, directly medial of the oculomotor nerve III. Inferior to the optic nerve, the ophthalmic artery is seen, followed in ventral direction by the trochlear nerve IV, the ophthalmic nerve V_1, and the maxillar nerve V_2.

Photograph by MedizinFotoKöln, Cologne, Germany. Sammlung des Zentrums Anatomie der Universität zu Köln

Box 9.4 Petroclinoid ligament [Ligamentum petroclinoideum]/Dorello's canal [Canalis nervi abducentis, Abducens nerve canal]

The Ligamentum petroclinoideum (also sphenopetrosum superius) is a fibrous structure connecting the lateral margin of the dorsum sellae (posterior clinoid process) and the upper margin of the Pars petrosa ossis temporalis (petrous ridge) (FIPAT Endnote 4). The petroclinoid ligament of Gruber (Wenzel Gruber, Russian anatomist, 1814–1890) is the roof of Dorello's fibroosseous canal [Canalis nervi abducentis] (Primo Dorello, Italian anatomist, 1872–1963) (FIPAT Endnote 5). They then pass through the cavernous sinus (Box 9.5, Figs. 9.8 and 9.9) in close association with the internal carotid artery and penetrate the orbita via the superior orbital fissure, within the common tendinous ring [Anulus tendineus communis (of Zinn)], to innervate the ipsilateral lateral rectus muscle. The annulus of Zinn (Johann Gottfried Zinn, German anatomist and botanist, 1727–1759) is a fibrous structure, at the apex of the orbita, in which the four rectus muscles of the eye are posteriorly inserted or anchored. Through it, optic nerve, and ophthalmic vessels, as well as other minor nerves and vessels, pass from the orbit to the eye bulb. It is strictly adherent to the optic nerve dural sheath and to the surrounding periosteum (Zampieri et al. 2015).

Box 9.5 Cavernous sinus [Sinus cavernosus]

The cavernous sinus belongs to a group of intracranial venous vessels, the so-called Dural venous sinuses [Sinus durae matris] (see Fig. 3.11). They are venous channels located intracranially between the endosteal and meningeal layer of the cranial dura mater [Dura mater cranialis]. Unlike other veins in the body, they do not run parallel to arteries and are valveless, allowing for bidirectional blood flow in intracranial veins (Al Kabbani et al. n.d.).

The cavernous sinuses, one on each side, surround the internal carotid arteries [Aa. carotides internae] and are situated on the body of the sphenoid bone (Danesh-Meyer 2012, Jacob 2007; see Figs. 9.8, 9.9, and 3.11). It extends from the superior orbital fissure [Fissura orbitalis superior] to the apex of the petrous temporal bone. Medially, the cavernous sinus is related to the pituitary gland and the sphenoidal sinus [Sinus sphenoidalis] (see Figs. 9.8 and 9.9). Laterally, it is related to the temporal lobe of the brain (Jacob 2007). The lateral wall has two dural layers between which the cranial nerves III, IV, and the ophthalmic division of the trigeminal nerve (CNV_1) (see Sect. 12.2.3.2.3) travel downwards. The maxillary division of the trigeminal (CNV_2) (see Sect. 12.2.3.2.3) and part of the trigeminal ganglion (see Sect. 12.2.3.2.3) may lie in the inferolateral wall of cavernous sinus (see Fig. 9.9), the CNV_2 sometimes just outside the sinus. The abducens nerve runs free in the cavernous sinus close to the internal carotid artery (Danesh-Meyer 2012). The endothelial lining separates these structures from the cavity of the sinus. The sinus is connected with the transverse/sigmoid sinus [Sinus transversus/sigmoideus]—which eventually drains into the internal jugular vein [Vena jugularis interna]—via the superior and inferior petrosal sinus [Sinus petrosus superior/inferior] (Fig. 3.11). Rostrally, it is in contact with the ophthalmic veins [V. ophthalmica] of the face. The cavernous sinuses of either side are connected to each other by parts of the sinuses rostral and caudal of the pituitary gland (for details, see Jacob 2007). A detailed demonstration of the basic and surgical anatomy of the cavernous sinus and its contents has been provided by Dolenc (1989).

After leaving the cavernous sinus, CNVI becomes visible between the optic nerve on its medial side and the lateral rectus muscle on the lateral side (Fig. 9.10).

A three-dimensional reconstruction of the course of the three oculomotor nerves based on a sliced anatomical specimen of the human head has been provided by Park et al. (2015).

9.2.1.1.4 Functional considerations

In addition to the innervation of the ipsilateral lateral rectus muscle of the eye, the abducens nucleus is involved in gaze holding and conjugate horizontal gaze during the vestibulo-ocular reflex and in the generation of saccades (Somani and Adesina 2019) (see Box 9.6).

Box 9.6 Basic eye movements

There are four basic types of eye movements: (1) vestibulo-ocular movements, (2) saccades, (3) smooth pursuit movements, and (4) vergence movements (Purves et al. 2001, see also for details).

1. Vestibulo-ocular movements (VOM)
   The function of the VOMs is to maintain the orientation of the eyes in space during head movements so that the visual image remains stable on the retina, accomplished by the generation of eye movements that are equal and opposite to head movements. The vestibulo-ocular reflex is generated by sensory signals from the labyrinthine canals and otoliths (see Figs. 5.4 and 5.5), relayed through the vestibular nuclei (see Sect. 5.3.1), where the signals are modulated by the cerebellum, then fed to the extraocular muscle motoneurons in the oculomotor, trochlear, and abducens nucleus via the medial longitudinal fasciculus (MLF) [Fasciculus longitudinalis medialis] (see Box 9.2) or the superior cerebellar peduncle. The VOM is a slow compensatory eye movement, and during a period of continuous stimulation it is interrupted by fast resetting saccades, giving rise to the typical fast and slow phases of vestibular nystagmus (involuntary oscillation of one or both eyes about one or more axes) (Horn and Adamczyk 2012). In the absence of such compensatory eye movements as mediated by the VOMs, the visual image would shift wildly, particularly during nonvoluntary head movements and visual orientation would be extremely difficult if not impossible.
2. Saccades
   Saccades are rapid, ballistic movements of the eyes that abruptly change the point of fixation. They range in amplitude from the small movements made while reading, for example, to the much larger movements made while gazing around a room. Saccades can be elicited voluntarily, but occur reflexively whenever the eyes are open, even when fixated on a target (Purves et al. 2001).
3. Smooth pursuit movements
   Smooth pursuit movements are much slower tracking movements of the eyes designed to keep a moving stimulus on the fovea. Such movements are under voluntary control in the sense that the observer can choose whether or not to track a moving stimulus (Purves et al. 2001).
4. Vergence movements
   Vergence movements align the fovea of each eye with targets located at different distances from the

observer. Unlike other types of eye movements in which the two eyes move in the same direction (conjugate eye movements), vergence movements are disconjugate (or disjunctive); they involve either a convergence or divergence of the lines of sight of each eye to see an object that is nearer or farther away. Convergence is one of the three reflexive visual responses elicited by interest in a near object. The other components of the so-called near reflex triad are accommodation of the lens (see CNIII Sect. 15.3.1.2), which brings the object into focus, and pupillary constriction (CNIII), which increases the depth of field and sharpens the image on the retina (Purves et al. 2001).

9.2.1.1.5 Living anatomy and clinical implications

A simple clinical test to get information about the motility of the extraocular eye muscles and conjugate eye movements consists in having the proband or patient look in the six cardinal positions of gaze bi- and monocularly (see Figs. 9.11 and 9.12) (Kheradmand and Zee 2012).

Congenital absence or hypoplasia of the abducens nucleus may occur with Möbius syndrome (see Table 8.2), a rare neurological disorder involving weakness or paralysis of multiple cranial nerves, most often CNVI and CNVII. Patients present with facial diplegia along with absent abduction or absent horizontal gaze bilaterally. Hypoplasia or atrophy of cranial nerve nuclei may be visible on neuroimaging, and most cases are thought to be due to a prenatal vascular insult to the brainstem.

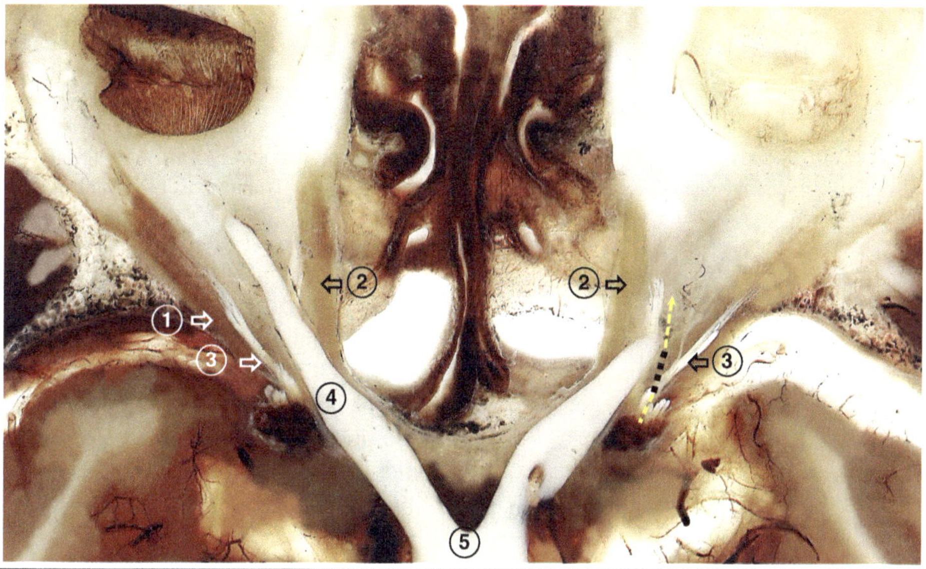

①	Lateral rectus muscle CNVI	---	Course of the inferior branch of CNIII --- ventral of CNII
②	Medial rectus muscle CNIII	④	Optic nerve CNII
③	Abducens nerve CNVI	⑤	Optic chiasm CNII

Fig. 9.10 Horizontal section through the human head at the level of the optic chiasm ⑤. Plastinated specimen. Dorsal view. The abducens nerve ③ is visible upon its entry into the orbita lateral of the optic nerve ④. It lies medial of the lateral rectus muscle ① into which the nerve ramifies. The medial rectus muscle ②—innervated by the oculomotor nerve (CNIII)—is the functional antagonist of the lateral rectus muscle leading to adduction of the eyeball upon contraction. Sammlung des Zentrums Anatomie der Universität zu Köln

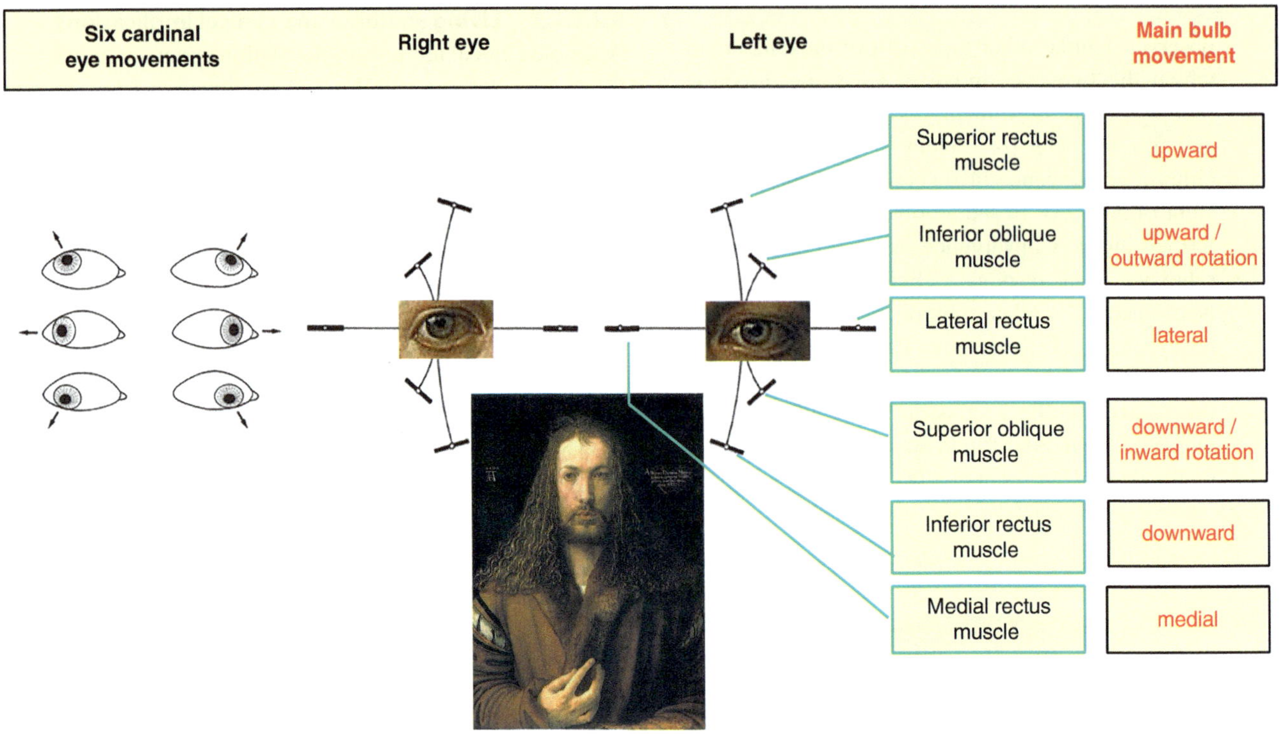

Fig. 9.11 Schematic representation of the six cardinal movements (here right eye) (see also Fig. 9.12) and the main directions into which the individual eye muscles move the eye bulb. Representations of the eyes taken from Albrecht Dürer: Selbstbildnis im Pelzrock (Self-Portrait with Fur-Trimmed Robe), 1500, Alte Pinakothek, Munich, Germany

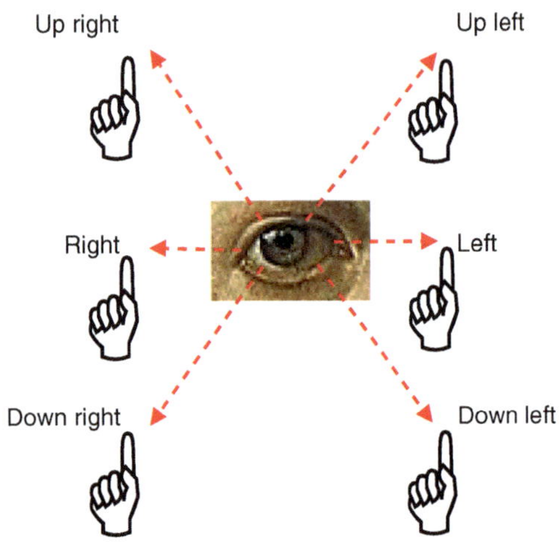

Fig. 9.12 Clinical examination of the six cardinal movements by following the examiner's index finger (see also Fig. 9.11)

9.2.1.2 Abducens nucleus, retractor bulbi part

While rodents and primates dispose of a functioning retractor bulbi muscle which retracts the ocular bulb into the orbita, in man, there are only certain muscular slips originating from some of the rectus muscles or the common tendinous ring (for details, see Haladaj et al. 2018).

9.2.1.3 Multiple sclerosis

We deal with multiple sclerosis in this place since you have seen in Box 9.3 that bilateral internuclear ophthalmoplegia is almost pathognomonic for multiple sclerosis.

Multiple sclerosis (MS) is an inflammatory, immune-mediated, demyelinating disorder that affects more than two million individuals worldwide (Reich et al. 2018). In its classic form, the clinical presentation is that of a relapsing-remitting disease with dissemination of lesions in time and space.

The mean age of onset is approximately 30 years, but the disease may first be manifest in patients over the age of 50 years and under the age of 15 years (Moore and Stadelmann-Nessler 2015). More than 90% of multiple sclerosis patients have onset before 50 years of age (Noseworthy et al. 2000). There seems to be an increase in incidence of multiple sclerosis in women (Orton et al. 2006). Males more frequently develop progressive disease than do females (Coyle 2005).

The type and severity of neurological impairment in individual patient is highly variable and the symptoms relate to sites of lesion predilection and include the visual system (optic neuritis) and spinal cord, the latter resulting in limb paresthesias and paralysis and bladder and bowel disturbances. Incoordination and gait abnormalities due to brainstem and cerebellar lesions are common manifestations (Moore and Stadelmann-Nessler 2015). Typically, patients can show episodes of visual loss in one eye, limb weakness, or sensory loss that can recover partly or completely. Many other symptoms can occur, including ataxia, double vision, loss of bladder control, and also cognitive decline (Richards et al. 2002).

Three different clinical forms of MS can be defined: primary progressive (10–15%), relapsing remitting (55–70%) and secondary progressive MS (15–30%) (Moore and Stadelmann-Nessler 2015). The disease can protract for decades with, in the end, a secondary progressive form. Occasionally, an aggressive, rapidly progressive form is seen that can affect almost all white matter (tumefactive MS or Marburg syndrome) (Nagappa et al. 2013).

The presence of lymphocytes and oligoclonal bands in the cerebrospinal fluid, the effect of plasmaphereses and the effect of steroids suggest involvement of autoantibodies. Family members have a higher risk for MS and the most significant gene involved is the human leukocyte antigen DRB1*1501 haplotype 3, suggesting indeed an autoimmune disorder.

The typical MS lesions can be detected by MRI (Thompson et al. 2018). Diagnostic criteria should include MR imaging of the spinal cord (Polman et al. 2011). New medicine, especially effective immunotherapy, can be effective and can slow down the inflammation (Kalincik et al. 2017).

Pathologically, myelinated fibers in both white and gray matter of the brain, cerebellum, and spinal cord can be affected. Furthermore, the optic nerve is often involved (Popescu and Lucchinetti 2012). The typical lesions, called MS plaques, are macroscopically sharply demarcated gray to brown lesions in the white matter that can extend into the gray matter or can even totally be located in the gray matter. They are situated particularly in the periventricular region (see Fig. 9.13).

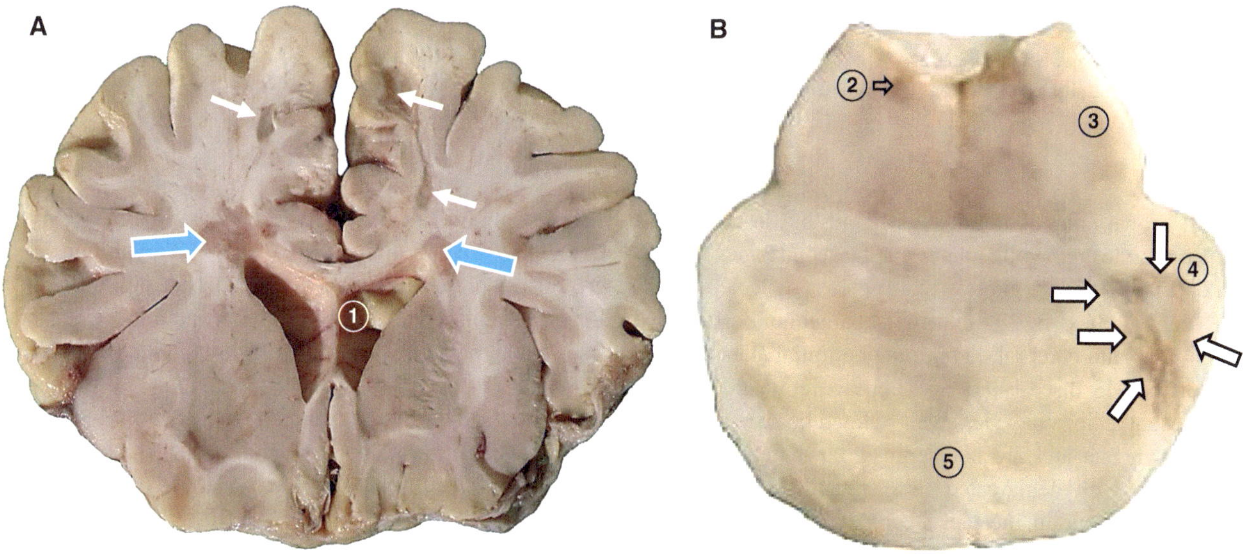

1	Lateral ventricle	4	Middle cerebellar peduncle
2	Locus caeruleus	5	Pons with corticospinal tract
3	Superior cerebellar peduncle		

Fig. 9.13 (**A**) Coronal section of a brain of an MS patient, multiple plaques in the white matter (white arrows). Note the sharp demarcation of these light brown MS lesions. Blue arrows indicate periventricular lesions. (**B**) Pons, slightly depressed lesion (white arrows). Collection A. Rozemuller

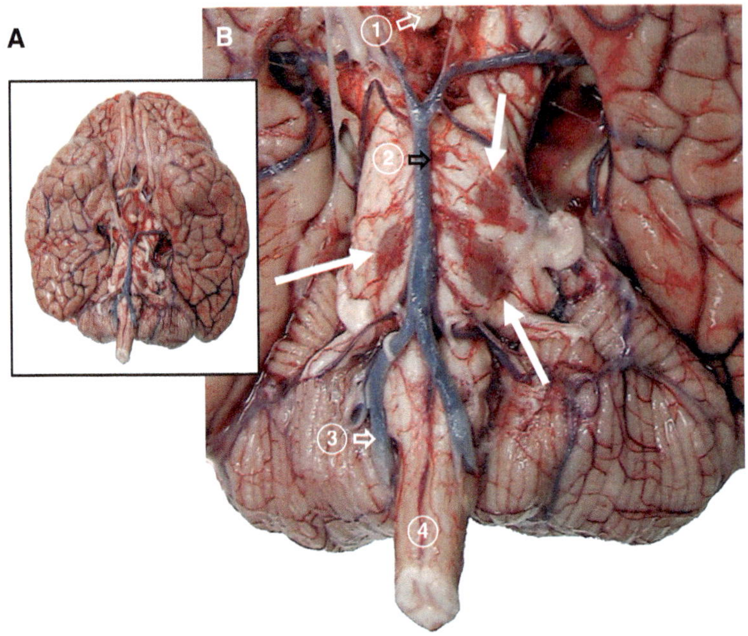

①	Mammillary body	③	Vertebral artery
②	Basilar artery	④	Medulla oblongata

Fig. 9.14 Ventral view of the brain of an MS patient (fresh specimen) (A) Overview, (B) Detail. External appearance showing several, brownish MS plaques on the surface of the pons (arrows). Collection A. Rozemuller

The brown color is not always found, but in some case just firmness of the white matter can be seen (Popescu et al. 2013) (Fig. 9.14).

Recent studies suggest that the subarachnoid space (see Sect. 1.2.1) and cerebral cortex may be initial sites and targets of the MS disease process (Popescu et al. 2013). The more recent lesions have less well-defined borders and more of a brown hue and are not depressed, whereas the older lesions are well demarcated, grayer, and depressed. MS lesions can occur in the mesencephalon, brainstem, especially in the pons, and in the spinal cord.

Microscopically, inactive MS plaques in the white matter show sharply delineated loss of myelin and secondary axonal loss with reactive astrocytes surrounding the lesion (occasionally with clumped eosinophilic protein aggregates, the so-called Rosenthal fibers, within their processes) but practically no microglia/macrophages. Active lesions show accumulation of blood-borne macrophages (containing degradation products of myelin) and activated microglia. The cellular, active lesions can be divided into early active lesions with clusters of activated microglia and with macrophages containing both major (PLP = myelin proteolipid protein, MBP = myelin basic protein) and minor (MOG = myelin oligodendrocyte glycoprotein, MAG = Myelin-associated glycoprotein, CNPase = 2′,3′-cyclic-nucleotide 3′-phosphodiesterase) myelin degradation products, in late active lesions (with macrophages containing only major myelin degradation proteins) and in "smoldering" plaques (Frischer et al. 2009).

The latter are slowly expanding plaques that typically show an inactive center with no or few macrophages, surrounded by a rim of activated microglia. These plaques occur later in the course (Frischer et al. 2015) (Figs. 9.15, 9.16, and 9.17).

Furthermore, accumulation of some lymphocytes is found in the rim of active plaques and cuffing of lymphocytes around nearby arterioles. Both T-cells and B-cells play a role in the pathogenesis. The inflammatory changes can also be detected in the leptomeninx (Popescu et al. 2013). Completely remyelinating lesions can be found as shadow plaques (Barkhof et al. 2003). Using immunohistochemistry with antibodies directed against myelin proteolipid protein (PLP), the myelin loss in MS plaques can be well visualized. In double stains with anti-PLP antibodies and antibodies against activated microglia, such as HLA-DR, both the lesions and the activity of the lesions can be seen in one view (Fig. 9.15B, C).

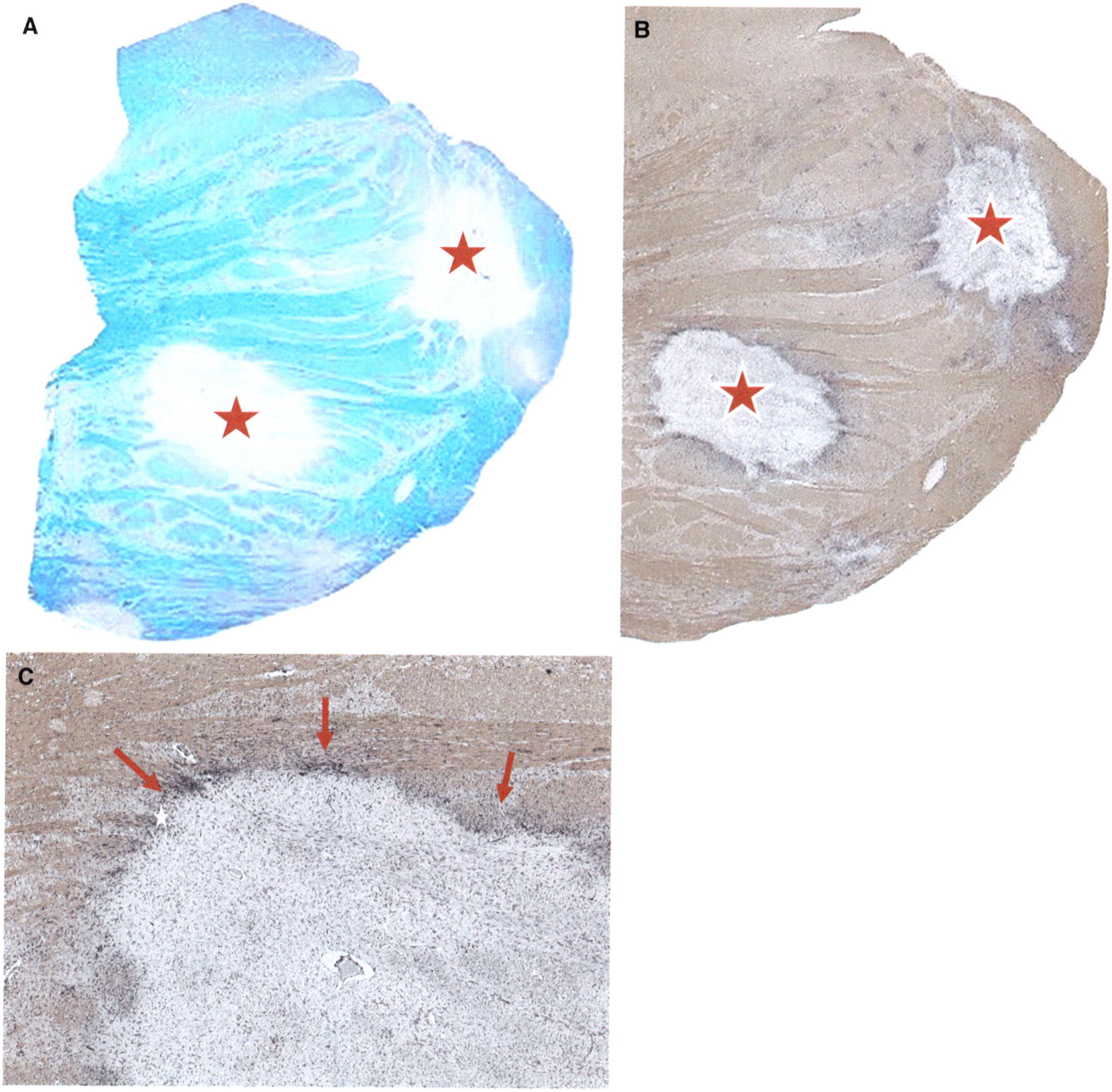

Fig. 9.15 Microscopical sections of part of the pons, showing demyelinated areas of smoldering MS plaques (red stars) in a Klüver-Barrera stain in (**A**). The same area in (**B** and **C**) (detail) with a double stain for PLP (myelin proteolipid protein) (brown) and HLA-DR (showing activated microglia) (dark blue). Note slightly cellularly populated margins of the MS plaques with microglia (red arrows). Collection A.Rozemuller

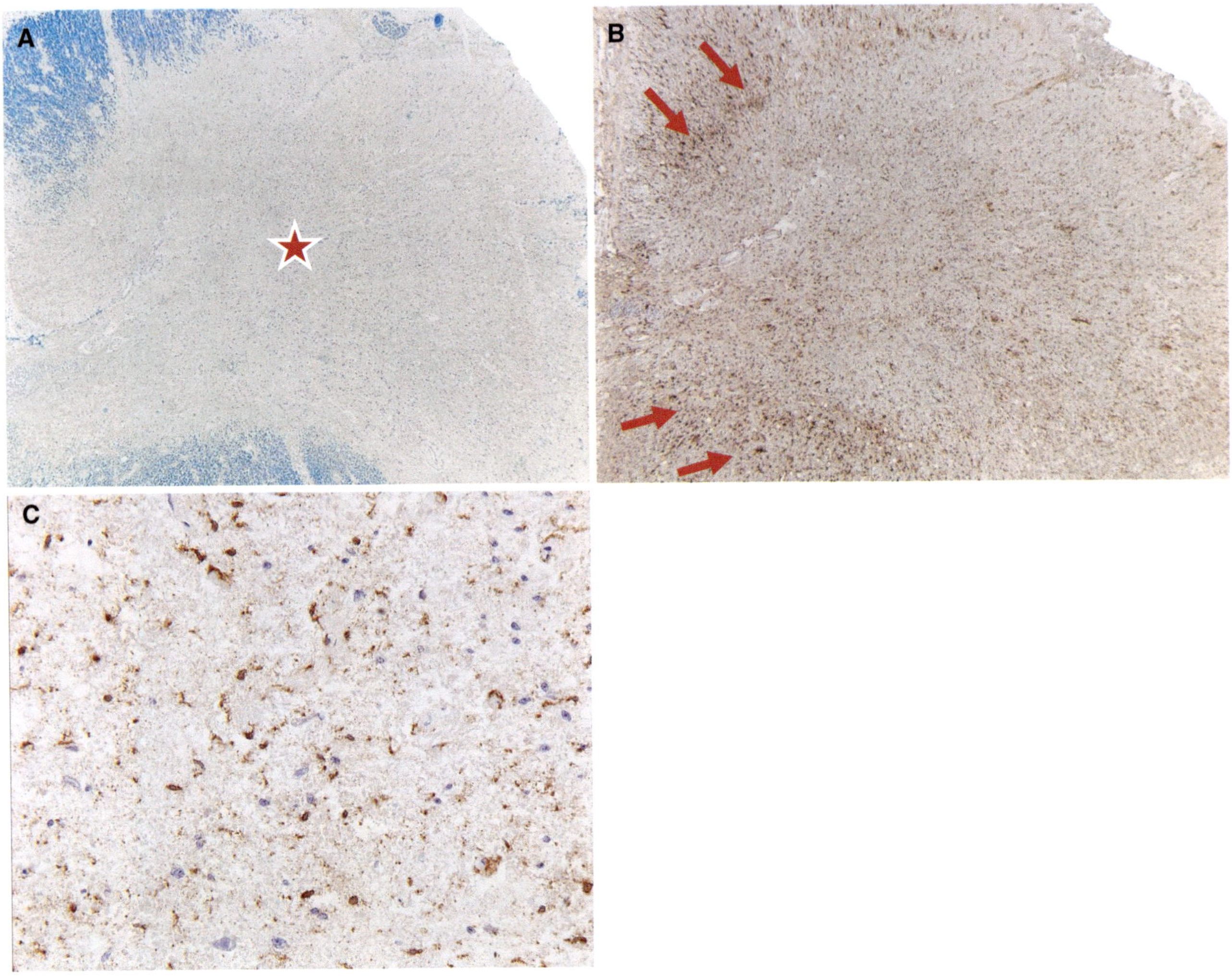

Fig. 9.16 In (**A**) and (**B**) smoldering (chronic active) lesion (red star) in the spinal cord, stained for Klüver-Barrera and anti-HLA-DR for microglia. Note the slightly more cellular margin (red arrows). In (**C**) detail of inactive center of the MS plaque stained for microglial cells (brown) using antibodies against HLA-DR. Collection A. Rozemuller

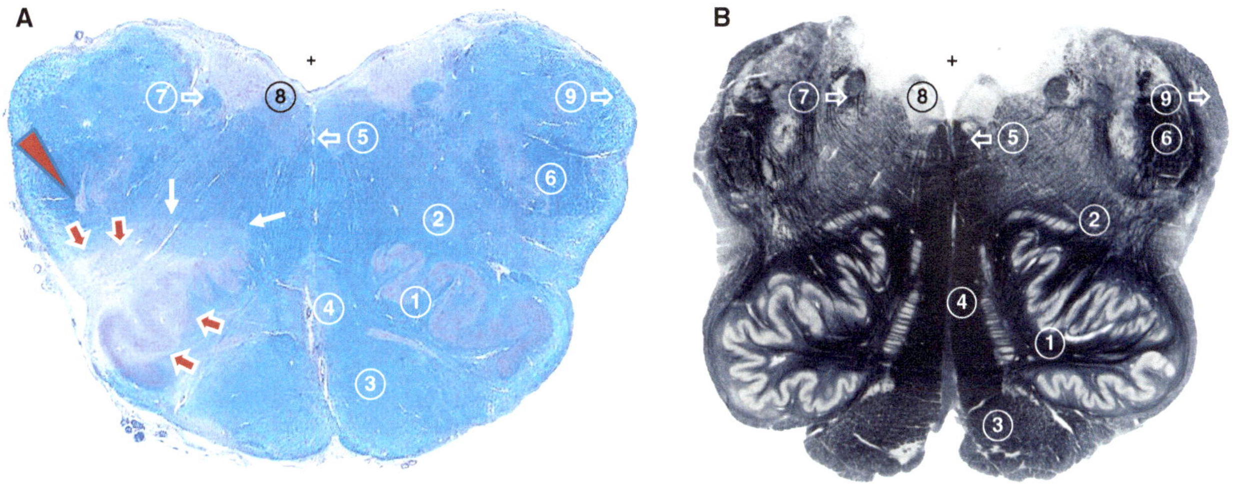

①	Inferior olivary complex	⑥	Spinal tract of trigeminal nerve CNV
②	Olivocerebellar tract	⑦	Solitary tract
③	Corticospinal tract	⑧	Ncl. of hypoglossal nerve CNXII
④	Medial lemniscus	⑨	Inferior cerebellar peduncle
⑤	Medial longitudinal fasciculus	+	4th ventricle

Fig. 9.17 (**A**) Sharply demarcated area (short red arrows) in a Klüver-Barrera myelin stain showing demyelination in the medulla oblongata, involving the inferior olivary complex ①. Also note the shadow of remyelination (lighter margin) (white arrows) and a second small lesion (red triangle). Compare to the medulla oblongata from a control case in (**B**) (modified Heidenhain stain for myelin). Collection A.Rozemuller and LabPON Twente

9.2.2 Branchial motor r5

9.2.2.1 Medioventral periolivary nucleus (MVPO)/Ventral nucleus of the trapezoid body [Ncl. ventralis corporis trapezoidei]

The term medioventral periolivary nucleus refers to one of four cell groups identified by Nissl stain in the periolivary region of the macaque.

http://braininfo.org/CentralDirectory.aspx?ID=1854&questID=574

In mice, this nucleus, together with the lateral superior olive, gives rise to cholinergic cochleopetal fibers (olivocochlear bundle) (Campbell and Henson 1988).

9.2.3 Visceral motor r5

9.2.3.1 Inferior salivatory nucleus

For detailed information on this nucleus, see Sect. 7.3.1.1.

9.2.4 Raphe nuclei r5

9.2.4.1 Interposed raphe nuclues [Ncl. raphes interpositus]

This nucleus was described in monkeys and humans by Büttner-Ennever and coworkers (1988) and a detailed cytoarchitectonic description has been provided by Rüb et al. (2003) (Fig. 9.18).

It is located in the midline of the caudal pons confined to the level where the rootlets of the abducens nerve—a landmark for the interpositus nucleus—pass through the brainstem (see Fig. 9.6B). The dorsal neighbor of the nucleus is the medial longitudinal fasciculus (Box 9.2), the ventral one the raphe magnus nucleus (see Fig. 9.18 ②) (see Sect. 7.3.3.2). The neurons of the nucleus raphe interpositus show a linear arrangement in a narrow band (for details, see Rüb et al. 2003). Although the nucleus is designated as raphes—and indeed is located at the seam of the pons—it is not serotonergic.

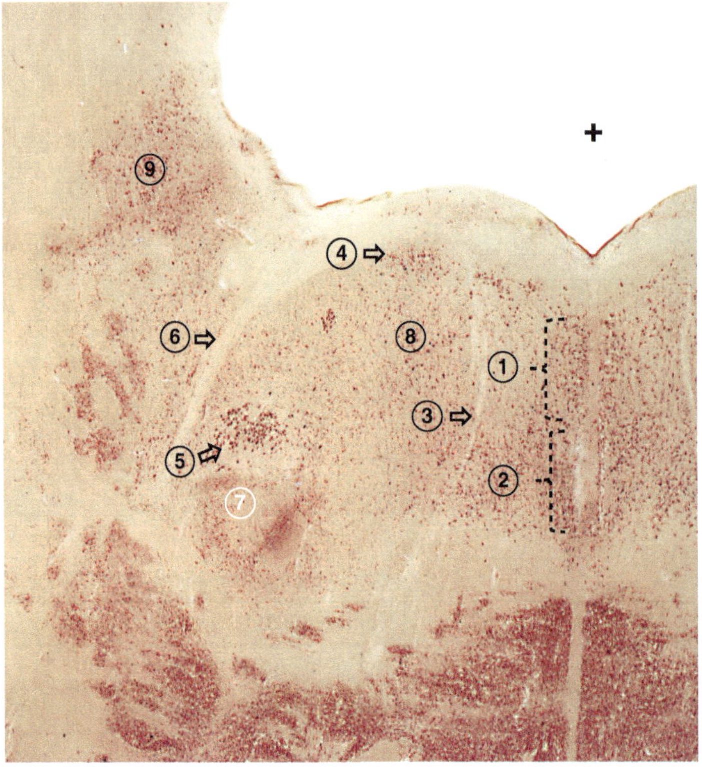

①	Interposed raphe ncl.	⑥	Facial nerve CNVII, descending part
②	Raphe magnus ncl.	⑦	Superior olivary complex
③	Abducens nerve CNVI	⑧	Pontine reticular ncl., caudal part
④	Ncl. of abducens nerve CNVI	⑨	Superior vestibular ncl. CNVIII
⑤	Motor ncl. of facial nerve CNVII	+	4th ventricle

Fig. 9.18 Horizontal section through the human pons at the level the interposed raphe nucleus ①. Darrow red staining. LabPON Twente

Functionally, the vast majority of the neurons are omnipause neurons related to saccades (see Box 9.6) in all directions. They are premotor neurons for the generation of saccades.

This nucleus shows a conspicous loss of neurons in spinocerebellar ataxia type 3 (SCA3), reactive astrocytes, and activated microglia (see Sect. 7.2.1.3.3).

In terms of function, it was suggested that lesions of the nucleus raphe interpositus may contribute to slowing of horizontal saccades but are not associated with saccadic oscillations (Rüb et al. 2003).

9.2.5 Basal tegmentum r5

9.2.5.1 Dorsomedial tegmental area

For detailed information on this structure, see Sect. 7.3.4.1.

9.2.6 Raphe nuclei r5

Information on the reticular nuclei listed can be found as indicated:

9.2.6.1 Parvocellular reticular nucleus

For detailed information on this nucleus, see Sect. 6.5.5.2.

9.2.6.2 Intermediate reticular nucleus

For detailed information on this nucleus, see Sect. 3.3.2.1.

9.2.6.3 Pontine reticular nucleus, caudal part

For detailed information on this nucleus, see Sect. 7.3.5.3.

9.2.6.4 Rostroventrolateral reticular nucleus

For detailed information on this nucleus, see Sect. 7.3.5.4.

9.2.6.5 Gigantocellular reticular nucleus

For detailed information on this nucleus, see Sect. 4.3.4.2.

9.2.6.6 Lateral paragigantocellular nucleus

For detailed information on this nucleus, see Sect. 7.3.5.6.

References

Al Kabbani A, Gaillard F et al (n.d.). Dural venous sinuses. https://radiopaedia.org/articles/dural-venous-sinuses

Ambekar S, Sonig A et al (2012) Dorello's canal and Gruber's ligament; Historical perspective. J Neurol Surg B 73:430–433

Barkhof F, Bruck W et al (2003) Remyelinated lesions in multiple sclerosis: magnetic resonance image appearance. Arch Neurol 60:1073–1081

Biazoli CE, Goto M et al (2006) The supragenual nucleus: a putative relay station for ascending vestibular signs to head direction cells. Brain Res 1094:138–148. [rat]

Buresch N (2005) Neuroanatomische Charakterisierung blickstabilisierender Neurone an der Hirnstammmittellinie der Primaten, einschließlich des Menschen. [Dissertation]. Munich, Germany: Ludwig-Maximilians Universität München

Büttner-Ennever JA, Büttner U et al (1982) Vertical gaze paralysis and the rostral interstitial nucleus of the medial longitudinal fasciculus. Brain 105:125–149

Büttner-Ennever JA, Cohen B et al (1988) Raphe nucleus of the pons containing omnipause neurons of the oculomotor system in the monkey, and its homologue in man. J Comp Neurol 267: 307–321

Campbell JP, Henson MM (1988) Olivocochlear neurons in the brainstem of the mouse. Hear Res 135:271–274. [white laboratory mice (ICR)]

Coyle PK (2005) Gender issues. Neurol Clin 23:39–60

Cullen KE (2019) Vestibular processing during natural self-motion: implications for perception and action. Nat Rev Neurosci 20:346–363

Danesh-Meyer HV (2012) Neuro-ophthalmology of brain tumors. In: Kaye AH, Laws ER Jr (eds) Brain tumors, 3rd edn. Elsevier, pp 214–236

Dietert SE (1965) The demonstration of different types of muscle fibers in human extraocular muscle by electron microscopy and cholinesterase staining. Investig Ophthalmol 4:51–63

Dolenc VV (1989) Anatomy and surgery of the cavernous sinus. Springer Wien New York

Fenichel GM (2009) Clinical pediatric neurology (sixth edition). Elsevier

Frischer JM, Bramow S et al (2009) The relation between inflammation and neurodegeneration in multiple sclerosis brains. Brain 132:1175–1189

Frischer JM, Weigand SD et al (2015) Clinical and pathological insights into the dynamic nature of the white matter multiple sclerosis plaque. Ann Neurol 78:710–721

Frohman TC, Galetta S et al (2008) Pearls & Oysters: the medial longitudinal fasciculus in ocular motor physiology. Neurology 70:e57–e67

Glover JC (2004) Encyclopedia of neuroscience. Elsevier

Haladaj R, Wysiadecki G et al (2018) Bilateral muscular slips between superior and inferior rectus muscles: case report with discussion on classification of accessory rectus muscles within the orbit. Surg Radiol Anat 40:855–862

Horn AKE, Adamczyk C (2012) The human nervous system (third edition). Elsevier

Horn AKE, Horng A et al (2018) Identification of functional cell groups in the abducens nucleus of monkey and human by perineuronal nets and choline acetyltransferase immunolabeling. Front Neuroanat 12:45

Huggenberger S, Moser N, Schröder H, Cozzi B, Granato A, Merighi A (2019) Neuroanatomie des Menschen. Springer, Heidelberg

Jacob S (2007) Human anatomy (first edition). Churchill Livingstone

Kalincik T, Brown JWL et al (2017) Treatment effectiveness of alemtuzumab compared with natalizumab, fingolimod, and interferon beta in relapsing-remitting multiple sclerosis: a cohort study. Lancet Neurol 16:271–281

Kheradmand A, Zee DS (2012) The bedside examination of the vestibulo-ocular reflex (VOR): an update. Rev Neurol (Paris) 168:710–719

Langer T, Fuchs AF et al (1985) Afferents to the flocculus of the cerebellum in the rhesus macaque as revealed by retrograde transport of horseradish peroxidase. J Comp Neurol 235:1–25

McGee S (2012) Evidence-based physical diagnosis (third edition). Elsevier Saunders

McGee S (2018) Evidence-based physical diagnosis (3rd edition). Elsevier Saunders

Moore GRW, Stadelmann-Nessler C (2015) Chapter 23: Demyelinating diseases. In: Love S, Perry A, Ironside J, Budka H (eds) Greenfield's neuropathology, 9th edn

Nagappa M, Taly AB et al (2013) Tumefactive demyelination: clinical, imaging and follow-up observations in thirty-nine patients. Acta Neurol Scand 128:39–47

Noseworthy JH, Lucchinetti C et al (2000) Multiple sclerosis. N Engl J Med 343:938–952

Olszewski J, Baxter D (1982) Cytoarchitecture of the human brain stem. Karger, Basel

Orton S-M, Herrera BM et al (2006) Sex ratio of multiple sclerosis in Canada: a longitudinal study. Lancet Neurol 5:932–936

Park HS, Chung MS et al (2015) Whole courses of the oculomotor, trochlear, and abducens nerves, identified in sectioned images and surface models. Anat Rec (Hoboken) 298:436–443

Polman CH, Reingold SC et al (2011) Diagnostic criteria for multiple sclerosis: 2010 revisions to the McDonald criteria. Ann Neurol 69:292–302

Popescu BF, Lucchinetti CF (2012) Pathology of demyelinating diseases. Annu Rev Pathol 7:185–217

Popescu BF, Pirko I et al (2013) Pathology of multiple sclerosis: where do we stand? Continuum (Minneap Minn) 19:901–921

Purves D, Augustine GJ et al (eds) (2001) Neuroscience, 2nd edn. Sinauer Associates, Sunderland, MA

Rea P (2015) Essential clinical anatomy of the nervous system. Elsevier

Reich DS, Lucchinetti CF et al (2018) Multiple sclerosis. N Engl J Med 378:169–180

Richards RG, Sampson FC et al (2002) A review of the natural history and epidemiology of multiple sclerosis: implications for resource allocation and health economic models. Health Technol Assess 6:1–73

Rüb U, Brunt ER et al (2003) The nucleus raphe interpositus in spinocerebellar ataxia type 3 (Machado-Joseph disease). J Chem Neuroanat 25:115–127

Somani AN, Adesina O-o (2019) Neuroanatomy, abducens nucleus. StatPearls [Internet]

Thompson AJ, Banwell BL et al (2018) Diagnosis of multiple sclerosis: 2017 revisions of the McDonald criteria. Lancet Neurol 17: 162–173

Zampieri F, Marrone D et al (2015) Should the annular tendon of the eye be named 'annulus of Zinn' or 'of Valsalva'? Acta Ophthalmol 93:97–99

Web Links

Terminologia neuroanatomica FIPAT, http://fipat.library.dal.ca/TNA/
http://braininfo.rprc.washington.edu/centraldirectory.aspx?ID=1981

Rhombomere 4 r4

10

Contents

H. Schröder et al., *The Human Brainstem*, https://doi.org/10.1007/978-3-030-89980-6_10

Abstract

This chapter deals with two important nuclei which provide the transfer of signals from the cerebral cortex to the cerebellum, the pontine nuclei, and from the mammillary region to the cerebellum, the reticulotegmental nucleus of the pons.

The third vestibular nucleus in the row, the lateral vestibular nucleus, is a derivative of rhombomere 4, too. It is strongly related to the cerebellum.

The pontine raphe nucleus belongs to the rostral group of raphe nuclei. It corresponds to the B5 serotonergic cell group and gives rise to serotonergic projections to the cerebellum.

The dorsal parts of the pontine reticular nucleus correspond to the horizontal gaze center or pontine paramedian reticular formation (PPRF).

Related to a paradigmatic lesion in the corticopontine tracts, which targets the pontine nuclei, this chapter deals with the topic of vascular dementia.

10.1 Rhombic lip r4

10.1.1 Precerebellar nuclei r4

10.1.1.1 Reticulotegmental nucleus of the pons/Reticulotegmental ncl. [Ncl. reticularis tegmenti pontis] (Bechterew)

The reticulotegmental nucleus of the pons is located in the tegmentum of the pons (see Box 10.1) at the level of the motor nucleus of trigeminal nerve (Fig. 10.1A).

Box 10.1 The pons

Figure 10.1B shows a typical traditional picture of the pons which has been challenged based on genoarchitectural studies by Watson et al. 2019. As Watson states, in many mammals, the pontine nuclei (Fig. 10.1A, B) as well as the reticulotegmental nucleus (Fig. 10.1A ①) and the nucleus of facial nerve (Fig. 10.1B ④, for comparison with the mouse brain see Fig. 6.32 in Schröder et al. 2020) are located at the ventral parts of rhombomeres 3 and 4 and the pontine bulge (see Inset Fig. 10.2) is restricted to this region (Watson et al. 2019).

Correlative with the expansion of the human cerebral cortex the "pons" has undergone a massive enlargement in humans, hiding thereby much of the prepontine hindbrain (from rhombomere 0, isthmus, to rhombomere 2) as well as the abducens nucleus, the superior olivary complex (see Fig. 10.1B ⑤, for comparison with the mouse brain see Figs. 6.34 and 6.35 in Schröder et al. 2020) and the facial nucleus in the retropontine region.

Watson suggests refraining from the use of the word "pons" as a topographical descriptor in all mammals, restricting the use of this term to the basilar pontine formation in rhombomeres 3 and 4. Though perfectly reasonable, it will probably get hard to change the terminology of an introduced anatomical term particularly since it is also constantly used for clinical purposes.

Studies that revealed the projection of different human cortical regions to the pontine nuclei (see here below Sect. 10.1.1.2) also revealed cortical targeting of the reticulotegmental nucleus of the pons (called papillioform nucleus by Olszewski and Baxter 1954).

The reticulotegmental nucleus of the pons receives suprabulbar input from the mammillary bodies [Corpora mammillaria] via the mammillotegmental fasciculus [Fasciculus mammillotegmentalis] and projects back to the mammillary bodies via the mammillary peduncle [Pedunculus mammillaris] (see Fig. 10.2, for details see ten Donkelaar 2011).

The reticulotegmental nucleus of the pons (eponym Bechterew, see Box 16.1 "Eponyms: Bechterew") has mainly efferent cerebellar connections. Furthermore, it is well known that the reticulotegmental nucleus of the pons (see Sect. 10.1.1.1) is involved in the generation of smooth pursuit eye movements (see Box 9.6 (3)).

In mice, the rostral part of the reticulotegmental nucleus is separated from the pontine nuclei by the medial lemniscus. It is known to be closely associated with the pontine nuclei in functional terms. Upon tracer injection into the cerebellum, many labeled cells were found in all parts of the nucleus (Fu et al. 2011). The mouse nucleus contains serotonin-immunoreactive neurons which are most prominent at the caudal pole of the interpeduncular nucleus (Sect. 14.5.4.4) (VanderHorst and Ulfhake 2006).

10.1.1.2 Pontine nuclei [Nuclei pontis]

The pontine nuclei (Fig. 10.1A ②, Fig. 10.1B①) are a group of nuclei scattered in the basis of the pons. They belong to the central gray of the pons and consist of small, oval, triangular, or elongated cells with indistinct Nissl granules (Olszewski and Baxter 1982).

Corticopontine projections in the human brain have been studied in the presence of large hemispheric, stroke-induced lesions by means of postmortem degeneration studies (com-

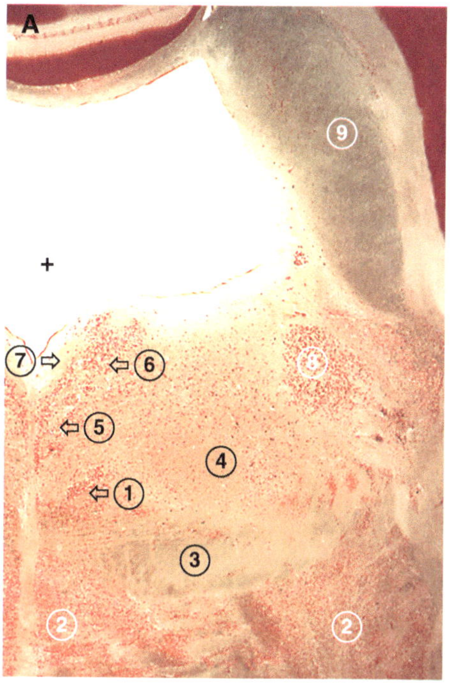

①	Reticulotegmental ncl. of the pons	⑥	Dorsal raphe ncl.
②	Pontine ncll.	⑦	Medial longitudinal fasciculus
③	Medial lemniscus	⑧	Motor ncl. of trigeminal nerve
④	Pontine reticular ncl., oral part	⑨	Superior cerebellar peduncle
⑤	Facial nerve CNVII	+	4th ventricle

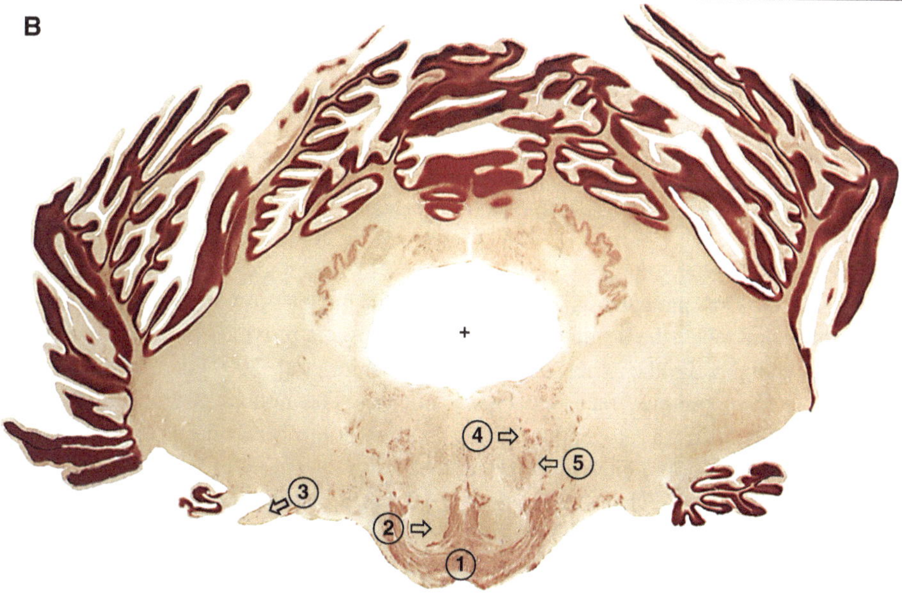

①	Pontine ncll.	④	Motor ncl. of facial nerve CNVII
②	Pyramidal (corticospinal) tract	⑤	Superior olivary complex
③	Facial nerve CNVII	+	4th ventricle

Fig. 10.1 (**A**) Horizontal section through the human pons at the level of the motor nucleus of trigeminal nerve ⑧. Darrow red stain. The reticulotegmental nucleus ① is located near to the midline in the tegmentum. It is separated from the pontine nuclei ② (see **B** ①) (see Sect. 10.1.1.2) by the medial lemniscus ③. Its dorsal neighbor is the dorsal raphe nucleus ⑤, its lateral one is the pontine reticular nucleus, oral part ④. LabPON Twente. (**B**) Horizontal section through the human pons at the level of the vestibulocochlear nerve ③. Darrow red stain. The microphotography clearly shows the large extension of the pontine nuclei ① at the basis of the pons. LabPON Twente

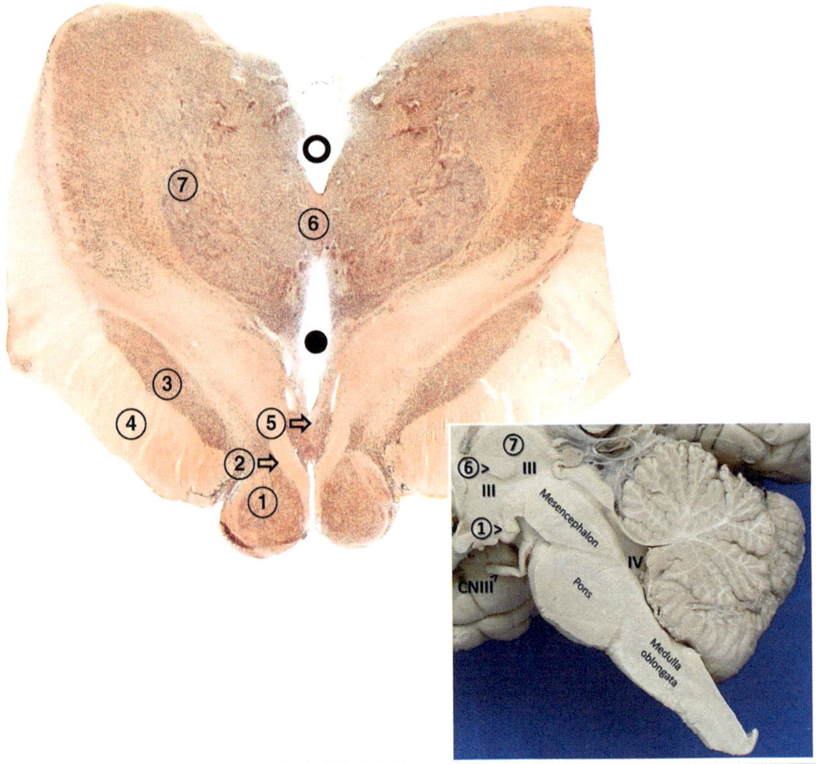

①	Optic chiasm	⑥	Interthalamic adhesion
②	Principal mammillary fasciculus	⑦	Thalamus
③	Subthalamic ncl.	○	3rd ventricle, dorsal part
④	Cerebral peduncle	●	3rd ventricle, ventral part
⑤	Fornix		

Fig. 10.2 Coronal section through the human brain at the level of the mammillary bodies ① (see inset for the location of the mammillary bodies in a sagittal section; III = third ventricle, IV = fourth ventricle, CNIII = oculomotor nerve). At this site, the mammillary bodies form part of the floor of the third ventricle. The main input structure to the mammillary bodies—from the hippocampus—is the fornix ⑤, its main efferent tract is the principal mammillary fasciculus ②. LabPON Twente

piled by ten Donkelaar 2011). One patient had extended lesions in the frontal and temporal lobes, the insula, the internal capsule, and the basal ganglia. In the other one, the lesions were confined to the frontopontine tract (see Figs. 3.14F–G). A third patient died with an otogenous brain abscess in the center of the temporal lobe, interrrupting the temporopontine tract (see Figs. 3.14F–G). The assessment of the degeneration signs showed a topotonic ending of the corticopontine tracts in different parts of the pontine nuclei.

As shown in detail in the rat, the pontine nuclei receive—as in man—input from the cerebral cortex (Leergard and Bjaalie 2007) and project to the cerebellum via the middle cerebellar peduncle. The pontine nuclei located in the ventral hindbrain (r3/r4) are—together with the inferior olivary complex in r10—the largest of the precerebellar nuclei in the mouse (Fu et al. 2011). Cholinergic input into the mouse pontine nuclei—as well as to the reticulotegmental nucleus of the pons—is provided by the laterodorsal tegmental nucleus (Tsutsumi et al. 2007) and the pedunculopontine tegmental nucleus. The authors assume an inhibitory effect on the neocortico-ponto-cerebellar projections by both nuclei.

10.1.1.3 Vascular dementia

In the vein of corticopontine connections, we here deal with the neuropathological basis of vascular dementia showing a patient with a lesion in the corticopontine tracts (Fig. 10.5). The most common cause of dementia worldwide is Alzheimer's disease (see Sect. 13.2.2.1.4). Many patients with dementia, however, now that the population is getting older, have a combination of abnormalities in their brains associated with AD, that is senile plaques, neurofibrillary tangles, and cerebral amyloid angiopathy (see Sect. 1.4.2.5)

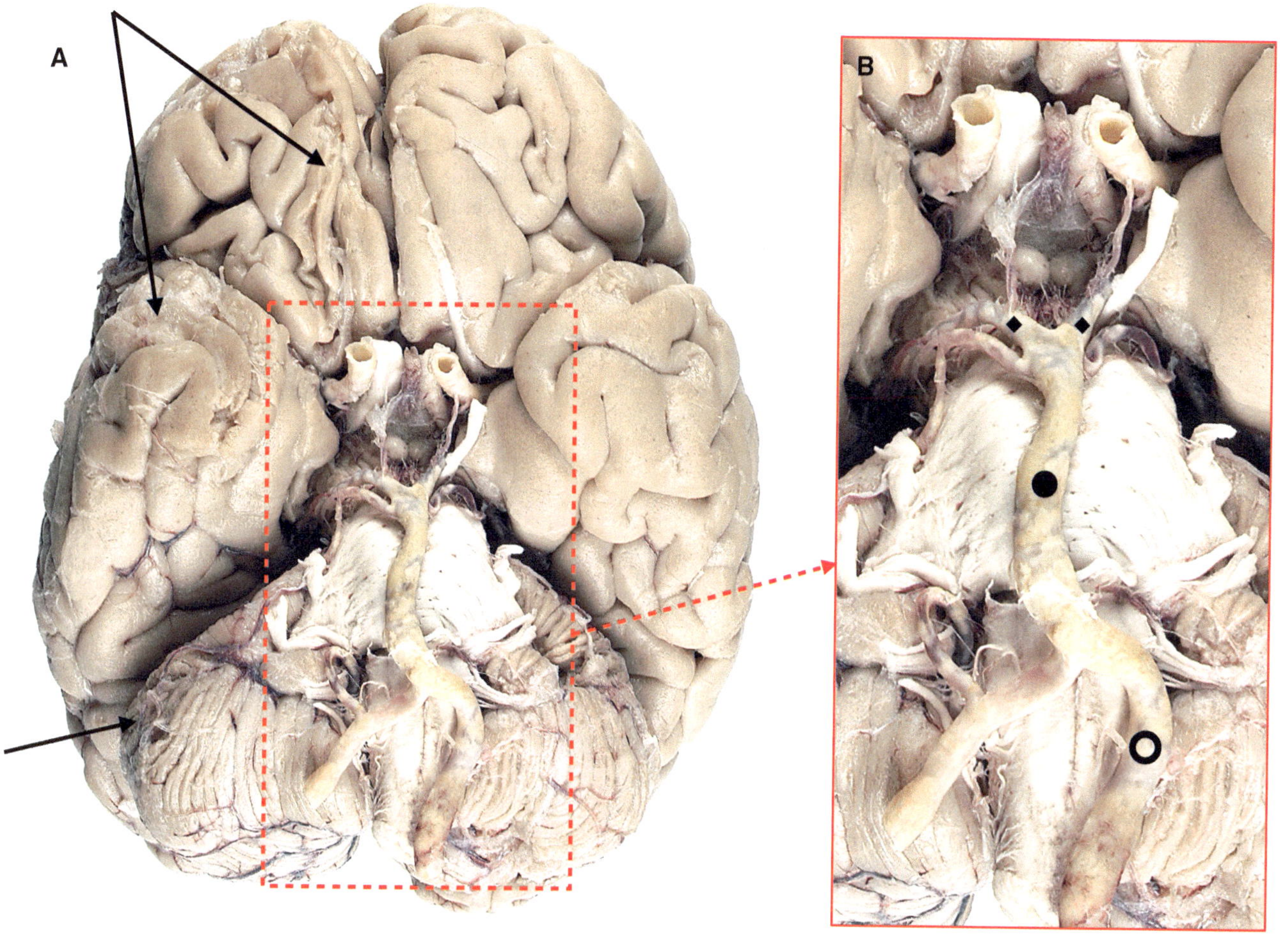

Fig. 10.3 (**A**) ventral view of the brain from a patient with vascular dementia (VD). Severe atherosclerosis of the circle of Willis [Circulus arteriosus cerebri] (detail in **B**). Stiffened, tortuous, yellowish aspect of the left vertebral artery (**O**), basilar artery (●), and the posterior cerebral arteries (◆). Multiple infarcts of the right hemisphere, frontotemporal, and the right cerebellar hemisphere (black arrows). LabPON Twente

as well as varying degrees of ischemic and/or hemorrhagic cerebrovascular diseases (see Sect. 1.4.2.7), commonly referred to as "mixed dementias" (Vinters et al. 2018; Neal 2012; Esiri et al. 1997). However, pure cerebrovascular disease in neuropathologically confirmed cohorts is a less frequent cause of dementia (McAleese et al. 2016).

Early studies identified cortical and subcortical infarcts, assumed to be the result of small diameter blood vessels disease (small vessel disease, SVD) as important factors in dementia (Esiri et al. 1997), but more recent evidence has shown that large vessel atherosclerosis (see Sect. 1.4.2.4) could also be important in the etiology of these changes in mainly elderly people (Markus 2008; Neal 2012) (Fig. 10.3).

Intracranial atherosclerosis has been identified as a risk factor for dementia that is independent of cerebral infarction and other risk factors, such as peripheral atherosclerosis, systemic hypertension, and age (Neal 2012; Troncosco et al. 2008). White matter lesions detected by imaging have also been shown to have a strong influence on progression of cognitive decline and are regarded as further surrogate markers of SVD (Neal 2012; Patel and Markus 2011). There still remains a lack of accepted neuropathological criteria to validate clinically identified vascular dementia (Wiederkehr et al. 2008), and there is controversy over the exact nature of the relation between cerebrovascular pathology and cognitive impairment (O'Brien and Thomas 2015). Small vessel disease is not a single process, as it includes arteriosclerosis, amyloid deposition, perivascular space and venous wall thickening, implying a range of different etiologies (Neal 2012). Comparisons across different studies is often difficult due to the wide range of pathology included in SVD. CT is sufficient to show established infarcts and extensive white matter lesions although MRI is highly preferable to show more precisely the degree, location, and extent of cerebrovascular disease (O'Brien and Thomas 2015). Genetic studies have mainly been on rare familial syndromes such as cerebral autosomal dominant arteriopathy with subcortical infarcts and leukoencephalopathy (CADASIL) and cathepsin A-related arteriopathy with strokes and leukoencephalopathy (CARASAL) (see Sect. 3.2.3.5). The majority of patients with vascular dementia showed multiple infarcts with mean loss of brain volume of 39 ml (Neal 2012) (Figs. 10.3, 10.4, and 10.5).

The location of the infarcts with regard to dementia was significant, 96% had bilateral temporal infarcts, over 80% subcortical infarcts in basal ganglia, globus pallidus,

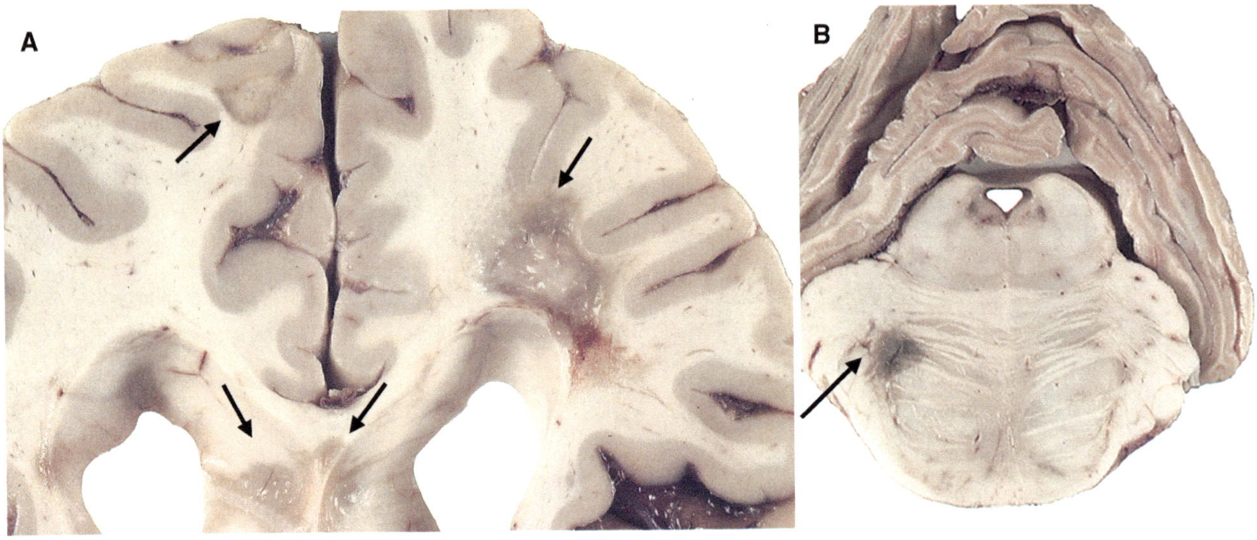

Fig. 10.4 Vascular dementia. In (**A**) coronal slice showing several infarcts in subortical areas and white matter (arrows). In (**B**) infarct of the pons (arrow). LabPON Twente

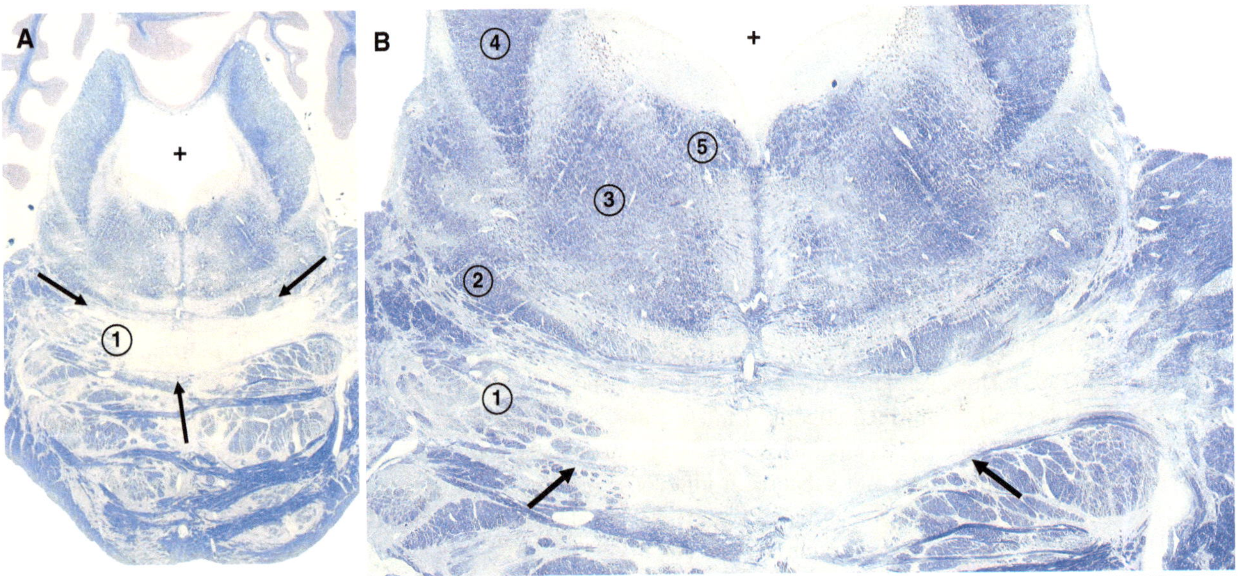

①	Corticopontine tracts	④	Superior cerebellar peduncle
②	Medial lemniscus	⑤	Medial longitudinal fasciculus
③	Central tegmental tract	+	4th ventricle

Fig. 10.5 Vascular dementia. (**A**) Infarct in the pons (arrows), region of the dorsal transverse pontine (pontocerebellar) fibers and pontine nuclei. Detail in (**B**) Klöver-Barrera stain for myelin. See also atlas part Campbell 13 for normal situation. LabPON Twente

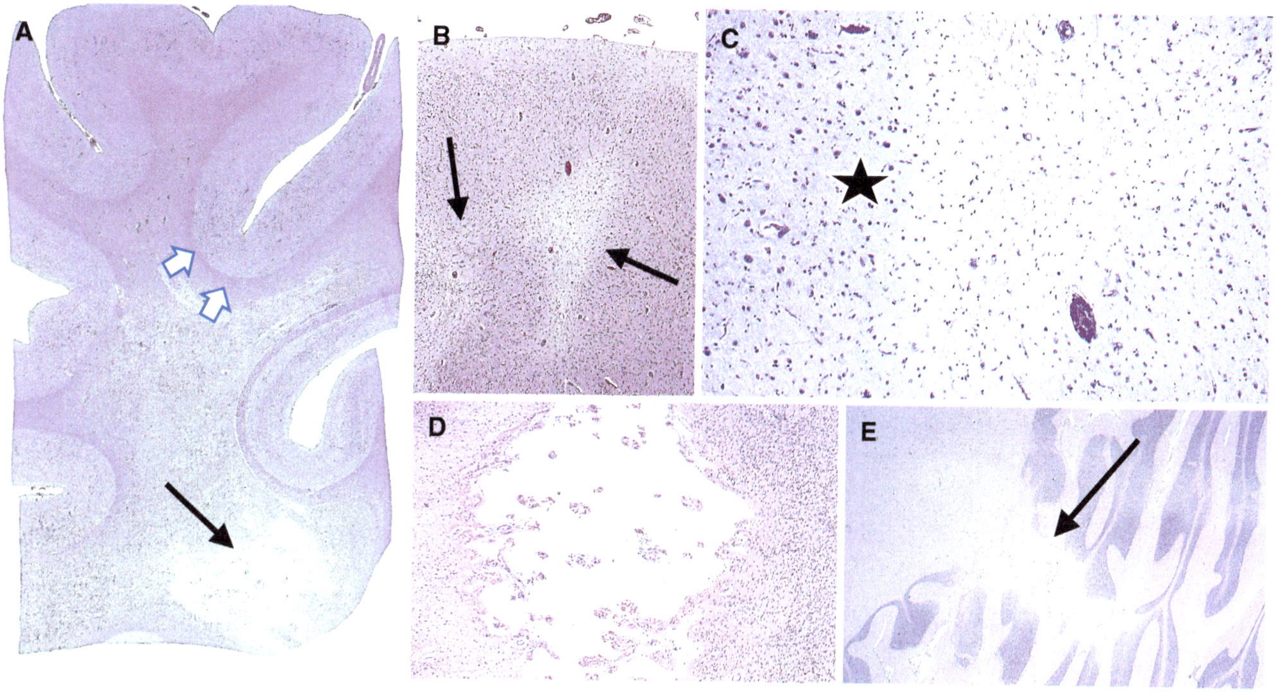

Fig. 10.6 Small vessel disease (SVD), Coronal section of the human brain, overview (**A**), demyelination with sparing of the subcortical U fibers (white arrows), lacunar infarct (black arrow) in the white matter, with associated pallor. (**B**) Microinfarcts in the frontal cortex (black arrows), detail in (**C**), showing reactive astrocytes (star). (**D**) Lacunar infarct in the basal ganglia, (**E**) infarct in the cerebellar cortex. Hematoxylin eosin stain. LabPON Twente

and caudate nucleus (Neal 2012; Erkinjuntti et al. 1988). Despite all these data, there is still no consensus on the diagnostic criteria for vascular cognitive impairment. Neuropathologists from seven UK centers developed a set of vascular cognitive impairment neuropathology guidelines (VCING). They found significant associations of cognitive impairment with microinfarcts, lacunar infarcts, large infarcts, arteriolosclerosis, perivascular space dilatation, myelin loss, and leptomeningeal cerebral amyloid angiopathy (CAA). Neuropathologists can use a combination of the three main determinants—moderate/severe occipital leptomeningeal, at least one large infarct, and moderate/severe arteriolosclerosis in the occipital white matter—to assign low, intermediate, or high likelihood that cerebrovascular disease contributed to cognitive impairment in an individual case (Skrobot et al. 2016). Therefore, a study was designed, the Vascular Impairment of Cognition Classification Consensus Study, to achieve a broader consensus on the conceptualization of impairment in cognition contributed by vascular pathology for clinical diagnosis and research (Skrobot et al. 2017a, b). In their model, atherosclerosis of the basal arteries was one of the variables not included. There is, however, evidence that atherosclerosis is associated with cerebral infarcts that are themselves a risk factor for vascular cognitive impairment (Dolan et al. 2010). It remains to be established whether the circle of Willis atherosclerosis is an independent determinant of vascular cognitive impairment (Skrobot et al. 2017a, b).

SVD is restricted by definition to small diameter vessels and is associated with pathological and radiological evidence of lacunar infarcts and white matter changes variously called leukoariosis (LA) and white matter hyperintense lesions (WML). Arteriolosclerosis is found in arterioles and also includes fibrinoid necrosis, microatheroma and microaneurysm formation (Neal 2012; Grinberg and Thal 2010) (Fig. 10.6).

The pathology of SVD is characterized by age-related loss of smooth muscle cells and replacement by deposits of fibro-lipohyalinosis material producing lumen occlusion and vessel wall thickening, including basement membrane structural changes (Lammie 2005). Venous collagenosis is a pathological thickening of veins, closely located to the lateral ventricles (Pantoni 2010). Thickening of the vessel wall can result in occlusion of the lumen, with a reduction of local venous return, increasing the likelihood of WML in periventricular areas (Neal 2012) (Fig. 10.7C and D).

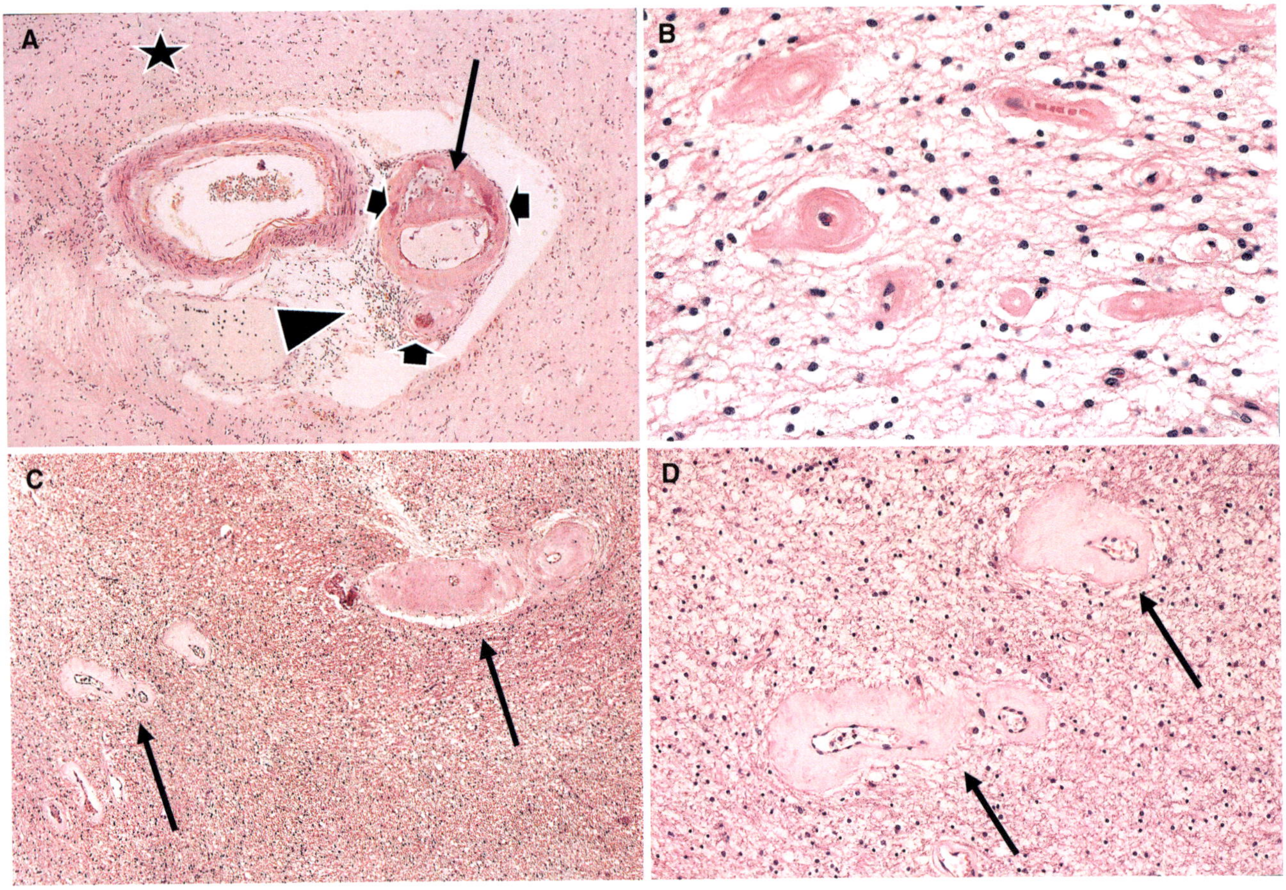

Fig. 10.7 Small vessel disease (SVD). In (**A**) marked hyalinized arterial wall (long arrow) with fibrinoid necrosis (short arrows), recanalization and narrowing of the lumen. Perivascular mononuclear infiltrate (lymphocytes, macrophages, hemosiderophages) (triangle). Note pallor of the surrounding white matter with reactive changes (star). Typical venous collagenosis, related to the lateral ventricle in **B**, **C**, and **D**. Collagenous thickening in varying density, causing partial occlusion of the vessels, pallor of the surrounding white matter (arrows in **C** and **D**). Hematoxylin eosin stain. LabPON Twente

10.2 Alar r4

10.2.1 Cochlear nuclei r4

See Chap. 8 r6.

10.2.1.1 Dorsal cochlear nucleus

See Chap. 8 r6.

10.2.1.2 Ventral cochlear nucleus

For detailed information on the cochlear nuclei, see Sect. 8.1.1.

10.2.2 Vestibular nuclei r4

10.2.2.1 Lateral vestibular nucleus [Ncl. vestibularis lateralis] r4 Eponym Deiters

The lateral vestibular nucleus (LVN) (Fig. 10.8) is the third in the row of vestibular nuclei (see Sect. 5.3.1), the second most cranial between the medial and the superior vestibular nuclei (see Sect. 11.2.2.1).

Kern et al. (2009) have provided a detailed comparative anatomical study on the lateral vestibular nuclei of man and Cetacea. The most important projection of the lateral vestibular nucleus—besides its role as a precerebellar nucleus—is the lateral vestibulospinal tract which contacts the anterior/ventral motor horn of the spinal cord (Fig. 10.8 ⑨), thus providing a major source for spinal activation and control (Kern et al. 2009; Oelschläger et al. 2008; Cozzi et al. 2016). The neurons forming this tract receive strong cerebellar input (Voogd 2016).

As shown by Kern et al. (2009), the interspecific volume of the lateral vestibular nucleus/body mass ratio, set to 1.0 for humans is 3.1 for dolphins (*Delphinus delphis*) and 5.2 for the La Plata dolphin (*Pontoporia blainvillei*). This allows some functional conclusions. In both dolphin species investigated, the growth of the LVN is in strong contrast to the remarkable reduction of the genuine vestibular nuclei. The usual habitat of *Pontoporia* is estuarine/coastal, while

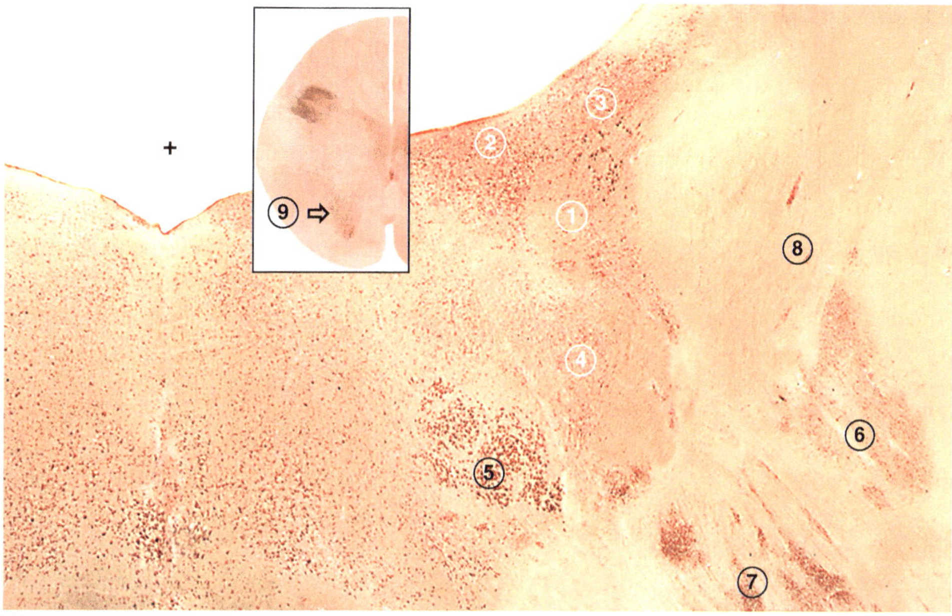

①	Lateral vestibular ncl. CNVIII	⑥	Anterior / ventral cochlear ncl. CNVIII
②	Medial vestibular ncl. CNVIII	⑦	Pontobulbar body
③	Superior vestibular ncl. CNVIII	⑧	Middle cerebellar peduncle
④	Spinal ncl. of trigeminal nerve CNV	⑨	Anterior / ventral horn of spinal cord
⑤	Motor ncl. of facial nerve CNVII	+	4th ventricle

Fig. 10.8 Horizontal section through the human pons at the level of the motor nucleus of facial nerve CNVII ⑤. Darrow red stain. At this level, three vestibular nuclei, the medial ②, the lateral ①, and the superior nucleus ③ (see Sect. 11.2.2.1) are visible. The inferior most, the spinal (inferior) one (see Sect. 5.3.1.1) is located caudal of this level. Inset: detail of spinal cord including anterior/ventral horn ⑨. Control case. LabPON Twente

Delphinus is a pelagic species (for details, see Kern et al. (2009). While the small size of the majority of the dolphin's vestibular nuclei correlates well with miniaturization of the semicircular canals (see Sect. 5.3.1), the size of the lateral vestibular nucleus [Nucleus vestibularis lateralis] seems to support its relative independence from the vestibular system and a close functional relationship with the cerebellum. In comparison with findings in humans and other terrestrial mammals, both of these aspects seem to be related to the physical conditions of aquatic life and locomotion in three dimensions (Kern et al. 2009). The larger LVN together with the lateral spinovestibular tract in *Delphinus* might thus be necessary for complicated three-dimensional high-speed maneuvers as well as for active hunting (Kern et al. 2009).

As in man, in mice, the lateral vestibular nucleus is limited dorsolaterally by the superior vestibular nucleus. Historically, the lateral vestibular nucleus in mice has been defined by the presence of very large neurons (Liang et al. 2014).

10.2.2.2 Medial vestibular nucleus

For detailed information on this nucleus, see Sect. 7.2.1.1.

10.2.3 Monoamine nuclei

10.2.3.1 A5 noradrenaline cells (NA5)

For detailed information on these cells, see Sect. 8.1.3.1.

10.2.4 Trigeminal sensory nuclei r4

10.2.4.1 Spinal trigeminal nucleus

For detailed information on this nucleus, see Sect. 3.2.2.2.

10.2.5 Superior olive nuclei r4

10.2.5.1 Dorsal periolivary region

For detailed information on these nuclei, see Sect. 8.1.1.5.

10.2.5.2 Lateral superior olive

For detailed information on these nuclei, see Sect. 8.1.1.5.

10.2.5.3 Medial superior olive

For detailed information on these nuclei, see Sect. 8.1.1.5.

10.2.6 Lateral lemniscus nuclei r4

10.2.6.1 Ventral nucleus of the lateral lemniscus (VLL)

For detailed information on these nuclei, see Sect. 8.1.1.5.

10.2.7 Central gray

For detailed information on this structure, see Sect. 13.2.7.

10.3 Basal r4

10.3.1 Visceral motor r4

10.3.1.1 Superior salivatory nucleus

For detailed information on this structure, see Sect. 7.3.1.1.1.

10.3.2 Branchial motor r4

10.3.2.1 Medioventral periolivary nucleus

For detailed information on this nucleus, see Sect. 9.2.2.1.

10.3.3 Raphe nuclei r4

10.3.3.1 Pontine raphe nucleus [Ncl. raphes pontis] B5

The pontine raphe nucleus belongs to the rostral group of raphe nuclei (see Sect. 3.2.1.3). It corresponds to the B5 serotonergic cell group and gives rise to serotonergic projections to the cerebellum.

In mice, at the level of the trigeminal motor nucleus (r3) 5-HT-immunoreactive neurons are present in the pontine raphe nucleus immediately ventral to the pontine gray (VanderHorst and Ulfhake 2006).

Detailed information on the following nuclei is provided as indicated below:

10.3.4 Basal tegmentum r4

10.3.4.1 Dorsomedial tegmental area

For detailed information on this area, see Sect. 7.3.4.1.

10.3.5 Reticular nuclei r4

10.3.5.1 Intermediate reticular nucleus

For detailed information on this nucleus, see Sect. 3.3.2.1.

10.3.5.2 Parvocellular reticular nucleus

For detailed information on this nucleus, see Sect. 6.5.5.2.

10.3.5.3 Pontine reticular nucleus, caudal part

For detailed information on this nucleus, see Sect. 7.3.5.3.

10.3.5.4 Oral pontine reticular nucleus/Pontine reticular nucleus, ventral part [Ncl. reticularis pontis oralis]

This nucleus is located in the pontine tegmentum at the level of the motor nucleus of trigeminal nerve (see Fig. 10.1A ④) between the trigeminal motor nucleus lateral and the reticulotegmental nucleus of the pons medial.

While the caudal part of this nucleus (see Sect. 7.3.5.3) contains—among other types—giant cells, this is not the case in the ventral/oral part.

The dorsal parts of the pontine reticular nucleus correspond to the horizontal gaze center or pontine paramedian reticular formation (PPRF) (ten Donkelaar 2011).

The pontine reticular nuclei receive input from the superior colliculus (see Sect. 15.1.3). In turn, the PPRF projects to the ipsilateral abducens nucleus (see Sect. 9.2.1.1) via internuclear neurons in the contralateral medial longitudinal fasciculus (see Box 9.2, see Fig. 3.14A–G) (Somani and Adesina 2020). The interneuronal axons synapse on the medial rectus subnucleus of the oculomotor nucleus (see Sect. 15.3.1.1) which innervates the medial rectus muscle (see Table 9.1, Fig. 9.4B ③).

This pathway guarantees equal innervation of horizontal eye muscles, the ipsilateral lateral rectus, and contralateral medial rectus to maintain conjugate gaze via coordinated activity. This coordination of input allows for conjugate horizontal gaze during saccades (see Box 9.6) via the parietal eye fields and the frontal eye fields targeting the superior colliculi. It is also relevant during voluntary eye movements.

10.3.5.5 Rostroventrolateral reticular nucleus

Information on this nucleus—not present in humans—is provided under Sect. 7.3.5.4.

References

Cozzi B, Huggenberger S et al (2016) Anatomy of dolphins insights into body structure and function. Academic Press

Dolan H, Crain B et al (2010) Atherosclerosis, dementia, and Alzheimer's disease in the Baltimore Longitudinal Study of Aging cohort. Ann Neurol 68:231–240

Erkinjuntti T, Halatia M et al (1988) Accuracy of the clinical diagnosis of vascular dementia: a prospective clinical and post mortem neuropathological study. J Neurol Neurosurg Psychiatry 51: 1037–1044

Esiri MM, Wilcock GK et al (1997) Neuropathological assessment of the lesions of significance in vascular dementia. J Neurol Neurosurg Psychiatry 63:749–753

Fu Y, Tvrdik P et al (2011) Precerebellar cell groups in the hindbrain of the mouse defined by retrograde tracing and correlated with cumulative Wnt1-Cre genetic labeling. Cerebellum 10:570–584

Grinberg LT, Thal DRF (2010) Vascular pathology in the aged human brain. Acta Neuropathol 119:277–290

Kern A, Seidel K et al (2009) The central vestibular complex in dolphins and humans: functional implications of Deiters' nucleus. Brain Behav Evol 73:102–110

Lammie GA (2005) Small vessel diease. In: Kalimo H (ed) Cerebrovascular disease (pathology and genetics). ISN Neuropath Press, Basel

Leergard TB, Bjaalie JG (2007) Topography of the complete corticopontine projection: from experiments to principal maps. Front Neurosci 1:211–223

Liang H, Bácskai T et al (2014) Projections from the lateral vestibular nucleus to the spinal cord in the mouse. Brain Struct Funct 219:805–815

Markus H (2008) Small vessel versus large vessel vascular dementia. Risk factors and MRI findings. J Neurol 255:1644–1651; discussion 1813–14

McAleese K, Alafuzoff I et al (2016) Post-mortem assessment in vascular dementia: advances and aspirations. BMC Med 14:129

Neal JW (2012) Vascular dementia: why pathology is still important. Rev Clin Geront 22:35–51

O'Brien JT, Thomas A (2015) Vascular dementia. Lancet 386: 1698–1706

Oelschläger HHA, Haas-Rioth M et al (2008) Morphology and evolutionary biology of the dolphin (*Delphinus* sp.) brain—MR imaging and conventional histology. Brain Behav Evol 71:68–86

Olszewski J, Baxter D (1954) The cytoarchitecture of the human brain stem. Karger, Basel

Olszewski J, Baxter D (1982) Cytoarchitecture of the human brain stem, 2nd edn. Karger, Basel

Pantoni I (2010) Cerebral small vessel disease: from pathogenesis and clinical characteristics to therapeutic challenges. Lancet Neurol 9:689–701

Patel B, Markus HS (2011) Magnetic resonance imaging in cerebral small vessel disease and its use as a surrogate disease marker. Int J Stroke 6:47–59

Schröder H, Moser N et al (2020) Neuroanatomy of the mouse. Springer

Skrobot OA, Attems J et al (2016) Vascular cognitive impairment neuropathology guidelines (VCING): the contribution of cerebrovascular pathology to cognitive impairment. Brain 139:2957–2969

Skrobot OA, O'Brien J et al (2017a) The vascular impairment of cognition classification consensus study. Alzheimers Dement 13:624–633

Skrobot OA, Atterms J et al (2017b) Reply: atherosclerosis and vascular cognitive impairment neuropathological guideline. Brain 140:e13

Somani AN, Adesina O-o (2020) Neuroanatomy, abducens nucleus. In: StatPearls [Internet]. StatPearls, Treasure Island, FL

ten Donkelaar HJ (2011) Clinical neuroanatomy: brain circuitry and its disorders. Springer

Troncosco JC, Zonderman AB et al (2008) Effects of infarcts on dementia in the Baltimore longitudinal study of aging. Ann Neurol 64:168–176

Tsutsumi T, Houtani T et al (2007) Vesicular acetylcholine transporter-immunoreactive axon terminals enriched in the pontine nuclei of the mouse. Neuroscience 146:1869–1878

VanderHorst VG, Ulfhake B (2006) The organization of the brainstem and spinal cord of the mouse: relationships between monoaminergic, cholinergic, and spinal projection systems. J Chem Neuroanat 31:2–36

Vinters HV, Zarow C et al (2018) Review: vascular dementia: clinicopathologic and genetic considerations. Neuropathol Appl Neurobiol 44:247–266

Voogd J (2016) Deiters' nucleus. Its role in cerebellar ideogenesis: the Ferdinando Rossi Memorial Lecture. Cerebellum 15:54–66

Watson C, Bartholomaeus C et al (2019) Time for radical changes in brain stem nomenclature: applying the lessons from developmental gene patterns. Front Neuroanat 13:10

Wiederkehr S, Simard M et al (2008) Validity of the clinical diagnostic criteria for vascular dementia: a critical review. Part II. J Neuropsychiatry Clin Neurosci 20:162–177

Rhombomere 3 r3

11

Contents

H. Schröder et al., *The Human Brainstem*, https://doi.org/10.1007/978-3-030-89980-6_11

Abstract

In this chapter, the row of vestibular nuclei is continued with the cranial most one, the superior vestibular nucleus. From the group of monoamine nuclei, the noradrenergic A7 group is dealt with. A large part of this chapter is occupied by the motor trigeminal nucleus; that part of the trigeminal complex innervates the masticatory (chewing) muscles. Finally, two other raphe nuclei, the median and paramedian raphe nuclei are subjects of this chapter.

11.1 Rhombic lip

11.1.1 Precerebellar nuclei r3

11.1.1.1 Reticulotegmental nucleus of the pons r3

For detailed information on this nucleus, see Sect. 10.1.1.1.

11.1.1.2 Pontine nuclei

For detailed information on the pontine nuclei, see Sect. 10.1.1.2.

11.2 Alar r3

11.2.1 Cochlear nuclei r3

For details on the cochlear nuclei, see Sect. 8.1.1.

11.2.1.1 Ventral cochlear nucleus

For detailed information, see Sect. 8.1.1.1.

11.2.2 Vestibular nuclei r3

11.2.2.1 Superior vestibular nucleus [Ncl. vestibularis superior]

The rostrocaudal extension of the superior vestibular nucleus (SVN)—exclusively at pontine levels—stretches about 4 mm from the rostral pole of the principal sensory trigeminal nucleus to the rostral third of the nucleus of abducens nerve (see Sect. 9.2.1.1). It is bordered ventrally by the medial and lateral vestibular nuclei, dorsally by the cerebellum. The presence of vertical bundles of myelinated vestibulocerebellar fibers [Fibrae vestibulocerebellares] (Fig. 11.1 ④) is characteristic for the superior vestibular nucleus (see Fig. 11.1).

The vestibular organ and the general connectivity of the vestibular nuclei are dealt with under Sect. 5.3.1 ff. The SVN projects to the contralateral mesencephalon via the ventral mesencephalic tract [Tractus vestibulomesencephalicus]. (Büttner-Ennever and Gerrits 2004).

In mice, the superior vestibular nucleus is a large area starting cranially at the level of the caudal one third of the locus caeruleus and is replaced at the level of the facial colliculus/internal knee of the facial nerve by the medial and lateral vestibular nuclei (Kovac and Denk 1968).

Detailed information on the lateral vestibular nucleus is available under the indicated number:

11.2.2.2 Lateral vestibular nucleus

For detailed information on this nucleus, see Sect. 10.2.2.1.

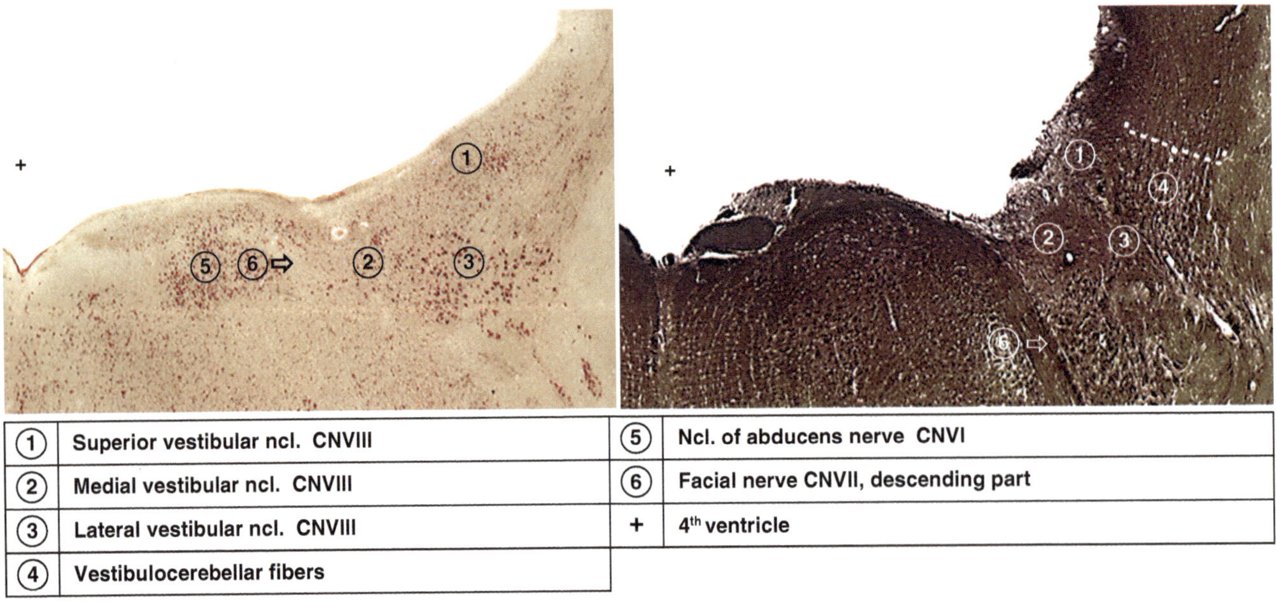

①	Superior vestibular ncl. CNVIII	⑤	Ncl. of abducens nerve CNVI
②	Medial vestibular ncl. CNVIII	⑥	Facial nerve CNVII, descending part
③	Lateral vestibular ncl. CNVIII	+	4th ventricle
④	Vestibulocerebellar fibers		

Fig. 11.1 Horizontal sections through the human brainstem at the level of the inner knee of the facial nerve/abducens nucleus ⑤. Left hand photograph Darrow Red (see also chapter 17 and atlas part Darrow red 19A), right hand Campbell fiber staining (see also atlas part Campbell 9, 9A, 9B). For details, see text. LabPON Twente

11.2.3 Monoamine nuclei r3

11.2.3.1 A7 noradrenaline cells (NA7)

The A7 cell group belongs to the pontine group of noradrenergic cells together with the A5 group (see Sect. 3.2.1). They are distributed from the level of the rostral facial nucleus up to the parabrachial nuclei (see Sect. 13.2.3). The cell groups A1, A2, A5, and A7 project in the ventral ascending bundle to the hypothalamus, the amygdala, the hippocampus, and the olfactory bulb (ten Donkelaar 2011).

A4 is projecting to the cerebellum together with A6, which provides noradrenergic input to the thalamus and forebrain (for details, see Sect. 13.2.2.1.3).

The group A3 is absent in the human brain.

The unequivocal identification of the brainstem adrenergic cell groups requires the use of immunocytochemical techniques (tyrosine hydroxylase, TH) as a survey marker for catecholaminergic cells, dopamine-ß-hydroxylase (DBH) for noradrenaline, and phenylethanolamine-N-methyltransferase (PNMT) for adrenaline (see also Fig. 3.3).

In mice, the TH-immunoreactive cell groups of the mesencephalon and pons (lateral tegmentum) form a more or less continuous column without distinct borders (VanderHorst and Ulfhake 2006), where A7 is located more rostral than A5. A5 neurons project to the spinal cord via a predominantly contralateral course (VanderHorst and Ulfhake 2006). A7 cell groups are involved in the control of nociception. Among other noradrenergic cell groups studied in rats, NA7 is mainly associated with motor control (Bruinstroop et al. 2012).

Detailed information on the following nuclei is available under the indicated numbers:

11.2.4 Trigeminal sensory nuclei r3

For detailed information, see Sect. 3.2.2

11.2.4.1 Spinal trigeminal nucleus, oral part

For detailed information, see Sect. 3.2.2.2

11.2.5 Lateral lemniscus r3

11.2.5.1 Ventral nucleus of the lateral lemniscus

For detailed information, see Sect. 8.1.1.5

11.2.6 Central gray r3

For detailed information on the central gray, see Sect. 13.2.7

11.3 Basal r3

11.3.1 Motor trigeminal complex r3

11.3.1.1 Motor nucleus of trigeminal nerve [Ncl. motorius nervi trigemini] Motor trigeminal nucleus

11.3.1.1.1 Location and cytoarchitecture of the motor trigeminal nucleus

In cranio-caudal direction, this nucleus appears medioventral of the superior cerebellar peduncle approximately at that site which is occupied further cranially by the locus caeruleus. It is bordered laterally by the principal sensory trigeminal nucleus, medially by the reticular formation, ventrally by the superior olivary complex and dorsal by the mesencephalic trigeminal nucleus (Fig. 11.2).

The neurons of the trigeminal motor nucleus are large and multipolar, as typical for somatomotor perikarya. Choline acetyltransferase-immunostaining renders the human trigeminal motoneurons as that of other human brainstem motor neurons positive (Oda and Nakanishi 2000). The axons of the trigeminal motoneurons run with the mandibular nerve (see Sect. 12.2.3.2.3) and give off different peripheral branches for the masticatory muscles (Figs. 11.3, 11.4, and 11.5, see Table 11.1) which can be distinguished roughly into jaw opening and closing muscles (see Table 11.1).

The development of the human motor nucleus of trigeminal nerve has been described in detail by Hamano et al. (1988).

In mice, the trigeminal motor nucleus—as well as the facial nucleus—displays a myotopic organization (Box 11.1, for details see, Terashima et al. 1994).

In a mouse model of tau protein hyperphosphorylation (pR5), mimicking one important aspect of Alzheimer pathology (see Sect. 13.2.2.1.4), i.e., neurofibrillary tangles, a massive infestation of the nucleus with AT8- and AT180-immunoreactive neurons with age was observed while the total number of neurons did not change. Interestingly, the body weight of pR5 mice assessed between 6, 14, and 17 months was consistently lower than that of in age-matched non-transgenic littermate controls (Huggenberger et al. 2021). Antibodies AT8 and AT180 are markers for the so-called pretangle stages of the tau protein. Possible changes in the trigeminal motor nucleus of AD patients have not been studied in detail (Wai et al. 2009) although weight loss in elderly humans has been suggested to be a marker for preclinical AD/mild cognitive impairment (Alhurani et al. 2016,

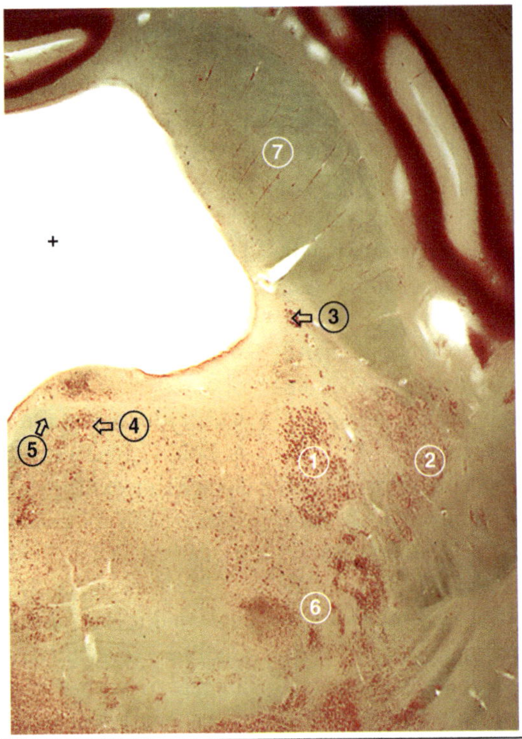

①	Motor ncl. of trigeminal nerve CNV	⑤	Genu of facial nerve
②	Principal sensory ncl. of trigeminal nerve CNV	⑥	Superior olivary complex
③	Mesencephalic ncl. of trigeminal nerve CNV	⑦	Superior cerebellar peduncle
④	Ncl. of abducens nerve CNVI	+	4th ventricle

Fig. 11.2 Horizontal section through the human brainstem at the level of the inner knee of the facial nerve ⑤/abducens nucleus. Darrow Red stain. (See also chapter 17 and atlas part Darrow red 23. LabPON Twente)

Jimenez and Pegueroles 2017) and full-blown AD (Mathys et al. 2017).

Box 11.1 Myotopy
Myotopy is the central nervous representation of muscles in a topographically ordered fashion.

Related phenomena are tonotopy in the auditory system (representation by individual frequencies), somatotopy in the sensory system (representation by peripheral regions), and retinotopy in the optic nerve, tract, and the visual cortex (representation by central, peripheral, and marginal regions of the retina).

11.3.1.1.2 Targets of the motor portion of the trigeminal nerve

The human trigeminal motor nucleus provides efferent innervation of the masticatory (chewing) muscles and part of the digastric muscle (floor of the oral cavity) (see Table 11.1 and following Figs. 11.3, 11.4, 11.5, and 11.6).

In Fig. 11.4, temporal and masseter muscles (compare with Fig. 11.3) have been removed as well as the coronoid process [Processus coronoideus] and parts of the ramus of mandible [Ramus mandibulae] to render visible the deep masticatory muscles (cf. with Figs. 11.6, 11.7, and 11.8): The lateral pterygoid muscle is composed of two heads. The superior one originates from the sphenoidal bone and runs to the lateral discus of the temporomandibular joint. The inferior head is running from the pterygoid process to the coronoid process of the mandibula. The superior head is the only masticatory muscle for jaw opening. The inferior one shifts the mandibula to the contralateral side; bilateral contraction protrudes the mandibula. The medial pterygoid originates in the pterygoid fossa and inserts at the masseteric tuberosity. It is involved in jaw closure (for details see, Table 11.1).

Finally, the anterior belly of the digastric muscle, innervated by the trigeminal motor nucleus is shown in Fig. 11.6.

11.3.1.1.3 Course of the motor portion of the trigeminal nerve

The motor root of the trigeminal nerve leaves the brainstem together with the other portions of the trigeminal nerve (see Sect. 12.2.3.2.3, Figs. 12.3, 12.4, and 12.12).

It bypasses the trigeminal ganglion [Ggl. trigeminale] (Figs. 12.3 and 12.7) and joins the mandibular nerve (V_3) which leaves the cranial cavity via the foramen ovale into the infratemporal fossa [Fossa infratemporalis] (Fig. 12.4).

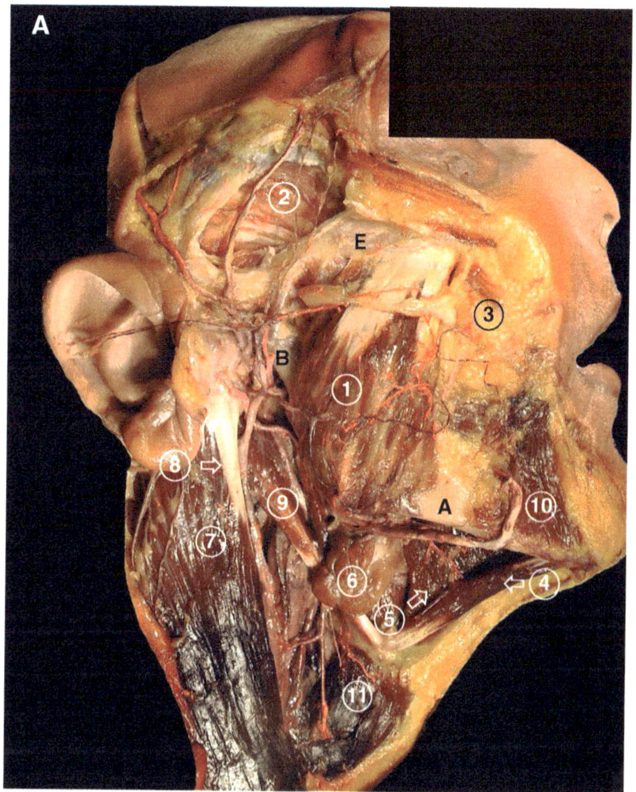

①	Masseter muscle CNV_3	⑩	Levator labii superioris alaeque nasi muscle CNVII
②	Temporalis / Temporal muscle CNV_3	⑪	Omohyoid muscle C_1-C_3
③	Buccal fat pad [Bichati]	A	Body of mandible
④	Digastric muscle, anterior belly CNV_3	B	Ramus of mandible
⑤	Mylohyoid muscle CNV_3	C	Coronoid process
⑥	Submandibular gland	D	Mandibular condyle
⑦	Sternocleidomastoid muscle CNXI	E	Zygomatic arch
⑧	Sternocleidomastoid muscle, tendo	F	Mandibular fossa
⑨	Digastric muscle, posterior belly CNVII		

Fig. 11.3 Lateral view of the human skull. (**A**) Skin and subcutaneous fat tissue partly removed. The buccal fat pad ⑤ covers the buccinator muscle (see Fig. 11.5 ⑬) which is innervated by the facial nerve (CNVII). The most superficial masticatory muscles visible are the masseter ① and ② temporal muscles. (**B**) The site of origin of the masseter muscle is the zygomatic arch E. The muscle is running to the masseteric tuberosity at the angle of the mandible [Angulus mandibulae] (**C**). The temporalis muscle originates in the temporal/parietal region of the head and inserts at the coronoid process of the mandibula. Both muscles are involved in jaw closure (see Table 11.1). (**A**) Courtesy Prof. Georg Feigl, Institute of Anatomy and Clinical Morphology, Witten/Herdecke University, Germany. Skull preparation: Sammlung des Zentrums Anatomie der Universität zu Köln. (**B**, **C**) Modified acc. to Lang 2004, Fig. 4 with permission

B

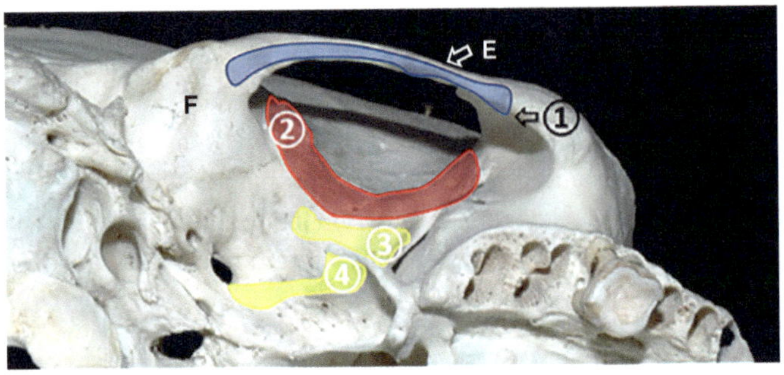

C

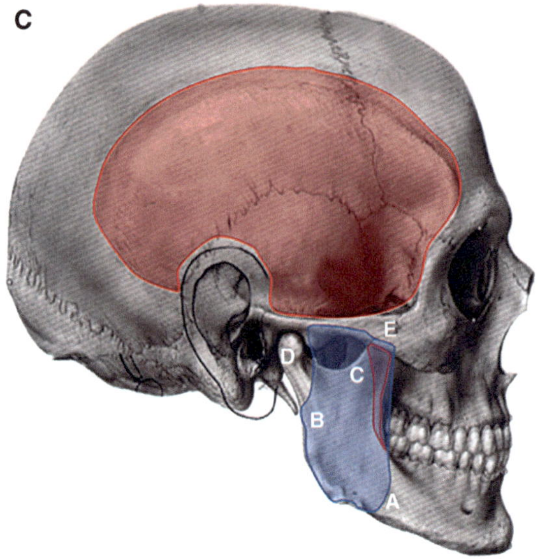

①	Masseter muscle CNV_3 (blue)	A	Body of mandible
②	Temporalis / Temporal muscle CNV_3 (red)	B	Ramus of mandible
③	Lateral pterygoid muscle CNV_3 (yellow)	C	Coronoid process
④	Medial pterygoid muscle CNV_3 (yellow)	D	Mandibular condyle
		E	Zygomatic arch
		F	Mandibular fossa

Fig. 11.3 (continued)

11.3.1.1.4 Central connectivity of the motor trigeminal nucleus

The anatomical knowledge about the cerebrocortical control of brainstem motor nuclei is based on the investigation by Kuypers in the 1950s (Kuypers 1958) on the brains of patients who died after an anatomically defined stroke. The thorough investigation of the specimens (here cited for a unilateral subtotal obstruction of the middle cerebral artery, acc. to ten Donkelaar 2011) revealed that degenerating corticobulbar (corticonuclear) fibers are distributed to both halves of the brainstem, the hypoglossal nuclei (particularly contralaterally), the ambiguus nucleus, the spinal trigeminal nucleus, the facial nucleus (in particular, contralaterally), and the motor trigeminal nucleus. The abducens nucleus was free of degenerating terminals.

11.3.1.1.5 Living anatomy and pathological aspects

Among the masticatory muscles, the temporalis muscle and the masseter muscle are accessible to a simple bedside testing. The examiner puts its palms on the masseter muscles of both sides. The patient is asked to tightly close the mouth. Under normal conditions, the examiner should feel the contraction of the masseter muscles on either side. The same procedure is used for the temporalis muscle accordingly.

11.3.2 Raphe nuclei r3

11.3.2.1 Median raphe nucleus B6 [Ncl. raphes medianus]

The median raphe nucleus belongs to the rostral group of raphe nuclei which accounts for 85% of all serotonergic neurons in the human brain (Hornung 2003, 2012). In addition to the median raphe nucleus, this group comprises the pontine raphe nucleus (B5), caudal linear nucleus (see Sect. 14.5.3.1), and the dorsal raphe nucleus (B7) (see Sect. 13.3.4.2). Formerly, the median raphe nucleus has also been called central raphe nucleus or central superior raphe nucleus (see e.g., Ohm et al. 1989, Rüb, personal communication).

For the caudal group (approx. 15% of 5-HT neurons), see the raphe magnus nucleus (B3) (see Sect. 7.3.3.2), the raphe pallidus nucleus (B1) (see Sect. 4.3.3.2), the raphe obscurus nucleus (B2) (see Sect. 4.3.3.1), and the serotonergic cells of vestibular area.

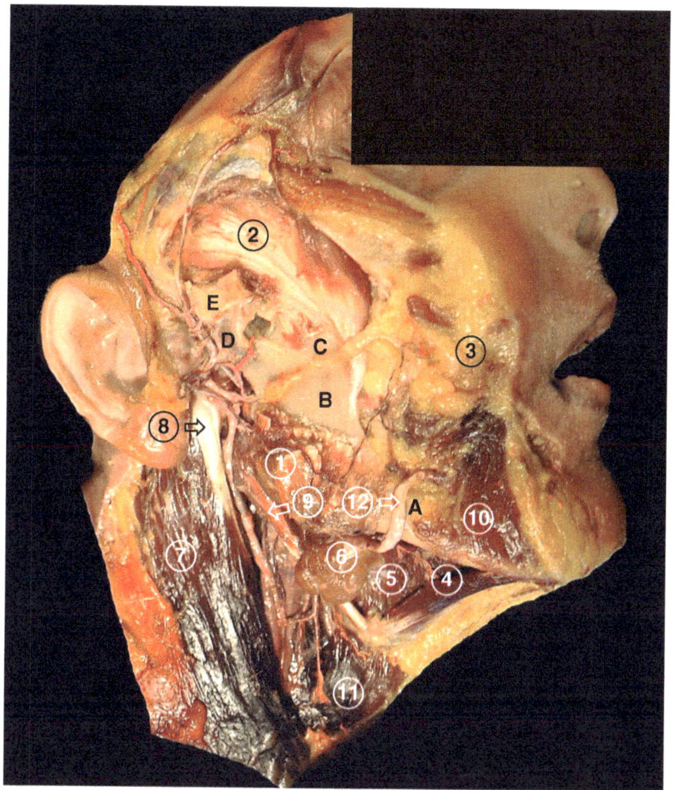

①	Masseter muscle CNV_3	⑩	Levator labii superioris alaeque nasi muscle CNVII
②	Temporalis / Temporal muscle CNV_3	⑪	Omohyoid muscle C_1-C_3
③	Buccal fat pad [Bichati] covering the buccinator muscle	⑫	Facial artery
④	Digastric muscle, anterior belly CNV_3	A	Body of mandible
⑤	Mylohyoid muscle CNV_3	B	Ramus of mandible
⑥	Submandibular gland	C	Coronoid process
⑦	Sternocleidomastoid muscle CNXI	D	Mandibular condyle
⑧	Sternocleidomastoid muscle, tendo	E	Zygomatic arch
⑨	Digastric muscle, posterior belly CNVII		

Fig. 11.4 Lateral view of the human skull. In addition to the dissection status in Fig. 11.3, the masseter muscle has been removed with the exception of origin and insertion as well as most of the zygomatic arch. This allows a view on the temporalis muscle ② down to its insertion at the coronoid process of the mandible C. Courtesy Prof. Georg Feigl, Institute of Anatomy and Clinical Morphology, Witten/Herdecke University, Germany. Skull preparation: Sammlung des Zentrums Anatomie der Universität zu Köln

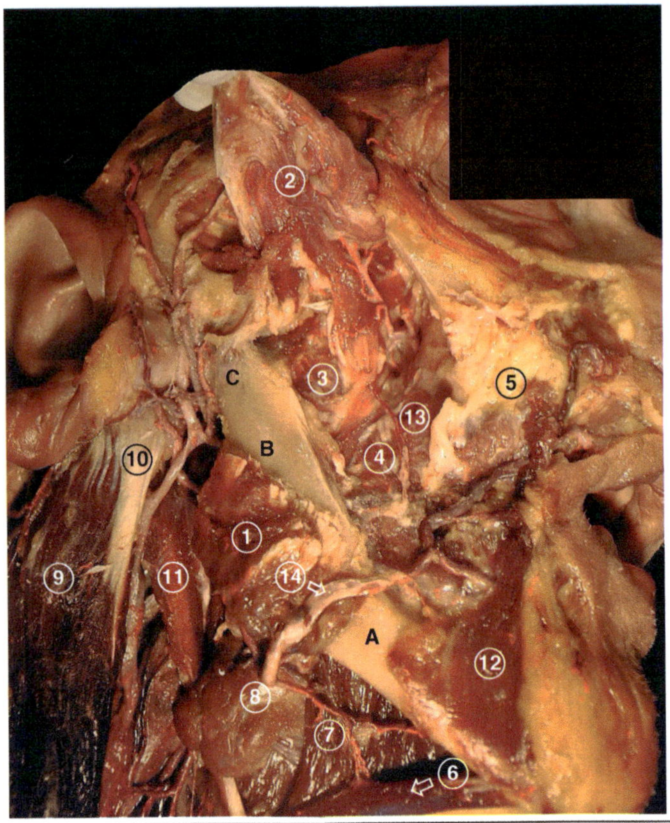

①	Masseter muscle CNV_3	⑩	Sternocleidomastoid muscle, tendo
②	Temporalis / Temporal muscle CNV_3	⑪	Digastric muscle, posterior belly CNVII
③	Lateral pterygoid muscle CNV_3	⑫	Levator labii superioris alaeque nasi muscle CNVII
④	Medial pterygoid muscle CNV_3	⑬	Buccinator muscle CNVII
⑤	Buccal fat pad [Bichati] covering the buccinator muscle	⑭	Facial artery
⑥	Digastric muscle, anterior belly CNV_3	A	Body of mandible
⑦	Mylohyoid muscle CNV_3	B	Ramus of mandible
⑧	Submandibular gland	C	Mandibular condyle
⑨	Sternocleidomastoid muscle CNXI		

Fig. 11.5 The masseter muscle has been removed completely; the tendon of the temporalis muscle from the coronoid process [Processus coronoideus] and the muscle belly are folded upwards. This allows a view onto the deep masticatory muscles, (cf. with Fig. 11.7A, B). The lateral pterygoid muscle is composed of two heads. The superior one originates from the sphenoidal bone and runs to the lateral discus of the temporomandibular joint. The inferior head is running from the pterygoid process to the coronoid process of the mandibula. The superior head is the only masticatory muscle for jaw opening. The inferior one shifts the mandibula to the contralateral side; bilateral contraction protrudes the mandibula. The medial pterygoid originates in the pterygoid fossa and inserts at the masseteric tuberosity (see also Fig. 11.8). It is involved in jaw closure (for details see, Table 11.1). Courtesy Prof. Georg Feigl, Institute of Anatomy and Clinical Morphology, Witten/Herdecke University, Germany. Skull preparation: Sammlung des Zentrums Anatomie der Universität zu Köln

Table 11.1 Human masticatory muscles innervated by mandibular nerve branches acc. to the classification by Stockstill and Mohl (2015) and https://m.thieme.de/viostatics/media/vio-2/final/de/media/pruefung/kiefergelenk-kaumuskeln-zaehne-nina-engel.pdf

Muscle	Heads	Origin	Insertion	Innervation	Function
Temporalis [M. temporalis]		Temporal line (Temporal/parietal bone)	Coronoid process of mandibula	Deep temporal nerves	Jaw closure, mandibula is drawn dorsally
Masseter [M. massetericus]	Superficial Deep	Zygomatic arch	Angle of mandible (masseteric tuberosity)	Masseter nerve	Jaw closure, muscle loop around the mandibula
Medial pterygoid [M. pterygoideus medialis]	Lateral Medial	Pterygoid fossa	Angle of mandible (masseteric tuberosity)	Medial pterygoid nerve	Moves the mandibula sideways
Lateral pterygoid [M. pterygoideus lateralis]	Superior	Infratemporal crest (sphenoid bone)	Discus lateralis (temporomandibular joint)	Lateral pterygoid nerve	Start of jaw opening
	Inferior	Pterygoid process, lateral plate	Condylar process	Lateral pterygoid nerve	Unilateral: Shifts mandibula to the contralateral side Bilateral: Protrusion
Mylohyoid [M. mylohyoideus]		Mylohyoid line (medial surface of the mandible)	Midline raphe (extends from mandibular symphysis to hyoid bone (Fig. 11.6))	Mylohyoid nerve	Elevates floor of mouth (e.g., in swallowing or protruding the tongue)
Digastric, anterior belly [M. digastricus, Venter ant.]		Intermediate tendon between anterior and posterior belly	Digastric fossa	Mylohyoid nerve	Part of the mouth floor

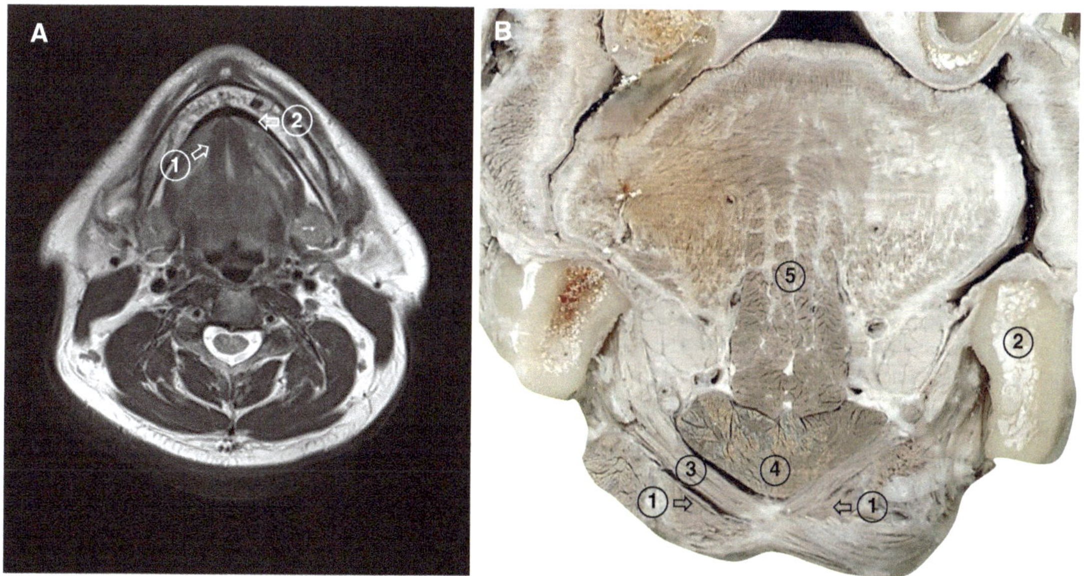

①	**Digastric muscle**	④	**Genioglossal muscle**
②	**Mandible**	⑤	**Tongue**
③	**Mylohyoid muscle**		

Fig. 11.6 View of the mandible with the anterior belly of the digastric muscle A ① (gaster Latin = belly) and the mylohyoid muscle B ③. Anterior and posterior bellies form an important constituent of the floor of the mouth together with the mylohyoid muscle. The anterior belly inserting at the mandible is innervated by the trigeminal motor nucleus, the posterior belly by the facial nerve. Sammlung des Zentrums Anatomie der Universität zu Köln

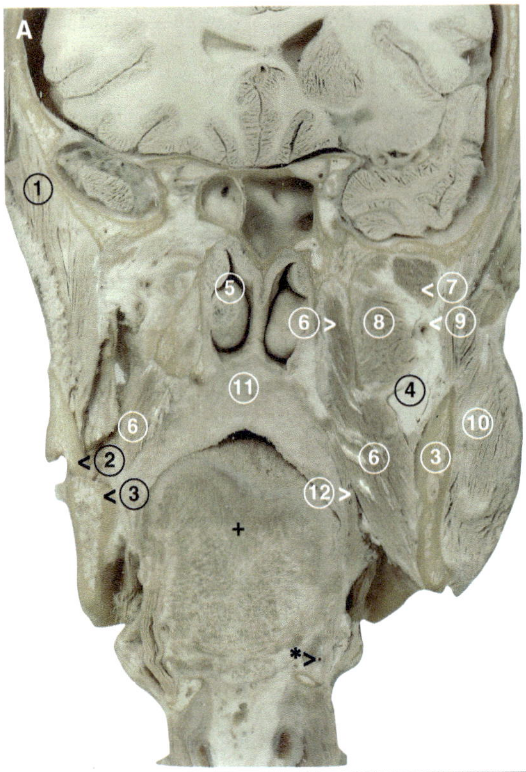

①	**Temporal muscle**	⑧	**Lateral pterygoid muscle, Inferior head**
②	**Mandible, coronoid process**	⑨	**Maxillary artery**
③	**Ramus of mandible**	⑩	**Masseter muscle**
④	**Buccal fat pad (Bichati)**	⑪	**Palate**
⑤	**Nasal cavity**	⑫	**Buccinator muscle**
⑥	**Medial pterygoid muscle**	+	**Tongue**
⑦	**Lateral pterygoid muscle, Superior head**	*	**Lingual artery**

Fig. 11.7 (**A**) Masticatory muscles in a coronal image (slight oblique cut) which can also be generated in vivo by means of MRT. The temporalis muscle ① is a flat muscle running parallel to the lateral parts of the skull to the head of the mandibula and the zygomatic bone (see also Fig. 11.3). It is covered by the galea and the skin of the head. The superior belly ⑦ of the lateral pterygoid muscle is seen directly ventral of the skull base. It is followed by the inferior belly ⑧ of the lateral pterygoid muscle. At the external surface of the mandibula, the masseter muscle ⑩ is located. The medial pterygoid muscle ⑥ is found medial of the lateral counterpart (see also Fig. 11.8). Sammlung des Zentrums Anatomie der Universität zu Köln. (**B**) Coronal section through the human head at the level of the pterygoid fossa. All masticatory muscles are hit in this section (see also Fig. 11.8). At higher magnification, it becomes obvious that the medial pterygoid muscle ① has its origin in the pterygoid fossa (hence the name) between the lateral ⑧ (red dotted line) and the medial plate ⑨ of the sphenoid bone. The right photograph shows the osteological situation in an oblique caudal view. The vomer bone (Latin for plow share, forming large part of the nasal septum) ⑩ indicates the midline. Sammlung des Zentrums Anatomie der Universität zu Köln

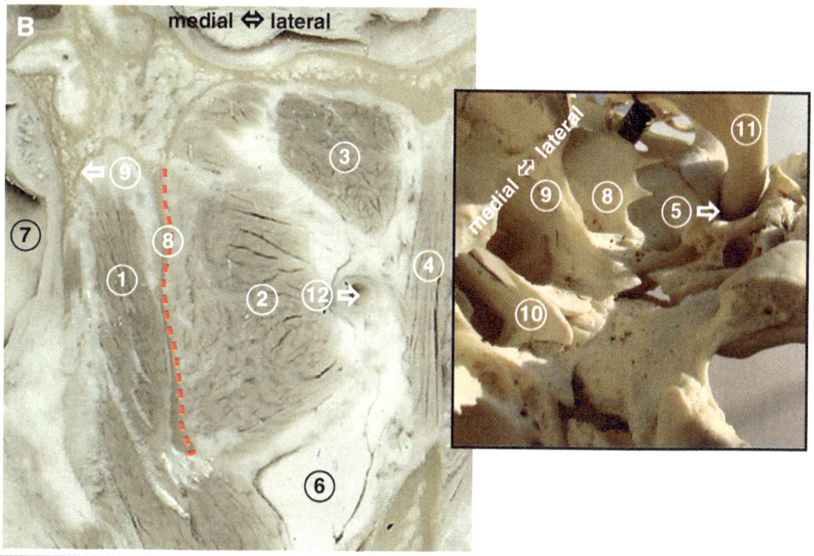

①	Medial pterygoid muscle	---	Lateral surface of lateral plate of sphenoid bone
②	Lateral pterygoid muscle, inferior head	⑧	Lateral plate of sphenoid bone
③	Lateral pterygoid muscle, superior head	⑨	Medial plate of sphenoid bone
④	Temporal muscle	⑩	Vomer
⑤	Temporomandibular joint	⑪	Ramus of mandible
⑥	Buccal fat pad (Bichati)	⑫	Maxillary artery
⑦	Nasal cavity		

Fig. 11.7 (continued)

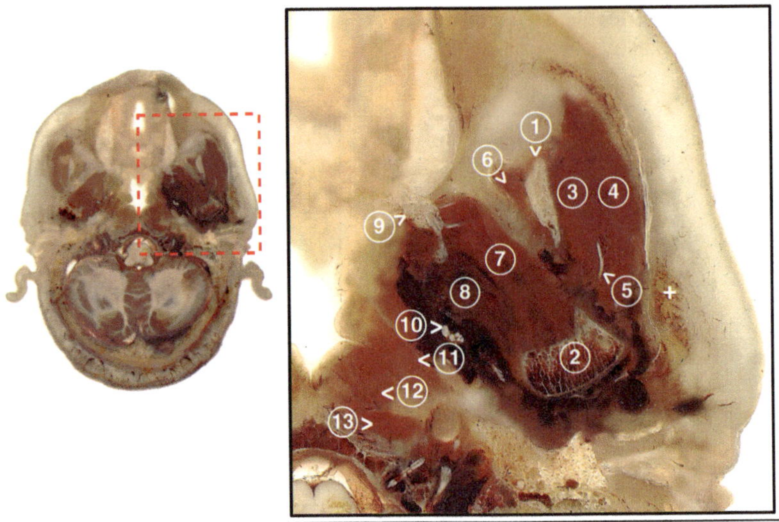

①	Mandible, coronoid process	⑧	Medial pterygoid muscle
②	Mandibular condyle	⑨	Pterygoid process
③	Masseter muscle, profound part	⑩	Mandibular nerve V_3
④	Masseter muscle, superficial part	⑪	Tensor veli palatini muscle
⑤	Masseteric nerve V_3	⑫	Levator veli palatini muscle
⑥	Temporalis muscle	⑬	Longus capitis muscle
⑦	Lateral pterygoid muscle	+	Parotid gland

Fig. 11.8 Masticatory muscles in a horizontal plane at the level of the confluence of the vertebral arteries. From lateral to medial the masseter muscle ③ ④, the lateral ⑦ and medial pterygoid muscles ⑧ are seen. The temporal muscle is hit in a very anterior position medially of the mandible ⑥. The caudal view of the skull shows the bony areas of insertion (temporalis) and origin (masseter, pterygoids) of the masticatory muscles. Mezey et al. (2022) recently claimed the existence of a deep or coronoid part of the masseter muscle thought to stabilize and retract the anterior-superior part of the mandible. In our material, we could not confirm the existence of this part of the masseter muscle. Sammlung des Zentrums Anatomie der Universität zu Köln

11.3.2.1.1 Location and cytoarchitecture of the median raphe nucleus

The B6 serotonergic cell group extends from the level of the trochlear nucleus (anular subnucleus) to the level of the locus caeruleus (principal subnucleus) (Rüb et al. 2000) (Fig. 11.9).

11.3.2.1.2 Connectivity of the median raphe nucleus

As known from animal studies, the median raphe as well as the dorsal raphe (see Sect. 13.3.4.2) are targeted from the lateral habenula, the interpeduncular nucleus (IPN) (see Sect. 14.5.4.4), several hypothalamic nuclei, the ventral tegmental area (VTA) (see Sect. 15.5.1.1), the laterodorsal tegmental nuclei (see Sect. 13.1.1.2), and the cingulate cortex which all are glutamatergic. Furthermore, the median nucleus receives input from the medial septum and the diagonal band of Broca (Box 11.2). Ascending input originates from the raphe magnus (see Sect. 7.3.3.2) and pallidus nuclei and the prepositus hypoglossi nucleus (see Sect. 6.2.1.5). The main efferent projections of the median raphe are via a ventral tract to the VTA (see Sect. 15.5.1.1) and IPN (see Sect. 14.5.4.4).

> **Box 11.2 Paul Broca (1824–1880)**
> The diagonal band of Broca is a basal forebrain structure first described by the French neuroanatomist Paul Broca. Today, he is well known for his discovery of a specific cortical region—today known as Broca's area—in post mortem brains of aphasic patients. These patients were suffering from an expressive aphasia and a damage of this specific region in the left telencephalic hemisphere in common.
> https://en.wikipedia.org/wiki/Paul_Broca

Possibly, the function most closely associated with the median raphe and sleep is the modulation of the hippocampal EEG or the theta rhythm. Specifically, the MR suppresses (or

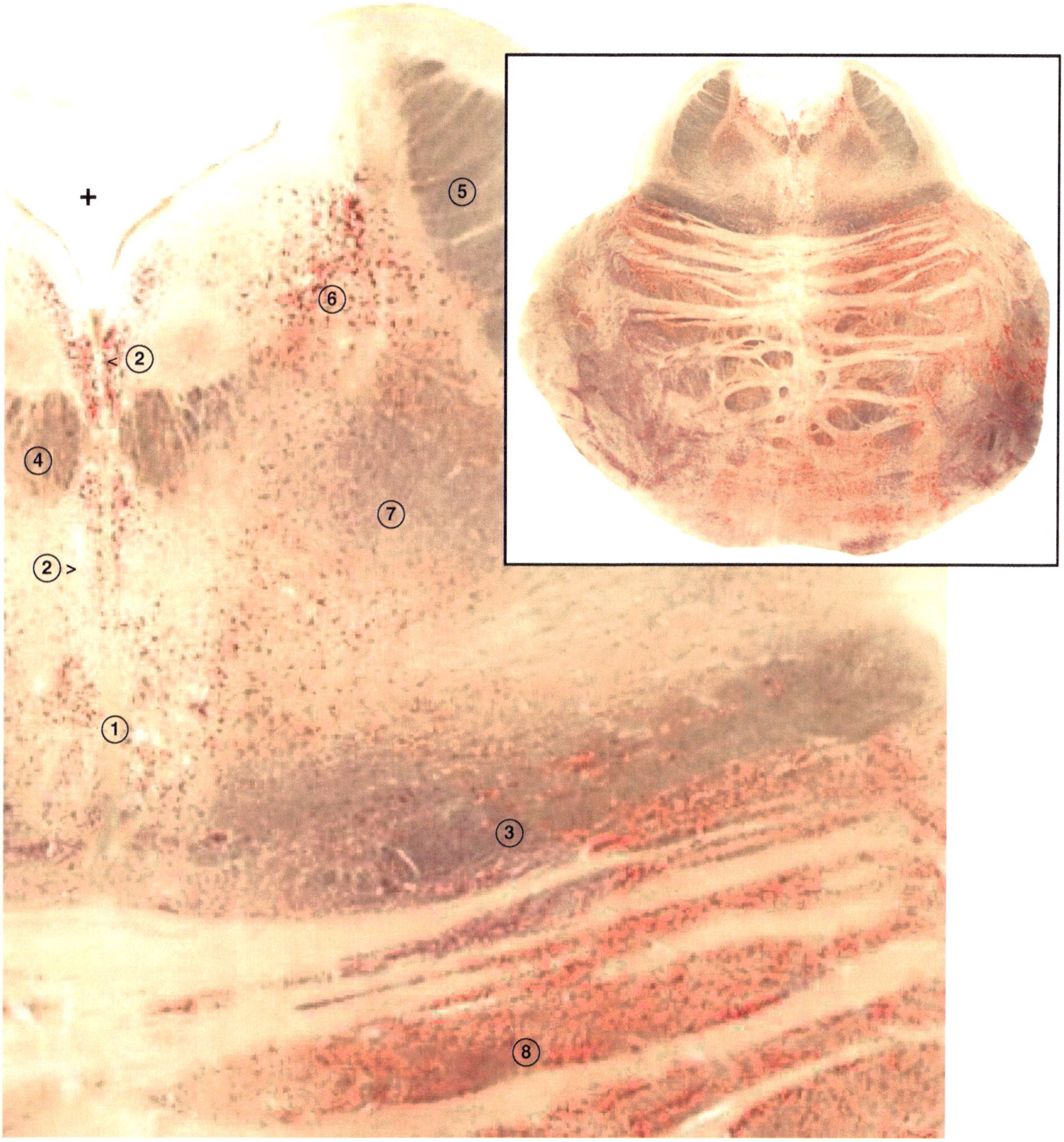

①	Median (central) raphe ncl., principal subncl.	⑤	Superior cerebellar peduncle
②	Dorsal raphe ncl., compact caudal subncl.	⑥	Locus caeruleus
③	Medial lemniscus	⑦	Central tegmental tract
④	Medial longitudinal fasciculus	⑧	Pontine ncll.
		+	4th ventricle

Fig. 11.9 Position of the median raphe nucleus ① in its superior part at the level of the Locus caeruleus ⑥ (see Sect. 13.2.2.1). Horizontal section. Darrow Red staining. The median raphe nucleus takes over a strictly midline position. At this level, it is bordered dorsally by the dorsal raphe nucleus ② and ventrolaterally by the medial lemniscus ③ (see Fig. 3.14A–G). See also atlas part Darrow red 31, 31A. LabPON Twente

blocks) theta in non-REM (NREM) sleep, while MR silencing in REM sleep (see Box 11.3) releases theta during that state (permissive role) (for details, see Linley and Vertes 2019).

> **Box 11.3 REM sleep**
> REM sleep (rapid eye movement sleep) is the fourth stage of sleep in which the eyes can be seen to be moving rapidly behind the eyelids. It is associated with increased brain activity and dreaming but decreased muscle tone.

11.3.2.2 Paramedian raphe nucleus [Ncl. raphes paramedianus]

The term paramedian raphe nucleus refers to a narrow layer of cells located lateral to the median raphe nucleus in the pontine reticular formation. It is found in man, the macaque, the rat, and the mouse. In man, but not in the rodent, it contains a substantial number of serotonergic cells.

http://braininfo.rprc.washington.edu/centraldirectory.aspx?ID=839

11.3.3 Basal tegmental nuclei r3

11.3.3.1 Dorsomedial tegmental area r3

For detailed information, see Sect. 7.3.4.1.

11.3.4 Reticular nuclei r3

11.3.4.1 Pontine reticular nucleus, oral part

For detailed information on this nucleus, see Sect. 10.3.5.4.

11.3.4.2 Parvocellular reticular nucleus

For detailed information on this nucleus, see Sect. 6.5.5.2.

References

Alhurani RE, Vassilaki M et al (2016) Decline in weight and incident mild cognitive impairment: mayo clinic study of aging. JAMA Neurol 73:439–446

Bruinstroop E, Cano G et al (2012) Spinal projections of the A5, A6 (locus coeruleus), and A7 noradrenergic cell groups in rats. J Comp Neurol 520:1985–2001

Büttner-Ennever JA, Gerrits NM (2004) Vestibular system. In: The human nervous system, 2nd edn, pp 1212–1240

Hamano S et al (1988) Development of the human motor trigeminal nucleus. Pediatr Neurosci 14:230–235

Hornung J-P (2003) The human raphe nuclei and the serotonergic system. J Chem Neuroanat 26:331–343

Hornung J-P (2012) The human raphe nuclei and the serotonergic system. In: The human nervous system (third edition)

Huggenberger S, Paterno M et al (2021) Longitudinal assessment of tau phosphorylation in the brainstem of P301L tau-transgenic pR5 mice. J Alzheimers Neurodegener Dis 7:051

Jimenez A, Pegueroles J (2017) Alzheimer's Disease Neuroimaging Initiative. Weight loss in the healthy elderly might be a non-cognitive sign of preclinical Alzheimer's disease. Oncotarget 8:104706–104716

Kovac W, Denk H (1968) Der Hirnstamm der Maus. Topographie, Cytoarchitektonik und Cytologie. Springer, Wien–New York

Kuypers HGJM (1958) Corticobulbar connections to the pons and lower brain stem in man. Brain 81:364–388

Lang J (2004) Praktische Anatomie, Band 1—Kopf, Teil A: Übergeordnete Systeme. Founded by Titus Ritter von Lanz, Werner Wachsmuth. Springer

Linley SB, Vertes RP (2019) Serotonergic systems in sleep and waking. In: Handbook of behavioral neuroscience, vol 30, pp 101–123

Mathys J, Gholamrezaee M et al (2017) Decreasing body mass index is associated with cerebrospinal fluid markers of Alzheimer's pathology in MCI and mild dementia. Exp Gerontol 100:45–53

Mezey SE, Müller-Gerbl M et al (2022) The human masseter muscle revisited: First description of its coronoid part. Ann Anat – Anat Anz 240:15178

Oda Y, Nakanishi I (2000) The distribution of cholinergic neurons in the human central nervous system. Histol Histopathol 15:825–834

Ohm TG, Heilmann R et al (1989) The human oral raphe system. Architectonics and neuronal types in pigment.Nissl preparations. Anat Embryol (Berl) 180:37–43

Rüb U, Del Tredici K et al. (2000) The evolution of Alzheimer's disease-related cytoskeletal pathology in the human rapke nuclei. Neuropathol Appl Neurobiol 26:553–567

Stockstill JW, Mohl ND (2015) Static and functional anatomy of the human masticatory system. In: Kandasamy S, Greene C, Rinchuse D, Stockstill J (eds) TMD and orthodontics. Springer

ten Donkelaar HJ (2011) Clinical neuroanatomy. Brain circuitry and its disorders. Springer

Terashima T, Kishimoto Y et al (1994) Musculotopic organization in the motor trigeminal nucleus of the reeler mutant mouse. Brain Res 666:31–42

VanderHorst VG, Ulfhake B (2006) The organization of the brainstem and spinal cord of the mouse: relationships between monoaminergic, cholinergic, and spinal projection systems. J Chem Neuroanat 31:2–36

Wai MSM, Liang C et al (2009) Co-localization of hyperphosphorylated tau and caspases in the brainstem of Alzheimer's disease patients. Biogerontology 10:457–469

Web Links

http://braininfo.rprc.washington.edu/centraldirectory.aspx?ID=839

https://m.thieme.de/viostatics/media/vio-2/final/de/media/pruefung/kiefergelenk-kaumuskeln-zaehne-nina-engel.pdf

https://en.wikipedia.org/wiki/Paul_Broca

Rhombomere 2 r2

12

Contents

H. Schröder et al., *The Human Brainstem*, https://doi.org/10.1007/978-3-030-89980-6_12

Abstract

In this neuromere, the second trigeminal nucleus following the spinal one (see Chap. 3) develops, i.e., the principal sensory trigeminal nucleus. By contrast to the protopathic properties of the spinal nucleus of the trigeminal nerve, this nucleus is the epicritic component of the trigeminal system providing sensibility to the skin of the face, the cranial mucosa of mouth, nasal cavities, teeth, and sinuses. This distribution pattern—alongside that of the spinal trigeminal nucleus—indicates the clinical disciplines that potentially deal with the epicritic properties of the trigeminal nerve like neurology, neurosurgery, ENT, neuroradiology but also dentistry. The principal sensory trigeminal nucleus gets input from three major branches, the ophthalmic, the maxillar, and the mandibular nerves. In addition to the sensory function, branches of the trigeminal nucleus subserve the guidance of parasympathetic and sympathetic fibers to their target organs, for example, to the lacrimal gland and the minor salivary glands. The trigeminal nerve plays an important role in neurological examination and for neurological disorders as, for example, trigeminal neuralgia and trigeminal herpes zoster.

The interpeduncular nuclei are the target structure of the habenulo-interpeduncular tract (formerly Fasciculus retroflexus) from the epithalamic habenular nuclei, one of the largest cholinergic tracts of the brain. Data from animal studies point to a role in the regulation of several behavioral patterns.

Barrington's nucleus—a derivative of rhombomere 2—is an important integration center for the regulation of the urinary bladder function.

The interpeduncular fossa, exit site of the oculomotor nerve, is an important anatomical and neuroradiological landmark. The close anatomical relationship between the oculomotor nerve and the vessels of the circle of Willis (see Chaps. 1 and 15) in the region of the interpeduncular fossa is the basis of aneurysm-related disturbances of the oculomotor nerve.

Detailed information on the following nuclei can be found as indicated:

12.1 Rhombic lip r2

12.1.1 Precerebellar nuclei r2

For detailed information, see Sect. 10.1.1.

12.1.1.1 Reticulotegmental nucleus of the pons

For detailed information on this nucleus, see Sect. 10.1.1.1.

12.2 Alar r2

12.2.1 Cochlear nuclei r2

For detailed information, see Sect. 8.1.1.

12.2.1.1 Dorsal cochlear nucleus

For detailed information on this nucleus, see Sect. 8.1.1.1.

12.2.2 Vestibular nuclei r2

For detailed information on the vestibular nuclei, see Sect. 11.2.2.

12.2.2.1 Superior vestibular nucleus

For detailed information on this nucleus, see Sect. 11.2.2.1.

12.2.3 Trigeminal sensory nuclei r2

12.2.3.1 Supratrigeminal nucleus r3 [Ncl. supratrigeminalis]

The supratrigeminal nucleus belongs to the group of the so-called premotor nuclei which send their axons to the hypoglossal nucleus, facial nucleus, and trigeminal motor nucleus. It is located dorsal and rostrodorsal to the mouse trigeminal motor nucleus. Tracing experiments in mice have shown reciprocal connections of this nucleus to the spinal trigeminal nucleus and the trigeminal motor nucleus. Li et al. (2005) conclude that some of the supratrigeminal neurons may represent premotor neurons to the trigeminal motor nucleus

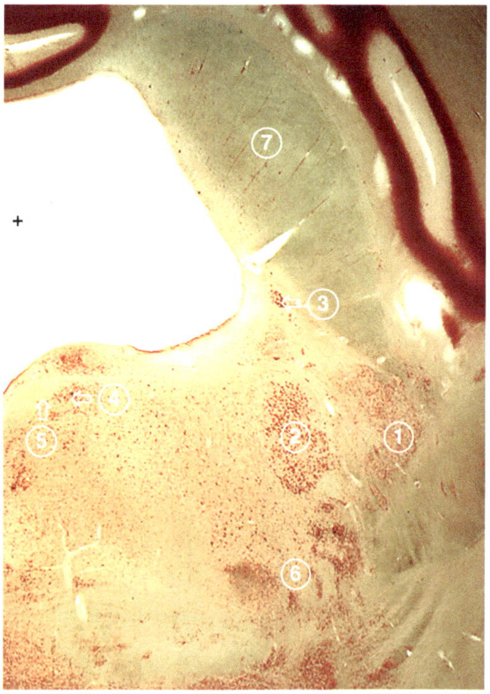

①	Principal sensory ncl. of trigeminal nerve CNV	⑤	Genu of facial nerve
②	Motor ncl. of trigeminal nerve CNV	⑥	Superior olivary complex
③	Mesencephalic ncl. of trigeminal nerve CNV	⑦	Superior cerebellar peduncle
④	Ncl. of abducens nerve CNVI	+	4th ventricle

Fig. 12.1 Horizontal section through the human pontine region at the level of the facial genu ⑤ and the three trigeminal nuclei ①–③ mentioned above under Sect. 12.2.3.2. The section is located cranial of the cranial end of the spinal ncl. of trigeminal nerve. Darrow red staining. The most conspicuous of the trigeminal nuclei is the motor ncl. of trigeminal nerve (for details, see Sect. 11.3.1.1) due to its large, densely packed neuronal perikarya ②. Directly lateral of the motor ncl. the principal sensory ncl. of trigeminal nerve is seen ① which is characterized by much smaller nerve cells. The mesencephalic ncl. of trigeminal nerve ③ is a band of large round loosely distributed neurons (see Sect. 13.2.6.1). See atlas part Darrow red 23. LabPON Twente

that receive nociceptive input from the spinal trigeminal nucleus to elicit jaw-opening reflexes by inhibiting jaw-closing trigeminal motor neurons. Human-specific data are not available (not listed in FIPAT Ch. 1).

12.2.3.2 Principal sensory trigeminal nucleus/ Principal sensory nucleus of trigeminal nerve r2 [Ncl. principalis n. trigemini]

The sensory trigeminal system (spinal nucleus of trigeminal nerve see Sect. 3.2.2.2, principal sensory nucleus of trigeminal nerve, mesencephalic nucleus of trigeminal nerve (see Sect. 13.2.6.1)) provides the sensory innervation for most of the head region. This is true for nociception (spinal trigeminal nucleus), epicritic sensory innervation (principal trigeminal nucleus), and proprioception (mesencephalic trigeminal nucleus). In the following, we will first deal with the principal trigeminal nucleus and then with the widespread branches of the trigeminal nerve.

12.2.3.2.1 Location and cytoarchitecture of the principal sensory trigeminal nucleus

The nucleus is located in the pontine tegmentum close to the superior cerebellar peduncle (see Fig. 12.1). Caudally, it is continuous with the spinal trigeminal nucleus (see Sect. 3.2.2.2). Cranially, it terminates 1 mm caudal to the oral pole of the motor trigeminal nucleus (see Sect. 11.3.1.1, Fig. 12.1). Its craniocaudal extension is about 5 mm (Olszewski and Baxter 1982).

On the histological level, the principal trigeminal nucleus is characterized by small oval or round cells with relatively large nuclei and diffusely distributed Nissl substance (Olszewski and Baxter 1982). The perikarya tend to congregate in irregular groups or clusters.

12.2.3.2.2 Target regions of the principal sensory trigeminal nerves

The trigeminal nerve has three major branches which collect sensory information from a plethora of smaller ramifications

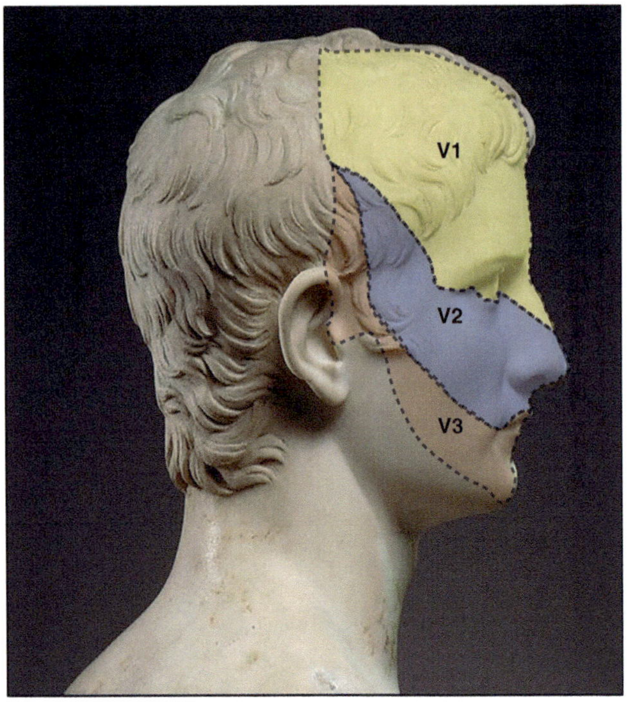

Fig. 12.2 Lateral view of the human face showing the peripheral cutaneous sensory supply zones of the three major trigeminal branches. The main access to examine the sensory part of the trigeminal nerve clinically is provided by the skin of the face (see also Sect. 12.2.3.2.4). V_1 = Ophthalmic nerve; V_2 = Maxillary nerve; V_3 = Mandibular nerve. The incomplete covering of the mandibular region by V_3 is due to the fact that the posterior region of the lower jaw is supplied by sensory branches originating in the cervical plexus [Plexus cervicalis]. Marble portrait bust of the Roman emperor Gaius, known as Caligula A.D. 37–41. The Metropolitan Museum of Art, State of New York (Public domain)

in the periphery and the inner surfaces of the head region. These three nerves are:

- The ophthalmic nerve [N. ophthalmicus] V_1
- The maxillary nerve [N. maxillaris] V_2
- The mandibular nerve [N. mandibularis] V_3

The anatomical terms provide a rough idea about the different supply areas (Fig. 12.2):

- **V_1:** ὁ ὀφθαλμός [ho ophthalmos] Greek = the eye
  Structures of the orbita, skin surrounding the orbita, conjunctiva, cornea, skin of the frontal bone
- **V_2:** Maxilla = upper jaw
  Inferior parts of the orbita, maxilla with upper teeth, gingiva, and mucosa, skin covering the maxillary bone, nose, upper lip
- **V_3:** Mandibula = lower jaw
  Mandibula with lower teeth, gingiva, and mucosa, partly skin covering the mandibula, lower lip, chin, meninges (see Sect. 1.2.1)

What you can see in schematic representations like in Fig. 12.2 are the sensory supply areas in the face of the three main trigeminal branches. What is not visible and would need anatomical dissection in anatomy or imaging techniques for experimental and clinical purposes is the vast network of trigeminal fibers below the surface and in the inner surfaces of the head ("tip of the iceberg"). A small proportion of the inner surfaces can be observed directly by means of inspection (oral cavity) or by specialized diagnostic otolaryngology equipment (e.g., endoscopy of nasal cavity, maxillary, or frontal sinuses).

This means that in addition to the skin innervation the trigeminal nerve innervates most of the mucosal surfaces of the head.

These are present in:

- The oral cavity
- The nasal cavity
- The maxillary, frontal, ethmoidal, and sphenoid sinuses (see also Fig. 12.13)

Other non-mucosal surfaces or coverings innervated by the trigeminal nerves are the sclera, the cornea, conjunctiva, the meninges, and the tympanic cavity. Finally, all sensory and nociceptive information from the teeth is mediated to the brainstem via trigeminal branches.

The proprioceptive signals from the masticatory muscles (Sect. 11.3.1.1.2) and most likely from the mimic muscles (see Sect. 8.2.1.6.1) are conveyed to the mesencephalic nucleus of trigeminal nerve (see Sect. 13.2.6.1).

The different sensory qualities are not segregated to own nerves in the periphery, but they share sensory branches. In addition, branches of the trigeminal nerve are "used" by other cranial nerves to get specific branches to their targets (mainly parasympathetic and sympathetic fibers, see below).

12.2.3.2.3 Course of the principal sensory trigeminal nerve branches

In the following, we will deal separately with the three main branches and their peripheral ramifications. We will base this mainly on anatomical specimens to give the reader an impression of the real spatial, imaging-related circumstances. Very tiny terminal branches will be mentioned but in several cases not shown.

In Fig. 12.3, you see a horizontal plastinated section displaying the exit of the trigeminal nerve ② from the pons ⑦. The thick nerve is running rostrally through the prepontine cistern and then continues into the trigeminal ganglion ① located on the inner surface of the skull base. The trigeminal ganglion is considered to be an analog of the spinal ganglia of the spinal cord.

In the periphery of the body, sensory perikarya are located inside the spinal ganglia. In the simplest case, a free nerve ending (mainly nociceptive, cf. Fig. 3.7) is the most proximal

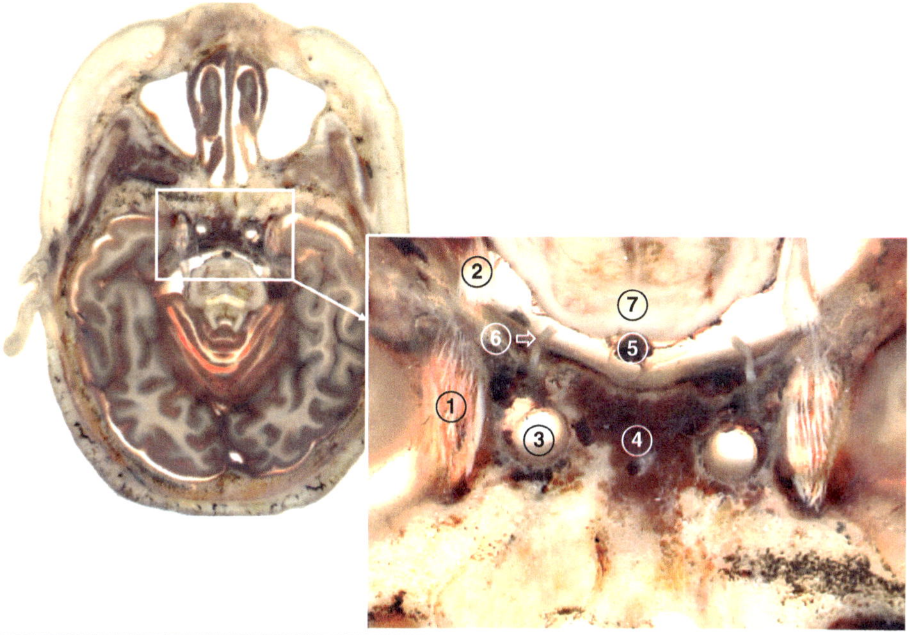

①	Trigeminal ganglion	⑤	Basilar artery
②	Trigeminal nerve CNV	⑥	Abducens nerve CNVI
③	Internal carotid artery inside the cavernous sinus	⑦	Pons
④	Cavernous sinus		

Fig. 12.3 Horizontal section through the human head at the level of the trigeminal ganglion ①. Plastinated specimen. The continuation to the brainstem is provided by the trigeminal nerve ② via the prepontine cistern. Note the vicinity to the internal carotid artery ③ and the cavernous sinus ⑤. The first (ophthalmic) and second (maxillary) main trigeminal branches are running through the cavernous sinus (see Fig. 12.5) to reach their target regions (for details, see text). Sammlung des Zentrums Anatomie der Universität zu Köln

part of the peripheral portion of the spinal ganglionic neurons (Fig. 3.8). The central portion would enter the spinal cord and reach the brainstem and thalamus via different ways (see Fig. 3.6B). The central sensory processes in their entirety is what we can see here as the trigeminal nerve. The neuronal perikarya are located in the trigeminal ganglia. The peripheral processes run with the three trigeminal branches.

On the histological level, Eftekhari et al. (2010) described a histological picture comparable to that of a spinal ganglion with ganglionic perikarya of various sizes and staining intensity (hematoxylin-eosin). Most of the cells in the trigeminal ganglion in their material were glial, satellite cells wrapped around the neuronal perikarya. Some ganglionic perikarya contained lipofuscin (see Box 4.4).

In addition, the authors studied the distribution of calcitonin gene-related peptide (CGRP) and the respective receptors in the human trigeminal ganglion. CGRP is supposed to play an important role in migraine (see below mandibular nerve V_3). They could show that more than almost half of the human trigeminal ganglionic neurons were immunopositive for CGRP. More than 30% of the neurons contained one or the other receptor marker while coexistence of CGRP and receptor-immunoreactivity was rare. The glial cells were immunonegative for CGRP but contained the CGRP receptor components. The findings indicate the possibility of CGRP signaling in the human trigeminal ganglion.

An ultrastructural study of the human trigeminal ganglion has been provided by Krastev et al. (2008).

The peripheral portion of the ganglionic neurons is what we are going to look for in the following. Remarkably close to the trigeminal ganglia, the separation into ophthalmic, maxillary, and mandibular nerve takes place. By contrast to the direction of sensory signals in vivo from the periphery to the center, for reasons of clarity and tradition, we will describe the trigeminal branches from their exit into the leptomeningeal space; in this case, the prepontine cistern (see Box 3.3) to the periphery as far as possible.

Ophthalmic nerve V_1

The ophthalmic nerve reaches its supply territory, the orbita, via the superior orbital fissure [Fissura orbitalis superior] (Fig. 12.4) and the cavernous sinus [Sinus cavernosus] (Fig. 12.5).

The superior ophthalmic vein (see Fig. 12.4 ④ [V. ophthalmica superior]) reaches the cavernous sinus (see Fig. 12.5) via the superior orbital fissure. The vein is

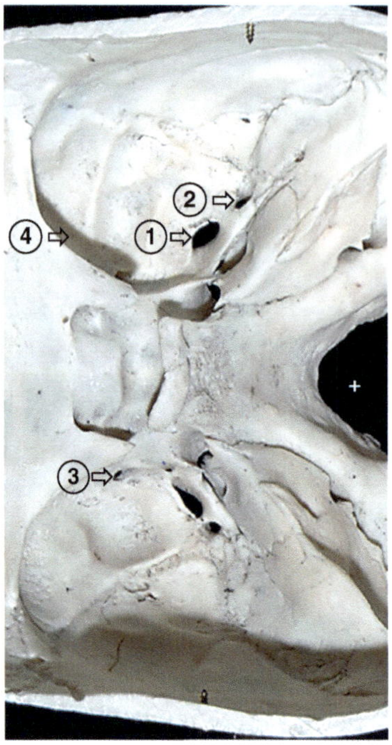

①	**Foramen ovale: Mandibular nerve V_3** **> Infratemporal fossa** Venous plexus of Foramen ovale	③	**Foramen rotundum: Maxillary nerve V_2** **> Pterygopalatine fossa**
②	**Foramen spinosum (from infratemporal fossa):** **Meningeal branch of mandibular nerve V_3** Middle meningeal artery	④	**Superior orbital fissure: Ophthalmic nerve (V1)** **Lacrimal, frontal, nasociliary nerves** **> Orbita** Superior ophthalmic vein
		+	**Foramen magnum**

Fig. 12.4 Dorsal view on the skull base with the exit foramina of the three main trigeminal branches. The ophthalmic nerve leaves the cranial cavity via the superior orbital fissure ④ [Fissura orbitalis superior] (not visible in this view) in direction to the orbita (see also Fig. 12.5). Sammlung des Zentrums Anatomie der Universität zu Köln

connected to the veins of the face via the angular vein (interior corner of the eye) [V. angularis]. Infections of the face can reach the cranial cavity (Cavernous sinus see Fig. 12.5) via this way.

Since the nerve immediately ramifies, its visualization in a topographic-anatomical specimen is difficult (see Figs. 12.6 and 12.7). Keep in mind that together with the ophthalmic nerve, three of its major branches, the lacrimal, frontal, and nasociliary nerves leave the middle cranial fossa. By removing the roof of the orbita and the intraorbital fat, it is possible to identify these branches in dorsal (Fig. 12.6) and lateral views (Fig. 12.7).

The maxillary nerve runs through the Foramen rotundum (round) to the pterygopalatine fossa [Fossa pterygopalatina] (see Figs. 12.4 and 12.16).

The mandibular nerve reaches the infratemporal fossa [Fossa infratemporalis] via the foramen ovale (see Fig. 12.4 ①).

Via the foramen spinosum the middle meningeal artery [A. meningea media]—branch of the external carotid artery—reaches the cranial cavity. If lesioned by, for example, a skull fracture, the artery may rupture and cause the so-called epidural hematoma which preferentially compresses the oculomotor nerve (see Sect. 15.3.1).

As shown in Fig. 12.6, a great number of different structures are contained in the orbita within a tight space. Visualizing the orbita as a pyramid with the apex directed to the superior orbital fissure and the base at the entrance of the orbita, most of the structures run more or less parallel from the apex toward the base / optic bulb (+). The ophthalmic nerve as an individual structure is not visible here (but see Fig. 12.7) since it has already split off into the frontal ①, lacrimal ③, and nasociliary nerves ⑫ [Nn. frontalis, lacrimalis, nasociliaris] during entering the orbita.

Before entering the fissure, V_1 has given off a recurrent meningeal nerve for the sensory supply of the tentorium cerebelli and falx cerebri (see Box 3.5). The thick frontal nerve ① is running in rostral direction on the levator palpebrae superioris muscle [M. levator palpebrae superioris] ② (see also Fig. 12.14A ①) although this muscle is innervated by the oculomotor nerve. After reaching the posterior circum-

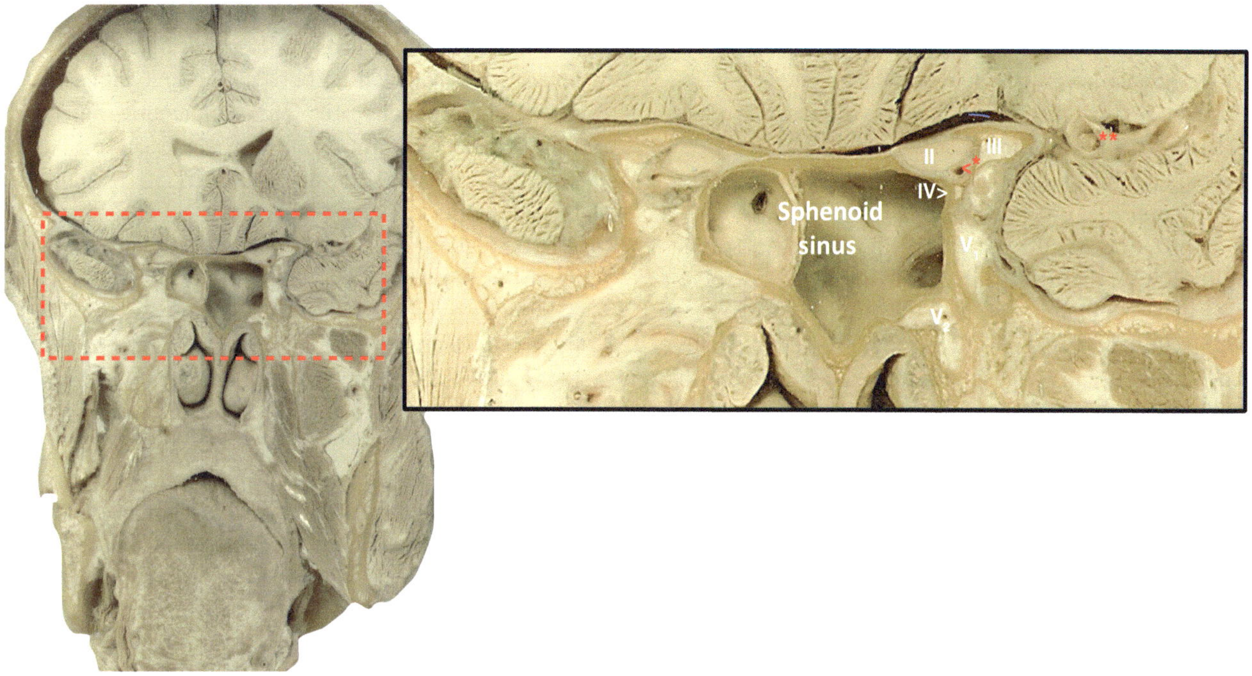

II	Optic nerve CNII	V_2	Maxillar nerve CNV
III	Oculomotor nerve CNIII	*	Ophthalmic artery
IV	Trochlear nerve CNIV	**	Middle cerebral artery in lateral sulcus
V_1	Ophthalmic nerve CNV		

Fig. 12.5 Coronal section (slightly oblique) through the human head at the level of the sphenoid sinus. Lateral of the sphenoid sinus, the cavernous sinus with CNII, III, IV, and VI and the ophthalmic artery is seen close to the apex of the orbita, i.e., the internal carotid artery (ICA) is not hit here. The ophthalmic nerve (V_1) lies ventral most in the cavernous sinus—ventral to the oculomotor (III) and the trochlear nerves (IV)—and leaves the skull via the superior orbital fissure (see Figs. 12.4, 12.6, and 12.7). The maxillary nerve (V_2) lies outside the cavernous sinus from where it will enter the inferior orbital fissure (see Fig. 12.11 ②). Photograph by MedizinFotoKöln, Cologne, Germany. Sammlung des Zentrums Anatomie der Universität zu Köln

ference of the optic bulb, the frontal nerve splits into the supratrochlear [N. supratrochlearis] ⑨ and the supraorbital nerves [N. supraorbitalis] ⑥ (see Fig. 12.11).

The supraorbital nerve splits again in a medial ⑧ and lateral ⑦ branch and supplies the conjunctiva of the eye [Tunica conjunctiva], the frontal sinus [Sinus frontalis], and the skin of the forehead (see Fig. 12.2 for skin territory). The medial and lateral branches reach the skin via the supraorbital notch (see Fig. 12.11 ③).

The supratrochlear nerve ⑨ reaches the medial corner of the eye where it splits in an ascending and a descending branch for the innervation of the skin.

The lacrimal nerve (Fig. 12.6 ③) runs parallel to the lacrimal artery [A. lacrimalis] (Fig. 12.6 ⑤). Both reach the lacrimal gland [Glandula lacrimalis] (Fig. 12.6 ④) (lacrima Latin = tear).

Via the so-called communicating branch with lacrimal gland of the zygomatic nerve (V_2) [R. communicans lacrimalis] the lacrimal nerve is provided with parasympathetic postganglionic fibers from the pterygopalatine ganglion [Ggl. pterygopalatinum] (see Box 12.1 for details, see Fig. 12.16 ①) for the lacrimal gland. The lacrimal nerve also has sensory portions for the conjunctiva and the lateral upper lid.

Sympathetic fibers just cross the ganglion, then continue to run along the zygomatic nerve (V_1) to blood vessels and secretomotor elements in the lacrimal gland, nose, and oral cavity (Goosmann and Dalvin 2018) (see Box 12.1 for details).

The nasociliary nerve (Fig. 12.6 ⑫) [N. nasociliaris] is located medial most in the orbita and has a number of branches. The first of them are the posterior (not visible in this preparation) and anterior ethmoidal nerves ⑪ (see also Fig. 12.10) [Nn. ethmoidalis posterior et anterior]. The posterior branch enters the mucosa of the sphenoid sinus [Sinus sphenoidalis] (Fig. 12.5) via the posterior ethmoidal foramen [Foramen ethmoidale posterius] (innervation of the meninges in the rostral parts of the anterior cranial fossa) while the anterior branch ⑪ runs into the anterior cranial fossa [Fossa cranii anterior] via the anterior ethmoidal foramen [Foramen

Box 12.1 Functional and topographic anatomy of the pterygopalatine and the submandibular ganglion

The pterygopalatine ganglion is one of the parasympathetic head ganglia, together with the ciliary ganglion (see here above), otic ganglion (see Fig. 12.8), and the submandibular one (see here below). While the general organization of the head ganglia is rather simple (Fig. 12.8), the topographical situation can be somewhat challenging.

In all cases, we have two different efferents, parasympathetic and sympathetic, and one kind of sensory afferents (Fig. 12.8). The parasympathetic afferents to the ganglia are preganglionic fibers from one of the salivatory nuclei. The sympathetic afferents are postganglionic fibers from the superior cervical ganglion which pervade the parasympathetic ganglia on route to the peripheral targets. Sensory afferents also pervade the ganglia but in centripetal direction and eventually reach the sensory trigeminal nuclei. For the sake of clarity, they have not been included in Fig. 12.8.

The anatomical relationships are rather simple to visualize by anatomical specimens for the ciliary ganglion (see Fig. 12.7), but this is not the case for the remaining ganglia. Using the semi-schematic illustration in Fig. 12.9, we will explain in the following the anatomy of the pterygopalatine and submandibular ganglia.

Pterygopalatine ganglion (see Fig. 12.16 ①)

The supply territories of the autonomic and sensory fibers related to the pterygopalatine ganglion are (visceromotor) lacrimal gland, glands of the nose and palate (Fig. 12.9).

The parasympathetic preganglionic fibers (GPN in Fig. 12.9, greater petrosal nerve [N. petrosus major]—formerly and better—called the parasympathetic root of pterygopalatine ganglion) join the facial nerve with which they enter the facial canal and split off the latter close to the geniculate ganglion (Fig. 12.9) and eventually reach the middle cranial fossa running in anterior direction (for details, see Box 8.5).

The postganglionic sympathetic fibers—now luckily called the sympathetic root of the pterygopalatine ganglion—reach the middle cranial fossa via a perivascular plexus (internal carotid artery) and meet with the parasympathetic root (greater petrosal nerve) at the entrance into the pterygoid canal. The latter connects the middle cranial fossa with the pterygopalatine fossa which harbors the homonymous ganglion. Anatomically, the parasympathetic and the sympathetic roots form one nerve, the nerve of the pterygoid canal (PC in Fig. 12.9) [N. canalis pterygoidei].

After entering the pterygopalatine fossa, the sympathetic fibers pervade the ganglion (see Fig. 12.16 ⑤) without synaptic contacts and reach the target organs by running along the zygomatic nerve (V_2) (see below) to blood vessels and secretomotor elements in the lacrimal gland, nose, and oral cavity (Goosmann and Dalvin 2018).

The preganglionic fibers of the parasympathetic root synapse on ganglionic neurons. Their axons reach the lacrimal gland and the mucosal glands of the nasal cavity and the palate by a detour via the zygomatic nerve (see below V_2) which has an anastomosis with the lacrimal nerve (Fig. 12.6 ③) (Nturibi and Bordoni 2020).

Finally, there is a sensory root of the ganglion. Fibers of the maxillary nerve supply the orbital periosteum and pass through the ganglion to reach the trigeminal ganglion. These fibers—as well as the sympathetic ones—are not shown in Fig. 12.8, but the course to the brainstem of the sensory fibers is similar to the gustatory fibers from the tongue to the solitary nucleus, labeled in yellow in Fig. 12.9.

Submandibular ganglion

The parasympathetic innervation of the "small" salivary glands, sublingual and submandibular, starts with preganglionic fibers in the superior salivatory nucleus (see Fig. 12.9). They run together with the facial nerve (for details, see Box 8.5) from which they split off in the facial canal. By traveling with the Chorda tympani (see Figs. 3.25A and 8.17), the secretomotor fibers eventually reach the lingual nerve (V_3) [N. lingualis] (Fig. 12.16 ④) from where they enter the submandibular ganglion. The postganglionic fibers innervate the sublingual (Fig. 12.14B ⑤) and submandibular glands [Gld. sublingualis/Gld. submandibularis] Fig. 12.15 ⑤).

There is also a sympathetic root of submandibular ganglion [Radix sympathica ganglii submandibularis].

The otic ganglion and the inferior salivatory nucleus are dealt with under Sect. 5.4.2.1.3.

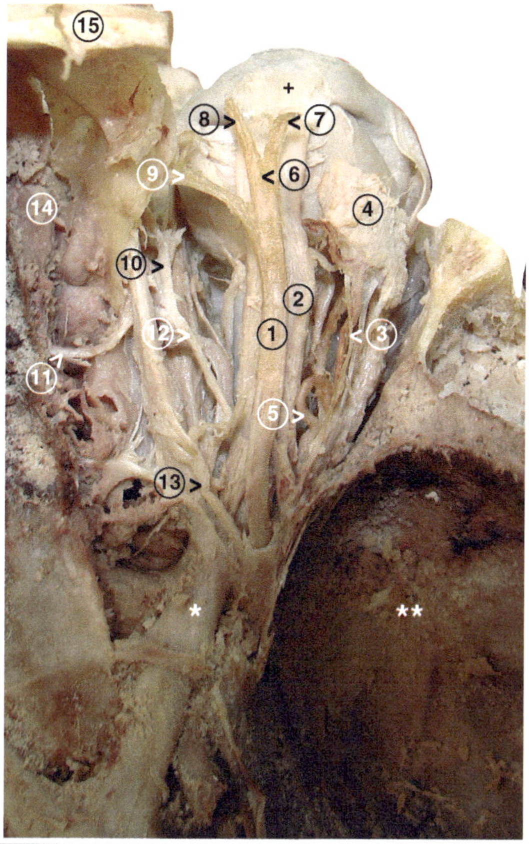

①	Frontal nerve V_1 lying on the levator palbebrae superioris muscle ②	⑪	Anterior ethmoidal nerve V_1
③	Lacrimal nerve V_1	⑫	Nasociliary nerve V_1
④	Lacrimal gland	⑬	Trochlear nerve
⑤	Lacrimal artery	⑭	Nasal cavity
⑥	Supraorbital nerve V_1	⑮	Frontal bone
⑦	Supraorbital nerve, lateral branch	+	Optic bulb
⑧	Supraorbital nerve, medial branch	*	Optic nerve
⑨	Supratrochlear nerve V_1	**	Middle cranial fossa
⑩	Infratrochlear nerve V_1		

Fig. 12.6 Dorsal view of the right human orbita. The brain has been removed as well as the roof and the medial and lateral walls of the orbita. In addition, the retroorbital fat [Corpus adiposum orbitae] has been excised. Thus, the view is free onto the neural, vascular, and muscular components of the orbita. For details, see text. Müller-Thomsen fecit. Sammlung des Zentrums Anatomie der Universität zu Köln

ethmoidale anterius]: it remains in an extradural position and eventually reaches the nasal cavity [Cavitas nasi] via the cribriform plate [Lamina cribrosa] (Fig. 12.10 ①) (cribrum Latin = sieve) giving off several nasal branches.

Note that the FIPAT directory shows the posterior ethmoidal nerve in brackets.

The infratrochlear nerve [N. infratrochlearis] (Fig. 12.6 ⑩) reaches the inner corner of the eye—crossing below the trochlea (see Box 12.2)—and supplies the lacrimal sac [Saccus lacrimalis], the lacrimal caruncle [Caruncula lacrimalis], and the surrounding skin including the medial parts of the upper and lower eye lids. The lacrimal caruncle

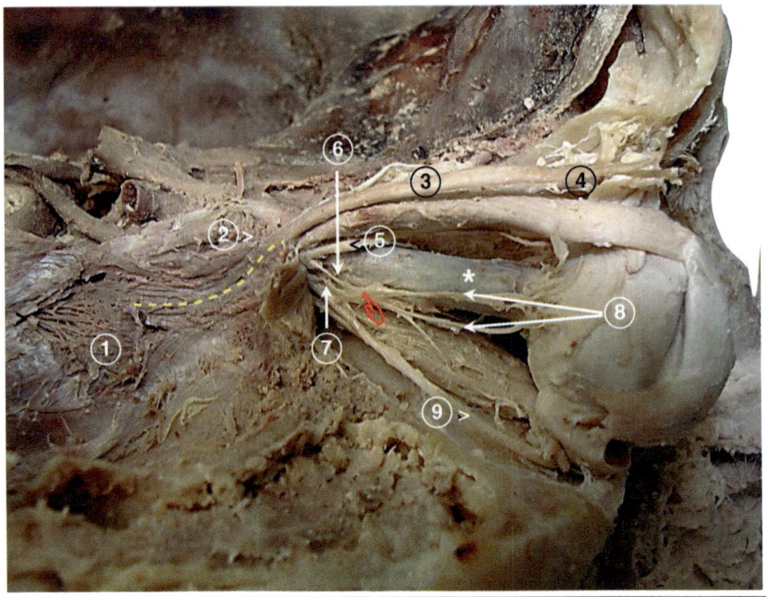

①	Trigeminal ganglion	(red outline)	Ciliary ganglion
- - -	Course of the ophthalmic nerve V_1	⑥	Branch to ciliary ganglion
②	Region of superior orbital fissure	⑦	Sympathetic root of ciliary ganglion
③	Frontal nerve V_1	⑧	Long ciliary nerves
④	Supraorbital nerve with branches at the inner corne of the eye V_1	⑨	Oculomotor nerve CNIII, inferior branch
⑤	Nasociliary nerve V_1 carrier of the branch to ciliary ganglion (parasympathetic) ⑥	*	Optic nerve

Fig. 12.7 Lateral view of the human orbita. The brain has been removed as well as the roof and the medial and lateral walls of the orbita and the orbital fat has been excised. Thus, the view is free onto the neural, vascular, and muscular components of the orbita. For details, see text. Müller-Thomsen fecit. Sammlung des Zentrums Anatomie der Universität zu Köln

Box 12.2 Etymology of the branches of the ophthalmic nerve (in alphabetic order)

1. Ethmoidal, from ancient Greek ἠθμοειδής (ethmoeides = perforated), from ἠθμός (ethmos = sieve) + -ειδής (-eides, "-form, -like"). The ethmoidal sinuses are part of the paranasal sinuses.
2. Frontal from Latin frons, frontis = forehead.
3. Infra/supratrochlear, related to trochlea = supporting point (hypomochlion) of the tendon of the superior oblique muscle (see Sect. 14.5.2.1.2); below/above the trochlea.
4. Lacrimal related to the lacrimal gland, from Latin lacrima = tear.
5. The nasociliary nerve has its name from the fact that its branches supply sensory innervation to the nasal cavity and with its terminal branch, the supratrochlear nerve, to the skin of the nose and the superior palpebra with its eyelashes, Latin ciliae (Sg. cilia).
6. Ophthalmicus, from ancient Greek ὁ ὀφθαλμός (ho ophthalmos) = the eye.
7. Pterygoid from *pterygoides*, from Greek *pterygoeides*, literally, shaped like a wing, from *pteryg-*, *pteryx* wing; akin to Greek to *pteron* (τὸ πτερόν) wing (pterygoid process of the sphenoid bone).

is the pink globular nodule at the very medial edge of the inner eye corner.

In addition to structures ③, ⑥, and ⑫ of Fig. 12.6, Fig. 12.7 shows the parasympathetic ganglion of the orbita, the ciliary ganglion [Ggl. ciliare] (see Box 12.1).

The parasympathetic preganglionic fibers for the ciliary ganglion are originating in the Edinger-Westphal nucleus (central cell group of the accessory nuclei of oculomotor nerve) (see Sect. 15.3.1.2). They join the oculomotor nerve (see Fig. 12.5). With the latter they leave the mesencephalon

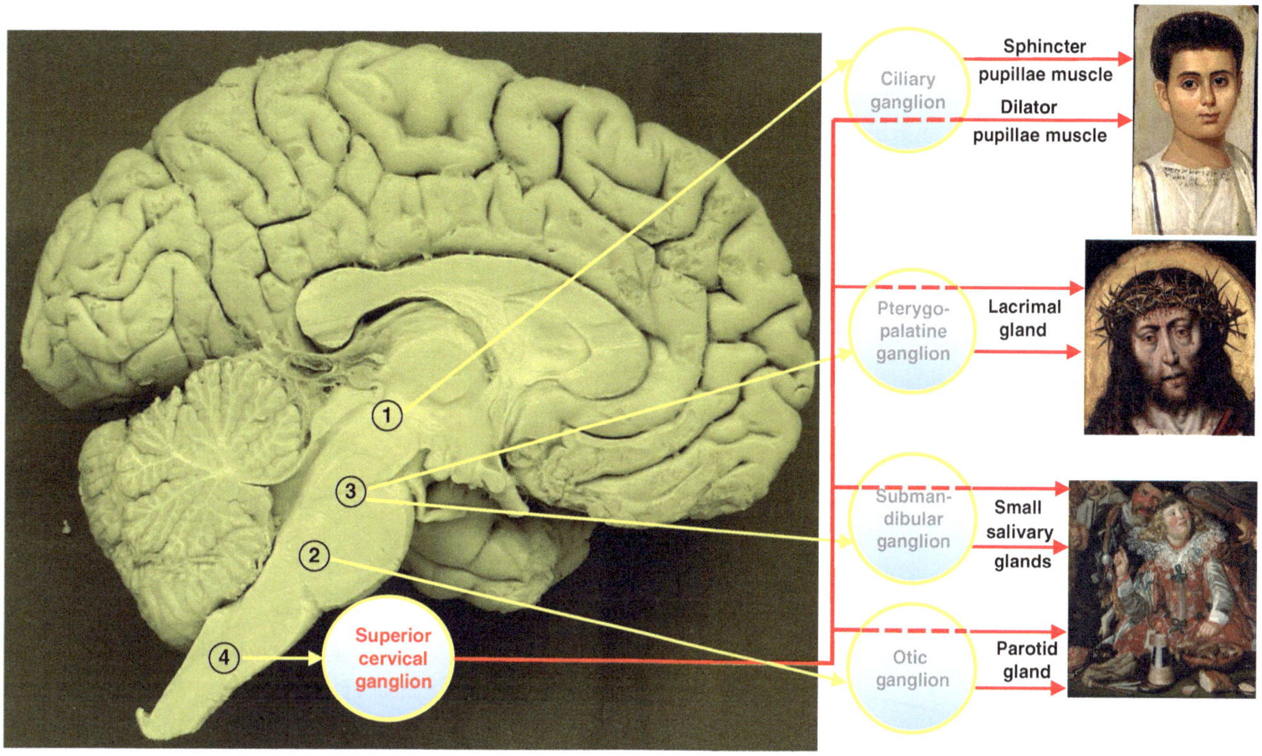

Fig. 12.8 Schematic survey of the parasympathetic and sympathetic innervation of smooth muscles and glands of the head region. Yellow arrows represent preganglionic, red ones postganglionic fibers. ① Edinger-Westphal nucleus [Ncl. accessorius N. oculomotorii, see Sect. 15.3.1.2]. Preganglionic fibers for the ciliary ganglion [Ggl. ciliare]. Postganglionic fibers to the ciliary [M. ciliaris] and the sphincter pupillae [M. sphincter pupillae] muscles of the eye. ② Superior salivatory nucleus. Preganglionic fibers for the pterygopalatine ganglion [Ggl. pterygopalatinum], postganglionic fibers for the lacrimal gland [Gld. lacrimalis]. ③ Superior salivatory nucleus (see Sect. 7.3.1.1). Preganglionic fibers for the submandibular ganglion [Ggl. submandibulare], postganglionic fibers for the submandibular [Gld. submandibularis] and sublingual glands [Gld. sublingualis]. ④ Inferior salivatory nucleus (see Sect. 7.3.1.1). Preganglionic fibers for the otic ganglion [Ggl. oticum], postganglionic fibers for the parotid gland [Gld. parotidea]. ⑤ Intermediolateral nucleus [Ncl. intermediolateralis] of the spinal cord. Preganglionic fibers for the superior cervical ganglion, postganglionic fibers for the dilator pupillae muscle [M. dilatator pupillae], the lacrimal gland [Gld. lacrimalis], submandibular [Gld. submandibularis], and sublingual glands [Gld. sublingualis]. Anonymous (Egypt, A.D. 100–150, Roman Period) Portrait of the Boy Eutyches. Detail. From the Metropolitan Museum of Art, New York (Public domain), with permission. Workshop of Aelbert Bouts (ca. 1451/54–1549) The Man of Sorrows. Detail. From the Metropolitan Museum of Art, New York (Public domain), with permission. Frans Hals (1582–1666) Merrymakers at Shrovetide. Ca. 1616–1617. Detail. From the Metropolitan Museum of Art, New York (Public domain), with permission. Brain specimen: Sammlung des Zentrums Anatomie der Universität zu Köln

in the interpeduncular fossa (Fig. 12.23), reach the orbita via the cavernous sinus (see Fig. 12.5) and the ciliary ganglion as branch of the nasociliary nerve [Ramus ganglionaris ciliaris] (Fig. 12.7 ⑥) (formerly—better from a functional perspective—parasympathetic root of ciliary ganglion). After having synapsed with the postganglionic parasympathetic ganglion cells, the postganglionic fibers form the Nn. ciliares longi (Fig. 12.7 ⑧) which perforate the sclera and reach the interior of the eye bulb. These parasympathetic fibers innervate the ciliary muscle [M. ciliaris] (accommodation, changes of focus from distant to near images by change in lens shape) and the sphincter pupillae [M. sphincter pupillae] (miosis, constriction of the pupil) (see also Sect. 15.3.1.2).

The sympathetic root of ciliary ganglion [Radix sympathica ganglii ciliaris] (Fig. 12.7 ⑦) enters the skull via the internal carotid plexus (see Fig. 12.8) and enters the orbita together with the ophthalmic artery.

The sympathetic fibers innervate the dilator pupillae [M. dilatator pupillae] (mydriasis, dilatation of the pupil).

Eventually, the ganglion is pervaded by a sensory root—running with the long ciliary nerves—with afferents from the iris, ciliary body [Corpus ciliare] and cornea (Branch to ciliary ganglion, formerly—better—sensory root of ciliary ganglion [Ramus ganglionaris ciliaris]) joining the nasociliary nerve in centripetal direction.

Maxillary nerve V_2

The maxillary nerve passes the Foramen rotundum (Fig. 12.12) to reach the pterygopalatine fossa and from there via the inferior orbital fissure (see Fig. 12.11) the orbita. The first branch released before entering the Foramen rotundum is the menin-

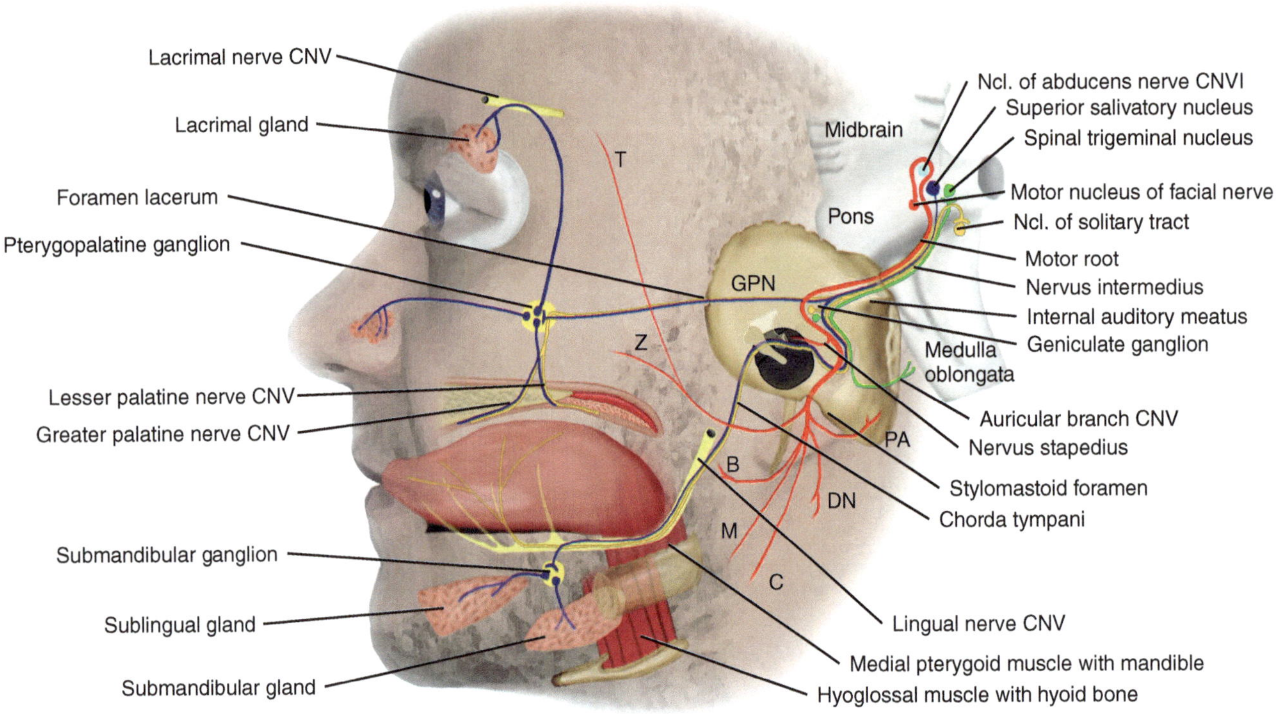

Fig. 12.9 Schematic illustration of the target organs of the pterygopalatine and submandibular ganglia

1. Branches of the trigeminal nerve (not shown in figure): Lacrimal nerve V_1, Greater/lesser palatine nerves V_2, Lingual nerve V_3
2. Branches of the facial nerve:
   - Mimic muscles: B—Buccal, C—Cervical, M—Mandibular, T—Temporal, Z—Zygomatic branches
   - DN—Digastric nerve
   - PA—Posterior auricular nerve
3. Vegetative branches:
   - GPN—Greater petrosal nerve
   - PC—Nerve of the pterygoid canal

Leander E. Huggenberger fecit Modified acc. to Takezawa et al. 2018, Fig. 2 with permission

geal branch [R. meningeus] which supplies the dura mater in the frontal territory of the middle meningeal artery. Figure 12.12 shows the main openings for the mandibular nerve and branches of the maxillar nerve on their way to the inferior surface of the skull base and further into the periphery.

Also in the pterygopalatine fossa, the maxillary nerve releases (1) sensory branches to the pterygopalatine ganglion (see also Box 12.1: Pterygopalatine ganglion, Fig. 12.16) and (2) the parasympathetic root of the pterygopalatine ganglion (greater petrosal nerve) (see also Box 12.1). The sensory root contains fibers from the orbital periost.

The greater petrosal nerve contains parasympathetic fibers for the lacrimal gland (see above ophthalmic nerve, Fig. 12.6 ④) and the mucosal glands of the nasal cavity and the pharynx.

The next branches are the Rr. orbitales which reach the orbita via the inferior orbital fissure and from there provide sensory innervation for the posterior ethmoidal cellulae and the sphenoid sinus.

The sympathetic postganglionic fibers run through the pterygopalatine ganglion, the sympathetic root of pterygopalatine ganglion [Radix sympathica ganglii pterygopalatini]. These fibers originate from the internal carotid plexus and run through the pterygopalatine ganglion to provide the sympathetic, secreto- and vasomotor innervation of the lacrimal gland, the oral and nasal mucosa (Goosmann and Dalvin 2018). These are followed by two groups of nasal branches for the nasal cavity and the ethmoidal sinus.

The nasopalatine nerve descends between the periost and mucosa of the nasal septum to the incisive foramen [Foramen incisivum] of the hard palate and from there supplies the anterior parts of the palatine mucosa and the gingiva of the upper incisor teeth.

The pharyngeal nerve [N. pharyngeus] provides sensory supply to the pharynx.

The parasympathetic innervation of the palatine mucosa is provided by the descending lesser palatine nerves [Nervi palatini minores] (soft palate) and the greater palatine nerve (hard palate) which reach their targets via the homonymous foramina (see Figs. 12.9 and 12.12 ① ②).

Also with its origin in the pterygopalatine fossa, the zygomatic nerve [N. zygomaticus] reaches the lateral wall of the

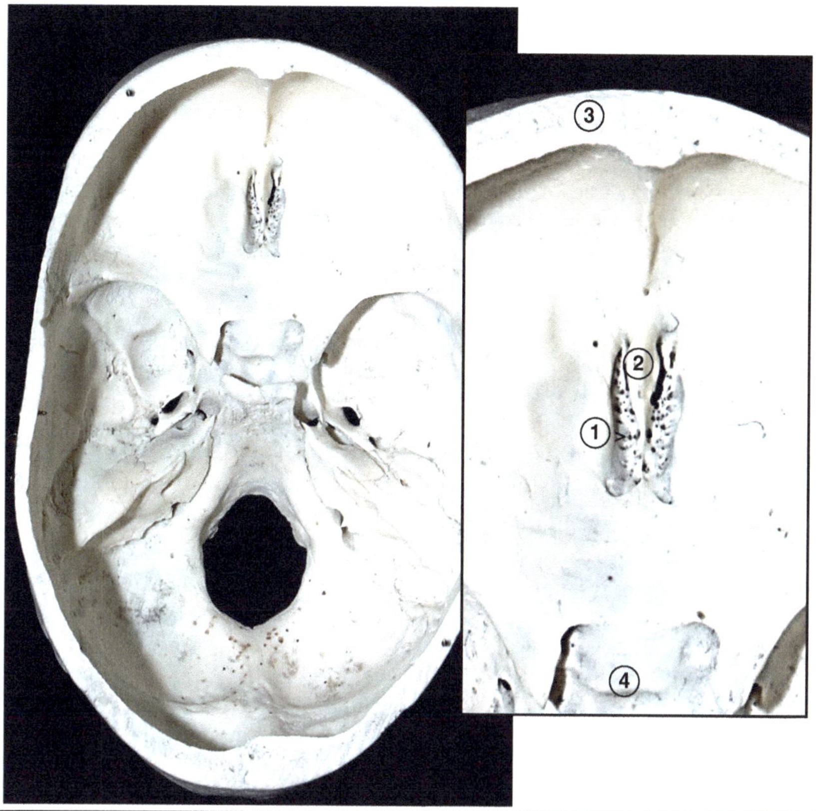

①	Cribriform plate	③	Frontal bone
②	Crista galli	④	Sella turcica

Fig. 12.10 By contrast to the other two cranial fossae, the anterior fossa is rather poor in openings. The main foraminal structure is the cribriform plate (cribris latin = sieve) which provides a connection between the nasal cavity and the interior of the skull. It is mainly occupied by the olfactory nerves of the first cranial nerve (CNI). The centripetal fibers from the olfactory epithelium in the nasal cavity pass through the cribriform plate to reach the olfactory bulb. In addition, the cribriform plate is the exit structure of the anterior ethmoidal nerve (V_1) (see Fig. 12.6 ⑪) which reaches the anterior cranial fossa via the anterior ethmoidal foramen, remains in an extradural position, and eventually reaches the nasal cavity via the cribriform plate. Anatomische Sammlung des Zentrums Anatomie

orbita via the inferior orbital fissure (Fig. 12.11 ②) and has an anastomosis with the lacrimal nerve (V_1) (Fig. 12.6 ③). Two small branches of the zygomatic nerve provide sensory innervation of the temporal and zygomatic skin.

The maxillary nerve supplies the upper teeth in the maxillary bone—as the mandibular nerve does with the lower teeth (mandibula)—via the superior alveolar nerve [N. alveolaris superior] with several superior alveolar branches [Rami alveolares superiores]. They form the superior dental plexus [Plexus dentalis superior] with superior dental branches for individual teeth (Fig. 12.15). In addition, the superior gingival nerves provide the sensory innervation of the local gingiva (Fig. 12.13).

The terminal branch of V_2 is the infraorbital nerve (see Fig. 12.14A ②) passing from the pterygopalatine fossa via the infraorbital groove and canal (Fig. 12.11) and the homonymous foramen to the skin of the lower eyelid, nose, upper lip, and cheeks which are innervated by its smaller branches (Fig. 12.2) (Box 12.3).

Box 12.3 Etymology of the branches of the maxillary nerve (V_2)

1. Alveolar, from latin alveolus, pl. alveoli diminutive of alveus = hollow. Here, the tooth sockets in maxilla and mandibula.
2. Dentalis, from Latin dens, dentis = tooth
3. Gingiva Latin = gum
4. Labial, from Latin labium = lip
5. Meningeus from ancient Greek ἡ μῆνιγξ, Gen. μήνιγγος he meninx, meningos = linings of the brain, meninges
6. Nasopalatinus, Palatinus, from Latin palatum = roof of the mouth, palate
7. Palpebral, from Latin palpebra = lid
8. Pterygo- from ancient Greek τό πτερύγιον to pterygion = small wing.

9. Pterygoides: pterygion + -ειδής (-eides, "-form, -like"), like a small wing. Alludes to the pterygoid process of the sphenoid bone
10. Zygomaticus, a, um meaning equivocal. Probably from ancient Greek τό ζυγόν to zygon = yoke. Os zygomaticum = cheekbone. Maybe because the cheekbone stands out prominently

Mandibular nerve V_3

The mandibular nerve (V_3) leaves the cranial cavity via the foramen ovale (Fig. 12.4 ① and Fig. 12.12 ③) into the infratemporal fossa [Fossa infratemporalis] where it starts giving off branches (for the branches of the trigeminal motor root see Sect. 11.3.1.1.3, for the etymology of the branches of the mandibular nerve see Box 12.4).

The first one is the meningeal branch, running back into the cranial cavity, supplies the meninges and part of the mastoid cells via the foramen spinosum (Fig. 12.4 ② and Fig. 12.12 ④).

A very frequent complaint that brings patients to either self-medication or to visit a physician is headache, one of the most prominent types being migraine. It is considered a neurovascular disorder involving local vasodilation of intracranial, extracerebral blood vessels and simultaneous stimulation

Box 12.4 Etymology of the branches of the mandibular nerve

1. Alveolar, from latin alveolus, pl. alveoli diminutive of alveus = hollow. Here the tooth sockets in maxilla and mandibula.
2. Auricularis, from Latin auris = ear, referring to the ear
3. Chorda tympani Greek ἡ χορδή, he chorde = string and τό τύμπανον, to tympanon = drum, here the middle ear. The chorda tympani is running through the cavum tympani to the mandibular region.
4. Dentalis, from Latin dens, dentis = tooth
5. Gingiva Latin = gum
6. Labial from Latin labium = lip
7. Lingualis, from Latin lingua = tongue
8. Mentalis from Latin mentum = chin
9. Spinosus, a, um Latin = thorny, spiny. The original meaning referred to the spinous process of the greater wing of the sphenoid bone (correctly foramen spinae, but later incorrectly changed to an adjective) (Fig. 12.12 ④).
10. Sublingual = Located below the tongue like the homonymous gland
11. Submandibular = Located below the mandible like the homonymous gland

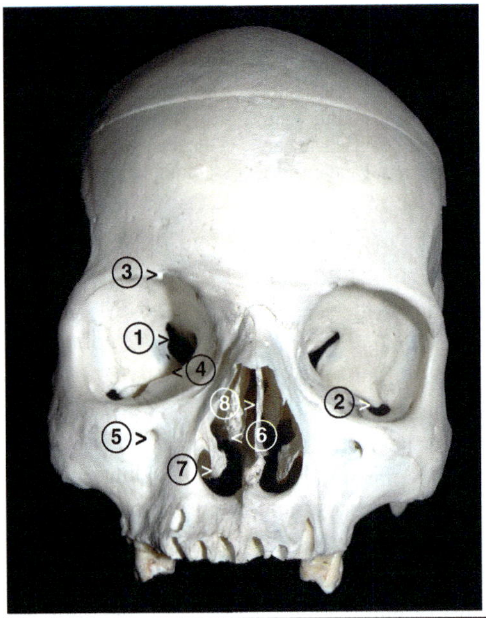

①	Superior orbital fissure	⑤	Infraorbital foramen / Infraorbital nerve V_2
②	Inferior orbital fissure	⑥	Nasal septum
③	Supraorbital notch / Supraorbital nerve V_1	⑦	Inferior nasal concha
④	Infraorbital groove	⑧	Middle nasal concha

Fig. 12.11 Frontal view of the human skull. Mandibula removed. The exit foramina of the terminal branches of V_1 (ophthalmic/supraorbital nerve) ③ and V_2 (maxillary/infraorbital nerve) ⑤ can be seen. The other major trigeminal branches, V_2 and V_3, leave the middle cranial fossa via proper foramina. Sammlung des Zentrums Anatomie der Universität zu Köln

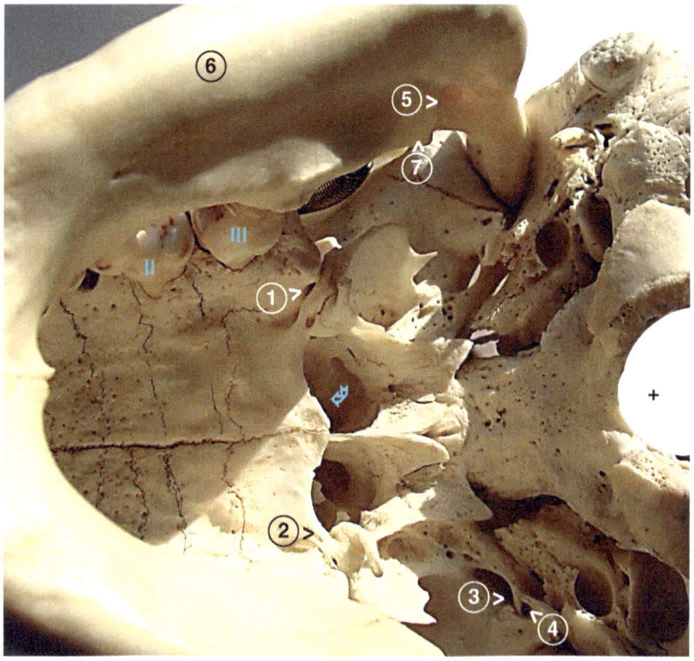

1	Greater palatine foramen V_2: Greater palatine nerve > Greater palatine canal Greater palatine artery	6	Body of mandible
2	Lesser palatine foramina V_2: Lesser palatine nerves > Lesser palatine canals Lesser palatine arteries	7	Lingula
3	Foramen ovale: Mandibular nerve V_3 > Infratemporal fossa Venous plexus of foramen ovale	+	Foramen magnum
4	Foramen spinosum: Meningeal branch of mandibular nerve > Middle cranial fossa Middle meningeal artery		Choana; Posterior nasal aperture
5	Aperture of mandible: Inferior alveolar nerve V_3 > Mandibular canal Inferior alveolar artery	II/III	Second / third molar tooth

Fig. 12.12 Ventral view of the human skull. The illustration shows the most important openings through which the trigeminal nerve branches (for sensory supply territories, see Fig. 12.2) reach the maxillary, palatine, and mandibular regions (for details, see text). Sammlung des Zentrums Anatomie der Universität zu Köln

of surrounding trigeminal sensory nervous pain pathways that results in headache (Aggarwal et al. 2012). The release of various vasodilators, especially calcitonin gene-related peptide (CGRP) (see above Sect. 12.2.3.2, trigeminal ganglion) induces pain response (for details, see Aggarwal et al. 2012, Buture et al. 2016).

After giving off a number of branches for masticatory and some other muscles (see Sect. 11.3.1.1.2), the next sensory branch is the buccal nerve [N. buccalis] for the buccal mucosa and gingiva followed by the auriculotemporal nerve [N. auriculotemporalis]. The latter usually encompasses the middle meningeal artery (see Figs. 12.4 and 12.12), has a small branch for the temporomandibular joint, and eventually supplies the temporal skin (see Fig. 12.2). In addition, it releases some small branches for the external acoustic meatus [Meatus acusticus externus], the auricle (see Fig. 7.37), the tympanic membrane [Membrana tympanica], and the parotid gland (see Sect. 7.3.1.1.3). Eventually, there is a communication with the facial nerve with parasympathetic fibers from the otic ganglion (see Sect. 5.4.2.1.3) via the facial nerve to the parotid gland.

The next branch of V_3 is the lingual nerve [N. lingualis]. It is running between the lateral and medial pterygoid muscle (see Fig. 11.7A) in an arched manner frontally to the floor of the mouth where it lies on the mylohyoid muscle. It has branches for the oropharyngeal isthmus [Isthmus faucium] and

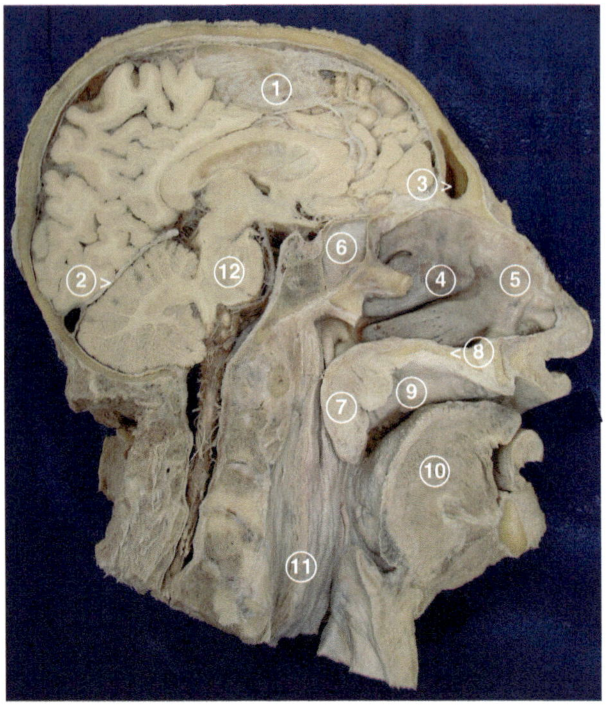

①	Falx cerebri as part of the meninges	⑦	Soft palate
②	Tentorium cerebelli as part of the meninges	⑧	Bony palate
③	Frontal sinus	⑨	Oral cavity
④	Nasal cavity	⑩	Tongue
⑤	Nasal septum	⑪	Pharynx
⑥	Sphenoid sinus	⑫	Brainstem

Fig. 12.13 Midsagittal section through the human head. Formalin-fixed specimen. The specimen displays some sensory supply territories of the three large trigeminal branches

①② Meninges: V_1: Tentorial branch for tentorium cerebelli (see Fig. 3.10) and Falx cerebri. Anterior meningeal branch for the meninges in the rostral parts of the anterior cranial fossa. V_2: Meningeal branch supplies the dura mater in the frontal territory of the middle meningeal artery. V_3: Meningeal branch accompanies the two branches of the middle meningeal artery.

Notabene: There is also a meningeal branch [Ramus meningeus of the vagus nerve] (see Sect. 7.3.2.1.2).

③ Frontal sinus: V_1: Posterior ethmoidal nerve, supraorbital nerve (Fig. 12.6)

④ Nasal cavity: V_1/V_2: Various nasal branches

⑤ Nasal septum: V_1/V_2: Anterior ethmoidal branch (Fig. 12.6), nasal branches, nasopalatine nerve

⑥ Sphenoid sinus: V_1: Posterior ethmoidal nerve. V_2: Orbital branches

⑦ Soft palate: V_2: Lesser palatine nerve (see Fig. 12.12)

⑧ Bony palate: V_2: Greater palatine nerve (see Fig. 12.12)

⑨ Oral cavity: V_1/V_2: Buccal nerve, gingival branches, lingual/sublingual nerve (Fig. 12.16)

⑩ Tongue: V_3: Lingual nerve (Fig. 12.16)

⑪ Pharynx: V_2: Pharyngeal nerve

Photograph by MedizinFotoKöln, Cologne, Germany. Sammlung des Zentrums Anatomie der Universität zu Köln

the tonsils, a communicating branch with the hypoglossal nerve.

The sublingual nerve (Fig. 12.14B) innervates the mucosa of the floor of the mouth and the gingiva of the anterior lower teeth and lingual branches supply the anterior two third of the tongue mucosa.

The Chorda tympani (for details, see Sect. 3.2.4.1.2) transfers signals from the taste receptors of the anterior two third of the tongue to the lingual nerve and then via the tympanic cavity and the petrotympanic fissure (see Fig. 3.26). In addition, efferent parasympathetic fibers from the superior salivatory nucleus run with the Chorda tympani to the submandibular ganglion [Ggl. submandibulare]. Postganglionic fibers (Fig. 12.8) reach the submandibular and sublingual glands (small salivatory glands) (Fig. 12.15) (see Eneroth et al. 1969).

One problem—less related to clinical examination but to the scientific study of taste (see Sect. 3.2.4.1.2 "Gustation")—is the fact that during the application of tastants to the proband's tongue not only the taste receptors but also adjacent trigeminal mechanoreceptors are stimulated unwillingly. That may lead to the misinterpretation of data obtained by

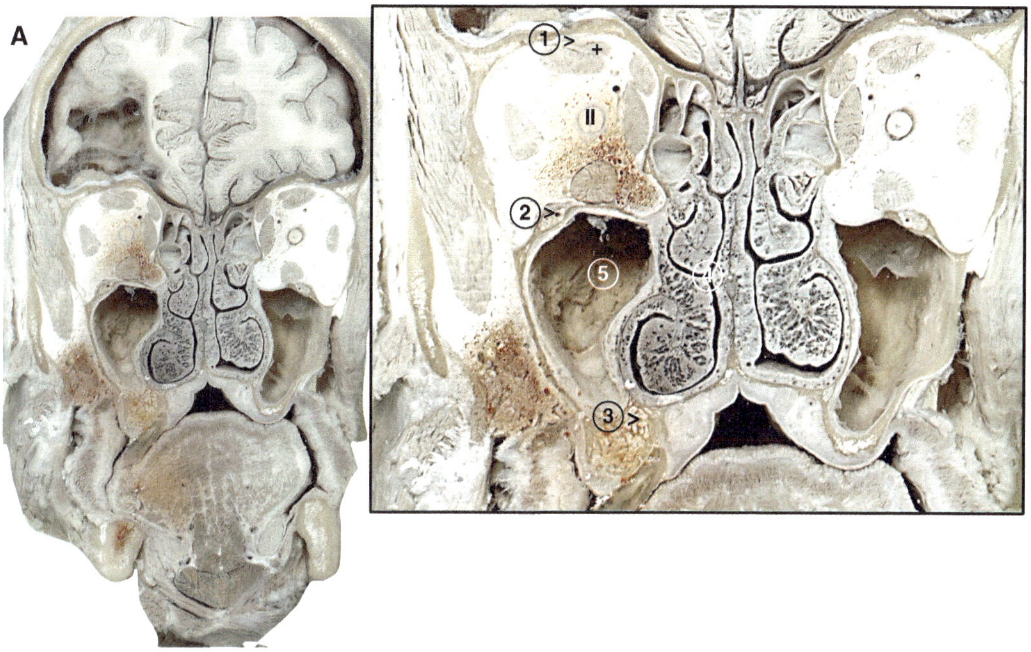

①	**Supraorbital nerve V_1 on levator palpebrae superioris (+)**	④	**Nasal cavity**
②	**Infraorbital nerve V_2**	⑤	**Maxillary sinus**
③	**Superior dental plexus V_2**	II	**Optic nerve CNII**

Fig. 12.14 (**A**) Coronal section through the human head at the level of the maxillary sinus ⑤ (as in **B**. see following page). Formalin-fixed specimen. In the magnified inset, three trigeminal branches (V_1/V_2) can be seen. The cystic structure in the right frontal lobe may be the consequence of a cerebral infarction. The supraorbital nerve ① originates from the ophthalmic nerve (V_1). It reaches the skin via the supraorbital notch (Fig. 12.11) and is the terminal peripheral branch of V_1 (see Fig. 12.2). The infraorbital and superior dental plexus are branches of the maxillary nerve (V_2). The infraorbital nerve ② is the peripheral terminal branch of V_2 and reaches the skin via the infraorbital foramen (Fig. 12.11). The superior dental plexus ③ as a branch of the superior alveolar nerve (V_2) provides the sensory innervation of the upper teeth. Photograph by MedizinFotoKöln, Cologne, Germany. Sammlung des Zentrums Anatomie der Universität zu Köln. (**B**) Coronal section through the human head at the level of the maxillary sinus (as in **A**). Formalin-fixed specimen. In the magnified inset, two trigeminal branches (V_2/V_3) can be seen. The superior dental plexus ① is a derivative of the superior alveolar nerve originating from the maxillary nerve (V_2). The inferior alveolar nerve ③ and the sublingual nerve ② are branches of the mandibular nerve (V_3). Photograph by MedizinFotoKöln, Cologne, Germany. Sammlung des Zentrums Anatomie der Universität zu Köln

functional MRI or electrophysiology as a readout of the stimulation experiments.

Furthermore, the trigeminal nerve is chemosensitive, in particular, to some spices like as, for example, chili peppers horseradish, wasabi roots or Szechuan pepper (sharp), the coolness of peppermint, and the tingle of carbonated drinks (Viana 2011). The trigeminal nerve can perceive chemical irritants like ammonium chloride (salmiac) or acetic acid even when olfactory perception (olfactory nerve) is not present. The typically evoked responses include salivation, tearing, coughing, respiratory depression, and sneezing. These may be seen as protective responses (Viana 2011).

The penultimate and strongest branch of V_3 is the inferior alveolar nerve [N. alveolaris inferior]—lateral of the lingual nerve—for the sensory innervation of the lower tooth (see Fig. 12.16). It runs via the aperture of the mandible (Fig. 12.12 ⑤) into the mandibular canal where it forms a dental plexus (inferior dental plexus [Plexus dentalis inferior]) from which—like in the maxilla—the inferior dental branches [Rr. dentales inferiores] run to the individual teeth (see Fig. 12.15). The most part of the mandibular buccal gingiva is supplied by the lower gingival branches.

The terminal branch of the mandibular/inferior alveolar nerve is the mental (mentum Latin = chin) nerve [N. mentale] which reaches the skin of the mental region via the mental foramen.

A good knowledge of local anatomy is a prerequisite for a successful anesthesia of the teeth in dental medicine. The conventional inferior alveolar nerve block is the most commonly used nerve block technique in dentistry (Kim et al. 2018). The anatomical target is the inferior alveolar nerve at the aperture of mandible (Fig. 12.12). Since this foramen cannot be palpated, anatomical imaginativeness is in particu-

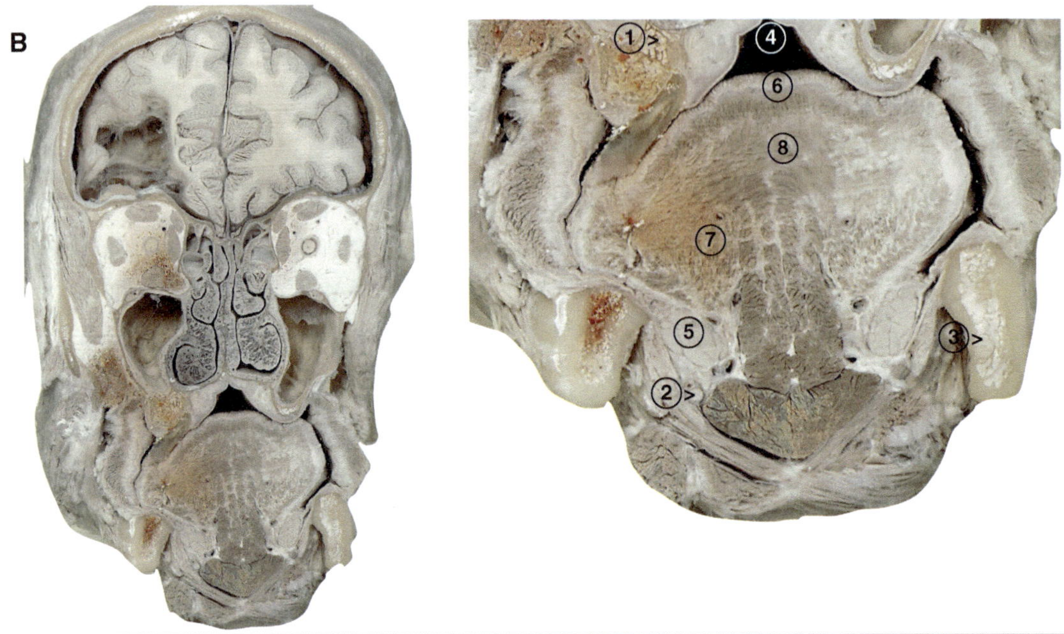

①	Superior dental plexus V_2	⑤	Sublingual gland
②	Sublingual nerve V_3	⑥	Superior longitudinal muscle of the tongue
③	Inferior alveolar nerve V_3	⑦	Inferior longitudinal muscle of the tongue
④	Oral cavity	⑧	Transverse muscle

Fig. 12.14 (continued)

lar required for this approach. The crucial clinical landmarks of this technique are the mandibular notch [Incisura mandibulae]—the greatest concavity on the anterior border of the mandibular ramus—and the pterygomandibular raphe. The pterygomandibular raphe is a connective tissue structure stretching from the sphenoid bone to the mandibula (Shimada and Gasser 1989).

The insertion point is located three fourth down the line drawn from the deepest part of the pterygomandibular raphe to the mandibular notch. The needle must be advanced until the bone is contacted. Aspiration is mandatory prior to administration of the local anesthetics and administration should be done very slowly. Ipsilaterally, the mandibular teeth to the midline, the body of the mandible, the lower part of the mandibular ramus, buccal periosteum, and mucous membrane to the premolars, anterior two third of the tongue, oral floor, lingual soft tissue, and the periosteum are all anesthetized (Kim et al. 2018).

In mice, the principal sensory trigeminal nucleus is the site of origin of ascending fibers carrying information from the vibrissae (whiskers) to the thalamus and subsequently to the so-called barrel field cortex (Woolsey and Welker 1975).

The whisker-barrel pathway starts with the infraorbital branch of the trigeminal ganglion which innervates the whisker pad on the mouse snout (Erzurumlu and Kind 2001). The peripheral spatial pattern of the vibrissae is reflected along the pathway in different arrangements. The vibrissae are extremely important for mice spatial orientation (for details, see Schröder et al. 2020).

Humans do not dispose of barrels.

12.2.3.2.4 Ascending connectivity of the trigeminal nucleus

The trigeminal afferents running to the spinal (see Sect. 3.2.2.2) and the principal trigeminal nucleus are organized comparable to the dorsal roots in the periphery of the body. The peripheral processes are the stimulus-perceiving part conveying signals to the pseudounipolar perikarya in the trigeminal ganglia (see Fig. 12.17). Pseudounipolar neurons are variations of bipolar neurons in that they have two processes which fuse during their development into one short common axon (Jennes 2017). The central processes form the trigeminal nerve and end in the corresponding nuclei. Exception is the mesencephalic trigeminal nucleus where the perikarya of the primary order neurons are ganglionic perikarya relocated into the brainstem (for details, see Sect. 13.2.6.1). The second-order neurons of the principal trigeminal nucleus release their axons into the trigeminothalamic tracts [Tractus trigeminothalamici], cross the midline, and reach the ventromedial posterior nucleus [Nucleus ventralis

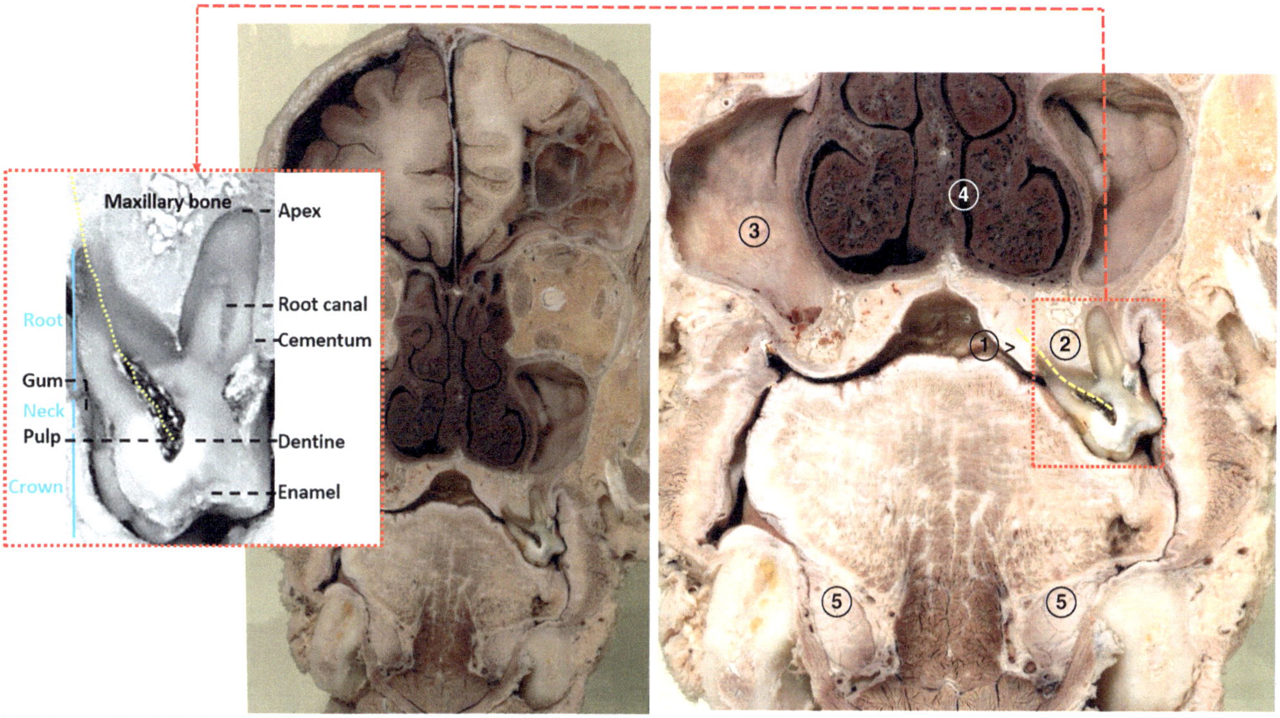

①	Superior alveolar nerve	④	Nasal septum
②	Maxillary bone	⑤	Submandibular gland
③	Maxillary sinus		

Fig. 12.15 Coronal section through the human head at the level of the maxillary sinus. Formalin-fixed specimen. The cystic structure in the right frontal lobe may be the consequence of a cerebral infarction as in Fig. 12.14A, B. The specimen shows the superior alveolar nerve ① together with one upper molar tooth. The yellow dotted line schematically represents the course of a branch from the alveolar nerve via the dental plexus into the root canal (---- see inset). At higher magnification (inset), the main structures of a human tooth can be seen. Enamel—the hardest and most resilient tissue of the human body—covers the outer surface of teeth. It can withstand biting forces of several 100 Newton. Dentine is the main supporting structure of teeth. Only enamel is harder than dentine. Cementum is the calcified tissue that forms the outer covering of the tooth root. Teeth are not massive structures but have cavities inside. At the apex, the supplying nerves and blood vessels enter the root canal which widens into the pulp. What cannot be seen here is the fact that tiny nerve fibers reach far into the dental cavity. Each tooth is anchored within the jaw's bone by collagen fibers (Sharpey fibers). Due to free nerve endings of the superior and inferior alveolar nerves, respectively, these fibers sense subtle changes of chewing pressures. This means that the anatomical substrate of tooth ache is the peripheral dental branches of the trigeminal nerve. Photograph by MedizinFotoKöln, Cologne, Germany. Sammlung des Zentrums Anatomie der Universität zu Köln

posteromedialis] of the thalamus. The last connection is that from the thalamus to the postcentral gyrus of the telencephalon, the cortical sensory representation of the face.

12.2.3.2.5 Living anatomy and pathological implications

Clinical testing of epicritic sensory function in the face

As for the periphery of the body, epicritic sensory function is mainly checked for by use of a tuning fork (normally 128 Hz) placed on predefined bony prominences (e.g., forehead, zygomatic bone, chin) as well as on the skin. The eyes of the patient should be shut. The resulting vibration signal is perceived mostly by Pacinian and Meissner corpuscles (see Fig. 3.7) and transferred as a neural signal to the central nervous system (Klingner and Witte 2018). Furthermore, gentle touches with the fingers or with a Q-tip should be applied. In both cases, the patient is asked to describe the sensations and about any differences in sensation on one side compared to the other side. This should be performed in the whole area of sensory supply of the trigeminal nerve and separately for the three main trigeminal branches. For testing of nociception function, see Sect. 3.2.2.2.3.

Clinical testing of trigeminal reflexes

There are several involuntary reflexes that can be elicited in the trigeminal supply territories.

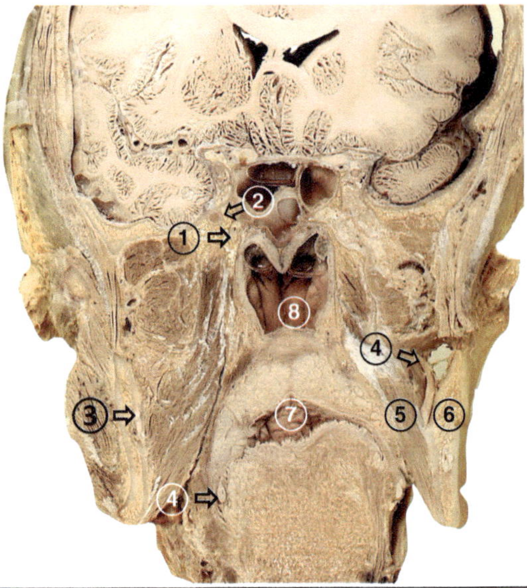

①	**Pterygopalatine ganglion**	⑤	**Medial pterygoid muscle**
②	**Maxillary nerve V_2**	⑥	**Mandible**
③	**Inferior alveolar nerve V_3**	⑦	**Oral cavity**
④	**Lingual nerve V_3**	⑧	**Nasal cavity**

Fig. 12.16 Coronal section through the human head at the level of posterior nasal aperture (see Fig. 12.12) in frontal view. Formalin-fixed specimen. The maxillary nerve ② is visible on the right side of the specimen—the section is somewhat shifted so that the right side is further posterior than the left side—and its main branches, the lingual nerve ④ and the inferior alveolar nerve ③ are best visible on the left right side. A rather long segment of the lingual nerve ④ is running on the medial pterygoid muscle ⑤. The latter is only visible after the buccal fat pad has been removed. The pterygopalatine ganglion can be seen on the right side ① ventromedial to the maxillary nerve ②, thereby indicating the location of the pterygopalatine fossa. Because of the advanced age of most body donors the maxillae and mandibulae often are either partly or completely toothless.

Photograph by MedizinFotoKöln, Cologne, Germany. Sammlung des Zentrums Anatomie der Universität zu Köln

Corneal reflex: CNVII

The cornea is extremely well equipped with trigeminal nerve endings. Stimulation by tactile (reflex test by the application of a cotton wisp), painful or thermal stimuli leads to the propagation of signals via the branches of the ophthalmic nerve (see V_1) (afferent limb of the reflex) and interneurons to the facial nucleus (see Sect. 8.2.1.1). The facial nerve (efferent limb) causes eyelid closure via activation of the orbicularis oculi muscle (see Fig. 8.22) (Ogino and Tadi 2019).

Blink reflex:

This reflex is elicited by the—unexpected for the patient—movement of an object into his peripheral visual field. Under normal conditions, the patient's reaction would be eyelid closure. The afferent limb of this reflex is the visual system (retina, optic tract, superior colliculi, see Sect. 15.1.3). As with the corneal reflex, the facial nucleus/nerve (efferent limb) causes eyelid closure via activation of the orbicularis oculi muscle (see Fig. 8.25) (Ogino and Tadi 2019).

Lacrimal reflex:

As with the corneal reflex, the cornea and/or conjunctiva is irritated by a cotton wisp. The afferent limb of the reflex is based on the ciliary fibers of the nasociliary nerve (V_1). In this case, the efferent fibers of the superior salivatory nucleus (see Fig. 8.12) running with the facial nerve as greater petrosal nerve (see Box 8.5) eventually synapse on postganglionic neurons of the pterygopalatine ganglion. These fibers join the zygomatic nerve and form an anastomosis with the lacrimal nerve, secretomotor for the lacrimal gland and the secretion of tears. It can be seen as a protective measure to flush out debris from the conjunctival sac [Saccus conjunctivalis].

Trigeminal Herpes zoster

A very painful, clinically rather impressive, and anatomically remarkably interesting entity are shingles or trigeminal Herpes zoster (see Fig. 12.18). Before the rash appears—potentially several days previously—patients often have pain, itching, or tingling in the area where it will develop.

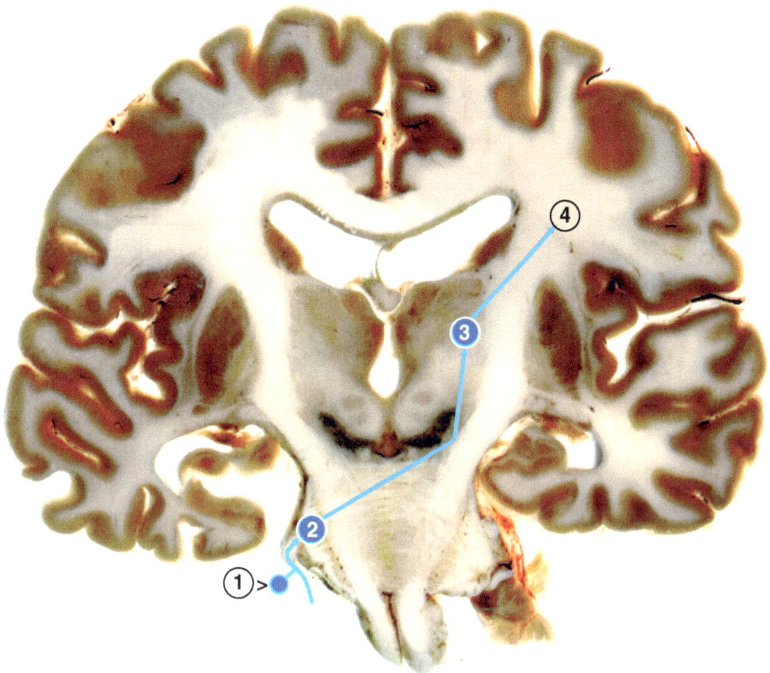

①	Pseudounipolar neuron of trigeminal ganglion	③	Ventral posteromedial nucleus (VPM)
②	Principal sensory nucleus of trigeminal nerve	④	To postcentral gyrus

Fig. 12.17 Central connectivity of the principal sensory trigeminal nucleus. The trigeminal sensory fibers have their perikarya located as pseudounipolar neurons in the trigeminal ganglion ① (see Fig. 12.3). The central processes enter the brainstem as the trigeminal nerve and synapse onto neurons of the principal sensory trigeminal nucleus ② (see Fig. 12.1). Their axons form the trigeminothalamic tracts [Tractus trigeminothalamici] (see also Fig. 3.6B) which cross to the contralateral ventral posteromedial nucleus [Nucleus ventralis posteromedialis] of the thalamus ③. The last neurons of the chain project to the postcentral gyrus ④. Do not confuse ventral posteromedial and posterolateral nuclei. The latter is the thalamic relay station for sensory signals from the periphery of the body (see Fig. 3.13B). Sammlung des Zentrums Anatomie der Universität zu Köln

The biologically interesting issue about this phenomenon is the relation of Herpes zoster with a preceding Varicella Zoster infection (chickenpox), potentially dating back decades. The virus that causes the well-known symptoms of chickenpox "hibernates" in the spinal and/or trigeminal ganglia after the acute infection.

Under certain circumstances (e.g., immunosuppression), the virus gets reactivated and moves from the ganglia retrogradely into the periphery of the supply territory down to the peripheral branches in the skin (for survey, see Klasser and Ahmed 2014). This causes an inflammation in the affected sensory supply area (see Fig. 12.18).

In case of trigeminal Herpes zoster, the dermatological changes topographically depend on which of the three main branches is affected. When you compare Fig. 12.18 with Fig. 12.2, you will immediately understand that in the diseased child the inflammation is located in the ophthalmic nerve territory.

Interestingly, reactivation of the virus in spinal ganglia may lead to the same kind of skin lesions in the peripheral dermatome supplied by the sensory fibers of the inflamed ganglion.

The disease is treated with an antiviral therapy and pain management. A feared complication is that the acute condition progresses in a prolonged phase, the so-called postherpetic neuralgia (Klasser and Ahmed 2014). See also trigeminal neuralgia under Sect. 3.2.2.2.3.

Detailed information on the following nuclei is provided as indicated below.

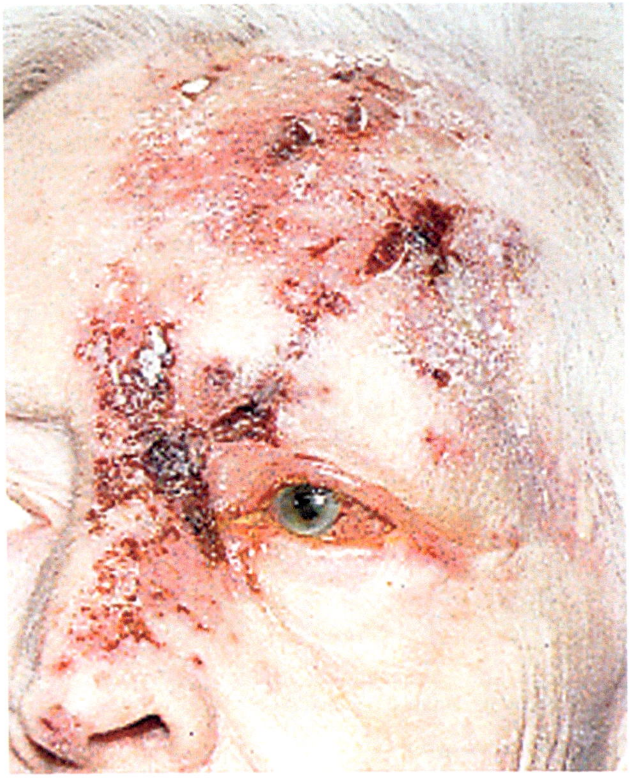

Fig. 12.18 Patient suffering from Varicella Zoster infection (shingles) of the face. The signs are present unilaterally more or less in the supply area of the ophthalmic nerve while those of the maxillary and mandibular nerves are spared (compare with Fig. 12.2). From Kaye 2018, Fig. 4.4 with permission

12.2.4 Monoamine nuclei r2

12.2.4.1 A7 noradrenaline cells (NA7)

For detailed information on these cells, see Sect. 11.2.3.1.

12.2.4.2 Subcaerulean nucleus

For detailed information on this nucleus, see Sect. 13.2.2.2.

12.2.5 Lateral lemniscus r2

12.2.5.1 Ventral nucleus of the lateral lemniscus

For detailed information, see Sect. 8.1.1.5.

12.2.5.2 Perilemniscal nucleus [Ncl. perilemniscalis]

FIPAT Ch. 1 does not list a perilemniscal nucleus but a paralemniscal nucleus (rhombomere 1). According to the data by Olszewski and Baxter (1982) and Nieuwenhuys et al. (1998), this nucleus is located near to the mesencephalic lateral surface between the lateral lemniscus and the brachium of the inferior colliculus (Fig. 12.19). The neurons are known to be small, triangular, or fusiform with long dendrites. Their appearance is similar to that of the sagulum nucleus (see Sect. 15.4.2) which is supposed to be connected with several auditory areas.

In mice, tracer studies (Liang et al. 2014) have shown this nucleus to project to the entire spinal cord mainly to the contralateral side.

Only very few specific data is available for the human paralemniscal nucleus [Ncl. paralemniscalis]. The name alludes to its location just lateral of the lateral lemniscus (see Fig. 12.19). A first functional attribution has been made by Kohnstamm (1912). His idea was that the paralemniscal nucleus is an acoustic reflex nucleus subserving acoustic defense movements and accommodation reaction. In addition to the input from the auditory system, the paralemniscal nucleus is supposed to receive retinal afferents (Ruddock 1995).

The paralemniscal region of the rat has been shown to get afferent innervation from the auditory cortex which also reaches the ipsi- and contralateral superior olive as well as the cochlear nuclei of either side (Saldaña 2015). In addition, the sagulum nucleus and the paralemniscal regions are contacted by these corticofugal tracts (for details see Saldaña 2015).

Detailed information on the following nuclei is provided as indicated below:

12.2.6 Central gray r2

For detailed information, see Sect. 13.2.7.

12.3 Basal r2

12.3.1 Motor trigeminal complex r2

For detailed information, see Sect. 11.3.1.

12.3.1.1 Motor trigeminal nucleus

For detailed information, see Sect. 11.3.1.

12.3.2 Raphe nuclei r2

12.3.2.1 Median raphe nucleus

For detailed information, see Sect. 11.3.2.1.

12.3.2.2 Paramedian raphe nucleus

For detailed information, see Sect. 11.3.2.2.

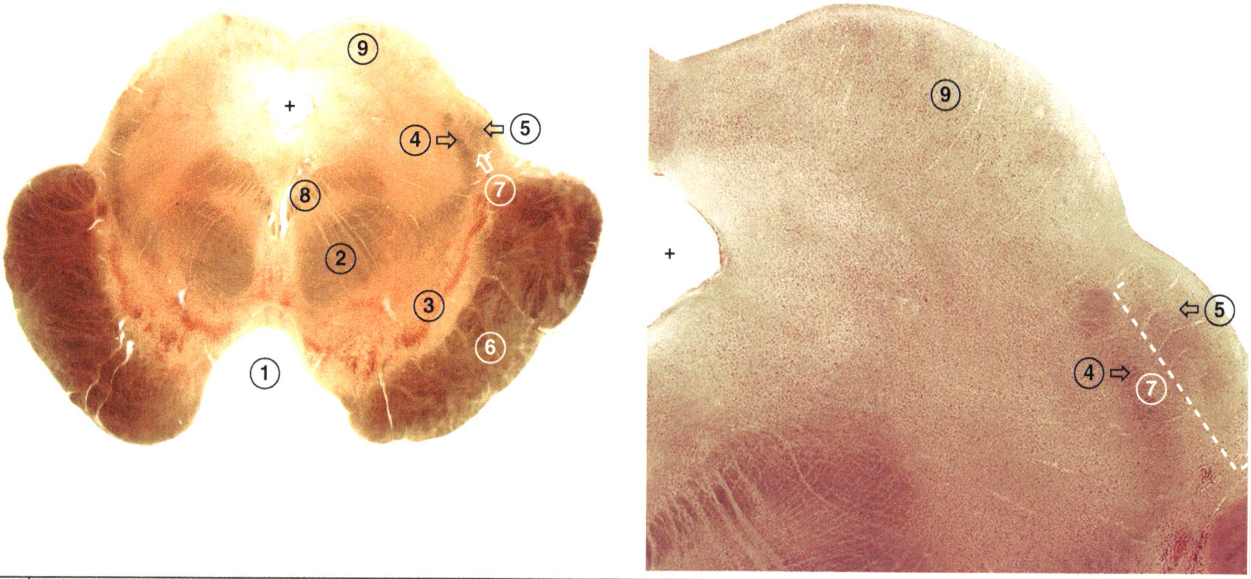

①	Interpeduncular fossa	⑥	Cerebral peduncle
②	Superior cerebellar peduncle / Red ncl.	⑦	Paralemniscal ncl.
③	Substantia nigra	⑧	Medial longitudinal fasciculus
④	Medial lemniscus	⑨	Superior colliculus
⑤	Brachium of inferior colliculus	+	Mesencephalic aqueduct

Fig. 12.19 Horizontal section through the human mesencephalon at the level of the decussation of the superior cerebellar peduncles. The paralemniscal nucleus ⑦—as the name says—is located, laterally, alongside the medial lemniscus ④ and directly medial to the brachium of the inferior colliculus ⑤. See atlas part Darrow red 38. LabPON Twente

12.3.3 Interpeduncular nuclei r2 [Nucleus interpeduncularis]

12.3.3.1 Location and cytoarchitecture of the interpeduncular nucleus

The interpeduncular nucleus (IPN) is an unpaired, approximately triangular structure located across the mesencephalic midline (see Fig. 12.20 ①). The name is derived from the interpeduncular fossa (see below), between the cerebral peduncles (Fig. 12.20 ⑤) (between the peduncles = inter pedunculos, hence the name). While in other mammals, a clear distinction of a medial and a lateral subnucleus can be made. This is not the case in the human brain (Olszewski and Baxter 1982).

The base of the triangle is neighbored by the intrapeduncular fossa (see also Sect. 12.4.1, Fig. 12.23). Laterally, the paranigral nuclei (between IPN and substantia nigra) are located. As visible in the inset, the neurons of the IPN are rather small (for details, see Olszewski and Baxter 1982). A Golgi impregnation study of the human IPN has been provided by Kemali and Casale (1982).

12.3.3.2 Connectivity of the interpeduncular nucleus

The main afferent input to the IPN in humans and other mammals is the cholinergic habenulo-interpeduncular tract [Tractus habenulo-interpeduncular tract (formerly Fasciculus retroflexus) from the medial habenular nucleus in the epithalamus (Naidich et al. 2009).

This massive tract has been described in developing human fetuses of 8–12 weeks (Cho et al. 2014). It starts on the dorsal surface of the prospective thalamus running dorsally and then ventrally around the thalamus, coming close to the red nucleus (see Fig. 12.20 ⑦) and then entering the IPN. Its identification in routine-stained histological specimens of the adult brain is difficult.

The habenula is thought to connect the forebrain to the ventral midbrain (Boulos et al. 2016). Based on neuroimaging studies, this appears to hold true for humans as well as for rodents. It regulates midbrain monoaminergic systems, notably serotonin and dopamine. It integrates cognitive with emotional and sensory processing (Boulos et al. 2016) based on the efferent connections of the IPN with the dopaminergic ventral tegmental area (see Sect. 15.5.1.1) and the serotonergic dorsal raphe nuclei (see Sect. 13.3.4.2) as shown in rodents.

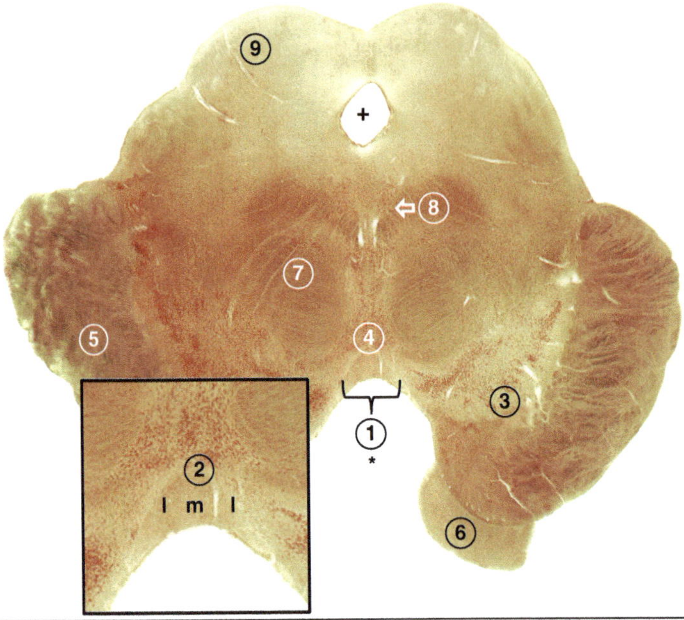

①	Interpeduncular ncl.	⑦	Superior cerebellar peduncles / Red ncl.
②	Interpeduncular ncl. m = medial / l = lateral	⑧	Medial longitudinal fasciculus
③	Substantia nigra	⑨	Superior colliculus
④	Caudal linear raphe ncl.	+	Mesencephalic aqueduct
⑤	Cerebral peduncle	*	Interpeduncular fossa
⑥	Pontine basis		

Fig. 12.20 Horizontal section through the human mesencephalon at the level of the Intrapeduncular fossa * and the superior colliculi ⑨. The interpeduncular nucleus (IPN) is located at the apex of the interpeduncular fossa * (see Sect. 12.4.1). Inset: At higher magnification, a lateral (l) and a medial subnucleus (m) can be distinguished (for details, see text). See atlas part Darrow red 37. LabPON Twente

In transgenic mice, the selective postnatal ablation of the medial habenula in expectedly leads to reduced acetylcholine levels in the interpeduncular nucleus accompanied by abnormalities in a wide range of behavioral domains (Kobayashi et al. 2013). Animals tended to be hyperactive during the early night period and are maladapted v repeatedly exposed to new environments.

Detailed information on the following nuclei is provided as indicated below:

12.3.4 Basal tegmental nuclei r2

12.3.4.1 Dorsomedial tegmental area

For detailed information, see Sect. 7.3.4.1.

12.3.5 Reticular r2

12.3.5.1 Pontine reticular nucleus, oral part

For detailed information on this nucleus, see Sect. 10.3.5.4.

12.3.5.2 Parvocellular reticular nucleus

For detailed information on this nucleus, see Sect. 6.5.5.2.

12.3.5.3 Barrington's nucleus/Nucleus of Barrington r2 (Pontine micturition center) [Centrum micturitionis, Regio M]

Frederick James Fitzmaurice Barrington (1925) was a surgeon–physiologist, founder of what today is called neurourology. His paper from 1925 on "The effect of lesions of the hind- and midbrain on micturition in the cat" laid ground to the idea of a brainstem structure controlling the lower urinary tract (see Morrison 2008).

12.3.5.3.1 Location and cytoarchitecture of Barrington's nucleus

Human-specific data on Barrington's nucleus is scarce. Most of what we know in terms of anatomy and function has been obtained from studies in the cat in the tradition of Frederick Barrington.

Blanco et al. (2013) have used corticotropin-releasing factor (CRF) as a marker for Barrington's nucleus in human

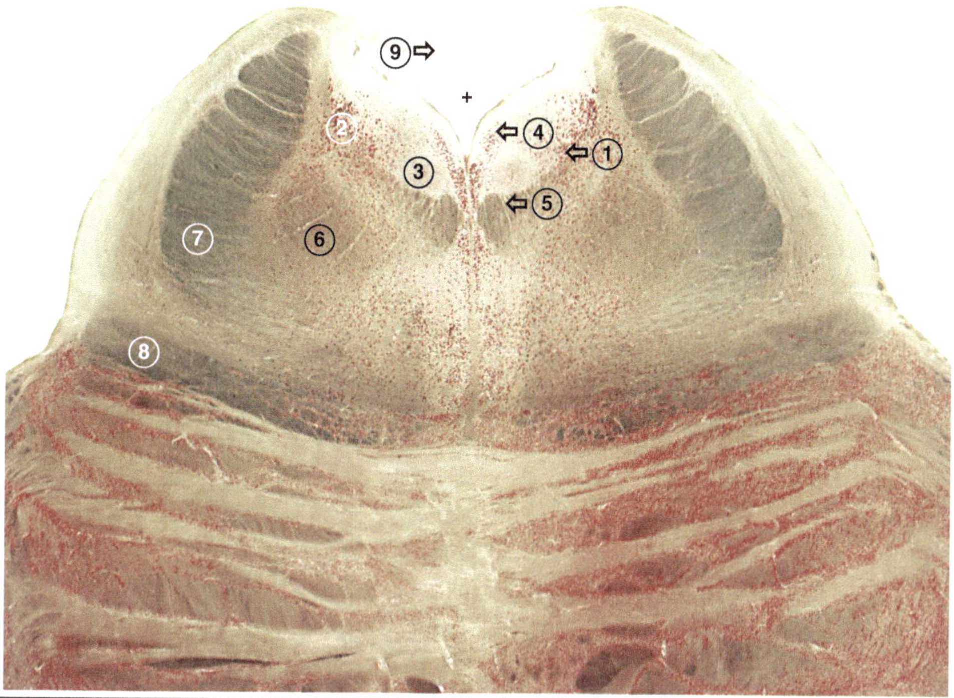

①	Barrington's ncl.	⑥	Central tegmental tract
②	Locus caeruleus	⑦	Superior cerebellar peduncle
③	Dorsal tegmental ncl.	⑧	Medial lemniscus
④	Dorsal raphe ncl.	⑨	Superior medullary velum
⑤	Medial longitudinal fasciculus	+	4th ventricle

Fig. 12.21 Horizontal section through the human pons at the level of the laterodorsal tegmental nucleus. Darrow red staining. Barrington's nucleus ① is located between the Locus caeruleus (formerly coeruleus) ② and the laterodorsal tegmental nucleus ③. For details, see text and atlas part Darrow red 31A. LabPON Twente

autopsy specimens. In conventional histological sections, it is difficult to identify the nucleus, however, the neighboring locus caeruleus (lateral, see Sect. 13.2.2.1) and laterodorsal tegmental nucleus (medial, Sect. 13.1.1.2) can be used to locate Barrington's nucleus (see Fig. 12.21).

12.3.5.3.2 Target regions of Barrington's nucleus

The current knowledge and ideas on human micturition—mainly derived from animal data and experiments—can be summarized in the simplified scheme on the next page (Fig. 12.22).

The lower urinary tract is supplied by a triple innervation by parasympathetic, sympathetic, and somatic motor (Onuf's nucleus in the ventral horn of the lumbosacral spinal cord) sources (see Fig. 12.22). These efferents are under control of brainstem and telencephalic centers (for details, see De Groat 2017). Feedback from the viscera of the pelvis to Barrington's nucleus is provided via the pelvic nerves and the spinal cord (for details, see Blanco et al. 2014).

The normal voiding cycle of the urinary bladder starts with its filling under conditions of low storage pressures (Zderic 2007). At a certain volume, the sensation of fullness is noted. Barrington's nucleus relinquishes its inhibitory influence on the detrusor. The sacral reflex arc (for details, see Uher and Swash 1998) is activated and with detrusor contraction the intravesical pressure rises. This is accompanied by funneling of the bladder neck and relaxation of the striated external sphincter (Zderic 2007) (see Fig. 12.22).

From studies in rodents, it can be inferred that Barrington's nucleus as the pontine micturition center (PMC) is an important structure for control of bladder, urethra, and urethral sphincter besides the medullary raphe nuclei, the Locus cae-

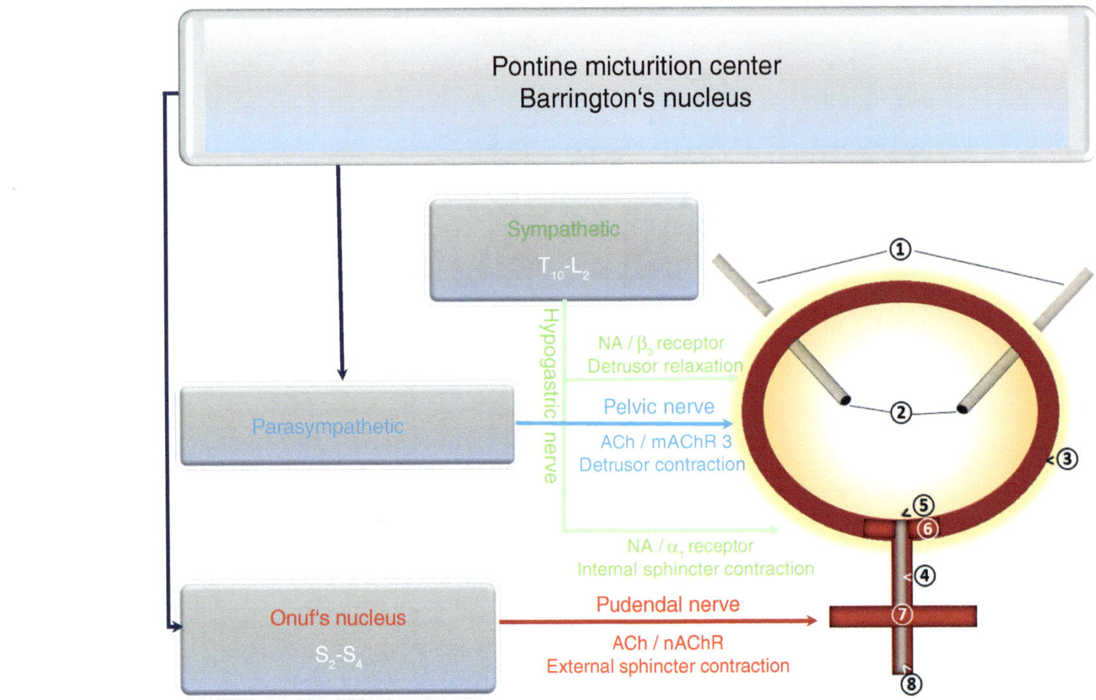

①	**Ureters**	⑤	**Internal urethral orifice**
②	**Ureteral openings**	⑥	**Internal urethral sphincter**
③	**Detrusor muscle**	⑦	**External urethral sphincter**
④	**Urethra**	⑧	**External urethral orifice**

Fig. 12.22 Schematic illustration of the human urinary bladder with the muscular effectors and their innervation. In terms of physiology, the urine is flowing from the kidneys via the paired ureters ① and their openings ② into the bladder. Under resting/filling conditions, the sympathetic innervation via the hypogastric nerve provides a relaxation (noradrenaline β2-adrenoreceptor) of the detrusor muscle ③ [M. detrusor vesicae] which forms a circular spiraling layer of smooth muscle around most of the bladder. Simultaneously, the sympathetic outflow leads to a contraction of the internal urethral sphincter ⑥ [M. sphincter urethrae internus] which prevents micturition (Blanco et al. 2013). The parasympathetic limb is under the control of Barrington's nucleus. In the activated state, the signals transmitted to the detrusor via the pelvic nerve lead to an interaction between acetylcholine (ACh) and the muscarinic ACh receptors (mAChRs) of the detrusor which results in contraction of the detrusor, the typical counterbalance scheme of sympathetic/parasympathetic innervation. Finally, the last gate in the urinary pathway, the external urethral sphincter (voluntary) ⑦ [M. sphincter externus] occludes the urethra below the level of the internal sphincter via cholinergic innervation and nicotinic ACh receptors (nAChRs). Modified acc. to Roy and Green 2019, Fig. 1 with permission

ruleus (see Sect. 13.2.2.1), the periaqueductal area (PAG) (see Sect. 13.2.7.1), and the A5 noradrenergic cell group (see Sect. 8.1.3.1). Most likely the hypothalamus and cortical areas are superordinate structures in this framework (De Groat 2017).

Efferent fibers from the hypothalamus project to (i) the sacral parasympathetic nucleus and the sphincter motor nucleus and (ii) the PMC. The latter targets the sacral parasympathetic nucleus, the lateral edge of the dorsal horn, and the dorsal commissure.

All these data can be summarized as follows:

1. On the afferent level, signals from the pelvic viscera converge directly or indirectly on Barrington's nucleus.
2. From there onward signals are most likely transferred to cortical areas via the Locus caeruleus (see Sect. 13.2.2.1) and the PAG (see Sect. 13.2.7.1).
3. Psychogenic stimuli may be mediated to the bladder and rectum from the PAG to Barrington's nucleus and from there to the spinal cord.

12.3.5.3.3 Pathological implications

Multiple system atrophy (MSA) is a progressive neurodegenerative disorder characterized by a combination of symptoms that affect both the autonomic nervous system (e.g., bladder function) and movement (see Sect. 3.3.2.3 for details).

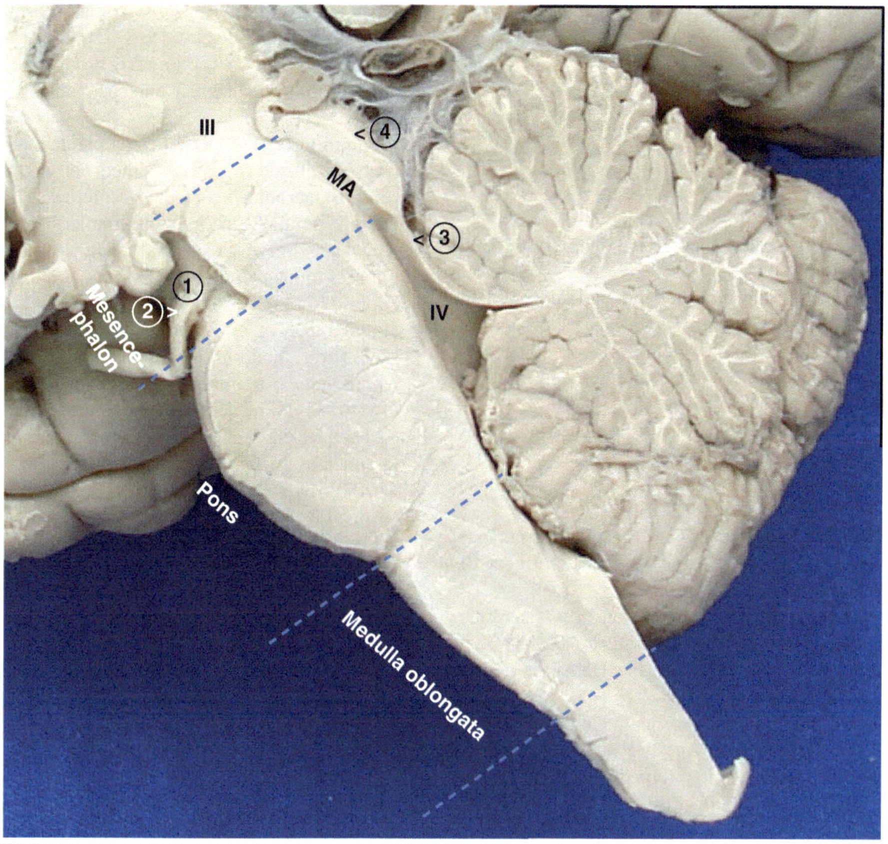

①	Interpeduncular fossa	III	3rd ventricle
②	Oculomotor nerve	IV	4th ventricle
③	Superior medullary velum	MA	Mesencephalic aqueduct
④	Quadrigeminal plate		

Fig. 12.23 Midsagittal section through the human brain. Formalin-fixed specimen. The interpeduncular fossa ① is located between superior and inferior borders of the mesencephalon. The oculomotor nerve ② exits from the brainstem in the interpeduncular fossa. Sammlung des Zentrums Anatomie der Universität zu Köln

The vast majority of individuals with MSA will experience parkinsonism symptoms. Two thirds of patients have a tremor or jerky movements (myoclonus) of the hands or fingers, and impaired bladder and bowel control including involuntary urination or defecation (urinary and rectal incontinence).

Benarroch and Schmeichel (2001) could show a massive loss of CRF neurons (see above Sect. 12.3.5.3.1) in the area of Barrington's nucleus in MSA patients who had suffered from urinary incontinence and retention.

Interestingly, in a transgenic mouse model of multiple system atrophy (MSA) Barrington's nucleus showed massive degeneration in 12-month-old animals (Boudes et al. 2013).

12.4 Floor plate r2

12.4.1 Interpeduncular fossa [Fossa interpeduncularis]

The interpeduncular fossa (Fig. 12.23) is a clearly visible depression at the ventral surface of the brain. In the interpeduncular fossa, the oculomotor nerve ② (see Sect. 15.3.1.1) (CN III) leaves the brainstem.

The fossa is bordered laterally by the cerebral peduncles, dorsally by the interpeduncular nucleus (see here above; Figs. 12.19 and 12.20).

References

Aggarwal M, Puri V et al (2012) Serotonin and CGRP in migraine. Ann Neurosci 19:88–94

Barrington F (1925) The effect of lesions of the hind- and mid-brain on micturition in the cat. J Exp Physiol 15:81–102

Benarroch EE, Schmeichel AM (2001) Depletion of corticotrophin-releasing factor neurons in the pontine micturition area in multiple system atrophy. Ann Neurol 50:640–645

Blanco L, Yuste J, et. (2013) Critical evaluation of the anatomical location of the Barrington nucleus: relevance for deep bran stimulation surgery of pedunculopontine tegmental nucleus. Neuroscience 247:351–363

Blanco L, Ros CM et al (2014) Functional role of Barrington's nucleus in the micturition reflex: relevance in the surgical treatment of Parkinson's disease. Neuroscience 266:150–161

Boudes M, Uvin P et al (2013) Bladder dysfunction in a transgenic mouse model of multiple system atrophy. Mov Disord 28:347–355

Boulos L-J, Darcq E et al (2016) Translating the Habenula—from Rodents to Humans. Biol Psychiatry 81:296–305

Buture A, Gooriah R et al (2016) Current understanding on pain mechanism in migraine and cluster headache. Anesth Pain Med 6:e35190

Cho KH, Mori S et al (2014) The habenulo-interpeduncular and mammillothalamic tracts: early developed fiber tracts in the human fetal diencephalon. Childs Nerv Syst 30:1477–1484

de Groat WC (2017) Autonomic nervous system: urogenital control. In: Reference module in neuroscience and biobehavioral psychology. Elsevier

Eftekhari S, Salvatore CA et al (2010) Differential distribution of calcitonin gene-related peptide and its receptor components in the human trigeminal ganglion. Neuroscience 169:683–696

Eneroth CM, Hökfelt T et al (1969) The role of the parasympathetic and sympathetic innervation for the secretion of human parotid and submandibular glands. Acta Otolaryngol 68:369–375

Erzurumlu RS, Kind PC (2001) Neural activity: sculptor of barrels in the neocortex. Trends Neurosci 24:589–595

Goosmann MM, Dalvin M (2018) Anatomy, head and neck, deep petrosal nerve. In: StatPearls [Internet]. Treasure Island, FL: StatPearls Publishing

Jennes L (2017) Cytology of the central nervous system: dendrites and axons. In: Conn's translational neuroscience. Elsevier

Kaye KM (2018) Herpes Zoster. In MSD Manual. Ausgabe für Medizinische Fachkreise. https://www.msdmanuals.com/de-de/profi/infektionskrankheiten/herpesviren/herpes-zoster

Kemali M, Casale E (1982) The morphology of the interpeduncular nucleus of man: a Golgi observation. Z Mikrosk Anat Forsch 96:591–599

Kim C, Hwang K-G et al (2018) Local anesthesia for mandibular third molar extraction. J Dent Anesth Pain Med 18:287–294

Klasser GD, Ahmed AS (2014) How to manage acute herpes zoster affecting trigeminal nerves. J Can Dent Assoc 80:e42

Klingner CM, Witte OW (2018) Somatosensory deficits. Handb Clin Neurol 151:185–206

Kobayashi Y, Sano Y et al (2013) Genetic dissection of medial habenulo-interpeduncular nucleus pathway function in mice. Front Behav Neurosci 7:17

Kohnstamm O (1912) Der Nucleus paralemniscalis inferior als akustischer Reflexkern und als Glied der zentralen Hörleitung (nebst einer Bemerkung über den Bechterew'schen Kern und den nucl. lateralis pontis). Eur Arch Otorhinolaryngol 89:59–60

Krastev D, Paloff A et al (2008) Ultra-structure of trigeminal ganglion in human. Journal of IMAB—Annual Proceeding (Scientific Papers) Book 1

Li JL, Wu SX et al (2005) Efferent and afferent connections of GABAergic neurons in the supratrigeminal and the intertrigeminal regions. An immunohistochemical tract-tracing study in the GAD67-GFP knock-in-mouse. Neurosci Res 51:81–91

Liang H, Bácskai T et al (2014) Projections from the lateral vestibular nucleus to the spinal cord in the mouse. Brain Struct Funct 219:805–815

Morrison JFB (2008) The discovery of the pontine micturition centre by F.J.F. Barrington Exp Physiol 93:742–745

Naidich TP, Duvernoy HM et al (2009) Duvernoy's atlas of the human brain stem and cerebellum. Springer

Nieuwenhuys R, Voogd J et al (1998) The human central nervous system. Springer

Nturibi E, Bordoni B (2020) Anatomy, head and neck, greater petrosal nerve. In: StatPearls. StatPearls Publishing, Treasure Island, FL

Ogino MH, Tadi P (2019) Neuroanatomy, trigeminal reflexes. Last Update: December 10, 2019. StatPearls [Internet]. Treasure Island, FL: StatPearls

Olszewski J, Baxter D (1982) Cytoarchitecture of the human brain stem, 2nd edn. Karger, Basel

Roy HA, Green AL (2019) The central autonomic network and regulation of bladder function. Front Neurosci 13:535

Ruddock KH (1995) Residual visual function in the absence of the human striate cortex. In: Robbins JG, Djamgoz MB, Taylor A (eds) Basic and clinical perspectives in vision research. Springer, pp 211–223

Saldaña E (2015) All the way from the cortex: a review of auditory corticosubcollicular pathways. Cerebellum 14:584–596

Schröder H, Moser N et al (2020) Neuroanatomy of the mouse. An introduction. Springer

Shimada K, Gasser RF (1989) Morphology of the pterygomandibular raphe in human fetuses and adults. Anat Rec 224:117–122

Takezawa K, Townsend G et al (2018) The facial nerve: anatomy and associated disorders for oral health professionals. Odontology 106:103–116

Uher E, Swash M (1998) Sacral reflexes: physiology and clinical application. Dis Colon Rectum 41:1165–1177

Viana F (2011) Chemosensory properties of the trigeminal system. ACS Chem Neurosci 2:38–50

Woolsey TA, Welker C (1975) Comparative anatomical studies of the SmL face cortex with special reference to the occurrence of "barrels" in layer IV. J Comp Neurol 164:79–94

Zderic SA (2007) Pediatric voiding function and dysfunction. In: Penn clinical manual of urology

Rhombomere 1 r1

13

Contents

H. Schröder et al., *The Human Brainstem*, https://doi.org/10.1007/978-3-030-89980-6_13

Abstract

This chapter is one of the most voluminous of this book. This is due to the fact that a number of important structures are derivatives of rhombomere 1 and that among those are nuclei which are intensely connected with neuropathology. This holds in particular true for the locus caeruleus and Alzheimer's disease. Therefore, this chapter contains the main contribution to this disease.

Other nuclei infested with Alzheimer pathology and dealt with here are the subpeduncular nucleus, the dorsal tegmental nucleus (Gudden), the parabrachial nuclei, and the dorsal raphe nucleus. In particular, the degeneration of the centripetal noradrenergic locus caeruleus and the serotonergic dorsal raphe nucleus are involved in the development of cognitive impairment since the alerting influence of these transmitter systems on the forebrain disappears in the course of the disease.

A meaningful relay structure involved with the handling of threat and responsible for pain modulation and psychopathological phenomena is the periaqueductal gray.

The pedunculopontine tegmental nucleus and the laterodorsal tegmental nucleus are important tegmental cholinergic nuclei.

Eventually, the mesencephalic trigeminal nucleus is responsible for processing of muscle spindle signals originating in the masticatory muscles.

13.1 Rhombic lip r1

13.1.1 Tegmentum r1

13.1.1.1 Alar cholinergic nuclei r1

13.1.1.1.1 Pedunculopontine tegmental nucleus/ Pedunculotegmental nucleus (PTg) [Ncl. tegmentalis pedunculopontinus]

This nucleus is a prominent cholinergic and NOS (nitric oxide synthase)-positive cell group, present in man, monkey, rat, and mouse (Watson et al. 2019). It has been renamed pedunculotegmental since it is not a pontine structure and has no relationship with the pontine nuclei in rhombomeres 3 and 4. As many other prepontine nuclei, it has been designed as pontine since it is covered by the rostrally extended human pons (Watson et al. 2019, see Box 1.1).

Location and morphology of the PTg

The pedunculopontine tegmental nucleus (PTg) is situated in the dorsolateral portion of the ponto-mesencephalic tegmentum. Its main mass is positioned at the trochlear nucleus level, and it is part of the mesencephalic locomotor region (MLR) in the upper brainstem (French and Muthusamy 2018) (see Box 13.1).

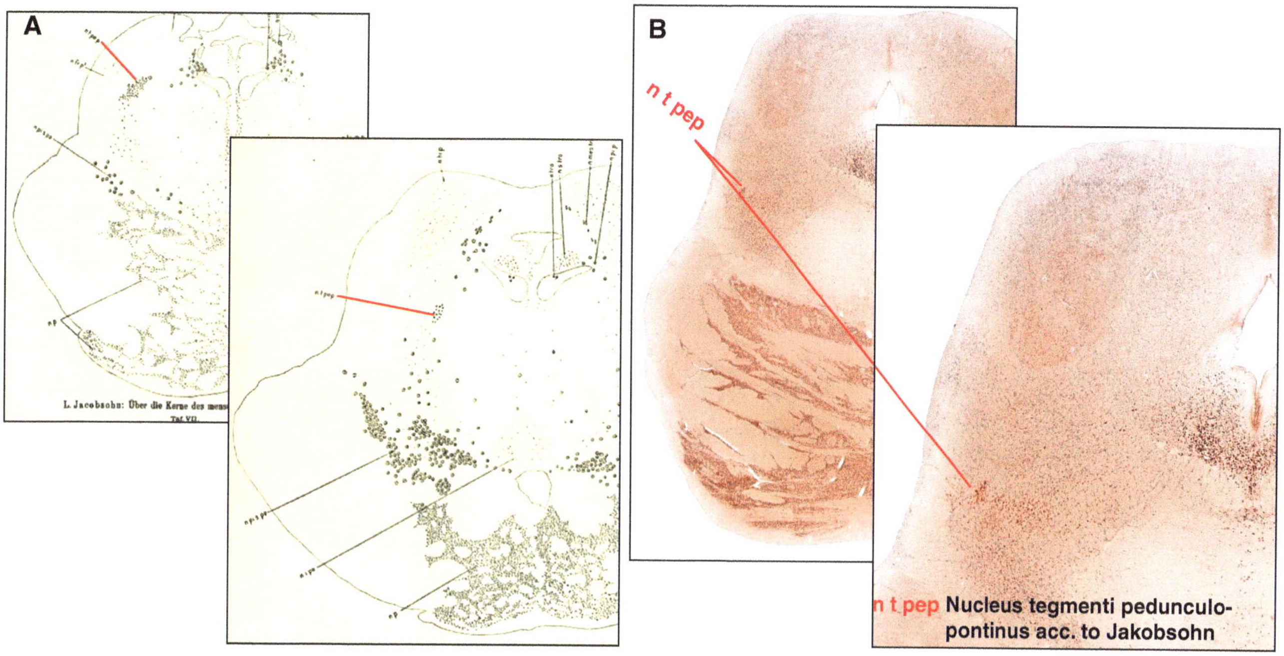

Fig. 13.1 (**A**) shows tables VII and VIII from Jacobsohn's work about the nuclei of the human brainstem with the first description of the "Nucleus tegmenti pedunculo-pontinus (n t pep)" (Jacobsohn 1909). (**B**) Corresponding histological section from our Darrow red series (see also atlas part Darrow red 36, 36A). (**A**) From Jacobsohn (1909) under: (Biodiversity Heritage library) https://www.biodiversitylibrary.org/ia/abhandlungenderk1909kn/#page/8/mode/1up. (**B**) LabPON Twente

The pedunculopontine tegmental nucleus corresponds to the cholinergic cell group Ch5 acc. to Mesulam et al. (1983) and is assumed to project to the thalamus.

It has been subdivided in a cholinergic, compact part [Pars compacta] and a diffuse, non-cholinergic dissipated part [Pars dissipata] (see Figs. 13.1 and 13.2) (for details, see ten Donkelaar 2011) (see also Sect. 13.2.4). The pars dissipata in primates is supposed to be the retrorubral nucleus of rodents (for details, see ten Donkelaar 2011).

Interestingly, even an extended search failed to find a current histological representation of the human PTg although this nucleus had already been described by Jacobsohn in 1909 under the Latin name Nucleus tegmenti pedunculopontinus (see Fig. 13.1A, B). Louis Jacobsohn (later Jacobsohn-Lask) (*1863 in Bromberg; †1940 in Sevastopol, emigrated under Nazi rule) was a German neurologist and neuroanatomist.

The compact part of the PTg is located lateral of the superior cerebellar peduncle. The reticular cells at the level of the decussation of the superior cerebellar peduncle aggregate in the dorsolateral part, which Jacobsohn named the nucleus tegmenti pedunculo-pontinus (see Fig. 13.1).

According to his description, the nucleus is small and elongated and shows a lower density in the ventral part. Its cells are medium-sized, acute-triangular, or oblong-blunt. The longitudinal extension of the nucleus reaches to the inferior colliculi.

The PTg contains cholinergic (see Fig. 13.2), GABAergic, glutamatergic neurons, and substance P-containing cells (Bennaroch 2013).

Connectivity and functional aspects of the PTg

The PTg receives direct input from the cerebral cortex, is reciprocally connected with the basal ganglia, and provides inputs to the thalamus and motor areas of the brainstem and spinal cord. Via these connections, the PTg is involved in mechanisms of cortical arousal and behavioral state control and participates in control of locomotion and muscle tone (Bennaroch 2013). These processes allow the PTg to sustain the background activity necessary for generating movement, so that disturbances in PTg function result in a number of devastating movement disorders (Garcia-Rill 2019) (see below). Mori et al. (2016) have suggested that the PTg also acts as an interface between the basal ganglia and the cerebellum.

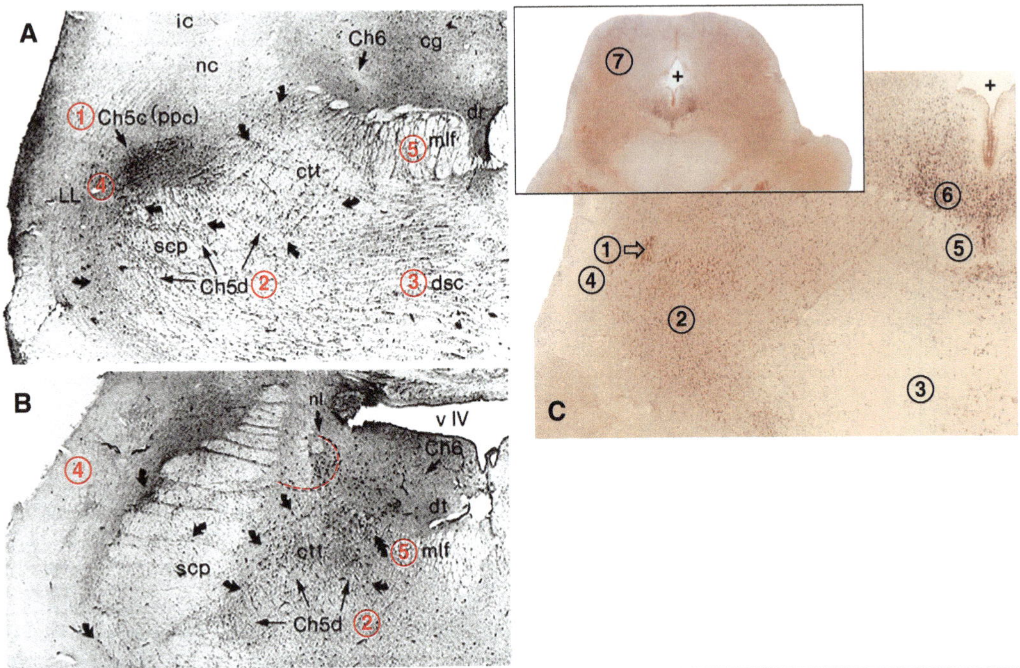

①	**Pedunculopontine tegmental ncl., compact part (Ch5c)**	⑤	**Medial longitudinal fasciculus**
②	**Pedunculopontine tegmental ncl., dissipate part (Ch5d)**	⑥	**Dorsal raphe ncl.**
③	**Decussation of the superior cerebellar peduncles**	⑦	**Inferior colliculus**
④	**Lateral lemniscus**	+	**Mesencephalic aqueduct**

Fig. 13.2 Horizontal sections through the pontomesencephalic transition zone. (**A**, **B**) ChAT immunohistochemistry from Mesulam et al. (1989). (**C**) Darrow red staining. LabPON Twente. Both parts of the PTg ① ② are visible in (**A**) and (**C**). Single curved arrows in (**A**) and (**B**) point to interstitial ChAT-positive dissipate part (Ch5d) neurons. Ch5d extends further caudally than the compact part (Ch5c). The broken red line in (**B**) is an arbitrary boundary for the locus caeruleus (nl). (**C**) shows the location of the PTg in one of our Darrow red brainstem sections at the level of the caudal inferior colliculi. (**A**, **B**): cg = central gray, Ch6 = laterodorsal tegmental nucleus (see Sect. 13.1.1.2), ctt = central tegmental tract, dsc = decussation of the superior cerebellar peduncles ③, dt = dorsal tegmental nucleus, LL = lateral lemniscus ④, ic = inferior colliculus, mlf = medial longitudinal fasciculus ⑤, nc = cuneiform nucleus, scp = superior cerebellar peduncle. (**A**, **B**) From Mesulam et al. 1989, Fig. 4 with permission. Compare with atlas Darrow red 36, 36A and Campbell 15, 16, 17, 17A

Pathological aspects of the PTg

The PTg is affected in Parkinson's disease (PD) (see Sect. 16.6.1.5) and atypical parkinsonian syndromes such as progressive supranuclear palsy (PSP, see Sect. 4.1.1.3.3) and multiple system atrophy (MSA, see Sect. 3.3.2.3). Involvement of the PTg may have an important role in gait impairment in these disorders (Bennaroch 2013). In PD patients, GABAergic basal ganglia output levels are abnormally increased, and gait disturbances are produced via abnormal increases in inhibition of the mesencephalic motor region (MLR) (see Box 13.1) by the reticular part of the Substantia nigra (see Sect. 16.6.1) (French and Muthusamy 2018).

Box 13.1 The mesencephalic locomotor region

It has been known for long that the transition zone between the mesencephalon and the hindbrain is involved in control of walking and running in a number of vertebrates including monkeys (for details, see Ryczko and Dubuc 2013). Historically, the cuneiform nucleus and the PTg are thought to be the anatomical correlate of the MLR (Chang et al. 2020). Humans asked to imagine that they are walking show an increased activity in those nuclei that are supposed to belong to the MLR, the PTg (see here above), the cuneiform (see Sect. 14.4.1.1) and the subcuneiform nuclei.

Since the PTg is vastly connected with the basal ganglia and the brainstem, dysfunction within these systems leads to advanced symptomatic progression in PD, including sleep and cognitive issues. The current therapy of choice for PTg-related disorders is deep brain stimulation (DBS) (see Box 13.2) by treating pathological circuitries within the parkinsonian brain (French and Muthusamy 2018) involving possible modulation of the cerebellum (Mori et al. 2016).

Box 13.2 Deep brain stimulation

Deep brain stimulation (DBS) is based on delivering constant electrical stimulation via electrodes implanted into certain defined target regions controlled by a subcutaneously placed pacemaker-like device. Best known is the use of DBS for the treatment of Parkinson's disease in patients in which drug therapy has ended to work. In Parkinson therapy, the electrodes are placed in the subthalamic nucleus (see Figs. 16.5 and 16.6). Currently, besides the treatment of Parkinson's disease or the motor impairments described under "Pathological aspects of the PTg," DBS is used routinely for the therapy of epilepsy, essential tremor, and obsessive-compulsive disorder (for review, see Gardner 2013).

https://www.mayoclinic.org/tests-procedures/deep-brain-stimulation/about/pac-20384562

13.1.1.2 Laterodorsal tegmental nucleus [Ncl. tegmentalis dorsolateralis]

The laterodorsal tegmental nucleus is a cholinergic nucleus (cell group Ch6) located ventral to the ventrolateral periaqueductal gray or pontine central gray (see Fig. 13.2). It is best identified by use of choline acetyltransferase immunohistochemistry (Mesulam et al. 1989, see Fig. 13.2).

Mostly from animal studies it is known that the cholinergic neurons of the laterodorsal tegmental nucleus are critical for the regulation of wakefulness and REM (rapid eye movement) sleep (Van Dort et al. 2015). Bueno et al. (2019) have shown the laterodorsal tegmental nucleus of the rat to be linked with the habenular nuclei pointing to the involvement of the laterodorsal tegmental nucleus in the regulation of aversive behaviors.

13.2 Alar r1

13.2.1 Vestibular r1

13.2.1.1 Superior vestibular nucleus

For detailed information on the superior vestibular nucleus, see Sect. 11.2.2.1.

13.2.2 Monoamine nuclei r1

13.2.2.1 Locus caeruleus/Caerulean nucleus [Locus caeruleus] (LC) A6

13.2.2.1.1 Location and morphology of the LC

The LC is the largest noradrenergic cell group (A6) in the brainstem (Fig. 13.3).

Due to its high density of neuromelanin-containing (see Box 13.5) neurons, it is easily detectable even at low magnification. A specific identification of the nucleus can be achieved using tyrosine hydroxylase as an immunohistochemical marker (see Fig. 3.3). The LC is located in the pontine tegmentum in the laterodorsal corner of the fourth ventricle and the mesencephalic aqueduct. Neighboring structures (see Fig. 13.3) are, laterally, the superior cerebel-

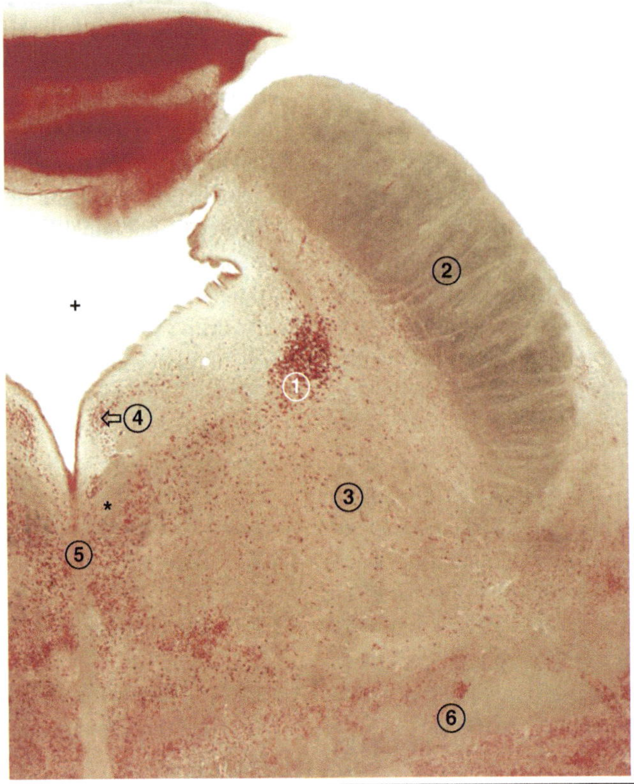

①	**Locus caeruleus**	⑤	**Dorsal raphe ncl.**
②	**Superior cerebellar peduncle**	⑥	**Medial lemniscus**
③	**Central tegmental tract**	*	**Medial longitudinal fasciculus**
④	**Dorsal raphe ncl.**	+	**4th ventricle**

Fig. 13.3 Horizontal section through the human pons at a mid-pontine level. Darrow red staining. The heavily stained neurons of the locus caeruleus are easily detected. In the more inferior parts, the LC looks like a boomerang with one wing oriented laterodorsally—in direction to the superior cerebellar peduncle and the other one oriented ventrolaterally. See also atlas part Darrow red 28. LabPON Twente

lar peduncle (Fig. 13.3 ②), more rostrally, the cuneiform nucleus (see Sect. 14.4.1.1), medially, the dorsal tegmental nucleus (see Sect. 13.3.1.1) and the periaqueductal gray (PAG) (see Sect. 13.2.7.1), dorsolaterally, the mesencephalic trigeminal nucleus (see Sect. 13.2.6.1), ventrolaterally, the medial parabrachial nucleus (see Sect. 13.2.3) and, ventromedially, the medial longitudinal fasciculus and the oral pontine reticular nucleus (see Sect. 10.3.5.4). The LC extends from the rostral pole of the trigeminal motor nucleus (see Sect. 11.3.1) to the caudal border of the cuneiform nucleus (see Sect. 14.4.1.1).

Older reports have it that the location of the locus caeruleus in the floor of the fourth ventricle/rhomboid fossa can be identified on the macroscopic level due to the neuromelanin-induced dark staining of the region (see Fig. 13.4).

Chan-Palay and Asan (1989a) have shown the human LC cells to belong to large multipolar neurons with round or multiangular somata, large elliptical "bipolar" neurons, small multipolar neurons, and small ovoid "bipolar" neurons. Most of the neurons contain neuromelanin pigment, but some larger neurons lack pigmentation. Dendritic arborization of all neurons is extensive.

The LC extends from the rostral pole of the trigeminal motor nucleus (see Sect. 11.3.1) to the caudal border of the cuneiform nucleus (see Sect. 14.4.1.1).

The Latin word caeruleus (coeruleus, ceruleus) means dark blue (see Fig. 13.4).

13.2.2.1.2 Functional aspects of the LC

Based on the connectivity data (see Sect. 13.2.2.1.3) (Figs. 13.5 and 13.6), the noradrenergic innervation plays a pivotal role in arousal, attention, stress response, and emo-

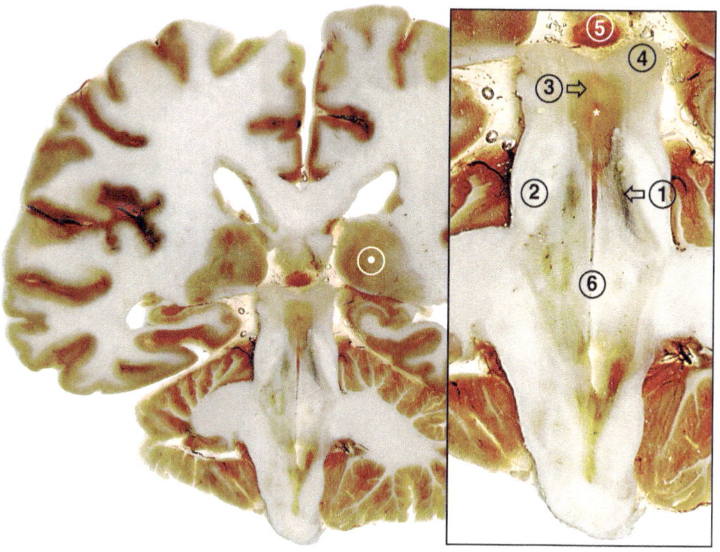

①	**Locus caeruleus**	⑤	**Pineal gland**
②	**Superior cerebellar peduncle**	⑥	**Rhomboid fossa / 4th ventricle**
③	**Periaqueductal gray**	*	**Mesencephalic aqueduct**
④	**Superior colliculus**	⊙	**Pulvinar**

Fig. 13.4 Coronal section through the human brain at the level of the pulvinar of the thalamus ⊙. Plastinated section. Due to the high neuromelanin (see Box 13.5) content of the LC, its location can be seen even in macroscopic specimen and its longitudinal extension can be estimated. Compare this anatomical specimen with the MRI data in Fig. 13.29. Sammlung des Zentrums Anatomie der Universität zu Köln

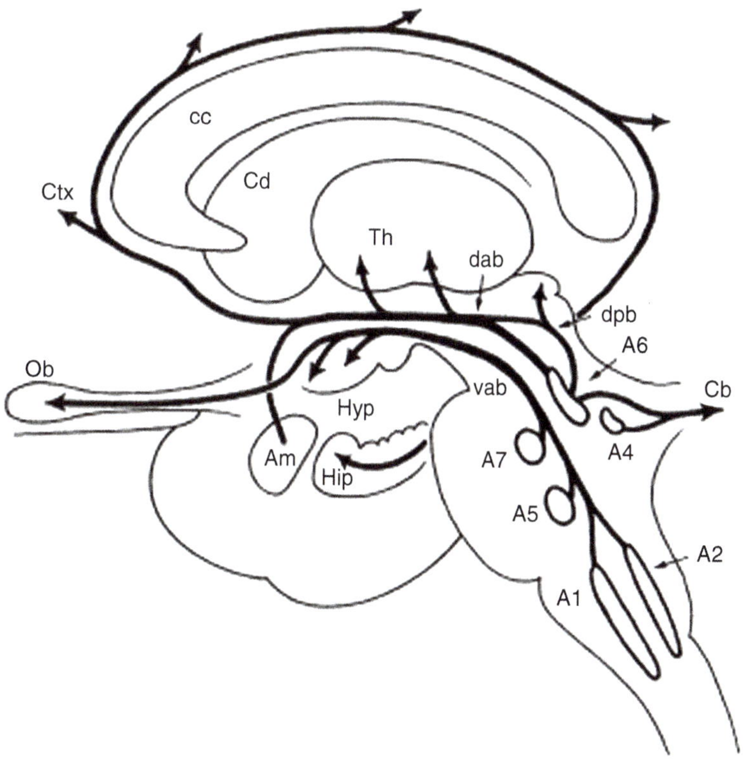

Fig. 13.5 Schematic representation of the ascending noradrenergic projections. The locus caeruleus (A6) projects via the dorsal ascending bundle (dab) to the thalamus (Th), the cerebral cortex (Ctx) and the amygdala (Am). A descending bundle is projecting to the cerebellum (Cb). Noradrenergic cells in medulla oblongata (A1, A2); Noradrenergic cells in caudolateral pons (A4); Noradrenergic cells of superior cerebellar peduncle (A5); Caerulean nucleus (A6); Noradrenergic cells in nucleus of lateral lemniscus (A7); Dopaminergic cells in retrorubral area (A8); Dopaminergic cells in compact part of substantia nigra (A9); Dopaminergic cells in ventral tegmental area (A10); Dopaminergic cells of periaqueductal gray (A11); *cc* Corpus callosum, *Cd* caudate nucleus, *dpb* dorsal periventricular bundle, *Hip* hippocampus, *Hyp* hypothalamus, *Ob* olfactory bulb, *vab* ventral ascending bundle. From ten Donkelaar 2011, Fig. 5.11 with permission

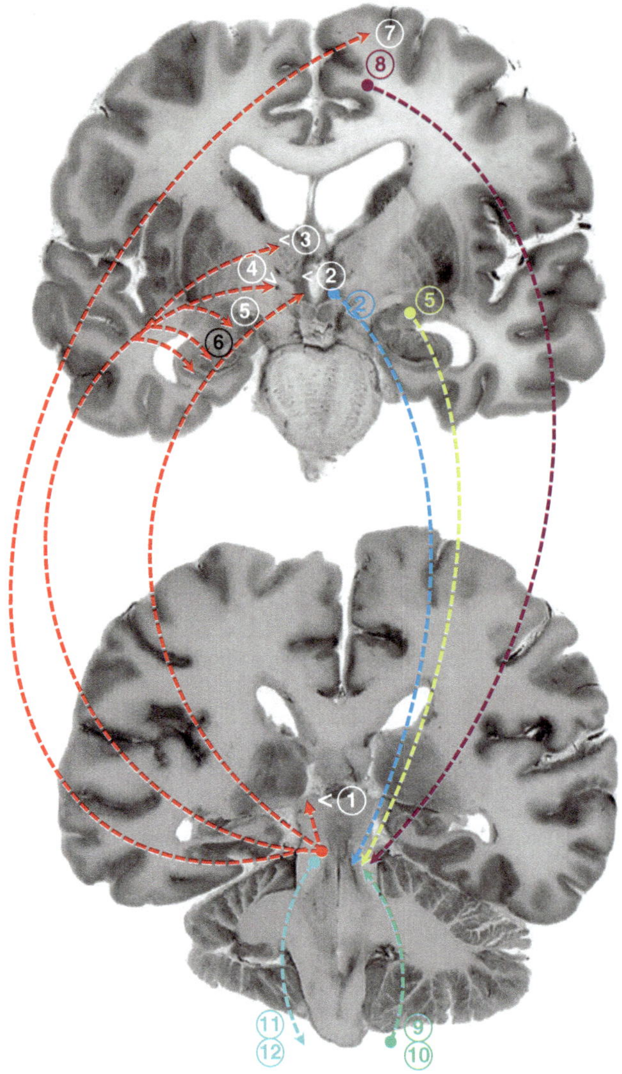

①	Colliculus superior	⑦	Prefrontal / frontal / parietal cortices
②	Hypothalamus	⑧	Medial prefrontal cortex
③	Thalamus	⑨	Rostral ventrolateral medulla C1
④	Midbrain dopaminergic groups	⑩	Dorsal horn lamina I
⑤	Amygdala	⑪	Medulla oblongata
⑥	Hippocampus / Parahippocampal cortex	⑫	Spinal cord

Fig. 13.6 Schematic representation of the most important efferent and afferent connections of the human locus caeruleus. Most of the data were derived from animal studies but the functionally and pathologically important innervation of the cerebral cortex has been corroborated for the human brain (Gaspar et al. 1989). The degeneration of the LC in Parkinson's and/or Alzheimer's disease results in the loss of noradrenergic innervation of the structures ①–⑦ in this graphic. H. Schröder fecit, modified acc. to Bennaroch 2018, Fig. 1 with permission

tional memory. Descending fiber bundles are involved in pain inhibition, motor control, and autonomic modulation (for details, see Bennaroch 2018).

13.2.2.1.3 Connectivity of the LC

The projections of its neurons innervate noradrenergically many forebrain structures including the cerebral cortex.

Most of the knowledge on efferent and afferent projections (see Fig. 13.5 and 13.6) is derived from animal studies.

In man, the somatosensory and motor cortices receive the most extensive noradrenergic innervation (Gaspar et al. 1989, Morrison and Foote 1982).

After having dealt with the other noradrenergic nuclei of the brainstem (A1, A2, A4, A5) in the previous chapters, Fig. 13.5 here provides a summary illustration of these nuclei and their target structures.

13.2.2.1.4 Pathological aspects of the LC/Alzheimer's disease

Loss of LC neurons with age has been described (Chan-Palay and Asan 1989a; Manaye et al. 1995) with the forebrain-projecting neurons undergoing loss to a greater extent than spinal cord-projecting ones (Manaye et al. 1995). Cell loss holds also true for chronically depressed patients without dementia (Chan-Palay and Asan 1989a). The most relevant disorder with regard to the locus caeruleus, however, is Alzheimer's disease (see Boxes 13.3 and 13.4).

Box 13.3 Alois Alzheimer

Alois Alzheimer was born on June 14th, 1864 in Franconia/Germany. He was a psychiatrist and neuropathologist and a collaborator of Emil Kraepelin, who coined the name "Alzheimersche Krankheit" (Alzheimer's disease). In 1907, he published the clinical and pathohistological findings on his famous patient, Auguste D. She had been hospitalized at the age of 51 years because of paranoia and extreme jealousy. He meticulously documented the course of her disease and she died several years later with full-blown dementia. Upon autopsy, Alzheimer detected the changes that still today are the pathological hallmarks of the disease (see Sect. 13.2.2.1.4). Macroscopically, he found an atrophy of the brain (see Figs. 13.7 and 13.8). Histologically, he described in Bielschowsky-stained sections (see Fig. 13.12C) a thickening and aggregation of neurofibrils, neurofibrillary tangles' (see Figs. 13.17–13.20). In addition, he described numerous miliary aggregations ("miliare Herdchen"), the changes that today are called amyloid plaques (see e.g., Figs. 13.10, 13.11, and 13.12) as well as endothelial changes (Fig. 13.15). In 1912, he was appointed Professor of Psychiatry at the University of Breslau (today Wrocław/Poland). On the journey, there he contracted a massive infection and died of pneumonia in Breslau on the 19th of December 1915 at the age of only 51 years (Schuchart 2017).

Box 13.4 Important facts about Alzheimer's disease (AD)

Alzheimer's disease is a variable disorder where genetic forms can start in the presenium and the risk on sporadic forms increases with age. The characteristic neuropathological changes are deposits of amyloid β proteins (plaques) and aggregates of hyperphosphorylated protein tau which are thought to impair normal neuronal and/or vascular function, ultimately resulting in neuronal loss. Several diagnostic algorithms for the pathologic changes can be used, such as Braak staging for tau, NIA-Reagan criteria for neuritic plaques, Thal phases for amyloid beta, and a combination of these (ABC-criteria). Although AD is predominantly a disease of the cerebral cortex, several groups suggest that the earliest manifestation of the disease starts in the lower brainstem. A definite diagnosis of Alzheimer's disease depends upon neuropathological evaluation. Diagnostic criteria such as biomarkers, including imaging and PET scanning, are increasingly used for the preclinical diagnosis of AD. CSF-biomarkers t-tau, p-tau, and Aβ42 reflect the pathological features of AD and evaluation of proteomic techniques to establish blood biomarkers are ongoing. Highly sensitive immunoassay techniques measuring plasma p-tau are promising techniques in diagnostic research.

The most common cause of dementia worldwide is Alzheimer's disease (AD). Many patients with dementia, however, now that the population is getting older, have a combination of abnormalities in their brains associated with AD, that is senile plaques, neurofibrillary tangles and cerebral amyloid angiopathy, as well as varying degrees of ischemic and/or hemorrhagic cerebrovascular disease. These are commonly referred to as "mixed dementias" (Vinters et al. 2018; Neal 2012; Esiri et al. 1997; Snowdon 1997). Pure vascular dementia is an infrequent diagnosis in neuropathological studies (McAleese et al. 2016).

The majority of patients with AD presents with disturbances of recent memory for 1–3 years and with the progression of the disease visuospatial, language, and attention disturbances can be seen. Many patients develop agitation or depression.

Diagnostic criteria to be used for AD have been developed. A workgroup presented criteria for all-cause dementia and for AD dementia. Biomarker evidence was also integrated into the diagnostic formulations for probable and possible AD for use in research settings (McKhann et al. 2011). A few years later, in 2014, recommendations have been made for the clinical diagnostic criteria. The term symptomatic AD is used to refer to all clinically expressed phases of

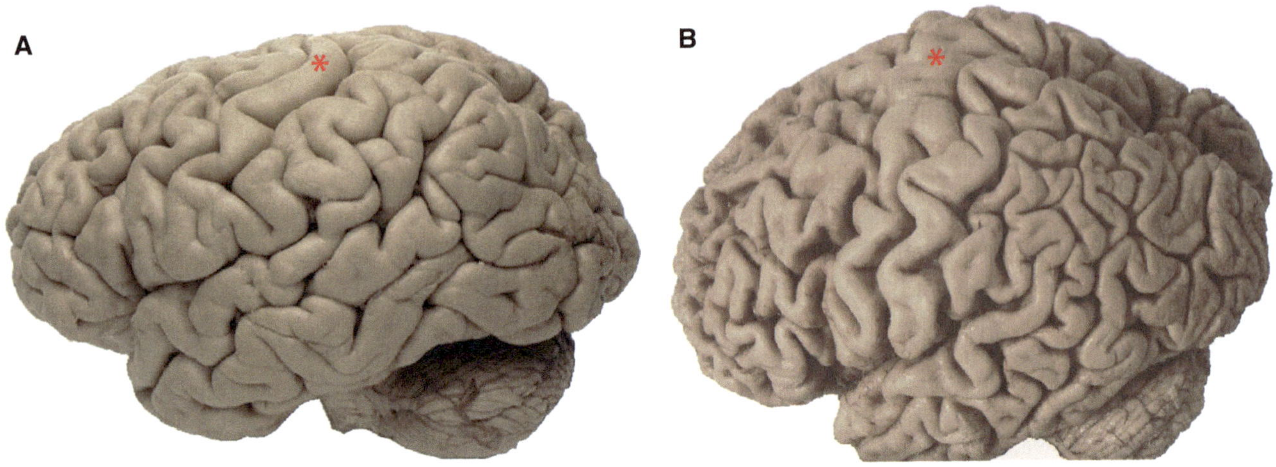

Fig. 13.7 Alzheimer's disease (AD), cortical atrophy. Lateral view of a control brain (**A**) and a patient with Alzheimer's disease (**B**). Relative sparing of primary motor and sensory areas (*). Severe atrophy is especially seen in familial cases. In the end stage of the disease, there is marked atrophy with enlargement of the sulci and flattening of the gyri. LabPON Twente

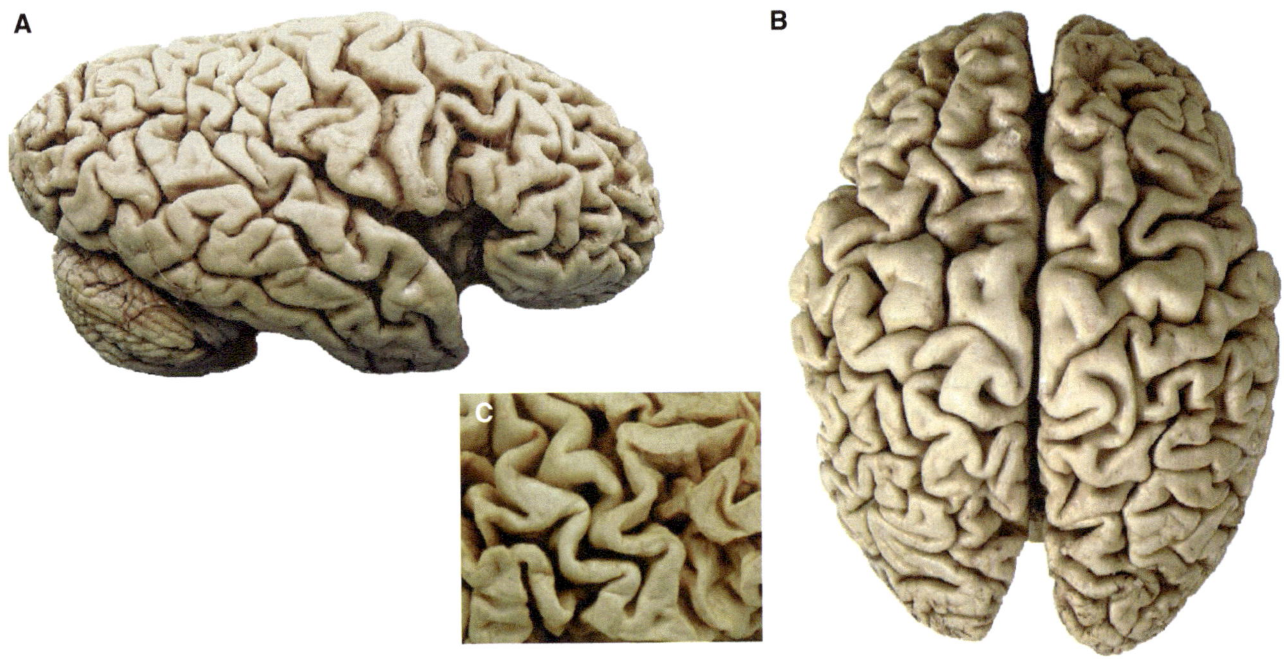

Fig. 13.8 Familial Alzheimer's disease: brain seen from lateral (**A**) and above (**B**) Detail cortex: enlargement of the sulci, flattened gyri (**C**). LabPON Twente

the disease, including its prodromal stages and also to consider incorporating biomarkers into diagnostic algorithms for AD and to allow non-amnestic, atypical presentations to be included as symptomatic AD, especially when there is supportive biomarker evidence (Morris et al. 2014).

AD biomarkers

Despite promising neuroimaging studies, definite diagnosis of AD depends upon neuropathological evaluation. Clinical diagnostic criteria are increasingly being interpreted based on biomarkers which could also be used for the preclinical diagnosis of AD (Lowe and Kalaria 2015). Two groups of biomarkers have been proposed (Jack and Barrio 2013; Rosen et al. 2013; Lowe and Kalaria 2015):

1. Biomarkers of brain Aβ deposition: amyloid PET scanning and reduced CSF Aβ42 level.
2. Biomarkers of neurodegeneration: atrophy on structural MRI-scanning, elevated levels of CSF total tau (t-tau) and phosphorylated tau, and hypometabolism on FDG-PET scans. The CSF biomarkers t-tau, p-tau, and Aβ42 reflect the pathological features of AD.

This framework for assessment of AD biomarkers has resulted in the AT(N) classification (Jack et al. 2018). Evaluations of a range of potential markers for aspects of AD development, such as Aβ, neuroinflammation, and neuronal dysfunction as well as evaluation of proteomic techniques to establish biomarkers in the blood are ongoing (Lowe and Kalaria 2015; Fagan and Perrin 2012; Bazenet and Lovestone 2012). Recently, the measurement of p-tau in plasma was shown to be a promising noninvasive and cost-effective AD biomarker. Likewise, highly sensitive immunoassays measuring plasma pT181-tau and pT217-tau effectively predicted tau and Aβ pathologies and could discriminate AD from other neurodegenerative disorders with similar accuracy as CSF or PET biomarkers (Karikari et al. 2020; Barthélemy et al. 2020; Palmqvist et al. 2020).

Genetic factors in AD

Genetic factors play a role in at least 80% of cases. Early-onset Alzheimer's disease (≤60 years) may rarely have autosomal dominant mutations in the amyloid precursor gene on chromosome 21. Patients with Down's syndrome have a highly increased risk to develop AD, probably reflecting the impact of the APP-gene on amyloid formation. Interestingly, a mutation in the APP gene (A673T) that decreased Aβ production was found to be protective against AD (Martiskainen et al. 2017; Jonsson et al. 2012). Other familial forms are linked to mutations on chromosome 14, whose gene is called Presenilin 1 (40% of all cases of early-onset Alzheimer's disease). This gene is also involved in amyloid β42 production, the more toxic amyloid form. In a few families, mutations in a similar gene, called Presenilin 2, are found on chromosome 1. APP, Presenilin 1 and Presenilin 2 mutations play a role in only 5% of all cases.

The majority of Alzheimer cases is late-onset (older than 60 years) and associated with a growing range of gene polymorphisms without any known cause and inheritance pattern (Gerrish et al. 2012). An increased risk of late-onset AD is associated with inheritance of either the apolipoprotein E4 (APOE4) allele, linked to chromosome 19, increasing the risk 12 times in homozygous individuals (Liao et al. 2017). Recent Genome Wide Association Studies (GWAS) have found more than 20 polymorphic genes, involved in immune response, lipid and APP metabolism, endocytosis, and synaptic function (Kunkle et al. 2019). Next to apolipoprotein E, the triggering receptor expressed on myeloid cells 2 (TREM2) was found as a major risk factor. This protein triggers APOE signaling which induces a microglia phenotypic switch from a homeostatic to a disease-associated phenotype. Furthermore, bridging integrator-1 (BIN1) was found as risk factor. This protein plays an important role in the pathogenesis of AD, probably by promoting tau pathology.

Neuroimaging in AD

Even though magnetic resonance imaging (MRI) provides useful information, it remains difficult to distinguish between the changes associated with normal aging and those that may cause or be related to the dementing process. In normal aging, which begins around the fifth to sixth decades, the ventricles become enlarged due to age-related neuronal loss. This process of atrophy does not occur uniformly throughout the brain. For example, the earliest changes are seen in the posterior fossa, especially in the vermis. By the early sixth decade, the third ventricle begins to enlarge, with enlargement of the cortical sulci and cisterns. The frontal cortex atrophies with age more rapidly than do the cerebral hemispheres as a whole. MRI studies indicate that, compared with normal control subjects, patients with AD exhibit global brain atrophy that is most pronounced in the anterior temporal lobe and the hippocampal formation. Hippocampal atrophy with dilatation of the lateral ventricles, is an early characteristic of AD but also occurs in FTLD (Frontotemporal lobar degeneration, see Sect. 4.3.1.1.5) (van de Pol et al. 2006). 8F-Fluorodeoxyglucose positron emission tomography (FDG-PET) commonly shows reduced glucose metabolism in the posterior cingulate and precuneus and in the parietotemporal regions. These changes may appear early in the disease course, and in patients with a diagnosis of mild cognitive impairment and earlier than hippocampal atrophy on morphological imaging (Petrella et al. 2003; Haller et al. 2013). Imaging studies using amyloid agents and other innovative techniques such as tau imaging, mass spectroscopy for CSF biomarker analysis, and others can detect early abnormalities in AD (Cohen and Klunk 2014; Weiner et al. 2017).

The characteristic neuropathological changes are deposits of amyloid and aggregates of hyperphosphorylated protein tau (see below). These changes are thought to impair normal neuronal and/or vascular function ultimately resulting in neuronal loss.

Macroscopic features of AD

Although the macroscopic abnormalities can vary between individuals, in AD disease, the brain is frequently smaller than average, with moderate to marked atrophy of the medial temporal lobe on cut surface (Fig. 13.9). Severe atrophy is especially seen in familial cases. In the end stage of the disease, there is marked atrophy with enlargement of the sulci and flattening of the gyri (Figs. 13.8 and 13.9). The brain may weigh up to 150–600 g less than that of a mentally "normal" age-matched individual (the normal weight is about 1400 g). Atrophy is especially present in the frontal, temporal, and parietal lobes while the occipital cortex is relatively spared.

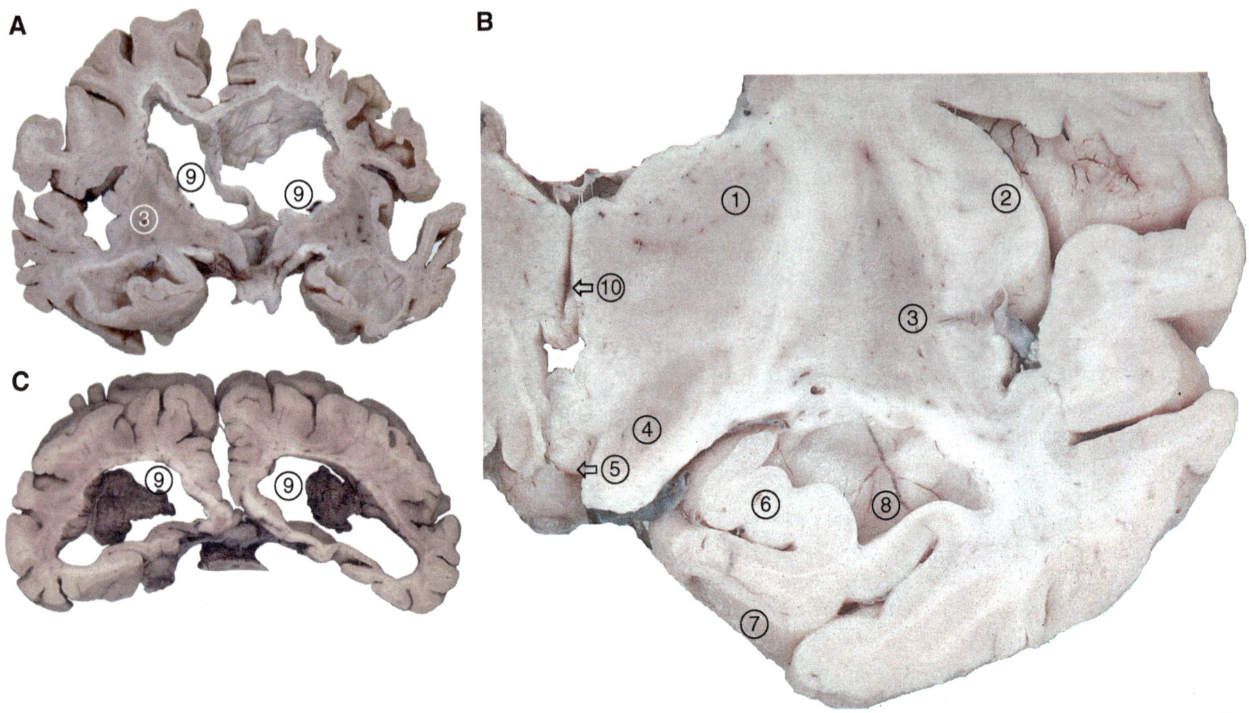

①	Thalamus	⑥	Hippocampus
②	Insula	⑦	Entorhinal cortex
③	Putamen	⑧	Temporal horn of lateral ventricle
④	Subthalamic ncl.	⑨	Lateral ventricle
⑤	Mammillary body	⑩	3rd ventricle

Fig. 13.9 Coronal sections. Frontal (**A**) and parietal level (**C**) and detail (**B**) at the level of the mammillary bodies ⑤. Atrophy with resultant dilatation of the ventricle system in Alzheimer's disease. Atrophy of the entorhinal cortex ⑦, shrinkage of the hippocampus ⑥ and widened temporal horn of the lateral ventricle ⑧. LabPON Twente

Microscopic features of AD

The microscopic hallmarks of AD are:

- **Amyloid deposits (senile plaques),** composed of Aβ peptide aggregates, derived by proteolytic breakdown from a neuronal membrane protein, the ß-amyloid precursor protein (APP) (Fig. 13.10).

The length of the Aβ protein can vary between 38 and 42 amino acids. Senile plaques show a morphologically heterogenous pattern in different brain regions. All types of Aβ deposits are referred to as senile plaques regardless of their morphology and can also be seen in cognitively normal individuals. The major difference between deposits in controls and those found in AD is their distribution. There are many different plaque types (Thal et al. 2006), for practical purposes, to be divided into diffuse, cored, and neuritic subtypes (Figs. 13.11 and 13.12). Of plaque-associated phagocytic microglia, a few can be seen in diffuse non-neuritic plaques (5%), but most are found in diffuse neuritic plaques (62%) and dense-core neuritic plaques (33%) (Griffin et al. 1995) (Fig. 13.13).

Coarse grained plaques and cotton wool plaques are rarer and often related to genetic mutation (Boon et al. 2020) (Fig. 13.12E, F).

Distribution of Aβ over the brain follows a specific pattern (Thal phases) (Thal et al. 2002) (Fig. 13.13B). The morphological appearance of distinct plaque types is also related to their anatomical distribution.

So-called lake-like amyloid occurs exclusively in the presubicular region (Fig. 13.14). Subpial band-like amyloid occurs in the subpial zone in all parts of the isocortex as well as in the entorhinal region and the cingulate gyrus in phase 3 (Figs. 13.10–13.14).

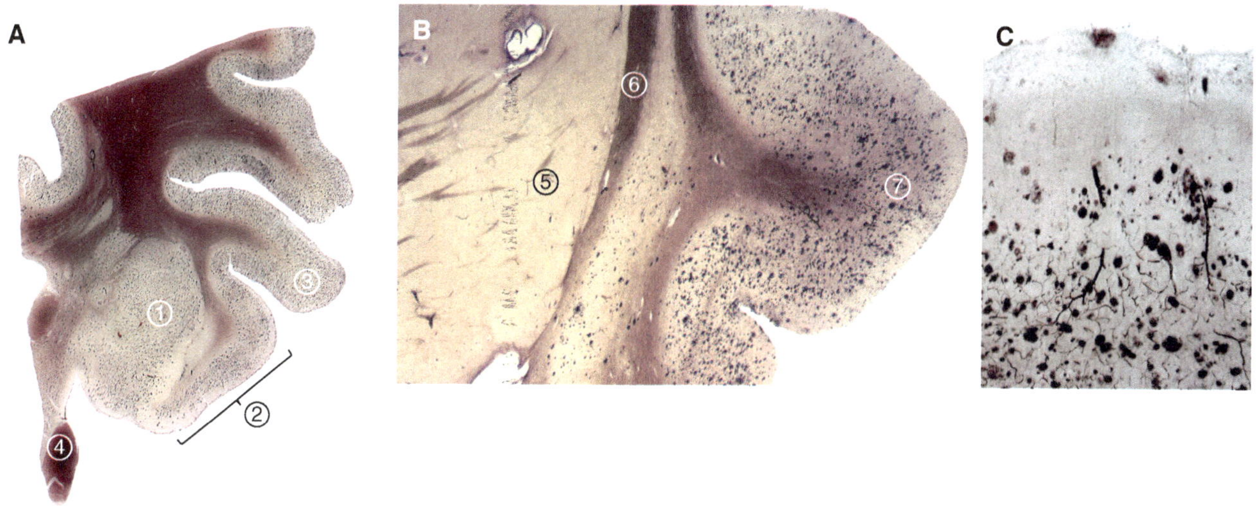

①	Amygdala	⑤	Putamen
②	Entorhinal cortex	⑥	Claustrum
③	Parahippocampal gyrus	⑦	Insular cortex
④	Optic tract		

Fig. 13.10 Alzheimer's disease. Aβ-deposits in all fields of the iso- and allocortex (**A**–**C**). PEG sections 100 μm, Campbell-Switzer staining the plaques against a pale background. LabPON Twente

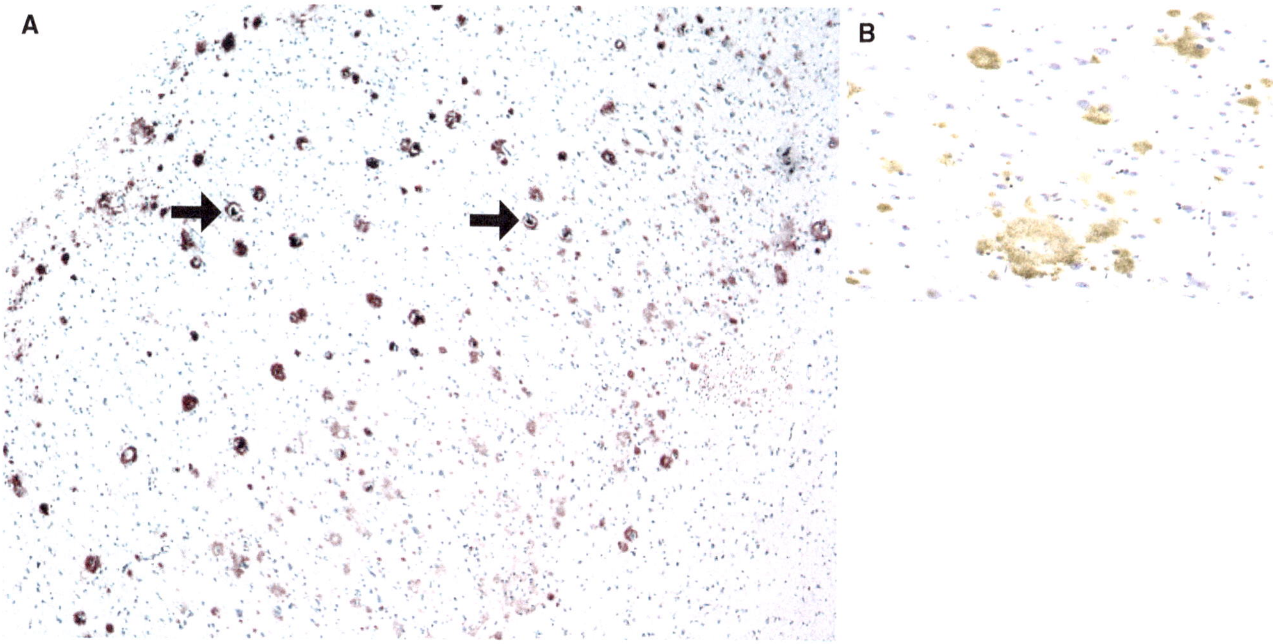

Fig. 13.11 Alzheimer's disease. Isocortex, Aβ-deposits in all layers (**A**), Aβ immunohistochemistry with Nissl counterstain. Cored plaques (arrows) and diffuse plaques (**B**). LabPON Twente

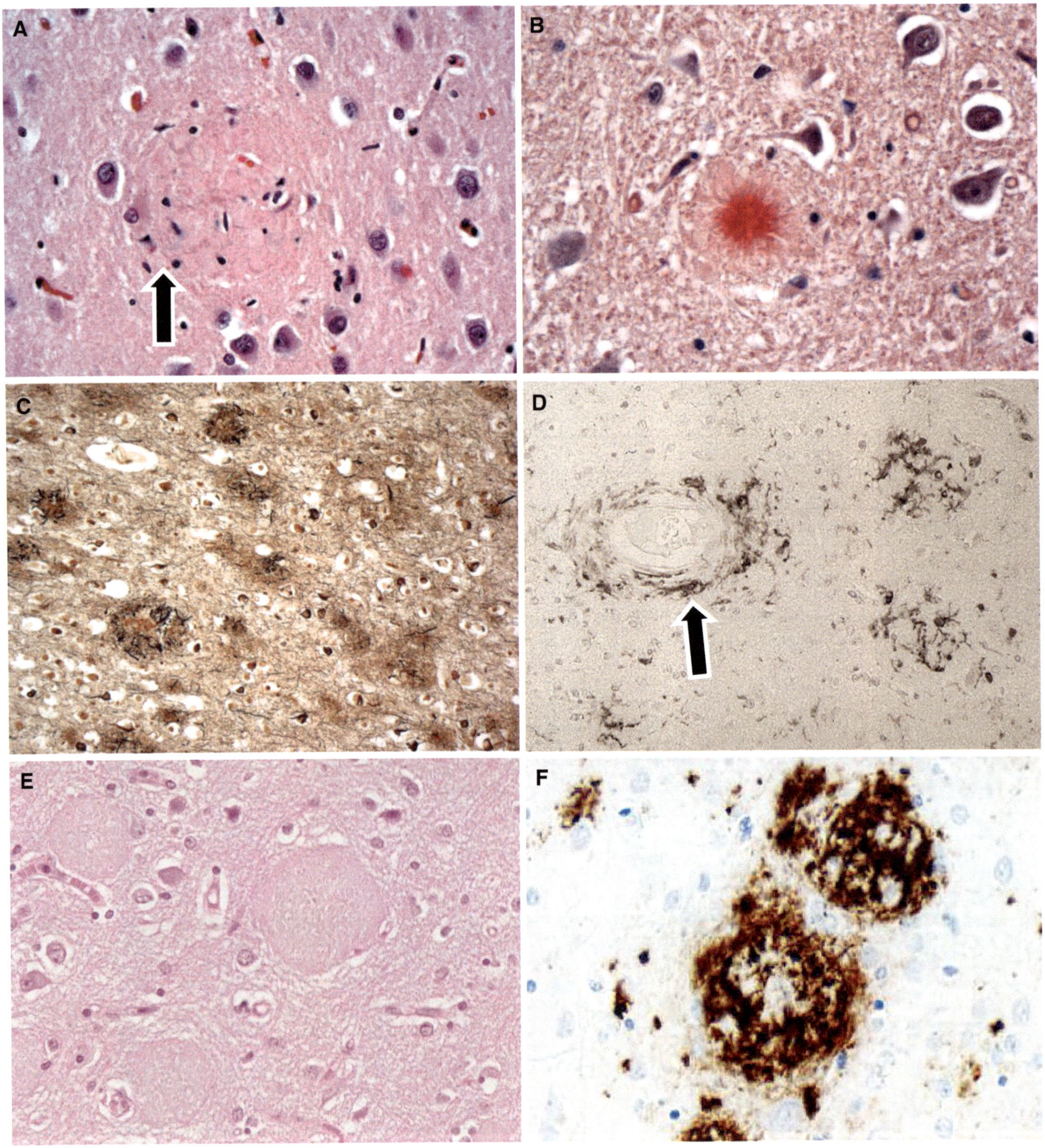

Fig. 13.12 Alzheimer's disease. In sections stained with hematoxylin and eosin (H & E), plaques are ill-defined areas of condensation in the neuropil, and some surrounding activated microglial cells (arrow) are seen in (**A**), plaques with a central core in (**B**). In a silver staining (Bielschowski), deposits can appear as cored plaques (**C**) and in CR3/43-immunostaining (**D**), a marker for activated microglia, cells are present surrounding neuritic plaques (arrow). Typical so-called cotton wool plaques in a H & E stain in (**E**). Coarse-grained plaques in (**F**) in an Aβ-immunostain (Nissl counterstain). Coarse-grainy deposits with the appearance of multiple cores. LabPON Twente

A

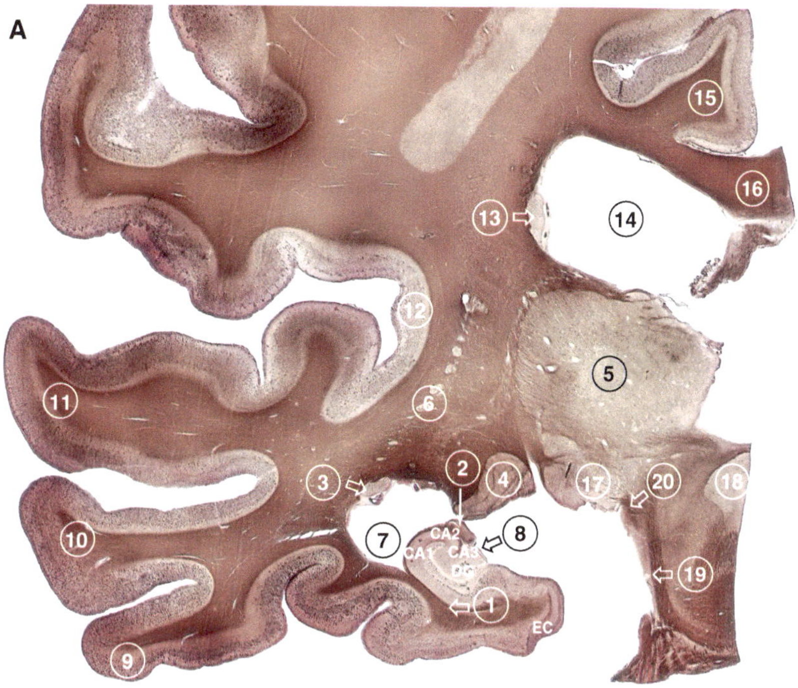

①	Subiculum	⑪	Superior temporal gyrus
②	Fimbria hippocampi	⑫	Insular cortex
③	Caudate ncl., cauda	⑬	Caudate ncl., cauda
④	Lateral geniculate body (LGB)	⑭	Lateral ventricle, central part
⑤	Thalamus	⑮	Cingulate gyrus
⑥	Putamen	⑯	Corpus callosum
⑦	Lateral ventricle, temporal horn	⑰	Medial geniculate body (MGB)
⑧	Fluffy material	⑱	Periaqueductal gray
⑨	Inferior temporal gyrus	⑲	Lateral lemniscus, entry into MGB
⑩	Middle temporal gyrus	⑳	Medial lemniscus

Fig. 13.13 Alzheimer's disease. (**A**) Left hemisphere in a Campbell Switzer staining. β-amyloidosis with black appearing Aβ-deposits (dots) in all fields of the isocortex, in the entorhinal region, the hippocampal formation, presubiculum, and the thalamus. Some fluffy material ⑧ can be seen along the free surface of the dentate gyrus (DG). Region of the hippocampus (CA, DG) and entorhinal cortex (EC) representing Thal phase 3 (see "Pathological staging—Diagnostic criteria of AD" and "Brainstem pathological changes in AD" here below). (**B**) Phases of β-amyloidosis acc. to Thal criteria (Thal et al. 2002; PEG 100 μm): phase 1 is characterized by exclusively isocortical Aβ deposits (Isocortex: black). Phase 2 shows additional allocortical Aβ deposits (red arrows), phase 3 additional Aβ deposits in diencephalic nuclei (red arrows) and the striatum (not shown), phase 4 additional Aβ deposits in distinct brainstem nuclei (red arrows), and phase 5 in the cerebellum and additional brainstem nuclei (red arrows) (for details, see "Pathological staging—Diagnostic criteria of AD" below). (**A**) LabPON Twente. (**B**) From Thal et al. 2002, Fig. 4 with permission

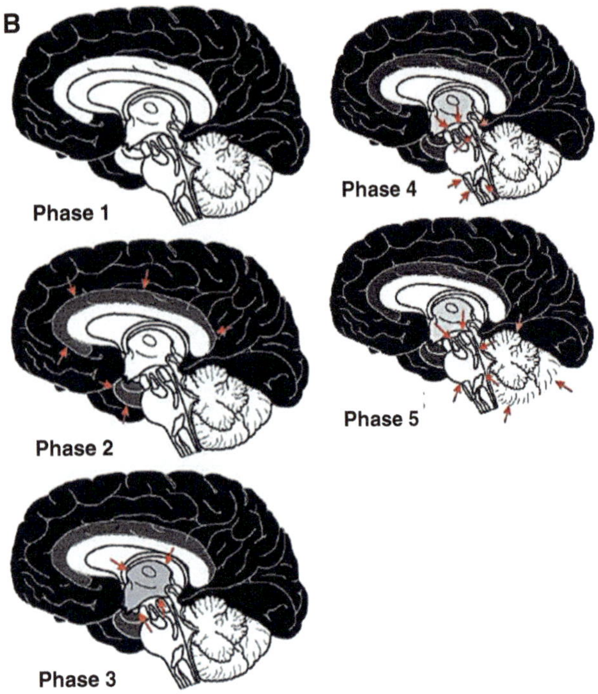

Fig. 13.13 (continued)

- **Cerebral amyloid angiopathy** (CAA)
  In many cases, amyloid deposits are present in vessel walls in the cerebral and cerebellar cortex and the leptomeninges. CAA can occur in all types of vessels including capillaries (Fig. 13.15).

Secondary changes may develop in affected microvessel walls. Capillary Aβ-deposits consist of Aβ deposits that are attached to the capillary basement membrane and can protrude into the neuropil in a plaque-like manner, the so-called dyshoric angiopathy (Surbek 1961) (Fig. 13.16).

CAA can lead to destruction of the vessel wall, focal fragmentations (double-barreled pattern) and the development of microaneurysms (see Fig. 13.15B) (Mandybur 1986). Amyloid plaques and CAA are not specific for symptomatic AD and can be seen in non-demented individuals. CAA (and Aβ plaques) develop first in isocortical areas and then expand in a characteristic sequence into further brain regions (Thal et al. 2015).

- **Neurofibrillary changes** can be found in neuritic plaques, neurofibrillary tangles, and neuropil threads. Different types of neurons develop different forms of

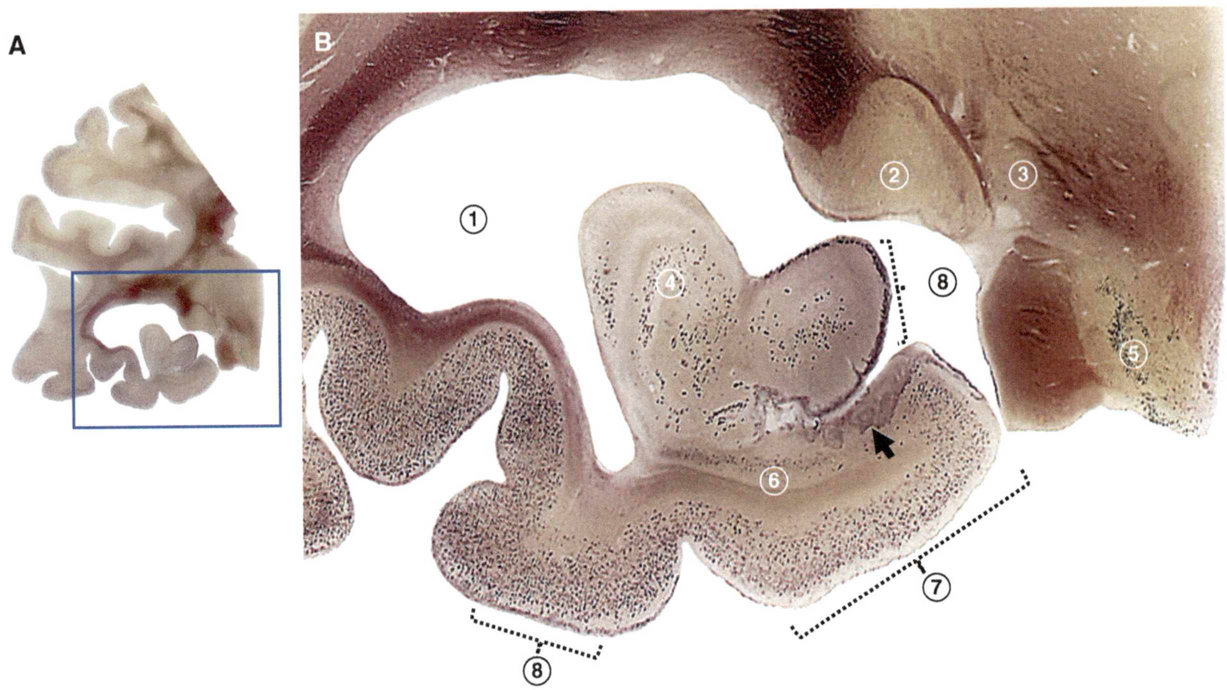

①	Temporal horn of lateral ventricle	⑤	Substantia nigra
②	Lateral geniculate body	⑥	Subiculum / CA1
③	Medial geniculate body	⑦	Entorhinal cortex
④	Dentate gyrus	⑧	Subpial band-like amyloid

Fig. 13.14 Alzheimer's disease. (**A**) Part of the left temporal lobe in a Campbell Switzer staining, Thal phase 3 (see "Pathological staging—Diagnostic criteria of AD" and "Brainstem pathological changes in AD" here below). (**B**) Hippocampal formation in detail. Presubicular lake-like amyloid (arrow) and subpial band-like amyloid ⑧. LabPON Twente

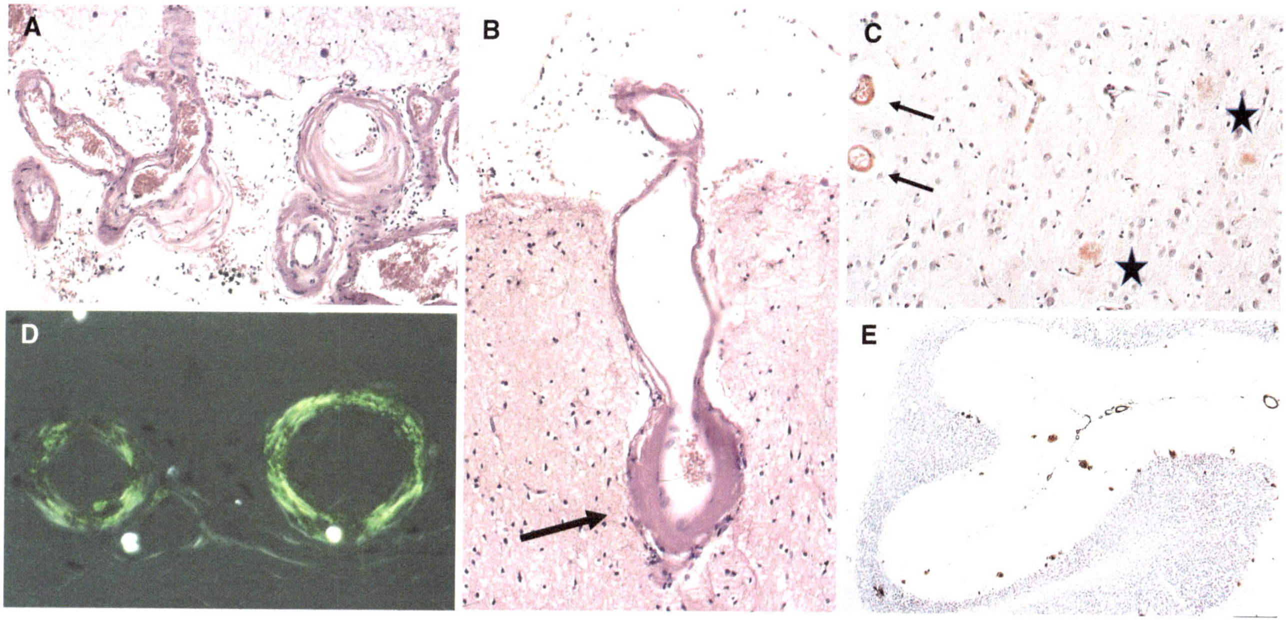

Fig. 13.15 Cerebral amyloid angiopathy. Thick-walled homogenous eosinophilic appearance of meningeal vessel walls (**A**), aneurysmatic dilatation (arrow, **B**), amyloid deposits in the neuropil (stars, **C**) as well as deposits in vessel walls in a Congo red staining (arrows, **C**). Amyloid in vessel walls shows apple-green birefringe in a Thioflavine S staining (**D**), amyloid in vessel walls in the cerebellum, immunostaining for Aβ peptide (**E**). LabPON Twente

neurofibrillary tangles (globular, flame-shaped, ghost) (Figs. 13.17 and 13.18).

The neurofibrillary tangles are composed of paired helical filaments and straight filaments which are formed from abnormally hyperphosphorylated tau proteins. Tau belongs to the family of microtubule-associated proteins and is involved in the physiological assembly and stabilization of microtubules. In the human brain, tau is produced from a single gene, located on chromosome 17q21. Six isoforms exist with either 3 or 4 microtubule-binding sites (3R/4R) and with or without one or 2 extra inserts in the N-terminal part. In neurodegenerative disorders, hyperphosphorylated tau proteins aggregate into intracellular filamentous inclusions, such as paired helical filaments in AD. Immunoblotting reveals the presence of a triplet of proteins at 55, 64 and 69 kDa. Neuropil threads are thread-like structures in the neuropil and form the earliest changes in AD (Fig. 13.19). Their density has been found to correlate with the severity of dementia (McKee et al. 1991). Neurons in the pre-tangle state show diffuse intracellular accumulation of hyperphosphorylated tau in the cytoplasm, which can be visualized immunohistochemically with antibodies such as monoclonal AT8. Tau starts to oligomerize when two identical tau motives form dimers (von Bergen et al. 2000), which can later form prefibrils and fibrils as seen in classical neurofibrillary tangles. Classical tangles, neuritic plaques, and neuropil threads can be visualized with the Gallyas silver staining (Figs. 13.18 and 13.19).

Ghost tangles (see Fig. 13.18C) are observed at late disease stages and reflect the accumulation of extracellular neurofibrils following the death of the tangle-bearing neuron (Uchihara 2014).

The tau molecule in AD has a unique 3D structure in cryogenic electron microscopy that is different from other tauopathies, each showing its unique tau conformation (Mukrasch et al. 2007; Saito et al. 2014). These different tau strains are probably the result of distinct post-translational modifications (Scheres et al. 2020; Falcon et al. 2018, 2019; Zhang et al. 2020).

- **Increased numbers of astrocytes and microglia**, especially surrounding cored and neuritic plaques (see Fig. 13.12D).
- **Granulovacuolar degeneration and Hirano bodies** (Fig. 13.20)

Although these changes are commonly found in AD, they are related more to aging in general than to a specific disease. Granulovacuolar degeneration (GVD) can be found before tangle formation, is present in the cell, and shows lysosomal structures that contain cytoskeletal proteins and proteins involved in the unfolded protein response.

Recently, Puladi et al. (2021) could show in the brains of elderly subjects - with Braak stages ranging from 0 to VI - that about 29% of AT8- (pre-tangle marker) immunoreactive neurons in the dorsal raphe nucleus and 31% in the locus caeruleus contained granulovacuolar bodies (GVB).

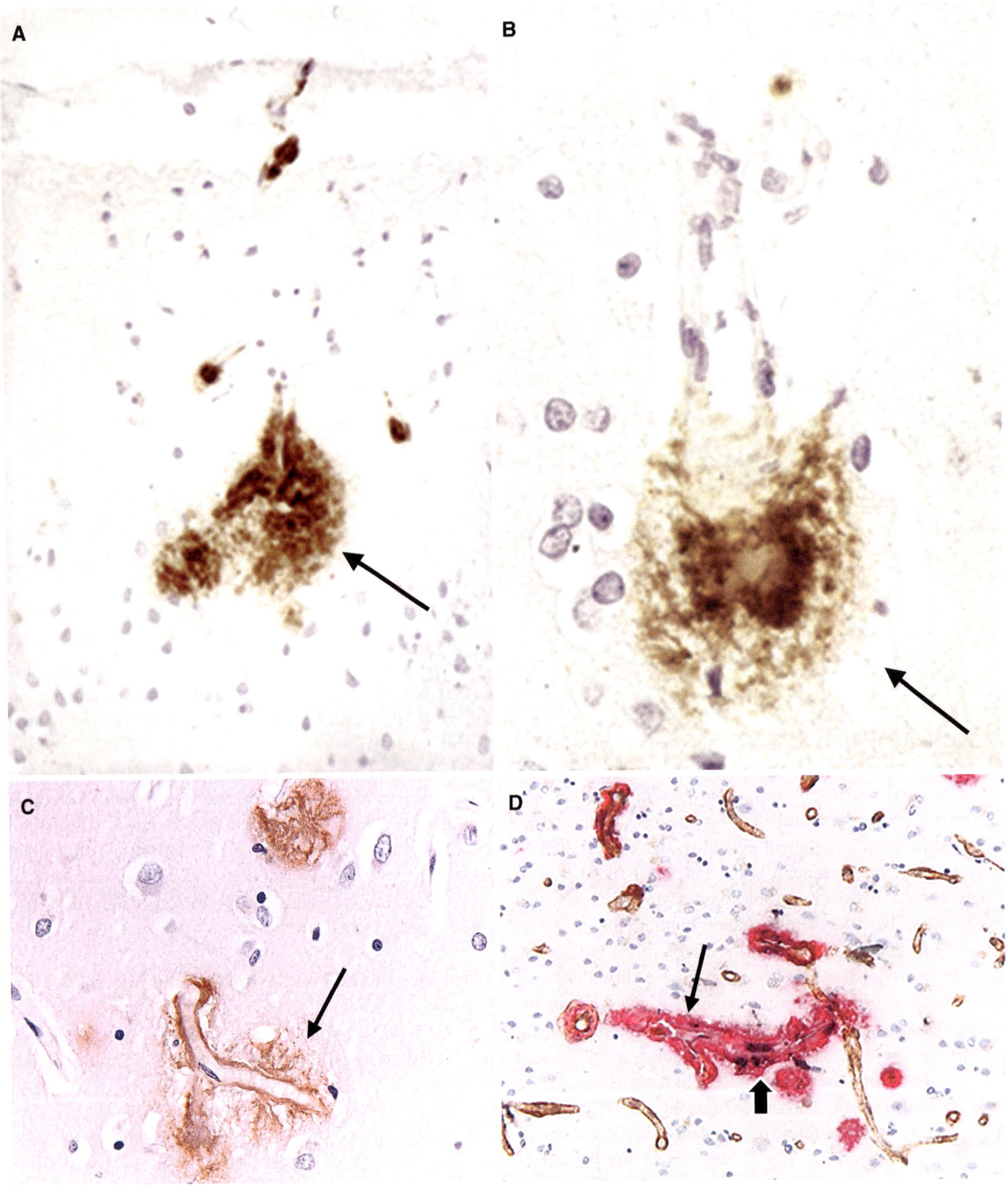

Fig. 13.16 Capillary amyloid angiopathy in capillaries in the leptomeninx (**A**) and in capillaries in the neuropil (**B** and **C**) displaying the typical picture of "dyshoric angiopathy," clearly showing amyloid, protruding into the neuropil (arrow) (Aβ-immunostain, Nissl counterstain). In (**D**), a triple stain Aβ (red), collagen IV (brown), and microglia (short thick arrow, blue) is seen. LabPON Twente

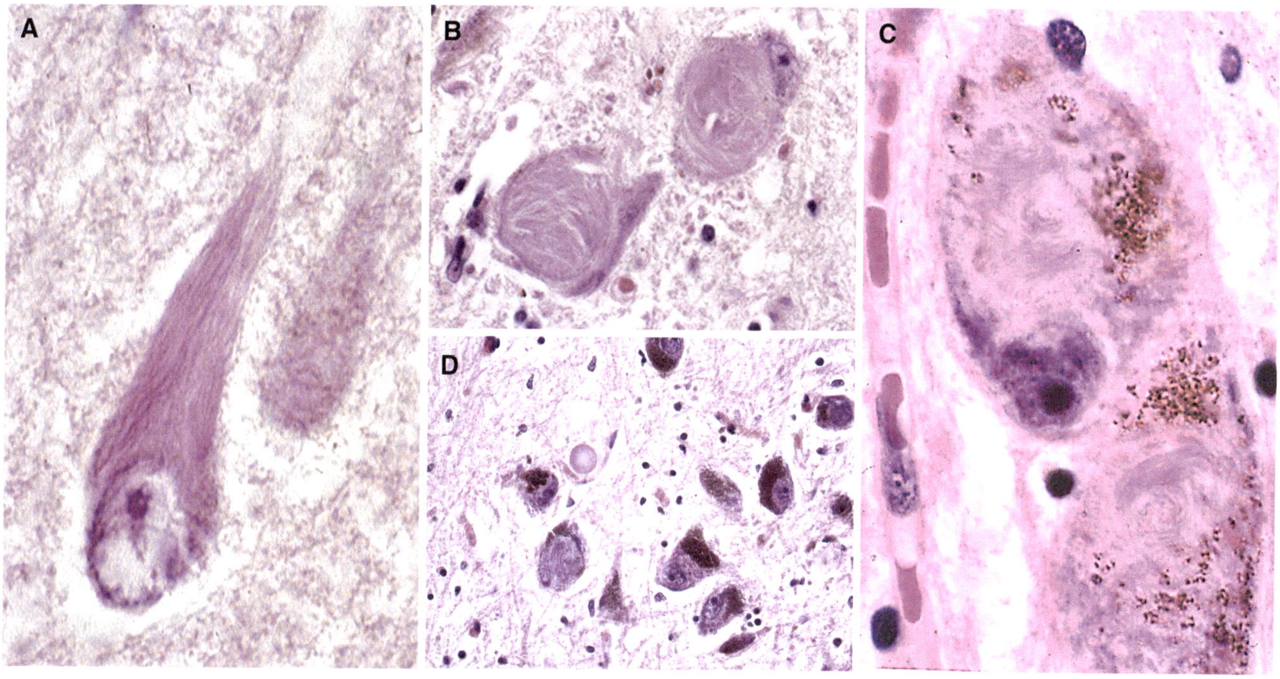

Fig. 13.17 Alzheimer's disease. H & E stain showing flame-shaped (**A**) and globosal-like older (**B**) neurofibrillary tangles in hippocampus and locus caeruleus, and globosal neurofibrillary tangles in the substantia nigra (**C** and **D**). LabPON Twente

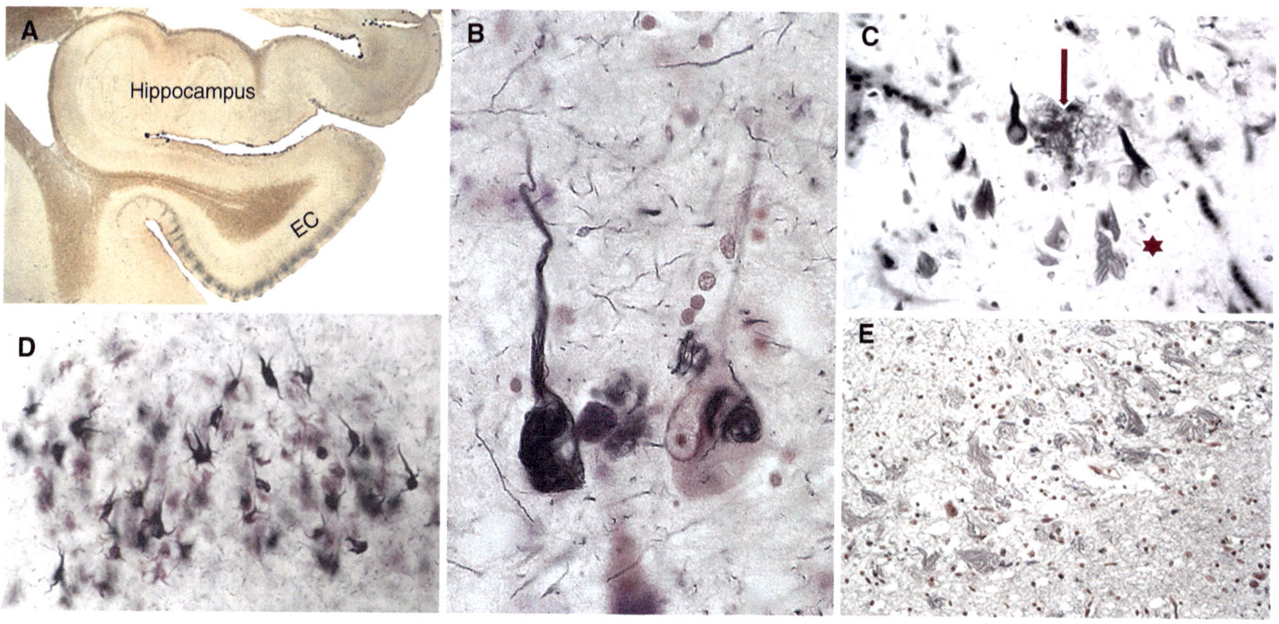

Fig. 13.18 Alzheimer's disease. (**A**) Overview of the entorhinal cortex (EC) (PEG 100 μm). Neurofibrillary tangles (NFTs) are virtually restricted to the transentorhinal and entorhinal region in the superficial pre-α layer (see Fig. 13.21), a transition zone between the entorhinal and temporal cortices. In (**B**), NFT in the subiculum, in (**C**) NFTs and a neuritic plaque (arrow) and a ghost tangle (star) are seen. In (**D**), NFTs in the pre-α entorhinal cell layer, in (**E**) ghost tangles in the entorhinal region are present. Gallyas silver staining. LabPON Twente

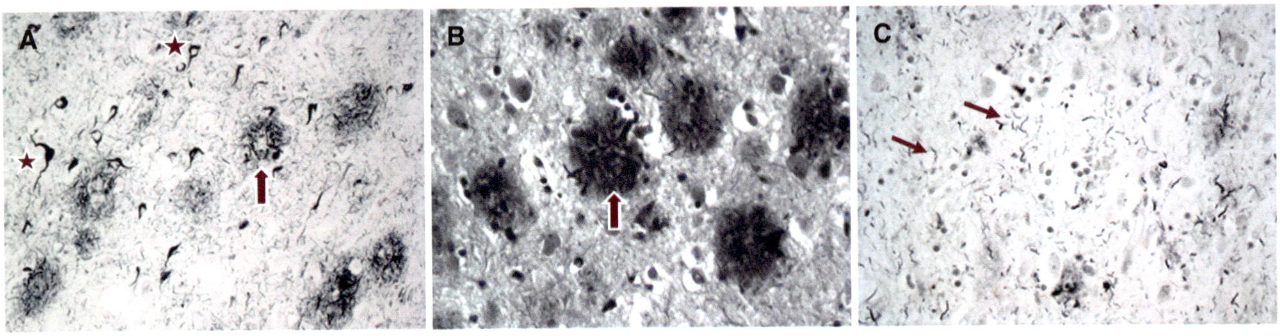

Fig. 13.19 (**A**–**C**) show the isocortex with neuritic plaques (arrow), neurofibrillary tangles (stars) and neuropil threads (thin arrows). Gallyas silver staining. LabPON Twente

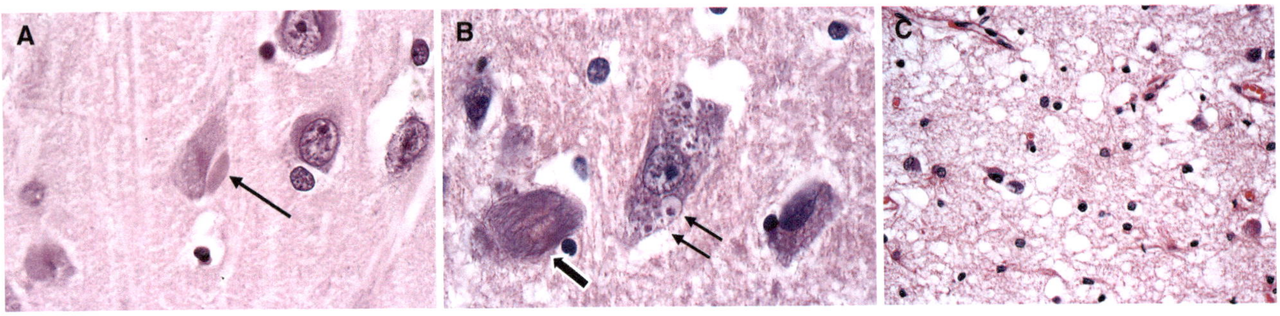

Fig. 13.20 Hirano body (arrow in **A**) and granulovacuolar degeneration (double arrows, **B**). The neuron toward the bottom left of (**B**) contains a neurofibrillary tangle (arrow). Irregular coarse vacuolation (spongiform change) in (**C**). H & E stain. LabPON Twente

Hirano bodies are eosinophilic, rod-shaped, ovoid, or spherical structures in dendrites. They are, like granulovacuolar degeneration, mainly located in the hippocampus (CA1) (see Fig. 13.20).

- **Synaptic dysfunction and neuronal loss**
  Synaptic density correlates with cognitive impairment in AD (Terry et al. 1991).
- **Spongiform change**
  A spongiform appearance usually involves the superior and inferior temporal and entorhinal cortices (Fig. 13.20C). It can also be seen in the amygdala.

Frequently found additional pathology

In Alzheimer's disease, additional α-synucleinopathy (abnormal accumulation of aggregates of α-synuclein protein in neurons, nerve fibers, or glial cells) can be found, especially in older patients. In those cases, the superior, inferior temporal cortex, (trans)entorhinal cortex, and amygdala can show spongiform changes. Even in young genetic AD patients, some Lewy bodies (see Sect. 16.6.1.6) can be found in the amygdala. Accumulation of the RNA-binding protein TDP-43 forms cytoplasmic inclusions in up to 70% AD cases (Josephs et al. 2016). Argyrophilic grains, spindle, or comma-shaped lesions are small 4-repeat tau-positive deposits in neuronal processes and often coexist with other neurodegenerative diseases, such as progressive supranuclear palsy (PSP see Sect. 4.1.1.3.3), Parkinson's disease (PD see Sect. 16.6.1.5), corticobasal degeneration (CBD, see Sect. 16.6.1.9), and AD. The frequency of argyrophilic grains in AD cases was reported up to 25% (Yokota et al. 2018).

Pathologic staging—Diagnostic criteria of AD

The definite diagnosis of AD remains to be a neuropathological one, as reliable clinical and radiological parameters do not predict the diagnosis with certainty. To determine the pathological alterations in AD, it is necessary to realize that amyloid and tau deposits can also be found in healthy non-demented elderly persons. Therefore, the diagnosis of Alzheimer's dementia must be related to the amount of changes present in the brain, age, and clinical picture. In order to search for a diagnostic (quantitative) cut-off level of pathological changes, several diagnostic algorithms were developed:

1. CERAD (Consortium to Establish a Registry for Alzheimer's Disease) criteria (Mirra et al. 1991; Khachaturian 1985, 2011; Powers 1997). These involve semi-quantitative scoring (sparse, moderate, or frequent) of the occurrence of neuritic plaques in three defined isocortical areas (superior temporal gyrus, medial frontal gyrus, and inferior parietal lobule). These age-related

plaque scores are used in combination with the clinical history of dementia to form criteria for possible, probable, or definite AD.

2. The Braak staging (Braak and Braak 1991) is based on the distribution pattern of neurofibrillary pathological changes: *Entorhinal (I and II)*, *Limbic (III and IV)*, and *Isocortical (V and VI)* (Fig. 13.21)
3. A combination of CERAD and Braak criteria is reflected in the NIA-Reagan criteria (Newell et al. 1999).
4. Thal criteria (Thal et al. 2002). Aβ-deposition follows a distinct sequence in which the regions are hierarchically involved (Fig. 13.13). Aβ-deposition expands anterogradely into regions that receive neuronal projections from regions already exhibiting Aβ (see "Brainstem pathological changes in AD"). In phase 1, there are Aβ deposits in the isocortex, for example, the frontal, parietal, temporal, or occipital cortex. In phase 2, there is an additional involvement of the allocortex, for example, entorhinal region, CA1, and in the insular cortex. In phase 3, Aβ deposits occur in the caudate nucleus, putamen, claustrum, basal forebrain nuclei, thalamus and hypothalamus, and white matter. In phase 4, several brainstem nuclei become involved, such as inferior olivary complex (see Sect. 4.1.1.3), reticular formation and the substantia nigra (see Sect. 16.6.1, see Fig. 13.28A), the central gray of the midbrain (see Sect. 13.2.7), the superior and inferior colliculi (see Fig. 13.22) (see Sects. 15.1.3 and 15.1.4), and the red nucleus (see Sect. 15.3.4). Phase 5 is characterized by Aβ deposits in the cerebellum.
5. The "ABC" criteria defined by NIA-AA (National Institute on Aging and Alzheimer's Association) as a practical approach, are a combination of Thal-criteria for amyloid, Braak staging for tau, and CERAD staging for neuritic plaques, each with a score of 1–3. Thal phase 1 and 2 are combined as A1 and Thal phase 4 and 5 are combined as A3. Likewise, Braak stage 1 and 2 become B1, Braak stage 3 and 4 become B2 and Braak stage 5 and 6 become B3 (Montine et al. 2012).

In the following, an explanation to the Braak staging is provided. AD is predominantly a disease of the cerebral cortex but also involves a specific set of subcortical nuclei including brainstem/hindbrain structures. The complex pattern of changes associated with AD develops gradually and more-or-less symmetrically in both hemispheres. In a study, Braak et al. (2011), Braak and Del Tredici (2011, 2012) found that younger age groups showed abnormal tau in the projection neurons of the locus caeruleus (see here above). They proposed a pretangle stage and assumed that the AT8-positive lesions in the brainstem are the earliest manifestations of the disease. Their data suggest that the tauopathy associated with sporadic AD begins earlier than previously thought and probably in the lower brainstem earlier than in the entorhinal region.

Previous investigations also found hyperphosphorylated tau during normal aging as well as indications that this finding becomes significantly more prominent in patients with mild cognitive impairment and early AD. It was concluded that the locus caeruleus is a brain region with a very high susceptibility to neurofibrillary degeneration (Grudzien et al. 2007). These data were confirmed in a more recent study analyzing the molecular characteristics of the locus caeruleus at early stages of neurofibrillary tangle pathology (Andrés-Benito et al. 2017) and in a quantitative study probing the correlation of neuronal loss, neurofibrillary tangles, and cell death markers across the Braak stages (Theofilas et al. 2018). Numerous studies already demonstrated that neuronal loss and synaptic loss are the best predictors of cognitive decline. Caspase-dependent autophagy pathways have been implicated in AD pathophysiology. The autophagy-lysosome pathway is one of two major systems responsible for degrading the cellular material and misfolded proteins, the other being the ubiquitin-proteasome system (Nijholt et al. 2011; Choi et al. 2013). Progressive changes in active caspase-6 and macroautophagy positivity during Braak stage progression and their co-occurrence with NFT burden was seen (Theofilas et al. 2018).

The original Braak staging of Alzheimer-related changes was published in 1991 (Braak and Braak 1991). The first two stages, the transentorhinal stages I and II, were characterized by mild or severe alterations of the transentorhinal layer Pre-α, a transition zone located between the entorhinal region and the adjoining temporal isocortex. Only mild onvolvement of the hippocampus (CA1) is seen (Fig. 13.21). The entorhinal region, hippocampal formation, and the olfactory areas are included in the allocortex. Recent studies however have argued (see above) that the pathologic process possibly begins with intraneuritic formation of pretangle material in the lower brainstem rather than in the transentorhinal region (Braak et al. 2011; Braak and Del Tredici 2011) (see Fig. 13.21). Stages III and IV, the limbic stages, are marked by a conspicuous involvement of the superficial layers of both the transentorhinal and the entorhinal cortex, with modest involvement of the first Ammon's horn sector (Figs. 13.21 and 13.23).

The presence of ghost tangles (see Fig. 13.18E) can be observed in the transentorhinal layer pre-α. The isocortex remains virtually devoid of changes or is only mildly affected. In most patients, there are minor changes in the magnocellular forebrain nuclei, anterodorsal nucleus of the thalamus, and amygdala. In stage IV, NFTs may also appear in large neurons located in basal portions of the putamen and the nucleus accumbens. The nucleus reuniens and the tuberomammillary nucleus are slightly more intensely affected,

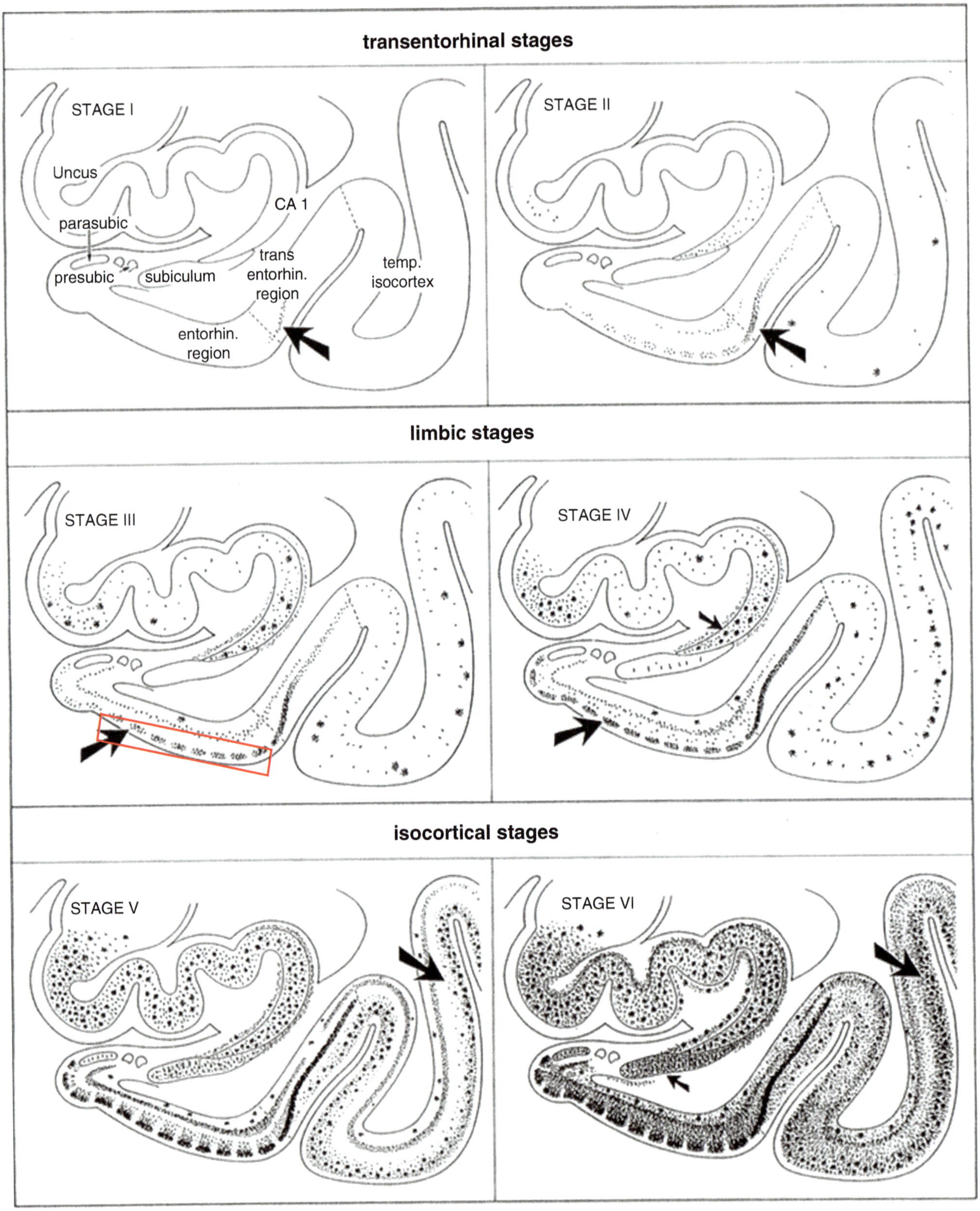

Fig. 13.21 Diagrams of cortical NFT-changes in stages I–VI of Alzheimer's disease according to Braak. Leading characteristics are indicated by arrows. The transentorhinal stages (I and II) are thought to represent clinically silent periods of AD with neurofibrillary tangle (NFT) involvement confined to transentorhinal layer pre-alpha. Limbic stages (III and IV) correspond with clinically incipient AD, and NFT involvement of hippocampal CA1. Isocortical stages (V and VI) represent fully developed AD, with NFT involvement of all areas of association cortex. The red frame indicates the cellular islands of the pre-α entorhinal layer. After Braak and Braak 1992, Fig. 8 with permission

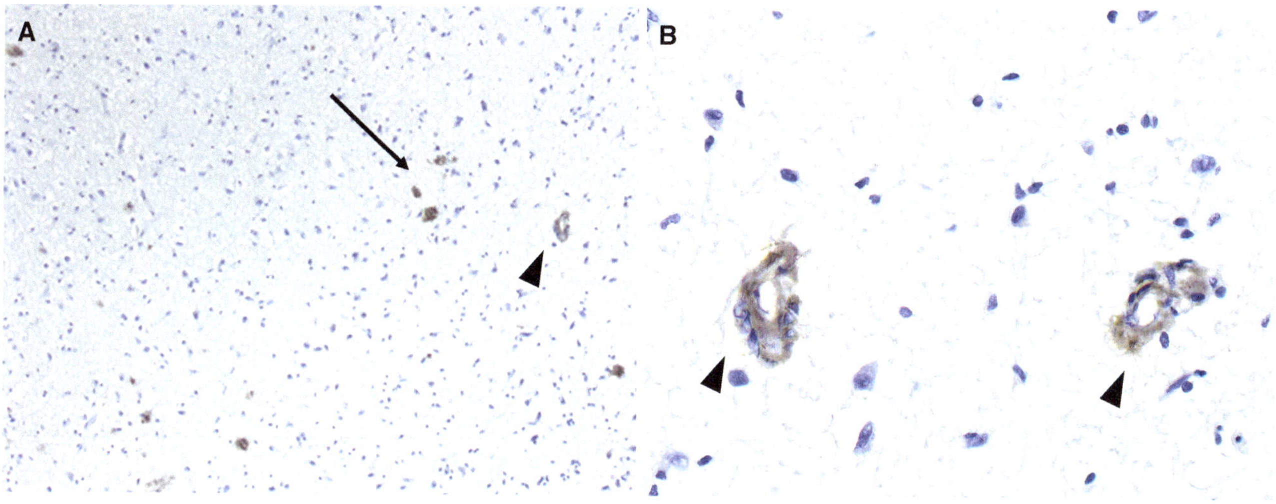

Fig. 13.22 Alzheimer's disease. Amyloid in arterial vessel walls (triangles; A, B) and some amyloid deposits (arrow, B) in the superior colliculus in an Aβ immunostain. Courtesy Prof. D.R. Thal

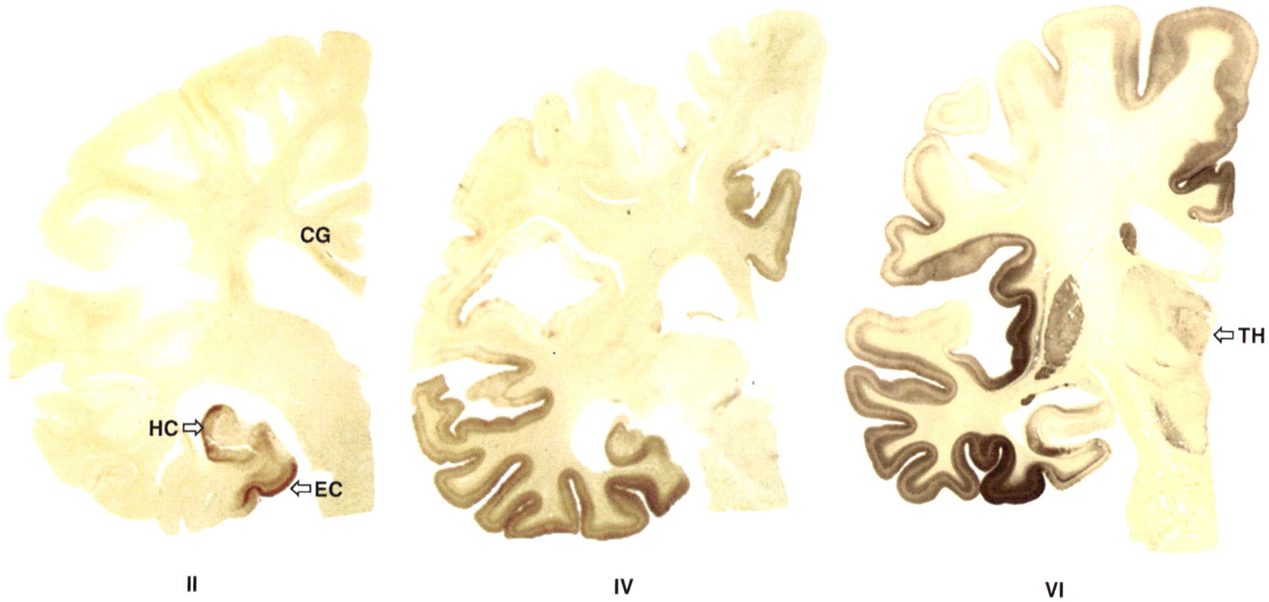

Fig. 13.23 Alzheimer's disease. Coronal sections through the human brain at the level of the hippocampus/entorhinal cortex. AD transentorhinal stage II, limbic stage IV, and isocortical stage VI in 100 μm PEG hemisphere sections immunostained for abnormal tau (AT8). *EC* entorhinal cortex, *CG* cingulate gyrus, *HC* hippocampus, *TH* thalamus. LabPON Twente

while the anterodorsal thalamic nucleus is densely filled with neurofibrillary tangles and neuropil threads.

The hallmark of the two isocortical stages, isocortical stages V-VI, is the destruction of virtually all isocortical association areas (see Fig. 13.23). In stage V, virtually all components of the hippocampal formation are involved. The development of particularly large numbers of neuropil threads (NT) and neurofibrillary tangles (NFT) is a feature of the subicular pyramidal cell layer (see Fig. 13.18B). The pathology includes all the changes noted in earlier stages. The primary sensory areas are still more or less resistant to the development of pathological changes, showing only modest numbers of neuritic plaques in layer III, with layer V being only slightly affected. The primary motor field shows sparse numbers of neuritic plaques (layer III) and is the last component of the isocortex to be affected by the pathological process. In stage VI, the primary sensory areas show a dense network of neuropil threads, in contrast to the small numbers of NFTs in layer V. The primary motor field remains almost devoid of neurofibrillary changes. There is an involvement of

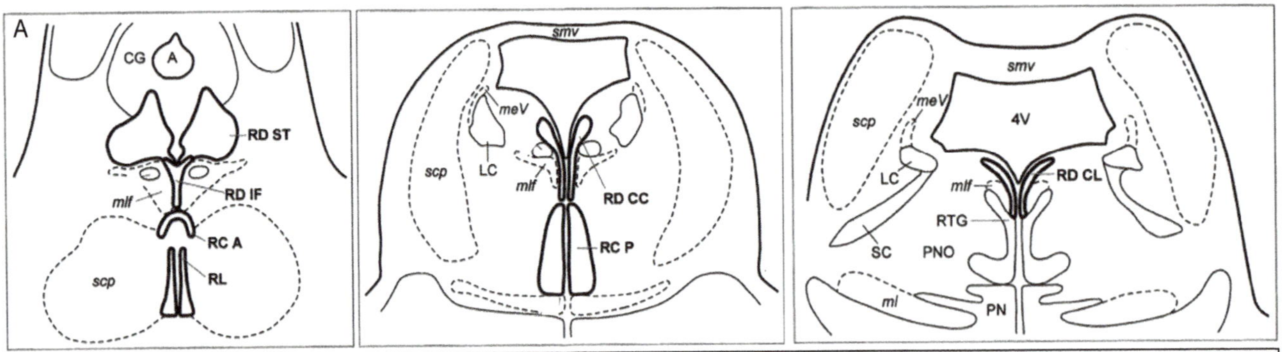

RD CC	Dorsal raphe, Compact caudal subncl.	CG	Central gray
RD CL	Dorsal raphe, Caudal lamellar subncl.	LC	Locus caeruleus
RD IF	Dorsal raphe, Interfascicular subncl.	*meV*	Mesencephalic trigeminal tract
RD ST	Dorsal raphe, Supratrochlear subncl.	*mlf*	Medial longitudinal fasciculus
RC A	Central raphe, Anular subncl.	RTG	Reticulotegmental ncl.
RC P	Central raphe, Principal subncl.	SC	Subcaerulean ncl.
RL	Linear raphe	*scp*	Superior cerebellar peduncle
A	Aqueduct	*smv*	Superior medullary velum

B

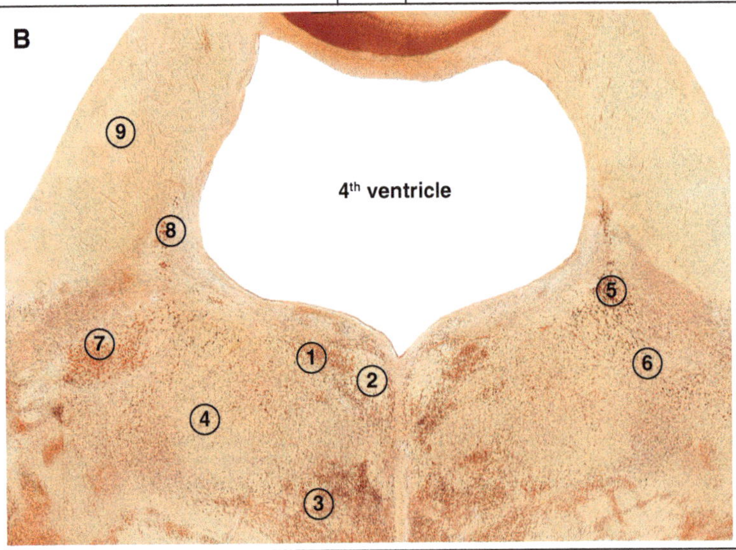

①	Dorsal raphe RD CL	⑥	Subcaerulean ncl. SC
②	Medial longitudinal fasciculus *mlf*	⑦	Trigeminal motor ncl.
③	Reticulotegmental ncl. RTG	⑧	Mesencephalic trigeminal tract *meV*
④	Central tegmental tract	⑨	Superior cerebellar peduncle *scp*
⑤	Locus caeruleus LC		

Fig. 13.24 (**A**) Schematical sections showing the nuclei of the raphe system, outlined by bold lines. (**B**) Darrow-red stained section, approx. corresponding to drawing C in (**A**). (**A**) After Rüb et al. 2000, Fig. 3 with permission. (**B**) LabPON Twente

the extrapyramidal system. Most of the nerve cells of the striatum contain NFTs. NFTs can also be observed in the substantia nigra (see Sect. 16.6.1.4).

The extent of brain destruction in stages III and IV often leads to the appearance of the first clinical symptoms. In stages V and VI (Figs. 13.21 and 13.23), there is fully developed AD.

Brainstem pathologic changes in AD

Aβ-depositions are first seen in the inferior olivary complex (see Sect. 4.1.1.3), the reticular formation of the medulla oblongata (see Sect. 3.3.2) and the substantia nigra (see Sect. 16.6.1) in Thal phase 4 (Thal et al. 2002). In addition, the central gray of the midbrain (see Sect. 13.2.7), the superior (see Fig. 13.22), and inferior colliculi (see Sects. 15.1.3 and 15.1.4) and the red nucleus (see Sect. 15.3.4) exhibit Aβ deposits. In Thal phase 5, plaques are also seen in the reticular formation of the pons, the pontine nuclei, the central and dorsal raphe nuclei (see Sect. 13.3.4.2), the locus caeruleus (Sect. 13.2.2.1), the parabrachial nuclei (Sect. 13.2.3), the dorsal tegmental nucleus (Gudden) (see Sect. 13.3.1.1), the reticulotegmental nucleus (Bechterew), and the cerebellum.

Abnormally phosphorylated tau is first seen in the dorsal raphe nucleus (see Rüb et al. 2000, 2001a, b). The raphe nucleus consists of an oral and caudal part.

The supratrochlear subnucleus (see Fig. 13.24), one of the components of the oral raphe, and located dorsally from the medial longitudinal fasciculus, is affected by neurofibrillary changes in the early stages of the disease (Rüb et al. 2000; Grinberg et al. 2009; Simič et al. 2009) (Fig. 13.24).

In Braak stage III and IV, the severity of the neurofibrillary pathology in the dorsal raphe increases. The central and linear raphe nuclei are now slightly affected. In the final stages V and VI, the dorsal raphe nucleus displays the most severe cytoskeletal pathology within the raphe system, followed by the central and linear raphe nuclei, whereas the changes in the caudal raphe nuclei are only slight. The developing damage within the nuclei of the raphe system correlates with the stages I–VI (Fig. 13.25).

In the early stages of AD, neurofibrillary changes in the brainstem are also seen in the locus caeruleus (see above), the pontine parabrachial region (see Sect. 13.2.3), the intermediate zone of the medullary reticular formation (see Sect. 3.3.2.1), and the reticulotegmental nucleus of the pons (formerly tegmentopontine reticular nucleus, nucleus of Bechterew). The development of the neurofibrillary pathology in the medial parabrachial nucleus (Fig. 13.26) (see Sect. 13.2.3), the subpeduncular tegmental nucleus (see Sect. 13.3.1.6), and the intermediate reticular zone (see Fig. 13.28) begins already in stage I, followed by the lateral parabrachial

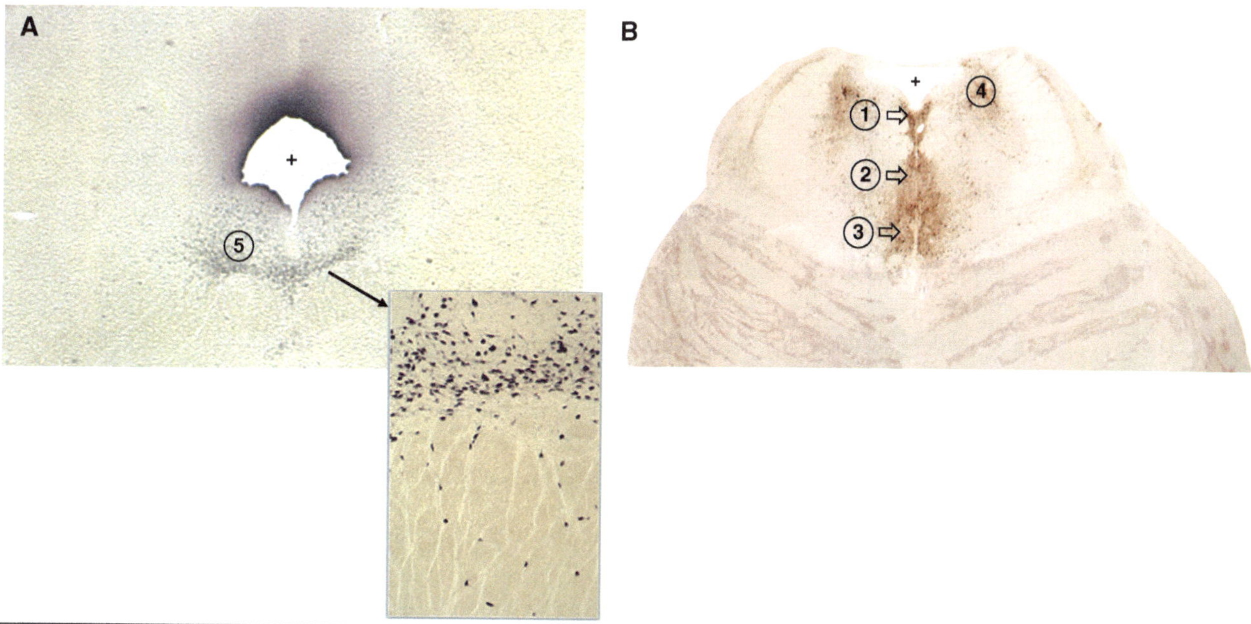

1	Dorsal raphe ncl., compact caudal subncl.	4	Locus caeruleus
2	Median (central) raphe ncl., principal subncl. PC	5	Dorsal raphe ncl., supratrochlear subncl.
3	Reticulotegmental ncl. RTg	+	Mesencephalic aqueduct

Fig. 13.25 Cytoskeletal changes in the raphe system. (**A**) Gallyas silver stain in isocortical stage V and detail (inset). (**B**) Pigment Nissl/AT8 stain shows AT8-immunoreactive changes in locus caeruleus and raphe system, stage V (see Fig. 13.21 and Fig. 13.24A). Compare with Atlas Darrow red 31A. LabPON Twente

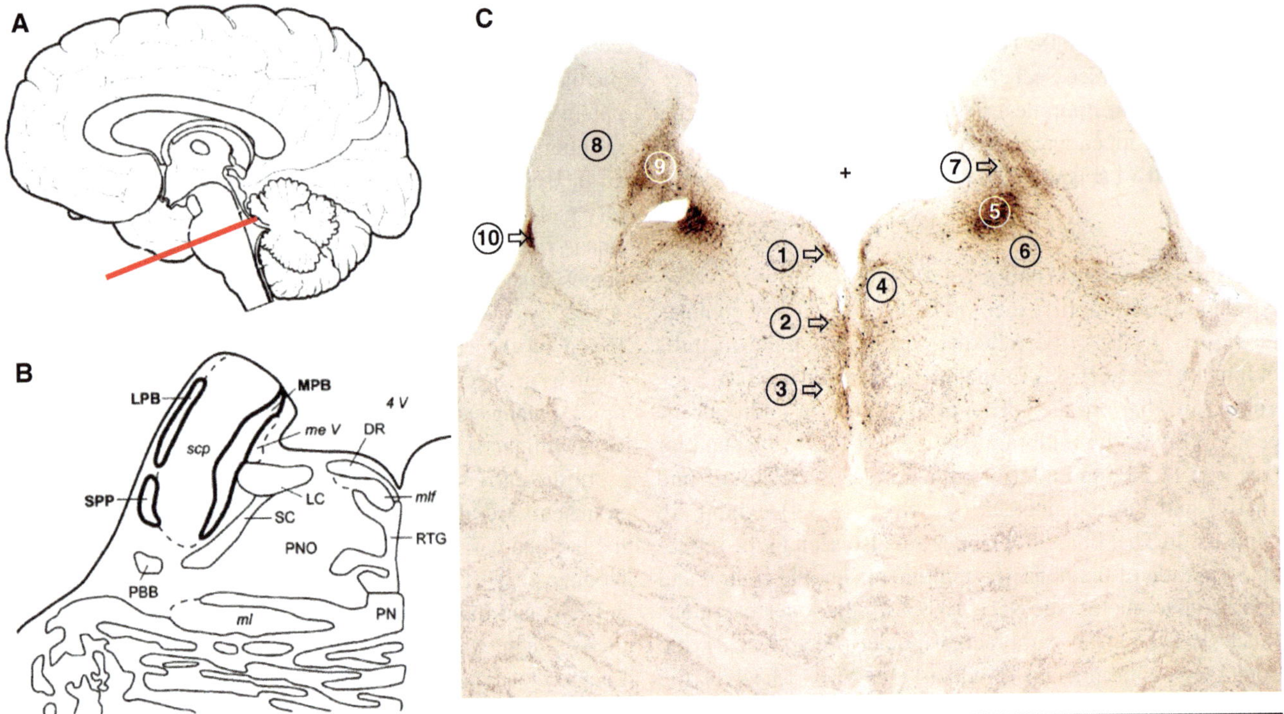

①	Dorsal raphe DR	⑦	Mesencephalic trigeminal tract *meV*
②	Median (central) raphe, principal subncl. PC	⑧	Superior cerebellar peduncle *scp*
③	Reticulotegmental ncl. RTG	⑨	Medial parabrachial ncl. MPB
④	Medial longitudinal fasciculus *mlf*	⑩	Subpeduncular pigmented ncl. SPP
⑤	Locus caeruleus LC	+	4th ventricle
⑥	Subcaerulean ncl. SC		

Fig. 13.26 In (**A**), a schematical median view of a midsagittal section through the brain (red line) at the level of the reticulotegmental nucleus ③ (RTG) (see also (**B**) and Fig. 13.24A). (**C**) shows a PEG section in Nissl stain and AT8-immunoreaction displaying severe cytoskeletal changes in the central raphe nucleus ②, the locus caeruleus ⑤, the medial parabrachial nucleus, and the subpeduncular pigmented nucleus. (**A**, **B**) After Rüb et al. 2001a, Fig. 2 with permission. (**C**) LabPON Twente

nucleus. The progression of the lesions correlates well with the NFT/NT cortical staging sequence I to VI (Rüb et al. 2001a, b; Ohm and Braak 1988). In the later stages, the severity of these changes increases and the intermediate reticular zone and the nucleus of Bechterew especially are severely affected (Fig. 13.27) (Rüb et al. 2001a, b). The dorsal motor vagus complex (today posterior ncl. of vagus nerve) as well as the substantia nigra (Fig. 13.28) (see Sect. 16.6.1) are also affected.

In Parkinson's disease (see Sect. 16.6.1.5) not only the dopaminergic neurons of the substantia nigra (see Sect. 16.6.1) undergo a significant loss but also the noradrenergic neurons of the LC (McMillan et al. 2011). It is a heavily affected brainstem nucleus in Parkinson's disease (for details see Sect. 16.6.1.5). The decrease in noradrenergic innervation of the cerebral cortex and the hippocampus most likely is important for the cognitive deficits observed in PD. Likewise, LC neurons degenerate in the course of Alzheimer's disease (Chan-Palay and Asan 1989b) in particular in those patients who simultaneously suffer from depression. The same association was found in Parkinson patients (Chan-Palay and Asan 1989b). In this vein, nor-

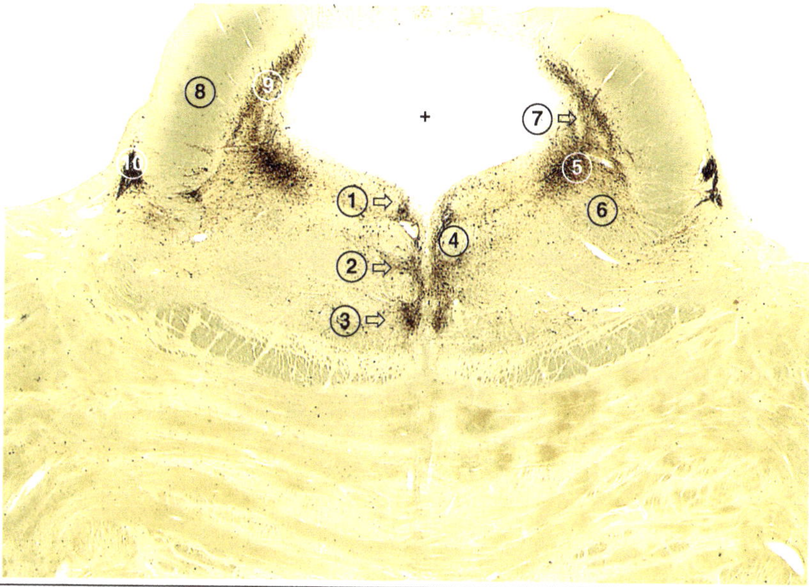

①	**Dorsal raphe DR CC, caudal compact subncl. CC**	⑦	**Mesencephalic trigeminal tract *meV***
②	**Median (central) raphe, principal subncl. PC**	⑧	**Superior cerebellar peduncle *scp***
③	**Reticulotegmental ncl. RTG**	⑨	**Medial parabrachial ncl. MPB**
④	**Medial longitudinal fasciculus *mlf***	⑩	**Subpeduncular pigmented ncl. SPP**
⑤	**Locus caeruleus LC**	+	**4th ventricle**
⑥	**Subcaerulean ncl. SC**		

Fig. 13.27 Detailed view of the upper pontine region (PEG section in Nissl stain and AT8-immunoreaction) showing severe cytoskeletal changes in the central raphe nucleus, the dorsal raphe nucleus, the locus caeruleus, the medial parabrachial nucleus, the subpeduncular pigmented nucleus, and the reticulotegmental nucleus. Abbreviations correspond to Figs. 13.24, 13.25, and 13.26. LabPON Twente

adrenaline uptake inhibitors are used together with serotonin uptake inhibitors (see Sect. 3.2.1.3.1) (Bradley and Lenox-Smith 2013) in the treatment of major depression.

In a review article, Galgani et al. (2021) summarized the MRI findings in normal and aged subjects as well as of those suffering from Alzheimer's or Parkinson's diseases. The authors claim that the neuroimaging data were largely in agreement with neuropathological evidence. LC signal intensity obviously corresponded to cognitive performance while, however, normal aging did not significantly affect the signal which is in contrast to the losses describes here above. A marked reduction of LC signal was observed in patients suffering from Alzheimer's and Parkinson's diseases. The signal corresponds well with the 3D model developed by Fernandes et al. (2012). A detailed discussion of the anatomical implications of these findings is beyond the scope of this book. Since this review article does not provide any illustration, we just mention two further examples by giving a view into the two papers by Betts et al. (2017, 2019). In their 2017 paper, they describe the MRI visualization of the LC in young and older adults (see Fig. 13.29) and in the 2019 paper a comparison between a healthy subject and an Alzheimer patient is shown. In the latter, the LC signal is no longer visible.

At various sites, the authors of this book have uttered their doubts about an exact anatomical localization of MRI or fMRI signals to structures in the human brainstem. This is not the case for the LC neuroimaging for two reasons: the size of the LC and the use of a neuromelanin-sensitive (see Box 13.5) procedure (see Betts et al. 2017). Betts et al. want the paper to be considered as a consensus on how non-invasive in vivo assessment of locus caeruleus integrity can be used for clinical research in neurodegenerative diseases.

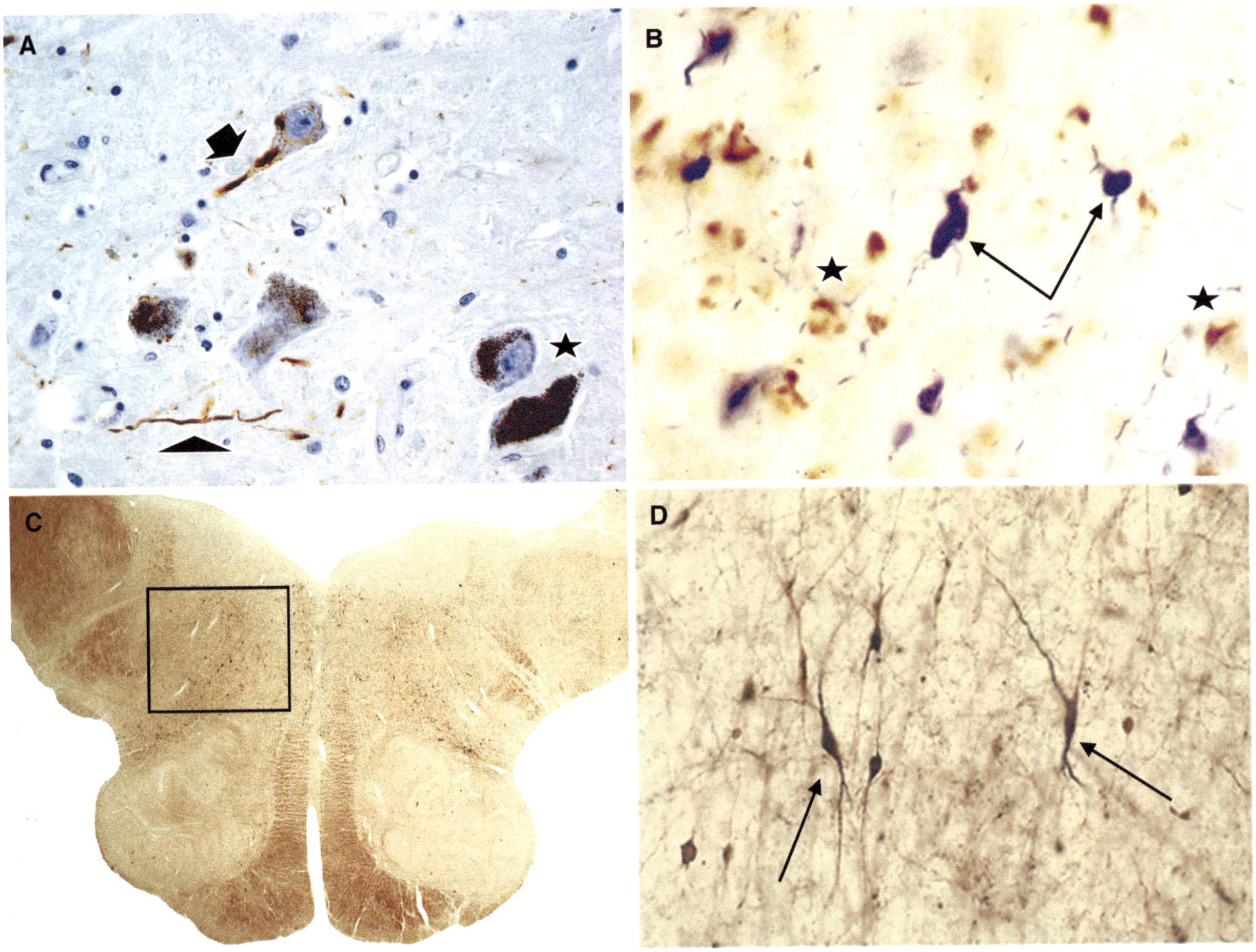

Fig. 13.28 Neurofibrillary tangles (NFTs) (arrows) and neuropil threads (NT) (triangle) in the substantia nigra (see Sect. 16.6.1) immunostained for AT8 in (**A**) (star = neuromelanin pigment) and in (**B**) (PEG 100 μm, Nissl counterstain in **A**). (**C**) Survey of the medulla oblongata at the level of the inferior olive. NFTs in the intermediate reticular zone (see Sect. 3.3.2.1) (rectangle) in AT8-immunostain (PEG 100 μm). (**D**) Detailed view of AT-8 immunoreactive neurons (arrows). LabPON Twente

Box 13.5 Neuromelanin

Neuromelanin is a pigmented polymer which results from the oxidation of catecholamines. Neuromelanin itself is an endogenous MR contrast agent and by binding, for example, iron and copper ions its relaxivity is strongly enhanced. This paramagnetic property shortens T1 time in MR images and is used to visualize neuromelanin in the locus caeruleus and the substantia nigra (see Sect. 16.6.1). The use of a 3D gradient echo T1-weighted sequence with MT pulses allows to increase the contrast between these neuromelanin-containing structures vs. brain background, even when using a 1.5T scanner (see Balážová et al. 2019).

13.2.2.2 Subcaerulean nucleus [Ncl. subcaeruleus]

The subcaerulean nucleus (Fig. 13.24) is located ventral of the locus caeruleus.

Its projections primarily go to preganglionic sympathetic neurons and somatic cranial nerve nuclei and as those of the LC to the brainstem reticular formation, the dorsal horn of the spinal cord, and the region surrounding the central canal and the ventral horn (see Bennaroch 2018).

In man, as an important issue the subcaerulean nucleus is involved in the so-called "idiopathic rapid eye movement sleep behavior disorder" (Ehrminger et al. 2016). The disease is characterized by nocturnal violence and an increased muscle tone during rapid eye movement sleep. It can precede Parkinson's disease (PD) disease by several years (Ehrminger

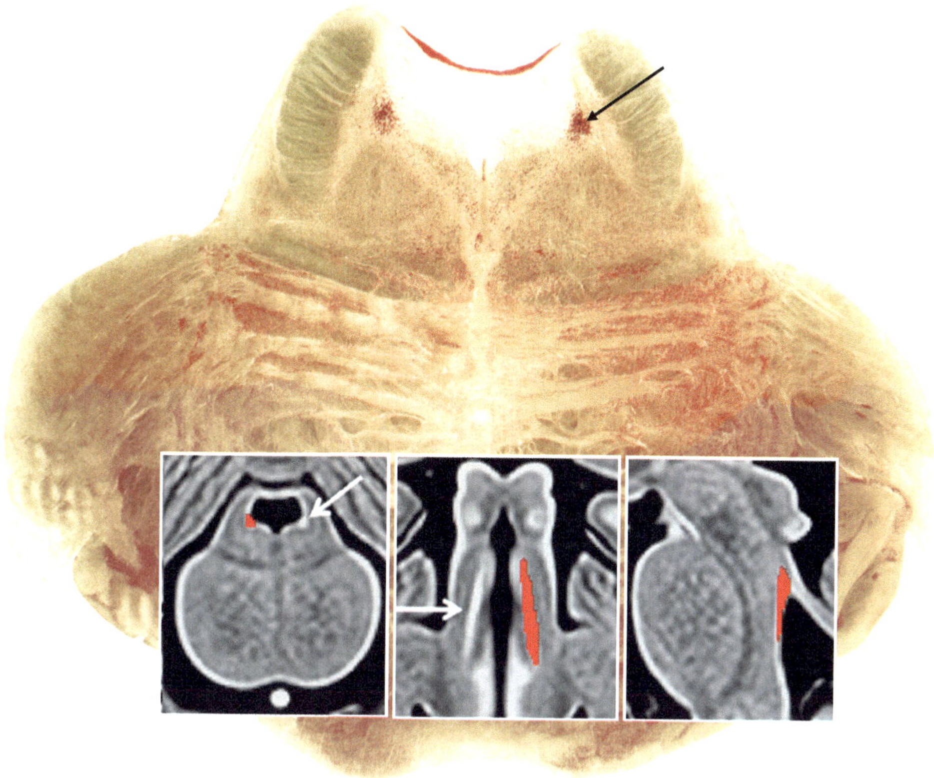

Fig. 13.29 Horizontal section through the human pons at a cranial pontine level showing the LC (arrow). Darrow red staining. In the foreground: Neuromelanin-sensitive MRI procedure clearly showing the LC in—from left to right—a horizontal, a coronal, and a sagittal plane (see Box 13.5). MRI images from Betts et al. 2017, Fig. 2 with permission

et al. 2016). In terms of neuroimaging, the PD patients and those with idiopathic rapid eye movement sleep behavior disorder show a reduced signal intensity in neuromelanin-sensitive MRI sequences. In a second study with a similar design, age- and gender-matched individuals were compared with the two disorders described above using a broad spectrum of investigative tools (for details, see Ehrminger et al. 2016). The patients more frequently had preclinical markers of α-synucleinopathies (see Sect. 16.6.1), including constipation, olfactory deficits, orthostatic hypotension, and subtle motor impairment. Reduced neuromelanin signal intensity was observed in the "idiopathic rapid eye movement sleep behavior disorder patients" in LC and subcaerulean nucleus. The authors suggest neuromelanin-sensitive imaging provides an early marker of non-dopaminergic α-synucleinopathy that can be detected on an individual basis.

13.2.2.3 A7 noradrenaline cells (A7)

For details, see Sect. 11.2.3.1 and Fig. 13.5.

13.2.2.4 B9 serotonin cells (B9) not listed in FIPAT Ch. 1

For details, see Sect. 3.2.1.3 and Fig. 13.38.

13.2.3 Parabrachial nuclei r1 (PBN) [Ncll. parabrachiales] (The medial PBN formerly called subpeduncular nucleus)

The lateral pontine tegmentum consists of the parabrachial nuclei and, as a ventral extension, the Kölliker-Fuse nucleus (see Sect. 13.2.4) (ten Donkelaar 2011).

The name "parabrachial nuclei" goes back to the old name for the superior cerebellar peduncle "brachium conjunctivum" (brachium [Latin] = arm, conjunctivum [Latin] = conjunctive, connectional). The parabrachial nuclei are arranged around the superior cerebellar peduncle. They are the second taste relay in rodents (Tokita et al. 2010) but not in humans where the ascending fibers from the solitary nucleus (Sect. 3.2.4.1.2) directly reach the parvocellular part of the ventral posteromedial nucleus of the thalamus (Fig. 13.30).

13.2.3.1 Location and morphology of the PBN

One can distinguish between the lateral (LPB) and medial (MPB) parabrachial nuclei (Fig. 13.30).

The LPB is a group of small neurons located at the lateral aspect of the superior cerebellar peduncle in the mesenceph-

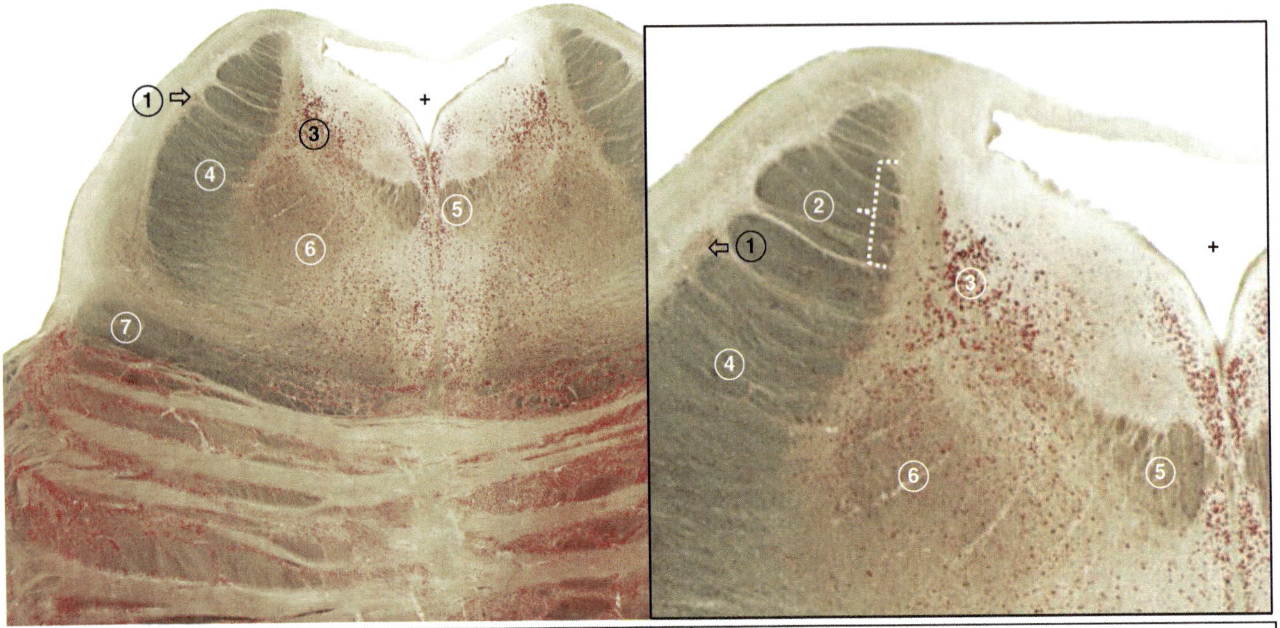

①	**Lateral parabrachial ncl.**	⑤	**Medial longitudinal fasciculus**
②	**Medial parabrachial ncl.**	⑥	**Central tegmental tract**
③	**Locus caeruleus**	⑦	**Medial lemniscus**
④	**Superior cerebellar peduncle**	+	**4th ventricle**

Fig. 13.30 Horizontal level through the human pons at the level of the parabrachial nuclei ①, ②. Darrow red staining. Both parabrachial nuclei are rather unequivocally identifiable only at higher magnification. The medial PBN ② lies in between the superior cerebellar peduncle ④ and the locus caeruleus ③. Approximately at that level, but at the lateral margin of the superior cerebellar peduncle, the lateral PBN is located ①. Compare with Figs. 13.26 and 13.27. See also atlas part Darrow red 31. LabPON Twente

alon, while the MPB is located dorsomedially of the superior cerebellar peduncle.

The LPB is most easily identified as a narrow strip of cells at the lateral edge of the superior cerebellar peduncle at that level where the subpeduncular (pigmented) tegmental nucleus (SPP) (Ohm and Braak 1988, see Sect. 13.3.1.6) is no longer globular in shape and starts to protrude dorsally. The largest diameter of the LPB is found at the caudal pole of the dorsal tegmental nucleus. Rostrally, it is displaced by the pedunculotegmental nucleus (see Sect. 13.1.1.1.1).

In the LPB, the cells are densely arranged smaller and more darkly stained compared to the MPB (Olszewski and Baxter 1982). The cells of the MPB are small, oval, or spindle shaped. Some larger cells may contain melanin (Olszewski and Baxter 1982).

The MPB precedes the LPB. It is bordered laterally by the superior cerebellar peduncle—just opposite to the subpeduncular (pigmented) tegmental nucleus (SPP) which is located at the lateral circumference of the peduncle (see below)—and medially by the nucleus of the mesencephalic trigeminal tract (see Sect. 13.2.6.1) and the locus caeruleus (see Sect. 13.2.2.1). The rostroventral end of the nucleus is indicated by the central tegmental tract before the rostral end of the LPB. The MPB contains small, oval, or spindle-shaped cells, some large cells, maybe containing melanin (Olszewski and Baxter 1982). It displays a strong AChE reactivity (see Box 3.10). Interestingly, the MPB can be identified very easily by the high amount of AT8-immunoreactivity in the brains of patients who suffered from Alzheimer's disease (pretangle stadium) (see Figs. 13.26 and 13.27) while its identification in conventionally stained sections is difficult.

13.2.3.2 Functional aspects of the PBN

Although it is tempting to assume that the human parabrachial nuclei are a gustatory relay station as in other mammals this is not the case.

Tegmental damage is known to cause coma. Parvizi and Damasio (2003) found in an MRI study on patients with brainstem strokes that in those with post-stroke coma—mostly bilateral lesions—the maximum overlap territory of the lesions coincided with the areas containing the parabrachial nuclei, the rostral raphe nuclei, the laterodorsal tegmental nuclei, the oral pontine nuclei, and the white matter between these nuclei.

13.2.3.3 Connectivity of the PBN

A number of ascending pain-transmitting pathways have been described for the human brain like the spinoreticular [Tractus spinoreticularis], the spinomesencephalic [Tractus spinomesencephalicus], the spinohypothalamic [Tractus spinohypothalamicus], the spinolimbic (not listed in FIPAT Ch. 1), and the spinoparabrachial tracts [Tractus spinoparabrachialis] (ten Donkelaar 2011). In the summary and interpretation by ten Donkelaar (2011) of a degeneration study by Mehler (1962) on a patient with bilateral anterolateral cordotomy and a postoperative survival time of 12 days it reads: "... *Extensive terminal degeneration was observed ventral to the brachium conjunctivum and lateral to the central tegmental tract in the subcoeruleus area, which might form part of the parabrachial nuclei*...". Similar findings were made in rodents (for details, see ten Donkelaar 2011).

Although unlike to rodents, primate parabrachial nuclei are not part of the ascending gustatory pathways (see Sect. 3.2.4.1) they nevertheless receive input from the solitary nucleus, however, only from its caudal and viscerosensory regions (for details, see ten Donkelaar 2011).

Human parabrachial projections to the thalamus, the amygdala, the bed nucleus of the stria terminalis, and the insular cortex have been made likely by the presence of CGRP-like immunoreactive terminals and perikarya (for details, see de Lacalle and Saper 2000).

The parabrachial nuclei are also known to be part of the central autonomic network and for projecting to the ambiguus and the posterior nucleus of vagus nerve. In conclusion, the parabrachial nuclei appear to be an important relay station of a viscerosensory pathway from the brainstem via the thalamus to the viscerosensory cortex.

13.2.3.4 Pathological aspects of the PBN

The parabrachial nuclei are a site of neurofibrillary pathology in Alzheimer's disease (for details, see Sect. 13.2.2.1.4, see Figs. 13.26 and 13.27).

13.2.4 Kölliker-Fuse nucleus (KF) [Syn: Ncl. subparabrachialis Kölliker-Fuse/ Subparabrachial nucleus Kölliker-Fuse]

13.2.4.1 Location and morphology of the KF

The localization and identification of this nucleus is not trivial in the human brainstem. Corresponding structures in the isthmic region were first described in 1896 by von Koelliker and later by Fuse (1913). Rudolf Albert von Koelliker (1817–1905) was a Swiss histologist and embryologist, working at the University of Würzburg, Gennosuke Fuse (1880–1946) was a Japanese embryologist.

The name KF was coined first by Castaldi (1923) (nucleo di Koelliker-Fuse) (for details, see Petrovický 1989).

Petrovický (1989) has traced the history of the name and restudied the location of the nucleus in human embryos and adult individuals. He pointed to the puzzling situation that in the literature two nuclei are referred to as the Kölliker-Fuse nucleus.

He came to the conclusion that the original descriptions by von Koelliker (1896) and by Fuse (1913) were identical with the compact part of the pedunculotegmental nucleus (PTg) (see Sect. 13.1.1.1.1) (see also Young et al. 1992, next paragraph). The "original" KF nucleus, however, was rather related to the principal nucleus of trigeminal nerve (see Sect. 12.2.3.2) according to Petrovický.

The nucleus today known as KF nucleus did not attract much medical attention until Young et al. (1992) could show that chronic stimulation of the Kölliker-Fuse nucleus region can relieve intractable pain in patients and that the effect is comparable to that of PAG/periventricular stimulation (see also Ren and Dubner 2008). Although the KF nucleus was not indicated in the then widely used stereotactic atlas by Schaltenbrand and Wahren (1977), the authors calculated a target point based on the note by Olszewski and Baxter (1982) that the KF nucleus was equivalent to the Pars compacta of the pedunculopontine tegmental nucleus (see here above see Sect. 13.1.1.1.1) and its topographical relationship with the locus caeruleus (see Sect. 13.2.2.1). By stimulation of this target, they successfully treated several patients suffering from otherwise intractable pain. This fits in with the finding that in cats (Stevens et al. 1982) the KF nucleus is the major source of spinopetal noradrenergic fibers.

Based on the study of Lavezzi et al. (2004), Ren and Dubner (2008) claimed that the KF nucleus is more developed in humans than in other animal species. The description by Lavezzi et al. (2004), however, provided illustrations that let it appear rather difficult to use them as a guide for identifying the nucleus in other section series.

In mice, VanderHorst and Ulfhake (2006) could show that significant populations of TH-immunoreactive neurons which project to the spinal cord were present in the KF region as well as the LC region.

13.2.4.2 Functional aspects of the KF nucleus

In addition to the obvious involvement of the nucleus in human nociception (see Sect. 13.2.4.1), based on clinical considerations and animal studies, the idea evolved that the KF nucleus has also a respiratory role as well as in cardiovascular regulation and cardiorespiratory coupling (Song et al. 2012).

13.2.4.3 Connectivity of the KF nucleus

By means of tract-tracing, Song et al. (2012) could show for the rat projections from the KF nucleus to among others the facial nucleus, trigeminal sensory nucleus, and various parts of ambiguus nucleus and its surrounding areas.

In conclusion, in any case, when dealing with anatomical and/or functional aspects of the so-called Kölliker-Fuse nucleus in planning studies or reading their results it should be determined whether authors studied the "original" nucleus or those that have been introduced later into the literature (Petrovický 1989).

13.2.5 Lateral lemniscus nuclei r1 [Nuclei leminisci lateralis]

13.2.5.1 Intermediate nucleus of the lateral lemniscus

The term nuclei of the lateral lemniscus refers to groups of cells located within or adjacent to the lateral lemniscus in the pontine tegmentum.

For details on the ascending auditory system, please see Sect. 8.1.1.5.

In the rat and the mouse, the structures include, between the dorsal and ventral nuclei, a narrow horizontal nucleus of the lateral lemniscus. Some authors subdivide the ventral nucleus of the rat and the mouse into an intermediate nucleus of the lateral lemniscus and a ventral nucleus of the lateral lemniscus. Functionally, the nuclei of the lateral lemniscus are part of the subcortical auditory system.

http://braininfo.rprc.washington.edu/centraldirectory.aspx?ID=1841

13.2.6 Trigeminal sensory nuclei r1

13.2.6.1 Mesencephalic trigeminal nucleus/Mesencephalic nucleus of trigeminal nerve [Ncl. mesencephalicus n. trigemini]

13.2.6.1.1 Location and morphology of the mesencephalic trigeminal nucleus

The mesencephalic trigeminal nucleus is the most rostral of the trigeminal nuclei (Fig. 13.31). The neurons of the mesencephalic trigeminal nucleus are unique in that they are derived from the neural crest like other neurons in the peripheral nervous system but enter secondarily into the central nervous system. The neural crest is a cell group developing from the wall of the neural tube but splitting off from the CNS (O'Rahilly and Müller 2007). Important examples for neural crest derivatives are the Schwann cells, meninges, sympathetic nerve ganglia, parts of the adrenal gland, melanocytes of the skin, and odontoblasts (anlage of teeth).

The cells are arranged in a band of large, scattered cells extending over 22–24 mm from the level of the principal sensory trigeminal nucleus up to the midbrain along the lateral border of the PAG (Olszewski and Baxter 1982).

It extends from pontine levels (level of facial nucleus, see Sect. 8.2.1) to the mesencephalon up to the superior colliculi commissure (Usunoff et al. 1997). Depending on the individual levels, the location changes.

In the most caudal part, the nucleus is dorsal of the locus caeruleus (see Sect. 13.2.2.1) (see Fig. 13.31B). The rostroventral end of the nucleus is indicated by the central tegmental tract, medial to the superior cerebellar peduncle (see Box 14.1) and lateral of the ependyma of the fourth ventricle.

In the mesencephalon, the neurons of the mesencephalic trigeminal nucleus delineate the lateral circumference of the periaqueductal gray (Sect. 13.2.7.1) (Fig. 13.31C) like pearls on a necklace. The perikarya are ovoid with regular outlines. The majority of the cells has a diameter of 40–50 µm (for details, see Usunoff et al. 1997).

In mice, combined lesioning-behavioral studies suggested that the mesencephalic trigeminal nucleus is involved in the control of feeding and exploratory behavior without modulation of the emotional state (Ishii et al. 2005).

Interestingly, a comparison between wild-type mice and Brn-3a-deficient animals revealed many proprioceptor cells in the mesencephalic trigeminal nucleus that contained parvalbumin immunoreactivity. In knockout mice, however, such neurons could not be detected (Ichikawa et al. 2005).

13.2.6.1.2 Connectivity and functional aspects of the mesencephalic trigeminal nucleus

Peripheral input to the mesencephalic trigeminal nucleus originates mainly in the masticatory muscles (see Sect. 11.3.1.1.2) which are equipped with muscle spindles (Szentagothai 1948). The main output of the nucleus—expectedly—reaches the trigeminal motor nucleus (Fig. 13.31A).

This connection is the basis for the masseter reflex (also called mandibular or jaw jerk reflex), the only important supraspinal monosynaptic reflex (Moini and Piran 2020)). Holding the mouth of the patient slightly open, the examiner puts its index finger below and parallel to the lips. The finger is then hit with the reflex hammer. The normal response is an upward movement of the mandible induced by the masseter muscles (see Sect. 11.3.1.1.2) mostly very slightly. A pronounced answer may be seen in patients with upper motor neuron lesions (impairment of the corticospinal tract) (Moini and Piran 2020). Usunoff et al. (1997) mention studies by Hassler who stated that proprioceptive efferents of the mesencephalic trigeminal nucleus reach six functionally different thalamic regions. A detailed account on the connections of the mesencephalic trigeminal nucleus has been provided by Usunoff et al. (1997).

13.2.6.2 Supratrigeminal nucleus

For detailed information on the supratrigeminal nucleus, see Sect. 12.2.3.1.

13.2.6.3 Principal sensory trigeminal nucleus

For detailed information on the principal sensory trigeminal nucleus, see Sect. 12.2.3.2.

13.2.7 Central gray, not subdivided

The central gray is an anatomically ill-defined aggregation of neurons near to the ventricular system, i.e., surrounding the mesencephalic aqueduct in the midbrain and located in the floor of the fourth ventricle at pontine levels. It is thought as the prolongation of the hypothalamus into the brainstem.

13.2.7.1 Periaqueductal gray (PAG) [Substantia grisea periaqueductalis, formerly also Substantia grisea centralis]

According to Watson et al. (2019), the tectal gray formation has not attracted much interest due to its misleading interpretation as a "posterior pretectal nucleus" (see Chap. 16, Sect. 16.4), a vaguely defined rostral part of the midbrain. This may explain why FIPAT Ch. 1 lists a posterior pretectal nucleus but not the tectal gray formation.

According to Puelles (2013), the gray matter of the hindbrain as a periventricular stratum has traditionally been named the periaqueductal gray or the central gray. Again, he states that this formation is usually assigned to the mesencephalon, based on the identification of the aqueduct as mesencephalic. This structure, however, extends from the pretectum (see Chap. 16) to rhombomere 2 (see Chap. 12).

Derivatives of the dorsolateral histogenetic alar zone are the tectal gray, the superior colliculus (see Sect. 15.1.3), the inferior colliculus and the preisthmic gray.

The superficial stratified tectal gray is a retinorecipient structure analogous to the superior colliculi (formerly called the ncl. of the optic tract [ncl. tractus optici]), while its central and intermediate parts were misidentified as the posterior pretectal nucleus.

The periaqueductal gray comprises a number of hindbrain nuclei like the locus caeruleus (Sect. 13.2.2.1), the mesence-

A

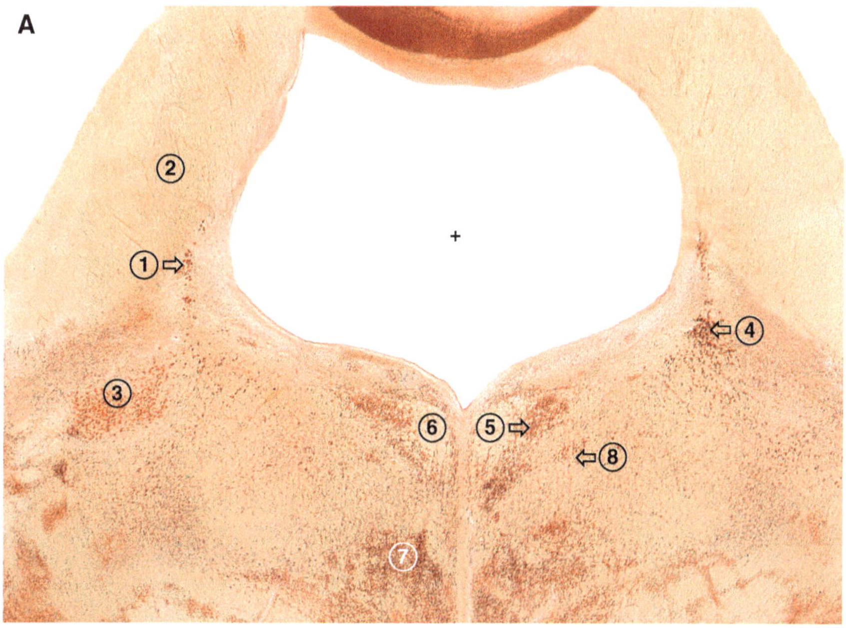

①	Mesencephalic trigeminal ncl.	⑥	Medial longitudinal fasciculus
②	Superior cerebellar peduncle	⑦	Reticulotegmental ncl.
③	Trigeminal motor ncl.	⑧	Posterodorsal tegmental ncl.
④	Locus caeruleus	+	4th ventricle
⑤	Dorsal raphe ncl.		

Fig. 13.31 Horizontal sections of the human brainstem at the level of the trigeminal motor nucleus (**A**), the locus caeruleus (**A**, **B**) and (**C**) the superior colliculi. Darrow red staining. In the more caudal sections (pons), the mesencephalic trigeminal nucleus is located near to the ependyma of the fourth ventricle and laterally flanked by the superior cerebellar peduncle (**A**) and ventrally by the locus caeruleus (**B** ③). With the expansion of the periaqueductal gray and the relocation of the superior cerebellar peduncles, the nucleus can be found at the lateral edge of the PAG (**C** ③), not easy to recognize because of its cell scarcity, but the size of its cells is helpful in identifying (see inset in **C**). (**A**) see also atlas part Darrow red 25, (**B**) Darrow red 26 and (**C**) Darrow red 39, 39A. LabPON Twente

B

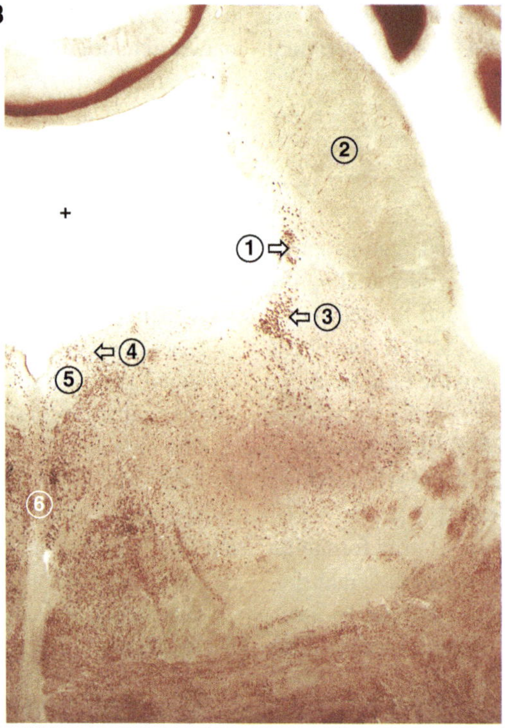

①	Mesencephalic trigeminal ncl.	⑤	Medial longitudinal fasciculus
②	Superior cerebellar peduncle	⑥	Dorsal raphe ncl.
③	Locus caeruleus	+	4th ventricle
④	Dorsal raphe ncl.		

C

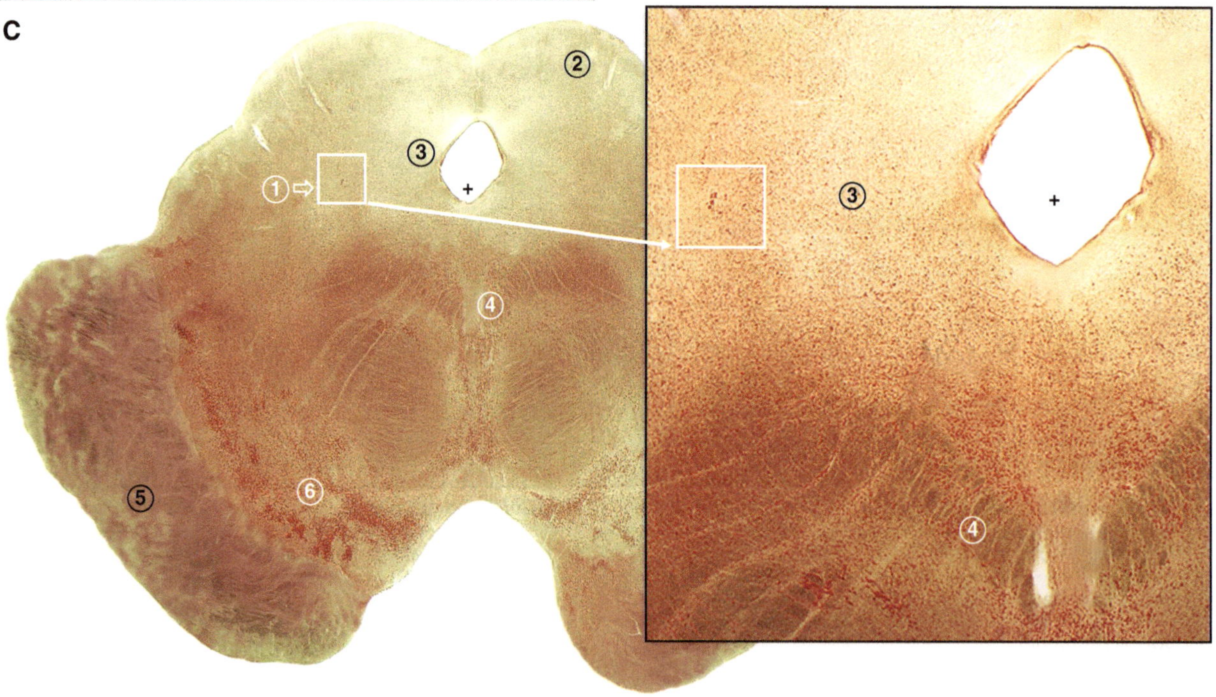

①	Mesencephalic trigeminal ncl.	⑤	Cerebral peduncle
②	Superior colliculus	⑥	Substantia nigra
③	Periaqueductal gray	+	Mesencephalic aqueduct
④	Medial longitudinal fasciculus		

Fig. 13.31 (continued)

phalic trigeminal nucleus (Sect. 13.2.6.1) and the laterodorsal tegmental nucleus (Sect. 13.1.1.2) (Puelles 2013). In addition, extending along the whole brainstem, the alar plate elements form the parvocellular reticular formation (Puelles 2013).

13.2.7.1.1 Location and morphology of the human PAG

As the name indicates, the PAG is a mass of gray substance located around the mesencephalic aqueduct (see Fig. 13.32). FIPAT lists this structure without any subdivision. In the classical work by Olszewski and Baxter (1982), the PAG was subdivided in a dorsal, a medial, and a lateral subnucleus. In animal research, the PAG is divided along its rostrocaudal axis into dorsomedial, dorsolateral, lateral, and ventrolateral columns (see Schröder et al. 2019).

The PAG consists of cells which envelop the dorsal and lateral parts of the aqueduct in the caudal half of the mesencephalon and surround the mesencephalic aqueduct in the oral half. It is directly continuous with the central gray of the pons and orally with the central gray of the third ventricle. Dorsally, the PAG is related to the cuneiform nucleus (see Sect. 14.4.1.1), ventrally to the dorsal raphe nucleus (Sect. 13.3.4.2), the oculomotor complex (see Sect. 15.3.1), the interstitial nucleus of Cajal (see Sect. 16.4.2), and the nucleus of Darkschewitsch/Elliptic nucleus (see Sect. 16.4.1). The subnuclei (PAGm—innermost portion, PAGl—outermost portion, PAGd—immediately dorsal to the AC, related to PAGl and on both sides to the PAGm) are distinguished based on their cytoarchitectonic characteristics (for details, see Olszewski and Baxter 1982).

13.2.7.1.2 Connectivity and function of the human PAG

The current opinion is that the PAG is a downstream regulatory station involved in a circuit including the prefrontal cortex (PFC), amygdala and hippocampus, exerting a descending control on the spinal cord (for details, see Cacciola et al. 2019) (see Fig. 13.33). For long it is known that stimulation of the dorsal PAG in humans undergoing surgery has produced reports of intense fear associated with autonomic reactions (e.g., tachycardia and hyperventilation) reminiscent of a full-blown panic attack (Nashold Jr et al. 1969).

In the meantime, a plethora of mainly animal studies has revealed that in this vein the main task of neurons in the PAG integrates negative emotions (threat) with the autonomic, neuroendocrine, and immune systems to facilitate adaptive

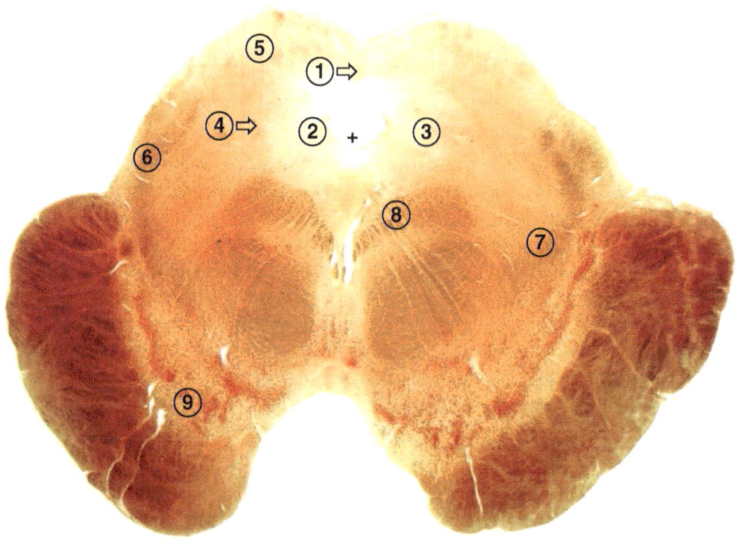

①	Periaqueductal gray, dorsal subncl.	⑥	Lateral lemniscus
②	Periaqueductal gray, medial subncl.	⑦	Medial lemniscus
③	Periaqueductal gray, lateral subncl.	⑧	Medial longitudinal fasciculus
④	Mesencephalic ncl. of trigeminal nerve CNV	⑨	Substantia nigra
⑤	Superior colliculus	+	Mesencephalic aqueduct

Fig. 13.32 Horizontal section through the human mesencephalon at the level of the superior colliculi ⑤. Darrow red staining. The PAG is arranged around the mesencephalic aqueduct (+) and—according to the classical description by Olszewski and Baxter (1982)—can be subdivided in a dorsal ①, a medial ② and a lateral ③ subnucleus. Note that the FIPAT Ch. 1 terminology only knows the undivided PAG. See also atlas part Darrow red 38. LabPON Twente

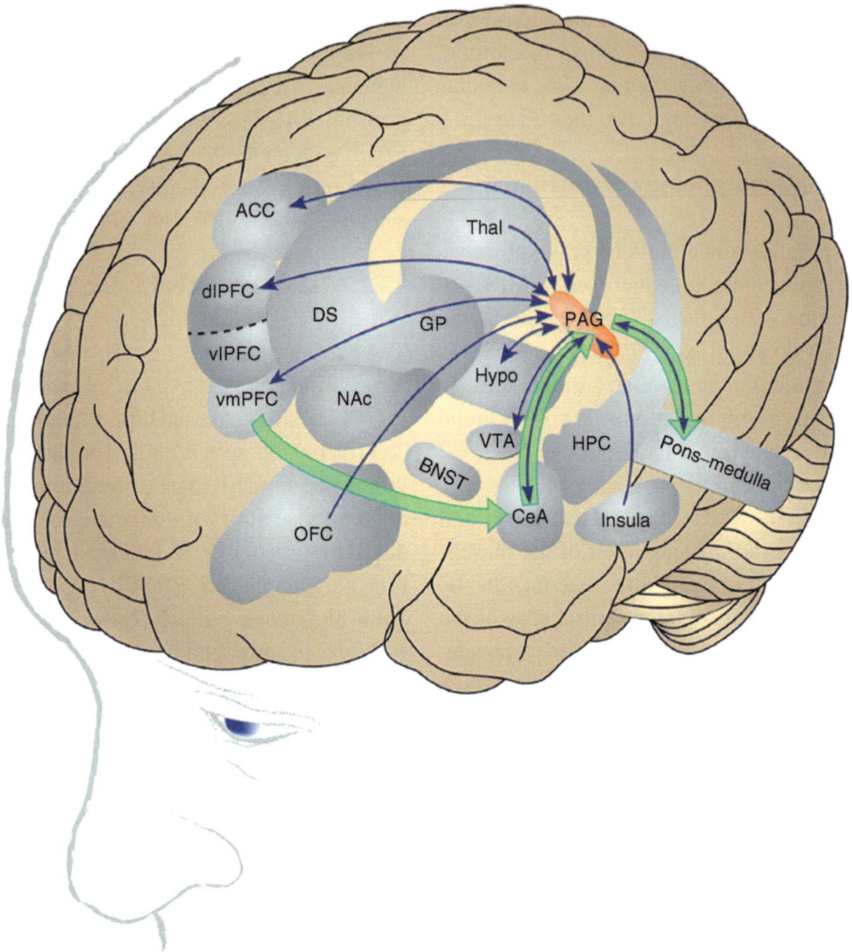

Fig. 13.33 Psychopathology-relevant connectivity of the human PAG. The connectivity of the PAG can be divided into reciprocal (**A**) and non-reciprocal connections (**B**). (**A**) Reciprocal connections of the PAG exist with the anterior cingulate cortex (ACC), dorsolateral prefrontal cortex (dlPFC), ventrolateral prefrontal cortex (vlPFC), ventromedial prefrontal cortex (vmPFC), nucleus accumbens (NAc), dorsal striatum (DS), globus pallidus (GP), central nucleus of the amygdala (CeA), ventral tegmental area (VTA), hippocampus (HPC), thalamus (Thal), and hypothalamus (Hypo). There are reciprocal connections between the PAG and brain stress and emotional systems in the hypothalamus, CeA, pons–medulla, and frontal cortex. (**B**) Afferents to the PAG without reciprocal connection include the ones from the thalamus (Thal), orbitofrontal cortex (OFC) and insula. Connections with the ACC and PFC are involved in top-down control of the function of the basal ganglia and extended amygdala (BNST and CeA). The highlighted broad arrows reflect the emphasis on the role of the PFC to CeA to PAG to pons–medulla pathways that are implicated in the pathophysiology of negative emotions. From George et al. 2019, Fig. 1 with permission

responses to threat (for review, see George et al. 2019). This is reflected by the schematic representation of connections shown in Fig. 13.33.

Highest in the respective hierarchy, the prefrontal cortex (PFC) projects mainly to the dorsal PAG driving social defeat response and the central nucleus of the amygdala (CeA) mediates innate fear responses (George et al. 2019). The ventral PAG is the mediator for learned fear-like behaviors, predatory hunting (important in animals), and depression-like behaviors.

In recent years, human-specific connectivity data for the PAG have been presented based on functional neuroimaging techniques. They are mainly concerned with connections of the PAG with other hindbrain regions. Cacciola et al. (2019) could show by means of high-resolution, multi-shell data of 100 healthy subjects from the open-access Human Connectome Project repository combined with constrained spherical deconvolution probabilistic tractography. Connectivity could be demonstrated between the PAG and the cerebellar nuclei—in particular the fastigial nucleus, vermal and hemispheric regions. This connection may provide one potential way the PAG coordinates some complex motor responses, such as freezing behavior.

Faull et al. (2015) could show the involvement of lateral and dorsomedial PAG in the network of conscious respiratory control by means of ultra-high field functional MRI (7T) (see also Faull et al. 2019).

Involving suprabulbar signalling, Satpute et al. (2013) could show by means of high-field strength fMRI and the use of emotionally aversive images that the activity was different depending on subregions of the PAG. The authors suggest that each subregion participates in a distinct functional circuitry. As several times pointed out in this book, again here, the attribution of signals to rather small subregions has, however, to be considered with caution.

Given that survival is dependent on the organism's ability to switch from prosocial to defensive behaviors, it seems feasible that the dysregulation of such mechanisms may lead to psychopathology (see George et al. 2019).

In mice, there is a midbrain complex of different subnuclei arranged around the cerebral aqueduct in a fashion reminiscent of a Gingko biloba leaf. It is divided along its rostrocaudal axis into different columns. Based on investigations in other mammals, it has been proposed that the dorsal columns of the PAG integrate behavioral and autonomic expression of defensive reactions, for example, following the exposure to a predator, i.e., animals that live by killing and eating other animals (prey) (see also Han et al. 2018). Recently, a connection between the hypothalamus and the PAG has been detected in mice (Celio et al. 2013) where a solitary cluster of parvalbumin-positive neurons (see Box 13.6)—the parvalbumin-positive (PV1) nucleus—has been shown to project predominantly ipsilaterally to the PAG (for details, see Celio et al. 2013).

Box 13.6 Parvalbumin (PARV, PV)

Parvalbumins are calcium (Ca^{2+})-binding proteins of low molecular weight. Like many other Ca^{2+}-binding proteins, they belong to the EF-hand family characterized by helix-loop-helix (HLH)-binding motifs (two helices packed together at an angle of ~90°, separated by a loop region where calcium binds). In the parvalbumin HLH structural motif, calcium is coordinated through one carbonyl oxygen atom and the oxygen-containing sidechains of 5 amino acid residues, or 4 residues and a water molecule.

http://www.ebi.ac.uk/interpro/entry/IPR008080 (InterPro entry—InterPro—EMBL-EBI)

Due to its colocalization in GABAergic neurons, PARV originally has been thought to be a marker for inhibitory interneurons. Recent data show that PARV neurons constitute excitatory pathways (Shang et al. 2015).

13.3 Basal r1

13.3.1 Basal tegmental nuclei r1

13.3.1.1 Dorsal tegmental nucleus [Ncl. tegmentalis dorsalis, Ncl. of von Gudden]

See Box 13.7, Fig. 13.34 and Sect. 13.3.1.3.

13.3.1.1.1 Location of the dorsal tegmental nucleus

A comprehensive study of the human dorsal, posterodorsal (see Sect. 13.3.1.2) and ventral (see Sect. 13.3.1.3) tegmental nuclei by means of AChE-histochemistry and cytoarchitectonics has been performed by Huang et al. (1992). Hayakawa and Zyo (1983) studied the comparative anatomy of Gudden's tegmental nuclei in several mammals. In conventional stains, this nucleus is easily to identify versus its neighbors (see Fig. 13.34), the dorsal raphe ⑨, the medial longitudinal fasciculus ④, and the locus caeruleus ③ by a bright, cell-poor rim around its spherical shape.

13.3.1.1.2 Connectivity and functional aspects of the dorsal tegmental nucleus

Besides the reticular formation, the ventral tegmental area (see Sect. 15.5.1.1) and the periaqueductal gray (PAG), the dorsal tegmental nucleus provides input to the hippocampal formation (ten Donkelaar 2011).

In rats, this nucleus receives massive input from the supragenual nucleus and therefore is probably involved in the head-direction cell-related circuitry (Biazoli et al. 2006).

13.3.1.1.3 Pathological aspects of the dorsal tegmental nucleus

The dorsal tegmental nucleus is known to display amyloid plaques in Alzheimer's disease (see Sect. 13.2.2.1.4).

Box 13.7 Gudden and the King of Bavaria

The psychiatrist Bernhard Aloys von Gudden (1824–1886) had been commissioned to write a report on the state of mind of King Ludwig II of Bavaria, famous for his architectonic projects like the Neuschwanstein Castle. Although he had never talked to the King himself for this report, he certified Ludwig suffering of "incurable mental weakness and paranoia." When he eventually met the King on June 13th, 1886, they went on a walk at Lake Starnberg. Both were found drowned shortly thereafter. A re-evaluation of the case by the German psychiatrist Heinz Häfner (2008) including the results of the King's autopsy presents evidence for the idea that Ludwig was not psychiatrically ill but the victim of an intrigue at the Bavarian court.

13.3.1.2 Posterodorsal tegmental nucleus [Ncl. tegmentalis posterodorsalis]

The posterodorsal tegmental nucleus (Fig. 13.31A ⑧) has its caudal end at the level of the rostral part of the trigeminal motor nucleus (Fig. 13.31A ③). It is composed of diffusely arranged small to medium neurons (Huang et al. 1992). AChE-histochemistry or substance-P like immunoreactivity is better suited to identify the nucleus unequivocally (Huang et al. 1992).

The nucleus is not listed in FIPAT Ch. 1.

13.3.1.3 Anterior/ventral tegmental nucleus [Ncl. tegmentalis anterior/ventralis, Ncl. of von Gudden]

In their comparative-anatomical study Hayakawa and Zyo (1983) claim that by contrast to all other mammals studied incl. monkeys, in man there is no ventral tegmental nucleus nor is a further subdivision of the dorsal tegmental nucleus possible. Huang et al. (1992) describe a human ventral tegmental nucleus which, however, based on their photographic material is not really comprehensible.

Interestingly, a ventral tegmental nucleus is listed in FIPAT Ch. 1.

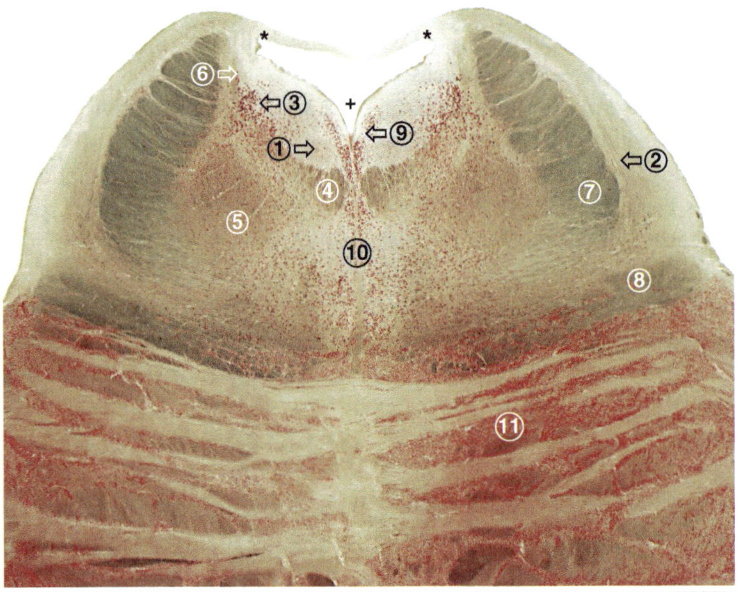

①	Dorsal tegmental ncl. (Gudden)	⑧	Medial lemniscus
②	Lateral lemniscus	⑨	Dorsal raphe ncl., compact caudal subncl.
③	Locus caeruleus	⑩	Median (central) raphe ncl., principal subncl.
④	Medial longitudinal fasciculus	⑪	Pontine ncll.
⑤	Central tegmental tract	*	Superior medullary velum
⑥	Mesencephalic trigeminal ncl.	+	4th ventricle
⑦	Superior cerebellar peduncle		

Fig. 13.34 Horizontal section through the human cranial pons at the level of the dorsal tegmental nucleus ①. Darrow red staining. The dorsal tegmental nucleus is located directly dorsal of the medial longitudinal fasciculus ④ and lateral of the compact caudal subnucleus of the dorsal raphe ⑨. Because of its low cell density in humans, it is hard to identify. A ventral tegmental nucleus is lacking in man (Hayakawa and Zyo 1983). See also atlas part Darrow red 31, 31A. LabPON Twente

13.3.1.4 Dorsomedial tegmental area [Area tegmentalis dorsomedialis]

Detailed information on this area is provided in Chap. 7 under Sect. 7.3.4.1.

13.3.1.5 Rhabdoid nucleus [Ncl. rhabdoideus]

The term rhabdoid nucleus refers to a group of cells in the superior central nucleus raphe of the mouse. The rhabdoid nucleus receives input from the habenula (Broms et al. 2014). Rhabdoid means rod-shaped. This nucleus is not listed in FIPAT Ch. 1.

13.3.1.6 Subpeduncular tegmental nucleus [Ncl. tegmentalis subpeduncularis]

The subpeduncular (pigmented) tegmental nucleus has been described for the human brain (Ohm and Braak 1988). It is located at the lower circumference of the superior cerebellar peduncle and characterized by neuromelanin-containing neurons. In Alzheimer-type dementia, the neuromelanin-containing neurons show severe neurofibrillary pathology (see Figs. 13.26 and 13.27). This nucleus is not listed in FIPAT Ch. 1.

13.3.2 Retrorubral nucleus A8

For the human retrorubral nucleus, Sect. 13.1.1.1.1.

The retrorubral nucleus corresponds to the TH-immunoreactive cell group A8 (Triarhou et al. 1988) in mice.

13.3.3 Interpeduncular nuclei r1

13.3.3.1 Interpeduncular nucleus

Detailed information on the interpeduncular nucleus is provided in Chap. 12 under Sect. 12.3.3.

13.3.4 Raphe nuclei r1

13.3.4.1 Median raphe nucleus

Detailed information on the median raphe nucleus is provided in Chap. 11 under Sect. 11.3.2.1.

13.3.4.2 Dorsal raphe nucleus (DR) [Ncl. raphes dorsalis] B6, B7

13.3.4.2.1 Location and morphology of the dorsal raphe nucleus

Among the oral raphe nuclei, the DR is the one with the most caudal extension. It is located mainly between the medial longitudinal fasciculus and the ependyma (Figs. 13.35, 13.36, and 13.37). The morphology of this nucleus is complicated; four subnuclei can be distinguished (Caudal lamellar subnucleus, in Figs. 13.35, 13.36, and 13.37 compact caudal subnucleus, interfascicular subnucleus, supratrochlear subnucleus) (cf. Rüb et al. 2000). Surprisingly, although the nature of these subnuclei is evident, FIPAT Ch. 1 only lists one dorsal raphe nucleus.

The DR has a longitudinal extension of about 20–25 mm, its caudal end situated at the level of the caudal end of the locus caeruleus (LC; see above Sect. 13.2.2.1). Rostrally, the nucleus reaches the nuclei of the oculomotor nerve. Most of the neurons in the DR are represented by medium-sized multipolar neurons with large amounts of lipofuscin.

80% of neurons in the dorsal raphe nucleus contain serotonin (Hornung 2003) but neurons also contain catecholamines and peptides (see Hornung 2003). Most of the serotonergic neurons are located in the rostral half of the DR.

The caudal lamellar subnucleus [Ncl. raphes dorsalis, subnucleus lamellaris caudalis] contains medium-sized fusiform neurons with clearly stained lipofuscin granula.

The compact caudal subnucleus [Ncl. raphes dorsalis, subnucleus compactus caudalis] is occupying almost all the space between the medial longitudinal fasciculus and the ependyma. Laterally, this part forms a border with the dorsal tegmental nucleus (Gudden) (see Sect. 13.3.1.1).

The interfascicular subnucleus [Ncl. raphes dorsalis, subnucleus interfascicularis], by contrast to the other raphe nuclei, in its oral part becomes a single structure (Fig. 13.37).

It is connected with the supratrochlear subnucleus [Ncl. raphes dorsalis, subnucleus supratrochlearis] by two lateral extensions. The nucleus is bordered dorsally by the lateral sulcus of the aqueductus mesencephali, ventrally by the medial longitudinal fasciculus. Most of the neurons in the DR are represented by medium-sized multipolar neurons with large amounts of lipofuscin.

13.3.4.2.2 Connectivity of the dorsal raphe nucleus

Parent et al. (2011) have shown in a detailed immunocytochemical study that the human substantia nigra (pars reticulata et compacta) (see Sect. 16.6.1), striatum, subthalamic nucleus and internal pallidum receive a serotonergic innervation. The serotonergic axons run in the transtegmental system and ascend with the medial forebrain bundle from where they reach their targets. The authors claim the existence of a direct contact of the serotonergic neurons with the dopaminergic nigrostriatal pathway, the main source of ascending serotonergic fibers being the dorsal raphe nucleus.

The human cerebral cortex has been shown to receive serotonergic afferents (Fig. 13.38) as revealed by Hornung (2003) in human cortical specimen obtained at neurosurgery. Using combined immunohistochemistry for 5-HT and calbindin—as an interneuron marker—he could show the associa-

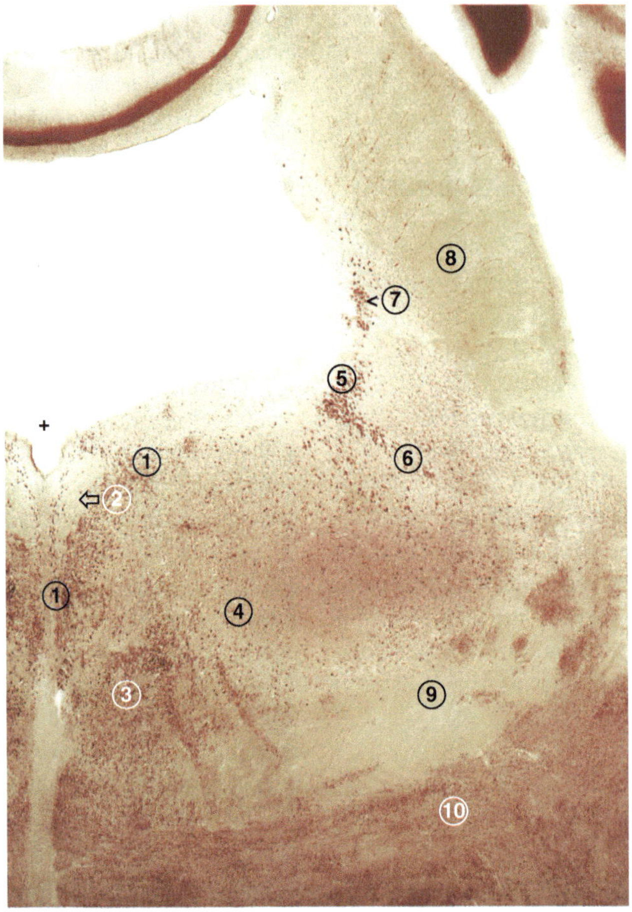

①	**Caudal lamellar subncl. of the dorsal raphe nucleus**	⑦	**Mesencephalic ncl. of trigeminal nerve CNV**
②	**Medial longitudinal fasciculus**	⑧	**Superior cerebellar peduncle**
③	**Reticulotegmental ncl. of the pons**	⑨	**Medial lemniscus**
④	**Pontine reticular ncl., oral part**	⑩	**Pontine ncll.**
⑤	**Locus caeruleus**	+	**4th ventricle**
⑥	**Subcaerulean ncl.**		

Fig. 13.35 Horizontal section through the cranial pons at the level of the caudal lamellar subnucleus of the DR ①. Darrow red staining. This subnucleus is located in the medial tegmentum near to the floor of the fourth ventricle, lateral of the medial longitudinal fasciculus ②. See also atlas part Darrow red 26. LabPON Twente

tion of the large varicose serotonin fibers with interneurons in the human frontal but not in the lateral temporal cortex. Thus, by targeting cortical interneurons, 5-HT could exert indirect influence on cortical projection neurons. Several 5-HT-receptors have been shown in the primate cerebral cortex (for details, see Hornung 2003).

Postmortem tract tracing in human brains has revealed projections from the solitary nucleus to the dorsal raphe nucleus (Ruggiero et al. 2000).

The afferent projections to the dorsal raphe nucleus—and to the median raphe—are mainly of limbic system origin and are derived mainly from animal studies (for details—also on additional connectivity data obtained in animal experiments see Hornung 2003).

13.3.4.2.3 Pathological aspects of the dorsal raphe nucleus

Matthews and Harrison (2012) have found modest alterations in the morphology of the dorsal raphe nucleus in suicide victims and patients having suffered from major depression (see Sect. 3.2.1.3.1) and bipolar disorder.

The most important disease affecting the dorsal raphe nucleus, however, is Alzheimer's disease (see Sect. 13.2.2.1.4).

In mice, located at the level of the trigeminal motor nucleus (r2/r3), the dorsal raphe nucleus (DR) neurons contains serotonergic neurons in particular in the inferior part of the nucleus. The DR provides the main serotonergic innerva-

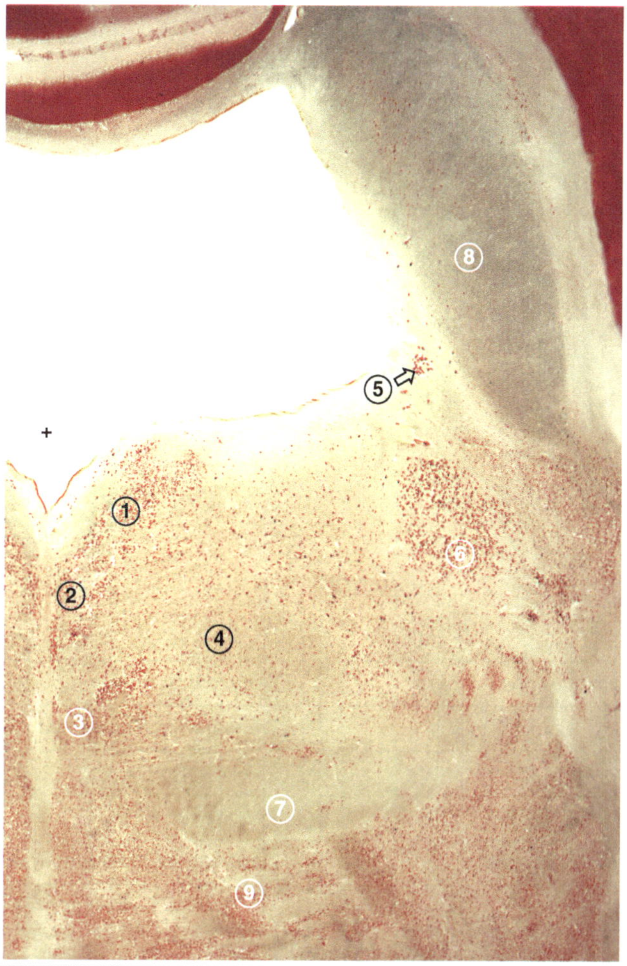

①	Dorsal raphe ncl., caudal lamellar subncl.	⑥	Motor ncl. of trigeminal nerve CNV
②	Dorsal raphe ncl., caudal lamellar subncl.	⑦	Medial lemniscus
③	Reticulotegmental ncl. of the pons	⑧	Superior cerebellar peduncle
④	Pontine reticular ncl., oral part	⑨	Pontine ncll.
⑤	Mesencephalic ncl. of trigeminal nerve CNV	+	4th ventricle

Fig. 13.36 Horizontal section through the cranial pons at the level of the motor nucleus of trigeminal nerve ⑥, caudal of the level shown in Fig. 13.35. Darrow red staining. The caudal lamellar subnucleus ①, ② is located medially of the oral pontine reticular nucleus ④. For the compact caudal subnucleus see Fig. 13.34. See also atlas part Darrow red 24. LabPON Twente

tion of the forebrain (VanderHorst and Ulfhake 2006; Muzerelle et al. 2016).

13.4 Reticular nuclei r1

13.4.1 Pontine reticular nucleus, oral part

For detailed information on the pontine reticular nuclei, see Sect. 10.3.5.4.

13.5 Floor plate r1

13.5.1 Interpeduncular fossa

For detailed information on the interpeduncular fossa, see Sect. 12.4.1.

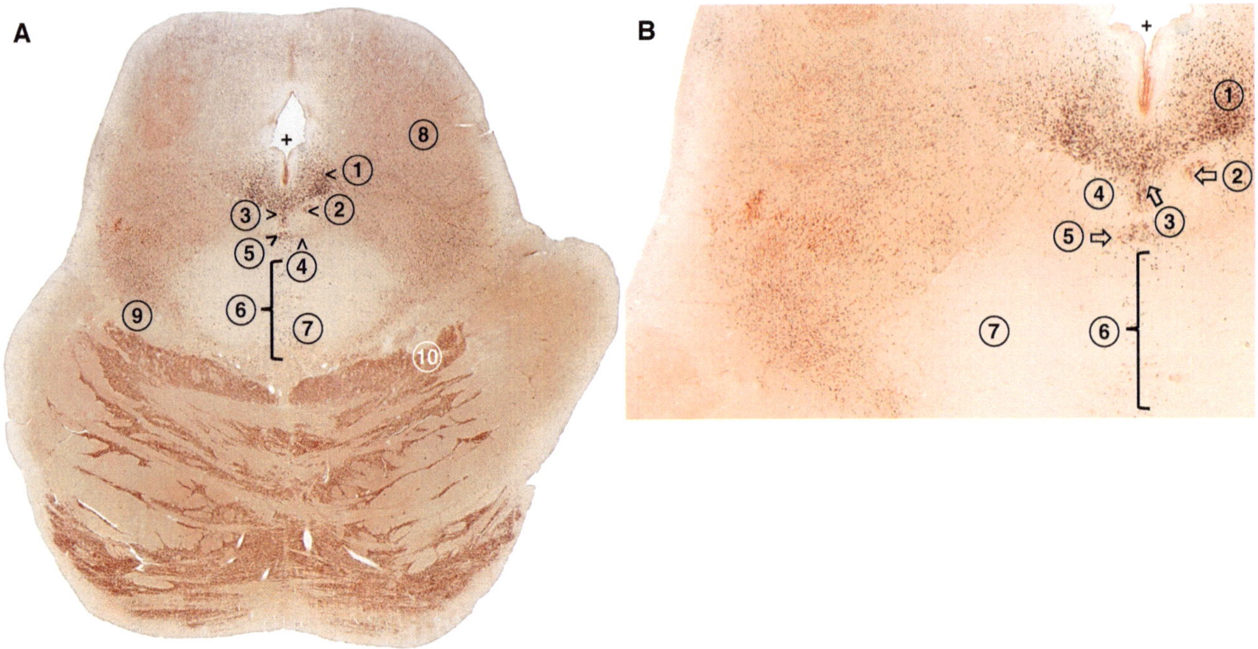

①	Supratrochlear subncl. of the dorsal raphe nucleus	⑦	Decussation of the superior cerebellar peduncles
②	Ncl. of trochlear nerve CNIV	⑧	Transition zone between inferior and superior colliculi
③	Interfascicular subncl. of the dorsal raphe nucleus	⑨	Medial lemniscus
④	Medial longitudinal fasciculus	⑩	Pontine ncll.
⑤	Anular subncl. of the median (central) raphe ncl.	+	Mesencephalic aqueduct
⑥	Linear raphe ncl.		

Fig. 13.37 (**A**) Horizontal section through the mesencephalon at the level of the nucleus of the trochlear nerve CNIV ② (**B**, detail). Darrow red staining. The supratrochlear subnucleus ① is located dorsal to the nucleus of trochlear nerve CNIV ② (hence the name!). The interfascicular subnucleus ③ lies in between the medial longitudinal fasciculi of either side ④ (hence the name!). Compare with Figs. 13.25, 13.26, and 13.27. See also atlas part Darrow red 36, 36A

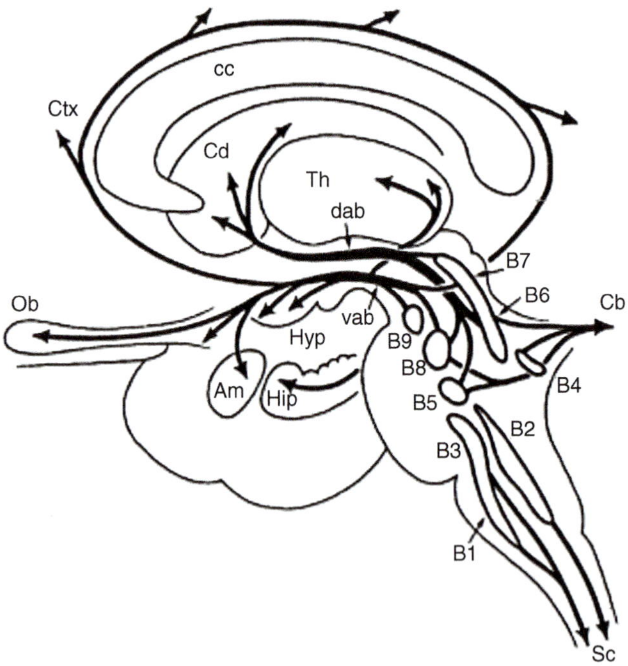

Fig. 13.38 Graphic representation of the serotonergic projections from the brainstem. B1: Raphe pallidus nucleus (see Sect. 4.3.3.2); B2: Raphe obscurus nucleus (see Sect. 4.3.3.1); B3: Raphe magnus nucleus (see Sect. 7.3.3.2); B4: Serotonergic cells of vestibular area; B5: Pontine raphe nucleus; B6: Median raphe nucleus (see Sect. 11.3.2.1); B7: Dorsal raphe nucleus (see here); B8: Linear raphe nucleus (see Sect. 15.5.1.4); B9: not listed in FIPAT Ch. 1. *Am* amygdala, *Cb* cerebellum, *cc* corpus callosum, *Cd* caudate nucleus, *Ctx* cerebral cortex, *dab* dorsal ascending bundle, *Hip* hippocampus, *Hyp* hypothalamus, *Ob* olfactory bulb, *Sc* spinal cord, *Th* thalamus, *vab* ventral ascending bundle. From ten Donkelaar 2011, Fig. 5.10 with permission

References

Andrés-Benito P, Fernandéz-Duenas V et al (2017) Locus coeruleus at asymptomatic early and middle Braak stages of neurofibrillary tangle pathology. Neuropathol Appl Neurobiol 43:373–392

Balážová Z, Nováková M et al (2019) Ch. 4: Structural and functional magnetic resonance imaging of dementia with Lewy bodies. In: Politis M (ed) International review of neurobiology, vol 144. Academic Press

Barthélemy NR, Horie K et al (2020) Blood plasma phosphorylated-tau isoforms track CNS-change in Alzheimer's disease. J Exp Med 217:e20200861

Bazenet C, Lovestone S (2012) Plasma biomarkers for Alzheimer's disease: much needed but tough to find. Biomark Med 6:441–454

Bennaroch EE (2013) Pedunculopontine nucleus. Functional organization and clinical implications. Neurology 80:1148–1155

Bennaroch EE (2018) Locus coeruleus. Cell Tissue Res 373:221–232

Betts MJ, Cardenas-Blanco A et al (2017) *In vivo* assessment of the human locus coeruleus along its rostrocaudal extent in young and older adults. NeuroImage 163:150–159

Betts MJ, Kirilina E et al (2019) Locus coeruleus imaging as a biomarker for noradrenergic dysfunction in neurodegenerative diseases. Brain 142:2558–2571

Biazoli CE, Goto M et al (2006) The supragenual nucleus: a putative relay station for ascending vestibular signs to head direction cells. Brain Res 1094:138–148

Boon BD, Bulk M et al (2020) The coarse-grained plaque: a divergent Aβ plaque-type in early-onset Alzheimer's disease. Acta Neuropathol 140:811–830

Braak H, Braak E (1991) Neuropathological stageing of Alzheimer-related changes. Acta Neuropathol 82:239–259

Braak H, Braak E (1992) The human entorhinal cortex: normal morphology and lamina-specific pathology in various diseases. Neurosci Res 15:6–31

Braak H, Del Tredici K (2011) The pathological process underlying Alzheimer's disease in individuals under thirty. Acta Neuropathol 121:171–181

Braak H, del Tredici K (2012) Where, when, and in what form does sporadic Alzheimer's disease begin? Curr Opin Neurol 6:708–714

Braak H, Thal DR et al (2011) Stages of the pathologic process in Alzheimer's disease: age categories from 1 to 100 years. J Neuropathol Exp Neurol 11:960–969

Bradley AJ, Lenox-Smith AJ (2013) Does adding noradrenaline reuptake inhibition to selective serotonin reuptake inhibition improve efficacy in patients with depression? A systematic review of meta-analyses and large randomised pragmatic trials. J Psychopharmacol 8:740–758

Broms J, Antolin-Fontes B et al (2014) Conserved expression of the GPR151 receptor in habenular axonal projections of vertebrates. J Comp Neurol 523:359–380

Bueno D, Lima LB et al (2019) Connections of the laterodorsal tegmental nucleus with the habenular-interpeduncular-raphe system. J Comp Neurol 527:3046–3072

Cacciola A, Bertino S et al (2019) Mapping the structural connectivity between the periaqueductal gray and the cerebellum in humans. Brain Struct Funct 224:2153–2165

Castaldi L (1923) Studi sulla struttura e sullo sviluppo del mesencefalo. Arch Ital Anat e Embr 20:23–225

Celio MR, Babalian A et al (2013) Efferent connections of the parvalbumin-positive (PV1) nucleus in the lateral hypothalamus of rodents. J Comp Neurol 521:3133–3153

Chang SJ, Cajigas I et al (2020) Dissecting brainstem locomotor circuits: converging evidence for cuneiform nucleus stimulation. Front Syst Neurosci 14:64

Chan-Palay V, Asan E (1989a) Quantitation of catecholamine neurons in the locus coeruleus in human brains of normal young and older adults and in depression. J Comp Neurol 287:357–372

Chan-Palay V, Asan E (1989b) Alterations in catecholamine neurons of the locus coeruleus in senile dementia of the Alzheimer type and in Parkinson's disease with and without dementia and depression. J Comp Neurol 287:373–392

Choi AM, Ryter SW et al (2013) Autophagy in human health and disease. N Engl J Med 368:651–662

Cohen AD, Klunk WE (2014) Early detection of Alzheimer's disease using PiB and FDG PET. Neurobiol Dis 72(Pt A):117–122

de Lacalle S, Saper CB (2000) Calcitonin gene-related peptide-like immunoreactivity marks putative visceral sensory pathways in human brain. Neuroscience 100:115–130

Ehrminger M, Latimier A et al (2016) The coeruleus/subcoeruleus complex in idiopathic rapid eye movement sleep behavior disorder. Brain 139:1180–1188

Esiri MM, Wilcock GK et al (1997) Neuropathological assessment of the lesions of significance in vascular dementia. JNNP 63:749–753

Fagan AM, Perrin RJ (2012) Upcoming candidate cerebrospinal fluid markers of Alzheimer's disease. Biomark Med 6:455–476

Falcon B, Zhang W et al (2018) Structures of filaments from Pick's disease reveal a novel tau protein fold. Nature 561:137–140

Falcon B, Zivanov J et al (2019) Novel tau filament fold in chronic traumatic encephalopathy encloses hydrophobic molecules. Nature 568:420–423

Faull OK, Jenkinson M et al (2015) Functional subdivision of the human periaqueductal grey in respiratory control using 7 tesla fMRI. NeuroImage 113:356–364

Faull OK, Subramanian HH et al (2019) The midbrain periaqueductal gray as an integrative and interoceptive neural structure for breathing. Neurosci Biobehav Rev 8:135–144

Fernandes P, Regala J et al (2012) The human locus coeruleus 3-D stereotactic anatomy. Surg Radiol Anat 34:879–885

French IT, Muthusamy KA (2018) A review of the pedunculopontine nucleus in Parkinson's disease. Front Aging Neurosci 10:99

Fuse G (1913) Die Randgebiete des Pons und des Mittelhirns, Arb Hirnanat Inst Zürich (herausgegeben von Prof. Dr. C. von Monakow). 7:211–253

Galgani A, Lombardo F et al (2021) Locus coeruleus magnetic resonance imaging in neurological diseases. Curr Neurol Neurosci Rep 21:2

Garcia-Rill E (2019) Arousal and movement disorders. In: Arousal in neurological and psychiatric diseases, pp 179–193

Gardner J (2013) A history of deep brain stimulation: technological innovation and the role of clinical assessment tools. Soc Stud Sci 43:707–728

Gaspar P, Berger B et al (1989) Catecholamine innervation of the human cerebral cortex as revealed by comparative immunohistochemistry of tyrosine hydroxylase and dopamine-beta-hydroxylase. J Comp Neurol 279:249–271

George DT, Ameli R et al (2019) Periaqueductal gray sheds light on dark areas of psychopathology. Trends Neurosci 42:349–360

Gerrish A, Russo R et al (2012) The role of variation at AßPP, PSEN1, PSEN2, and MAPT in late onset Alzheimer's disease. J Alzheimers Dis 28:377–387

Griffin WST, Sheng JG et al (1995) Interleukin-1 expression in different plaque types in Alzheimer's disease in plaque evolution. J Neuropathol Exp Neurol 54:276–281

Grinberg LT, Rüb U et al (2009) The dorsal raphe nucleus shows phosphor-tau neurofibrillary changes before the transentorhinal region in Alzheimer's disease. A precocious onset? Neuropathol Appl Neurobiol 35:406–416

Grudzien A, Shaw P et al (2007) Locus coeruleus neurofibrillary degeneration in aging, mild cognitive impairment and early Alzheimer's disease. Neurobiol Aging 28:327–335
Häfner H (2008) Ein König wird beseitigt, Ludwig II. von Bayern. (A king is eliminated, Ludwig II of Bavaria) C.H. Beck, München
Haller S, Garibotto VE et al (2013) Neuroimaging of dementia in 2013: what radiologists need to know. Eur Radiol 23:3393–3404
Han W, Tellez LA et al (2018) Integrated control of predatory hunting by the central nucleus of the amygdala. Cell 168:311–324
Hayakawa T, Zyo K (1983) Comparative cytoarchitectonic study of Gudden's tegmental nuclei in some mammals. J Comp Neurol 216:233–244
Hornung J-P (2003) The human raphe nuclei and the serotonergic system. J Chem Neuroanat 26:331–343
Huang X-F, Törk I et al (1992) The dorsal, posterodorsal, and ventral tegmental nuclei: a cyto- and chemoarchitectonic study in the human. J Comp Neurol 318:117–137
Ichikawa H, Qiu F et al (2005) Brn-3a is required for the generation of proprioceptors in the mesencephalic trigeminal tract nucleus. Brain Res 1053:203–206
Ishii T, Furuoka H et al (2005) The mesencephalic trigeminal sensory nucleus is involved in the control of feeding and exploratory behavior in mice. Brain Res 1048:80–86
Jack CR, Barrio JR (2013) Cerebral amyloid PET imaging in Alzheimer's disease. Acta Neuropathol 126:643–657
Jack CR, Bennett DA et al (2018) NIA-AA Research framework: toward a biological definition of Alzheimer's disease. Alzheimers Dement 14:535–562
Jacobsohn L (1909) Über die Kerne des menschlichen Hirnstamms. Abh. der Königl. Preuss. Akad. der Wiss. Jg. 1909. Phys.-Math. Classe. 1–70
Jonsson T, Atwal JK, e al. (2012) A mutation in APP protects against Alzheimer's disease and age-related cognitive decline. Nature 488:96–99
Josephs KA, Murray ME et al (2016) Updated TDP-43 in Alzheimer's disease staging scheme. Acta Neuropathol 131:571–585
Karikari TK, Pascoal TA et al (2020) Blood phosphorylated tau 181 as a biomarker for Alzheimer's disease: a diagnostic performance modelling study using data from four prospective cohorts. Lancet Neurol 19:422–433
Khachaturian ZS (1985) Diagnosis of Alzheimer's disease. Arch Neurol 42:1097–1105
Khachaturian ZS (2011) Revised criteria for diagnosis of Alzheimer's disease: National Institute on Aging-Alzheimer's Association diagnostic guidelines for Alzheimer's disease. Alzheimers Dement 7:253–256
Kunkle BW, Grenier-Boley B et al (2019) Genetic meta-analysis of diagnosed Alzheimer's disease identifies new risk loci and implicates Aβ, tau, immunity and lipid processing. Nat Genet 51:414–430
Lavezzi AM, Ottaviani G et al (2004) Cytoarchitectural organization of the parabrachial/Kölliker-Fuse complex in man. Brain and Development 26:316–320
Liao F, Yoon H et al (2017) Apolipoprotein E metabolism and functions in brain and its role in Alzheimer's disease. Curr Opin Lipidol 28:60–67
Lowe J, Kalaria R (2015). In: Love S, Perry A, Ironside J, Budka H (eds). Greenfield's Neuropathology Ninth ed. 85
Manaye KF, McIntire DD et al (1995) Locus coeruleus cell loss in the aging human brain: a non-random process. J Comp Neurol 358:79–87
Mandybur TI (1986) Cerebral amyloid angiopathy: the vascular pathology and complications. J Neuropathol Exp Neurol 45:79–90
Martiskainen H, Herukka S-K et al (2017) Decreased plasma β-amyloid in the Alzheimer's disease APP A673T variant carriers. Ann Neurol 82:128–132
Matthews PR, Harrison PJ (2012) A morphometric immunohistochemical, and in situ hybridization study of the dorsal raphe nucleus in major depression, bipolar disorder schizophrenia, and suicide. J Affect Disord 137:125–134
McAleese KE, Alafuzoff I et al (2016) Post-mortem assessment in vascular dementia: advances and aspirations. BMC Med 14:129
McKee AC, Kosik KS et al (1991) Neuritic pathology and dementia in Alzheimer's disease. Ann Neurol 30:156–165
McKhann GM, Knopman DS et al (2011) The diagnosis of dementia due to Alzheimer's disease: recommendations from the National Institute on Aging-Alzheimer's association workgroups on diagnostic guidelines for Alzheimer's disease. Alzheimers Dement 7:263–269
McMillan PJ, White SS et al (2011) Differential response of the central noradrenergic nervous system to the loss of locus coeruleus neurons in Parkinson's disease and Alzheimer's disease. Brain Res 1373:240–252
Mehler WR (1962) The anatomy of the so-called "pain tract" in man: an analysis of the course and distribution of the ascending fibres of the fasciculus anterolateralis. In: French JD, Porter RW (eds) Basic research in paraplegia. Thomas, Springfield, IL, pp 26–55
Mesulam M-M, Mufson EJ et al (1983) Central cholinergic pathways in the rat: an overview based on an alternative nomenclature (Ch1-Ch6). Neuroscience 10:1185–1201
Mesulam M-M, Geula C et al (1989) Human reticular formation: cholinergic neurons of the pedunculopontine and laterodorsal tegmental nuclei and some cytochemical comparisons to forebrain cholinergic neurons. J Comp Neurol 281:611–633
Mirra S, Heyman A et al (1991) The consortium to establish a registry for Alzheimer's disease (CERAD). Part II: standardization of the neuropathologic assessment of Alzheimer's disease. Neurology 41:479–486
Moini J, Piran P (2020) Trigeminal and facial nerves. In: Functional and clinical neuroanatomy: a guide for health care professionals. Academic Press
Montine TJ, Phelps CH et al (2012) National Institute on Aging-Alzheimer's Association guidelines for the neuropathologic assessment of Alzheimer's disease: a practical approach. Acta Neuropathol 123:1–11
Mori F, Okada K-I et al (2016) The pedunculopontine tegmental nucleus as a motor and cognitive interface between the Cerebellum and Basal Ganglia. Front Neuroanat 10:109
Morris JC, Blennow K et al (2014) Harmonized diagnostic criteria for Alzheimer's disease: recommendations. J Intern Med 275:204–213
Morrison JH, Foote SL (1982) Laminar, tangential and regional organization of the noradrenergic innervation of monkey cortex: dopamine-beta-hydroxylase immunohistochemistry. Brain Res Bull 9:309–319
Mukrasch MD, von Bergen M et al (2007) The "jaws" of the tau-microtubule interaction. J Biol Chem 282:12230–12239
Muzerelle A, Scotto-Lomassese S et al (2016) Conditional anterograde tracing reveals distinct targeting of individual serotonin cell groups (B5-B9) to the forebrain and brainstem. Brain Struct Funct 221:535–561
Nashold BS Jr, Wilson WP et al (1969) Sensations evoked by stimulation in the midbrain of man. J Neurosurg 30:14–24
Neal JW (2012) Vascular dementia: why pathology is still important. Rev Clin Geront 22:35–51
Newell KL, Hyman BT et al (1999) Application of the National Institute on Aging (NIA)—Reagan Institute criteria for the neuropathological diagnosis of Alzheimer's disease. J Neuropathol Exp Neurol 58:1147–1155
Nijholt AT, De Kimpe L et al (2011) Removing protein aggregates: the role of proteolysis in neurodegeneration. Curr Med Chem 18:2459–2476

Ohm TG, Braak H (1988) The pigmented subpeduncular nucleus: a neuromelanin-containing nucleus in the human pontine tegmentum. Morphology and changes in Alzheimer's disease. Acta Neuropathol 77:26–32

Olszewski J, Baxter D (1982) Cytoarchitecture of the human brain stem, 2nd edn. Karger, Basel, pp 1–199

O'Rahilly R, Müller F (2007) The development of the neural crest in the human. J Anat 211:335–351

Palmqvist S, Janelidze S et al (2020) Discriminative accuracy of plasma phospho-tau 217 for Alzheimer's disease vs other neurodegenerative disorders. JAMA 324:772–781

Parent M, Wallmann M-J et al (2011) Serotonin innervation of basal ganglia in monkeys and humans. J Chem Neuroanat 41:256–265

Parvizi J, Damasio AR (2003) Neuroanatomical correlates of brainstem coma. Brain 126:1524–1536

Petrella JR, Coleman RE et al (2003) Neuroimaging and early diagnosis of Alzheimer's disease: a look to the future. Radiology 226: 315–336

Petrovický P (1989) The nucleus Koelliker-Fuse (K-F) and parabrachial nuclear complex (PBNC) in man. Location, cytoarchitectonics and terminology in embryonic and adult periods, and comparison with other mammals. J Hirnforsch 30:551–563

Powers JM (1997) Diagnostic criteria for the neuropathologic assessment of Alzheimer's disease. Neurobiol Aging 84:S53–S54

Puelles J (2013) Ch. 10: Plan of the developing vertebrate nervous system: relating embryology to the adult nervous system. In: Rubenstein J, Rakic P (eds) Comprehensive developmental neuroscience: patterning and cell type specification in the developing CNS and PNS. Elsevier

Puladi B, Dinekov M et al (2021) The relation between tau pathology and granulovacuolar degeneration of neurons. Neurobiol Dis 147:105–138

Ren K, Dubner R (2008) The Kölliker-Fuse nucleus. In: The senses: a comprehensive reference, 2008

Rosen C, Hansson O et al (2013) Fluid biomarkers in Alzheimer's disease—current concepts. Mol Neurodegener 8:20

Rüb U, Del Tredici K et al (2000) The evolution of Alzheimer's disease-related cytoskeletal pathology in the human raphe nuclei. Neuropathol Appl Neurobiol 26:553–567

Rüb U, Schultz C et al (2001a) Early involvement of the tegmentopontine reticular nucleus during the evolution of Alzheimer's disease-related cytoskeletal pathology. Brain Res 908:107–112

Rüb U, Del Tredici K et al (2001b) The autonomic higher order processing nuclei of the lower brain stem are among the early targets of the Alzheimer's disease-related cytoskeletal changes. Acta Neuropathol 101:555–564

Ruggiero DA, Underwood MD et al (2000) The human nucleus of the solitary tract: visceral pathways revealed with an "in vitro" postmortem tracing method. J Auton Nerv Syst 79:181–190

Ryczko D, Dubuc R (2013) The multifunctional mesencephalic locomotor region. Curr Pharm Des 19:4448–4470

Saito T, Matsuba Y et al (2014) Single App knock-in mouse models of Alzheimer's disease. Nat Neurosci 17:661–663

Satpute AB, Wager TD et al (2013) Identification of discrete functional subregions of the human periaqueductal gray. Proc Natl Acad Sci U S A 110:17101–17106

Schaltenbrand G, Wahren W (1977) Atlas for stereotaxy of the human brain, 2nd edn. Thieme

Scheres SH, Zhang W et al (2020) Cryo-EM structures of tau filaments. Curr Opin Struct Biol 64:17–25

Schröder H, Moser N et al (2019) Neuroanatomy of the mouse. Springer

Schuchart S (2017) Berühmte Entdecker von Krankheiten: Alois Alzheimer hat nicht geirrt. (Famous discoverers of diseases: Alois Alzheimer was not at fault). Dtsch Arztebl 114: [120]

Shang C, Liu Z et al (2015) BRAIN CIRCUITS. A parvalbumin-positive excitatory visual pathway to trigger fear responses in mice. Science 348:1472–1477

Simič G, Stanic G et al (2009) Does Alzheimer's disease begin in the brainstem? Neuropathol Appl Neurobiol 35:532–554

Snowdon D (1997) Brain infarction and the clinical expression of Alzheimer's disease: findings from the Nun study. JAMA 277:813–817

Song G, Wang H et al (2012) Kölliker–Fuse neurons send collateral projections to multiple hypoxia-activated and nonactivated structures in rat brainstem and spinal cord. Brain Struct Funct 217:835–858

Stevens RT, Hodge CJ Jr et al (1982) Kolliker-Fuse nucleus: the principal source of pontine catecholaminergic cells projecting to the lumbar spinal cord of cat. Brain Res 239:589–594

Surbek B (1961) L'angiopathie dyshorique (Morel) de l'écorce cérébrale. Etude anatomo-clinique et statistique, aspect génétique. Acta Neuropathol 1:168–197

Szentagothai J (1948) Anatomical considerations on monosynaptic reflex arcs. J Neurophysiol 11:445–454

ten Donkelaar HJ (2011) Clinical neuroanatomy: brain circuitry and its disorders. Springer

Terry RD, Masliah E et al (1991) Physical basis of cognitive alterations in Alzheimer's disease: synapse loss is the major correlate of cognitive impairment. Ann Neurol 30:572–580

Thal DR, Rüb U et al (2002) Phases of Aß-deposition in the human brain and its relevance for the development of AD. Neurology 58:1791–1800

Thal DR, Capetillo-Zarate E et al (2006) The development of amyloid ß protein deposits in the aged brain. Sci Aging Knowl Environ 6(re1)

Thal DR, Walter J et al (2015) Neuropathology and biochemistry of Aβ and its aggregates in Alzheimer's disease. Acta Neuropathol 129:167–182

Theofilas P, Ehrenberg AJ et al (2018) Probing the correlation of neuronal loss, neurofibrillary tangles and cell death markers across the Alzheimer's disease Braak stages: a quantitative study in humans. Neurobiol Aging 61:1–12

Tokita K, Inoue T et al (2010) Subnuclear organization of parabrachial efferents to the thalamus, amygdala and lateral hypothalamus in C57BL/6J mice: a quantitative retrograde double labeling study. Neuroscience 171:351–365

Triarhou LC, Norton J et al (1988) Mesencephalic dopamine cell deficit involves areas A8, A9 and A10 in weaver mutant mice. Exp Brain Res 70:256–265

Uchihara T (2014) Pretangles and neurofibrillary changes: similarities and differences between AD and CBD based on molecular and morphological evolution. Neuropathology 34:571–577

Usunoff KG, Marani E et al (1997) The trigeminal system in man. Adv Anat Embryol Cell Biol 136:1–126

Van de Pol LA, Hensel A et al (2006) Hippocampal atrophy on MRI in frontotemporal lobar degeneration and Alzheimer's disease. J Neurol Neurosurg Psychiatry 77:439–442

Van Dort CJ, Zachs DP et al (2015) Optogenetic activation of cholinergic neurons in the PPT or LDT induces REM sleep. Proc Natl Acad Sci U S A 112:584–589

VanderHorst VG, Ulfhake B (2006) The organization of the brainstem and spinal cord of the mouse: relationships between monoaminergic, cholinergic, and spinal projection systems. J Chem Neuroanat 31:2–36

Vinters HV, Zarow C et al (2018) Review: vascular dementia: clinicopathologic and genetic considerations. Neuropathol Appl Neurobiol 44:247–266

Von Bergen M, Friedhoff P et al (2000) Assembly of tau protein into Alzheimer paired helical filaments depends on a local sequence motif (306VQIVYK311) forming β structure. PNAS 97: 5129–5134

Von Koelliker A (1896) Handbuch der Gewebelehre des Menschen (6. Aufl.) Vol. II Engelmann Verlag Leipzig

Watson C, Bartholomaeus C et al (2019) Time for radical changes in brain stem nomenclature: applying the lessons from developmental gene patterns. Front Neuroanat 13:10

Weiner MW, Veitch DP et al (2017) The Alzheimer's disease neuroimaging initiative 3: continued innovation for clinical trial improvement. Alzheimers Dement 13:561–571

Yokota O, Miki T et al (2018) Neuropathological comorbidity associated with argyrophilic grain disease. Neuropathology 38: 82–97

Young RF, Tronnier V et al (1992) Chronic stimulation of the Kölliker-Fuse nucleus region for relief of intractable pain in humans. J Neurosurg 76:979–985

Zhang W, Tarutani A et al (2020) Novel tau filament fold in corticobasal degeneration. Nature 580:283–287

Web Links

https://www.biodiversitylibrary.org/ia/abhandlungenderk1909kn/#page/8/mode/1up

http://braininfo.rprc.washington.edu/centraldirectory.aspx?ID=1841

http://www.ebi.ac.uk/interpro/entry/IPR008080(InterPro/entry—InterPro—EMBL-EBI)

https://www.mayoclinic.org/tests-procedures/deep-brain-stimulation/about/pac-20384562

Isthmus r0

14

Contents

H. Schröder et al., *The Human Brainstem*, https://doi.org/10.1007/978-3-030-89980-6_14

Abstract

In this neuromere, the second trigeminal nucleus following the spinal one (see Chap. 3) develops, i.e., the principal sensory trigeminal nucleus. By contrast to the protopathic properties of the spinal nucleus of the trigeminal nerve this nucleus is the epicritic component of the trigeminal system providing sensibility to the skin of the face, the cranial mucosa of mouth, nasal cavities, teeth, and sinuses. This distribution pattern—alongside that of the spinal trigeminal nucleus—indicates the clinical disciplines that potentially deal with the epicritic properties of the trigeminal nerve like neurology, neurosurgery, ENT, neuroradiology but also dentistry. The principal sensory trigeminal nucleus gets input from three major branches, the ophthalmic, the maxillar and the mandibular nerves. In addition to the sensory function, branches of the trigeminal nucleus subserve the guidance of parasympathetic and sympathetic fibers to their target organs, for example, to the lacrimal gland and the minor salivary glands. The trigeminal nerve plays an important role in neurological examination and for neurological disorders as, for example, trigeminal neuralgia and trigeminal herpes zoster.

The interpeduncular nuclei are the target structure of the habenulo-interpeduncular tract (formerly Fasciculus retroflexus) from the epithalamic habenular nuclei, one of the largest cholinergic tracts of the brain. Data from animal studies point to a role in the regulation of several behavioral patterns.

Barrington's nucleus—a derivative of rhombomere 2—is an important integration center for the regulation of the urinary bladder function.

The interpeduncular fossa, exit site of the oculomotor nerve, is an important anatomical and neuroradiological landmark. The close anatomical relationship between the oculomotor nerve and the vessels of the circle of Willis (see Chaps. 1 and 15) in the region of the interpeduncular fossa is the basis of aneurysm-related disturbances of the oculomotor nerve.

14.1 Isthmus

The isthmus of the mammalian brain is a distinct but relatively thin segment of the neural tube, separating the caudal mesencephalon from the first rhombomere. Usually, a small tissue bridge between two anatomical structures, in case of the central nervous isthmus a constriction between midbrain and the caudal rhombomeres, is meant.

According to recent findings, the isthmus is the rostral-most part of the rhombencephalon (for details, see Sect. 2.4 "Segmental development of the brainstem").

It is of Cre *Fgf8* lineage (Fgf8 = fibroblast growth factor, a developmental organizer for the midbrain-hindbrain boundary) and comprises the trochlear nucleus/nerve, the parabigeminal nucleus, the microcellular tegmental nucleus, and the decussation of the superior cerebellar peduncles (Watson et al. 2017).

14.2 Isthmus roof plate

14.2.1 Superior medullary velum (SMV) [Velum medullare superius]

The plate-like SMV (see Figs. 1.5, 1.9, and 13.34) is the rostral counterpart of the inferior medullary velum. Together (see Fig. 1.5) they form the roof of the fourth ventricle versus the cerebellum.

14.3 Isthmus rhombic lip

From the isthmus rhombic lip parts of the cerebellar nuclei and the cerebellar cortex develop (see Sect. 2.4).

14.4 Isthmus alar

14.4.1 Isthmus alar tegmentum

14.4.1.1 Cuneiform nucleus [Ncl. cuneiformis]

The cuneiform nucleus (cuneus Latin = wedge, not to confuse with the cuneate nucleus, see Sect. 3.2.3.3) is located ventrally of the inferior colliculus, laterally bordered by the lateral lemniscus, and medially by the periaqueductal gray (see atlas Darrow red 37, Fig. 14.12).

It is mainly composed of relatively densely packed small cells, intermingled with some medium-sized and large cells (ten Donkelaar 2011).

The cuneiform and the pedunculopontine nucleus (see Sect. 13.1.1.1.1) are assumed to form the physiologically defined mesencephalic locomotor region (MLR) in the midbrain reticular formation. The MLR is involved in locomotor rhythm generation and control of postural tone (MacKinnon 2018) (see also Sect. 13.1.1.1).

In addition, the cuneiform nucleus is part of the human descending pain modulatory pathway according to Sprenger et al. (2018).

In mice, the stimulation of the cuneiform nucleus has been suggested to induce antiepileptic effects similar to that elicited by stimulation of the part of the pedunculopontine tegmental nucleus (see Sect. 13.1.1.1.1) (Hong et al. 2014).

14.4.1.2 Dorsal nucleus of lateral lemniscus (DNLL) [Nucleus dorsalis lemnisci lateralis]

The dorsal nucleus of lateral lemniscus (DNLL) is part of the ascending auditory pathways. The lateral lemniscus consists of ascending auditory fibers from the brainstem to the inferior colliculus (see Sects. 15.1.4 and 8.1.1.5).

In mice, it appears histologically as a group of neurons scattered within the dorsal portion of the lateral lemniscus (see Iwahori 1986). By means of Golgi impregnation, Iwahori (1986) could distinguish three different neuron types in this nucleus which is considered to be one of the commissural nuclei of the ascending auditory system.

Bajo et al. (1999) have provided a detailed tracing study of the connection between DNLL and inferior colliculus in the cat. They concluded that the laminar structure of the DNLL may be the structural basis for its tonotopical organization and that this structural pattern might be common among mammals.

https://link.springer.com/referenceworkentry/10.1007%2F978-3-540-29678-2_4058

14.4.1.3 Microcellular tegmental nucleus [Ncl. tegmentalis microcellularis]

This nucleus is not listed in FIPAT and human-specific data are not available.

It is a cell group partially overlapping the lateral lemniscus in the pontine tegmentum of the mouse. Some authors regard this nucleus to be part of the sagulum nucleus (see Sect. 15.4.2).

http://braininfo.rprc.washington.edu/centraldirectory.aspx?ID=1825

14.4.1.4 Parabigeminal nucleus [Nucleus parabigeminalis]

The cholinergic neurons of the parabigeminal nucleus are called Ch8 (cholinergic) group (Mesulam et al. 1989). According to these authors, the parabigeminal nucleus with its small and hypochromic neurons is located at the lateral surface of the pontomesencephalic region with the lateral lemniscus as medial and the inferior colliculus as dorsal neighbor. 80–90% of the Nissl-stained neurons are moderately ChAT-immunoreactive (Mesulam et al. 1989).

The same group could show that, in rodents, the Ch8 neurons are the major source of cholinergic projections to the superior colliculi and that part of the cholinergic thalamic innervation also originates in Ch8.

In macaque monkeys, the parabigeminal nucleus projects to the dorsal lateral geniculate nucleus (Wilson et al. 1995).

In cats, Ma et al. (2013) have shown that the parabigeminal nucleus encodes estimations about target trajectories of momentarily invisible moving objects by combining sensory representations of target location, extrapolated positions of briefly obscured targets, and eye position information.

14.4.1.5 Paralemniscal nucleus

Detailed information on the para/perilemniscal nucleus is provided under Sect. 12.2.5.2.

14.5 Basal isthmus

14.5.1 Retrorubral field (RRF)

14.5.1.1 Dopamine cell group A8 (DA8)

14.5.1.1.1 Retrorubral field and A8 dopamine cells (RRF/A8)

Here, we encounter a terminology problem since two structures of the brainstem have been baptized "retrorubral" (Watson et al. 2019): the retrorubral dopaminergic field/A8 dopamine cell group (located in m2) and the retrorubral tegmental or reticular nucleus (r1). This has led to much confusion about the actual nuclei dealt with in many papers (for details, see Watson et al. 2019). In order to correct this problem, Paxinos and Watson (2014) have renamed the retrorubral nucleus as the retroisthmic nucleus. As described in detail by Watson et al. (2019) "… it lies immediately caudal to the caudal boundary of the isthmus. The retroisthmic nucleus is therefore defined as an area in rhombomere 1 between the pedunculotegmental nucleus medially, and the lateral lemniscus and its nuclei laterally. Rostrodorsal to it appears the microcellular tegmental nucleus of the isthmus, and rostral to it is the caudal (isthmic) pole of the substantia nigra. …..”

It is probable that the cholinergic retroisthmic nucleus is the rodent homolog of the pars dissipata of peduncular tegmental nucleus of primates (Watson 2012).

14.5.2 Isthmus trochlear complex

14.5.2.1 Trochlear nucleus/Ncl. of trochlear nerve [Ncl. n. trochlearis]

The trochlear nucleus is located partially inside the medial longitudinal fasciculus at the level of the caudal end of the mesencephalic aqueduct (Fig. 14.1).

14.5.2.1.1 Location and fine structure of the trochlear nucleus

The nuclei of the trochlear nerve are located in the mesencephalic tegmentum at the level of the red nucleus and the interpeduncular fossa (see Fig. 14.1). Mesulam et al.

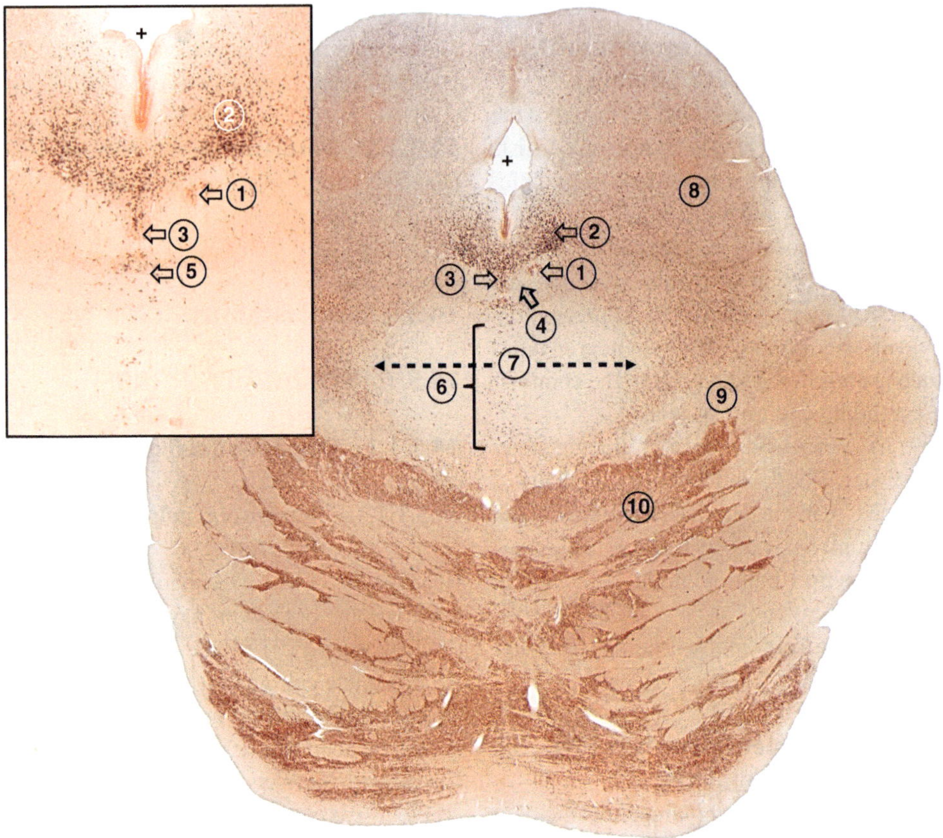

①	Ncl. of trochlear nerve CNIV	⑦	Decussation of the superior cerebellar peduncles
②	Supratrochlear subncl. of the dorsal raphe ncl.	⑧	Transition zone between inferior and superior colliculi
③	Interfascicular subncl. of the dorsal raphe ncl.	⑨	Medial lemniscus
④	Medial longitudinal fasciculus	⑩	Pontine nuclei
⑤	Anular subnucleus of the median (central) raphe ncl.	+	Mesencephalic aqueduct
⑥	Linear raphe ncl.		

Fig. 14.1 Location of the nucleus of trochlear nerve CNVI. Horizontal section through the human mesencephalon at the level of the supratrochlear subnucleus of the dorsal raphe ②. Darrow red staining. The reference point for the identification of the trochlear nucleus is the medial longitudinal fasciculus ④. At its dorsal margin, close to the dorsal raphe ③, the trochlear nucleus is located ①. For details on the decussation of the superior cerebellar peduncles ⑦, see Box 14.1. See also Atlas part Campbell 18. LabPON Twente

(1989) have shown that all of the large neurons in the oculomotor and trochlear nerve nuclei were ChAT-positive. Dendrites and perikarya remained confined to nuclear boundaries. The axons formed bundles that exited the nuclei as intensely ChAT-positive rootlets (Mesulam et al. 1989).

Olszewski and Baxter (1982) described that there is an almost seamless continuation of the oculomotor nucleus (see Sect. 15.3.1.1) to the main part of the trochlear nucleus. The latter, however, has some smaller groups in a more caudal position which are embedded into the medial longitudinal fasciculus. This is the situation which is shown in Fig. 14.1.

14.5.2.1.2 Target muscle of the trochlear nerve

The target muscle of the trochlear nerve is the superior oblique muscle [M. obliquus superior] (Fig. 14.2). This muscle originates at the tendinous sheath of superior oblique muscle [Vagina tendinis m. obliqui superioris] on the lesser wing of the sphenoid bone medial to the optic canal (Remington 2012). From there, the muscle runs close to the medial wall of the orbita rostrally until it reaches the trochlea [Latin for pulley, hence the name of the nerve]. The trochlea is a U-shaped piece of cartilage at the orbital plate of the frontal bone (Remington 2012). Before entering the trochlea, the muscle passes into its tendon of insertion, passes through the trochlea and then lies inferior to the superior rectus muscle (see Fig. 14.3). From its

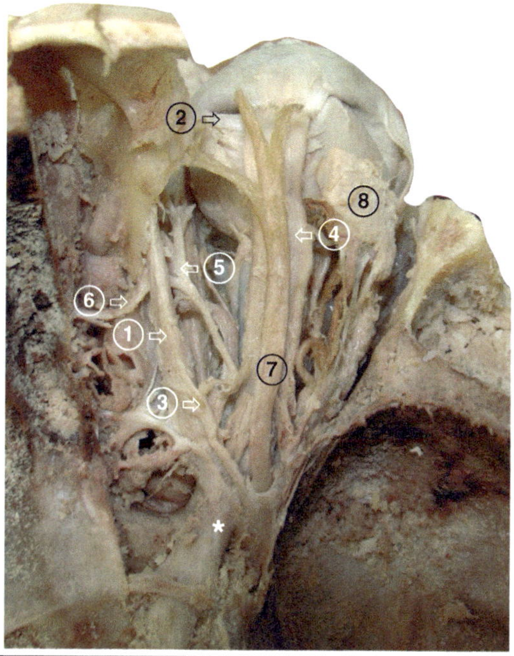

①	Superior oblique muscle CNIV, muscle belly	⑥	Anterior ethmoideal nerve CNV_1
②	Superior oblique muscle CNIV, tendon	⑦	Frontal nerve CNV_1
③	Trochlear nerve CNIV	⑧	Lacrimal gland
④	Superior rectus muscle CNIII	*	Optic nerve
⑤	Infratrochlear nerve CNV_1		

Fig. 14.2 Dorsal view onto the human orbita. Liquid-preserved specimen. The dorsal roof of the orbita, the levator palpebrae superioris muscle (CNIII) and the orbital fat have been removed to show the intraorbital structures Fig. 14.3. The superior oblique muscle ① is the medial most one. The trochlear nerve ③, the belly of the superior obliquus muscle ① and its tendon ② are visible. The insertion zone at the eye bulb below the superior rectus muscle (CNIII) is shown in Fig. 14.3. At the periorbita, the muscle tendon is running through the trochlea (name of CNVI!) which changes its course from a rostral to a lateral direction ② (for details, see Remington 2012). Müller-Thomsen fecit. Sammlung des Zentrums Anatomie der Universität zu Köln

originally rostral direction the course of the tendon is changed by the trochlea to a lateral direction.

14.5.2.1.3 Course of the trochlear nerve

The trochlear nerve is the only cranial nerve that leaves the brainstem at its dorsal surface and contralateral to its nucleus. It crosses the midline inside the brainstem (Decussation of trochlear nerves [Decussatio fibrarum nervorum trochlearium] Fig. 14.5 ②) and exits at the dorsal surface caudal to the inferior colliculi (Figs. 14.4 and 14.5).

The position of the nucleus in comparison to the other cranial nerve nuclei is shown in Fig. 14.4.

The nerve is then traversing the cerebellopontine cistern (see Box 3.3) and enters the cavernous sinus (Fig. 14.6). It runs in the wall of the sinus caudal of the oculomotor nerve (see Fig. 14.9) and enters the orbita together with the other oculomotor nerves via the superior orbital fissure.

14.5.2.1.4 Functional considerations, living anatomy, and clinical implications of the trochlear nerve

In simplified terms, the superior oblique muscles move the eyeball upon contraction downwards (depression) and medially toward the nose (intorsion) (Scully 2014).

The specific clinical investigation of the eye motility related to the trochlear nerve is shown in Fig. 14.10. The palsy of the trochlear nerve (isolated loss of superior oblique muscle) results in problems with normal adduction of the eyes. Patients suffer from diplopia with diplopic images which are shifted upwards and laterally to each other. In daily life, this causes main problems for reading and going downstairs. To compensate for the diplopia, the patients adopt a chin-down and contralateral head-tilt position (Eggenberger and Pula 2014).

Figure 14.10 shows an example of a right trochlear palsy (Abkur 2017). When testing the eye motility as shown in this figure, the patient will complain about the diplopia being

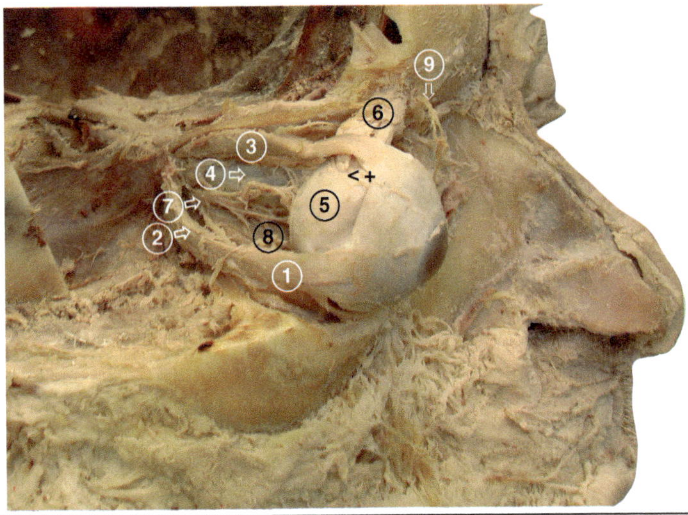

①	Lateral rectus muscle CNVI	⑥	Superior oblique muscle CNIV Tendon part
②	Abducens nerve CNVI	+	Superior oblique muscle CNIV Insertion site
③	Medial rectus muscle CNIII	⑦	Ciliary ganglion
④	Optic nerve	⑧	Ciliary nerves
⑤	Ocular bulb	⑨	Supraorbital nerve CNV_1

Fig. 14.3 Lateral view onto the human orbita. Liquid-preserved specimen. The lateral wall of the orbita, its roof and the A. and N. lacrimalis have been removed to allow for the visualization of the structures inside the orbita. Compare with Fig. 14.2. Müller-Thomsen fecit. Sammlung des Zentrums Anatomie der Universität zu Köln

worst when looking down and to the left (head fixed). This is an indication for the weak muscles to be either the left inferior rectus (CNIII) (see Sect. 15.3.1.1.2) or the right superior oblique muscle (Abkur 2017).

The causes of a trochlear palsy are manifold but traumatic injury is the most common cause of a trochlear nerve palsy because of the length of the nerve, as well as its emergence from the dorsum of the brainstem. Other causes are ischemic or vascular disease as in the case described here and congenital cases, such as maldevelopment of the trochlear nerve (Morillon and Bremner 2017).

14.5.2.2 Trochlear nucleus shell region

The trochlear nucleus shell region is a small area at the dorsomedial surface of the trochlear nucleus in mice. No human-specific data are available.

14.5.2.3 Paratrochlear nucleus [Ncl. paratrochlearis]

The term paratrochlear nucleus refers to a group of cells adjacent to the medial longitudinal fasciculus of the midbrain and the trochlear nucleus in the mouse. No human-specific data are available.

http://braininfo.rprc.washington.edu/centraldirectory.aspx?ID=1109

14.5.3 Isthmus Raphe

14.5.3.1 Caudal linear nucleus of the raphe/Caudal linear nucleus (CLi) [Ncl. linearis caudalis]

The serotonergic caudal linear nucleus of the raphe—located in the midline of the brainstem—is one of two parts (second

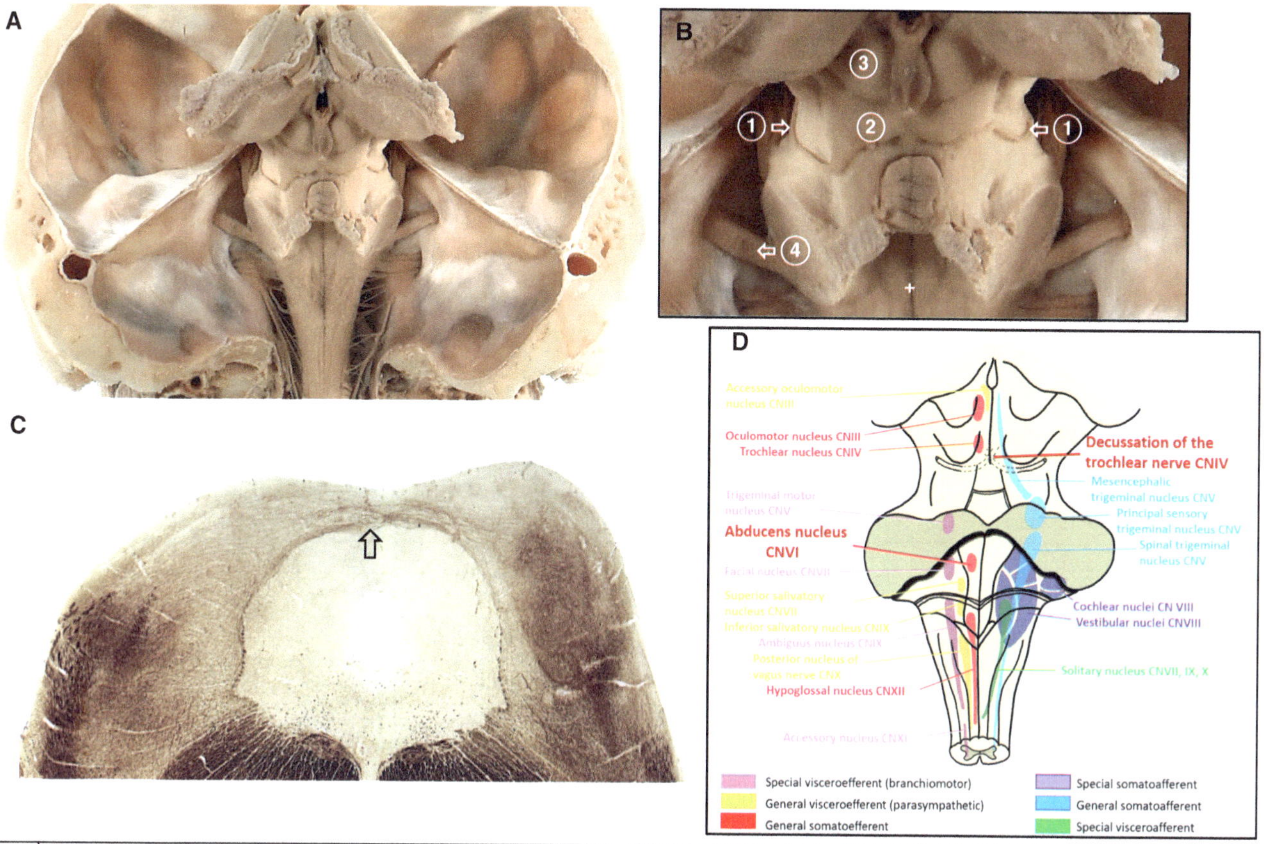

①	**Trochlear nerve CNIV**	④	**Vestibulocochlear nerve CNVIII**
②	**Inferior colliculus**	+	**4th ventricle / Rhomboid fossa**
③	**Superior colliculus**	⇨	**Decussation of CNIV**

Fig. 14.4 (**A**) Dorsal view of the human brainstem. Formalin-fixed specimen. At higher magnification (**B**), the dorsal exit of the trochlear nerve ① caudal of the inferior colliculus ② can be seen. The nerves cross to the contralateral side inside the brainstem (compare with the Campbell stain in (**C**) and with Fig. 14.5). After exiting, they run around the brainstem ventral of the inferior colliculi to become visible at the ventrolateral surface of the mesencephalon (see also Fig. 14.5). The schematic illustration (**D**) shows the intracerebral location of the nucleus of trochlear nerve (see Fig. 14.1) in relation to the exit of the nerve. (**A**, **B**) Sammlung des Zentrums Anatomie der Universität zu Köln. (**C**) LabPON Twente. (**D**) Modified after Huggenberger et al. 2019, Fig. 15.5 with permission

one: rostral linear nucleus of the raphe) of the linear nucleus of the raphe in the ventral mesencephalic tegmentum at the level of the oculomotor nuclear complex (Halliday and Törk 1986, see Figs. 14.11 and 14.12). It is the most rostral serotonergic nucleus in the brainstem. In detail, it is located in the ventral tegmental area, dorsally and caudally of the interpeduncular nucleus, flanked by the dentato-thalamic axons (see Box 14.1, Fig. 14.8) in the superior cerebellar peduncle (Halliday and Törk 1986). In addition to the 5-HT neurons—ten times less numerous than in the dorsal (see Sect. 13.3.4.2) or median raphe—the CLi also contains tyrosine hydroxylase-containing (catecholamine synthesis, see Fig. 3.3) and substance P-containing neurons (Halliday et al. 1990).

FIPAT Ch. 1 has the group of ventral tegmental nuclei [Nuclei tegmentales ventrales] comprising the caudal linear nucleus, the rostral linear nucleus, the paranigral nucleus (see Sect. 14.6.1), the interfascicular nucleus (see Sect. 14.5.4.3), the parabrachial pigmented (see Sect. 15.5.1.2), and the parapeduncular nucleus.

FIPAT Ch. 1 lists a caudal (#1430) and a rostral linear nucleus (#1431) [Ncl. linearis caudalis/rostralis] and a linear raphe nucleus (#1412) [Ncl. raphes linearis], obviously identical with the serotonergic cell group B8. The corresponding FIPAT endnote to the latter, #64, states that, because the Ncl. linearis inferior (All terms in the endnote in Latin) contains some serotonergic neurons, Nieuwenhuys et al. 2008 called

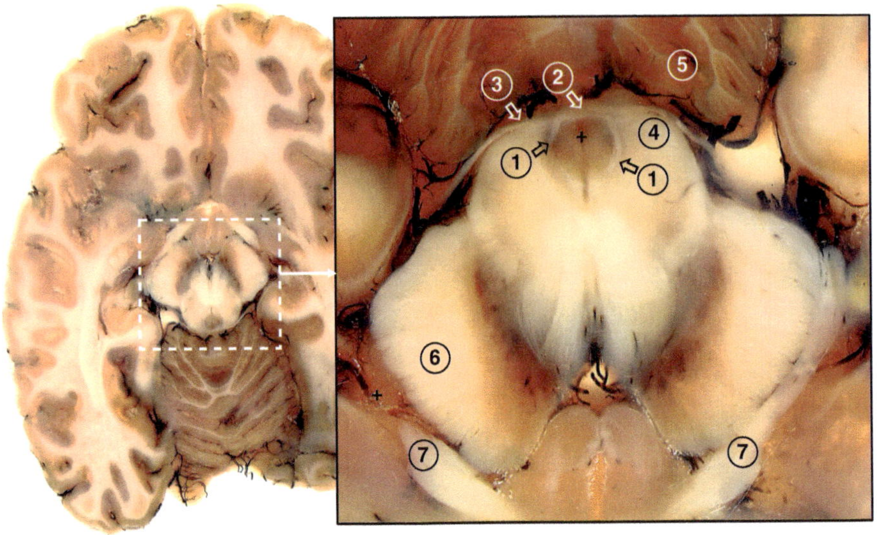

①	Trochlear nerve CNIV inside the brainstem	⑤	Cerebellum
②	Decussation of the trochlear nerves	⑥	Cerebral peduncle
③	Exit of the trochlear nerve	⑦	Optic tract
④	Inferior colliculus	+	Mesencephalic aqueduct

Fig. 14.5 Horizontal section through the human brain at the level of the exit of the trochlear nerves. Plastinated specimen. Note that the overview (left) is dorsoventrally inverted. The trochlear nerves ① run from the trochlear nuclei (see Fig. 14.1) in laterodorsal direction, lateral to the periaqueductal gray around the mesencephalic aqueduct +, and cross dorsal of the latter inside the mesencephalon to the other side. Shortly thereafter, they leave the brainstem ③ and start running in ventrorostral direction (see Figs. 14.4 and 14.6). Sammlung des Zentrums Anatomie der Universität zu Köln

Box 14.1 Cerebellar peduncles

1. The superior cerebellar peduncles [Pedunculus cerebellaris superior] contain a compact, sickle-shaped fiber bundle, the brachium conjunctivum, which consists of fibers from all the ipsilateral cerebellar nuclei (Voogd and Rugirok 2012) (Figs. 14.7 and 14.8). They are the main efferent pathway of the cerebellum and decussate at the level of the inferior colliculi (Fig. 14.5 ④ and Fig. 14.6 ⑤).

   The dorsomedial portion of the peduncles and their ventral fibers are the first to decussate, the middle part decussates at a more rostral level (Kweldam et al. 2014).

   The term decussation/decussatio comes from Latin decim = X = ten. The X is thought as a symbol for a crossing.

   In detail, the superior cerebellar peduncle comprises the fiber bundles entering or leaving the cerebellum rostrally that include the ventral spinocerebellar and uncinate tracts, the rostral-most fibers of the inferior cerebellar peduncle (Fig. 14.7), and a bundle of nucleo-olivary fibers (see Fig. 14.8) from the dentate and interposed nuclei of the cerebellum (Voogd and Rugirok 2012). The hooked bundle (uncinate tract/fasciculus of the cerebellum [Fasciculus uncinatus cerebelli], tract/bundle of Russell, not to confuse with the uncinate fasciculus of the telencephalon) is an efferent pathway from the fastigial cerebellar nucleus, crossing within the cerebellum and reaching the vestibular nuclei (see Sect. 5.3.1) and the pontine and bulbar reticular formation. Cerebellar input into the dorsal thalamus has been shown for a number of mammals. In primates, cerebellar input into the thalamus reaches the ventrolateral and the ventroposterior motor nuclei (cf. Tlamsa and Brumberg 2010).

   Voogd and Rugirok (2012) demonstrated one of the few cases in the literature where the superior cerebellar peduncle was traced in a human brain with Nauta's silver impregnation of degenerated axons after a lesion of the cerebellar nuclei which shows the targeting of the thalamic motor nuclei (dentato-rubro-thalamic tract) (see Fig. 14.8). Nauta's technique is based on the detection of degenerating axons to the point of terminal boutons by an impregnation with metallic silver applicable in animal experimental studies as well as in post-mortem examination of the human brain (Saleeba et al. 2019, 2020).

Using tractography, the existence of a crossing dentate-thalamic tract could be shown in vivo in humans (Law et al. 2012). Via an efferent connection to the frontal cortex, this tract is an important component of a motor regulatory loop. Disruption of this tract may contribute to speech-language problems in children (Law et al. 2012).

Axons from vestibulooculomotor neurons of the superior vestibular nucleus (see Sect. 11.2.2.1), and possibly from the group Y, that transmit information from the anterior semicircular canal (see Sect. 5.3.1.1.2), make up the extreme ventrolateral pole of the superior cerebellar peduncles and join its decussation (Voogd and Rugirok 2012).

2. The middle cerebellar peduncle [Pedunculus cerebellaris medius] (Brachium pontis), a purely afferent system, contains the pontocerebellar (now transverse pontine fibers) mossy fibers from the pontine nuclei to the cerebellum.
3. The outer portion of the inferior cerebellar peduncle [Pedunculus cerebellaris inferior] is the afferent, so-called restiform body [Corpus restiforme], which primarily contains spinocerebellar, trigeminocerebellar, olivocerebellar, and arcuate fibers (ten Donkelaar 2011). The medial, partially efferent, juxtarestiform body contains the dorsal spinocerebellar tract (Nieuwenhuys et al. 2008).

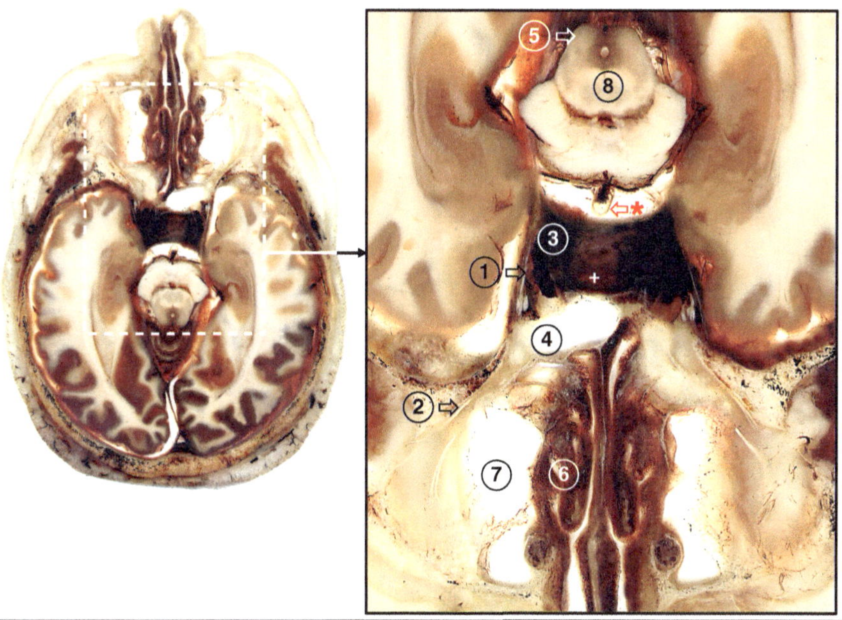

①	Trochlear nerve CNIV in the wall of the cavernous sinus	⑥	Nasal cavity
②	Trochlear nerve CNIV, entry into the orbita	⑦	Maxillary sinus
③	Cavernous sinus	⑧	Decussation of superior cerebellar peduncles
④	Sphenoid sinus	+	Pituitary gland
⑤	Quadrigeminal plate	*	Basilar artery

Fig. 14.6 Horizontal section through the human brain at the level of the decussation of superior cerebellar peduncles ⑧ [Decussatio pedunculorum cerebellarium superiorum] (see Box 14.1). Note that the overview (left) is dorsoventrally inverted. Sammlung des Zentrums Anatomie der Universität zu Köln

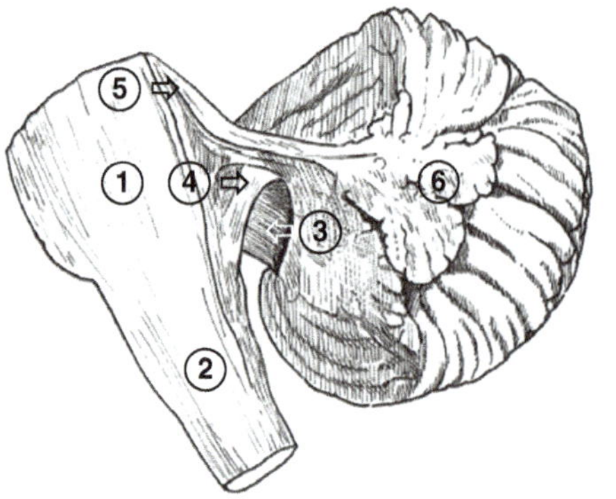

①	**Pons**	④	**Middle cerebellar peduncle**
②	**Medulla oblongata**	⑤	**Superior cerebellar peduncle**
③	**Inferior cerebellar peduncle**	⑥	**Cerebellum**

Fig. 14.7 Drawing of a human brainstem specimen in which the connection of the brainstem with the cerebellum via the three cerebellar peduncles was dissected. The topographical situation is in so far complicated as the three peduncles do not target the cerebellum in a clear-cut longitudinal order (for details see Nieuwenhuys et al. 2008). Ribbers fecit

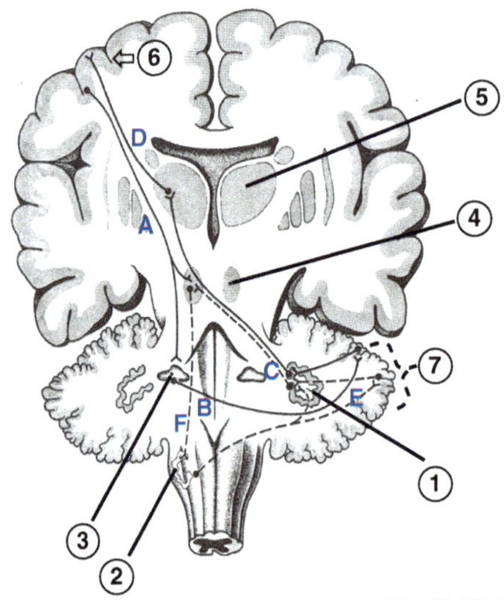

①	**Dentate nucleus**	A	**Corticopontine fibers**
②	**Inferior olivary complex**	B	**Transverse pontine / Pontocerebellar* fibers**
③	**Pontine nuclei**	C	**Dentate-rubro-thalamic tract***
④	**Red nucleus**	D	**Thalamocortical fibers***
⑤	**Thalamus / Ventral lateral complex**	E	**Olivocerebellar tract**
⑥	**Cerebral cortex**	F	**Central tegmental tract**
⑦	**Cerebellar cortex / Purkinje cells**		

Fig. 14.8 Schematic representation of the cortico-cerebellar-thalamocortical loop. It starts in the frontal cortex ⑥ (**A**), runs through the internal capsule between thalamus ⑤ and basal ganglia and mainly targets the pontine nuclei ③ (see Sect. 10.1.1.2) with collaterals to the red nucleus ④ (see Sect. 15.3.4). From the pontine nuclei the transverse pontine (or formerly and better pontocerebellar) fibers (**B**) cross to the contralateral half of the cerebellum where they make synaptic contact with the Purkinje cells in the cerebellar cortex ⑦. All cerebellar output runs via the Purkinje axons to the cerebellar nuclei, in particular the dentate nucleus ①. From there the next axon of the loop (**C**) crosses the midline and reaches the ventrolateral nucleus of the thalamus ⑤ after decussating and sending fibers to the contralateral red nucleus ④. Eventually, the loop is closed by a thalamocortical connection (**D**). Functionally related to the loop are (1) the central tegmental tract (**F**) (see Sect. 4.1.1.3.2 and Fig. 3.14B–G) from the red nucleus to the inferior olive (see Sect. 4.1.1.3) and (2) from there the climbing fiber olivocerebellar tract (**E**) with collaterals to the dentate nucleus ① (see Fig. 3.14B ⑦). Ribbers fecit. * These terms are common but not listed in FIPAT Ch. 1

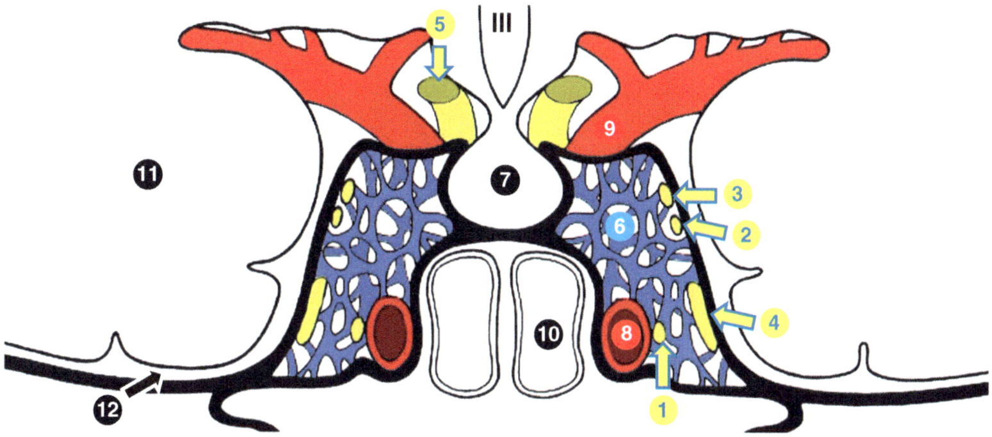

(1)	Abducens nerve CNVI	(8)	Internal carotid artery, cavernous part
(2)	Trochlear nerve CNIV	(9)	Middle cerebral artery
(3)	Oculomotor nerve CNIII	(10)	Sphenoid sinus
(4)	Ophthalmic nerve CNV$_1$	(11)	Telencephalon
(5)	Optic nerve	(12)	Leptomeningeal space
(6)	Cavernous sinus / trabeculae	▬	Dura mater
(7)	Pituitary gland	III	3rd ventricle

Fig. 14.9 Schematic representation of a frontal/coronal section through the human head at the level of the pituitary gland (7) and the cavernous sinus (6). The trochlear nerve (2) is located in the wall of the cavernous sinus between the oculomotor nerve dorsal and the ophthalmic nerve ventral of it. Ribbers et S. Huggenberger fecerunt

it Ncl. raphes linearis. Furthermore, the terminology introduced by Paxinos (Mai and Paxinos 2012) is mentioned which calls the median serotonergic part "azygos part of CLi" and the paramedian parts as "zygos parts of CLi". Dealing this way with the different parts of the CLi, in our opinion, does not really contribute to the necessary clarity.

14.5.3.2 Dorsal Raphe nucleus (DR)

For details on this nucleus, see Sect. 13.3.4.2.

14.5.4 Isthmus basal tegmental nuclei

14.5.4.1 Isthmic reticular formation

The isthmic reticular formation, together with the mesencephalic reticular formation, the microcellular tegmental nucleus, and the ventral tegmental area (VTA) (see Sect. 15.5.1.1), is a region involved in arousal, sleep, and reward (for details, see Singh et al. 2021).

14.5.4.2 Isthmus interpeduncular nuclei

For details on this nucleus, see Sect. 12.3.3.

14.5.4.3 Interfascicular nucleus (IF) [Ncl. interfascicularis]

The interfascicular nucleus is located dorsal to the interpeduncular nucleus (see Fig. 14.11E (9)). In contrast to rats, cats, and monkeys, in which the interfascicular nucleus is a median cluster, the human IF is small and consists of two paramedian clusters connected only by cell bridges (Halliday and Törk, 1986). Compared to other nuclei of the ventral mesencephalic tegmentum, it has significantly smaller cells (Halliday and Törk, 1986). Both the cells and the neuropil are densely AChE positive.

The interfascicular nucleus is part of the A10 cell group (Yamaguchi et al. 2011). It is located between the tractus habenulointerpeduncularis, formerly fasciculi retroflexi.

http://braininfo.rprc.washington.edu/centraldirectory.aspx?ID=1191

14.5.4.4 Interpeduncular nucleus

For detailed information on the interpeduncular nucleus, see Sect. 14.5.4.2.

Detailed information is given in Chap. 12 under Sect. 12.3.3.

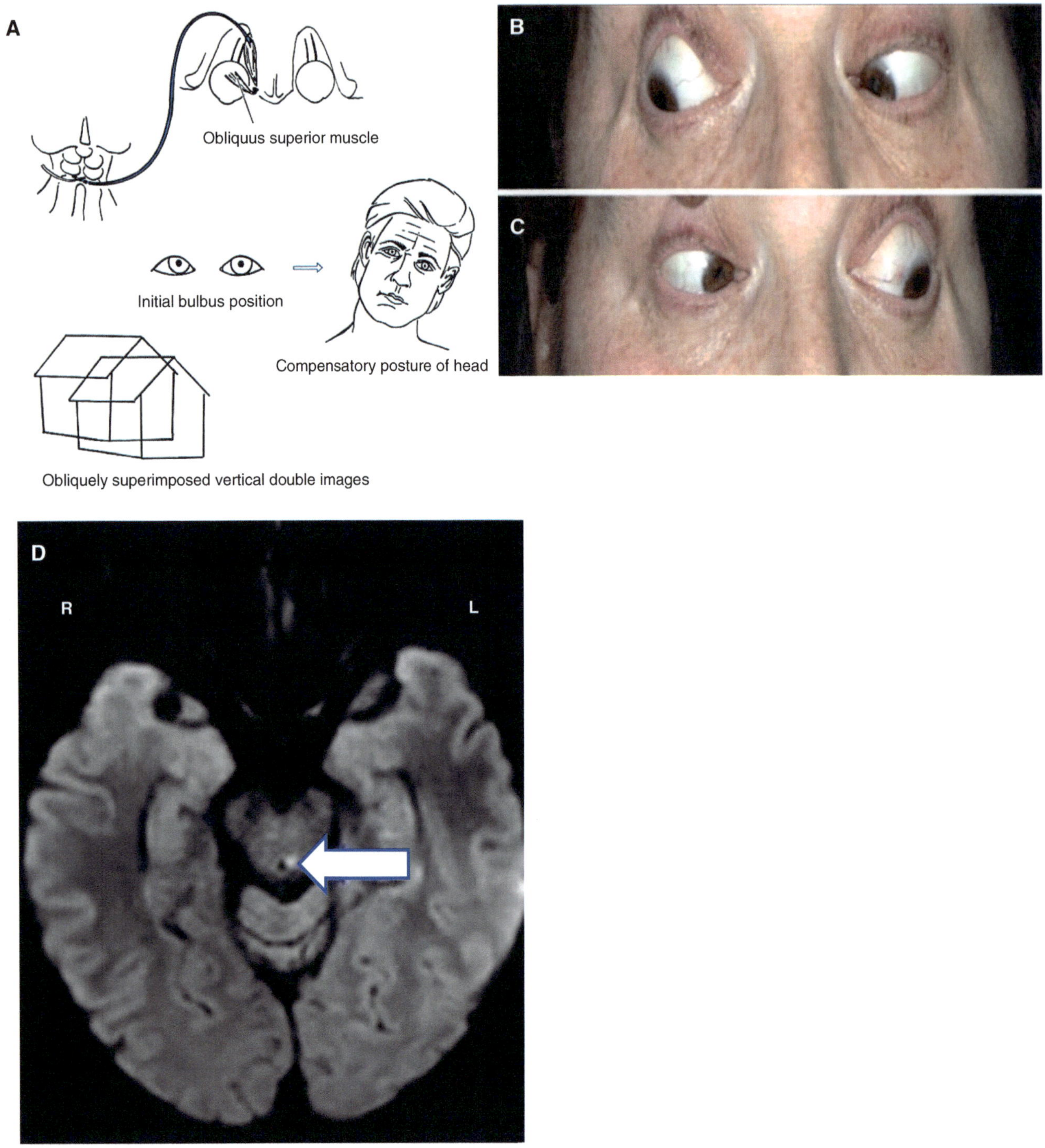

Fig. 14.10 Trochlear palsy. (**A**) Schematic drawing of the trochlear nerve (CNIV) and the superior oblique muscle. On inspection (upright head position), the right eye is shifted to the left. Patients try to minimize the disturbing diplopia by a compensatory tilting of the head away from the affected superior oblique muscle (and thus toward the lesion, see **D**). From Berlit 2014, Fig. 5.6 with permission. (**B**, **C**) Neuroophthalmological findings in a 53-year-old patient with a history of hypertension and hyperlipidemia who acutely developed a diplopia on looking downwards or to the left. (**B**) Normal left superior oblique action. (**C**) Impaired right eye depression in adduction. (**D**) MR imaging showed a left-sided hyperintense lesion—interpreted as infarction (see Sect. 1.4.2)—lateral to the mesencephalic aqueduct, anatomically corresponding to the intracerebral course of the trochlear nerve (see Fig. 14.5). From Abkur 2017, Figs. 1 and 2 with permission

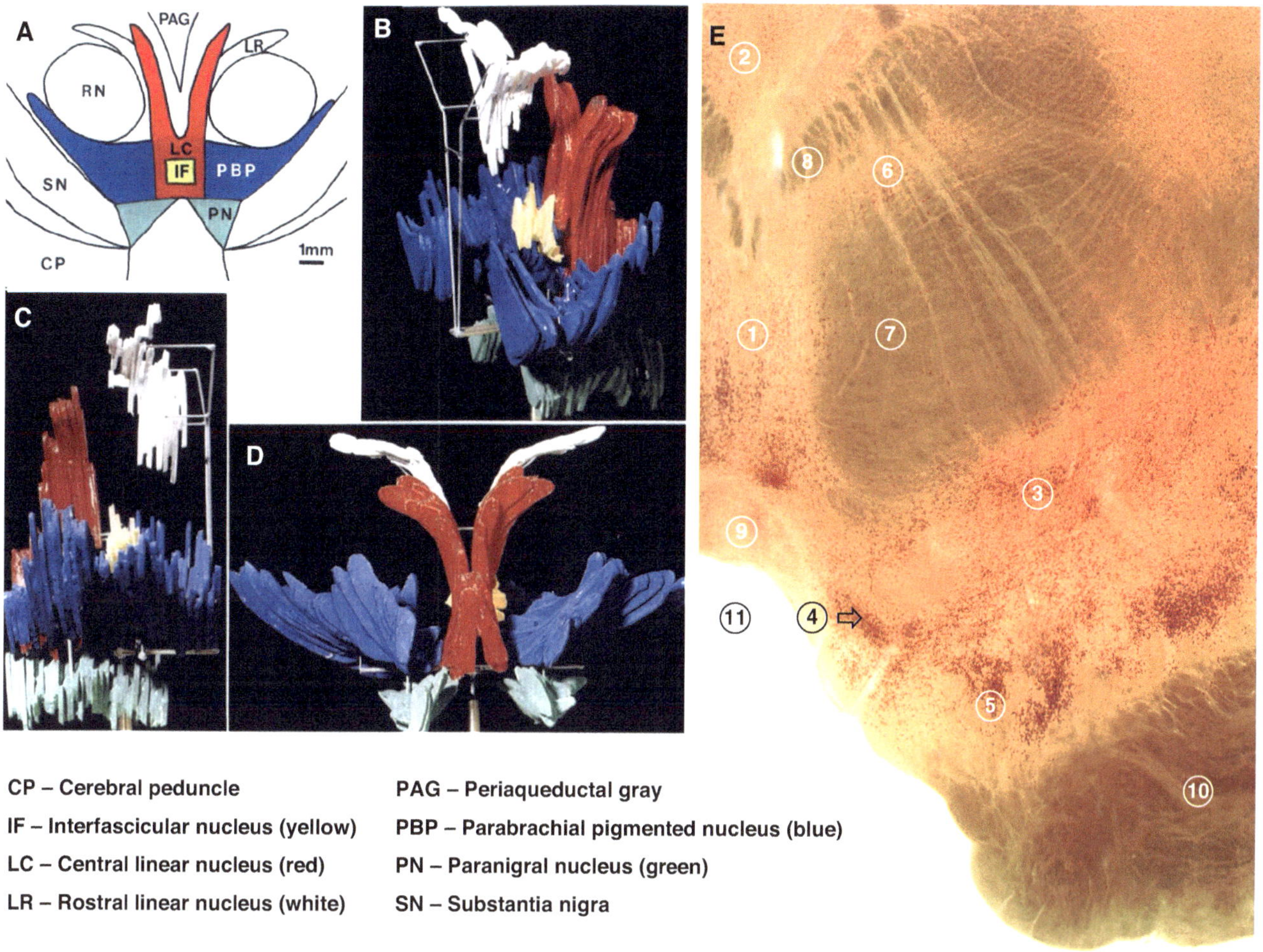

①	Linear raphe ncl.	⑦	Red ncl.
②	Dorsal raphe ncl.	⑧	Medial longitudinal fasciculus
③	Parabrachial pigmented ncl.	⑨	Interpeduncular ncl.
④	Paranigral ncl.	⑩	Cerebral peduncle
⑤	Substantia nigra, compact part	⑪	Interpeduncular fossa
⑥	Rostral linear ncl.		

Fig. 14.11 The ventromedial mesencephalic tegmentum in a schematic representation (**A**) and 3D reconstruction (**B–D**). Compare with the right-hand histological section of the nuclei of the ventromedial mesencephalic tegmentum (**E**) (Darrow red stain) (see also 3.2.1.3). (**A–D**) Modified acc. to Halliday and Törk 1986, Fig. 10. (**E**) LabPON Twente

14.6 Isthmus floor plate

14.6.1 Paranigral nucleus of the VTA/ Paranigral nucleus [Ncl. paranigralis]

The paranigral nucleus is part of the ventral tegmental area (VTA) (see Sect. 15.5.1.1 and Fig. 14.12 ⑫). It contains pigmented, dopaminergic cells associated with the medial extremity of the substantia nigra (SN). It extends from the level of the caudal pole of the SN to the most rostral rootlets of the oculomotor nerve (5 mm). Caudally, it is medially bordered by the interpeduncular nucleus (see Sect. 12.3.3), ventrally by the subpial layers of the interpeduncular fossa (see Sect. 12.4.1), laterally by the SN (see Sect. 16.6.1) and dorsally by the parabrachial pigmented nucleus (see Sect. 15.5.1.2). The nuclei of both sides bridge at the caudal pole of the oculomotor complex. Orally, the paranigral nucleus is related laterally to the SN, the other portions are in contact with the posterior perforate substance.

The neurons are small to medium-sized, slender, multipolar with long dendrites and contain melanin (Olszewski and Baxter 1982).

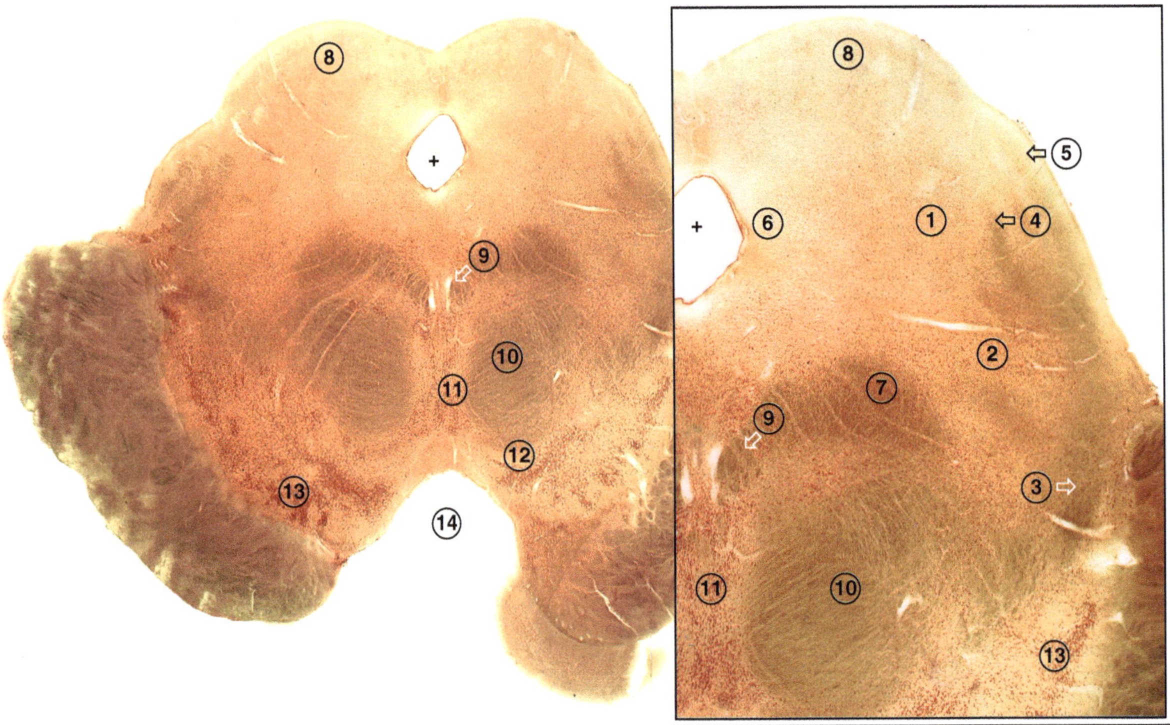

①	Cuneiform ncl.	⑨	Medial longitudinal fasciculus
②	Subcuneiform ncl.	⑩	Superior cerebellar peduncle / Red ncl.
③	Medial lemniscus	⑪	Linear raphe ncl. / Caudal (inferior) linear ncl.
④	Medial lemniscus	⑫	Paranigral ncl.
⑤	Brachium of the inferior colliculus	⑬	Substantia nigra
⑥	Periaqueductal gray	⑭	Interpeduncular fossa
⑦	Central tegmental tract	+	Mesencephalic aqueduct
⑧	Superior colliculus		

Fig. 14.12 Horizontal section through the human cranial mesencephalon. Darrow red staining. See also atlas part Campbell 19. LabPON Twente

References

Abkur TM (2017) Trochlear nerve palsy. Pract Neurol 17:474–475

Bajo VM, Merchán MA et al (1999) Topographic organization of the dorsal nucleus of the lateral lemniscus in the cat. J Comp Neurol 407:349–366

Berlit P (2014) Erkrankungen der Hirnnerven. In: Basiswissen Neurologie. Springer-Lehrbuch. Springer, Berlin, Heidelberg

Halliday GM, Törk I (1986) Comparative anatomy of the ventromedial mesencephalic tegmentum in the Rat, Cat, Monkey and Human. J Comp Neurol 252:423–445

Halliday GM, Gal WP et al (1990) Substance P-containing neurons in the pontomesencephalic tegmentum of the human brain. Neuroscience 39:81–96

Hong Q, Ke B et al (2014) Cuneiform nucleus stimulation as adjunct treatment for intractable epilepsy: a virally mediated transsynaptic tracing study in spinally transected transgenic mice. Epilepsy Behav 33:135–137

Huggenberger S, Moser N et al (2019) Neuroanatomie des Menschen. Springer

Eggenberger ER, Pula JH (2014) Neuro-ophthalmology in medicine. In: Aminoff's neurology and general medicine (fifth edition)

Iwahori N (1986) A Golgi study on the dorsal nucleus of the lateral lemniscus in the mouse. Neurosci Res 3:196–212

Kweldam CF, Gwynn H et al (2014) Undecussated superior cerebellar peduncles and absence of the dorsal transverse pontine fibers: a new axonal guidance disorder? Cerebellum 13:536–540

Law N, Greenberg M et al (2012) Clinical and neuroanatomical predictors of cerebellar mutism syndrome. Neuro-Oncology 14:1294–1303

Ma R, Cui H et al (2013) Predictive encoding of moving target trajectory by neurons in the parabigeminal nucleus. J Neurophysiol 108:2029–2043

Mai JK, Paxinos G (eds) (2012) The human nervous system, 3rd edn. Elsevier, Amsterdam

MacKinnon CD (2018) Balance, gait, and falls. Handbook of clinical neurology. Elsevier

Mesulam M-M, Geula C et al (1989) Human reticular formation: cholinergic neurons of the pedunculopontine and laterodorsal tegmental nuclei and some cytochemical comparisons to forebrain cholinergic neurons. J Comp Neurol 281:611–633

Morillon P, Bremner F (2017) Trochlear nerve palsy. Br J Hosp Med (Lond) 78:C38–C40

Nieuwenhuys R, Voogd J, van Huijzen C (2008) The human central nervous system, 4th edn. Springer

Olszewski J, Baxter D (1982) Cytoarchitecture of the human brain stem, 2nd edn. Karger, Basel

Paxinos G, Watson C (2014) Paxinos and Watson's the rat brain in stereotaxic coordinates, 7th edn. Elsevier Academic Press, San Diego, CA

Remington LA (2012) Clinical anatomy and physiology of the visual system, 3rd edn. Elsevier

Saleeba C, Dempsey B et al (2019) A student's guide to neural circuit tracing. Front Neurosci 13:897

Saleeba C, Dempsey B et al (2020) Corrigendum: a student's guide to neural circuit tracing. Front Neurosci 14:177. Erratum for: Front Neurosci. 2019 Aug 27;13:897

Scully C (2014) Scully's medical problems in dentistry (seventh edition)

Singh K, García-Gomar MG et al (2021) Probabilistic atlas of the mesencephalic reticular formation, isthmic reticular formation, microcellular tegmental nucleus, ventral tegmental area nucleus complex and caudal-rostral linear raphe nuclei complex in living humans from 7 Tesla MRI. Brain Connect. Epub ahead of print

Sprenger C, Eichler IC et al (2018) Altered signaling in the descending pain modulatory system after short-term infusion of the μ-opioid agonist remifentanil. J Neurosci 38:2454–2470

ten Donkelaar HJ (2011) Clinical neuroanatomy: brain circuitry and its disorders. Springer

Tlamsa AP, Brumberg JC (2010) Organization and morphology of thalamocortical neurons of mouse ventral lateral thalamus. Somatosens Mot Res 27:34–43

Voogd J, Rugirok TJH (2012) Cerebellum and precerebellar nuclei. In: The human nervous system. (Third Edition) Elsevier

Watson C (2012) Hindbrain: cholinergic nuclei. In: Watson C, Paxinos G, Puelles L (eds) The mouse nervous system. Academic Press

Watson C, Shimogori T et al (2017) Mouse Fgf8-Cre-LacZ lineage analysis defines the territory of the postnatal mammalian isthmus. J Comp Neurol 525:2782–2799

Watson C, Bartholomaeus C et al (2019) Time for radical changes in brain stem nomenclature: applying the lessons from developmental gene patterns. Front Neuroanat 13:10

Wilson JR, Hendrickson AE et al (1995) Sources of subcortical afferents to the macaque's dorsal lateral geniculate nucleus. Anat Rec 242:566–574

Yamaguchi T, Wang HL et al (2011) Mesocorticolimbic glutamatergic pathway. J Neurosci 31:8476–8490

Web Links

http://braininfo.rprc.washington.edu/centraldirectory.aspx?ID=1109
http://braininfo.rprc.washington.edu/centraldirectory.aspx?ID=1825
http://braininfo.rprc.washington.edu/centraldirectory.aspx?ID=1191
https://link.springer.com/referenceworkentry/10.1007%2F978-3-540-29678-2_4058

Mesencephalon m1/m2

15

Contents

H. Schröder et al., *The Human Brainstem*, https://doi.org/10.1007/978-3-030-89980-6_15

Abstract

Two important structures of the visual/oculomotor system originate from this neuromere, the superior colliculus, and the oculomotor nucleus/nerve CNIII.

The superior colliculus is an important mesencephalic relay station of the visual system.

The auditory counterpart of the superior colliculus is the inferior colliculus. It collects the signals from the ascending auditory system and propagates them via the brachium of the inferior colliculus to the medial geniculate body, the thalamic relay station of the auditory system. This makes the region of the colliculi—the so-called quadrigeminal plate—an important anatomical issue in neuroophthalmology (tumors of the nearby pineal gland may exert pressure on the superior colliculi leading to disturbances in the coordination of external eye muscles) as well as audiology.

The red nucleus is an interesting structure in terms of evolution. It first appears in the vertebrate brain. As an example for red nucleus affection, we have included in this chapter the prion diseases which may affect the brainstem. In humans, it is an important motor relay station with efferent connections to the inferior olive (central tegmental tract) and afferent input from the cerebellum in the framework of the dentate-rubro-thalamic tract.

In conjunction with findings in the red nucleus demonstrated in this chapter, the prion diseases are dealt with in this here.

The ventral tegmental area VTA—besides the Substantia nigra which, however, is of pretectal origin—is an important source of dopaminergic input into the forebrain. Due to its important role in drug addiction and the central reward system, this region is not only relevant for neurology and neuropharmacology but also for psychiatry.

15.1 Alar mesencephalon

15.1.1 Tectal gray formation

Detailed information on the tectal gray formation can be found in Chap. 13 under Sect. 13.2.7.

15.1.2 Mesencephalic periaqueductal gray [Substantia grisea periaqueductalis, formerly also Substantia grisea centralis]

Detailed information on the mesencephalic periaqueductal gray (PAG) is provided in Chap. 13 under Sect. 13.2.7.1.

15.1.3 Superior colliculus (SC) [Colliculus superior] (Colliculus Latin = small hill)

The name of the superior (and the inferior) colliculi becomes evident in a macroscopic view of the mesencephalon (see Figs. 1.5, 1.6B, and 15.3): The quadrigeminal plate consists of four small bulges, the two superior (see Fig. 15.1) and two inferior colliculi.

15.1.3.1 Location and morphology of the superior colliculi

The superior colliculi are the rostral most part of the mesencephalic tectum. Their caudal neighbors are the inferior colliculi (Fig. 15.3). The human superior colliculi have a laminar appearance that becomes obvious at higher magnification (Fig. 15.2). Unfortunately, the delineation of the individual layers is not as simple as with other CNS structures like the layering of the cerebral or cerebellar cortices. In man—as in many other mammals—the superior colliculi are part of the

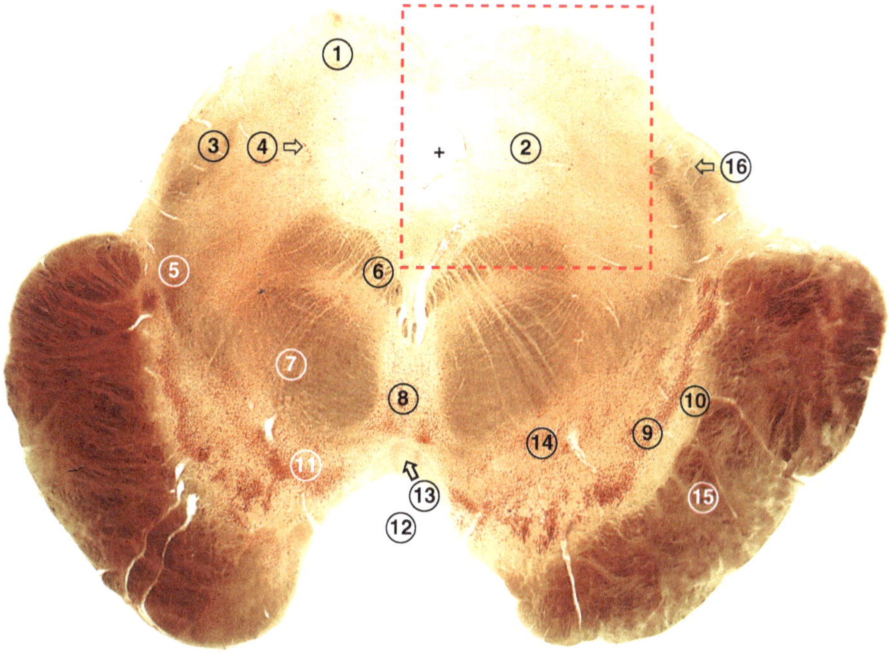

①	Superior colliculus	⑩	Substantia nigra reticular part
②	Periaqueductal gray	⑪	Paranigral ncl.
③	Medial lemniscus	⑫	Interpeduncular fossa
④	Mesencephalic trigeminal ncl.	⑬	Interpeduncular ncl.
⑤	Medial lemniscus	⑭	Parabrachial pigmented ncl.
⑥	Medial longitudinal fasciculus	⑮	Cerebral peduncle
⑦	Superior cerebellar peduncles	⑯	Brachium of inferior colliculus
⑧	Linear raphe ncl.	+	Mesencephalic aqueduct
⑨	Substantia nigra compact part		

Fig. 15.1 Horizontal section through the human mesencephalon at the level of the superior colliculi. Darrow red stain. The frame indicates the magnified area in Fig. 15.2. See also atlas part Darrow red 38, 38A, Campbell 19. LabPON Twente

visual system but also receive somatosensory (mainly whisker-derived) and auditory input (Dräger and Hubel 1975).

The superficial three layers (Lamina I/zonal layer [Stratum zonale], Lamina II/superficial gray layer [Stratum griseum superficiale], Lamina III/optic layer [Stratum opticum]) receive visual input.

The deeper layers (Lamina IV/intermediate gray layer [Stratum griseum intermedium], Lamina V/intermediate white layer [Stratum medullare intermedium], Lamina VI/deep gray layer [Stratum griseum profundum]) are the target of somatosensory and auditory but also of visual input. Finally, there is also a Lamina VII/deep white layer [Stratum medullare profundum].

15.1.3.2 Connectivity and function of the superior colliculi

In general terms, the superior colliculi belong to the visual system as the inferior colliculi to the auditory system. Using various tracers in postmortem human brains, Tardif et al. (2005) suggest that intrinsic connections within superficial layers and radial superficial layers to deep layers exist in human SC.

15.1.3.3 Pathological aspects of the superior colliculi

In vivo the pineal gland (of diencephalic origin) "lies" on the quadrigeminal plate (see Fig. 15.3 ⑧). An enlarge-

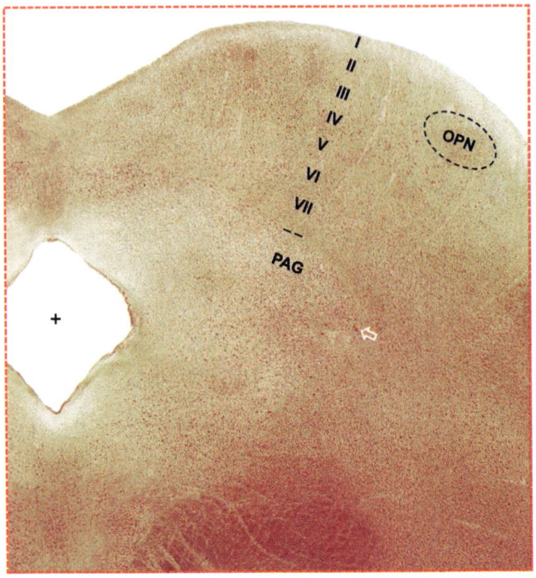

I	Zonal layer	VI	Deep gray layer
II	Superficial layer	VII	Deep white layer
III	Optic layer	OPN	Olivary pretectal ncl.
IV	Intermediate gray layer	PAG	Periaqueductal gray
V	Intermediate white layer	+	Mesencephalic aqueduct

Fig. 15.2 Horizontal section through the human midbrain at the level of the superior colliculi (SC). Darrow red stain. Tectum and part of the tegmentum (see Fig. 15.1 for overview). At higher magnification, it becomes obvious that the superior colliculi have a layered structure. The superficial three layers receive visual input, the deeper layers are targets of somatosensory and auditory but also of visual input. Finally, there is also a deep white layer at the border with the periaqueductal gray. ⇦ Neuronal perikaryon of the mesencephalic trigeminal nucleus at the border between PAG and superior colliculus. See also atlas part Darrow red 38B. LabPON Twente

ment of the pineal gland, for example, by a tumor or a cyst, may exert pressure on the region of the colliculus superior with compression of the rostral midbrain and pretectum at the level of the superior colliculi. This may manifest clinically in the so-called Parinaud syndrome (Henri Parinaud, French ophthalmologist, 1844–1905) (Feroze and Patel 2021).

The characteristic symptom of Parinaud syndrome is limited conjugate upgaze. This is due to the involvement of the adjacent vertical gaze center like, for example, the interstitial nuclei of Cajal (see Sect. 16.4.2) and the rostral interstitial nucleus of the medial longitudinal fasciculus (riMLF) (see Sect. 16.4.5) (see also Figs. 16.1 and 16.2) (Feroze and Patel 2021).

Armstrong et al. (2017) studied the human superior colliculi in chronic traumatic encephalopathy compared to neurologically healthy brains. They could show decreased neuron densities in the encephalopathy cases mainly in intermediate and lower layers. Based on this distribution pattern, the authors concluded that these changes may rather compromise motor function than visual sensory analysis.

In Alzheimer's disease, amyloid deposits have been shown in the superior as well as in the inferior colliculi (see Sect. 13.2.2.1.4 Brainstem pathologic changes in AD, see Fig. 13.22.)

In mice, the superior colliculi are part of the visual system, too, but also receive somatosensory (mainly whisker-derived) and auditory input (Dräger and Hubel 1975). Recently, an important role of the mice SC in fear responses (Box 15.1) has been detected (Shang et al. 2015). The authors could show that parvalbumin-containing neurons, located predominantly in the layer II in the mouse superior colliculi, play a key role in mediating fear responses. They get input from retinal ganglion cells (optic nerve/tract) and project to the parabigeminal nucleus (see Sect. 14.4.1.4). The latter projects to the central nucleus of the amygdala. Thus, a pathway (retina ⇨ SC ⇨ parabigeminal nucleus ⇨ amygdala) has

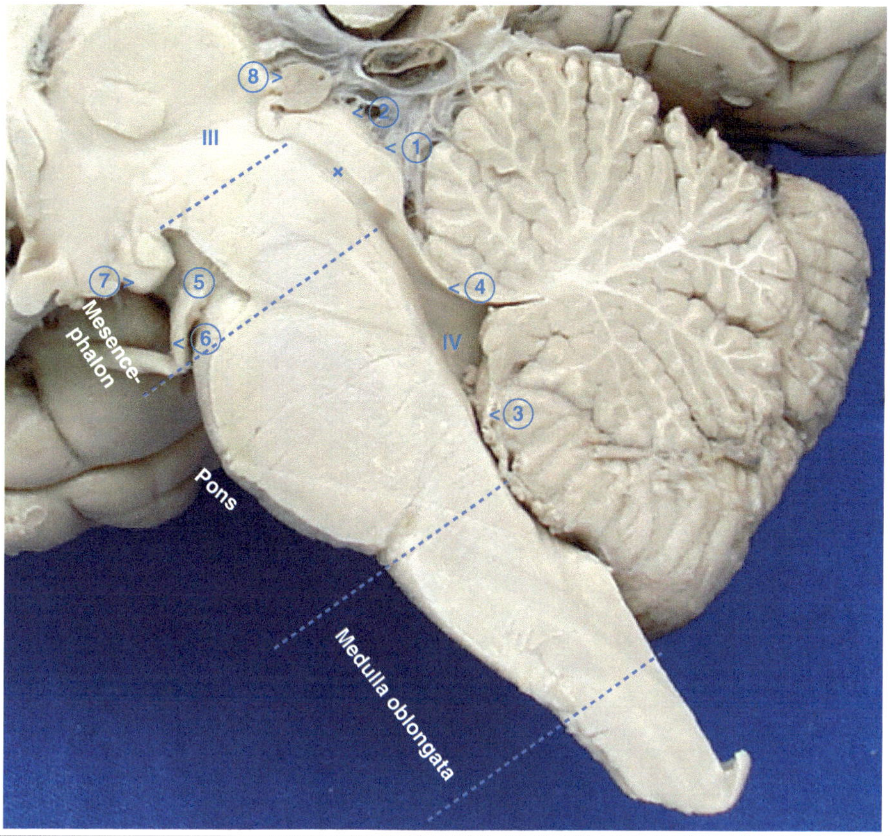

①	Inferior colliculus	⑦	Mammillary body
②	Superior colliculus	⑧	Pineal body
③	Inferior medullary velum	III	3rd ventricle
④	Superior medullary velum	IV	4th ventricle
⑤	Interpeduncular fossa	+	Mesencephalic aqueduct
⑥	Oculomotor nerve CNIII		

Fig. 15.3 Sagittal section through the human brain. Formalin-fixed specimen. The SC is in close contact with the epithalamic pineal gland ⑧. The fourth ventricle (covered by the medullary vela ③ ④) narrows into the mesencephalic aqueduct (+), dorsally overcast by the quadrigeminal plate consisting of inferior ① and superior colliculi ②. Sammlung des Zentrums Anatomie der Universität zu Köln

been identified that subserves the processing of threatening visual signals.

Box 15.1 Fear responses

When facing threats, for example, the presence of a predator, animals have the option of flight or fight. The flight-or-fight response is preceded by the so-called freezing, a phase of increased alertness accompanied by a temporary immobility. This phase is considered as an attempt to feigning death.

15.1.4 Inferior colliculus (IC) [Colliculus inferior]

The inferior colliculi are an important relay station between the cochlear nuclei (see Sect. 8.1.1) and the medial geniculate body in the thalamus.

All auditory signals reach the IC via the lateral lemniscus (see Fig. 15.5) (for functional considerations of the IC and the ascending auditory pathways see Sect. 8.1.1.5). The IC are macroscopically visible caudal of the SC (Fig. 15.3 ① and Fig. 15.4).

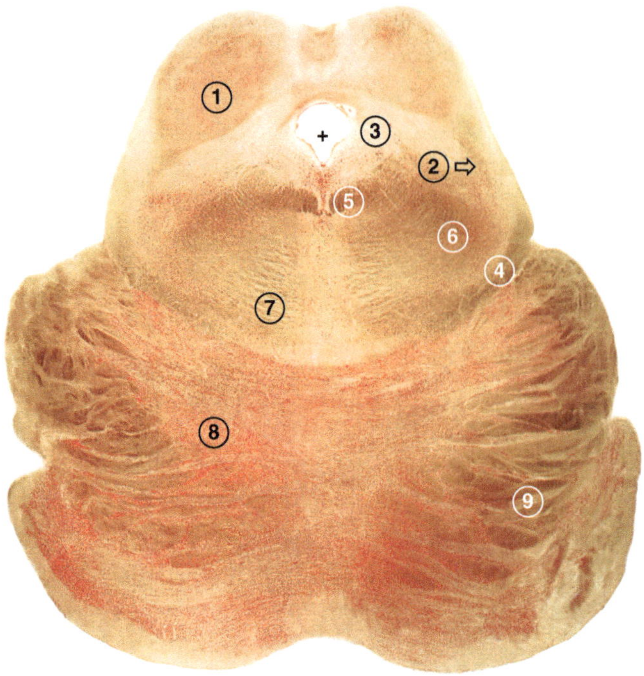

①	Inferior colliculus	⑥	Central tegmental tract
②	Lateral lemniscus	⑦	Superior cerebellar peduncle
③	Periaqueductal gray	⑧	Pontine ncll.
④	Medial lemniscus	⑨	Pyramidal (corticospinal) tract
⑤	Medial longitudinal fasciculus	+	Mesencephalic aqueduct

Fig. 15.4 Horizontal section through the human mesencephalon at the level of the inferior colliculi. Darrow red stain. See atlas part Darrow red 33, Campbell 17, 17A. LabPON Twente

15.2 Liminal midbrain

15.2.1 Ventrolateral periaqueductal gray

For detailed information on the periaqueductal gray, see Sect. 15.1.2.

15.3 Mesencephalic basal plate

15.3.1 Oculomotor complex

15.3.1.1 Nucleus of oculomotor nerve (CNIII)/Oculomotor nucleus [Nucleus n. oculomotorii]

The CNIII nucleus (Figs. 15.6 and 15.7) gives rise to the axons of the oculomotor nerve which traverse the red nucleus (Fig. 15.7) and exit in the interpeduncular fossa.

15.3.1.1.1 Location and morphology of the oculomotor nucleus

The delineation of motoneuron subgroups for the different oculomotor (see Box 15.2) muscles (medial, inferior, superior rectus; inferior oblique; levator palpebrae superioris) has been achieved for the monkey (Che Ngwa et al. 2014; Che Ngwa 2016) (see Figs. 15.8, 15.9, and 15.10).

Box 15.2 Term "oculomotor"

The adjective oculomotor in the stricter sense is assigned to the nucleus of oculomotor nerve and the oculomotor nerve. It is, however, used in broader sense for all structures related to movement of the eyes, like the other oculomotor nuclei (CNIV, CNVI), their muscles (obliquus superior/lateral rectus), etc.

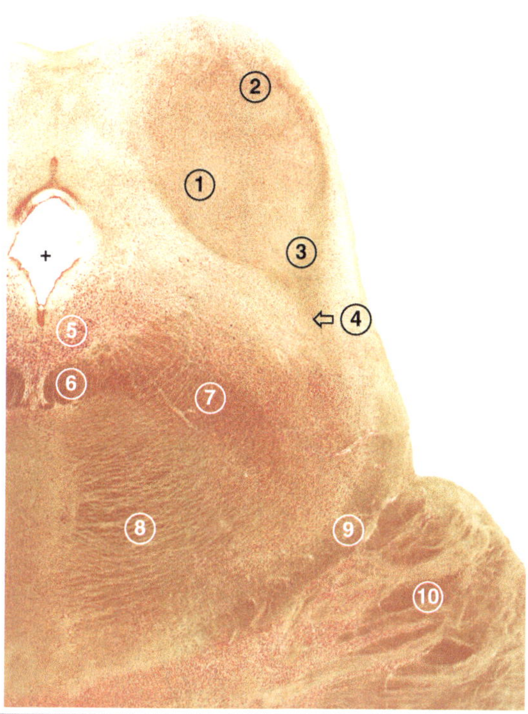

①	Inferior colliculus, central ncl.	⑦	Central tegmental tract
②	Inferior colliculus, pericentral ncl.	⑧	Superior cerebellar peduncle
③	Inferior colliculus, external ncl.	⑨	Medial lemniscus
④	Lateral lemniscus	⑩	Cerebral peduncle
⑤	Supratrochlear subncl. of the dorsal raphe ncl.	+	Mesencephalic aqueduct
⑥	Medial longitudinal fasciculus		

Fig. 15.5 Horizontal section through the human mesencephalon at the level of the inferior colliculi [Colliculus inferior/Colliculi inferiores]. Darrow red stain. Tectum and tegmentum. Three different nuclei, the central nucleus [Ncl. centralis], pericentral [Ncl. pericentralis] (with four different layers), and the external nucleus [Ncl. externus] have been described (for details FIPAT Ch. 1 endnote # 67and # 1500 ff.). See also atlas part Darrow red 34A, Campbell 17A. LabPON Twente

The CNIII nucleus contains the motoneurons for all external ocular muscles except of those for the superior oblique (CNIV) and lateral rectus muscle (CNVI) (see Table 15.1).

15.3.1.1.2 Target muscles of the oculomotor nucleus/nerve

The muscles innervated by the oculomotor nerve are listed in Table 15.1. Figure 15.11 shows a schematic representation of the general possibilities of the eye bulb to move around the main axes and Fig. 15.12 displays the anatomy of the oculomotor (CNIII, CNIV) muscles in a coronal section through the human head.

The eyeball can be imagined as a sphere with three main axes, arranged perpendicular to each other, similar to the main body axes (see Figs. 1.1 and 1.2). There is a sagittal or anterior-posterior (A-P) axis through the pupil of the eye, a horizontal axis running from lateral to medial (L-M) and a vertical axis running from dorsal to ventral (D-V) (see Fig. 15.11). Rotation around the individual axes results in elevation/depression, adduction/abduction (see Table 15.1). The rectus muscles subserve only one movement (Table 15.1). The obliquus inferior draws the eye upward and outward, the obliquus superior downwards and inwards (Fig. 15.8, Table 15.1).

Porter and Donaldson (1991) could show that in cats the cells innervating the spindles of the extraocular muscles are located mainly in the trigeminal ganglion (see Fig. 12.3).

The four rectus muscles of the eye are posteriorly anchored at the annulus of Zinn (German anatomist and botanist, 1727–1759) (see Fig. 15.8B ④), a fibrous structure at the apex of the orbita. Through it, optic nerve and ophthalmic vessels, as well as other minor nerves and vessels, pass from the orbit to the eye bulb. It is strictly adherent to the optic nerve dural sheath and the surrounding periosteum

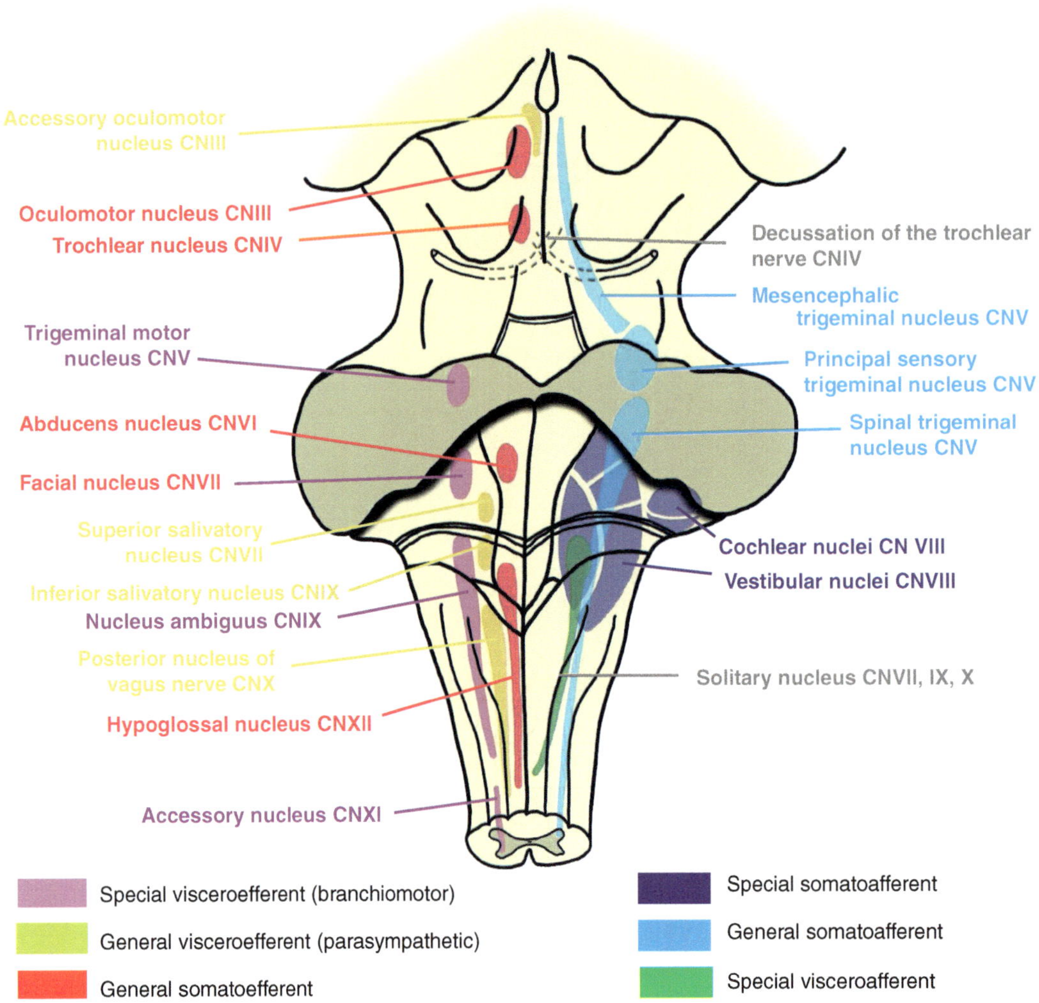

Fig. 15.6 Schematic representation of the nucleus of oculomotor nerve and the accessory nuclei of oculomotor nerve (Edinger-Westphal nucleus see Sect. 15.3.1.2). From Huggenberger et al. 2019, Fig. 15.5 with permission

(Zampieri et al. 2015). Starting with Fig. 15.15 seven photographs of horizontal plastinated sections will be presented through the human head in craniocaudal order in a schematic view and a fluid-preserved anatomical specimen.

Since these sections also hit CNIII and other cranial nerves, we will deal with them also under Sect. 15.3.1.1.3 "Course of the oculomotor nerve." For a better understanding, the plastinated sections will be preceded by two macroscopic specimens which show the contents of the orbita from above/dorsal (see Figs. 15.13 and 15.14).

15.3.1.1.3 Course of the oculomotor nerve

The axons of the oculomotor neurons leave the nucleus at its ventral side and enter the red nucleus. Even in routinely stained sections, the fiber bundles of the oculomotor nerve leaving the nucleus and running to the interpeduncular fossa through the red nucleus can be seen (see Fig. 15.21).

The fibers then traverse the interpeduncular cistern (see Box 3.3) and run to the cavernous sinus [Sinus cavernosus]. Here, the CNIII comes in contact with the petroclinoid ligament (see Fig. 9.7, Box 9.4) being potential subject to pressure-induced lesions (see Sect. 15.3.1.2.4). Figure 15.20 shows a horizontal section through the human head at the level of the CNIII exit into the interpeduncular fossa. Luckily, the saw cut hit the nerve almost in its entirety from exit to the entry in the orbita.

The oculomotor nerve is located upper most of the other sinus-related nerves (CN IV, V_1, VI) in the lateral wall of the sinus (see Fig. 15.23). After leaving the sinus, the oculomotor nerve enters the orbita via the superior orbital fissure within the common tendinous ring [Annulus tendineus communis (of Zinn) see Fig. 15.8B]. It then divides into a superior branch for the levator palpebrae superioris and the superior rectus muscles (see Figs. 15.8, 15.16, and 15.18)

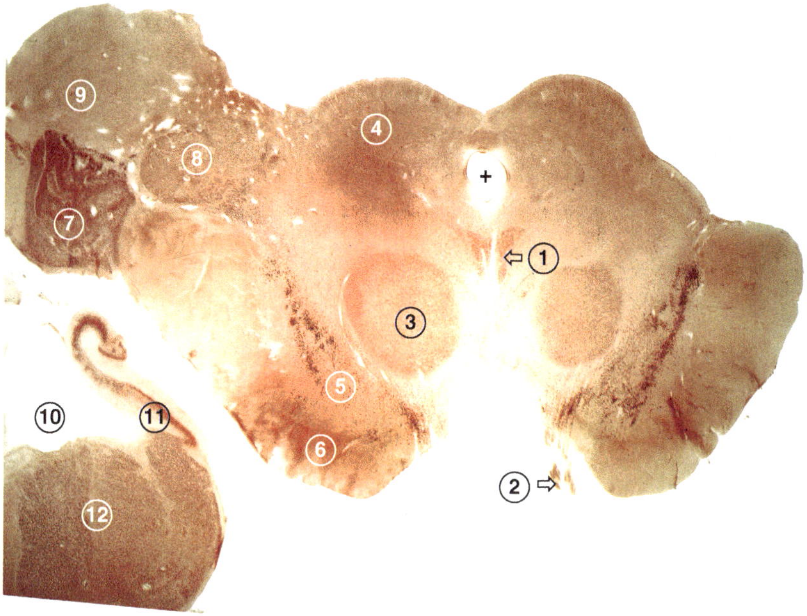

①	Oculomotor complex	⑧	Medial geniculate body
②	Oculomotor nerve CNIII	⑨	Pulvinar
③	Red ncl.	⑩	Inferior (temporal) horn of lateral ventricle
④	Superior colliculus	⑪	Hippocampus
⑤	Substantia nigra	⑫	Amygdala
⑥	Cerebral peduncle	+	Mesencephalic aqueduct
⑦	Lateral geniculate body		

Fig. 15.7 Survey of the human oculomotor complex. Horizontal section through the human midbrain at the level of the SC. Darrow red staining. The oculomotor complex ① with the oculomotor and the Edinger-Westphal nucleus, located at either side of the midline, can be identified even at low magnification due to their relatively high cell density. See also atlas part Darrow red 41, Campbell 20. LabPON Twente

and an inferior branch (see Figs. 15.8, 15.13, 15.14, 15.19, and 15.20) for the other CNIII-innervated muscles (medial rectus, inferior rectus, inferior obliquus).

Box 15.3 The Josephinum

The Josephinum, established in 1785 by the German emperor Joseph II in Vienna, Austria, as a medicosurgical military academy, was equipped with a vast collection of wax models of all organs including the brain. They are inspired by the wax models in La Specola in Florence, Italy. The Vienna models were produced under supervision of Felice Fontana, director of La Specola (anatomist and physicist) and Paolo Mascagni (anatomist). In addition to their use in teaching they were, even then, intended for the general public. The majority of the models survived to this day and is on display. A visit is absolutely worthwhile as well as that of La Specola.

https://www.josephinum.ac.at/en/

https://www.naturacollecta.unifi.it/vp-126-the-ceroplastic-workshop.html

15.3.1.1.4 Connectivity of the oculomotor nucleus CNIII

As to the pretectal control of oculomotor nuclei Onodera and Hicks (1996) had proposed a direct projection from the elliptic nucleus (see Sect. 16.4.1) for the cat. As described by Horn and Büttner-Ennever (1998) upon tracer injection into vertically pulling eye muscles, the elliptic nucleus did not show transsynaptically labeled neurons which speaks against a direct projection to the oculomotor nucleus.

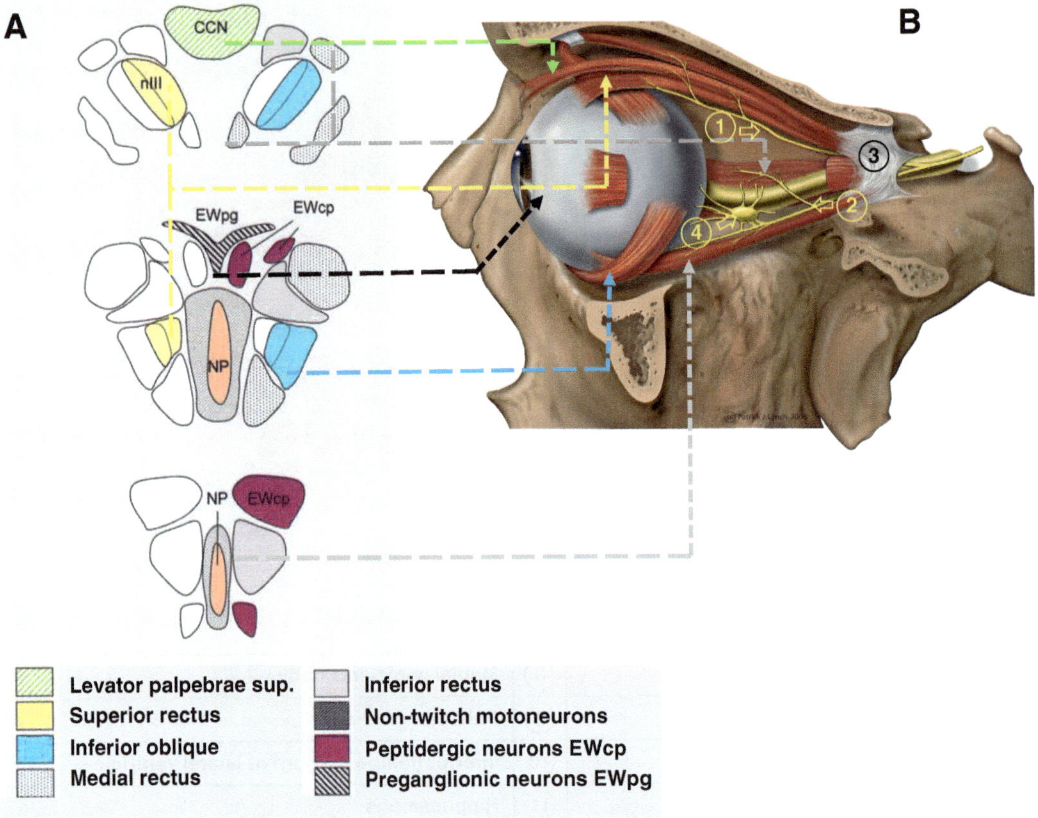

①	Oculomotor nerve CNIII Superior branch	CCN	Central caudal ncl.
②	Oculomotor nerve CNIII Inferior branch	EWcp	Edinger Westphal ncl., centrally projecting neurons
③	Annulus of Zinn	EWpg	Edinger Westphal ncl., preganglionic neurons
④	Ciliary ganglion	nIII	Central ncl.
		NP	Ncl. of Perlia

Fig. 15.8 Musculotopic Innervation of the CNIII muscles. Warwick (1953) studied the musculotopic pattern of subnuclei in the CNIII nucleus of *Cercocebus fuliginosus* (Sooty mangabey) and *Macaca mulatta* combining electrophysiological and anatomical techniques. A schematic coronal representation of the different subnuclei of the oculomotor nucleus and their muscular targets is shown in (**A**) and (**B**). The nucleus of Perlia so far has only been described in humans. It had been considered as center for convergence, but its exact function is still not known (Zeeh and Horn 2012). For reasons of clarity not all of the subnuclei have been connected by colored lines with the innervated muscles. (**A**) Modified after Zeeh and Horn 2012, Fig. 5 with permission. (**B**) Patrick J. Lynch (2006), https://upload.wikimedia.org/wikipedia/commons/archive/6/60/20100101235241%21Lateral_orbit_nerves_chngd.jpg with permission

As shown in monkeys, the oculomotor nucleus contains a population of internuclear neurons projecting to the contralateral abducens nucleus (Ugolini et al. 2006).

Using diffusion tensor tractography, Jang et al. (2018) could show a high degree of connectivity of the oculomotor nucleus with the vestibular nuclei.

The nuclei of CNIII, CNVI and CNIV are connected via the medial longitudinal fasciculus (MLF) (see Box 9.2) (see Fig. 3.14A–G) serving to coordinate conjugate eye movements, especially lateral gaze (Link and Sloan 2003). Impulses relayed to the right abducens nucleus (for the right lateral rectus muscle) are also transmitted via the MLF to the left oculomotor nucleus (for the left medial rectus muscle) so the eyes look to the right in a conjugate fashion (Link and Sloan 2003). A lesion of the MLF—most often seen in multiple sclerosis (see Sect. 9.2.1.3 and Box 9.2)—results in the inability of the eye on the affected side to adduct past midline and nystagmus of the abducting eye (Link and Sloan 2003).

15.3.1.1.5 Living anatomy and clinical aspects of oculomotor nucleus/nerve

The examination of the motility of the oculomotor muscles (see also Sect. 15.3.1.2.4) is a compulsory part of any physical examination. It can be carried out simply and fast following the schemes provided in Fig. 15.24A and B.

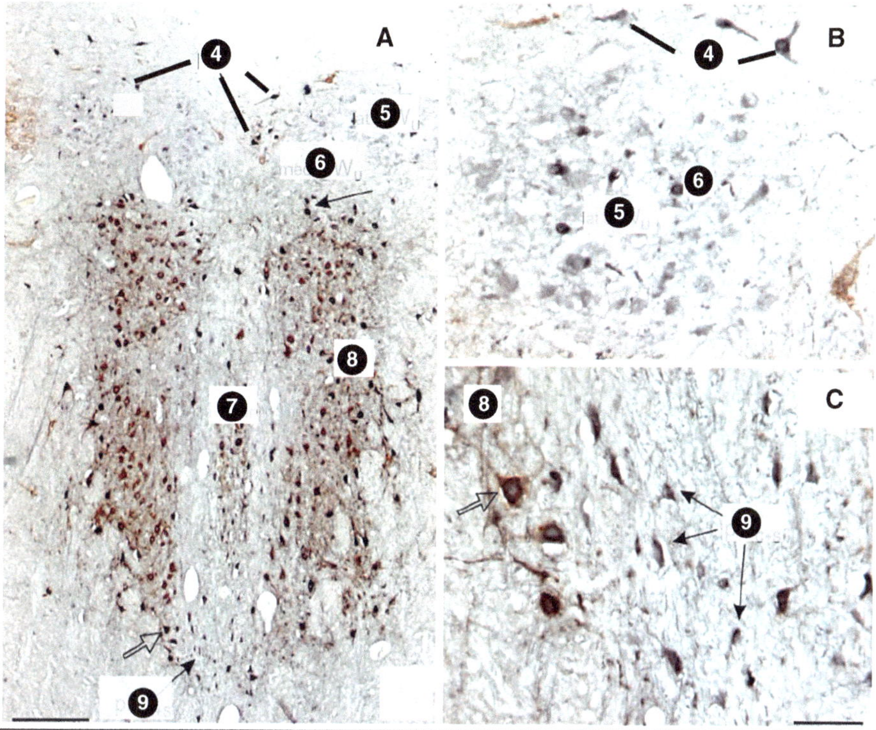

④	Perioculomotor preganglionic neurons in human $pIII_{PG}$	⑧	Oculomotor ncl. nIII
⑤	Lateral Edinger-Westphal urocortin neurons Lat EW_U	⑨	Perioculomotor motoneurons of multiply innervated eye muscle fibers in human $pIII_{C+S}$
⑥	Medial Edinger-Westphal urocortin neurons Med EW_U		
⑦	Perioculomotor ncl. of Perlia in human $pIII_{NP}$		

Fig. 15.9 The histological pictures (**A**–**C**) show horizontal sections through the CNIII nucleus using antibodies to choline acetyltransferase (ChAT) (black) and chondroitin sulfate proteoglycan (CSPG) (brown). The neurons of the oculomotor nucleus are typically somatomotor, i.e., large with a clearly visible nucleolus inside the cell nucleus. As all cranial motor nuclei, they can be labeled by means of ChAT-immunoreactivity. Almost all cholinergic neurons in ⑦ and ⑧ are ensheathed by CSPG-ir perineuronal nets (open arrows ⇨). Neurons indicated by black solid arrows (**A**, **C** ⑨ →)—lacking CSPG-immunoreactivity—are supposed to be the multiply innervated fiber (MIF) motoneuron population of extraocular eye muscles (for details, see Box 9.1). ④ indicates the cholinergic preganglionic parasympathetic neurons projecting to the ciliary ganglion (see Sect. 15.3.1.2 Edinger-Westphal nucleus pg) with postganglionic innervation of the sphincter pupillae muscle (miosis) and the ciliary body (accomodation) (for details, see Sect. 15.3.1.2.3 and Fig. 15.28). ⑤ and ⑥ label the regions of the Edinger-Westphal nucleus with non-cholinergic, urocortin-containing and centrally projecting neurons (for details, see Sect. 15.3.1.2.1, see also). Scale bars (**A**) 0.5 mm (**B**, **C**) 100 µm. From Horn et al. 2008, Fig. 4 with permission

The examiner carefully observes whether the eye movements smoothly follow the movements of the examiner's index finger or whether the eye bulb(s) lag behind in certain directions (for an example, see Fig. 15.32).

The oculomotor nerve is subject of lesions, for example, either by increased intracranial pressure or by the local impact of an aneurysm of the circle of Willis.

More information on this topic is provided in Sect. 15.3.1.2.4 (here below).

15.3.1.2 Edinger-Westphal nucleus (EW)/ Accessory nuclei of oculomotor nerve [Nuclei accessorii n. oculomotorii]

15.3.1.2.1 Location and morphology of the Edinger-Westphal nucleus (EW)

This nuclear complex has a rather complicated history. The nucleus has first been described by Ludwig Edinger (1855–1918, German neurologist and neuroscientist, University of Frankfurt) (see Fig. 15.25) in the human brain (1885).

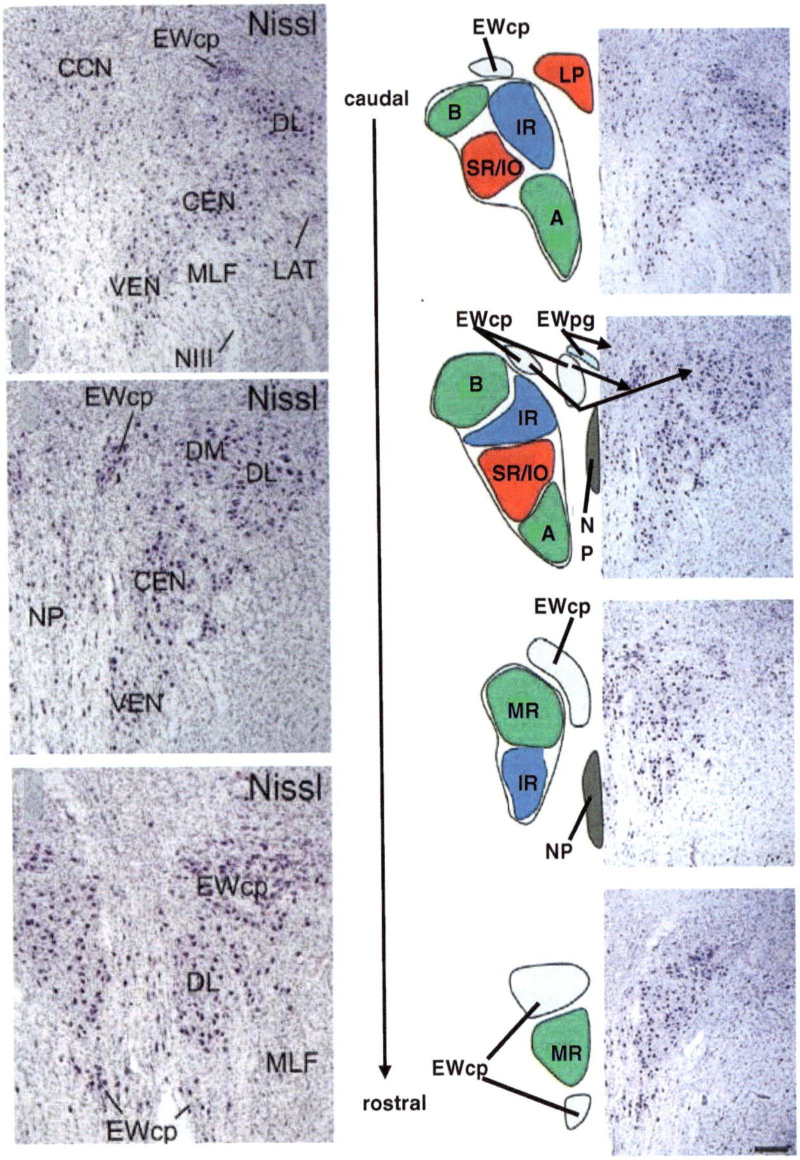

A	A group of medial rectus muscle	IR	Inferior rectus muscle
B	B group of medial rectus muscle	LAT	Lateral subgroup
CCN	Central caudal ncl.	LP	Levator palpebrae muscle
CEN	Central subgroup	MLF	Medial longitudinal fasciculus
DL	Dorsolateral subgroup	MR	Medial rectus muscle
DM	Dorsomedial subgroup	NP	Ncl. of Perlia
EWcp	Edinger-Westphal ncl., centrally projecting neurons	NIII	Oculomotor nerve
EWpg	Edinger-Westphal ncl., preganglionic neurons	SR	Superior rectus muscle
IO	Inferior oblique muscle	VEN	Ventral subgroup

Fig. 15.10 The musculotopy of the human CNIII nucleus has been studied in detail by Che Ngwa et al. (2014), Che Ngwa (2016). The left column in the figure shows transverse sections from caudal to rostral. The right column of the figure shows schematic representations of the subnuclei presented in the photo to the right of it. The muscles innervated by the subnuclei are labeled accordingly. Calibration bar 500 μm. The CNIII subnuclei are not listed in FIPAT Ch. 1. Modified after Che Ngwa et al. 2014, Figs. 1, 2, 3, 8 with permission

Table 15.1 Human external/extraocular eye muscles (see Figs. 15.8, 15.19, and 15.20)

Muscle Innervating nerve	Main functions/muscle moves eye
Lateral rectus muscle [M. rectus lateralis] Abducens nerve CNVI	Laterally
Medial rectus muscle [M. rectus medialis] Oculomotor nerve CNIII	Medially
Superior rectus muscle [M. rectus superior] Oculomotor nerve CNIII	Upwards
Inferior rectus muscle [M. rectus inferior] Oculomotor nerve CNIII	Downwards
Inferior oblique muscle [M. obliquus inferior] Oculomotor nerve CNIII	Upwards/outwards
Levator palpebrae superioris muscle [M. levator palpebrae superioris] Oculomotor nerve CNIII	Raising of the upper lid
Superior oblique muscle [M. obliquus superior] Trochlear nerve CNIV	Downwards/inwards

By contrast to, for example, rodents, humans lack the retractor bulbi muscle which retracts the bulb into the orbita. For the internal eye muscles, see Sect. 15.3.1.2

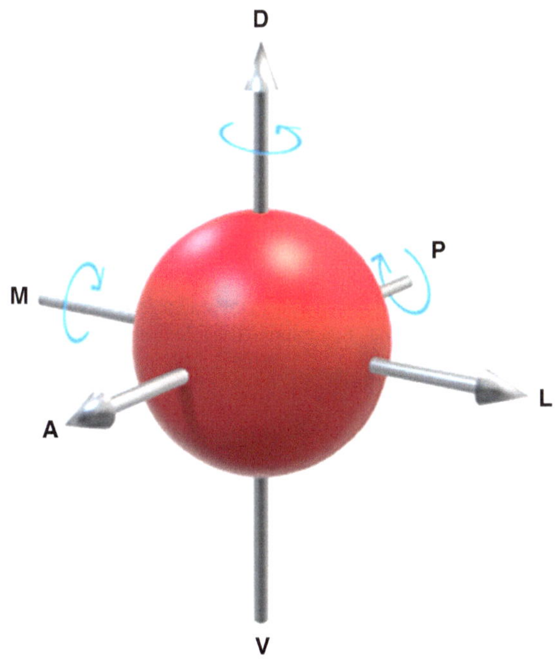

Fig. 15.11 Schematic visualization of the movement possibilities of the eye bulb around the three main axes. The anterior-posterior or sagittal axis—which would enter the bulb at the pupil—allows for rotation of the eyeball outwards (obliquus inferior) or inwards (obliquus superior). The horizontal axis—perpendicular to the sagittal and the vertical axis—runs from medial (nasal) to lateral (temporal) and allows for elevation (rectus superior, obliquus inferior) or depression (rectus inferior, superior oblique). The vertical axis runs from ventral (floor of the orbita) to dorsal (roof of the orbita) and allows for abduction (lateral rectus muscle) or adduction (medial rectus muscle). H. Schröder fecit

In his paper of 1885, Edinger wrote: *"Dorsal und ventral von jedem Oculomotoriuskern lieg (sic), medial zum Hauptkern, je ein kleiner Kern spindelförmiger Zellen, dessen Beziehungen zum Nerv nicht sicher sind,"* i.e., dorsal and ventral of either oculomotor nucleus lies medial to the main nucleus a small nucleus with fusiform cells, the relation of which to the nerve is not unequivocal.

This short communication did not contain any illustrations.

Carl Friedrich Otto Westphal (1833–1890, German psychiatrist and neurologist, Charité, Berlin) complemented the investigation of this nucleus by a clinico-anatomical study (1887). Westphal described a patient with a chronic-progressive palsy of the external eye muscles (Ophthalmoplegia externa). In addition, this patient suffered from unresponsive pupils while the accommodation was undisturbed.

Upon autopsy histological sections of the midbrain were prepared. Westphal wrote: *"Dorsalwärts von diesen atrophischen Oculomotoriuskernen sieht man …. an dem oberen Theile der Säule des Oculomotoriuskerns, …. unmittelbar neben der Raphe eine eigenthümliche (mediale) Gruppe von Ganglienzellen, die noch nirgends beim (erwachsenen) Menschen beschrieben zu sein scheint; die Gruppe hat die Form eines Ovals, …. (vergl. Taf. XVIII. Fig. I, II, IV) …."*

Dorsal of the atrophied oculomotor nuclei you see …. at the upper part of the column of the oculomotor nucleus, … directly besides the midline (raphe) a peculiar (medial) group of ganglion cells, which appears not to have been described anywhere in (adult) humans, the group is of oval shape, … (compare plate XVIII, Fig. I, II, IV) (see Fig. 15.26).

Welchem Zwecke könnten nun aber die gesund gebliebene mediale und laterale Gruppe dienen? Eine Antwort darauf liegt sehr nahe; es könnten möglicherweise die inneren glatten Augenmuskeln von diesen Gruppen aus innervirt (sic) werden.

What purpose could the non-affected (healthy) medial and lateral group subserve? An answer is obvious, possibly the internal smooth eye muscles could be innervated by these groups.

As described by Westphal himself (see here above), the patient suffered from unresponsive pupils while the autopsy had not revealed an atrophy of the optic nerve. This probably should have raised some doubts about a visceromotor role of the nuclei described by Westphal but he continued:

Die Pupillenstarre lässt sich mit grosser Wahrscheinlichkeit als ein Symptom der gleichzeitig vorhanden gewesenen Tabes resp. Allgemeinen Paralyse deuten; aber wir kennen bei beiden Krankheiten resp. ihrer Combination die Stellen des Nervensystems, von denen die Pupillenveränderungen ausgehen, nicht.

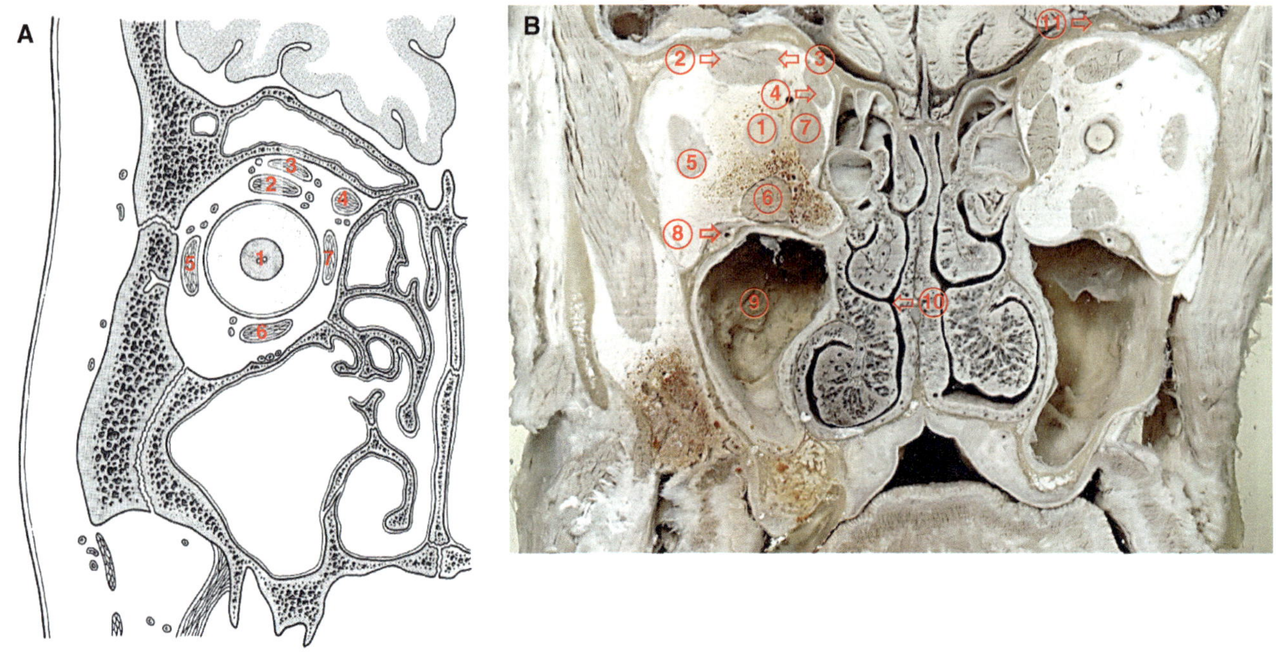

①	**Optic nerve CNI**	⑦	**Rectus medialis muscle CNIII**
②	**Rectus superior muscle CNIII**	⑧	**Infraorbital nerve V_2**
③	**Levator palpebrae superioris muscle CNIII**	⑨	**Maxillary sinus**
④	**Obliquus superior muscle CNIV**	⑩	**Nasal cavity**
⑤	**Rectus lateral muscle CNIII**	⑪	**Dura mater**
⑥	**Rectus inferior muscle CNIII**		

Fig. 15.12 Coronal sections through the human head at the middle third of the orbita. The external eye muscles are arranged in a circle around the optic nerve ①. In clockwise direction, starting with the superior rectus muscle ②, we see the superior oblique ④, the medial rectus ⑦, and the inferior rectus ⑥ muscle. The levator palpebrae superioris muscle ③ is located on top of the rectus superior ②. The inferior oblique has crossed from medial to lateral posterior to the section seen here (see Fig. 15.8). The relatively small bony floor of the orbita may fracture upon violent impact and the rectus inferior muscle might be trapped. Photograph by MedizinFotoKöln, Cologne, Germany. (**A**) Ribbers fecit. (**B**) Sammlung des Zentrums Anatomie der Universität zu Köln

The unresponsive pupil most likely can be interpreted as a symptom of the simultaneous Tabes or general paralysis, but for both disorders and their combination, respectively, we do not know the sites of the nervous system, which are the origin of the pupil changes.

As we know today, most likely Edinger and Westphal described that part of the eponymous nucleus (see Fig. 15.26) that today is classified as its centrally projecting part (EWcp) (see also Figs. 15.8, 15.9, and 15.10). It is not involved in the motor regulation of the internal smooth muscles but its urocortin-containing neurons project to several central nervous sites (see Kozicz et al. 2011a for details).

The visceromotor cholinergic sites are scattered dorsal of the EWcp nucleus, now called the EWpg (preganglionic) nucleus (see Kozicz et al. 2011a for details). Given the scarcity of these neurons in the human brain, not to speak of the lack of any suited markers in the nineteenth century, it is not surprising that they eluded the attention of Edinger as well as Westphal. It later turned out that EWpg and EWcp are present in non-human primates, rodents, and even birds (Kozicz et al. 2011a) (Fig. 15.27).

FIPAT Ch. 1 endnote 62 recommends that *the name Edinger-Westphal nucleus should be restricted to the cytoarchitectonically defined central cell group traditionally considered as the location of preganglionic neurons. The lateral, nonganglionic part of the nucleus contains urocortin-positive neurons which project to the lateral septum, the raphe nuclei, and the spinal cord.*

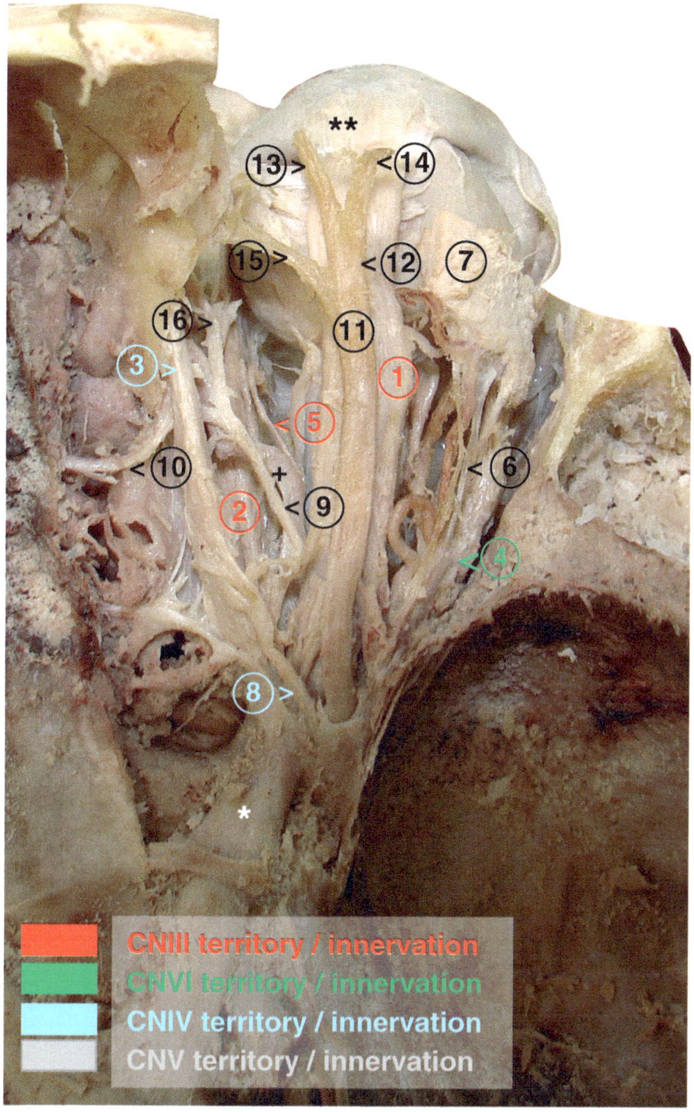

(1)	Levator palpebrae superioris muscle CNIII	(10)	Anterior ethmoideal nerve V_1
(2)	Medial rectus muscle CNIII	(11)	Frontal nerve V_1
(3)	Obliquus superior muscle CNIV	(12)	Supraorbital nerve V_1
(4)	Lateral rectus muscle CNVI	(13)	Supraorbital nerve V_1, medial branch
(5)	Oculomotor nerve CNIII, inferior branch	(14)	Supraorbital nerve V_1, lateral branch
(6)	Lacrimal nerve V_1	(15)	Supratrochlear nerve V_1
(7)	Lacrimal gland	(16)	Infratrochlear nerve V_1
(8)	Trochlear nerve CNIV		*Optic nerve **Optic bulb + Ophthalmic artery
(9)	Nasociliary nerve V_1		

Fig. 15.13 Dorsal view of the human orbita. Roof, medial and lateral wall of the orbita have been removed as well as a part of the frontal bone. The intraorbital fat was excised to show the intraorbital structures. Those that can also be identified in the plastinated sections Figs. 15.15, 15.16, 15.17, 15.18, 15.19, and 15.20 are either given in red (CNIII and muscles), light blue (CNIV) (as in Fig. 15.14), or green (CNVI). Structures given in black are not visible in the plastinated sections. The frontal nerve (V_1) (11) is running anterior on the levator palpebrae muscle (1) thus hiding the superior rectus muscle (CNIII). The medial rectus muscle (2) is innervated by the inferior branch of CNIII which in addition innervates the inferior rectus (see Fig. 15.14 (2)) and inferior oblique muscles. For details on CNIV, see Sect. 14.5.2.1, on CNVI Sect. 9.2.1.1 and on the trigeminal branches Sect. 12.2.3.2. Sammlung des Zentrums Anatomie der Universität zu Köln. Müller-Thomsen fecit

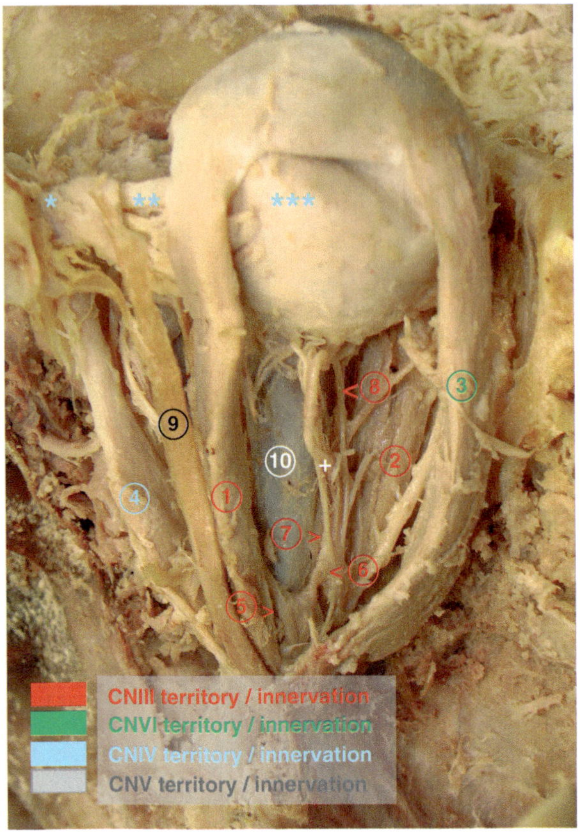

①	Medial rectus muscle CNIII	⑤	Oculomotor nerve, inferior branch CNIII
②	Inferior rectus muscle CNIII	⑥	Ciliary ganglion
③	Lateral rectus muscle CNVI	⑦	Short ciliary nerves
④	Superior oblique muscle CNIV, muscle belly * Approximate site of the trochlea ** Tendon *** Approximate insertion site at the eye bulb	⑧	Long ciliary nerves
		⑨	Frontal nerve, relocated medially
		⑩	Optic nerve CNII
		+	Short posterior ciliary arteries

Fig. 15.14 Dorsal view of the human orbita. In addition to the dissection shown in Fig. 15.13 levator palpebrae and superior rectus muscles have been removed enabling the view onto the CNIII-innervated muscles medial rectus ① and inferior rectus ②. The inferior oblique (CNIII) is hidden by the medial rectus and superior oblique muscles. The ciliary nerves ⑦ ⑧ are running from the ciliary ganglion ⑥ to the eyeball where they provide (1) parasympathetic innervation (Edinger-Westphal nucleus) for the sphincter pupillae muscle (miosis, constriction of the pupil) and the ciliary muscle (accommodation, i.e., changes of focus from distant to near images by changing lens shape) (see Fig. 15.29), (2) sympathetic innervation (superior cervical ganglion via vessel plexus) for the dilator pupillae muscle [Note that it is M. "dilatator" pupillae in Latin] (mydriasis, dilation of the pupil), and (3) sensory innervation of cornea, conjunctiva, sclera, iris, choroidea, and ciliary body. Sammlung des Zentrums Anatomie der Universität zu Köln. Müller-Thomsen fecit

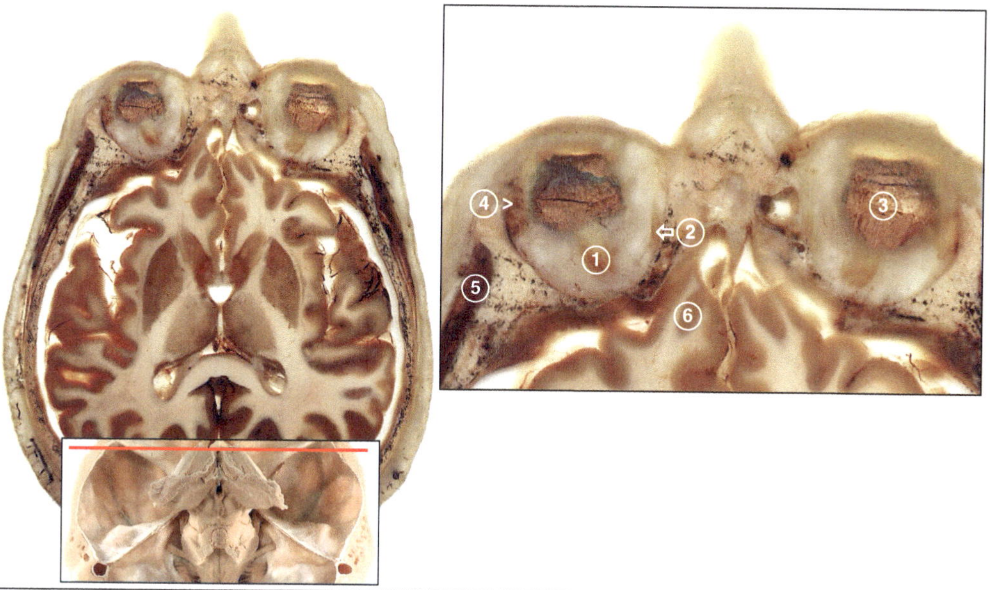

①	Superior rectus muscle CNIII	④	Lacrimal gland
②	Superior oblique muscle CNIV	⑤	Temporal muscle
③	Dorsal view on the pigmented layer of the retina	⑥	Frontal lobe

Fig. 15.15 Plate I dorsal view. This is the cranial most of the sections shown in this series. It hits the superior surface of the retina (pigmented layer) but is too far caudal to show the levator palpebrae superioris muscle (see Figs. 15.12 and 15.13). The cranial most muscle present here is the superior rectus muscle (CNIII) ① and parts of the superior oblique muscle (CNIV) ② (compare with Figs. 15.13 and 15.14). Because located rather cranial in the orbita, the lacrimal gland [Glandula lacrimalis] ④ is visible in this section (compare with Fig. 15.13). The small rectangular insets show for each section the approximate position in dorsal view of an open skull with the brainstem. Sammlung des Zentrums Anatomie der Universität zu Köln

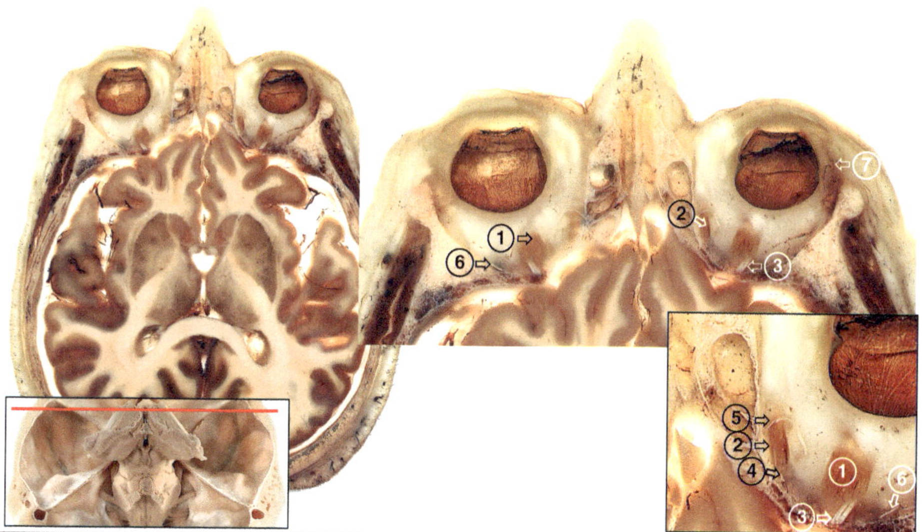

①	Superior rectus muscle CNIII	⑤	Anterior ethmoidal nerve V_1
②	Superior oblique muscle CNIV	⑥	Lacrimal nerve V_1
③	Oculomotor nerve, superior branch CNIII	⑦	Lacrimal gland
④	Trochlear nerve CNIV		

Fig. 15.16 Plate I ventral view. This photograph shows the ventral side of plate I. Therefore, it is not surprising that the structures visible are not much different from those seen in the dorsal view of plate I. The main difference is the existence of the cranial nerves innervating the rectus superior muscle (CNIII) and the superior obliquus muscle (CNIV) (see Figs. 15.13, 15.14). The lacrimal nerve (V_1) ⑥, secretomotor for the lacrimal gland ⑦, and the anterior ethmoidal nerve (V_1) ⑤ providing sensory innervation of the nasal mucosa (see Sect. 12.2.3.2.3), are visible. Sammlung des Zentrums Anatomie der Universität zu Köln

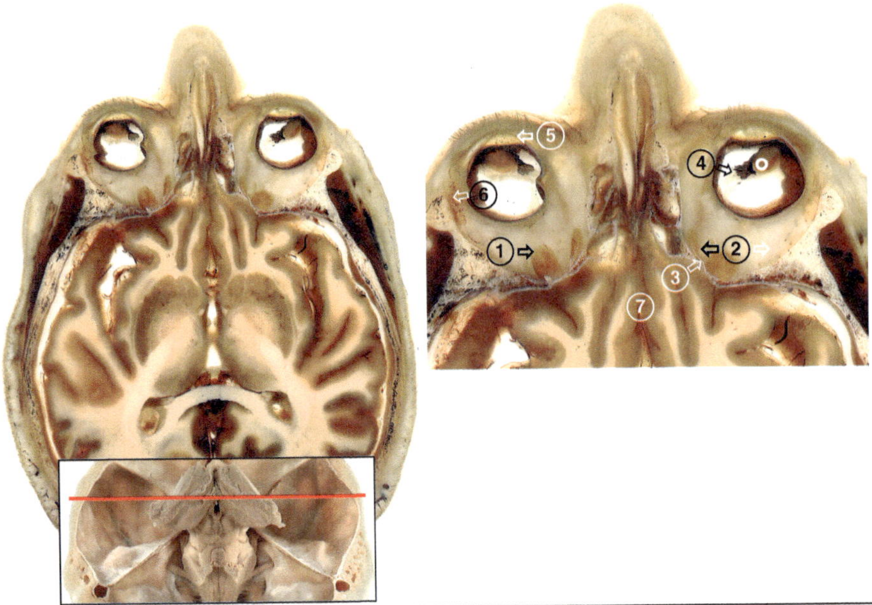

①	Superior rectus muscle CNIII	⑤	Cornea
②	Superior oblique muscle CNIV	⑥	Lacrimal gland
③	Trochlear nerve CNIV	⑦	Frontal lobe
④	Lens **O**, zonular fibers medially detached		

Fig. 15.17 Plate II dorsal view. The next eye muscle encountered here is the superior oblique (CNIII) ②. The partially detached zonular fibers plus ciliary body (see Fig. 15.28) are of interest as to innervation by the Edinger-Westphal nucleus (Sect. 15.3.1.2). Sammlung des Zentrums Anatomie der Universität zu Köln

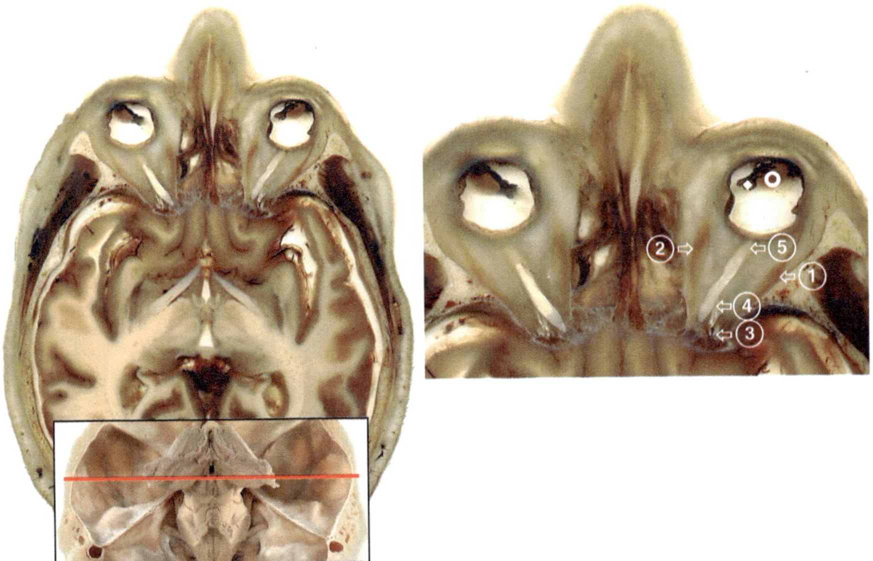

①	Lateral rectus muscle CNVI	④	Nascociliary nerve V_1
②	Medial rectus muscle CNIII	⑤	Optic nerve CNII
③	Oculomotor nerve, superior branch CNIII	**O**	Lens with medial zonular fibers ◆ detached

Fig. 15.18 Plate II ventral view. The anatomical situation gets more complicated. The most impressive structure is the optic nerve (CNII) ⑤ running from the eyeball/retina to the canal of CNII inside the annulus of Zinn, the common site of origin of all extraocular eye muscles including lateral and medial rectus muscles ① ②. The medial rectus muscle is innervated by the inferior branch of CNIII (not visible here), the superior branch ③ (see Fig. 15.8B) innervates the superior rectus and the levator palpebrae superioris muscles. Sammlung des Zentrums Anatomie der Universität zu Köln

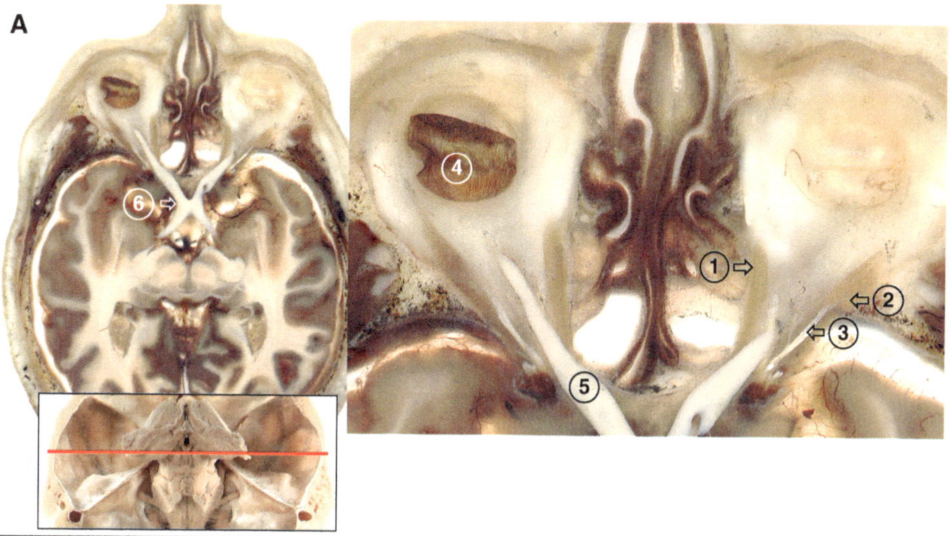

①	Medial rectus muscle CNIII	④	Retina, partly detached
②	Lateral rectus muscle CNVI	⑤	Optic nerve CNII
③	Abducens nerve CNVI	⑥	Optic chiasm CNII

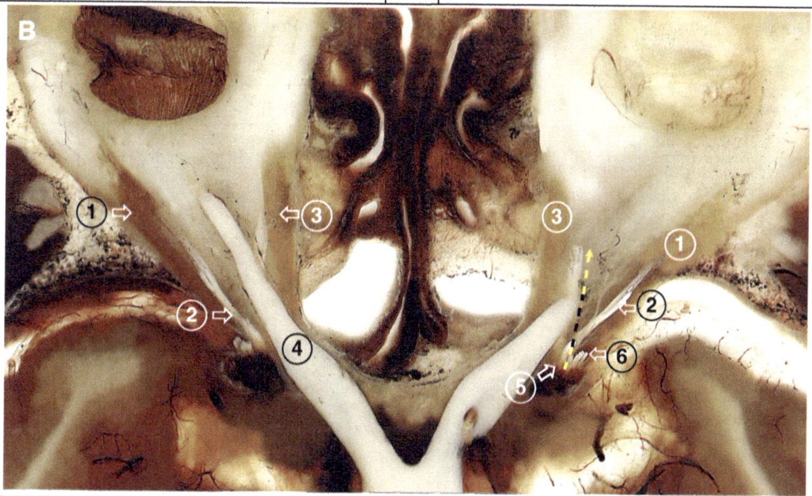

①	Lateral rectus muscle CNVI	④	Optic nerve CNII
②	Abducens nerve CNVI	⑤	Nasociliary nerve V_1
③	Medial rectus muscle CNIII	⑥	Superior branch of the oculomotor nerve CNIII
---	Course of the inferior branch of CNIII --- ventral of CNII		

Fig. 15.19 (**A**) Plate III dorsal view. The optic chiasm ⑥ is now visible. As in Fig. 15.18 the medial ③ and lateral ① rectus muscles are visible into which the innervating fibers of the abducens nerve (CNVI) ② (see Sect. 9.2.1.1) enter. For the other nerves hit, see the magnification in (**B**). Sammlung des Zentrums Anatomie der Universität zu Köln. (**B**) Plate III dorsal view at higher magnification. Besides the abducens nerve (CNVI) ②, from medial to lateral the inferior branch of CNIII ----, the nasociliary nerve ⑤ and the superior branch of CNIII ⑥ are visible. Sammlung des Zentrums Anatomie der Universität zu Köln

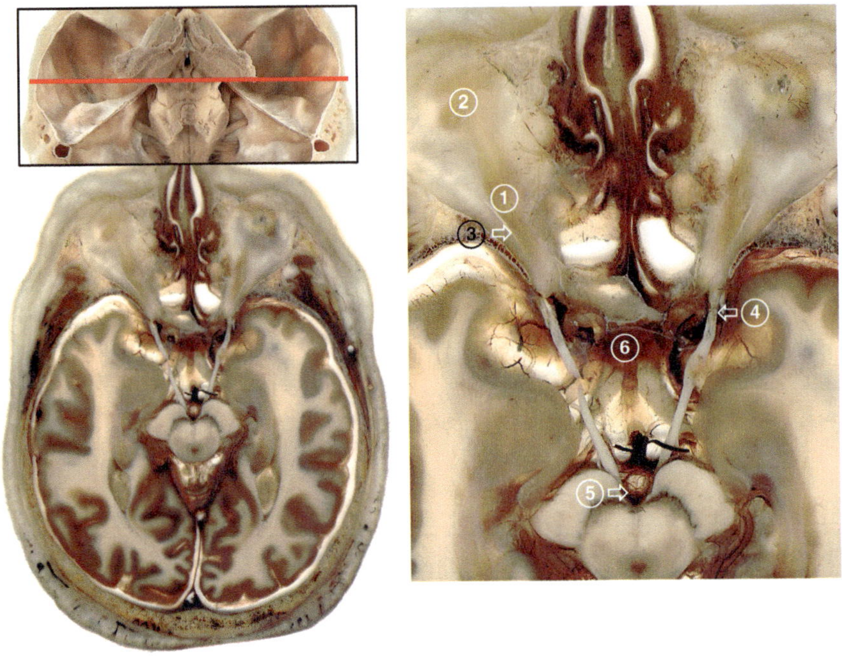

①	Inferior rectus muscle CNIII	④	Oculomotor nerve CNIII
②	Inferior oblique muscle CNIII	⑤	Interpeduncular fossa
③	Oculomotor nerve, inferior branch	⑥	Pituitary gland

Fig. 15.20 Plate III ventral view. The caudal most muscles, inferior rectus ① and inferior oblique ②, are visible. The pituitary gland ⑥ with its stalk at the posterior surface of the optic chiasm marks the midline. Lateral of it almost the complete course of the CNIII from its exit into the interpeduncular fossa (see Figs. 15.21 and 15.22) via the cavernous sinus (see Fig. 15.23) to the superior orbital fissure is visible. Sammlung des Zentrums Anatomie der Universität zu Köln

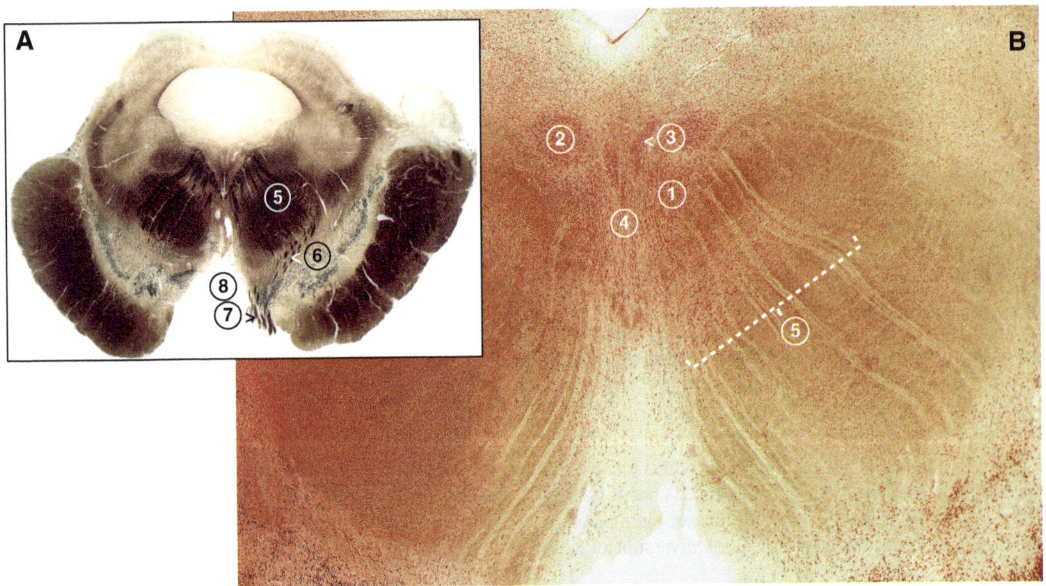

①	Ncl. of oculomotor nerve CNIII, central part	⑤	CNIII fibers traversing the red nucleus
②	Ncl. of oculomotor nerve CNIII, dorsal part	⑥	CNIII fibers running dorsomedial of the Substantia nigra
③	Accessory ncl. of CNIII, nonganglionic projecting part	⑦	CNIII entering the interpeduncular fossa
④	Oculomotor nucleus CNIII, Ncl. Perlia	⑧	Interpeduncular fossa

Fig. 15.21 Horizontal sections through the human mesencephalon at the level of the superior colliculi (see Sect. 15.1.3). The survey picture (Campbell stain) shows the intracerebral course of CNIII by the dark brownish label of the descending fiber bundles. In the high-power microphotograph (Darrow red staining), the oculomotor complex is visible. Due to the less intensive staining of the traversing CNIII fibers compared to the parenchyma of the red nucleus, they can be clearly seen ⑤. LabPON Twente

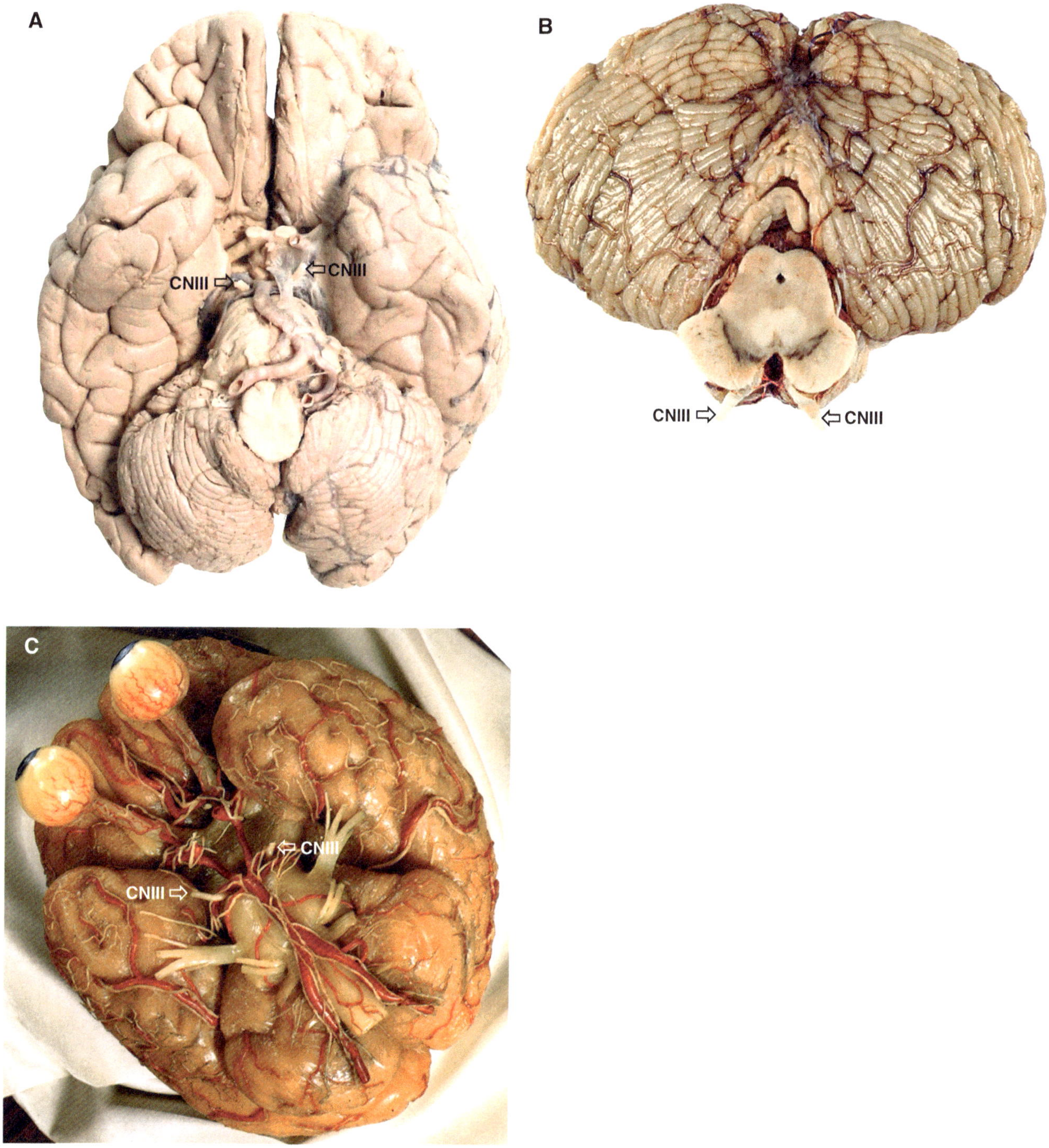

Fig. 15.22 Topographical situation of the CNIII exit into the interpeduncular fossa in (**A**) a liquid-preserved human brain (ventral view), (**B**) a formalin-fixed human brain with the brainstem cut horizontally at the exit level of CNIII with a ventral view of the cerebellum and (**C**) in one of the famous wax moulages (ventral view of the brain) from the Josephinum, Vienna, Austria with permission (see Box 15.3). (**A**) and (**B**) LabPON Twente

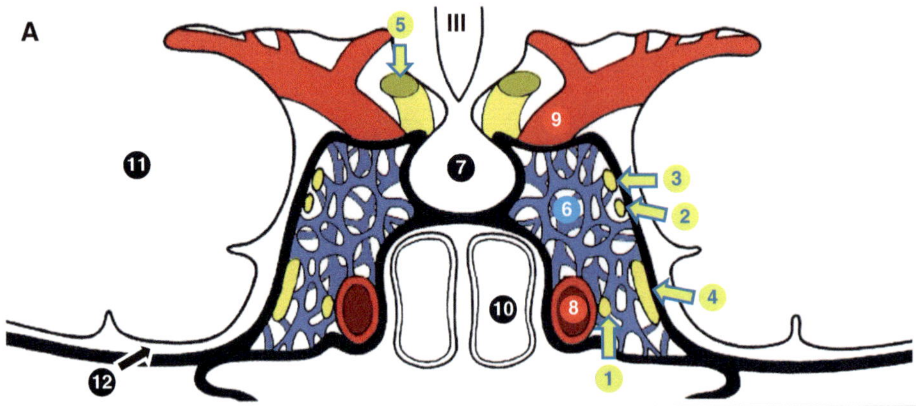

①	Abducens nerve CNVI	⑧	Internal carotid artery, cavernous part
②	Trochlear nerve CNIV	⑨	Middle cerebral artery
③	Oculomotor nerve CNIII	⑩	Sphenoid sinus
④	Ophthalmic nerve CNV$_1$	⑪	Telencephalon
⑤	Optic nerve	⑫	Leptomeningeal space
⑥	Cavernous sinus / trabeculae	▬	Dura mater
⑦	Pituitary gland	III	3rd ventricle

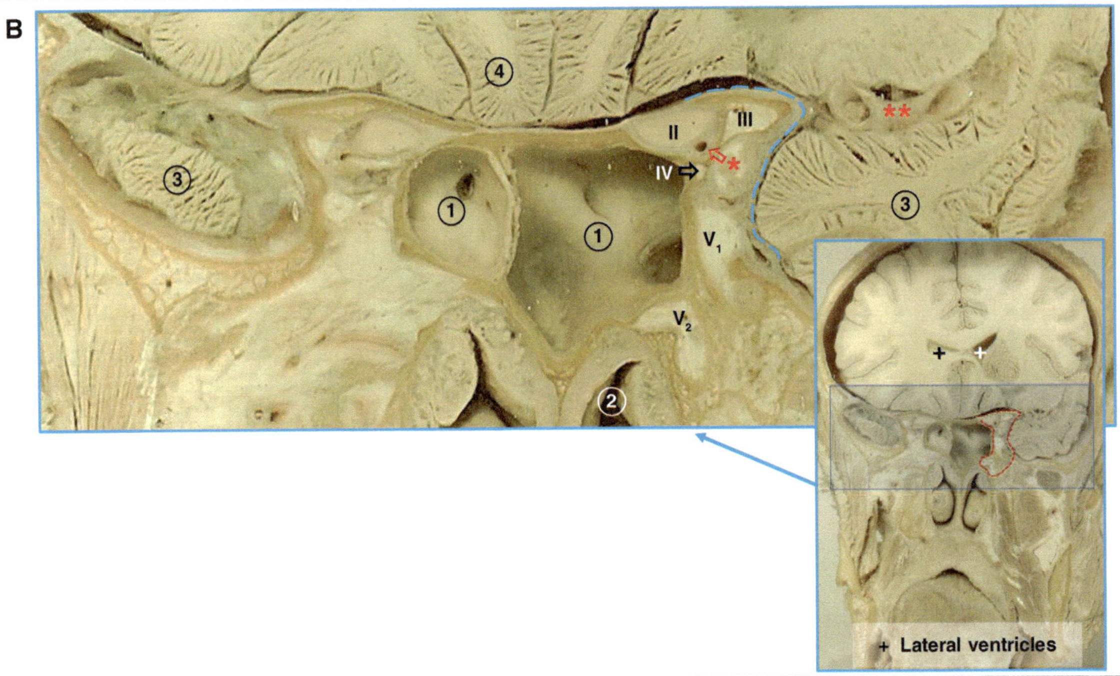

---	Dorsolateral border of cavernous sinus	④	Frontal lobe
①	Sphenoid sinus	*	Ophthalmic artery
②	Nasopharynx	**	Middle cerebral artery in lateral sulcus
③	Temporal lobe		

Fig. 15.23 (**A**) Schematic representation of a coronal section through the cavernous sinus. The position of the three oculomotor nerves (CNIII, CNIV, CNVI) in relation to the cavernous sinus can clearly be seen (Compare with **B**). Ribbers et Huggenberger fecerunt. (**B**) Coronal section through the human head at the level of the cavernous sinus. Fluid-preserved specimen. The large cavity in the middle ① is the sphenoid sinus, the posterior most of the human nasal sinuses with its posterior wall being the anterior wall of the Sella turcica (see Fig. 12.10). CNIII and CNIV as well as the ophthalmic nerve (V_1) and the maxillar nerve (V_2) can be seen in the wall of the cavernous sinus lateral of the sphenoid sinus. Photograph by MedizinFotoKöln, Cologne, Germany. Sammlung des Zentrums Anatomie der Universität zu Köln

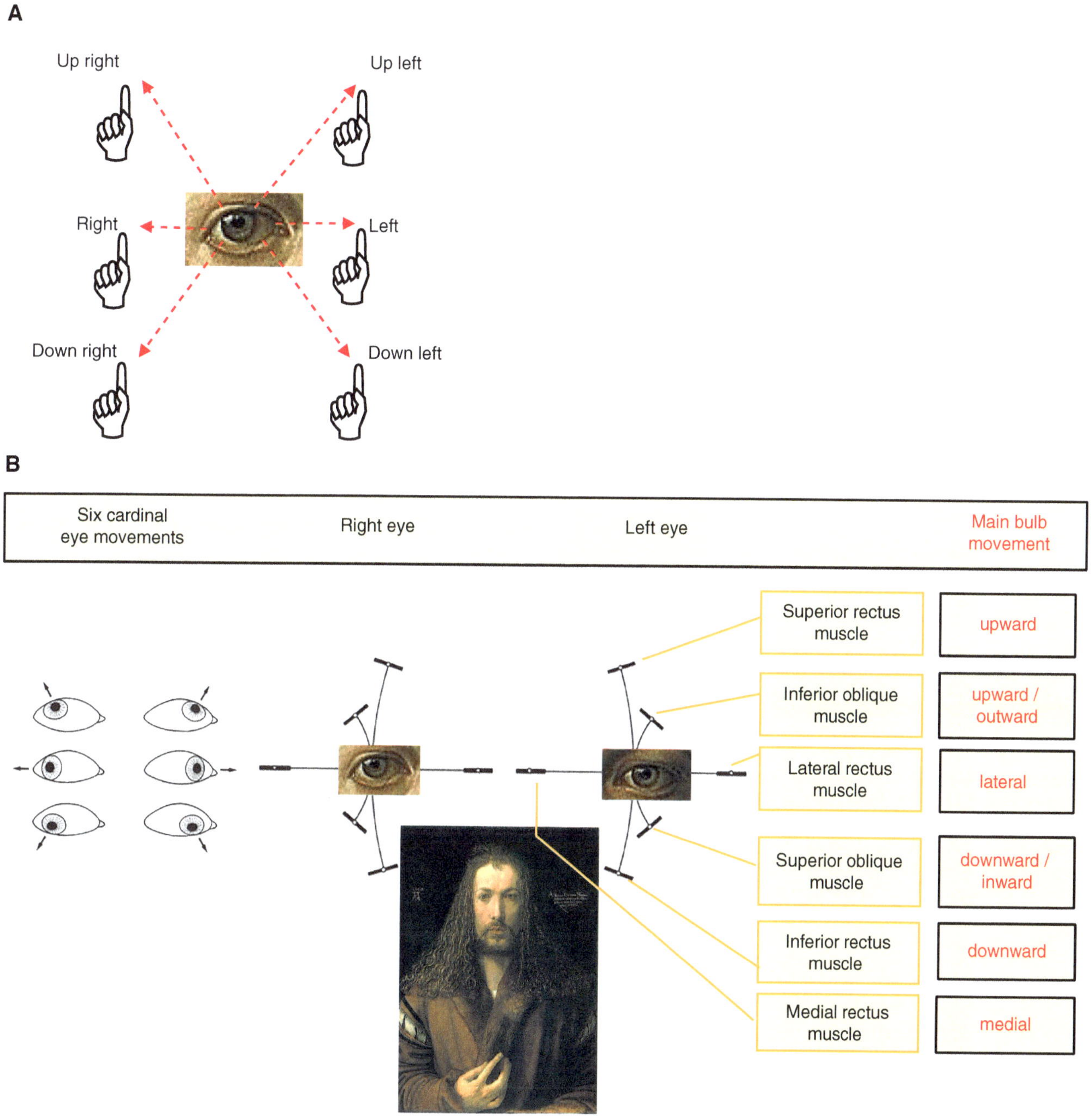

Fig. 15.24 (**A**) In this simple orientating test, the examiner moves its index finger in the directions shown here and asks the patient to follow the finger's movement while the examiner observes the movement of the patient's eyes. H. Schröder fecit. (**B**) In essential, this is an extended illustration to (**A**). It shows the six cardinal eye movements that can be observed when the test shown in (**A**) is executed and correlates them with the potential extraocular eye muscle movements. H. Schröder fecit. Albrecht Dürer Selbstbildnis im Pelzrock (Self-portrait with fur-trimmed robe) (1500). Alte Pinakothek, München, Germany with permission

Fig. 15.25 Lovis Corinth (1909), Ludwig Edinger beim Sezieren eines Gehirns (Ludwig Edinger dissecting a brain). © Historisches Museum Frankfurt, photographer: Horst Ziegenfusz, HMF.B.1956.01. Lovis Corinth (1858–1925) was one of the best-known German impressionists and a sought-after portraitist. For an account of the history of this portrait, see Mann (1974). Stahnisch (2008) writes in his biographic sketch of Ludwig Edinger: *"As Edinger was of Jewish origin, his plans to lecture as a Privatdozent* (i.e., private lecturer; the authors) *in Giessen were interrupted when the head of the university barred him from exercising his Venia legendi. Consequently, in 1883 Edinger established himself as the first Nervenarzt* (neurologist) *in Frankfurt. Fortunately, the Senckenberg Foundation gave him full status as a scientific member which allowed him to pursue his academic career."* He was one of the founders of the University of Frankfurt with his Institute of Neurology which later was named in his honor Edinger-Institute. The merciless irony of history has it that the infamous Nazi neuropathologist Julius Hallervorden (see Box 16.4) spent the last years of his work in the Edinger-Institute in Frankfurt/M. Edinger's daughter Tilly (1897–1967) published the founding work of paleoneurology, Die Fossilen Gehirne (Fossil Brains) in 1929. She discovered that the imprints of brains on the internal surface of fossil skulls could be used to produce endocasts. These enable to study size and surface of brains of extinct animal species (for details, see Buchholtz and Seyfarth 1999). Working at the Senckenberg Institute till 1938, she had to leave Germany because of Nazi rule, her brother died in the Holocaust. She eventually emigrated to the United States where she worked at the Museum of Comparative Zoology at Harvard

It should be noted that the designation *"central cell group"* is not precise for the location of the EWpg (see Figs. 15.8A and 15.27A "Human").

Very recently, May et al. (2020) could identify the EWpg in Macaque monkeys by injection of retrograde tracer in the ciliary ganglion.

Summarized by Kozicz et al. (2011b), after 15 years of urocortin research it appears clear that midbrain urocortin 1 (Ucn1) is important for food intake, the homeostatic equilibrium, and the regulation of mood under stress. Stress reaction may be initially under the influence of corticotrophin-releasing factor, while Ucn1 seems to be crucial for management of the later adaptive phase and imbalance between the two players may be of pathogenetic importance. The fact that Ucn1 neurons project to, among other nuclei, the dorsal raphe nucleus (see Sect. 13.3.4.2) involves the Ucn1 neurons in EWcp in the control of dorsal raphe serotonergic activity, thereby modulating the stress response and stress-related behaviors (for details, see Kozicz et al. 2011b).

15.3.1.2.2 Target muscles of the Edinger-Westphal nucleus EWpg

The EWpg innervates the ciliary and sphincter pupillae muscles (Fig. 15.28).

15.3.1.2.3 Course of the parasympathetic fibers of the Edinger-Westphal nucleus

The parasympathetic, preganglionic fibers originating from the cholinergic neurons of EWpg join CNIII with which they reach the orbita. Here they leave CNIII and enter the ciliary ganglion (see Figs. 15.8, 15.9, 15.10, and 15.14 ⑥) as branch to ciliary ganglion of the nasociliary nerve (Fig. 15.13 ⑨) [Ramus ganglionaris ciliaris] (formerly—better—parasympathetic root of ciliary ganglion). After having synapsed with the postganglionic parasympathetic ganglion cells, the postganglionic fibers form the Nn. ciliares longi (Fig. 15.14 ⑧) which perforate the sclera and reach the interior of the eye bulb. The parasympathetic fibers innervate the ciliary muscle [M. ciliaris] (accommodation (Remington 2012), changes of focus from distant to near images by change in lens shape) and the sphincter pupillae [M. sphincter pupillae] (miosis, constriction of the pupil).

The sympathetic root of ciliary ganglion [Radix sympathica ganglii ciliaris] enters the skull via the internal carotid plexus. The sympathetic fibers innervate the dilator pupillae [M. dilatator pupillae] (mydriasis, dilatation of the pupil).

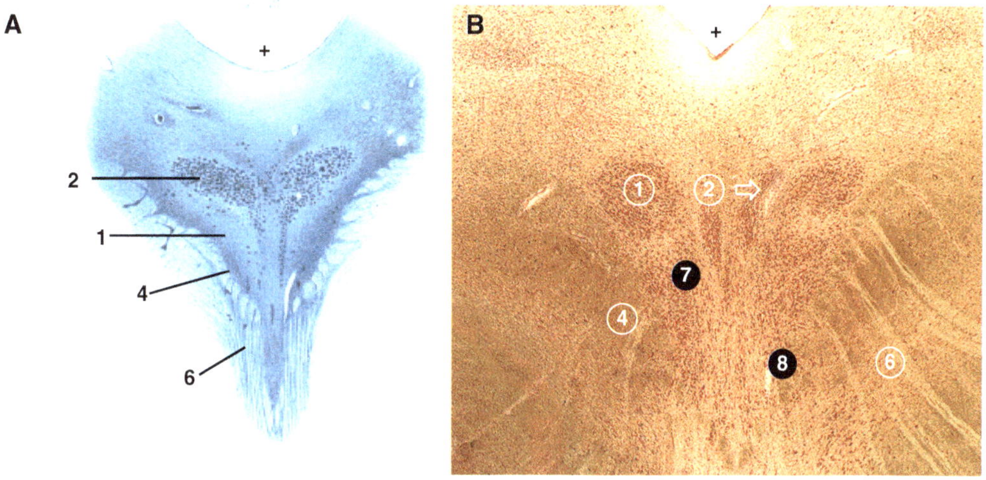

①	Atrophischer Oculomotoriuskern / Atrophic ncl. of oculomotor nerve
②	Mediale Zellengruppe / Medial cell group
④	Hinteres Längsbündel / Posterior longitudinal fasciculus
⑥	Atrophische Wurzelfasern / Atrophic root fibers
+	Aqueductus Sylvii / Aqueduct of Sylvius / Mesencephalic aqueduct

①	Ncl. of oculomotor nerve CNIII, dorsal part
②	Accessory ncl. of oculomotor nerve, nonganglionic projecting part (Edinger-Westphal ncl., centrally projecting part)
④	Medial longitudinal fasciculus
⑥	Oculomotor nerve CNIII
❼	Ncl. of oculomotor nerve CNIII, central part
❽	Ncl. of oculomotor nerve CNIII, ventral part

Fig. 15.26 (**A**) From Westphal 1887, Fig. 1 with permission. The horizontal detail drawing shows that part of the mesencephalic tegmentum which is bordered ventrally by the—posterior in his time, now called—medial longitudinal fasciculus **4** and dorsally by the mesencephalic aqueduct (+) from the case described by Westphal. (**B**) Normal histology of the same region in a Darrow red staining. Note the well-developed nucleus of oculomotor nerve ① is almost lacking in Westphal's case. Since the patient described had a paralysis of external but not of the internal eye muscles the conclusion Westphal drew was that the so-called medial **2** and lateral cell groups were responsible for the innervation of the smooth muscles of the eye. LabPON Twente

15.3.1.2.4 Living anatomy and clinical aspects of the Edinger-Westphal nucleus

The most common clinical test to examine the function of the Edinger-Westphal nucleus and its efferent fibers is the pupillary reaction/light reflex (Fig. 15.29).

Human-specific data on the afferent part of the light reflex and the pretectal path to the EW are not available. For the macaque brain, however, an extensive antero- and retrograde tracing study has revealed the connectivity of pretectal nuclei (May and Warren 2020). Within the pretectum, the vast majority of neurons projecting to the preganglionic Edinger-Westphal nucleus was found within the olivary pretectal nucleus (OPN) [Ncl. pretectalis olivaris]. The olivary pretectal nucleus is the first CNS nucleus involved in the pupillary light reflex pathway, the circuit that adjusts the diameter of the pupil in response to ambient light levels. Retinal terminals were concentrated within the borders of the olivary pretectal nucleus. This was supposed be due to the fact that each macaque retina provides nearly equal density projections to the ipsilateral and contralateral olivary pretectal nucleus (May and Warren 2020).

The human pretectal olivary nucleus has been described by Kuhlenbeck and Miller (1949), using a non-vertebrate terminology. It is easiest to recognize laterally of the superior layers of the superior colliculi (see Fig. 15.2 OPN).

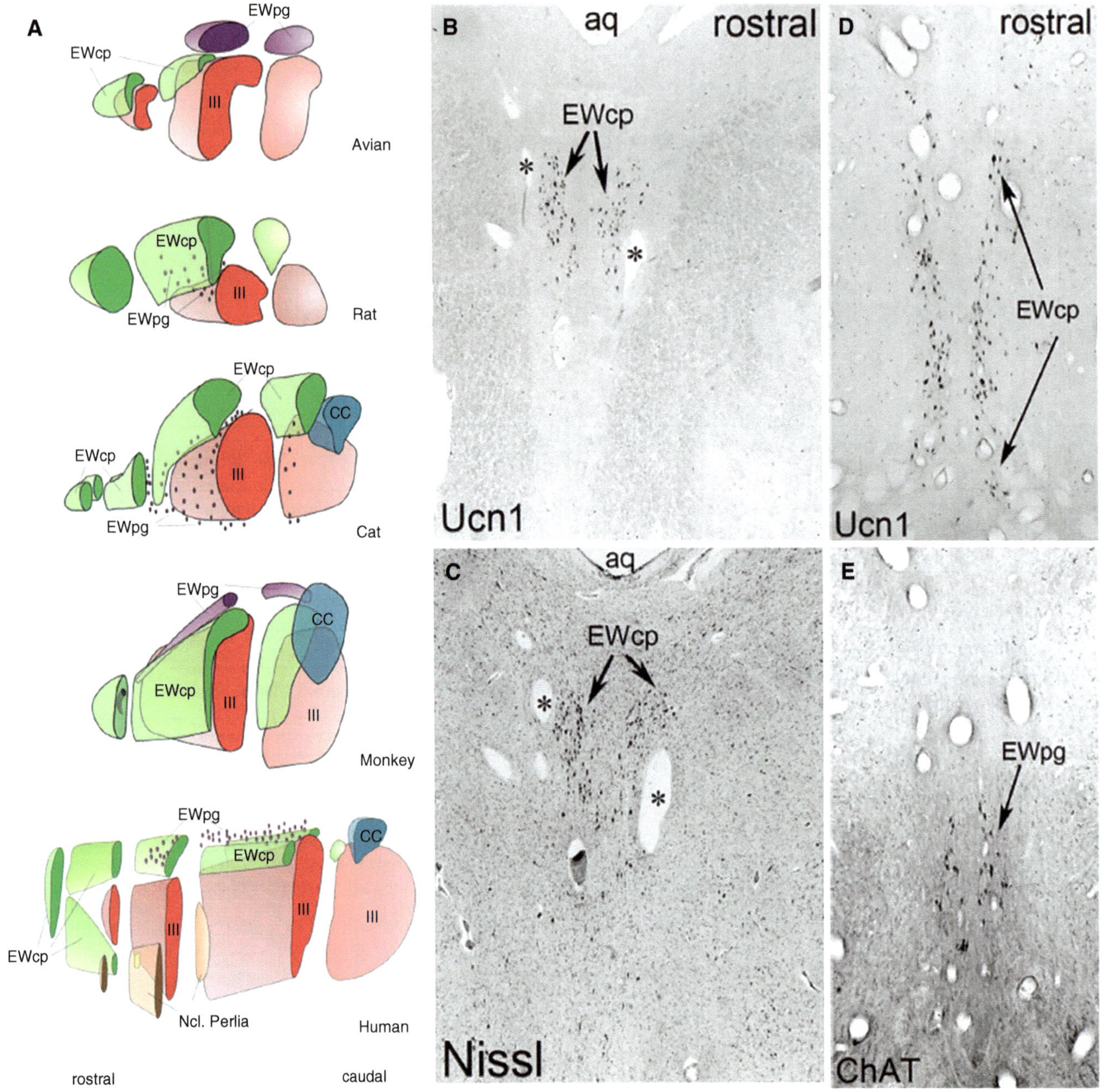

Fig. 15.27 Comparative anatomy of the Edinger-Westphal nucleus. (**A**) Schematic summary of the oculomotor/EW region in different species. The parasympathetic part of EW (EWpg) is a compact structure in birds and monkeys but scattered in rat, cat, and man. This is the reason why this part of the nucleus eludes identification with conventional staining as used, for example, by Westphal (1887), Nissl or also Darrow red. It can be visualized as all preganglionic parasympathetic nuclei using choline acetyltransferase (ChAT)-immunohistochemistry (**E**, monkey). (**C**) The compact structure described by Westphal is what today is called the EWcp, clearly visible in man in a Nissl stain. Furthermore, it can be identified unequivocally by use of urocortin-immunohistochemistry in man (**B**) and monkey (**D**). * = Corresponding vessels. Modified from Kozicz et al. 2011a, Figs. 1, 2, 7 with permission

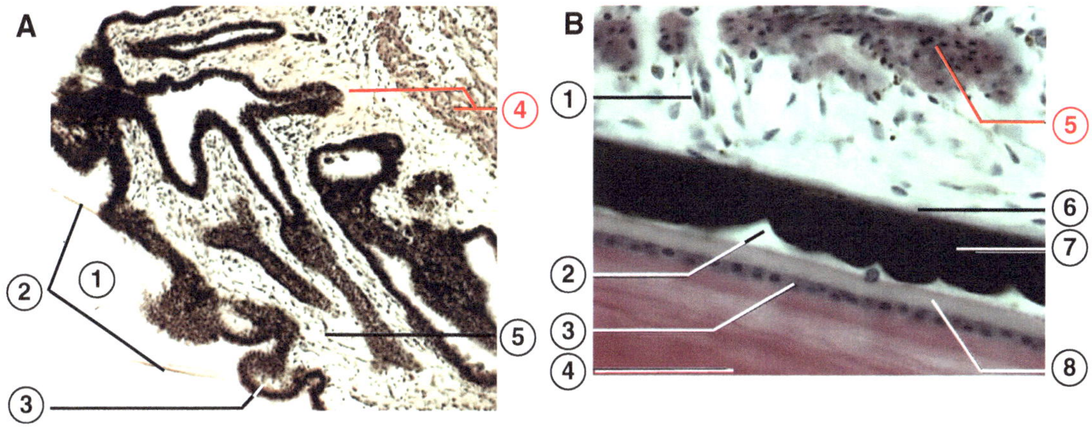

	A		B
①	Posterior eye chamber	①	Iris (Stroma)
②	Zonular fibers	②	Connection between anterior and posterior eye chamber
③	Pigment epithelium	③	Lens epithelium
④	Ciliary muscle	④	Lens ncl.
⑤	Ciliary body (Stroma)	⑤	Sphincter pupillae
		⑥	Dilator pupillae
		⑦	Iris epithelium
		⑧	Lens capsule

Fig. 15.28 Target muscles of the preganglionic Edinger-Westphal nucleus. Two smooth muscles of the eye are innervated by postganglionic fibers from the ciliary ganglion (see Sect. 15.3.1.2.3), the ciliary muscles (**A** ④), and the sphincter pupillae (**B** ⑤). The ciliary muscles ④ are connected with the lens via the zonular fibers (see also Fig. 15.17, mostly destroyed during tissue processing) and are thereby involved in accommodation (see Sect. 15.3.1.2.3). The sphincter pupillae muscle ⑤ is a very small longitudinal band of smooth muscles close to the pigmented iris epithelium. Its contraction leads to a decrease of the pupil diameter (miosis), i.e., less photons/light reach the retina. Miosis is the reflex reaction in the light reflex (see Sect. 15.3.1.2.3). H & E stain. Figures modified from www.anatomiedesmenschen.de/Sammlung des Zentrums Anatomie der Universität zu Köln

A similar neuronal pathway is used by the **accommodation-convergence reaction** (near point reaction, also accommodation reflex). It is not a true reflex but rather a synkinesis or an association of three occurrences: convergence, accommodation, and miosis (Remington 2012).

As an object is brought near along the midline, the medial rectus muscles (see Figs. 15.12, 15.13, 15.14, 15.18, and 15.19) contract to move the image onto each fovea; the ciliary muscle (see Fig. 15.28) contracts to keep the near object in focus; and the sphincter muscle (see Fig. 15.28) constricts to decrease the size of the pupil, thereby improving depth of field (Remington 2012).

By contrast to the light reflex, the afferents of this reaction follow the visual pathway beyond the midbrain (see Fig. 15.29) to the visual cortex which transfers signals to the frontal eye fields. These communicate with the oculomotor nucleus (see Sect. 15.3.1.1) and the Edinger-Westphal nucleus through a pathway that passes through the internal capsule. The efferent pathway, via CNIII, innervates the medial rectus muscle, and the parasympathetic pathway (EW) innervates the ciliary muscle and iris sphincter (Remington 2012).

When dealing in emergency medicine with unconscious patients, testing the light reflex is of utmost importance to check an increase in intracranial pressure since it does not require the cooperation of the patient. Supratentorial pressure increase (see Fig. 15.30) is propagated to the tentorial notch where the oculomotor nerve passes over the petroclinoid ligament. The pressure exerted on the nerve may lead to first reversible, later irreversible lesions. In the latter case, autopsy findings have shown a sanguineous indentation of the oculomotor nerve (Fig. 15.31).

As long as the patient is conscious, the motility of the eye muscles can be examined as shown in Figs. 15.24 and 15.32. In case of an oculomotor (CNIII) palsy, the clinical picture can be deduced from the fact that usually only the abducens nerve (lateral rectus) and the trochlear nerve (superior obliquus) are intact. All the remaining muscles are paralytic. The predominance of lateral rectus and superior obliquus drives the eyeball of the lesioned side laterally (lateral rectus) and downwards (obliquus superior). Furthermore, usually the levator palpebrae superioris does not work anymore resulting in the closure of the eyelid (ptosis) on the lesioned side.

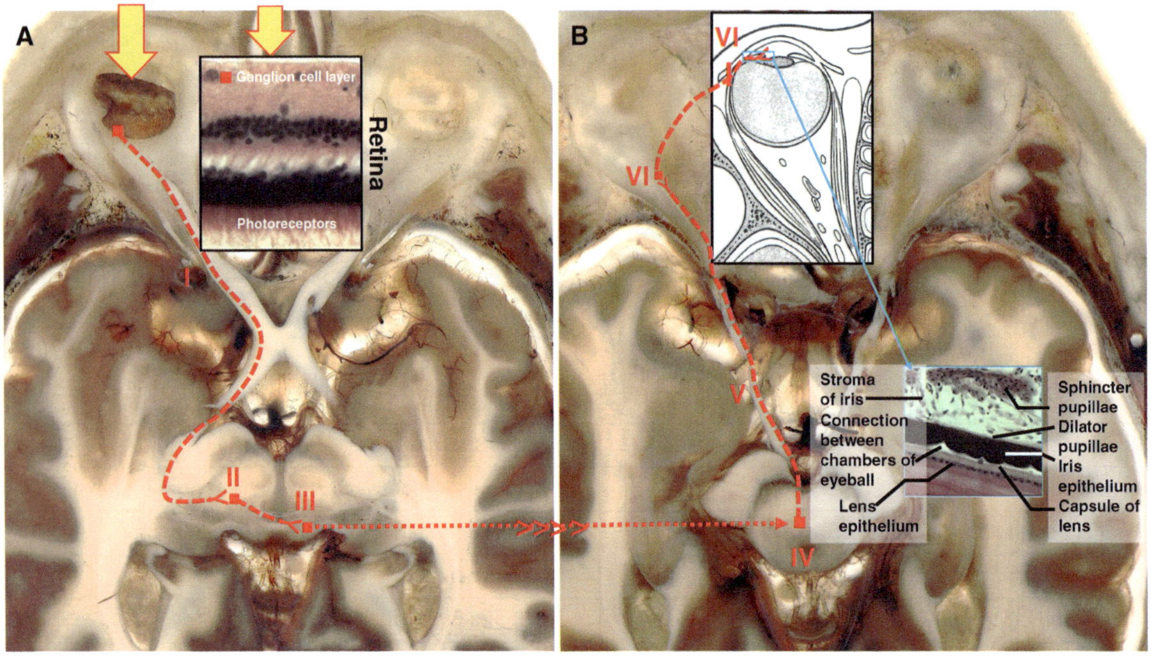

Fig. 15.29 Anatomy of the light reflex. (**A**) The afferent part of the reflex starts with lighting one eye while covering the other. The light passes the cornea, lens, hyaloid body, and the retina. At the most peripheral part of the retina, the photoreceptors are activated. The signal is propagated from the receptors via the bipolar cells to the ganglion cells. The axons of the ganglion cells form the optic nerve ([I] CNII) and those axons involved in the light reflex reach via interneurons [II] the olivary pretectal nucleus ([II] see Fig. 15.2) which project [III] **bilaterally** to the Edinger-Westphal nucleus [IV]. (**B**) From there the axons [V] run with the oculomotor nerve to the ciliary ganglion [VI] in the orbita. Eventually, the postganglionic fibers innervate the sphincter pupillae [VII] muscle in the iris which leads to the observable effect of the reflex, i.e., a bilateral decrease in pupillary diameter (miosis). Sammlung des Zentrums Anatomie der Universität zu Köln. Inset photographs from www.anatomiedesmenschen.de/Inset drawing. Ribbers fecit

Fig. 15.30 Semi-schematic model of the sequels of a tumor (T)—as well as of intracranial hematomas (see Fig. 15.36)—of the forebrain. Superior arrow: The cingulate gyrus is moved to the contralateral side. Middle arrows: Parts of the temporal lobe are herniated lateral of the brainstem into the tentorial notch (see Box 3.5). Inferior arrows: The brainstem is moved caudally and the cerebellar tonsils are pressed into the Foramen magnum. From Zülch 1959, Fig. 32 with permission

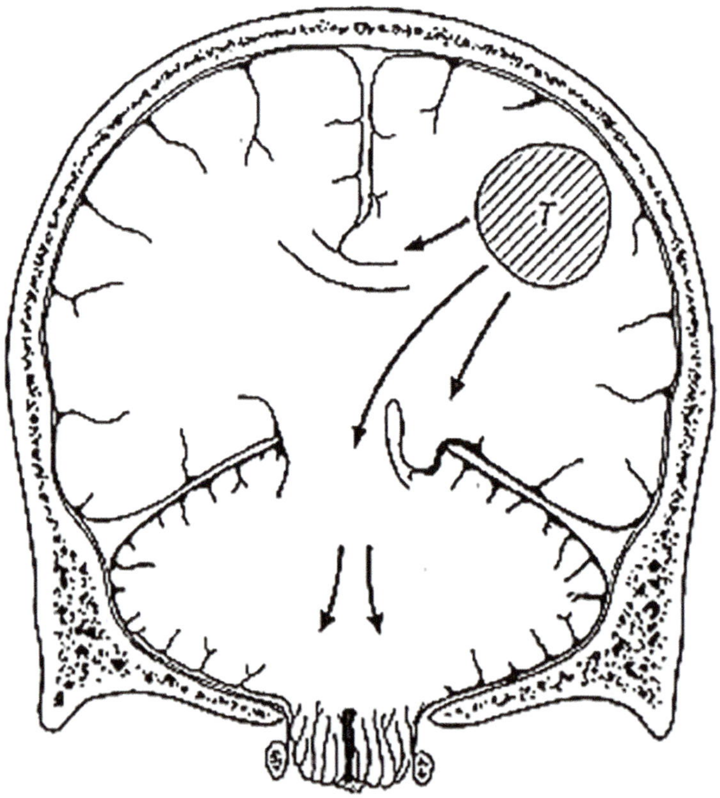

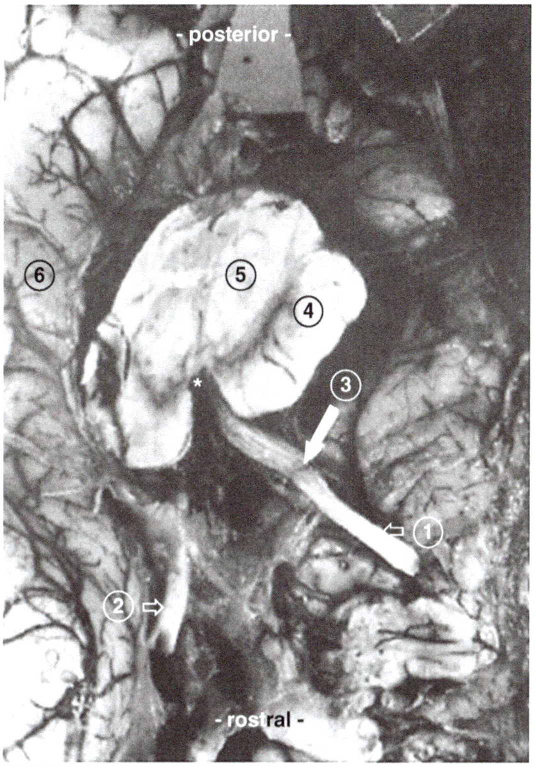

①	Left oculomotor nerve CNIII	⑤	Mesencephalic tegmentum
②	Right oculomotor nerve CNIII	⑥	Temporal lobe, ventral surface
③	Hemorrhagic furrow	*	Interpeduncular fossa
④	Cerebral peduncle		

Fig. 15.31 Autopsy specimen showing the right oculomotor nerve ① CNIII after entrapment into the tentorial notch (dura mater removed) with compression in the region of the superior edge of the clivus by the petroclinoid ligament (see Fig. 9.7). The caudal view of the removed brain (brainstem sectioned horizontally at the level of the mesencephalon, see interpeduncular fossa *, see Figs. 15.3 and 15.7) shows the course of both CNIII from their exit (at the right hand better visible than left) to the cavernous sinus. The right CNIII shows a dark, stripe-shaped discoloration ⇨, approximately half-way from its exit which represents a hemorrhagic furrow ③ caused by a caudal directed space-occupying lesion. The latter compresses the nerve at the region of the petroclinoid ligament (see Fig. 9.7). From Zülch 1959, Fig. 81 with permission

A clinical example in a conscious patient is shown in Figs. 15.32 ff. She complained about an episode of headache and elevated blood pressure (almost 200 mmHg) ca. one and a half months before admission, 5 days before she experienced diplopia. As obvious from mere inspection, the patient presented with a left ptosis (center picture). The patient is asked to look to the left side where the intact lateral rectus muscle cooperates with the left medial rectus to produce a conjugate movement. Looking to the right, however, reveals the inability of the left medial rectus to move the left eye towards the nose while the right lateral rectus moves the eye completely to the right. When looking upwards, the left eye lags behind compared to the right one due to the impaired function of the left superior rectus. Finally, when looking downwards again, the left eye lags behind the right one since the left eye lacks the action of the obliquus and rectus inferior muscles. All other parameters of the ophthalmological examination were normal (Slit lamp examination—except a mid-dilated pupil left—fundoscopy, perimetry).

In case a traumatic event can be excluded like here (see, however, Fig. 15.34), the most likely cause of an oculomotor palsy is an aneurysm of the cerebral circulation in a vessel close to the extracranial course of CNIII. For this reason, a carotid angiography was performed (see Fig. 15.33A) revealing a lobulated protruding component (1 cm) at the junction

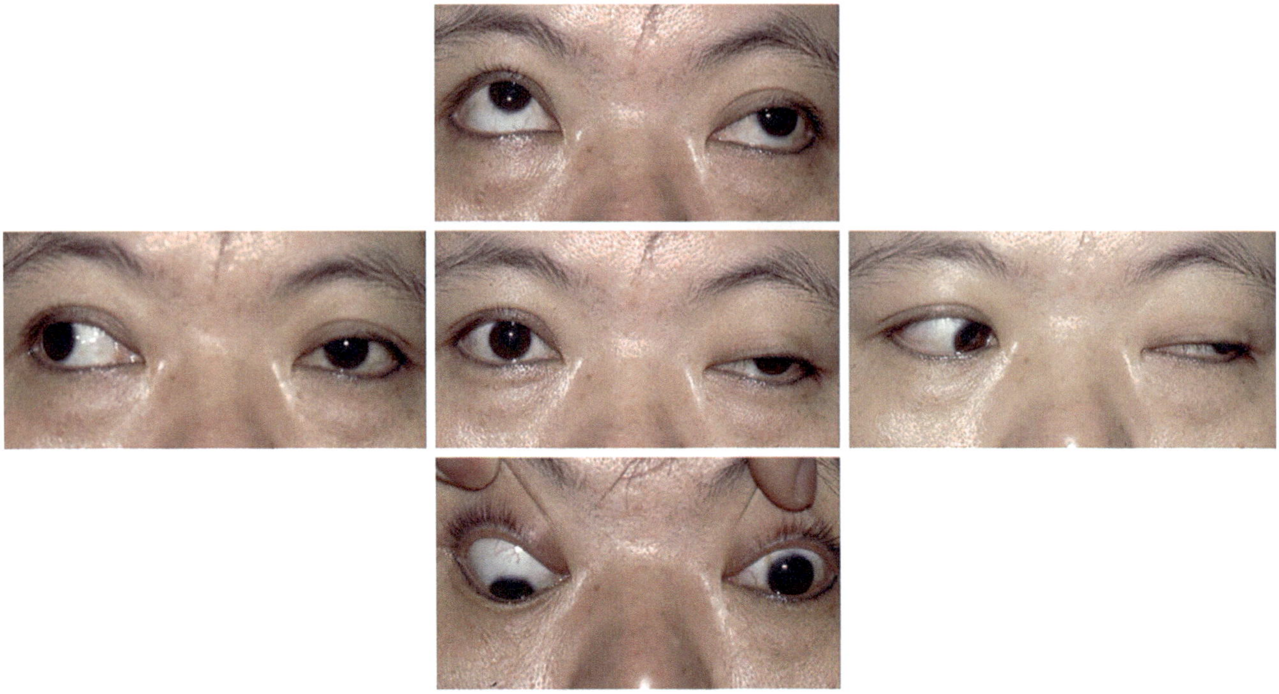

Fig. 15.32 A 42-year-old female had an episode of headache with elevated blood pressures to 197 mmHg about 1.5 months ago. She underwent an EEG evaluation at a local hospital which came back normal. Diplopia occurred 5 days prior. The patient's visual acuity was 6/6.7 in the right eye and 6/8.6 in the left. There was left ptosis and 25 prism diopters of exotropia in the left eye. Extraocular movements revealed moderate limitation of adduction, supraduction, and infraduction of the left eye. Slit lamp examination showed a normal anterior segment OU, except for a mid-dilated pupil in the left eye. Fundoscopic examination and visual field examination were normal. She was admitted under the impression of left oculomotor nerve palsy with pupil involvement. From Wang 2018, Fig. 31.1 with permission

of the left internal carotid artery and the posterior communicating artery.

The patient underwent a craniotomy and the aneurysm was treated by clipping (see Box 15.4). Two months later the patient showed a nearly normal eye motility (Fig. 15.33C).

Box 15.4 Surgical clipping of aneurysms

The classical neurosurgical approach to the treatment of an aneurysm is the application of the so-called clip onto the base of the aneurysm (see Fig. 15.34) after surgical exposure of the region where the aneurysm is located (Trepanation under general anesthesia).

The goal of the procedure is to isolate the aneurysm from the normal circulation without blocking off any small perforating arteries nearby.

The clip (see Fig. 15.34) works like a tiny coil-spring clothespin, in which the blades of the clip remain tightly closed until pressure is applied by a clip applier to open the blades. Clips are made of titanium and remain on the artery permanently.

https://mayfieldclinic.com/pe-clipping.htm

Another site in the brainstem where the oculomotor nerve may be compressed—eliciting an oculomotor palsy—based on the anatomical features is the passage of CNIII between the posterior cerebral artery (PCA) and the superior cerebellar artery (SCA) (see Fig. 15.35). An aneurysm of one of these arteries may lesion the oculomotor nerve.

Finally, a head trauma may be the cause of an oculomotor palsy, such as an intracranial hematoma, most frequently the so-called epidural hematoma (Fig. 15.36). Usually, in a supratentorial position, these hematomas develop very quickly after a skull fracture which traverses and thereby lacerates the middle meningeal artery [A. meningea media]. As described above (see Fig. 15.30), any increment in intracranial volume, like a tumor or, as here, an arterial bleeding, is trapped inside the bony cage of the skull. The only direction of pressure propagation is to the larger dural (tentorial notch) or bony openings (Foramen magnum) of the skull. The first location causing problems during this shifting process is the tentorial notch with the sphenoidal ridge and the petroclinoid ligament (see Fig. 9.7). Regardless of the cause of pressure increase, eventually, it will result in a lesion of the oculomotor nerve. The faster a pressure increase develops, the faster the clinical signs become obvious. In case of

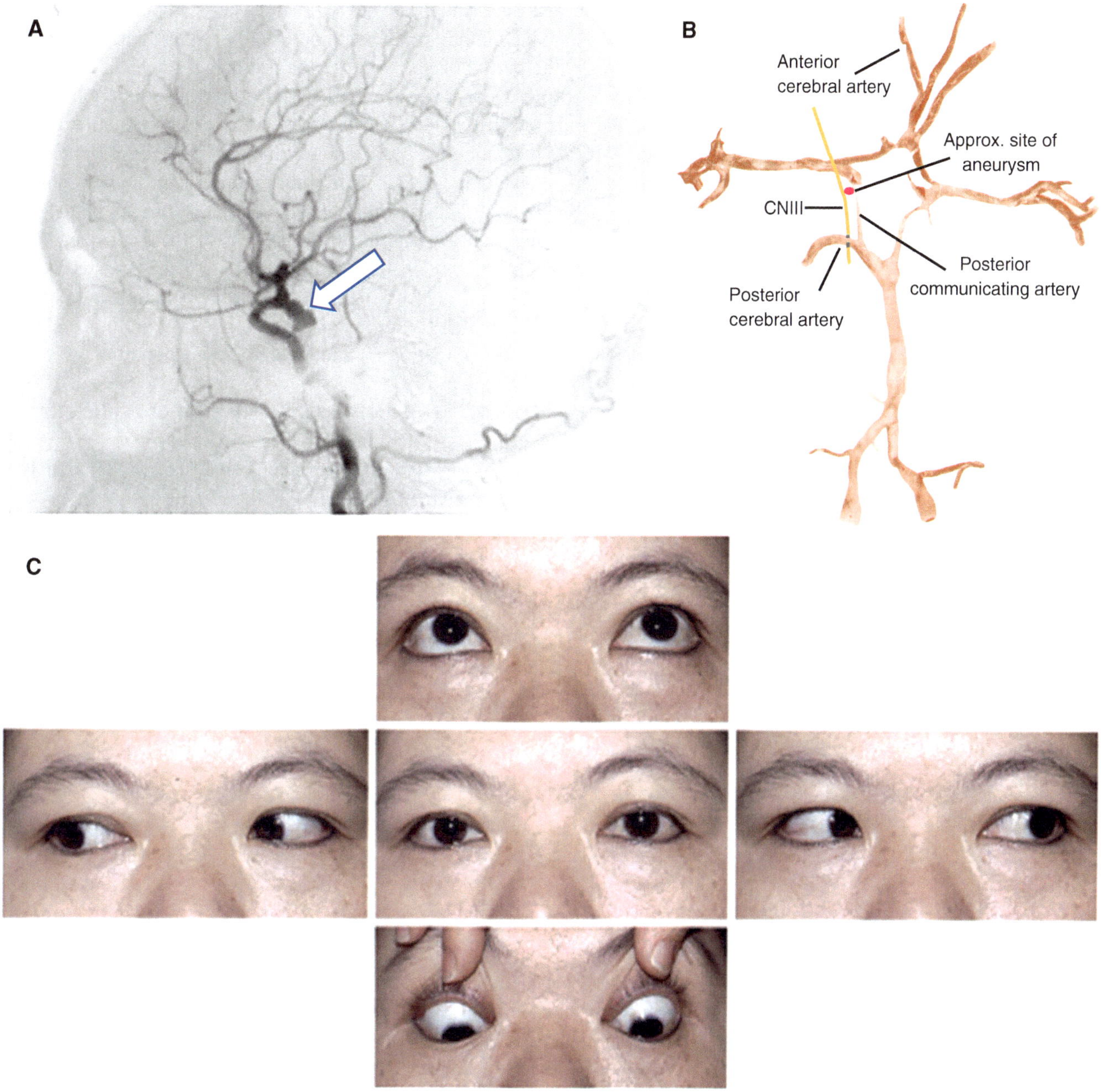

Fig. 15.33 (**A**) Carotid angiography. Lateral view. The arrow points to a lobulated protruding structure, an aneurysm approx. 1 cm in size, located at the junction of the left internal carotid and the posterior communicating artery (compare with Fig. 15.31B). The pathoanatomical situation is shown in a three-dimensional fashion in Fig. 15.33B. From Wang 2018, Fig. 31.2 with permission. (**B**) Dorsal view on the human circle of Willis (see Figs. 1.16–1.18 and Sect. 1.4.1.1). The left posterior communicating artery shows a ventral bulge in its anterior one third (red ovoid, schematic visualization) which comes in close contact with the oculomotor nerve potentially lesioning the somatomotor portions and the preganglionic fibers from the Edinger-Westphal nucleus pg (see Sect. 15.3.1.2.2). The latter are supposed to be located at the dorsal medial surface of CNIII. CNIII is crossing the posterior cerebral artery below (----) and lies lateral of the internal carotid artery (see also Fig. 15.23A). LabPON Twente. H. Schröder fecit. (**C**) Postoperative examination (2 months) of eye motility (compare with Fig. 15.32). From Wang 2018, Fig. 31.3 with permission

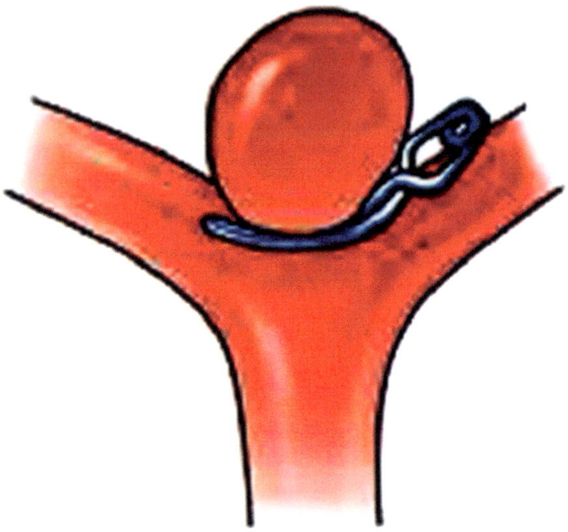

Fig. 15.34 As an example of aneurysm clipping, here, the graphical representation of an aneurysm located at a branching point of two cerebral vessels (different from the situation of the case report shown in Fig. 15.33). The general approach of clipping is to separate the aneurysm with its fragile vessel wall from the general circulation by a clip (blue) which is applied to the so-called neck of the aneurysm. From Huang et al. 2018, Fig. 4 with permission

a tumor this may be years, in case of an aneurysm months and an epidural hematoma usually elicits symptoms within a period of 30 min after the traumatic impact. The patients lose consciousness which impedes the evaluation of eye motility as described above. However, no collaboration of the patient is required to test the pupillary (light) reflex. The development of a wide, subsequently fixed pupil—unreactive to light—is a reliable sign of an oculomotor nerve damage, i.e., an imminent entrapment of the brainstem in the tentorial notch. In case no therapy will be initiated at that point the pressure increase would reach the medulla oblongata which becomes entrapped together with the cerebellar tonsils in the foramen magnum, a lethal condition.

15.3.1.3 Medial accessory oculomotor nucleus [Ncl. accessorius medialis]

For information on this nucleus, see Sect. 16.4.4 (see Box 16.1 Eponyms: Bechterew).

15.3.1.4 Oculomotor nucleus, parvocellular part

Oculomotor nucleus, parvocellular part, or parvocellular oculomotor nucleus is synonymous with the accessory

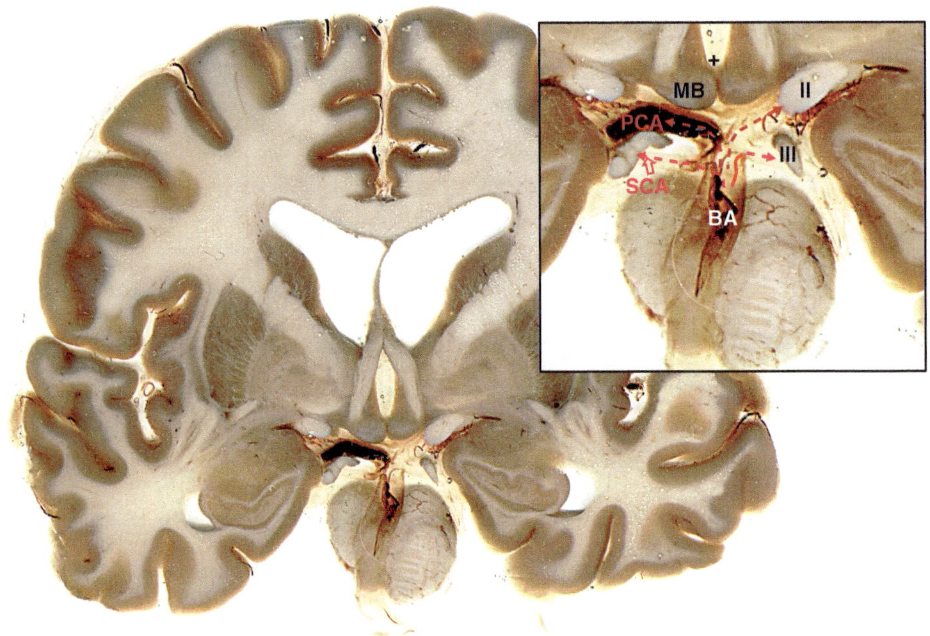

II	Optic tract CNII	PCA	Posterior cerebral artery
III	Oculomotor nerve CNIII	MB	Mammillary body
BA	Basilar artery	+	3rd ventricle
SCA	Superior cerebellar artery		

Fig. 15.35 Coronal section through the human brain at the level of the mammillary bodies (MB). Plastinated section. The specimen shows a ventral view on the surface of the pons where the basilar artery (BA) is running to its terminal branching point into the superior cerebellar artery (SCA) and the posterior cerebral artery (PCA). The latter is running dorsally around the mesencephalon. The oculomotor nerve (III) on its way to the cavernous sinus (i.e., towards the beholder) traverses the space between the SCA and the PCA. In case of expansion of one of the vessels (e.g., cerebral aneurysm, see Box 15.4), the oculomotor nerve may be subject to compression at this site. Sammlung des Zentrums Anatomie der Universität zu Köln

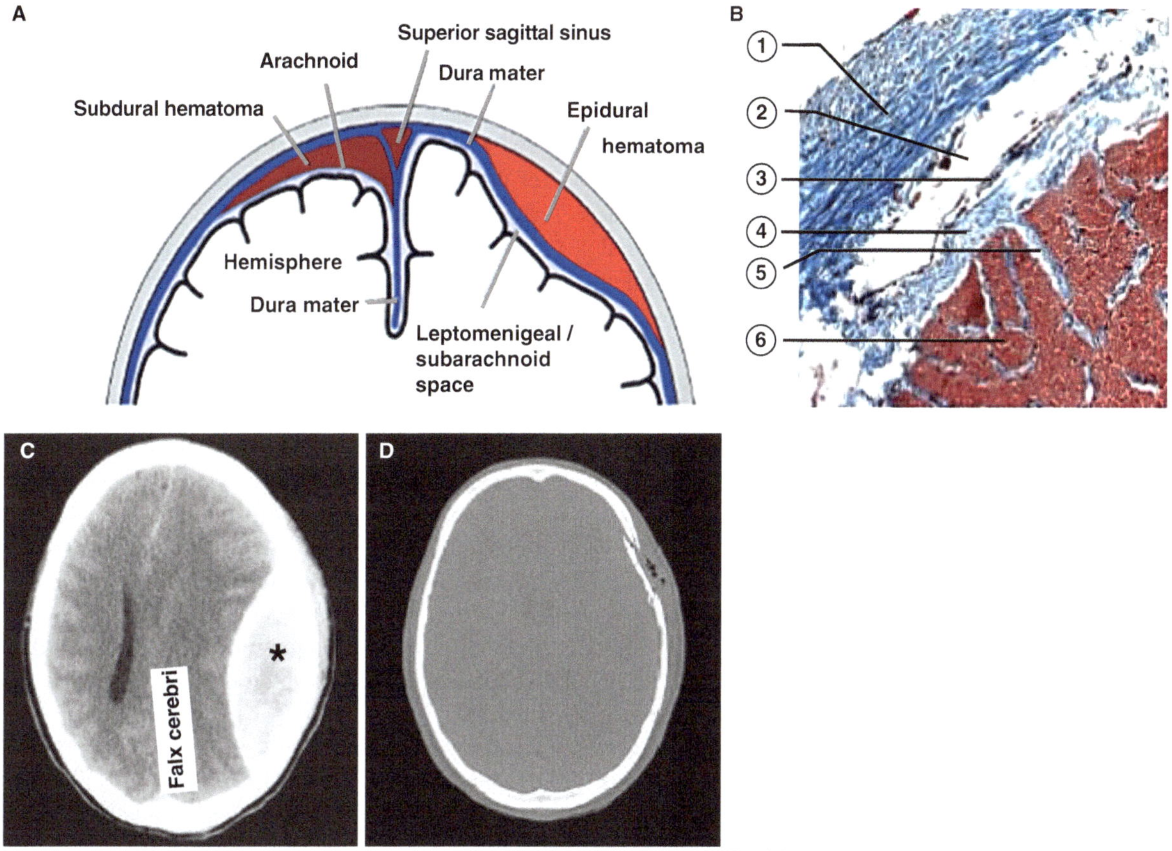

①	Dura mater	④	Pia mater
②	Leptomeningeal space	⑤	Septa
③	Arachnoid	⑥	Optic nerve

Fig. 15.36 Topography of intracranial hematomas. (**A**) An epidural hematoma is an arterial bleeding from a meningeal artery. These arteries are located upon the outer surface of the dura mater. During rupture, the blood is collecting upon (Greek epi) the dura mater. A subdural hematoma is due to the rupture of dural veins on the inner surface of the dura: the blood is collecting under (Latin sub) the dura. (**B**) Histological section of the dura mater surrounding the optic nerve ⑥. The dura ① is the outer most and thickest part of the meninges. Between the dura and the arachnoidea ③ the leptomeningeal space ② (see Fig. 1.8)—artificially enlarged—is extending, filled with cerebrospinal fluid. Below the inner layer of the arachnoid the pia mater ④ is seen which extends in septa ⑤ into the optic nerve (red) ⑥. (**C**) As indicated in (**A**), the neuroradiological sign of an epidural hematoma (*) is its convex surface versus the brain parenchyma. The increased intracranial pressure has led to a complete compression of the left lateral ventricle and a displacement of the Falx cerebri to the contralateral side. (**D**) The CT bone scan shows a discontinuity of the skull on the left side with impression of the bone. Together, these findings suggest a left-sided epidural hematoma due to a skull fracture. The neurosurgical therapy consists of trepanation above the fracture, removing the bone and the hematoma and, if possible, ligation of the lesioned artery. From Huggenberger et al. 2019, Figs. 15.18 and 15.19 with permission

oculomotor nucleus, Edinger-Westphal nucleus (see above Sect. 15.3.1.2).

15.3.1.5 Supraoculomotor cap

The supraoculomotor cap is a small band of neurons in the periaqueductal gray below the level of the aqueduct and directly above the supraoculomotor gray in the middle of the antero-posterior extent of the PAG which in rat, rabbit, cat, monkey, and man contains the enzyme NADPH diaphorase. NADPH diaphorase is a nitric oxide synthase (Carrive and Paxinos 1994).

15.3.2 Basal tegmentum

15.3.2.1 Pararubral nucleus [Ncl. pararubralis]

The term pararubral nucleus refers to a group of cells located dorsal and lateral to the red nucleus in the mouse.

15.3.3 Mesencephalic reticular formation [Formatio reticularis mesencephali]

In the latest edition of the Paxinos et al. (2020), brainstem book the mesencephalic reticular formation (MRF) is identified as the homolog of the rodent and non-human primate retroparafascicular nucleus. It is located lateral of the periaqueductal gray with no anatomically defined borders.

Recently, May and Warren (2020) could show that in monkeys the central mesencephalic reticular formation contains premotor neurons controlling lens accommodation (see Sect. 15.3.1.2.3). This is in line with previous findings that this part of the MRF also has premotor neurons for medial rectus motoneurons (see Figs. 15.8, 15.9, and 15.10) and plays an important role for near response and disjunctive saccades when viewers look between targets located at different distances (for details, see May and Warren 2020). It has been known to be involved in the control of both, horizontal and vertical eye movements.

15.3.4 Red nucleus [Ncl. ruber]

15.3.4.1 Location and morphology of the red nucleus

The spherical red nucleus [Ncl. ruber] got its name from its reddish appearance which is obvious on unfixed brain sections (Fig. 15.37). This coloring is due to a high content of iron which is also the basis of its visibility in T2-weighted MRI images (Cacciola et al. 2019) (see Fig. 15.37B). It can be subdivided in a rostral parvo- and a caudal magnocellular part.

15.3.4.2 Connectivity and function of the red nucleus

The parvocellular part exclusively projects ipsilaterally to the inferior olive via the central tegmental tract (see Fig. 3.14A–G) (ten Donkelaar 2011). The magnocellular part is the origin of the rubrospinal tract, compared to rats and monkeys of small size (ten Donkelaar 2011). In humans, Nathan and Smith (1982) could show the existence and course of the rubrospinal and the central tegmental tract by means of neurodegeneration studies. The rubrospinal tract was shown to contain only few fibers and not to extend beyond the upper cervical segments. The main descending motor tract of the human brain is the corticospinal tract.

To understand the function of the red nucleus, it is useful to consider its development during evolution which is well studied (ten Donkelaar 1988). Comparative anatomical studies have shown that the red nucleus plays an important role in the transition from aquatic to terrestrial life and/or the development of limbs (for details, see review by Basile et al. 2021).

The development of function during the evolutionary process has been summarized by Basile et al. (2021) as follows:

- The first part to appear is the magnocellular one related to simple locomotor pattern while the parvocellular part appears later in evolution and is involved in more complex motor activity.
- The functional specialization for upper limb movements of the magnocellular part become more evident with the evolution of bipedalism.
- What is still unclear is the question whether the magnocellular red nucleus undergoes a complete regression in the adult human brain. It could be possible, however, that it plays a role during human ontogenesis and in the recovery of pyramidal lesions.
- The parvocellular red nucleus may be situated at a hierarchically higher level of motor control with the olivocerebellar and basal ganglia systems.
- As already suggested for rodents (see here below), the red nucleus may be involved in mediating antinociceptive responses to pain stimulation.

An overview about our current knowledge on red nucleus (RN) connectivity as shown by Basile et al. 2021 is given in Fig. 15.38.

By contrast to man, mainly the parvocellular region gives rise to the rubrospinal tract (Liang et al. 2012) in mice. The fibers of this tract terminate in the contralateral spinal laminae 5, 6, 7, and 9. In the latter, fibers were close to the hand motor neurons (C8-T1) and foot motor neurons (L5-L6). This is consistent with the view of rubrospinal fibers to play a role in distal limb movements in rodents. The input from the interstitial system of the trigeminal spinal tract to the red nucleus has been studied in detail in mice (Pinto et al. 2007). The projection from the interstitial system of the trigeminal spinal tract to the red nucleus may mediate modulation of the facial muscles by pain and other sensory information (Olyntho-Tokunago et al. 2008).

The human red nucleus as well as the substantia nigra receives a cholinergic innervation (Mesulam et al. 1992).

15.3.4.3 Pathology of the red nucleus

Over more than 100 years, a number of symptoms have been ascribed to the dysfunction of the red nucleus such as certain forms of tremor (for details, see Basile et al. 2021). Nevertheless, the role of RN in tremor generation is not yet completely understood. An interesting aspect, based on rodent as well as primate studies, is the idea of a possible role of the rubrospinal tract to replace the corticospinal tract after pyramidal lesions. The general outcome of motor deficits after, for example, stroke, however, sheds doubt on the possibility of a replacement sensu strictiori.

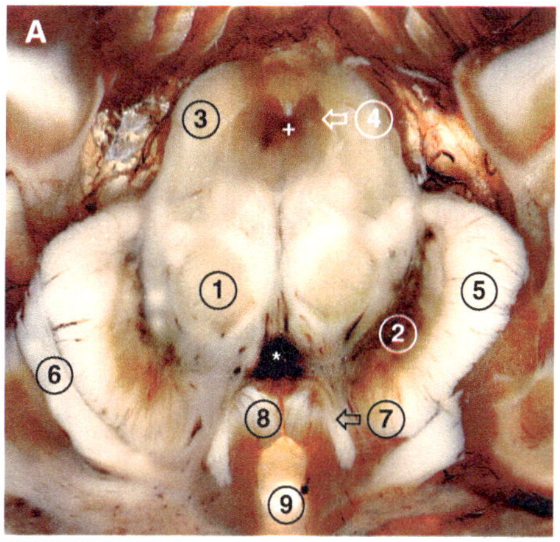

①	Red ncl.	⑦	Mammillothalamic fasciculus
②	Substantia nigra	⑧	Column of fornix
③	Inferior colliculus	⑨	3rd ventricle
④	Periaqueductal gray	+	Mesencephalic aqueduct
⑤	Cerebral peduncle	*	Interpeduncular fossa
⑥	Optic tract CNII		

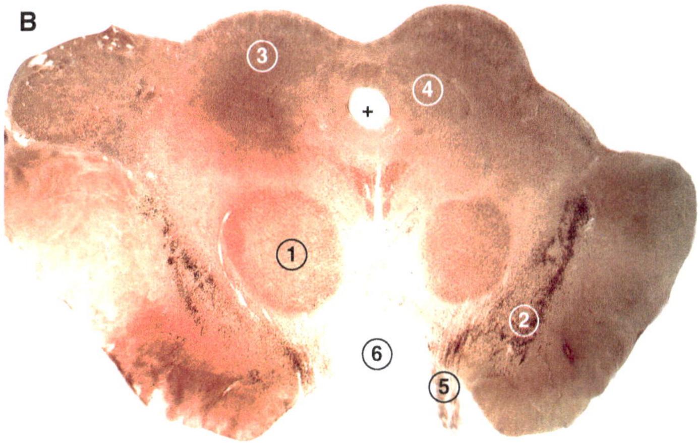

①	Red ncl. parvocellular part	⑤	Oculomotor nerve CNIII
②	Substantia nigra	⑥	Interpeduncular fossa
③	Superior colliculus	+	Mesencephalic aqueduct
④	Periaqueductal gray		

Fig. 15.37 Horizontal sections through the human midbrain at the level of the red nucleus. (**A**) Fixed specimen. The eponymous red coloring of the red nucleus is clearly visible. (**B**) Darrow red staining. In the histological section, the red nucleus is intensely stained. See atlas part Darrow red 41, Campbell 19. LabPON Twente

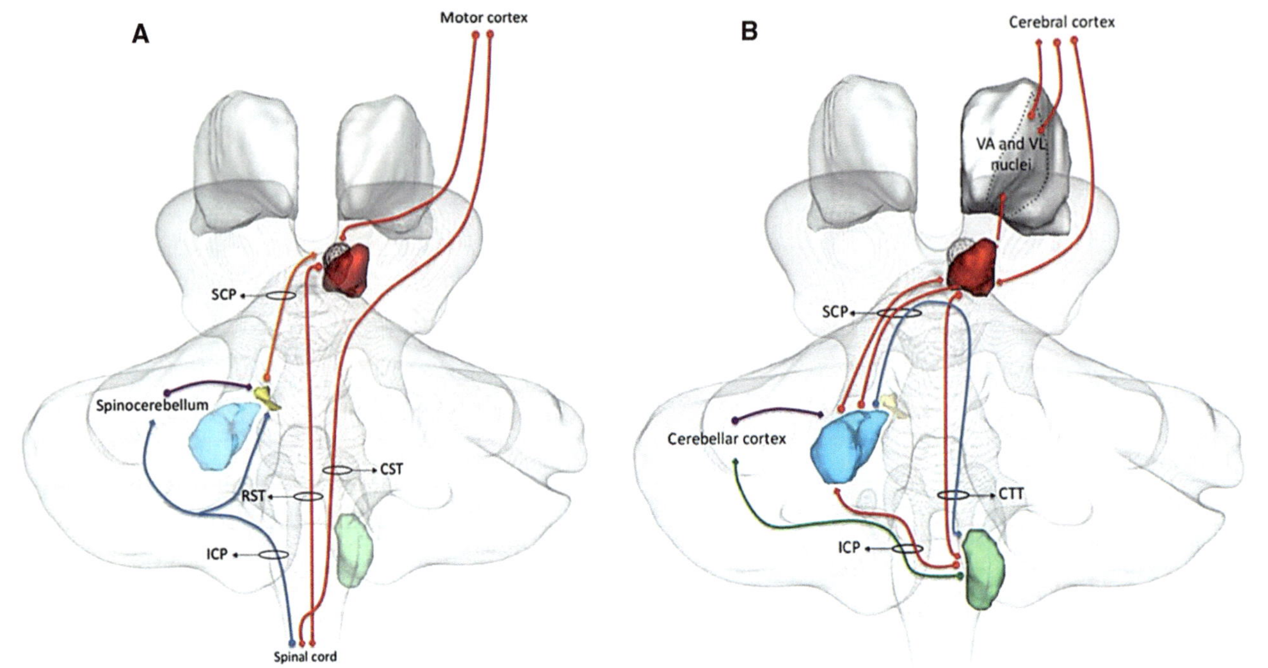

Fig. 15.38 Schematic drawing of the main connectivity of the red nucleus (red). (**A**) Connectivity of the magnocellular part of the RN (RNmc). The RNmc receives input from the motor cortex. It is the origin of a direct descending connection to the spinal cord (RST). Ascending spinal projections reach the RNpc via the cerebellum (inferior and superior cerebellar peduncles) and the interposed nucleus (yellow). (**B**) Connectivity of the parvocellular part of the RN (RNpc). As the RNmc, it is targeted by the cerebral cortex. Its efferent descending fibers, the rubroolivary fibers, reach the inferior olive (green) via the CTT. From there, the contralateral cerebellum (cerebellar cortex and dentate nucleus = blue) is reached. The ascending afferent input to the red nucleus is provided from the contralateral dentate nucleus via the superior cerebellar peduncle. Eventually, the loop is closed via the ventral anterior and ventral lateral nuclei of the thalamus to the cerebral cortex. *CST* corticospinal tract, *CTT* central tegmental tract, *ICP* inferior cerebellar peduncle, *RST* rubrospinal tract, *SCP* superior cerebellar peduncle. From Basile et al. 2021, Fig. 2 with permission

Related to the possible role of RN in anti/nociception (see above), it is interesting that fMRI studies indicate an involvement in migraine attacks (see Basile et al. 2021).

Although a correlation with clinical symptoms is difficult, in the following paragraph Fig. 15.44 shows a neuropathological example for an affection of the red nucleus.

15.3.4.4 Prion diseases

15.3.4.4.1 General considerations

Prion diseases (or transmissible spongiform encephalopathies) exemplified in humans by Creutzfeldt-Jakob disease (CJD) are a heterogeneous group of invariably fatal neurodegenerative disorders. Although they can be located at any site in the CNS, we have placed this section here because of an impressive example in the red nucleus (see Fig. 15.44) and the substantia nigra. Prion diseases have been recognized in animals, for example, in sheep (as scrapie) and in cows (as bovine spongiform encephalopathy) (Prusiner 1982, 1998). In humans, 85% of prion diseases occur sporadically, as in sporadic CJD (sCJD), sporadic Fatal Insomnia (sFI) (Mastrianni et al. 1999), and the protease-sensitive prionopathy (PSPr) (Masters et al. 1979). Around 10–15% of cases arise due to dominantly inherited mutations in the prion protein gene (PRNP) on chromosome 20, as in genetic CJD (gCJD), Gerstmann-Sträussler-Scheinker (GSS) disease and Fatal Familial Insomnia (FFI).

What sets prion diseases apart from other neurodegenerative disorders is that they are *infectious*: spontaneous transmission from one individual to another has occurred outside an experimental setting (in less than 1% of cases) (Prusiner 1998). The disease Kuru is exemplary, attributed to the consumption of brain tissue during cannibalistic rituals in Papua New Guinea (Collinge et al. 2006). In modern days, most cases of CJD transmission between humans were traced to prion-contaminated corneal transplants, dura mater grafts and intramuscularly injected pituitary-derived growth hormone (Brown et al. 2000). A more recent example is that of the new variant CJD (vCJD) in humans, which is linked to bovine spongiform encephalitis (BSE) through the introduction of contaminated meat into the human food chain (Will et al. 1996). The incidence of sporadic CJD is two in a million; the incidence of vCJD is far rarer and has dropped since stringent measurements have been taken after the outbreak of mad cow disease in the 1970s and 1980s.

The expression pattern of the normal (cellular) prion molecule (PrP^c) in mammals is broad and includes skeletal muscle, kidney, heart, secondary lymphoid organs, and the central nervous system (CNS). Within the CNS, high PrP^C levels are found in the synaptic membranes of neurons and on astrocytes. In the periphery, PrP^C is expressed on lymphocytes, follicular dendritic cells, and erythrocytes. However, in prion disorders the prion protein is misfolded as PrP^{sc} and aggregates, which makes it toxic, resulting in the typical spongiform changes and/or prion amyloid deposits. These characteristics of misfolding and self-assemblage of prions are now used for the new diagnostic so-called RT-QuIC test in cerebrospinal fluid of suspected patients during life (Schmitz et al. 2016; Green 2018). This test has greatly improved the recognition and surveillance of CJD.

The enormous clinical and pathological heterogeneity of prion diseases can be explained by the phenotypical variability of the prion molecule, determined by polymorphism of the prion gene in codon 129 (methionine (M) or valine (V)) and physicochemical characteristics of the prion protein (type 1 protein of 21 kDa or type 2 molecule of 19 kDa). This gives six possible isoforms (MM1, MV1, MM2, MV2, VV1, and VV2). In their pure forms, almost all these isoforms have different histopathological substrates ("histotypes") and clinical consequences as described in the new consensus classification (Parchi and Saverioni 2012; Parchi et al. 2012). However, MM1 and MV1 are indistinguishable and have merged into one type MM1/MM variant. Next to these variables, the prion molecule can be glycosylated at none, one or two sites which also influences its misfolding and propagation. Variant CJD (vCJD or MM 2 V) represents an additional very distinctive CJD phenotype. It is characterized by a bovine type 2 PrP^{SC} form that differs from the type 2 found in sporadic CJD due to a relatively high representation of the diglycosylated PrP^{SC} glycoform (Collinge et al. 1996).

The macroscopy in rapidly progressive disorders is often normal as there is not enough time for atrophy. In CJD cases with relative long duration, atrophy of the anterior part of the vermis can be found. In sporadic FI, thalamus atrophy can be detected. An exception is the panencephalopathic CJD form with extreme atrophy and loss of white matter ("walnut brain"). This form is not related to a specific histotype (Jansen et al. 2009).

The histopathology of prion diseases shows, in general, loss of neurons, gliosis and PrP^{sc} deposits. Spongiosis is often seen and is characterized by miliary, sharply demarcated vacuoles in the parenchyma of the involved gray matter. Fine vacuoles of about 2–4 μm are seen in MM1, MV1, VV2, and MV2, while medium-sized vacuoles of 4–6 μm are found in VV1 and conflating, relatively large vacuoles (10 μm) can be found in the MM2C (cerebral) histotype but not in MM2T (thalamic) subtype. After long duration, the vacuolization can disappear, while the gliosis and neuronal loss increase.

Antibodies against prions (such as mab 3F4) are most useful for studying the pattern and distribution of the abnormal prion molecule. The staining pattern can be synaptic, perivacuolar, perineuronal, or compact/plaque-like with (congophilic) prion plaques (Parchi and Saverioni 2012, Parchi et al. 2012). The flow chart developed in 2012 (Parchi et al. 2012) can help to distinguish the different histotypes, including combinations of two types (having type 1 and type 2 prion molecules in one case), which occurs in about 10% (Parchi and Saverioni 2012) (Fig. 15.39).

15.3.4.4.2 Sporadic CJD (sCJD)

CJD is a rapidly progressive disease that is fatal in a few weeks to months, occasionally takes a few years, and is clinically characterized by fast cognitive decline, pyramidal- and extrapyramidal signs, cerebellar signs, visual disorders, and myoclonics. Triphasic spikes in the EEG can be found (Parchi and Saverioni 2012). The RT-QuIC test on CSF is essential for the diagnosis and has a specificity of 100% and sensitivity of 92%. The most frequent occurring, classic type of CJD has histotype MM1/MV1 and especially shows involvement of all layers of the cerebral cortex, caudate, putamen, thalamus, and cerebellar cortex. The brainstem is slightly involved, especially the substantia nigra and pons nuclei (Fig. 15.40).

The second most common phenotype is the VV2 histotype, seen in about 11–15% of sporadic cases. The mean age of onset and clinical duration are 60 years and 6.5 months, respectively. Clinically, VV2 patients show prominent gait ataxia at onset, while dementia occurs later in the course of the illness. In this type, especially the cerebellar cortex of the vermis, the brainstem, the deeper layers of the limbic cortex, striatum, thalamus, and hypothalamus are concerned, while the deep layers in the neocortex are less affected (Parchi and Saverioni 2012).Next to synaptic prion staining, perineuronal staining and plaque-like deposits are seen. In the brainstem, all the gray areas are severely involved (Figs. 15.41 and 15.42).

The rarer CJD form MV2K, affects about 8% of cases. The mean age of onset and clinical duration are 59 years and 18 months, respectively. This subtype has a significantly lower progression rate, usually more than 1 year. Clinical duration of the disease exceeding 2 years is not infrequent in this type and could raise diagnostic difficulties at the clinical stage. Clinically, both ataxia and dementia are prominent (Parchi and Saverioni 2012). It is characterized by Kuru type of plaques, especially in the cerebellar cortex. The cerebral

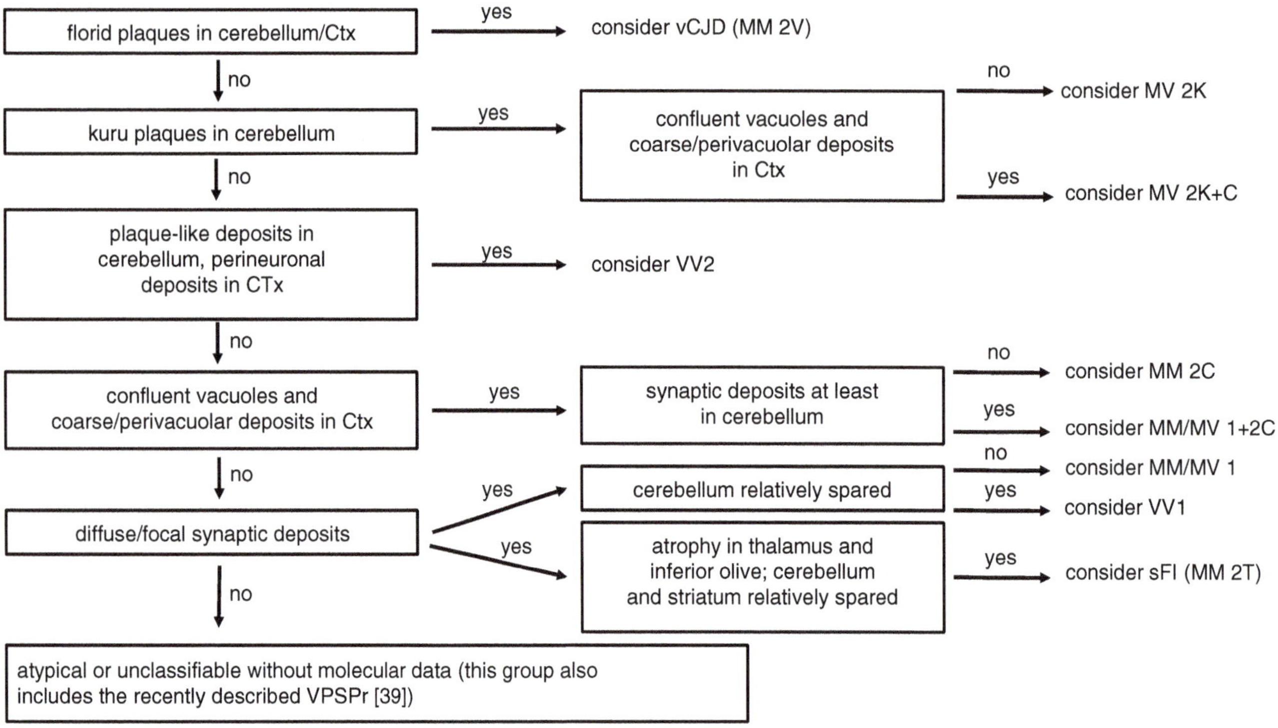

Fig. 15.39 Recommended diagnostic flowchart for diagnosing human sporadic prion disease histotypes and vCJD. Modified from Parchi et al. 2012, Fig. 1 with permission

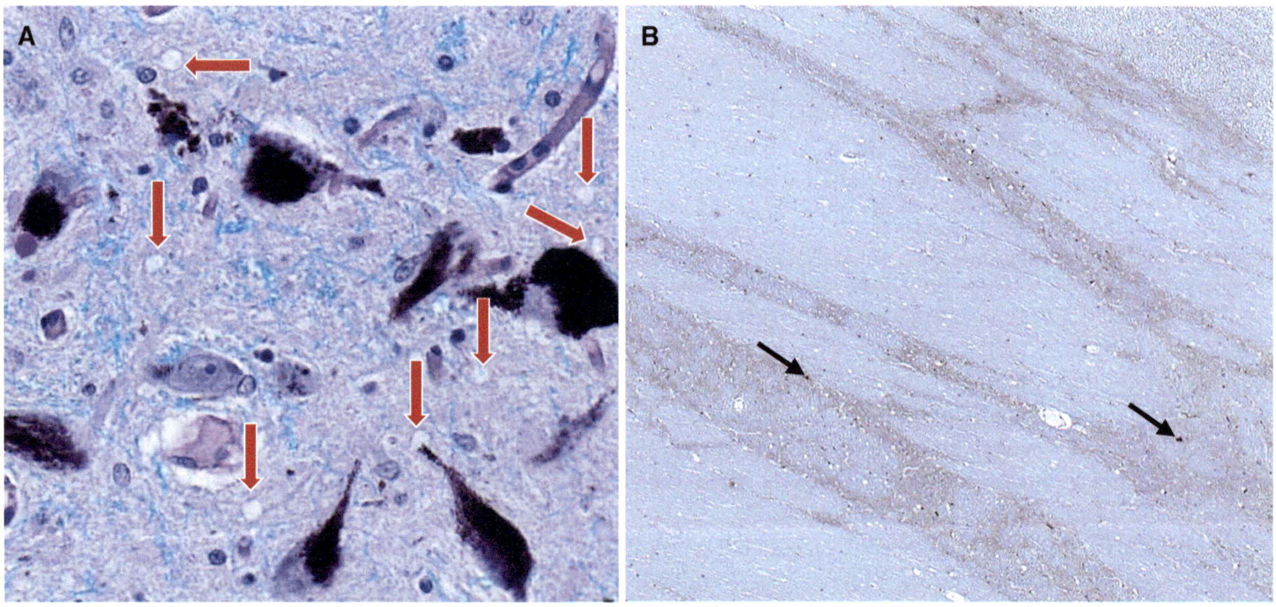

Fig. 15.40 (**A**) Microscopical section of the substantia nigra (pars compacta) of a sCJD patient with MM/MV1 subtype, showing spongiosis in a Luxol & PAS stain. Note the fine, sharply demarcated, and round vacuoles in the parenchyma (red arrows). (**B**) Microscopical section of the pons of an sCJD patient (MM/MV1 type) in a synaptic prion staining (weak brown color), using immunohistochemistry with mab 3F4. Next to faint synaptic staining also some coarse prion deposits are seen (arrows). Collection A. Rozemuller

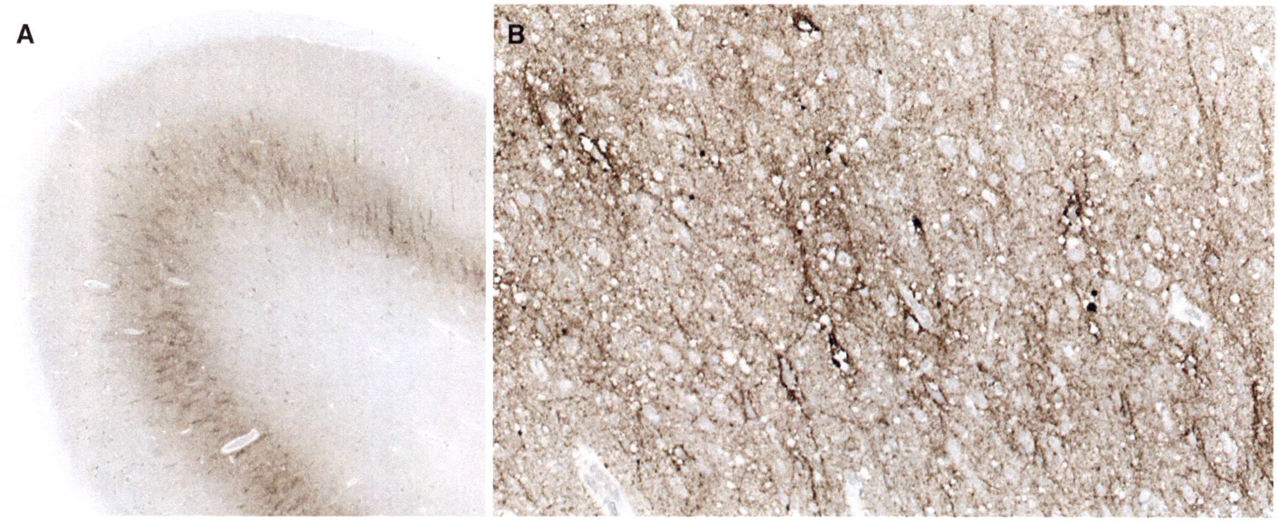

Fig. 15.41 (**A**) Limbic cortex of a sCJD patient, VV2 subtype, showing intense prion staining of the deeper layers (both perineural and synaptic staining). (**B**) Detail of deeper layers. Mab 3F4 immunostain. Collection A. Rozemuller

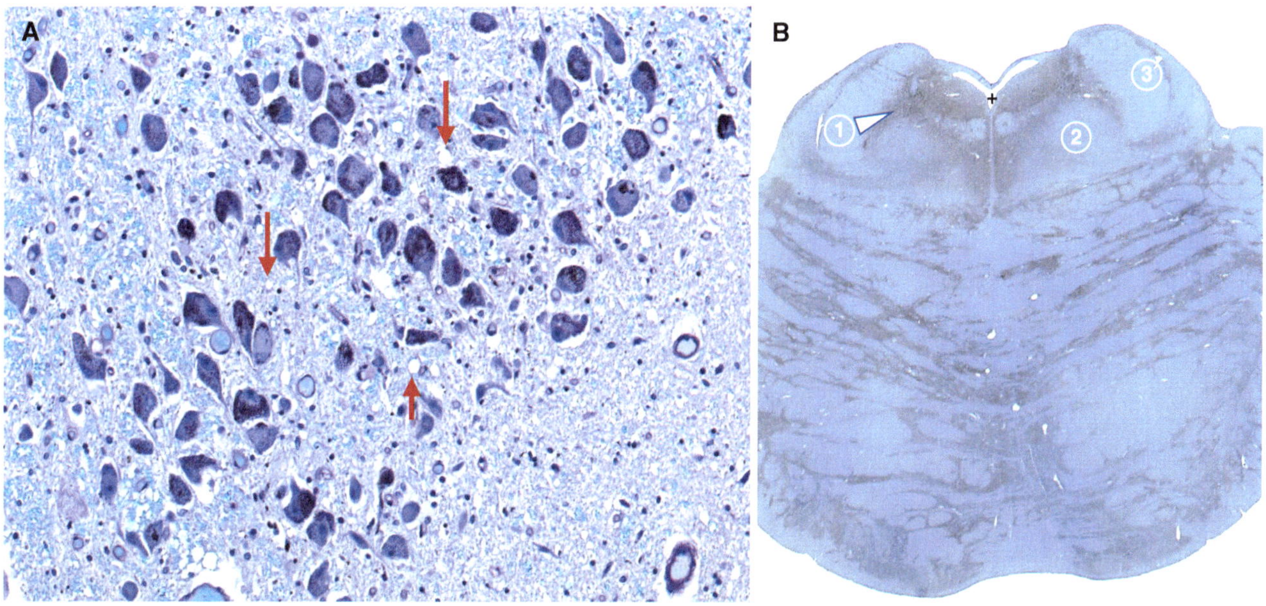

①	**Locus caeruleus**	③	**Superior cerebellar peduncle**
②	**Central tegmental tract**	+	**4th ventricle**

Fig. 15.42 (**A**) Locus caeruleus (see Sect. 13.2.2.1) of an sCJD patient with subtype VV2, fine vacuolar spongiosis, i.e., multiple rounded small vacuoles (red arrows). Luxol & PAS stain. (**B**) Horizontal section through the pons of an sCJD patient with VV2 subtype. Diffuse (brown) synaptic prion staining through the gray areas of the pons, including pontine nuclei and locus caeruleus. Mab 3F4 immunostain. Collection A. Rozemuller

cortex (most often frontal lobes and cingulate gyri), basal ganglia and thalamus also show spongiosis and Kuru type of plaques (Fig. 15.43).

Furthermore, in the white matter some prion plaques can be detected. In the nuclei of the mesencephalon, especially in the substantia nigra, a perineuronal prion staining is seen, but also synaptic staining, just as in the pons, raphe nuclei, and inferior olives (Fig. 15.44).

Moreover, prion plaques can be found in the periaqueductal gray and in the raphe nuclei. There is no obvious cell loss in the nuclei of vagus nerve or hypoglossal nerve although some synaptic prion deposits can be found in the nucleus of

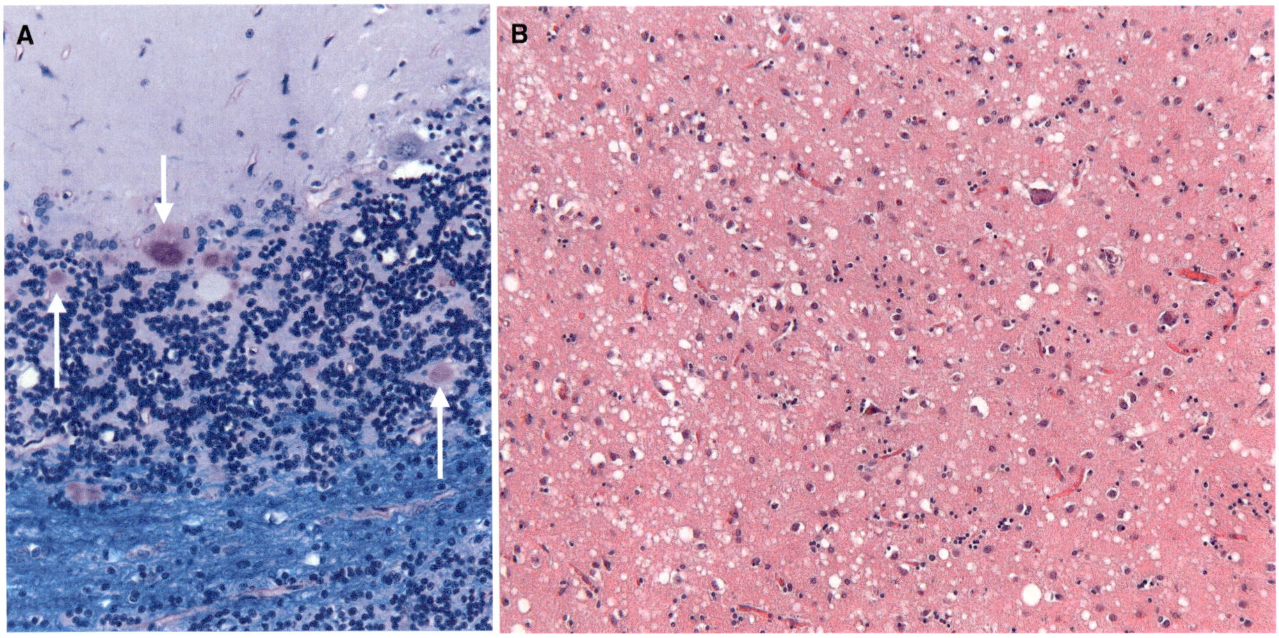

Fig. 15.43 (**A**) Cerebellum with Kuru type of plaques in a patient with sCJD subtype MV2K. Larger and smaller PAS-positive purple plaques in the granular layer (white arrows). Luxol & PAS stain. (**B**) Frontal cortex showing extensive spongiosis and gliosis. H & E stain. Collection A. Rozemuller

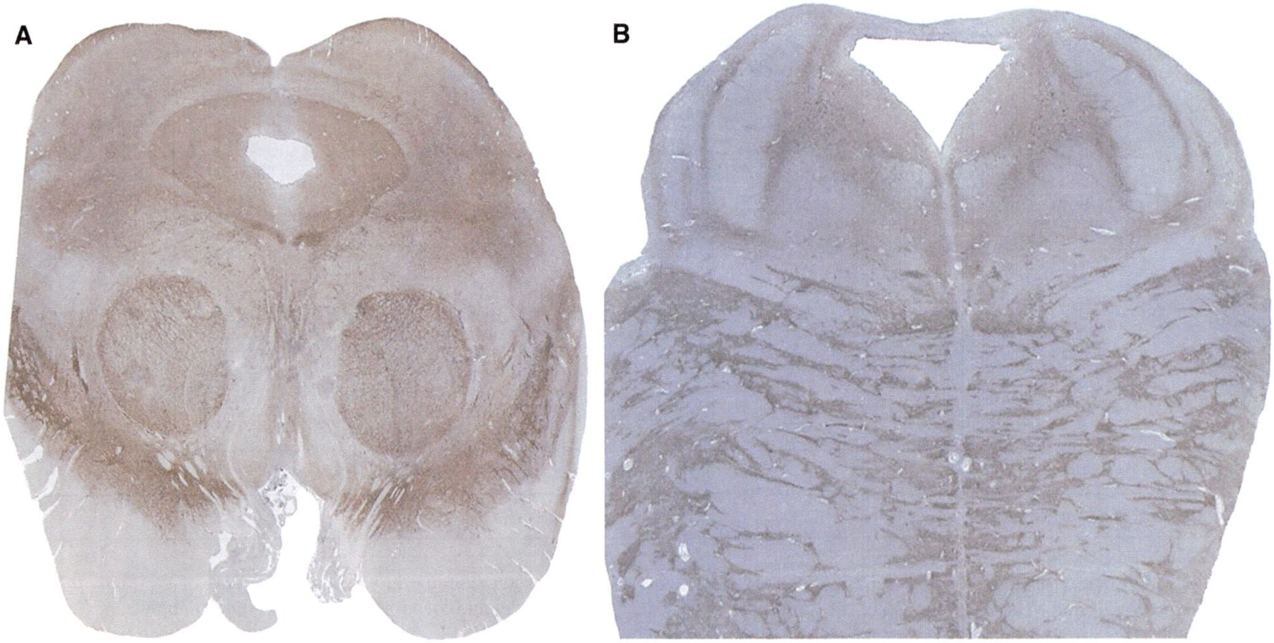

Fig. 15.44 (**A**) Horizontal section through the mesencephalon of an sCJD patient with MV2K subtype showing diffuse synaptic prion staining especially in the red nucleus, substantia nigra, and periaqueductal gray. (**B**) Diffuse brown prion staining in all gray areas of the pons. Mab 3F4 immunostain

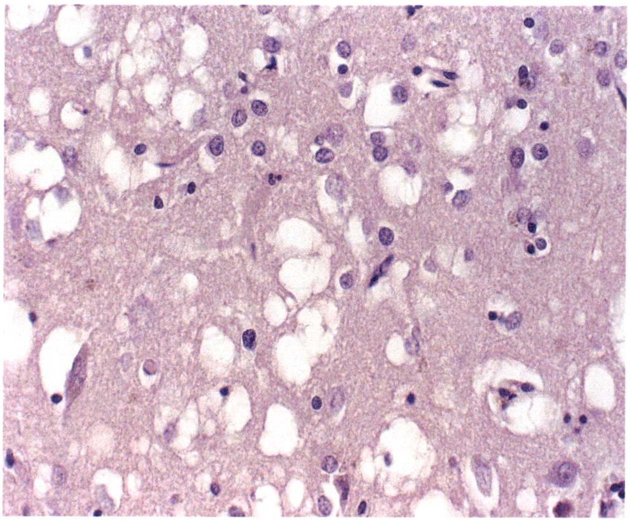

Fig. 15.45 Cortex showing conflating vacuoles (grape-like clusters) in an sCJD patient with MM2 subtype. H & E stain. Collection A. Rozemuller

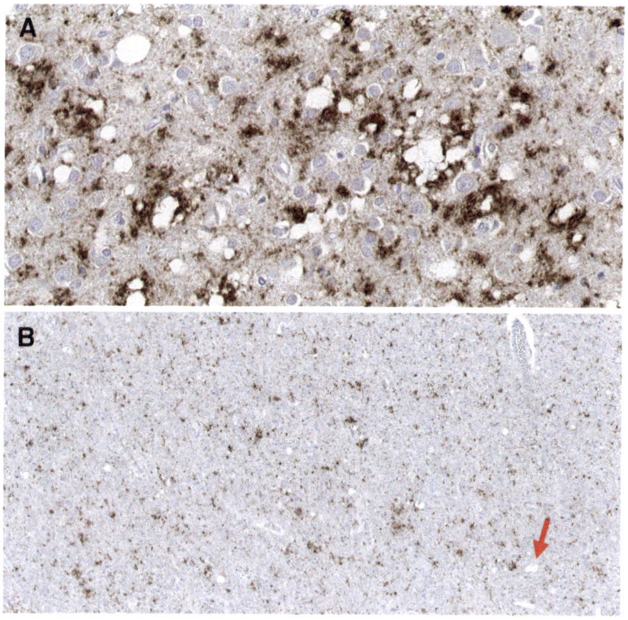

Fig. 15.46 (**A**) Frontal cortex of an sCJD patient with MM2 subtype, showing conflating, rather large vacuoles surrounded by brown prion deposits ("perivacuolar staining"). In addition, weak synaptic staining ("punctuate staining") is seen in the parenchyma. (**B**) Midbrain section (tegmentum) of a sCJD patient, MM2C subtype, showing small dispersed brown prion deposits and a few fine vacuoles in the parenchyma that occasionally conflate (red arrow). Mab 3F4 immunostain. Collection A. Rozemuller

vagus nerve. A rare phenotype that affects about 1% of the sCJD population is the VV1 subtype. The VV1 subjects show a mean age at onset of about 40 years, which is by far the youngest among sCJD subtypes, and a mean duration of 15 months (Parchi and Saverioni 2012). The rare VV1 histotype shows ballooned neurons, spongiosis, and faint synaptic prion deposits in the frontal cortex but does not affect the brainstem.

Patients with MM2C (cortical) subtype show conflating medium-sized vacuoles as grape-like clusters dispersed in the cortex (Fig. 15.45).

The brainstem and cerebellum are relatively spared but can show spongiosis and small deposits (Figs. 15.46 and 15.47).

In sCJD MM 2T, which is clinically characterized by sleep disorders (sporadic FI) and cognitive decline, severe neuronal loss and gliosis is seen in the medial thalamic and inferior olives and relative sparing of anterior striatum. The cortex and the rest of the brainstem are not involved.

15.3.4.4.3 Variant CJD (VCJD)

Variant CJD, related to consumption of meat of BSE (bovine spongiform encephalopathy) cows, has almost been eradicated. It affects especially younger people, always with an MM genotype, and has no specific test (RT-QuIC is negative). The duration is often more than 2 years with psychiatric symptoms. The PrP^{SC} is found in dendritic cells and in the typical "florid" prion plaques that are congophilic and show radiating amyloid fibrils in the core surrounded by vacuoles. Multiple florid plaques are seen in the cerebral cortex (all lobes) and cerebellar cortex (Fig. 15.48).

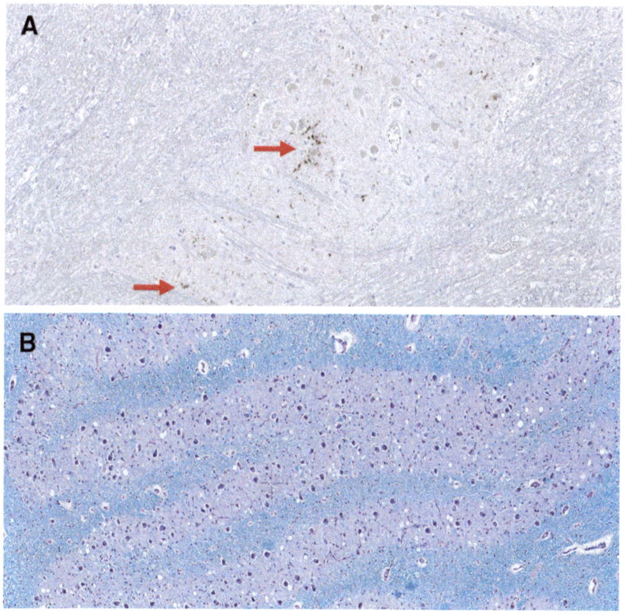

Fig. 15.47 (**A**) Inferior olive nucleus of the same patient with a few coarse and synaptic brown prion deposits (red arrows). Mab 3F4 immunostain. (**B**) Inferior olive nucleus with spongiosis. Note the large number of small vacuoles. Luxol & PAS stain. Collection A. Rozemuller

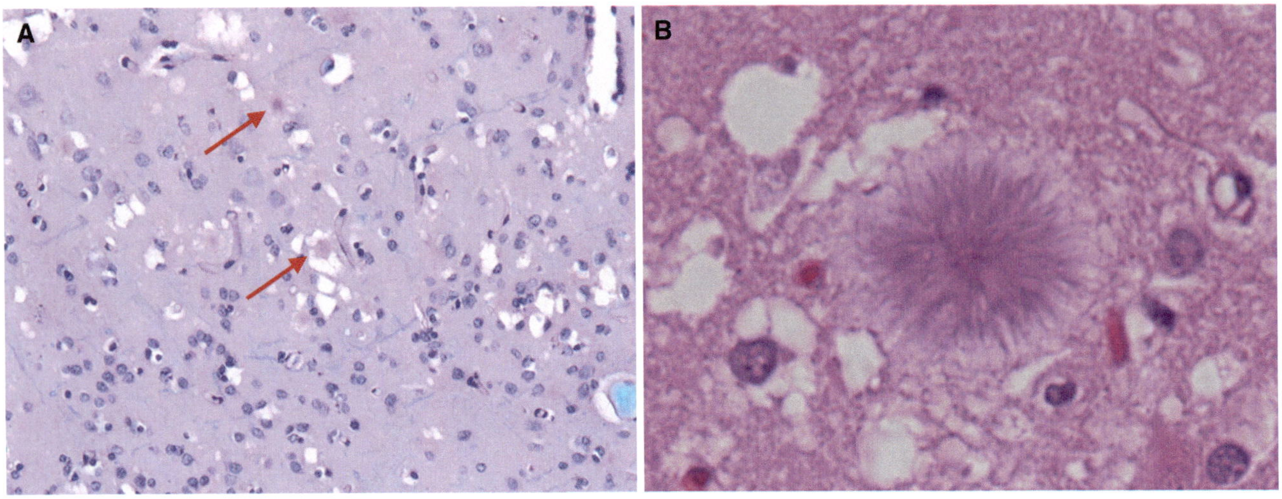

Fig. 15.48 (**A**) Frontal cerebral cortex of a vCJD patient, showing typical "florid" plaques (red arrows). Luxol & PAS stain. (**B**) Typical amyloid plaque in variant CJD. H & E staining. Collection A. Rozemuller

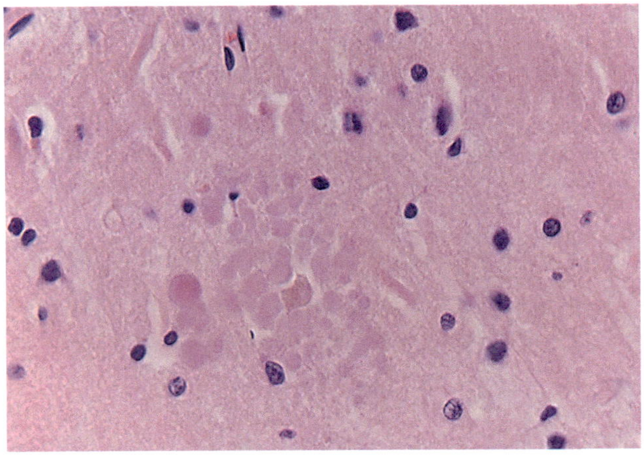

Fig. 15.49 Cerebellar cortex (molecular layer) of a patient with Gerstmann-Sträussler-Scheinker disease (GSS), showing the typical multicentric prion plaque. H & E stain. Collection A. Rozemuller

The spongiform encephalopathy is most marked in the caudate nucleus and putamen. Severe neuronal loss and gliosis is present in the thalamus, especially the pulvinar. Next to PrP^C-positive florid plaques, small cluster plaques are detected as well as amorphous pericellular and pericapillary deposits. Furthermore, prion-positive dendritic cells are seen. The distribution of the spongiosis is widespread throughout the brain, including the brainstem nuclei but prion plaques are not detected in the brainstem nuclei. The substantia nigra is severely affected and shows prion-positive neuronal extension.

15.3.4.4.4 Hereditary forms of prion diseases

More than 50 different mutations can be found in the PRNP gene leading to either Gerstmann-Sträussler-Scheinker disease (GSS), FFI, or hereditary CJD. However, also modifying genetic factors have been described (Mead et al. 2019). In GSS, typical amyloid fibrils are formed of 7–10 KDa, often in the cerebellum with clinical ataxia. Typical for GSS are multicentric plaques (Fig. 15.49).

Some mutations in the prion gene lead to prion amyloid plaques localized especially in the cerebral cortex and, clinically, to a frontotemporal dementia. Diffuse and compact prion plaques can also be found in the cerebellum. The duration can be up to 17 years (Fig. 15.50).

In cases with prion amyloidosis and long duration, a secondary tauopathy is found. The brainstem is relatively spared. Rarely, prion amyloid angiopathy is found (Jansen et al. 2010) (Fig. 15.51). In FFI, a mutation is found in codon 178 of PRNP and the thalamus and inferior olives are most affected like in the sporadic variant.

15.3.5 Retrorubral field

For detailed information on the retrorubral field, see Sect. 14.5.1.

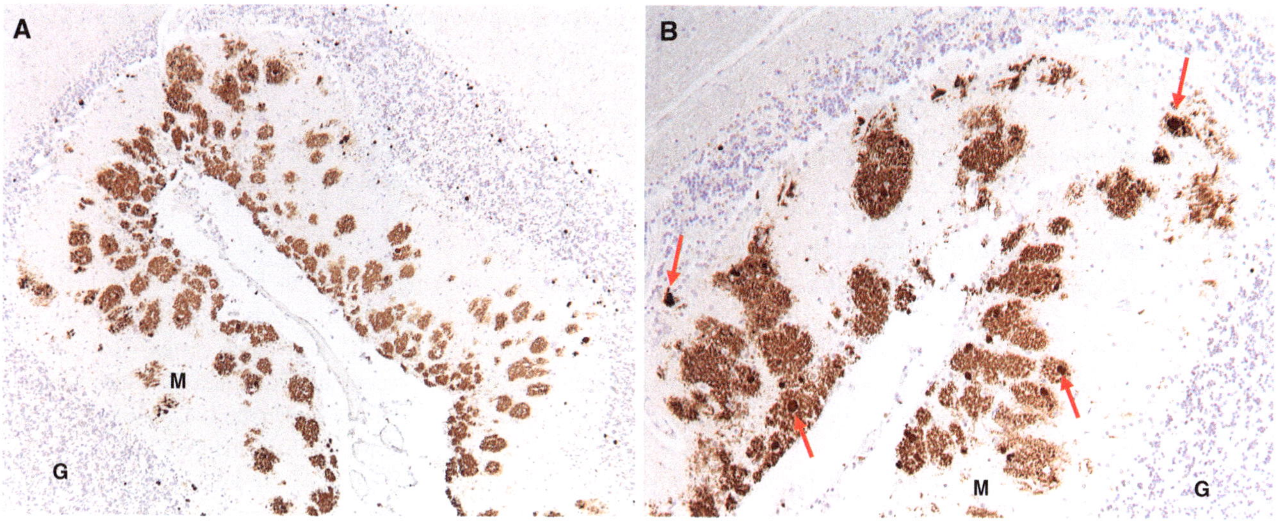

Fig. 15.50 Cerebellar cortex (overview and detail) of a hereditary prion disease with G131V mutation in PRNP, which leads to widespread prion amyloid deposits in the whole brain. More diffuse prion plaques in the molecular layer and small compact prion deposits (red arrows) in molecular as well as granular layer. Mab 3F4 immunostain. M = molecular layer, G = granular layer. Collection A. Rozemuller

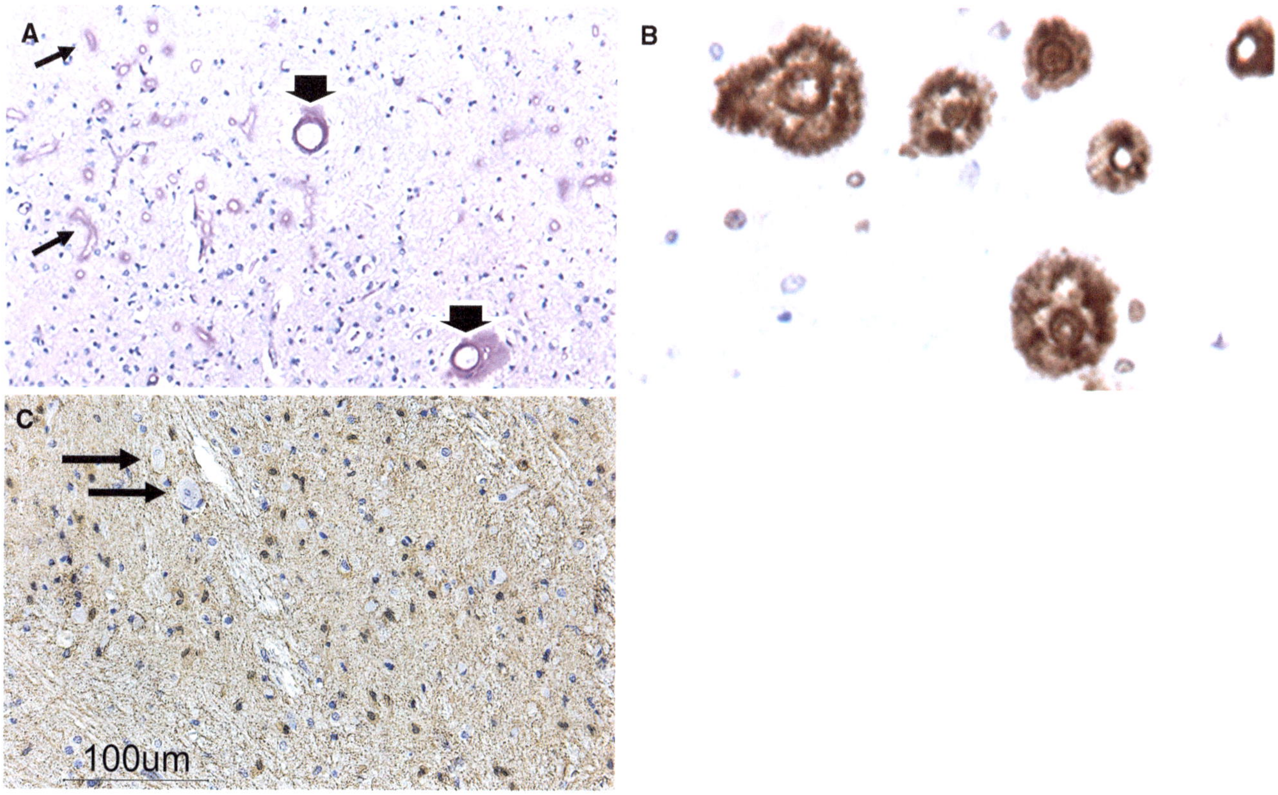

Fig. 15.51 (**A**) Frontal cortex of a cerebral prion amyloid angiopathy in a patient with hereditary CJD (Y226X mutation). PAS-positive purple prion amyloid in and around the wall of capillaries (small black arrows) and larger vessels (thick arrows). Luxol & PAS stain. (**B**) Frontal cortex with prion stain. Mab 3F4 immunostain. (**C**) Inferior olive nucleus of an FFI patient. Severe loss of neurons, only a few neurons are left (arrows) and an increase of glial cells (gliosis) in a GFAP immunostain. Collection A. Rozemuller

15.4 Pre-isthmus midbrain

15.4.1 Cuneiform nucleus

For detailed information on this nucleus, see Sect. 14.4.1.1.

15.4.2 Sagulum/Sagulum nucleus [Ncl. saguli] (Sagulum = small military cloak)

This nucleus is located rostral to the microcellular tegmental nucleus and caudal to the inferior colliculus. As shown for other mammals, it is connected with several auditory areas.

15.4.3 Mesencephalic trigeminal nucleus

For detailed information on this nucleus, see Sect. 13.2.6.1.

15.4.4 Precuneiform area

For detailed information on the precuneiform area, see Sect. 14.4.1.1.

15.4.5 Tectal gray group

See Sect. 15.1.1.

15.4.5.1 Posterior pretectal nucleus

With regard to the tectal gray group and the posterior pretectal nucleus and the related terminology problems, see Sect. 13.2.7.1 in Chap. 13.

15.5 Mesencephalic floor plate

15.5.1 Ventral tegmental nuclei [Ncll. tegmentales ventrales]

FIPAT Ch. 1 has the group of ventral tegmental nuclei [Nuclei tegmentales ventrales] comprising the caudal linear nucleus 14.5.3.1, the rostral linear nucleus 15.5.1.4, the paranigral nucleus (see Sect. 14.6.1), the interfascicular nucleus (see Sect. 14.5.4.3), the parabrachial pigmented (Sect. 15.5.1.2), and the parapeduncular nucleus (not dealt with here).

15.5.1.1 Ventral tegmental area (VTA)

The VTA (see Sect. 15.5.1), according to ten Donkelaar (2011), is made up of

- the rostral linear nucleus (Sect. 15.5.1.4 and Fig. 14.11 ⑥)
- the caudal linear nucleus (see Sect. 14.5.3.1)
- the interfascicular nucleus (see Sect. 14.5.4.3 and Fig. 14.11 ②)
- the paranigral nucleus (see Sect. 14.6.1 and Fig. 14.11 ④)
- the parabrachial pigmented nucleus (Sect. 15.5.1.2 and Fig. 14.11 ③)

Generally spoken, the VTA is the origin of the mesolimbic dopamine pathway which targets the accumbens nucleus [Ncl. accumbens] and the prefrontal cortex (Robeck and Mudra 2021). In a recent diffusion tensor study, Coenen et al. (2018) could identify and trace the medial forebrain bundle [Fasciculus prosencephali medialis] as the main efferent connection of the VTA. Most VTA efferents bilaterally reach the superior and middle frontal gyrus and the lateral orbitofrontal region corresponding to Brodmann areas (BA) 10, 9, 8, 11, and 11m (Coenen et al. 2018).

In terms of function, the mesolimbic dopamine system provides reward for behaviors necessary for survival of the individual and the species such as food, water, nurturing, and sex (Robeck and Mudra 2021).

The flip side of these functional properties is the central role of the mesolimbic dopamine system in addiction where, for example, drugs substitute these behaviors. To mention just one example, nicotine abuse, based on the overall central nervous existence of nicotinic acetylcholine receptors, results in a rapid increase in dopamine release in the VTA and the accumbens nucleus occurring within 10 s of smoking a cigarette. The normal reward system renumerating food consumption, social affiliation, and sexual activity is "abused" by the taking of drugs (Engelmann et al. 2020).

In mice, the VTA [Area tegmentalis ventralis] corresponds to the dopaminergic cell group A10 (see VanderHorst and Ulfhake 2006).

Based on the TH-immunoreactive neurons, the border between the VTA and the compact part of the substantia nigra cannot be drawn clearly (VanderHorst and Ulfhake 2006). The VTA is involved in several higher order cognitive processes such as reward seeking and working memory. In this regard, the VTA is an extremely important structure for drug abuse-related behavior, for example, concerning nicotine and cocaine. The anatomical substrate of reward behavior in animals is the mesocortico-accumbens pathway which originates in the VTA and projects to the telencephalic nucleus accumbens.

15.5.1.2 Parabrachial pigmented nucleus of the VTA [Ncl. parabrachialis pigmentosus]

This nucleus is located laterally of the central linear and the interfascicular nuclei dorsal of the substantia nigra (Figs. 14.11 and 14.12). Together with the paranigral nucleus and the caudal linear raphe nucleus, it belongs to the three main dopamine containing cell groups of the VTA.

15.5.1.3 Interfascicular nucleus

Detailed information on this nucleus is provided in Chap. 14 under Sect. 14.5.4.3.

15.5.1.4 Rostral linear nucleus of the raphe [Ncl. linearis rostralis]

FIPAT Ch. 1 lists a caudal (#1430) and a rostral linear nucleus (#1431) [Ncl. linearis caudalis/rostralis] and a linear raphe nucleus (#1412) [Ncl. raphes linearis], obviously identical with the serotonergic cell group B8. The corresponding endnote to the latter, # 64 states that, because the Ncl. linearis inferior (all terms in the endnote in Latin) contains some serotonergic neurons, Nieuwenhuys et al. (2008) called it Ncl. raphes linearis. Furthermore, the terminology introduced by Paxinos (Mai and Paxinos 2012) is mentioned, which calls the median serotonergic part "azygos part of caudal linear nucleus of the raphe (CLi)" and the paramedian parts as "zygos parts" & of CLi (Azygos from Greek ἄζυξ, azyx, -ygos = unpaired, zygos ≈ paired). Dealing this way with the different parts of the CLi, in our opinion, does not really contribute to the necessary clarity.

References

Armstrong RA, McKee AC et al (2017) Pathology of the superior colliculus in chronic traumatic encephalopathy. Optom Vis Sci 94:33–42

Basile GA, Quartu M et al (2021) Red nucleus structure and function: from anatomy to clinical neurosciences. Brain Struct Funct 226:69–91

Brown P, Preece M et al (2000) Iatrogenic Creutzfeldt-Jakob disease at the millennium. Neurology 55:1075–1081

Buchholtz EA, Seyfarth EA (1999) The gospel of the fossil brain: Tilly Edinger and the science of paleoneurology. Brain Res Bull 48:351–361

Cacciola A, Bertino S et al (2019) Mapping the structural connectivity between the periaqueductal gray and the cerebellum in humans. Brain Struct Funct 224:2153–2165

Carrive P, Paxinos G (1994) The supraoculomotor cap: a region revealed by NADPH diaphorase histochemistry. Neuroreport 5: 2257–2260

Che Ngwa E (2016) Identification of motoneurons innervating individual extraocular muscles within the oculomotor nucleus in human. Dissertation, LMU München: Medizinische Fakultät

Che Ngwa E, Zeeh C et al (2014) Delineation of motoneuron subgroups supplying individual eye muscles in the human oculomotor nucleus. Front Neuroanat 8:2

Coenen VA, Schumacher LV et al (2018) The anatomy of the human medial forebrain bundle: ventral tegmental area connections to reward-associated subcortical and frontal lobe regions. Neuroimage Clin 18:770–783

Collinge J, Sidle KCL et al (1996) Molecular analysis of prion strain variation and the aetiology of 'new variant' CJD. Nature 383:685–690

Collinge J, Whitfield J et al (2006) Kuru in the 21st century—an acquired human prion disease with very long incubation periods. Lancet 367:2068–2074

Dräger UC, Hubel DH (1975) Responses to visual stimulation and relationship between visual, auditory, and somatosensory inputs in mouse superior colliculus. J Neurophysiol 38:690–713

Edinger L (1885) Über den Verlauf der centralen Hirnnervenbahnen mit Demonstrationen von Präparaten. Arch Psychiatr Nervenkr 16:858–859

Edinger T (1929) Die fossilen Gehirne. Anat EntwGesch 28:1–249

Engelmann JM et al (2020) Nicotine dependence. In: Abeloff's clinical oncology (sixth edition)

Feroze KB, Patel BC (2021) Parinaud syndrome. In: StatPearls [Internet]. StatPearls Publishing, Treasure Island, FL

Green AJE (2018) RT-QuIC: a new test for sporadic CJD. Pract Neurol 19:49–55

Horn AKE, Büttner-Ennever JA (1998) Premotor neurons for vertical eye movements in the rostral mesencephalon of monkey and human: histologic identification by parvalbumin immunostaining. J Comp Neurol 392:413–427

Horn AK, Eberhorn A et al (2008) Perioculomotor cell groups in monkey and man defined by their histochemical and functional properties: reappraisal of the Edinger-Westphal Nucleus. J Comp Neurol 507:1317–1335

Huang CQ, Kang DZ et al (2018) The classification of intracranial aneurysm neck: a single center research experience. Chin Neurosurg Jl 4:39

Huggenberger S, Moser N, Schröder H, Cozzi B, Granato A, Merighi A (2019) Neuroanatomie des Menschen. Springer, Heidelberg

Jang SH, Lee MY et al (2018) Structural neural connectivity of the vestibular nuclei in the human brain: a diffusion tensor imaging study. Neural Regen Res 13:727–730

Jansen C, Head MW et al (2009) Panencephalopathic Creutzfeldt-Jakob disease in the Netherlands and the UK: clinical and pathological characteristics of nine patients. Neuropathol Appl Neurobiol 35:272–283

Jansen C, Parchi P et al (2010) Prion protein amyloidosis with divergent phenotype associated with 2 novel nonsense mutations in PRNP. Acta Neuropathol 119:189–197

Kozicz T, Bittencourt JC et al (2011a) The Edinger-Westphal nucleus: a historical, structural, and functional perspective on a dichotomous terminology. J Comp Neurol 519:1413–1434

Kozicz T, Sterrenburg L et al (2011b) Does midbrain urocortin 1 matter? A 15-year journey from stress (mal)adaptation to energy metabolism. Stress 14:376–383

Kuhlenbeck H, Miller RN (1949) The pretectal region of the human brain. J Comp Neurol 91:369–407

Liang H, Paxinos G et al (2012) The red nucleus and the rubrospinal projection in the mouse. Brain Struct Funct 217:221–232

Link MJ, Sloan CY (2003) Oculomotor nucleus (third cranial nerve) In: Encyclopedia of the neurological sciences

Mai J, Paxinos G (2012) The human nervous system, 3rd edn. Elsevier

Mann G (1974) Das Porträt des Neuroanatomen Ludwig Edinger von Lovis Corinth (1909). Med J 9:325–328

Masters CL, Harris JO et al (1979) Creutzfeldt-Jakob disease: patterns of worldwide occurrence and the significance of familial and sporadic clustering. Ann Neurol 5:177–188

Mastrianni JA, Nixon R et al (1999) Prion protein conformation in a patient with sporadic fatal insomnia. N Engl J Med 340:1630–1638

May PJ, Warren S (2020) Pupillary light reflex circuits in the Macaque Monkey: the olivary pretectal nucleus. Brain Struct Funct 225:305–320

May PJ, Sun W et al (2020) Pupillary light reflex circuits in the macaque monkey: the preganglionic Edinger-Westphal nucleus. Brain Struct Funct 225:403–425

Mead S, Lloyd S et al (2019) Genetic factors in mammalian prion diseases. Annu Rev Genet 53:117–147

Mesulam M-M, Mash L et al (1992) Cholinergic innervation of the human striatum, globus pallidus, subthalamic nucleus, substantia nigra and red nucleus. J Comp Neurol 323:252–268

Nathan PW, Smith MC (1982) The rubrospinal and central tegmental tracts in man. Brain 105:223–269

Nieuwenhuys et al (2008) The human central nervous system, 4th edn. Springer

Olyntho-Tokunago HH, Pinto ML et al (2008) Projections from the anterior interposed nucleus to the red nucleus diminish with age in the mouse. Anat Histol Embryol 37:438–441

Onodera S, Hicks TP (1996) A projection linking motor cortex with the LM-suprageniculate nuclear complex through the periaqueductal gray area which surrounds the nucleus of Darkschewitsch in the cat. Prog Brain Res 112:85–98

Parchi P, Saverioni D (2012) Molecular pathology, classification, and diagnosis of sporadic human prion disease variants. Folia Neuropathol 50:20–45

Parchi P, de Boni L et al (2012) Consensus classification of human prion disease histotypes allows reliable identification of molecular subtypes: an inter-rater study among surveillance centres in Europe and USA. Acta Neuropathol 124:517–529

Paxinos G, Furlong TM et al (2020) Human brainstem: cytoarchitecture, chemoarchitecture, myeloarchitecture. Revised edition. Academic Press

Pinto ML, Olyntho-Tokunaga HH et al (2007) The interstitial system of the trigeminal spinal tract projects to the red nucleus in mice. Somatosens Mot Res 24:221–225

Porter JD, Donaldson IM (1991) The anatomical substrate for cat extraocular muscle proprioception. Neuroscience 43:473–481

Prusiner SB (1982) Novel proteinaceous infectious particles cause scrapie. Science 216:136–144

Prusiner SB (1998) Prions. Proc Natl Acad Sci U S A 95:13363–13383

Remington L-A (2012) Autonomic innervation of ocular structures. Accommodation-convergence reaction (near-point reaction). In: Clinical anatomy and physiology of the visual system (third edition)

Robeck I, Mudra SA (2021) Pain and addiction. Pathophysiology. In: Pain care essentials and innovations. Elsevier

Schmitz M, Cramm M et al (2016) The real-time quaking-induced conversion assay for detection of human prion disease and study of other protein misfolding diseases. Nat Protoc 11:2233–2242

Shang C, Liu Z et al (2015) A parvalbumin-positive excitatory visual pathway to trigger fear responses in mice. Science 348:1472–1477

Stahnisch FW (2008) Ludwig Edinger (1855–1918). J Neurol 255:147–148

Tardif E, Delacuisine B et al (2005) Intrinsic connectivity of human superior colliculus. Exp Brain Res 166:316–324

ten Donkelaar HJ (1988) Evolution of the red nucleus and rubrospinal tract. Behav Brain Res 28:9–20

ten Donkelaar HJ (2011) Clinical neuroanatomy: brain circuitry and its disorders. Springer

Ugolini G, Klam F et al (2006) Horizontal eye movement networks in primates as revealed by retrograde transneuronal transfer of rabies virus: differences in monosynaptic input to "slow" and "fast" abducens motoneurons. J Comp Neurol 498:762–785

VanderHorst VG, Ulfhake B (2006) The organization of the brainstem and spinal cord of the mouse: relationships between monoaminergic, cholinergic, and spinal projection systems. J Chem Neuroanat 31:2–36

Wang AG (2018) Pcom aneurysm with Oculomotor Nerve Palsy (ONP). In: Emergency neuro-ophthalmology. Springer, Singapore

Warwick R (1953) Representation of the extraocular muscles in the oculomotor nuclei of the monkey. J Comp Neurol 98:449–503

Westphal CFO (1887) Ueber einen Fall von chronischer progressiver Lähmung der Augenmuskeln (Ophthalmoplegia externa) nebst Beschreibung von Ganglienzellengruppen im Bereiche des Oculomotoriuskerns. Arch Psychiat und Nervenkr 18:846–871

Will RG, Ironside JW et al (1996) A new variant of Creutzfeldt-Jakob disease in the UK. Lancet 347:921–925

Zampieri F, Marrone D et al (2015) Should the annular tendon of the eye be named 'annulus of Zinn' or 'of Valsalva'? Acta Ophthalmol 93:97–99

Zeeh C, Horn AKE (2012) Der Okulomotoriuskern und seine Subnuklei beim Menschen/The Subnuclei of the Oculomotor Nucleus in Humans. Klin Monatsbl Augenheilkd 229:1083–1089

Zülch KJ (1959) Störungen des intrakraniellen Druckes. In: Olivecrona H, Tönnis W (eds) Handbuch der Neurochirurgie Erster Band/Erster Teil Grundlagen I. Springer

Web Links

www.anatomiedesmenschen.de/Insetdrawing

https://www.josephinum.ac.at/en/

https://mayfieldclinic.com/pe-clipping.htm

https://www.naturacollecta.unifi.it/vp-126-the-ceroplastic-workshop.html

https://www.neurosurgicalatlas.com/volumes/cerebrovascular-surgery/aneurysms/posterior-communicating-artery-aneurysm

https://upload.wikimedia.org/wikipedia/commons/archive/6/60/20100101235241%21Lateral_orbit_nerves_chngd.jpg

Pretectum p1 (Prosomere 1)

16

Contents

H. Schröder et al., *The Human Brainstem*, https://doi.org/10.1007/978-3-030-89980-6_16

Abstract

Structures dealt with in this chapter according to genoarchitectonical reasons no longer can be summarized under the term brainstem (Watson et al., Front Neuroanat. 13:10, 2019). They are included here for historical reasons and their use in clinical settings in close relationship to the "classical" brainstem. They comprise structures of enormous clinical importance like the substantia nigra (see Sect. 16.6.1), the ventral tegmental area (see Sects. 16.7.1 and 15.5.1.1), and nuclei important for visual coordination as the nucleus of Darkschewitsch/Elliptic nucleus, the interstitial nucleus of Cajal, the medial accessory nucleus of Bechterew, and the rostral interstitial nucleus of the medial longitudinal fasciculus.

The substantia nigra is a hotspot for various neurodegenerative diseases. Besides well-known Parkinson's disease (PD), this chapter will deal with dementia with Lewy bodies (DLB), dementia with argyrophilic grains (DAG), neurodegeneration with brain iron accumulation (NBIA), and corticobasal degeneration (CBD).

16.1 Precommissural pretectum

The precommissural pretectum is not dealt with in this book

16.2 Justacommissural pretectum

The justacommissural pretectum is not dealt with in this book

16.3 Commissural pretectum

The commissural pretectum is not dealt with in this book

16.4 Liminal pretectum

The pretectum has its name from its position between the tectum (quadrigeminal plate, see Fig. 1.5) and the diencephalon. In terms of classical neuroanatomy, it does not belong to the brainstem. According to the genoarchitectonical considerations by Watson et al. (2019) (see Box 1.1) certain structures, formerly considered as part of the mesencephalon, cannot be regarded that way anymore. This holds true in particular for the Substantia nigra. They will be dealt with, therefore, not under neuromere m1/m2 (Chap. 15) but here.

For reasons of function, in particular for the control of oculomotor activity, certain nuclei of the pretectum are closely related to the brainstem and will be included here.

In this vein, the rostral interstitial nucleus of medial longitudinal fasciculus has been included here. This nucleus (Sect. 16.4.5) together with the elliptic nucleus (see Sect. 16.4.1) and the interstitial nucleus (see Sect. 16.4.2) are listed in FIPAT Ch. 1 under "Prerubral tegmentum" [Tegmentum prerubrale, formerly Tegmentum diencephali].

The rest of the pretectal structures (Sects. 16.1–16.3), however, will not be dealt with here.

16.4.1 Elliptic nucleus/Nucleus of Darkschewitsch [Ncl. ellipticus]

The designation "elliptic" originally comes from Cetacea and Proboscidea (see here below). It is located in the ventral wall of the mesencephalic aqueduct (Fig. 16.1). The FIPAT terminology lists this nucleus under the name elliptic nucleus.

FIPAT Ch 1. endnote 72 *Nucleus commissurae posterioris* states that the elliptic nucleus or nucleus of Darkschewitsch originally was considered as the ventral division of the Nucleus commissurae posterioris which also has a principal and magnocellular part. The ventral part or Ncl. ellipticus is not considered part of the Ncl. of posterior commissure anymore (for details, also see Horn and Adamczyk 2012).

The elliptic nucleus plays an important role in dolphins where it is—together with the inferior olive (see Sect. 4.1.1.3)—prominent for its size (Cozzi et al. 2017; Huggenberger et al. 2019). The elliptic nucleus projects to the inferior olive via the medial tegmental tract. Input to the elliptic nucleus comes from the limbic system, the extrapyramidal system, the pyramidal tract, and the auditory system. The nucleus ellipticus is the origin and target of a loop to the inferior olive, the cerebellum and back. In primates, the analogous system originates in the red nucleus, from there to the inferior olivary complex, the cerebellum and back to the red nucleus (see Box 14.1).

As described by Horn and Büttner-Ennever (1998), upon tracer injection into vertically pulling eye muscles, the elliptic nucleus did not show transsynaptically labeled neurons in primates and humans which speaks against a direct projection of this nucleus to the nucleus of oculomotor nerve (see Sect. 15.3.1) which has been proposed by Onodera and Hicks (1996) for the cat.

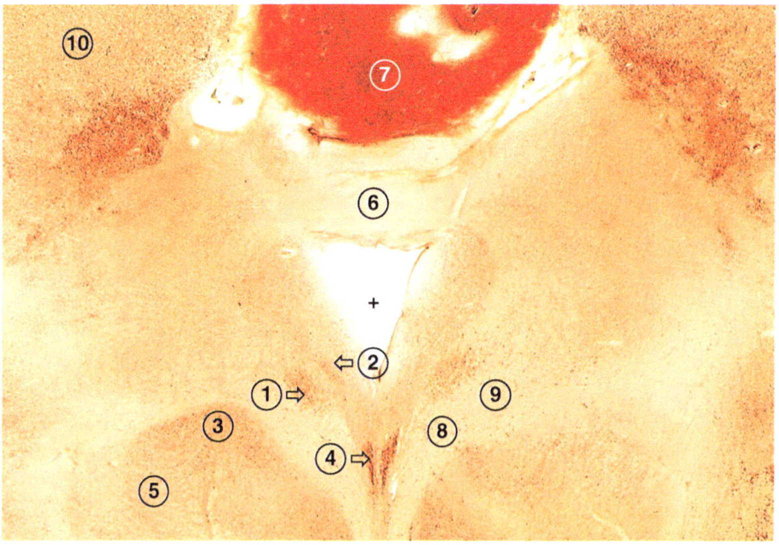

①	Interstitial ncl. (Cajal)	⑦	Pineal gland
②	Elliptic ncl. (Darkschewitsch)	⑧	Medial longitudinal fasciculus
③	Medial accessory ncl. of Bechterew	⑨	Medial tegmental tract
④	Edinger-Westphal ncl., centrally projecting	⑩	Pulvinar
⑤	Red ncl., parvocellular part	+	Mesencephalic aqueduct
⑥	Posterior commissure		

Fig. 16.1 Horizontal section through the transition zone mesencephalon/pretectum/thalamus at the level of the posterior commissure ⑥. Darrow red stain. The posterior commissure forms the roof of the mesencephalic aqueduct (+). On top of the commissure, the epithalamic pineal gland ⑦ is located. The nucleus of Darkschewitsch ② lies lateral of the ependyma on either side of the aqueduct. Lateral of the nucleus of Darkschewitsch the medial accessory nucleus of Bechterew ③ and the nucleus interstitialis Cajal ① are located. Ventromedial of the latter the Edinger-Westphal nucleus EWcp ④ can be seen, which projects centrally (Sect. 15.3.1.2). The parvocellular part of the red nucleus ⑤ (see Sect. 15.3.4) is separated from the aforementioned nuclei by the medial tegmental tract ⑨ and the medial longitudinal fasciculus ⑧ (see also atlas part Darrow red 42, Campbell part 20). LabPON Twente

16.4.2 Interstitial nucleus (Cajal) [Ncl. interstitialis]

This nucleus is located ventrolaterally of the elliptic nucleus (see Figs. 16.1 and 16.2, see Box 4.2).

Functionally, it acts as an integrator for the eye-velocity signals to eye-position signals (Horn and Büttner-Ennever 1998) and is involved in eye-head coordination. The interstitial nucleus projects to the vertical extraocular muscles (superior rectus, inferior rectus, superior oblique, inferior oblique muscles, see Sect. 14.5.2.1.2 and Sect. 15.3.1.1.2), the vestibular nuclei (see Sect. 5.3.1), and the neck muscle motoneurons in the spinal cord (for details, see Horn and Büttner-Ennever 1998).

The interstitial nucleus is most easily identified via its parvalbumin (PARV)-immunoreactivity (see Box 13.6) which visualizes the nucleus as an ovoid structure at the border of the periaqueductal gray (see Sect. 13.2.7.1) directly ventral of the nucleus of Darkschewitsch (Horn and Büttner-Ennever 1998). In terms of cytoarchitecture, these authors ascribe the latter more elongated cells with their long axis parallel to the border of the PAG.

Horn and Büttner-Ennever (1998) describe unilateral lesions of the interstitial nucleus to result in contralateral head tilt, torsion of the eyes to the contralateral side as for the rostral interstitial nucleus of the medial longitudinal fasciculus but a torsional spontaneous nystagmus to the ipsilateral side (in contrast to riMLF) (Halmagyi et al. 1994).

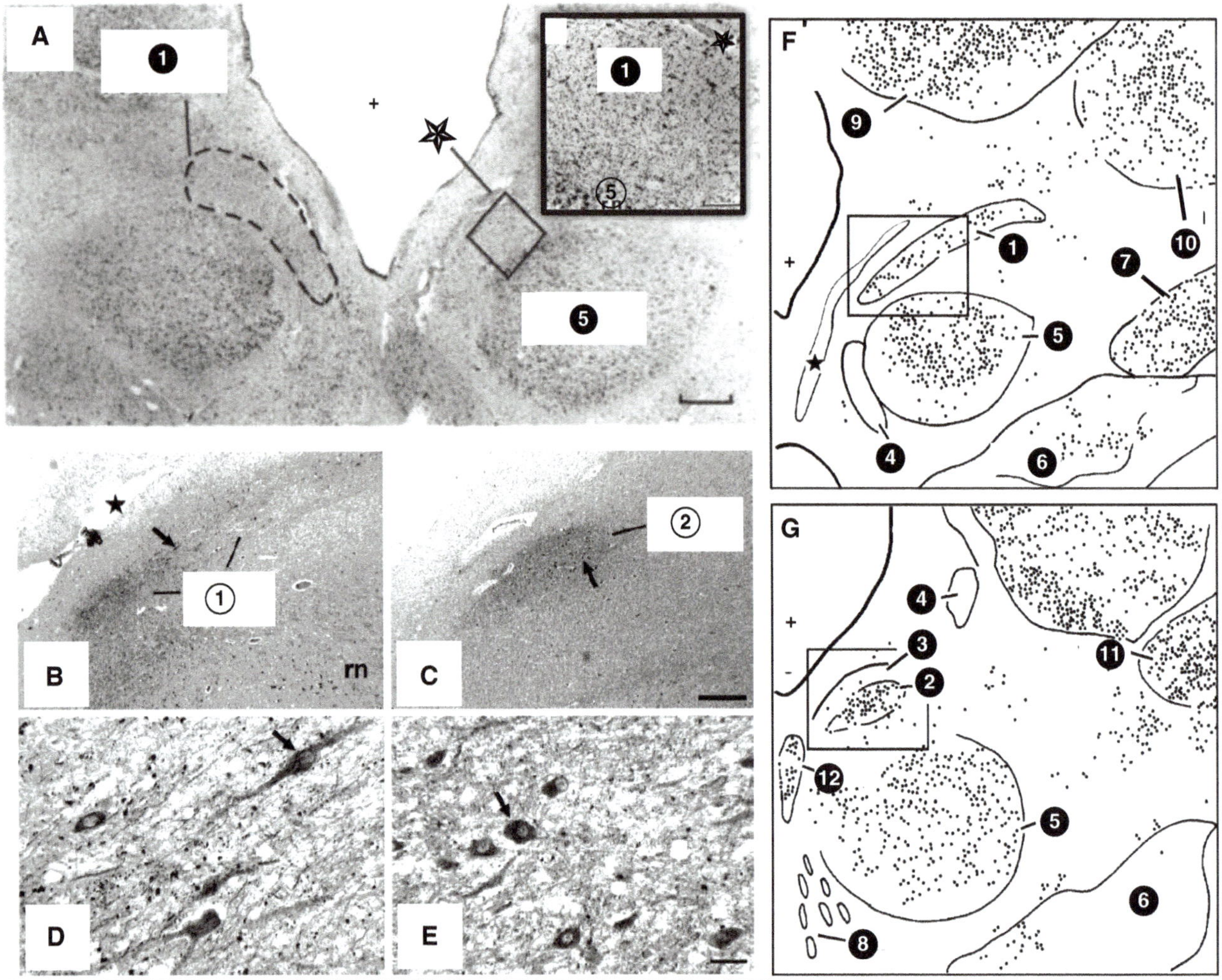

①	Rostral interstitial ncl. of medial longitudinal fasciculus	⑧	Oculomotor nerve
②	Interstitial ncl. (Cajal)	⑨	Centromedian ncl. of thalamus
③	Elliptic ncl.	⑩	Ventral lateral thalamic ncl.
④	Habenulo-interpeduncular tract	⑪	Ventral posterior medial thalamic ncl., parvocellular part
⑤	Red ncl.	⑫	Oculomotor ncl.
⑥	Substantia nigra	✯	Thalamo-subthalamic paramedian artery
⑦	Subthalamic ncl.	+	3rd ventricle

Fig. 16.2 Frontal/coronal sections through the human mesencephalon at the level of the rostral red nucleus ⑤. Note that in Büttner-Ennever et al. (1982) as well as in Horn and Büttner-Ennever (1998) the authors speak of "transverse" sections. However, when you look at Fig. 6 of Horn and Büttner-Ennever 1998 you will see that the cutting planes shown are frontal. (**A**) Nissl (cresyl violet)-stain. The site of the nucleus is indicated by the dashed line and a microphotograph is shown in the rectangle. Conventional identification is not trivial but use of parvalbumin (PARV)-immunohistochemistry clearly shows the rostral interstitial nucleus of the medial longitudinal fasciculus ① (**B, D**) (see Sect. 16.4.5) as well as the interstitial nucleus (Cajal) ② (**C**). The distribution of PARV-ir neurons is shown in the schematic illustrations (**F**) and (**G**). Note that the medial accessory nucleus (Bechterew) (see Sect. 16.4.4) is almost devoid of PARV-immunopositivity. The terminology of the thalamic nuclei shown here has been adapted to that used by Rüb et al. (2016). Without embarking on the mind-blowing discussion of thalamic nuclei nomenclature, it should be mentioned that FIPAT Ch 1 has for: ⑩ a "Ventral lateral complex" Nr. 1829 (VL) and for; ⑪ a "Ventromedial posterior nucleus, parvocellular part" Nr. 1842 (VPMpc). The centromedian nucleus ⑨ is listed as such under Nr. 1819 (CMn). (**A**) modified acc. to Büttner-Ennever et al. 1982, Fig. 2 with permission. (**B–G**) modified acc. to Horn and Büttner-Ennever 1998, Figs. 8, 9 with permission

16.4.3 Pre-Edinger-Westphal nucleus

This nucleus is not listed in FIPAT Ch. 1 nor do exist any references in the common scientific search machines.

16.4.4 Medial accessory nucleus of Bechterew [Ncl. accessorius medialis]

This nucleus is not listed in FIPAT Ch. 1. It was originally called the "Medial accessory nucleus" by Bechterew (1909) (Box 16.1) and later been termed as the "Nucleus of Bechterew" by Fuse (1937) in various mammals. Leichnetz (1982) and Leichnetz et al. (1984) have studied this nucleus and its connectivity (see below) in monkeys.

A recent description of the human medial accessory nucleus has been provided by Onodera and Hicks (2009).

Box 16.1 Eponyms: Bechterew

The use of the eponym Bechterew is paradigmatic for the use of eponyms in general.

Probably, the best known use of his name is in conjunction with Bechterew's disease (Morbus Bechterew) a chronic inflammatory rheumatic disorder of the vertebral column, due to the simultaneous studies of the German internist Adolf von Strümpell and the French neurologist Pierre Marie also known as Bechterew-Strümpell-Marie disease. However, there are certain structures of the human brain that also bear the eponym Bechterew:

1. Superior vestibular nucleus (see Sect. 11.2.2.1)
2. Reticulotegmental nucleus (see Sect. 10.1.1.1)
3. Medial accessory nucleus of Bechterew (see here above Sect. 16.4.4)
4. Band of Kaes-Bechterew (A horizontal band of myelinated fibers in the cerebral cortex)

Furthermore, a number of reflexes historically bears the name Bechterew.

Beyond his scientific merits, Bechterew's life is of interest in relation to the century of the great dictators (Schuchart 2020). **Vladimir Mikhailovich Bekhterev** was born in 1857 in Sorali/Ural in Russia. He started studying medicine at the age of 16 in St. Petersburg and finished his studies in 1878, worked as a psychiatrist and wrote his dissertation on the body temperature in mental diseases. In 1884, he won a scientific competition associated with a stipendium that allowed him to get in contact with the leading researchers in his field in Paris, Vienna, Berlin, and Leipzig. After his return to Russia, he became Professor in St. Petersburg and was the founder of the first institute for experimental psychology in Kazan. In 1913, he was dismissed because of being critical of the government. After the Russian Revolution, he was the founder of the Institute of Brain Research (which will later bear his name) and was responsible for the treatment of critically ill Lenin (Crome 1971). In 1928, he was asked for medical advice by the personal physician of **Stalin**. After exploring Stalin for 3 h, Bechterew told his colleague his diagnosis, severe paranoia. Two days later, he was dead after "enjoying" a light meal at a visit to a theater after he had examined Stalin. After the snack, he suffered from stomach pain and vomiting. He was "cared for" till his death by two physicians who were members of the secret service. Without autopsy and against the will of his family he was cremated, becoming a non-person for half a century. Members of his family were persecuted and even killed (see also Box 3.10).

By use of tract tracing techniques in monkeys, Leichnetz (1982) and Leichnetz et al. (1984) could show that the rostral interstitial nucleus of the medial longitudinal fasciculus, the nucleus of Darkschewitsch, the medial accessory nucleus of Bechterew, and the dorsomedial parvocellular red nucleus receive projections from several cortical areas including the frontal eye field. The interstitial nucleus Cajal had only light projections from the frontal cortex if at all (Leichnetz et al. 1984).

16.4.5 Rostral interstitial nucleus of the medial longitudinal fasciculus [Ncl. interstitialis rostralis fasciculi medialis longitudinalis]

This nucleus (see Fig. 16.2) was first detected in monkeys as involved in vertical eye movements (Büttner-Ennever et al. 1982). It realizes this function in concert with the interstitial nucleus Cajal. The eponym Cajal was useful to avoid confusions between both nuclei. Lesions of the rostral interstitial nucleus of the medial longitudinal fasciculus (riMLF) lead to vertical gaze paralysis. Based on the anatomical data obtained in monkeys a homologous structure was described in man (Büttner-Ennever et al. 1982; Horn and Büttner-Ennever 1998). The rather difficult conventional localization of this nucleus is based on transsynaptic labeling studies in monkeys using the injection of tetanus toxin fragment C into vertical-pulling muscles (i.e., superior rectus, inferior rectus, superior oblique, inferior oblique muscles, see Sect. 15.3.1.1.2, note that these muscles are innervated by different cranial nerves) (Horn and Büttner-Ennever 1998). The tracer was found in the riLMF mainly ipsilateral and bilateral in the interstitial nucleus Cajal. Combining the tracing with parvalbumin (PARV)-immunohistochemistry, the authors could show that all premotor neurons of the riMLF and almost all of them in the interstitial nucleus were immunopositive for PARV (see Fig. 16.2). The riLMF was identified in man as a wing-shaped nucleus dorsomedial of the red nucleus, rostral to the traversing habenulo-interpeduncular tract (formerly Fasciculus retroflexus).

16.4.6 Substantia nigra, lateral part

For detailed information on the Substantia nigra, lateral part, see Sect. 16.6.1.

16.5 Ventral pretectum

16.5.1 Red nucleus, parvocellular part

For detailed information on the red nucleus incl. the parvocellular part, see Sect. 15.3.4.

16.5.2 p1 reticular formation (p1Rt)

This structure has been described as "basal reticular formation of p1" structure in the Developing Mouse Brain Atlas/ p1BRt, Developing Mouse Brain Atlas, Mus musculus Atlas: Developing Mouse P56.

http://search.brain-map.org/search/index.html?query=basal%20reticular%20formation%20of%20p1

Adult Mouse Brain Tissue Gene Expression Profiles dataset under: formation+of+p1/Allen+B/Harmonizome/gene_set/basal+reticular+formation+of+p1/Allen+Brain+Atlas+Adult+Mouse+Brain+Tissue+Gene+Expression+Profiles

This nucleus is not listed in FIPAT Ch. 1 nor do any human-specific references exist in the common scientific search machines.

16.6 Basal pretectum

16.6.1 Substantia nigra, compact part (SNC)

16.6.1.1 The mesodiencephalic SN/VTA complex

According to Watson et al. (2019), it is widely assumed that the Substantia nigra and the ventral tegmental area (VTA) (see Sect. 15.5.1.1) are integral part of the midbrain but only their caudal parts are located in the compressed true ventral midbrain (for details, see Watson et al. 2019). The rostral parts of SN and VTA are derived from prosomeres 1–3 and lie in the diencephalon. Therefore, for these dopaminergic cell groups the term "mesodiencephalic SN/VTA complex" (not indicated in FIPAT Ch.1) is recommended (Watson et al. 2019).

16.6.1.2 Location and morphology of the Substantia nigra (SN)

The SN takes up a large part of the ventral mesencephalic tegmentum. Its ventral neighbor are the cerebral peduncles, dorsally it is covered by the decussation of the superior cerebellar peduncles in the more caudal parts, further rostrally by the red nucleus. The longitudinal extension of the SN comprises the whole extension of the midbrain. It starts rostrally approximately at the level of the superior colliculi (see Fig. 16.4B) stretching orally up to the subthalamic nucleus (STN) (Olszewski and Baxter 1982). Figure 16.3 shows a typical horizontal midbrain section at the level of the superior colliculi (see Sect. 15.1.3). The dark pigmentation of the SN is usually seen in neurologically healthy individuals. In Fig. 16.4, the SN is shown at three different levels, (A) at the transition between pons and midbrain, (B) at the level of the superior colliculi, and (C) at the most cranial section of our atlas series.

The SN can be subdivided into a compact and a reticular part. This terminology indicates a rather high density of neurons in the compact [Pars compacta] and a higher percentage of neuropil in the reticular part [Pars reticulata]. The compact part can be further subdivided into two parts with individual subnuclei (for details, see Braak and Braak 1986, van Domburg and ten Donkelaar 1991). As indicated in endnote 65 of FIPAT Ch 1, there is a partial correspondence of the subnuclei to the subdivision of the SN into nigrosomes and matrix as shown by Damier et al. (1999).

16.6.1.3 Connectivity and function of the Substantia nigra

The connection between SN and caudate putamen (CP)—the most important one in terms of pathology—belongs to those for which relatively many human-specific data are available. This is also due to the fact that because of the important role of this system in diseases like Parkinson's a lot of scientific work has been invested into elucidation of the system.

In terms of anatomy the connection between the SN and the CP relies on the nigrostriatal fibers [Fibrae nigrostriatales], formerly called the nigrostriatal tract. It is formed by the axons of dopaminergic neurons of the SN. Inside the CP, they make synaptic contact with at least two neuron populations with different neurochemistry. One group of neurons is equipped with dopamine (DA) D1 receptors which inhibit the propagation of signals from the nigrostriatal terminals.

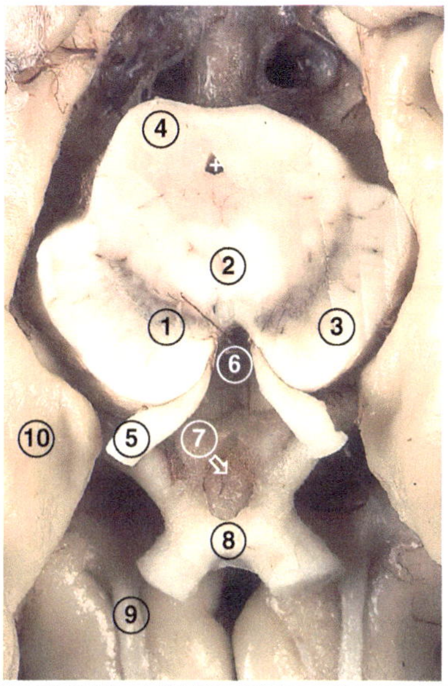

①	**Substantia nigra**	⑦	**Pituitary stalk**
②	**Decussation of the superior cerebellar peduncles**	⑧	**Optic chiasm**
③	**Cerebral peduncle**	⑨	**Olfactory nerve CNI**
④	**Superior colliculus**	⑩	**Ambient gyrus / Entorhinal cortex**
⑤	**Oculomotor nerve CNIII**	+	**Mesencephalic aqueduct**
⑥	**Interpeduncular fossa**		

Fig. 16.3 Horizontal section through the human mesencephalon at the level of the exit of the oculomotor nerve. Unfixed specimen. LabPON Twente

The second group bears DA D2 receptors which facilitate the transmission of signals. This is an interesting example for two completely different effects of one neurotransmitter depending on the kind of receptors they interact with (see Fig. 16.6).

You should, however, keep in mind that most of the connectivity data on the human SN are derived from animal studies. An interesting exception is a Marchi stain study (degenerating myelin) by Larson et al. (1982) showing a reciprocal connection of the globus pallidus with the ventral lateral nucleus of the thalamus.

Recently, Zhang et al. (2017) studied anatomical and functional aspects of the human SN using diffusion and functional MRI data from the Human Connectome Project (https://www.humanconnectome.org). They postulate a tripartite connectivity based on the parcellation of the SN with a limbic, cognitive, and motor arrangement. According to their data, the medial SN connects with limbic striatal and cortical regions and encodes value, the ventral SN connects with associative cortical regions and striatum, encoding salience. The lateral SN connects with cortical and striatal somatomotor regions, also encoding salience (Zhang et al. 2017).

16.6.1.4 Pathological aspects of Substantia nigra connectivity

For any brain structure, it is crucial to know how it behaves with normal aging to understand pathological changes. This task has been executed for the human brain stem catecholaminergic nuclei, i.e., SN, central gray substance, ventral tegmental area, peri- and retrorubral area and locus caeruleus (LC) by Kubis and Faucheux (2000). Quantitative evaluation of tyrosine hydroxylase (TH) immunolabeled cells revealed that in none of the nuclei studied was a statistically significant decrease of the TH-neurons nor a global neuron loss. This points to Parkinson's disease (PD) most likely not to be related to accelerated degeneration with age (Kubis and Faucheux 2000). The neuropathological process will be reviewed in detail under Sect. 16.6.1.5. Here, the anatomy-related consequences of a dopaminergic cell loss are considered based on the schemes presented in Fig. 16.6 (see legend for detailed explanation).

In Parkinson's disease, the dopaminergic SN neurons undergo degeneration, the axons to the CP get lost and eventually the CP is completely deprived of its dopaminergic input. From the scheme in Fig. 16.6 it becomes obvious that the direct pathway loses its excitatory input while the indi-

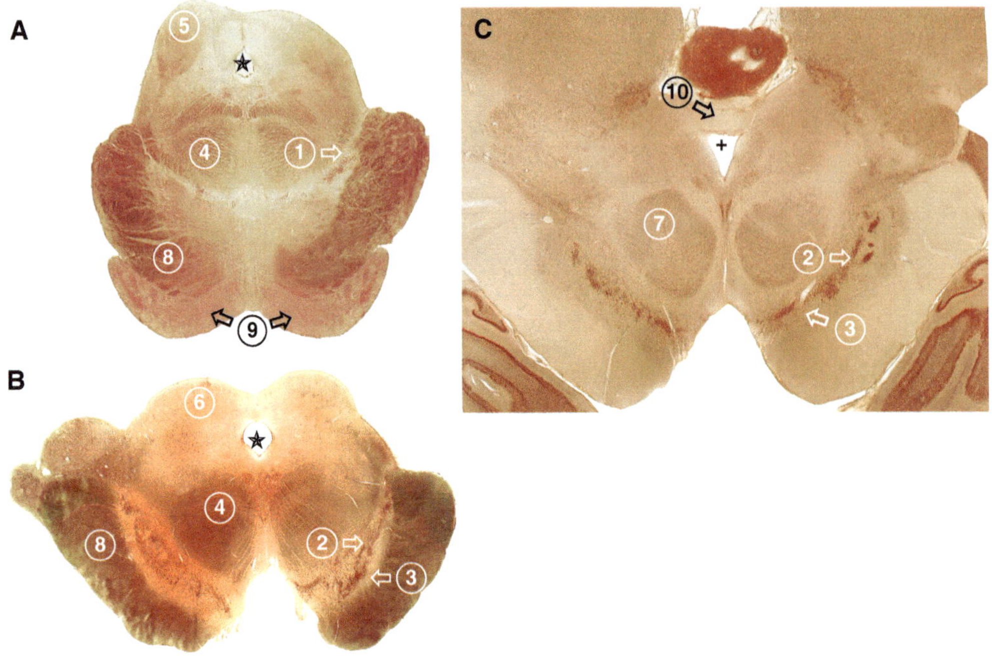

①	Substantia nigra	⑦	Red ncl.
②	Substantia nigra, compact part	⑧	Cerebral peduncle
③	Substantia nigra, reticular part	⑨	Pontine basis
④	Superior cerebellar peduncle	⑩	Posterior commissure
⑤	Inferior colliculus	★	Mesencephalic aqueduct
⑥	Superior colliculus	+	3rd ventricle

Fig. 16.4 Horizontal histological sections (Darrow red stain) showing the human Substantia nigra at different levels (**A–C**). In terms of connectivity, the two different caudate putamen (CP) dopamine-sensitive neuron types stand at the beginning of two different pathways from the CP via several stations eventually reaching the cerebral cortex (Figs. 16.5 and 16.6) where they again elicit different effects. (See also atlas part Darrow red 35, 40, 42, Campbell 17, 19, 20). LabPON Twente

rect pathway is deprived from the inhibitory dopaminergic input. This finding may explain at least one of the three major clinical symptoms already described by James Parkinson (1817) (see Box 16.2) in:

- Bradykinesia (Greek βραδύς bradys = slow, clumsy ἡ κίνησις he kinesis = movement
- Resting tremor
- Rigor

Two of these symptoms are obvious even to laymen who are confronted with a person suffering from Parkinson's disease, i.e., reduced motor activity (bradykinesia) and the shaking (tremor). The third symptom, rigor, is experienced by the examining physician when he/she passively moves the forearm of the patient in the elbow joint which mediates the impression that the joint is a cogwheel with discrete steps of movement instead of a smooth movement under normal circumstances.

While up to today, there are no convincing explanations for tremor and rigor the development of bradykinesia is relatively easy to deduce from Fig. 16.6, considering the loss of dopaminergic modulation of nigrostriatal connectivity.

PD also is one of the neurological diseases serving as a classical paradigm for rational drug therapy. As you can see in Fig. 16.7, biosynthesis of catecholamines as, for example, DA is an enzymatically controlled series of defined steps. The third of these steps transforms L-DOPA to DA under the

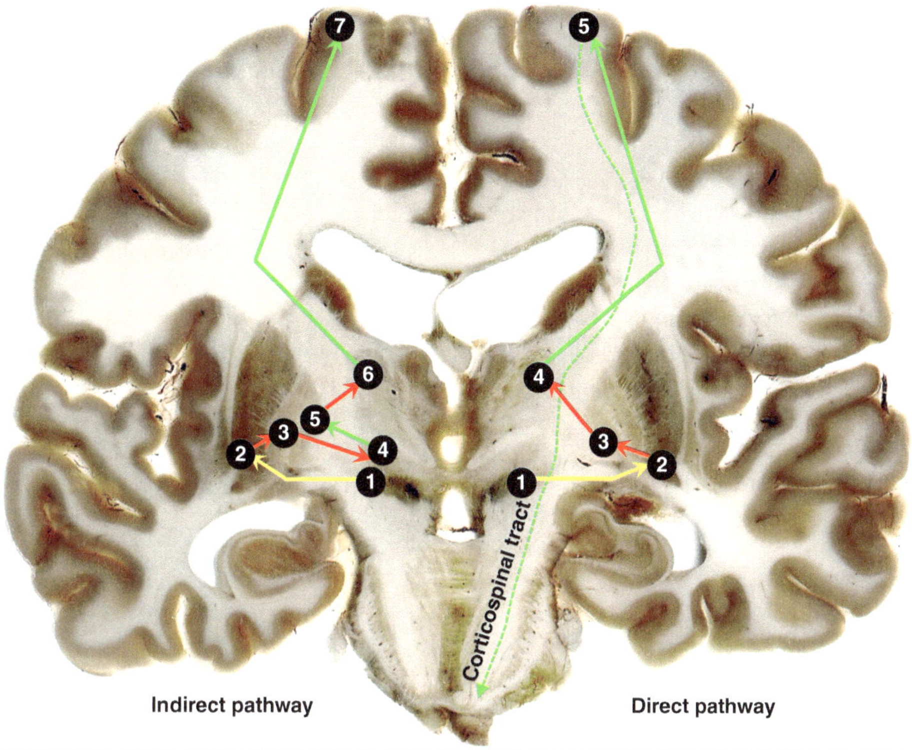

①	Substantia nigra		Direct pathway
②	Caudate putamen	③	Globus pallidus, medial segment
	Indirect pathway	④	Lateral ventroanterior nucleus of thalamus *
③	Globus pallidus, lateral segment	⑤	Primary motor cortex BA4
④	Subthalamic nucleus		Dopamine / Glutamate / GABA
⑤	Globus pallidus, medial segment		
⑥	Lateral ventroanterior nucleus of thalamus *		
⑦	Primary motor cortex BA4		

Fig. 16.5 Both pathways start in the Substantia nigra (SN) ① and reach as first target the caudate putamen (CP) ②. From there, in the indirect pathway the neuronal chain is continued to the globus pallidus lateral segment [Globus pallidus lateralis/externus] ③. Projections of the latter reach the subthalamic nucleus ④ which projects back to the globus pallidus medial segment [Globus pallidus medialis/internus] ⑤. Next station is the lateral anterior thalamic nucleus, one of the thalamic motor nuclei ⑥. The direct pathway is continued from the CP directly to globus pallidus medial segment ③, then reaches the lateral ventroanterior thalamic nucleus (* terminology/identification acc. to Mai and Majtanik 2017) ④. The hierarchical highest part of both pathways is the primary motor cortex (BA4 = Brodmann area 4) ⑤ ⑦. The integrity or lesions of the pathways (see Fig. 16.6) are reflected in movements mediated by the corticospinal tract. D1 receptor neurons form the so-called direct pathway while D2 receptor neurons are the first part of the so-called indirect pathway. The term "indirectly" refers to the fact that this connection has one more station—the subthalamic nucleus—as compared to the direct pathway. As you can see in Fig. 16.6, the different pharmacological effects observed rely on the action of two different classical transmitters and their receptors. On the excitatory side, there is glutamate (Glu), on the inhibitory side γ-aminobutyric acid (GABA) (see Fig. 16.6). H. Schröder fecit. Sammlung des Zentrums Anatomie der Universität zu Köln

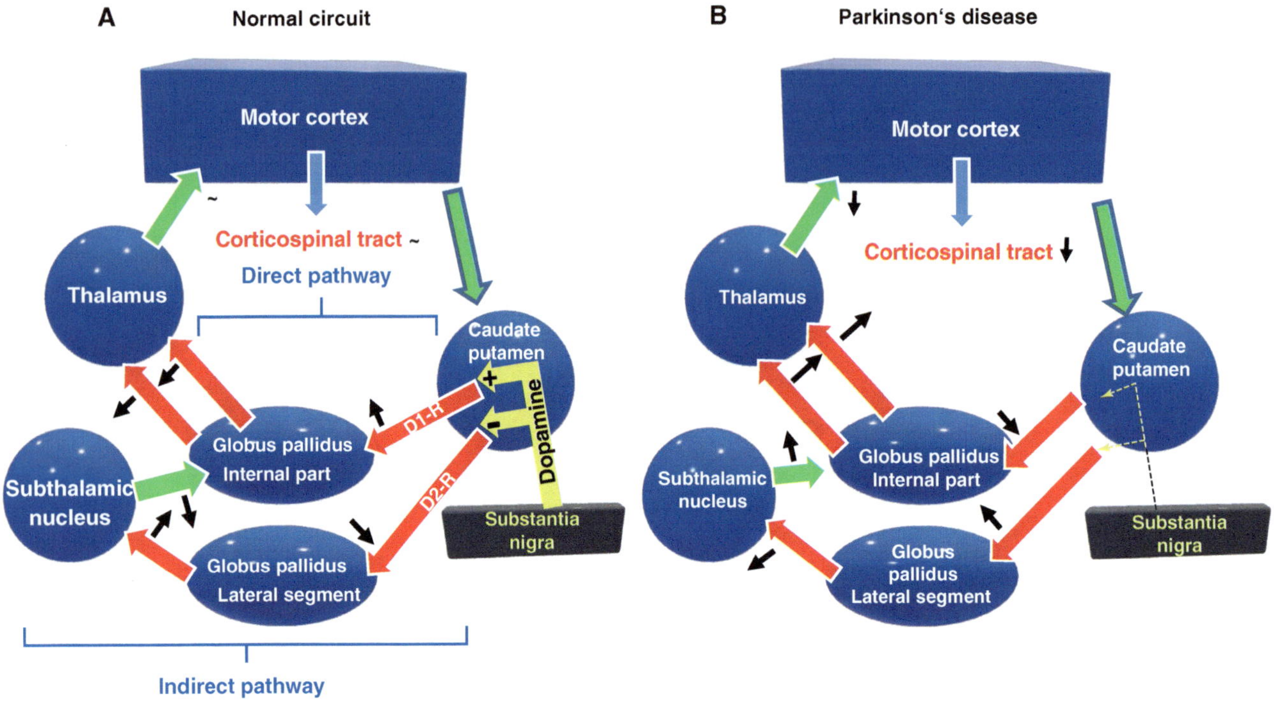

Fig. 16.6 This schematic representation is based on an illustration by Albin et al. (1989) and the transformation in a better intelligible form by one of the authors (FH). As shown in Fig. 16.5, there are two different circuits under the control of dopaminergic (Fig. 16.7) projections from the SN. In simplified terms of neurochemistry, in addition to dopamine (DA) (yellow arrows), there are only two more transmitters involved, the inhibitory transmitter **GABA** (γ-aminobutyric acid) (red arrows) and the excitatory transmitter **glutamate** (GLU) (green arrows). Note that the majority of synapses in the circuits are GABAergic inhibitory. Only the connection from the subthalamic nucleus (STN) (indirect pathway) to the globus pallidus medial segment and from the motor nuclei of the thalamus to the primary motor cortex are glutamatergic (excitatory). The understanding of the circuit-mediated effects therefore is mainly based on the inhibitory and/or disinhibitory effects of GABA. (**A**) Under normal conditions, two different types of caudate putamen (CP) neurons with different DA receptors each are the starting point of the circuits. D1 receptors mediate an excitatory effect of DA, D2 receptors an inhibitory effect. Please note that the segregation into two DA receptor subtypes also segregates the circuits in the direct and the indirect pathway. **Direct** and **indirect** refers to the way the GABAergic neurons of either pathway reach the globus pallidus medial segment. The **direct pathway** has only one synapse, formed presynaptically by a GABAergic projection from the CP to postsynaptic GABA receptors on neurons of the globus pallidus medial segment. The indirect pathway, too, eventually reaches the globus pallidus medial segment but performs kind of a detour via globus pallidus lateral segment and STN. From the globus pallidus medial segment a "united" pathway runs to thalamic motor nuclei (see Fig. 16.5) which then contact the primary motor cortex. The motor readout—transferred via the corticospinal tract—allows for a clinical evaluation. Arrows symbolize increase (⇧) or decrease (⇩) of transmitter release at the individual presynaptic sites. Under normal conditions, DA stimulates the first neuron of the direct pathway resulting in a maintained high GABA release at the synapse with the globus pallidus medial segment. Its activity is thus inhibited resulting in reduced GABA release at the pallido-thalamic junction. This leads to disinhibition of the GLUergic thalamocortical projection and a stimulation of the motor cortex. In the **indirect pathway**, D2 receptors mediate an inhibitory effect on the first GABAergic neuron which synapses on neurons of the globus pallidus lateral segment which likewise are GABAergic. They are disinhibited and mediate an inhibition of the glutamatergic STN, i.e., again a decreased excitation of the globus pallidus medial segment as with the direct pathway and identical effects for the thalamus, motor cortex, and pyramidal tract. (**B**) In PD, with loss of DAergic CP innervation (degeneration of SN neurons and axons) a reversal of the quantitative effects on the synaptic transmission in both circuits takes place, eventually resulting in inhibition of thalamus, cortex, and activity levels of the corticospinal tract, clinically assessable as bradykinesia. H. Schröder fecit

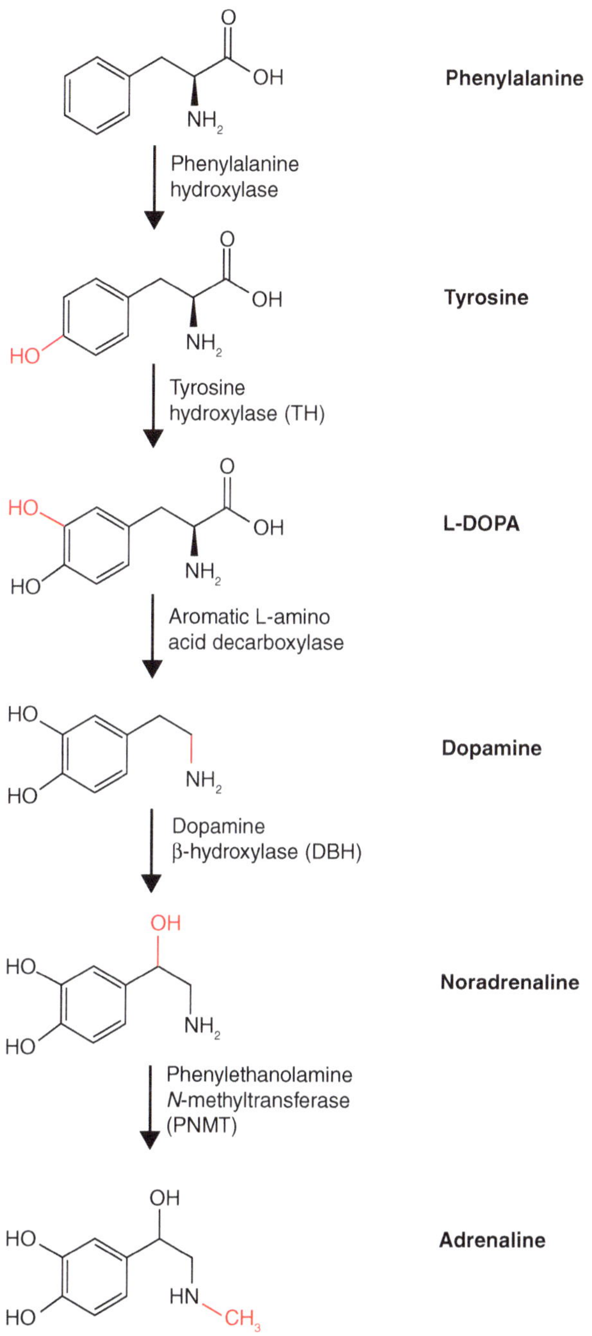

Fig. 16.7 Biosynthesis of catecholamines. Dopamine (DA) forms by decarboxylation of L-DOPA. L-DOPA, by contrast to DA, can cross the blood–brain barrier (see Sect. 1.4.1.1). L-DOPA is taken up by DAergic neurons and converted to DA. This approach gets useless when a critical number of SN DAergic neurons has undergone neurodegeneration. Sophia Schröder fecit

influence of the enzyme L-amino acid decarboxylase. Under normal conditions, this happens in the neurons of the SN. The more of these neurons degenerate during the course of PD (see Sect. 16.6.1.5) the more the biosynthesis of DA decreases—accompanied by movement restriction.

In this situation, administration of L-DOPA bypasses the synthesis problem.

16.6.1.5 Neuropathological aspects of Substantia nigra/Parkinson's disease

Parkinson's disease (PD) is a frequent neurodegenerative movement disorder, characterized clinically by rigidity, akinesia, resting tremor, and postural instability due to progressive degeneration of the DAergic nigrostriatal system and other neuronal networks mainly caused by loss of pigmented neurons in the substantia nigra (SN) (Jellinger and Mizuno 2004) (also see Sect. 16.6.1.4) (see Box 16.2).

Box 16.2 Parkinson's disease (PD)/Dementia with Lewy bodies (DLB)

Parkinson's disease (PD) is the second most common neurodegenerative disease in the world after Alzheimer's disease (see Sect. 13.2.2.1.4). Genetic forms have revealed much about the molecular pathogenesis of PD, but most cases, however, represent the idiopathic form of the disease. PD, PD with dementia, and Dementia with Lewy bodies (DLB) are clinical syndromes characterized by the neuropathological accumulation of α-synuclein in the CNS that represent a clinicopathological known as Lewy body disorders (see Box 16.3). Available therapies for PD only treat the symptoms of the disease (see Sect. 16.6.1.4). Since the underlying causes are heterogeneous, selected molecular pathways should be targeted for treatment of the symptoms of the disease. Drugs that enhance intracerebral DA concentrations remain the mainstay of treatment for motor symptoms. Deep brain stimulation (see Box 13.2) is a well-established treatment for the motor symptoms of the disease.

PD is a multisystem disorder in which predisposed neuronal types in specific regions of the peripheral, enteric, and central nervous systems become progressively involved. It has been suggested that the onset of the disease could be localized in the peripheral organs, such as the gastrointestinal tract and the olfactory bulb, before the neuropathological changes occur in the CNS.

Within the spectrum of Lewy body disorders, there is a group of elderly demented people who have a consistent pattern of clinical features, such as fluctuation in cognition, visual hallucinations, REM sleep disorders as well as spontaneous motor features of parkinsonism. This entity is called Lewy body dementia (DLB). Important realizing is that Lewy bodies (LBs) and Lewy neurites (LNs), which are essential for the diagnosis of PD, are also the morphological hallmark of DLB, which means that the two diseases cannot be distinguished solely on the grounds of neuropathological changes. Although Alzheimer-type changes and vascular pathology often are found in DLB patients, the pattern of Lewy pathology in DLB/PD is different from the distribution of neurofibrillary pathology in AD (see Sect. 13.2.2.1.4).

Box 16.3 Charles Parkinson and Friedrich Heinrich Lewy

Charles Parkinson (1755–1824) was an English physician who published "An Essay on the Shaking Palsy" in 1817 describing six personally observed cases. The term "shaking palsy" refers to the resting tremor patients display together with reduced motor activity (hypokinesia) and rigor (muscle stiffness). Friedrich Heinrich Lewy (1885–1950) was a German neurologist who worked in Alois Alzheimer's laboratory. He was the first to describe the later the so-called Lewy bodies (see Fig. 16.14ff.) in Parkinson's disease and dementia with Lewy bodies (DLB). He left Germany because of the Nazi regime, emigrating to the USA (Frederic Henry Lewey).

PD with a prevalence of approx. 2% is the second most common neurodegenerative disease in the world after Alzheimer's disease (AD). The incidence increases rapidly with age. In sporadic cases, causes and etiology are largely unknown. Older age and smoking habits are the only risk factors for PD that have consistently been found across studies (de Lau and Breteler 2006). Converging evidence from studies from neuropathology and biomarkers determined in the CSF (Sect. 1.2.1), neuroimaging and genetic studies, suggests that β-amyloid and tau pathology as well as cerebrovascular disease and TDP-43 (43-kDa TAR DNA-binding protein) are likely to influence clinical features and progression (van Rumund et al. 2019; Coughlin et al. 2020; Buchman et al. 2019). Most cases of PD represent the idiopathic forms of the disease (de Lau and Breteler 2006; Funayama et al. 2015). PD is typically a sporadic disease but, only in a small proportion of patients, mutation of a dominantly or recessively inherited gene can cause PD (de Lau and Breteler 2006; Funayama et al. 2015; Kalia and Lang 2015). Although rare, these genetic forms of PD have revealed much about the molecular pathogenesis of sporadic PD (Funayama et al. 2015, Kalia and Lang 2015). The first gene associated with PD was SNCA, which encodes α-synuclein and the discovery of SNCA mutations as a cause of autosomal dominant PD led to the finding of α-synuclein as primary component of Lewy pathology. PD, PD with dementia, and Dementia with Lewy Bodies (DLB) are clinical syndromes characterized by the neuropathological accumulation of α-synuclein in the CNS representing a clinicopathological spectrum known as Lewy body disorders (Coughlin et al. 2020). Since the discovery of SNCA mutations, five additional genes have been proposed to mediate autosomal dominant forms of PD (Kalia and Lang 2015). There are also three genes that now are established as associated with autosomal recessive forms of PD. However, not all genetic forms of PD are associated with Lewy body pathology (Kalia and Lang 2015; Poulopoulos et al. 2012). Misfolding of α-synuclein is the key to pathology and used for new biomarkers in CSF (van Rumund et al. 2019). Available therapies for PD only treat symptoms (Kalia and Lang 2015). Drugs that enhance intracerebral dopamine concentrations (see Sect. 16.6.1.4, see Fig. 16.7) or stimulate dopamine receptors remain the mainstay of treatment for motor symptoms. However, since the underlying causes of PD are heterogeneous, and multiple cellular processes are variably involved, a more effective strategy might be to target selected molecular pathways in specific patients and with several drugs. Potential pharmacological targets for disease modification in PD include neuroinflammation, mitochondrial dysfunction and oxidative stress, calcium channel activity, LRRK2 kinase activity, as well as α-synuclein accumulation, aggregation, and cell-to-cell transmission (including immunotherapy techniques) (Kalia and Lang 2015; Aldakheel et al. 2014; Tran et al. 2014). Deep brain stimulation (DBS) (see Box 13.2) is a well-established treatment for PD motor symptoms. The STN is the preferred DBS target in advanced PD (Figs. 16.8 and 16.9).

Unilateral pallidotomy (Fig. 16.10) is an optional treatment for PD patients in remote areas where the mandatory frequent follow-up for the adjustment of stimulation parameters is impossible (Esselink et al. 2007) or in well-informed patients who decide against stimulation. Thalamic DBS is also an option for the treatment of tremor, but its success is rather limited (Esselink et al. 2007; Kalia and Lang 2015).

Macroscopically, the external surface of the brain in PD is unremarkable. On cut surface, pallor of the substantia nigra

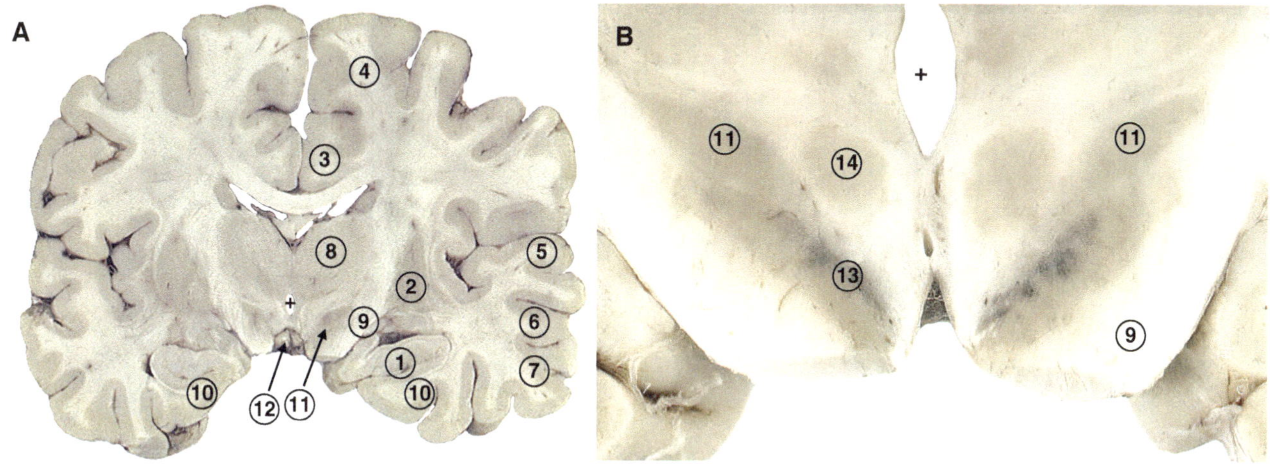

①	Hippocampus	⑨	Cerebral peduncle
②	Putamen	⑩	Entorhinal cortex
③	Cingulate gyrus	⑪	Subthalamic ncl.
④	Precentral gyrus	⑫	Mammillary bodies
⑤	Superior temporal gyrus	⑬	Substantia nigra
⑥	Middle temporal gyrus	⑭	Red ncl.
⑦	Inferior temporal gyrus	+	3rd ventricle
⑧	Thalamus		

Fig. 16.8 Coronal section of the normal human brain at the level of the thalamus ⑧ and subthalamic nucleus ⑪ in (**A**) and in higher magnification of the subthalamic nucleus in (**B**) ⑪. Note the dark-stained, neuromelanin pigment-containing substantia nigra ⑬ dorsomedial of the cerebral peduncle ⑨. LabPON Twente

and locus caeruleus is evident (Figs. 16.12 and 16.13). The appearances of the globus pallidus and CP are normal. The loss of specific subsets of melanin-containing projection cells in the substantia nigra pars compacta (Fig. 16.14) as well as of melanin-containing noradrenergic neurons in the locus caeruleus (Fig. 16.15, see Fig. 13.29, Box 13.5) is a characteristic finding. The formation of intraneuronal inclusions—pale and Lewy bodies in cell somata, and of Lewy neurites in cellular processes of a few vulnerable cell types—is pathognomonic for PD (Fig. 16.14). However, characterizing PD as a loss of nigral dopaminergic neurons overemphasizes a single feature among the many facets of this disorder. PD is a multisystem disorder in which predisposed neuronal types in specific regions of peripheral, enteric, and central nervous systems become progressively involved.

Brainstem LBs show a classic morphology comprising an eosinophilic hyaline core and a pale-staining peripheral halo. α-synuclein-immunoreactive LBs and LNs can be found in the posterior nucleus of vagus nerve (Figs. 16.14, 16.16, and 16.17, see Sect. 7.3.2.1), locus caeruleus (Fig. 16.15) (see Sect. 13.2.2.1), SN (Fig. 16.19) (see Sect. 16.6.1), ventral tegmental area (see Sect. 15.5.1.1), and pedunculopontine nucleus (see Sect. 13.1.1.1.1) (Braak et al. 2003; Seidel et al. 2015; Kalia and Lang 2015; Tubert et al. 2019) and the intermediate reticular zone (Fig. 16.18) (see Sect. 3.3.2.1). α-synuclein immunoreactive coiled bodies (CBs) are smaller and more difficult to detect than LBs and LNs. They surround the lucent nucleus and are twisted in the small cytoplasmic rim of the affected oligodendrocytes. The distribution and extent of the CBs show no differences between PD and DLB patients (Seidel et al. 2015) (Fig. 16.19).

Neuronal loss and LBs in these brainstem regions are often accompanied by the presence of macrophages (filled with pigment) and gliosis (Fig. 16.19). Results of a study of cases with incidental LB pathology indicate that PD commences with the formation of LBs and LNs not only in these brainstem nuclei but also in the olfactory system (del Tredici et al. 2002). The first pathological changes appear in the anterior olfactory nucleus and olfactory bulb (Fig. 16.20).

The amygdala undergoes severe pathological changes during the course of PD (Braak et al. 1994). The most prominent changes occur in the central and accessory cortical nuclei of the amygdala (Fig. 16.20).

LBs can also be found in both the transentorhinal and entorhinal regions. Typically, long LNs, extending through-

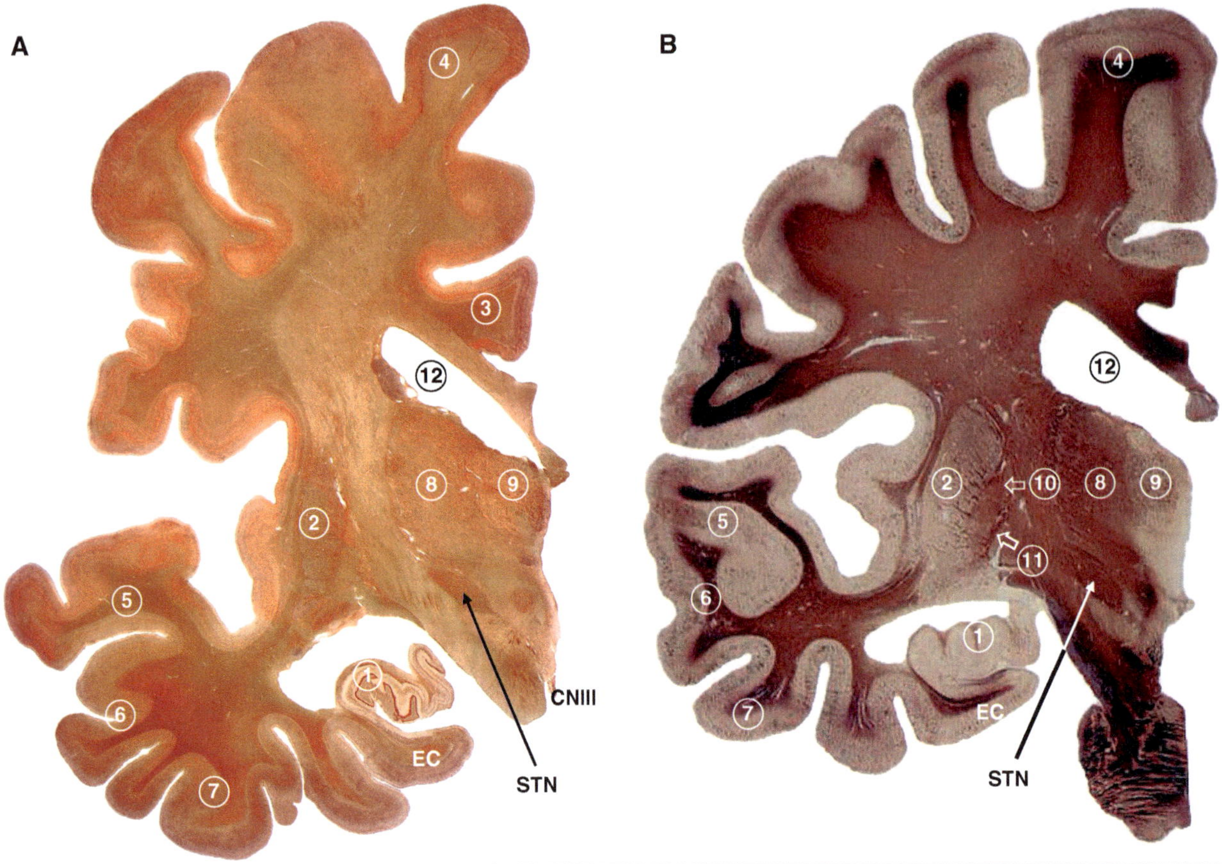

①	Hippocampus	⑦	Inferior temporal gyrus
②	Putamen	⑧	Ventrolateral thalamus
③	Cingulate gyrus	⑨	Mediodorsal thalamus
④	Precentral gyrus	⑩	Globus pallidus, lateral segment
⑤	Superior temporal gyrus	⑪	Globus pallidus, medial segment
⑥	Middle temporal gyrus	⑫	Lateral ventricle

Fig. 16.9 Coronal section of the hemisphere at the same level as in Fig. 16.8 in a Darrow red stain (**A**) and a Campbell Switzer stain (**B**). (PEG 100 μm). Arrows points to the subthalamic nucleus (STN). In this case, amyloid β-deposits are present in neo- and allocortex. EC = entorhinal cortex, CNIII = oculomotor nerve. LabPON Twente

out the second sector of the Ammon's horn, are often present (de Vos et al. 1995) (see Fig. 16.21).

LBs and LNs are also often present in the magnocellular nuclei of the forebrain (medial septal nucleus, both vertical and horizontal limbs of the diagonal band), the basal nucleus of Meynert, the tuberomammillary nucleus, and the midline nuclei of the thalamus. They are also found in areas of the neocortex, especially in anterior cingulate (Fig. 16.21), insular, and temporal cortices. Less frequently, LBs are found in the occipital cortex. In the cortex, they are much less distinct in conventional stains (H & E), appearing as diffuse eosinophilic spheroids, often without halo. In the spinal cord, involvement of parasympathetic and sympathetic neurons has been shown (van de Berg et al. 2007; Braak et al. 2007a) (Fig. 16.21). The main constituents of LBs are—apart from α-synuclein (Spillantini et al. 1997)—neurofilament proteins, dardarin, ubiquitin, and proteins involved in ubiquitin metabolism.

Ultrastructurally, the halo of classical LBs consists of radially arranged intermediate filaments (7–20 nm) associated with a granular electron-dense coating material and vesicular structures. They are easily visualized by staining for α-synuclein and by ubiquitin, P62, or silver staining, i.e., the Campbell-Switzer silver staining method. Ubiquitin and P62 staining are less specific and sensitive than an α-synuclein staining because they stain not only LBs and LNs but also

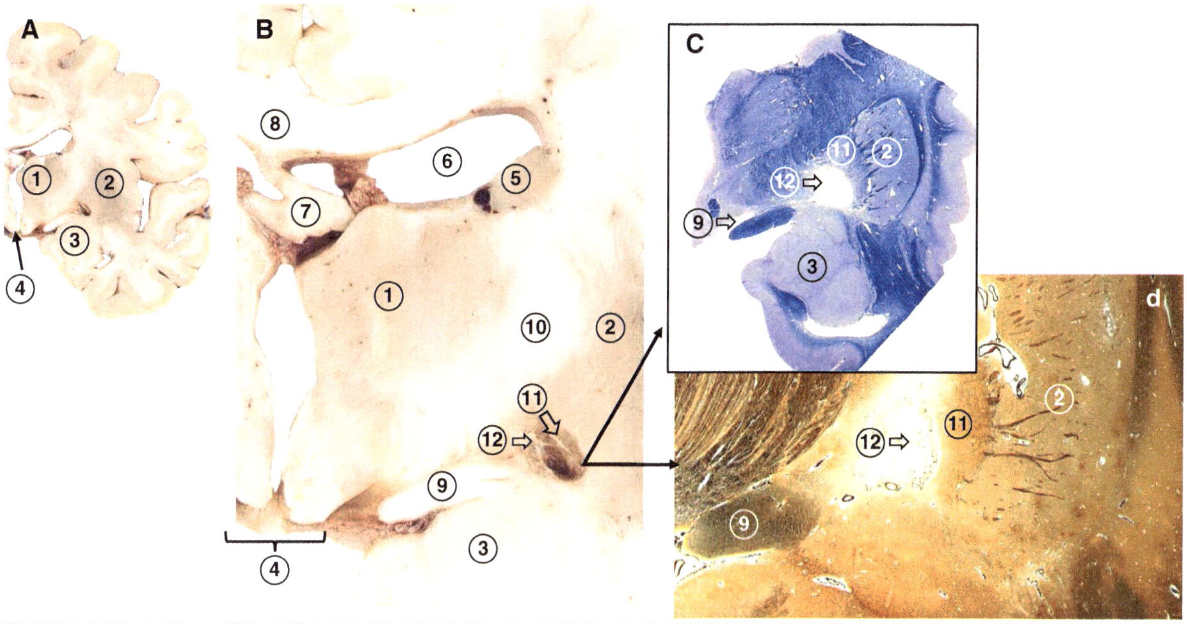

①	Thalamus	⑦	Fornix
②	Putamen	⑧	Corpus callosum
③	Amygdala	⑨	Optic tract
④	Mammillary bodies	⑩	Internal capsule
⑤	Caudate nucleus	⑪	Globus pallidus, lateral segment
⑥	Lateral ventricle	⑫	Globus pallidus, medial segment

Fig. 16.10 Coronal view of the right hemisphere from a patient who underwent unilateral pallidotomy (**A**), in detail shown in (**B**). The surgical defect in the globus pallidus medial segment (see Fig. 16.9 ⑪) is clearly visible in a Klüver-Barrera myelin (**C**) and a Bielschowsky silver stain (**D**). Compare with Fig. 16.11. LabPON Twente

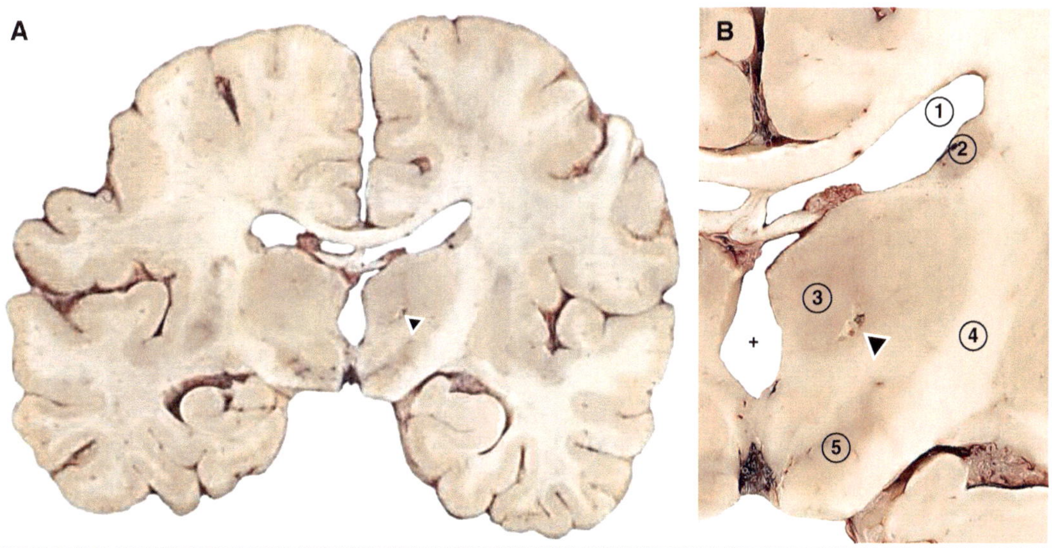

①	Lateral ventricle	④	Internal capsule
②	Caudate ncl.	⑤	Substantia nigra
③	Thalamus	+	3rd ventricle

Fig. 16.11 Coronal view of a hemisphere of a PD patient upon unilateral thalamotomy (**A**). Detail (**B**) shows the surgical defect (arrowhead, see also survey in (**A**)). Note the loss of pigmentation in the substantia nigra ⑤ (Compare with Figs. 16.3, 16.4, 16.5, and 16.12). LabPON Twente

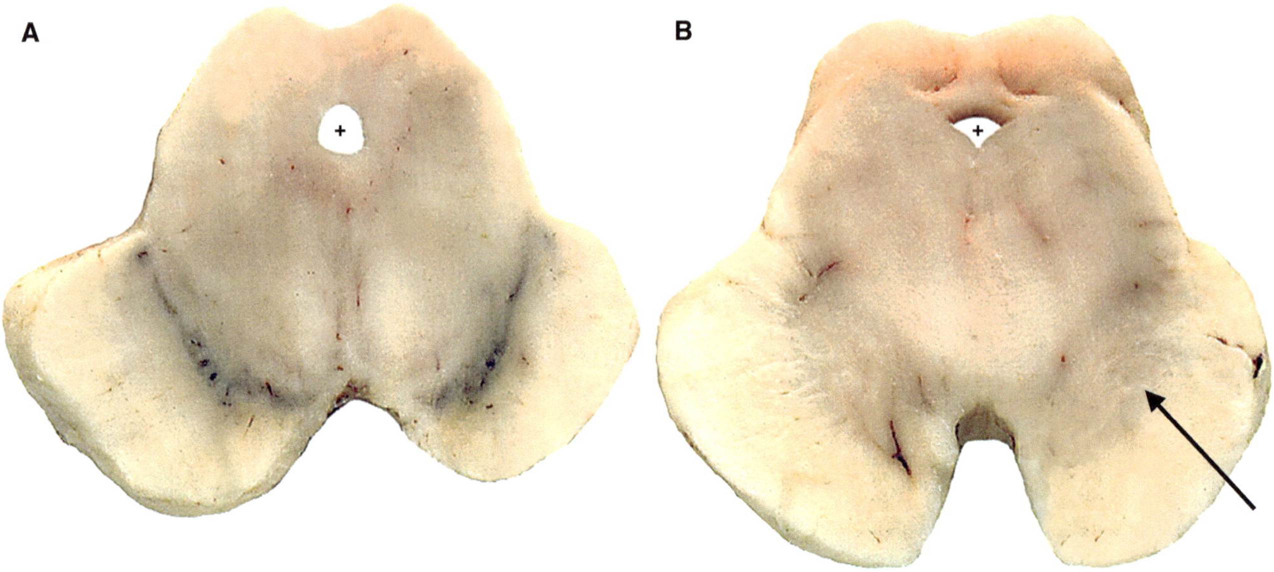

Fig. 16.12 Comparison between control (**A**) and depigmentation of substantia nigra in a PD patient (**B**). Arrow points to the SN. + = mesencephalic aqueduct. LabPON Twente

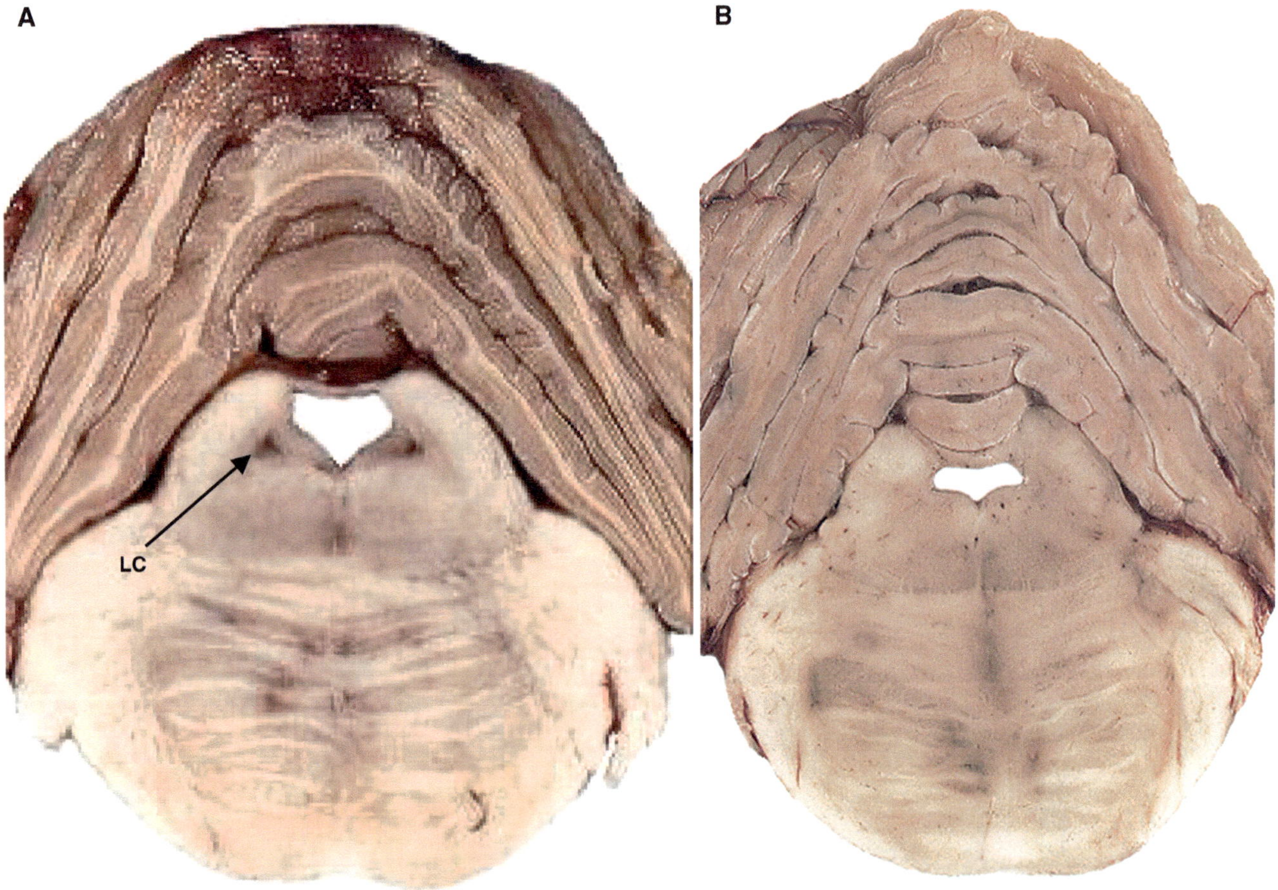

Fig. 16.13 Depigmentation of the locus caeruleus (LC) in the rostral pons in a PD patient (**B**) compared to a control case (arrow) in (**A**) (see also Sect. 13.2.2.1). LabPON Twente

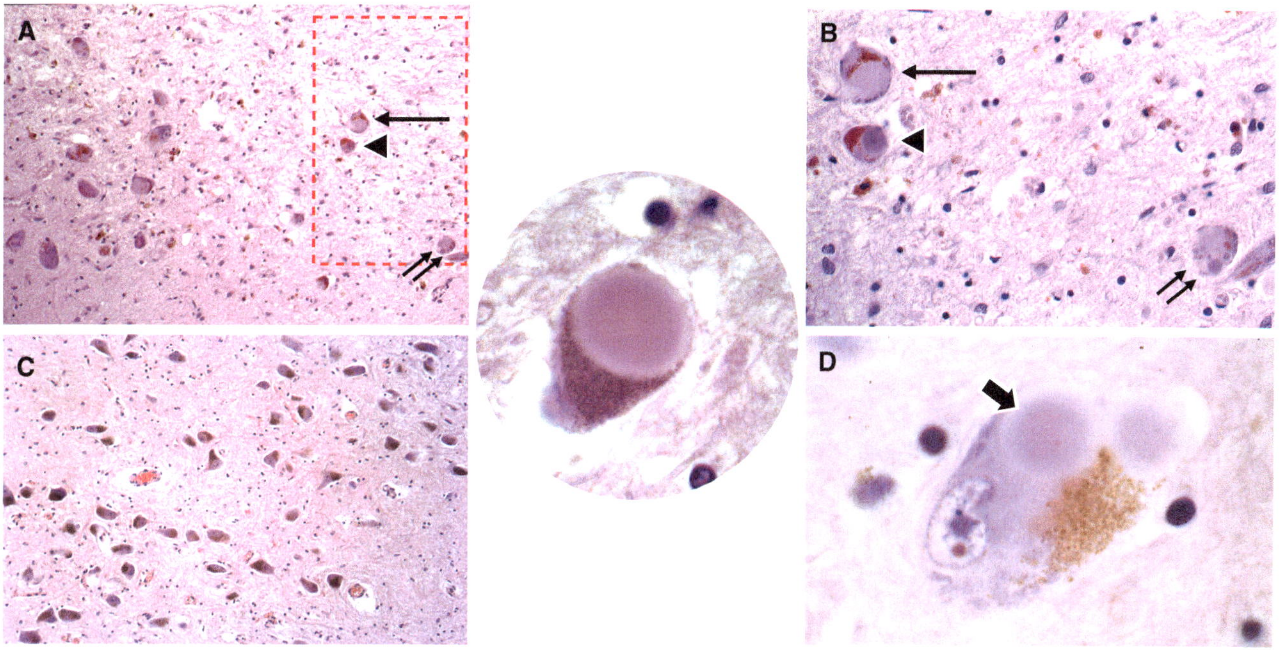

Fig. 16.14 PD. Loss of neuromelanin-containing cells in the SN, pigment incontinence in a PD patient compared to a control (**C**). In (**B**)—compare with the approximately corresponding red quadrangle in (**A**)—a pale body (arrow), a single Lewy body (arrowhead) and a single neuron containing several Lewy bodies (double arrows) are seen. In the center and in (**D**) LBs with a hyaline eosinophilic core and a typical halo (arrow) are shown. H & E stain. LabPON Twente

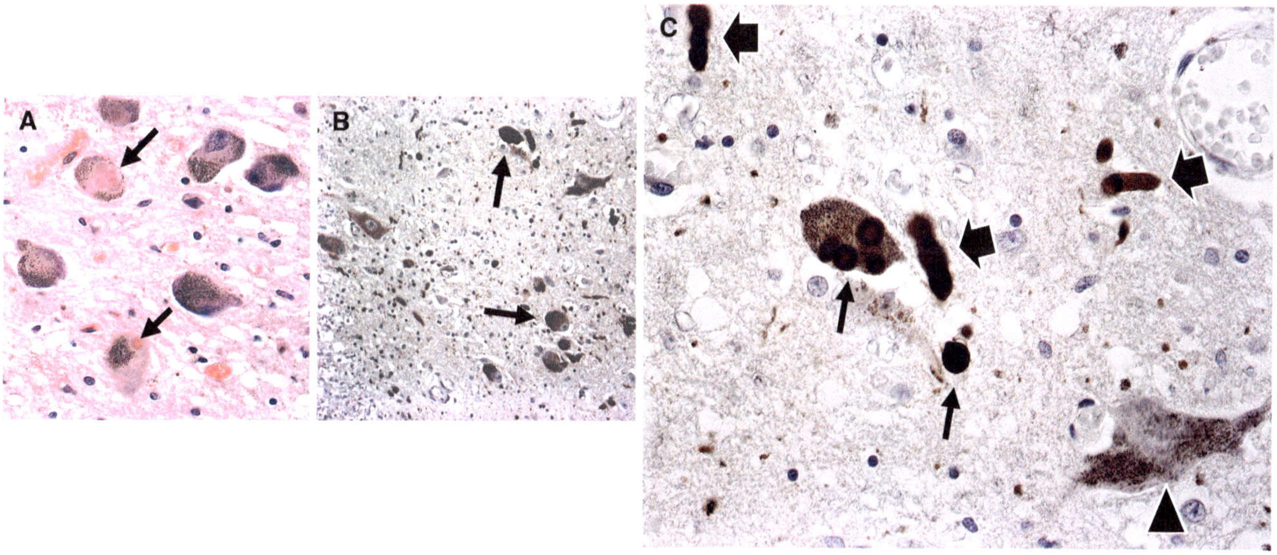

Fig. 16.15 Locus caeruleus in a PD patient. (**A**) Neuromelanin-pigmented neurons containing one or more LBs (H & E stain, arrows). (**B**) Loss of neurons containing LBs (arrows) and LNs in an α-synuclein-immunostain. In (**C**) (detail) LBs (arrows) and locally "varicose" appearance of the LNs (thick arrows) and neuromelanin-containing neuron (arrow head). LabPON Twente

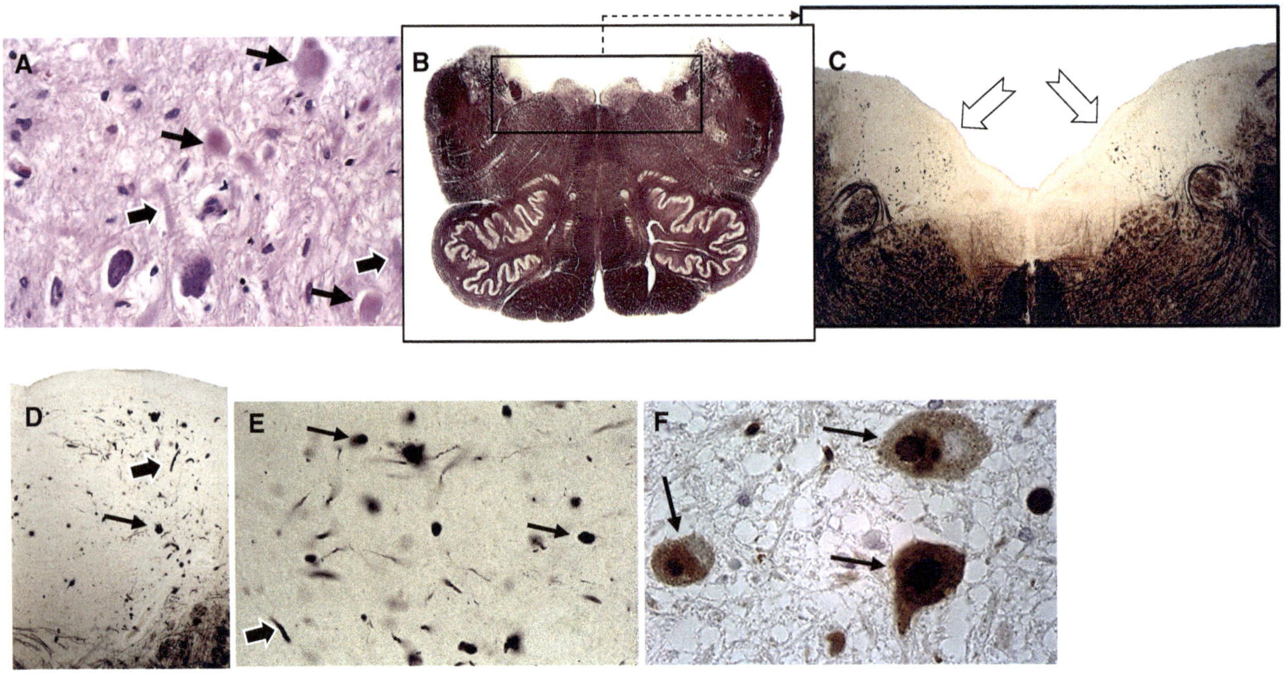

Fig. 16.16 PD. Neuronal loss in the region of the posterior nucleus of the vagus nerve. (**A**) LBs (small arrows) and LNs (thick short arrows) (H & E stain). (**B**) Campbell Switzer stain (PEG 100 µm) of an overview of the normal medulla oblongata and its detail (**C**). Open arrows point to the position of the posterior nucleus of the vagus nerve. (**D–F**) details of the region of the posterior nucleus of the vagus nerve including LBs (arrows) and LNs (thick short arrows). In (**F**) an α-synuclein immunostain shows LBs (arrows) and some small LNs (not labeled). LabPON Twente

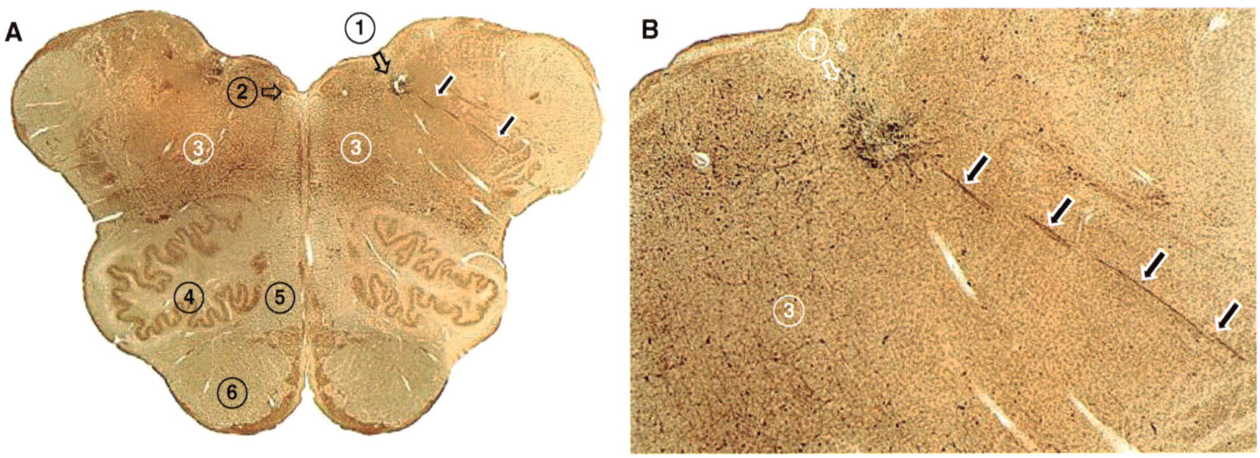

①	Region of posterior ncl. of CNX	④	Inferior olivary complex
②	Hypoglossal ncl.	⑤	Medial lemniscus
③	Intermediate reticular zone	⑥	Pyramidal (corticospinal) tract

Fig. 16.17 PD. Medulla oblongata at the level of the posterior nucleus of the vagus nerve (CNX) in an α-synuclein immunostain (**A**, PEG 100 µm). Immunoreactivity in the CNX area ① and in the intermediate reticular zone/nucleus ③. In (**B**) Lewy neurites in axons, following the course of the vagus nerve (CNX, arrows). LabPON Twente

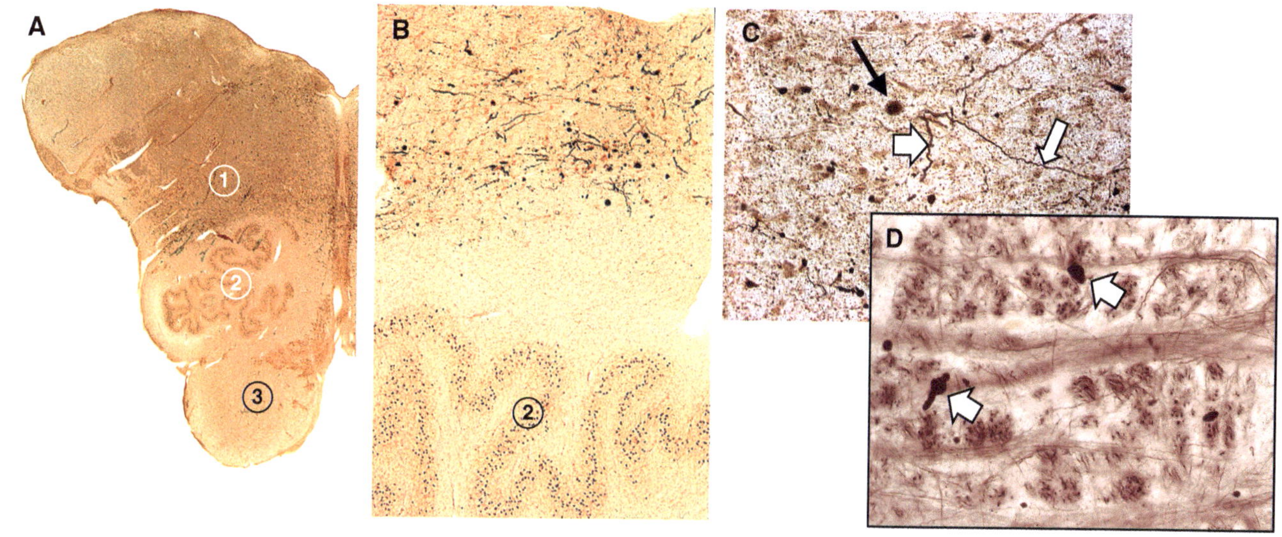

①	Intermediate reticular zone	③	Pyramidal (corticospinal) tract
②	Inferior olivary complex		

Fig. 16.18 PD. Medulla oblongata. (**A**) Overview and (**B**) Detail of the intermediate reticular zone, Lewy bodies and Lewy neurites, α-synuclein immunostain, PEG 100 μm. In (**C**) and (**D**) pigment Nissl/α-synuclein stain showing LBs (black arrow), thread-like LNs (small white arrow) and thick irregular varicose Lewy neurites (thick white arrows). LabPON Twente

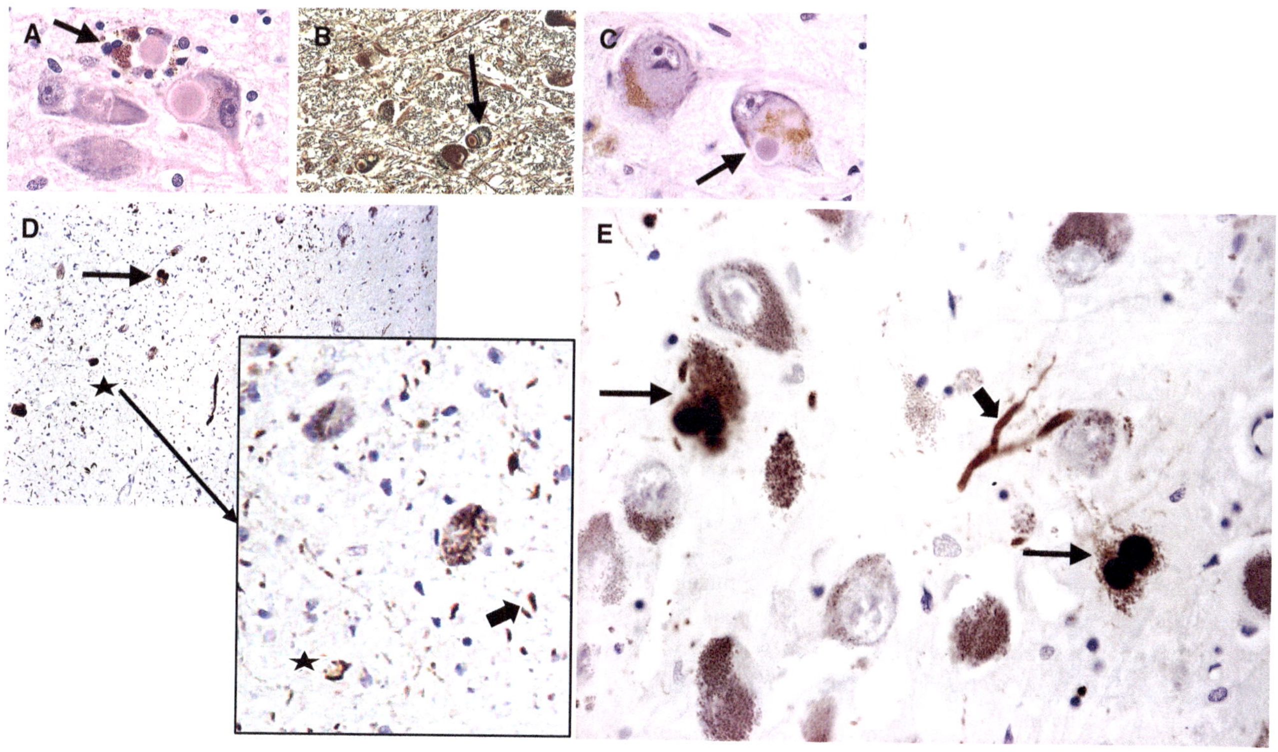

Fig. 16.19 PD. LBs (arrows) in the Substantia nigra (**A**) and (**C**) H & E stain, (**B**) Bielschowsky stain. (**A**) LB in a dying neuron with microglia reaction and pigment incontinence. In (**D** + inset) and (**E**) LBs, thread-like, varicose and branching Lewy neurites (thick short arrows) and a coiled body (star) in an α-synuclein immunostain. LabPON Twente

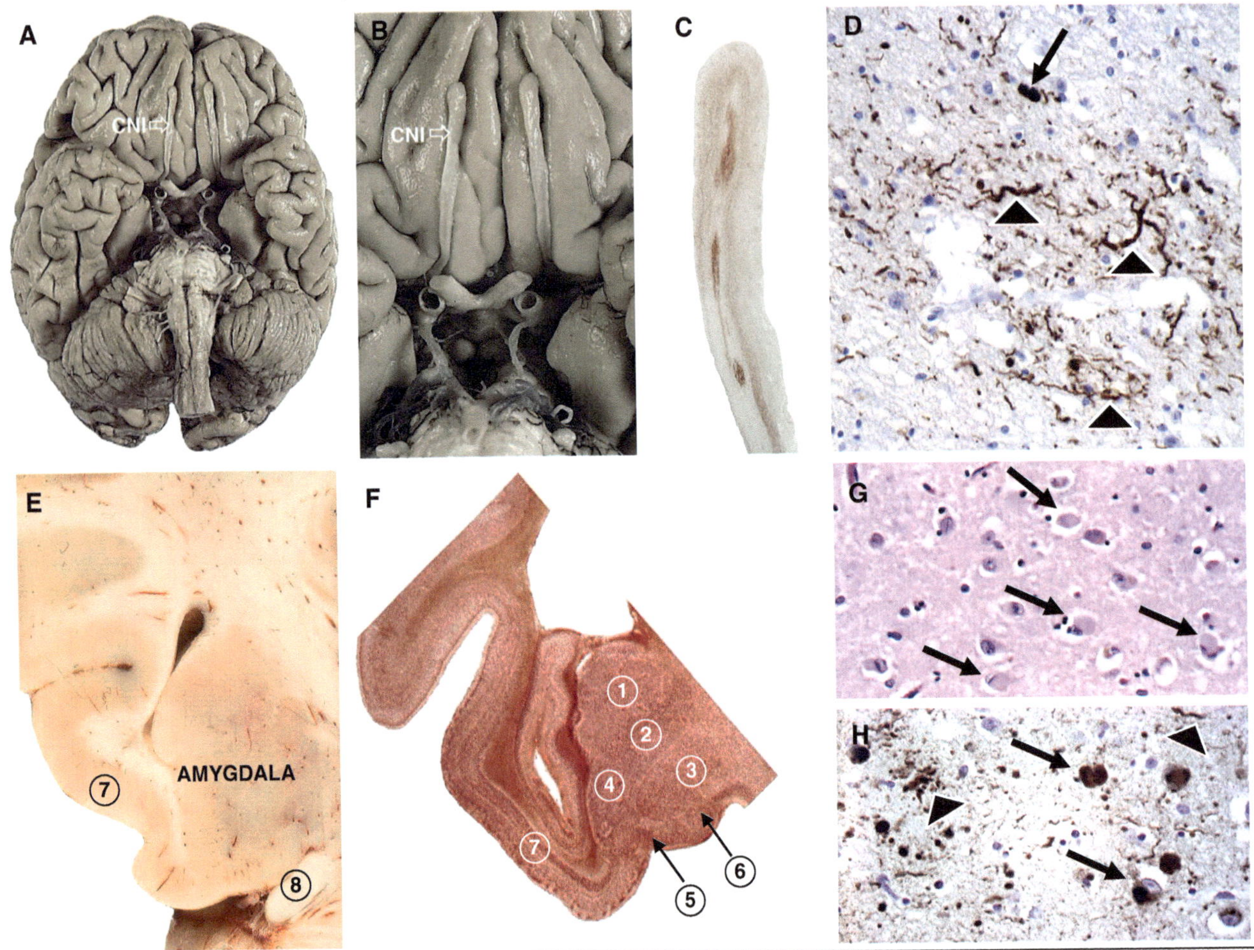

①	Lateral amygdaloid ncl.	⑤	Accessory cortex
②	Basal amygdaloid ncl.	⑥	Periamygdaloid cortex
③	Accessory basal ncl.	⑦	Entorhinal cortex
④	Transitory zone	⑧	Optic tract CNII

Fig. 16.20 PD. Ventral view (**A**, **B**) of the brain showing the olfactory tract and bulb (CNI). (**C**) The olfactory tract contains islands of the anterior olfactory nucleus. A dense network of LBs (arrow) and LNs (arrowheads) is seen in the anterior olfactory nucleus. α-synuclein immunostain (**C** and **D**). (**E**) Anteromedial part of the temporal lobe with the amygdala. In (**F**) different nuclei of the amygdala in a pigment Nissl stain are displayed. In (**G**) and (**H**) LBs (arrows) and LNs (arrowheads) in the accessory cortical nucleus (H & E stain and α-synuclein immunostain). LabPON Twente

corpora amylacea and neurofibrillary tangles. A recent study showed that the core of LBs consists of crowded organelles and lipid membranes (Shahmoradian et al. 2019). LNs consist of intraneuritic proteinaceous material and are not detected by conventional H & E staining but are visualized by staining for ubiquitin and α-synuclein.

Pale bodies (see Fig. 16.14B) are different from LBs but are also seen in neurons of the substantia nigra and locus caeruleus (see Sect. 13.2.2.1). These structures have been proposed as precursors of LBs and have a similar immunocytochemical profile. Because of their close association with true LBs, the presence of pale bodies alone is a reason to look for LBs. In PD, Lewy inclusions can also be found in astroglial cells in the prosencephalic regions in advanced cases (Braak et al. 2007b).

Outside the CNS, LBs and LNs may be seen in sympathetic and parasympathetic neurons, as well as in the enteric nervous system (myenteric and submucosal plexuses) (Braak et al. 2006; van de Berg et al. 2007) and cardiac sympathetic nerves (Iwanaga et al. 1999) and sympathetic trunk (van de

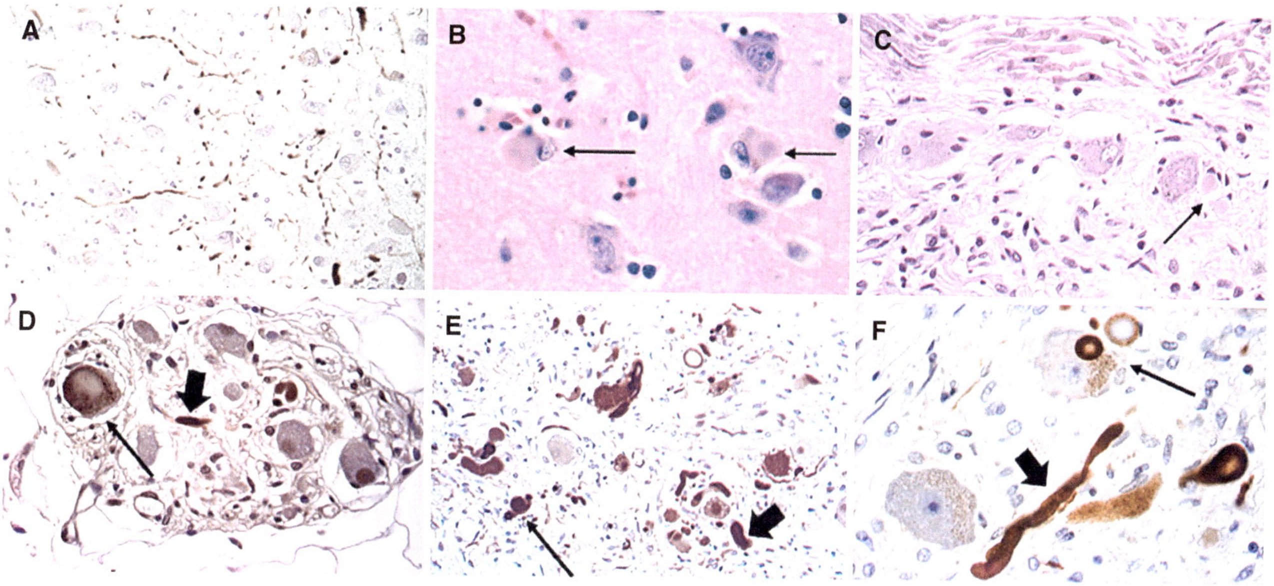

Fig. 16.21 PD. (**A**) CA2 sector of the hippocampal Ammon's horn showing the typical long and varicose LNs in an α-synuclein immunostain. In (**B**) cortical LBs in the cingulate gyrus (arrows) in a H & E stain, homogeneous eosinophilic structures, pushing the nucleus to one side of the cell. Cortical LBs usually lack a halo but in this picture an indication of halo formation is seen. (**C**) LBs (arrow) in the enteric nervous system: myenteric (Auerbach) plexus of the esophagus (H & E stain). (**D**) LBs (arrows) and LNs (thick arrows) in the postganglionic sympathetic neurons in the myocard, in the autonomic ganglia of the sympathethic trunk (**E**, **F**) in an α-synuclein immunostain. LabPON Twente

Berg et al. 2007; Kalia and Lang 2015) (Fig. 16.21C–F). It has been suggested that the onset of the disease is to be localized to peripheral organs, particularly the gastrointestinal tract (GIT) and the olfactory bulb soon before neuropathological changes occur in the CNS (Harsanyiova et al. 2020). It is not yet possible to clearly determine the primary inititation site of the onset of the pathology within the GIT and where or to which organs the pathology subsequently spreads (Harsanyiova et al. 2020; Cersosimo 2015).

The distribution pattern of α-synuclein-positive LBs and LNs in PD is not random. A recent staging procedure for the inclusion body pathology associated with PD proposes that, in the brain, the pathological process (formation of proteinaceous intraneuronal LBs and LNs) begins at two sites and continues in a topographically predictable sequence in six stages, during which components of the olfactory, autonomic, limbic, and somatomotor systems become progressively involved (Braak et al. 2003). In the first two stages, α-synuclein pathology is confined to the posterior nucleus of the vagal nerve (Sect. 7.3.2.1), the olfactory bulb (see Fig. 16.20), the caudal raphe nuclei (see Sect. 3.2.1.3), the gigantocellular reticular nucleus (see Sect. 4.3.4.2), and the locus caeruleus (see Fig. 16.15). As the disease progresses, other areas develop pathological changes, such as the substantia nigra and the forebrain (stages 3–4) (Figs. 16.14 and 16.21B). Melanin-laden nerve cells of the posterior nucleus of the vagus nerve begin to develop LBs. In the end stages 5–6, the pathological process reaches the sensory association cortex, prefrontal cortex and in the last stage, the entire neocortex. Thread-like LNs are typical elements at the level of the posterior vagal complex (Fig. 16.17). The origin of these LNs in this area, their size, topographical orientation, and connection to proximal portions of the CNIX/CNX permit their identification as intra-axonal deposits (Braak et al. 2001, 2003) (Fig. 16.17).

The first two stages may represent the prodromal phase of the disease in which the classical motor symptoms of PD are not yet visible. At stages 3–4, the symptomatic phase of the disease starts. Interestingly, Jellinger (2004) has shown that approximately 30% of aged individual cases—with no clinical reference to neuropsychiatric disorders and/or extrapyramidal signs—show incidental α-synuclein pathology in the substantia nigra pars compacta and other brainstem areas. As Jellinger (2004) states, however, it is unclear whether these patients represent pre- or subclinical forms of dementia with Lewy bodies (DLB) (see here below Sect. 16.6.1.6) with no or little clinical references.

In approximately 10–20% of aged unimpaired individuals, α-synuclein pathology has been observed in the olfactory bulb, posterior nucleus of the vagal nerve and spinal cord (Bloch et al. 2006; Klos et al. 2006).

Demented individuals with PD exhibit more severe cortical Alzheimer's disease-associated neurofibrillary pathology and amyloid angiopathy (see Fig. 13.15) than non-demented PD patients. β-Amyloid depositions (diffuse plaques) and amyloid angiopathy are prevalent in PD patients with con-

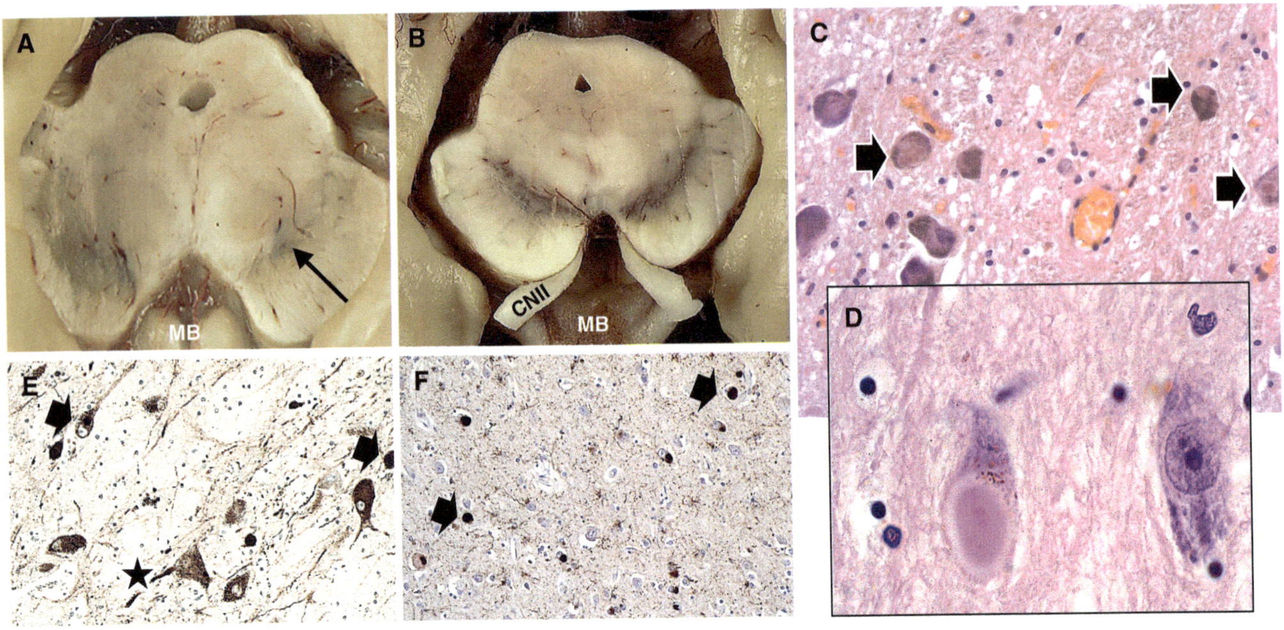

Fig. 16.22 (**A**) Depigmentation of the Substantia nigra (SN) (arrow) in a patient with DLB compared to a control case (**B**). MB = mammillary bodies (**C**) LBs in neuromelanin-containing neurons in the locus caeruleus, detail of LB in the SN (**D**) (H & E stain). LBs and LNs in SN (**E**) and in the cingulate gyrus (**F**) in an α-synuclein immunostain. LBs (thick arrows), LNs (star). LabPON Twente

comitant AD. This does not rule out the possibility that other causes may be responsible for the appearance of dementia in PD (co-occurrence of argyrophilic grains, cortical LBs, multiple infarctions). An increase in the number of α-synuclein-positive LBs and LNs in the limbic cortices or neocortex has been described in PD cases with dementia, but not in PD cases without dementia (Braak et al. 2005). As dementia also occurs in PD patients with only few LBs and LNs in these brain regions, loss of cortical input might also contribute to the cognitive deficits. The coexistence of neurofibrillary tangles may exacerbate cognitive deficits in Parkinson's disease with dementia.

LBs and LNs are not specific for PD; they can also be found in DLB, NBIA (see Sect. 16.6.1.8), neuroaxonal dystrophy, and Ataxia telangiectasia.

16.6.1.6 Dementia with Lewy bodies (DLB)

Within the spectrum of Lewy body disorders there is a group of elderly demented people who have a consistent pattern of clinical features, such as fluctuation in cognitive function, visual hallucinations, REM sleep disorders, and spontaneous motor features of parkinsonism. Memory function is relatively spared (Mori 2017; Barrett and Armstrong 2017). Although parkinsonism is one of the three major core features for the diagnosis of dementia with Lewy bodies (DLB) the frequency of tremor, wearing off, and dyskinesia seem to be lower than those of PD (Mizuno 2017).

In patients with DLB as well as in PD, Lewy bodies (LBs) and Lewy neurites (LN) are observed in not only the central nervous system (Fig. 16.22) but also the peripheral autonomic nervous system (Orimo 2017) (see Fig. 16.21). Other supportive features including repeated falls, syncope, and transient loss of consciousness can also be partly attributed to the presence of autonomic dysfunction (Orimo 2017).

These findings may highlight a problem for pathologists, namely that LBs and LNs, which are essential for the diagnosis of Parkinson's disease (PD), are also the morphological hallmark of Dementia with Lewy Bodies (DLB) meaning that the two diseases cannot be distinguished solely on the grounds of pathological changes. Given the overlap between the morphological changes of DLB and PD, some investigators consider PD, PD with dementia (PDD), and DLB to represent a phenotypic spectrum of the same disease process (Lewy body disease). Thus, it seems likely that it is not possible to reliably distinguish the neuropathological features of PD and DLB patients in the advanced clinical stages by means of histological and immunocytochemical postmortem evaluation (Weinstein et al. 2000; van de Berg et al. 2007; Seidel et al. 2015).

It was originally thought that the presence of cortical LBs formed a separate entity, known as diffuse Lewy body disease (Kosaka et al. 1984). Numerous terms have been used in the past to describe the same or similar entities, senile dementia of Lewy body type (Perry et al. 1990), Lewy body dementia (Gibb et al. 1987), dementia with Lewy bodies (Bergeron et al. 1996), and Lewy body variant of Alzheimer's disease (Hansen et al. 1990). These terms describe in fact similar groups of patients with a degree of neuropathological and clinical variation probably attributable to referral-based sampling biases. DLB was eventually agreed as a term to include all of these within one set of operationalized consensus criteria (McKeith et al. 1999). The Consortium on Dementia with Lewy bodies published consensus guidelines for not only clinical but also pathological diagnosis, recommending use of dementia with Lewy bodies as a generic term, because it acknowledges the presence of LBs without specifying their relative importance in symptom function.

The interpretation of the neuropathologic data is made difficult by the fact that patients can have AD-type pathology and or vascular pathology (McKeith et al. 1999). Since then, many reports on DLB have been published. DLB has been reported to be the second most frequent dementia following AD (Aarsland et al. 2008; Mc Keith et al. 2004; Chin et al. 2019; Outeiro et al. 2019). In the latest revision of the DSM5 (Diagnostic and Statistical Manual of Mental Disorders, 5th ed), the neurocognitive disorder with Lewy bodies (NCDLB) was therefore included (McKeith 2017).

In vitro and in vivo experimental models demonstrate that amyloid-like alpha-synuclein fibrils show prion-like properties and are able to convert normal alpha-synuclein to an abnormal form. It seems that this prion-like propagation of alpha-synuclein through neuronal networks is the key mechanism of formation of alpha-synuclein pathology (Hasegawa 2017).

Many patients with Lewy body disease (DLB, PD) have abnormalities similar to those seen in Alzheimer's disease (AD) (see Sect. 13.2.2.1.4), such as amyloid deposits (diffuse plaque formation, amyloid angiopathy) and neurofibrillary changes, although the latter ones are seen in only a minority of the cases.

It also should be remembered that vascular pathology is found not only in patients with AD but also in at least 30% of patients with Lewy body disease. Spongiform change is a feature of some DLB cases and occurs mainly in temporal cortex.

Diagnosing DLB is, however, still challenging as symptoms are heterogeneous and not specific to the disease (Vergouw et al. 2020; Chin et al. 2019). DLB and PDD differ from each other, and from AD, in a cognitive domain-specific manner (Smirnov et al. 2020). DLB and PDD declined more rapidly than AD in the visuospatial domain (Smirnov et al. 2020). Early identification of Lewy body disease could be of help in identifying individuals at risk of progression to dementia. Criteria might be made at a stage when an individual has at least one clinical feature suggestive of prodromal DLB, for example, cognitive or neuropsychiatric features (McKeith 2017). Although structural and functional neuroimaging can be of help in the diagnosis of DLB, there is limited evidence regarding early diagnosis (Kasanuki and Iseki 2017).

The interplay between cognitive, psychiatric, motor, and autonomic symptoms of DLB makes the management of these symptoms complex and challenging. The pharmacological management of DLB is usually polypharmacy and is complicated by the risk of adverse reactions to medication (Ikeda 2017; Boot 2015). All treatments for DLB are still symptomatic for the time being (Postuma and Walker 2017). Mitochondria-related proteins have been described in LBs, and this could be of interest for future treatment strategies for PD and DLB (Kawamoto et al. 2020).

There are no characteristic or specific macroscopic abnormalities for DLB. Variable pallor of the SN is visible (Fig. 16.22). The locus caeruleus is usually depigmented. Brainstem or cortical LBs and LNs are the only features considered essential for a pathological diagnosis of DLB (Fig. 16.22) although AD pathology and spongiform changes may also be seen. Coiled bodies (CB) are smaller and more difficult to detect than LBs and LNs, surrounding the lucent nucleus and twisted in the small cytoplasmic rim of the affected oligodendrocytes. Distribution and extent of the CBs show no differences between PD and DLB patients (Seidel et al. 2015).

AD pathology is frequently present in DLB patients, usually in the form of diffuse amyloid-beta plaques or as neuritic plaques, with neocortical neurofibrillary tangles being much less common. Spongiform changes in the temporal and entorhinal cortex, as mentioned before, have also been described (Brown et al. 1998). The Consortium on Dementia with Lewy bodies published revised criteria in 2005 for the clinical and pathologic diagnosis of DLB incorporating new information about the core clinical features and suggesting improved methods to assess them (McKeith et al. 2005).

The revised consensus pathological guidelines were formulated to provide morphological guidelines to score severity and anatomical distribution pattern of LBs in five cortical regions (McKeith et al. 2005) but do not provide a protocol for the diagnosis of DLB. The authors proposed a new scheme for the pathologic assessment of LBs and LNs using α-synuclein immunohistochemistry and semi-quantitative grading of lesion density, with the pattern of regional involvement being more important than the total LB count. The Consortium on Dementia with Lewy bodies proposed a semi-quantitative grading of the severity of LB pathology (0 = none; 1 = mild; 2 = moderate; 3 = severe; 4 = very severe). Depending on the anatomical distribution pattern, three different categories can be assigned: 1. brainstem dominant; 2. limbic (or transitional); 3. diffuse neocortical type. While brainstem nuclei are affected in virtually every case of LB disease, the severity of brainstem pathology is highly variable. The consortium proposed diffuse neocortical DLB cases with low Braak stages most likely to be clinically diagnosed with DLB. This approach is based on studies demonstrating that the clinical diagnostic accuracy for DLB is higher in patients with a low burden of AD-type pathology. This proposal assumes that it is necessary to distinguish DLB from PD with dementia (PDD) or at least the different variants. It is likely, however, that DLB and PDD represent a pathological spectrum in which anatomical distribution of Lewy pathology overlaps. The pattern of LB distribution in DLB and PD is different from that of neurofibrillary tangles described for AD and is the same as the pattern of LB distribution described for PD (Braak et al. 1995; McKeith et al. 2005; Harding et al. 2000). Vascular pathology is not uncommon in DLB, and approximately 30% of the DLB cases show some vascular abnormalities (McKeith et al. 1999). The clinical significance of these vascular abnormalities in DLB is, however, unknown.

16.6.1.7 Argyrophilic Grain Disease (AgD) (Dementia with argyrophilic grains)

Argyrophilic Grains (ArG) were first described in 8 out of a series of 56 brains of patients suffering from adult onset dementia by Braak and Braak (1987). Later, the same authors described ArG as main histological feature in 125 cases in a study of 2661 non-selected brains and called it Argyrophilic grain disease (AgD). They also supported the view that AgD is not a rare disorder (Braak and Braak 1998). In later years, more studies have recognized AgD as a frequent neuropathological entity among degenerative tauopathies (Tolnay and Probst 2008; Lee et al. 2001). Based on further studies, it was concluded that ArG are not regularly present in brains from cognitively unimpaired subjects and therefore cannot be regarded as age-related change. On the other hand, the presence of ArG in human brain is not necessarily associated with cognitive decline (Tolnay and Probst 2008). AgD can coexist with various other degenerative diseases such as progressive supranuclear palsy (PSP) (see Sect. 4.1.1.3.3) and corticobasal degeneration (CBD) (see Sect. 16.6.1.9) which have distinct disease-specific glial inclusions (Ikeda et al. 2016; Yokota et al. 2018; Buée and Delacourte 1999). Argyrophilic inclusions were also described in cases of PD (see Sect. 16.6.1.5) (Hishikawa et al. 2001), DLB (see Sect. 16.6.1.6), and amyotrophic lateral sclerosis (Soma et al. 2012) (see Sect. 4.3.1.1.6). Mild neurofibrillary pathology (Braak stages I–III) is common in AgD. Tau-immunoreactive neurons with granular or diffuse cytoplasmic staining consistent with pretangles are often seen (Togo et al. 2002; Ikeda et al. 2000; Martinez-Lage and Munoz 1997; Ferrer et al. 2008). Tau and αB-crystallin-positive ballooned cells (swollen neurons containing the stress protein αB-crystallin) can be found in limbic areas, especially the amygdala (Tolnay et al. 2002; Ferrer et al. 2003). Furthermore, AgD may be a pathological base in some patients with mild cognitive impairment, late-onset psychosis, bipolar disorder, delusions and hallucinations, schizophrenia, and depression (Yokota et al. 2018; Asaoka et al. 2010; Nagao et al. 2014). AgD share the pathological tau doublet (64 and 69 kDa) and predominance of 4R tau isoforms with CBD and PSP (Tolnay et al. 2002) whereas AD is characterized by simultaneous presence of both 3R and 4R tau proteins. Macroscopically, AgD brains appear unchanged or show mild frontotemporal atrophy. In most cases, there is no obvious atrophy of the hippocampal formation and amygdala (Tolnay and Probst 2008). AgD cases clinically suggestive of frontotemporal dementia (FTD) (see Sect. 4.3.1.1.5) have been described with atrophy of the medial temporal lobe (Tsuchiya et al. 2001; Maurage et al. 2003; Ishihara et al. 2005).

The histopathological hallmark of AgD are the ArGs, small, oval to spindle-shaped structures easily detectable by conventional silver stain (e.g., Gallyas stain). ArG predominate in the limbic system, mainly the CA1 region, subiculum, entorhinal, and transentorhinal cortices (Fig. 16.23), furthermore in amygdala and hypothalamus. ArG have been described in midbrain and pons (Ishihara et al. 2005; Ikeda et al. 2016; Yokota et al. 2018; Ferrer et al. 2008; Asaoka et al. 2010). The lower brainstem occasionally shows a few ArG in the periaqueductal gray matter (see Sect. 13.2.7.1), anterior raphe nuclei (see Sect. 3.2.1.3), dorsal vagus area (see Sect. 7.3.2.1), and locus caeruleus (Fig. 16.24) (see Sect. 13.2.2.1) (Tolnay and Probst 2008). ArG can be easily visualized by markers for abnormally phosphorylated tau protein such as antibody AT8. Like in other 4R tauopathies, such as PSP (see Sect. 4.1.1.3.3) and CBD, coiled bodies (CBs) are a consistent finding, largely restricted to white matter close to the cortical areas especially in the limbic cortex. Ultrastructurally, ArG differ from neurofibrillary lesions consisting of aggregates of straight filaments (9–18 nm) and bundles of 25 nm smooth tubules (Tolnay and Probst 2008). Neuronal cytoplasmic tau-positive, Gallyas-negative inclusions, the so-called pretangles (compare Sect. 13.2.2.1.4.5), are distributed similarly to ArG. No familial form of AgD is known so far. Recent studies describe two MAPT mutations causing pathological features consistent with AgD (Kovacs et al. 2008; Ronnback et al. 2014).

16.6.1.8 Neurodegeneration with brain iron accumulation (NBIA)

Neurodegeneration with brain iron accumulation (NBIA), a subform formerly referred to as Hallervorden-Spatz syndrome (Box 16.4), is a heterogeneous group of disorders with excessive CNS iron deposition, particularly in globus pallidus and substantia nigra, recognizable as decreased routine T2w MRI signal with sometimes a central small hyperintense area (the "eye of the tiger" sign) (Dusek et al. 2015). The NBIA group comprises distinct genetic metabolic diseases in some of which iron accumulates only in a subgroup of affected subjects (Di Meo and Tiranti 2018).

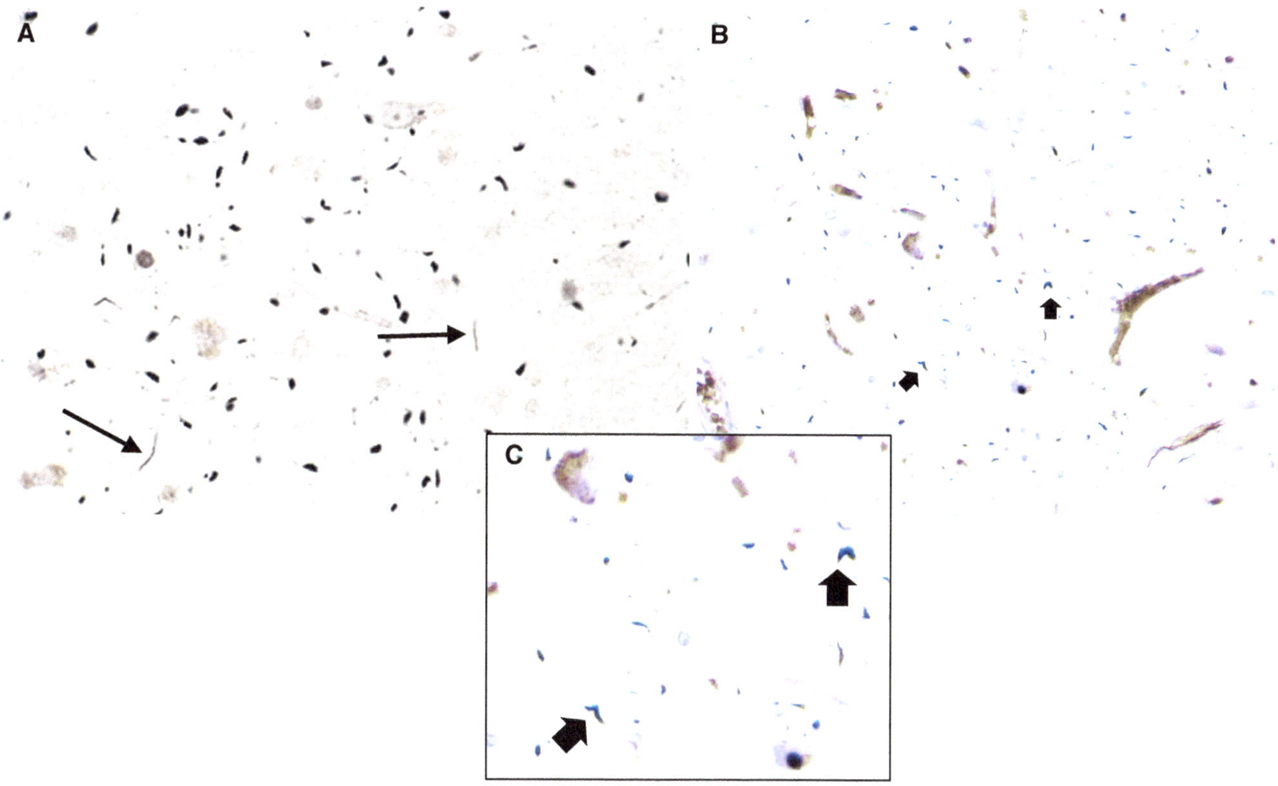

Fig. 16.23 Argyrophilic Grain Disease (AgD). (**A**) CA1 region in a Gallyas silver stain. Argyrophilic grains (spindle- and comma-shaped), neuropil threads (long arrows), and coiled bodies (short thick arrows, **B**, and higher magnification, **C**). Brown colored erytrocytes, in the lumina of vessels. LabPON Twente

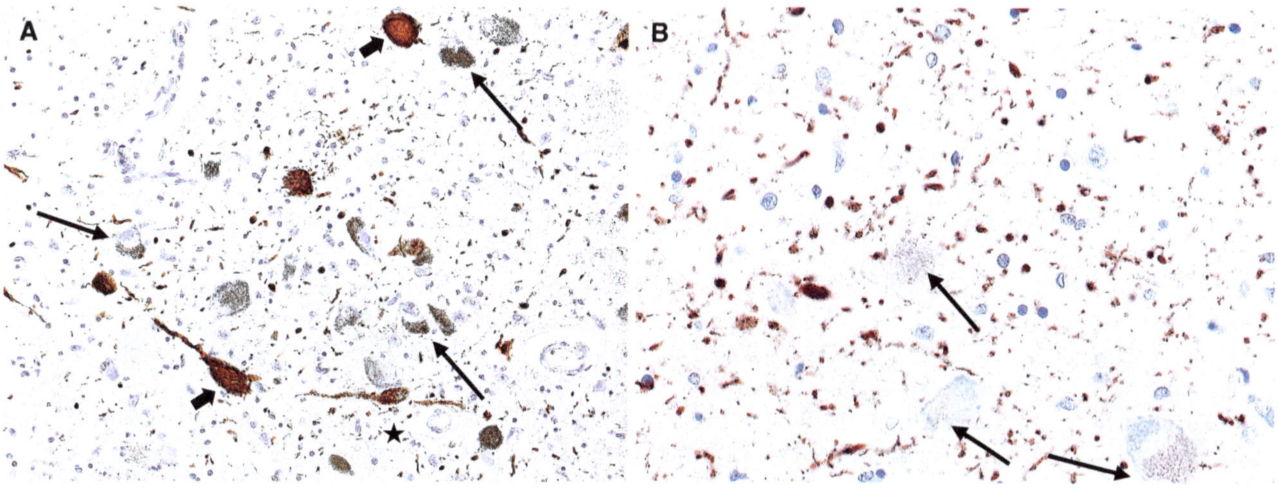

Fig. 16.24 Argyrophilic Grain Disease (AgD). Locus caeruleus (LC) in an AT8-immunostain. Many argyrophilic grains, a few neurofibrillary tangles (short thick arrows) and a pretangle (star) are present (**A**). Preexisting neuromelanin-containing neurons of the LC are indicated by long small arrows (**B**). Courtesy and copyright: Prof. Y. Saito

Box 16.4 Hugo Spatz (1888–1969)/Julius Hallervorden (1882–1965)

Hugo Spatz succeeded Oskar Vogt as head of the Kaiser-Wilhelm-Institut (KWI) für Hirnforschung (Brain Research) in Berlin-Buch (Germany). He was responsible for brain autopsy of executed prisoners ("Begleitforschung" = accompanying research) performed by Julius Hallervorden. The latter also participated in the action T4 euthanasia program. Both are credited with the discovery of Hallervorden-Spatz syndrome (now referred to as Pantothenate kinase-associated neurodegeneration). After WWII, Hallervorden became President of the German Neuropathological Society, continuing his research as head of Max Planck Institute (MPI) in Giessen. He relocated to the MPI in Frankfurt/M. in 1958, working at the Edinger-Institute in Frankfurt/M. Franz Seitelberger (1916–2007) (see below), a former SS member, worked on brains obtained by Spatz and Hallervorden after the war in Hallervorden's institute. Due to the infamous activities of Spatz and Hallervorden this eponym—and a number of other neurological ones (Kondziella 2009)—has to be used with demanded caution.

Pantothenate-kinase-associated neurodegeneration (PKAN), the most common form of NBIA (NBIA type 1) is caused by autosomal recessive mutations in the gene encoding pantothenate-kinase 2 (PANK2), which maps for chromosome 20p13. COASY protein-associated neurodegeneration (CoPAN) is an early onset rare autosomal recessive form of NBIA due to mutations in COASY, a gene located on human chromosome 17q21. PANK2 and COASY code for enzymes directly involved in biosynthesis of coenzyme A that, functioning as high-energy carrier of acetyl and acyl moieties, represents a central metabolic cofactor in energy production and fatty acid metabolism (Di Meo and Tiranti 2018). Mutations in PLA2G6 (chromosome 22q13.1) encoding for calcium-independent phospholipase A2 group VI has been associated to PLA2G6-associated neurodegeneration (PLAN), also called NBIA type 2 or infantile neuroaxonal dystrophy, previously Seitelberger's disease (Box 16.4). It is an autosomal recessive heterogeneous group of related neurodegenerative conditions comprising infantile neuroaxonal dystrophy associated neurodegeneration with brain iron accumulation. PLA2G6, FA2H (fatty acid hydroxylase 2) and C19orf12 genes code for proteins of lipid metabolism, membrane integrity, and mitochondrial function. Other forms of lipid metabolism-associated neurodegeneration with brain iron metabolism are SCP2 and leukoencephalopathy with dystonia and motor neuropathy. WDR-45 and ATP13A2 are implicated in autophagosome formation and lysosomal activity, respectively. SCP2 codes for a peroxisomal enzyme with lipid transfer activity (Di Meo and Tiranti 2018). Neuroferritinopathy (NFP) due to mutations in the ferritin light-polypeptide (FTL1) gene and aceruloplasminemia (ceruloplasmin gene mutations) affect the iron metabolic pathway (Dusek et al. 2015).

MRI findings suggestive of iron deposits also have been described in rare disorders, such as Woodhouse-Sakati and Kufor-Rakeb syndrome and in lysosomal disorders. Although iron deposits were not confirmed histopathologically, typical MRI findings lump these diseases together with NBIA.

PKAN is the most prevalent disorder from the NBIA group, accounting for 50% of patients. PKA manifests as a classic phenotype in 75% of patients. The classic phenotype may begin with nonspecific symptoms, later accompanied by postural instability, lower limb spasticity, and dystonia. Visual symptoms due to retinopathy are common (70%) (Dusek et al. 2015). Cognitive dysfunction is common, but highly variable (Kurian and Hayflick 2013). Chelating therapy is the best option for aceruloplasminaemia but the clinical experience with chelation therapy in other NBIA subforms has yielded mixed results so far (Suzuki et al. 2013).

The macroscopic appearance in cases of NBIA syndrome is a striking rust-brown pigmentation and atrophy of the globus pallidus and reticular part of the substantia nigra. In some cases, such as the lipid metabolism-associated neurodegeneration cases, cortical and/or cerebellar atrophy is present (Figs. 16.25 and 16.26).

Histology shows a diffuse loss of neurons and myelinated fibers, gliosis, and accumulation of iron-containing pigment in these regions, which also contain the so-called spheroids. Spheroids are thought to represent swollen axonal terminals. Spheroids can also be seen in the cerebral cortex, brainstem nuclei, and spinal cord. The largest number are in the most disrupted areas, and usually more abundant in the pallidum than the substantia nigra. Spheroids are strongly immunopositive for neurofilament protein. Spongiform change (prominent microvacuolation of the neuropil) together with an increase in number of reactive astrocytes can be seen (Fig. 16.27). Pigment is present in surviving neurons, within spheroids, and in astrocytes. The pigment contains lipofuscin (see Box 4.4), neuromelanin (see Box 13.5), and iron. Clusters of iron-laden macrophages are seen as well as extracellular iron pigment in a perivascular distribution and as free deposits (Fig. 16.28).

Lewy body pathology (see Box 16.2) has also been mentioned in NBIA cases but has mainly been described in the lipid metabolism-associated neurodegeneration cases in which widespread α-synuclein-positive Lewy pathology can be found (Dooling et al. 1974; Olanow 1994; Paisan-Ruiz et al. 2012; Li et al. 2013) (Fig. 16.29).

In some cases of NBIA, type2 tau pathology (neurofibrillary tangles and neuropil threads) can also be found. They are composed of both 3-repeat and 4-repeat tau isoforms (LI et al. 2013) (Fig. 16.30).

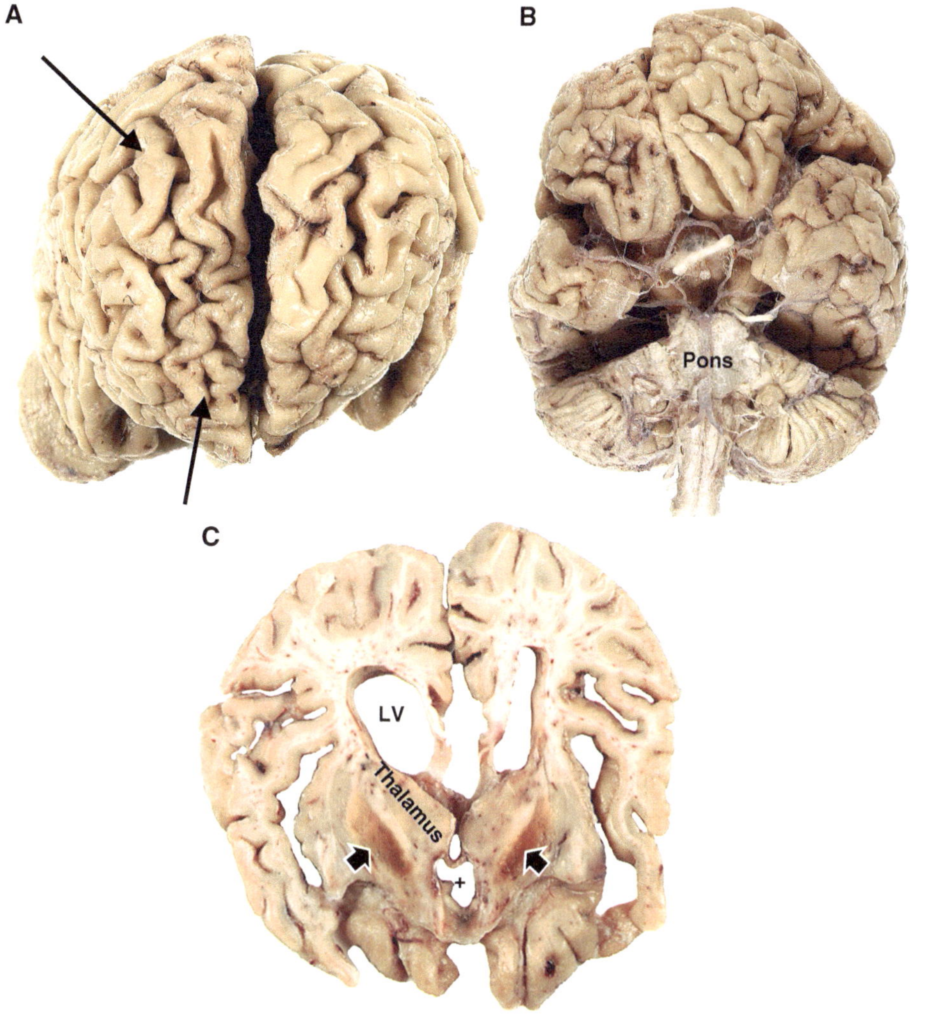

Fig. 16.25 Brain from a patient with NBIA syndrome, probably lipid metabolism-associated neurodegeneration with brain iron accumulation, type C19orf12 mutation. (**A**, **B**) Frontal and frontoventral views of the brain showing cerebral and pontocerebellar atrophy (Arrows point to severe atrophy, widened sulci, narrow gyri). (**C**) Coronal section (level of the thalamus) showing a typical rusty brown discoloration in the atrophic globus pallidus (thick arrows). Dilatation of the ventricular system. LV = lateral ventricle, + = third ventricle. LabPON Twente

The clinical and pathological features could represent a link between PLA2G6 and parkinsonian disorders (Paisan-Ruiz et al. 2012). A variable depletion of cerebral and cerebellar cortical neurons with astrocytosis can be present, especially in NBIA type 2 cases.

16.6.1.9 Corticobasal degeneration (CBD)

Corticobasal degeneration was first described by Rebeiz et al. (1968), called corticodentonigral degeneration with neuronal achromasia. CBD can present with various clinical phenotypes including Richardson's syndrome (RS). The term corticobasal syndrome is used to describe one of the characteristic clinical syndromes associated with CBD pathology and is the classic clinical presentation typically showing asymmetric parkinsonism with a variable combination of ideomotor apraxia, rigidity, myoclonus and dystonia often associated with the alien limb phenomenon (Bayram et al. 2020; Grijalvo-Perez and Litvan 2014; Kouri et al. 2011; Saranza et al. 2019). Speech apraxia is also a well-known symptom (Ruggeri et al. 2020).

The distribution of neuronal loss and tau pathologies determines the clinical presentation (Saranza et al. 2019). CBD typically presents in the sixth decade, men and women being equally affected. It is a relentless disease with poor prognosis, having a mean disease duration of 6.6 years (Saranza et al. 2019; Ling et al. 2010). Less than 50% of patients receive a CBD diagnosis and there is phenomenological overlap with several neurodegenerative disorders (Svenningsson 2019). Patients with parkinsonian features, especially without tremor, that are not responsive to levodopa, usually have one of these three major neurodegenerative disorders rather than PD: progressive supranuclear palsy (see Sect. 4.1.1.3.3), multiple system atrophy (see Sect. 3.3.2.3), or CBD.

Pathological criteria for CBD diagnosis have been formulated by a working group supported by the office of rare diseases of the NIH and subsequently validated by an independent group of neuropathologists (Dickson et al. 2002). Minimal pathologic features for CBD are cortical and striatal tau-positive neuronal and glial lesions.

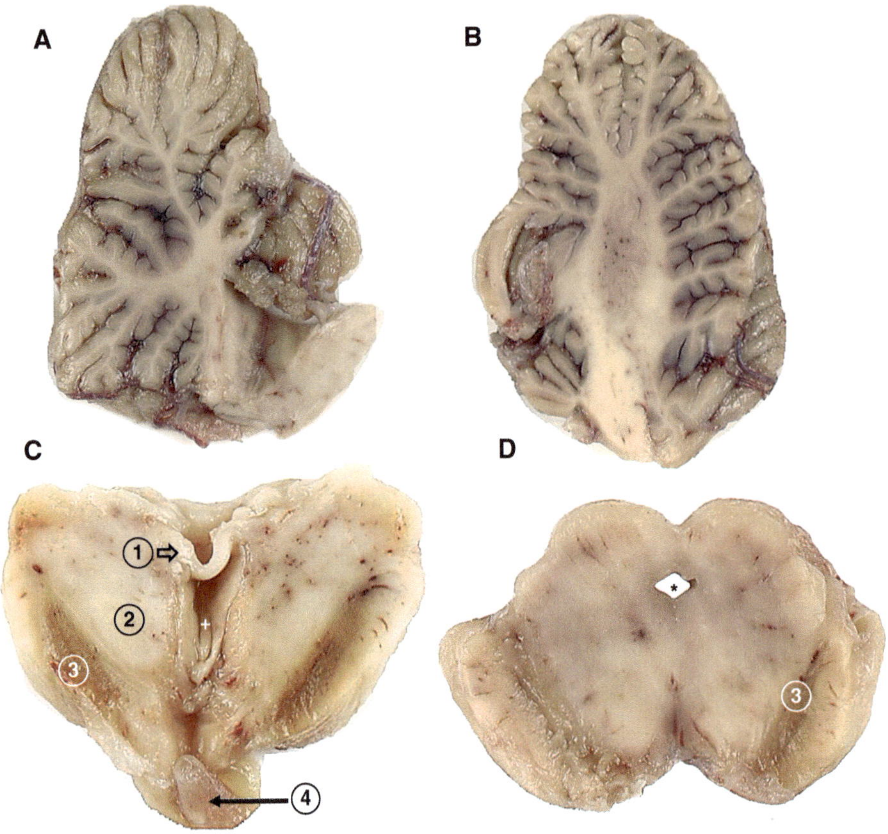

①	**Interthalamic adhesion**	④	**Pituitary stalk**
②	**Thalamus**	+	**3rd ventricle**
③	**Substantia nigra**	*	**Mesencephalic aqueduct**

Fig. 16.26 Cerebellar atrophy in a brain from a patient with NBIA syndrome (**A**, **B**) in sagittal section and (**C**) in transverse section at the thalamic level and the pituitary stalk. The latter points to the observer. (**D**) shows the rusty brown discoloration of the substantia nigra ③ in a horizontal section at the level of the mesencephalon. LabPON Twente

CBD is a 4R-tauopathy, a class of disorders in which the tau protein forms insoluble inclusions in the brain. The H1 haplotype of MAPT (the tau gene) is present in cases of CBD at a higher frequency than in controls (Zhang et al. 2020). There is no CSF biomarker that can differentiate CBD from PSP (see Sect. 4.1.1.3.3) and MSA (see Sect. 3.3.2.3) yet (Svenningsson 2019). An effective treatment is not yet available and therapy is symptomatic, based on similarity with other diseases (Caixeta et al. 2020).

Macroscopically, there is a typically asymmetrical cortical atrophy of the posterior frontal, parietal, and pre- and postcentral gyri contralateral to the limbs that are most severely affected in life. There is a relative sparing of the occipital lobes. A marked atrophy of the anterior corpus callosum is often seen (Kouri et al. 2011). The atrophy is associated with a variable dilatation of the ventricular system. The substantia nigra may show depigmentation, especially in cases with motor symptoms (Figs. 16.31 and 16.32).

A gray discoloration of the cerebellar dentate nucleus can be seen and the tegmentum is slightly atrophic and the aqueduct enlarged (Dickson 1999).

CBD is characterized by neuronal loss, astrocytosis, and microglia proliferation. In addition, it is associated with four main histopathological features: swollen cortical neurons containing neurofilament protein, 4-repeat tau-containing neuropil threads, 4-repeat tau-containing astrocytes, and filamentous inclusions composed of 4-repeat tau in basal neurons and rarely in cortical neurons.

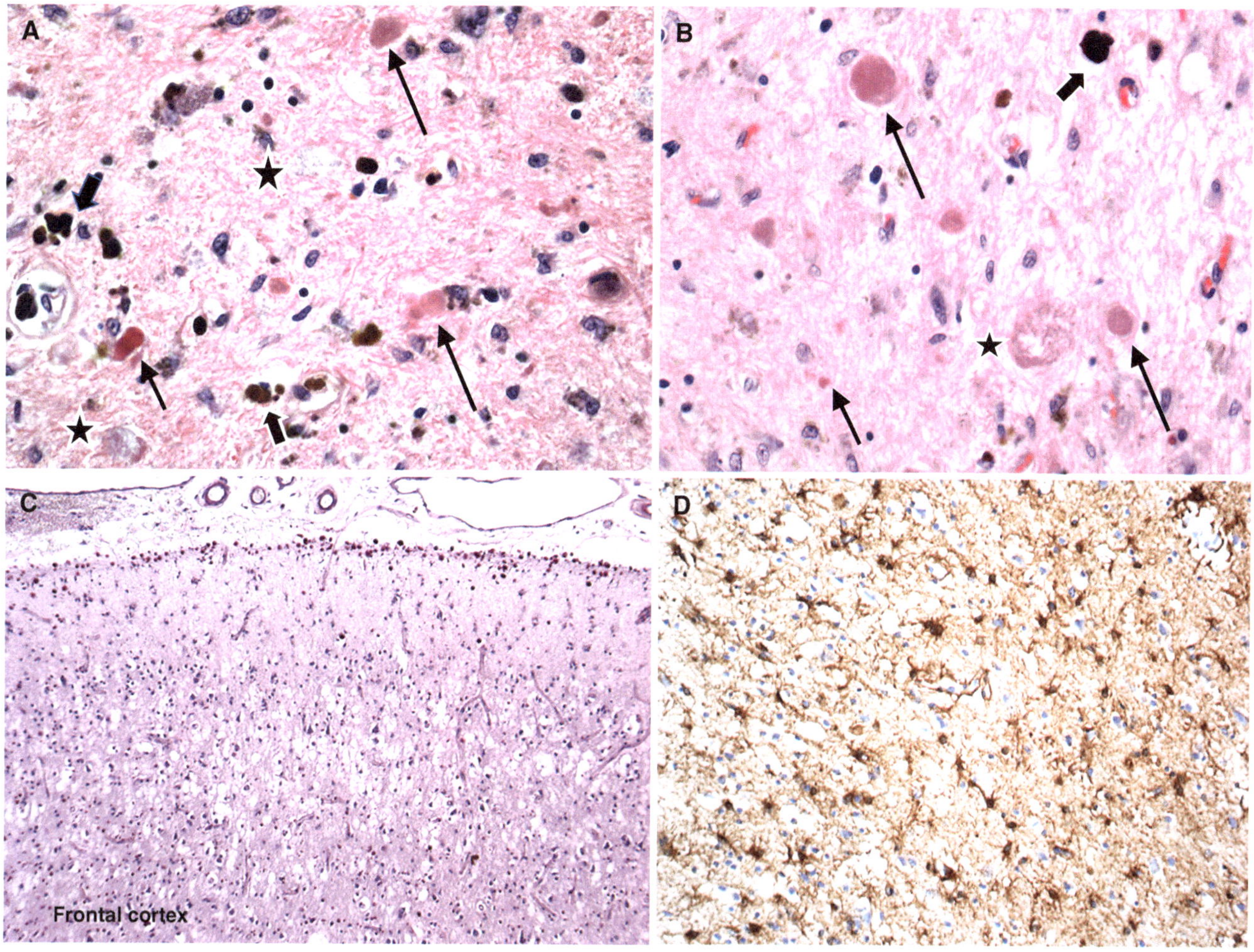

Fig. 16.27 NBIA. Neuroaxonal dystrophy in the globus pallidus (**A**, **B**, H & E stain). Axonal swellings (eosinophilic axonal spheroids) varying in size from small to large (arrows). Spheroids are present in all NBIA types. Vacuoles are seen in close vicinity of spheroids, possibly due to resorption of swellings (stars). Deposits of iron pigment (brownish, small thick arrows). (**C**, **D**) Neocortex with spongiform change (**C**, H & E stain) and astrogliosis (**D**, GFAP immunostain). LabPON Twente

A key pathological feature is the presence of swollen cortical neurons in affected areas (also termed achromatic or ballooned neurons) (Fig. 16.33). They are most frequent in cortical layers III, V, and VI and are apparent in H & E-stained sections as pale-stained large neurons, often with a surrounding artifactual area of vacuolation. These swollen neurons are immunoreactive for phosphorylated neurofilament as well as αB-crystallin, a stress protein. Tau (4-repeat)-positive neuronal perinuclear fibrillary inclusions are more frequently seen in small neurons. They are less well demarcated and less round than Pick bodies and, in contrast to Pick's disease (see Sect. 4.3.1.1.5), they are not found in the dentate gyrus.

Astrocytic plaques, as collections of abnormal tau in the distal processes of astrocytes, are a characteristic finding in CBD. These plaques can be found especially in precentral and premotor gyri but also in putamen, thalamus, and subthalamic nucleus (Komori 1999) (Fig. 16.33B).

Neuropil threads (dystrophic neurites) are numerous in CBD found throughout gray and white matter of the affected cortical areas and the deep gray matter (Fig. 16.33D). In contrast to AD, the density of neuropil threads can be as great in white matter as in cortical gray matter. Their number is larger than in PSP, but both disorders have the same tau isoforms with 4-repeat of the microtubule-binding region.

Oligodendroglial 4-repeat tau inclusions called coiled bodies (CBs) (Fig. 16.33C) and threads are (Fig. 16.33D) frequently found in degenerating areas like the frontoparietal cortex, deep gray and white matter, mesencephalon, and pontine base (Komori 1999). The tau-positive inclusions have an annular and fibrillar appearance and are distinct both morpho-

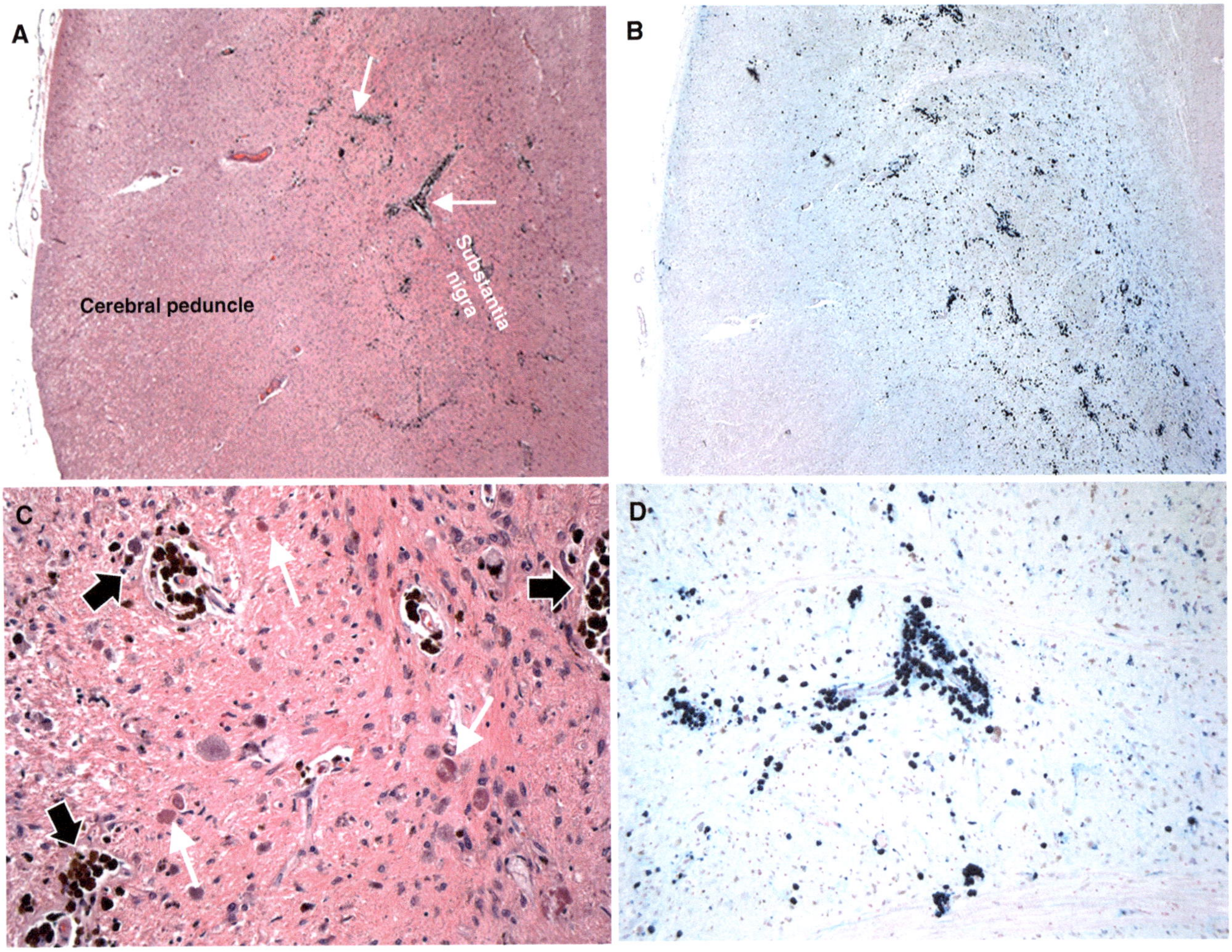

Fig. 16.28 NBIA. (**A**) Conspicuous accumulation of iron pigment in the substantia nigra, reticular part, partially extracellularly around blood vessels (white arrows), numerous spheroids (white arrows) and iron pigment (small thick black arrows) in a H & E stain (**C**, high power) and in a Perls' stain (blue) in (**B** and **D**, high power). LabPON Twente

logically and antigenically from the round or crescent-shaped, ubiquitinated, oligodendroglial inclusions of MSA (see Sect. 3.3.2.3). Thick, comma-like CBs are often observed in CBD. In nearly all patients, there is neuronal loss, pigment incontinence, and gliosis of the subcortical nuclei. There is also almost always cell loss and astrocytosis in the lateral portion of the substantia nigra. A characteristic feature is the presence of slightly basophilic filamentous inclusions in residual nigral neurons (originally termed corticobasal bodies), which can mimic globose tangles, characteristic of PSP (see Sect. 4.1.1.3.3) (Gibb et al. 1989) (Fig. 16.34).

Mild-to-moderate neuronal loss, gliosis, and/or neurofibrillary tangles can be found in other subcortical and brainstem structures such as thalamus, subthalamic nucleus, dentate nuclei, red nuclei, pons, reticular formation, and inferior olivary nuclei. The dentate nucleus shows eosinophilic degeneration like in PSP (see Sect. 4.1.1.3.3). However, in contrast with PSP, most changes are seen in the cortex and subcortical areas.

16.6.2 Red nucleus, parvocellular part

For detailed information on the parvocellular part of the red nucleus, see Sect. 15.3.4.

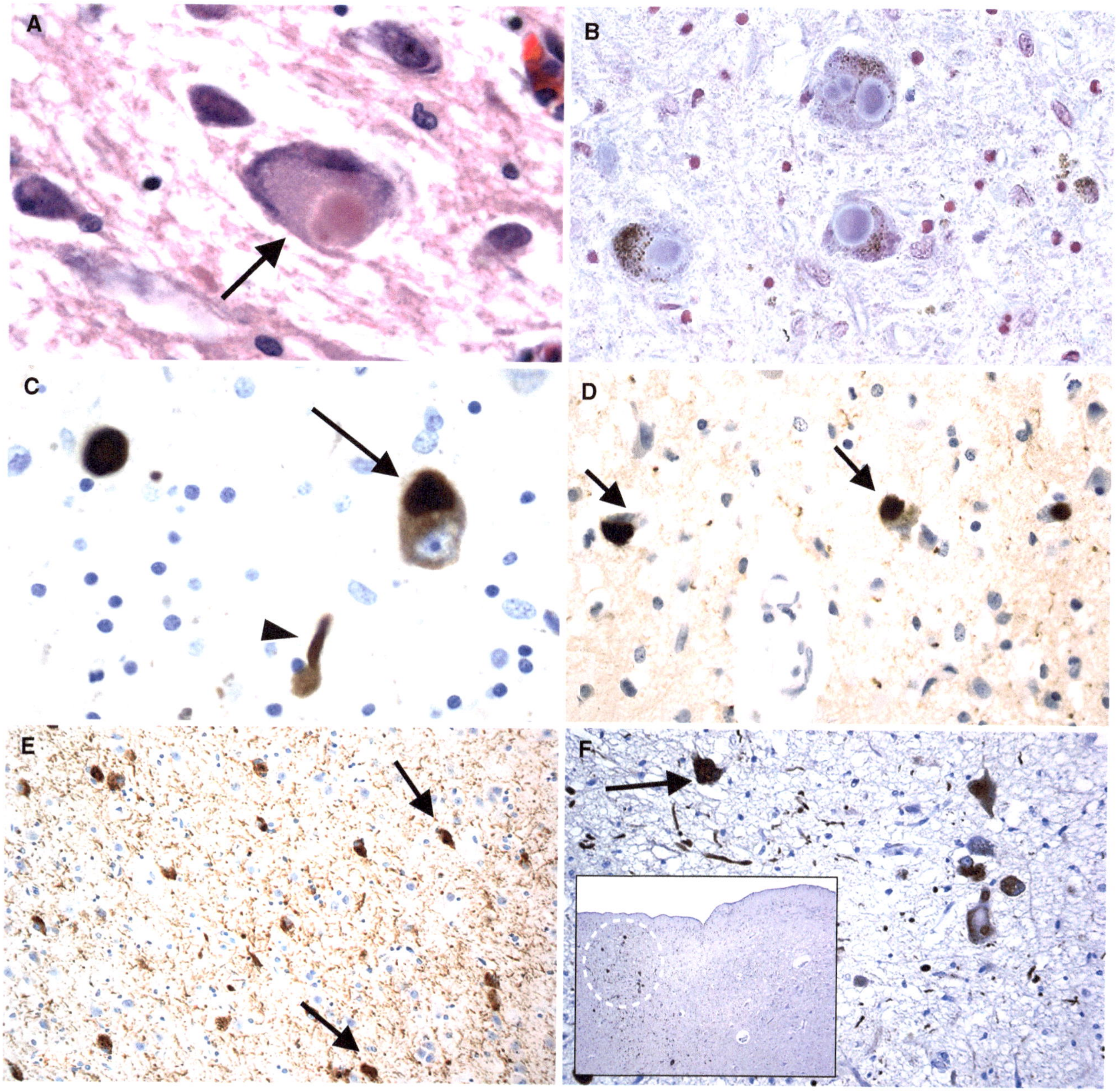

Fig. 16.29 In NBIA type 2 (infantile neuroaxonal dystrophy), widespread Lewy pathology can be found as in this case. (**A**–**C**) LBs (arrows) and LNs (arrow head) in the substantia nigra (**A**: H & E stain, **B**: Alcian Blue stain, **C**: α-synuclein immunostain). Lewy pathology in neo- and allocortex (**D**, **E**, α-synuclein immunostain) and in the medulla oblongata (**F**) with loss of neurons in the region of the posterior nucleus of the vagus nerve (white circle), a preferential location of LBs (arrow) and LNs. Inset: low power view of (**F**). LabPON Twente

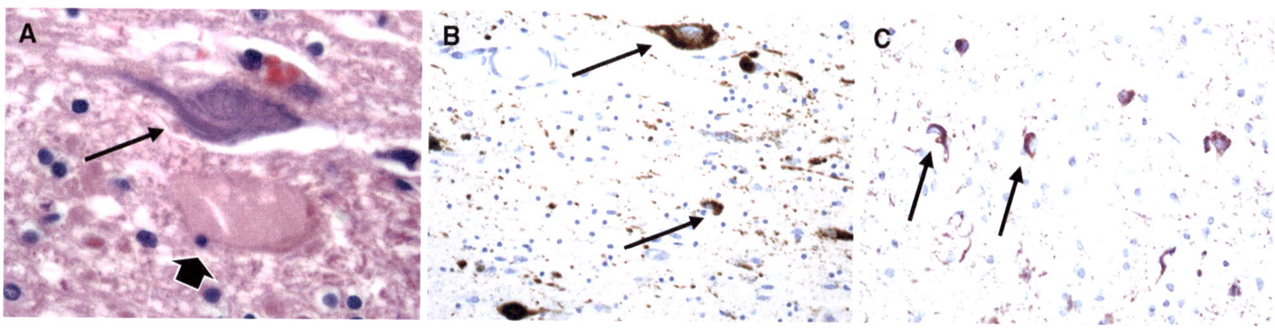

Fig. 16.30 Tau pathology in NBIA type 2. (**A**) Neurofibrillary tangle (NFT, arrow) and spheroid (thick arrow) in the substantia nigra (SN) (H & E stain). NFTs (arrows) and neuropil threads in the SN (**B**) and frontal cortex (**C**) (α-synuclein immunostain). LabPON Twente

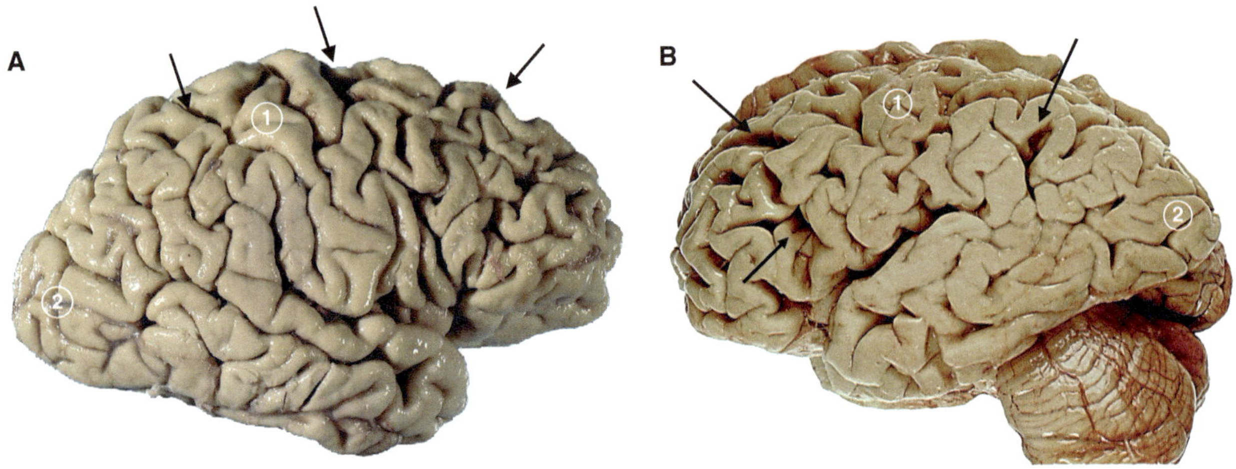

Fig. 16.31 Lateral views of brains from two CBD patients. Frontoparietal atrophy (arrows) in the peri-Rolandic area ①, surrounding the central sulcus, with relative sparing of the occipital lobes ② (**A** and **B**). B unfixed specimen. Collection A. Rozemuller and LabPON Twente

Fig. 16.32 Coronal sections at (**A**) the level of the thalamus/subthalamic nucleus showing atrophy of the superior frontal gyrus ① and thinning of the corpus callosum (arrow), in (**B**) at the level of the here pale, depigmented Substantia nigra, in (**C**) at the level of the pale locus caeruleus (arrows). Collection A. Rozemuller

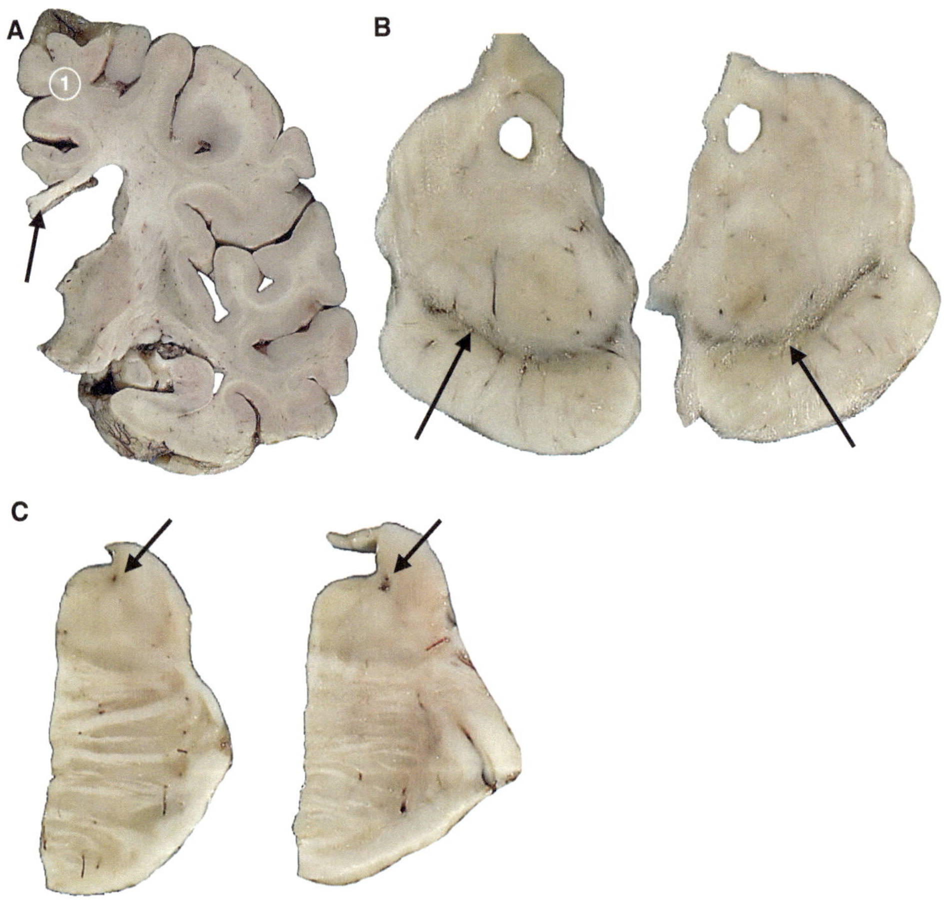

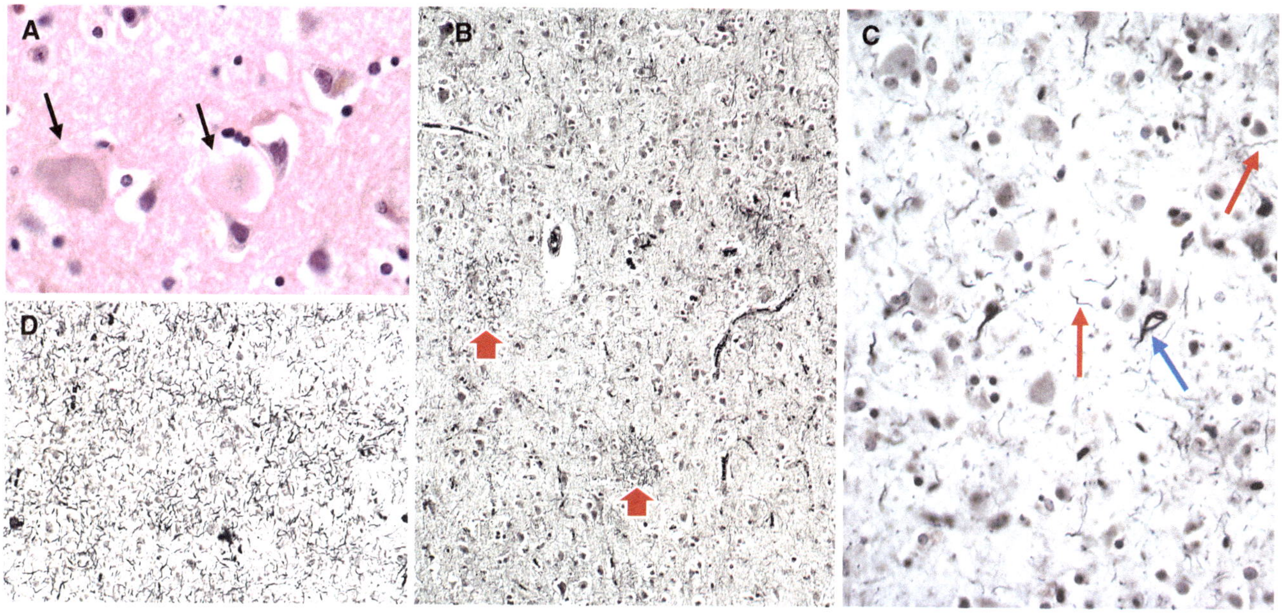

Fig. 16.33 Corticobasal degeneration (CBD). In (**A**) characteristic ballooned neurons (arrows) in the frontal cortex in a H & E stain. In (**B**) astrocytic plaques, a typical finding in CBD (red arrows), Gallyas silver stain. In (**C**) oligodendroglial coiled bodies (blue arrow) and threads (red arrows). (**D**) shows numerous diffusely spread neuropil threads in a Gallyas silver stain. LabPON Twente

16.6.3 p1 reticular formation

See Sect. 16.5.2.

16.6.4 Pararubral nucleus

For detailed information on the Pararubral nucleus, see Sect. 15.3.2.1.

16.6.5 Substantia nigra, reticular part (SNR)

For details on SNR, see Sect. 16.6.1.

16.7 Pretectal floor plate

16.7.1 Ventral tegmental area (VTA)

For detailed information on this area, see Sect. 15.5.1.1.

16.7.2 Parabrachial pigmented nucleus of the VTA

For detailed information on this nucleus, see Sect. 15.5.1.2.

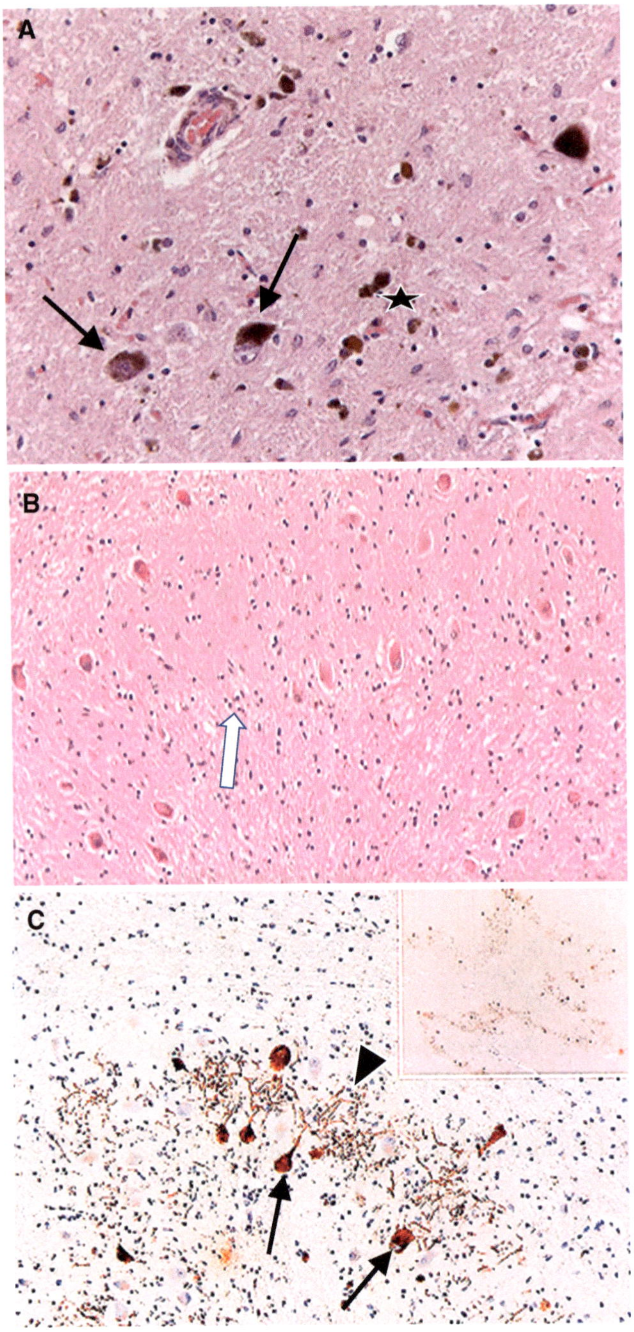

Fig. 16.34 Corticobasal degeneration (CBD). In (**A**) loss of neuromelanin-containing cells in the Substantia nigra, pigment incontinence (star). Preserved neuromelanin-containing neurons (arrows, H & E stain). In (**B**) loss of neurons in the dentate nucleus and slight gliosis (white arrow) in a H & E stain and AT8 tau immunostain in (**C**), showing globose tangles (corticobasal bodies, arrows) and threads (triangle) (inset: overview). Collection A. Rozemuller

References

Aarsland D, Rongve A et al (2008) Frequency and case identification of dementia with Lewy bodies using the revised consensus criteria. Dementia Geriatr Cogn Disord 26:445–452

Albin RL, Young AB et al (1989) The functional anatomy of basal ganglia disorders. Trends Neurosci 12:366–375

Aldakheel A, Kalia LV, Lang AE (2014) Pathogenesis-targeted, disease-modifying therapies in Parkinson's disease. Neurotherapeutics 11:6–23

Asaoka T, Tsuchiya K et al (2010) Argyrophilic grain disease with delusions and hallucinations: a pathological study. Psychogeriatrics 10:69–76

Barrett M, Armstrong MJ (2017) Dementia with Lewy bodies advances: a new consensus report. Neurology 89:18–19

Bayram E, Dickson DW et al (2020) Pathology-proven corticobasal degeneration presenting as Richardson's syndrome. Mov Disord Clin Pract 7:267–272

Bechterew W v (1909) Die Funktionen der Nervencentra, Heft II. Verlag von Gustav Fischer, Jena, pp 1018–1024

Bergeron C, Burns A et al (1996) Consensus guidelines for the clinical and pathological diagnosis of dementia with Lewy bodies; report of the consortium on DLB International workshop. Neurology 47:1113–1124

Bloch A, Probst A et al (2006) Alpha-synuclein pathology of the spinal and peripheral autonomic nervous system in neurologically unimpaired elderly subjects. Neuropathol Appl Neurobiol 32:284–295

Boot MP (2015) Comprehensive treatment of dementia with Lewy bodies. Alzheimers Res Ther 7:45

Braak H, Braak E (1986) Nuclear configuration and neuronal types of the nucleus niger in the brain of the human adult. Human Neurobiol 5:71–82

Braak H, Braak E (1987) Argyrophilic grains: characteristic pathology of cerebral cortex in cases of adult onset dementia without Alzheimer changes. Neurosci Lett 76:124–127

Braak H, Braak E (1998) Argyrophilic grain disease: frequency of occurrence in different age categories and neuropathological diagnostic criteria. J Neural Transm (Vienna) 105:801–819

Braak H, Braak E et al (1994) Amygdala pathology in Parkinson's disease. Acta Neuropathol 88:493–500

Braak H, Braak E et al (1995) Nigra and extranigral pathology in Parkinson's disease. J Neural Transm Suppl 46:15–31

Braak E, Sandmann-Keil D et al (2001) Alpha-synuclein immunopositive Parkinson's disease-related inclusion bodies in the lower brainstem nuclei. Acta Neuropathol 101:195–201

Braak H, del Tredici K et al (2003) Staging of brain pathology related to sporadic Parkinson's disease. Neurobiol Aging 24:197–211

Braak H, Rüb U et al (2005) Cognitive status correlates with neuropathologic stage in Parkinson's disease. Neurology 64:1404–1410

Braak H, de Vos RAI et al (2006) Gastric alpha-synuclein immunoreactive inclusions in Meissner's and Auerbach's plexuses in cases staged for Parkinson's disease-related brain pathology. Neurosci Lett 396:67–72

Braak H, Sastre M et al (2007a) Parkinson's disease: lesions in the dorsal horn layer I, involvement of parasympathetic and sympathetic neurons. Acta Neuropathol 113:421–429

Braak H, Sastre M, del Tredici K et al (2007b) Development of alpha-synuclein immunoreactive astrocytes in the forebrain parallels stages of intraneuronal pathology in sporadic Parkinson's disease. Acta Neuropathol 114:231–241

Brown DF, Dababo MA et al (1998) Neuropathologic evidence that the Lewy body variant of Alzheimer's diease represents coexistence of Alzheimer's disease and idiopathic Parkinson disease. J Neuropathol Exp Neurol 57:39–46

Buchman AS, Yu L et al (2019) Progressive Parkinsonism in older adults is related to the burden of mixed brain pathologies. Neurology 92:e1821–e1830

Buée L, Delacourte A (1999) Comparative biochemistry of tau in progressive supranuclear palsy, corticobasal degeneration, FTDP-17 and Pick's disease. Brain Pathol 9:681–693

Büttner-Ennever JA, Büttner U et al (1982) Vertical gaze paralysis and the rostral interstitial nucleus of the medial longitudinal fasciculus. Brain 105:125–149

Caixeta L, de Melo Caixeta V et al (2020) Pharmacological interventions in corticobasal degeneration: a review. Dement Neuropsychol 14:243–247

Cersosimo MG (2015) Gastrointestinal biopsies for the diagnosis of alpha-synuclein pathology in Parkinson's disease. Gastroenterol Res Pract 2015:476041

Chin KS, Theodorczuk A et al (2019) Dementia with Lewy bodies: challenges in the diagnosos and management. Aust N Z J Psychiatry 53:291–303

Coughlin DG, Hurtig HI et al (2020) Pathological influences on clinical heterogeneity in Lewy body disorders. Mov Disord 35:5–19

Cozzi B, Huggenberger S et al (2017) Anatomy of dolphins: insights into body structure and function. Chapter 6. Elsevier

Crome L (1971) The medical history of V. I. Lenin. History of Medicine 4:20–22

Damier P, Hirsch EC et al (1999) The substantia nigra of the human brain. I. Nigrosomes and the nigral matrix, a compartmental organization based on calbindin D28k immunohistochemistry. Brain 122:1421–1436

de Lau LM, Breteler MM (2006) Epidemiology of Parkinson's disease. Lancet Neurol 5:525–535

de Vos RAI, Jansen ENH et al (1995) "Lewy body disease": clinicopathological correlations in 18 consecutive cases of Parkinson's disease with and without dementia. Clin Neurol Neurosurg 97:13–22

del Tredici K, Rüb U et al (2002) Where does Parkinson disease pathology begin in the brain. J Neuropathol Exp Neurol 61:413–426

Di Meo I, Tiranti V (2018) Classification and molecular pathogenesis of NBIA syndromes. Eur J Paediatr Neurol 22:272–284

Dickson DW (1999) Neuropathologic differentiation of progressive supranuclear palsy and corticobasal degeneration. J Neurol 246(Suppl 2):S6–S15

Dickson DW, Bergeron C et al (2002) Office of rare diseases neuropathologic criteria for corticobasal degeneration. J Neuropath Exp Neurol 61:935–946

Dooling EC, Schoene WC et al (1974) Hallervorden-Spatz syndrome. Arch Neurol 30:70–83

Dusek P, Litwin T et al (2015) Wilson disease and other neurodegenerations with metal accumulations. Neurol Clin 33:175–204

Esselink RAJ, de Bie RMA et al (2007) Parkinson's disease-relates stereotactic functional neurosurgery. In: Wolters ECH, van Laar T, Berendse HW (eds) Parkinsonism and related disorders, vol 12, pp 241–249

Ferrer I, Barrachina M et al (2003) Phosphorylated protein kinases associated with neuronal and glial tau deposits in argyrophilic grain disease. Brain Pathol 13:62–78

Ferrer I, Santpere G et al (2008) Argyrophilic grain disease. Brain 131:1416–1432

Funayama M, Ohe K et al (2015) CHCHD2 mutations in autosomal dominant late-onset Parkinson's disease: a genome-wide linkage and sequencing study. Lancet Neurol 14:274–282

Fuse G (1937) Ein neuer Versuch zur Unterteilung des ruber tegmenti bei den Karnivoren und zur phylogenetischen Bewertung seiner Entwicklung unter Berücksichtigung der an Karnivoren und Affen gewonnenen Ergebnisse. Arb Anat Inst Kaiserl Jpn Univ Sendai 20:123–188

Gibb WRG, Esiri MM et al (1987) Clinical and pathological features of diffuse cortical Lewy body disease (Lewy body dementia). Brain 110:1131–1153

Gibb WR, Luthert PJ et al (1989) Corticobasal degeneration. Brain 112:1171–1192

Grijalvo-Perez AM, Litvan I (2014) Corticobasal degeneration. Semin Neurol 34:160–173

Halmagyi GM, Aw ST et al (1994) Jerk-waveform see-saw nystagmus due to unilateral meso-diencephalic lesion. Brain 117:789–803

Hansen L, Salmon D et al (1990) The Lewy vody variant of Alzheimer's disease: a clinical and pathological entity. Neurology 40:1–8

Harding HJ, Krill JJ et al (2000) Practical measures to simplify the Braak tangle staging method for routine pathological screening. Acta Neuropathol 99:199–208

Harsanyiova J, Buday T et al (2020) Parkinson's disease and the gut: future perspectives for early diagnosis. Front Neurosci 14:626

Hasegawa M (2017) Molecular biology of Dementia with Lewy bodies. In: Kosaka K (ed) Dementia with Lewy bodies. Clinical and biological aspects. Springer Japan, pp 41–55

Hishikawa N, Hashizume Y et al (2001) Widespread occurrence of argyrophilic glial inclusions in Parkinson's disease. Neuropathol Appl Neurobiol 27:362–372

Horn AKE, Adamczyk C (2012) Reticular formation: eye movements, gaze and blinks. (Nucleus of the posterior commissure (PCom)). In: Mai JK, Paxinos G (eds) The human nervous system, 3rd edn, pp 328–366

Horn AKE, Büttner-Ennever JA (1998) Premotor neurons for vertical eye movements in the rostral mesencephalon of monkey and human: histologic identification by parvalbumin immunostaining. J Comp Neurol 392:413–427

Huggenberger S, Oelschläger HA et al (2019) Atlas of the anatomy of dolphins and whales. Chapter 5. Elsevier

Ikeda M (2017) Pharmocotherapy in dementia with Lewy bodies. In: Kosaka K (ed) Dementia with Lewy bodies. Clinical and biological aspects. Springer Japan, pp 215–233

Ikeda K, Akiyama H et al (2000) Clinical aspects of argyrophilic grain disease. Clin Neuropathol 19:278–284

Ikeda C, Yokota O et al (2016) The relationship between development of neuronal and astrocytic tau pathologies in subcortical nuclei and progression of argyrophilic grain disease. Brain Pathol 26:488–505

Ishihara K, Araki S et al (2005) Argyrophilic grain disease presenting with frontotemporal dementia: a neuropsychological and pathological study of an autopsied case with presenile onset. Neuropathology 25:165–170

Iwanaga K, Wakabashi K et al (1999) Lewy body-type degeneration in cardiac plexus in Parkinson's and incidental Lewy body diseases. Neurology 52:1269–1271

Jellinger K (2004) Lewy body-related alpha-synucleinopathy in the aged brain. J Neural Transm (Vienna) 111:1219–1235

Jellinger K, Mizuno Y (2004) In: Dickson DW (ed) Neurodegeneration: the molecular pathology of dementia and movement disorders. ISS Neuropath Press, Basel, pp 17–21

Kalia LV, Lang AE (2015) Parkinson's disease. Lancet 386:896–912

Kasanuki K, Iseki E (2017) CT, MRI, SPECT, and PET in DLB. In: Kosaka K (ed) Dementia with Lewy bodies. Clinical and biological aspects. Springer Japan, pp 131–139

Kawamoto Y, Ayaki T et al (2020) Accumulation of HAX-1 and PARL in brainstem-and cortical-type Lewy bodies in Parkinson's disease and dementia with Lewy bodies. J Neurol Sci 415:116928
Klos K, Ahlskog JE et al (2006) Alpha-synuclein pathology in the spinal cords of neurologically asymptomatic aged individuals. Neurology 66:1100–1102
Komori T (1999) Tau-positive glial inclusions in progressive supranuclear palsy, corticobasal degeneration and Pick's disease. Brain Pathol 9:663–679
Kondziella D (2009) Thirty neurological eponyms associated with the Nazi era. Eur Neurol 62:56–64
Kosaka K, Yoshimura M et al (1984) Diffuse type of Lewy body disease: progressive dementia with abundant cortical Lewy bodies and senile changes of varying degree—a new disease? Clin Neuropathol 3:185–192
Kouri N, Murray ME et al (2011) Neuropathological features of corticobasal degeneration presenting as corticobasal syndrome or Richardson syndrome. Brain 134:3264–3275
Kovacs GG, Pittman A et al (2008) MAPT S3051 mutation: implications for argyrophilic grain disease. Acta Neuropathol 116:103–118
Kubis N, Faucheux BA (2000) Preservation of midbrain catecholaminergic neurons in very old human subjects. Brain 123:366–373
Kurian MA, Hayflick SJ (2013) Pantothenate kinase associated neurodegeneration (PKAN) and PLA2G6-associated neurodegeneration (PLAN): review of two major neurodegeneration with brain iron accumulation (NBIA) phenotypes. Int Rev Neurobiol 110:49–71
Larson SJ, Sances A Jr et al (1982) Cerebello-pallido-thalamic connections in man. Appl Neurophysiol 45:549–562
Lee VM-Y, Goedert M et al (2001) Neurodegenerative tauopathies. Ann Rev Neurosci 24:1121–1159
Leichnetz GR (1982) The medial accessory nucleus of Bechterew: a cell group within the anatomical limits of the rostral oculomotor complex receives a direct prefrontal projection in the monkey. J Comp Neurol 210:147–151
Leichnetz GR, Spencer RF et al (1984) Cortical projections to nuclei adjacent to the oculomotor complex in the medial dien-mesencephalic tegmentum in the monkey. J Comp Neurol 228:359–387
Li A, Paudel R et al (2013) Pantothenate kinase-associated neurodegeneration is not a synucleinopathy. Neuropathol Appl Neurobiol 39:121–131
Ling H, O'Sullivan SS et al (2010) Does corticobasal degeneration exist? A clinicopathological re-evaluation. Brain 133:2045–2057
Mai JK, Majtanik M (2017) Human brain in standard MNI space. A comprehensive pocket atlas. Academic Press
Martinez-Lage P, Munoz DG (1997) Prevalence and disease associations of argyrophilic grains of Braak. J Neuropathol Exp Neurol 56:157–164
Maurage C-A, Sergeant N et al (2003) Diffuse form of argyrophilic grain disease: a new variant of four-repeat tauopathy different form limbic argyrophilic grain disease. Acta Neuropathol 106:575–583
Mc Keith IG, Galasko D et al (2004) Dementia with Lewy bodies. Lancet Neurol 3:19–28
McKeith I (2017) Clinical diagnostic criteria for dementia with Lewy bodies. In: Kosaka K (ed) Dementia with Lewy bodies. Clinical and biological aspects. Springer Japan, pp 59–71
McKeith IG, Perry E et al (1999) Report of the second dementia with Lewy body international workshop: diagnosis and treatment. Consortium on dementia with Lewy bodies. Neurology 53:902–905
McKeith IG, Dickson DW et al (2005) Diagnosis and management of dementia with Lewy bodies: third report of the DLB-consortium. Neurology 65:1863–1872
Mizuno Y (2017) Parkinson symptoms in dementia with Lewy bodies. In: Kosaka K (ed) Dementia with Lewy bodies. Clinical and biological aspects. Springer Japan, pp 93–109
Mori E. (2017) Cognitive impairments of dementia with Lewy bodies. In: Dementia with Lewy bodies. Clinical and biological aspects. Kosaka, K (eEd). Springer, pp. 73-85
Nagao S, Yokota O et al (2014) Argyrophilic grain disease as a neurodegenerative substrate in late-onset schizophrenia and delusional disorders. Eur Arch Psychiatry Clin Neurosci 264:317–331
Olanow CW (1994) Hallervorden-Spatz syndrome: an iron-storage disease. In: Calne DB (ed) Neurodegenerative diseases. W.B. Saunders, pp 807–824
Olszewski J, Baxter D (1982) Cytoarchitecture of the human brain stem, 2nd edn. Karger, Basel
Onodera S, Hicks TP (1996) A projection linking motor cortex with the LM-suprageniculate nuclear complex through the periaqueductal gray area which surrounds the nucleus of Darkschewitsch in the cat. Prog Brain Res 112:85–98
Onodera S, Hicks TP (2009) A comparative neuroanatomical study of the red nucleus of the Cat, Macaque and Human. PLoS One 4:e6623
Orimo S (2017) Autonomic symptoms in dementia with Lewy bodies. In: Kosaka K (ed) Dementia with Lewy bodies. Clinical and biological aspects. Springer Japan, pp 111–127
Outeiro TF, Koss DJ et al (2019) Dementia with Lewy bodies: an update and outlook. Mol Neurodegener 14:5
Paisan-Ruiz C, Li A et al (2012) Widespread Lewy body and tau accumulation in childhood and adult onset dystonia-parkinsonism cases with PLA2G6 mutations. Neurobiol Aging 33:814–823
Parkinson J (1817) An essay on the shaking palsy. 1817. J Neuropsychiatry Clin Neurosci 14:223–236
Perry RH, Irving D et al (1990) Senile dementia of Lewy body type. A clinically and neuropathologically distinct form of Lewy body dementia in the elderly. J Neurol Sci 95:119–139
Postuma RB, Walker Z (2017) Dementia with Lewy bodies. Diagnosis is only skin deep? Neurology 89:310–311
Poulopoulos M, Levy OA et al (2012) The neuropathology of genetic Parkinson's disease. Mov Disord 27:831–842
Rebeiz JJ, Kolodny EH et al (1968) Corticodentatonigral degeneration with neuronal achromasia. Arch Neurol 18:20–33
Ronnback A et al (2014) Neuropatholoigical characterization of two siblings carrying the MAPT S305S mutation demonstrates features resembling argyrophilic grain disease. Acta Neuropathol 127:297–298
Rüb U, Stratmann K et al (2016) Hierarchical distribution of the tau cytoskeletal pathology in the thalamus of Alzheimer's disease patients. J Alzheimers Dis 49:905–915
Ruggeri M, Biagioli C et al (2020) Progressive aphasia, apraxia of speech and agraphia in corticobasal degeneration: a 12-case series clinical and neuropsychological descriptive study. Int J Lang Commun Disord 55:867–874
Saranza GM, Whitwell JL et al (2019) Corticobasal degeneration. Int Rev Neurobiol 149:87–136
Schuchart S (2020) Vom Leben und plötzlichen Sterben des Wladimir M. Bechterew. Ärzteblatt 117:40
Seidel K, Mahlke J et al (2015) The brainstem pathologies of Parkinson's disease and dementia with Lewy bodies. Brain Pathol 25:121–135
Shahmoradian SH, Lewis AL et al (2019) Lewy pathology in Parkinson's disease consists of crowded organelles and lipid membranes. Nat Neurosci 22:1099–1109
Smirnov DS, Galasko D et al (2020) Cognitive decline profiles differ in Parkinson disease dementia and dementia with Lewy bodies. Neurology 94:e2076–e2087

Soma K, Fu Y-J et al (2012) Co-Occurrence of argyrophilic grain disease in sporadic amyotrophic lateral sclerosis. Neuropathol Appl Neurobiol 38:54–60

Spillantini MG, Schmidt ML et al (1997) Alpha-synuclein in Lewy bodies. Nature 388:839–840

Suzuki Y, Yoshida K et al (2013) Effectiveness of oral iron chelator treatment with deferasirox in an aceruloplasminaemia patient with a novel ceruloplasmin gene mutation. Intern Med 52:1527–1530

Svenningsson P (2019) Corticobasal degeneration: advances in clinicopathology and biomarkers. Curr Opin Neurol 32:597–603

Togo T, Cookson N et al (2002) Argyrophilic grain disease: neuropathology, frequency in a dementia brain bank and lack of relationship with apolipoprotein E. Brain Pathol 12:45–52

Tolnay M, Probst A (2008) Argyrophilic grain disease. In: Duyckaerts C, Litvan I (eds) Handbook of clinical neurology. 3rd series. Elsevier, pp 553–564

Tolnay M, Sergeant N et al (2002) Argyrophilic grain disease and Alzheimer's disease are distinguished by their different distribution of tau protein isoforms. Acta Neuropathol 104:425–434

Tran HT, Chung C-Y et al (2014) alpha-synuclein immunotherapy blocks uptake and templated propagation of misfolded alpha-synuclein and neurodegeneration. Cell Rep 7:2054–2065

Tsuchiya K, Mitani K et al (2001) Argyrophilic grain disease mimicking temporal Pick's diease: a clinical, radiological, and pathological study of an autopsy case with a clinical course of 15 years. Acta Neuropathol 102:195–199

Tubert C, Galtieri D et al (2019) The pedunculopontine nucleus and Parkinson's disease. Neurobiol Dis 128:3–8

van de Berg WDJ, Rozemuller JM et al (2007) Neuropathology in movement disorders. In: Parkinsonism and related disorders. In: Wolters, van Laar, Berendse (eds), pp 97-128

van Domburg PHMF, ten Donkelaar HJ (1991) The human substantia nigra and ventral tegmental area. Adv Anat Embryol Cell Biol 121:1–132

van Rumund A, Green AJE et al (2019) a-synuclein real time quaking-induced conversion in the cerebrospinal fluid of uncertain cases of parkinsonism. Ann Neurol 85:777–781

Vergouw LJM, Marler LP et al (2020) Dementia with Lewy bodies. A clinicopathologic study series of false-positive cases. Alzheimer Dis Assoc Disord 34:178–182

Watson C, Bartholomaeus C et al (2019) Time for radical changes in brain stem nomenclature: applying the lessons from developmental gene patterns. Front Neuroanat 13:10

Weinstein HC, de Vos RAI et al (2000) Dementie met Lewy lichaampjes: klinische en neuropathologische aspecten. Ned Tijdschr Neurol 3:140–146

Yokota O, Miki T et al (2018) Neuropathological comorbidity associated with argyrophilic grain disease. Neuropathology 38:82–97

Zhang Y, Michel-Herve K et al (2017) Anatomical and functional organization of the human substantia nigra and its connections. elife 6:e26653

Zhang W, Tarutani A et al (2020) Novel tau filament fold in corticobasal degeneration. Nature 80:283–287

Web Links

Terminologia neuroanatomica FIPAT, http://fipat.library.dal.ca/TNA/

https://www.humanconnectome.org/

https://maayanlab.cloud/Harmonizome/gene_set/basal+reticular+formation+of+p1/Allen+Brain+Atlas+Adult+Mouse+Brain+Tissue+Gene+Expression+Profiles

http://search.brain-map.org/search/index.html?query=basal%20reticular%20formation%20of%20p1

Part II

Neuroanatomical Atlases of the Human Brainstem

Histology atlas of the human brainstem

17

Contents

17.1 Darrow red series (Visualization of perikarya)

17.1.1 Use of the atlas

17.1.1.1 Generation of the sections

This atlas was prepared in the Laboratorium Pathologie Oost-Nederland, Twente, The Netherlands. Horizontal PEG sections were prepared and stained either according to the Darrow red procedure or the Campbell procedure. Darrow red stained sections show the perikarya while Campbell sections are used for the demonstration of fiber tracts (for technical details, see section Materials and Methods in the Foreword).

17.1.1.2 Identification and labeling of structures

The labeling of structures in the present atlas was restricted to those nuclei and fiber tracts that a person with a moderate experience in microscopical work can retrace and use for his or her own work.

Certain nuclei of the human brainstem, in particular somatomotor motor nuclei like the ncl. of the hypoglossal nerve CNXII (see atlas part Darrow red 10A), the motor ncl. of the facial nerve CNVII with its subnuclei (see Darrow red 17A ff.) or the motor ncl. of the trigeminal nerve CNV (see atlas part Darrow red 23) are easy to identify because of the large diameter of their neurons and the mere size of the perikarya.

The delineation of nuclei with small diameter or small neuronal perikarya is mainly performed on the basis of their topographical relation with readily detectable nuclei or tracts like e.g. the posterior paragigantocellular reticular ncl. directly ventral of the hypoglossal ncl. in Darrow red 10A, or the tiny nucleus of trochlear nerve CNIV inside the medial longitudinal fasciculus (see Darrow red 36, 36A) or the parabrachial nuclei with small perikarya related to the superior cerebellar peduncle (see Darrow red 29A).

The identification of tracts is easy in cases where they display a considerable caliber like the pyramidal tract, the medial longitudinal fasciculus, and the medial lemniscus (see atlas part Campbell 1). Other functionally important but more diffuse tracts like the anterolateral tract (region of) (see atlas part Campbell 1) have to be identified based on their topographic relation with larger tracts or prominent nuclei.

17.1.1.3 Literature used

For the identification of the structures mainly the Olszewski and Baxter atlas of 1982 was used in addition to original papers that have been used in conjunction with the description of structures in the individual chapters of this book.

Partly the Brainstem atlas by Paxinos et al. (2020) was used, mainly the myelin stains in the rear of the book—keeping in

H. Schröder et al., *The Human Brainstem*, https://doi.org/10.1007/978-3-030-89980-6_17

mind that this atlas is hampered by the fact that the majority of sections has been processed in a distorted way, i.e., the photographs do not represent the geometry of the histological sections in that the photographs show a shortening with regard to the medial—lateral extension. The exception of this problem is the fiber stainings in the very end of the Paxinos atlas.

Darrow red 1

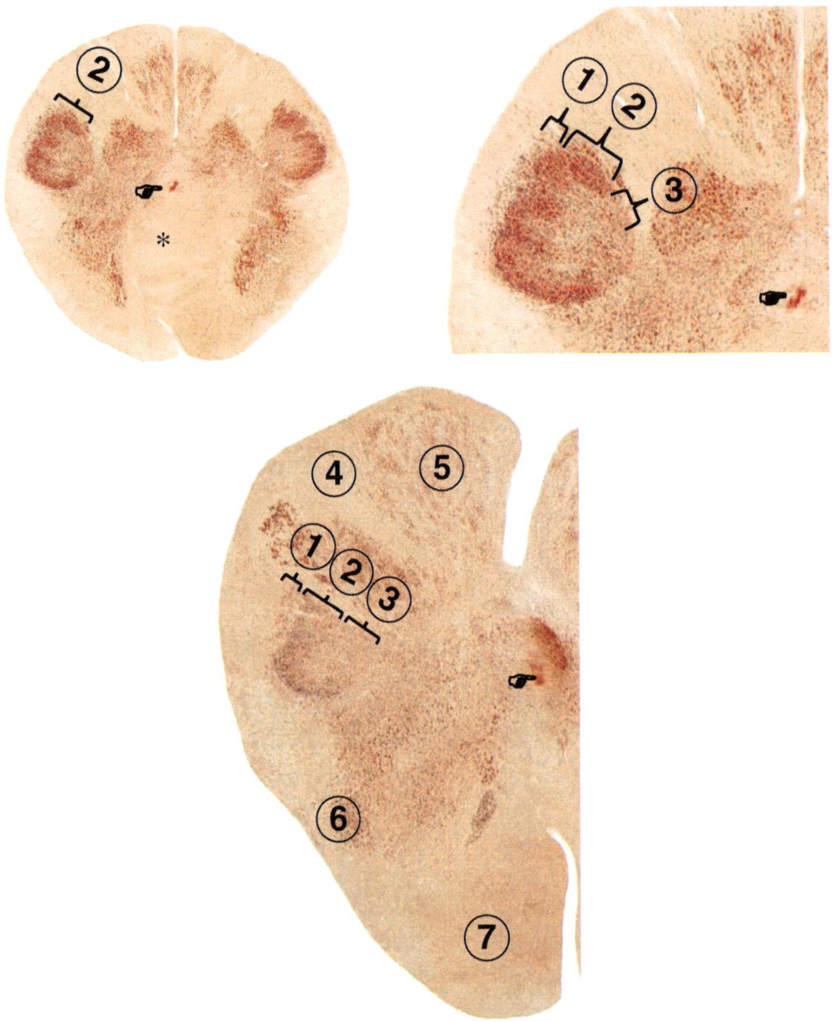

①	Trigeminal lamina I	⑥	Retroambiguus ncl.
②	Trigeminal lamina II	⑦	Pyramidal tract
③	Trigeminal lamina III	☛	Central canal
④	Cuneate ncl.	*	Decussation of pyramids
⑤	Gracile ncl.		

Darrow red 2

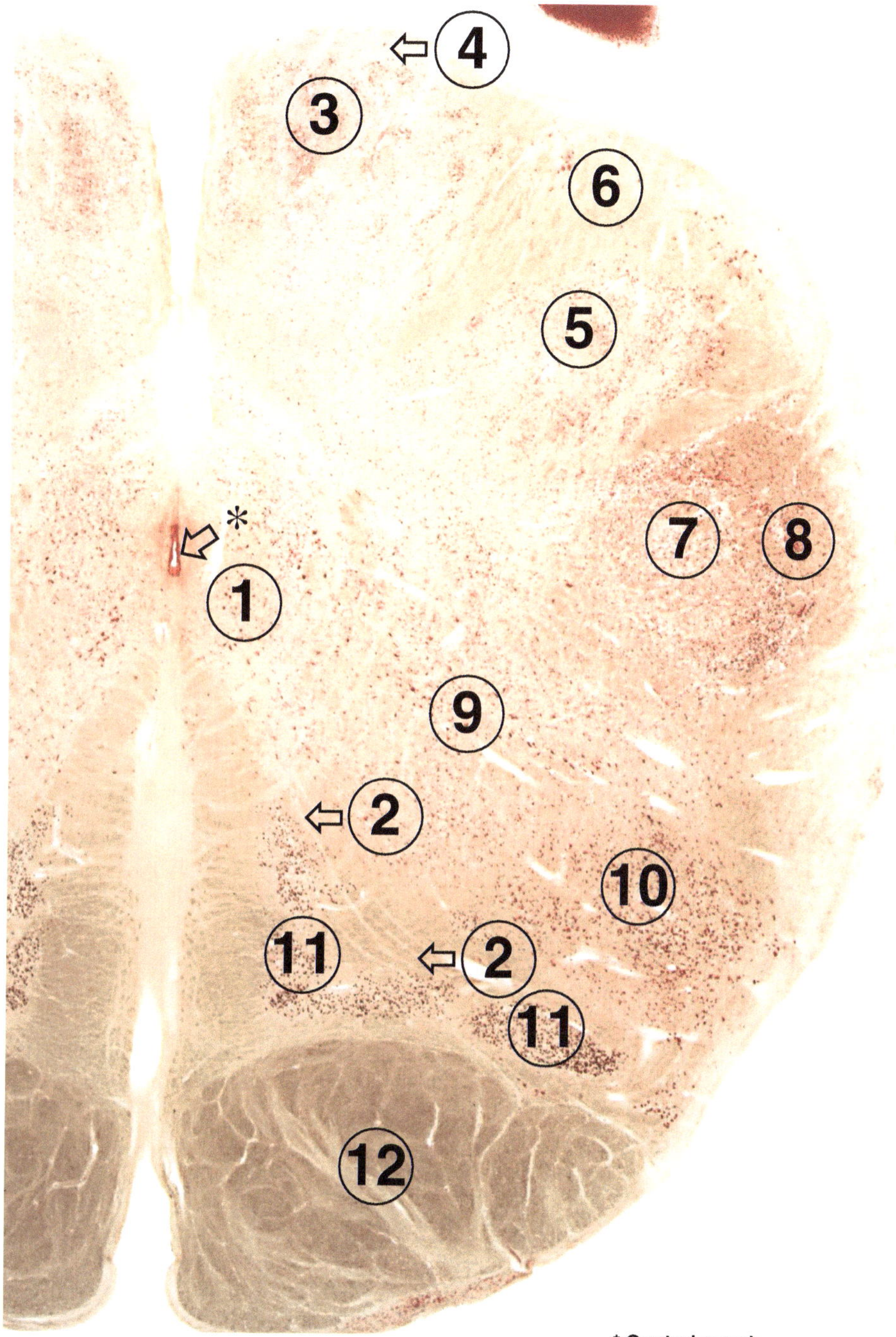

* Central canal

①	Ncl. of hypoglossal nerve CNXII	⑦	Spinal ncl. of trigeminal nerve CNV
②	Hypoglossal nerve CNXII	⑧	Spinal tract of trigeminal nerve
③	Gracile ncl.	⑨	Intermediate reticular zone
④	Gracile fasciculus	⑩	Lateral reticular ncl.
⑤	Cuneate ncl.	⑪	Inferior olivary complex
⑥	Cuneate fasciculus	⑫	Pyramidal (corticospinal) tract

Darrow red 2A

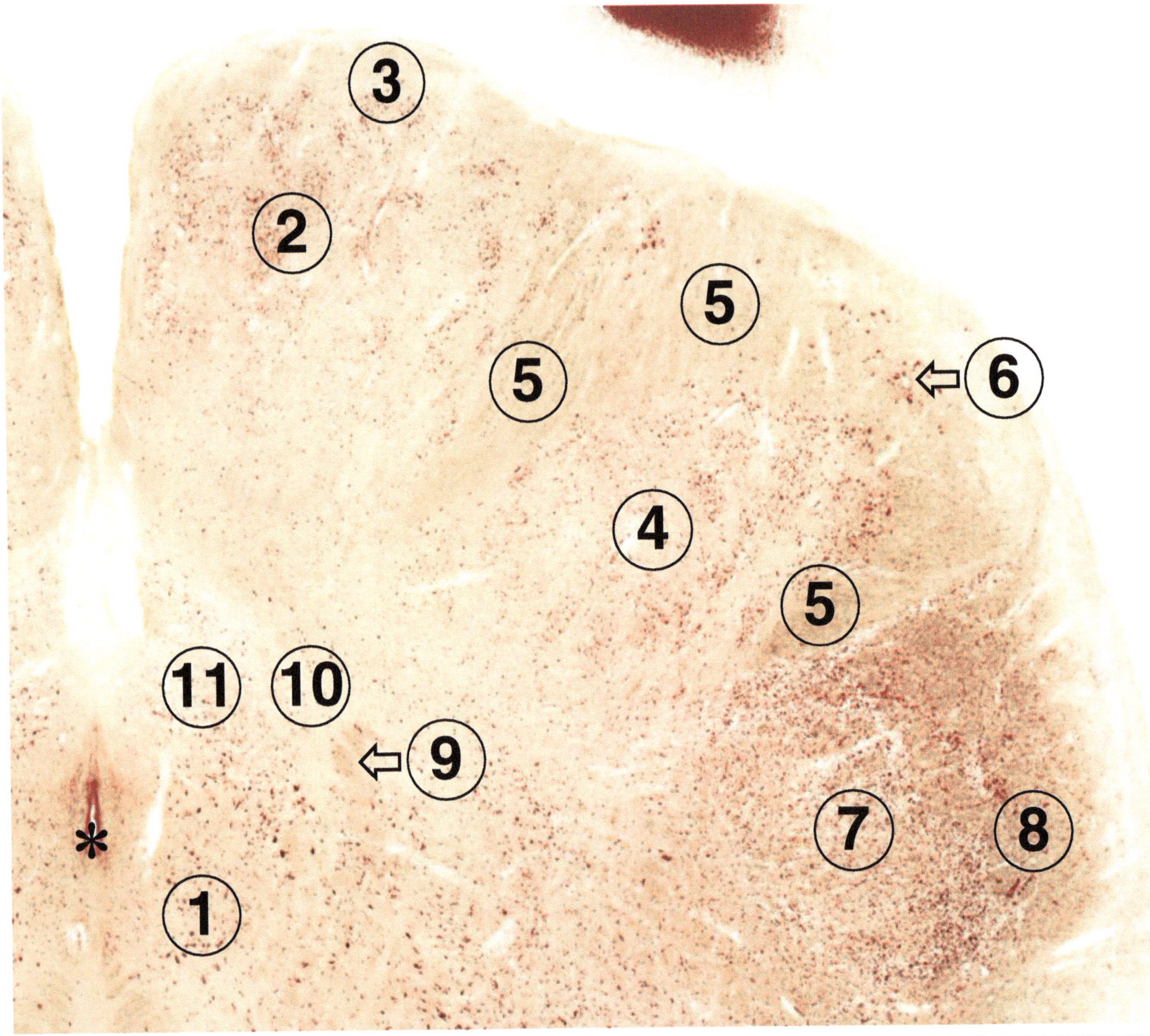

①	Ncl. of hypoglossal nerve	⑦	Spinal ncl. of trigeminal nerve
②	Gracile ncl.	⑧	Spinal tract of trigeminal nerve
③	Gracile fasciculus	⑨	Solitary tract
④	Cuneate ncl.	⑩	Ncll. of the solitary tract
⑤	Cuneate fasciculus	⑪	Posterior ncl. of vagus nerve CNX
⑥	Accessory cuneate ncl.	*	Central canal

Darrow red 2B

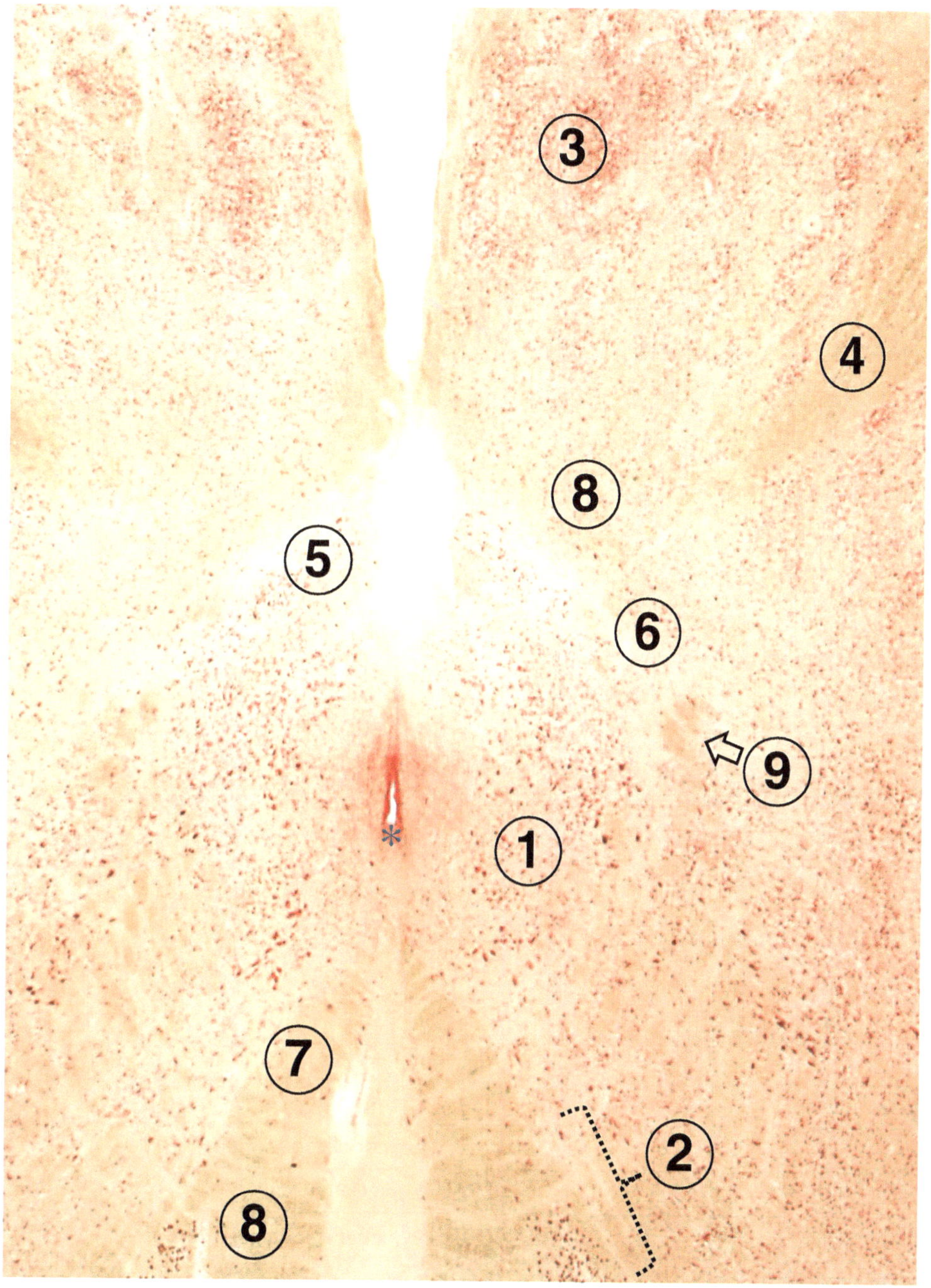

①	Ncl. of hypoglossal nerve	⑥	Ncll. of solitary tract
②	Hypoglossal nerve	⑦	Medial longitudinal fasciculus
③	Gracile ncl.	⑧	Region of the tectospinal tract
④	Cuneate fasciculus	⑨	Solitary tract
⑤	Posterior ncl. of vagus nerve	*	Central canal

Darrow red 3

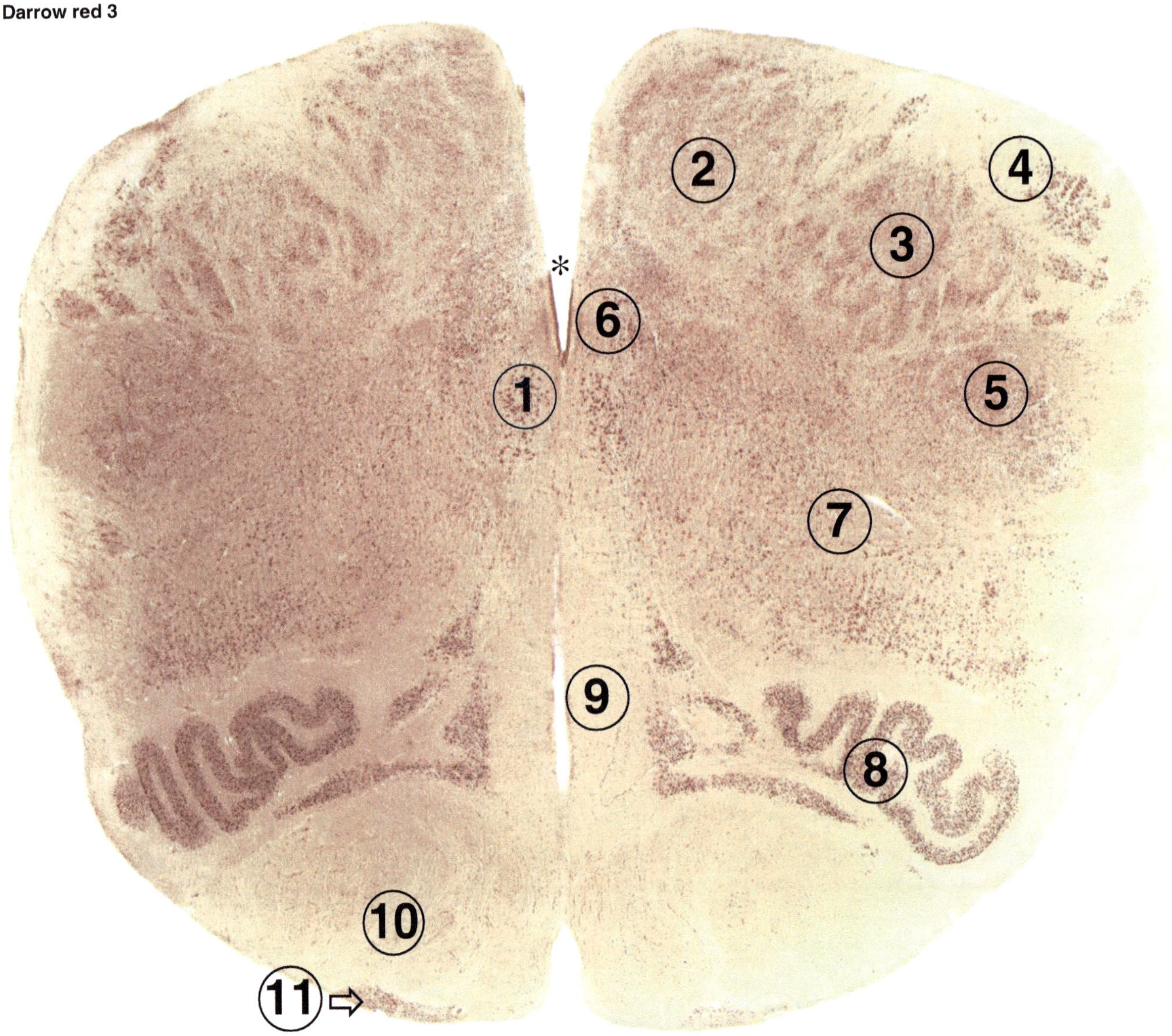

①	Ncl. of hypoglossal nerve	⑦	Intermediate reticular zone
②	Gracile ncl.	⑧	Inferior olivary complex
③	Cuneate ncl.	⑨	Medial lemniscus
④	Accessory cuneate ncl.	⑩	Pyramidal (corticospinal) tract
⑤	Spinal ncl. of trigeminal nerve	⑪	Arcuate ncl.
⑥	Posterior ncl. of vagus nerve	*	Transition central canal > 4th ventricle

Darrow red 3A

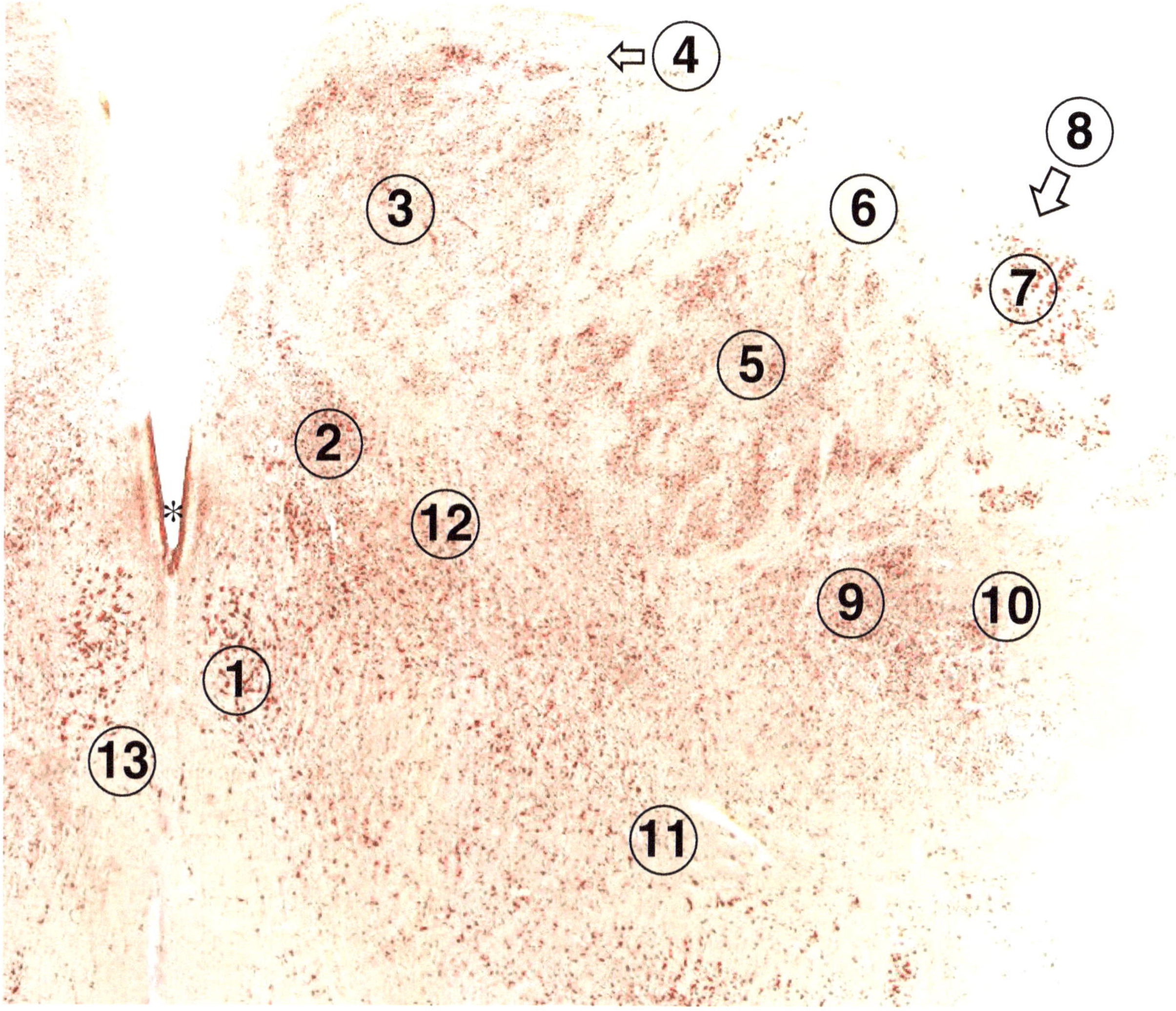

①	Ncl. of hypoglossal nerve	⑧	Inferior cerebellar peduncle
②	Posterior ncl. of vagus nerve	⑨	Spinal ncl. of trigeminal nerve
③	Gracile ncl.	⑩	Spinal tract of trigeminal nerve
④	Gracile fasciculus	⑪	Intermediate reticular zone
⑤	Cuneate ncl.	⑫	Ncll. of solitary tract
⑥	Cuneate fasciculus	⑬	Medial longitudinal fasciculus
⑦	Accessory cuneate ncl.	*	Transition central canal > 4th ventricle

Darrow red 4

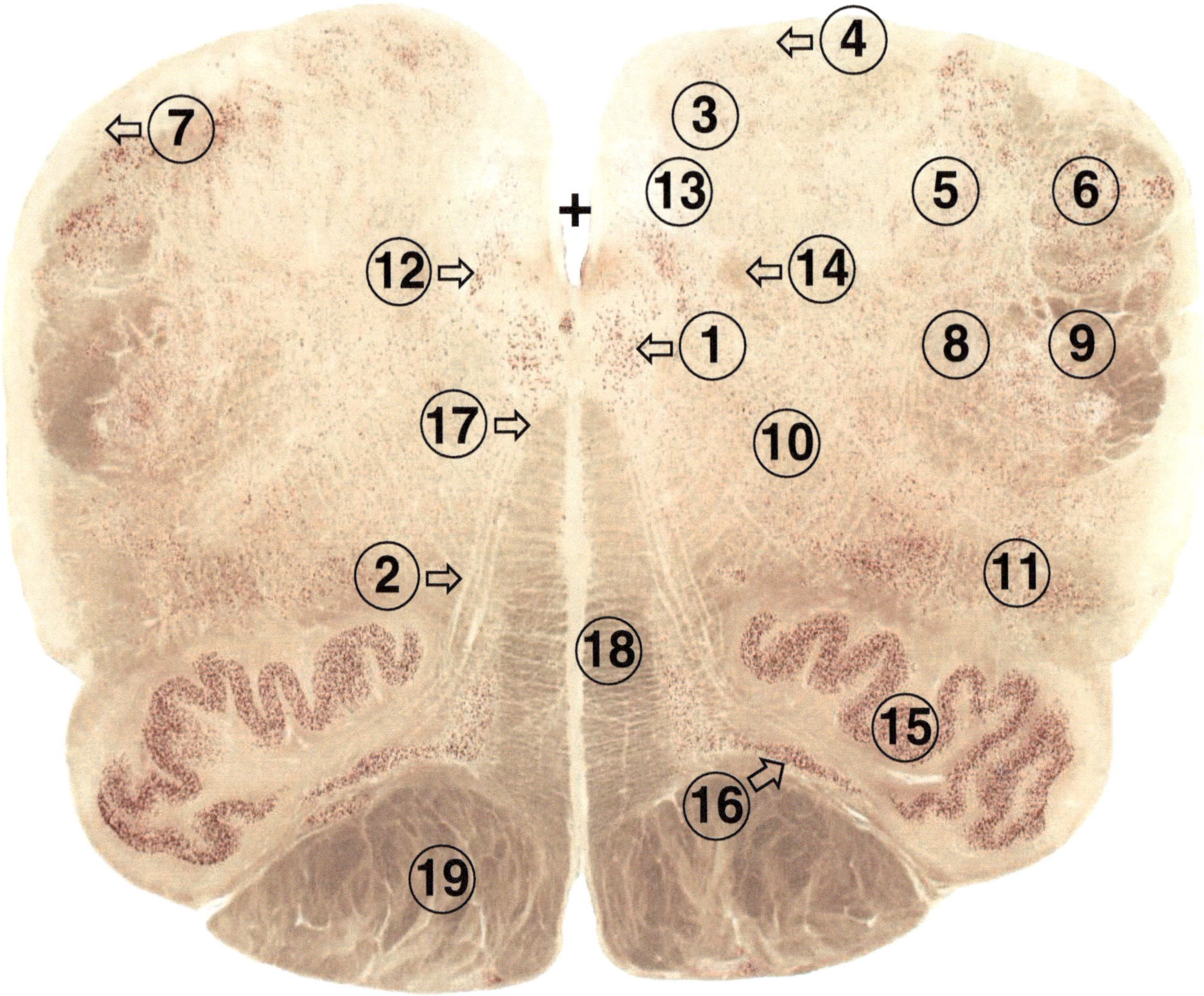

①	Ncl. of hypoglossal nerve	⑪	Lateral reticular ncl.
②	Hypoglossal nerve	⑫	Posterior ncl. of vagus nerve
③	Gracile ncl.	⑬	Ncll. of solitary tract CNVII, IX, X
④	Gracile fasciculus	⑭	Solitary tract
⑤	Cuneate ncl.	⑮	Principal olivary ncl. of inferior olivary complex
⑥	Accessory cuneate ncl.	⑯	Medial accessory olivary ncl.
⑦	Inferior cerebellar peduncle	⑰	Medial longitudinal fasciculus
⑧	Spinal ncl. of trigeminal nerve CNV	⑱	Medial lemniscus
⑨	Spinal tract of trigeminal nerve	⑲	Pyramidal (corticospinal) tract
⑩	Intermediate reticular zone	+	4th ventricle

Darrow red 5

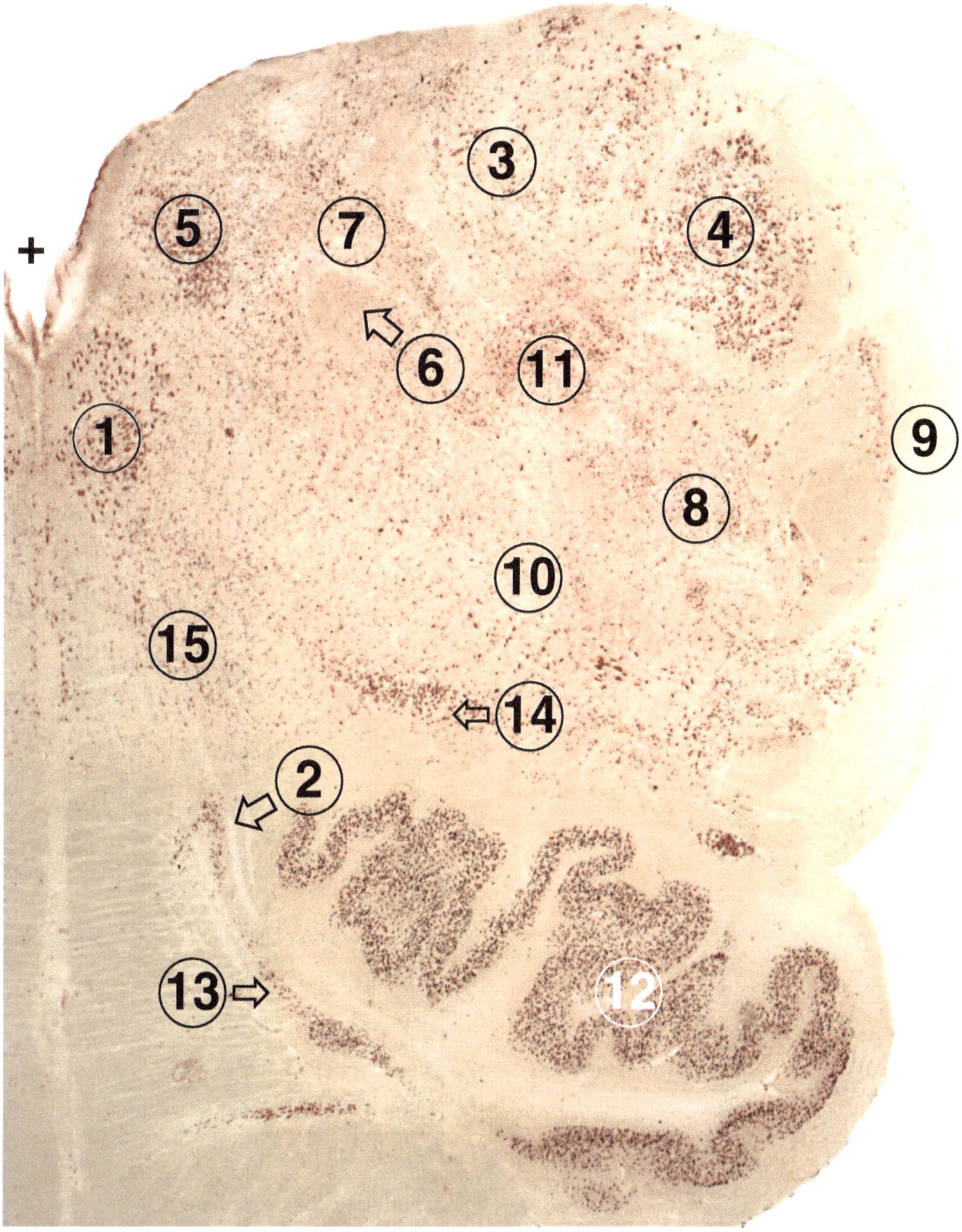

①	Ncl. of hypoglossal nerve	⑨	Inferior cerebellar peduncle
②	Hypoglossal nerve	⑩	Intermediate reticular zone
③	Cuneate ncl.	⑪	Parvocellular reticular ncl.
④	Accessory cuneate ncl.	⑫	Principal olivary ncl.
⑤	Posterior ncl. of vagus nerve	⑬	Medial accessory olivary ncl.
⑥	Solitary tract	⑭	Posterior accessory olivary ncl.
⑦	Ncll. of solitary tract	⑮	Gigantocellular reticular ncl.
⑧	Spinal ncl. of trigeminal nerve	+	4th ventricle

Darrow red 5A

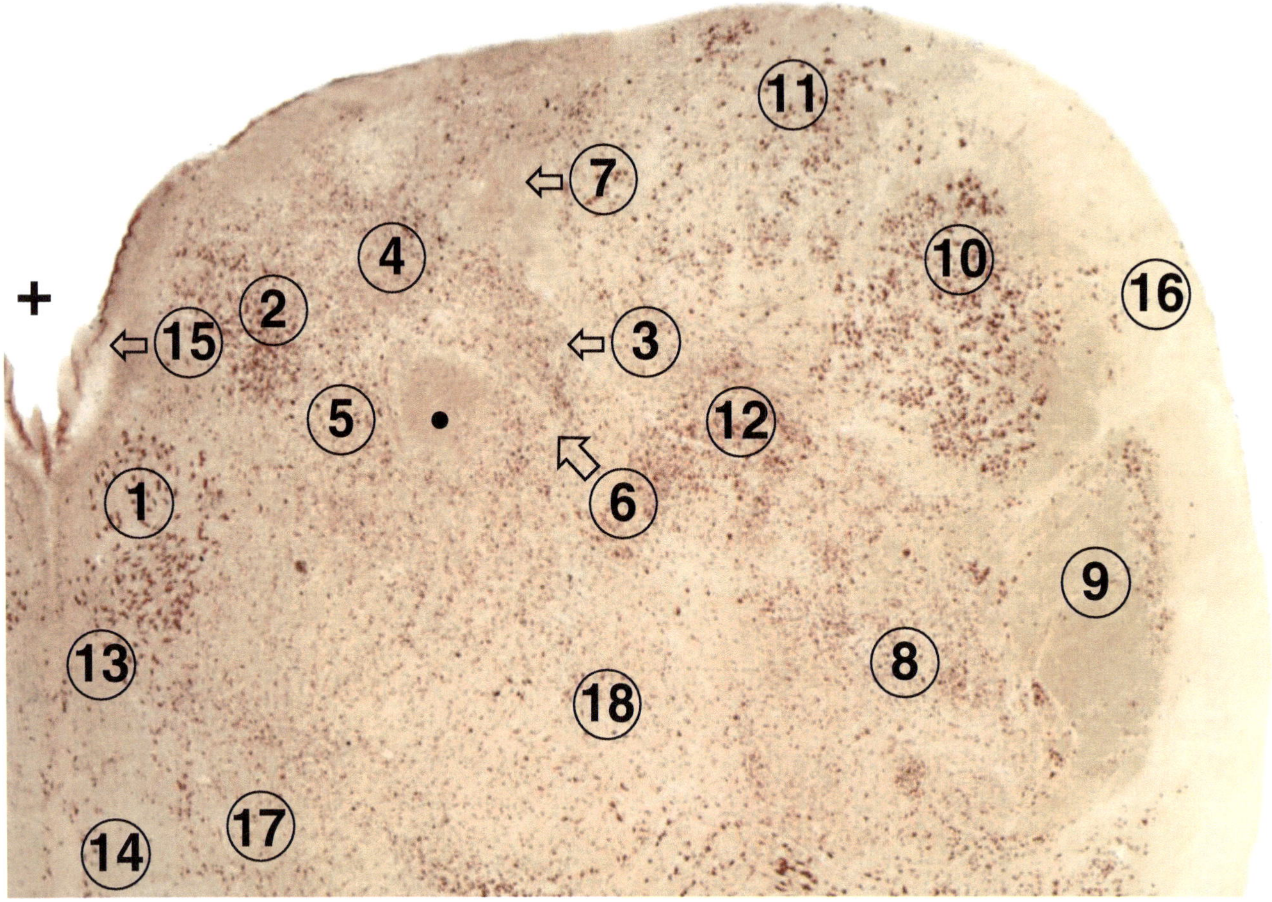

①	Ncl. of hypoglossal nerve	⑩	Accessory cuneate ncl.
②	Posterior ncl. of vagus nerve	⑪	Cuneate ncl.
•	Solitary tract	⑫	Parvocellular reticular ncl.
③	Posterolateral solitary ncl.	⑬	Medial longitudinal fasciculus
④	Medial solitary ncl.	⑭	Region of the tectospinal tract
⑤	Intermediate solitary ncl.	⑮	Area postrema
⑥	Anterolateral solitary ncl.	⑯	Inferior cerebellar peduncle
⑦	Gelatinous solitary ncl.	⑰	Gigantocellular reticular ncl.
⑧	Spinal ncl. of trigeminal nerve	⑱	Intermediate reticular zone
⑨	Spinal tract of trigeminal nerve	+	4th ventricle

Darrow red 6

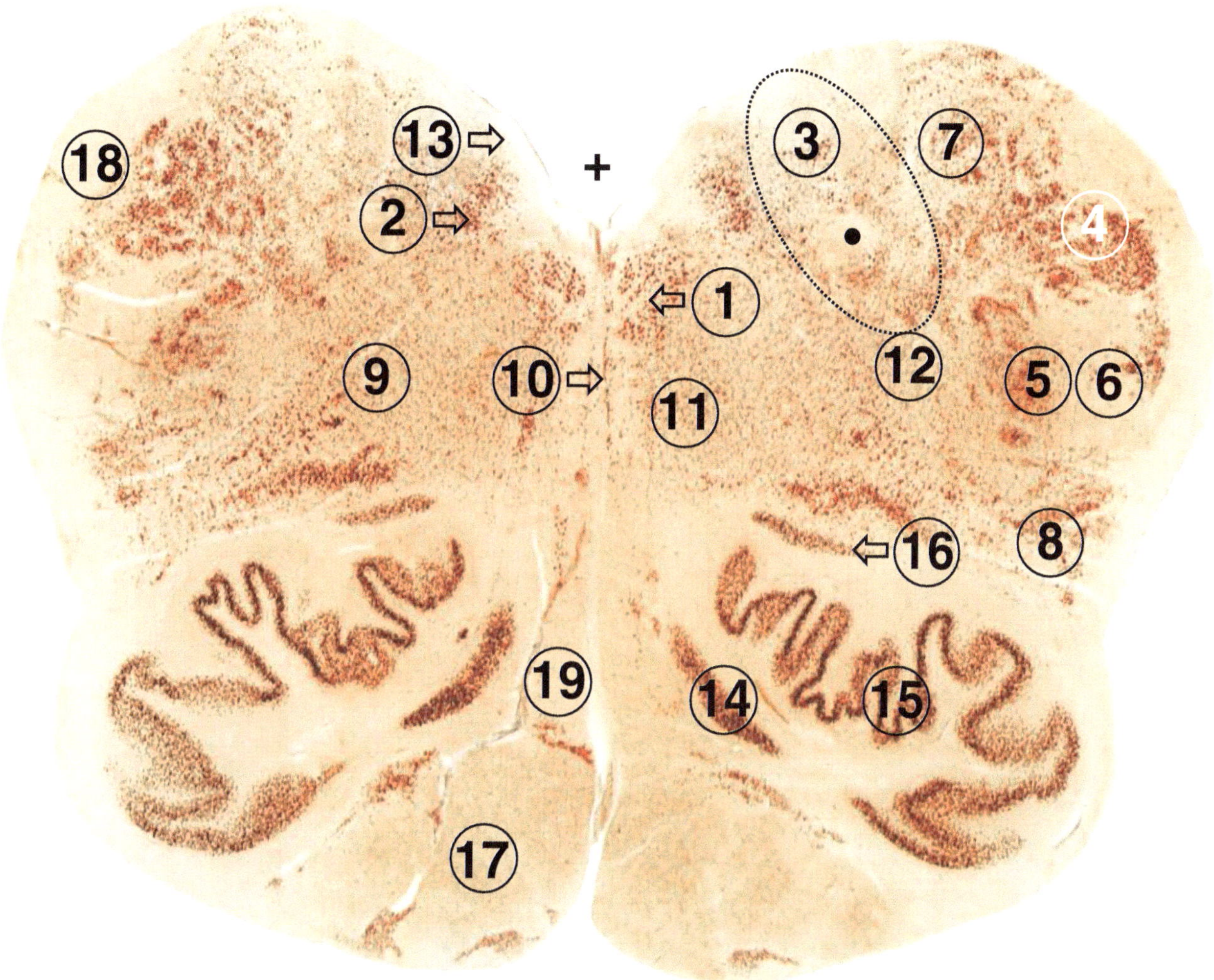

①	Ncl. of hypoglossal nerve		⑪	Gigantocellular reticular ncl.
②	Posterior ncl. of vagus nerve		⑫	Parvocellular reticular ncl.
•	Solitary tract		⑬	Area postrema
③	Ncll. of solitary tract --- For details see Darrow red 10A		⑭	Medial accessory olivary ncl.
④	Accessory cuneate ncl.		⑮	Principal olivary ncl.
⑤	Spinal ncl. of trigeminal nerve		⑯	Posterior accessory olivary ncl.
⑥	Spinal tract of trigeminal nerve		⑰	Pyramidal (corticospinal) tract
⑦	Inferior vestibular ncl. CNVIII		⑱	Inferior cerebellar peduncle
⑧	Lateral reticular ncl.		⑲	Medial lemniscus
⑨	Intermediate reticular zone		+	4th ventricle
⑩	Raphe obscurus ncl.			

Darrow red 7

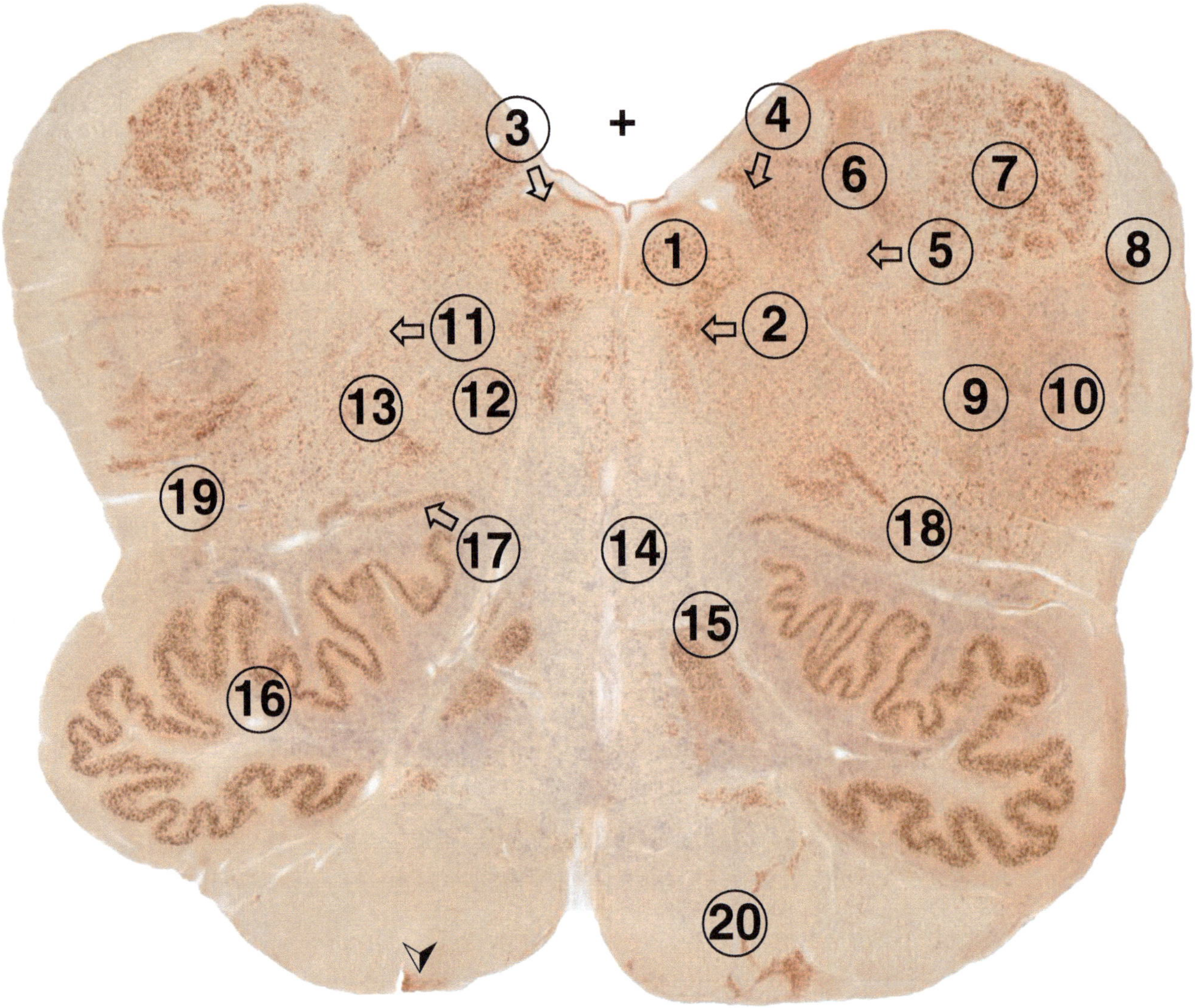

①	Ncl. of hypoglossal nerve	⑫	Gigantocellular reticular ncl.
②	Subhypoglossal ncl. Roller	⑬	Intermediate reticular zone
③	Intercalated ncl.	⑭	Medial lemniscus
④	Posterior ncl. of vagus nerve	⑮	Medial accessory olivary ncl.
⑤	Solitary tract	⑯	Principal olivary ncl.
⑥	Ncll. of solitary tract For details see Darrow red 10A	⑰	Posterior accessory olivary ncl.
⑦	Accessory cuneate ncl.	⑱	Lateral paragigantocellular ncl.
⑧	Inferior cerebellar peduncle	⑲	Lateral reticular ncl.
⑨	Spinal ncl. of trigeminal nerve	⑳	Pyramidal (corticospinal) tract
⑩	Spinal tract of trigeminal nerve	⮟	Arcuate ncl.
⑪	Parvocellular reticular ncl.	+	4th ventricle

Darrow red 8

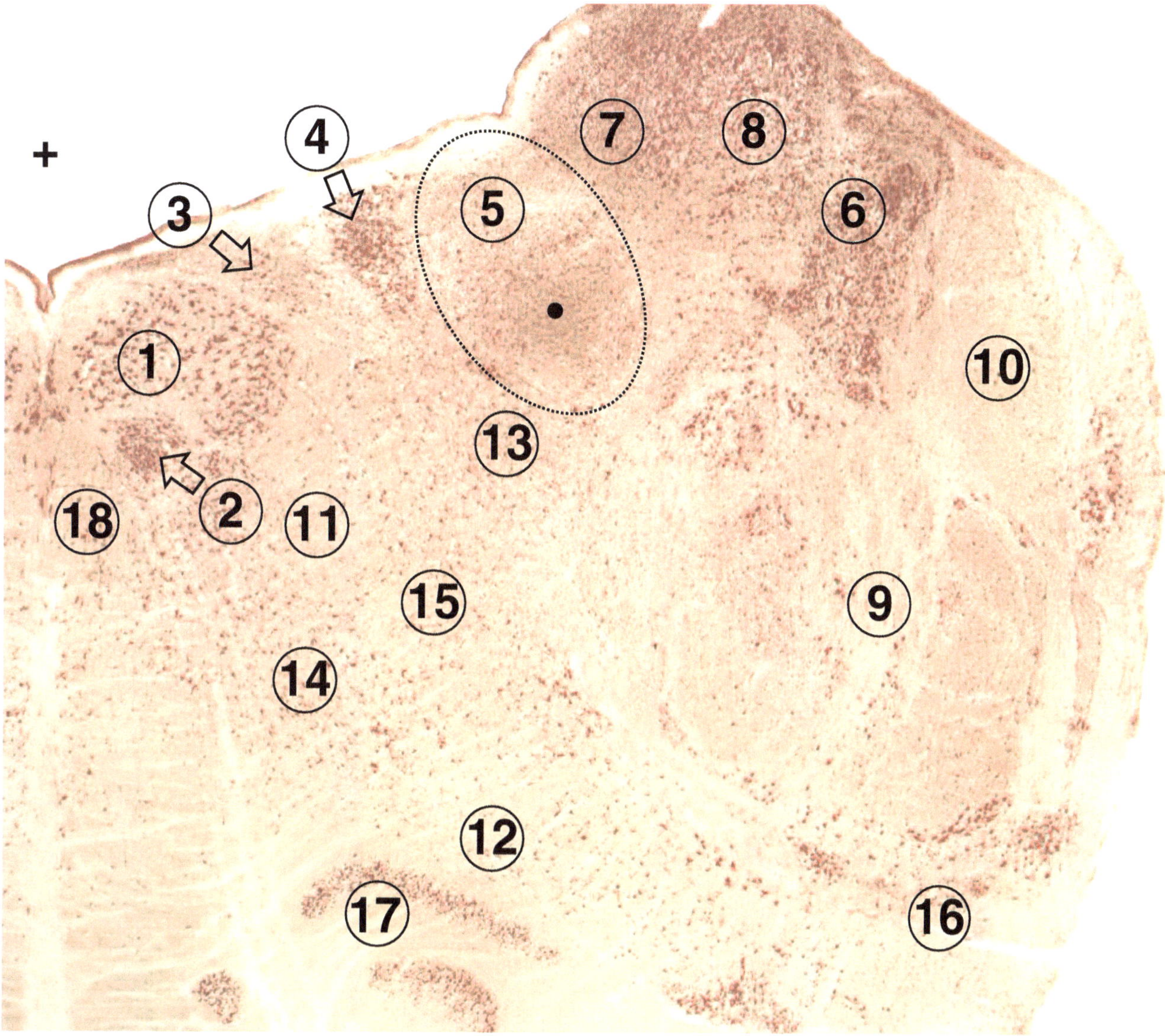

①	Ncl. of hypoglossal nerve	⑩	Inferior cerebellar peduncle
②	Subhypoglossal ncl. Roller	⑪	Posterior paragigantocellular reticular ncl.
③	Intercalated ncl.	⑫	Lateral paragigantocellular ncl.
④	Posterior ncl. of vagus nerve	⑬	Parvocellular reticular ncl.
•	Solitary tract	⑭	Gigantocellular reticular ncl.
⑤	Ncll. of solitary tract --- For details see Darrow red 10A	⑮	Intermediate reticular zone
⑥	Accessory cuneate ncl.	⑯	Lateral reticular ncl.
⑦	Medial vestibular ncl. CNVIII	⑰	Inferior olivary complex
⑧	Inferior vestibular ncl. CNVIII	⑱	Medial longitudinal fasciculus
⑨	Spinal ncl. / tract of trigeminal nerve	+	4th ventricle

Darrow red 9

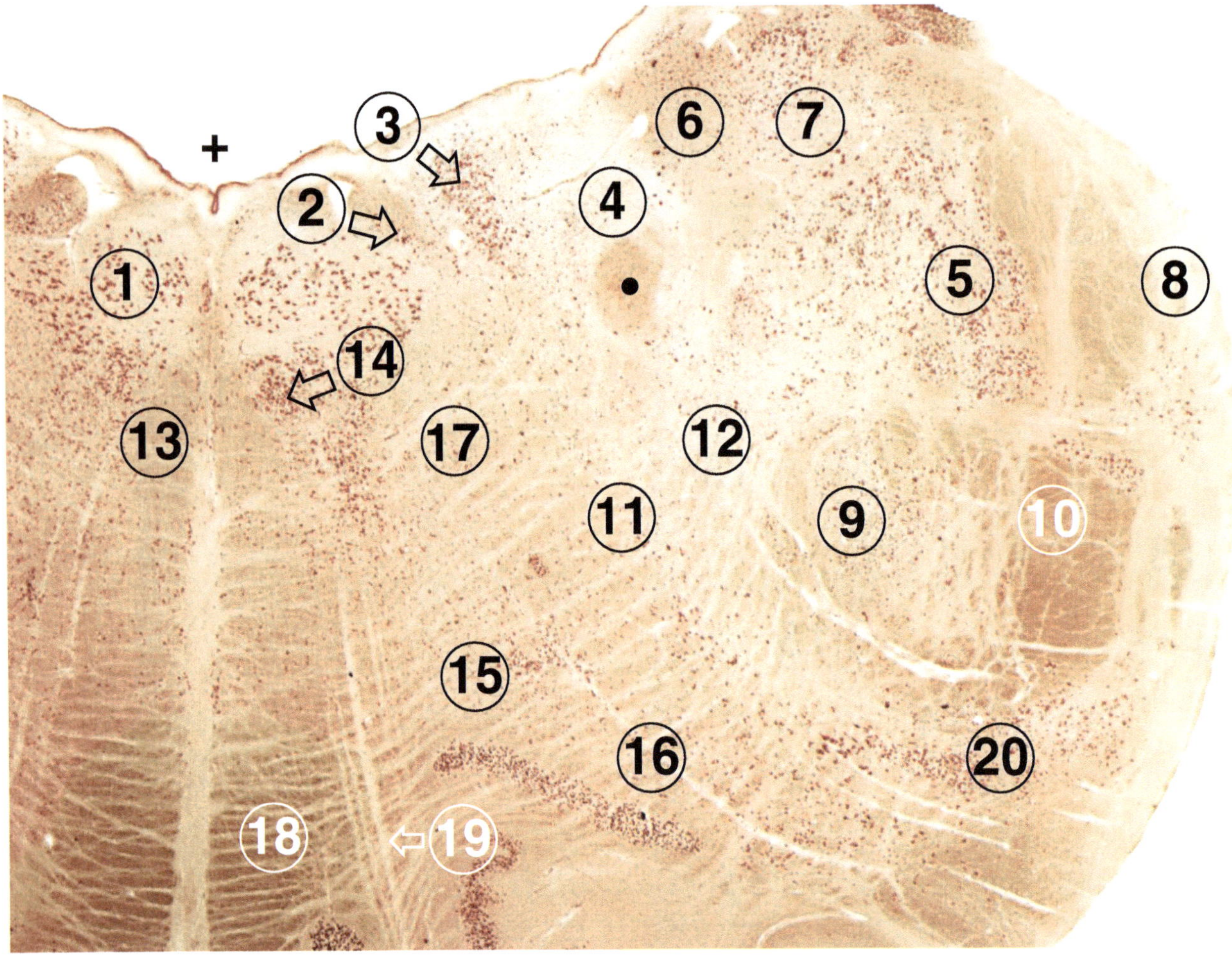

①	Ncl. of hypoglossal nerve	⑪	Intermediate reticular zone
②	Intercalated ncl.	⑫	Parvocellular reticular ncl.
③	Posterior ncl. of vagus nerve	⑬	Medial longitudinal fasciculus
•	Solitary tract	⑭	Subhypoglossal ncl. Roller
④	Ncll. of solitary tract For details see Darrow red 10A	⑮	Gigantocellular reticular ncl.
⑤	Accessory cuneate ncl.	⑯	Lateral paragigantocellular ncl.
⑥	Medial vestibular ncl.	⑰	Posterior paragigantocellular reticular ncl.
⑦	Inferior vestibular ncl.	⑱	Medial lemniscus
⑧	Inferior cerebellar peduncle	⑲	Hypoglossal nerve
⑨	Spinal ncl. of trigeminal nerve	⑳	Lateral reticular ncl.
⑩	Spinal tract of trigeminal nerve	+	4th ventricle

Darrow red 10

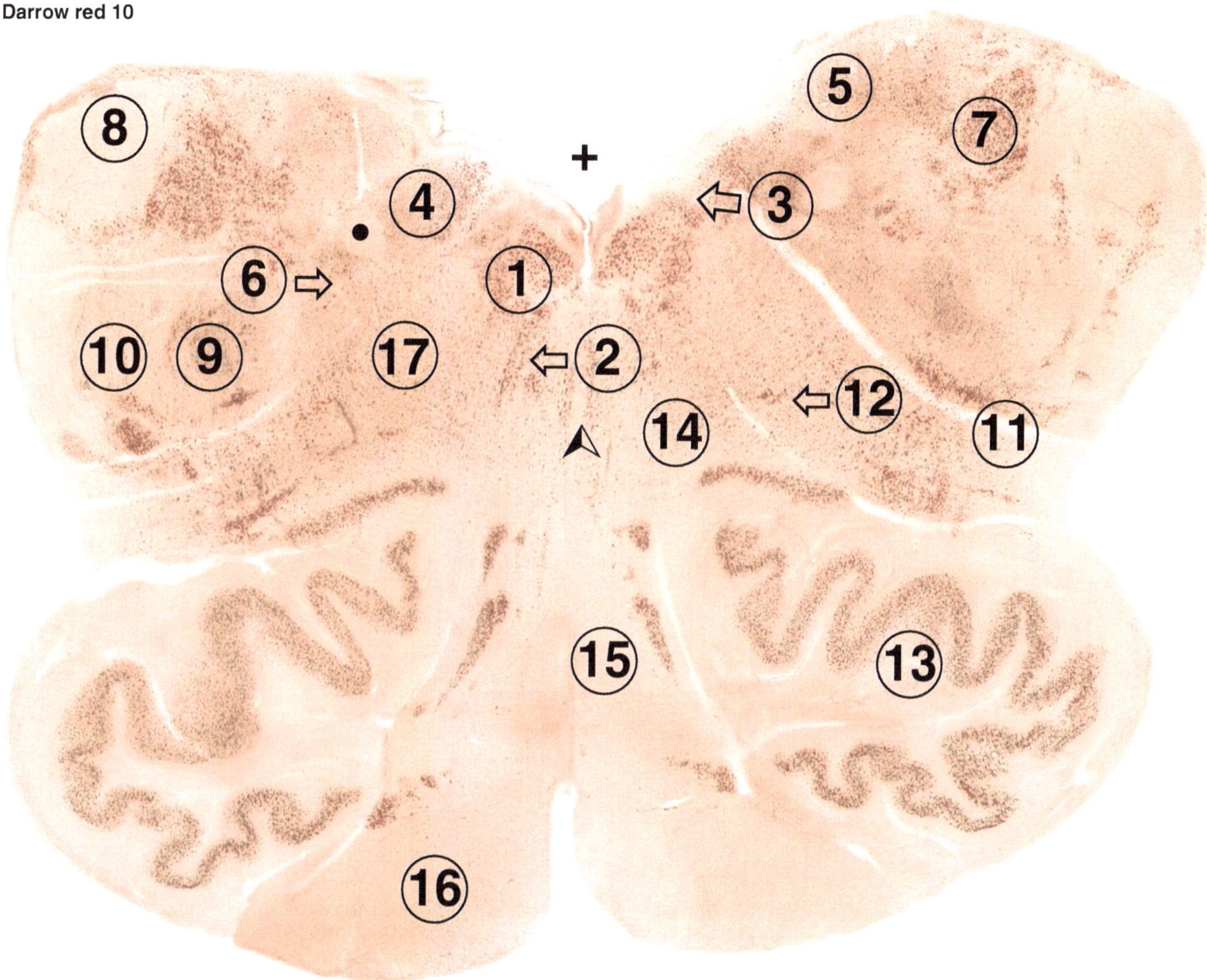

①	Ncl. of hypoglossal nerve	⑪	Lateral reticular ncl.
②	Subhypoglossal ncl. Roller	⑫	Ncl. ambiguus CNIX
③	Intercalated ncl.	⑬	Principal olivary ncl.
④	Posterior ncl. of vagus nerve	⑭	Gigantocellular reticular ncl.
⑤	Medial vestibular ncl.	⑮	Medial lemniscus
⑥	Ncll. of solitary tract For details see Darrow red 10A	⑯	Pyramidal (corticospinal) tract
⑦	Accessory cuneate ncl.	⑰	Intermediate reticular zone
⑧	Inferior cerebellar peduncle	⮝	Raphe obscurus ncl.
⑨	Spinal ncl. of trigeminal nerve	•	Solitary tract
⑩	Spinal tract of trigeminal nerve	+	4th ventricle

Darrow red 10A

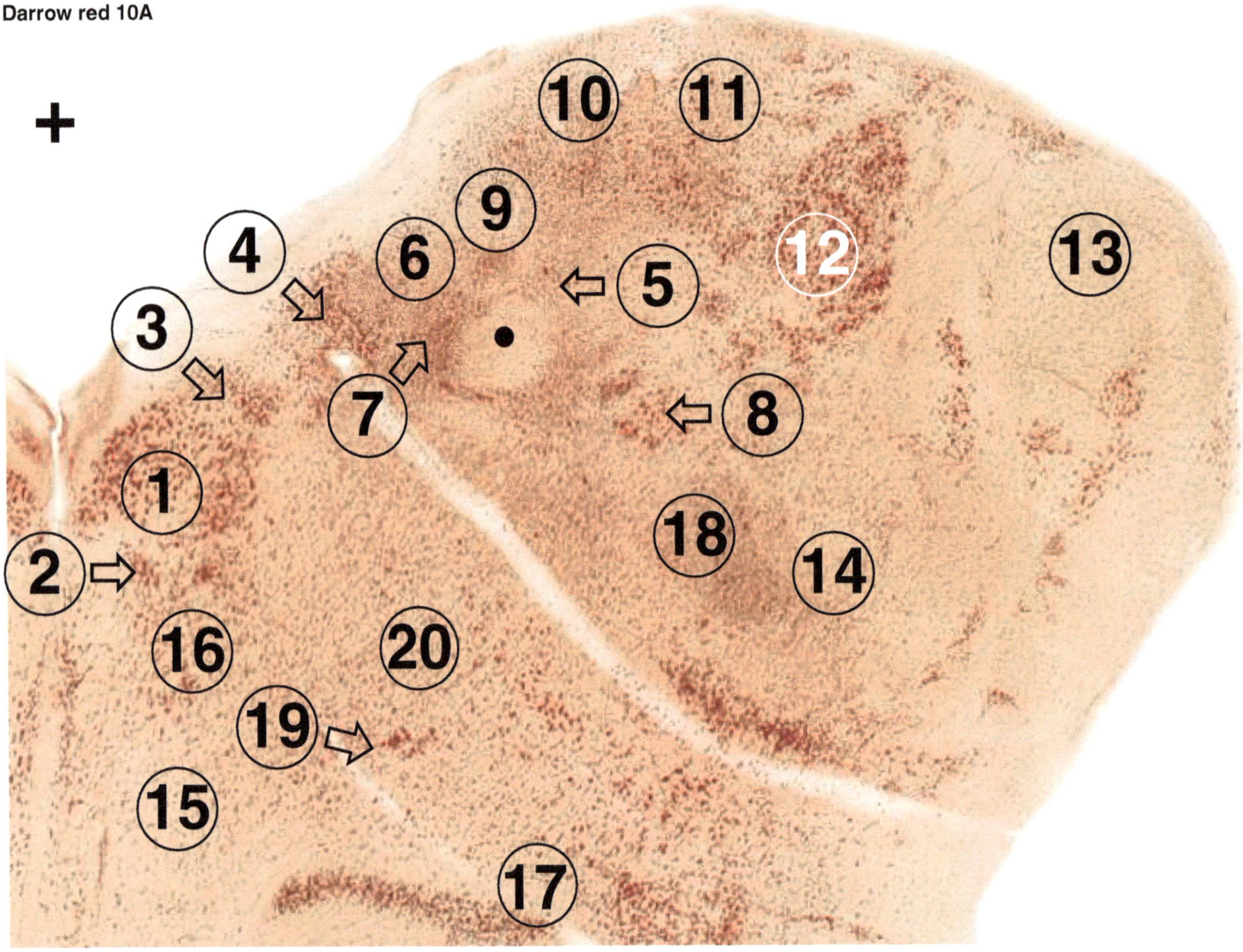

①	Ncl. of hypoglossal nerve	⑪	Inferior vestibular ncl.
②	Subhypoglossal ncl. Roller	⑫	Accessory cuneate ncl.
③	Intercalated ncl.	⑬	Inferior cerebellar peduncle
④	Posterior ncl. of vagus nerve	⑭	Spinal ncl. / tract of trigeminal nerve
•	Solitary tract	⑮	Gigantocellular reticular ncl.
⑤	Posterolateral solitary ncl.	⑯	Posterior paragigantocellular reticular ncl.
⑥	Medial solitary ncl.	⑰	Lateral paragigantocellular ncl.
⑦	Intermediate solitary ncl.	⑱	Parvocellular reticular ncl.
⑧	Anterolateral solitary ncl.	⑲	Ncl. ambiguus
⑨	Gelatinous solitary ncl.	⑳	Intermediate reticular zone
⑩	Medial vestibular ncl.	+	4th ventricle

Darrow red 11

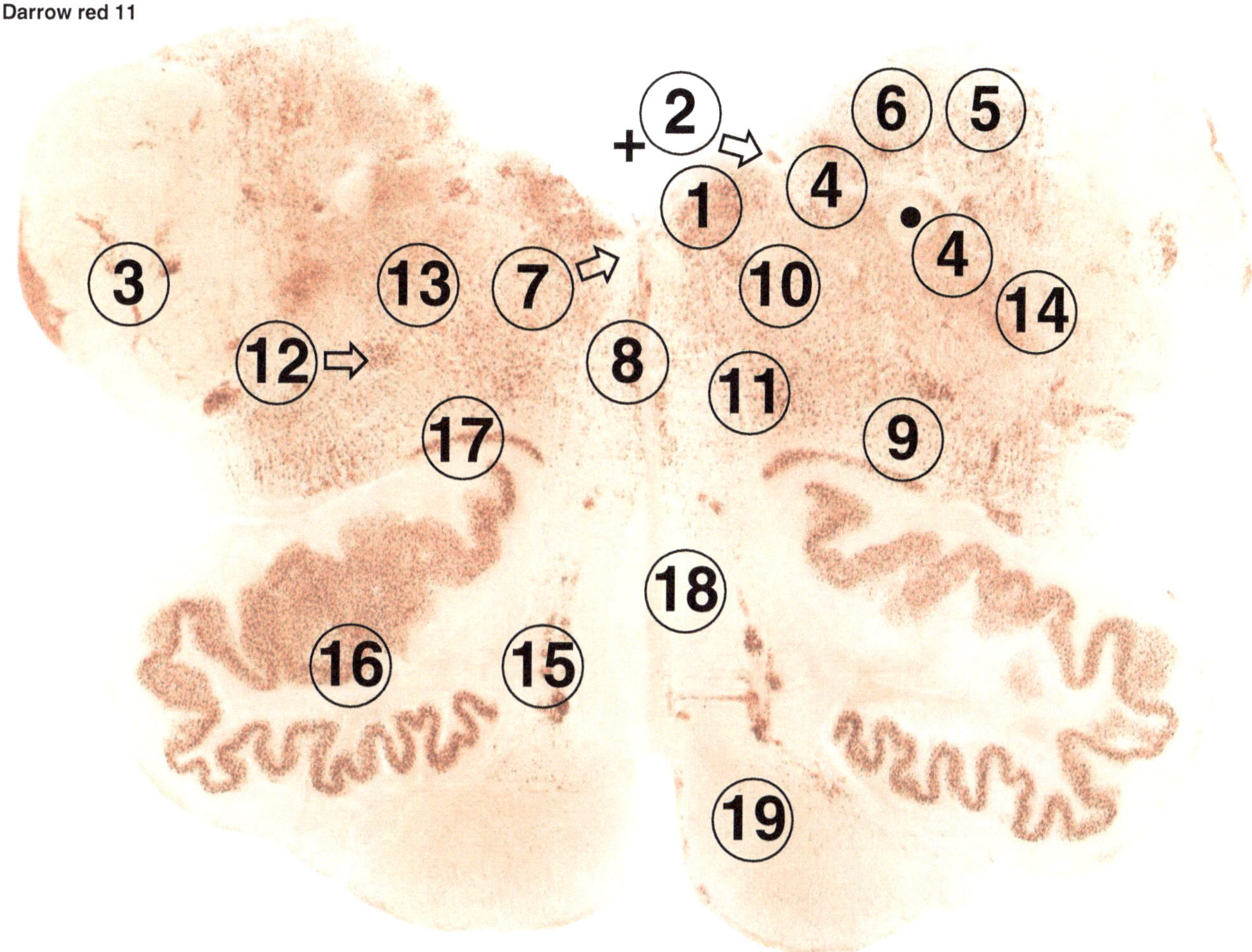

①	Prepositus hypoglossi ncl.	⑪	Gigantocellular reticular ncl.
②	Interpositus ncl.	⑫	Ncl. ambiguus
③	Inferior cerebellar peduncle	⑬	Intermediate reticular zone
④	Solitary ncll. For details see Darrow red 11A	⑭	Spinal ncl. / tract of trigeminal nerve
•	Solitary tract	⑮	Medial accessory olivary ncl.
⑤	Inferior (spinal) vestibular ncl.	⑯	Principal olivary ncl.
⑥	Medial vestibular ncl.	⑰	Posterior accessory olivary ncl.
⑦	Medial longitudinal fasciculus	⑱	Medial lemniscus
⑧	Raphe obscurus ncl.	⑲	Pyramidal (corticospinal) tract
⑨	Lateral paragigantocellular ncl.	+	4th ventricle
⑩	Posterior paragigantocellular reticular ncl.		

Darrow red 11A

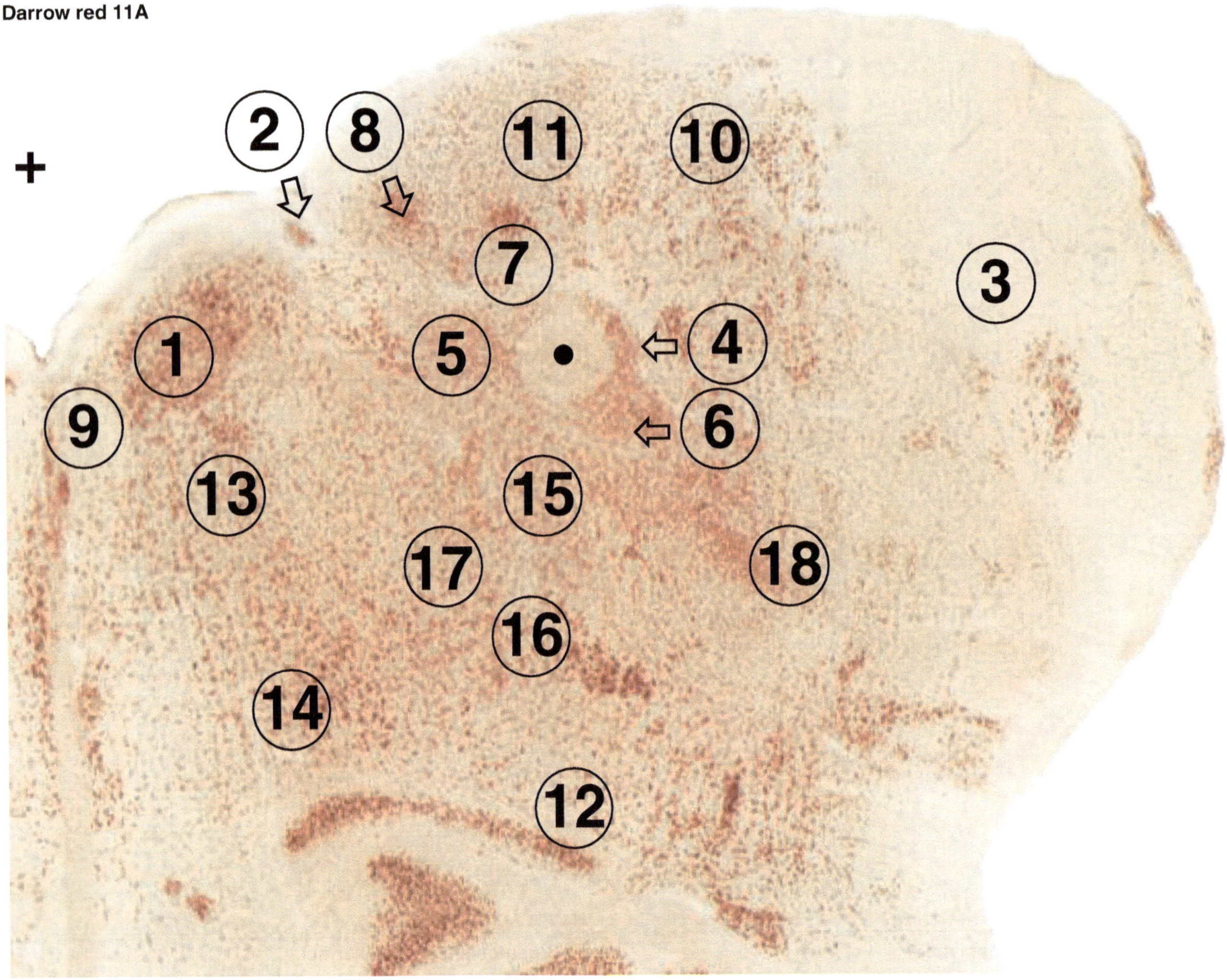

①	Prepositus hypoglossi ncl.	⑩	Inferior vestibular ncl.
②	Interpositus ncl.	⑪	Medial vestibular ncl.
③	Inferior cerebellar peduncle	⑫	Lateral paragigantocellular ncl.
•	Solitary tract	⑬	Posterior paragigantocellular reticular ncl.
④	Posterolateral solitary ncl.	⑭	Gigantocellular reticular ncl.
⑤	Intermediate solitary ncl.	⑮	Parvocellular reticular ncl.
⑥	Anterolateral solitary ncl.	⑯	Ncl. ambiguus
⑦	Medial solitary ncl.	⑰	Intermediate reticular zone
⑧	Posterior ncl. of vagus nerve	⑱	Spinal ncl. / tract of trigeminal nerve
⑨	Medial longitudinal fasciculus	+	4th ventricle

Darrow red 12

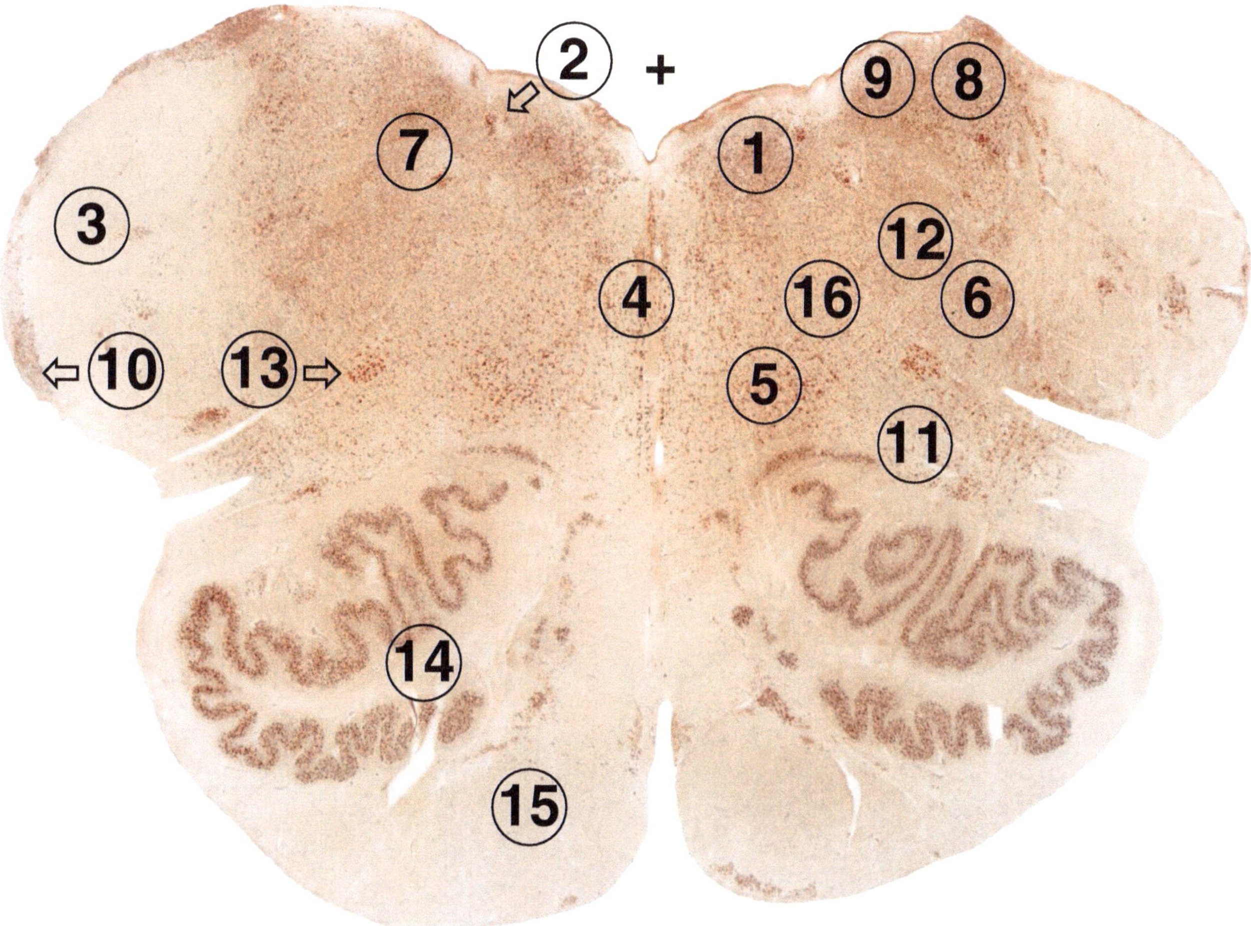

①	Prepositus hypoglossi ncl.		⑨	Medial vestibular ncl.
②	Interpositus ncl.		⑩	Anterior (ventral) cochlear ncl.
③	Inferior cerebellar peduncle		⑪	Lateral paragigantocellular ncl.
④	Raphe obscurus ncl.		⑫	Parvocellular reticular ncl.
⑤	Gigantocellular reticular ncl.		⑬	Ncl. ambiguus
⑥	Spinal ncl. / tract of trigeminal nerve		⑭	Inferior olivary complex
⑦	Solitary tract / NclI. of solitary tract		⑮	Pyramidal (corticospinal) tract
⑧	Inferior vestibular ncl.		⑯	Intermediate reticular zone
			+	4th ventricle

Darrow red 13

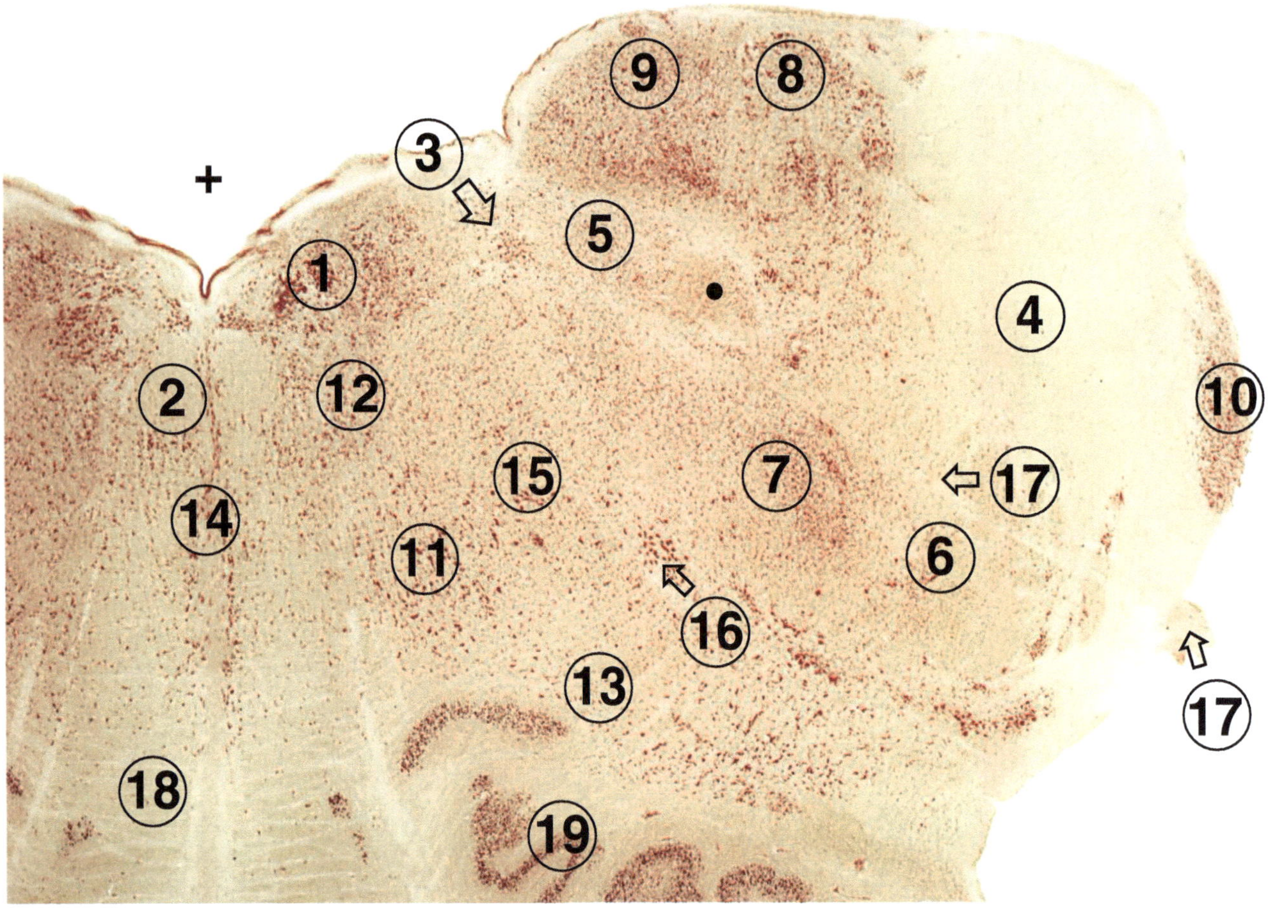

①	Prepositus hypoglossi ncl.	⑪	Gigantocellular reticular ncl.
②	Medial longitudinal fasciculus	⑫	Posterior paragigantocellular reticular ncl.
③	Posterior ncl. of vagus nerve	⑬	Lateral paragigantocellular ncl.
④	Inferior cerebellar peduncle	⑭	Raphe obscurus ncl.
•	Solitary tract	⑮	Intermediate reticular zone
⑤	NcII. of solitary tract	⑯	Ncl. ambiguus
⑥	Spinal ncl. / tract of trigeminal nerve	⑰	Glossopharyngeal nerve CNIX Vagus nerve CNX See Campbell 4 (CNX) 5 (CNIX)
⑦	Parvocellular reticular ncl.		
⑧	Inferior vestibular ncl. CNVIII	⑱	Medial lemniscus
⑨	Medial vestibular ncl. CNVIII	⑲	Inferior olivary complex
⑩	Posterior (dorsal) cochlear ncl.	+	4th ventricle

Darrow red 14

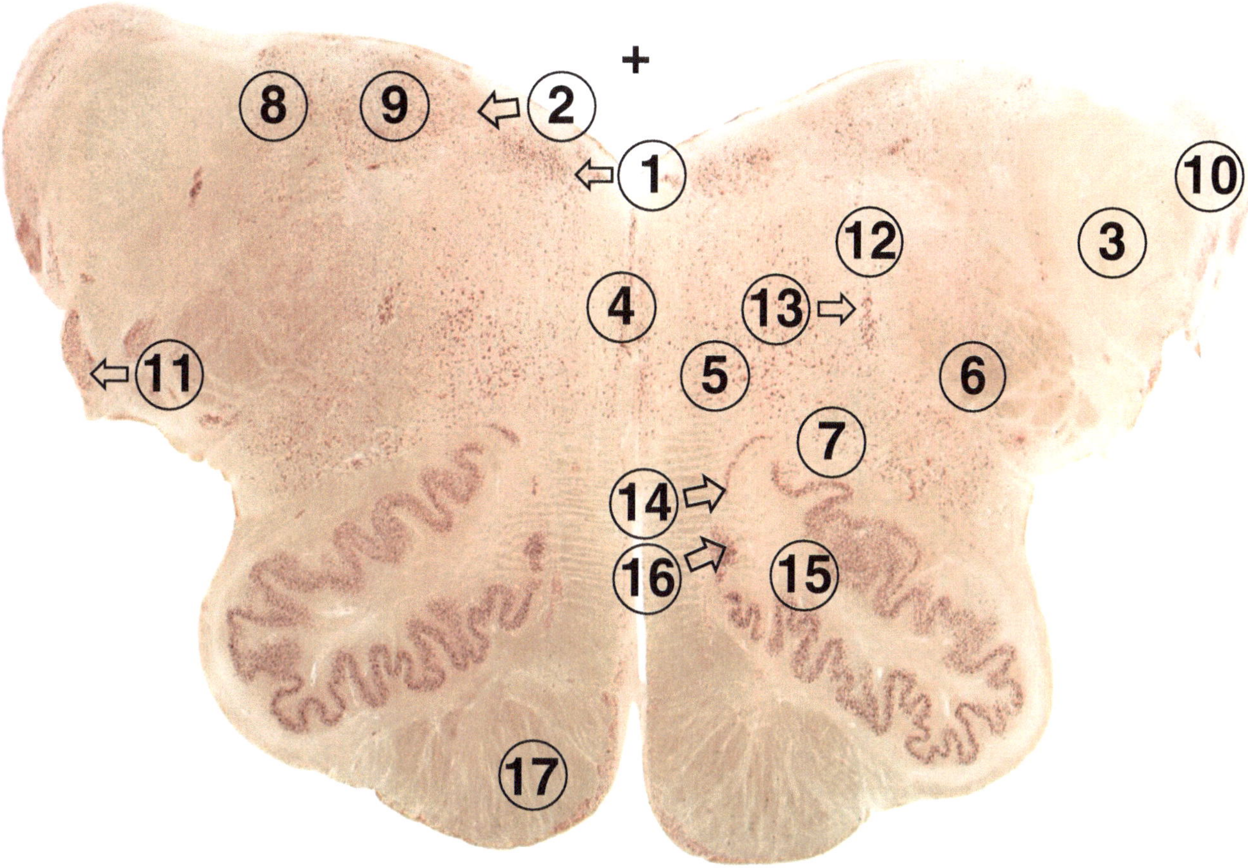

①	Prepositus hypoglossi ncl.	⑩	Posterior cochlear ncl. CNVIII
②	Interpositus ncl.	⑪	Anterior cochlear ncl. CNVIII
③	Inferior cerebellar peduncle	⑫	Parvocellular reticular ncl.
④	Raphe obscurus ncl. See also Fig. 4.35	⑬	Ncl. ambiguus
⑤	Gigantocellular reticular ncl.	⑭	Posterior accessory olivary ncl.
⑥	Spinal ncl. / tract of trigeminal nerve	⑮	Principal olivary ncl.
⑦	Lateral paragigantocellular ncl.	⑯	Medial accessory olivary ncl.
⑧	Inferior vestibular ncl.	⑰	Pyramidal (corticospinal) tract
⑨	Medial vestibular ncl.	+	4th ventricle

Darrow red 14A

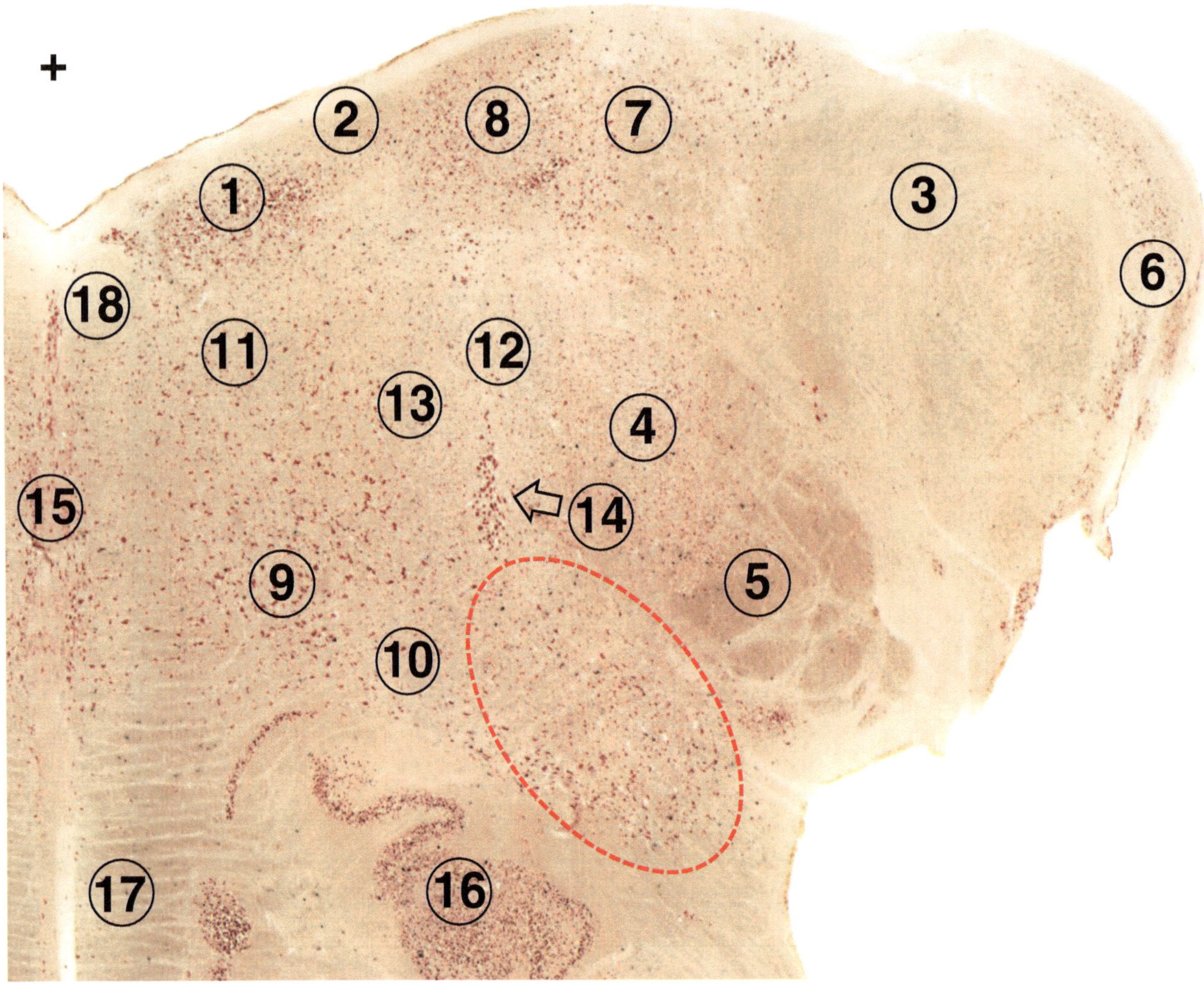

①	Prepositus hypoglossi ncl.	⑪	Posterior paragigantocellular reticular ncl.
②	Interpositus ncl.	⑫	Parvocellular reticular ncl.
③	Inferior cerebellar peduncle	⑬	Intermediate reticular zone
④	Spinal ncl. of trigeminal nerve	⑭	Ncl. ambiguus CNIX, CNX
⑤	Spinal tract of trigeminal nerve	⑮	Raphe obscurus ncl. See also Fig. 4.35
⑥	Cochlear ncll.	⑯	Inferior olivary complex
⑦	Inferior vestibular ncl.	⑰	Medial lemniscus
⑧	Medial vestibular ncl.	⑱	Medial longitudinal fasciculus
⑨	Gigantocellular reticular ncl.	--	Approximate region of the pre-Bötzinger complex
⑩	Lateral paragigantocellular ncl.	+	4th ventricle

Darrow red 15

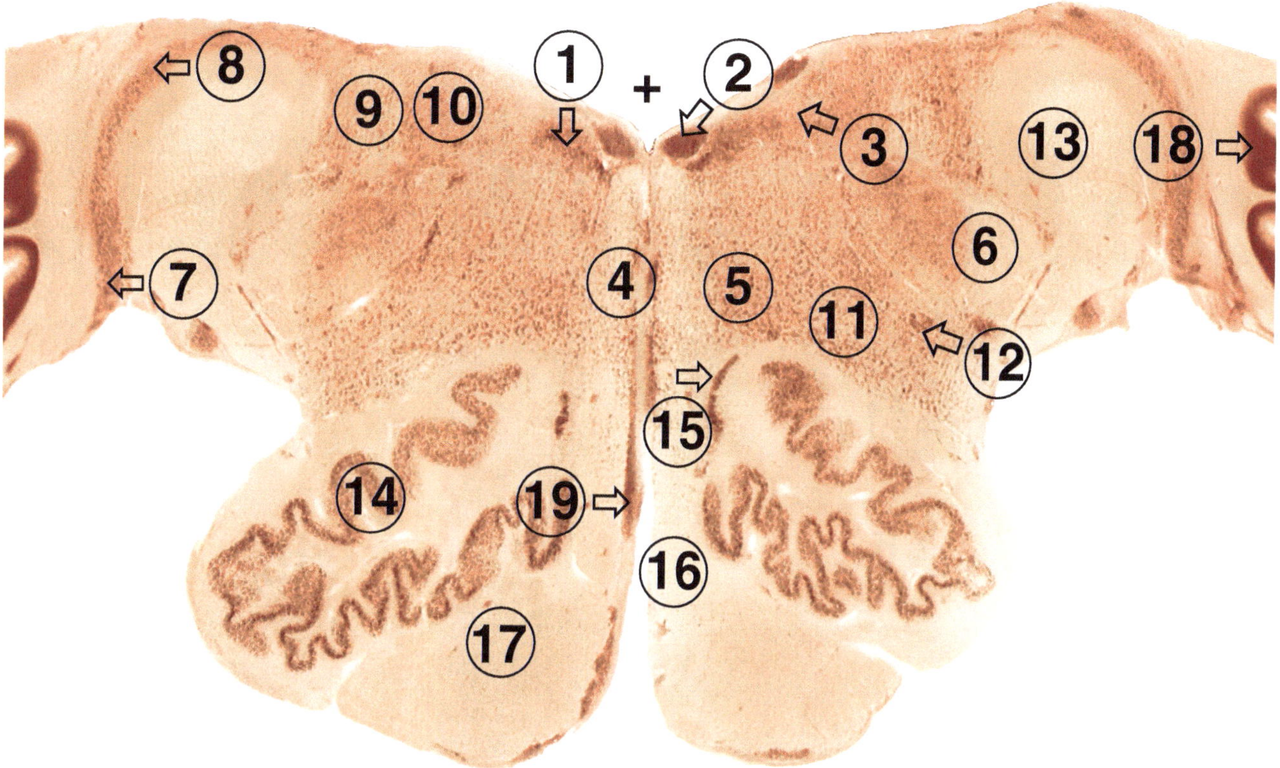

①	Prepositus hypoglossi ncl.	⑪	Lateral paragigantocellular ncl.
②	Posterior paramedian ncl.	⑫	Motor ncl. of facial nerve CNVII
③	Interpositus ncl.	⑬	Inferior cerebellar peduncle
④	Raphe obscurus ncl.	⑭	Principal olivary ncl.
⑤	Gigantocellular reticular ncl.	⑮	Posterior accessory olivary ncl.
⑥	Spinal ncl. / tract of trigeminal nerve CNV	⑯	Medial lemniscus
⑦	Anterior cochlear ncl.	⑰	Pyramidal (corticospinal) tract
⑧	Posterior cochlear ncl.	⑱	Cerebellum
⑨	Inferior vestibular ncl.	⑲	Raphe magnus ncl.
⑩	Medial vestibular ncl.	+	4th ventricle

Darrow red 15A

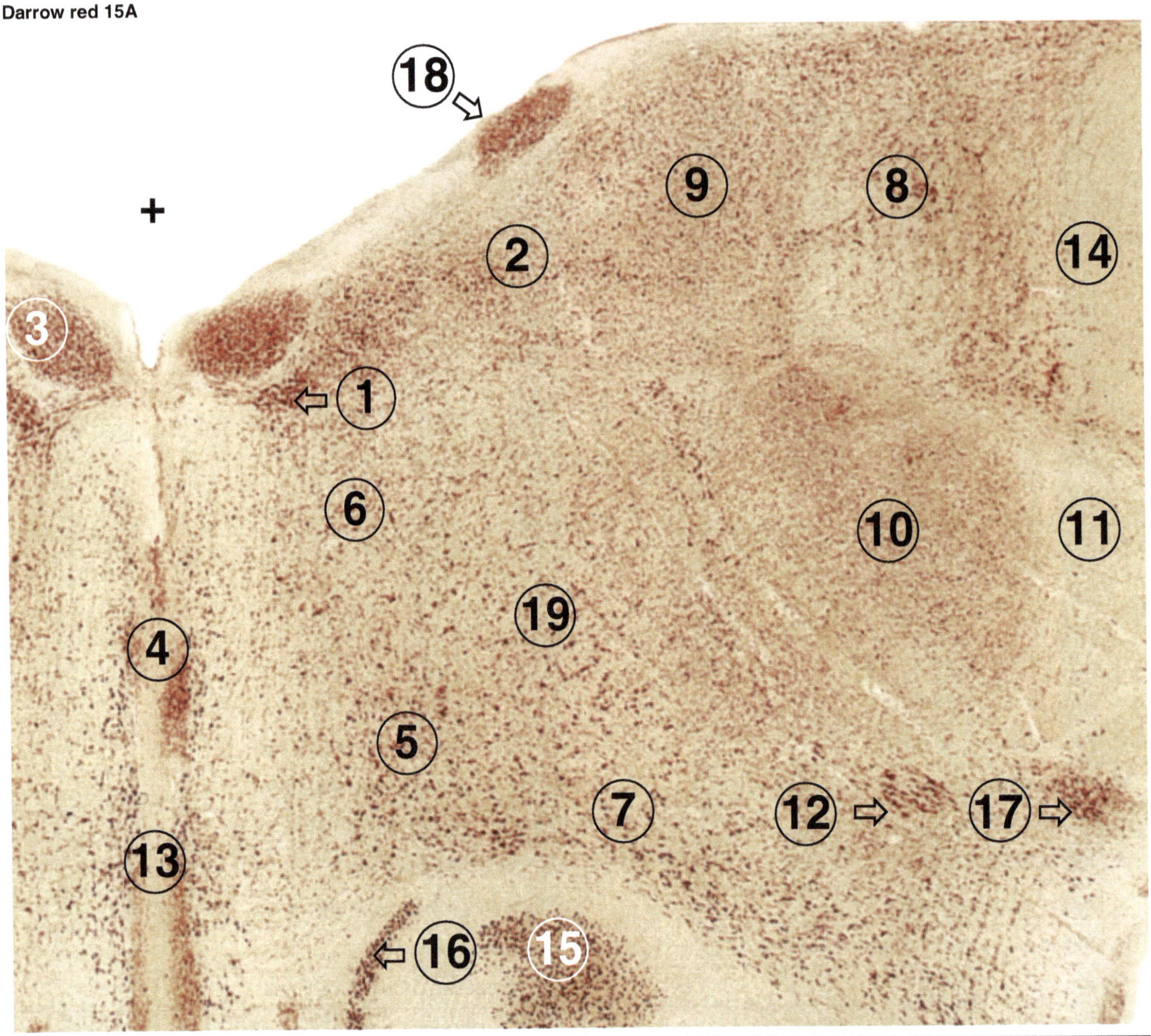

①	Prepositus hypoglossi ncl.		⑪	Spinal ncl. / tract of trigeminal nerve
②	Interpositus ncl.		⑫	Motor ncl. of facial nerve CNVII
③	Posterior paramedian ncl.		⑬	Raphe magnus ncl.
④	Raphe obscurus ncl.		⑭	Inferior cerebellar peduncle
⑤	Gigantocellular reticular ncl.		⑮	Principal olivary ncl.
⑥	Posterior paragigantocellular reticular ncl.		⑯	Posterior accessory olivary ncl.
⑦	Lateral paragigantocellular ncl.		⑰	Lateral reticular ncl.
⑧	Inferior vestibular ncl.		⑱	Subventricular ncl.
⑨	Medial vestibular ncl.		⑲	Intermediate reticular zone
⑩	Parvocellular reticular ncl.		+	4th ventricle

Darrow red 16

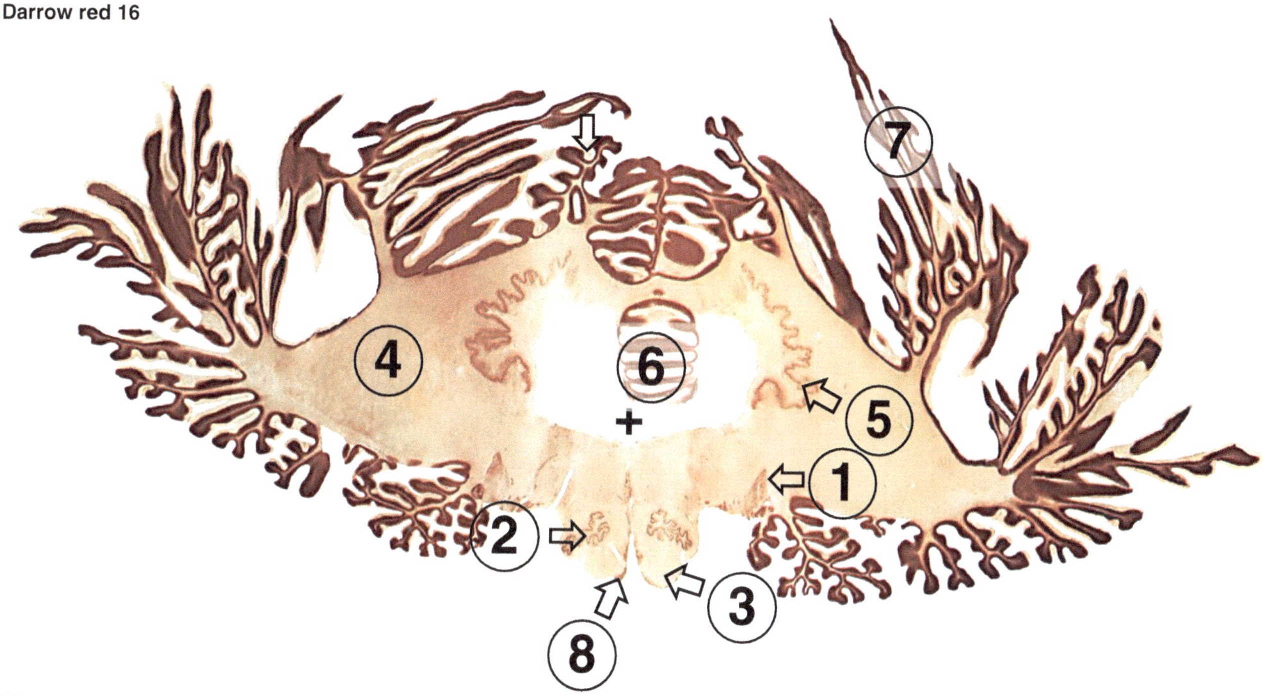

①	Anterior cochlear ncl.	⑤	Dentate ncl.
②	Inferior olivary complex	⑥	Nodule
③	Pyramidal (corticospinal) tract	⑦	Cerebellar cortex
④	White matter of cerebellum	⑧	Arcuate ncl.
		+	4th ventricle

Darrow red 16A

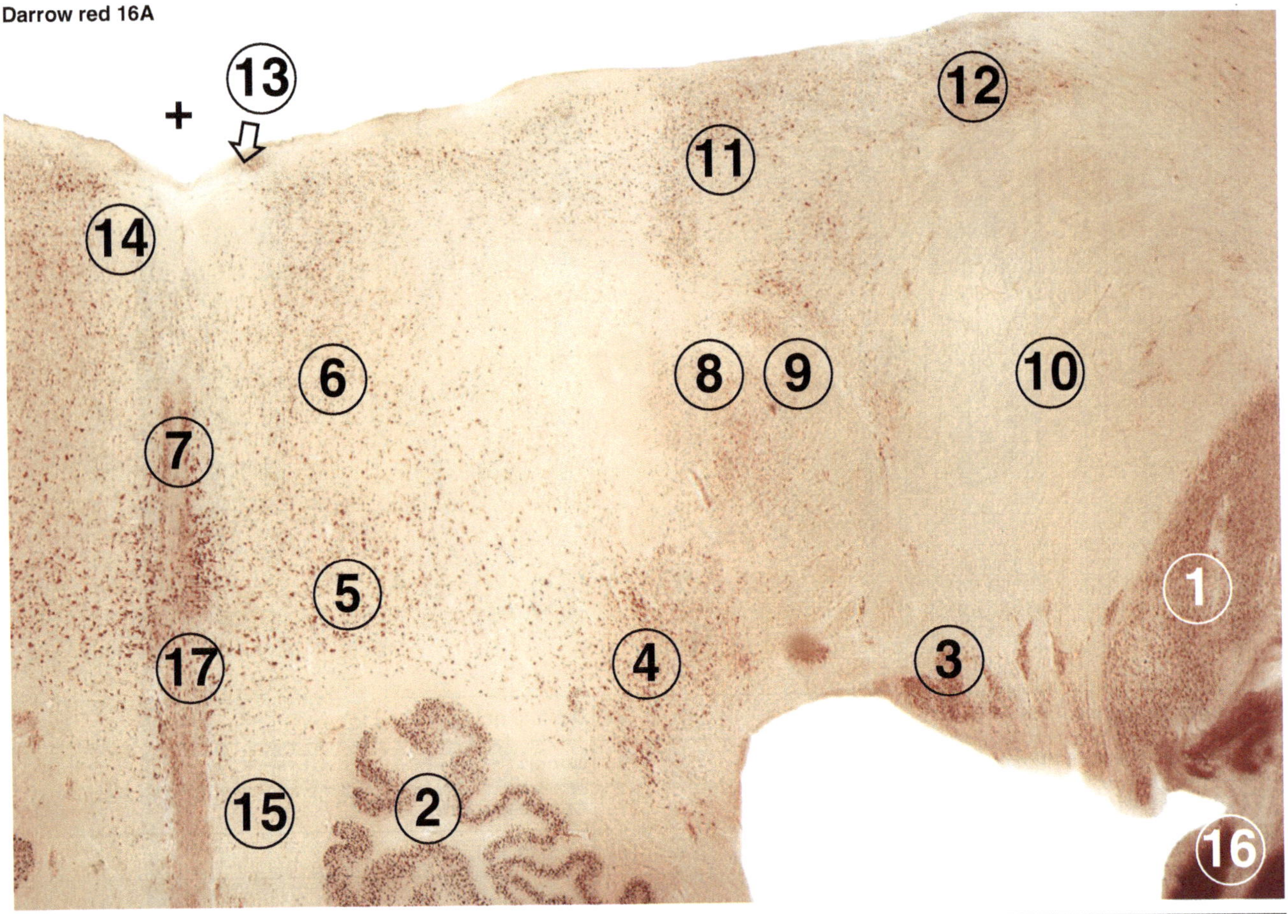

①	Anterior cochlear ncl.	⑩	Inferior cerebellar peduncle
②	Inferior olivary complex	⑪	Medial vestibular ncl.
③	Pontobulbar body	⑫	Inferior vestibular ncl.
④	Lateral paragigantocellular ncl.	⑬	Posterior paramedian ncl.
⑤	Gigantocellular reticular ncl.	⑭	Medial longitudinal fasciculus
⑥	Posterior paragigantocellular reticular ncl.	⑮	Medial lemniscus
⑦	Raphe obscurus ncl.	⑯	Cerebellum
⑧	Spinal ncl. of trigeminal nerve	⑰	Raphe magnus ncl.
⑨	Spinal tract of trigeminal nerve	+	4th ventricle

Darrow red 17

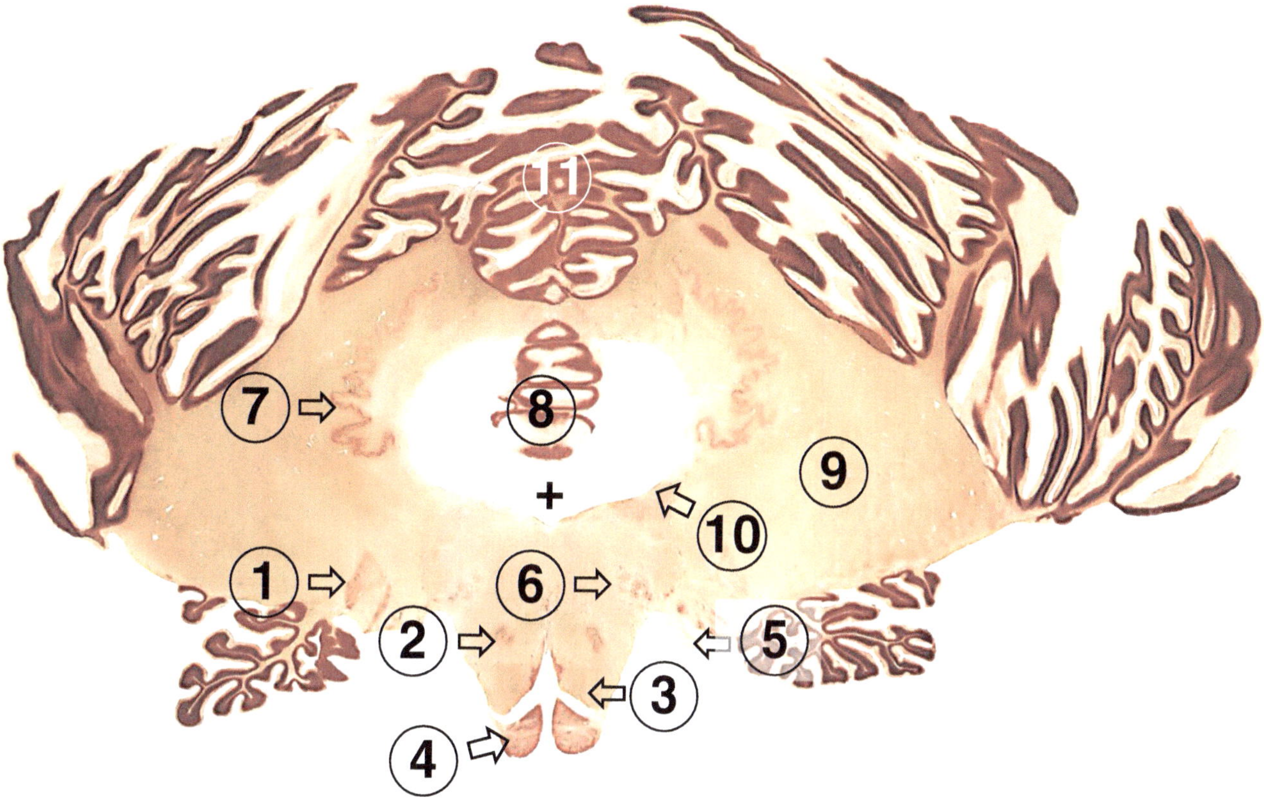

①	Anterior cochlear ncl.	⑦	Dentate ncl.
②	Inferior olivary complex	⑧	Nodule
③	Pyramidal (corticospinal) tract	⑨	Inferior cerebellar peduncle
④	Pontine basis / Pontine ncll.	⑩	Vestibular ncll.
⑤	Vestibulocochlear nerve CNVIII	⑪	Cerebellum
⑥	Motor ncl. of facial nerve CNVII	+	4th ventricle

Darrow red 17A

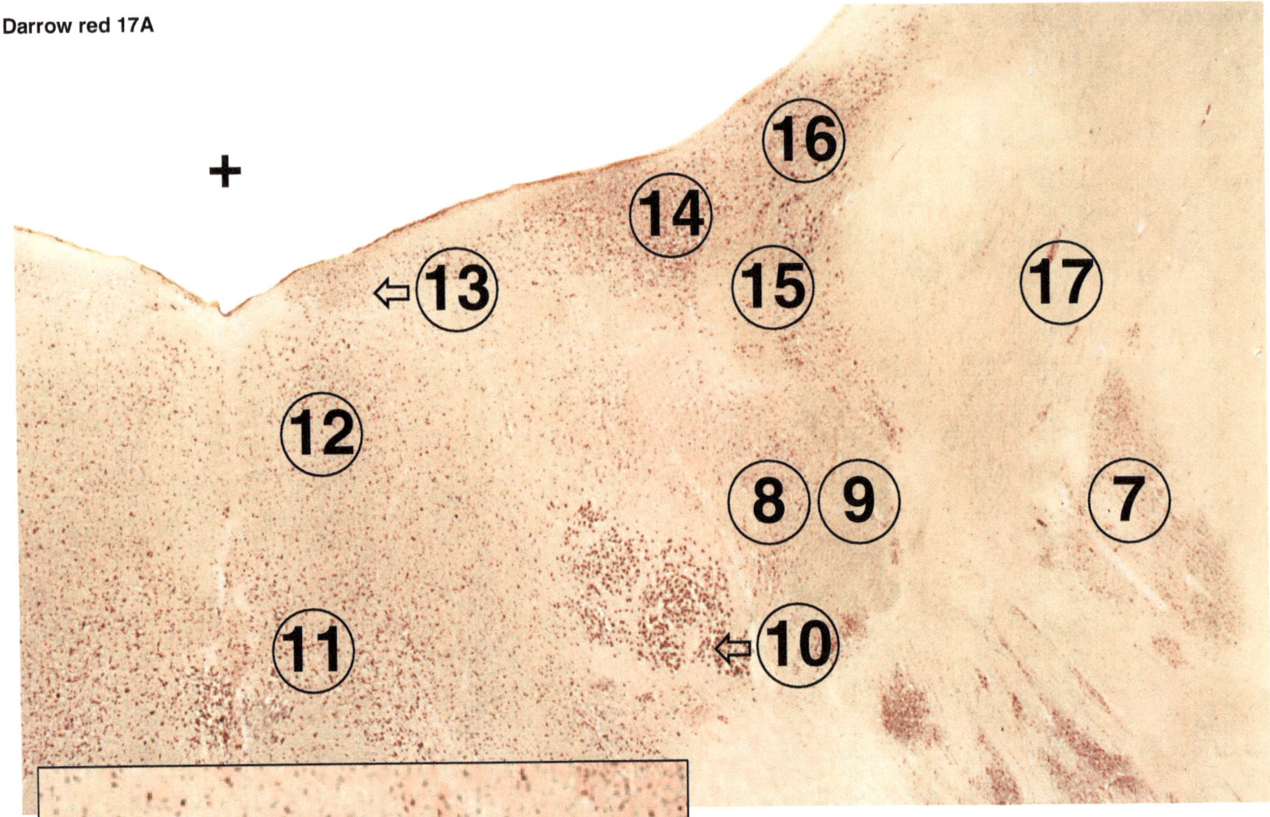

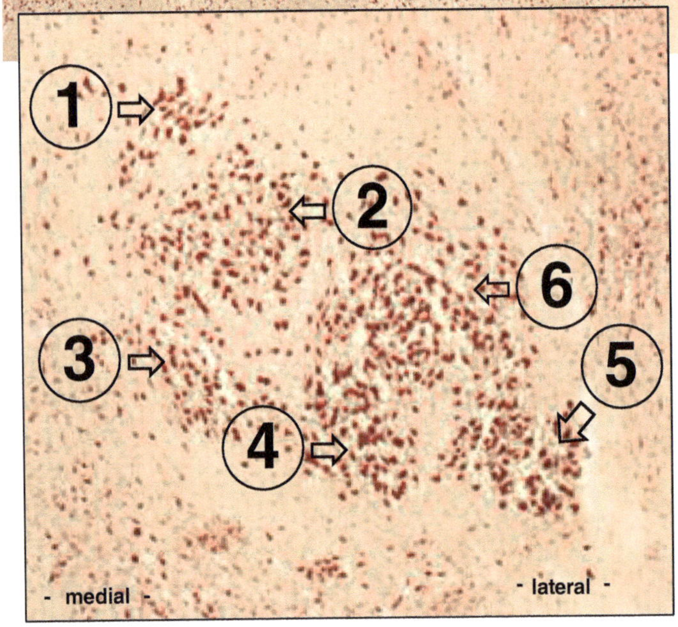

Ncll. of facial nerve CNVII	
①	Dorsal subnucleus
②	Intermediate subnucleus
③	Medial subnucleus
④	Ventromedial subnucleus
⑤	Ventrolateral subnucleus
⑥	Ventral subnucleus

⑦	Anterior cochlear ncl.	⑬	Posterior paramedian ncl.
⑧	Spinal ncl. of trigeminal nerve	⑭	Medial vestibular ncl.
⑨	Spinal tract of trigeminal nerve	⑮	Lateral vestibular ncl.
⑩	Motor ncl. of facial nerve CNVII	⑯	Superior vestibular ncl. CNVIII
⑪	Gigantocellular reticular ncl.	⑰	Middle cerebellar peduncle
⑫	Posterior paragigantocellular reticular ncl.	+	4th ventricle

Darrow red 18

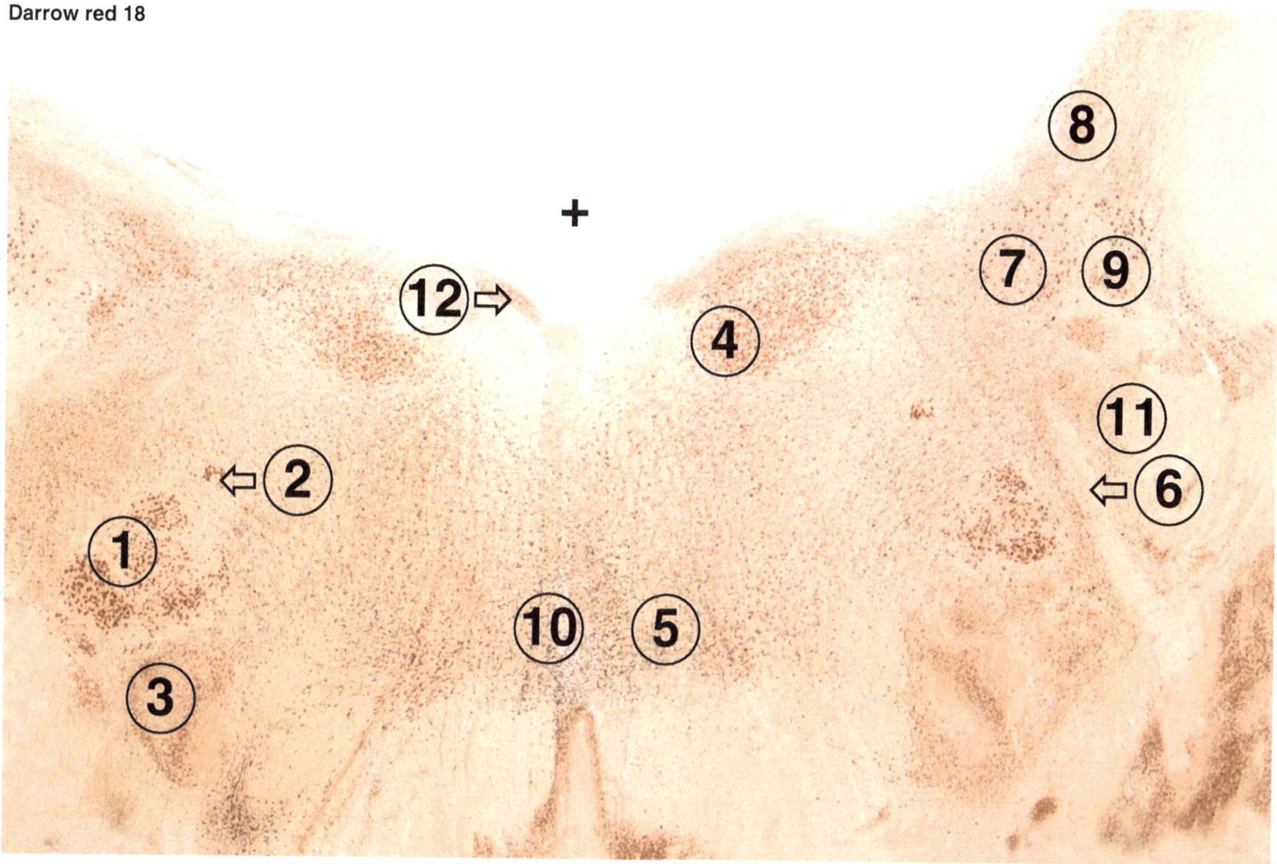

①	Motor ncl. of facial nerve	⑦	Medial vestibular ncl.
②	Retrofacial ncl.	⑧	Superior vestibular ncl. CNVIII
③	Superior olivary complex	⑨	Lateral vestibular ncl.
④	Ncl. of abducens nerve CNVI	⑩	Raphe magnus ncl.
⑤	Gigantocellular reticular ncl.	⑪	Spinal ncl. / tract of trigeminal nerve
⑥	Facial nerve, descending part See atlas part Campbell 8A, 9B	⑫	Posterior paramedian ncl.
		+	4th ventricle

Darrow red 18A

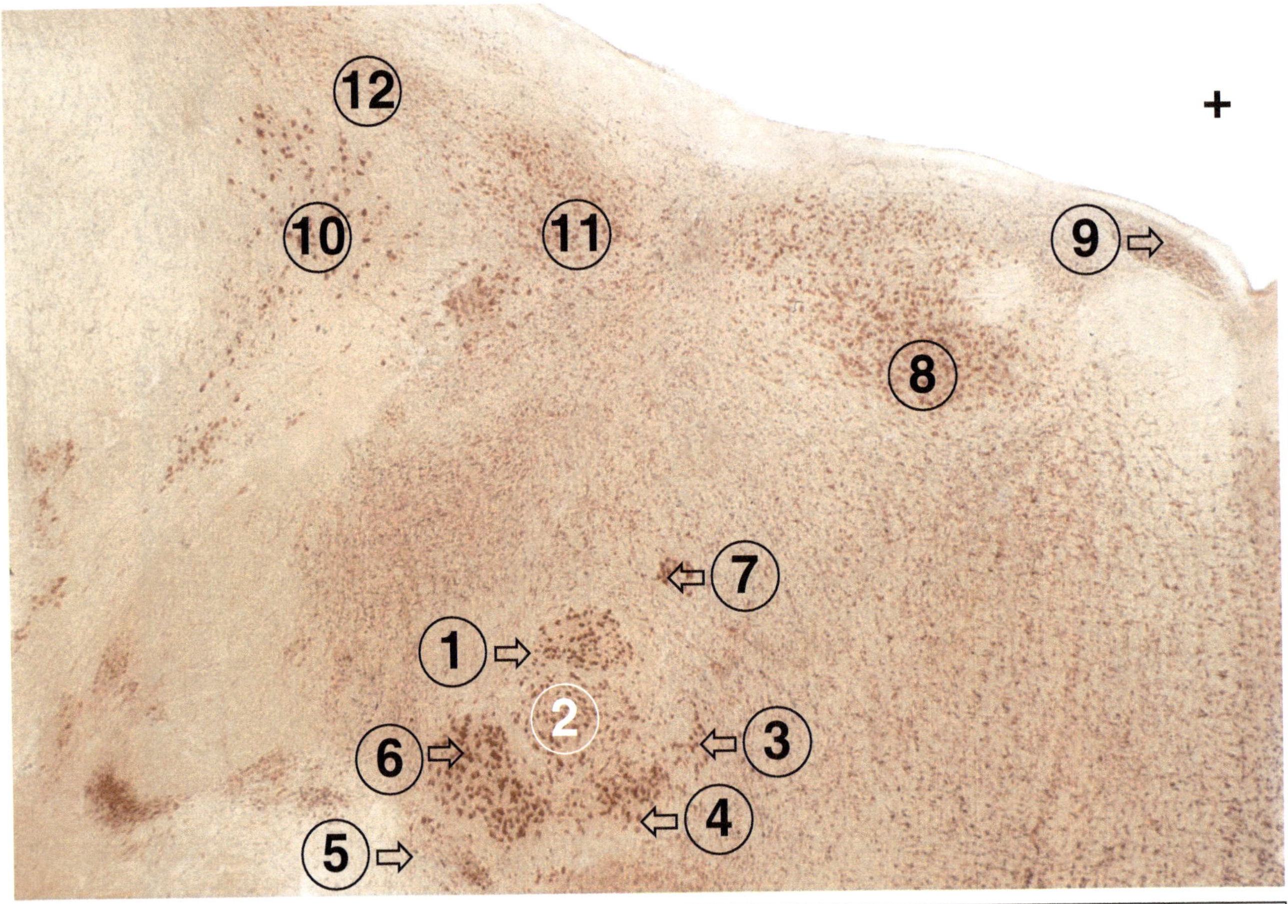

	Motor ncll. of facial nerve CNVII		
⑦	Retrofacial ncl.		
①	Dorsal subnucleus	⑧	Ncl. of abducens nerve
②	Intermediate subnucleus	⑨	Posterior paramedian ncl.
③	Medial subnucleus	⑩	Lateral vestibular ncl.
④	Ventromedial subnucleus	⑪	Medial vestibular ncl.
⑤	Ventrolateral subnucleus	⑫	Superior vestibular ncl.
⑥	Ventral subnucleus	+	4th ventricle

Darrow red 19

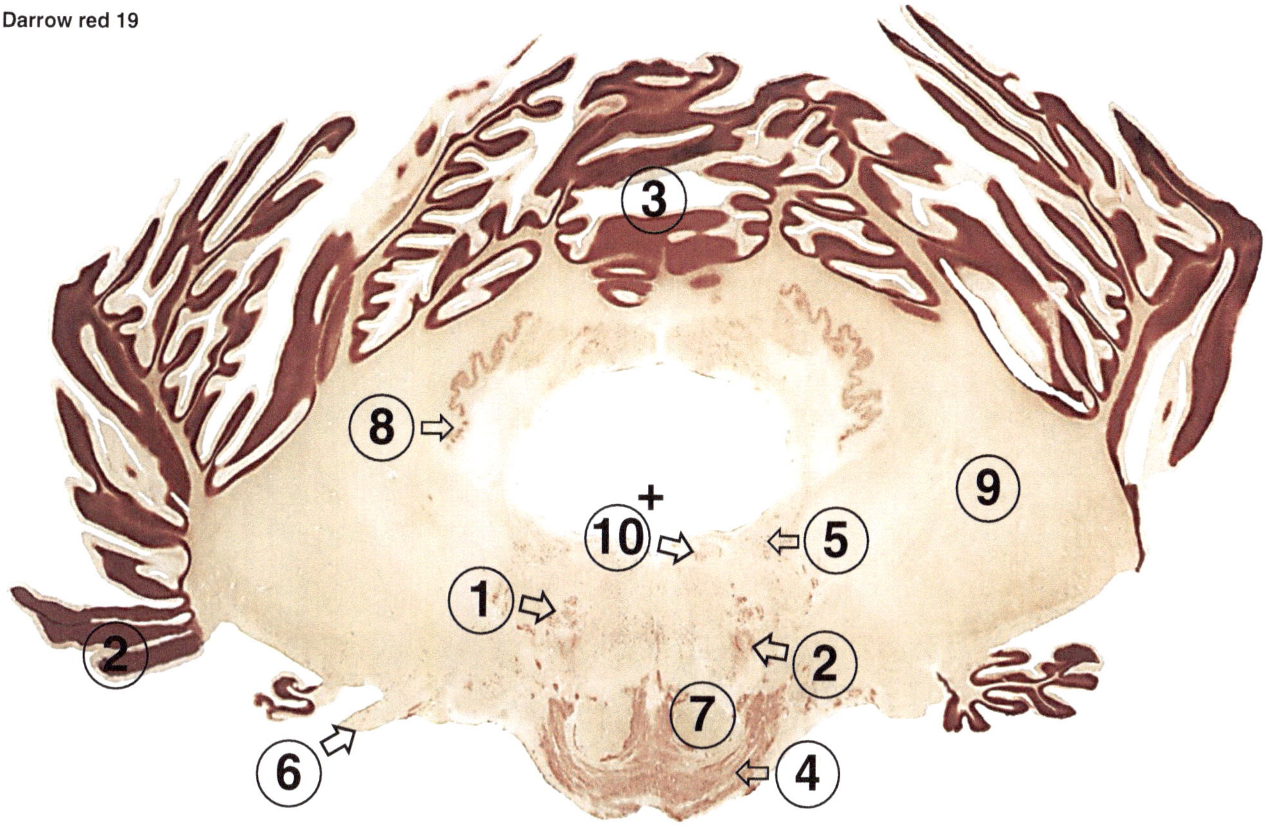

①	Motor ncl. of facial nerve
②	Superior olivary complex
③	Cerebellum
④	Pontine basis / Pontine ncll.
⑤	Vestibular ncll.
⑥	Vestibulocochlear nerve CNVIII
⑦	Pyramidal (corticospinal) tract
⑧	Dentate ncl.
⑨	Middle cerebellar peduncle
⑩	Ncl. of abducens nerve CNVI
+	4th ventricle

Darrow red 19A

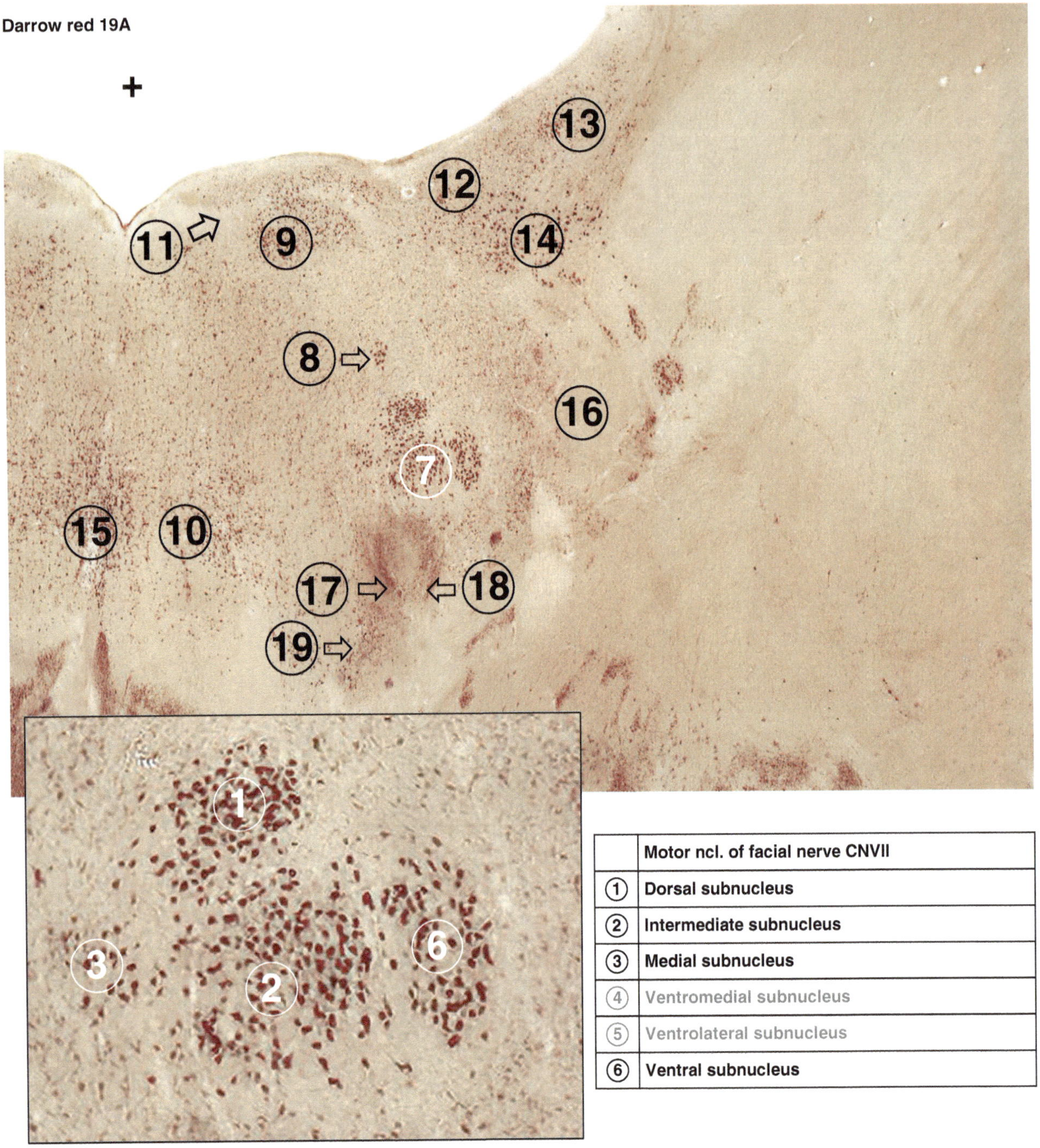

	Motor ncl. of facial nerve CNVII
①	Dorsal subnucleus
②	Intermediate subnucleus
③	Medial subnucleus
④	Ventromedial subnucleus
⑤	Ventrolateral subnucleus
⑥	Ventral subnucleus

⑦	Motor ncl. of facial nerve	⑭	Lateral vestibular ncl.
⑧	Retrofacial ncl.	⑮	Raphe magnus ncl.
⑨	Ncl. of abducens nerve	⑯	Spinal ncl. / tract of CNV
⑩	Gigantocellular reticular ncl.	⑰	Medial superior olivary ncl.
⑪	Genu of facial nerve	⑱	Lateral superior olivary ncl.
⑫	Medial vestibular ncl.	⑲	Ncl. of trapezoid body
⑬	Superior vestibular ncl.	+	4th ventricle

Darrow red 20

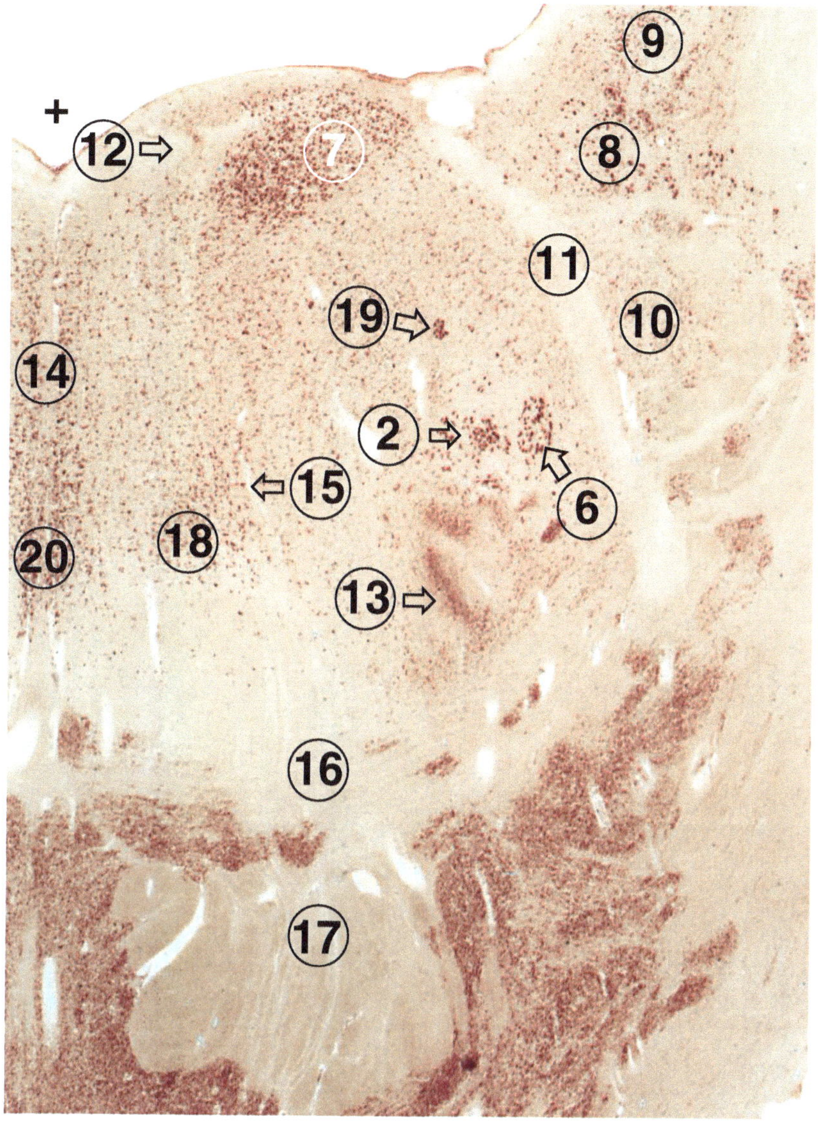

	Ncll. of facial nerve CNVII
①	Dorsal subnucleus
②	Intermediate subnucleus
③	Medial subnucleus
④	Ventromedial subnucleus
⑤	Ventrolateral subnucleus
⑥	Ventral subnucleus

⑦	Ncl. of abducens nerve
⑧	Lateral vestibular ncl.
⑨	Superior vestibular ncl.
⑩	Spinal ncl. / tract CNV
⑪	Facial nerve, descending part
⑫	Genu of facial nerve
⑬	Superior olivary complex
⑭	Raphe interpositus ncl.
⑮	Abducens nerve
⑯	Medial lemniscus
⑰	Pyramidal (corticospinal) tract
⑱	Gigantocellular reticular ncl.
⑲	Retrofacial ncl.
⑳	Raphe magnus ncl.
+	4th ventricle

Darrow red 21

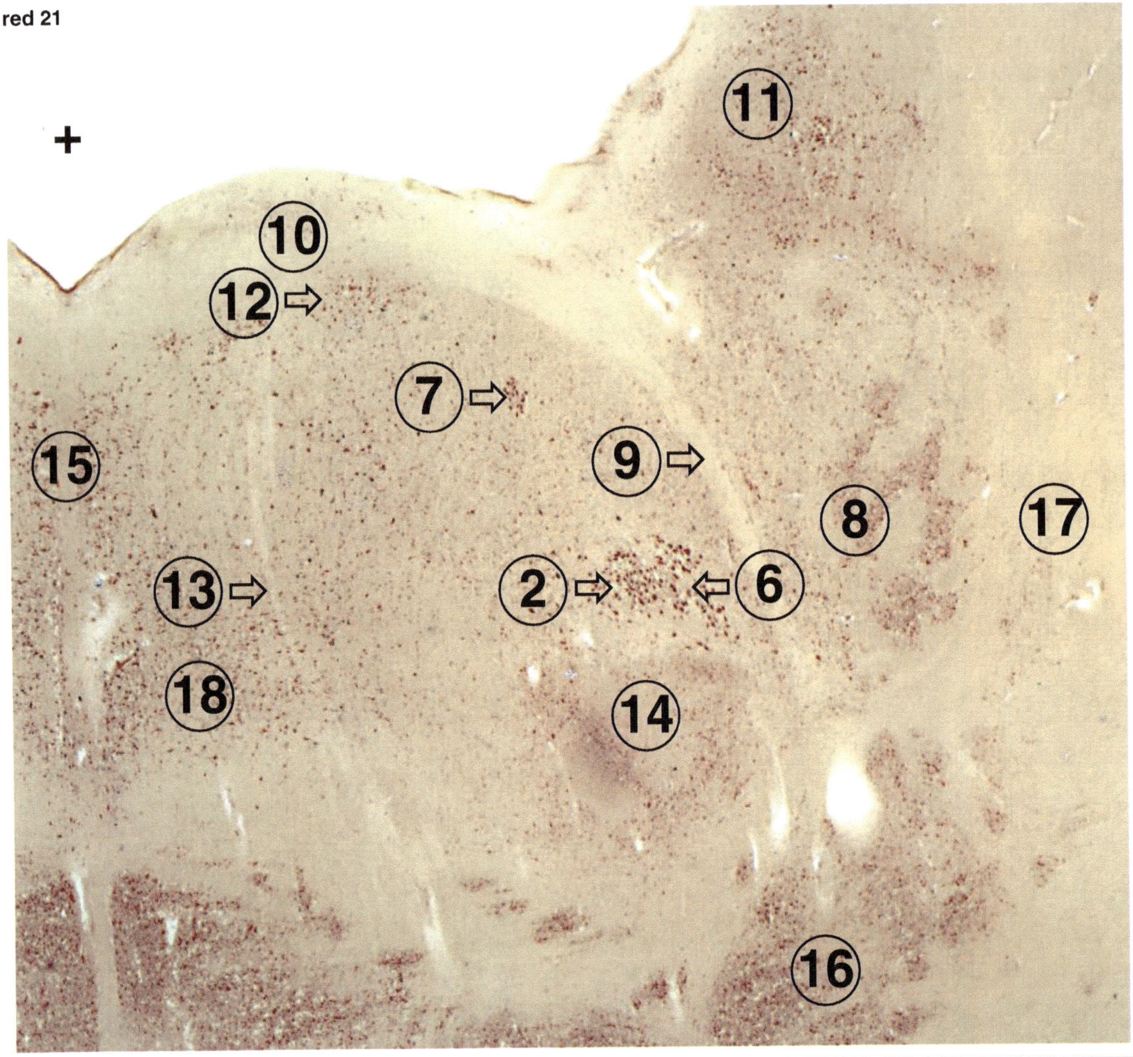

	Ncll. of facial nerve CNVII	④	Ventromedial subnucleus
①	Dorsal subnucleus	⑤	Ventrolateral subnucleus
②	Intermediate subnucleus	⑥	Ventral subnucleus
③	Medial subnucleus		

⑦	Retrofacial nucleus	⑭	Superior olivary complex
⑧	Spinal ncl. of trigeminal nerve	⑮	Raphe interpositus ncl.
⑨	Facial nerve / descending part	⑯	Pontine ncll.
⑩	Genu of facial nerve	⑰	Middle cerebellar peduncle
⑪	Vestibular ncll.	⑱	Gigantocellular reticular ncl.
⑫	Ncl. of abducens nerve	+	4th ventricle
⑬	Abducens nerve		

Darrow red 22

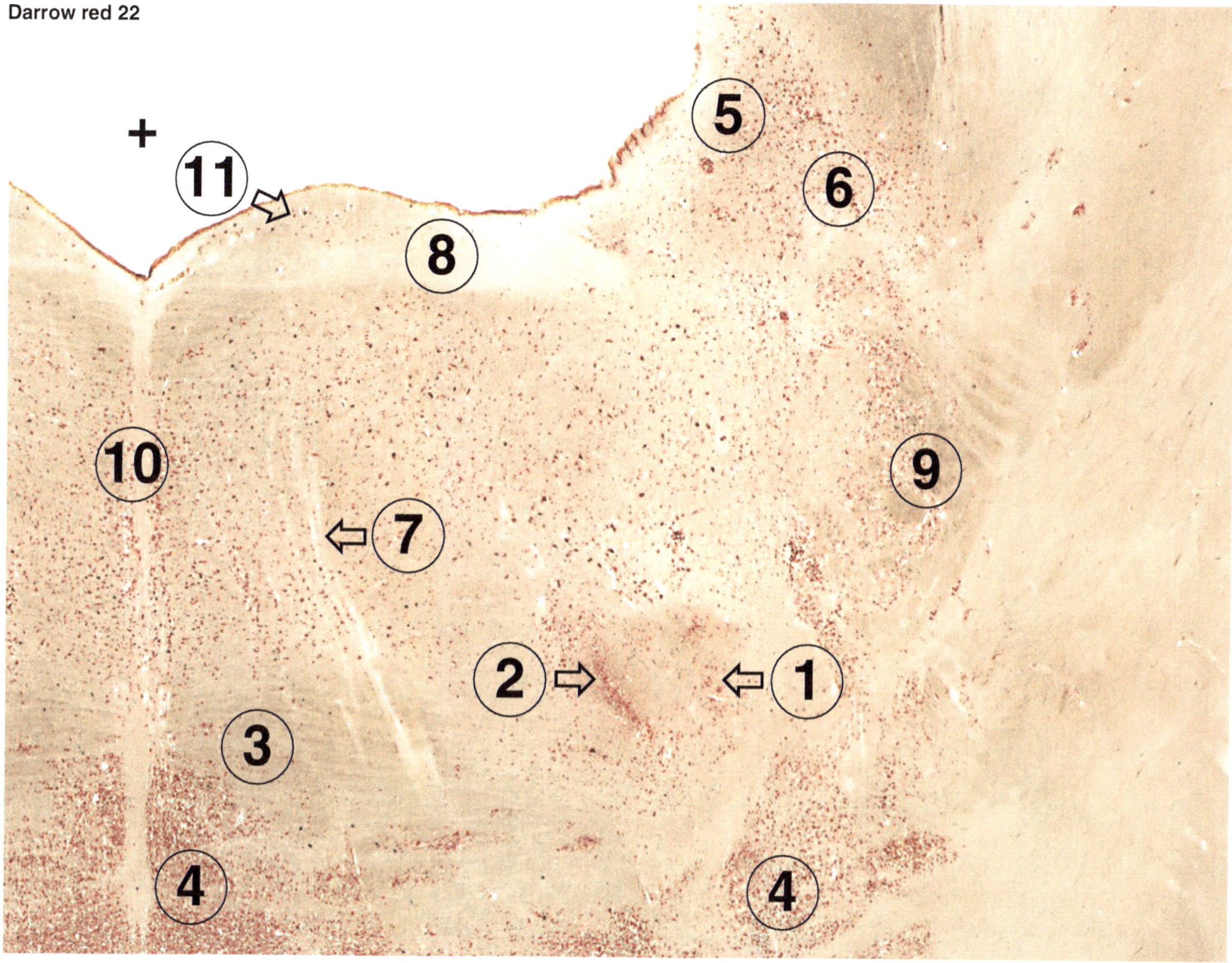

①	Lateral superior olivary ncl. CNVIII	⑦	Abducens nerve See atlas part Campbell 9B
②	Medial superior olivary ncl. CNVIII	⑧	Genu of facial nerve
③	Trapezoid body CNVIII See atlas part Campbell 9B	⑨	Spinal ncl. / tract of trigeminal nerve
④	Pontine ncll.	⑩	Raphe interpositus ncl.
⑤	Superior vestibular ncl.	⑪	Supragenual ncl.
⑥	Lateral vestibular ncl.	+	4th ventricle

Darrow red 23

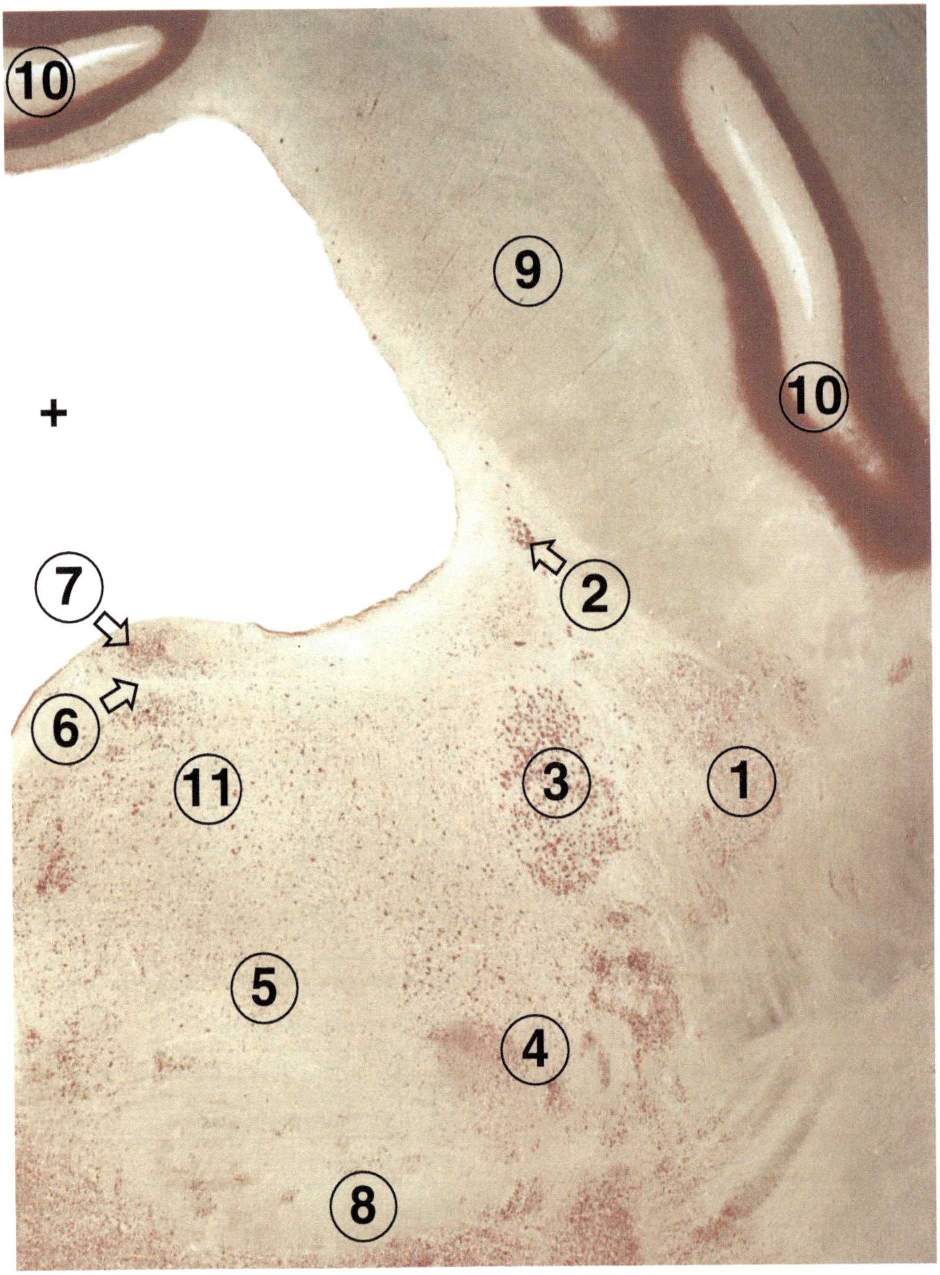

①	Principal sensory ncl. of trigeminal nerve CNV	⑦	Supragenual ncl.
②	Mesencephalic ncl. of trigeminal nerve CNV	⑧	Medial lemniscus
③	Motor ncl. of trigeminal nerve CNV	⑨	Superior cerebellar peduncle
④	Superior olivary complex	⑩	Cerebellum
⑤	Pontine reticular ncl., caudal part	⑪	Dorsomedial tegmental area
⑥	Genu of facial nerve	+	4th ventricle

Darrow red 24

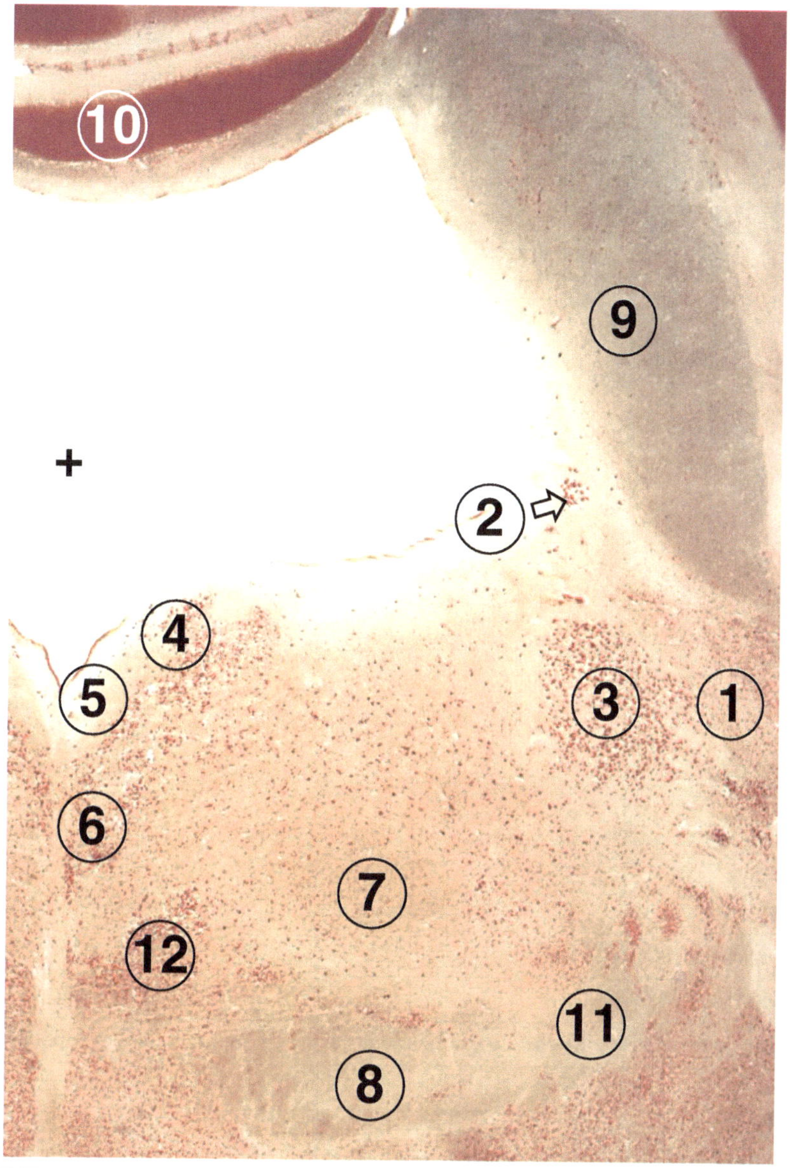

①	Principal sensory trigeminal ncl.	⑦	Pontine reticular ncl., oral part
②	Mesencephalic ncl. of trigeminal nerve	⑧	Medial lemniscus
③	Motor ncl. of trigeminal nerve	⑨	Superior cerebellar peduncle
④	Dorsal raphe ncl., caudal lamellar subnucleus	⑩	Cerebellum
⑤	Medial longitudinal fasciculus	⑪	Lateral lemniscus
⑥	Dorsal raphe ncl., caudal lamellar subnucleus	⑫	Reticulotegmental ncl.
		+	4th ventricle

Darrow red 25

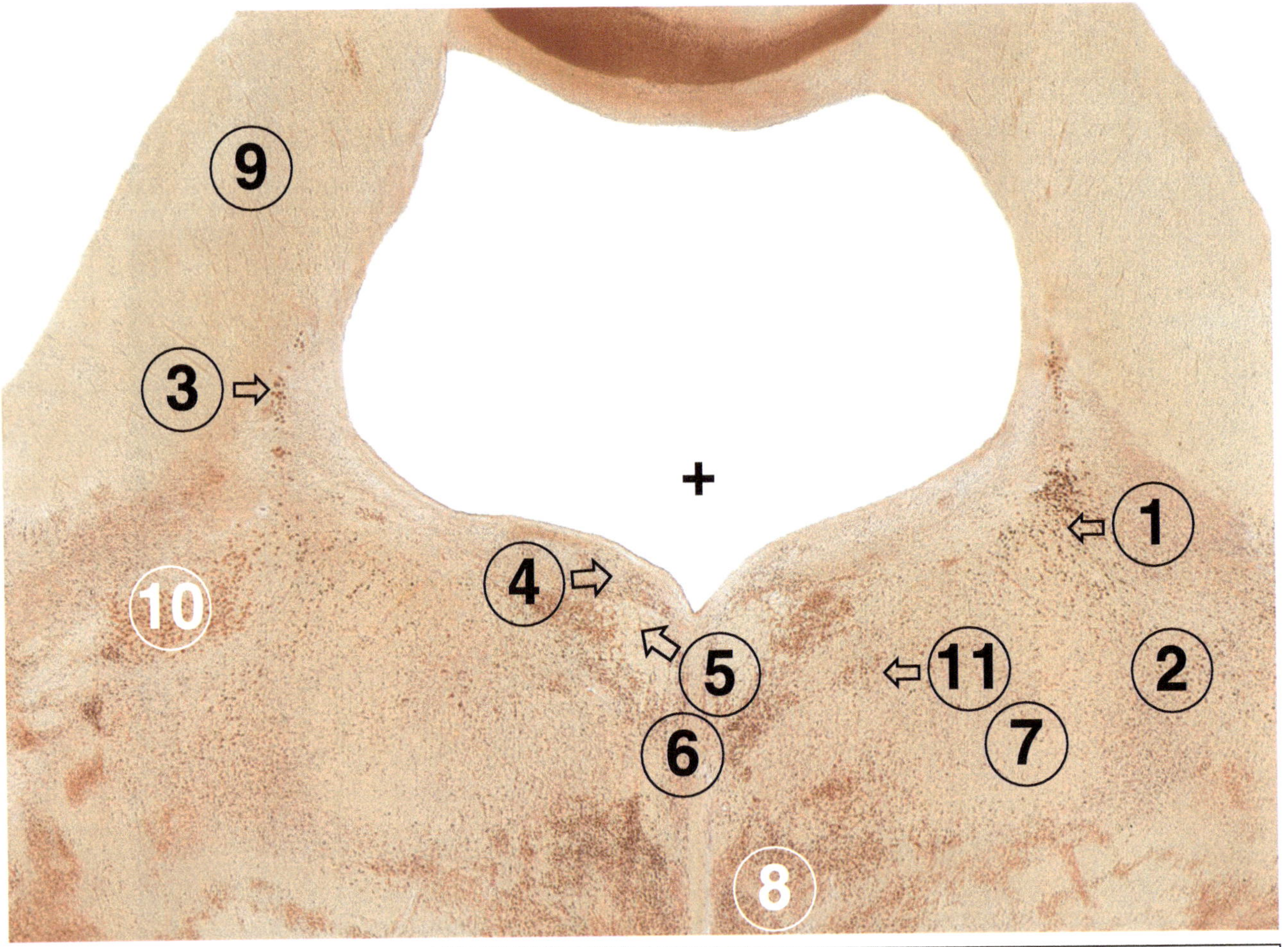

①	Locus caeruleus See atlas part Campbell 15	⑦	Central tegmental tract See atlas part Campbell 13
②	Subcaerulean ncl.	⑧	Reticulotegmental ncl.
③	Mesencephalic ncl. of trigeminal nerve See atlas part Campbell 13	⑨	Superior cerebellar peduncle
④	Dorsal raphe ncl., caudal lamellar subnucleus	⑩	Motor ncl. of trigeminal nerve
⑤	Medial longitudinal fasciculus See atlas part Campbell 13	⑪	Dorsomedial tegmental area
⑥	Dorsal raphe ncl., caudal lamellar subnucleus	+	4th ventricle

Darrow red 25A

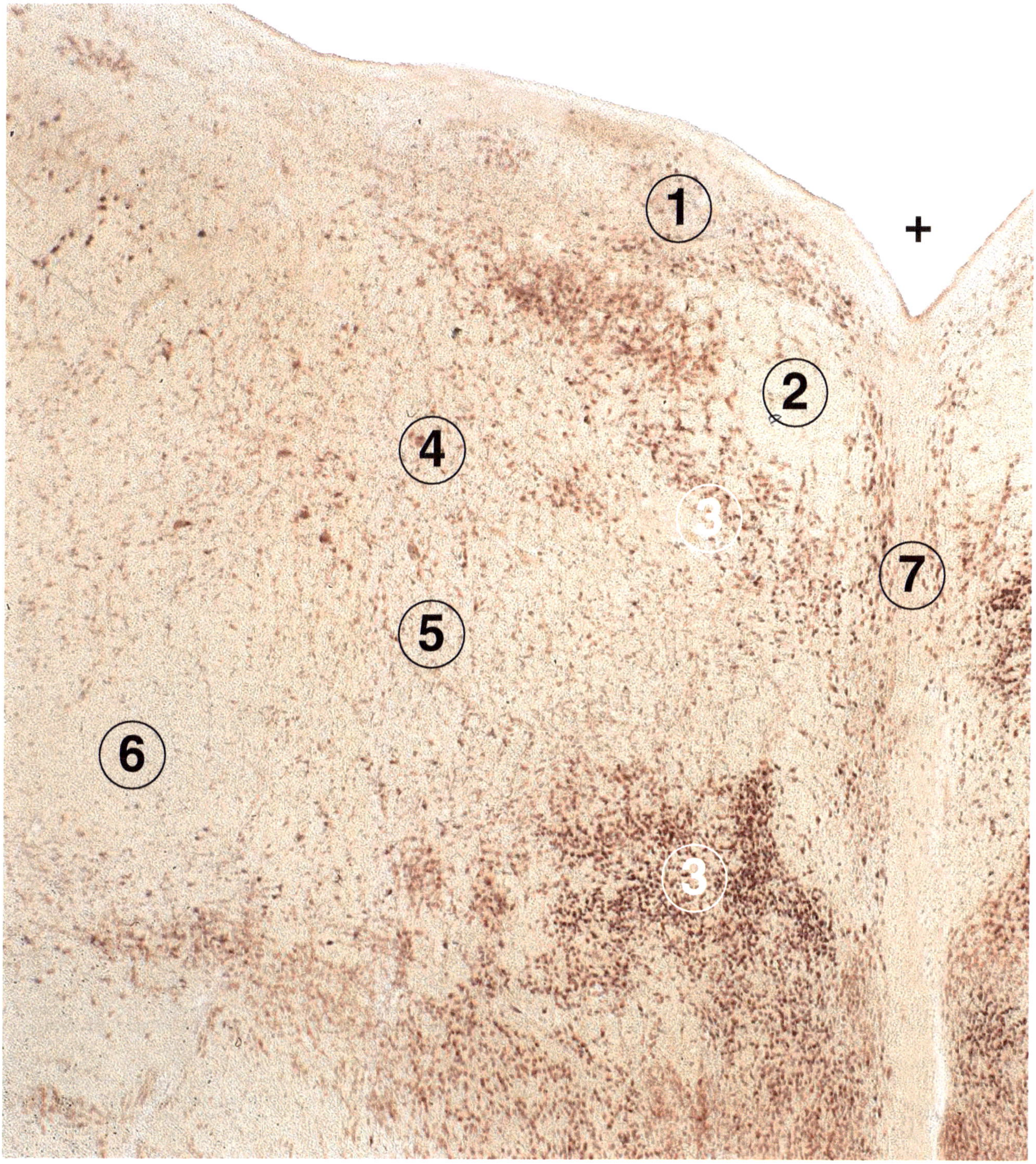

①	Dorsal raphe ncl., caudal lamellar subnucleus	⑤	Pontine reticular ncl., oral part
②	Medial longitudinal fasciculus See atlas part Campbell 15	⑥	Central tegmental tract See atlas part Campbell 15
③	Reticulotegmental ncl.	⑦	Dorsal raphe nucleus, caudal lamellar subnucleus
④	Dorsomedial tegmental area	+	4th ventricle

Darrow red 26

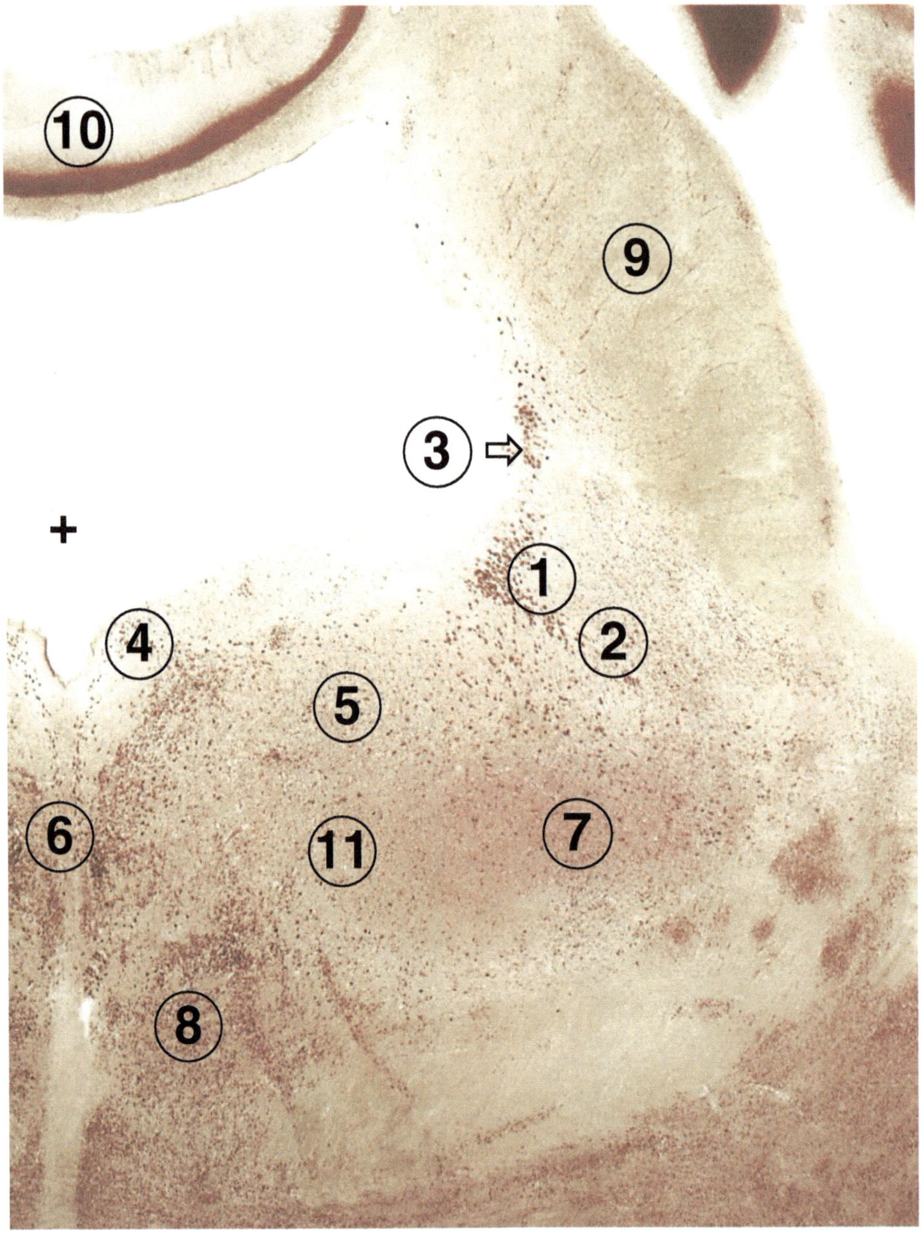

①	Locus caeruleus See atlas part Campbell 13-15	⑦	Central tegmental tract See atlas part Campbell 13-15
②	Subcaerulean ncl.	⑧	Reticulotegmental ncl.
③	Mesencephalic ncl. of trigeminal nerve	⑨	Superior cerebellar peduncle See atlas part Campbell 13-15
④	Dorsal raphe ncl., caudal lamellar subnucleus	⑩	Cerebellum
⑤	Dorsomedial tegmental area	⑪	Pontine reticular ncl., oral part
⑥	Dorsal raphe ncl., caudal lamellar subnucleus	+	4th ventricle

Darrow red 27

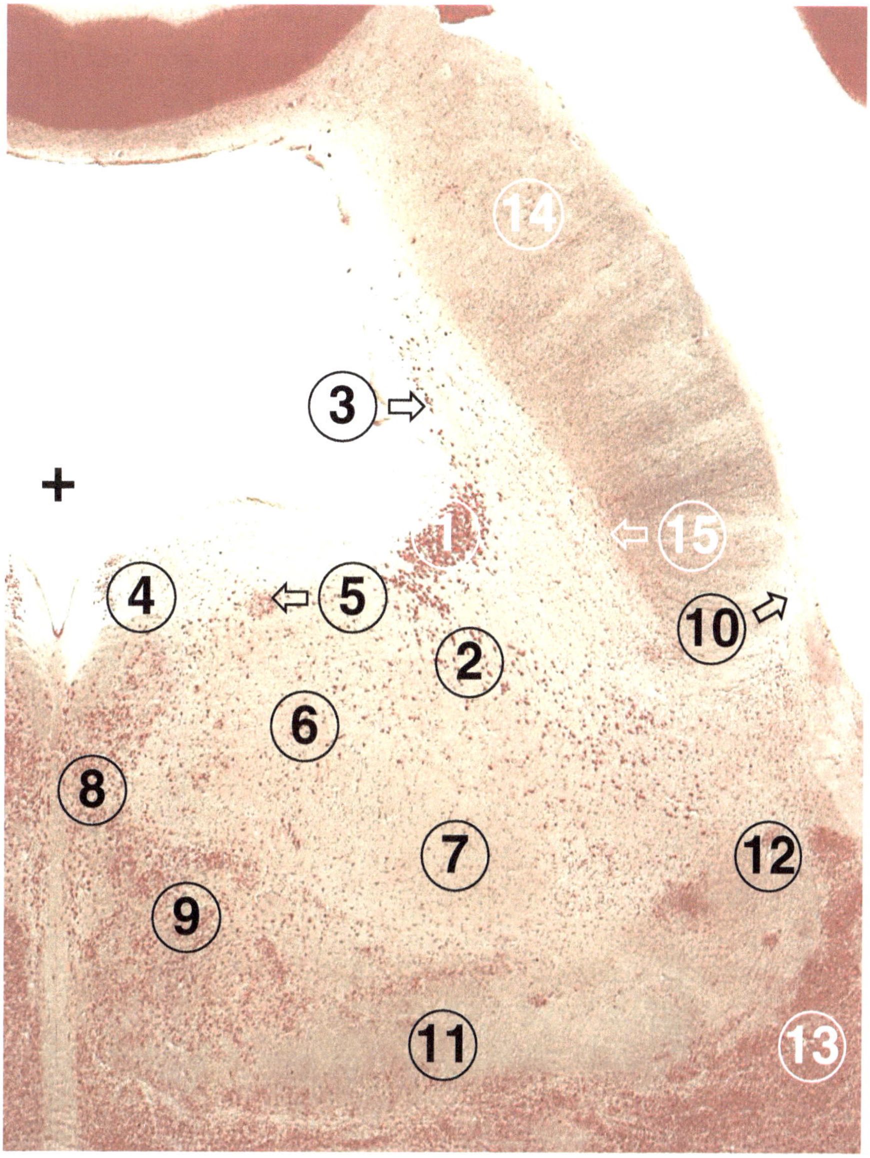

①	Locus caeruleus	⑨	Reticulotegmental ncl.
②	Subcaerulean ncl.	⑩	Subpeduncular tegmental (pigmented) ncl.
③	Mesencephalic ncl. of trigeminal nerve	⑪	Medial lemniscus
④	Dorsal raphe ncl., caudal lamellar subnucleus	⑫	Lateral lemniscus
⑤	Posterodorsal tegmental ncl.	⑬	Pontine ncll.
⑥	Pontine reticular ncl., oral part	⑭	Superior cerebellar peduncle
⑦	Central tegmental tract	⑮	Medial parabrachial ncl.
⑧	Dorsal raphe nucleus, caudal lamellar subnucleus	+	4th ventricle

Darrow red 28

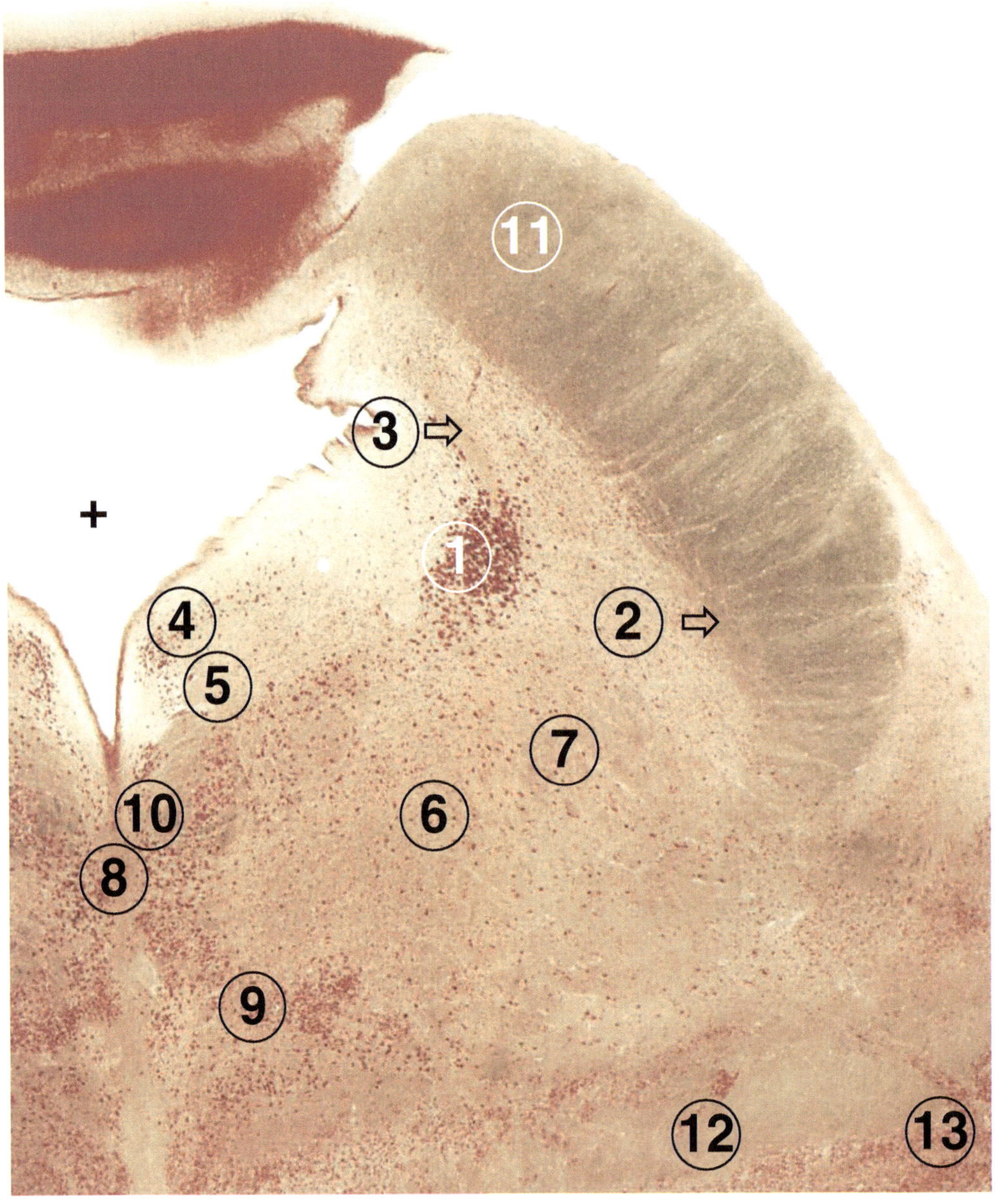

①	Locus caeruleus	⑧	Dorsal raphe nucleus, caudal lamellar subnucleus
②	Medial parabrachial ncl.	⑨	Reticulotegmental ncl.
③	Mesencephalic ncl. and tract of trigeminal nerve See atlas part Campbell 15	⑩	Medial longitudinal fasciculus See atlas part Campbell 15
④	Dorsal raphe ncl., caudal lamellar subnucleus	⑪	Superior cerebellar peduncle
⑤	Posterodorsal tegmental ncl.	⑫	Medial lemniscus
⑥	Pontine reticular ncl., oral part	⑬	Pontine ncll.
⑦	Central tegmental tract	+	4th ventricle

Darrow red 29

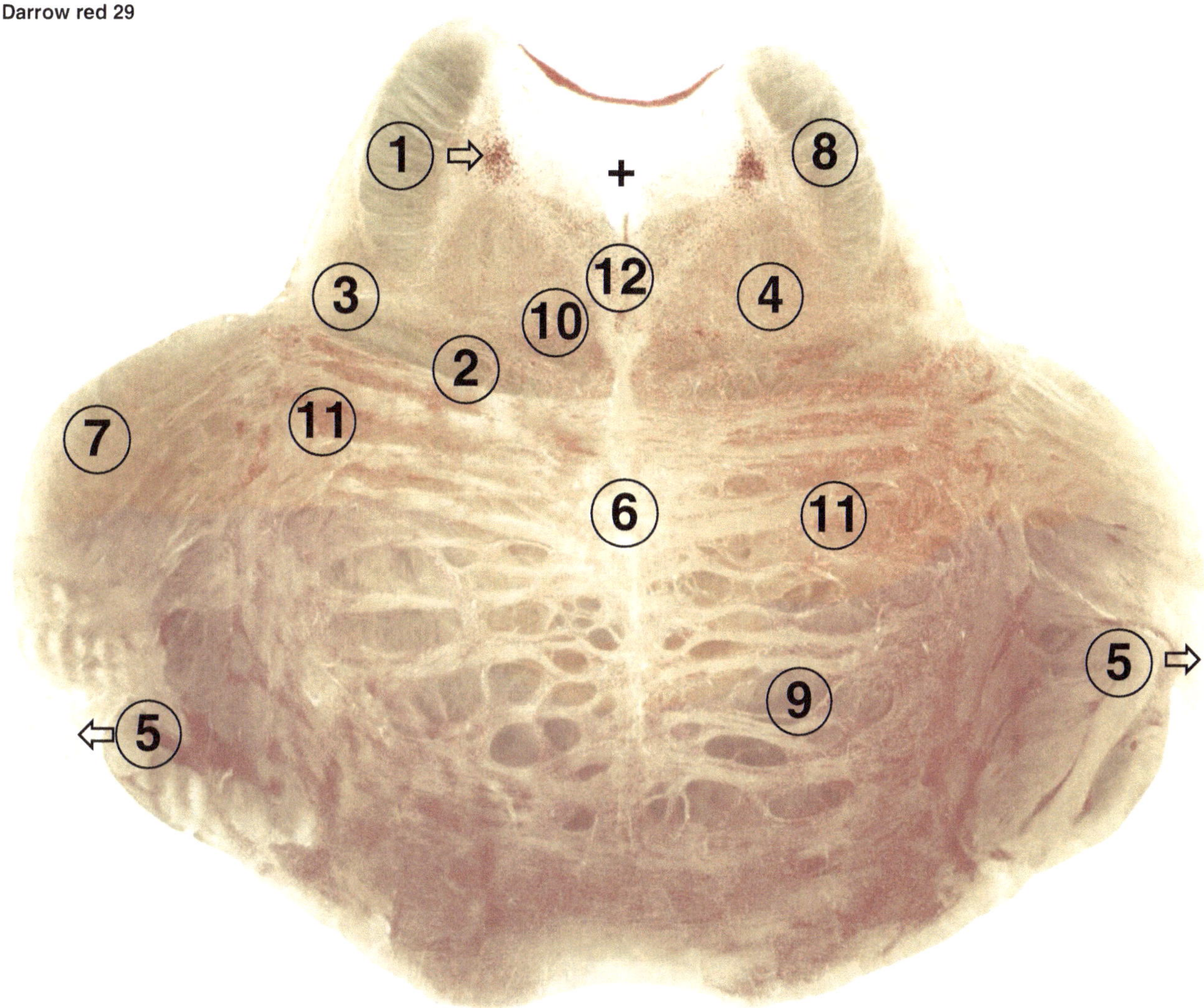

①	Locus caeruleus	⑧	Superior cerebellar peduncle
②	Medial lemniscus	⑨	Pyramidal (corticospinal) tract
③	Lateral lemniscus		②-⑨ See atlas part Campbell 13-15
④	Central tegmental tract	⑩	Reticulotegmental ncl.
⑤	Trigeminal nerve CNV	⑪	Pontine ncll.
⑥	Transverse pontine fibers	⑫	Dorsal raphe ncl., caudal lamellar subnucleus
⑦	Middle cerebellar peduncle	+	4th ventricle

Darrow red 29A

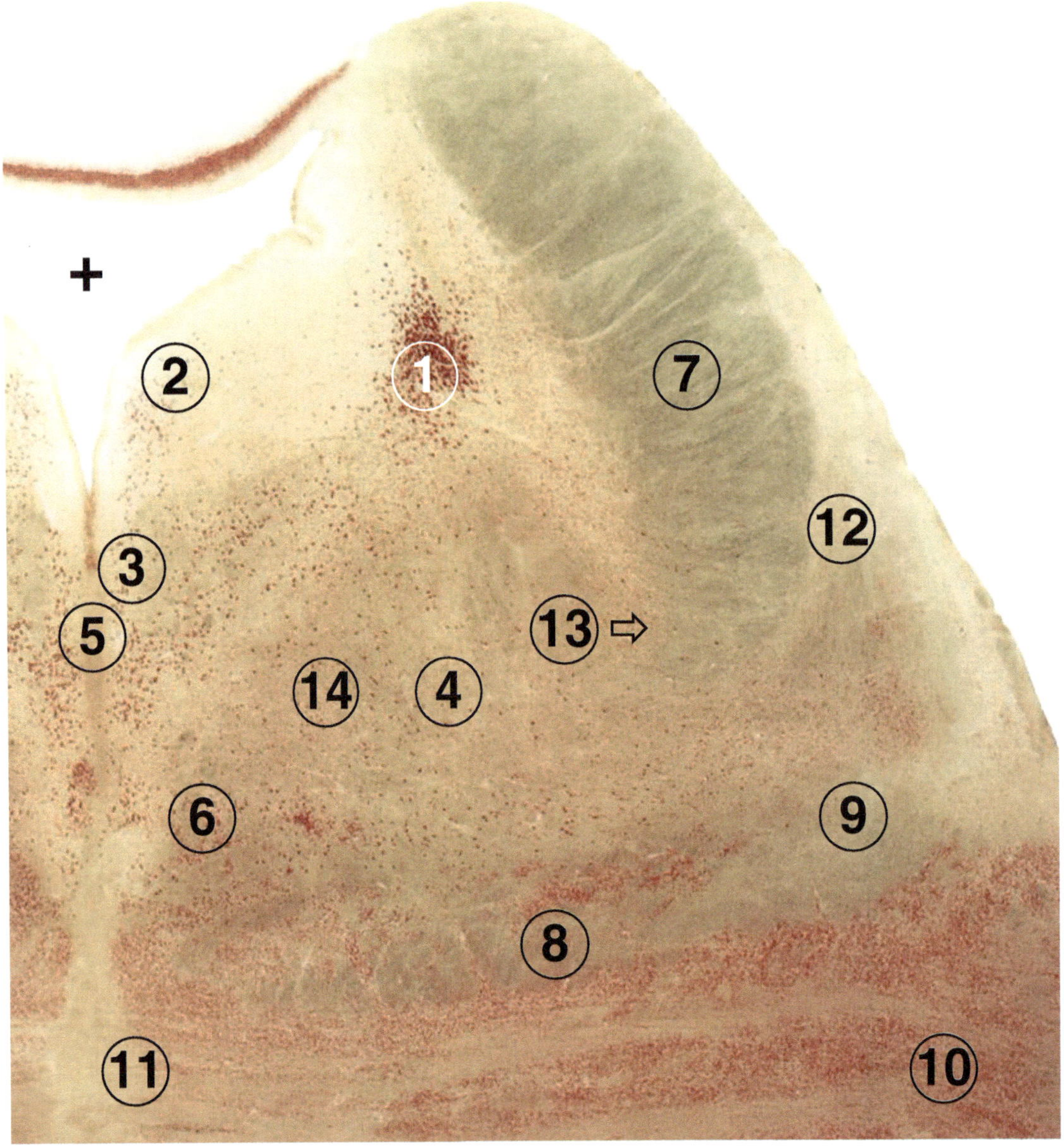

①	Locus caeruleus		⑨	Lateral lemniscus
②	Dorsal raphe ncl., caudal lamellar subnucleus		⑩	Pontine ncll.
③	Medial longitudinal fasciculus		⑪	Transverse pontine fibers See atlas part Campbell 13
④	Central tegmental tract		⑫	Subpeduncular tegmental (pigmented) ncl.
⑤	Dorsal raphe ncl., caudal lamellar subnucleus		⑬	Medial parabrachial ncl.
⑥	Reticulotegmental ncl.		⑭	Pontine reticular ncl., oral part
⑦	Superior cerebellar peduncle		+	4th ventricle
⑧	Medial lemniscus			

Darrow red 30

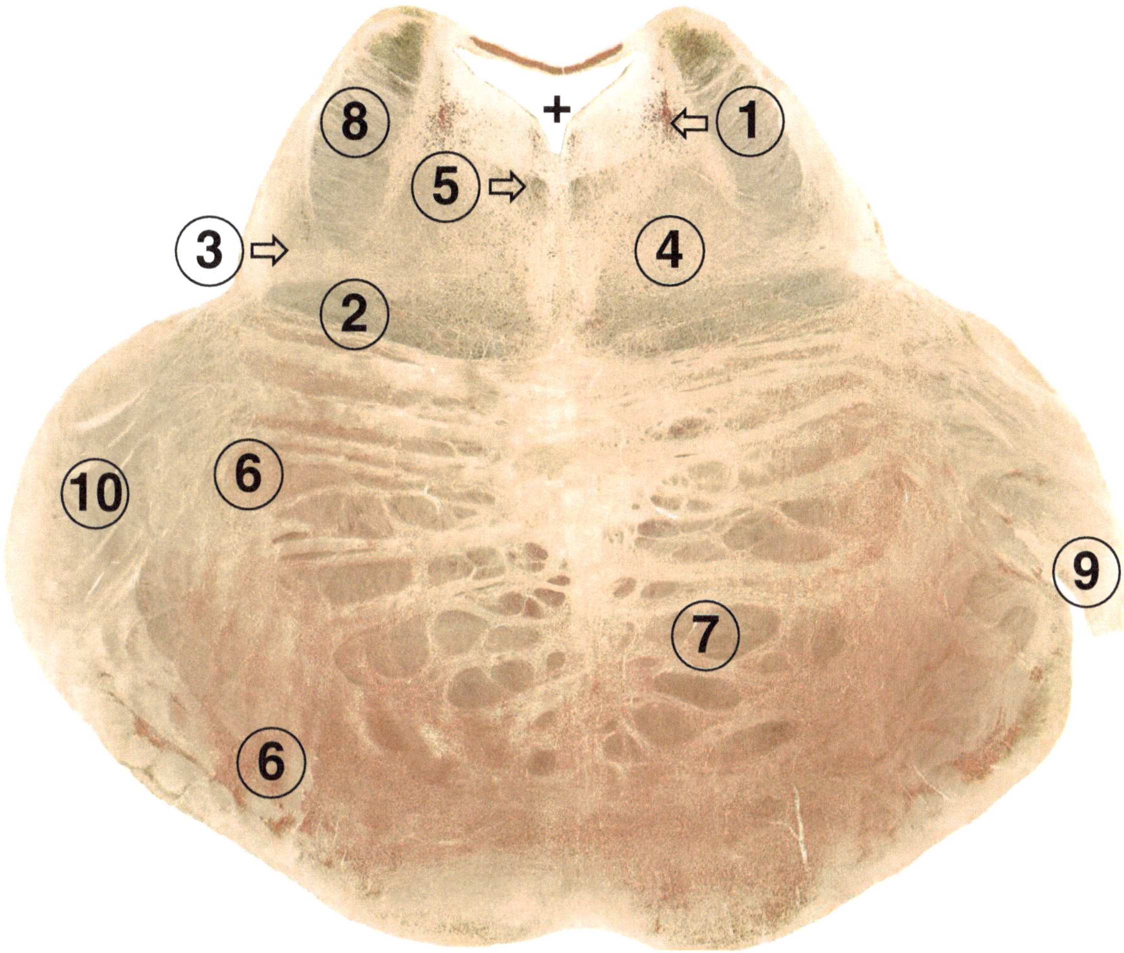

①	Locus caeruleus	⑦	Pyramidal (corticospinal) tract
②	Medial lemniscus	⑧	Superior cerebellar peduncle
③	Lateral lemniscus	⑨	Trigeminal nerve
④	Central tegmental tract	⑩	Middle cerebellar peduncle
⑤	Medial longitudinal fasciculus	+	4th ventricle
⑥	Pontine ncll.		

Darrow red 31

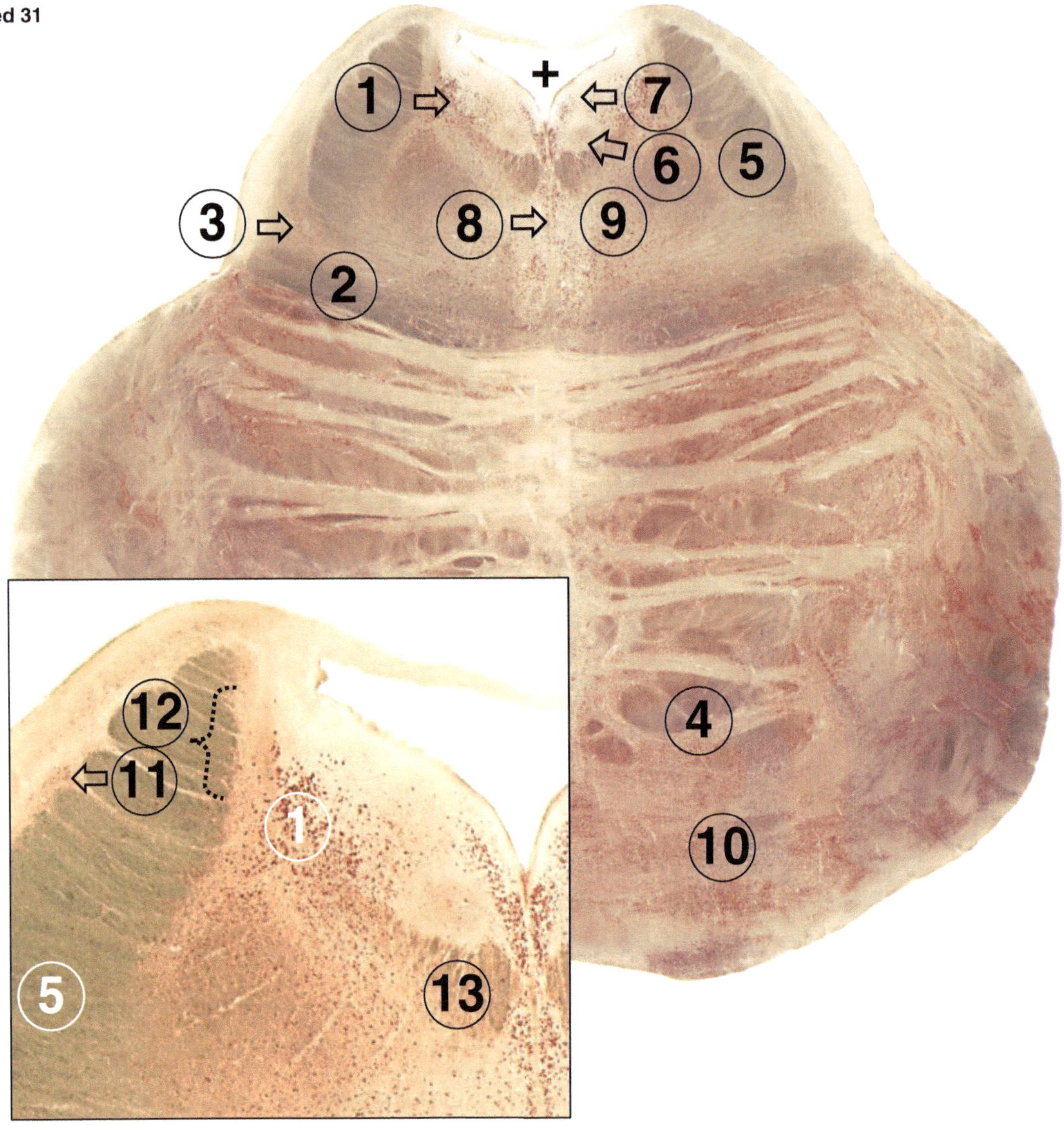

①	Locus caeruleus		⑧	Dorsal raphe ncl., compact caudal subnucleus
②	Medial lemniscus		⑨	Pontine reticular ncl., oral part
③	Lateral lemniscus		⑩	Pontine ncll.
④	Pyramidal (corticospinal) tract		⑪	Lateral parabrachial ncl.
⑤	Superior cerebellar peduncle		⑫	Medial parabrachial ncl.
	②-⑤ See atlas Campbell 14, 15		⑬	Medial longitudinal fasciculus
⑥	Dorsal tegmental ncl.		+	4th ventricle
⑦	Dorsal raphe ncl., compact caudal subnucleus			

Darrow red 31A

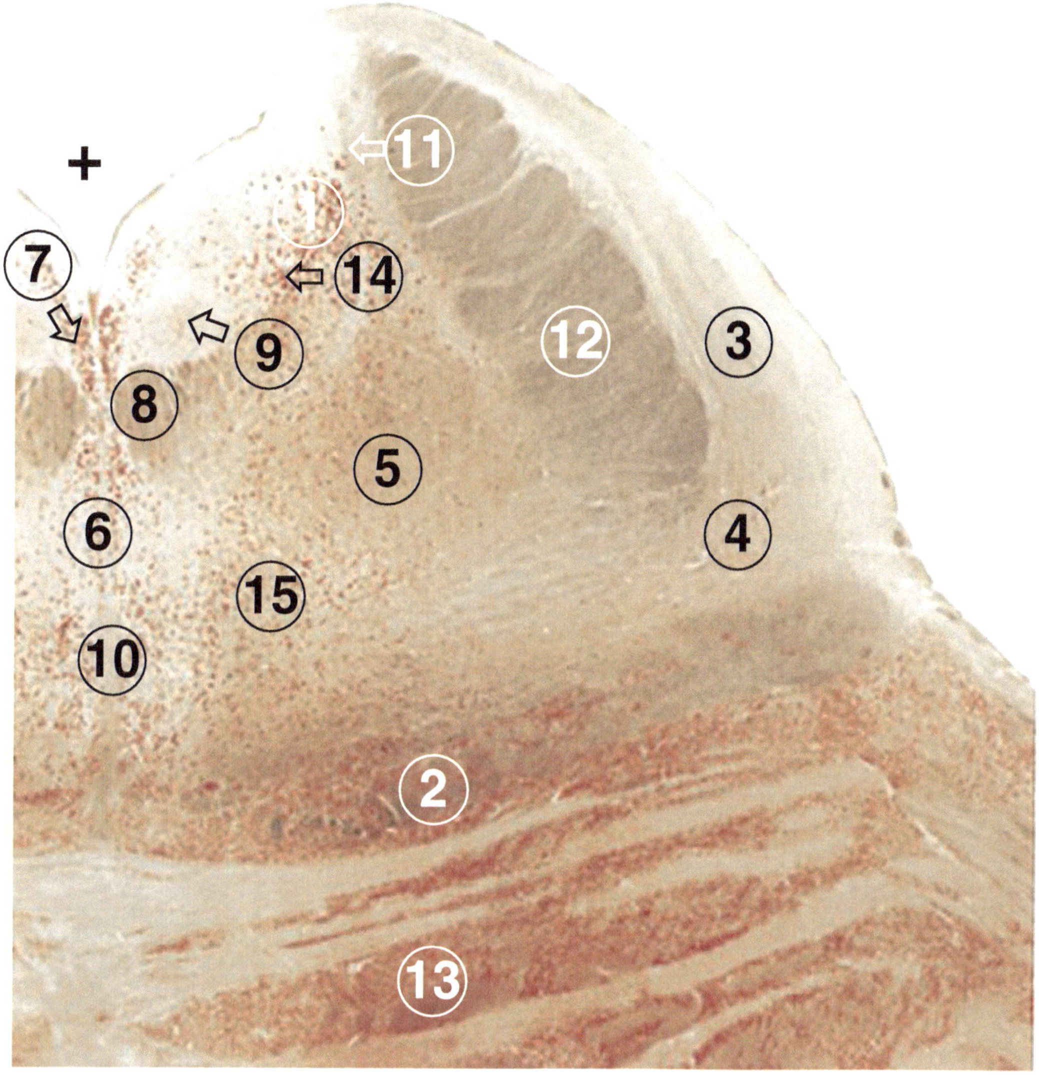

①	Locus caeruleus	⑨	Dorsal tegmental ncl.
②	Medial lemniscus	⑩	Median (central) raphe ncl., principal subnucleus
③	Lateral lemniscus	⑪	Mesencephalic ncl. and tract of trigeminal nerve
④	Ventral ncl. of lateral lemniscus	⑫	Superior cerebellar peduncle
⑤	Central tegmental tract	⑬	Pontine ncll.
⑥	Dorsal raphe ncl., compact caudal subnucleus	⑭	Barrington's ncl.
⑦	Dorsal raphe ncl., compact caudal subnucleus	⑮	Pontine reticular ncl., oral part
⑧	Medial longitudinal fasciculus	+	4th ventricle

Darrow red 32

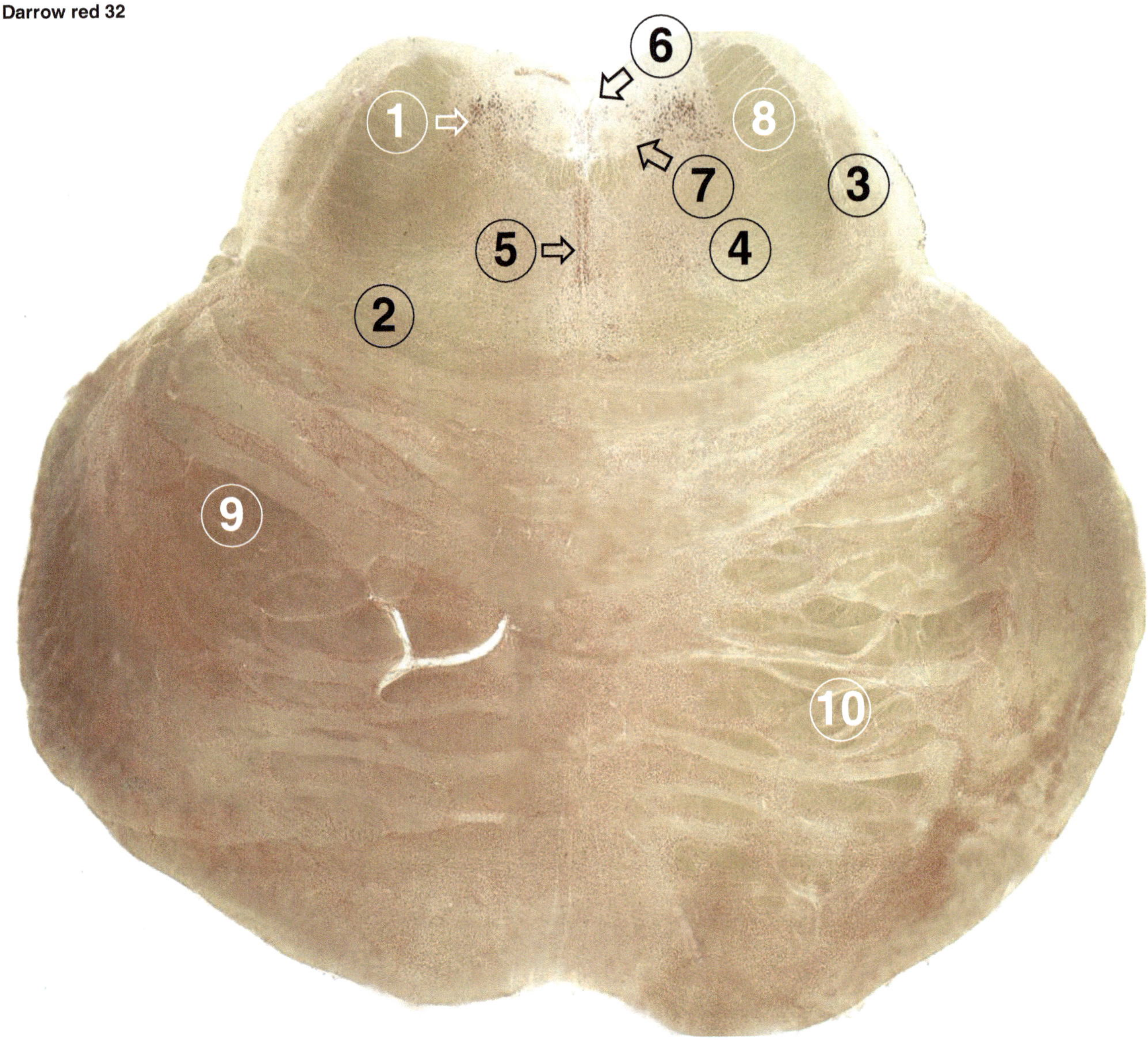

①	Locus caeruleus	⑥	Dorsal raphe ncl., compact caudal subnucleus
②	Medial lemniscus	⑦	Dorsal tegmental ncl.
③	Lateral lemniscus	⑧	Superior cerebellar peduncle
④	Central tegmental tract	⑨	Pontine ncll.
⑤	Median (central) raphe ncl., principal subnucleus	⑩	Pyramidal (corticospinal) tract

Darrow red 33

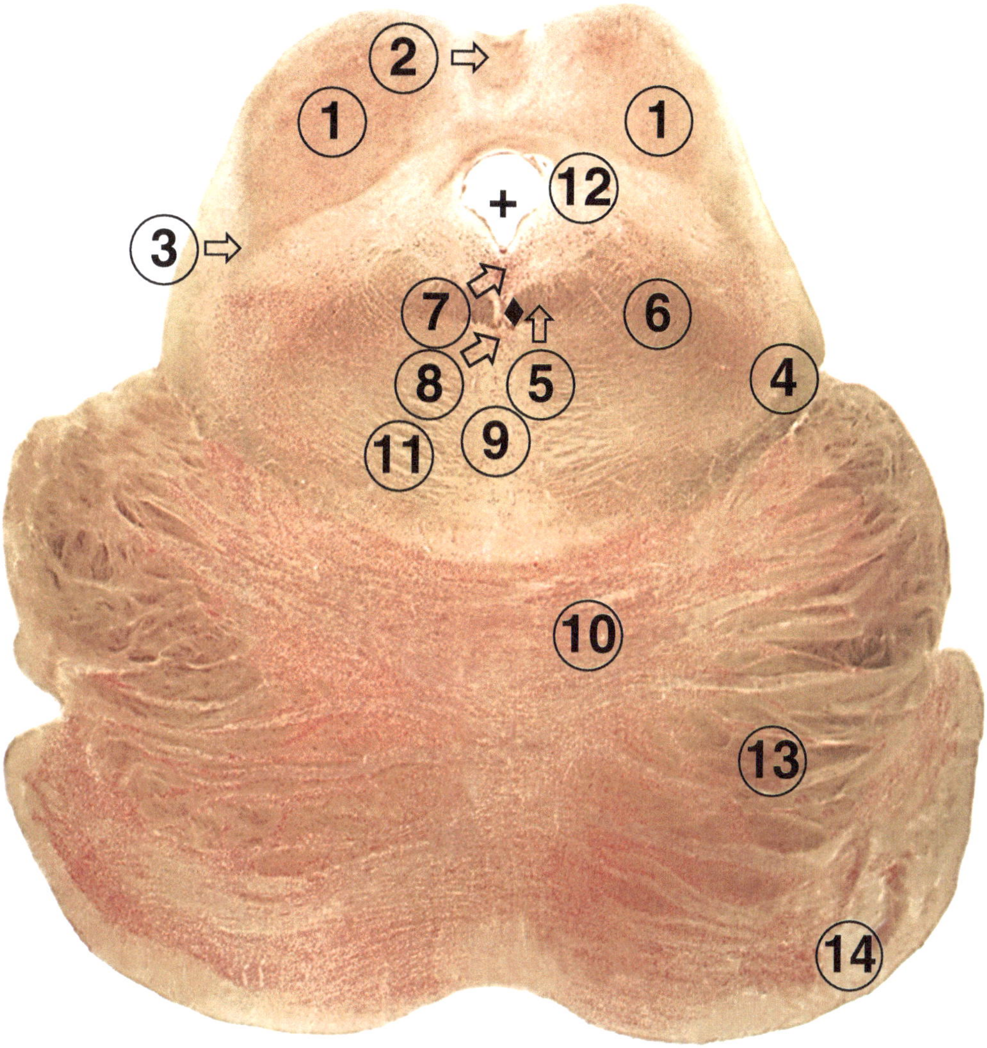

①	Inferior colliculus	⑧	Median (central) raphe ncl., anular part
②	Commissure of inferior colliculi See atlas part Campbell 17A	⑨	Linear raphe ncl.
③	Lateral lemniscus	⑩	Pontine ncll.
④	Medial lemniscus	⑪	Superior cerebellar peduncle
⑤	Medial longitudinal fasciculus	⑫	Periaqueductal gray
⑥	Central tegmental tract	⑬	Pyramidal (corticospinal) tract
⑦	Dorsal raphe ncl., supratrochlear part	⑭	Pontine basis
♦	Dorsal raphe ncl., interfascicular part	+	Mesencephalic aqueduct

Darrow red 34

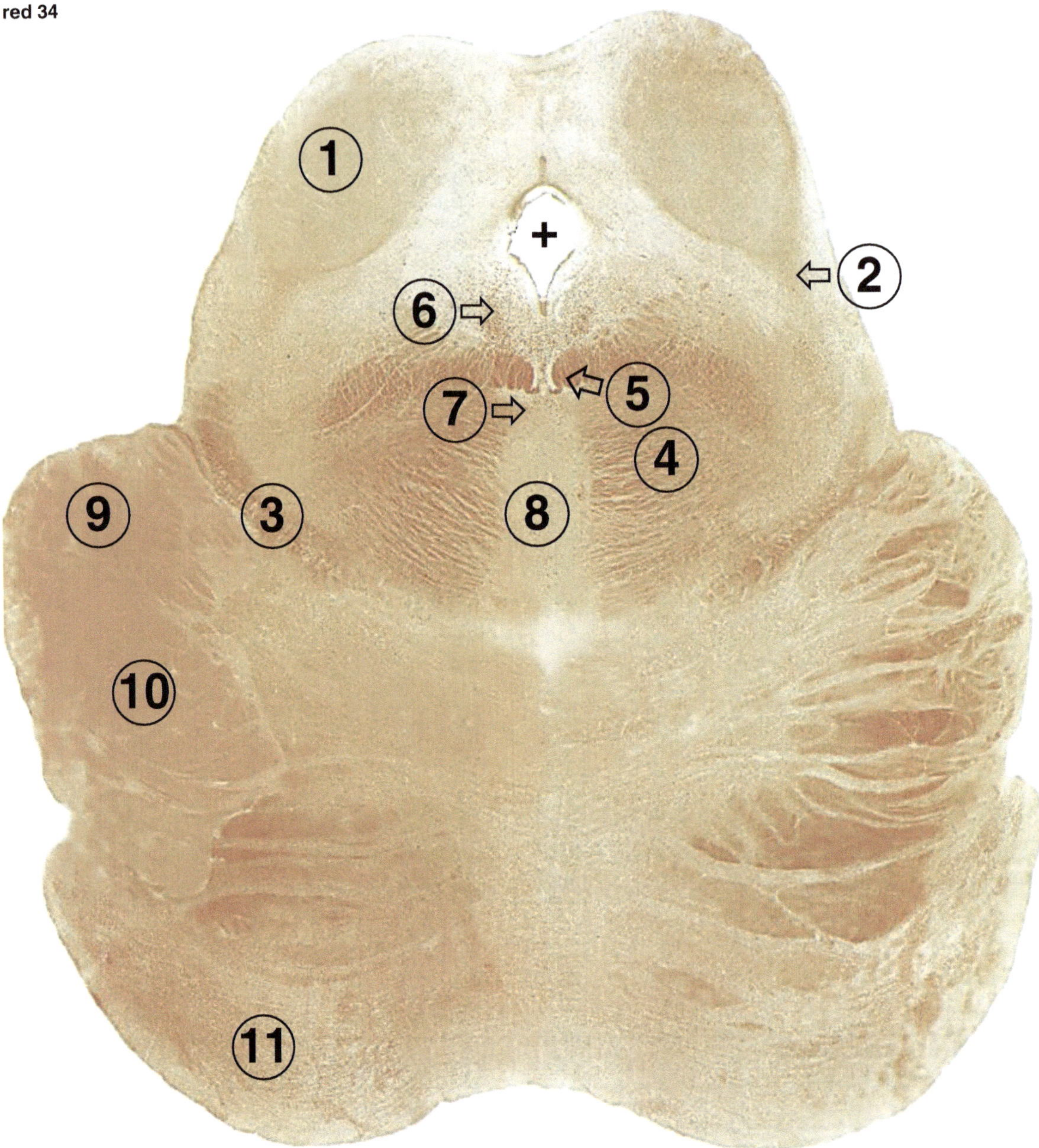

①	Inferior colliculus	⑦	Median (central) raphe ncl., anular part
②	Lateral lemniscus	⑧	Linear raphe ncl.
③	Medial lemniscus	⑨	Parietotemporopontine fibers
④	Superior cerebellar peduncle	⑩	Pyramidal (corticospinal) tract
⑤	Medial longitudinal fasciculus ①-⑤ See atlas part Campbell 17A	⑪	Pontine basis
⑥	Dorsal raphe ncl., supratrochlear part	+	Mesencephalic aqueduct

Darrow red 34A

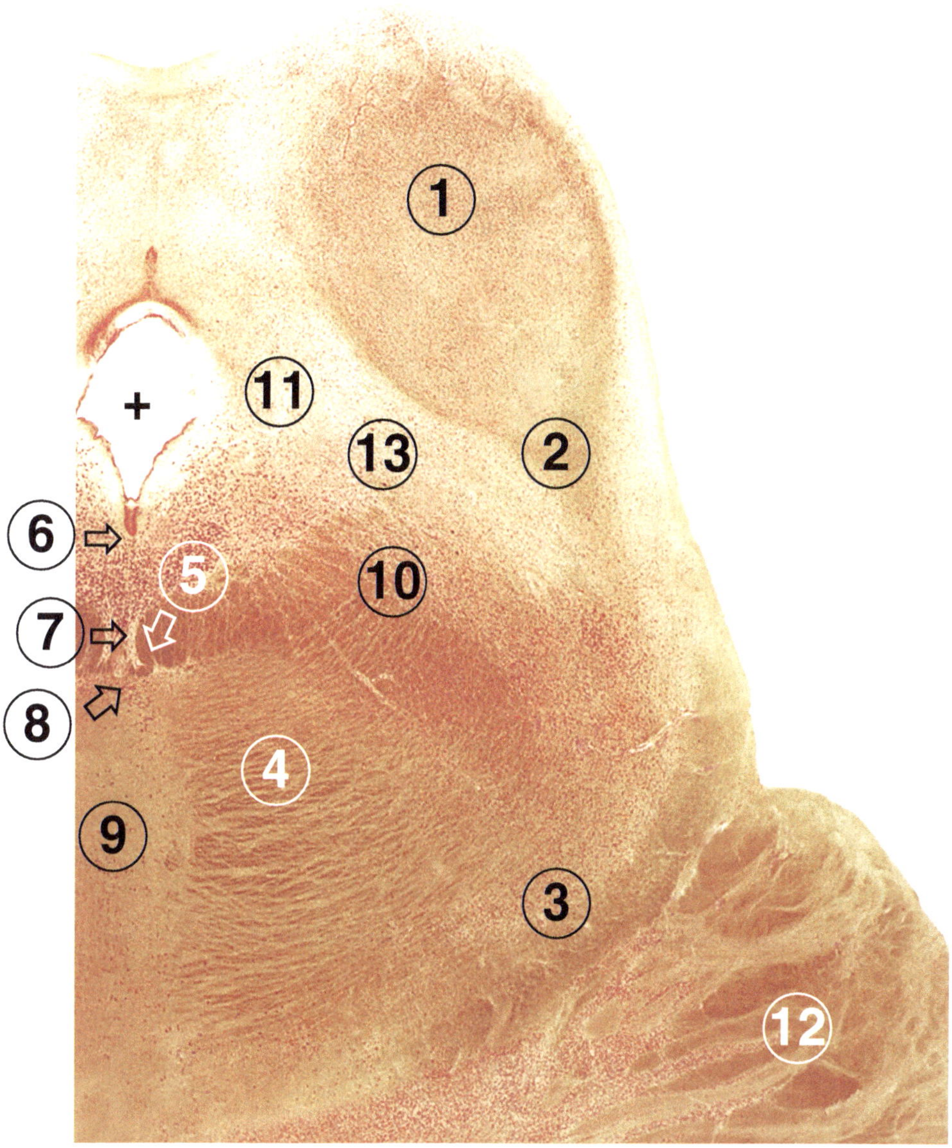

①	Inferior colliculus	⑧	Median (central) raphe ncl., anular part
②	Lateral lemniscus	⑨	Linear raphe ncl.
③	Medial lemniscus	⑩	Central tegmental tract
④	Superior cerebellar peduncle	⑪	Periaqueductal gray
⑤	Medial longitudinal fasciculus ①-⑤ See atlas part Campbell 17A	⑫	Cerebral peduncle
⑥	Dorsal raphe ncl., supratrochlear part	⑬	Cuneiform ncl.
⑦	Dorsal raphe ncl., interfascicular part	+	Mesencephalic aqueduct

Darrow red 35

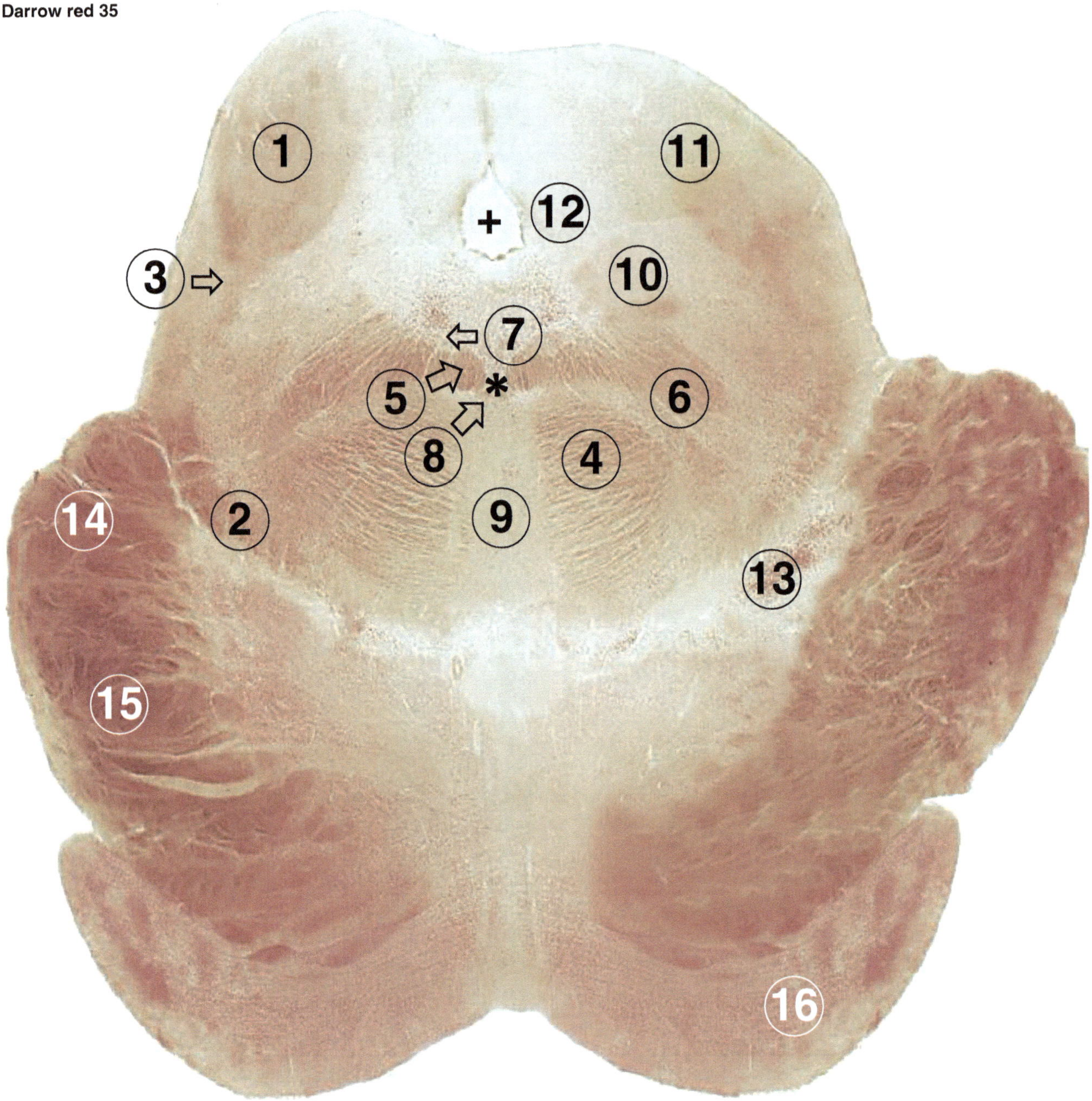

①	Inferior colliculus	⑨	Linear raphe ncl.
②	Medial lemniscus	⑩	Cuneiform ncl.
③	Lateral lemniscus	⑪	Transition inferior > superior colliculus
④	Superior cerebellar peduncle	⑫	Periaqueductal gray
⑤	Medial longitudinal fasciculus	⑬	Substantia nigra
⑥	Central tegmental tract	⑭	Parietotemporopontine tract
⑦	Dorsal raphe ncl., supratrochlear part	⑮	Pyramidal (corticospinal) tract
*	Dorsal raphe ncl., interfascicular part	⑯	Pontine basis
⑧	Median (central) raphe ncl., anular part	+	Mesencephalic aqueduct

Darrow red 36

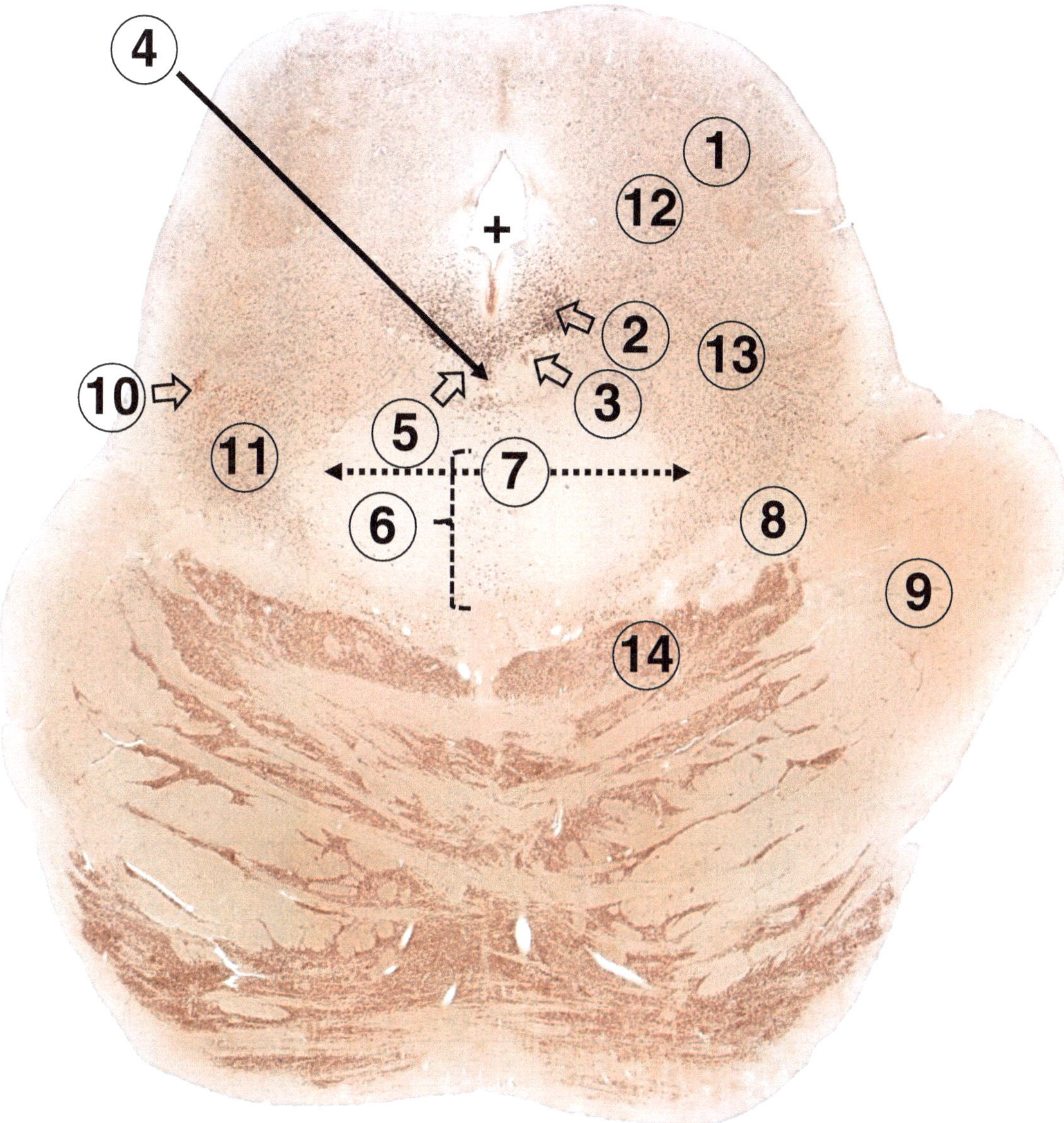

①	Transition inferior > superior colliculi	⑨	Cerebral peduncle
②	Supratrochlear subnucleus of the dorsal raphe ncl.	⑩	Pedunculopontine tegmental ncl., compact part (Ch5c)
③	Ncl. of trochlear nerve CNIV	⑪	Pedunculopontine tegmental ncl., dissipate part (Ch5d)
④	Interfascicular subnucleus of the dorsal raphe ncl.	⑫	Cuneiform ncl.
⑤	Medial longitudinal fasciculus	⑬	Subcuneiform ncl.
⑥	Linear raphe ncl.	⑭	Pontine ncll.
⑦	Decussation of the superior cerebellar peduncles	+	Mesencephalic aqueduct
⑧	Medial lemniscus ⑦-⑧ See atlas part Campbell 16, 17, 17A		

Darrow red 36A

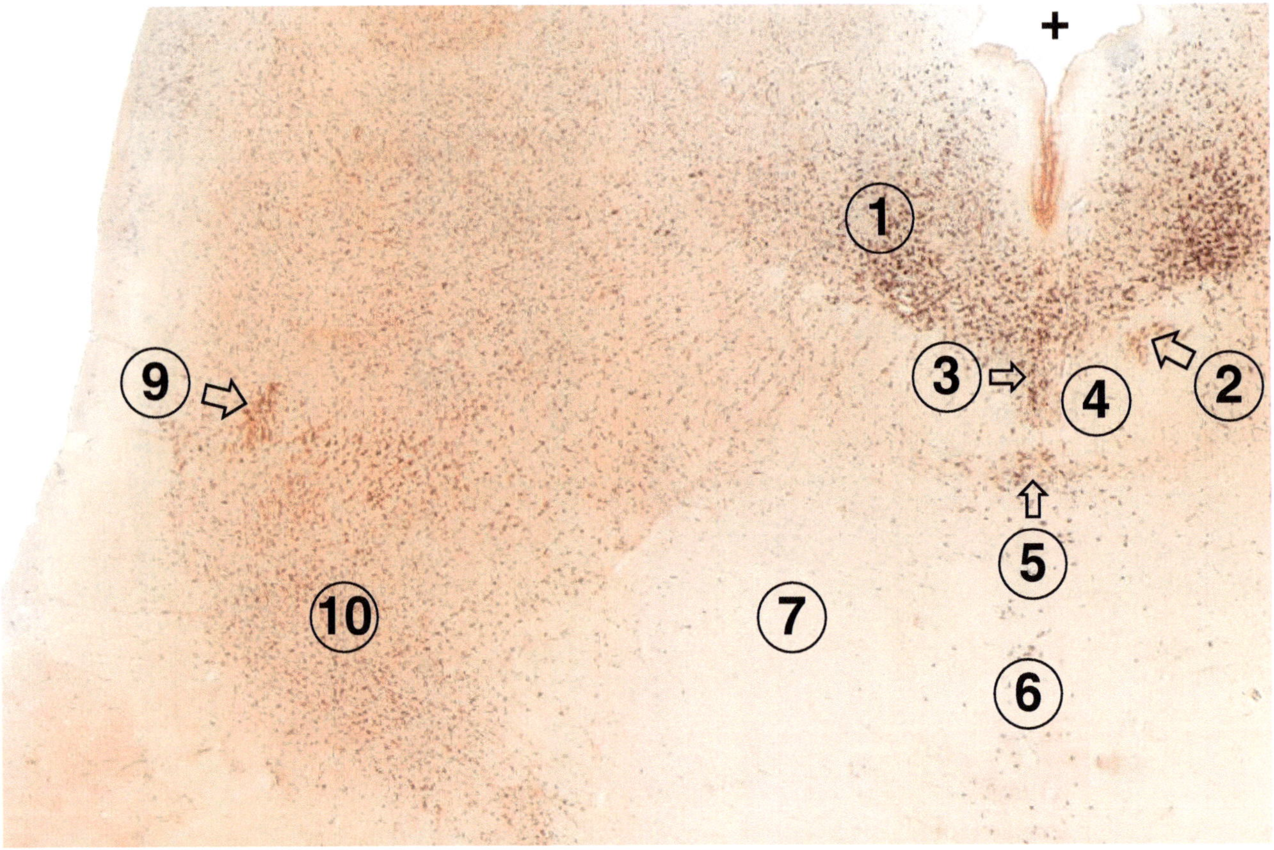

①	Supratrochlear subnucleus of the dorsal raphe ncl.	⑥	Linear raphe ncl.
②	Ncl. of trochlear nerve CNIV	⑦	Decussation of the superior cerebellar peduncles See atlas part Campbell 17
③	Interfascicular subnucleus of the dorsal raphe ncl.	⑨	Pedunculopontine tegmental ncl., compact part (Ch5c)
④	Medial longitudinal fasciculus See atlas part Campbell 17	⑩	Pedunculopontine tegmental ncl., dissipate part (Ch5d)
⑤	Anular subnucleus of the median (central) raphe nucleus	+	Mesencephalic aqueduct

Darrow red 37

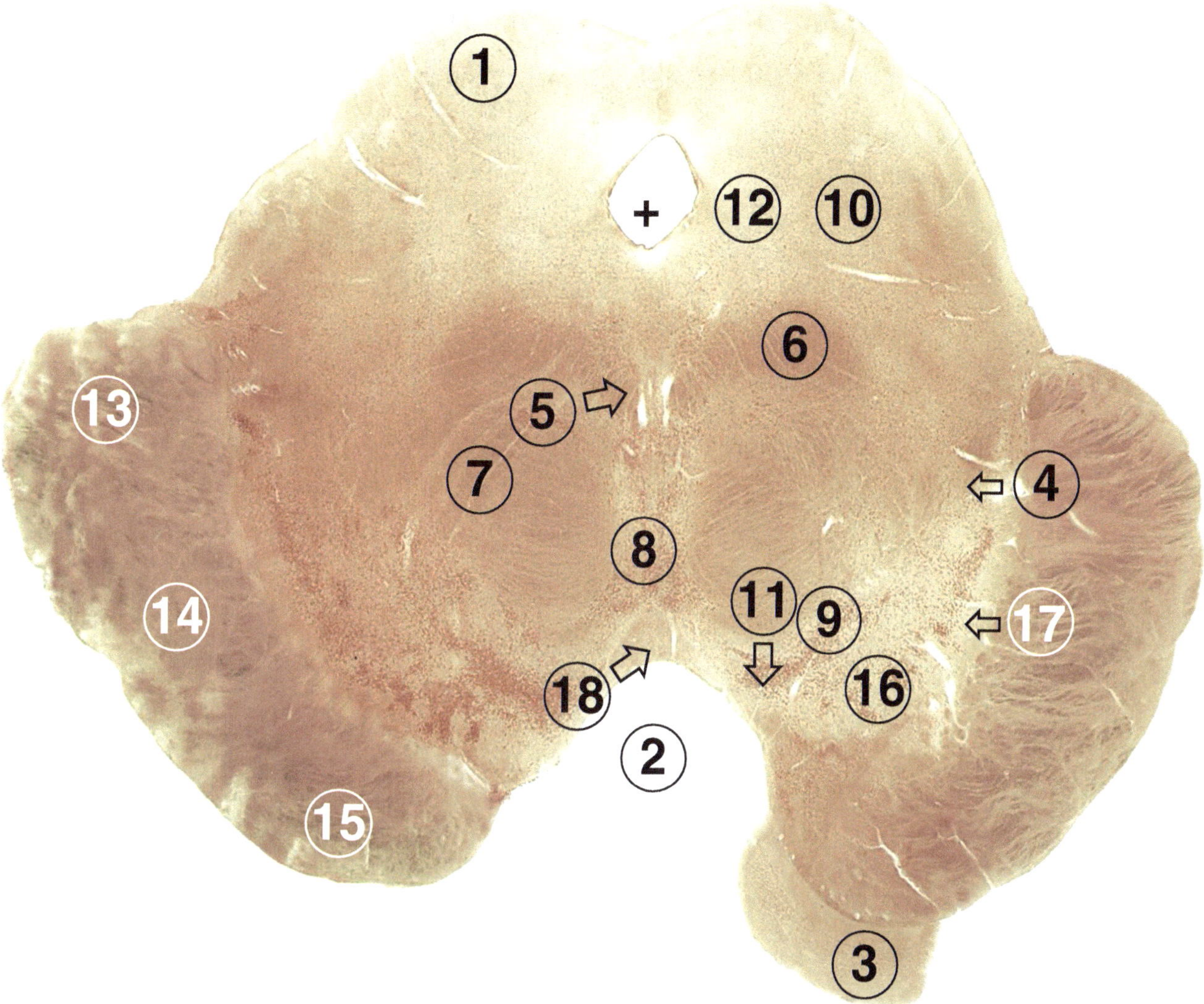

①	Superior colliculus	⑪	Paranigral ncl.
②	Interpeduncular fossa	⑫	Periaqueductal gray
③	Pontine basis	⑬	Parietotemporopontine tract
④	Medial lemniscus	⑭	Pyramidal (corticospinal) tract
⑤	Medial longitudinal fasciculus	⑮	Frontopontine tract
⑥	Central tegmental tract	⑯	Substantia nigra, compact part
⑦	Superior cerebellar peduncle ④-⑦ See atlas part Campbell 19	⑰	Substantia nigra, reticular part
⑧	Linear raphe ncl. / Caudal (inferior) linear ncl.	⑱	Interpeduncular ncl.
⑨	Parabrachial pigmented ncl.	+	Mesencephalic aqueduct
⑩	Cuneiform ncl.		

Darrow red 38

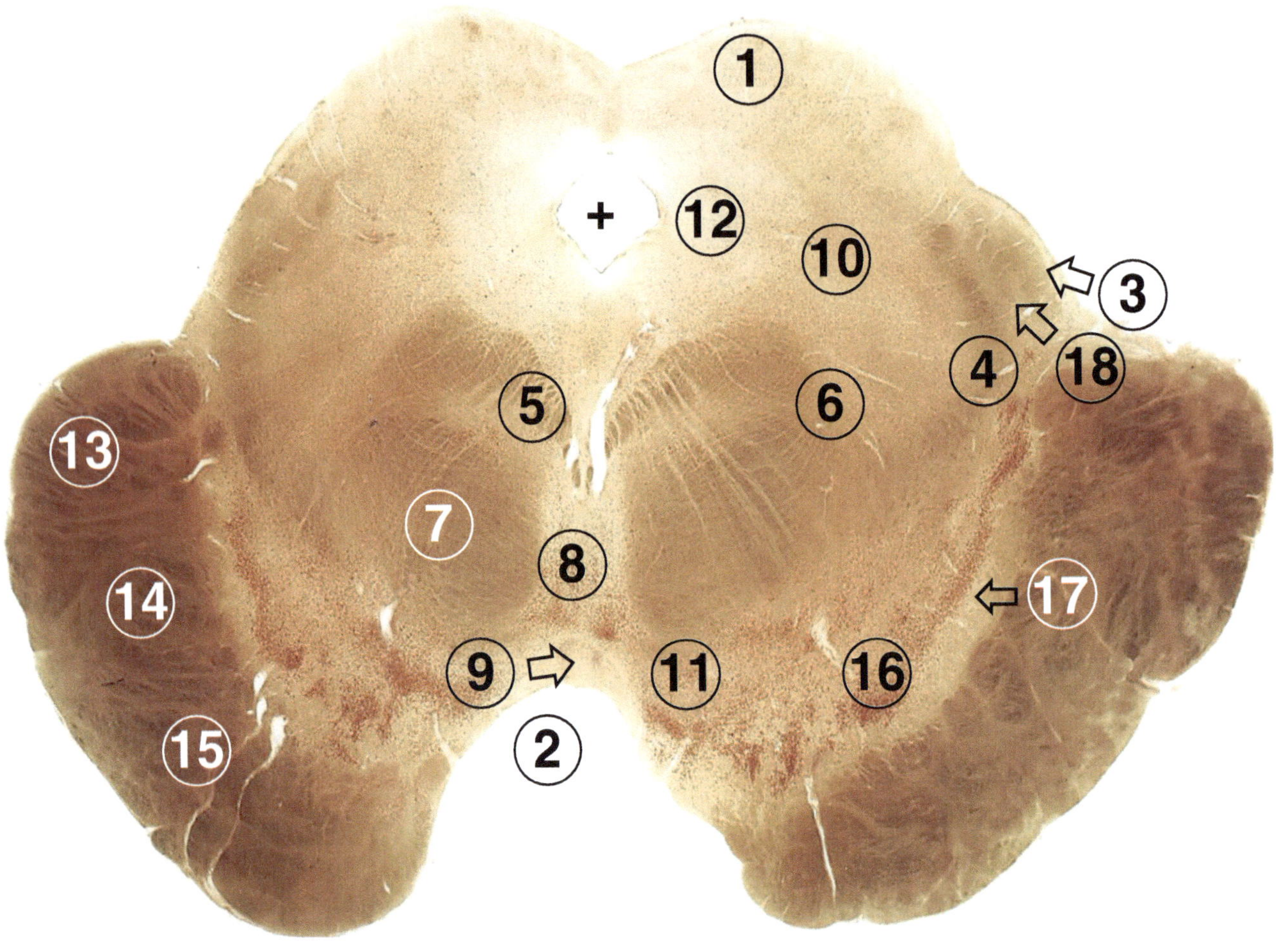

①	Superior colliculus	⑩	Precuneiform ncl.
②	Interpeduncular fossa	⑪	Paranigral ncl.
③	Brachium of inferior colliculus	⑫	Periaqueductal gray
④	Medial lemniscus	⑬	Parietotemporopontine tract
⑤	Medial longitudinal fasciculus	⑭	Pyramidal (corticospinal) tract
⑥	Central tegmental tract	⑮	Frontopontine tract
⑦	Superior cerebellar peduncle ③-⑦ See atlas part Campbell 19	⑯	Substantia nigra, compact part
⑧	Linear raphe ncl.	⑰	Substantia nigra, reticular part
⑨	Interpeduncular ncl.	⑱	Paralemniscal ncl.
		+	Mesencephalic aqueduct

Darrow red 38A

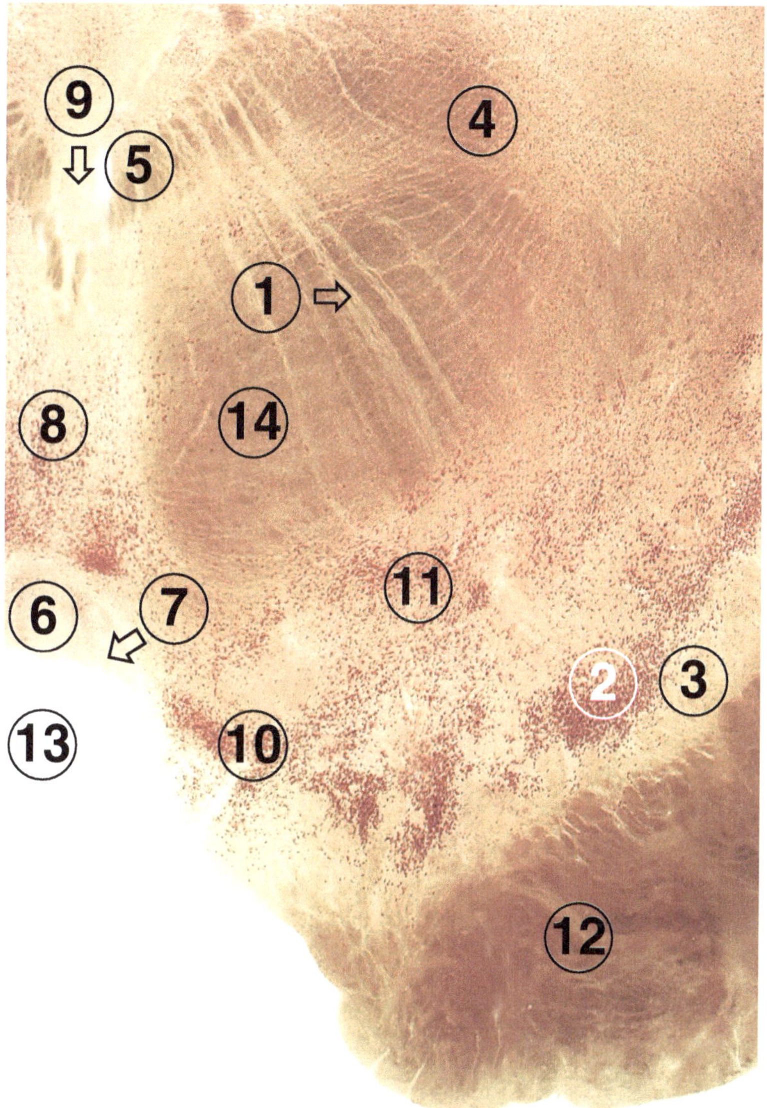

①	Oculomotor nerve CNIII	⑧	Caudal linear raphe ncl.
②	Substantia nigra, compact part	⑨	Dorsal raphe ncl., interfascicular part
③	Substantia nigra, reticular part	⑩	Paranigral ncl.
④	Central tegmental tract	⑪	Parabrachial pigmented ncl.
⑤	Medial longitudinal fasciculus	⑫	Cerebral peduncle
⑥	Interpeduncular ncl., medial part	⑬	Interpeduncular fossa
⑦	Interpeduncular ncl., lateral part	⑭	Superior cerebellar peduncle / Red ncl.

Darrow red 38B

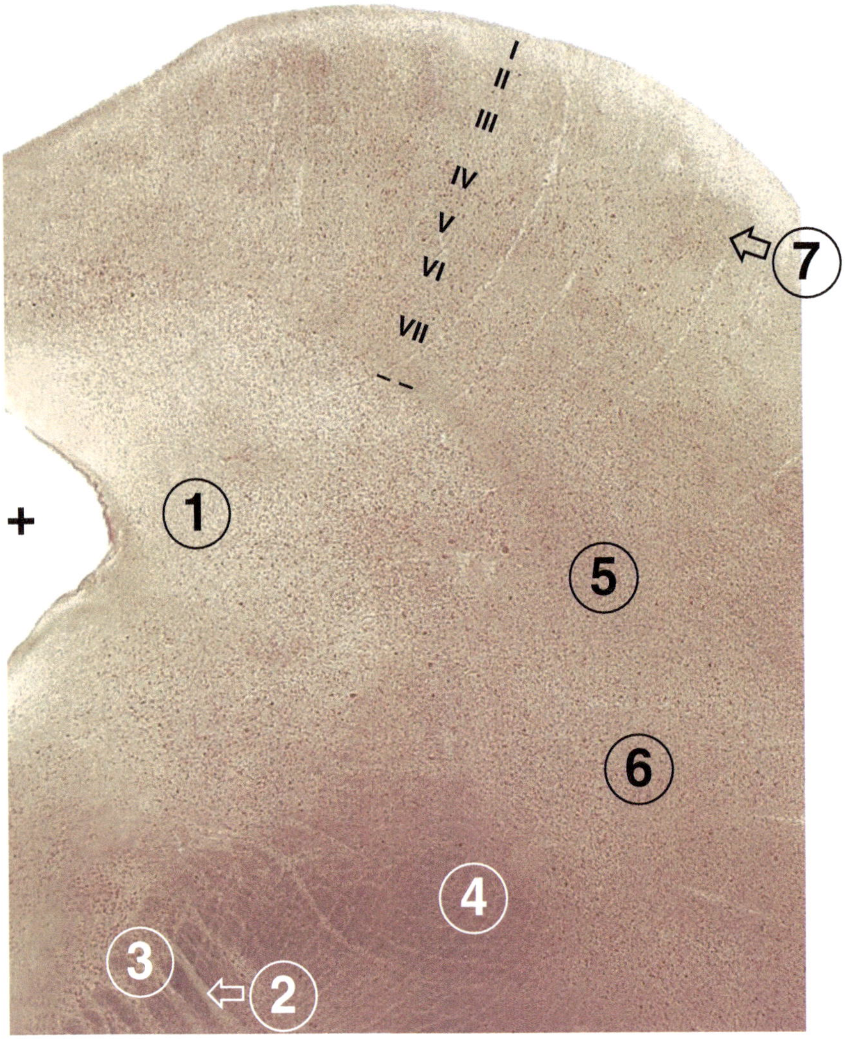

I -VII	Layers of superior colliculus	①	Periaqueductal gray
Ⓘ	Zonal layer	②	Oculomotor nerve CNIII
Ⓘ	Superficial gray layer	③	Medial longitudinal fasciculus
Ⓘ	Optic layer	④	Central tegmental tract
Ⓘ	Intermediate gray layer	⑤	Precuneiform ncl.
Ⓥ	Intermediate white layer	⑥	Subcuneiform ncl.
Ⓥ	Deep gray layer	⑦	Oval pretectal ncl.
Ⓥ	Deep white layer	+	Mesencephalic aqueduct

Darrow red 39

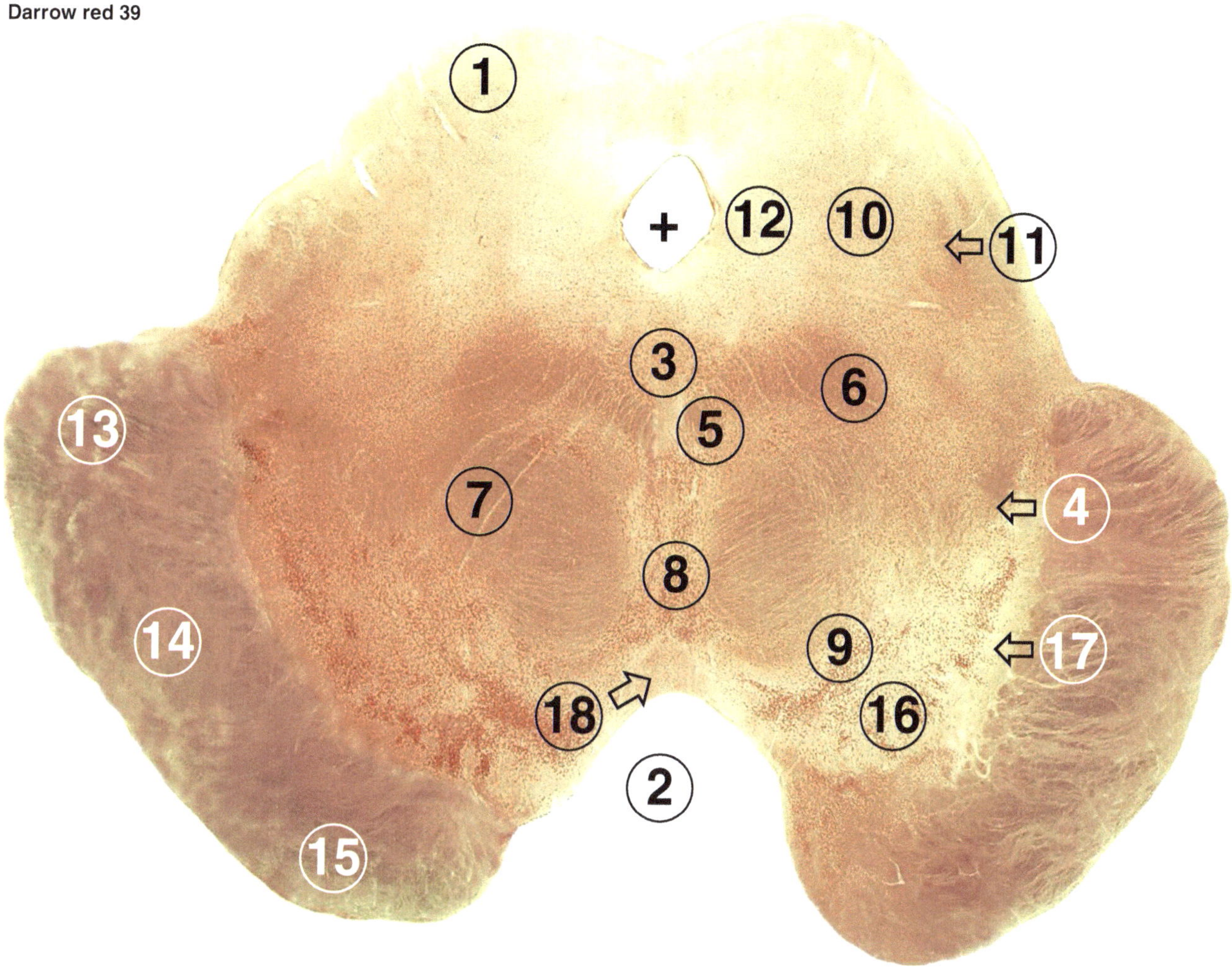

①	Superior colliculus
②	Interpeduncular fossa
③	Ncll. of oculomotor nerve CNIII / Oculomotor complex
④	Medial lemniscus
⑤	Medial longitudinal fasciculus
⑥	Central tegmental tract
⑦	Superior cerebellar peduncle / Red. ncl.
⑧	Linear raphe ncl. / Caudal (inferior) linear nucleus
⑨	Parabrachial pigmented ncl.
⑩	Precuneiform ncl.

⑪	Medial lemniscus
⑫	Periaqueductal gray
⑬	Parietotemporopontine tract
⑭	Pyramidal (corticospinal) tract
⑮	Frontopontine tract
⑯	Substantia nigra, compact part
⑰	Substantia nigra, reticular part
⑱	Interpeduncular ncl.
+	Mesencephalic aqueduct

Darrow red 39A

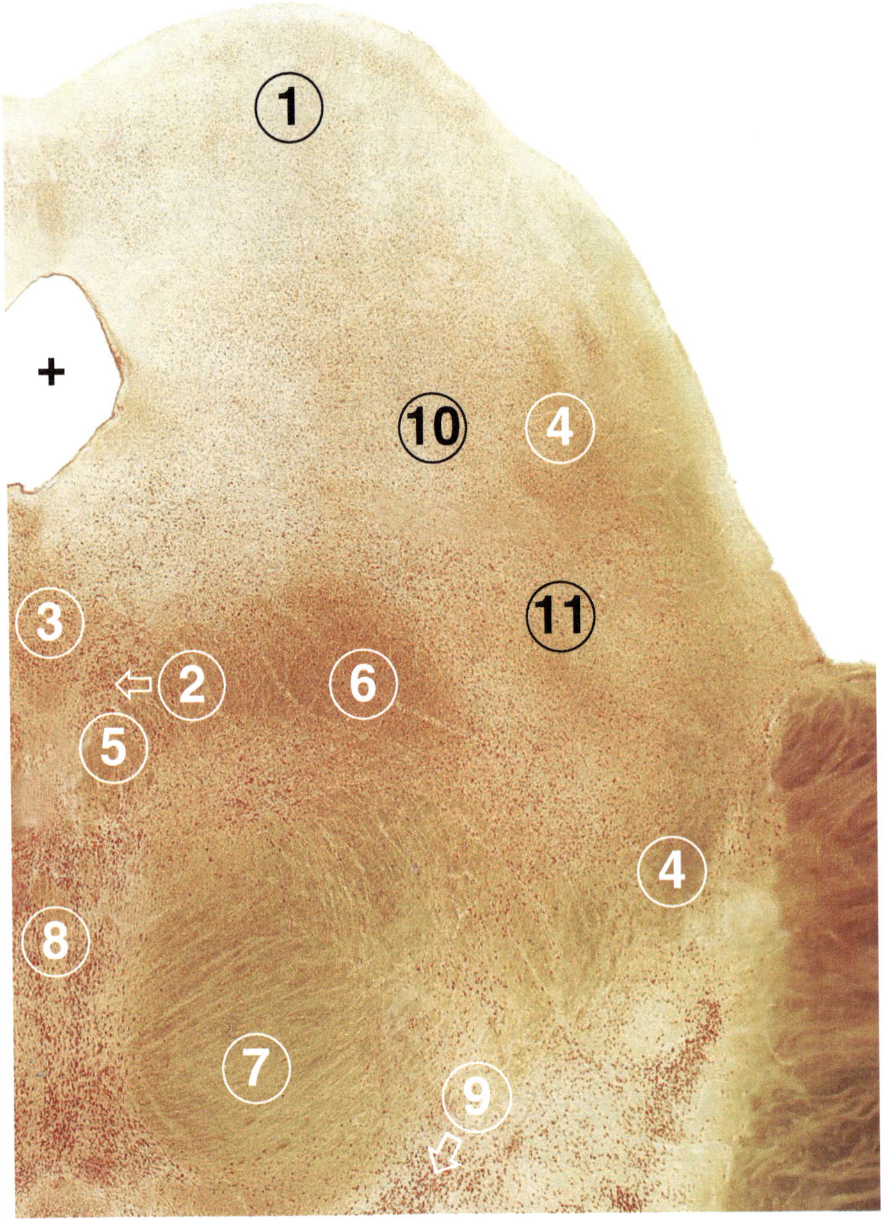

①	Superior colliculus	⑦	Superior cerebellar peduncle
②	Ncl. of oculomotor nerve CNIII, dorsal part	⑧	Linear raphe ncl.
③	Accessory ncl. of CNIII Edinger Westphal ncl., centrally projecting part	⑨	Parabrachial pigmented ncl.
④	Medial lemniscus	⑩	Precuneiform ncl.
⑤	Medial longitudinal fasciculus	⑪	Subcuneiform ncl.
⑥	Central tegmental tract	+	Mesencephalic aqueduct

Darrow red 40

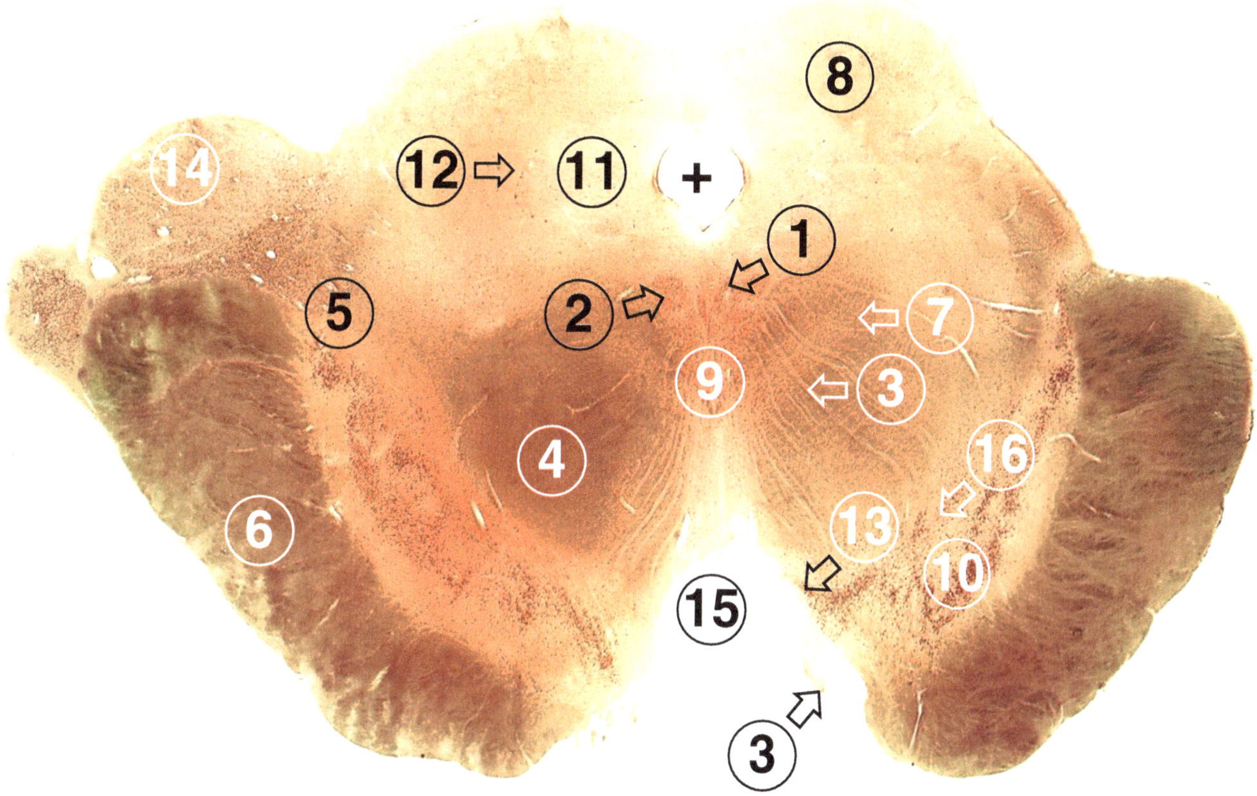

①	Accessory ncl. of oculomotor nerve, nonganglionic projecting part (Edinger-Westphal ncl., centrally projecting part)	⑨	Linear raphe ncl.
		⑩	Substantia nigra
②	Ncl. of oculomotor nerve CNIII	⑪	Periaqueductal gray
③	Oculomotor nerve	⑫	Mesencephalic ncl. of trigeminal nerve
④	Superior cerebellar peduncle / Red ncl.	⑬	Paranigral ncl.
⑤	Medial lemniscus	⑭	Medial geniculate body
⑥	Cerebral peduncle	⑮	Interpeduncular fossa
⑦	Central tegmental tract ③-⑦ See atlas part Campbell 19	⑯	Parabrachial pigmented ncl.
⑧	Superior colliculus	+	Mesencephalic aqueduct

Darrow red 40A

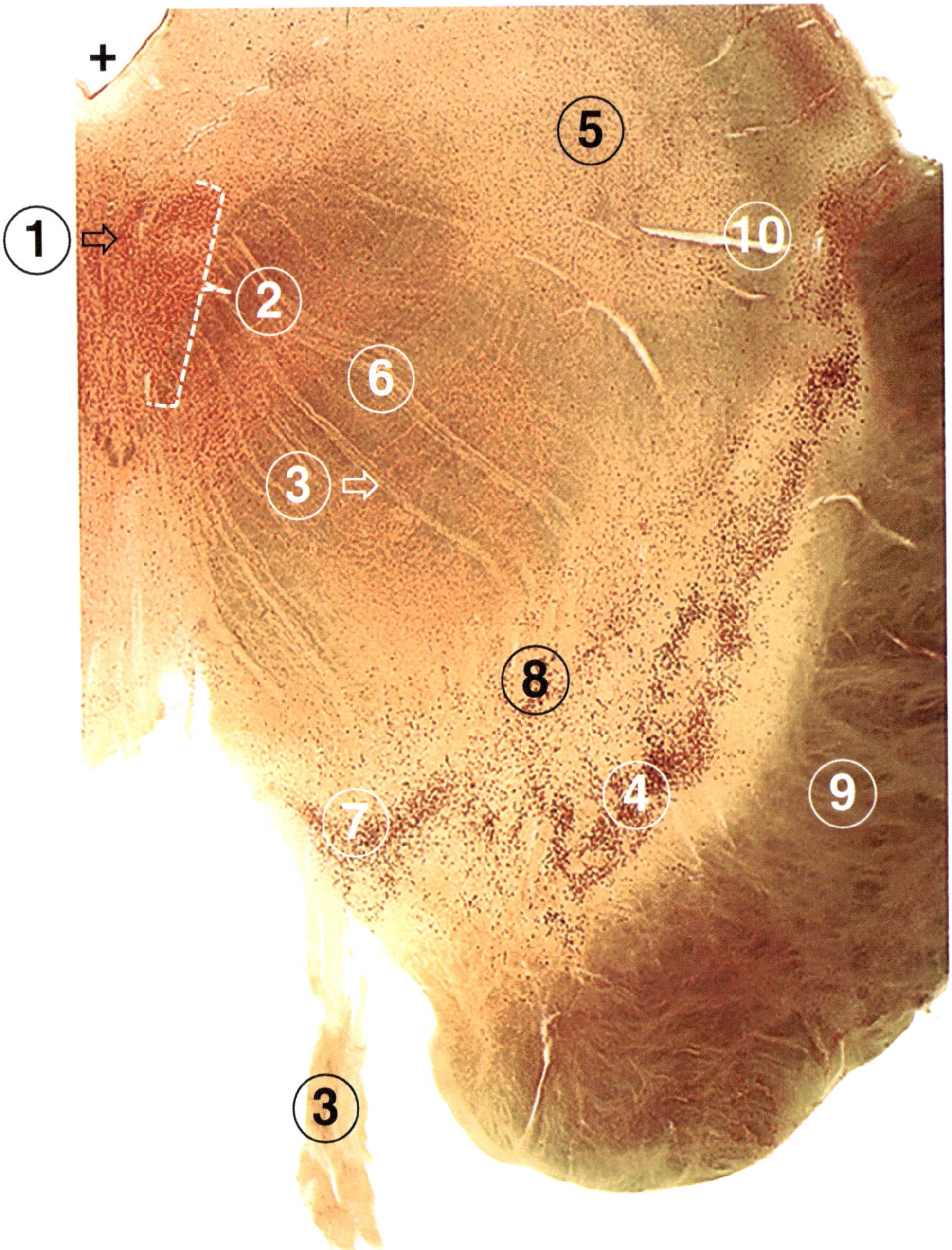

①	Accessory ncl. of oculomotor nerve, nonganglionic projecting part (Edinger-Westphal ncl., centrally projecting part)	⑥	Red ncl.
		⑦	Paranigral ncl.
②	Ncl. of oculomotor nerve	⑧	Parabrachial pigmented ncl.
③	Oculomotor nerve CNIII	⑨	Cerebral peduncle
④	Substantia nigra	⑩	Medial lemniscus
⑤	Precuneiform ncl.	+	Mesencephalic aqueduct

Darrow red 40B

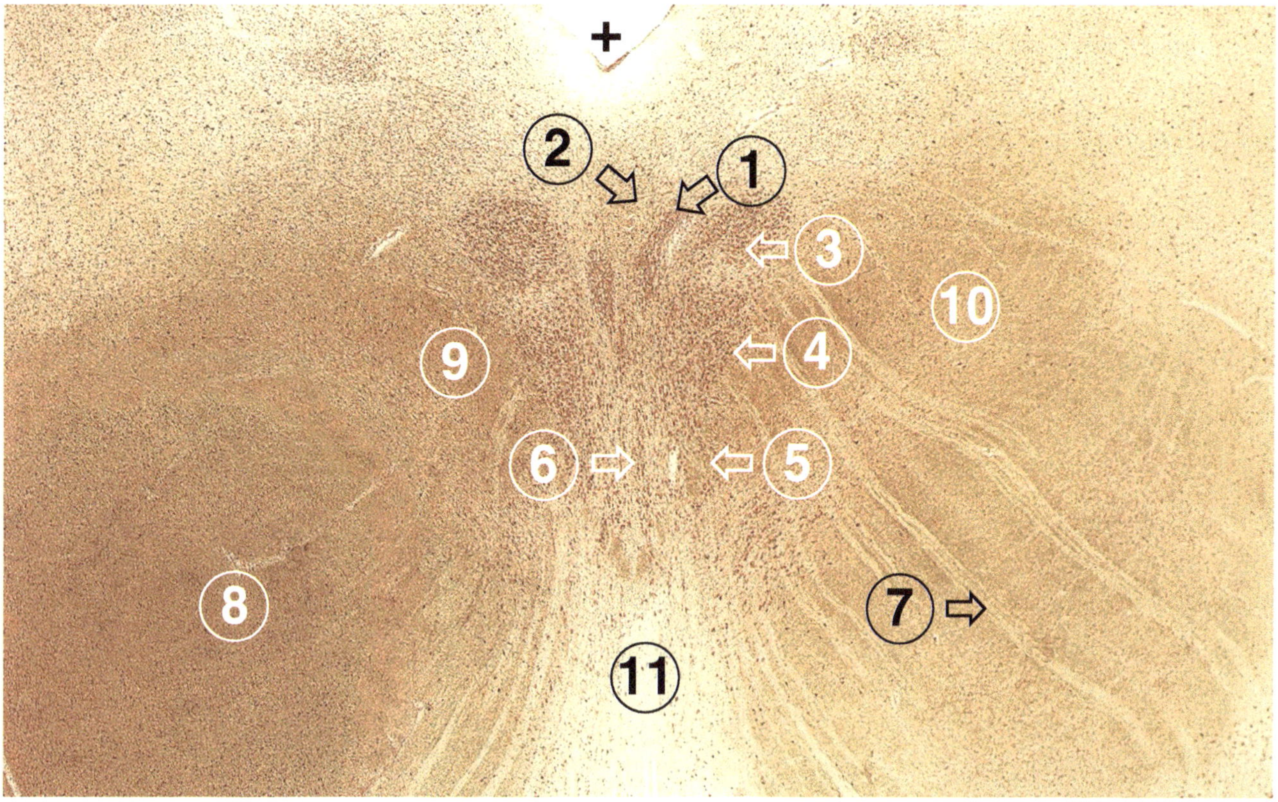

①	Accessory ncl. of oculomotor nerve, nonganglionic projecting part (Edinger-Westphal ncl., centrally projecting part)	⑥	Ncl. of oculomotor nerve, interoculomotor part (Ncl. of Perlia)
		⑦	Oculomotor nerve
②	Region of accessory ncl. of oculomotor nerve, preganglionic part (Edinger-Westphal ncl., preganglionic part)	⑧	Red nucleus, parvocellular part
		⑨	Medial longitudinal fasciculus
③	Ncl. of oculomotor nerve, dorsal part	⑩	Central tegmental tract
④	Ncl. of oculomotor nerve, central part	⑪	Linear raphe ncl.
⑤	Ncl. of oculomotor nerve CNIII, ventral part	+	Mesencephalic aqueduct

Darrow red 41

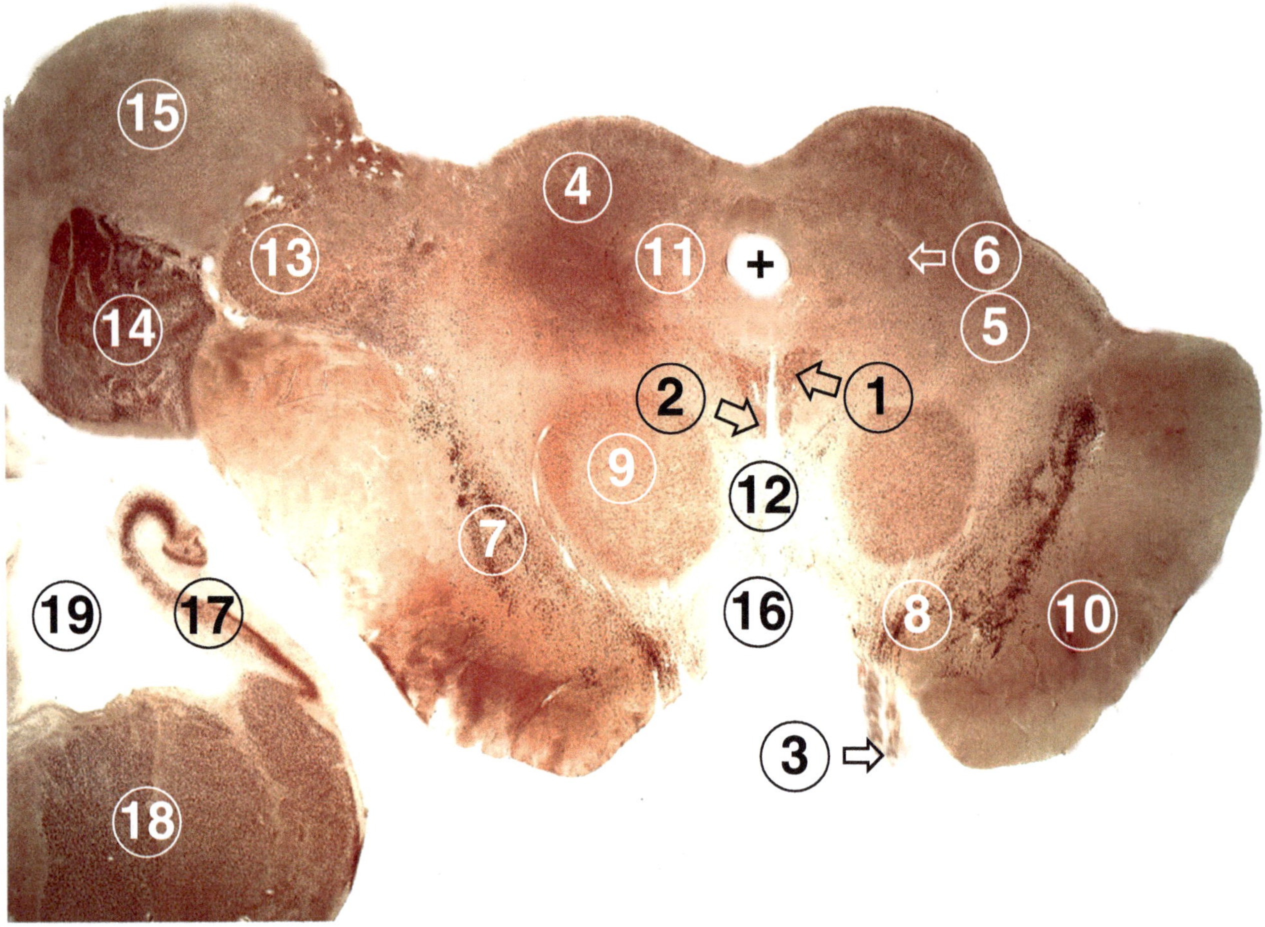

①	Accessory ncl. of oculomotor nerve, nonganglionic projecting part (Edinger-Westphal ncl., centrally projecting part)	⑪	Periaqueductal gray
		⑫	Linear raphe ncl.
②	Ncl. of oculomotor nerve CNIII	⑬	Medial geniculate body
③	Oculomotor nerve CNIII	⑭	Lateral geniculate body
④	Superior colliculus	⑮	Pulvinar
⑤	Precuneiform ncl.	⑯	Interpeduncular fossa
⑥	Mesencephalic ncl. of trigeminal nerve	⑰	Hippocampus
⑦	Substantia nigra	⑱	Amygdala
⑧	Paranigral ncl.	⑲	Temporal horn of lateral ventricle
⑨	Red nucleus, parvocellular part	+	Mesencephalic aqueduct
⑩	Cerebral peduncle See atlas part Campbell 19, 20		

Darrow red 42

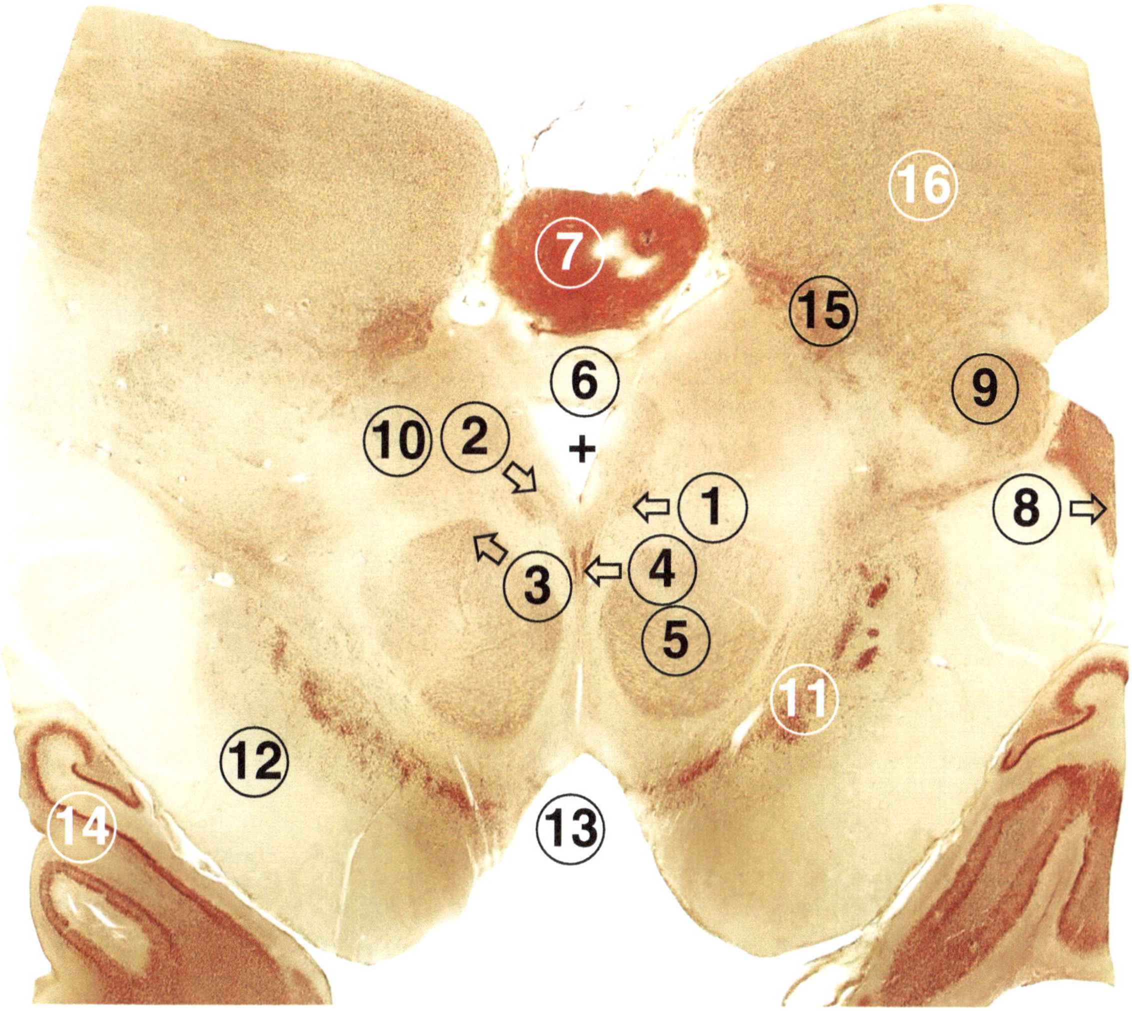

①	Interstitial nucleus (Cajal)	⑨	Medial geniculate body
②	Elliptic nucleus (Darkschewitsch)	⑩	Periaqueductal gray
③	Medial accessory nucleus of Bechterew	⑪	Substantia nigra
④	Accessory ncl. of oculomotor nerve, nonganglionic projecting part (Edinger-Westphal ncl., centrally projecting part)	⑫	Cerebral peduncle
		⑬	Interpeduncular fossa
⑤	Red nucleus, parvocellular part	⑭	Hippocampus
⑥	Posterior commissure	⑮	Caudal intralaminar group of the thalamus
⑦	Pineal gland	⑯	Pulvinar
⑧	Lateral geniculate body	+	Mesencephalic aqueduct

17.2 Campbell series (Visualization of fiber tracts)

Campbell 1

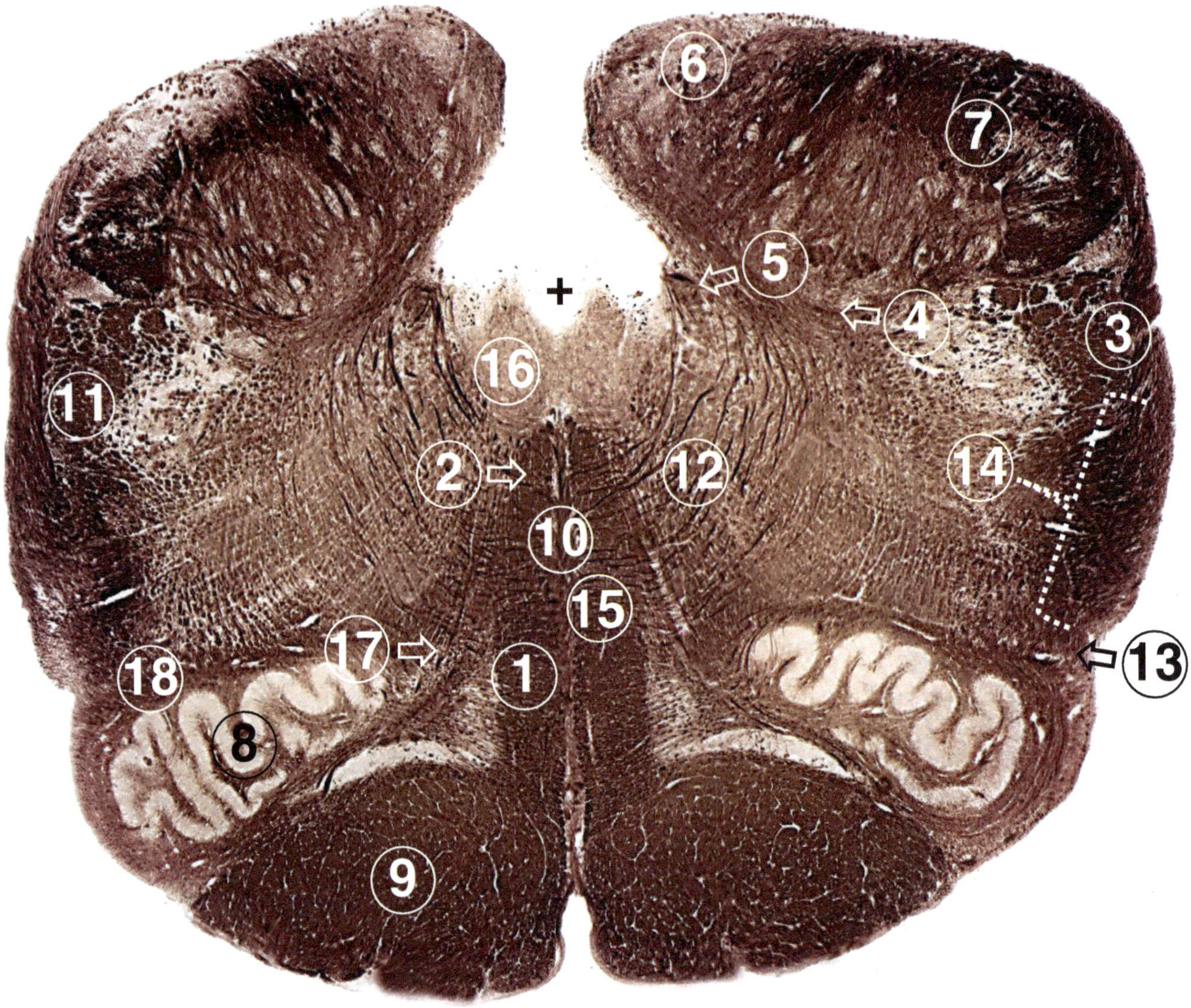

①	Medial lemniscus	⑪	Spinal tract of trigeminal nerve CNV
②	Medial longitudinal fasciculus	⑫	Internal arcuate fibers
③	Inferior cerebellar peduncle	⑬	Region of the spinothalamic fibers / Anterolateral tract
④	Bulbar root of accessory nerve CNXI	⑭	Spinocerebellar tracts
⑤	Solitary tract	⑮	Region of the tectospinal tract
⑥	Gracile fasciculus	⑯	Ncl. of hypoglossal nerve CNXII
⑦	Cuneate fasciculus	⑰	Hypoglossal nerve CNXII
⑧	Inferior olivary complex	⑱	Central tegmental tract
⑨	Pyramidal (corticospinal) tract	+	4th ventricle
⑩	Decussation of medial lemnisci		

Campbell 2

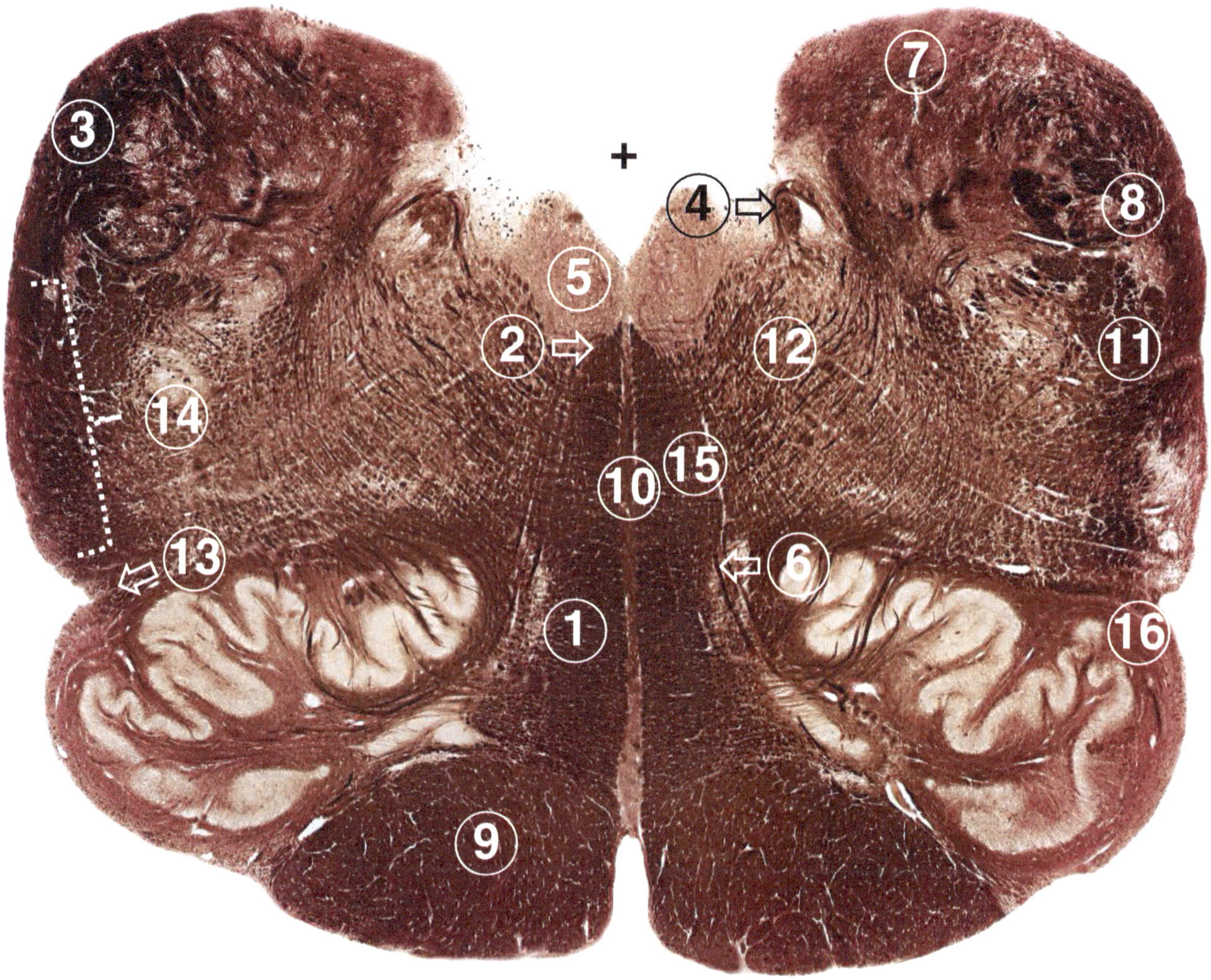

①	Medial lemniscus	⑩	Decussation of medial lemnisci
②	Medial longitudinal fasciculus	⑪	Spinal tract of trigeminal nerve
③	Inferior cerebellar peduncle	⑫	Internal arcuate fibers
④	Solitary tract	⑬	Region of the spinothalamic fibers / Anterolateral tract
⑤	Ncl. of hypoglossal nerve	⑭	Spinocerebellar tracts
⑥	Hypoglossal nerve	⑮	Region of the tectospinal tract
⑦	Gracile fasciculus	⑯	Central tegmental tract
⑧	Cuneate fasciculus	+	4th ventricle
⑨	Pyramidal (corticospinal) tract		

Campbell 3

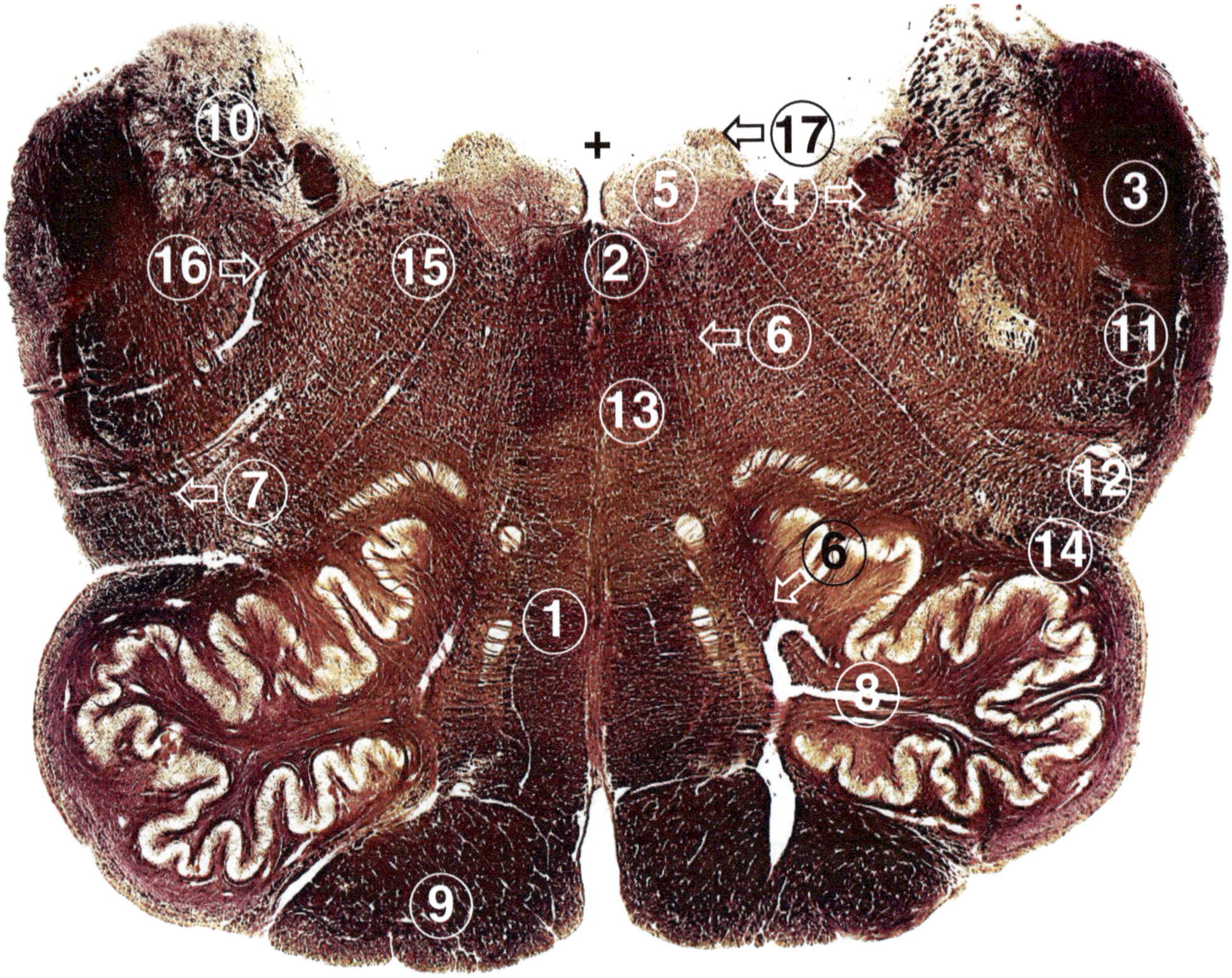

①	Medial lemniscus	⑩	Inferior (spinal) vestibular ncl. CNVIII
②	Medial longitudinal fasciculus	⑪	Spinal tract of trigeminal nerve
③	Inferior cerebellar peduncle	⑫	Region of the spinothalamic fibers / Anterolateral tract
④	Solitary tract	⑬	Region of tectospinal tract
⑤	Ncl. of hypoglossal nerve	⑭	Central tegmental tract
⑥	Hypoglossal nerve	⑮	Internal arcuate fibers
⑦	Olivocerebellar tract	⑯	Vagus nerve CNX
⑧	Inferior olivary complex	⑰	Posterior longitudinal fasciculus
⑨	Pyramidal (corticospinal) tract	+	4th ventricle

Campbell 4

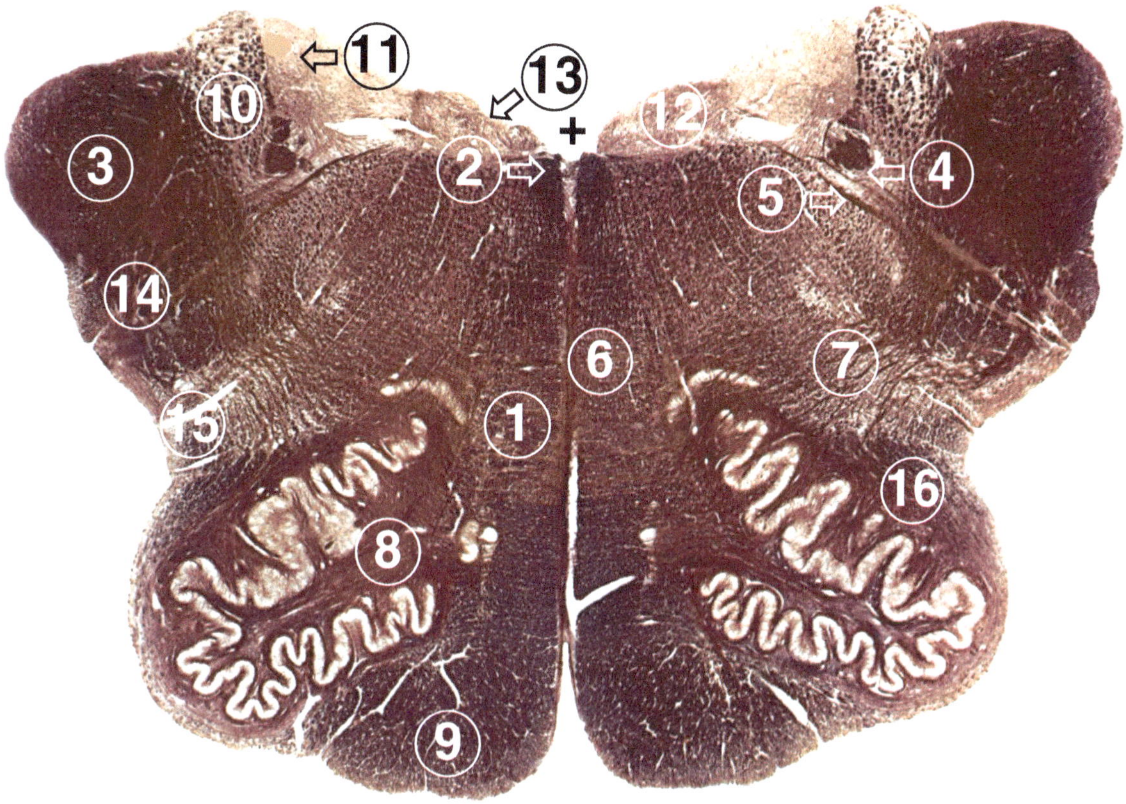

①	Medial lemniscus	⑩	Spinal (inferior) vestibular ncl.
②	Medial longitudinal fasciculus	⑪	Medial vestibular ncl. CNVIII
③	Inferior cerebellar peduncle	⑫	Prepositus hypoglossi ncl.
④	Solitary tract	⑬	Posterior longitudinal faciculus
⑤	Vagus nerve CNX See Darrow red 13	⑭	Spinal tract of trigeminal nerve
⑥	Region of tectospinal tract	⑮	Region of the spinothalamic fibers / Anterolateral tract
⑦	Olivocerebellar tract	⑯	Central tegmental tract
⑧	Inferior olivary complex	+	4th ventricle
⑨	Pyramidal (corticospinal) tract		

Campbell 5

①	Medial lemniscus	⑨	Pyramidal (corticospinal) tract
②	Medial longitudinal fasciculus	⑩	Central tegmental tract
③	Inferior cerebellar peduncle	⑪	Spinal trigeminal tract
④	Glossopharyngeal nerve CNIX See Darrow red 13	⑫	Decussation of medial lemnisci
⑤	Solitary tract	⑬	Region of the spinothalamic fibers / Anterolateral tract
⑥	Vestibular nerve CNVIII	⑭	Medial vestibular ncl.
⑦	Olivocerebellar tract	⑮	Inferior vestibular ncl.
⑧	Inferior olivary complex	+	4th ventricle

Campbell 6

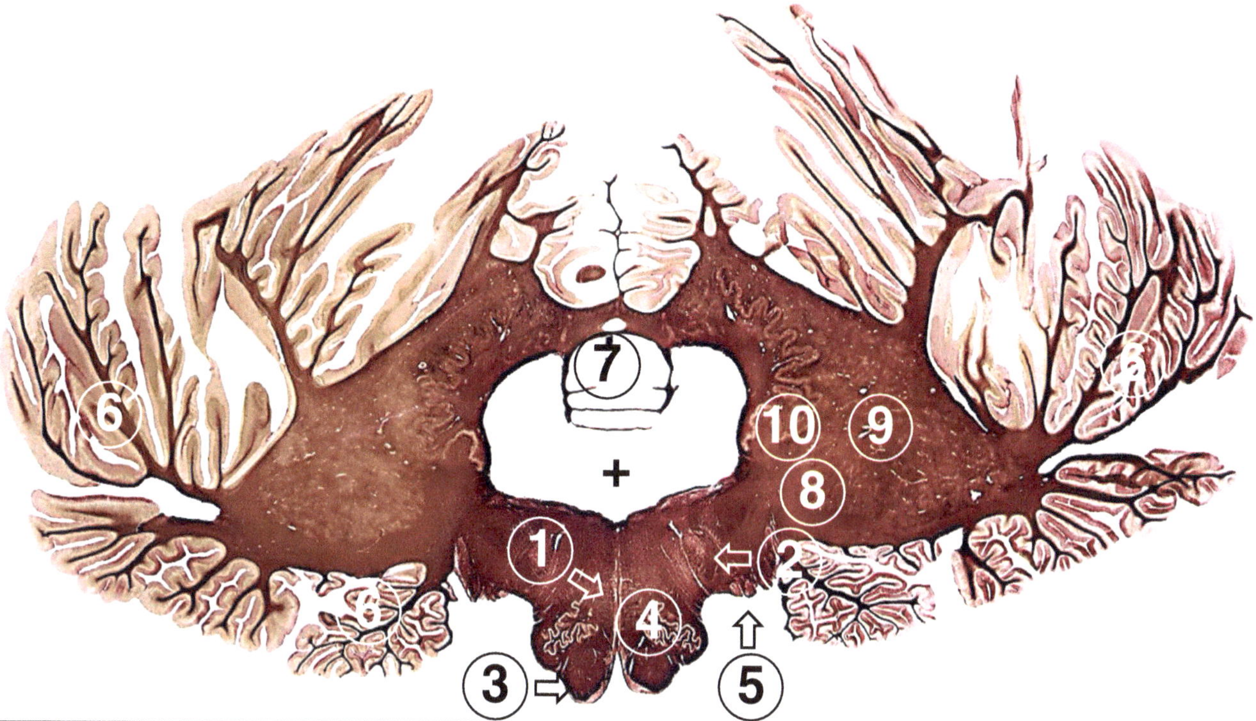

①	Medial lemniscus		⑦	Cerebellar nodule
②	Spinal tract / ncl. of trigeminal nerve		⑧	Inferior cerebellar peduncle
③	Pyramidal (corticospinal) tract		⑨	Middle cerebellar peduncle
④	Inferior olivary complex		⑩	Dentate ncl.
⑤	Cerebellopontine angle		+	4th ventricle
⑥	Cerebellum			

Campbell 6A

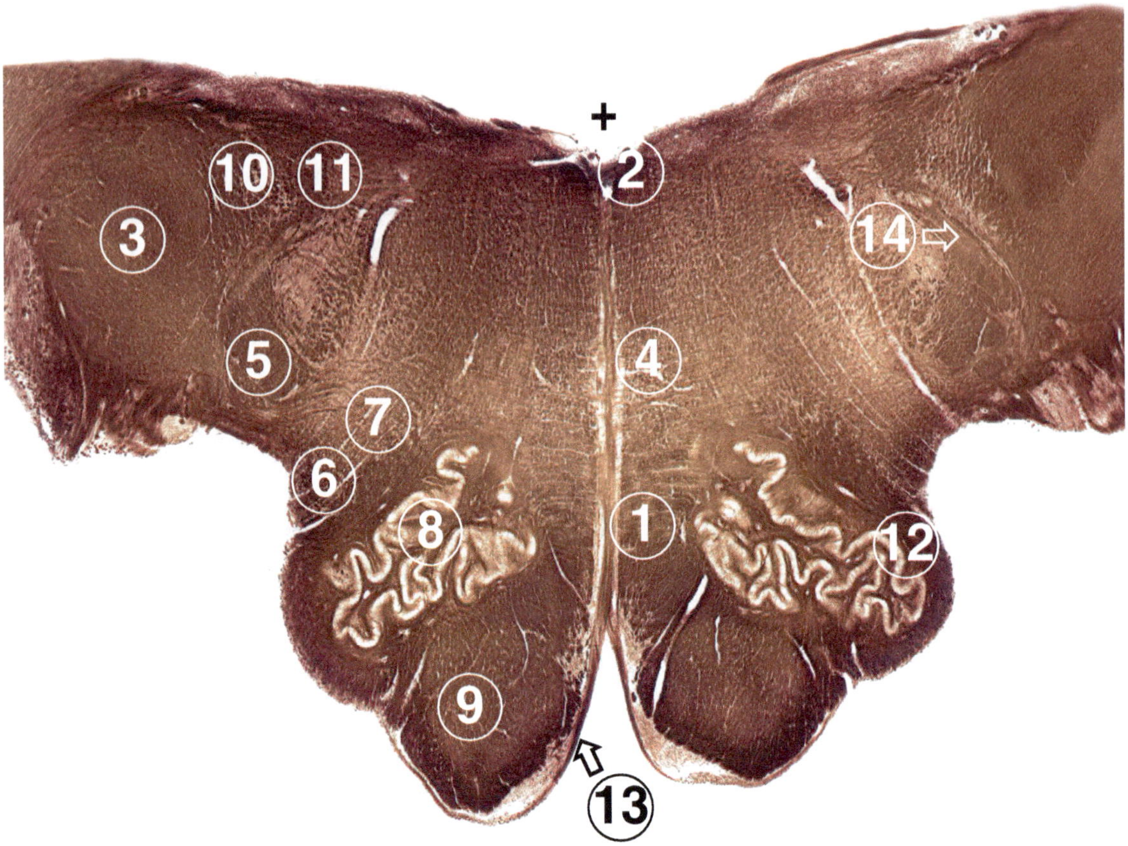

①	Medial lemniscus	⑨	Pyramidal (corticospinal) tract
②	Medial longitudinal fasciculus	⑩	Spinal vestibular ncl.
③	Inferior cerebellar peduncle	⑪	Medial vestibular ncl.
④	Region of tectospinal tract	⑫	Central tegmental tract
⑤	Spinal tract of trigeminal nerve	⑬	Anterior external arcuate fibers
⑥	Region of the spinothalamic fibers / Anterolateral tract	⑭	Glossopharyngeal nerve CNIX
⑦	Olivocerebellar tract	+	4th ventricle
⑧	Inferior olivary complex		

Campbell 7

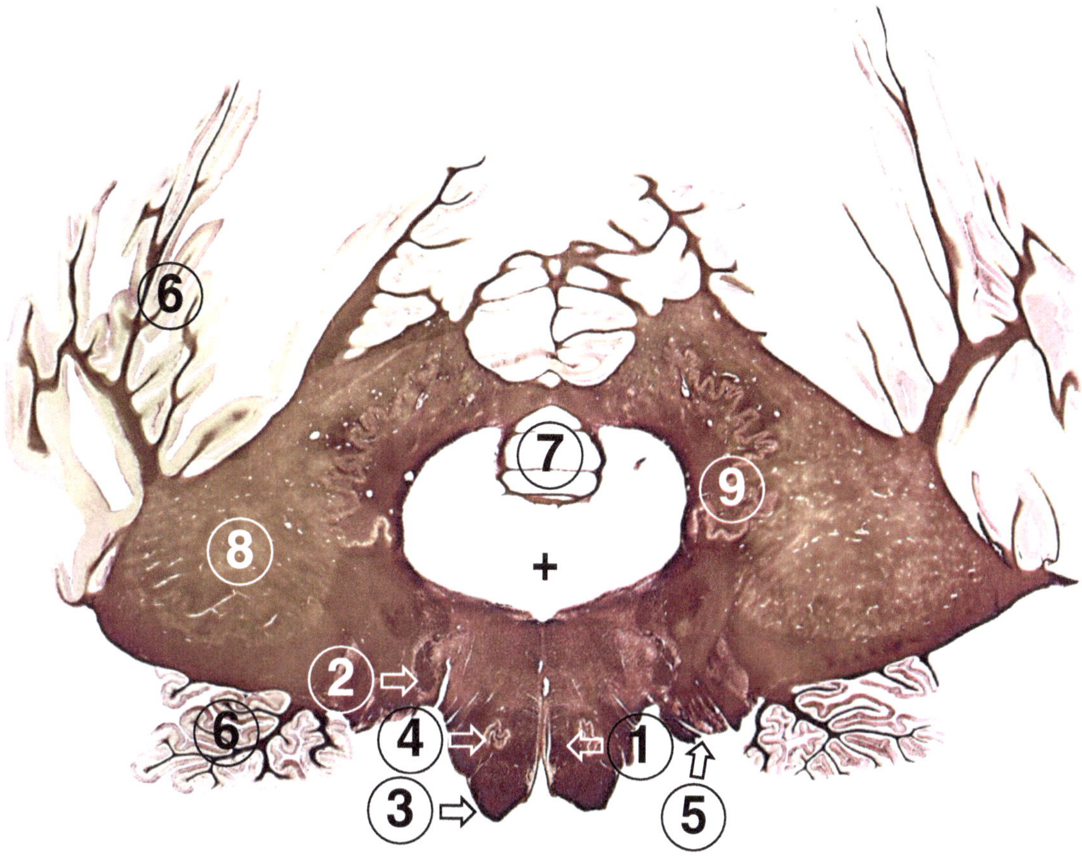

①	Medial lemniscus	⑥	Cerebellum
②	Spinal tract / ncl. of trigeminal nerve	⑦	Cerebellar nodule
③	Pyramidal (corticospinal) tract	⑧	Middle cerebellar peduncle
④	Inferior olivary complex	⑨	Dentate ncl.
⑤	Cerebellopontine angle	+	4th ventricle

Campbell 8

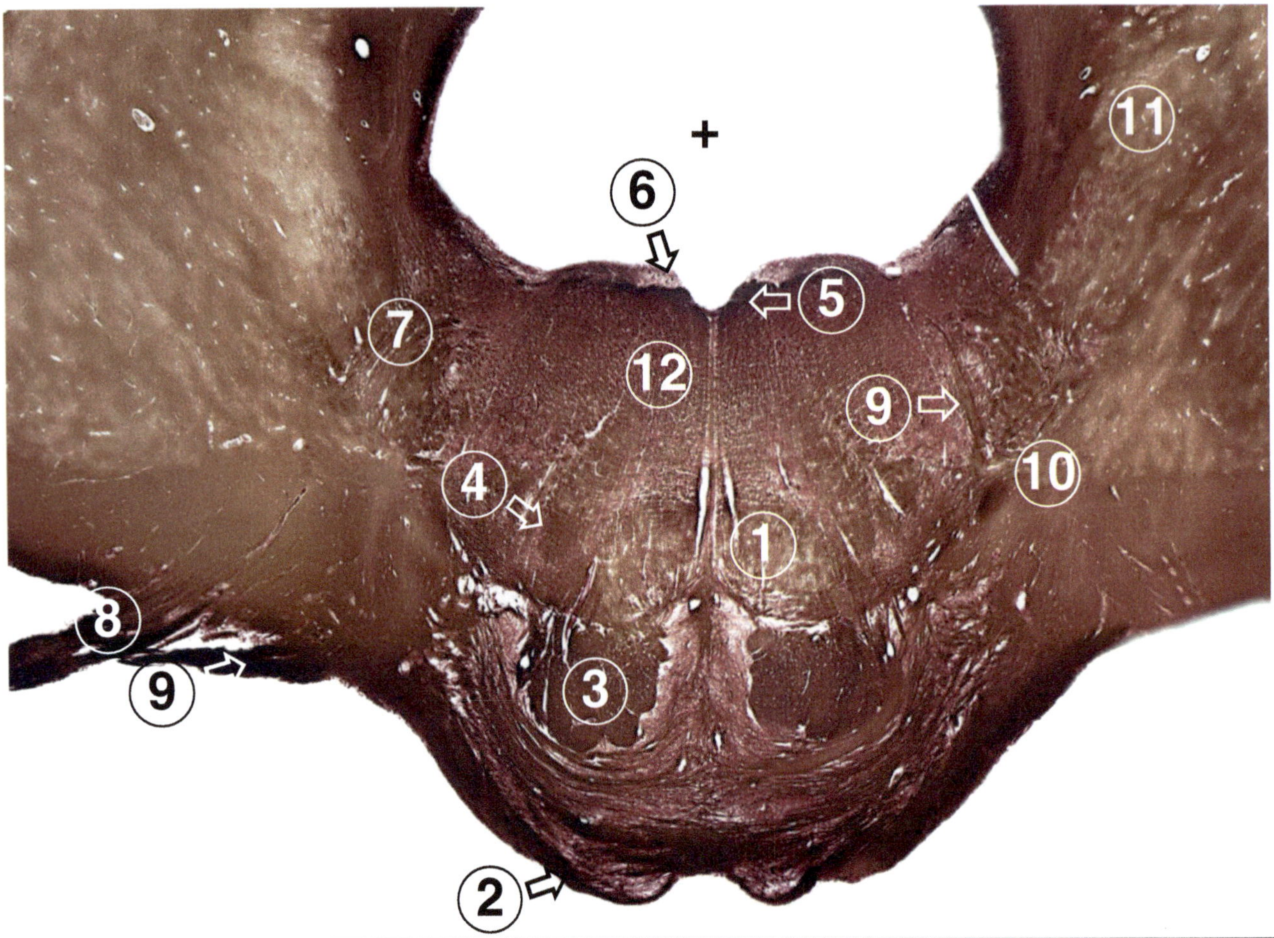

①	Medial lemniscus	⑦	Spinal tract of trigeminal nerve
②	Pontine basis	⑧	Vestibulocochlear nerve CNVIII
③	Pyramidal (corticospinal) tract	⑨	Facial nerve CNVII
④	Central tegmental tract	⑩	Inferior cerebellar peduncle
⑤	Medial longitudinal fasciculus	⑪	Middle cerebellar peduncle
⑥	Posterior longitudinal fasciulus	⑫	Region of tectospinal tract
		+	4th ventricle

Campbell 8A

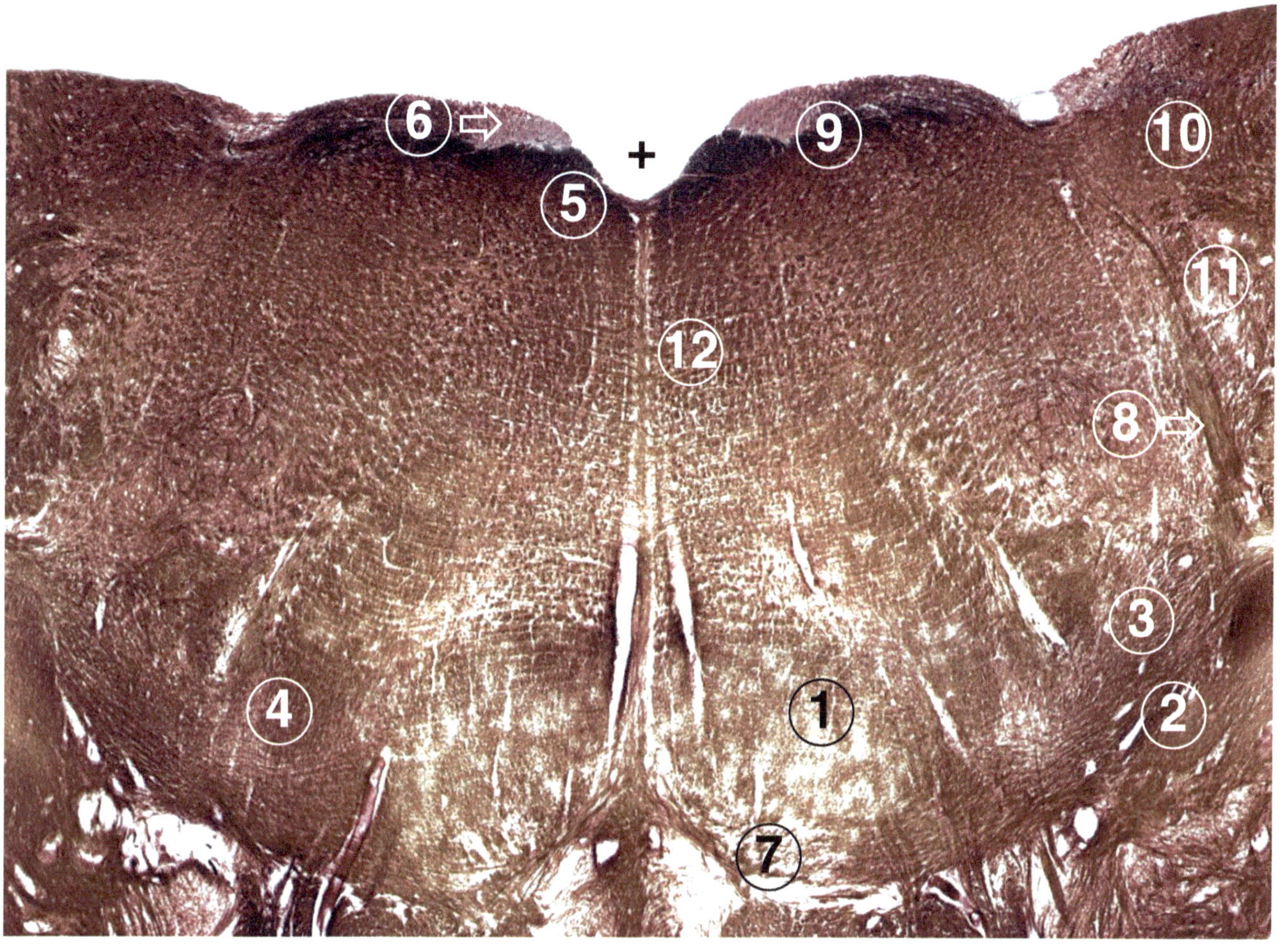

①	Medial lemniscus		⑦	Trapezoid body
②	Region of the spinothalamic fibers / anterolateral tract		⑧	Facial nerve, descending part
③	Lateral lemniscus		⑨	Genu of facial nerve
④	Central tegmental tract		⑩	Vestibular ncll.
⑤	Medial longitudinal fasciculus		⑪	Spinal ncl. of trigeminal nerve
⑥	Posterior longitudinal fasciulus		⑫	Region of tectospinal tract
			+	4th ventricle

Campbell 9

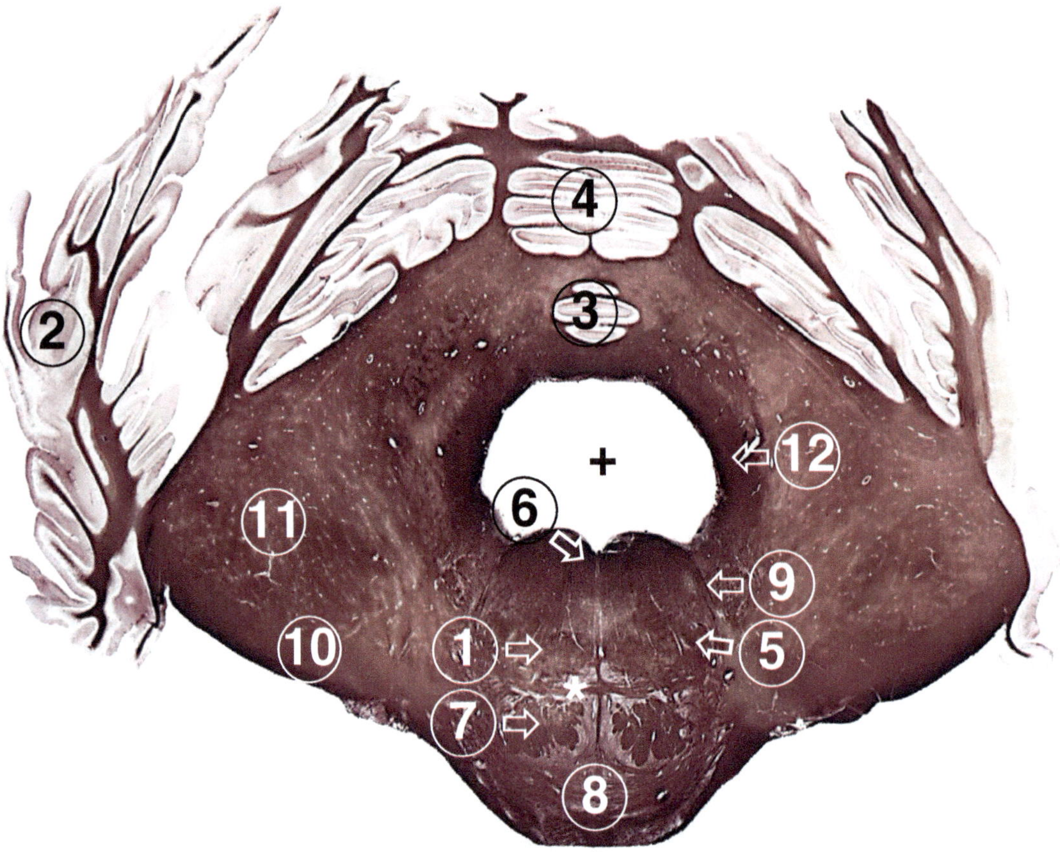

①	Medial lemniscus	⑧	Transverse pontine fibers (Pontocerebellar fibers)
②	Cerebellum	⑨	Facial nerve
③	Nodule	⑩	Inferior cerebellar peduncle
④	Uvula	⑪	Middle cerebellar peduncle
⑤	Superior olivary complex	⑫	Superior cerebellar peduncle
⑥	Medial longitudinal fasciculus	*	Trapezoid body
⑦	Pyramidal (corticospinal) tract	+	4th ventricle

Campbell 9A

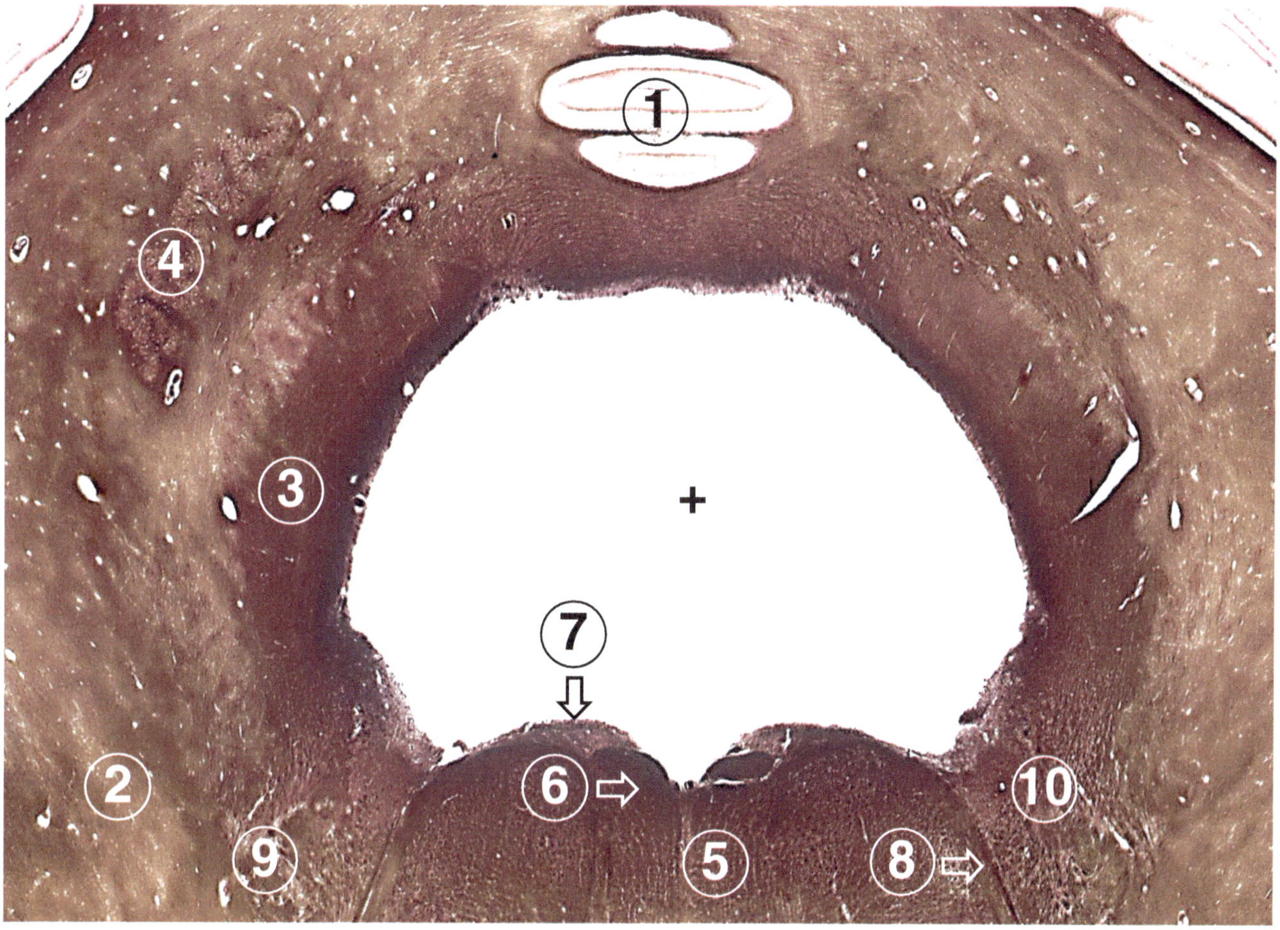

①	Nodule		⑦	Posterior longitudinal fasciculus
②	Middle cerebellar peduncle		⑧	Facial nerve, descending part
③	Superior cerebellar peduncle		⑨	Spinal tract of trigeminal nerve
④	Dentate ncl.		⑩	Vestibular ncll.
⑤	Region of tectospinal tract		+	4th ventricle
⑥	Medial longitudinal fasciculus			

Campbell 9B

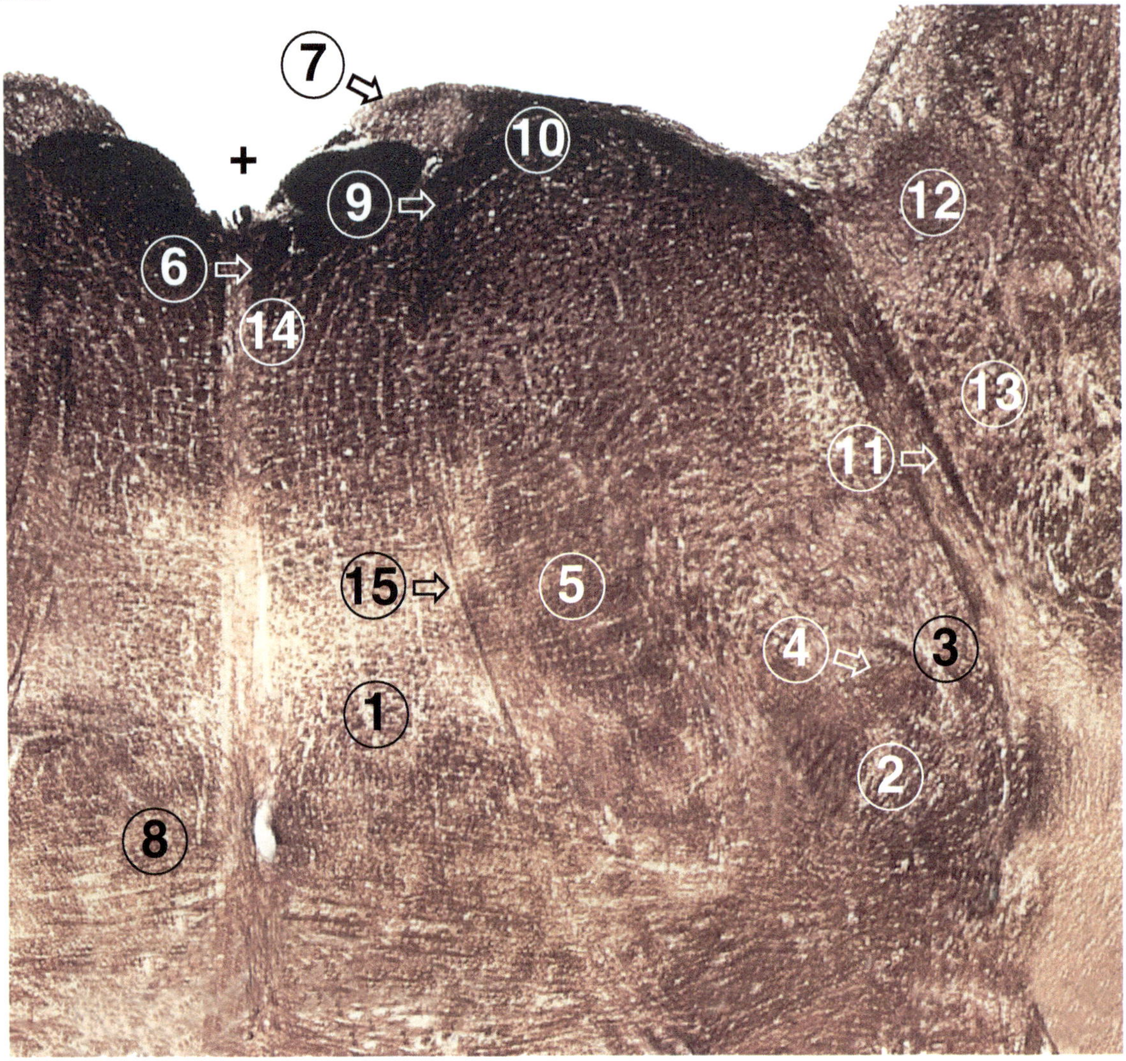

①	Medial lemniscus	⑨	Facial nerve, ascending part
②	Region of the spinothalamic fibers / anterolateral tract	⑩	Genu of facial nerve
③	Lateral lemniscus	⑪	Facial nerve, descending part
④	Superior olivary complex	⑫	Vestibular ncll.
⑤	Central tegmental tract	⑬	Spinal ncl. of trigeminal nerve
⑥	Medial longitudinal fasciculus	⑭	Region of tectospinal tract
⑦	Posterior longitudinal fasciculus	⑮	Abducens nerve CNVI
⑧	Trapezoid body	+	4th ventricle

Campbell 10

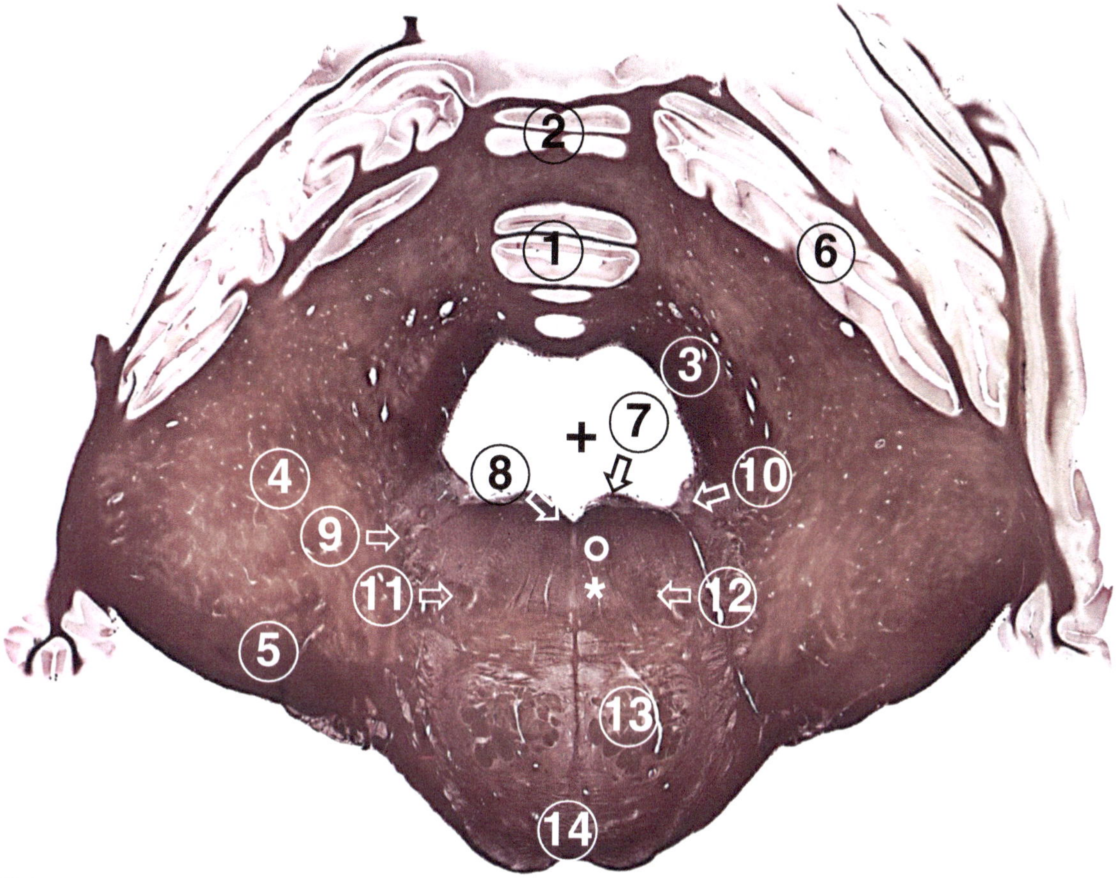

①	Nodule	⑩	Vestibular ncll.
②	Uvula	⑪	Superior olivary complex
③	Superior cerebellar peduncle	⑫	Central tegmental tract
④	Middle cerebellar peduncle	⑬	Pyramidal (corticospinal) tract
⑤	Inferior cerebellar peduncle	⑭	Transverse pontine fibers (Pontocerebellar fibers)
⑥	Cerebellum	*	Medial lemniscus
⑦	Posterior longitudinal fasciculus	∘	Region of tectospinal tract
⑧	Medial longitudinal fasciculus	+	4th ventricle
⑨	Spinal tract of trigeminal nerve		

Campbell 10A

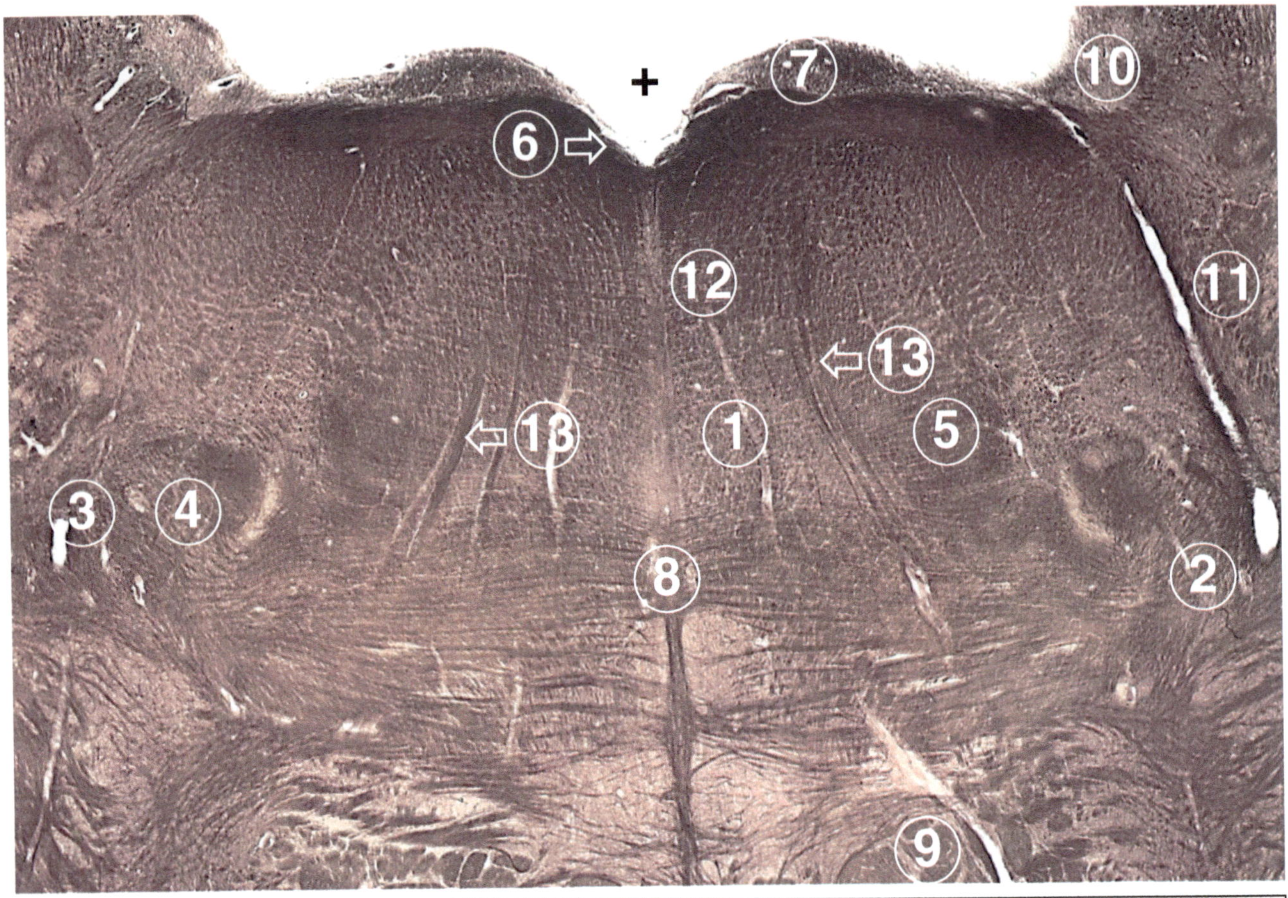

①	Medial lemniscus		⑧	Trapezoid body
②	Region of the spinothalamic fibers / anterolateral tract		⑨	Pyramidal (corticospinal) tract
③	Lateral lemniscus		⑩	Vestibular ncll.
④	Superior olivary complex		⑪	Spinal ncl. of trigeminal nerve
⑤	Central tegmental tract		⑫	Region of tectospinal tract
⑥	Medial longitudinal fasciculus		⑬	Abducens nerve CNVI
⑦	Posterior longitudinal fasciculus		+	4th ventricle

Campbell 11

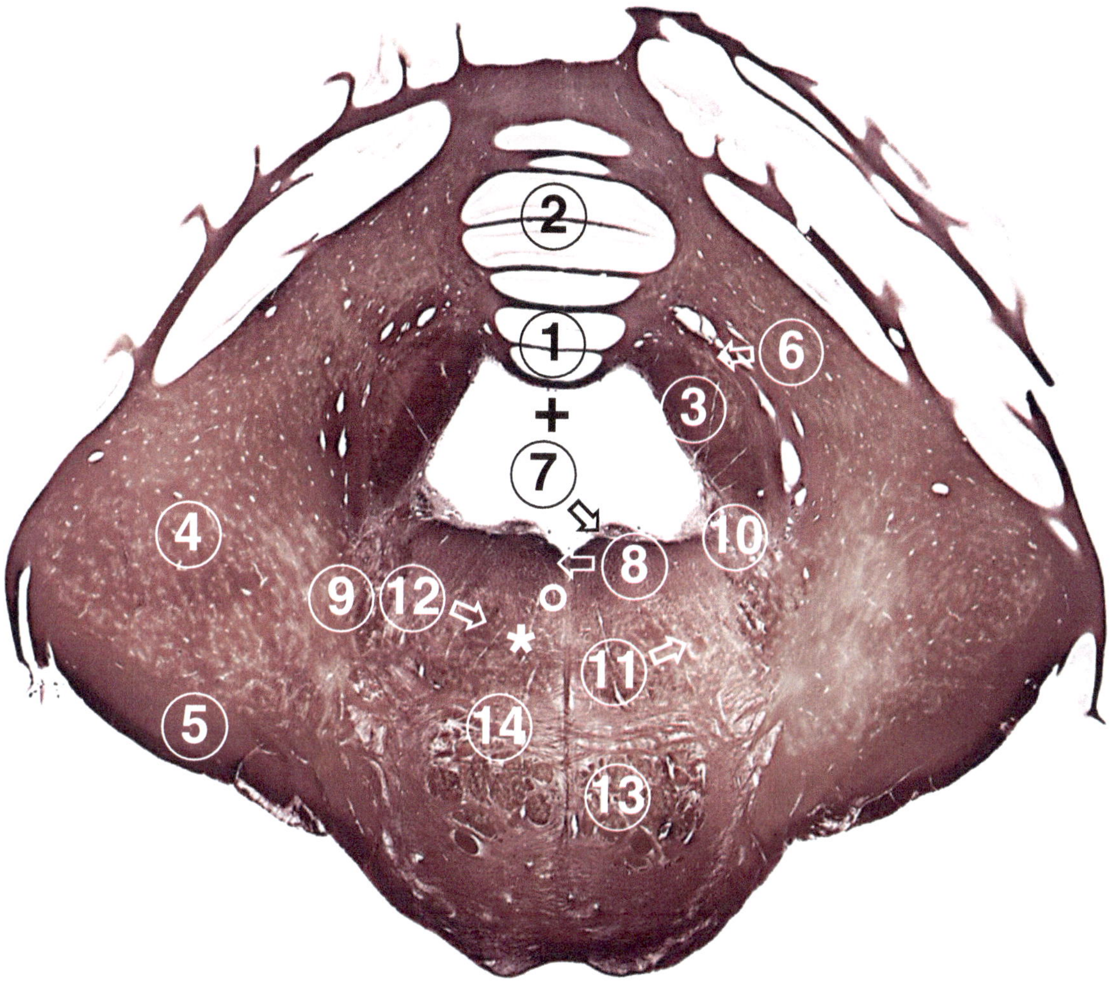

①	Nodule		⑩	Vestibular ncll.
②	Uvula		⑪	Superior olivary complex
③	Superior cerebellar peduncle		⑫	Central tegmental tract
④	Middle cerebellar peduncle		⑬	Pyramidal (corticospinal) tract
⑤	Inferior cerebellar peduncle		⑭	Transverse pontine fibers (Pontocerebellar fibers)
⑥	Dentate ncl.		*	Medial lemniscus
⑦	Posterior longitudinal fasciculus		∘	Region of tectospinal tract
⑧	Medial longitudinal fasciculus		+	4th ventricle
⑨	Spinal tract of trigeminal nerve			

Campbell 11A

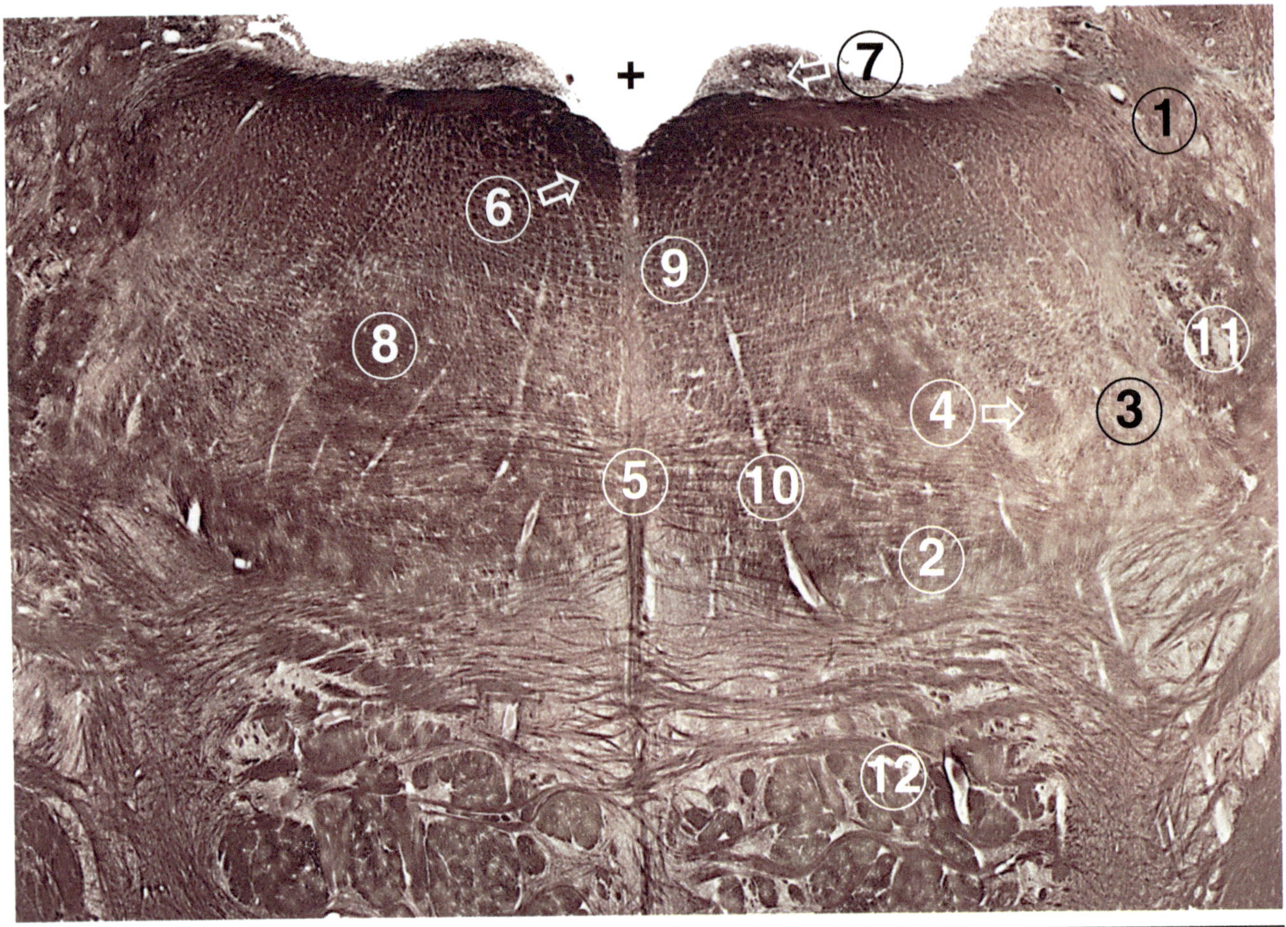

①	Vestibular ncll.	⑧	Central tegmental tract
②	Region of the spinothalamic fibers / anterolateral tract	⑨	Region of tectospinal tract
③	Lateral lemniscus	⑩	Medial lemniscus
④	Superior olivary complex	⑪	Spinal tract of trigeminal nerve
⑤	Trapezoid body	⑫	Pyramidal (corticospinal) tract
⑥	Medial longitudinal fasciculus	+	4th ventricle
⑦	Posterior longitudinal fasciulus		

Campbell 12

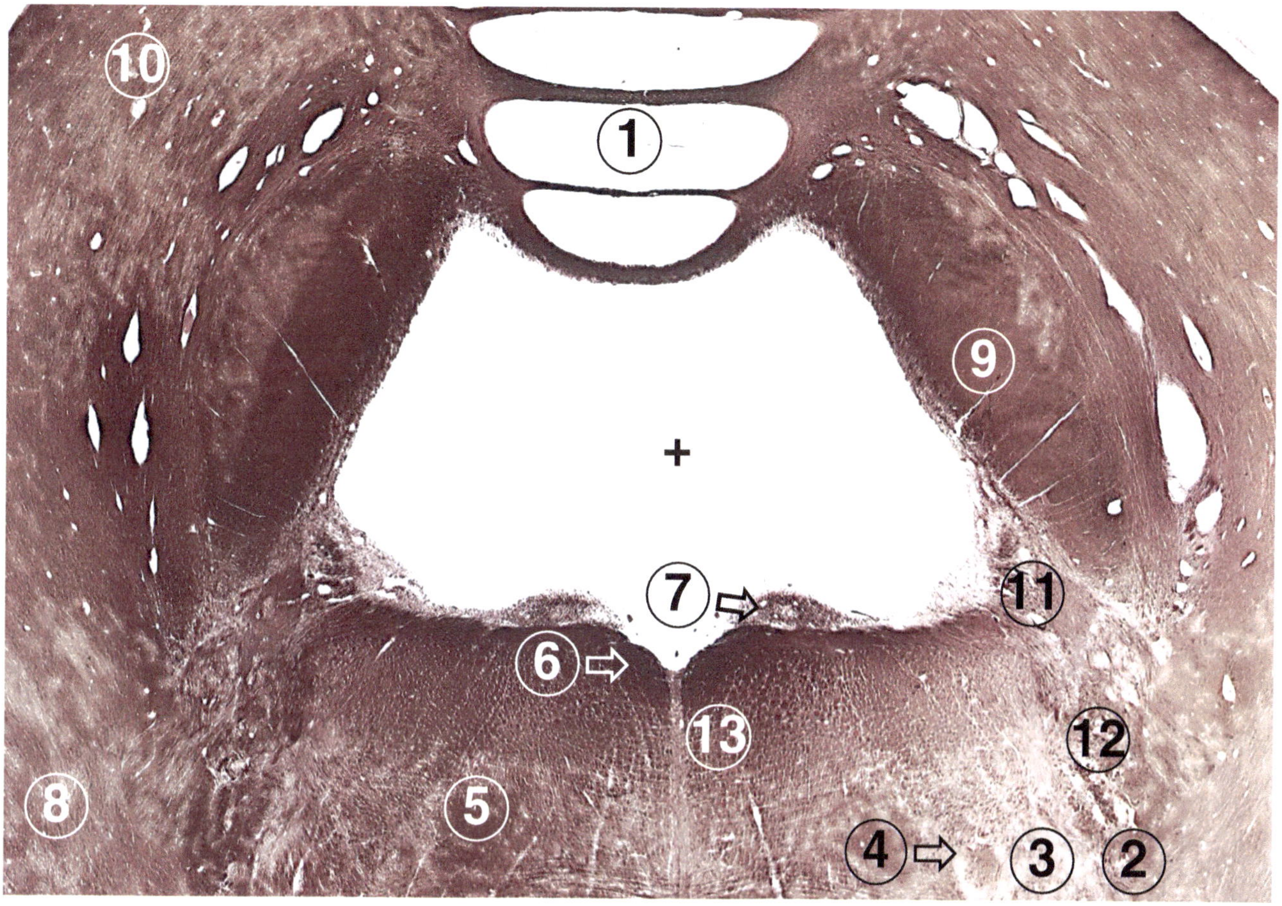

①	Nodule	⑧	Middle cerebellar peduncle
②	Region of the spinothalamic fibers / anterolateral tract	⑨	Superior cerebellar peduncle
③	Lateral lemniscus	⑩	Cerebellar white matter
④	Superior olivary complex	⑪	Vestibular ncll.
⑤	Central tegmental tract	⑫	Spinal tract of trigeminal nerve
⑥	Medial longitudinal fasciculus	⑬	Region of tectospinal tract
⑦	Posterior longitudinal fasciulus	+	4th ventricle

Campbell 13

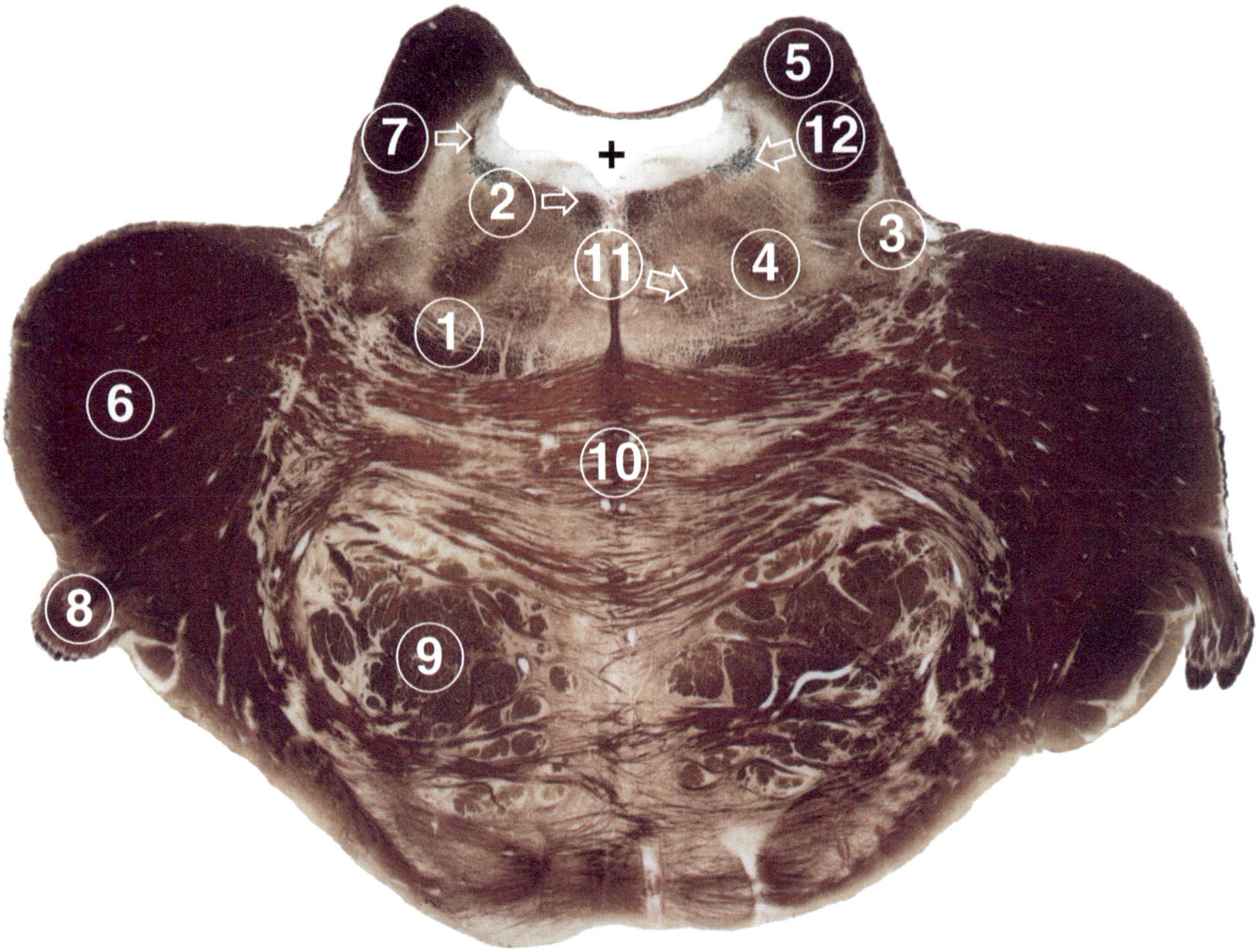

①	Medial lemniscus	⑦	Mesencephalic tract of trigeminal nerve
②	Medial longitudinal fasciculus	⑧	Trigeminal nerve CNV
③	Lateral lemniscus	⑨	Pyramidal (corticospinal) tract
④	Central tegmental tract	⑩	Transverse pontine fibers (Pontocerebellar fibers)
⑤	Superior cerebellar peduncle	⑪	Tectospinal tract
⑥	Middle cerebellar peduncle	⑫	Locus caeruleus
		+	4th ventricle

Campbell 14

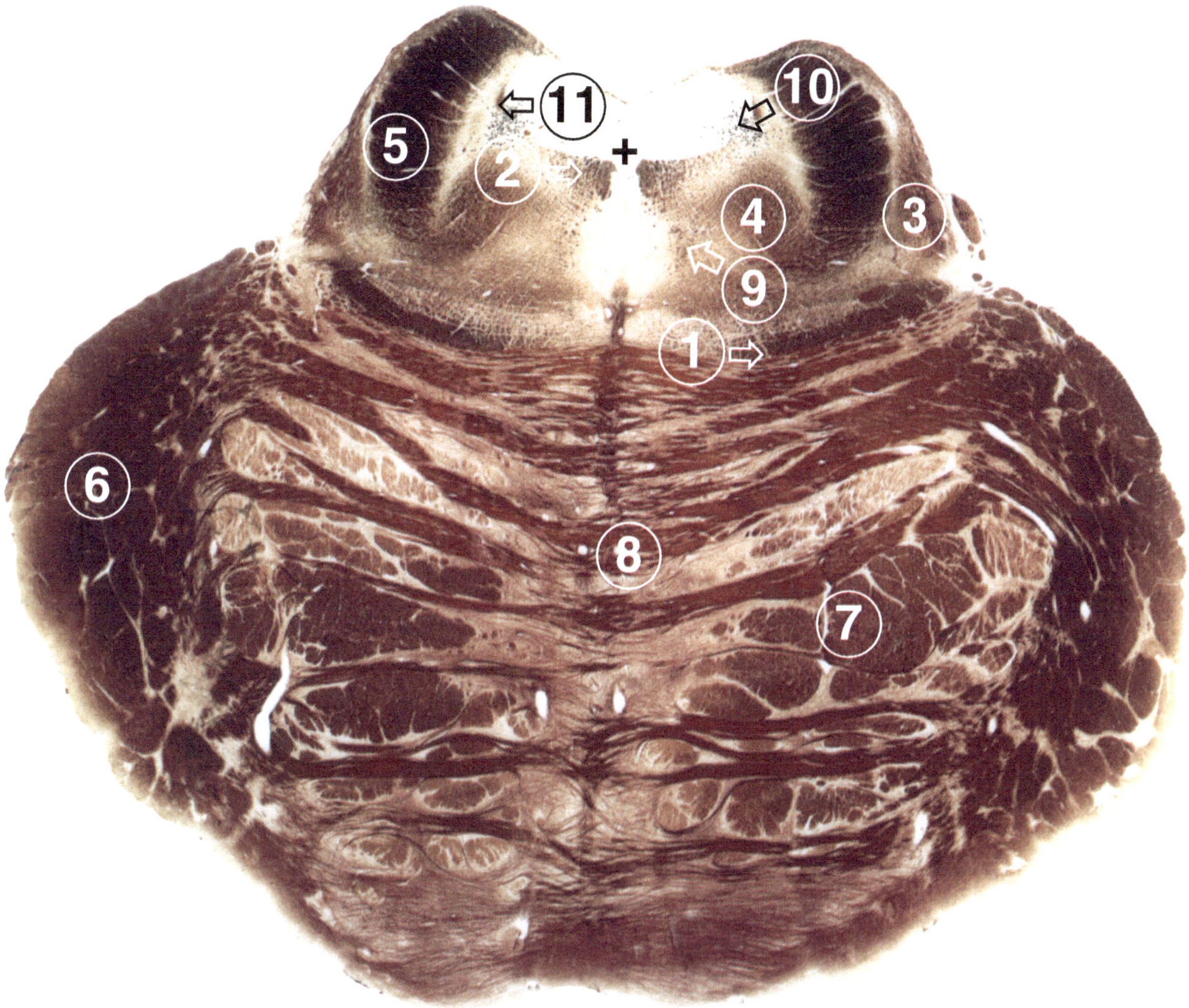

①	Medial lemniscus	⑦	Pyramidal (corticospinal) tract
②	Medial longitudinal fasciculus	⑧	Transverse pontine fibers (Pontocerebellar fibers)
③	Lateral lemniscus	⑨	Region of tectospinal tract
④	Central tegmental tract	⑩	Locus caeruleus
⑤	Superior cerebellar peduncle	⑪	Mesencephalic tract of trigeminal nerve
⑥	Middle cerebellar peduncle	+	4th ventricle

Campbell 15

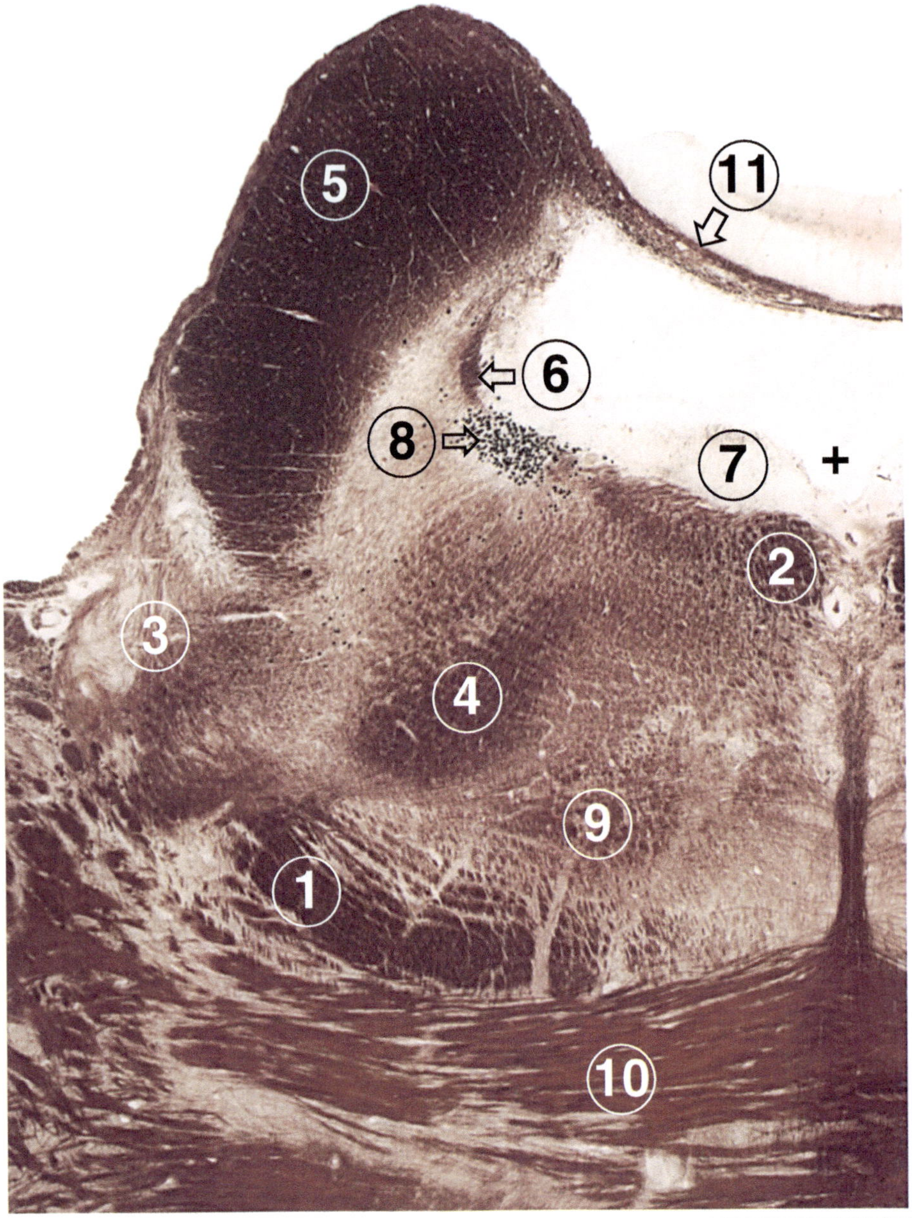

①	Medial lemniscus	⑦	Posterior longitudinal fasciculus
②	Medial longitudinal fasciculus	⑧	Locus caeruleus
③	Lateral lemniscus	⑨	Tectospinal tract
④	Central tegmental tract	⑩	Transverse pontine fibers (Pontocerebellar fibers)
⑤	Superior cerebellar peduncle	⑪	Superior medullary velum
⑥	Mesencephalic tract of trigeminal nerve	+	4th ventricle

Campbell 16

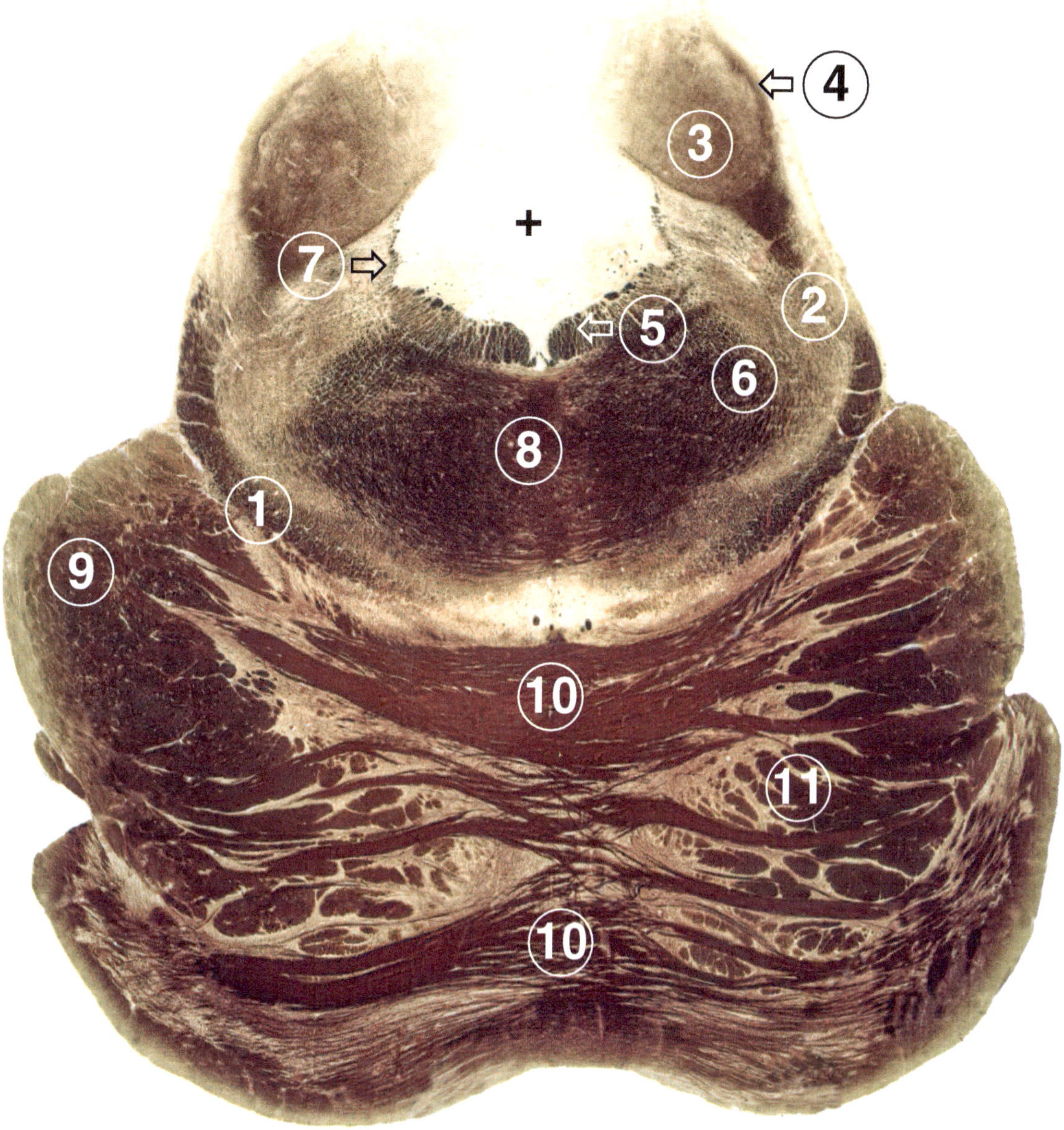

①	Medial lemniscus	⑦	Mesencephalic tract of trigeminal nerve
②	Lateral lemniscus	⑧	Decussation of superior cerebellar peduncles
③	Inferior colliculus	⑨	Parietotemporopontine fibers
④	Brachium of inferior colliculus	⑩	Transverse pontine fibers (Pontocerebellar fibers)
⑤	Medial longitudinal fasciculus	⑪	Pyramidal (corticospinal) tract
⑥	Central tegmental tract	+	Mesencephalic aqueduct

Campbell 17

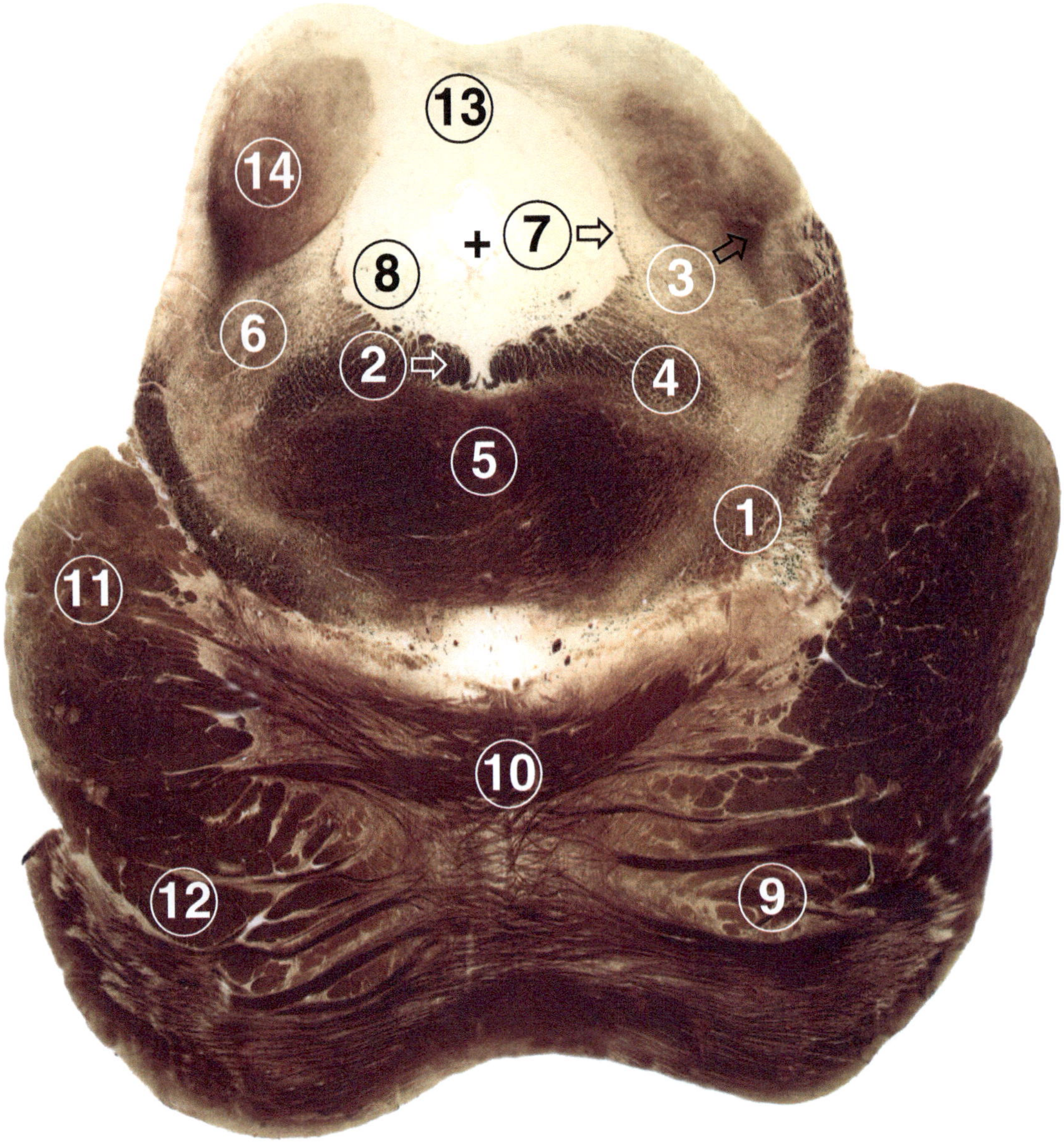

①	Medial lemniscus	⑧	Posterior longitudinal fasciculus
②	Medial longitudinal fasciculus	⑨	Pyramidal (corticospinal) tract
③	Lateral lemniscus	⑩	Transverse pontine fibers (Pontocerebellar fibers)
④	Central tegmental tract	⑪	Parietotemporopontine fibers
⑤	Decussation of superior cerebellar peduncles	⑫	Frontopontine fibers
⑥	Anterolateral tract	⑬	Periaqueductal gray substance
⑦	Mesencephalic tract of trigeminal nerve	⑭	Inferior colliculus
		+	Mesencephalic aqueduct

Campbell 17A

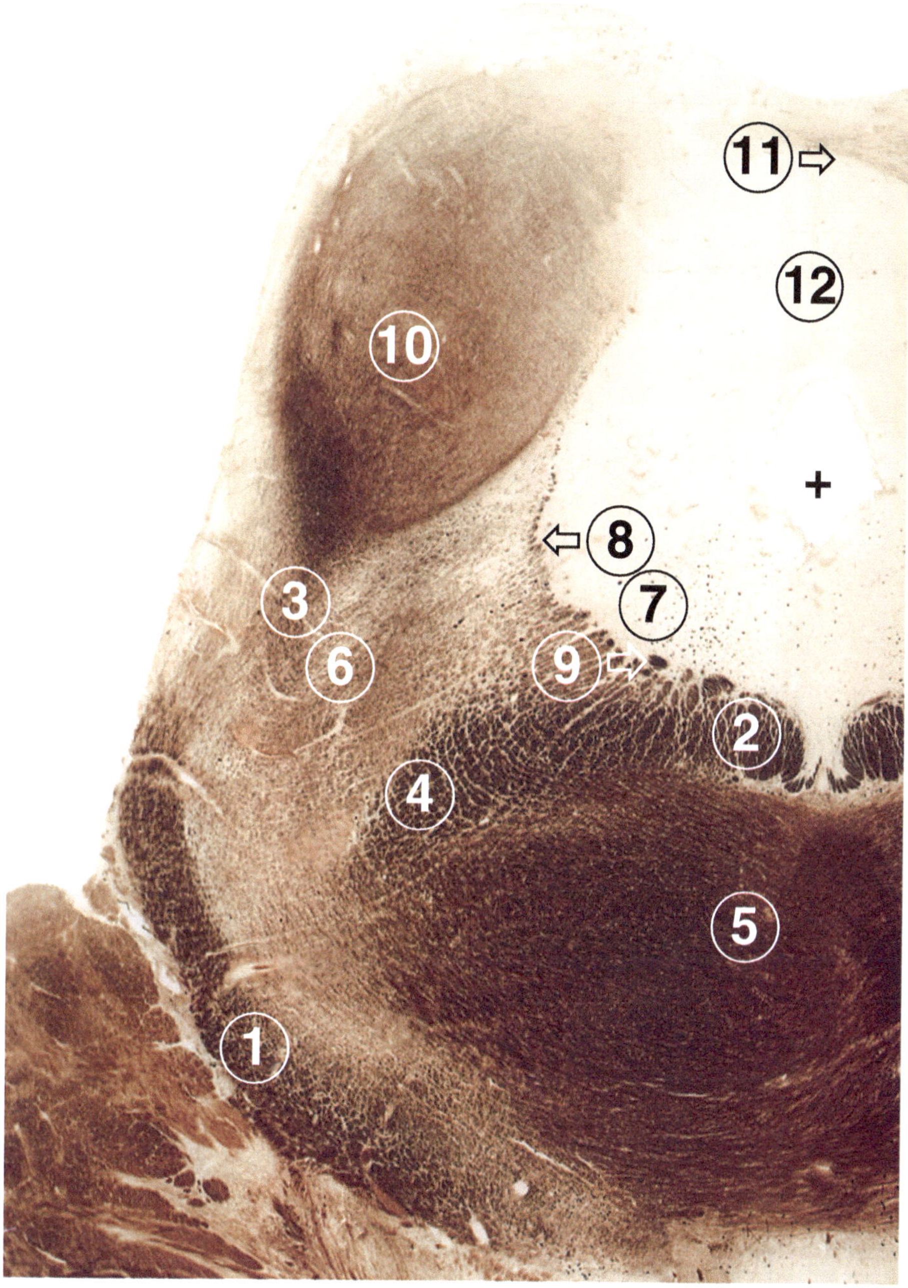

①	Medial lemniscus	⑦	Posterior longitudinal fasciculus
②	Medial longitudinal fasciculus	⑧	Mesencephalic tract of trigeminal nerve
③	Lateral lemniscus	⑨	Trochlear nerve CNIV
④	Central tegmental tract	⑩	Inferior colliculus
⑤	Decussation of superior cerebellar peduncles	⑪	Commissure of inferior colliculi
⑥	Anterolateral tract	⑫	Periaqueductal gray substance
		+	Mesencephalic aqueduct

Campbell 18

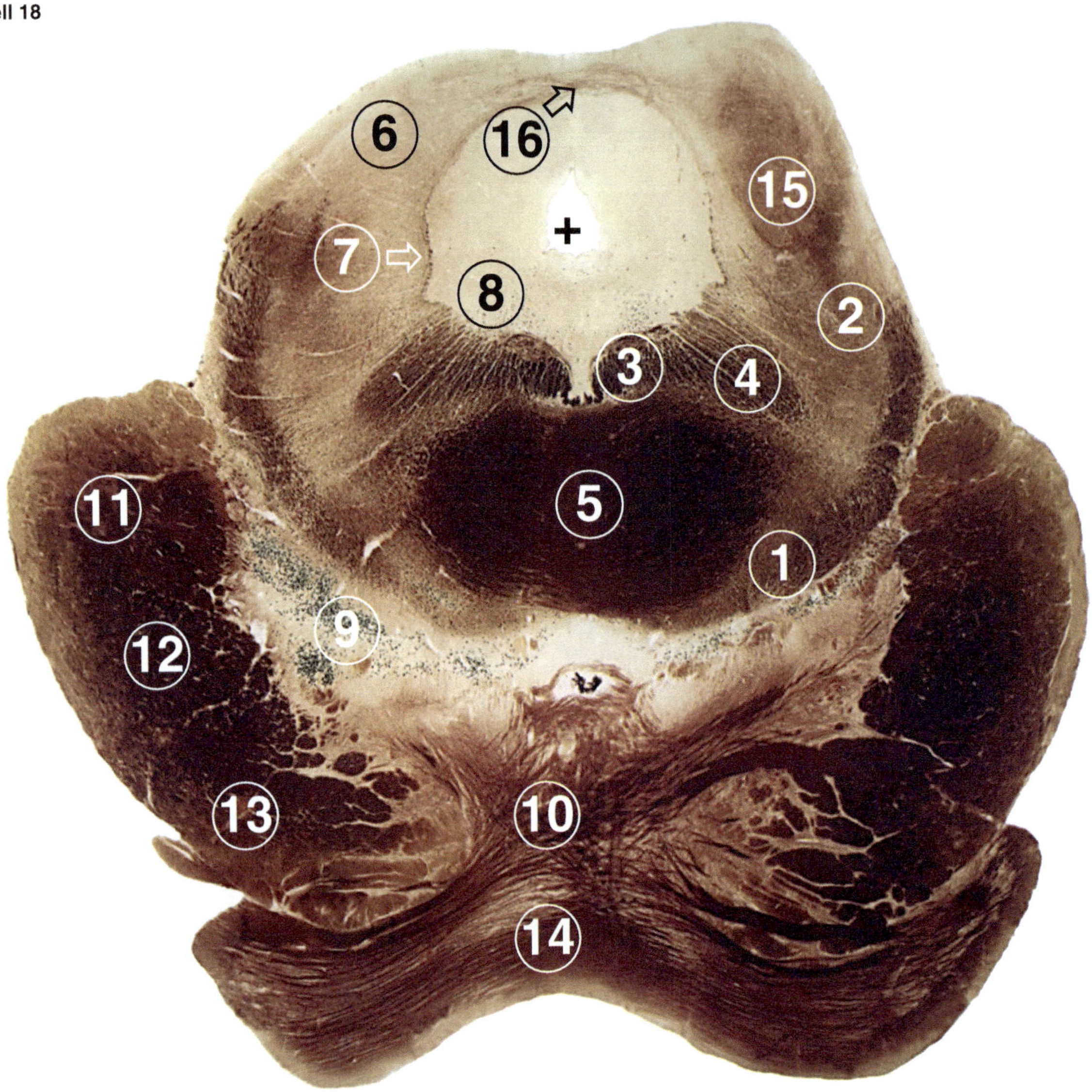

①	Medial lemniscus	⑨	Substantia nigra
②	Lateral lemniscus	⑩	Transverse pontine fibers (Pontocerebellar fibers)
③	Medial longitudinal fasciculus	⑪	Parietotemporopontine tract
④	Central tegmental tract	⑫	Pyramidal (corticospinal) tract
⑤	Decussation of superior cerebellar peduncles	⑬	Frontopontine tract
⑥	Transition inferior > superior colliculus	⑭	Pontine basis
⑦	Mesencephalic tract CNV	⑮	Inferior colliculus
⑧	Posterior longitudinal fasciculus	⑯	Decussation of trochlear nerves
		+	Mesencephalic aqueduct

Campbell 19

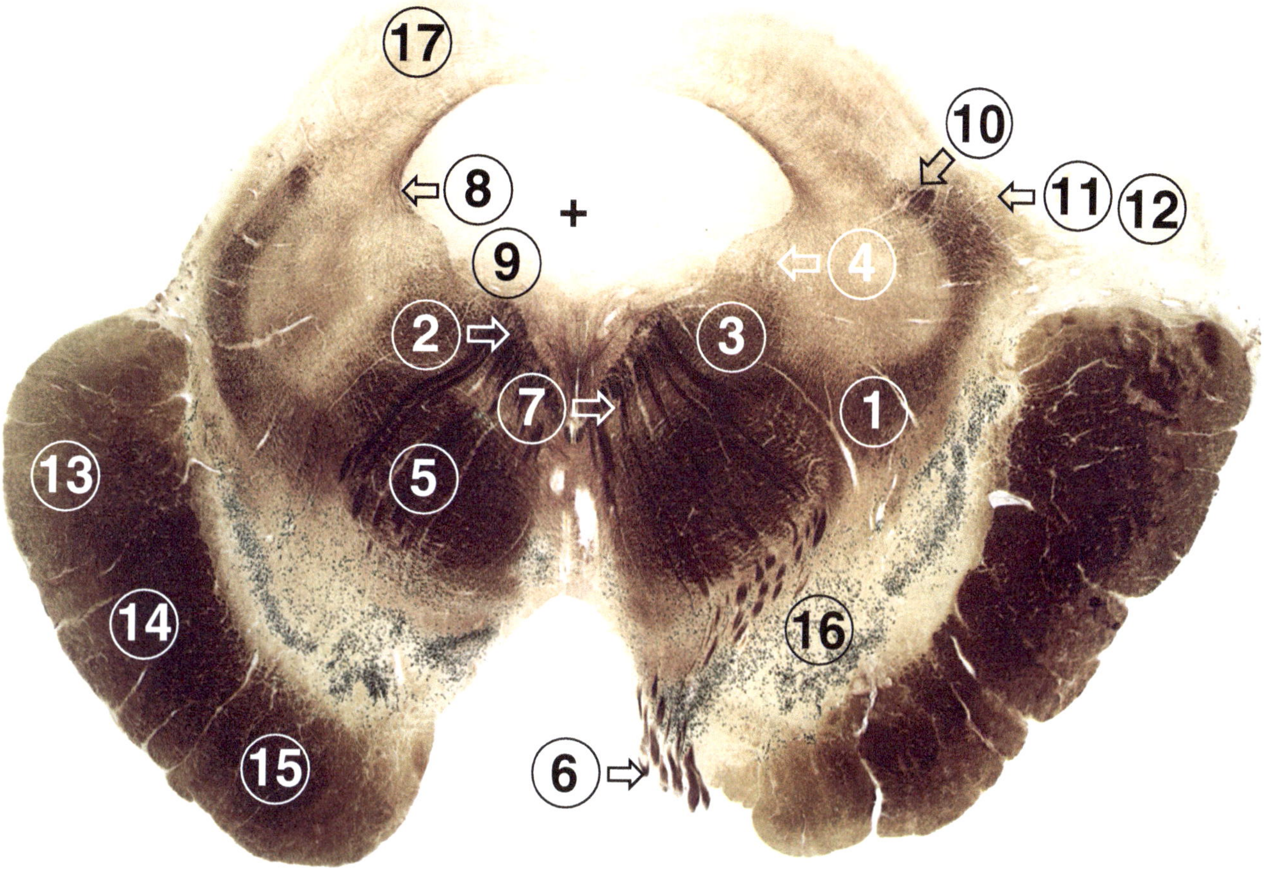

①	Medial lemniscus	⑩	Anterolateral tract
②	Medial longitudinal fasciculus	⑪	Brachium of inferior colliculus
③	Central tegmental tract	⑫	Medial geniculate body
④	Tectospinal tract	⑬	Parietotemporopontine tract
⑤	Superior cerebellar peduncle / Red ncl.	⑭	Pyramidal (Corticospinal) tract
⑥	Oculomotor nerve CNIII	⑮	Frontopontine tract
⑦	Oculomotor nerve inside red nucleus	⑯	Substantia nigra
⑧	Mesencephalic tract CNV	⑰	Superior colliculus
⑨	Posterior longitudinal fasciculus	+	Mesencephalic aqueduct

Campbell 20

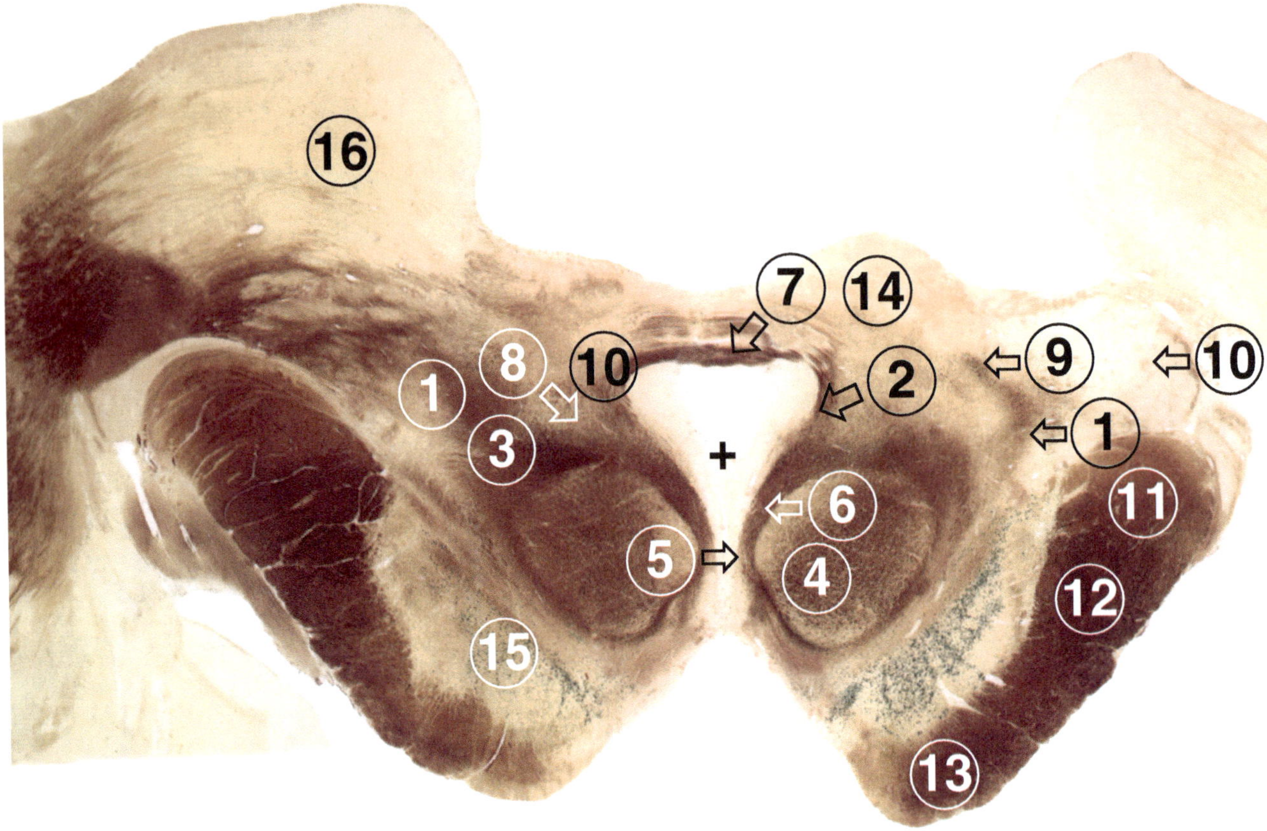

①	Medial lemniscus	⑨	Anterolateral tract
②	Posterior longitudinal fasciculus	⑩	Medial geniculate body
③	Central tegmental tract	⑪	Parietotemporopontine tract
④	Superior cerebellar peduncle (Red ncl.)	⑫	Pyramidal (corticospinal) tract
⑤	Habenulointerpeduncular tract	⑬	Frontopontine tract
⑥	Medial longitudinal fasciculus	⑭	Superior colliculus
⑦	Posterior commissure	⑮	Substantia nigra
⑧	Posterior trigeminothalamic tract	⑯	Pulvinar
		+	Mesencephalic aqueduct

References

Olszewski J, Baxter D (1982) Cytoarchitecture of the human brain stem, 2nd edn. Karger, Basel

Paxinos G, Furlong TM et al (2020) Human brainstem: cytoarchitecture, chemoarchitecture, myeloarchitecture. Revised edition. Academic Press

Atlas of the human brainstem: magnetic resonance imaging (MRI) 18

Contents

18.1 Annotations MRI Atlas

Although MRT technology and imaging quality has steadily progressed over the last years, in particular because of the small size of the brainstem and the structures contained, it is still difficult to obtain a direct spatial correlation with histological findings or even those in brain plastinates with the MR images. Only for large structures a direct identification is possible, like e.g. the inferior olivary complex, the cerebellar peduncles or the red nucleus. However, using identifiable structures and the histological atlas at least an educated guess on the location of smaller structures is possible which also means that a brainstem lesion can be correlated with the potential nuclei and/or fiber tracts concerned. For example, the location of the central tegmental tract—not directly visible—in the medulla oblongata can be assumed via the location of the inferior olivary complex (see atlas MRI 2). Another example is the location of the medial and lateral lemnisci in relation to the superior cerebellar peduncle (MRI 9).

The present primary use of T2 sequences with the cerebrospinal fluid-filled spaces appearing hyperintense (white) enables the identification of a number of cranial nerves as well as cerebral vessels in their course through the CSF spaces and a comparison with the topographical details provided by the plastinates. This holds also true e.g. for the organs of the inner ear (MRI 6). Cross references to the topographical atlas (see atlas P1 etc.) are provided. A MRI survey as well as a dissected specimen of the circle of Willis (for details see Chap. 1) is shown in the first section MRI 1 of this atlas part to enable the subsumption of individual structures in the MR image with the whole of the head and brain and the cerebrovascular circulation.

Finally, although most of the readers will know, skull bone in T2 sequences appears as a hypointense, difficult to detect structure. The hyperintense outermost structure in the posterior parts of the head of the present series is the scalp.

H. Schröder et al., *The Human Brainstem*, https://doi.org/10.1007/978-3-030-89980-6_18

18.2 MRI Atlas of the Human Brainstem

MRI 1

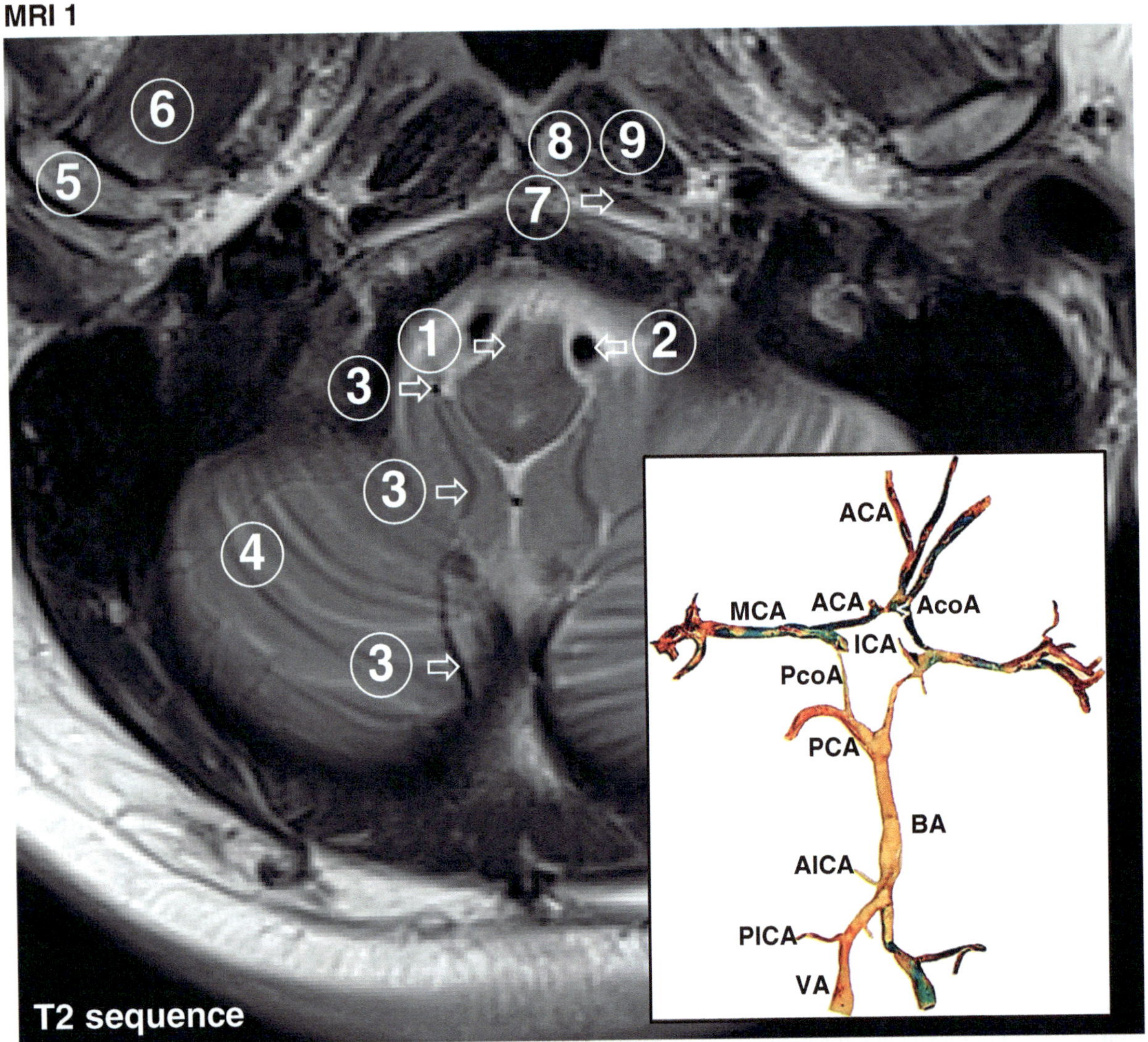

①	Pyramidal (corticospinal) tract
②	Vertebral artery (VA)
③	Branches of posterior inferior cerebellar artery (PICA)
④	Cerebellar hemisphere
⑤	Mandibular condyle
⑥	Pterygoid muscles
⑦	Longus capitis muscle
⑧	Levator veli palatini muscle
⑨	Tensor veli palatini muscle
	See atlas part plastinates P1, P1 A, P1 B

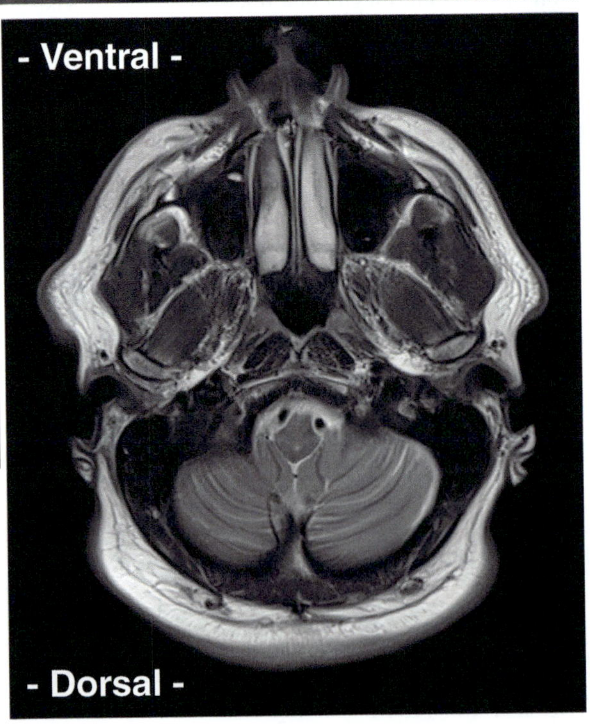

MRI 2

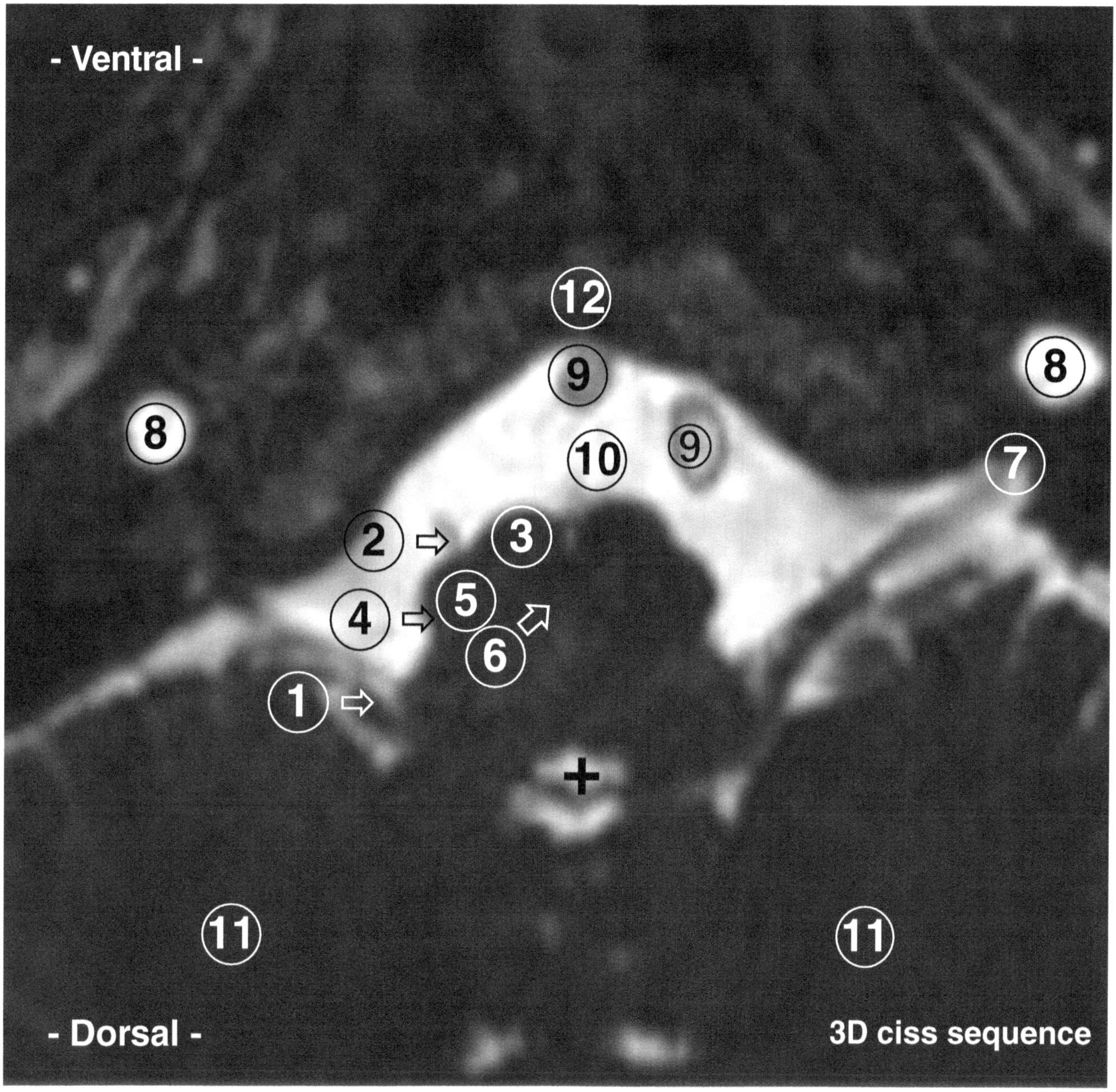

①	Vagus nerve CNX / glossopharyngeal CNX	⑧	Internal carotid arteries (ICA)
②	Hypoglossal nerve CNXII	⑨	Vertebral arteries (VA)
③	Pyramidal (corticospinal) tract	⑩	Prepontine cistern
④	Central tegmental tract	⑪	Cerebellum
⑤	Region of inferior olivary complex	⑫	Clivus
⑥	Region of the medial lemniscus	+	4th ventricle
⑦	Jugular foramen		See atlas part P1, P1 A

MRI 3

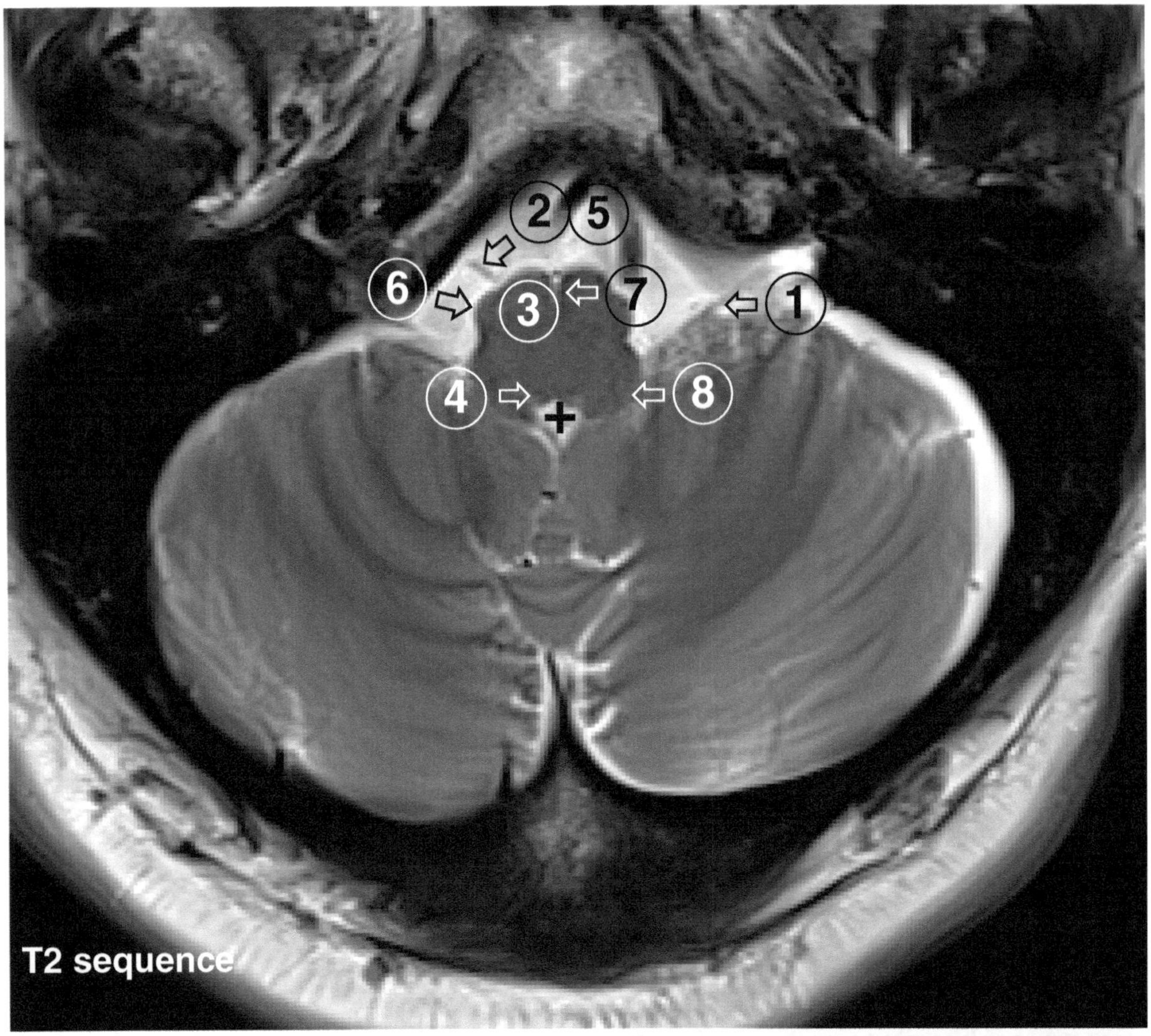

①	Vagus nerve CNX
②	Hypoglossal nerve CNXII
③	Pyramidal (corticospinal) tract
④	Ncl. of hypoglossal nerve
⑤	Vertebral arteries (VA) merging into basilar artery (BA)
⑥	Region of inferior olivary complex
⑦	Region of arcuate ncl.
⑧	Inferior cerebellar peduncle
+	4th ventricle
	See atlas part P1, P1 A

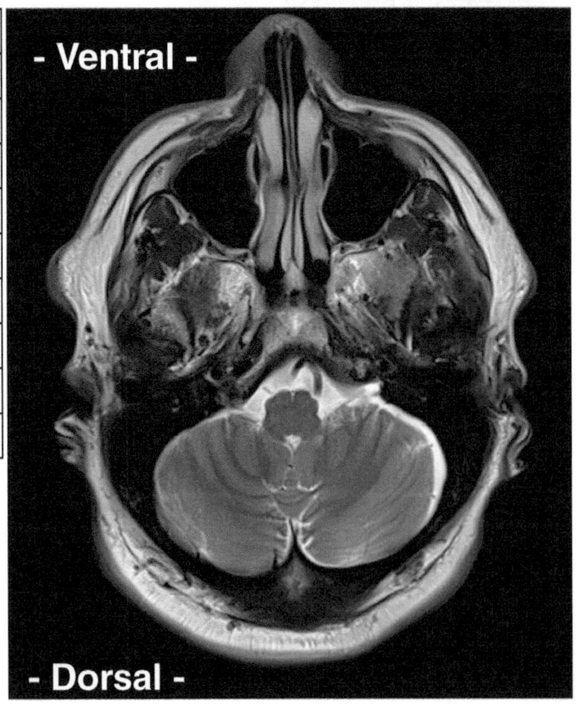

MRI 4

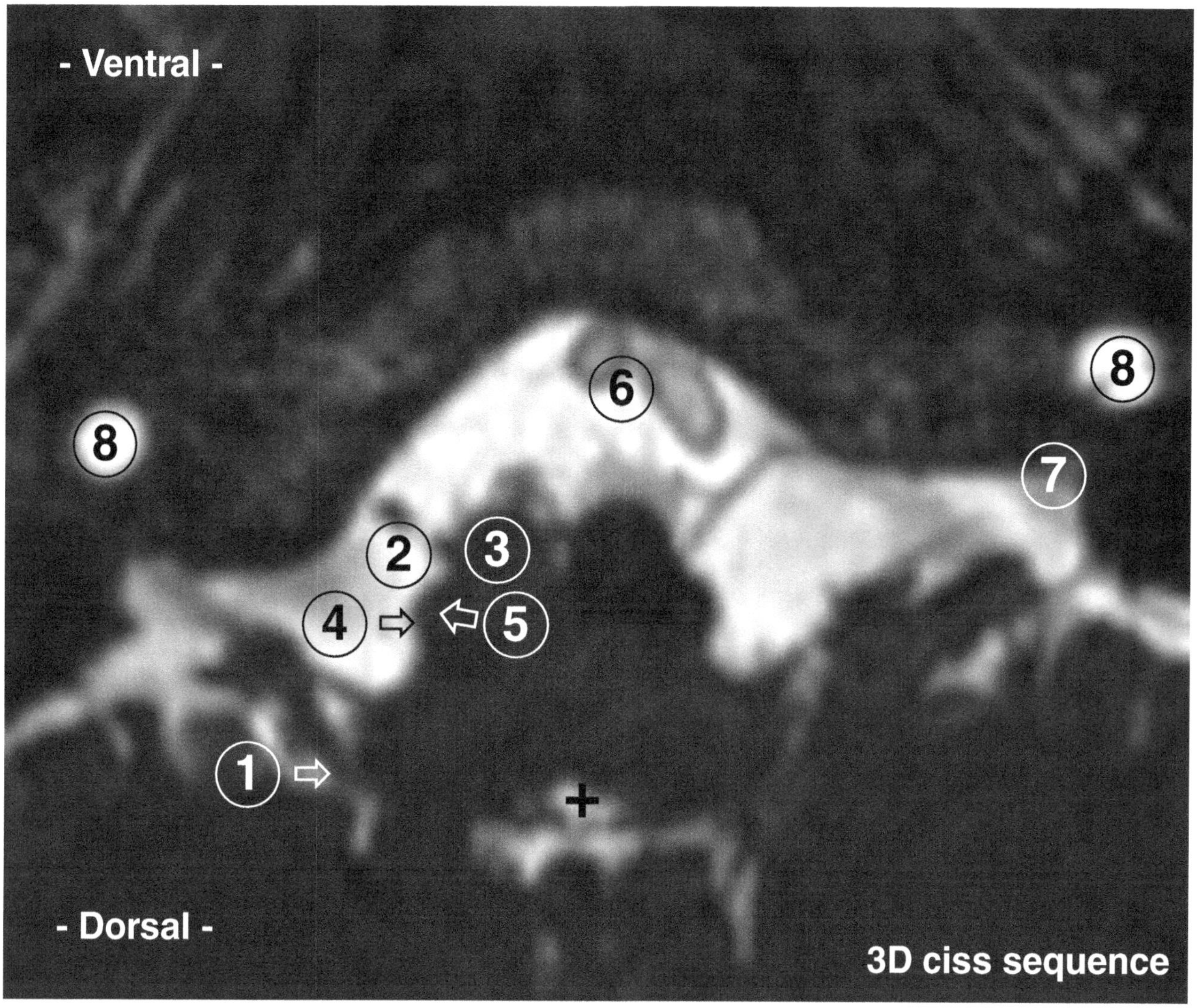

①	Glossopharyngeal nerve CNIX		⑥	Vertebral arteries (VA) merging into basilar artery (BA)
②	Hypoglossal nerve CNXII		⑦	Jugular foramen
③	Pyramidal (corticospinal) tract		⑧	Internal carotid arteries (ICA)
④	Region of the central tegmental tract		+	4th ventricle
⑤	Inferior olivary complex			

MRI 5

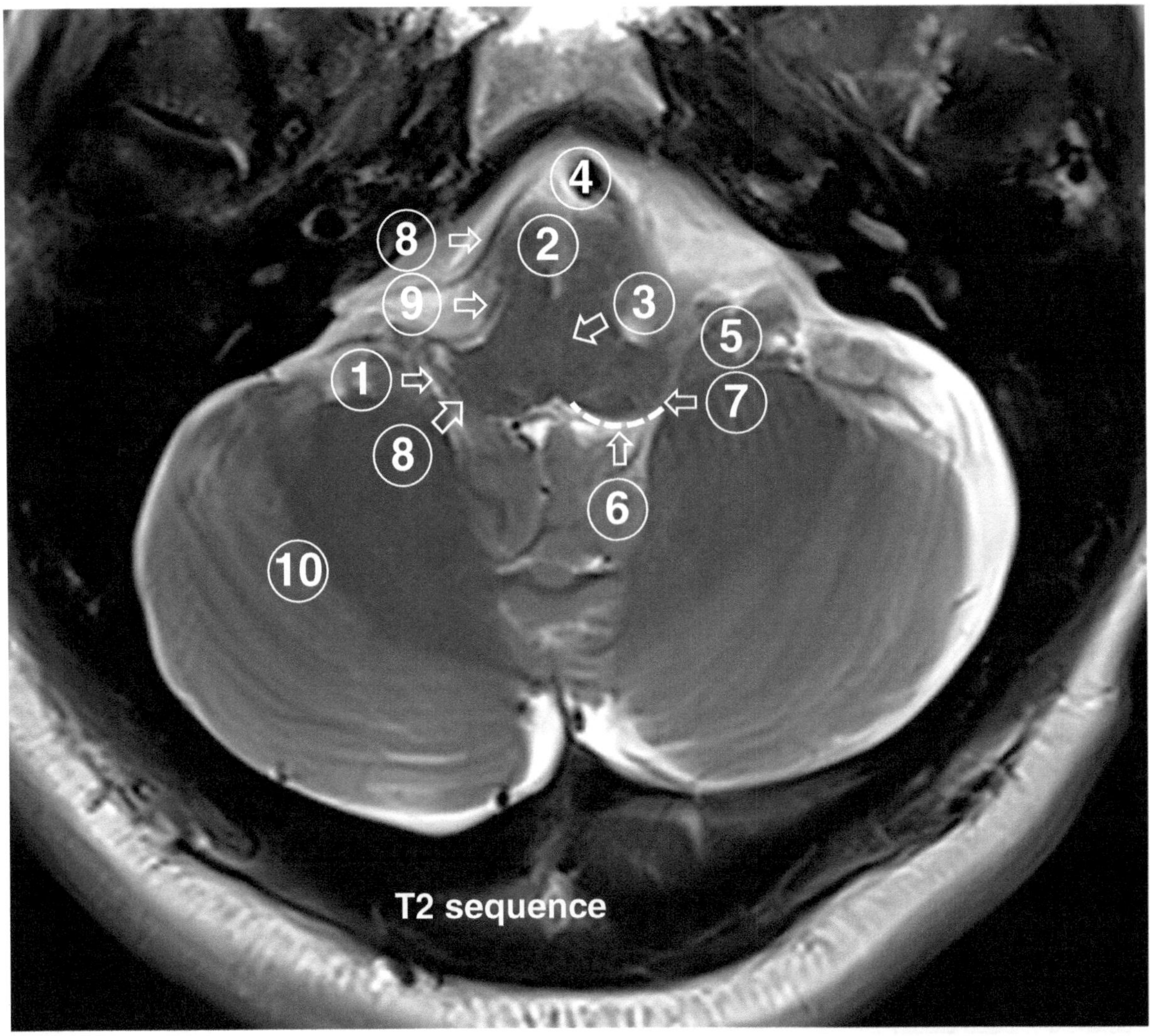

①	Glossopharyngeal nerve CNIX
②	Pyramidal (corticospinal) tract
③	Region of the medial lemniscus
④	Basilar artery (BA)
⑤	Flocculus
⑥	Region of the dorsal acoustic striae
⑦	Region of posterior cochlear ncl. CNVIII
⑧	Anterior inferior cerebellar artery (AICA)
⑨	Branch of vertebral artery (VA)
⑩	Cerebellar hemisphere
	See atlas part P3, P3 A

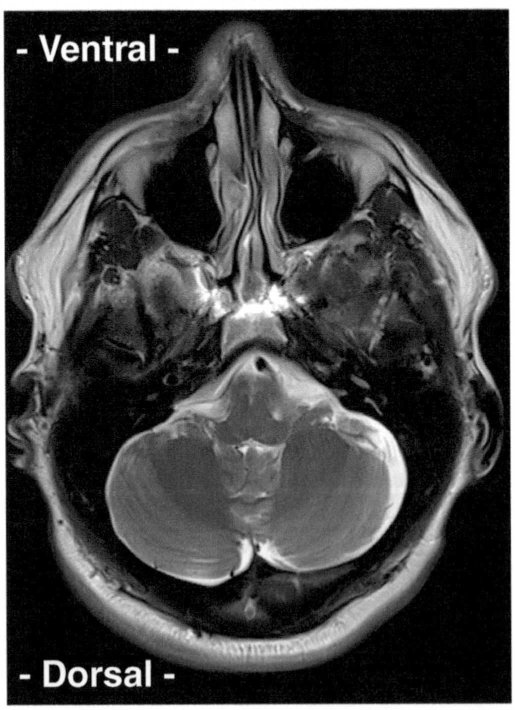

MRI 6

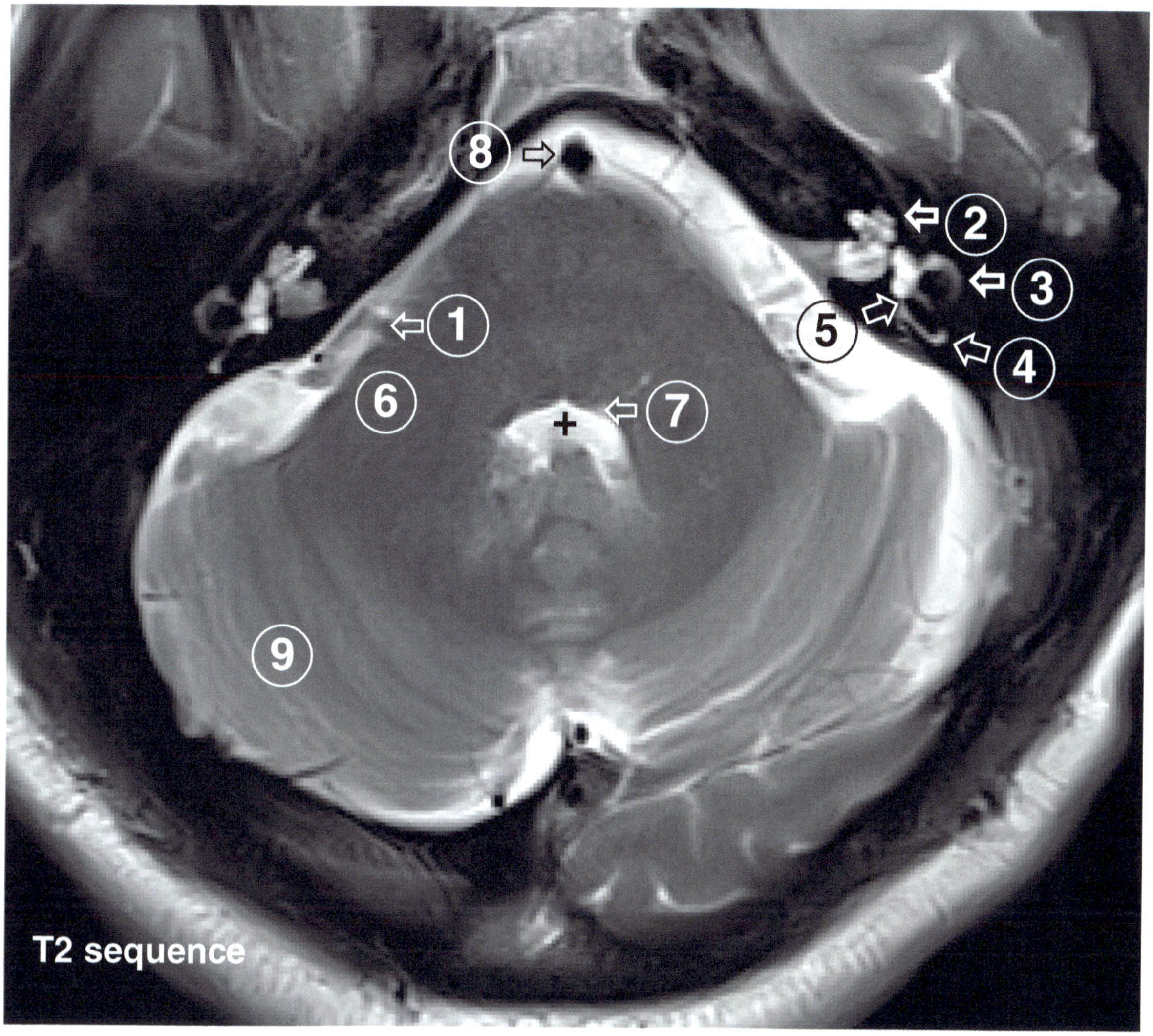

①	Facial nerve CNVII / vestibulocochlear nerve CNVIII
②	Cochlea
③	Lateral semicircular duct
④	Posterior semicircular duct
⑤	Anterior semicircular duct / utricle
⑥	Middle cerebellar peduncle
⑦	Facial colliculus
⑧	Basilar artery (BA)
⑨	Cerebellar hemisphere
+	4th ventricle
	See atlas part P3, P3 A, P4, P4 A

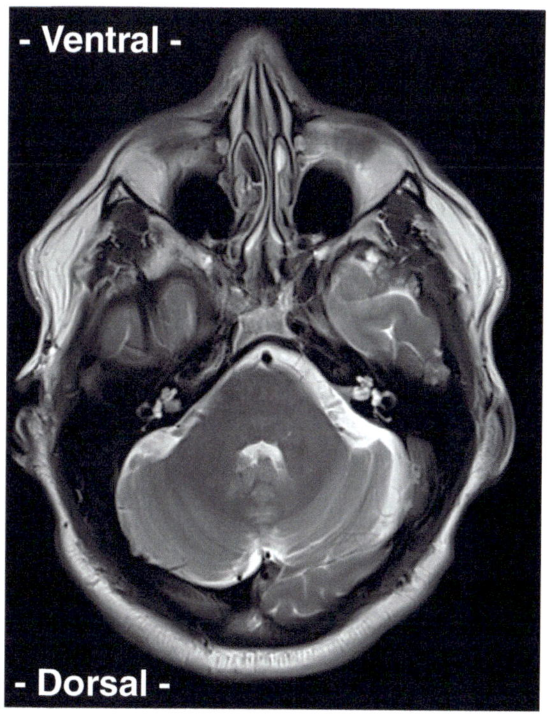

MRI 7

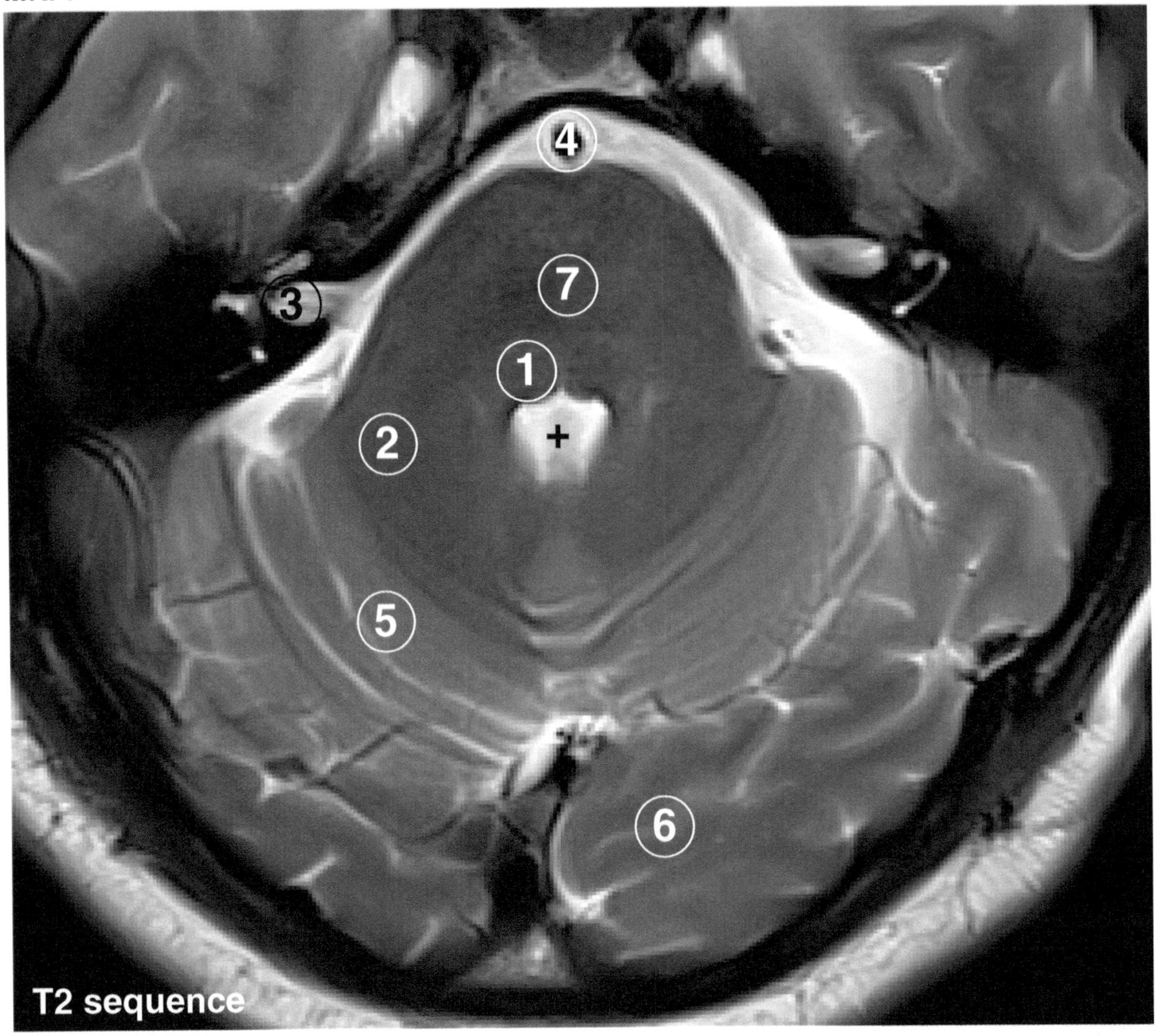

①	Region of abducens nerve ncl. CNVI
②	Middle cerebellar peduncle
③	Facial nerve CNVII / vestibulocochlear nerve CNVIII
④	Basilar artery (BA)
⑤	Cerebellar hemisphere
⑥	Occipital lobe
⑦	Pons
+	4th ventricle
	See atlas part P3, P3 A, P4, P4 A

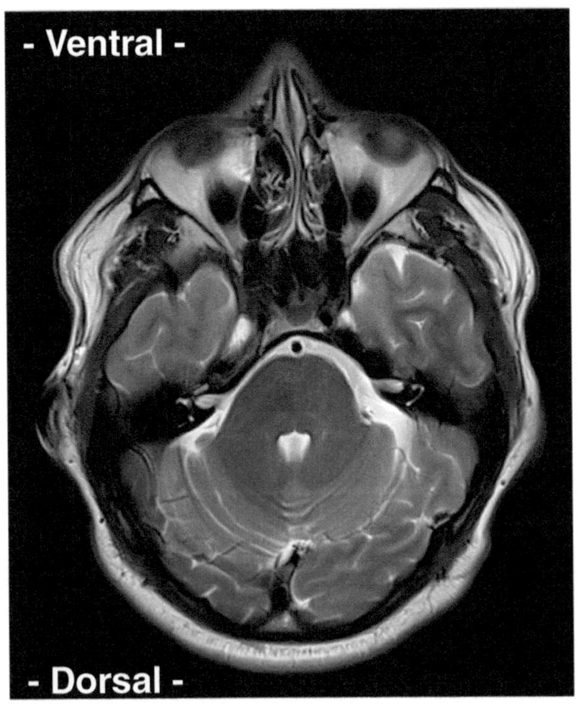

MRI 8

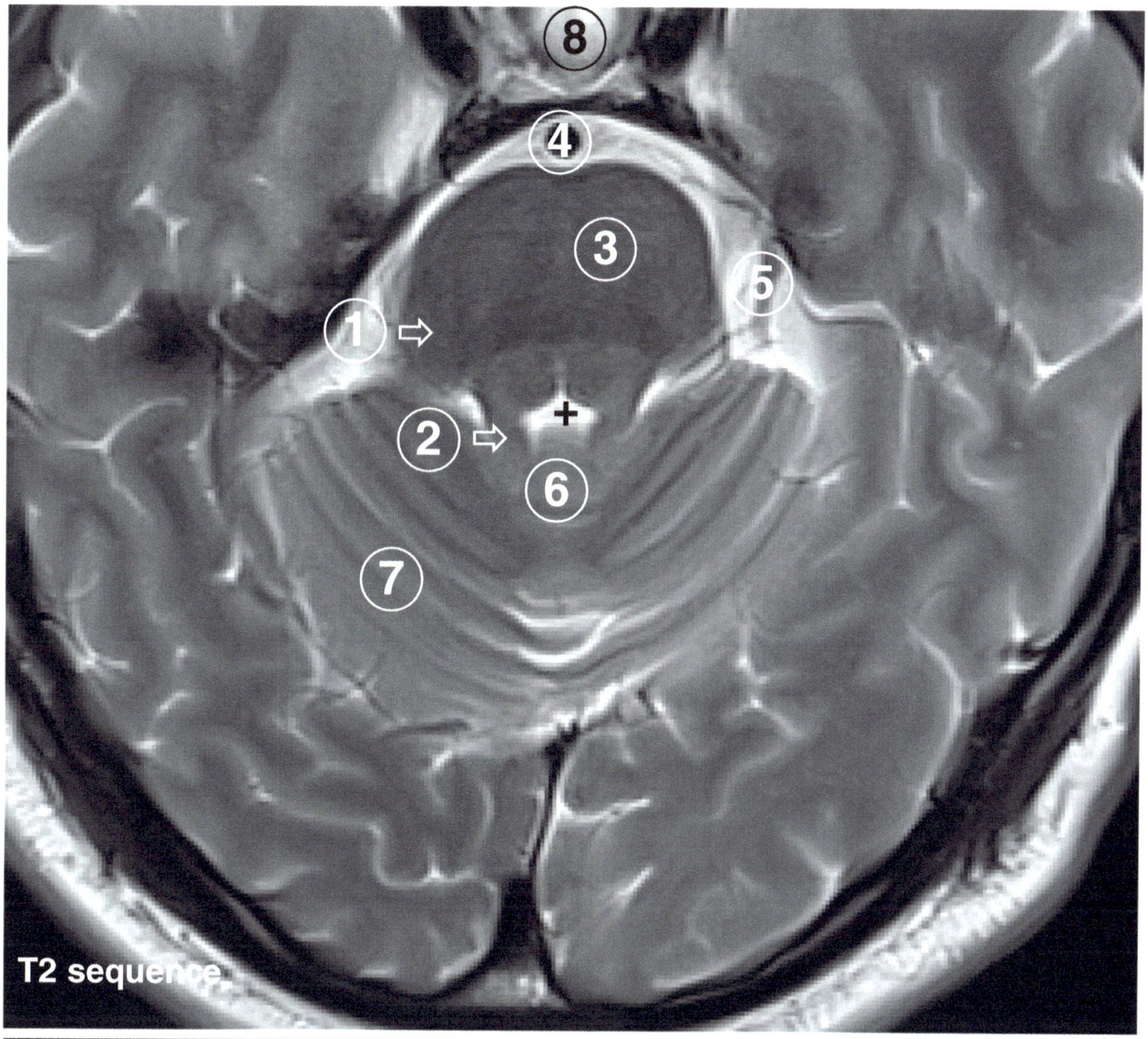

①	Region of medial lemniscus
②	Superior cerebellar peduncle
③	Transverse pontine (pontocerebellar) fibers and pontine nuclei
④	Basilar artery (BA)
⑤	Posterior inferior cerebellar artery (PICA)
⑥	Vermis
⑦	Cerebellar hemisphere
⑧	Pituitary gland
+	4th ventricle
	See atlas part P3, P3 A, P4, P4 A

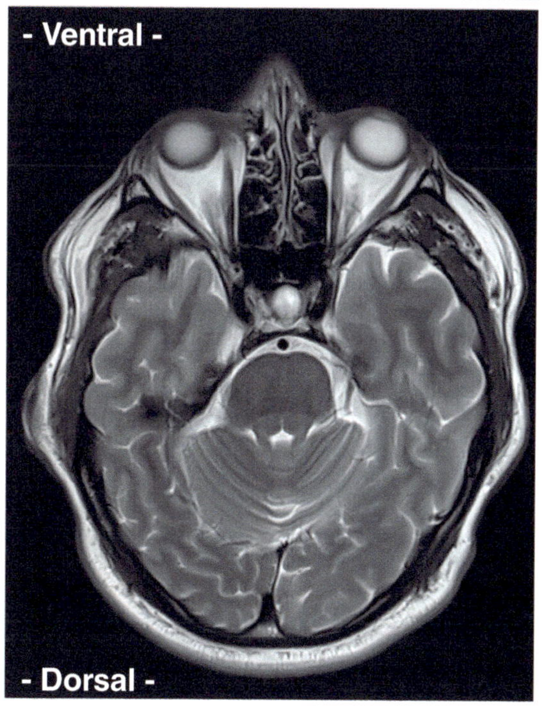

MRI 9

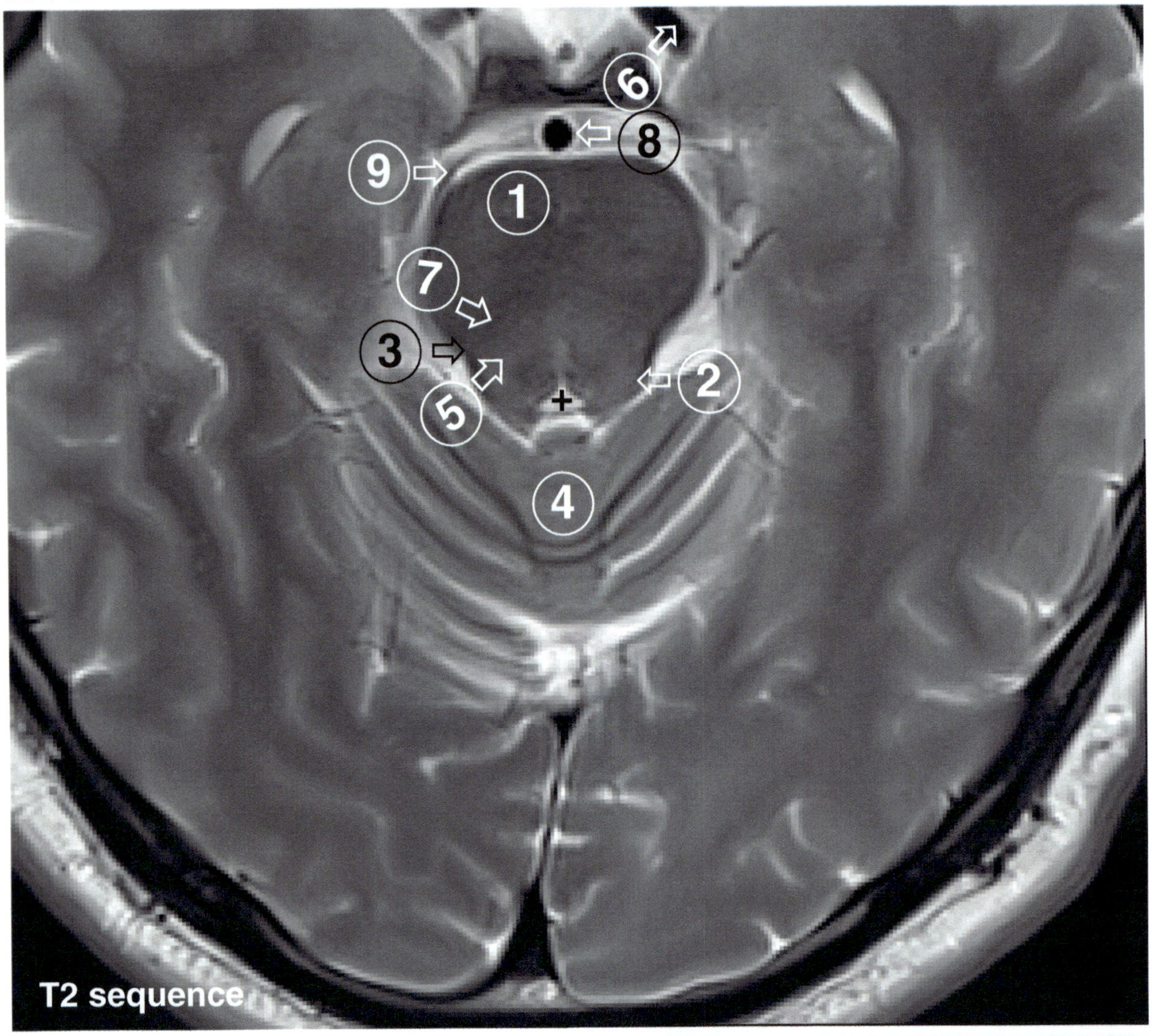

①	Pontine basis
②	Superior cerebellar peduncle
③	Region of the lateral lemniscus
④	Vermis
⑤	Region of the locus caeruleus
⑥	Internal carotid artery (ICA)
⑦	Region of the medial lemniscus
⑧	Basilar artery (BA)
⑨	Superior cerebellar artery (SCA)
+	4th ventricle
	See atlas part P3, P3 A, P4, P4 A

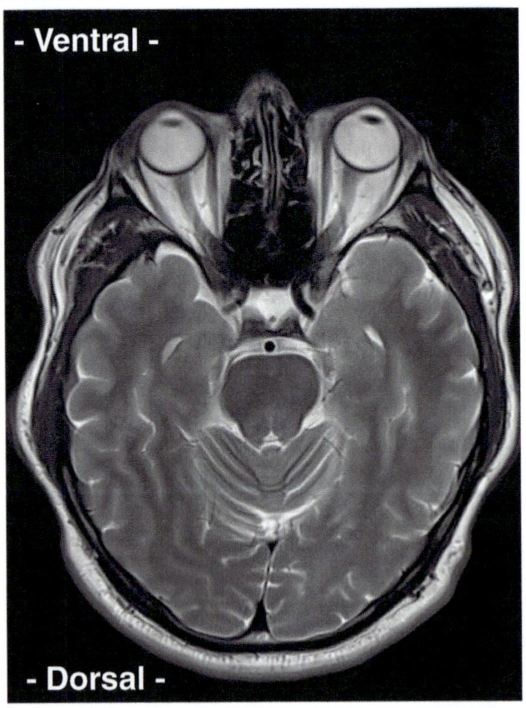

MRI 9 A (post mortem formalin-fixed specimen)

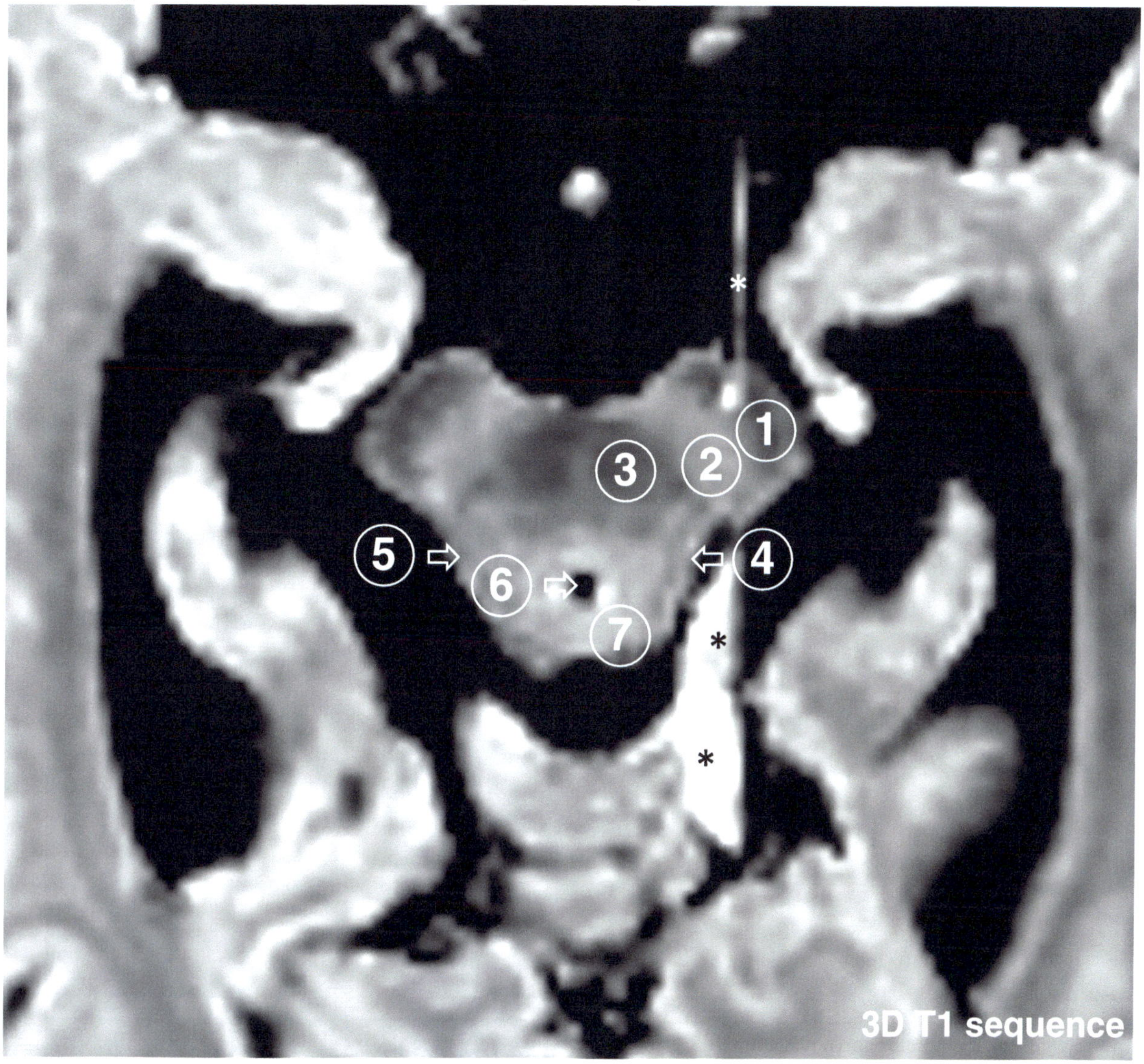

①	Cerebral peduncle
②	Substantia nigra
③	Red ncl.
④	Brachium of inferior colliculus
⑤	Region of the medial lemniscus
⑥	Mesencephalic aqueduct
⑦	Superior colliculus
*	This artefact was not reproducible by other scan methods of the same specimen. Note that this specimen was formalin fixed. The resulting MRI scans obviously are not of the same quality as in-vivo scans (compare with MRI 10)

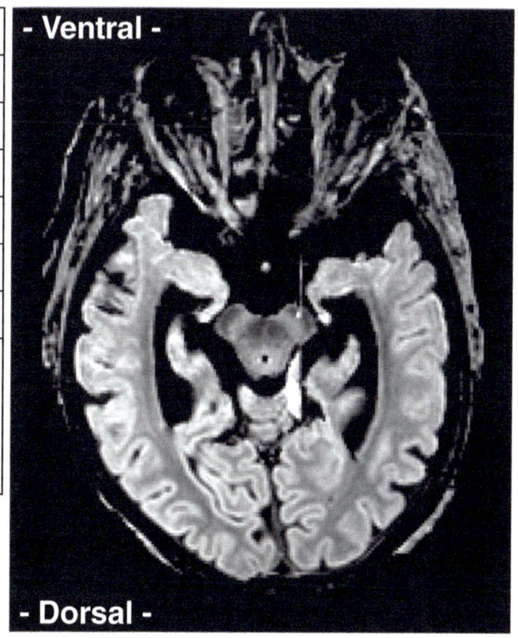

MRI 10

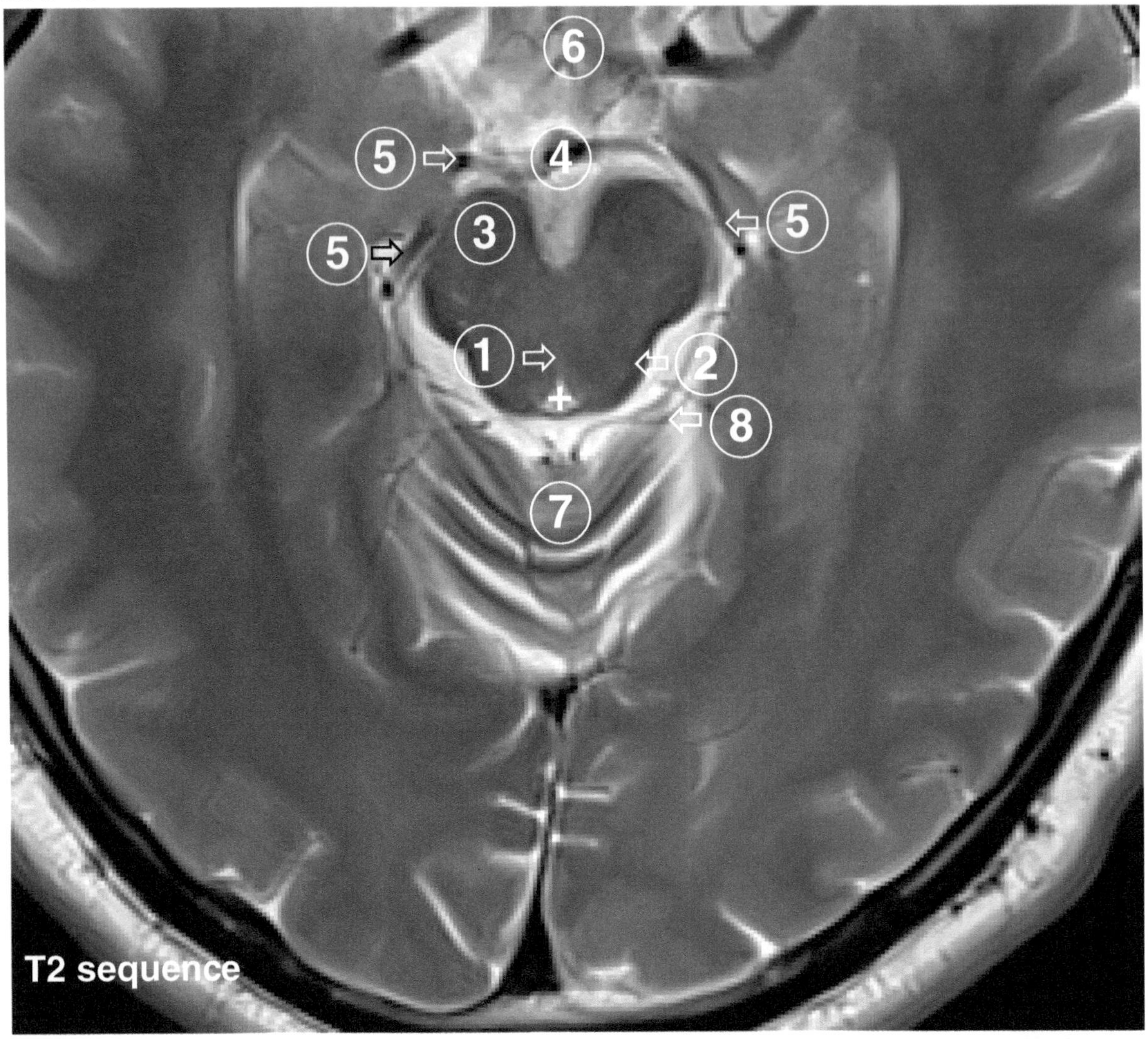

①	Region of oculomotor nerve ncl. CNIII
②	Superior cerebellar peduncle
③	Cerebral peduncle
④	Basilar artery (BA)
⑤	Posterior cerebral artery (PCA)
⑥	Optic chiasm
⑦	Cerebellum
⑧	Branches of superior cerebellar artery (SCA)
+	Mesencephalic aqueduct
	See atlas part P6 – P9

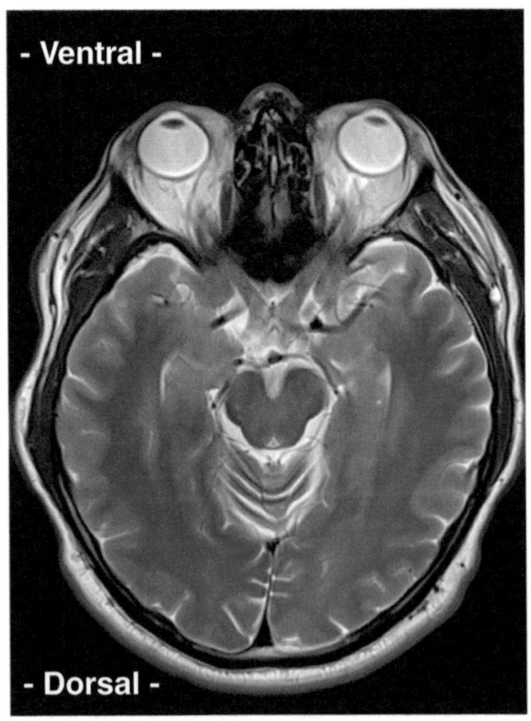

MRI 11

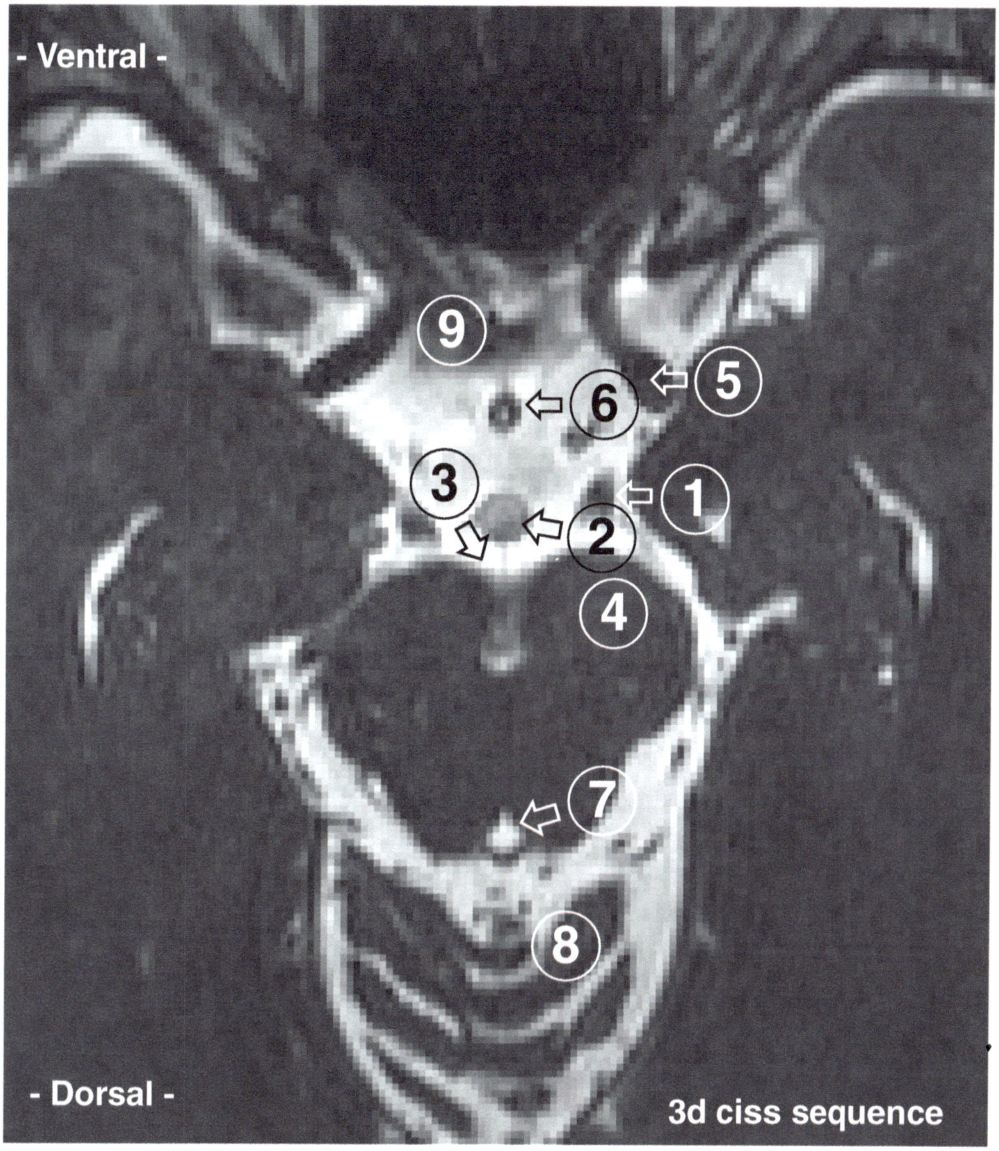

1	Oculomotor nerve CNIII
2	Basilar artery (BA)
3	Interpeduncular cistern
4	Cerebral peduncle
5	Internal carotid artery (ICA)
6	Pituitary stalk
7	Mesencephalic aqueduct
8	Cerebellum
9	Optic chiasm
	See atlas part P6 –P9

MRI 12

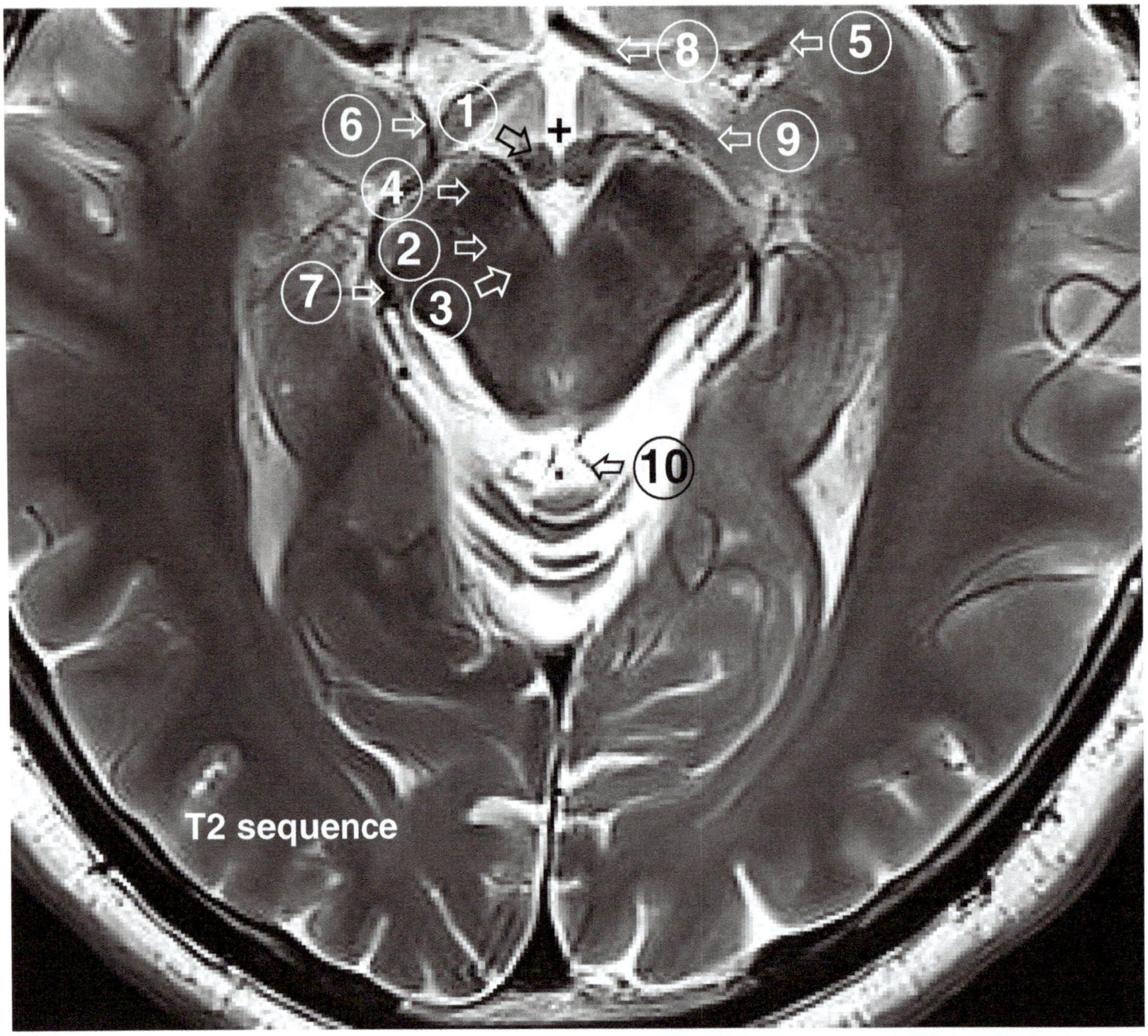

①	Corpus mammillare
②	Region of Substantia nigra
③	Red ncl.
④	Cerebral peduncle
⑤	Middle cerebral artery (MCA)
⑥	Posterior communicating artery (PCoA)
⑦	Posterior cerebral artery (PCA)
⑧	Anterior cerebral artery (ACA)
⑨	Optic tract
⑩	Branches of superior cerebellar artery (SCA)
+	3rd ventricle
	See atlas part P10 – P11

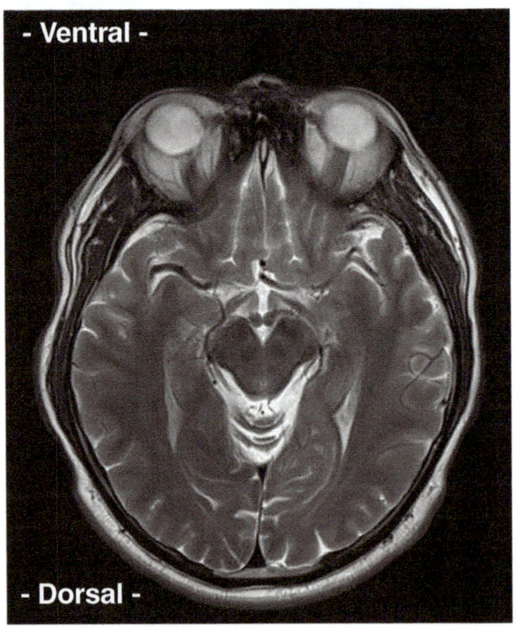

MRI 13 A, B

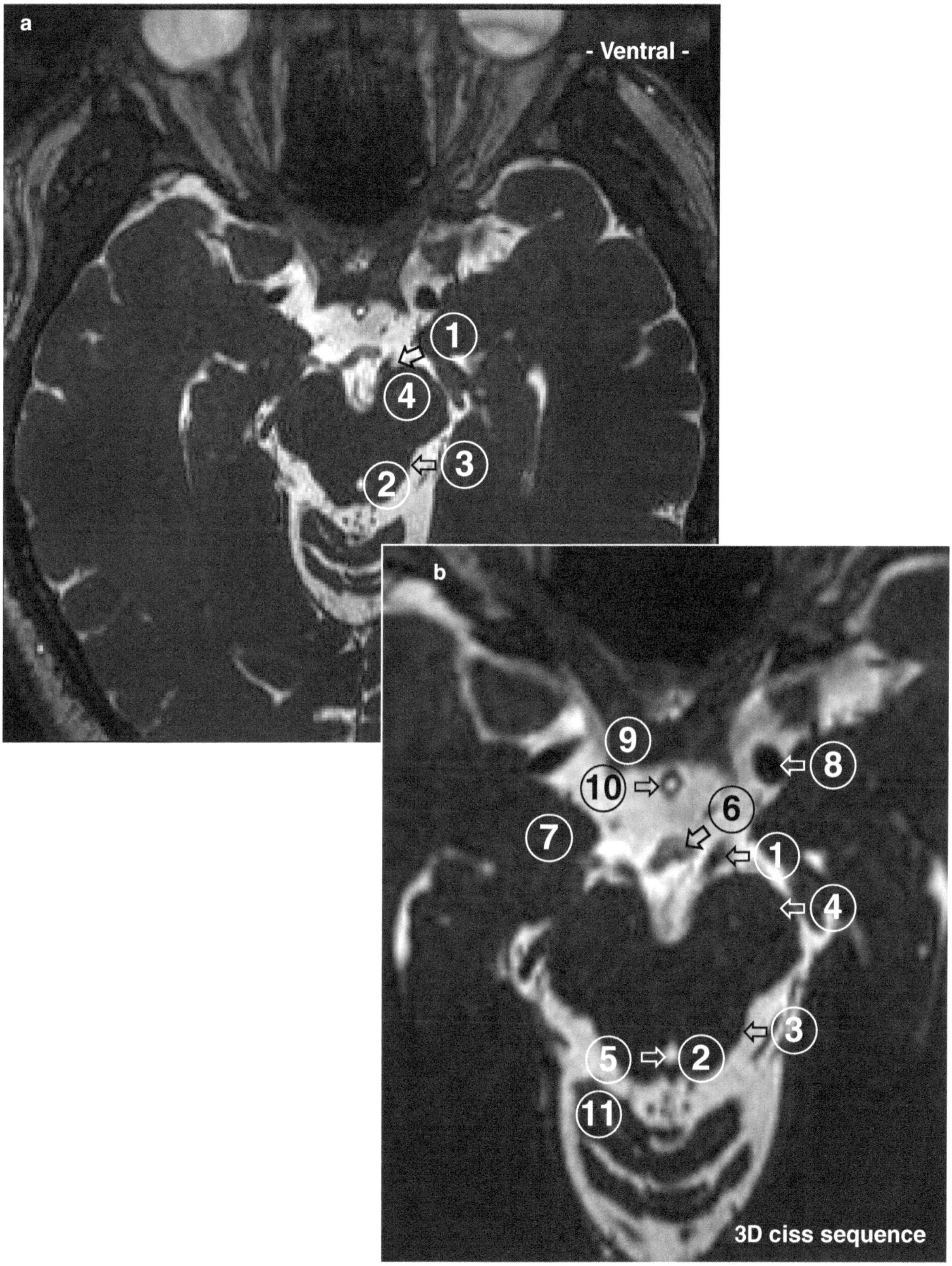

①	Oculomotor nerve CNIII	⑦	Uncus
②	Superior colliculus	⑧	Internal carotid artery (ICA)
③	Region of brachium of inferior colliculus	⑨	Optic chiasm
④	Cerebral peduncle	⑩	Pituitary stalk
⑤	Mesencephalic aqueduct	⑪	Cerebellum
⑥	Basilar artery		See atlas part P10 – P11

MRI 14

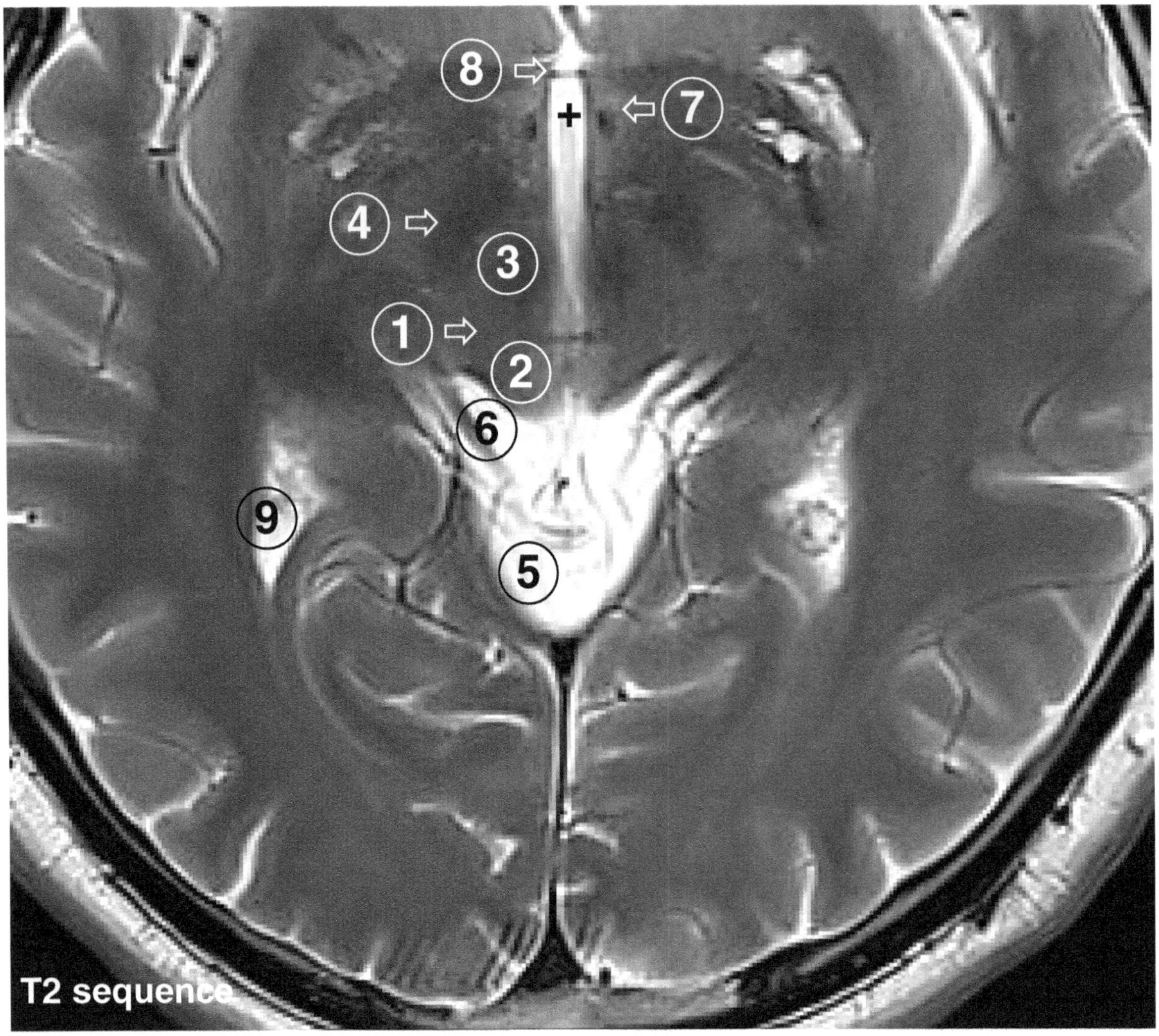

①	Pretectal area
②	Superior colliculus
③	Hypothalamus
④	Region of substantia nigra
⑤	Ambient cistern
⑥	Basal vein
⑦	Fornix, columna
⑧	Anterior commissure
⑨	Lateral ventricle, occipital horn
+	3rd ventricle
	See atlas part P10 –P11

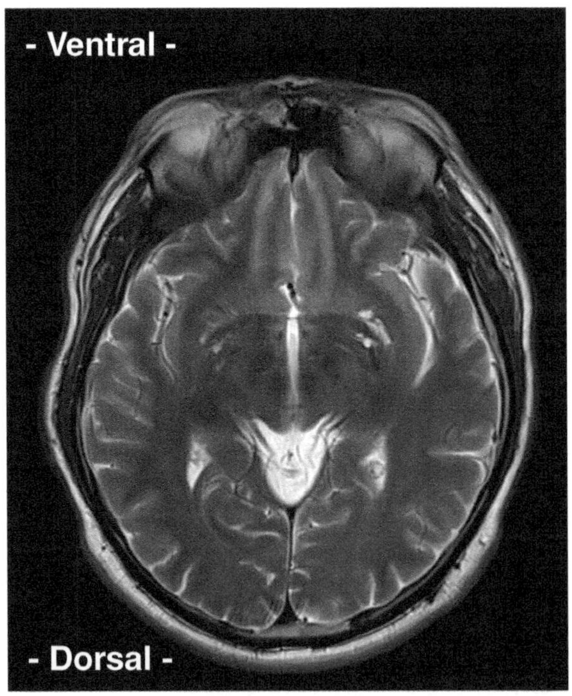

Topographical atlas of plastinated sections of the human brainstem and cranial nerve target structures

19

Contents

19.1 Annotations Atlas Plastinates and MRI

The generation of the plastinate specimens is described under Materials and Methods (Foreword p. 117). In this part of the atlas you will find a series of horizontal plastinated sections starting with the most caudal section P1. P1–P14 show horizontal brainstem sections, P15–18 show the brainstem in the coronal plane.

All specimens shown here belong to the collection of the Center of Anatomy at the University of Cologne (Sammlung des Zentrums Anatomie der Universität zu Köln).

The photographs of sections P1–P14 are endowed with

1. an overview illustration with a small inset showing the approximately corresponding MRI figure from the MRI atlas (see chapter 18). Please note that there is no complete correspondence between the plastinated sections and the MRI sections.
2. detail photographs of the individual sections which correspond to illustrations in the individual chapters of the book with an inset showing the approximate level of the illustrations related to the brainstem. This arrangement allows for a simple comparison of the survey illustrations (and MRI insets)—in which frames for the detail pictures are provided—with a detailed description (see e.g. p. 653. P1A For description see Fig. 5.15). Using the figure numbers information on the photographs can be found in the legends in the chapters and the chapter sections itself.

Photographs of sections P15–18 show surveys of the coronal sections. Brainstem level of the individual section is indicated in a midsagittal photograph of the brain.

H. Schröder et al., *The Human Brainstem*, https://doi.org/10.1007/978-3-030-89980-6_19

19.2 Topographical Atlas: Plastinates and MRI

P1 - Cranial view

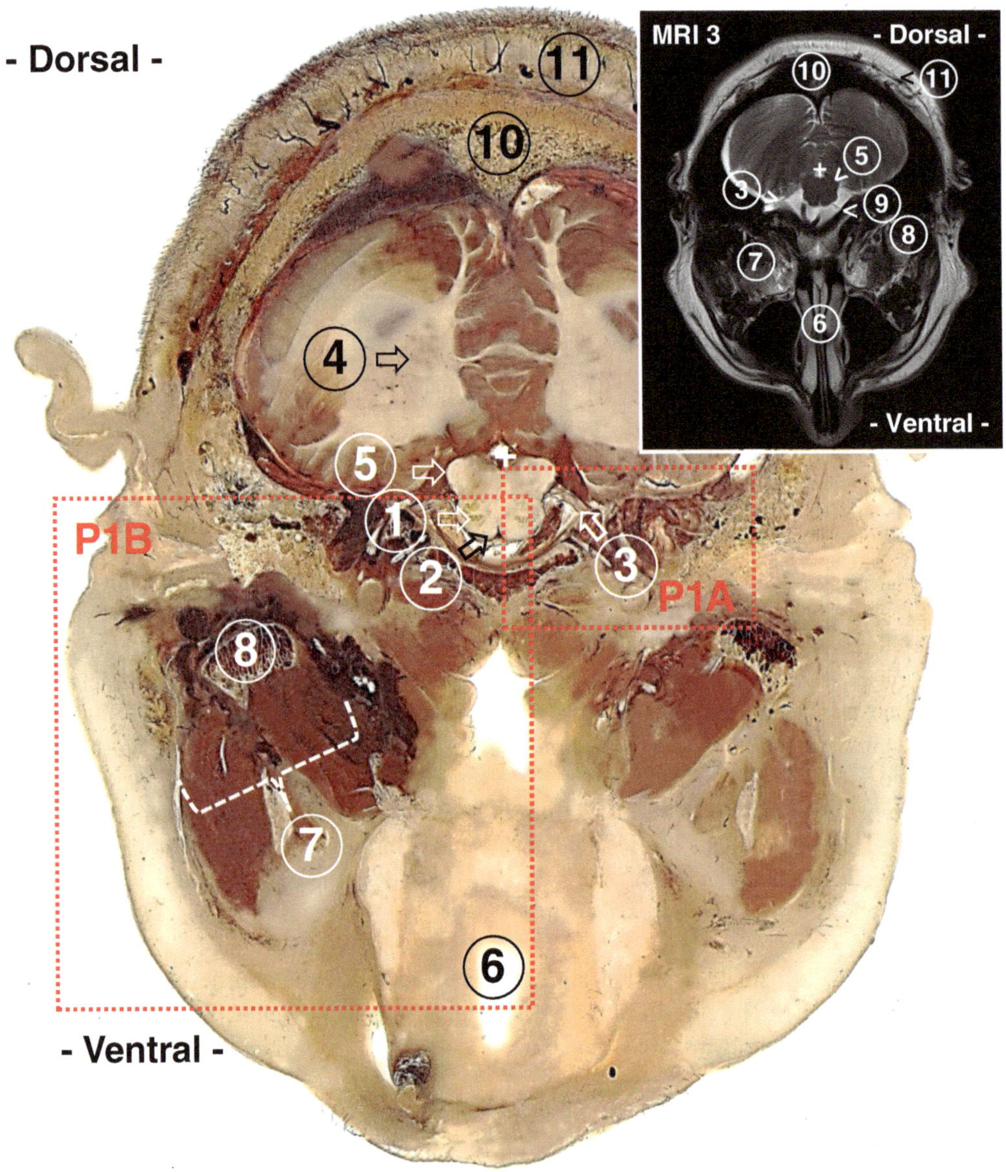

①	Inferior olivary complex	⑦	Masticatory muscles
②	Pyramidal (corticospinal) tract	⑧	Mandibular condyle
③	Vagus nerve CNX	⑨	Hypoglossal nerve CNXII
④	Dentate nucleus / cerebellum	⑩	Skull bone, calvarium
⑤	Inferior cerebellar peduncle	⑪	Scalp
⑥	Nasal cavity	+	4th ventricle

P1A Details see Fig. 5.15 – Cranial view

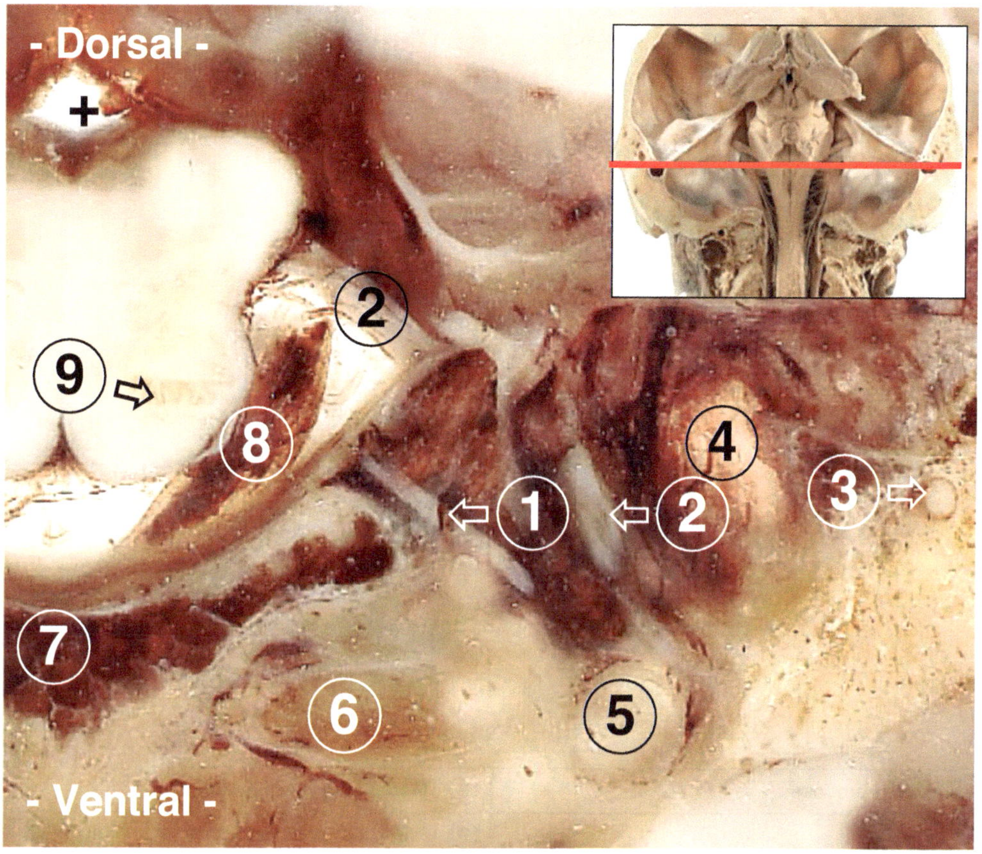

①	Glossopharyngeal nerve CNIX
②	Vagus nerve CNX
③	Facial nerve CNVII
④	Jugular bulb
⑤	Internal carotid artery (ICA)
⑥	Longus capitis muscle
⑦	Clivus
⑧	Vertebral artery (VA)
⑨	Inferior olivary complex
+	4th ventricle

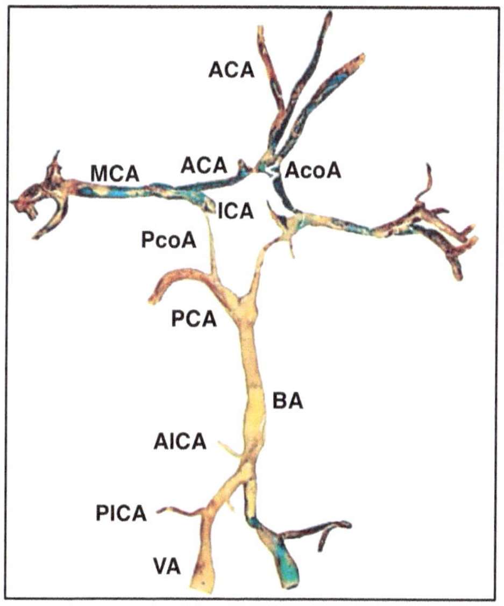

P1B Details see Fig. 11.8 – Cranial view

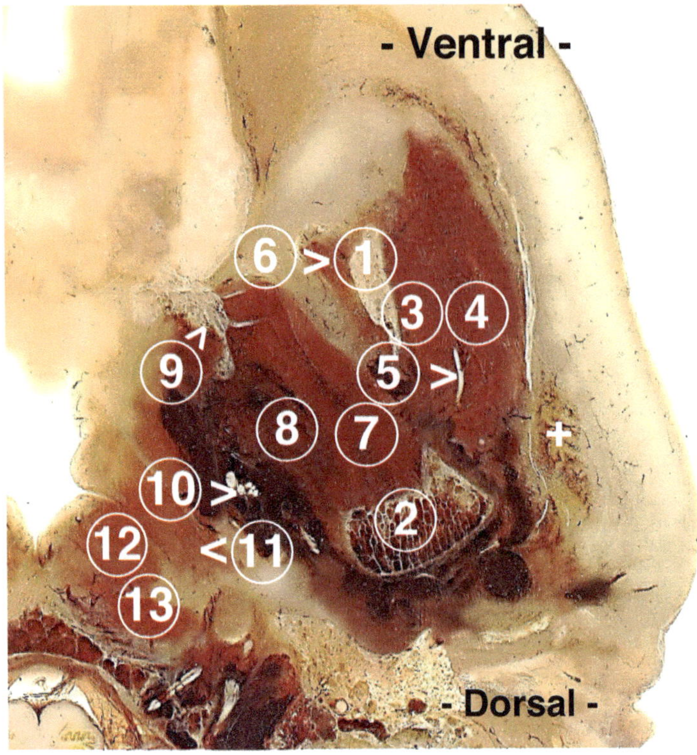

①	Mandible, coronoid process	⑧	Medial pterygoid muscle
②	Mandibular condyle	⑨	Pterygoid process
③	Masseter muscle, Profound part	⑩	Mandibular nerve V_3
④	Masseter muscle, superficial part	⑪	Tensor veli palatini muscle
⑤	Masseteric nerve V_3	⑫	Levator veli palatini muscle
⑥	Temporalis muscle	⑬	Longus capitis muscle
⑦	Lateral pterygoid muscle	+	Parotid gland

P2 – Caudal view

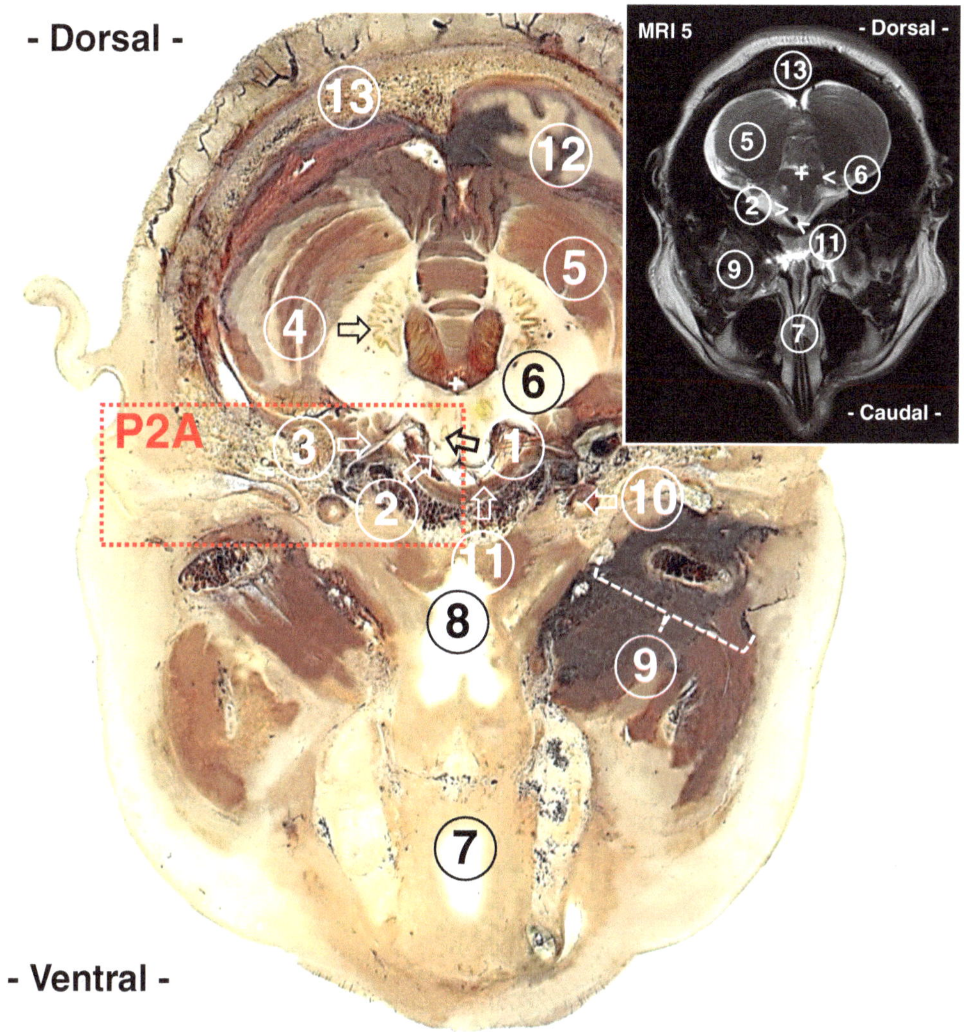

①	Inferior olivary complex	⑧	Pharnyx
②	Pyramidal (corticospinal) tract	⑨	Masticatory muscles
③	Vestibulocochlear nerve CNVIII	⑩	Internal carotid artery
④	Dentate ncl.	⑪	Vertebral artery
⑤	Cerebellum	⑫	Occipital lobe
⑥	Inferior cerebellar peduncle	⑬	Skull bone, calvarium
⑦	Nasal cavity	+	4th ventricle

P2A Description see Fig. 8.17C – Caudal view

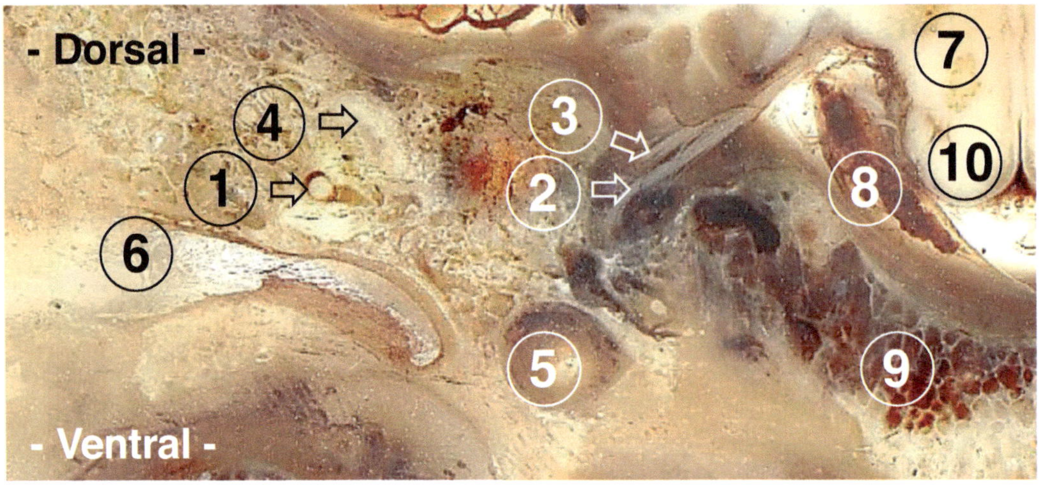

①	Facial nerve CNVII
②	Cochlear nerve
③	Vestibular nerve
④	Lateral semicircular canal
⑤	Internal carotid artery (ICA)
⑥	External acoustic meatus
⑦	Inferior olivary complex
⑧	Vertebral artery (VA)
⑨	Clivus
⑩	Pyramidal (corticospinal) tract

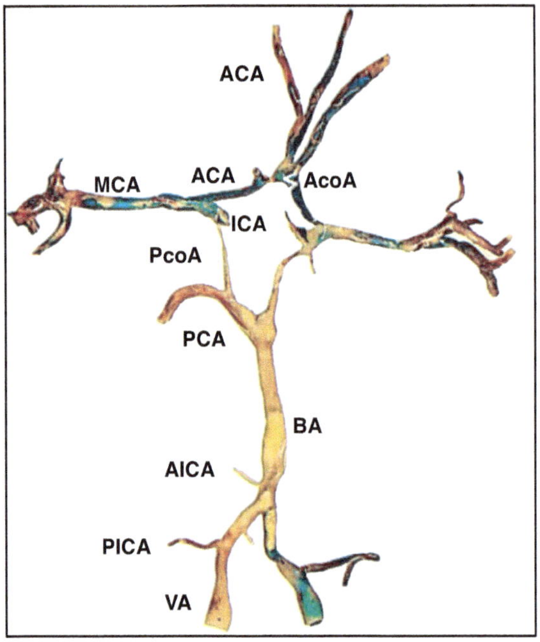

P3 - Cranial view

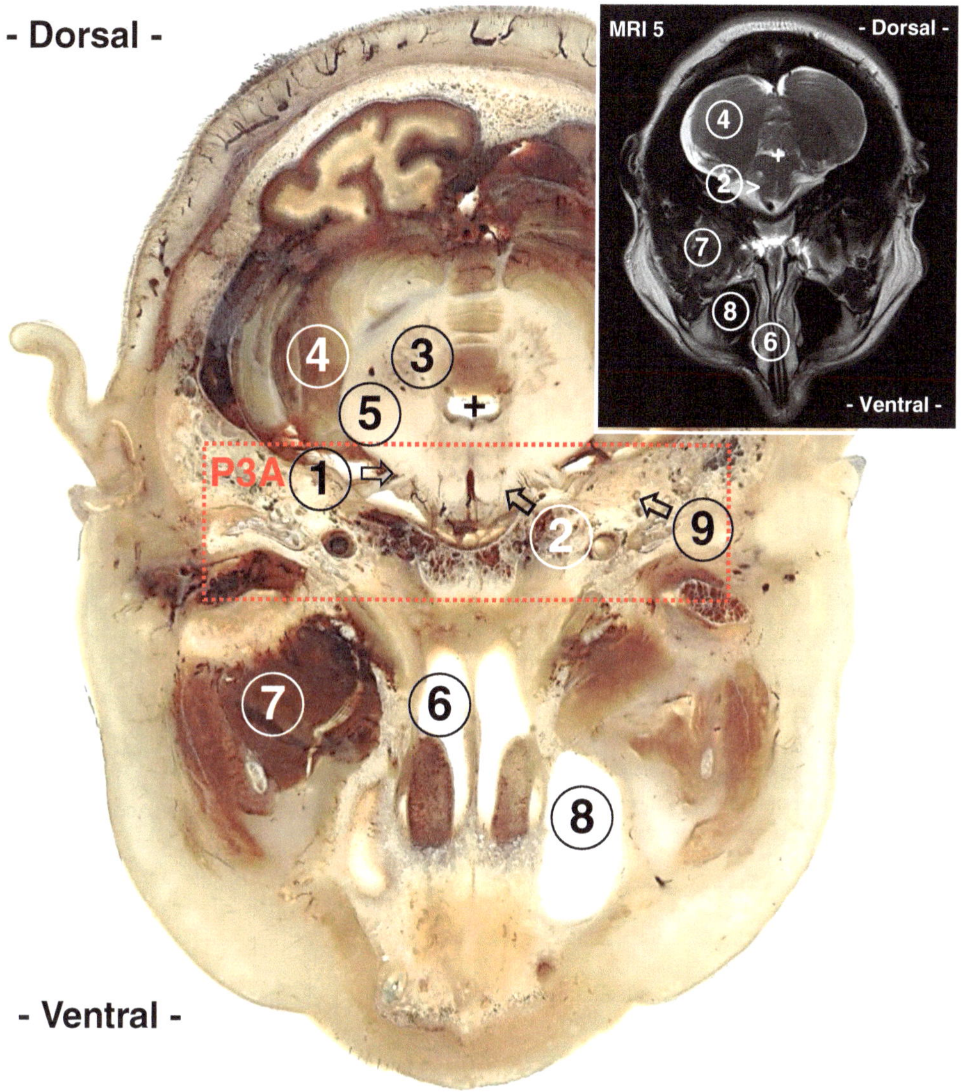

①	Vestibulocochlear nerve CNVIII / Facial nerve CNVII	⑥	Nasal cavity
②	Pyramidal (corticospinal) tract	⑦	Masticatory muscles
③	Dentate ncl.	⑧	Maxillary sinus
④	Cerebellum	⑨	Anterior semicircular canal
⑤	Inferior cerebellar peduncle	+	4th ventricle

P3A Detail Description see Fig. 8.17A - Cranial view

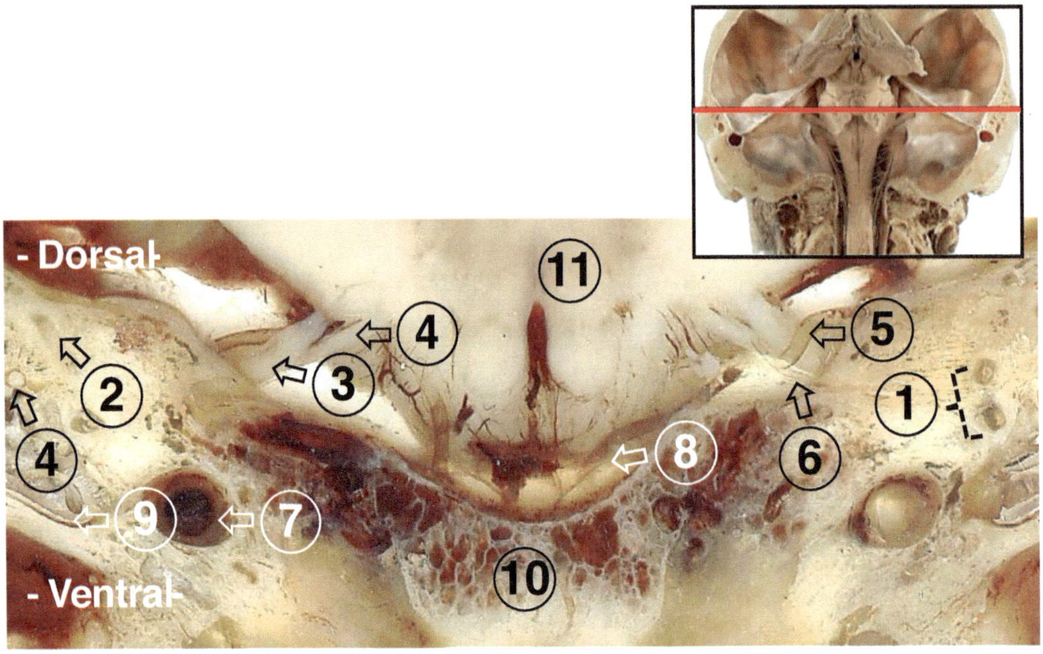

①	Anterior semicircular canal
②	Posterior semicircular canal
③	Vestibulocochlear nerve in prepontine cistern
④	Facial nerve CNVII
⑤	Anterior inferior cerebellar artery (AICA)
⑥	Labyrinthine artery
⑦	Internal carotid artery (ICA)
⑧	Vertebral artery (VA)
⑨	Tympanic cavity
⑩	Clivus
⑪	Pons

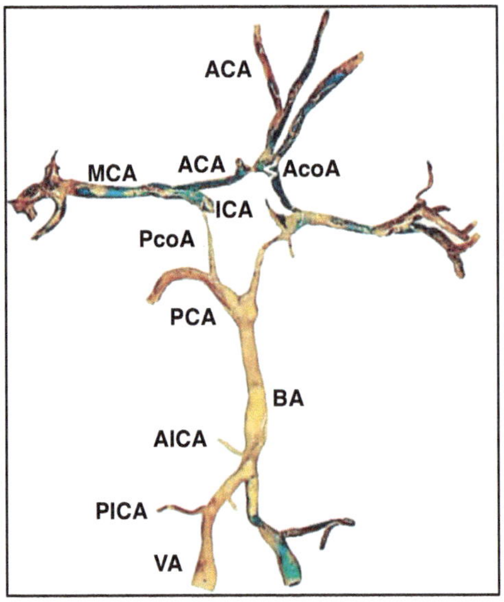

P4 – Caudal view

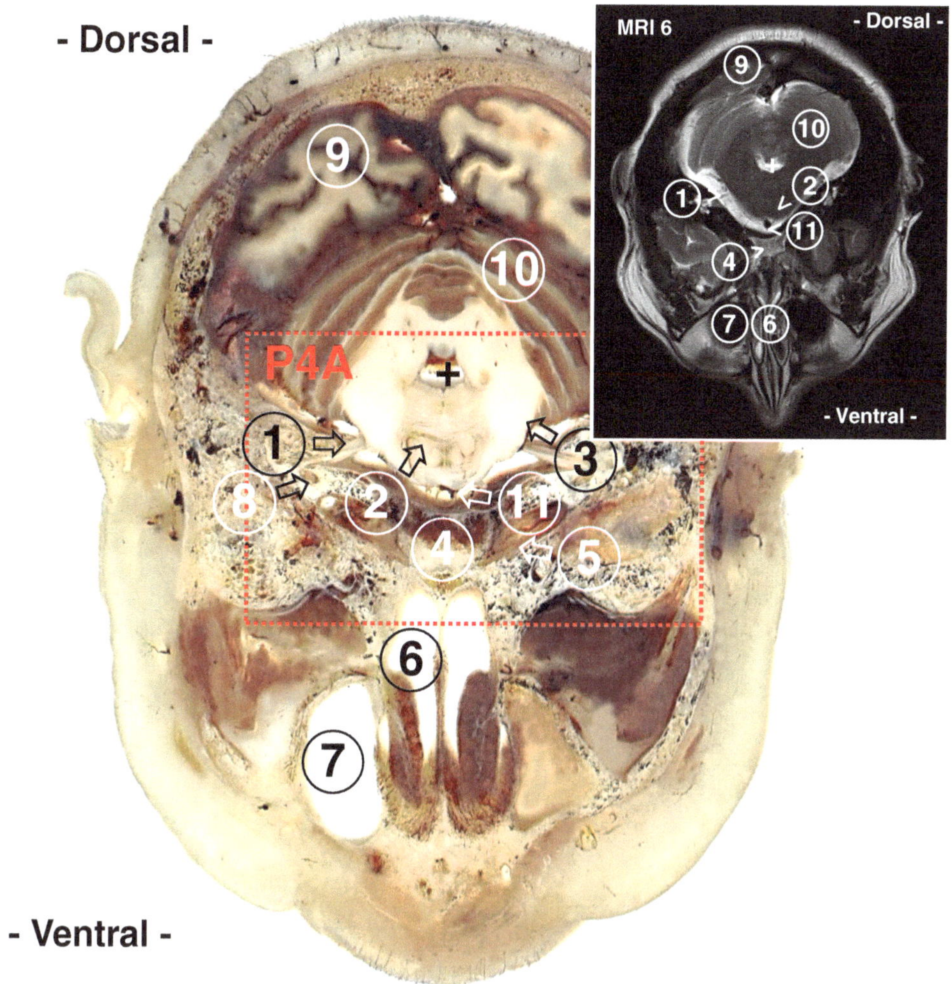

①	Vestibulocochlear nerve CNVIII / Facial nerve CNVII	⑦	Maxillary sinus
②	Pyramidal (corticospinal) tract	⑧	Cochlea
③	Inferior cerebellar peduncle	⑨	Occipital lobe
④	Clivus	⑩	Cerebellum
⑤	Internal carotid artery (ICA)	⑪	Basilar artery (BA)
⑥	Nasal cavity	+	4th ventricle

P4A Description see Fig. 8.7 – Caudal view

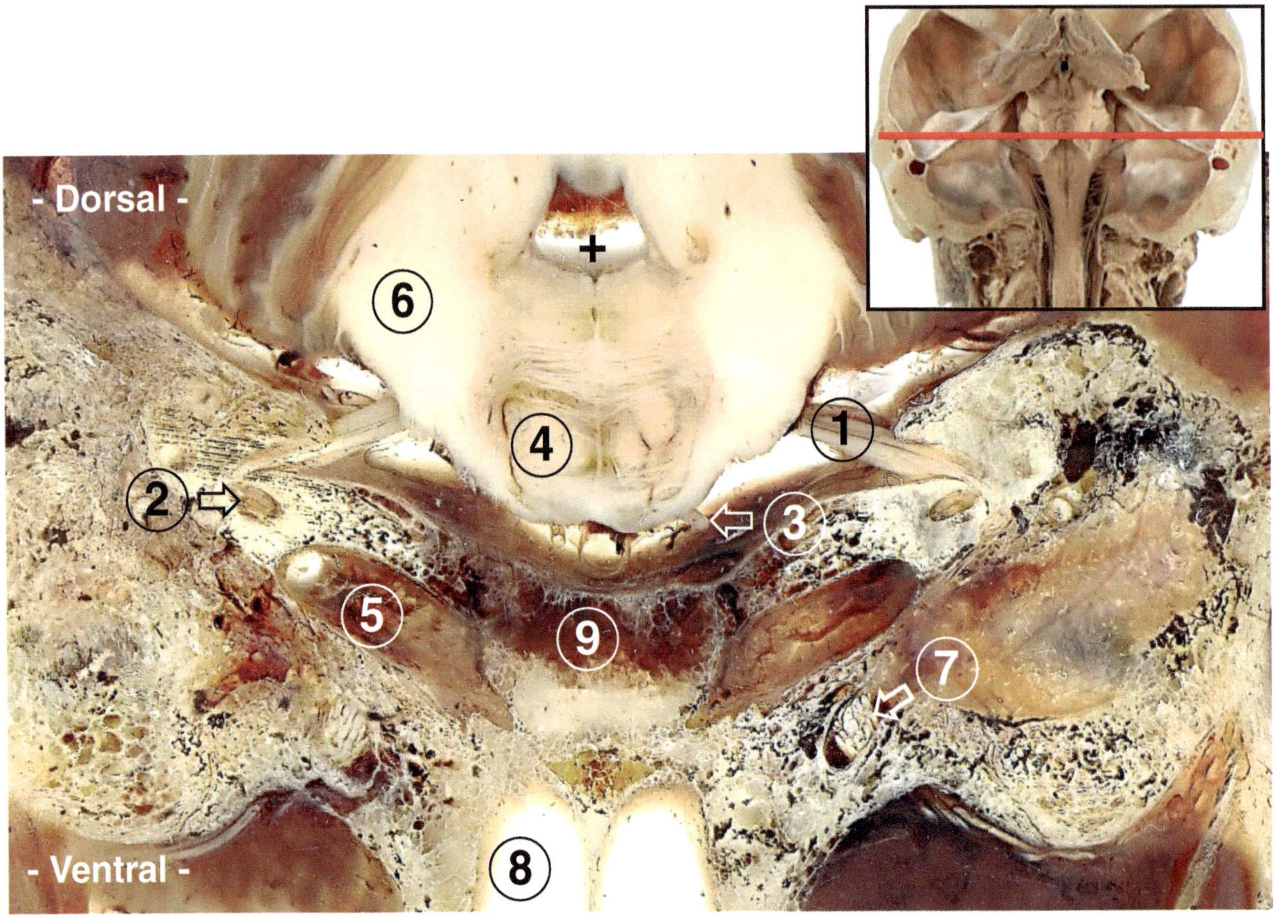

①	Vestibulocochlear nerve CNVIII / Facial nerve CNVII
②	Cochlea
③	Abducens nerve CNVI
④	Pyramidal (corticospinal) tract
⑤	Internal carotid artery (ICA)
⑥	Middle cerebellar peduncle
⑦	Trigeminal nerve CNV
⑧	Nasal cavity
⑨	Clivus
+	4th ventricle with choroid plexus

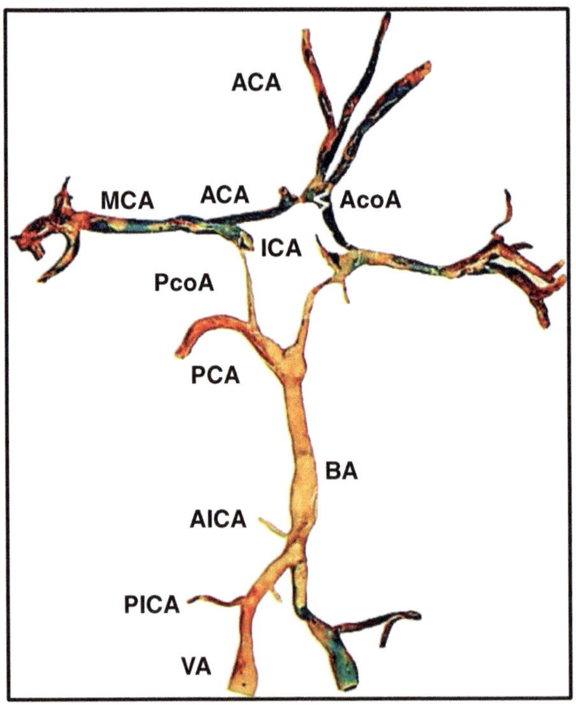

P5 – Caudal view

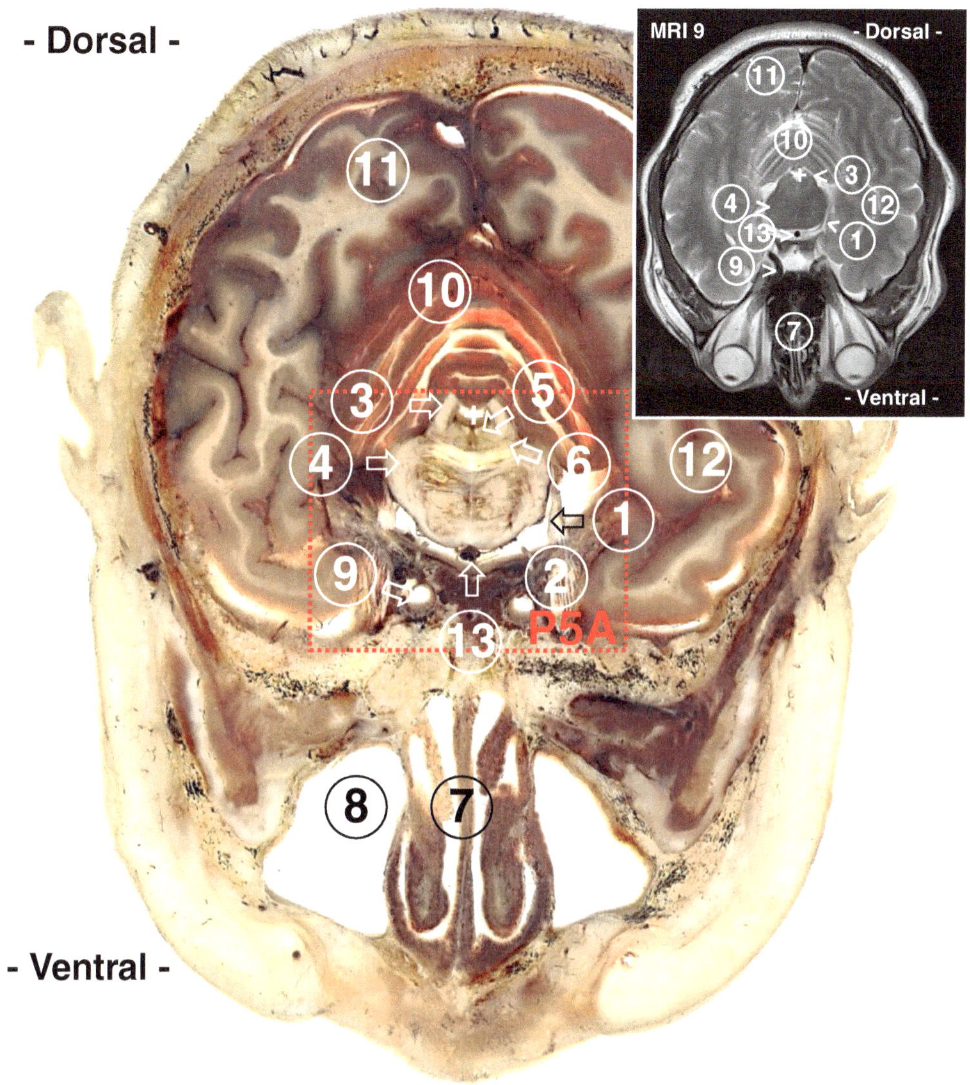

①	Trigeminal nerve CNV	⑧	Maxillary sinus
②	Trigeminal ganglion CNV	⑨	Internal carotid artery (ICA)
③	Superior cerebellar peduncle	⑩	Cerebellum
④	Middle cerebellar peduncle	⑪	Occipital lobe
⑤	Medial longitudinal fasciculus	⑫	Temporal lobe
⑥	Medial lemniscus	⑬	Basilar artery (BA)
⑦	Nasal cavity	+	4th ventricle

P5A Details see Fig. 3.8 – Caudal view

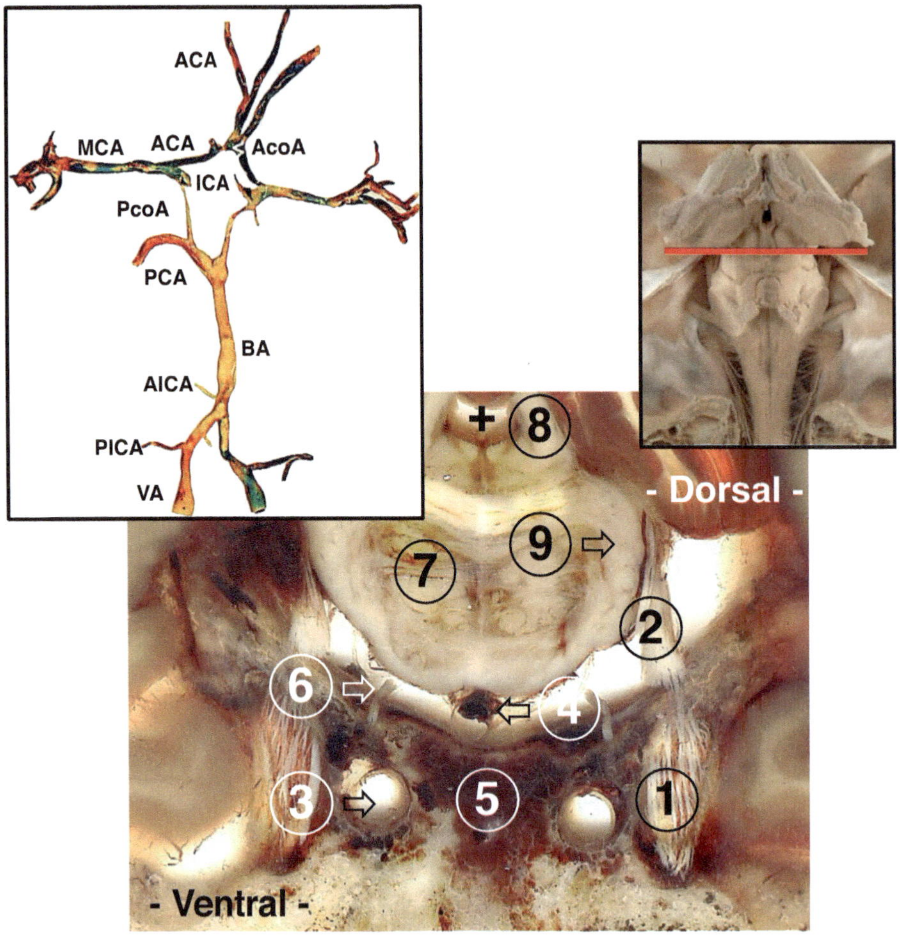

①	Trigeminal ganglion CNV	⑥	Abducens nerve CNVI
②	Trigeminal nerve CNV	⑦	Pons
③	Internal carotid artery inside ⑤ (ICA)	⑧	Superior cerebellar peduncle
④	Basilar artery (BA)	⑨	Middle cerebellar peduncle
⑤	Cavernous sinus	+	4th ventricle

P6 – Cranial view

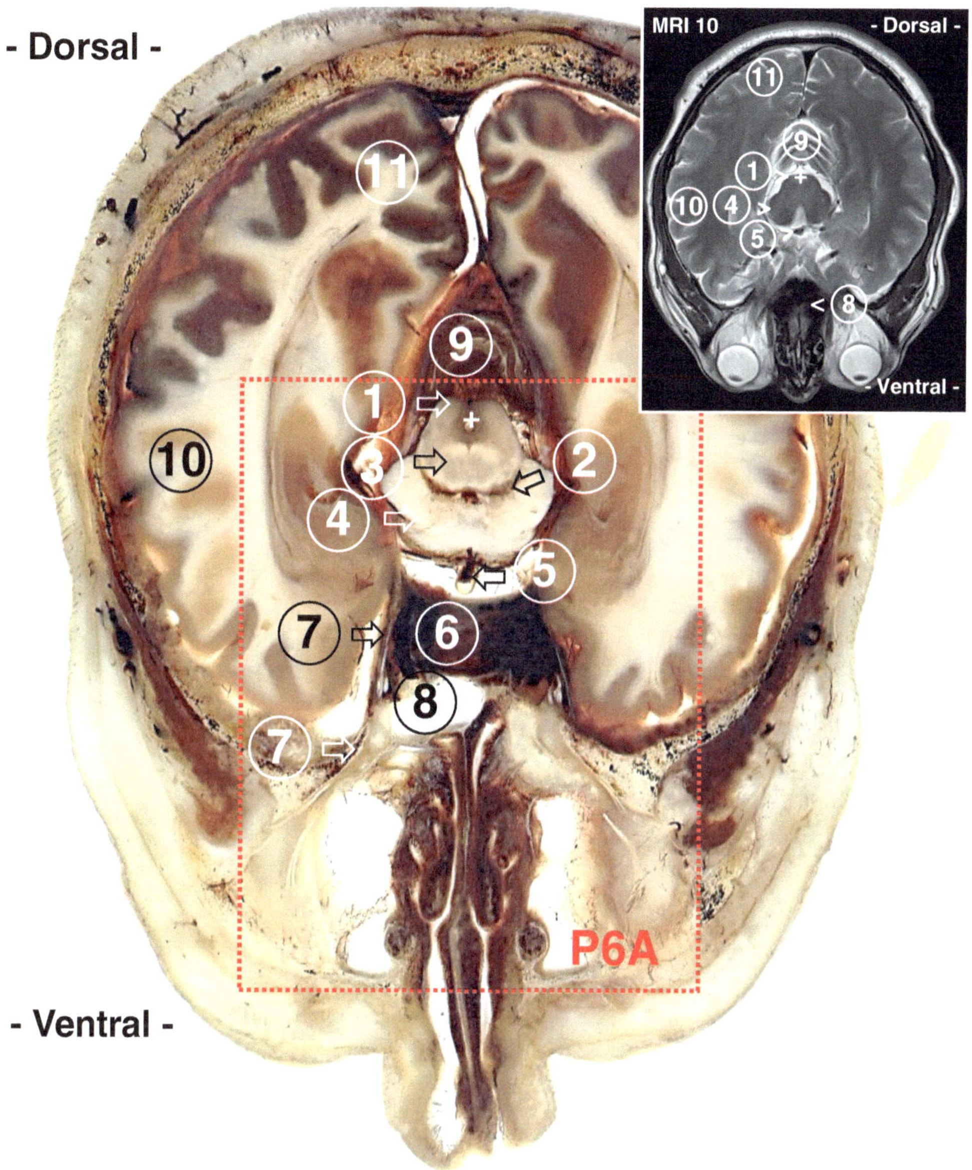

①	Inferior colliculus	⑦	Abducens nerve CNVI
②	Substantia nigra	⑧	Sphenoid sinus
③	Superior cerebellar peduncle	⑨	Cerebellum
④	Cerebral peduncle	⑩	Temporal lobe
⑤	Basilar artery (BA)	⑪	Occipital lobe
⑥	Cavernous sinus	+	Mesencephalic aqueduct

P6A Description see Fig. 14.6 – Cranial view

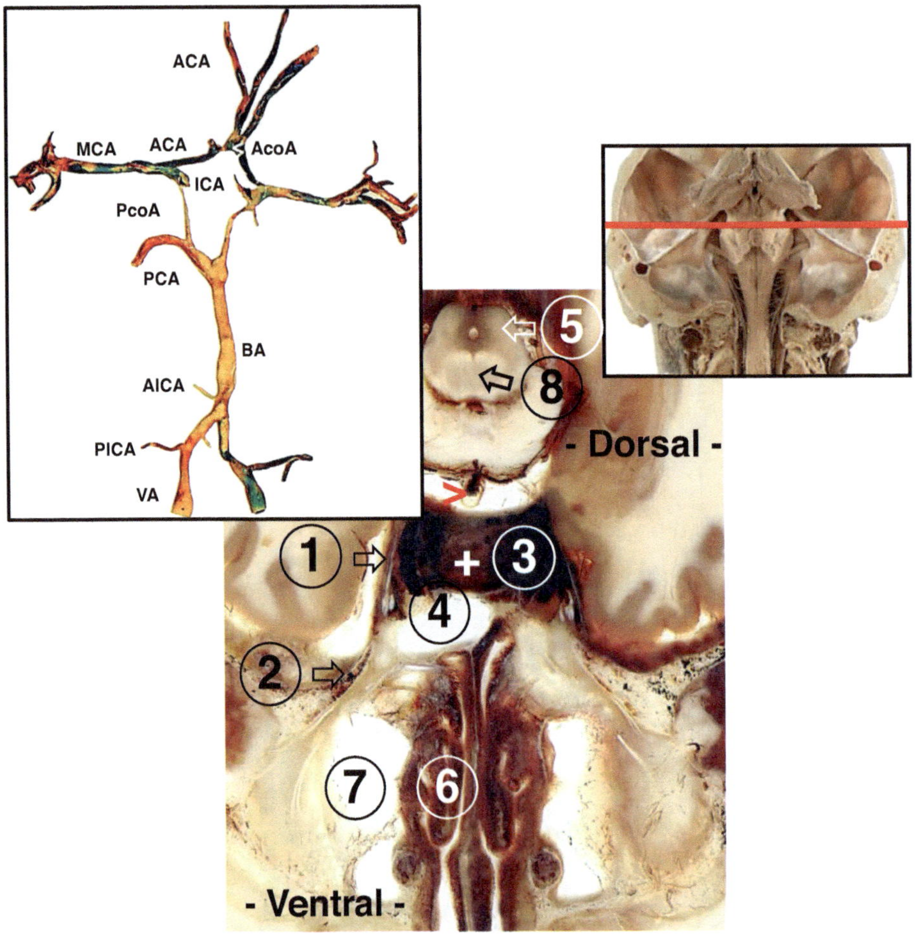

①	Trochlear nerve (CNIV) in the wall of the cavernous sinus	⑥	Nasal cavity
②	Trochlear nerve (CNIV), entry into the orbita	⑦	Maxillary sinus
③	Cavernous sinus	⑧	Decussation of superior cerebellar peduncles
④	Sphenoid sinus	+	Pituitary gland
⑤	Inferior colliculus	>	Basilar artery (BA)

P7

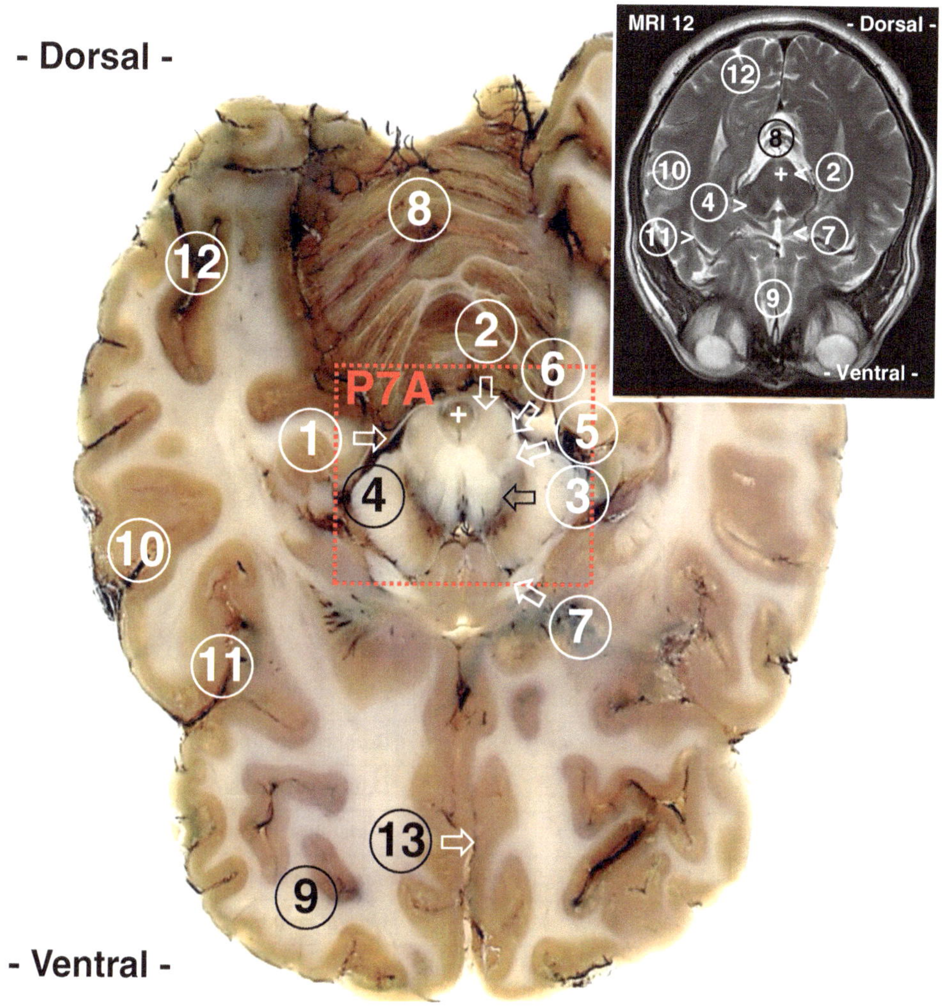

①	Trochlear nerve CNIV	⑧	Cerebellum
②	Inferior colliculus	⑨	Frontal lobe
③	Substantia nigra	⑩	Temporal lobe
④	Cerebral peduncle	⑪	Lateral sulcus
⑤	Medial lemniscus	⑫	Occipital lobe
⑥	Lateral lemniscus	⑬	Interhemispheric cleft
⑦	Optic tract	+	Mesencephalic aqueduct

P7A Description see Fig. 14.5

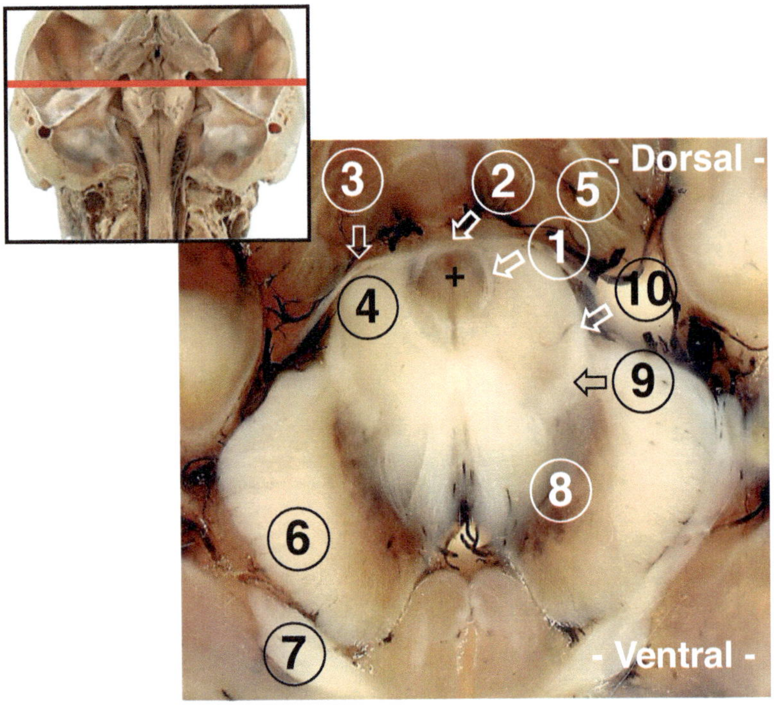

①	Trochlear nerve CNIV inside the brainstem	⑥	Cerebral peduncle
②	Decussation of the trochlear nerves CNIV	⑦	Optic tract CNII
③	Exit of the trochlear nerve CNIV	⑧	Substantia nigra
④	Inferior colliculus	⑨	Medial lemniscus
⑤	Cerebellum	⑩	Lateral lemniscus
		+	Mesencephalic aqueduct

P8

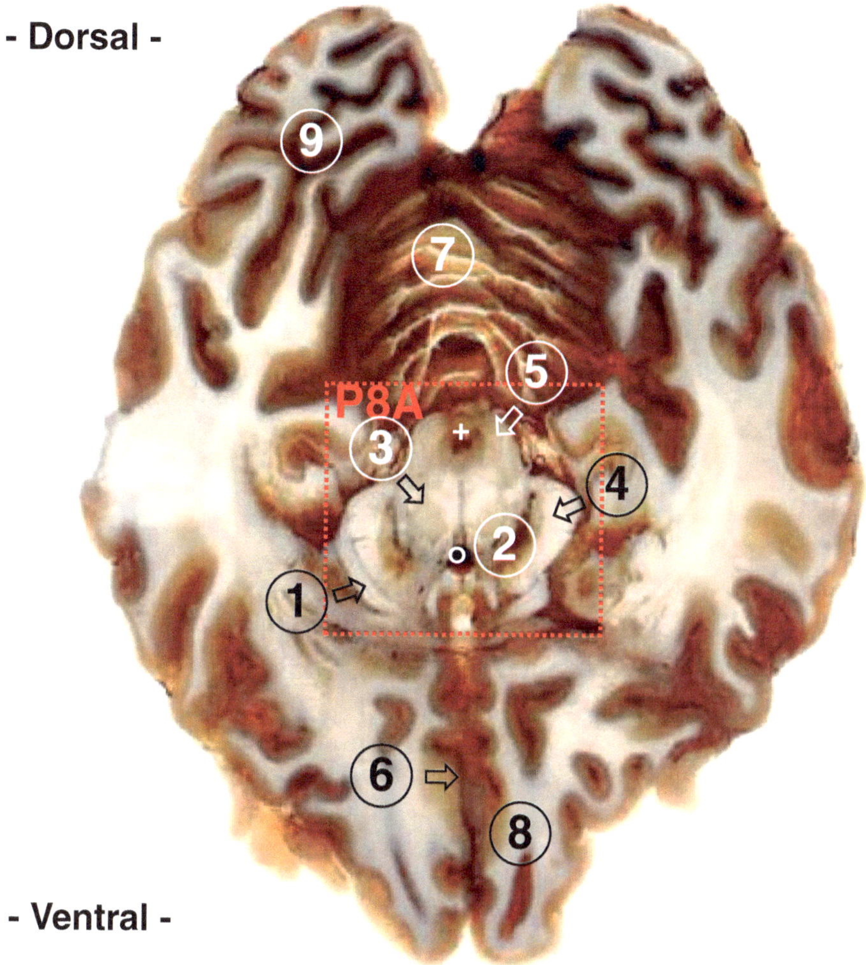

①	Optic tract CNII	⑥	Interhemispheric cleft
②	Substantia nigra	⑦	Cerebellum
③	Red ncl.	⑧	Frontal lobe
④	Cerebral peduncle	⑨	Occipital lobe
⑤	Superior colliculus	o	Interpeduncular fossa
		+	Mesencephalic aqueduct

P8A

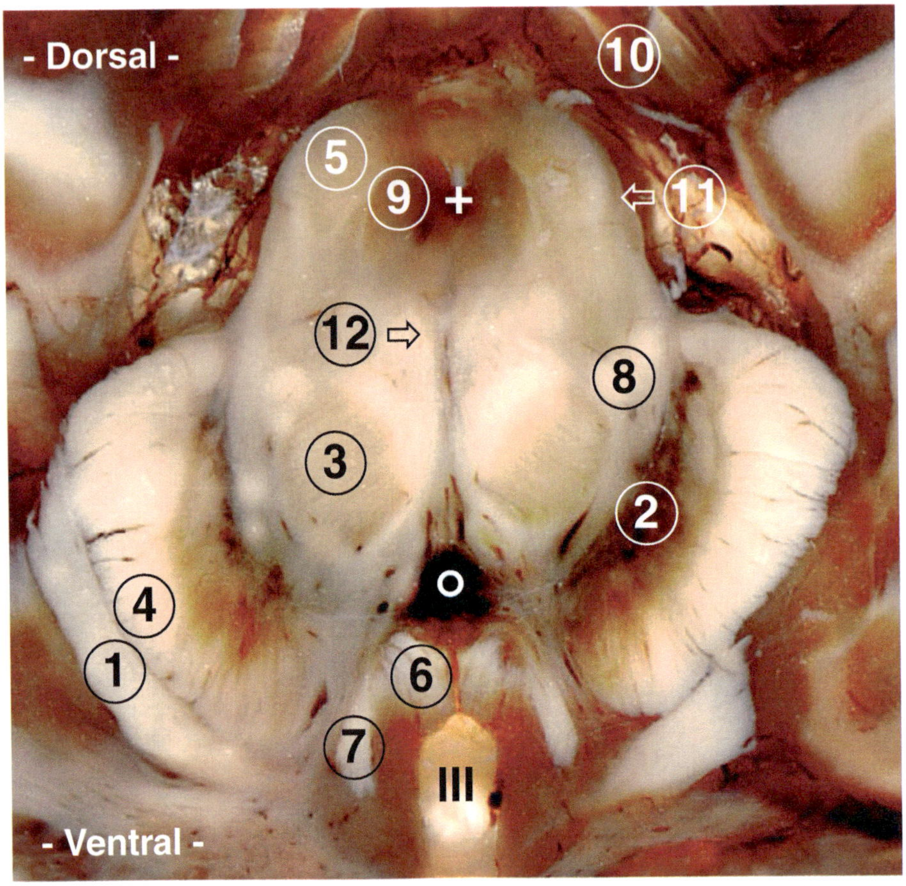

①	Optic tract CNII	⑧	Medial lemniscus
②	Substantia nigra	⑨	Periaqueductal gray
③	Red ncl.	⑩	Cerebellum
④	Cerebral peduncle	⑪	Brachium of inferior colliculus
⑤	Superior colliculus	⑫	Medial longitudinal fasciculus
⑥	Fornix, columna	III	3rd ventricle
⑦	Mammillothalamic tract	o	Interpeduncular fossa
		+	Mesencephalic aqueduct

P9 – Cranial view

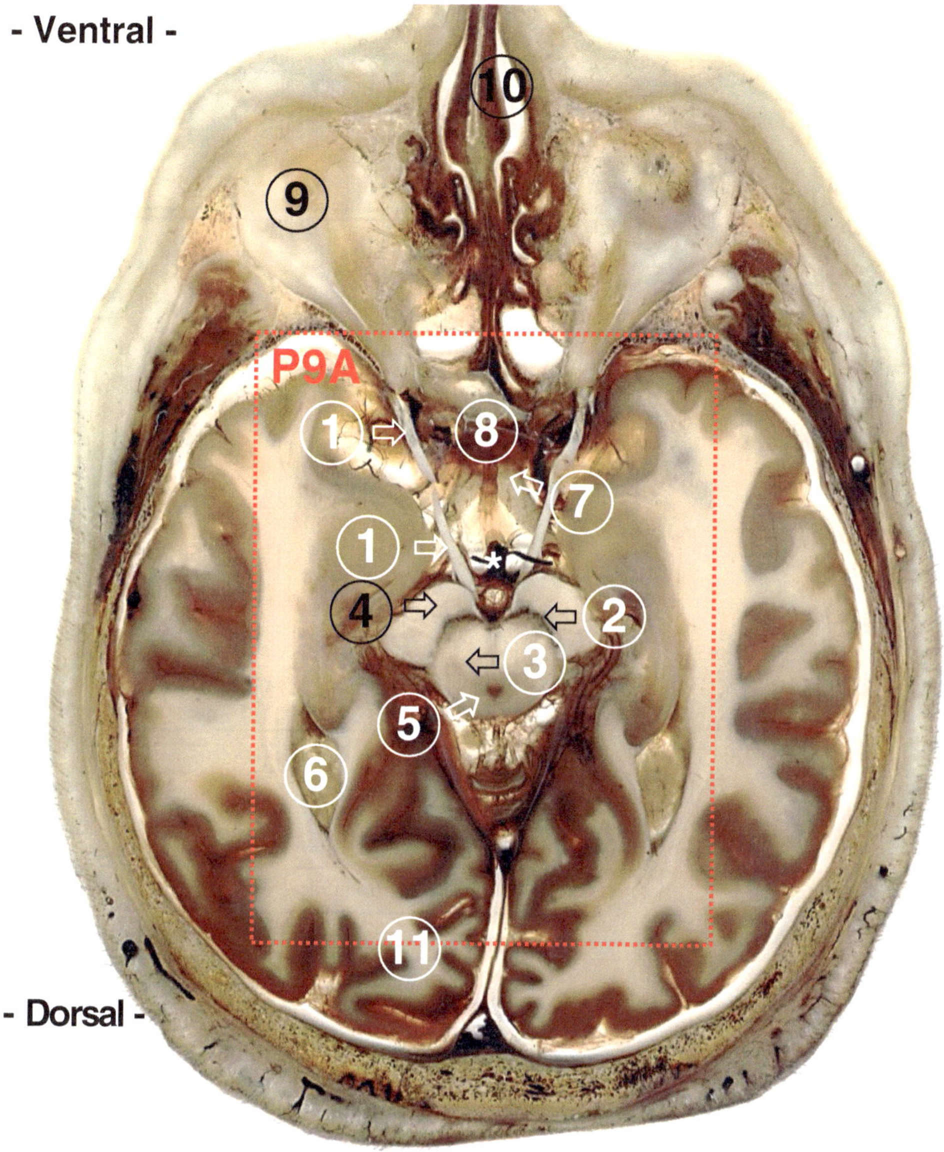

①	Oculomotor nerve CNIII	⑦	Optic chiasm
②	Substantia nigra	⑧	Pituitary gland
③	Superior cerebellar peduncles, decussation	⑨	Eye bulb
④	Cerebral peduncle	⑩	Nasal cavity
⑤	Superior colliculus	⑪	Occipital lobe
⑥	Lateral ventricle occipital horn	*	Basilar artery (BA)

P9A Details see Fig. 15.20 – Cranial view

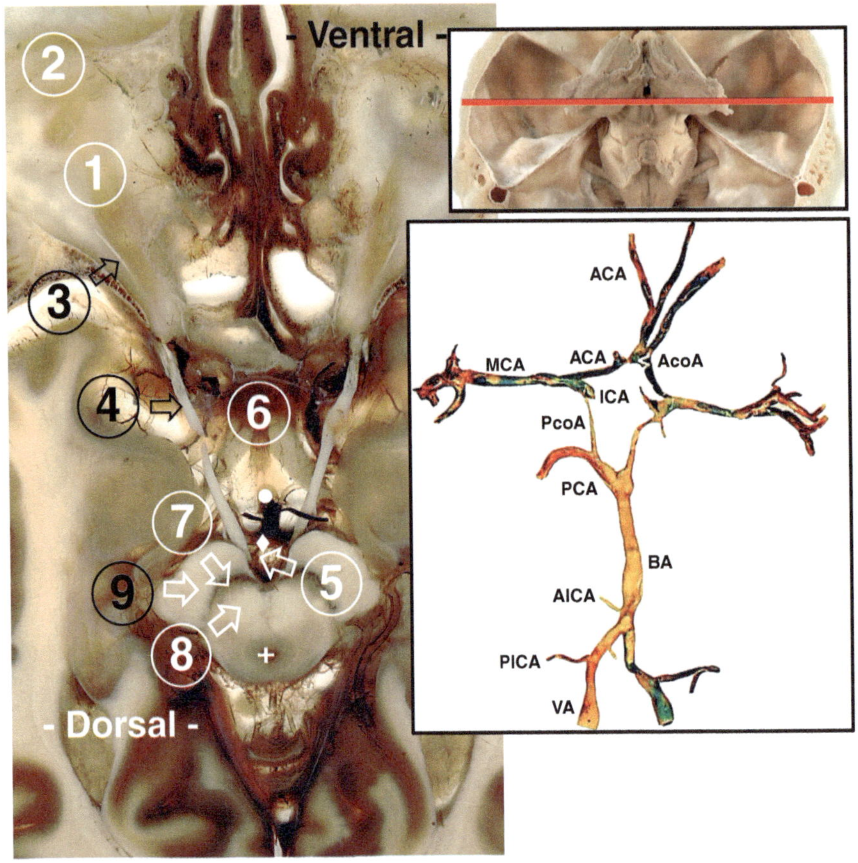

①	Inferior rectus muscle CNIII	⑦	Substantia nigra
②	Inferior oblique muscle CNIII	⑧	Superior cerebellar peduncles, decussation
③	Oculomotor nerve, inferior branch	⑨	Cerebral peduncle
④	Oculomotor nerve CNIII	•	Basilar artery (BA)
⑤	Interpeduncular fossa	♦	Bifurcation of BA into the posterior cerebral arteries (PCA)
⑥	Pituitary gland	+	Mesencephalic aqueduct

P10 – Cranial view

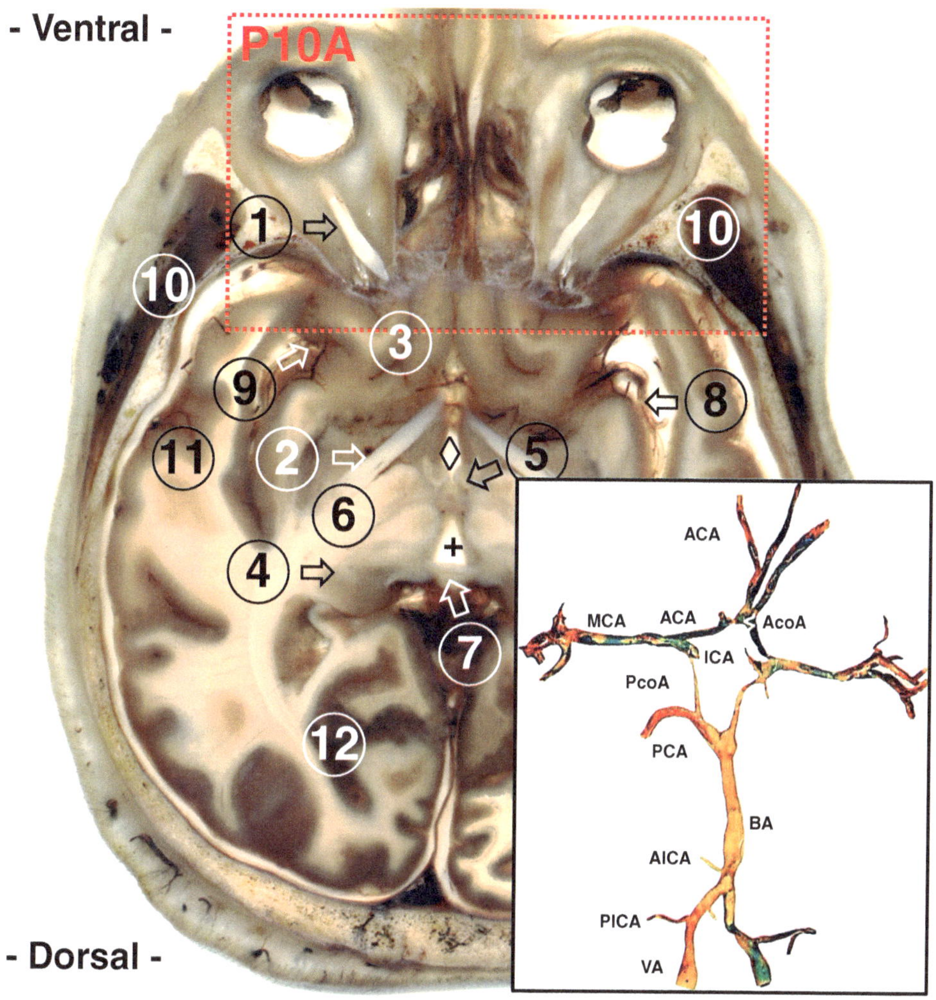

①	Optic nerve CNII	⑧	Middle cerebral artery (MCA)
②	Optic tract CNII	⑨	Lateral sulcus
③	Frontal lobe	⑩	Temporal muscle
④	Pulvinar	⑪	Temporal lobe
⑤	Mammillary bodies	⑫	Occipital lobe
⑥	Cerebral peduncle	◊	Interpeduncular fossa
⑦	Posterior commissure	+	3rd ventricle

P10A Details see Fig. 15.18

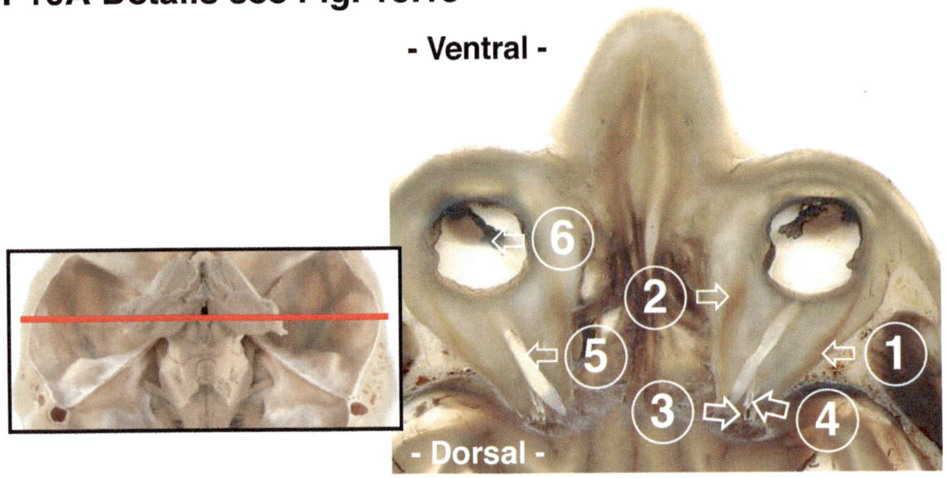

①	Lateral rectus muscle CNVI	④	Nascociliary nerve V_1 of trigeminal nerve CNV
②	Medial rectus muscle CNIII	⑤	Optic nerve CNII
③	Oculomotor nerve, superior branch CNIII	⑥	Lens with medial zonular fibers detached

P11 – Caudal view

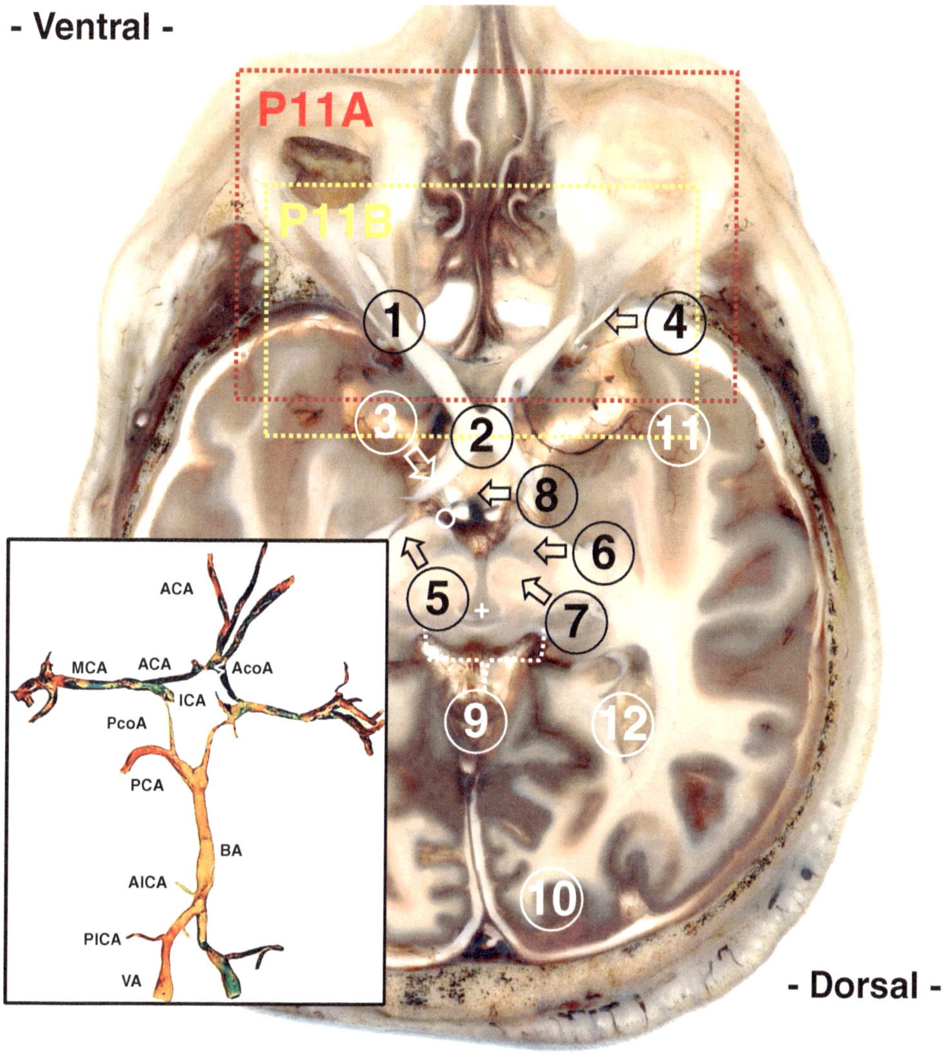

①	Optic nerve CNII	⑧	Basilar artery (BA)
②	Optic chiasm CNII	⑨	Quadrigeminal plate
③	Optic tract CNII	⑩	Occipital lobe
④	Abducens nerve CNVI	⑪	Temporal lobe
⑤	Cerebral peduncle	⑫	Lateral ventricle, occipital horn
⑥	Substantia nigra	+	Mesencephalic aqueduct
⑦	Red ncl.	o	Posterior cerebral artery (PCA)

P11A / P11B Details see Figs. 15.19 (A) and 15.19 (B) – Caudal view

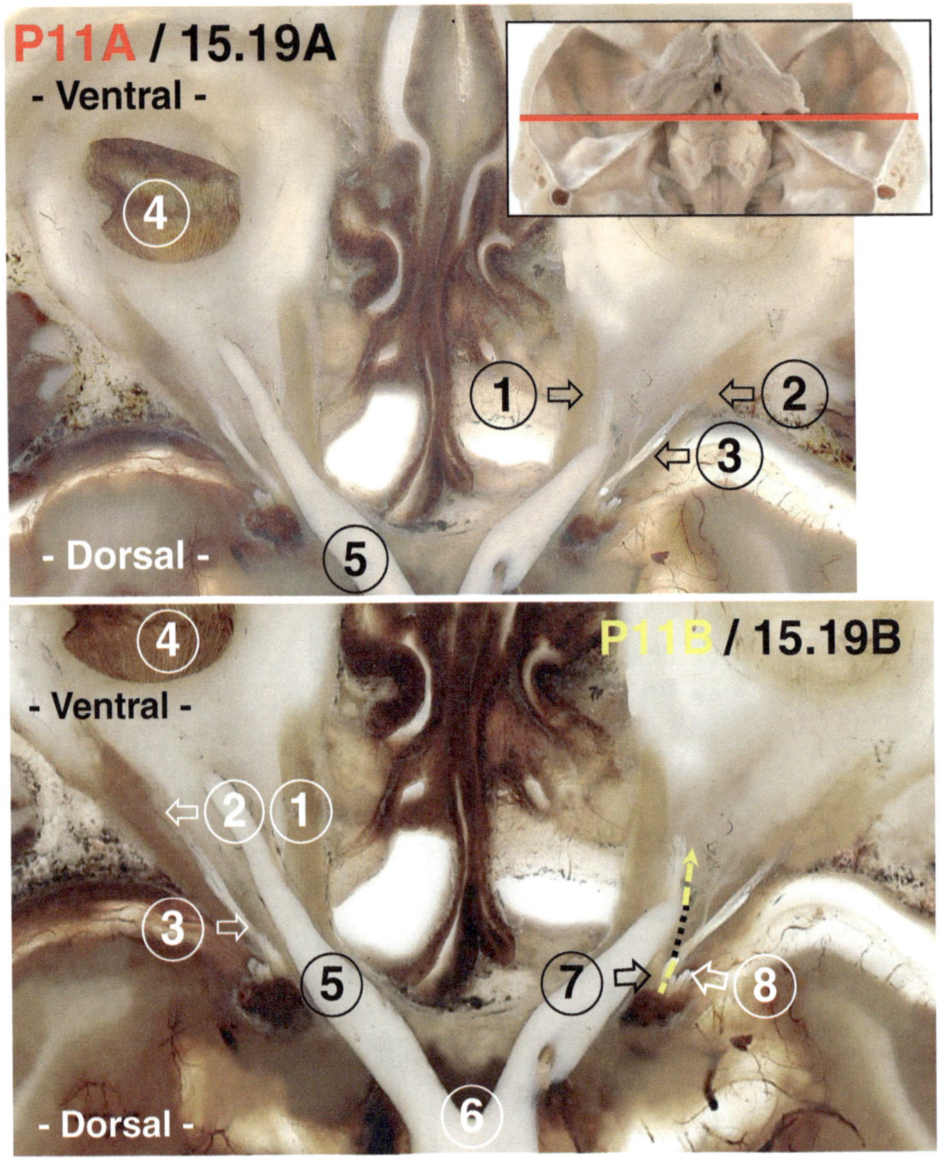

①	Medial rectus muscle	⑥	Optic chiasm CNII
②	Lateral rectus muscle	⑦	Nasociliary nerve V_1
③	Abducens nerve CNVI	⑧	Superior branch of CNIII
④	Retina, partly detached	---	Course of inferior branch of CNIII, ---ventral of CNII
⑤	Optic nerve CNII		

P12 – Caudal view

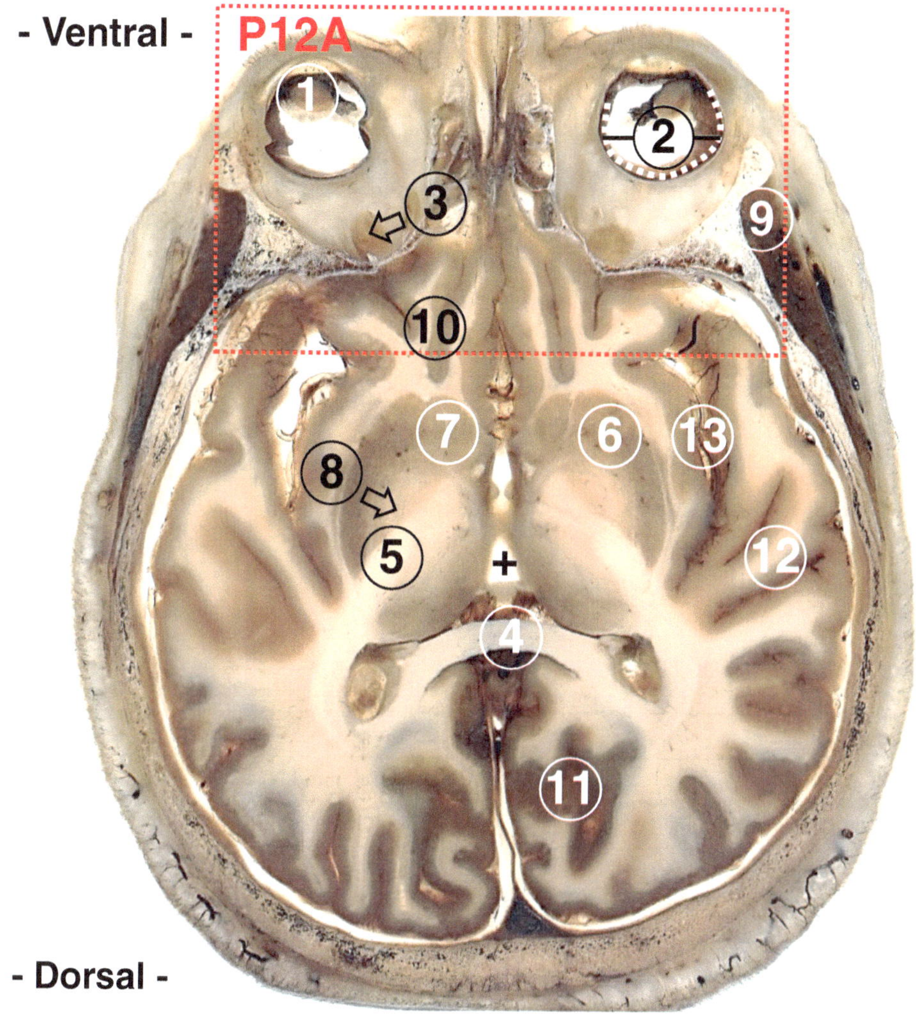

①	Lens	⑧	Globus pallidus
②	Retina ---	⑨	Temporal muscle
③	Superior rectus muscle CNIII	⑩	Frontal lobe
④	Corpus callosum	⑪	Occipital lobe
⑤	Internal capsule	⑫	Temporal lobe
⑥	Putamen	⑬	Lateral sulcus
⑦	Caudate ncl.	+	3rd ventricle

P12A Description see Fig. 15.17 – Cranial view

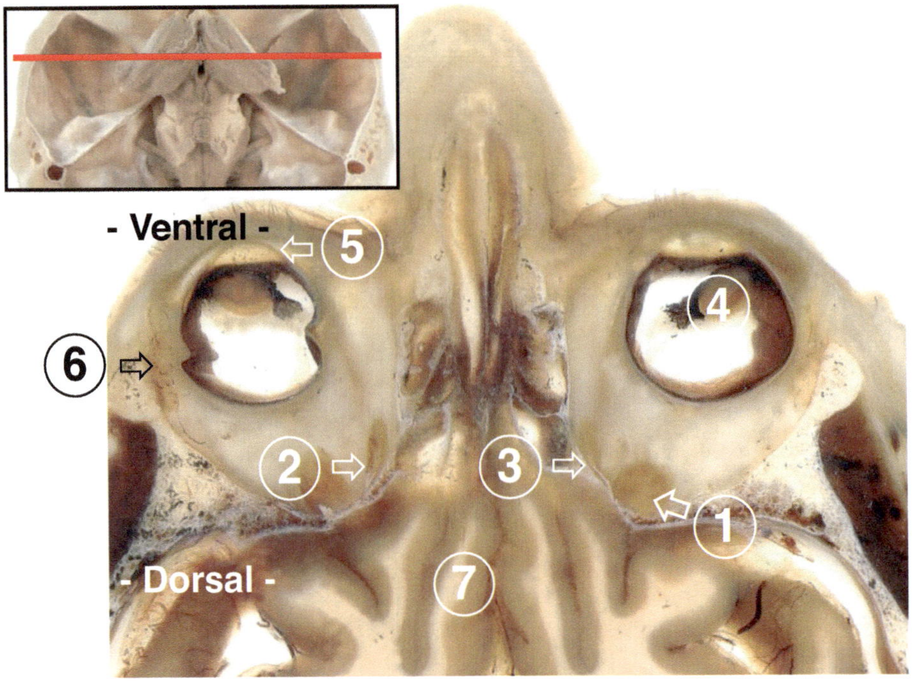

①	Superior rectus muscle CNIII	⑤	Cornea
②	Medial rectus muscle CNIII	⑥	Lacrimal gland
③	Oculomotor nerve CNIII, superior branch	⑦	Frontal lobe
④	Lens, zonular fibers medially detached		

P13 – Cranial view

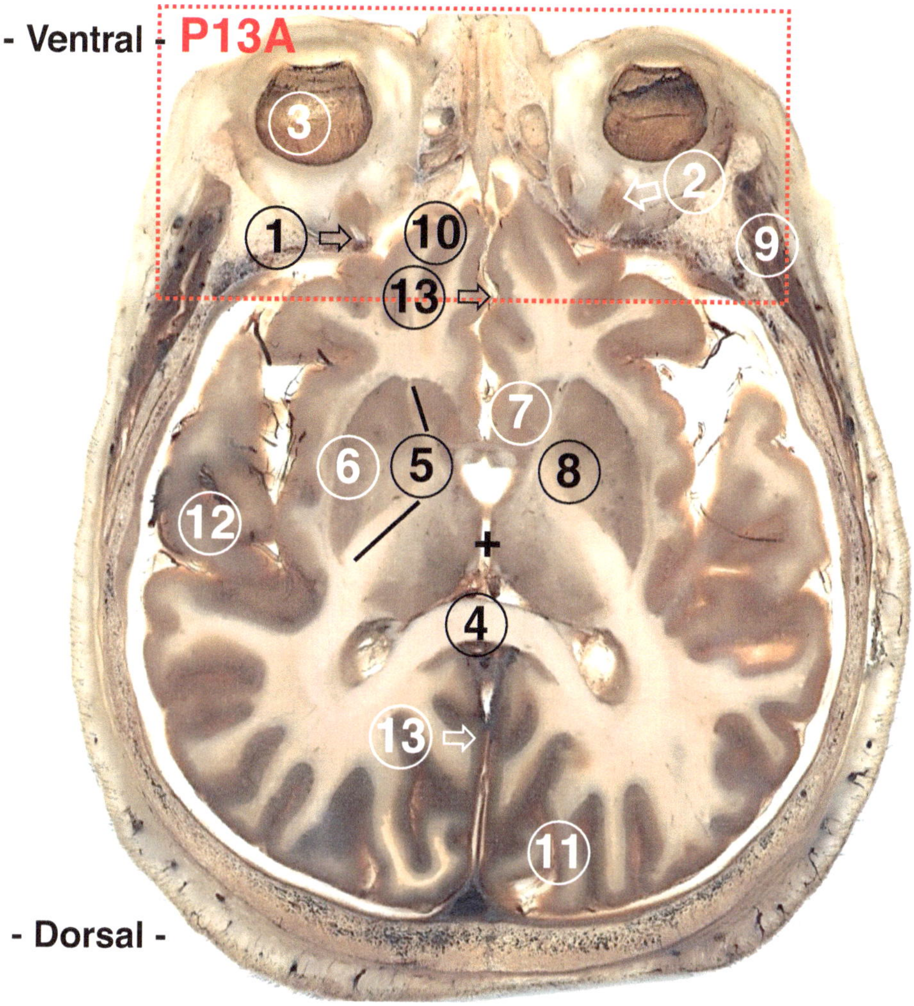

①	Oculomotor nerve CNIII, superior branch	⑧	Globus pallidus
②	Superior rectus muscle CNIII	⑨	Temporal muscle
③	Retina	⑩	Frontal lobe
④	Corpus callosum	⑪	Occipital lobe
⑤	Internal capsule	⑫	Temporal lobe
⑥	Putamen	⑬	Interhemispheric cleft
⑦	Caudate ncl.	+	3rd ventricle

P13A Description see Fig. 15.16 – Cranial view

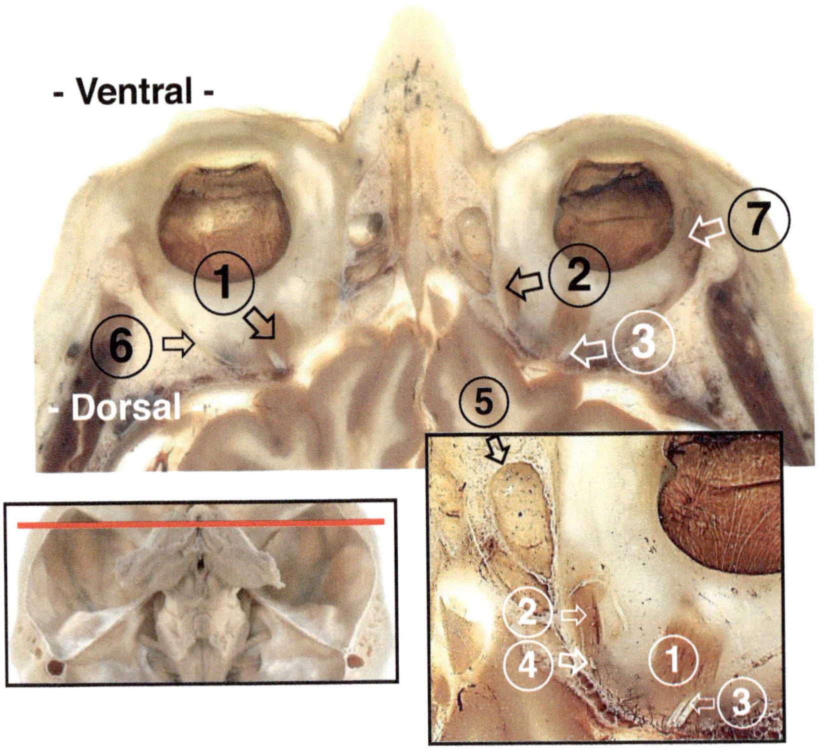

①	Superior rectus muscle CNIII	⑤	Anterior ethmoidal nerve V_1
②	Superior oblique muscle CNIV	⑥	Lacrimal nerve V_1
③	Oculomotor nerve, superior branch CNIII	⑦	Lacrimal gland
④	Trochlear nerve CNIV		

P14 – Caudal view

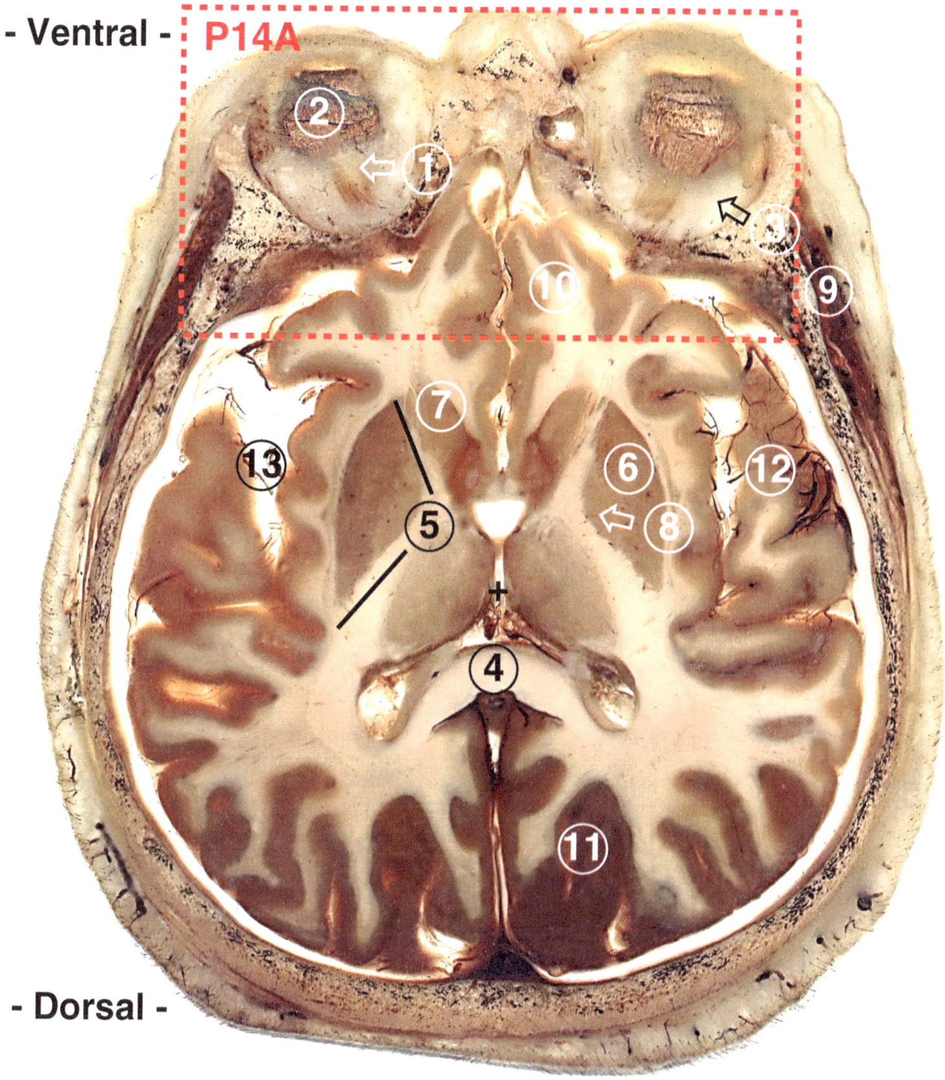

①	Superior rectus muscle CNIII	⑧	Globus pallidus
②	Retina	⑨	Temporal muscle
③	Retrobulbar fat body	⑩	Frontal lobe
④	Corpus callosum	⑪	Occipital lobe
⑤	Internal capsule	⑫	Temporal lobe
⑥	Putamen	⑬	Lateral sulcus
⑦	Caudate ncl.	+	3rd ventricle

P14A Description see Fig. 15.15 – Caudal view

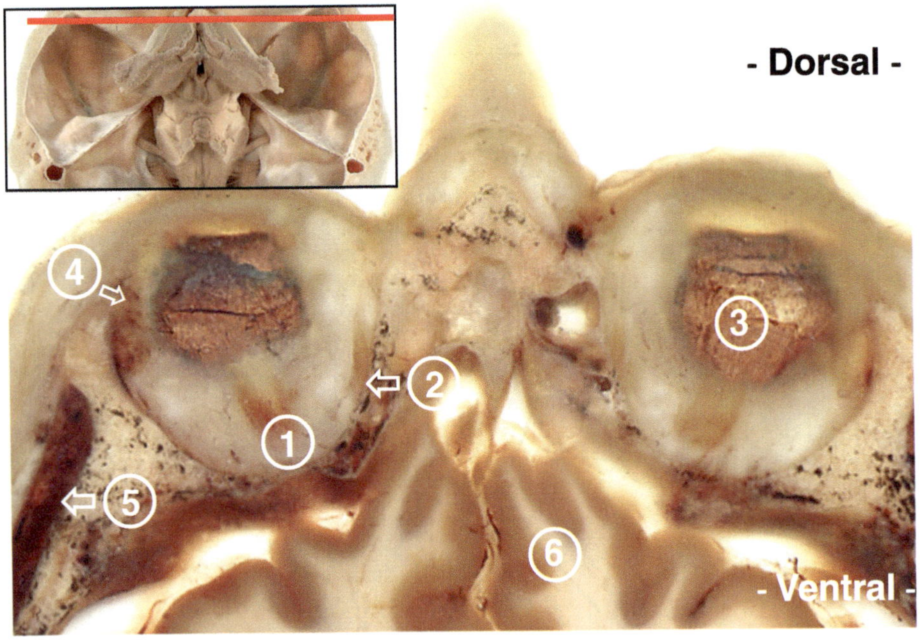

①	Superior rectus muscle CNIII	④	Lacrimal gland
②	Superior oblique muscle CNIV	⑤	Temporal muscle
③	Dorsal view on the pigmented layer of the retina	⑥	Frontal lobe

P15 Coronal section 1

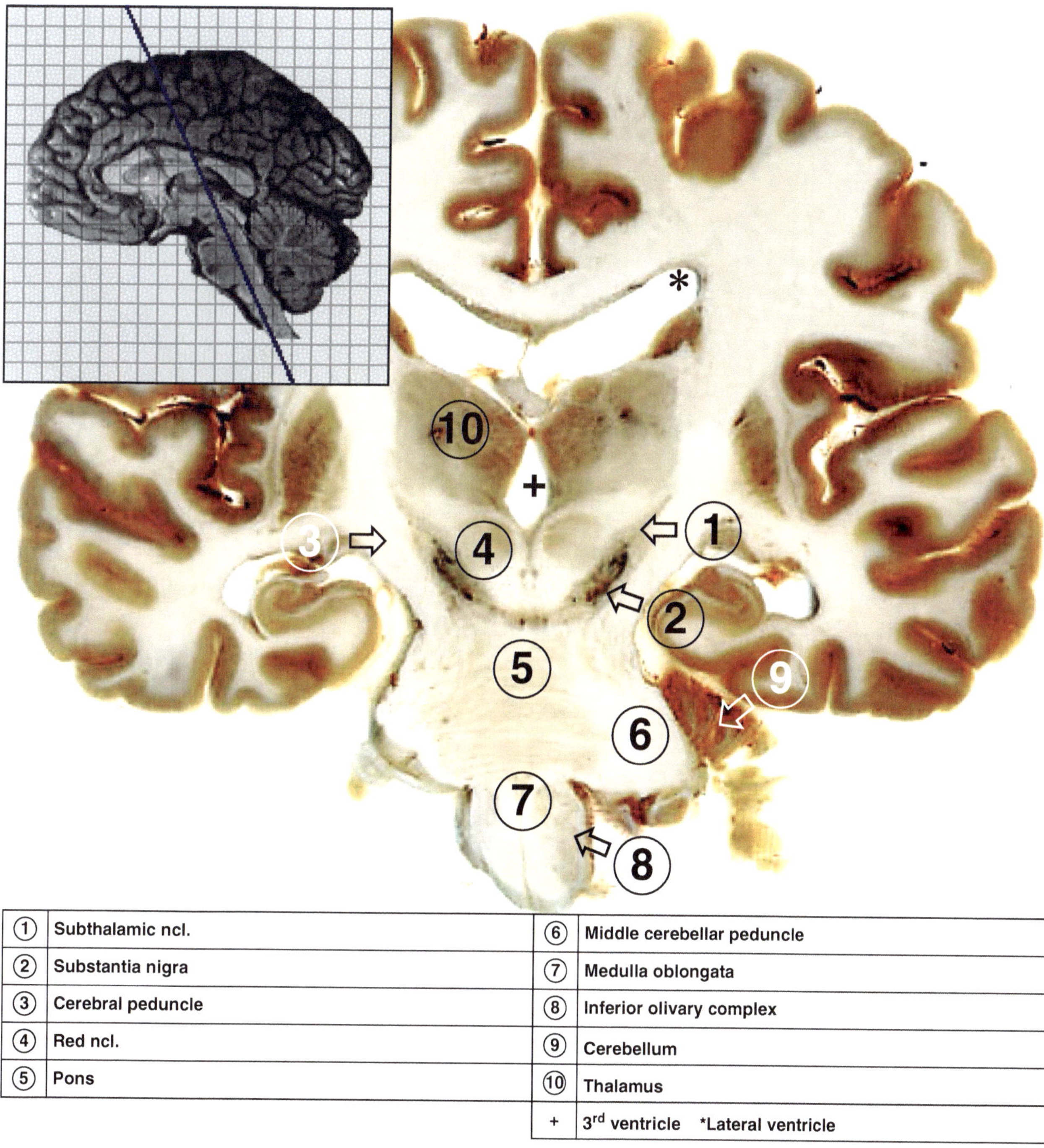

①	Subthalamic ncl.	⑥	Middle cerebellar peduncle
②	Substantia nigra	⑦	Medulla oblongata
③	Cerebral peduncle	⑧	Inferior olivary complex
④	Red ncl.	⑨	Cerebellum
⑤	Pons	⑩	Thalamus
		+	3rd ventricle *Lateral ventricle

P16 Coronal section 2

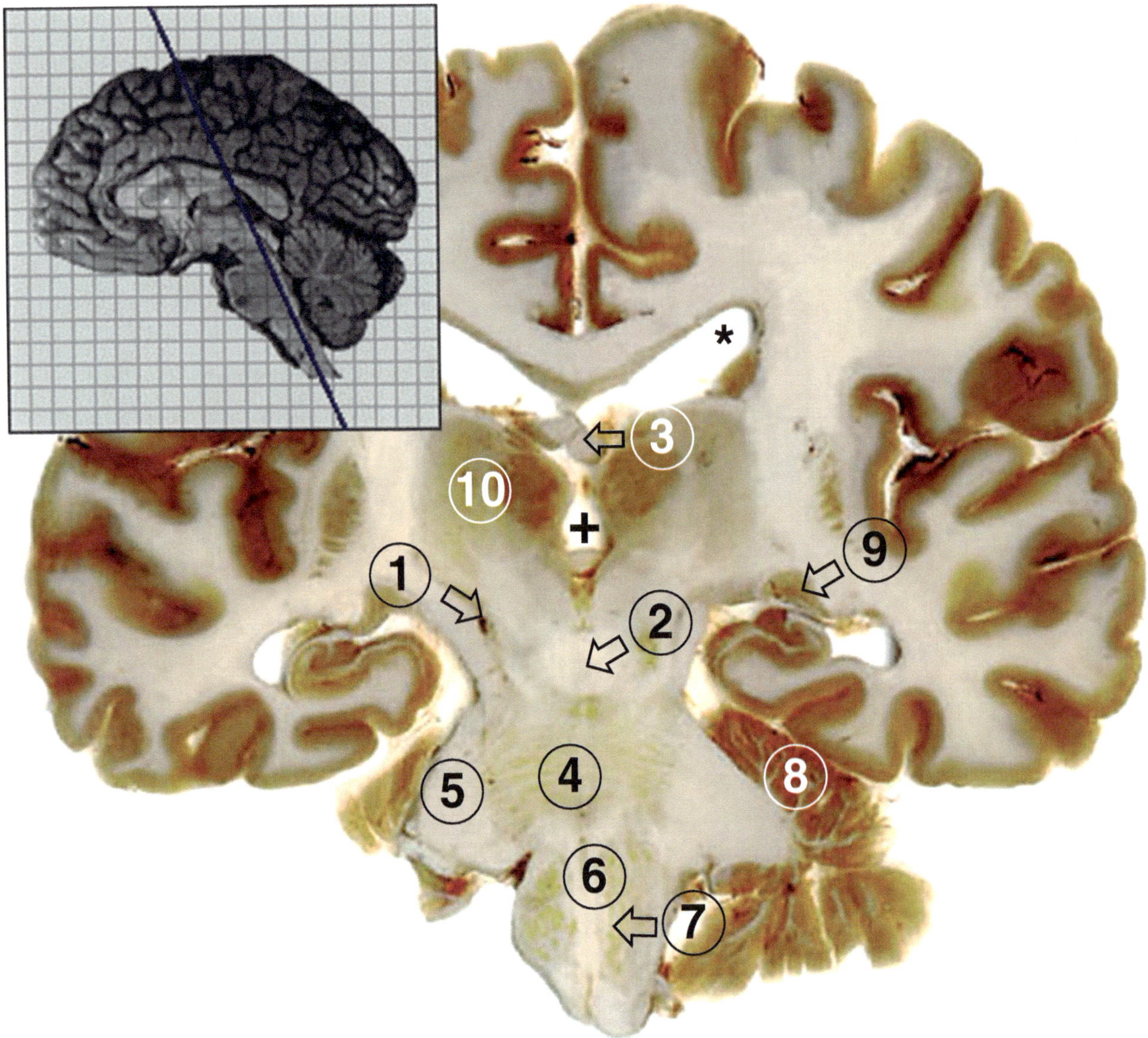

①	Substantia nigra	⑥	Medulla oblongata
②	Decussation of the superior cerebellar peduncles	⑦	Inferior olivary complex
③	Fornix	⑧	Cerebellum
④	Pons	⑨	Lateral geniculate body
⑤	Middle cerebellar peduncle	⑩	Thalamus
		+	3rd ventricle *Lateral ventricle

P17 Coronal section 3

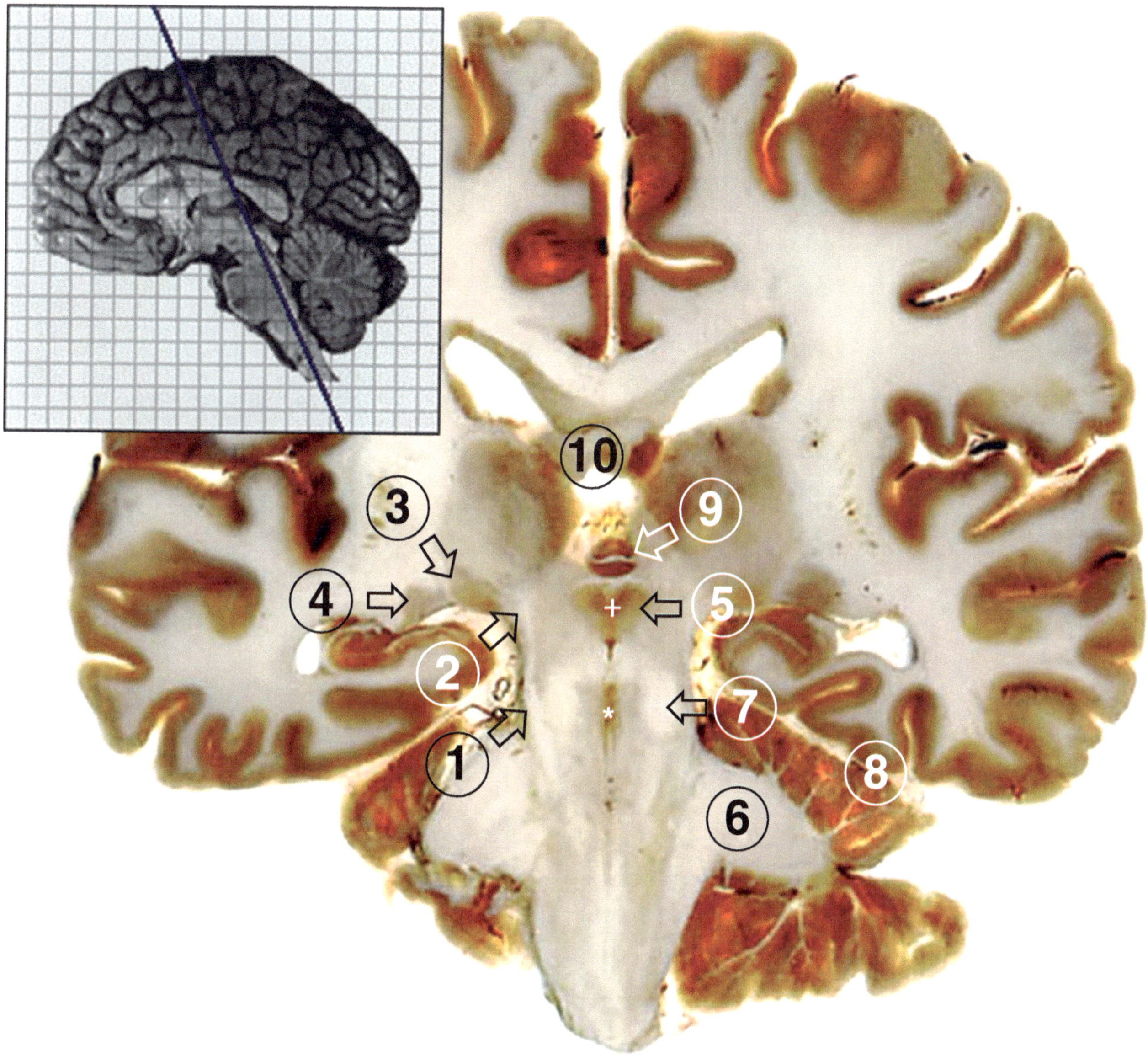

①	Lateral lemniscus	⑥	Middle cerebellar peduncle
②	Brachium colliculi inferioris	⑦	Superior cerebellar peduncle
③	Medial geniculate body	⑧	Cerebellum
④	Lateral geniculate body	⑨	Pineal gland with posterior commissure
⑤	Periaqueductal gray	⑩	Fornix
		+	Mesencephalic aqueduct *4th ventricle

P18 Coronal section 4

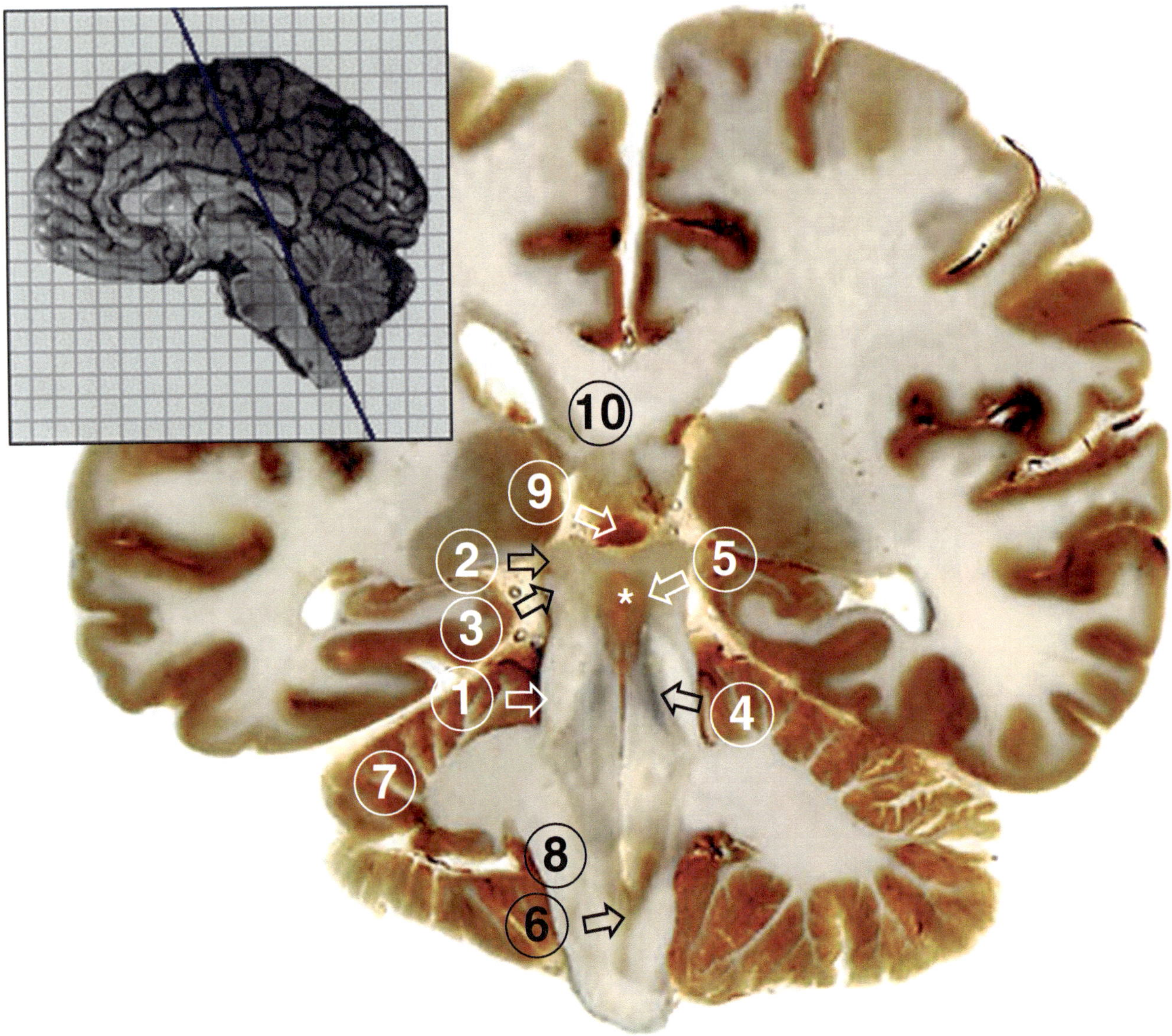

①	Superior cerebellar peduncle	⑥	Obex
②	Inferior colliculus	⑦	Cerebellum
③	Lateral lemniscus	⑧	Inferior cerebellar peduncle
④	Locus caeruleus	⑨	Pineal gland
⑤	Periaqueductal gray	⑩	Splenium of the corpus callosum
		*	Mesencephalic aqueduct + 4th ventricle / rhomboid fossa

Index

H. Schröder et al., *The Human Brainstem*, https://doi.org/10.1007/978-3-030-89980-6

B

C

M

N

O

Q

R

S

GPSR Compliance

The European Union's (EU) General Product Safety Regulation (GPSR) is a set of rules that requires consumer products to be safe and our obligations to ensure this.

If you have any concerns about our products, you can contact us on ProductSafety@springernature.com

In case Publisher is established outside the EU, the EU authorized representative is:

Springer Nature Customer Service Center GmbH
Europaplatz 3
69115 Heidelberg, Germany

Batch number: 10020312

Printed by Printforce, the Netherlands